The Electrical Engineering Handbook Series

Series Editor
Richard C. Dorf
University of California, Davis

Titles Included in the Series

The Electrical Engineering Handbook, Richard C. Dorf
The Biomedical Engineering Handbook, Joseph D. Bronzino
The Circuits and Filters Handbook, Wai-Kai Chen
The Transforms and Applications Handbook, Alexander D. Poularikas
The Control Handbook, William S. Levine
The Electronics Handbook, Jerry C. Whitaker
The Industrial Electronics Handbook, J. David Irwin
The Communications Handbook, Jerry D. Gibson
The Mobile Communications Handbook, Jerry D. Gibson
The Technology Management Handbook, Richard C. Dorf

THE

TECHNOLOGY

MANAGEMENT

HANDBOOK

Editor-in-Chief

Richard C. Dorf

 CRC PRESS

A CRC Press Handbook Published in Cooperation with IEEE Press

Acquiring Editor:	*Nora Konopka*
Project Editor:	*Susan Fox*
Marketing Manager:	*Jane Stark*
Cover design:	*Dawn Boyd*

Library of Congress Cataloging-in-Publication Data

The technology management handbook / editor-in-chief, Richard C. Dorf.
      p. cm.
    Includes bibliographical references and index.
    ISBN 0-8493-8577-6 (alk. paper)
    1. Technology—Management—Handbooks, manuals, etc. I. Dorf, Richard C.
  T49.5.T4454 1998
  658.5'14—dc21

98-22328
CIP

Preface

In a single volume *The Technology Management Handbook* provides a ready reference for the technology manager in industry, government, and academia. Useful for the project manager, lead engineer, section manager, or executive in a technology company, the book also will be a primary resource for the manager of technology applications in a service or manufacturing company. Divided into 7 sections consisting of 22 chapters, this comprehensive format encompasses the field of technology management – offering the most up-to-date information in economics, marketing, product development, manufacturing, finance, accounting, innovation, project management, human resources, and international business.

Written from the technical manager's perspective and written for the technologists who are managers, *The Technology Management Handbook* provides in-depth information on the science and practice of management.

Key features include:

- Contributed articles by more than 200 technical management experts
- Offers a technical manager's perspective for technologists who are managers
- Covers the true business end of engineering management
- Describes the fundamentals of running a business by addressing the subjects of economics, finance, accounting, project management, marketing, and manufacturing
- Outlines a historical perspective on engineering management
- Examines crisis management
- Includes nearly 200 articles and 1,000 defined terms — cross-referenced article to article
- Serves as a comprehensive reference for all engineers

Organization

The fundamentals of technology management have evolved to include a wide range of knowledge, substantial empirical data, and a broad range of practice. The focus of the handbook is on the key concepts, models, and methods that enable the manager to effectively manage the development and utilization of technologies. While data and formulae are summarized, the main focus is the provision of the underlying theories and concepts and the appropriate application of these theories to the field of technology management. Thus, the reader will find the key concepts defined, described, and illustrated in order to serve the needs of the reader over many years.

The level of conceptual development of each topic is challenging, but tutorial and relatively fundamental. Each article, of which there are nearly 200, is written to enlighten the expert, refresh the knowledge of the mature manager, and educate the novice.

The information is organized into 20 major sections which is further broken down into 22 chapters. Each section contains an historical vignette that serves to enliven and illuminate the history of the subject of that section.

Each article includes three important and useful categories: defining terms, references, and further information. *Defining terms* are key definitions and the first occurrence of each term defined is indicated in boldface in the text. The definitions of these terms are summarized as a list at the end of each chapter or article. The *references* provide a list of useful books and articles for follow-up reading. Finally, *further information* provides some general and useful sources of additional information on the topic.

Locating Your Topic

Numerous avenues of access to information contained in the handbook are provided. A complete table of contents is presented at the front of the book. In addition, an individual table of contents precedes each of the 7 sections. Finally, each chapter begins with its own table of contents. The reader should look over these tables of contents to become familiar with the structure, organization, and content of the book. A subject index is available at the end of the book, which also can be used to locate definitions.

Acknowledgments

This handbook is a testimony to the dedication of the Editorial Board, the publishers, and my editorial associates. I particularly wish to acknowledge at CRC Press Ron Powers, Publisher; Nora Konopka, Associate Editor, and Susan Fox, Project Editor. I also wish to acknowledge Kristen Maus, Consulting Editor. Finally, I am indebted to the assistance of Sara Hare, who served as Managing Editor.

Richard C. Dorf
Editor-in-Chief

Editor-in-Chief

Richard C. Dorf, professor of management and professor of electrical and computer engineering at the University of California, Davis, teaches graduate courses in the Graduate School of Management on technology management, innovation, and entrepreneurship. He earned a Ph.D. in electrical engineering from the U.S. Naval Postgraduate School, an M.S. from the University of Colorado, and a B.S. from Clarkson University.

Professor Dorf has extensive experience with education and industry and is professionally active in the fields of new ventures and innovation and entrepreneurship. He has served as a visiting professor at the University of Edinburgh, Scotland; the Massachusetts Institute of Technology; Stanford University; and the University of California, Berkeley.

A Fellow of The Institute of Electrical and Electronics Engineers, Dr. Dorf is widely known for his *Modern Control Systems,* 8th edition (Addison-Wesley, 1998), *The International Encyclopedia of Robotics* (Wiley, 1988), and *The Engineering Handbook* (CRC Press, 1996). Dr. Dorf has served as associate editor of the IEEE Transactions on Engineering Management since 1988.

Editorial Board

Contributors

Layek Abdel-Malek
New Jersey Institute of Technology
Newark, New Jersey

Jeffrey Alexander
Washington Core: International
    Technology and Policy
    Consultants
Washington, D.C.

Janet K. Allen
Georgia Institute of Technology
Atlanta, Georgia

Kwasi Amoako-Gyampah
University of North Carolina
Greensboro, North Carolina

Philip Anderson
Dartmouth College
Hanover, New Hampshire

Lynda M. Applegate
Harvard Business School
Boston, Massachusetts

Mark Atlas
Carnegie Mellon University
Pittsburgh, Pennsylvania

Tung Au
Carnegie Mellon University
Pittsburgh, Pennsylvania

B. Michael Aucoin
Texas A&M University
College Station, Texas

Constance E. Bagley
Stanford Business School
Stanford, California

A. Terry Bahill
University of Arizona
Tucson, Arizona

Catherine Banbury
Saint Mary's College
Moraga, California

Lynne F. Baxter
Heriot-Watt University
Edinburgh, Scotland

Robert G. Beaves
Robert Morris College
Moon Township, Pennsylvania

Paul A. Beck
Paul A. Beck & Associates
Pittsburgh, Pennsylvania

Roger J. Best
University of Oregon
Eugene, Oregon

S. Bhimjee
San Francisco State University
San Francisco, California

David Birchall
Henley Management College
Greenlands, Henley-on-Thames,
    Oxfordshire, England

Benjamin S. Blanchard
Virginia Polytechnic Institute &
    State University
Blacksburg, Virginia

Carol I. Bordas
Thorp, Reed & Armstrong
Pittsburgh, Pennsylvania

Gordon Brown
Digital Image Technology
Pittsburgh, Pennsylvania

George Bugliarello
Polytechnic University
Brooklyn, New York

Robert A. Burgelman
Stanford University
Stanford, California

Tom Byers
Stanford University
Stanford, California

Dick Campion
Rational Solutions
Calgary, Alberta, Canada

Elias G. Carayannis
The George Washington University
Washington, D.C.

Mario W. Cardullo
Virginia Polytechnic Institute and
    State University
Alexandria, Virginia

Kenneth E. Case
Oklahoma State University
Stillwater, Oklahoma

Alok K. Chakrabarti
New Jersey Institute of Technology
Newark, New Jersey

William L. Chapman
Hughes Aircraft Company
Tucson, Arizona

Clayton M. Christensen
Harvard Business School
Boston, Massachusetts

Kim B. Clark
Harvard Business School
Boston, Massachusetts

Peter K. Clark
University of California
Davis, California

Robert T. Clemen
Duke Universtiy
Durham, North Carolina

Adrienne J. Colella
Texas A&M University
College Station, Texas

James C. Collins
University of Virginia
Charlotteville, Virginia

Jay A. Conger
University of Southern California
Los Angeles, California

Noellette Conway-Schempf
Carnegie Mellon University
Pittsburgh, Pennsylvania

Robert G. Cooper
McMaster University
Hamilton, Ontario, Canada

Joseph J. Cordes
The George Washington University
Washington, D.C.

David M. Cottrell
Brigham Young University
Provo, Utah

Masako N. Darrough
University of California
Davis, California

Thomas H. Davenport
University of Texas at Austin
Austin, Texas

Jerry Dechert
University of Oklahoma
Stillwater, Oklahoma

L. Frank Demmler
Carnegie Mellon University
Pittsburgh, Pennsylvania

Angelo S. DeNisi
Texas A&M University
College Station, Texas

B. S. Dhillon
University of Ottawa
Ottawa, Ontario, Canada

Terry E. Dielman
Texas Christian University
Fort Worth, Texas

Robert J. Dolan
Harvard Business School
Boston, Massachusetts

Richard C. Dorf
University of California, Davis
Davis, California

Kathleen M. Eisenhardt
Stanford University
Stanford, California

Howard Eisner
The George Washington University
Washington, D.C.

Kimberly D. Elsbach
University of California
Davis, California

Wolter J. Fabrycky
Virginia Polytechnic Institute &
State University
Blacksburg, Virginia

Loren Falkenberg
University of Calgary
Calgary, Alberta, Canada

David Flath
North Carolina State University
Raleigh, North Carolina

Richard Florida
Carnegie Mellon University
Pittsburgh, Pennsylvania

Ron Franklin
University of Calgary
Calgary, Alberta, Canada

William M. Frix
John Brown University
Siloam Spring, Arkansas

Urs E. Gattiker
Aalborg University
Aalborg, Denmark

Marc S. Gerstein
Massachusetts Institute of
Technology
Cambridge, Massachusetts

Eitan Gerstner
University of California
Davis, California

James P. Gilbert
Rollins College
Winter Park, Florida

Thomas W. Gilligan
University of Southern California
Los Angeles, California

Paul A. Gompers
Harvard Business School
Boston, Massachusetts

Richard Goodman
University of California,
Los Angeles
Los Angeles, California

Paul E. Green
University of Pennsylvania
Philadelphia, Pennsylvania

Paul A. Griffin
University of California
Davis, California

Yash P. Gupta
University of Colorado
Denver, Colorado

Michael R. Hagerty
University of California
Davis, California

Robert W. Hall
Indiana University and Purdue
University
Indianapolis, Indiana

Robert B. Handfield
Michigan State University
East Lansing, Michigan

Andrew B. Hargadon
Stanford University
Stanford, California

Kathryn Rudie Harrigan
Columbia University
New York, New York

Francis T. Hartman
University of Calgary
Calgary, Alberta, Canada

David R. Henderson
Hoover Institution and Naval
Postgraduate School
Pacific Grove, California

Chris Hendrickson
Carnegie Mellon University
Pittsburgh, Pennsylvania

Amy E. Herrmann
Georgia Institute of Technology
Atlanta, Georgia

John Heskett
Illinois Institute of Technology
Chicago, Illinois

Paul Heyne
University of Washington
Seattle, Washington

Christopher T. Hill
George Mason University
Fairfax, Virginia

Robert D. Hisrich
Case Western Reserve University
Cleveland, Ohio

James E. Hodder
University of Wisconsin
Madison, Wisconsin

Clyde W. Holsapple
University of Kentucky
Lexington, Kentucky

Earl D. Honeycutt, Jr.
Old Dominion University
Norfolk, Virginia

Herman P. Hoplin
Syracuse University
Syracuse, New York

Cynthia Huffman
University of Pennsylvania
Philadelphia, Pennsylvania

Marco Iansiti
Harvard University
Boston, Massachusetts

Anil. B. Jambekar
Michigan Technological University
Houghton, Michigan

Daniel L. Jensen
The Ohio State University
Columbus, Ohio

Dana M. Johnson
Wayne State University
Detroit, Michigan

Barbara E. Kahn
University of Pennsylvania
Philadelphia, Pennsylvania

Jahangir Karimi
University of Colorado
Denver, Colorado

Waldemar Karwowski
University of Louisville
Louisville, Kentucky

Suleiman K. Kassicieh
University of New Mexico
Albuquerque, New Mexico

Ralph Katz
Northeastern University and
Massachusetts Institute
of Technology
Boston, Massachusetts

Robert Keeley
University of Colorado — Colorado
Springs
Colorado Springs, Colorado

Ralph L. Keeney
University of Southern California
San Francisco, California

Kevin Lane Keller
Dartmouth College
Hanover, New Hampshire

John Leslie King
University of California
Irvine, California

Heleen Kist
Stanford University
Stanford, California

Janice A. Klein
Massachusetts Institute
of Technology
Cambridge, Massachusetts

Joseph F. Kmec
Purdue University
West Lafayette, Indiana

Edward M. Knod, Jr.
Western Illinois University
Macomb, Illinois

Patrick N. Koch
Georgia Institute of Technology
Atlanta, Georgia

Timothy G. Kotnour
University of Central Florida
Orlando, Florida

Walter Kuemmerle
Harvard University
Boston, Massachusetts

Michael D. Kull
George Washington University
Washington, D.C.

Denis Lambert
IDX Corporation
Colorado Springs, Colorado

Peter LaPlaca
The University of Connecticut
Vernon, Connecticut

Lester Lave
Carnegie Mellon University
Pittsburgh, Pennsylvania

Chung-Shing Lee
The George Washington University
Washington, D.C.

Donald R. Lehmann
Columbia University
New York, New York

Dorothy Leonard
Harvard University
Boston, Massachusetts

Josh Lerner
Harvard Business School
Boston, Massachusetts

David M. Levy
George Mason University
Fairfax, Virginia

C. Richard Liu
Purdue University
West Lafayette, Indiana

John D. Lyon
University of California
Davis, California

Michael W. Maher
University of California
Davis, California

Larry A. Mallak
Western Michigan University
Kalamazoo, Michigan

Mike Markel
Boise State University
Boise, Idaho

Harold E. Marshall
National Institute of Standards and
   Technology
Gaithersburg, Maryland

Florence M. Mason
F. Mason and Associates
Dallas, Texas

Richard O. Mason
Southern Methodist University
Dallas, Texas

Thomas W. Mason
Rose-Hulman Institute of
   Technology
Terre Haute, Indiana

John H. Mather
Carnegie Mellon University
Pittsburgh, Pennsylvania

**Christopher M.
McDermott**
Rensselaer Polytech Institute
Troy, New York

Richard B. McKenzie
University of California
Irvine, California

Sue McNeil
Carnegie Mellon University
Pittsburgh, Pennsylvania

Ajay Menon
Colorado State University
Fort Collins, Colorado

Marc H. Meyer
Northeastern University
Watertown, Massachusetts

Ralph F. Miles, Jr.
University of Southern California
   and California Institute of
   Technology
Altadena, California

Stephen M. Millett
Battelle
Columbus, Ohio

Farrokh Mistree
Georgia Institute of Technology
Atlanta, Georgia

Robert P. Morgan
Washington University in St. Louis
St. Louis, Missouri

Timothy Morgan
The Automated Answer,
   Incorporated
San Juan Capistrano, California

Mary Munter
Dartmouth College
Hanover, New Hampshire

E. Lile Murphree, Jr.
The George Washington University
Washington, D.C.

Donald D. Myers
University of Missouri-Rolla
Rolla, Missouri

Steven Nahmias
Santa Clara University
Santa Clara, California

Prasad Naik
University of California
Davis, California

Hamid Noori
Wilfrid Laurier University
   and Hong Kong Polytechnique
   University
Waterloo, Ontario, Canada

Charles A. O'Reilly III
Stanford University
Stanford, California

Gerardo A. Okhuysen
The University of Texas
Richardson, Texas

Nick Oliver
University of Cambridge
Cambridge, England

Sharon M. Oster
Yale University School of
  Management
New Haven, Connecticut

Donald Palmer
University of California
Davis, California

David P. Paul, III
Monmouth University
Monmouth, New Jersey

E. Brian Peach
The University of West Florida
Pensacola, Florida

Terry Pearce
Leadership Communication
Novato, California
Boise, Idaho

Karol I. Pelc
Michigan Technological University
Houghton, Michigan

Jesse Peplinski
Georgia Institute of Technology
Atlanta, Georgia

Michael Pinedo
New York University
New York, New York

Jeffrey K. Pinto
The Pennsylvania State University
Erie, Pennsylvania

Ivan Png
National University of Singapore
Kent Ridge, Singapore

Jerry I. Porras
Stanford University
Stanford, California

Alan L. Porter
Georgia Institute of Technology
Atlanta, Georgia

Kevin P. Prykull
Senstar Capital Corporation
Pittsburgh, Pennsylvania

Kenneth J. Purfey
Executive Financial Consultant
Wexford, Ohio

Mansour Rahimi
University of Southern California
Los Angeles, California

Lenny Ralphs
Franklin Covey Company
Salt Lake City, Utah

Rama Ramakumar
Oklahoma State University
Stillwater, Oklahoma

Vithala R. Rao
Cornell University
Ithaca, New York

Richard Reeves
Cranfield University
Bedford, U.K.

Richard Reimer
The College of Wooster
Wooster, Ohio

Arnold Reisman
Reisman and Associates
Shaker Heights, Ohio

John L. Richards
University of Pittsburgh
Pittsburgh, Pennsylvania

Henry E. Riggs
Keck Graduate Institute
Claremont, California

L. James Ristas
Alix, Yale & Ristas, LLP
Hartford, Connecticut

Mary Lou Roberts
University of Massachusetts
Boston, Massachusetts

Ralph Roberts
The University of West Florida
Pensacola, Florida

Everett M. Rogers
University of New Mexico
Albuquerque, New Mexico

Peter S. Rose
Texas A&M University
Bryan, Texas

Richard S. Rosenbloom
Harvard Business School
Boston, Massachusetts

Frederick A. Rossini
Atlanta, Georgia

Liora Salter
York University
North York, Ontario, Canada

William Samuelson
Boston University
Boston, Massachusetts

Susan Walsh Sanderson
Rensselaer Polytechnic Institute
Troy, New York

Yolanda Sarason
University of New Mexico
Albuquerque, New Mexico

Terri A. Scandura
University of Miami
Miami, Florida

Stephen R. Schmidt
Air Academy Associates
Colorado Springs, Colorado

Richard J. Schonberger
University of Washington
Seattle, Washington

Jules J. Schwartz
Boston University
Boston, Massachusetts

Sridhar Seshadri
New York University
New York, New York

Walter D. Short
National Renewable Energy
  Laboratory
Golden, Colorado

Robert Simons
Harvard University
Boston, Massachusetts

Mark L. Sirower
New York University
New York, New York

Sim B. Sitkin
Duke University
Durham, North Carolina

Stanley Slater
University of Washington
Bothell, Washington

Hyrum W. Smith
Franklin Covey Company
Salt Lake City, Utah

Keith E. Smith
The George Washington University
Washington, D.C.

Marlene A. Smith
University of Colorado at Denver
Denver, Colorado

Robert P. Smith
University of Washington
Seattle, Washington

Timothy L. Smunt
Wake Forest University
Winston-Salem, North Carolina

Morten Steffensen
University of New Mexico
Albuquerque, New Mexico

Matthew P. Stephens
Purdue University
West Lafayette, Indiana

Karen Stephenson
University of California
Los Angeles, California

Kathleen M. Sutcliffe
University of Michigan Business
  School
Ann Arbor, Michigan

Robert I. Sutton
Stanford University
Stanford, California

Derby A. Swanson
Applied Marketing Science, Inc.
Waltham, Massachusetts

Hildy Teegen
The George Washington University
Washington, D.C.

Hans J. Thamhain
Bentley College
Waltham, Massachusetts

Robert J. Thomas
Georgetown University
Washington, D.C.

John P. Ulhøi
The Aarhus School of Business
Aarhus, Denmark

Mastafa Usumeri
Auburn University
Auburn, Alabama

Nikhil P. Varaiya
San Diego State University
San Diego, California

Donald E. Vaughn
Southern Illinois University
Carbondale, Illinois

Karl H. Vesper
University of Washington
Seattle, Washington

Mary Ann Von Glinow
Florida International University
Miami, Florida

Andrew Ward
Emory University
Atlanta, Georgia

Karl E. Weick
University of Michigan Business
  School
Ann Arbor, Michigan

Allen M. Weiss
University of Southern California
Los Angeles, California

Douglas C. West
Henley Management College
Henly-on-Thames, Oxfordshire,
  England

Steven C. Wheelwright
Harvard Business School
Boston, Massachusetts

Jerry Wind
University of Pennsylvania
Philadelphia, Pennsylvania

Russell S. Winer
University of California
Berkeley, California

Lan Xue
The George Washington University
Washington, D.C.

Xiaoping Yang
Purdue University
West Lafayette, Indiana

Kuang S. Yeh
Sun Yat-Sen University
Kaohsiung, Taiwan

Hui-Yun Yu
Sun Yat-Sen University
Kaohsiung, Taiwan

Lucio Zavanella
Universita degli Studi di Brescia
Brescia, Italy

Contents

Section I The Technology Manager and the Modern Context

Section II   Knowledge for the Technology Manager

Section III   Tools for the Technology Manager

Section IV Managing the Business Function

Section V   Strategy of the Firm

Section VII   Global Business Management

Appendixes

I

The Technology Manager and the Modern Context

1

Entrepreneurship and New Ventures

Tom Byers
Stanford University

Heleen Kist
Stanford University

Robert I. Sutton
Stanford University

Thomas W. Mason
Rose-Hulman Institute of Technology

L. Frank Demmler
Carnegie Mellon University

Paul A. Gompers
Harvard Business School

Josh Lerner
Harvard Business School

Hildy Teegen
The George Washington University

Karl H. Vesper
University of Washington

Elias G. Carayannis
The George Washington University

Jeffrey Alexander
Washington Core: International Technology and Policy Consultants

George Bugliarello
Polytechnic University

Everett M. Rogers
University of New Mexico

Morten Steffensen
University of New Mexico

1.1 Characteristics of the Entrepreneur: Social Creatures, Not Solo Heroes[1]

Tom Byers, Heleen Kist, and Robert I. Sutton

More than a dozen founders of successful companies received awards at the 1997 "Northern California Entrepreneurs of the Year" banquet. The first was given to Margot Fraser, president and founder of

[1] The authors wish to thank the Kauffman Center for Entrepreneurial Leadership for providing financial support to help us write this chapter.

Birkenstock Footprint Sandals, which is now in its third decade as the exclusive U.S. distributor of Birkenstock footwear. Fraser was lauded for her persistence and good business sense, which were portrayed as key causes of her firm's success. In an emotional acceptance speech, Fraser protested that she did not deserve the award alone; rather, it would be shared with her employees. In addition, Fraser said that her firm's success would have been impossible without so much support from her supplier in Germany and her family.

The last and most prestigious award for "Master Entrepreneur" was given to Dado Banatao, co-founder of several successful start-ups, including Chips and Technologies and S3, and current chairman of five high-technology companies. Like the other winners, Mr. Banatao was recognized for his personal qual-ities, including a competitive spirit, love of taking calculated risks, and technical knowledge. As in nearly all the other acceptance speeches, Mr. Banatao claimed that he was given too much individual credit and that he was just one of many people who enabled these start-ups to succeed. He emphasized that this success stemmed not just from him but from a group of close friends and associates who understood one another's strengths and how to combine them to launch a start-up.

This paradox, where the presenter focused on the entrepreneur's individual qualities but the winner attributed success to others, was evident in each of the dozen or so award presentations that night. It reflects widely held, but inaccurate, beliefs about what it takes to be a successful entrepreneur, and about the broader nature of the entrepreneurial process.

Numerous studies by psychologists suggest that, when a person does something, those who are watching ("observers") will say it is caused by a characteristic of the person (such as persistence or optimism), while the person being watched (the "actor") will say he or she is doing it because of something about the situation (e.g., "Mary told me to do it" or "I know this is what he expects me to do"). This pattern is called the "fundamental attribution error" because observers usually place too much credit or blame on personality characteristics and not enough on factors outside the person that drove him or her to action [Fiske and Taylor, 1991]. This attribution error occurs because observers notice the person who takes action but do not notice the other, external forces that often cause the behavior. Such perceptions are fueled by our culture, which glorifies heroes and rugged individualists, but scapegoats those who fail and overlooks the average team player. Similarly, leadership researcher James Meindl [1990] has shown that writings by business reporters and by scholars place too much emphasis on leaders' individual characteristics (especially personality) as causes of firm performance and not enough emphasis on factors outside the leader, such as other people or structural opportunities and constraints.

Popular and academic writings on **entrepreneurship** have been especially prone to romanticizing individual founders and CEOs when new firms are successful and vilifying them when such firms fail. Academic researchers, along with authors who write for broader audiences, such as journalists, venture capitalists, and entrepreneurs, have expended much time and text in a quest to predict who will succeed as an entrepreneur and who will fail. These diverse writings emphasize that personality, along with other individual characteristics, such as demographic and cultural background, will predict who will become an entrepreneur and which entrepreneurs will succeed. On the face of it, there are good reasons to give founders and CEOs the lion's share of credit and blame in young, small organizations. When the organization is small, the leader can devote more time to influencing each member, and when it is young, the force of a leader's personality is dampened less by organizational history and accepted procedures. Some evidence implies that a leader personality has a stronger impact on structure in small and young organizations than in large and old organizations [Miller and Dorge, 1986].

Traditional Views on the Characteristics of Entrepreneurs

As early as the 1950s, researchers began looking for personality factors that determine who is — and who is not — likely to become an entrepreneur. McClelland [1961] found that entrepreneurs had a higher need for achievement than nonentrepreneurs and were, contrary to popular opinion, only moderate risk takers. A great deal of research on the personality characteristics and sociocultural backgrounds of successful entrepreneurs was conducted in the 1980s and 1990s. Timmons' [1994] analysis of more than

50 studies found a consensus around six general characteristics of entrepreneurs: (1) commitment and determination, (2) leadership, (3) opportunity obsession, (4) tolerance of risk, ambiguity, and uncertainty, (5) creativity, self-reliance, and ability to adapt, and (6) motivation to excel. A related stream of research examines how individual demographic and cultural backgrounds affect the chances that a person will become an entrepreneur and be successful at the task. For example, Bianchi's [1993] review indicates these characteristics include: (1) being an offspring of self-employed parents, (2) being fired from more than one job, (3) being an immigrant or a child of immigrants, (4) previous employment in a firm with more than 100 people, (5) being the oldest child in the family, and (6) being a college graduate.

At the same time, the business press has devoted much attention to the backgrounds, personalities, and quirks of successful entrepreneurs, turning founders such as Bill Gates and Steve Jobs into names that are probably more familiar to most Americans than those of their representative in the U.S. Congress. This media coverage has fueled the belief that successful entrepreneurs are a breed apart. These press reports, and much academic literature, have reinforced the myth that the success of a new venture depends largely on the words and deeds of a brilliant and inspiring superstar, or perhaps two superstars. Although other people are involved, people in our culture seem to believe that they play far less important roles and are easily replaced.

Although this myth is propagated and accepted in many corners of America and other western nations, it is supported by weak evidence. Many scholarly studies conducted over the past three decades have found that the success of a start-up is not significantly affected by the personality or background of the founder or leader. Although the personality and sociocultural variables proposed to distinguish successful from unsuccessful entrepreneurs seem logical and often are gleaned from legends about successful start-ups, these variables explain only a small part of who will be a successful entrepreneur and which ventures will succeed.

Entrepreneurship as a Social Activity

We propose that a more accurate picture of entrepreneurship emerges when it is viewed as a social rather than an individual activity. Building a company entails hiring, organizing, and inspiring a group of people who typically need to get start-up funds from other people, buy things from other people, and ultimately, will flourish or fail together as a result of the ability to sell things to yet another group of people. The emphasis on rugged individualism is so prevalent in western culture that many of the lists of "characteristics of successful entrepreneurs" barely reflect that launching a start-up entails constant interaction with others. For example, five of the six characteristics of entrepreneurs identified in Timmons' [1994] review could just as easily be used for identifying which people are best suited for a solo activity that entails little or no interaction with others, such as racing a sailboat alone. Commitment, opportunity obsession, tolerance of risk, ambiguity and uncertainty, creativity, self-reliance, ability to adapt, and motivation to excel all seem to describe the kind of rugged individualist who struggles alone to win a contest under difficult circumstances. Only one of the categories identified by Timmons — leadership — refers to the social nature of entrepreneurship.

Recently, organizational sociologists, including Howard Aldrich [1986] and John Freeman [1996], have been developing theory and doing research on the implications of viewing entrepreneurship as a social process. Aldrich [1986] proposes that entrepreneurship is embedded in a social context and channeled and facilitated (or inhibited) by a person's position in a social network. Not only can social networks facilitate the activities of potential entrepreneurs by introducing them to opportunities they would otherwise have missed or not have pursued, but social networks are essential to providing resources to a new venture. This view suggests that success does not depend just on the initial structural position of the entrepreneur but also on the personal contacts he or she establishes and maintains throughout the process.

Freeman [1996] emphasizes that, as a result, successful entrepreneurs are especially skilled at using their time to develop relationships with people who are crucial to the success of their new venture. A new venture may start as the brainchild of one or very few people, but it takes many more people to put together the pieces of the puzzle that constitute a successful firm. The first few pieces of the puzzle usually come from and through the existing network of the entrepreneur or "insiders": friends, family, and co-founders. As the creation of the venture progresses, however, entrepreneurs need to reach beyond their individual social network and involve "outsiders" such as banks, venture capitalists, lawyers, accountants, strategic partners, customers, and industry analysts and influencers.

Key Characteristics of Entrepreneurs as Social Creatures

This view that entrepreneurship is a social process, and that entrepreneurs act largely as social rather than solo players, should not lead researchers and educators to conclude that entrepreneurs' individual knowledge and skill are irrelevant to the success of a new venture. Rather, the essentially social nature of establishing connections and fruitful relationships with both insiders and outsiders means that the study — and teaching — of entrepreneurship should focus more heavily on social behavior. This includes how people identify which relationships will be crucial to the success of their venture and how they develop and maintain the relationships that enable their firm to obtain the information, funds, legitimacy, and help needed for their firm to survive and flourish.

Our view that entrepreneurship is largely a social activity and that entrepreneurs are largely social creatures suggests not only that academics need to continue research on the social networks of new ventures but that aspiring entrepreneurs would benefit from learning the rudimentary tools of social network analysis [Burt, 1992]. These tools can be used to identify which key people and firms are connected to one another and which central players are crucial to a start-up's success. For example, as Freeman [1996] points out, although the money that venture capitalists provide to a start-up is critical to its success, the network of contacts provided by the venture capital firm may actually be more important to the success of a new start-up. Freeman quotes a promotional brochure from the Mayfield Fund (one of the most well-known venture capital firms of Silicon Valley) that states, "Mayfield has close working relationships with technology leaders, universities, other venture capital firms, financial institutions, consultants and corporations throughout the United States, as well an international network. Mayfield's contacts provide a key resource for developing the relationships critical to a growing technology-based company, including potential corporate partners, both here and abroad." A network perspective suggests that an entrepreneur may be placing his or her firm at considerable risk by accepting start-up funds from a venture capital firm that provides favorable financial terms, but lacks access to the network of contacts the firm needs to succeed.

Once an entrepreneur has determined which relationships are crucial to the success of his or her new venture, most of his or her time is spent building, negotiating, and maintaining these relationships. This implies that successful entrepreneurs are able to persuade others to enter relationships and to take actions that will help the new venture. A key part of any leader's role is persuading people to do things that they are unsure about or do not want to do at all. It is no accident that many of the most famous entrepreneurs are renowned for their interpersonal influence skills. Herb Kelleher is co-founder and CEO of Southwest Airlines, which provided shareholders with a yearly average return of 22% between 1972 and 1992, the single best performing stock during this 20-year period. Kelleher, who some consider unconventional and eccentric, has also been called America's best CEO by *Fortune* Magazine [Labich, 1994]. Our view of Kelleher's seemingly over-the-top style is that he is a master of persuasion. He charms, flatters, and jokes with employees and customers, he bullies competitors, and he battles with politicians, often all at once, to do what is best for Southwest.

Most people are not born with Kelleher's flair, but there are proven means of influence that people can learn to use. The fundamentals of interpersonal influence are studied and taught in many corners of academia, especially in psychology and management departments. Robert Cialdini's [1992] book *Influence: The Psychology of Persuasion* is complete and readable and is used in many courses on interpersonal persuasion.

In addition, entrepreneurs often influence others through negotiations. They may negotiate with venture capitalists over the terms of an investment in the firm, with employees over their salary and stock options, and with potential acquirers over a purchase price for the entire company. More research is needed on the prevalence and importance of negotiation skills as a characteristic of entrepreneurs. In addition to teaching interpersonal influence skills, perhaps entrepreneurship courses and programs should include the theory and practice of negotiation. Bazerman and Neale's [1992] *Negotiating Rationally* would be a useful text for such courses.

Finally, the social nature of entrepreneurship means that entrepreneurs spend a great deal of their time in groups. There is already compelling evidence that the characteristics of top management teams are important to the success of a new venture [Eisenhardt and Schoonhoven, 1990]. The myth that any given entrepreneur is a rugged individualist, who toils alone, can often be demolished by just asking to look at his or her calendar. Nearly always, the majority of his or her time will be spent as part of a group, at meetings with the management team, board meetings, project team meetings, and so on. As such, knowledge about leading a group, being a constructive group member, dynamics of healthy vs. destructive groups, and designing — or repairing — groups so that they function well should be a central component of entrepreneurship research and education. Robert Reich [1987], former U.S. Secretary of Labor, echoes this view in his article "Entrepreneurship Reconsidered: The Team as Hero." There are useful materials available about teams and how to manage them, such as the edited collections by Paul Goodman [1986] and Richard Hackman [1990]. Materials of this kind can help aspiring (and perhaps struggling) entrepreneurs learn a great deal about groups: how roles emerge, how their members jockey for influence, the functions and dysfunctions of group conflict, how to inspire creativity, and how to enhance group decision making, all of which can help entrepreneurs become more effective.

Conclusion

We began by asserting that individual entrepreneurs get too much credit and blame for the fate of new ventures. We also emphasized that successful entrepreneurs are those who can develop the right kinds of relationships with others inside and outside their firm. Our perspective suggests that, in trying to predict which entrepreneurs will succeed or fail, instead of turning attention to the characteristics of individual founders and CEOs, researchers and teachers would be wiser to turn attention to the *other* people with whom the entrepreneur spends time and how they respond. Our perspective also implies that the format of the "Entrepreneurs of the Year" competition described at the outset of this chapter ought to be changed. Rather than using such events to recognize individual CEOs or founders from successful start-ups, awards could be presented to recognize the intertwined group of people who made each start-up a success.

Defining Terms

Entrepreneurship: Although there is no official definition of entrepreneurship, the following one has evolved from work done at Harvard Business School and is now generally accepted by authors: "Entrepreneurship is the process of creating or seizing an opportunity and pursuing it regardless of the resources currently controlled" [Timmons, 1994].

References

Aldrich, H. and Zimmer, S. Entrepreneurship through social networks, In *The Art and Science of Entrepreneurship*, D. L. Sexton and R. W. Smilor, eds., pp. 3–23, Ballinger, Cambridge, MA, 1986.

Bazerman, M. H. and Neale, M. A. *Negotiating Rationally*, The Free Press, New York, 1992.

Bianchi, A. Who's most likely to go it alone?, *Inc.*, 15(5): 58, 1993.

Burt, R. S. *Structural Holes: The Social Structure of Competition*, Harvard University Press, Cambridge, MA, 1992.

Cialdini, R. *Influence: The Psychology of Persuasion*, William Morrow and Company, New York, 1993.

Eisenhardt, K. M. and Schoonhoven, C. B. Organizational growth: linking founding team, strategy, environment, and growth among U.S. semiconductor ventures, 1978–1988, *Admin. Sci. Q.*, 35: 504–529, 1990.

Fiske, S.T. and Taylor, S.E. *Social Cognition, 2nd ed.*, McGraw-Hill, New York, 1991.

Freeman, J. Venture Capital as an Economy of Time, Working paper, Haas Business School, University of California at Berkeley, 1996.

Goodman, P. A. et al. *Designing Effective Work Groups*, Jossey-Bass, San Francisco, 1986.

Hackman, J. R. *Groups that Work (and Those that Don't)*, Jossey-Bass, San Francisco, 1986.

Labich, K. Is Herb Kelleher America's best CEO?, *Fortune*, May 2: 44–52, 1994.

McClelland, D. A. *The Achieving Society*, Van Nostrand, Princeton, NJ, 1961.

Meindl, J. R. On leadership: an alternative to conventional wisdom, In *Research in Organizational Behavior*, B. M. Staw and L. L. Cummings, eds., pp. 159–204, JAI Press, Greenwich, CT, 1990.

Miller, D. and Dröge, C. Psychological and traditional determinants of structure, *Admin. Sci. Q.*, 31: 539–560, 1986.

Reich, R. Entrepreneurship reconsidered: the team as hero, Harv. Bus. Rev., May–June: 77–83, 1987.

Timmons, J. A. *New Venture Creation: Entrepreneurship for the 21st Century*, 4th ed., Irwin Press, Burr Ridge, IL, 1994.

1.2 Barriers to Entry

Thomas W. Mason

There is a "paradox of entrepreneurship" because of barriers to entry [Dollinger 1995]. If entry is easy, then many will enter and profits will be limited. If barriers to entry are high, it will be difficult to cover the start-up costs and profits will be limited. Yet, entrepreneurs do enter new markets and contest the established firms in old ones, especially when technology changes. The successful new venture thrives because decision makers have understood what the barriers to entry are, how they can be overcome, and how they can be used against others who would like to launch competing ventures. Bain [1956] showed that higher profits were associated with high barriers to entry due to scale economies, product differentiation, absolute cost advantages, and capital requirements. Porter [1980] discussed entry barriers as major structural components in the determining competitive strategies. In recent years, rapid change and technology have become more important for conditions of entry. However, there are still some basic rules that should be followed by those who wish to launch new ventures.

A Description of the Barriers

Economies of scale are commonly viewed as a barrier because new entrants to a market will usually have much lower sales than established producers, and average cost per unit often falls over a wide range of production. For example, the primary costs for a software product are likely to be for development. Costs of discs or CDs and printed material plus packaging and distribution are almost trivial. Therefore, the average cost per software license goes down as more and more licenses are sold. The sellers of large volumes can charge a low price and still exceed this average cost. That same price for a new entrant may be well below the average cost per unit of its product. In addition to high development costs, scale economies can arise from volume discounts for large-quantity purchases of inputs, effective use of production equipment, establishment of distribution channels, advertising, sales forces, service operations, and other parts of the production and sale of the product. Often, per-unit costs fall with higher output because the costs of one or more indivisible inputs are the same no matter how much gets produced. The new entrant must either spend large amounts of money to rapidly grow its business or endure a long period of losses until its output grows to a level that is sufficient to eliminate this disadvantage. This can certainly discourage new ventures.

Technology and rapidly changing markets can affect economies of scale either negatively or positively. The automation of processes will mean more of the total costs of production are fixed. While labor and materials costs can vary with output, the costs of robots and large computer systems do not. Moreover, the constant need to adapt to changing customer requirements leads to large expenditures on research and development. A small-volume supplier is at a greater disadvantage if these high fixed costs are spread over a smaller output. Arthur [1994] has argued that many of the new industries actually exhibit increasing returns to scale that never level off, let alone bring about any diseconomies of scale. This implies that the first firm to win the dominant position in the market for a product will just get more and more cost advantages as its output grows. This phenomenon has enormous, controversial implications for economic systems, but, to someone considering a new venture in such an industry, the lesson is clear. If you are not in the market in a big way early in the window of opportunity, you will have no chance to be successful.

While early dominance appears to be critical, technology and the growing role of outsourcing and applications of the concept of the virtual firm may reduce barriers due to scale economies. Historically, firms often had to be vertically integrated from raw materials through the stages of production and distribution to the final consumer. Today, firms are disintegrating and using outside suppliers for many nonstrategic functions. This enables a new entrant to contract out parts of the process to firms who are achieving scale economies by producing for multiple customers. For example, many relatively small producers of electronic components will contract out the manufacture of the hardware to facilities whose business with multiple such customers allows them to achieve the cost savings of large-volume operations.

Another way that technology can overcome economies of scale is when a new, more cost-effective approach is discovered. The power of personal computers has actually put large, established firms who were locked into high-cost main-frame systems at a cost disadvantage. An analogous situation has occurred in the telecommunications industry, where the huge embedded base of technology of the Regional Bell Operating Companies has enabled firms such as Metropolitan Fiber Systems (now part of Worldcom) to enter the local-access telecommunications markets and grow very rapidly because their up-to-date technology has both better quality and lower costs.

Product differentiation is another barrier to entry. Cumulative effects of advertising and other brand promotion give firms an advantage over new competitors. In consumer products, the brand name alone for Coca Cola is worth billions, and any entrant to the soft drink industry is at a serious disadvantage all over the world. In industrial markets, it is generally less expensive to keep an existing customer than it is to acquire a new one. This is especially true when there are **switching costs** for customers as well as issues of inertia and loyalty to overcome. Changing to a new computer system implies that a business must endure high training costs and inevitable delays and other problems. The advantages with a new system supplier must be great to get customers to switch. Service reputation and availability may also give existing firms advantages over potential entrants.

To the extent that the established brand name is associated with standardized, one- product-fits-all characteristics, a new firm may actually overcome product differentiation and other barriers by finding a niche that is not being served. Years ago, product differentiation barriers were thought to be very high in brewing. In the past decade, microbrewers have built successful businesses by offering something different and turning the lack of a national brand name into part of their image. Small-scale manufacturers of personal computers have been able to survive in markets dominated by IBM, Compaq, and other giants who have large advertising budgets and significant-scale economies. Rapidly changing technology can make customers dependent upon a personalized approach based upon knowledge of their particular needs [Ramstad, 1997]. As industries mature and products become more stable and better understood, such niches may not exist. However, substantial business can be won before that happens and sometimes small niche players can grow to be quite large, the way Dell and Gateway have in the PC business.

Entrepreneurs with new innovations can have some confidence that new products can enter a profitable niche in a market, but keeping the foothold may be difficult. New ventures that do not erect barriers or create switching costs for their newly won customers will have their markets quickly captured by other entrants. For example, Compression Engineering, a small, rapidly growing provider of rapid prototyping services has done well, but the firm will have to commit to a continuous and expensive plan of equipment and software upgrades if it is to keep its leading edge reputation. Such product differentiation is often the competitive advantage that brings sustained profitability.

Absolute cost advantages can also be a barrier to entry for a new venture or a source of protection for an established firm. While scale economy barriers hurt a new entrant until adequate volumes of sales occur, absolute cost advantages mean lower costs for the established firm even when the entrant produces the same amount. These barriers may be due to legal/institutional factors discussed below, but they can also exist because of control of the best source of raw materials or possession of a superior location or as many other sources as there are factors affecting production and distribution. Advantages also arise because economies of scope produce cost savings when two or more products can be produced and distributed together. An entrant with a single product has a disadvantage.

In technology-based business, knowledge may be the critical source of value. Intellectual "stars" with genius in particular technologies can provide real advantages to their firms. Moreover, the learning curve concept says that, as cumulative production goes up over time, the cost per unit of production goes down. Early participants in a market have employees who know more about the product and processes, and this cost advantage persists even if a later entrant matches their output level in any time period. The only solution for such advantages may be for an entrant to "leapfrog" to a new technology that makes that of existing firms obsolete.

Legal and institutional barriers can cause absolute cost advantages or more effective entry deterrents. Obviously, tariffs, quotas, and other measures to protect domestic producers can make international markets inaccessible. Government licensing requirements can completely block entry, and getting such licenses or permits can represent significant barriers, especially in international markets. Governments also enforce, or do not enforce, intellectual property protection such as patents, copyrights, trademarks, and trade secrets. Government regulatory procedures often have the side effect of protection of existing firms. For example, the lucrative pharmaceutical and medical device market of the United States cannot be entered by new ventures without several years and many millions of dollars spent on the approval processes required by the U.S. Food and Drug Administration. Firms with established products may be able to make continuous, marginal product improvements without the same magnitude of regulatory approval. Other regulations such as pollution control may inhibit entry by requiring more stringent and expensive emission limitations on new firms than those imposed on existing operations.

In addition to formal, legal restraints, there can be institutional barriers. Customs may prevent particular cultures from embracing a new product or even change in general. National loyalties may cause people to avoid imports, even if there are no legal trade barriers. Moreover, technical standards may differ, which make a potential entrant's new product completely incompatible with local systems. This is a key issue in various parts of the telecommunications industry.

If there were no other entry barriers, perhaps **capital requirements** would not be so great a problem. However, there is much uncertainty about overcoming barriers and establishing a business that will yield an adequate, sustained level of profitability. As a practical matter, financial markets are far from omniscient, and the needs to acquire facilities and equipment, do the research and development, get the necessary government approvals, hire and train the work force, establish distribution channels, and launch the advertising campaign mean entry will not occur unless substantial money is available. An established firm can innovate using less-expensive internally generated funds. Even if external debt or equity is required, there is likely to be less perceived risk, so funds at reasonable costs are more available, compared to those for a new entrant.

A Note on Barriers to Exit

Large capital requirements and expenditures to develop a new product are sunk costs, but responsibilities to provide returns on the investments that have been made can affect decisions to get out. Commitments to employees, customers, suppliers, and others can make it very difficult to leave a market once a business is launched. In some countries, there may even be serious legal restrictions that are intended to keep firms from abandoning employees and their communities. On the other hand, some businesses may be easy to exit. For example, the core technology or even the whole business may be an excellent prospect for acquisition by another firm. Such ease of exit will affect the prospects for entry. Sources of capital will be more readily available if there is a good scenario for recovering the funds. When plans are made to launch a venture, they should include the possibilities for exit, including the chances of recovery for investment in facilities and research and development. At the same time, firms should understand that long-term abnormal profitability is unlikely if there is easy entry and exit.

Some Practical Steps for Considering Entry Barriers

Based upon the description of the barriers to entry, there are some practical rules for new technological ventures. While the following are not adequate to cover all of the entry barrier issues, they are a good starting point.

1. Use available technical data and industry information to develop a good understanding of the costs, cost structure, and capital requirements of your own product and those of potential competitors. Existing industry prices may change rapidly in the face of entry, and it is important to know how the new venture's costs compare to the costs of established firms.
2. Understand the product and the stage of its life cycle to determine if a winner has been established or whether there is an opportunity to occupy a dominant position. Even a superior technology or product may not overcome the scale economy and absolute cost advantages of established firms.
3. Look for new technologies that overcome scale and absolute cost advantages.
4. Identify a niche or niches in the market within which competitive advantage and profitability can be established and sustained. Barriers caused by product differentiation can be overcome and then used to advantage with this type of approach.
5. Look for ways to form alliances, partnerships, and contractual relationships that take advantage of scale economy, absolute cost, and product differentiation advantages of other firms that are not direct competitors.
6. Plan for rapid expansion, including early entry into international business. If advantages of lower costs and more rapid product development accrue to the dominant player, one needs to prepare to take that market position.
7. Beware of legal and institutional barriers including intellectual property and standards issues, and be ready to in turn use such barriers to protect a foothold in the market.
8. Understand the barriers to exit, as well as entry, and include possibilities for exit in the development of the plan for a new venture. Investors, employees, and the entrepreneur are all well served by knowing what the exit possibilities might be, and it may facilitate entry.

An Example Scenario

The rules above can be illustrated by a hypothetical firm with a new technology. There is great excitement when a new technology works. However, the wise entrepreneur quickly moves to determine how much it is going to cost to manufacture, distribute, and support.

Are there expensive, specialized pieces of equipment to produce the new hardware? Will new distribution channels have to be established? Will the software involved require extensive development efforts, and will it be difficult to support? Suppose the answers to these questions imply significant economies of scale. This may not be bad news if the firm has a reasonable opportunity to dominate the market and have the scale advantages on its side. If the new technology is going to provide an order of magnitude improvement in the customers' cost, then growth may come very quickly, or there may be a niche in the market that is uniquely served by the new product. If this niche is either large enough or growing rapidly, the new venture can again achieve rapid growth. With this rapid growth, the firm can show the high returns needed to attract the money to overcome capital requirements barriers. This is especially true if there are likely opportunities to exit by selling the business and its technology to an established firm. Once the capital is obtained and the market is entered, there are ways to enhance growth. For example, the venture might form alliances with large, established firms who have appropriate distribution channels but need the new technology.

Such alliances or partnerships can also be used effectively to rapidly enter international markets for further advances in the volume of sales.

At the same time, the firm should be aware of patents, copyrights, and institutional barriers that can protect it from followers. These are unlikely to completely inhibit entry by themselves; patents, for example, are only as good as your ability to defend them. However, they can be potent additions to other barriers.

This example shows how barriers to entry can be positive sources of protection for a business based on new technology. Firms such as Microsoft, Netscape, Intel, MCI, and many more all have parts of their histories that illustrate both overcoming barriers to entry and using them as protection from subsequent entrants. Overcoming and using the paradox caused by entry barriers is essential to sustained profits from innovations.

Conclusion

Barriers to entry can appear formidable. However, there have been many cases of entrants establishing profitable positions in spite of them. Rapidly changing technology may present new difficulties for entrepreneurial ventures, but such changes also create new opportunities not only to capture business in existing markets, but to create new markets. The keys are to understand how the new technological venture provides value to its customers, and how it can continue to be the best source of that value.

Defining Terms

Absolute cost advantages: A situation where an existing firm has lower costs per unit than an entrant, even if their output levels are the same.

Capital requirements: This refers to the entry barrier created by the need for large amounts of money to successfully enter a market.

Economies of scale: Per unit cost decreases that occur with increasing size.

Legal and institutional barriers: These are laws, customs, and standards that make it difficult for a new firm to enter a market.

Product differentiation: The creation of perceived differences in products with similar functions through advertising and other brand promotion techniques.

Switching costs: Costs, such as training or retooling, that are incurred by a customer who changes suppliers.

References

Arthur, W. B. *Increasing Returns and Path Dependence in the Economy,* The University of Michigan Press, Ann Arbor, MI, 1994.

Bain, J. *Barriers to New Competition: Their Character and Consequences in Manufacturing Industries,* Harvard University Press, Cambridge, MA, 1956.

Dollinger, M. *Entrepreneurship: Strategies and Resources,* Irwin, Burr Ridge, IL, 1995.

Porter, M. E. *Competitive Strategy: Techniques for Analyzing Industries and Competitors,* Free Press, New York, 1980.

Ramstad, E. Defying the odds: despite giant rivals, many tiny PC makers are still doing well, *Wall Street J.,* January 8: 1, 1997.

1.3 Raising Money from Private Investors

L. Frank Demmler

It has been estimated that private investors invest more than ten times as much money in small businesses than institutional venture capital, which translates into about $50 billion per year. In order to raise private capital successfully, entrepreneurs need to understand why private investors invest in early-stage companies.

The Investor Perspective

A sophisticated investor looks at a business as a "black box" in terms of its attractiveness as an investment. He asks, "Can this business transform a little bit of my money into a large amount?" Private investors typically are seeking returns greater than 20% and often greater than 30% per year.

Investors will not invest in a business in which their perception of the risk is "too high". Place yourself in the investor's shoes and ask the following questions:

- Is the proposed business clearly described? Would your mother understand it?
- Are there customers who **care** about the product?
- Are there enough customers to make a viable business?
- Is the business model consistent with the investor's expectations?
- Does the management team have the right experience and skill set?
- Are the next critical steps identified and appropriate action plans developed?

How do you get an investor to meet with you? The answer lies primarily with your business plan. Within 20 seconds of looking at the plan, the experienced investor has probably made one of two decisions — "no" or "maybe". Your goal is to get a "maybe". The speed that an investor makes this initial decision may not seem fair to the entrepreneur, but that's the real world of raising money.

To simplify these issues, the risk factors can be clustered into two areas from the investor's perspective:

- The business — does it make sense?
- The people — can they succeed and provide my desired rate of return?

Identifying Potential Investors

You can now begin the process of identifying potential investors (Fig. 1.1). Let's start with the **people who know you**, including friends, family, neighbors, and relatives, i.e., individuals on your rolodex, your Christmas card list, etc. These are the people who know *you* and are willing to invest in *you*, perhaps even regardless of the business. Statistically, over 90% of all money invested in small business comes from this group.

Unfortunately, there are some drawbacks. If things don't go well, there may be more than just financial penalties to face. It can be awkward sitting across the Thanksgiving table from Uncle George if you lost his retirement nest egg.

Another group of potential investors are **people who are familiar with your industry** and can react quickly to your business proposal. They will understand the market dynamics and quickly decide whether they agree with the concept. Included in this group are suppliers, customers, channel members (your representatives and/or distributors), and successful entrepreneurs within each of these organizations. If they are impressed with the business concept, then the perceived risk is limited to their evaluation of you.

Figures 1.1 and 1.2 provide a framework for identifying potential investors, categorizing them, and prioritizing your efforts.

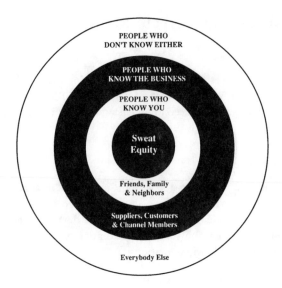

FIGURE 1.1

Risk Reduction Techniques

A variety of risk-reduction techniques exists that an entrepreneur may employ to lower an **investor's perception of risk.**

Demonstrate Your Ability to Perform against Plan

With advance planning, it is possible to use time as a strategic weapon in fundraising. For example, if you cite certain accomplishments that should occur while you are still in the fundraising process (a customer order, for example) and they happen, that can build significant credibility.

Unfortunately, the converse is also true.

Stage the Investment and Tie It to Milestones

If the investment is divided into two or more increments, funding can be tied to the successful accomplishment of major milestones.

Build In "Go/No-Go" Decision Points

In addition to milestones, "go/no-go" decision points can be identified in a particular project. With the identification of these, exposure of the investor's capital can be prudently staged. For example, FDA approval represents such a go/no-go decision point in the medical products area.

		THE BUSINESS	
		KNOW	DON'T KNOW
THE PEOPLE INVOLVED IN THE BUSINESS	KNOW	• quick decisions • great anchors • halo effect • can add value	• quick decision • lower economic cost • highest psychological cost
	DON'T KNOW	• moderate decision time • halo effect • higher return expectations • can add value	• long decision time • highest return requirement • costly & time intensive because of due diligence • largest pool of capital

FIGURE 1.2

Use Credible Intermediaries

Another risk-reduction technique involves using intermediaries who have independent credibility.

If you use **somebody to introduce you to the investor**, the quality of the person who provides the introduction says much about you. For example, the CEO of a successful company in which your targeted investor has placed money would be a very powerful intermediary for these purposes.

Similarly, using **professionals** (often called investment bankers) who have successfully raised money for similar businesses in the past can be an effective strategy. They already have a network of investors who will often invest in particular situations based primarily upon the recommendation of the financial professional.

Before moving on, there are two important points related to the use of professional intermediaries.

- You must ask the question, "May I please have the names and contact information for three of my peers for whom you have provided these services?"
- Increasingly, successful professional fundraisers use an upfront fee to screen prospective clients. Therefore, if the references check out well, you should not reject an upfront fee out of hand.

Halo Effect of Advisors

Just as intermediaries use their credibility to attract investors, those people that you can identify as being involved in your company as mentors, advisors, or board members can also be very effective in lowering risk perceptions.

There is one caution, however. If these individuals are not also investing in the company, a "red flag" will be raised with many investors and this may work to your disadvantage.

Third-Party Confirmation or Endorsement

Quite often an investor is going to be skeptical about the claims an entrepreneur makes. Therefore, independent sources, such as a trade journal article or an independent market study, that confirm critical elements of a business proposal can be a very effective tool in building credibility and investor interest.

Relevant Analogs

Comparing your business to one that is a known success can be a very effective risk-reduction technique and can help you describe your business in easily understood terms. For example, if your business is a restaurant chain that is anticipating franchising, describing it as an Olive Garden or Chi-Chi's with a menu of some particular theme will help the reader immediately understand the business concept.

Deal Structures

Deal structures and investor return expectations span a continuum from those who want to lock in higher-than-market rates with as little risk as possible to those willing to risk everything on the opportunity for a "home run".

Deal structures tend to reflect the risk profile of the investor. The following are some deal structures that investors commonly use:

- Notes with higher-than-market interest rates — Some relatively conservative investors have money and want to receive a higher return than what is available from high interest rate bonds. Therefore, they look for opportunities to lock in higher interest rates, often in the range of 18 to 30% per year. They often require personal guarantees.
- Notes with warrants — These are notes like those just described, but with one added feature — warrants to buy a specific amount of stock at a specific price. The warrants let the investor participate in the "upside" of equity ownership if the company is extremely successful.

- Convertible notes — In this case, the full principal amount of the note can be converted into common stock. Often, interest only is paid monthly or quarterly, with little, if any, principal pay down.
- Preferred stock — Preferred stock can be converted to common stock under certain conditions and has certain preference items, such as liquidation preference if the company is sold for some amount less than had been hoped for. This means that the investor gets his money first, before management. This form of stock is relatively complex and usually requires significant legal work and related fees. For these reasons, preferred stock is rarely used in smaller, private investor deals, although venture capitalists almost always invest in this form of equity.
- Common stock — As its name implies, this is a "plain vanilla" form of security that represents equity ownership but does not contain any special rights or protections.

Beyond these deal structures, many other creative relationships can be negotiated. For example, deals based upon royalties on sales are common.

In conclusion, deal structures are not set in stone. While it is likely that the experienced investor will have a limited number of deal "templates" that he will consider, deal structures are only limited by the imaginations of the involved parties. If you and an investor want to do a deal, it is likely you will be able to structure something that satisfies both sides.

Alternatives

Alternatives should always be considered. Raising money is rarely a linear path, so building in flexibility as well as developing viable options will work to the entrepreneur's advantage. It is not at all uncommon for final deals to look like a smorgasbord with "a little bit of this and a little bit of that". Here are some of the questions that you should ask yourself:

- Can I make meaningful progress with **less money**?
- Can I **stage the investment**? Can I close on a smaller amount of investment now and with the expectation that successful accomplishment of milestones will be sufficient to attract the next round of investment, hopefully at a higher valuation?
- Can I get development **funds from a customer or supplier** who will ultimately benefit from my success? Will my suppliers and customers consider special terms and conditions for me? Might my supplier give me goods on consignment? Might my customer place a deposit at the time of an order or at least agree to progress billings?
- Might money be available from an **economic development agency**? Typically, such agencies focus on jobs with loan availability being roughly $15,000 to $30,000 per job to be created or retained within 3 years.
- Are **grants** a possibility through state programs such as the Ben Franklin Partnership Program in Pennsylvania or any of the Federal Small Business Innovation Research (SBIR) Programs?
- Is there a **competitor** who might be better able to exploit the commercial potential of what I'm doing?
- Is there a **boot strap** alternative? Can I sell consulting or services to generate cash flow and create industry visibility and credibility?

Time to Raise Capital

The time it takes to raise capital can be excruciatingly long. As a rule of thumb, entrepreneurs should plan on any where from 6 to 12 months in order to be successful in raising a significant amount of capital. The primary steps are as follows:

1. Develop a business plan — 2 to 9 months
2. Initiate discussions with potential investors — 2 to 4 months
3. Respond to investor requests for additional information — 0 to 3 months

4. Bide time while potential investors do their due diligence — 0 to 6 months
5. Receive a term sheet and negotiate a deal — 1 to 6 months
6. Prepare legal work, negotiate the fine points, and close the deal — 1 to 3 months

It is unlikely that any company would be at the high end for all factors. As a practical matter, for most deals, if they haven't closed within a year, something else is going on and it is unlikely that the deal will ever close.

Conclusion

The primary barrier to raising money is the investor's perception of risk. It is the entrepreneur's job to understand that and prepare himself and his company appropriately.

Raising money is not an easy undertaking. Trying to raise money will take you away from building your business on a day-to-day basis. You must keep in mind that it takes time to raise capital, probably longer than you ever imagined, and there is no guarantee that you will be successful in your efforts. You will have to run as many parallel paths as possible until a deal begins take shape. During that period you will likely be in "boot strap" mode. You must make a conscious decision that the opportunity cost and the potential downside are worth risking.

Defining Terms

Bootstrap: The steps necessary to grow a company without outside investment capital.

Common stock: Stock that represents the class of owners who have claims on the assets and earnings of a corporation once all vendors', debt holders', and preferred stockholders' claims have been satisfied.

Convertible: Stock that is convertible into common stock. Usually used to describe preferred stock that is convertible into common stock.

Cumulative preferred stock: Preferred stock whose dividends, if not paid annually, accumulate as a commitment that must be paid before dividends to common shareholders can be declared.

Dilution: A potential reduction in earnings per share when stock options are exercised or other securities are converted into common stock.

Due diligence: The evaluation process an investor engages in before deciding to invest in a company. Typically, this process includes a thorough analysis of the company's historical financial statements and future financial projections; reference checks of vendors, customers, and personal references; discussions with outside experts familiar with the industry; and any other steps the investor feels are necessary to become comfortable with the investment opportunity.

Equity: A claim to assets. Usually refers to the stockholders or owners of a company. Also a term used to indicate the type of investment an investor is prepared to make. Compare and contrast equity to debt: someone who lends a company money.

Exit strategy: The strategy that defines how an investor will extract his money once invested in a company. This could be structured as a formal legal agreement to buy or sell stock at a predetermined price, a loan with specific payment terms, or an informal understanding between the entrepreneur and the investor.

Options: The right to buy something during a specified period at a specified price. Frequently refers to company stock that can be purchased or earned by management or investors.

Post-financing valuation: The value of a company **after** an investment is made.

Preferred stock: Capital stock with a claim to income or assets after creditors and bondholders but before common stockholders.

Pre-financing valuation: The value of a company **before** an investment is made.

Term sheet: A document that describes the business agreement between an entrepreneur and an investor. The document is not legally binding but is frequently used by attorneys to create a legally binding agreement.

Further Information

Entreprenet	www.enterprise.org/enet
Money Hunter	www.money hunter.com
SCOR-NET	www.scor-net.com
Inc.Online	www.inc.com
Finance Hub.Com	www.financehub.com
Venture Capital MetaResource	www.nvst.com
EntreWorld	www.entreworld.org

1.4 Venture Capital

Paul A. Gompers and Josh Lerner

Over the past 2 decades, there has been a tremendous boom in the venture capital industry. (**Venture capital** can be defined as equity or equity-linked investments in young, privately held companies, where the investor is a financial intermediary and is typically active as a director, advisor, or even manager of the firm.) The pool of venture partnerships has grown tenfold, from under $4 billion in 1978 to about $40 billion at the end of 1995. Venture capital's recent growth has outstripped that of almost every class of financial product.

While the growth in venture capital has been striking, the potential for future development is even more impressive: the private equity pool today remains relatively small. For every $1 of private equity in the portfolio of U.S. institutional investors, there are about $30 of publicly traded equities.

Both the demand for and supply of such capital are likely to expand. On the demand side, many studies suggest that privately held firms continue to face substantial problems in accessing the financing necessary to undertake profitable projects. These problems stem from the complex nature of many start-up firms and the difficulties most investors have in assessing these complexities. The value of due diligence and monitoring will be substantial. In addition, the specialized skills of the venture capitalists in terms of industry knowledge and experience will be important in reducing asymmetric information. The supply of private equity is also likely to continue growing. Within the past 2 years, numerous pension funds have invested in private equity for the first time. Many experienced investors have also decided to increase their allocations to venture capital and buyout funds. These increased allocations will take a number of years to implement. These patterns are even more dramatic overseas. While fundraising by other types of private equity funds — especially those active in buyouts — has climbed in recent years, venture capital activity in Europe and Asia has been many times less than in the United States.[2]

At the same time, the venture capital industry has been quite turbulent, passing through many peaks and valleys. Investments in U.S. venture funds between 1980 and 1995 would have yielded returns that were below those from investments in most public equity markets.[3] Due to the illiquidity and risk of private equity, one would expect instead a *higher* return. These poor returns largely stemmed from funds begun in the 1980s, when a large number of inexperienced venture capitalists raised first funds and established organizations aggressively expanded. Many of the new funds could not find satisfactory investments, while rapid growth created turmoil at some established organizations. The early 1990s saw far fewer funds raised and rising returns. With the recent growth in fundraising, it is unclear whether the high returns seen in the recent years can be sustained. This cycle of growth and disillusionment has created much instability in the industry.

---

[2] For a statistical overview of international venture capital activity, see European Venture Capital Association, *1996 EVCA Yearbook*, Zaventum, Belgium, European Venture Capital Association, 1996; and Asian Venture Capital Journal, *Venture Capital in Asia: 1995/96 Edition*, Hong Kong, Asian Venture Capital Journal, 1995.

[3] Venture Economics, *Investment Benchmarks Report: Venture Capital*, New York, Venture Economics, 1996, page 281.

We will review the distinct phases associated with venture capital investment: fundraising, investing, and exiting. The bibliography lists works on each topic.

Fundraising

Private equity funds often have complex features, and the legal issues involved are frequently arcane. The fund's structure, however, has a profound effect on the behavior of venture investors.

An example may help to illustrate this point. Almost all venture and buyout funds are designed to be "self-liquidating", i.e., to dissolve after 10 or 12 years. The need to terminate each fund imposes a healthy discipline, forcing private equity investors to take the necessary-but-painful step of terminating under-performing firms in their portfolios. (These firms are sometimes referred to as the "living dead" or "zombies".) However, the pressure to raise an additional fund can sometimes have less-pleasant consequences. Young private equity organizations frequently rush young firms to the public marketplace in order to demonstrate a successful track record, even if the companies are not ready to go public. This behavior, known as "grandstanding", can have a harmful effect on the long-run prospects of the firms dragged prematurely into the public markets.[4]

There are a wide array of actors in private equity fundraising. Investors — whether pension funds, individuals, or endowments — each have their own motivations and concerns. These investors frequently hire intermediaries, often referred to as "gatekeepers". Sometimes gatekeepers play a consultative role, recommending attractive funds to their clients. In other cases, they organize "funds-of-funds" of their own. Specialized intermediaries concentrate on particular niches of the private equity industry, such as buying and selling interests in limited partnerships from institutional investors. In addition, venture and buyout organizations are increasingly hiring placement agents who facilitate the fundraising process.

Investing in a private equity fund is, in some respects, a "leap of faith" for institutional investors. Most pension funds and endowments typically have very small staffs. At the largest organizations, a dozen professionals may be responsible for investing several billion dollars each year. Meanwhile, venture capital funds undertake investments that are in risky new firms pursuing new technologies that are difficult for outsiders to assess and whose true value may not be known for many years.

Many of the features of venture capital funds can be understood as responses to this uncertain environment, rife with many information gaps. For instance, the **carried interest** — the substantial share of profits that are allocated to the private equity investors — helps address these information asymmetries by ensuring that all parties gain if the investment does well. Similarly, pension funds hire **gatekeepers** to ensure that only sophisticated private equity funds with well-defined objectives get funded with their capital. Gatekeepers are specialized intermediaries who review and select venture capital funds for institutional investors.

At the same time, other features of venture capital funds can be seen as attempts to *transfer* wealth between parties rather than as efforts to increase the size of the overall amount of profits generated by these investments. An example was the drive by many venture capital funds in the mid-1980s — a period when the demand for their services was very strong — to change the timing of their compensation. Prior to this point, venture capital funds had typically disbursed all the proceeds from their first few successful investments to their investors until the investors had received their original invested capital back. The venture capitalists would then begin receiving a share of the subsequent investments that they exited. Consider a fund that had raised capital of $50 million, whose first three successful investments yielded $25 million each. Under the traditional arrangement, the proceeds from the first two offerings would have gone entirely to the institutional investors in their fund. The venture capitalists would have only begun receiving a share of the proceeds at the time that they exited the third investment.

In the mid-1980s, venture capitalists began demanding — and receiving — the right to start sharing in even the first successfully exited investments. The primary effect of this change was that the venture

---

[4]Paul A. Gompers, Grandstanding in the venture capital industry, *Journal of Financial Economics*, 43 (September 1996) pages 133–156.

capitalists began receiving more compensation early in their funds' lives. Put another way, the net present value of their compensation package increased considerably. It is not surprising that, as the inflow into venture capital weakened in the late 1980s, institutional investors began demanding that venture capitalists return to the previous approach of deferring compensation. This twin tension — between behavior that increases the size of the "pie" and actions that simply change the relative sizes of the slices — runs through the fundraising process.

Investing

The interactions between venture capitalists and the entrepreneurs that they finance are at the heart of the venture cycle. These complex interactions can be understood through the use of two frameworks.

Entrepreneurs rarely have the capital to see their ideas to fruition and must rely on outside financiers. Meanwhile, those who control capital — for instance, pension fund trustees and university overseers — are unlikely to have the time or expertise to invest directly in young or restructuring firms. Some entrepreneurs might turn to other financing sources, such as bank loans or the issuance of public stock, to meet their needs. However, because of four key factors, some of the most potentially profitable and exciting firms are likely to be unable to access these financing sources.

The first factor, uncertainty, is a measure of the array of potential outcomes for a company or project. The wider the dispersion of potential outcomes, the greater the uncertainty. By their very nature, young companies are associated with significant levels of uncertainty. Uncertainty surrounds whether the research program or new product will succeed. The response of firm's rivals may also be uncertain. High uncertainty means that investors and entrepreneurs cannot confidently predict what the company will look like in the future.

Uncertainty affects the willingness of investors to contribute capital, the desire of suppliers to extend credit, and the decisions of a firm's managers. If managers are adverse to taking risks, it may be difficult to induce them to make the right decisions. Conversely, if entrepreneurs are overoptimistic, then investors want to curtail various actions. Uncertainty also affects the timing of investment. Should an investor contribute all the capital at the beginning, or should he stage the investment through time? Investors need to know how information-gathering activities can address these concerns and when they should be undertaken.

The second factor, asymmetric information (or information disparities), is distinct from uncertainty. Because of his day-to-day involvement with the firm, an entrepreneur knows more about his company's prospects than investors, suppliers, or strategic partners. Various problems develop in settings where asymmetric information is prevalent. For instance, the entrepreneur may take detrimental actions that investors cannot observe, perhaps undertaking a riskier strategy than initially suggested or not working as hard as the investor expects. The entrepreneur might also invest in projects that build up his reputation at the investors' expense.

Asymmetric information can also lead to selection problems. The entrepreneur may exploit the fact that he knows more about the project or his abilities than investors do. Investors may find it difficult to distinguish between competent entrepreneurs and incompetent ones. Without the ability to screen out unacceptable projects and entrepreneurs, investors are unable to make efficient and appropriate decisions.

The third factor affecting a firm's corporate and financial strategy is the nature of its assets. Firms that have tangible assets, e.g., machines, buildings, land, or physical inventory, may find financing easier to obtain or may be able to obtain more favorable terms. The ability to abscond with the firm's source of value is more difficult when it relies on physical assets. When the most important assets are intangible, such as trade secrets, raising outside financing from traditional sources may be more challenging.

Market conditions also play a key role in determining the difficulty of financing firms. Both the capital and product markets may be subject to substantial variations. The supply of capital from public investors and the price at which this capital is available may vary dramatically. These changes may be a response to regulatory edicts or shifts in investors' perceptions of future profitability. Similarly, the nature of product markets may vary dramatically, whether due to shifts in the intensity of competition with rivals or in the nature of the customers. If there is exceedingly intense competition or a great deal of uncertainty about the size of the potential market, firms may find it very difficult to raise capital from traditional sources.

Venture capitalists have at least six classes of mechanisms at their disposal to address these changing factors. Careful crafting of financial contracts and firm strategies can alleviate many potential roadblocks.

The first set relates to the financing of firms. First, *from whom* a firm acquires capital is not always obvious. Each source — private equity investors, corporations, and the public markets — may be appropriate for a firm at different points in its life. Furthermore, as the firm changes over time, the appropriate source of financing may change. Because the firm may be very different in the future, investors and entrepreneurs need to be able to anticipate change.

Second, the *form* of financing plays a critical role in reducing potential conflicts. Financing provided by private equity investors can be simple debt or equity or involve hybrid securities such as convertible preferred equity or convertible debt. These financial structures can potentially screen out overconfident or underqualified entrepreneurs. The structure and timing of financing can also reduce the impact of uncertainty on future returns.

A third element is the *division* of the profits between the entrepreneurs and the investors. The most obvious aspect is the pricing of the investment: for a given cash infusion, how much of the company does the private equity investor receive? Compensation contracts can be written that align the incentives of managers and investors. Incentive compensation can be in the form of cash, stock, or options. Performance can be tied to several measures and compared to various benchmarks. Carefully designed incentive schemes can avert destructive behavior.

The second set of activities of venture capitalists relates to the strategic control of the firm. *Monitoring* is a critical role. Both parties must ensure that proper actions are taken and that appropriate progress is being made. Critical control mechanisms, e.g., active and qualified boards of directors, the right to approve important decisions, and the ability to fire and recruit key managers, need to be effectively allocated in any relationship between an entrepreneur and investors.

Private equity investors can also encourage firms to *alter the nature of their assets* and thus obtain greater financial flexibility. Patents, trademarks, and copyrights are all mechanisms to protect firm assets. Understanding the advantages and limitations of various forms of intellectual property protection and coordinating financial and intellectual property strategies are essential to ensuring a young firm's growth. Investors can also shape firms' assets by encouraging certain strategic decisions, such as the creation of a set of "locked-in" users who rely on the firm's products.

Evaluation is the final, and perhaps most critical, element of the relationship between entrepreneurs and venture capitalists. The ultimate control mechanism exercised by the private equity investors is to refuse to provide more financing to a firm. In many cases, the investor can — through direct or indirect actions — even block the firm's ability to raise capital from other sources.

Exiting

The final aspect of the venture cycle is the exiting of investments. Successful exits are critical to insuring attractive returns for investors and, in turn, to raising additional capital. However, private equity investors' concerns about exiting investments — and their behavior during the exiting process itself — can sometimes lead to severe problems for entrepreneurs.

Perhaps the clearest illustration of the relationship between the private and public markets was seen during the 1980s and early 1990s. In the early 1980s, many European nations developed secondary markets. These sought to combine a hospitable environment for small firms (e.g., they allowed firms to be listed even if they did not have an extended record of profitability) with tight regulatory safeguards. These enabled the pioneering European private equity funds to exit their investments. A wave of fund-raising by these and other private equity organizations followed in the mid-1980s. After the 1987 market crash, initial public offering activity in Europe and the United States dried up. However, while the U.S. market recovered in the early 1990s, the European market remained depressed. Consequently, European private equity investors were unable to exit investments by going public. They were required either to continue to hold the firms or to sell them to larger corporations, often at unattractive valuations. While U.S. private equity investors — pointing to their successful exits — were able to raise substantial amounts of new capital, European private equity fundraising during this period remained depressed. The influence of exits on the rest of the private equity cycle suggests that this is a critical issue for funds and their investors.

The ability to exit also affects the types of deals that are financed. In the European example, the reduced opportunities for initial public offerings has led many private equity firms to finance later-stage companies in industries that are amenable to corporate mergers or acquisitions. These tend to be firms in non-high-technology industries. The amount of early-stage financing in Europe is substantially lower, even as a percentage of the total venture capital pool, than it is in the United States.

The exiting of private equity investments also has important implications for entrepreneurs. As discussed above, the typical private equity fund is liquidated after 1 decade (though extensions of a few years may be possible). Thus, if a venture capitalist cannot foresee how a company will be mature enough to take public or to sell at the end of a decade, he is unlikely to invest in the firm. If it was equally easy to exit investments of all types at all times, this might not be a problem. However, interest in certain technologies by public investors seems to be subject to wide swings. For instance, in recent years "hot issue markets" have appeared and disappeared for computer hardware, biotechnology, multimedia, and Internet companies. Concerns about the ability to exit investments may have led to too many private equity transactions being undertaken in these hot industries. At the same time, insufficient capital may have been devoted to industries not in the public limelight.

Concerns about exiting may also adversely affect firms once they are financed by private equity investors. Less-scrupulous investors may occasionally encourage companies in their portfolio to undertake actions that boost the probability of a successful initial public offering, even if they jeopardize the firm's long-run health, e.g., increasing earnings by cutting back on vital research spending. In addition, many private equity investors appear to exploit their inside knowledge when dissolving their stakes in investments. While this may be in the best interests of the limited and general partners of the fund, it may have harmful effects on the firm and the other shareholders.

The exiting of private equity investments involves a diverse range of actors. Private equity investors exit most successful investments through taking them public.[5] A wide variety of actors are involved in the initial public offering. In addition to the private equity investors, these include the investment bank that underwrites the offering, the institutional and individual investors who are allotted the shares (and frequently sell them immediately after the offering), and the parties who end up holding the shares.

Few private equity investments are liquidated at the time of the initial public offering. Instead, private equity investors typically dissolve their positions by distributing the shares to the investors in their funds. These distributions usually take place 1 to 2 years after the offering. A variety of other intermediaries are involved in these transactions, such as distribution managers who evaluate and liquidate distributed securities for institutional investors.

Many of the features of the exiting of private equity investments can be understood as responses to many uncertainties in this environment. An example is the **lock up** provisions that prohibit corporate insiders and private equity investors from selling at the time of the offering. This helps avoid situations where the officers and directors exploit their inside knowledge that a newly listed company is overvalued by rapidly liquidating their positions.

At the same time, other features of the exiting process can be seen as attempts to transfer wealth between parties. An example may be the instances where private equity funds distribute shares to their investors that drop in price immediately after the distribution. Even if the price at which the investors

---

[5] A Venture Economics study finds that a $1 investment in a firm that goes public provides an average cash return of $1.95 in excess of the initial investment, with an average holding period of 4.2 years. The next best alternative, an investment in an acquired firm, yields a cash return of only 40¢ over a 3.7-year mean holding period. See Venture Economics, *Exiting Venture Capital Investments*, Wellesley, Venture Economics, 1988.

ultimately sell the shares is far less, the private equity investors use the share price *before* the distribution to calculate their fund's rate of return and to determine when they can begin profit sharing.

Defining Terms

Carried interest: The substantial share of profits, often 20%, that are allocated to the venture capitalists.
Gatekeepers: Specialized intermediaries who review and select venture capital funds for institutional investors.
Hot issue markets: Periods when investor demand for initial public offerings in particular industries, or offerings in general, appears to be particularly high.
Lock-up: Agreements between investment bankers, corporate insiders, and venture capitalists that prohibit the insiders and investors from selling at the time of or for a set period after a public stock offering.
Venture capital: Equity or equity-linked investments in young, privately held companies, where the investor is a financial intermediary and is typically active as a director, advisor, or even manager of the firm

References

Bartlett, J. W. *Venture Capital: Law, Business Strategies, and Investment Planning*, John Wiley & Sons, New York, 1988, and supplements (best overview of legal issues).
Fenn, G., Liang, N., and Prowse, S. T*he Economics of the Private Equity Industry*, Federal Reserve Board, Washington, DC, 1995 (best general overview of the industry and its workings).
VentureOne, National Venture Capital Association Annual Report, VentureOne, San Francisco, various years (for recent data on venture capital activity).

For More Information

For detailed bibliographies of academic, practitioner, legal, and case study materials about each portion of the venture investment process, the interested reader is referred to http://www.people.hbs.edu/jlerner/vcpe.html.

For detailed information on current trends and activity in the venture capital industry, the interested reader should consult Venture Economics' *Venture Capital Journal* or Asset Alternatives' *Private Equity Analyst*.

1.5   Valuation of Technology and New Ventures

Hildy Teegen

Firms considering new investments in technology ventures face the important challenge of determining the investment level appropriate for the venture. This determination is made by converting expectations of the technology's viability and potential commercial success into an R&D budget or reservation price for technology acquisition. Although objective measures of value exist, subjective measures of value based upon the perceptions of evaluators are a necessary component of the overall valuation of technology and new technology ventures. This is due largely to the intangible nature of technology and technology ventures and the different perspectives of providers/sellers and users/purchasers of the technology. Many factors are considered when establishing the value of technology and a new venture — financial, strategic, and environmental factors among them. Once the firm has determined the various dimensions of value for the technology or venture, these value sources are converted into an overall valuation through applying various valuation methods. Finally, the valuation is codified into a "score", which may reflect a price (when addressing external markets) or a budget (when addressing internal markets) for the technology or new venture.

The Technology Manager and the Modern Context

Dimensions of Value

When considering the value of a technology venture, it is important to distinguish between internal market considerations and external market considerations. Although new technology ventures are often entered into for the purpose of developing technology that will be deployed within the investing firm, it is increasingly recognized that the technology firm's performance can depend on the external commercialization of technology in markets that are peripheral, lateral, or additional to the company's own production [Adoutte, 1989]. Given shortened product life cycles in technology, the ability to quickly recoup R&D investments through early selling to other firms becomes critical [Neil, 1988].

Internal markets are within the firm and concern the decision-making criteria that are relevant there. **External markets** are those outside of the firm and include firms and investors independent of the technology venture firm as well as potential or actual alliance partners to the technology venture firm. In practice, the valuation considerations facing internal and external markets largely overlap; however, some considerations are unique to a given market.

Common to both internal and external market valuation considerations are three areas of prospective performance of the technology venture: technological, financial, and market performance. These three areas of performance largely define the ultimate value of a technology venture.

Technological performance corresponds to the viability of the proposed technology. Technology will be viable if it is relatively error-free in application, is demonstrably superior in application to competing (existing and in-development) technologies, and is compatible with existing technology platforms and standards. Technology ventures that have met the conditions of technological performance will be valued more highly than those that have not.

Financial performance evaluates the return on investment in the technology as viewed by enhanced incomes from applying the technology (or licensing/selling it to other firms) and/or reduced costs of production from applying the technology. For internal markets, the basis for the return calculation is the capital invested in the R&D effort; for external markets, the basis for the return calculation is the price paid for the technology. Financial performance entails also a firm's cost of capital, the risk associated with a particular technology venture investment, inflation, and tax implications of the investment decision.

No technology is created independent of market acceptability. Even in the case of technology intended exclusively for internal markets, end users of products produced with, or including, the technology must be considered in gauging the value of the technology. **Market performance** of a technology is captured by measures such as projected market share/installed user base and customer adoption rates. These measures will be impacted by factors such as the compatibility of the technology with existing platforms and standards, the availability of service and parts [Lloyd, 1986], stage of life cycle of the product and technology category, and perceived advantage over competing alternatives.

External market members evaluate technology ventures of independent firms based upon certain specific factors in addition to the technology, financial, and performance dimensions discussed above. Alternative technologies sourced from other firms or internally are a key consideration for external markets. Where alternatives are abundant and/or superior, external markets will degrade the value of this technology venture. The analog to this effect concerns the number and bargaining power of potential purchasers of the technology (or potential investors or potential allies). Where greater potential purchasers/investors/allies exist for a technology, the producer of the technology has greater bargaining power, and thus the value of the technology venture is enhanced [Emerson, 1962].

When the technology venture in question is more innovative in nature (and thus less "proven"), the risk for the external market's acquisition of the technology (or shares of the technology venture or partnering with the venture as an ally) is greater, thus degrading the value of the technology venture.

Control over the technology and its application is another consideration for external markets. The more the technology producer retains control over the technology and its commercialization (through restrictions on second-generation R&D, territory restrictions for commercialization, usage performance quotas, required investments in commercialization, etc.), the less valuable the technology becomes to external markets.

Internal market evaluators of technology ventures implicitly must take into account the various considerations of the external markets where the technology could potentially be commercialized. As mentioned previously, these external market sales can represent significant additional value for the technology venture. However, internal markets have unique considerations for valuing technology ventures. These considerations concern the competitive strategy of the firm.

Technology ventures that represent investments in areas that are central or core to the firm's strategy will be evaluated more favorably by internal markets. **Core areas** are those aspects of the firm that have deliberately been selected as the drivers of the firm's competitive ability in the market. These are the areas (product lines, technology families, etc.) that the firm has specialized in and that provide the firm's greatest contribution to overall competitive performance. Core activities are cornerstones of the firm's stated mission. Activities in the firm that are not the principal drivers of competitive performance are called peripheral activities.

By reflecting an investment in a core area, a technology venture will more easily demonstrate a valid use of scarce firm resources. The opportunity costs associated with the deployment of capital, human resources, productive capacity, and other firm resources will be lower for core areas than for peripheral activities. Investments with low associated opportunity costs will be valued more highly by the internal market decision maker.

Technology ventures that represent the best prospects for the firm in the future will be the most highly valued by internal markets. Thus, where a technology venture represents an early stage in a stream of R&D that can produce successive generations of technology, it will be more highly valued than those ventures geared toward later-stage innovations, as gains in early-stage innovations can be leveraged by the firm in later-stage technology development.

The riskiness of a technology venture further impacts its value on internal markets. Riskier ventures will be valued less than more certain ventures. Risk can be defined in terms of initial investment required; larger initial investments are more risky as more investment can be lost should the venture fail. The time frame for development of the technology is directly related to risk — longer ventures expose the firm to greater contingencies, which could adversely impact the venture. The greater the likelihood of a successful (technological, financial, and market acceptable) innovation, the lower the risk of the venture as its outcome is projected to be more favorable. Where many competing firms are active in similar technology ventures, the risk of the venture is increased as the probability of alternative technologies reaching the market increases, degrading the value of the technology venture. In terms of competitive response, however, internal markets must also consider the competitive ramifications of *not* pursuing a technology venture that may prove critical to their ability to compete for the future. Finally, those technology ventures dedicated to innovations that have wide market appeal (as indicated by a large existing user base, for instance) or that have potential for transfer among subsidiaries or partnerships of the investing firm are less risky in that the payoff to successful innovations will be greater; these ventures will be valued more highly by internal markets.

The final unique consideration for internal markets when valuing a technology venture concern the **"make vs. buy vs. ally"** decision. Firms seeking technological innovations will decide on investments in technology ventures within the firm (make) only when such an investment is valued more highly than acquiring comparable technology through purchases in the market (buy) and more highly than accessing comparable technology through an alliance partner such as a joint venture partner (ally). The relative values of each option consider simultaneously the various factors impacting value discussed in the preceding sections.

International Considerations for Valuing Technology Ventures

When firms are active in global markets, additional considerations become important for valuing technology. The economic development level of the target market will determine how appropriate a given technological innovation is for the factors of production prevalent in the market. For instance, technologies that favor capital intensity will be relatively less valuable in the developing nations that have labor as the more abundant (and thus less expensive) factor of production. Nations that do not provide adequate protection of intellectual property rights will be markets for which a given technology will be less valuable — its uniqueness does not provide it with a strong competitive advantage in a market where it can be readily copied and commercialized by unauthorized parties.

Valuation Methods

The most basic method of valuing a technology venture is the identification of *comparable technology ventures* in the market; the stated value of those ventures is then used to proxy the value of the technology venture in question. Thus, where a firm is considering the investment in a routing system software package, the value of existing routing system software packages available in the market would be used (and perhaps adjusted) to estimate the value of this technology venture. The availability of good proxies for novel technologies is a particularly daunting limitation to this valuation method, however.

Financial valuation methods based on an incomes approach are performed by projecting cash flows associated with the technology venture into the future. These estimated cash flows are then discounted to their present value by considering the timing of the cash flows and an appropriate discount rate. The discount rate considers such factors as the riskiness of the investment, inflation, cost of capital, and currency (monetary unit) of the transaction [Davidson, 1996]. Tax considerations are included in the estimation of cash flows over time. For publicly held firms, the investment in the technology venture will impact the share price in the firm. Typically, market prices of shares take into account the current assets of the firm as well as a growth component [Rotman, 1996]. Where the technology venture allows for enhanced growth opportunities (as perceived by the market), the share price of the firm will increase upon investing in the venture.

The most accepted method of accounting for the value of technology ventures to date relies on the **options pricing model**. Options pricing concerns the valuation of the right or option to purchase or sell an asset at some date in the future. Technology investments are seen as options in that investments today allow the firm to "go to the next stage" in an innovation stream [Williams, 1994]. The options pricing model typically produces a higher value for a technology investment than would a simple discounted cash flow method [Kahn, 1992]. It considers the current value of the underlying asset(s)/technology base, the volatility of the technology's price in the market, the exercise price of the option (the cost of continuing to innovate), the time to maturity (when the investment will "pay off"), and the interest rate (cost of capital to the investing firm).

The best valuation strategy considers multiple valuation methods for triangulation of results — the various estimates of value then produce estimate boundaries, and the decision maker can select a value within these boundaries. Additionally, all valuations should incorporate sensitivity analyses that adjust the valuation estimates for changes in the underlying assumptions. These sensitivity analyses serve to further define the boundaries on the value estimates and increase the utility of the final value estimate produced.

Value Codification

Once the technology venture has been valued, this value must be translated into a number, a budget, or a price. Technologies that will be sold in external markets to independent firms or to alliance partners will have clearly codified values representing the transfer of monetary or other resources that will accompany the transfer of the technology produced by the innovating firm.

Most typically, technology is transferred under a license that stipulates the usage conditions for the purchased technology. **License agreements** are granted in exchange for payment of one or a combination of two types: **lump sum payments** and **royalties** over sales. Lump sum payments are most usual upon the initial transfer of the technology and put more risk on the purchaser of the technology. Royalties are tied to performance (e.g., sales) or to time and thus put more risk on the seller of the technology. Since sellers and buyers of technology have different incentives to favor one or the other of these payment structures, a combination typically results in practice.

A derivative of these financial compensation systems are technology transfers that are compensated through acquisition of other technology from the purchasing firm (**reciprocal technology transfers**) or through products produced using the technology that can then be sold to third parties (**buy back arrangements**). Both such compensation types tend to more closely bind the incentives of the buying and selling firms and thus are used to facilitate smooth technology transfers.

Conclusion

The valuation of technology and new technology ventures requires the consideration of multiple factors that impact a technology's value. Various methodologies exist for incorporating the impacts of these factors into a value score for a technology venture. The technology score is then codified into a price for the venture, which can be expressed in terms of financial or technology/product exchanges.

Defining Terms

Buy back arrangements: A method for nonfinancially compensating the seller of technology where products produced with the transferred technology are received in exchange for the technology. The "seller" of the technology is then responsible for the disposal of the products received.

Core areas/activities: Aspects of the firm that drive its competitive ability. Opposite of peripheral areas/activities.

External markets: The group of decision makers located outside the firm, including firms and investors that are independent of the technology venture firm and can be potential or actual alliance partners.

Financial performance: Measure of the return on investment in the technology.

Internal markets: The group of decision makers located within the firm considering the investment in the technology venture.

License agreement: Contract stipulating usage of acquired technology.

Lump sum payments: A method for financially compensating the seller of technology where funds are provided as a single sum. The seller is guaranteed this specified sum of money in exchange for the technology.

Make vs. buy vs. ally: Decision by the firm as to where technology should best be accessed: within the firm, purchased from an independent firm, or accessed from an alliance partner, respectively.

Market performance: Measure of the acceptance by the market for the technology.

Options pricing model: Financial valuation method that accounts for the right to participate in future innovations when valuing technology.

Reciprocal technology transfers: A method for nonfinancially compensating the seller of technology where other technology is received in exchange for the technology "sold".

Royalty fees: A method for financially compensating the seller of technology where funds are provided based upon a percentage of sales or over time. The seller is not guaranteed a specific sum of money for the technology.

Technological performance: Measure of the viability of the technology.

References

Adoutte, R. High technology as a commercial asset, *Int. J. Technol. Mgt.*, 4(4): 397–406, 1989.

Davidson, A. S. A potpourri of valuation issues, *CA Mag.*, 129(7): 35–37, 1996.

Emerson, R. Power-dependence relations, *Am. Sociolog. Rev.*, 27: 31–41. 1962.

Kahn, S. Using option theory to value assets, *Glob. Fin.* 6(4): 82–85, 1992.

Neil, D. J. The valuation of intellectual property. *Int. J. Technol. Mgt.*, 3(1–2): 31–42, 1988.

Rotman, D. Do you know what your technology is worth?, *Chem. Week.*, 158(14): 50, 1996.

Williams, E. Modelling Merck: a conversation with CFO Judy Lewent, *CFO*, 10(8): 44–50, 1994.

Further Information

For a complete examination of the usage of options pricing and valuation, see Faulkner, T., Applying options thinking to R&D valuation, *Res. Technol. Mgt.,* 39(3): 50–56, 1996.

For insight into measuring the intellectual capital of a firm, see Stewart, T. A., Measuring company IQ, *Fortune,* 129(2): 24, 1994.

For further discussion on setting royalty rates, see Udell, G. G. and Potter, T. A., Pricing new technology, *Res.-Tech. Mgt.,* 32(4): 14–18, 1989.

1.6   Internal Ventures[6]

Karl H. Vesper

Corporate venturing is an alluring concept. It suggests the possibility of combining the established reputation, massive talent, and powerful resources of a major established company with the innovativeness, flexibility, aggressiveness, and frugality of a start-up to move the big company smartly ahead in sales and profits. Venturing connotes doing new things, departing from what is customary to pursue opportunities that might otherwise (1) not be detected, (2) not be appreciated even though detected, or (3) not be effectively exploited even though appreciated.

Internal ventures occasionally happen and may take different forms. They may be more or less radical, may be initiated at different organizational levels, and may emerge at different stages in a given project. Historically, departures have at times had major positive impact on the development of companies, such as 3M, IBM, and Tandy.

It would be nice to be able to institutionalize such occurrences. The argument advanced to support concepts of "corporate entrepreneurship", "internal entrepreneurship", or "intrapreneurship", is that it should help make corporations more innovative and more responsive to new market opportunities, while motivating employees through a greater sense of ownership in their intracorporate ventures to perform better [Pinchot, 1985].

This idea is neither new nor untried. The extent to which it should be possible for a venture to be created, develop, and operate inside a large organization, as opposed to simply in the open economy outside a corporation, is a question that has repeatedly been subject to academic discussion and corporate experimentation over the years (see Vesper and Holmdahl [1963]). Descriptions of experiments occasionally crop up [Bailey, 1987]. Missing, however, are reports of strings of significant successes. Instead, the corporate venture programs that are heralded in the press as were those of AT&T [Cohen et al., 1985] and Eastman Kodak [Chandler, 1986] over 10 years ago tend later to disappear without fanfare.

The most applauded examples of corporate ventures, however, have with regrettably few exceptions not been produced by systematic programs deliberately set up and managed in mature firms. In an extensive and insightful recent book on corporate ventures, scholars Block and MacMillan [1993] assert that "venturing has proved successful for many organizations," citing as among the more notable 3M, Raychem, and Woolworth. These companies certainly have ventured, but it is not clear that they did so as a result of mounting programs to become more entrepreneurial than they naturally were. 3M and Raychem have certainly been innovative companies, but as a matter of top management style and gradual evolution from the beginning, rather than from later plugging in programs to add emulation of independent venture creation internally.

It is easy to identify new products, such as nylon, transistors, and innumerable process improvements that were created in normal corporate development processes. There have also been, and doubtless continue to be, some corporate experiments designed and adopted for "cultivating" ventures. However,

---

[6] This chapter is adapted from a discussion entitled "Corporate Ventures", which appears in *New Venture Strategies* by Karl H. Vesper, Prentice Hall, Englewood Cliffs, NJ, 1990. With permission.

it is difficult to find many examples of strong performance by the latter. There have been reports of singular successes, such as Tandy's microcomputer, which was initiated by an unauthorized person, driven upstream against resistance by the company's traditions, and had a major positive impact on the corporation, but cascades of such examples were not systematically produced. So the quest for a "formula" that will produce ventures within corporations analogously to the way independent ventures arise in the open economy remains open.

Venture Forms

Some forms venture initiatives can take include the following:

1. *Strategically bold moves* intended to take the company out of its customary lines of business through internal initiatives rather than acquisitions. Established management provides the venturing initiative. These may include not only internal start-ups but also joint ventures where two companies form and co-own a third and strategic partnering where a large company and a smaller one team up to exploit a market or technology.

2. *Venture division operation* wherein the corporation sets up an organizational subunit whose job is to encourage, spawn, nurture, and protect innovative projects that do not fit elsewhere in the organization. They may take the form of separate "skunk works" where ventures can develop offbeat ideas, and/or "internal venture capital", offices that provide funding or grants for employees to pursue projects outside their accustomed jobs. The sponsoring entity may simply be one executive or a separate group or venture division.

3. *Task forces* would likely be part of a venture division's activities, but they may also be set up independently within the company or in a laboratory. Fluke Manufacturing, for instance, sometimes groups employees made redundant by project terminations into "Phoenix teams" with the mission essentially of coming up with new ventures to pursue, which they then propose to management. Xerox formed a group with the mission of inventing the "office of the future" in its Palo Alto Research Center, and from that effort came the microcomputer, the mouse, icons, word processing software, and the laser printer. Up to that point it might, in hindsight, be viewed as the most spectacularly successful of all industrial innovation projects. Unfortunately, the company effectively sank the project at that point, leaving most of the opportunity to be taken by others to create the microcomputer industry.[7]

4. *External venturing* may also be undertaken through a venture office or venture division. In this activity, the venture is started somewhere outside the corporation, and the venture office becomes involved in it by either helping to fund it or acquiring it. The funding may take the form of research contracts to university researchers, for instance, or venture capital investment, possibly in collaboration with other venture capital firms.

5. *Employee innovation programs,* often operated in conjunction with outside consultants, are activities undertaken to stimulate more venturing-type activities by more employees. Training sequences, incentive schemes, recognition awards, retreats, task forces, newsletters, and various kinds of hoopla may be undertaken to stimulate more innovative thinking by more employees.

6. *Management transformation* is likely to include employee programs, as above, but goes further into how the company is run from top to bottom and from environmental scanning for opportunities, both outside and inside, strategic definition, organizational structure, communications systems, reward systems, control systems, leadership, and cultural change. Cornwall and Perlman in a textbook nicely illustrated with vignettes, not including any spectacular innovations, state

---

[7] Two books nicely highlight the history of this spectacular venture story: *Fumbling the Future,* by two journalists, Douglas K. Smith and Robert C. Alexander, Morrow, New York, 1988, and *Prophets in the Dark,* by the CEO of Xerox, David T. Kearns, and David A. Nadler, Harper, New York, 1992.

that "organizational entrepreneurship does not require a different set of strategies, but rather a different approach to corporate strategies" that in turn affects virtually everything that goes on in the corporation.

7. *Initiative from below,* where an employee goes beyond normal expectations of a job to pioneer a product or service, often beginning sub rosa and "bootlegging" resources informally that were assigned to other purposes, is another form of venturing. If successful, the project may eventually pick up a sponsor and move forward. If it does not pick up a sponsor, the employee will probably have to choose between dropping the venture and pursuing it outside the company.

Exceptional autonomy plus a dedicated champion seem to be necessary if new and different projects or ventures are to succeed. If the venture is to become substantial relative to the company's other activities, then higher executive support or sponsorship also seems necessary. Championing behavior continues to be studied, but reports of it tend to lack examples of significant successes [Ohe et al., 1993]. The same can be said about studies of venture teams [Churchill and Muzyka, 1995] and attitudes of managers toward venturing [Weaver and Henderson, 1995]. Such studies also tend to lack statement of what they mean by entrepreneurship and venturing, whether one of the seven types suggested above or some other, and how corporate venturing is different from traditional R&D or occasionally unusual innovation sequences.

Venturing Variables

A host of variables can enter into the corporate venturing process, including the following:

1. Happenstance. Spontaneous ventures occur from time to time in the normal course of business without precalculation. It is tempting, but not easy, to seek ways of prompting more of them.
2. Organizational setting and rules. What elements of the organizational setting foster ventures? Corporate culture and history are claimed to be important.
3. Objectives of the venture activity. They may include only "bottom line" results of the venturing or they may include the overall impact of the venturing on the organization. How accomplishments are to be measured in the latter case can be a problem.
4. Organizational level of the prime mover. Particularly in small companies, it usually seems to be the CEO. In large corporations, it may be subordinate employees [Schon, 1963]. Sometimes teams are involved. Sometimes ventures shift from one champion to another.
5. Other players and their positions. Who else besides the prime mover is crucial to the process? Sponsors who provide support and protection, and gatekeepers who provide needed information, as well as team members have been seen as important [Roberts and Frohman, 1972]. Strategy and tactics used by the players. Some books have offered rules for both corporations [Brandt, 1986] and individuals. Pinchot [1985], for instance, has suggested "ten commandments" for entrepreneurs operating inside corporations.
6. Missions assigned. Sometimes venture task forces are assigned to find as a venture anything that will build on the strengths of the company and reach specified levels of sales and ROI within a specified time period. At Xerox the charter was to "invent the office of the future", a goal that appears in hindsight to have been incredibly powerful. Not to be overlooked, however, were also the choices made in terms of people chosen for the task, where they were located, and how they were supported.
7. Seeking major departures vs. small innovations. Lee Iacocca's turnaround of Chrysler was hailed by some as a major entrepreneurial feat, while Kanter [1983] characterized as entrepreneurial much smaller developments inside General Motors and other companies.
8. Risk and commitment levels are dimensions of these major and minor ventures that have been given almost no treatment in the literature. It has concentrated mainly on organizational processes.
9. Budgeting approach. Venture capitalists, for instance, tend to use achievement of milestones as a basis for capital infusions, whereas corporations traditionally have used calendar-based budgeting. Internal ventures may call for the former rather than the latter.

10. Venture outcome and its impact have also been given little analysis. The Strategic Planning Institute years ago published some data on variables correlating with corporate venture performance. For instance, Hobson and Morrison [1983] reported on correlates with success, as were mentioned in a study by Miller et al. [1987] and in another by MacMillan and Day [1987]. However these all were limited to statistics based on ratios due to the nature of the PIMS database and hence are abstract from other dimensions of the venture processes and limited.

Beyond these are two more general perspectives from which corporate venturing can be considered: the venture participant and the corporation. These two will mainly be used here, touching on the others as they may be involved.

Individual Perspective

The prime mover of a venture may be either the chief executive officer of the corporation or someone subordinate. Examples of the former would include the following:

- William Hewlett of the Hewlett-Packard company, who initiated the development of the company's pocket calculator because he wanted one, against recommendations of the company's market researchers, who predicted few would be sold.
- Kazuo Inamori, founder and CEO of Kyocera, a leading Japanese electronics company, who directed his researchers to undertake the development of artificial emeralds after he learned from a jeweler that such gems were in short and shrinking supply. When the established marketing channels would not accept his new gems, he had his company enter the retail jewelry business by establishing a store in Beverly Hills.

A venture leader can also occur lower in the organization and may be either appointed or a volunteer, or even a renegade. An appointed leader may be similar to simply a project manager except for other aspects that set venturing apart from typical projects. Kidder [1985], for instance, describes a project team, working on a project that it is not supposed to, inside a computer company where the project leader can be seen as a venturer with his own sub rosa enterprise. Or, it can involve an individual employee pursuing a venture solo, as illustrated by Don French at Tandy Corporation, who developed a prototype for the TRS-80 after management told him the company had no interest in computer products and he should not pursue them on work time [Swaine, 1981].

Aspects that may distinguish a venture from a conventional research or engineering project can include the following:

- A venture connotes a more complete business, including such elements as not only technical development but also profit responsibility, overall business planning, follow-through to market, production, selling, and servicing. In contrast, a project typically is more limited to particular functional specialties and lacks profit-and-loss responsibility.
- A venture usually involves a new direction for the firm or a more radical change in the way it does business, whereas a project connotes a more limited innovation, usually in line with the accustomed strategy and direction of the company.
- A venture needs greater autonomy than a project because it fits less well with the company's customary procedures. This autonomy may come about by "hiding" the venture, by separating it geographically, or by housing it in a special organizational unit capable of shielding it from the normal company activities.

Corporate vs. Independent Venturing

Autonomy can in some ways make the venture similar to an independent startup. But, from the prospective internal entrepreneur's point of view, some **major contrasts** between venturing inside a corporation and venturing independently may be the following:

Internal Entrepreneurship	Independent Entrepreneurship
The person in charge of the venture still reports to a boss who has power of dismissal and can overrule decisions	The one in charge of the venture has no superior officer, although subject to desires of customers, financers, and possibly directors and colleagues
Financial risk is all carried by the parent company	Financial risk is shared by the entrepreneur in charge, other shareholders, suppliers, and lenders
Financial capacity is determined by parent company; outside sources may not be used without parental consent	Financial capacity determined by venture itself; any sources can be used
Administrative formalities are decreed by the parent in such areas as accounting, personnel, contracts, public relations, advertising, and customer servicing	Administrative formalities are at management's discretion and very minimal
Success will not make a great amount of money for those in the venture; can mean promotion	The entrepreneur and founding investors may make millions; can mean financial independence
Failure will not put managers of the venture out of jobs; they can return to the parent	Failure will mean that everyone in the venture, including managers, will have to find new employers
Having managed an internal venture is likely to enhance career advancement if the venture succeeds and to retard career advancement if it fails	Having managed an independent company is likely to enhance career advancement whether or not the venture succeeds

Corporate Perspective

Almost all the discussion in the academic literature of corporate venturing is from the perspective of the corporation and corporate management, as opposed to that of the subordinate who carries out the venture. Some of it has been academic description of alternative venture forms. Schollhammer [1982], for instance, suggested as alternative modes administrative, opportunistic, initiative, acquisitive, and incubative. Much of it extols venturing as a kind of corporate elixer for rejuvenating the company and shifting its growth trajectory upward. The argument is generally based on one or more anecdotes of dramatic success in which a major product innovation occurred. The inference drawn is that, if it happened in such an instance, then it can be made to happen elsewhere by calculated management action.

Circumstances under which corporations typically tend to go into one form of venturing or another are when (1) they are afraid they will not grow enough from "business as usual", (2) they have been losing creative employees and promising ventures to outside backers, (3) profits are up and cash is available, (4) takeover is a threat and management wants to "show some action", and (5) management develops enthusiasm because the concept of venturing is new or it has read and heard "wonder stories" of venture exploits at companies such as 3M.

The argument against corporate cultivation of ventures is that corporations that have tried cultivating them have often written such efforts off as expensive and unsuccessful [Mueller, 1973]. To this the proponents respond that the efforts either were not properly applied or were not pursued long enough to succeed.

Fast [1979] observed that corporate venture divisions typically were set up when old products were maturing and profits and cash availability were high. Later, when cash became short, profits turned down, new products began to catch on, or some combination of such events occurred, management would decide the new ventures were costing a lot of money and could be dispensed with. The cycle of this phase-out might run for 3 or 4 years, which is probably less than the time needed for the ventures to mature and begin to prove their worth. Hence, they would be written off and venturing would appear in hindsight to have been a waste of time and money.

Certainly, goals of corporate venturing such as more effective innovation, faster and more economical paths from conception to market, and heightened employee productivity continue to be worth pursuing. Occasionally, companies mount programs that work at least for a while. For example, Signode Corporation launched a program in which teams of successful managers were invited to volunteer to seek, for a specified time period and within a specified budget, new product directions that would become consequentially contributive to the company within 5 years. This worked, but only until the company was acquired by another and the project was curtailed. A similar pattern of success and shutdown was reported for a program at Omark corporation, which, in contrast to Signode, mounted a program of training and incentives for large numbers of employees to seek new ideas.

Because some of these venture program initiatives have shown promise, however briefly, and because of the importance of their goals, more such experimentation, coupled with tracking and reporting of results seems called for. In each company it will almost certainly require strong initiative from the top and flexibility for accommodating the company's individuality to find the combination of bold and subtle elements that will work for it in particular both economically and reliably.

References

Bailey, J. E. Developing corporate entrepreneurs — three Australian case studies, In *Frontiers of Entrepreneurship Research*, Churchill et al., eds., p. 553, 1987.

Block, Z. and MacMillan, I. *Corporate Venturing*, Harvard Business School Press, Boston, 1993.

Brandt, S. C. *Entrepreneurship in Established Companies*, Dow Jones Irwin, Homewood, IL, 1986.

Chandler, C. Eastman Kodak opens window of opportunity, *J. Bus. Strat.*, 7(1): 5, 1986.

Churchill, N. C. and Muzyka, D. F. High performance entrepreneurial teams, In *Frontiers of Entrepreneurship Research*, p. 503, 1995.

Cohen, D. J., Graham, R. J., and Shils, E. La Brea Tar Pits revisited: corporate entrepreneurship and the AT&T dinosaur, In *Frontiers of Entrepreneurship Research*, Hornaday, J. A., Shils, E. B., and Timmons, J. A., eds., p. 621, 1985.

Fast, N. A visit to the new venture graveyard, *Res. Mgt.*, 22(2): 18–22, 1979.

Hobson, E. L. and Morrison, R. M. How do corporate start-up ventures fare?, In *Frontiers of Entrepreneurship Research*, Hornaday, J. X., Timmons, J. A., and Vesper, K. H., eds., Babson Center for Entrepreneurial Studies, Wellesley, MA, p. 390, 1983.

Kanter, R. M. *The Change Masters*, Simon & Schuster, New York, 1983.

Kidder, T. *Soul of a New Machine*, Little Brown, Boston, 1981; see also Computer engineers memorialized seek new challenges, *Wall Street J.*, September 20, p. 1, 1985.

MacMillan, I. C. and Day, D. L. Corporate ventures into industrial markets: dynamics of aggressive entry, *J. Bus. Ventur.*, 2(1): 29, 1987.

Miller, A., Wilson, R., and Gartner, W. B. Entry strategies of corporate ventures in emerging and mature industries, In *Frontiers of Entrepreneurship Research*, Churchill et al., eds., p. 481, 1987.

Mueller, R. K. Venture vogue: boneyard or bonanza?, *Columbia J. World Bus.*, Spring: 78, 1973.

Ohe, T. et al. Championing behavior, In *Frontiers of Entrepreneurship Research*, Churchill, N.C. et al., eds., Wellesley, p. 427, 1993.

Pinchot, G. *Intrapreneuring*, Harper & Row, New York, 1985.

Roberts, E. D. and Frohman, A. B. Internal entrepreneurship: strategy for growth, *Bus. Q.*, Spring: 71–78, 1972.

Scholhammer, H. Internal corporate entrepreneurship, In *Encyclopedia of Entrepreneurship*, Kent, C. A., Sexton, D. L., and Vesper, K. H., eds., Prentice-Hall, Englewood Cliffs, NJ, p. 209, 1982.

Schon, D. A. Champions for radical new inventions, *Harv. Bus. Rev.*, March–April: 84, 1963.

Swaine, M. How the TRS-80 was born, *Infoworld*, August 31, p. 40, 1981.

Weaver, R. Y. and Henderson, S. Entrepreneurship in organizations — the Avuiall Studies, a model for corporate growth, In *Frontiers of Entrepreneurship Research*, Bygrave, W. et al., eds., p. 652, 1995.

Vesper, K. H. and Holmdahl, T. G. Venture management in more innovative large industrial firms, *Res. Mgt.*, May 1986; Orenbuch, D. I. The new entrepreneur or venture manager, *Chem. Technol.*, October, p. 584, 1971; Hanan, M. *Venture Management*, McGraw-Hill, New York, 1976; Burgelman, R. A. Designs for corporate entrepreneurship in established firms, *Calif. Mgt. Rev.*, 26(3): 154, 1984; Costello, D. R. *New Venture Analysis*, Dow Jones-Irwin, Homewood, IL, 1985.

1.7   Technology-Driven Strategic Alliances: Tools for Learning and Knowledge Exchange in a Positive-Sum World

Elias G. Carayannis and Jeffrey Alexander

Abstract

Strategic alliances as a socioeconomic phenomenon, business process, and knowledge pooling and sharing protocol are profiled and studied in technology-driven environments.

The concepts of technological learning, co-opetition, and strategic technological capability options are also presented in the context of strategic alliance formation, evolution, and dissolution.

Lessons learned and critical success and failure factors in choosing how, why, and when to enter in a strategic partnership are identified and discussed.

Fundamentals

The term **strategic alliance** (SA) appears often in the business press and in press releases to describe new relationships formed between companies. Usage of the term is particularly popular for cooperative partnerships involving companies that normally compete with each other in the marketplace. Several databases tracking SAs in technology-intensive industries show that the rate of formation of such alliances has accelerated dramatically since the late 1970s.[8] Current conditions in the global economy make these alliances advantageous, perhaps necessary, for firm competitiveness in areas of advanced technology. The challenges for technology managers are to determine when to form SAs, whom to choose as partners, and how to ensure that such alliances provide adequate benefits to justify the substantial effort needed to create and maintain them.

The importance of technology development and sharing as a motivation for the formation and operation of alliances has been stated by Osborn and Baughn [1990]: "Many of the alliances made between firms with headquarters in developed nations are in high-tech areas, and many also involve joint research and development." Yoshino and Rangan [1995] identify strategic flexibility, the protection of core form assets, learning opportunities, and value-adding potential as strategic objectives motivating an alliance. Schmidt and Fellerman [1993] identify the following four kinds of benefits associated with forming an alliance:

1. Economies of scale of the static and dynamic kind and economies of scope
2. Quick and easy access to knowledge and markets
3. The reduction of the capital requirements and the risks involved in the development of new kinds of products and technologies
4. The possibility of influencing the structure of competition in the relevant markets

Placing Strategic Alliances in Context

A close analysis of the term strategic alliance shows that it has two facets. The word alliance describes a type of relationship between two or more companies, where the firms are allied toward some common outcome or direction. The word strategic refers to the significance of the relationship to its member firms. The alliance moves the firms into a position of greater competitive advantage in the marketplace than they could attain without the alliance.

SAs are of particular interest to academics and managers alike because they seem to counter the traditional assumption that competition produces the greatest gains to individual firms and to the economy as a whole. For an individual firm, "perfect" competition means that no firm earns a return

---

[8] Examples include the Cooperative Agreements and Technology Indicators (CATI) database developed at the Maastricht Economic Research Institute on Innovation and Technology (MERIT) and the Advanced Research Project on Agreements (ARPA) database at Milan Polytechnic.

that is above those earned by its competitors, and thus no firm has a true "competitive advantage". The real goal of business strategy is instead to find ways to earn "abnormal returns". In macroeconomic terms, interfirm alliances may produce greater overall returns than free competition. The application of game theory to business strategy has shown that total competition may lead to "no-win" situations, such as the classic "prisoners' dilemma", or "win-lose" propositions where only one firm can dominate the industry [Carayannis et al., 1998]. Pure competition often leads to a zero-sum game, where firms are simply seeking ways to divide up shares of a market with no growth potential. The true goal of strategy and policy should be to identify "positive sum" or "win-win" situations, where competitive conditions lead to market growth, benefiting multiple players in the market [Carayannis and Alexander, 1997]. Business researchers Brandenburger and Nalebuff [1996] call this range of competitive conditions "coopetition", reflecting a state between full cooperation and perfect competition (see Table 1.1).

TABLE 1.1 Dynamics and Outcomes of Competition, Coopetition, and Cooperation

	Perfect Competition	Coopetition	Perfect Cooperation
Dynamics	Winner-take-all Absence of coordination	Integration of complementary and competitive relations	Close coordination of actions
Focus	Emphasis on efficiency	Emphasis on gain sharing	Emphasis on maximizing rents
Outcomes	Zero-sum environment	Positive-sum environment	Monopoly or integration

Knowledge Exchange, Cooperation, and Competition in Strategic Alliances

Given that firms are unable to protect their knowledge absolutely, they may be able to preempt the misappropriation of their knowledge by sharing it with their competitors. The equal exchange of knowledge constitutes a "quid pro quo", which, in turn, reinforces a growing trust between the parties to that transaction [Carayannis and Alexander, 1997]. As long as the knowledge exchanged between the firms is perceived by the recipients to be of equal value, trust can be built. Many alliances and joint ventures fail because the participants are unwilling to share knowledge, on the faulty assumption that they are "parting with" that expertise.

Knowledge exchange does not rule out downstream competition between two firms. Again, knowledge is simply the foundation of intellectual capital, not its equal. Possession of knowledge is one thing; its application and control is much more significant. Hence, firms may be willing to cooperate to share and develop jointly "generic knowledge", which they then apply in their unique ways in differentiated products on the market. The availability of that pool of generic knowledge adds value to the products of both firms, yet still allows them the freedom to compete with each other at the market level.

SAs are a principal tool of coopetitive strategy, since they represent the opportunistic alignment of multiple firms for mutual benefit. This perspective reveals some important features of SAs, many of which were identified by Hamel et al. [1989]:

Interfirm collaboration is a form of competition. Although the optimal alliance will result in gains for all parties involved, each member must keep in mind that its partners' eventual objective is to attain market dominance, and thus they may try to manipulate the alliance for individual gain.

Strategic alliances, like all good things, will eventually come to an end. Although "strategic alliances" are often interpreted as long-term relationships, they almost always have finite usefulness to their members. Therefore, firms must realize that these alliances will eventually outlive their ability to generate strategic advantage.

Opportunities to form SAs require substantial up-front analysis and constant assessment. Since these are relatively new and fairly complex phenomena, firms should enter SAs only after careful consideration of their costs, benefits, risks, and opportunities. Furthermore, firms must monitor their alliances to ensure that they are extracting appropriate benefits without "giving away the store".

Carayannis et al. [1997] validated empirically the following attributes of technology-driven SAs:

1. The presence of a champion is highly correlated with the success of a technology-based, seed-capital-financed SA.
2. The use of funds derived from a SA is primarily aimed at acquiring and developing tangible strategic assets such as proprietary technology, secondarily for general working capital, and thirdly for acquiring and developing intangible strategic assets such as skills and know-how possessed by key managerial personnel.
3. The established firm in a SA receives primarily technology-related intellectual property rights and secondarily marketing rights more often than equity, manufacturing rights, etc. in exchange for their capital infusion.
4. SAs are set up based on word-of-mouth and informal networking among the alliance partner principals rather than on formal search processes, and SAs that provide seed capital do not tend to foster the creation of a new organizational entity to supplement the partners.
5. SAs are set up without the use of intermediaries and they are governed informally and with exchange of directors rather than through such formal means as joint management committees; most successful SAs are set up with domestic, rather than foreign, partners with a previous success record in such ventures.
6. The alternatives to a SA for SA partners are to become a competitor or have a licensing agreement rather than becoming a customer or supplier; SAs were the first choice for securing early-stage seed capital for SA partners, while venture capital funds were the last one.

Justifying Strategic Alliances

Botkin and Matthews [1992] and Niederkofler [1991] predict an acceleration of partnerships between large and small companies, which will become a substantial source of early-stage risk capital for embryonic firms. The evidence is not conclusive, however, that small firms typically benefit from alliances with established firms. In their examination of the information technology industry, Hagedoorn and Schakenraad [1994] determine that small firms in such an alliance often lack the capabilities to take advantage of the opportunities afforded by alliances. Nonetheless, Deeds and Hill [1996] show that SAs, up to a certain level at least, help increase the rate of new product development in the case of entrepreneurial biotechnology firms.

Lorange and Roos [1993] contend that the ability to form and successfully manage alliances will be a key factor in the future success of many firms. Also, evidence from a study by Booz Allen suggests that a surprising percentage of alliances are successful even though U.S. managers are behind Japanese and European companies in the alliance formation process [Bamford, 1994]. According to Bamford, Booz Allen found that the average return on investment (ROI) in alliances has been nearly 17%, which is significantly higher than normal ROI of a single company.

A comprehensive summary of the various motives for the formation of SAs in technology is provided by Hagedoorn [1993] (see Table 1.2). There are two components to these motives. First, the emergence of SAs is driven by changes in the nature of business competition and technology development resulting from several underlying global economic trends. Second, alliances are formed because they best attain specific objectives of their member firms.

The globalization of economic competition promotes the use of collaboration, as it puts a premium on the speed of innovation. The nature of innovation in today's market has also changed in the past 20 years. First, innovation now occurs through the fusion of technologies from different industries as well as the development of revolutionary technologies within a single industry [Kodama, 1992]. Second, product technologies themselves are becoming more complex, so that mastery of those technologies requires a larger and more diverse group of experts [Kash and Rycroft, 1994]. It is more and more difficult to assemble sufficient skills and technologies within a single firm, in the timeframe driven by competition, leading to the rise of alliances to accomplish various objectives:

TABLE 1.2   An Overview of Motives for (Strategic) Interfirm Technology Cooperation

I   Motives related to basic and applied research and some general characteristics of technological development
Increased complexity and intersectoral nature of new technologies, cross-fertilization of scientific disciplines and fields of technology, monitoring of evolution of technologies, technological synergies, access to scientific knowledge or complementary technology
Reduction, minimizing and sharing of uncertainty in R&D
Reduction and sharing of costs of R&D
II Motives related to concrete innovation processes
Capturing of partner's tacit knowledge of technology, technology transfer, technological leapfrogging
Shortening of product life cycle, reducing the period between invention and market introduction
III Motives related to market access and search for opportunities
Monitoring of environmental changes and opportunities
Internationalization, globalization, and entry to foreign markets
New products and market, market entry, expansion of product range

Source: Hagedoorn, J. *Str. Mgt. J.,* 14: 371–385, 1993. With permission.

Acquire access to a technology. As noted above, many products require the assembly of multiple technologies. If those technologies are not available on the open market, then firms will generally resort to an agreement such as a license or co-agreement to gain access to that technology.

Exchange technologies. Firms also find that they can be more competitive by offering products that bundle the complementary technologies of two companies. Although bundling is coordinated most easily within an integrated firm, forming an alliance allows the firms to bundle specific technologies with fewer complications than a merger.

Obtain access to specialized skills. Firms may also need very specific technical skills, but only for limited periods during technology development. In this case, it is much less costly to outsource those skills from a specialized firm than to hire in those skills.

Assemble cumulative resources. As noted in Table 1.2, one trend in technology development is that the scale of effort required to create complex technologies has increased dramatically. Companies therefore form alliances to pool their common resources, which includes financial, material, and human assets. Examples of such alliances include R&D corporations (such as Bellcore and Cable-Labs) and R&D consortia (such as SEMATECH).

Assemble complementary resources. Often, a smaller firm has a technology but not the complementary resources (e.g., a distribution network) to bring it to market [Teece, 1986]. SAs may be used to obtain those complementary resources from another firm, without the cost of building those resources internally. This is a common pattern in the biotechnology industry, where small start-ups develop new treatments but the large pharmaceutical firms take over manufacturing, marketing, and distribution.

Acquire a strategic resource. According to the resource-based view of the firm, a company attains sustained competitive advantage through the development of key strategic resources. These resources tend to be complex, meaning that they cannot be easily broken down into their component inputs, and are also subject to causal ambiguity, meaning that they cannot be imitated or reproduced by other firms. As demonstrated by Chi [1994], firms that want to use the strategic resource of another company must generally use an alliance, as acquisition of all or part of another firm tends to incur monitoring costs and other drawbacks that eliminate the strategic resource.

Obtain a strategic technological capability option. One relatively new idea in strategic management is the application of options theory to investment and resource allocation. Strategic options are initial investments that are made to retain future flexibility while controlling one's risk exposure under proper hedging schemes. Alliances are often entered as a strategic option because they create an opportunity to pursue more coordinated effort in the future or if needed end the collaboration without substantial sunk costs. Kogut [1991] shows xphow firms enter joint ventures to open up the option of acquiring their partner at a later date. Bleeke and Ernst [1995] also discuss how strategic alliances often end in the sale of one partner to the other. Carayannis et al. [1996] underscore the significance of strategic technological capability options as follows:

> Machine tool companies that adopted numerically-controlled production technology in the early 1970's, found it easier to switch to integrated computer-based manufacturing a decade later...These firms developed *capability options,* which could be "exercised" to reduce the cost of adopting subsequent, more advanced technology vintages. In this context, the *capabilities* developed in the first technology transition facilitated a subsequent transition to the more advanced technology. This implies that part of the benefit from the original investment was the *capability* to work with automated processing equipment, a capability that translated more easily into the adoption of CIM technology in the 1980's.

Create learning opportunities. From an organizational learning perspective, the "dynamic capabilities" theory developed by Teece et al. [1997] posits that the strategic advantage of a firm comes not only from its existing capabilities but also from its potential future abilities, reflected in the capacity and opportunities for corporate learning. SAs play a significant role in building dynamic capabilities, since they represent a common medium for exposing firms to new technologies and techniques and transferring those capabilities in-house (see Hamel [1991]).

The Structure of Strategic Alliances

Bamford [1994] comments on some "secrets of success" in setting up SAs:

> During alliance negotiations see if you can articulate the other side's objectives. . . Decide which potential partners are most important and make your own contacts... Don't enter an alliance thinking it will solve a weakness... a partner that is not sharing a strength will get gobbled up... Some of the longest-lasting and most lucrative alliances. . . started out simply as licensing agreements or arm's length supplier relationships.

Niederkofler [1991] also provides a list of key success factors in creating a strategic partnership:

1. A cooperative agreement should be based on a clear understanding of each partner's resources and interests, focused on specific goals.
2. Time pressures must not be allowed to take priority in the negotiation process.
3. The amount of actual cooperation should be limited to the amount necessary to achieve the goals of the cooperation.
4. Well-connected, veteran entrepreneurial managers with boundary-spanning skills should be assigned as liaisons by the larger firm.
5. Top management needs to legitimate and support the entrepreneurial actions of the liaison manager.
6. Cooperation management focused on the creation of trust and good will creates the best basis for a mutually beneficial relationship.
7. A sequential build-up of relationship intensity enables firms to get to know each other's interests and operating styles.
8. Flexibility is key in gaining benefits from cooperation.

An alliance is an organizational form involving two or more firms that creates a relationship between markets and hierarchies [Williamson, 1985]. This relationship is generally cemented with a formal agreement that sets out certain conditions under which the two firms will cooperate, including possibly the contributions of each partner, a timeframe for collaboration, and the organizational structure of the alliance.

Since these alliances are "strategic", firms should support their commitment to the alliance with some form of investment, such as an exchange of equity between firms or a contribution of resources.

Commitment may also be demonstrated by some agreement on how the firm will conduct its cooperation, such as technologies it will contribute or personnel assigned to the alliance's activities.

Although it is not necessary, most alliances are established for a specific timeframe. At a minimum, an alliance will involve a firm commitment to engage in a series of transactions or interactions, such as a strategic supplier agreement. In such a case, the transactions must somehow be interlinked, not discrete. Also, many alliance agreements include some mechanism for ending the cooperation, using such delimiters as a predetermined date, the accomplishment of a goal, or the decision to quit by one or more partners.

SAs take many shapes. Alliances arrange for some level of interaction between the member firms, for example, through meetings or the creation of a "virtual team" that spans firms. The organization of SAs will in part depend on the nature of the alliance members. For example, alliances between private firms can generally follow any of a wide range of structures, except as proscribed by regulations such as antitrust. However, if one or more members is a government organization or a university, the nature of the alliance may be constrained by the public duties or political expectations placed on such institutions.

Catalogs of formal modes of alliance and collaboration in technology have been created by Chesnais [1987] and Hagedoorn [1990], among others. Some typical modes for defining the interfirm relationship in an alliance include

Licensing: the exchange of access to a technology and perhaps associated skills from one company for a regular stream of cash flows from another.

Cross-licensing: an agreement between two firms to allow each other use of or access to specific technologies owned by the firms.

Strategic supplier agreement: a long-term supply contract, including guarantees of future purchases and greater integration of activity than a casual market relationship. One prominent example is the second-source agreements signed between semiconductor chip manufacturers.

Contract R&D: an agreement under which one company or organization, which generally specializes in research, conducts research in a specific area on behalf of a sponsoring firm.

Direct investment: the purchase of a substantial share of equity (generally defined as 10% ownership or more) in one company by another as a means of securing a strategic relationship.

Equity swap: an agreement for two companies to purchase equity stakes in each other's operations as a means of securing their relationship.

Joint R&D agreement: an agreement under which two or more companies agree to cooperate in a specific area of R&D or a specific project, coordinating research tasks across the partner firms and with sharing of research results.

R&D corporation: the establishment of a separate organization, jointly owned by two or more companies, that conducts research on behalf of its owners. A notable example is Bellcore, which originally was established by the seven Regional Bell Holding Companies of the United States and which would conduct research and set standards for the local telephone system.

Research consortium: any organization with multiple members that is formed to conduct joint research in a broad area, often in its own facilities and using personnel on loan from member firms and/or direct hires. The Microelectronics and Computer Technology Corporation (MCC) and SEmiconductor MAnufacturing TECHnology (SEMATECH) are examples of such organizations.

In the case of MCC, Gibson and Rogers [1994] identify learning about collaborating as one of the main benefits from forming this technology partnership:

It is in learning about the process of technology collaboration where MCC may have its greatest impact on the member companies in particular and enhanced US industrial competitiveness in general. MCC has served as an important real-life laboratory to study the barriers and facilitators of collaborative R&D leading to technology application and commercialization.

In the case of SEMATECH, Carayannis and Gover [1997] describe the consortium as a quasi-horizontal partnership that pursued win-win combinations of participant resources and capabilities:

SEMATECH has operated as an industry-wide, horizontal/quasi-vertical consortium focused on strengthening suppliers for their member companies and in this sense, the SEMATECH member companies were collaborating in supporting their suppliers to enable them to compete more vigorously... The quasi-vertical qualification is used to indicate that while SEMATECH's emphasis is on strengthening their U.S. suppliers, their suppliers are not members of SEMATECH. This collaboration provided the SEMATECH members with strategic technological options (access to a healthy semiconductor equipment manufacturing infrastructure) they could exercise at will in the future and thus in a win-win, or co-opetitive, game-theoretic context.

Each of these mechanisms has its own advantages and drawbacks. Also, certain structures are designed to be used only under specific conditions. For example, licensing is generally limited to situations where one firm needs to obtain access to a finished technology developed by another firm, rather than the acquisition of technology still under development. The above list merely outlines the broad range of formal structures for SAs, so it is important to analyze the nature and intent of these collaborations as well as their mechanisms.

Considerations in Partner Selection

Once a company has determined that some form of alliance is required for the development or management of strategic technologies, the next steps of critical importance are identifying a suitable partner for collaboration. Again, drawing on the resource-based view of the firm, a company must first understand its own capabilities and position to determine what it would need in an external partner. The company should conduct some form of what Ford [1988] calls a "technology audit". This is the process by which a company identifies the technologies that are critical to its current and future competitiveness in its businesses, benchmarks its position in those technologies, and decides the optimal means of achieving improvement in that position. Based on this audit, a company can analyze whether it needs to pursue an alliance and can generate a rough profile of what a potential partner firm should provide as part of that alliance.

Potential partners can be found among many areas, including competitors, current or potential suppliers, and customers. Each of these groups of firms can contribute strategic information, resources, or capabilities. For example, technology users are often a valuable source of ideas for technology improvement, and an alliance is one means of obtaining access to those ideas. Similarly, in many industries manufacturers are off-loading more research and development responsibility to their suppliers, with the expectation that the suppliers provide technical improvements that can be easily integrated into the assembled product.

Summary

While not panaceas by any means, various modalities of collaborating by pooling tangible resources and intangible assets such as individual and firm competencies (i.e., sets of skills and technologies [Prahalad and Hamel, 1994]), internal structures (i.e., management, legal structure, manual systems, attitudes, R&D, and software), and external structures (i.e., brands, customer, and supplier relations [Svelby, 1997]) can be powerful means for adding value long term and in a sustainable manner through leveraging technology and knowledge.

There are clear trends for the increasing use of SAs and other forms of firm collaboration such as joint ventures, consortia, and also government-university-industry partnerships as research budgets shrink, and the level of technological complexity and rate of technological innovation increase exponentially. Such forms of technology partnerships are effective risk management and uncertainty filtering as well as tangible and intangible resource-leveraging tools — they facilitate collaborating to compete more effectively, which is reflected in the emerging co-opetitive theory of the firm.

Technology and the ways people and firms work together evolve in a mutually interactive and increasingly influencing manner. As geographic and product or service markets, suppliers and customers, and competitors and complementors become increasingly intertwined, and knowledge generation and exchange becomes the core of economic endeavor, forming SAs is becoming almost unavoidable. Thus, knowing when to form and when to end such a collaborative relationship and with whom and under what terms to create it must be a core competence of the knowledge-driven firm.

Defining Terms

Allocentrism: As explained by Brandenburger and Nalebuff [1996], allocentrism is the practice of "putting yourself in the shoes of other players" to better understand their motives and hence be able to act more effectively in evaluating your own value in the relationship, predicting future moves by others, and understanding their perceptions.

Co-opetition: A concept created by Brandenburger and Nalebuff [1996], where they describe co-opetition as "a duality in every relationship — the simultaneous elements of cooperation and competition. War and peace. Co-opetition."

Intellectual capital: The tangible and intangible assets of a firm, encompassing human capital and pure "know-how", which enables the creation of a flow of new ideas.

Strategic alliance: A relationship between two companies, encompassed in some formal agreement, which initiates collaborative effort toward increasing mutual competitive advantage and where the results of that collaboration are greater than either firm could achieve independently.

Strategic technological capability option: As noted by Carayannis et al [1996], this concept refers to the fact that "investing in given technologies, firms create opportunities or 'options' for themselves, to make still additional investments in the future." These investments are necessary so that the firm can at least "stay on the learning curve", which allows it to retain the ability to adopt newer generations of technology as they emerge.

Technology audit: As defined by Ford [1988], this is a methodology by which a firm is able to identify key technologies essential to its competitive advantage, assess its relative position in those technologies within its industries, and identify how to acquire new technologies or to more fully exploit existing technologies.

References

Anonymous. *Wall Street J.* April 20, 1995.

Bamford, A. Not so dangerous liaisons, *Fin. World,* 163(25): 56–57, 1994.

Bleeke, J. and Ernst, D. Is your strategic alliance really a sale?, *Harv. Bus. Rev.* January-February: 97–105, 1995.

Botkin and Matthews. *Winning Combinations: The Coming Wave of Entrepreneurial Partnerships between Large and Small Companies,* John Wiley & Sons, New York, 1992.

Brandenburger, A. M. and Nalebuff, B. J. *Coopetition,* Doubleday, New York, 1996.

Carayannis, E. G. et al. Architectural Innovation, Technological Learning, and the Virtual Utility Concept, *Paper presented at the 1996 International Conference on Engineering and Technology Management,* IEEE Engineering Management Society, IEMC '96, Vancouver, Canada, August 18–20, 1996.

Carayannis, E. G. et al. Business — University Virtual Teaming for Strategic Planning and Knowledge Management, *Technological Forecasting and Social Change Journal,* March 1998.

Carayannis, E. G. and Alexander, J. M. The Role of Knowledge Exchange in Trust, Co-opetition and Post-Capitalist economics, *Proceedings of the European Institute for Advanced Studies in Management (EIASM) 1997 Conference,* Leuven, Belgium, June 4–6, 1997.

Carayannis, E. G. and Gover, J. "Co-opetition", Strategic Technology Options and Game Theory in Science and Technology Policy: The Case of Sematech, *Paper presented to the Portland International Conference on Management of Engineering and Technology,* Portland, Oregon, July 27–31, 1997.

Carayannis, E. G., Kassicieh, S., and Radosevich, R. Financing Technological Entrepreneurship: The Role of Strategic Alliances in Procuring Early Stage Seed Capital, Portland International Conference on Management of Engineering and Technology, Portland, Oregon, July 27–31, 1997.

Chesnais, F. Technical co-operation agreements between firms, *STI Rev.* pp. 51–119, 1987.

Chi, T. Trading in strategic resources: necessary conditions, transaction cost problems, and choice of exchange structure, *Strat. Mgt. J.,* 15: 271–290, 1994.

Deeds, M. and Hill, C. Strategic alliances and the rate of new product development: an empirical study of entrepreneurial biotechnology firms, *J. Bus. Ventur.* 11: 41–55, 1996.

Ford, D. Develop your technology strategy, *Long Range Plan.,* 21: 85–95, 1988.

Gibson, D. and Rogers, E. *R&D Collaboration on Trial,* Harvard Business School Press, Cambridge, MA, 1994.

Hagedoorn, J. Organizational modes of inter-firm cooperation and technology transfer, *Technovation,* 10(1): 17–30, 1990.

Hagedoorn, J. Understanding the rationale of strategic technology partnering: interorganizational modes of cooperation and sectoral differences, *Str. Mgt. J.* 14: 371–385, 1993.

Hagedoorn, J. and Schakenraad, J. The effect of strategic technology alliances on company performance, *Str. Mgt. J.,* 15(4): 291–309, 1994.

Hamel, G. Competition for competence and inter-partner learning within international strategic alliances, *Str. Mgt. J.,* 12: 83–103, 1991.

Hamel, G., Doz, Y. L., and Prahalad, C. K. Collaborate with your competitors — and win, *Harv. Bus. Rev.,* January–February: 133–139, 1989.

Kodama, F. Technology fusion and the new R&D, *Harv. Bus. Rev.,* July-August: 70–78, 1992.

Kogut, B. Joint ventures and the option to expand and acquire, *Mgt. Sci.,* 37(1): 19–33, 1991.

Lorange, P. and Roos, J. *Strategic Alliances: Formation, Implementation and Evolution,* Blackwell Publishers, Cambridge, MA, 1993.

Niederkofler, P. The evolution of strategic alliances: opportunities for managerial influence, *J. Bus. Ventur.,* 6: 237–257, 1991.

Osborn, R. and Baughn, C. C. Forms of interorganizational governance for multinational alliances, *Acad. Mgt. J.* 33(3): 503–519, 1990.

Prahalad, C. K. and Hamel, G. *The Core Competency of the Corporation,* Harvard Business School Press, Cambridge, MA, 1994.

Schmidt and Fellerman. On explaining strategic alliances, *J. Int. Theo. Econ.,* 148(9): 748–755, 1993.

Svelby, K. *The New Organizational Wealth: Managing and Measuring Knowledge-based Assets,* Berrett-Koehler, San Francisco, 1997.

Teece, D. J. Profiting from technological innovation: implications for integration, collaboration, licensing and public policy, *Res. Pol.,* 15(6): 285–305, 1986.

Teece, D. J., Pisano, G. and Shuen, A. Dynamic capabilities and strategic management. *Strat. Mgt. J.,* in press.

Williamson, O. E. *The Economic Institutions of Capitalism,* The Free Press, New York, 1985.

Yoshino, M. Y. and Rangan, U. S. *Strategic Alliances: An Entrepreneurial Approach to Globalization,* Harvard Business School Press, Boston, 1995.

Further Information

Contractor, F. J. and Lorange, P., eds. *Cooperative Strategies in International Business,* Lexington Books, Lexington, MA, 1988. The product of a colloquium on international alliances sponsored by Rutgers University and the Wharton School, this is a pioneering work on strategic alliances and the first compendium of alliance research.

Beamish, P.W. and Killing, J. P., eds. *Cooperative Strategies (three volumes),* The New Lexington Press, San Francisco, 1997. These three books contain the papers from a series of three conferences held in 1996 as a follow-up to the 1986 Rutgers/Wharton conference. Each volume presents research from and on alliance activity in a region of the Triad (North America, Europe, and Asia). Selected papers were published in the *Journal of International Business Studies, 1996 Special Issue on Global Perspectives on Cooperative Strategies,* 27(5).

Dodgson, M. *Technological Collaboration in Industry: Strategy, Policy and Internationalization in Innovation,* Routledge Press, London, 1993. This book provides an overview specifically on technology-based SAs, looking at both the positive and negative aspects of such ventures.

The Academy of Management Journal, Special Research Forum on Alliances and Networks, April, 40(2), 1997. This special issue provides current empirical studies of alliances and interfirm networks, especially in technology-intensive activities.

1.8 Knowledge Parks and Incubators

George Bugliarello

Knowledge parks and incubators are special physical environments for the creation — directly or indirectly — of economic value through the development, application, and transfer of knowledge and the creation of new enterprises. Most parks are planned, but some happen spontaneously.

The Parks are largely a post-World War II phenomenon, triggered by the increasingly large role played by knowledge (e.g., R&D) in industry and other value-added activities. Where successful, the parks become a powerful instrument of economic and social development and can profoundly influence regional and urban planning [Biggs, 1985; Bugliarello, 1996].

Typology

These knowledge-focused environments may be characterized by a variety of thrusts, different mixes of knowledge-generating and knowledge-applying entities, and different scales — from the vastness of a region, to the high density of an urban setting, to the microenvironment of an incubator.

Their typology — at times confusing in its terminology — ranges from research parks to university-industry parks, industry parks, and incubators as well as to broader and more complex environments, such as science cities or science regions. Sometimes more generic terms such as "science park" or "technopolis" are also used [AURRP, 1994; Scott, 1993]. Yet all these environments have certain common characteristics and requirements and are often created through similar processes.

The concept of a knowledge park can extend to knowledge institutions other than universities or research laboratories, e.g., specialized high schools, community colleges, or hospitals. In turn, industry can be intended in a broad sense to encompass, beyond manufacturing companies, other economic activities that depend on knowledge, such as the service industry (finance, merchandising, etc.).

- *Research parks* are environments in which a number of research laboratories — both public and private — are concentrated.
- *University-industry parks* — the most common, and often synonymous with research parks when the latter include a university — combine university and industry facilities to enhance their interaction.
- *Industry parks* are special areas in which concentrations of typically high-tech industry facilities are located. Many towns, large and small, povide areas for this purpose.
- *Incubators* are environments, not as large and demanding of resources as the parks, specifically designed to encourage entrepreneurs to bring to commercial realization scientific and technological ideas.
- *Science cities* are urban areas catalyzed by the presence of universities, industries, research laboratories, etc., such as Tsukuba in Japan.
- *Science regions* are broader environments, such as Route 128 in Boston, Silicon Valley, or the expansion of the Research Triangle Park in North Carolina, in which high-tech industry and research laboratories tend to concentrate for a variety of reasons, such as the relative closeness to universities and other research institutions, ease of communications, or the initial presence (as originally the case with Route 128) of available and inexpensive facilities.

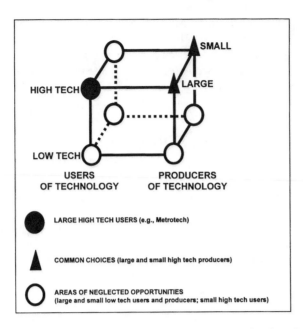

FIGURE 1.3   Strategic choices for knowledge parks. (Source: Bugliarello, G. Urban knowledge parks and economic and social development strategies, J. Urban Plan. Dev., 122 (2): 33–45, 1996.)

There is also an increasing opportunity for some parks and incubators, thanks to today's advances in telecommunications, to be conceived as "virtual" environments in which the common ground is not physical space but a network of supporting elements — universities, research laboratories, faculty, suppliers, etc. — all potentially capable of interacting in real time (including, for example, teaching or operating laboratories at a distance).

Often, a knowledge park does not fall into a single category, and a single type of park is unlikely to satisfy the complex economic and social development needs of a city or region. Thus, the aim of a farsighted modern development strategy is the creation of an intelligent mix.

Strategies

The basic strategic choices offered in the development of a knowledge park are summarized in Fig. 1.3. They entail deciding whether the park should be hi-tech or low-tech, whether it should be oriented toward the users or toward the producers of technology, and whether these should be large or small. Most common are parks oriented toward high-tech producers, large or small, Few parks, as of now, focus on the users of technology — whether large or small — as well as on users or producers of low-tech. Yet, these are all sectors with a glaring need for the support a knowledge park can offer them to enhance their technological skills and the value added of their products. Universities, being inherently high-tech, tend to favor an association with high-tech producers or, much less frequently, with high-tech users of technology (as in the case of Metrotech in Brooklyn, New York, which is focused on the financial industry). However, colleges and institutions involved primarily in teaching rather than research are often more suitable to focusing on low-tech users and producers and on their personnel education and training needs.

To create a park and assemble the necessary players requires linking the park to broader regional or local strategies and developing appropriate coalitions that will vary with the goals and strategies chosen for the park. Thus, if the goal is jobs creation, the most effective and rapid strategy is likely to be attraction of massive amounts of industry that, by necessity, become the key player in the coalition. The involvement of government, community, and the university are also essential. If the principal goal is generation of knowledge, the strategy must be the attraction of research laboratories and the strengthening of the university. Again, government and the community continue to be important elements of such a coalition, but university and industry must jointly play a key role. If the goal is creation of new enterprises, a logical strategy is development of incubators and research laboratories with the support of the university or other knowledge institution.

In any case, there is a need for a strong champion of the park and for a framework to relate goals, strategies, and coalitions. In practice, neither goals nor strategies are likely to be pure, and the coalitions may vary with the specifics of the situation. For example, it is always essential to involve the most powerful and influential leaders in the community, whoever they may be, regardless of affiliation.

Often, the goals may be multiple and differ for different components of the coalition. This is not detrimental as long as the goals are compatible and lead to a common action by all players. Problems arise when they do not.

Resources

The *physical resources* needed for a knowledge park include ground space and adequate infrastructure. To launch a park, a minimum of physical space must be available from the start for the potential occupants, including, if appropriate, some incubators. Some initial space is also needed for administration and for marketing the park. Additional space must be made available, when required. for further occupants.

The occupants of the park are often offered options such as renting, leasing of land and/or facilities, or outright ownership.

The necessary infrastructure includes

- Transportation, to give access to the park. The success of a park is dependent on ease of access. Roads, railroads, and air transportation are important as well as, for certain industrial parks, water transportation. Availability of parking is also a must.
- Global telecommunications (satellite and fiberoptic links and telephone networks).
- Hotels, restaurants, day care centers, schools, complexes for sports activities, shops, banks, government offices with which the activities in the park may interact (e.g., post office), etc.

The region where the park is located should offer the park other important resources such as amenities and industries that can supply the needs of the park and help create a critical mass in the region. In most cases, absence of these supporting elements, as when a park is in an isolated locality, reduces the park's attractiveness and viability.

The need for these infrastructural elements also varies with the dimensions of the park and can be satisfied by a mix of in-park and outside facilities. However, most of the elements should be in place from the very beginning.

The *financial resources* include

- An initial investment to plan, acquire, and clear the ground and build the first phase of the park.
- Resources to cover the initial phase of the operation of the park, before it can become self-sufficient (usually 3 to 7 years if the park is viewed as a commercial enterprise).
- Capital for future investments (buildings, infrastructure, incubators, etc.) or, at least, certainty that such capital will be available when needed.
- When appropriate, a venture capital fund to increase the capacity of the park and enhance the economy of its region through investment in promising enterprises attracted to it.
- Tax abatements and other incentives, to be provided by local or state governments.

The *human resources* potentially available, both in number and quality, are a key ingredient for the success of a knowledge park. Thus, for a research or a university-industry park, the number and areas of expertise of faculty in the university or Ph.D.s in the region as well as the overall quality of the university are a deciding factor. Similarly, for an industry park, the potential availability and characteristics of the work force (numbers, skills, quality, costs, union environment, etc.) are key.

The Process

The process by which a knowledge park is developed varies according to whether the park in is in an urban environment (such as Metrotech in Brooklyn or the University City Science Center in Philadelphia) or — as the majority are, for reasons of cost, convenience, and perceived attractiveness — in a less-dense suburban or rural environment. It also varies according to local socioeconomic and geopolitical conditions as well as the park goals. Almost invariably, however, it is a complex process, particularly if it occurs in an urban area where it is bound to involve the displacement of numerous occupants of the area or if it requires the cooperation of different jurisdictions, e.g., states or municipalities.

Typically, particularly for research and university-industry parks, the principal steps involved (not in strict chronological order) include

1. Development of the concept
2. Exploratory contacts with interested parties and development of initial plans
3. Securing of financing commitments (usually contingent upon achievement of certain benchmarks)
4. Search and conditional commitments by potential tenants
5. Environmental statements
6. Approvals by planning agencies
7. Establishment, where appropriate, of knowledge units that can serve as catalysts for the park, e.g., a specialized research laboratory
8. Commitments, by the local government and by the developer of the park, to infrastructural improvements and to incentives (tax elements, subsidized training for the workforce, creation of enterprise zones, reduced utility rates, etc.)
9. Alliances of sponsor, developer, local government, state government, the community, etc., for marketing the park
10. Consolidation of anchor tenants and search for additional tenants
11. Site acquisition and preparation (a step that may occur wherever appropriate in this sequence)
12. Construction (including, in case, refurbishing of existing facilities)
13. Attraction of subsidiary facilities (hotels, shops, restaurants, etc.)
14. Operational agreements among the park occupants and creation of a management structure

Organization

It is difficult to establish a clear precept for a park's organization, since it will depend on the nature of the park and the imperatives to which it responds. Usually, however, it may include

- A board constituted by the representatives of the key participants in the park who have a stake in its success and that, through committees and compacts, deals with various operational issues, such as transportation, personnel, security, health, waste disposal
- A manager of broad views and flexibility and with experience in fields relevant to the activities of the park
- A marketing and promotional office
- An office to manage the services necessary for the park (including maintenance, utilities, and security)
- An office of administration and finance
- A human resources office

Defining Terms

Incubator: A facility for the transformation of research results or inventive ideas into viable commercial enterprises.

Industry park: A knowledge park comprised of industry units.

Knowledge park: Special environment for the creation, directly or indirectly, of economic value through the development, application, and transfer of knowledge and the creation of new enterprises.

Research park: A knowledge park comprised of research laboratories.

Science city: A city — typically a new city — focused on science and/or its applications.

University-industry park: A knowledge park combining university and industry facilities.

References

Association of University Related Research Parks (AURRP). *Worldwide Research and Science Parks Directory,* Tempe, AZ, 1994.

Bugliarello, G. Urban knowledge parks and economic and social development strategies, *J. Urb. Plan. Dev.,* 122 (2): 33–45, 1996.

Gibbs, J. M. *Science Parks and Innovation Centers — Their Economic and Social Impact,* Elsevier, New York, 1985.

Scott, A. J. *Technopolis,* University of California Press, Berkeley, CA, 1993.

Further Information

Saxenian, A. *Regional Networks: Industrial Adaptation in Silicon Valley and Route 128,* Harvard University Press, Cambridge, MA, 1993.

1.9   Spin-Offs

Everett M. Rogers[9] and Morten Steffensen

This essay explains the spin-off process of new companies from research and development (R&D) organizations. We provide examples from our current research on spin-offs from federal R&D laboratories and from university-based research centers in New Mexico.

Technology transfer is becoming increasingly important for many R&D organizations. The technology-based spin-off company is an effective approach for transferring technology from an R&D organization to a commercial organization [Roberts and Malone, 1996]. The parent organization may be (1) a research university, (2) a government R&D laboratory, or (3) a private company. In this essay we look at the general nature of the spin-off process from any type of R&D organization.

Spin-Offs as Technology Transfer

Technology is information put into use in order to carry out some task. *Technology transfer* is the application of information into use [Eto et al., 1995]. Technology transfer ordinarily involves (1) a source of technology that possesses specialized technical skills and (2) the transfer of technology to receptors who do not posses these specialized skills and who cannot create the technology themselves [Williams and Gibson, 1990]. Many attempts have been made to transfer technology. Technology licensing and spin-off companies are two of the main techniques used to commercialize technology [Roberts and Malone, 1996].

---

[9] Everett M. Rogers is professor, Department of Communication and Journalism, University of New Mexico, and is known for his book, *Diffusion of Innovations,* published in its fourth edition in 1995. Morten Steffensen was a visiting research scientist in the Department of Communication and Journalism at the University of New Mexico in 1996 to 1997, while on leave from the University of Bergen, Norway.

The term "spin-off" means a new company that arises from a parent organization. Typically, an employee leaves the parent organization, taking along a technology that serves as the entry ticket for the new company in a high-tech industry. Spin-offs are also known as "start-ups" and "spin-outs". Some founders of spin-off companies react negatively to the term spin-off, which they feel emphasizes an indebtedness to the parent organization that does not fully recognize the sacrifices in money, time, and effort usually required to establish the spin-off company.

A *spin-off* is a new company that is (1) formed by individuals who were former employees of a parent organization and (2) a core technology that is transferred from the parent organization. This definition is based on that of Smilor et al. [1990] in their study of 23 spin-offs from the University of Texas at Austin and is generally similar to the definition used by most scholars. A spin-off is a mechanism of technology transfer because it is usually formed in order to commercialize a technology that originated in (1) a government R&D laboratory, (2) a university, or (3) a private R&D organization.

Roberts and Malone [1996] define a spin-off with emphasis on its source of funding. A spin-off company is usually formed with inputs of investment funding. Four principal entities are often involved in the spin-off process: (1) the *technology originator*, the person or organization that brings the R&D from basic concepts through the stages of the innovation development process to the point at which the transfer of technology can begin, (2) *the entrepreneur* (or the entrepreneurial team), who takes the technology created by the originator and attempts to create a new business venture, (3) the *R&D organization* that often is represented in the spin-off process by its technology licensing office, and (4) the *venture investor*, generally a venture capital organization that provides funding for the new venture in return for partial equity ownership in the new company.

A study of spin-offs from government R&D laboratories and research universities in the United States and Japan indicates that it is somewhat of an oversimplification to define a spin-off as a new company in which both the founders and the core technology are transferred from a parent organization [Caryannis et al., 1997]. Often only one or the other of these two factors is transferred. The parent organization may provide the spin-off with venture capital, building space, or other needed resources, or these resources may come from another organizations. Therefore, either (1) the definition of a spin-off should be expanded to include these other resource-transfers (thus defining a *spin-off* as a new company that is established by transferring its core technology, founders, or other resources from a parent organization) or (2) limit the concept of spin-off to a specific resource transfer, such as a technology spin-off, a founder spin-off, a venture capital spin-off, etc.

The Originator and the Entrepreneur

There are two kinds of entrepreneurs: (1) *inventor entrepreneurs*, who are or were employees in an R&D organization and who actively seek to commercialize their own technology, and (2) *surrogate entrepreneurs*, who are not the inventors but who acquire rights to the technology [Radosevitch, 1995].

The typical scientist is motivated by peer recognition and usually publishes scientific papers as a main means of technology transfer. For most researchers the business world is unknown, but a few researchers become entrepreneurs. Perhaps an entrepreneur becomes frustrated with his present employer. Smilor et al. [1990] found that "push" factors played a minor role in the decision to become an entrepreneur for their university spin-offs of study in Austin, Texas. Several entrepreneurs said they started their spin-off in order to make money by capitalizing on a market opportunity.

The Parent Organization

The success of new spin-off companies depends in part on whether they are supported actively or passively in the start-up process by their parent organization or whether they are opposed by the parent. Entrepreneurs spinning off from the University of Texas considered the university to be the most important organizational influence on their spin-off company and one that was generally encouraging.

Most individuals in R&D organizations realize that important differences exist between the scientific culture and the business culture. Often the commercialization effort of an R&D organization is organized through a technology licensing office in order to manage intellectual property rights. The basic function of a technology licensing office is to link technology with the marketplace. A technology licensing office evaluates the technological innovations that come from research, and, if favorable, the technology is licensed, a spin-off is encouraged, or other arrangements for technology transfer are made.

Many R&D organizations, especially research universities, have established business incubators to help spin-off new companies. A *business incubator* provides a nurturing environment in order to facilitate successful spin-offs. Usually a business incubator provides office space, legal consulting, and management services for the new business ventures.

The Agglomerative Effect of Spin-Offs

Certain regions have a concentration of spin-off companies. Examples are Boston's Route 128; Northern California's Silicon Valley; Austin, Texas; Cambridge, England; Tsukuba Science City in Japan; and Sofia Antiopolis, on the French Riviera. Why do spin-offs agglomerate in certain regions and not in others?

One reason is infrastructure in the form of venture capital, specialized legal services, and personnel with technical expertise. Another infrastructural factor is available technology and the presence of the R&D organizations that produce such technology.

An example of a single research university's impact on the economy by means of spin-offs is provided by MIT (the Massachusetts Institute of Technology). A study by BankBoston [1997] identified 4000 MIT-related spin-off firms that employed at least 1.1. million people and generated $232 billion in worldwide sales. Since 1990, 150 new firms per year have spun-out by MIT graduate students or faculty. This spin-off rate from MIT is exceptional. Why do not other research universities (and other R&D organizations) have as many spin-offs? The answer is the infrastructural support and services provided by research universities (or other R&D organizations), the availability of venture capital, and access to technological innovations. In addition, the local entrepreneurial climate for spin-offs (sometimes called "entrepreneurial fever") is very important in encouraging the rate of spin-offs in a region [Rogers and Larsen, 1983].

The effect of positive feedback from earlier spin-out successes can be an important factor on entrepreneurship and the rate of spin-offs [Roberts, 1991]. The glamorization of spin-off successes in the popular press triggers more spin-offs. When R&D workers read about the success of Bill Gates, Michael Dell, and Steve Jobs, they are influenced to become an entrepreneur. The net effect of infrastructural factors is the agglomeration of high-tech spin-offs in technopolises (technology cities). Once a critical mass of spin-offs is founded in a region, further spin-offs are most likely to occur in that same region.

The Technology Commercialization Environment

Technology transfer is regarded as a crucial resource for economic development in New Mexico. The state ranks 44th among the 50 U.S. states in per capita income, while it ranks 4th in federal and university R&D performance and 21st in R&D performance in the private technology sector [*State of New Mexico*, 1996]. In addition to 3 research-oriented universities, over 20,000 scientists work at three large federal R&D laboratories and their contractors. New Mexico's technology commercialization environment is unique, but it also has several weak points that limit the growth of a technopolis:

1. The abundance of available technologies is the result of publicly sponsored research (which is mainly defense related) and is difficult to commercialize.
2. Few large firms are headquartered in New Mexico.
3. Isolation from mass markets and suppliers.
4. A dependency culture (especially dependent upon federal government spending).
5. A shortage of entrepreneurs.
6. A rather fragmented, competitive infrastructure for technology commercialization.
7. Federal legislation that is increasingly supportive of technology transfer and commercialization [Radosevitch, 1995].

Two important factors limit the growth of new high-tech companies in New Mexico: (1) a lack of venture capital (only one venture capital company is located in Albuquerque) and (2) a lack of organized efforts to maximize the commercial potential of technological innovations. However, efforts are underway to remedy these shortcomings.

During the past decade, the New Mexico state government and several local governments have promoted technology commercialization through policy changes, incubators, advisory groups, public seed capital, collaborative agreements between federal R&D laboratories and private industry, and conferences.

The University of New Mexico (UNM) is the largest of the three research universities in New Mexico. Technology transfer from UNM was not of highest priority until the 1990s. Today, UNM has 55 research centers, in comparison to only 51 academic departments [Hashimoto et al., 1997]. The university's technology licensing and commercialization activity was not well organized until the Science and Technology Corporation (STC) was founded in 1995. STC is a nonprofit organization that conducts technology commercialization activities on behalf of UNM. It oversees all intellectual property management at the university, including promotion, developing of industry relationships, and evaluation of patents.

A Spin-Off from a University Research Center

One successful spin-off from UNM in recent years is Nanopore, founded in 1994 as a spin-off from the UNM Center of Micro-Engineered Ceramics (CMEC) by the then director and associate director. These two founders had different ideas about the development of the company, such as whether they should go public and if they should seek venture capital. One partner bought out the other after the company's 1st year of operation. The president of Nanopore decided not to seek venture capital. In 1997 Nanopore formed *its* first spin-off, Nanoglass, a joint venture with a California-based company.

One incentive for the entrepreneur to start this company was the pleasure of seeing the results of his research work put into commercial application. "What happens when you write a paper is that you send it away, and after three years you realize that either it resulted in nothing or that somebody else commercialized it." Nanopore's core technology is the production of a porous metal that meets an important market demand in the semiconductor industry, as it moves to smaller and smaller semiconductors with more and more circuitry on each chip. Nanopore licensed this core technology from UNM, but the University did not play a significant role in the company's formation. Nanopore has expanded through collaboration with large companies such as Texas Instruments, which provides them with worldwide sales of their products. The story of Nanopore shows that spin-offs can be successful without venture capital and other support structures.

References

BostonBank. *MIT: The Impact of Innovation,* Boston, 1997.

Caryannis, E. G., Rogers, E. M., Kurihara, K., and Allbritton, M. High-Technology Spin-Offs from Government R&D Laboratories and Research Universities, University of New Mexico, Albuquerque, 1997.

Eto, M., Rogers, E. M., Wierengo, D., and Albbritton, M. Technology Transfer from Government R&D Laboratories in the United States and Japan: Focus on New Mexico, University of New Mexico, Albuquerque,1995.

Hashimoto, M., Rogers, E. M., Hall, B.J., Steffensen, M., Speakman, K., and Timko, M. Technological Innovation and Diffusion Mechanisms in High-Technology Industry: Technology Transfer from University Research Centers, University of New Mexico, Albuquerque, 1997.

Radosevitch, R. A model for entrepreneurial spin-offs from public technology sources, *Int. J. Technol. Mgt.,* 10 (7/8): 879–893, 1995.

Roberts, E. B. *Entrepreneurs in High Technology: Lessons from MIT and Beyond,* Oxford University Press, New York, 1991.

Roberts, E. B. and Malone, D. E. Policies and structures for spinning off new companies from research and development organizations, *R&D Mgt.*, 26, (1): 17–48, 1996.

Rogers, E. M. and Larsen, J. K. *Silicon Valley Fever: The Impact of High-Tech Culture*, Basic Books, New York, 1983.

Smilor, R. W., Gibson, D. V., and Dietrich, G. B. University spin-out companies: technology start-ups from UT-Austin, *J. Bus. Ventur.*, 5: 63–76, 1990.

State of New Mexico, Economic Development Department. *The New Mexico Handbook,* Santa Fe, New Mexico, 1996.

Williams, F. and Gibson, D. V. *Technology Transfer: A Communication Perspective,* Sage, Newbury Park, CA, 1990.

TEN PRINCIPLES
OF ECONOMICS

1. People face tradeoffs.

2. The cost of something is what you give up to get it.

3. Rational people think at the margin.

4. People respond to incentives.

5. Trade can make everyone better off.

6. Markets are usually a good way to organize economic activity.

7. Governments can sometimes improve market outcomes.

8. A country's standard of living depends on its ability to produce goods and services.

9. Prices rise when the government prints too much money.

10. Society faces a short-run tradeoff between inflation and unemployment.

Source: Principles of Economics
N. Gregory Mankiw

2

Science and Technology Policy

Robert P. Morgan
Washington University in St. Louis

Joseph J. Cordes
The George Washington University

Christopher T. Hill
George Mason University

Hui-Yun Yu
Kuang S. Yeh
Sun Yat-Sen University

Yolanda Sarason
University of New Mexico

Suleiman K. Kassicieh
University of New Mexico

Keith E. Smith
The George Washington University

2.1 Government Policy

Robert P. Morgan

The U.S. government plays a significant role in the evolution of scientific and technological discovery and progress in the United States. That role greatly expanded during and after World War II, spurred on by military wartime needs and reinforced by the Cold War that followed. Today, the focus has shifted somewhat to trying to determine the proper role of government in fostering economic growth and competitiveness through support for science and technology. Issues of recent concern are the level of government support for research and the balance between government intervention and the role and responsibilities of the private sector in civilian research and development (R&D).

Science and technology policy is defined in a variety of ways. For our purposes here, science and technology policy means government plans, programs, and initiatives in support of science and technology. Sometimes distinctions are made between **science policy** and **technology policy**, depending upon where on the spectrum of science vs. technology one focuses. For example, the government is a major factor in the support of **basic research** in this country (see Fig. 2.1 and Defining Terms), generally viewed primarily as a "scientific" activity that falls within the domain of science policy. On the other hand, the government's role in the **development** of technology, which has been somewhat restricted to specific sectors in which the government is the primary customer, namely defense and space, fits within the technology policy rubric. A major technology policy debate raged in the first Clinton administration between the Democratic president and the Republican Congress over the proper government role in **applied research** and development directed towards the civilian sector, an activity thought by some to be the exclusive province of private industry in the United States, in contrast to the practice in other countries, e.g., Japan.

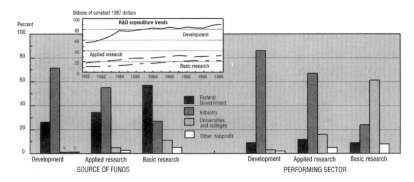

FIGURE 2.1   National R&D expenditures, funders, and performers, by character of work: 1996. Funds for federally funded R&D center performers are included in their affiliated sectors. (*Source:* National Science Foundation, Division of Science Resources Studies, *National Patterns of R&D Resources: 1996*, NSF 96-333.)

Sometimes additional distinctions are drawn. For example, the term **policies for science** generally includes R&D budgets, science advice to the president, and government support for science education, among other topics. *Science for policy* refers to the use of scientific approaches to aid in policy formulation, such as improved accuracy of environmental monitoring techniques and application of decision analysis to policy problems. A more encompassing practice is to include within science and technology policy all *policy issues with significant scientific and technological components*, which, these days, might include just about everything!

Historical Evolution

In the early days of the United States, government support for science and technology was sporadic and not of major importance [Dupre and Lakoff, 1962; Morin, 1993]. Some prominent individuals who founded the United States, such as Franklin, Washington, and Jefferson, had some scientific and technological skills and interests. Early government support was made available for the Lewis and Clark Expedition to explore the West and to establish the U.S. Military Academy. However, science and technology were not intense concerns of government; an early scientific institution, the Smithsonian, was started by a grant from an Englishman, James Smithson.

Gradually, however, with the industrialization of the country, government became more involved. Boiler explosions on steamships led to government intervention and regulation in the 19th century. With the outbreak of the Civil War, interest in using science and technology for military purposes received a boost, especially from the Navy, and the National Academy of Sciences was created. World War I led to much greater government interest in science and technology, involvement of leading inventors such as Thomas Edison, and creation of the National Research Council. Meanwhile, government departments and agencies with significant technological components were springing up, such as the Department of Agriculture, the first agency to provide extramural support for science, the Food and Drug Administration, the Public Health Service, and the National Advisory Committee on Aeronautics (NACA), the predecessor of NASA.

Government involvement in science and technology changed in a dramatic way during World War II following some initial attempts on the part of the Franklin Roosevelt Administration to involve science and technology in response to the Great Depression. A key element was the establishment of the Office of Scientific Research and Development, which coordinated the involvement of scientists and engineers in the war effort, fostered government-industry-university cooperation, and gave scientists and engineers unprecedented access to government in an advisory capacity. The successful development of the atomic bomb gave high visibility and influence to the scientific establishment.

A seminal event in the evolution of U.S. science and technology policy was the delivery of a report, "Science, The Endless Frontier" to President Truman in 1945 [Bush, 1945]. The report, written by Vanevar Bush, a prominent electrical engineer, made a strong case for science as essential not only to military strength but also to job creation, health, and economic growth. It argued that government support for science was of the utmost importance and that, if left to their own devices with minimal government intervention, scientists would produce results of enormous benefit to the nation. Some refer to this argument as "the social contract for science", which has been generally accepted until very recently when it has begun to be called into question.

The Post–World War II Period

Following World War II, the shape of government support for science and technology began to emerge in a form that has persisted, with some modifications, until the present. Government emerged as a principal supporter of basic research (see Fig. 2.1) through funding provided by a variety of federal agencies, including the National Institutes of Health, the National Science Foundation, NASA, the Office of Naval Research, and the Department of Energy. The main performer of basic research in the United States is the universities, supported by a system of grants and contracts that evolved during World War II. University support has enabled a close coupling to develop between support for research and the education of graduate students, who would go on to become the researchers and professors of the future. Government also provides support for applied R&D, particularly in the defense and space sectors. The principal performer of applied R&D is private industry (See Fig. 2.1).

Historically, government support for science has gone through some distinct periods since World War II [Smith, 1990]. First, there was a period of rapid growth, spurred on by the Cold War and spending on defense and space. In the mid-1960s, a period of retrenchment set in, with little if any growth, in which people questioned whether the benefits of science and technology were worth the costs (e.g., pollution, the nuclear arms race, and the Vietnam War), and calls were heard for more immediate positive returns from science and technology to help solve pressing societal problems, (e.g., the "war on cancer" and the "war on poverty"). Government spending on nonmilitary R&D grew to be equal to that of defense R&D (see Fig. 2.2). That situation began to change toward the end of the Carter administration in the late 1970s, and, with the advent of the presidency of Ronald Reagan, military spending grew rapidly and once again became the dominant focus of R&D. Government spending for civilian R&D was cut, under the rationale that such work should be left to private industry.

During the Bush administration in the late 1980s, with the demise of the Cold War and the collapse of the Soviet Union, momentum began to develop for government support for **generic, precompetitive**

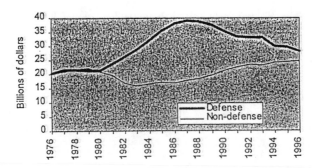

FIGURE 2.2   Federal budget authority for defense and nondefense R&D: fiscal years 1976 through 1996 (billions of constant 1987 dollars). (*Source:* National Science Foundation Report NSF 95-342.)

research on problems of interest to civilian industry. During this period, organizations such as the Semiconductor Research Corporation were created to foster such research and make the United States more competitive with other nations. What began as a modest effort in the Bush administration emerged as a major focus of technology policy in the Clinton administration. The Advanced Technology Program (ATP) of the National Institute of Standards and Technology (NIST) and the Technology Reinvestment Project of the Department of Defense were slated for rapid growth. With the election of a Republican Congress in 1994, these initiatives faltered. Although the ATP program survives, it is considerably smaller than projected. Other programs include Cooperative Research and Development Agreements (CRADAs) in which government laboratories (Federally Financed Research and Development Centers — FFRDCs — such as Los Alamos National Laboratory) cooperate with industry on projects that seek to transfer government know-how to the private sector.

Government involvement in applied R&D is not new. It has been happening in defense and space since World War II. What is at issue is involvement of the government in funding applied research directed at problems in the civilian sector. Early initiatives of this kind in the Johnson and Carter administrations faltered, some argued, because the government attempted to "pick winners" in commercial R&D. Current supporters of efforts such as ATP and the "Clean Car" Initiative argue that this danger is being avoided by heavy industry involvement in the formulation and execution of the projects.

Another significant shift took place in the 1980s. Prior to that, government-supported research outputs could generally not be privately patented or licensed. With the passage of the Bayh-Dole Act in 1980 and subsequent legislation, these restrictions were lifted, and research supported by government in universities and industry can now lead to patented outcomes for those performers.

In the mid-1990s, an overriding concern arose about government support for science and technology associated with the tight federal budget situation. It appeared as though support for R&D would fall over the next 5 years, in keeping with bipartisan desires to balance the federal budget. Research and development is in the discretionary part of the budget, that part most vulnerable to cuts. So far, at least one sector, health research, has done reasonably well, but overall we appeared to be in a no-growth or even reduced-funding mode. The effects could be particularly severe on universities, which are the primary performers of basic research. Industry support for R&D could conceivably make up some of this gap, particularly for engineering, although this is by no means certain, and the nature of the research performed in universities could very well change with a shift from government to industry support. Support for long-term, fundamental research is in jeopardy, as is industry's willingness to perform such research. At the start of 1998, there have been some indications that things may be beginning to turn around. Although countries go through phases in which such research may not be a significant part of their economic growth (e.g., Japan in the generation after World War II), fundamental research capability appears to be essential for the long term. Scientists and engineers are becoming more active and more vocal in making the case for support for R&D. Talk of federal R&D cuts has been replaced by talk of sizable R&D increases.

Recent Science and Technology Policy Issues

Within the past 5 years, numerous science and technology policy issues have received attention [Branscomb, 1993; Teich, 1997]. In one way or another these issues impact technology managers in industry as well as government. Some, such as the government's role in civilian R&D and the level of government support for basic research, have been discussed above. Other illustrative policy areas include the following:

Dual-use technologies: With the demise of the Cold War and limitations on defense budgets, interest arose in developing technologies that could be used to underpin developments both in the military and civilian sectors. Some significant steps have been made in this direction, driven both by the need to economize and to bring military technological developments to the level of recent civilian developments. This has resulted in the elimination of duplication, albeit at the expense of no longer keeping certain military and civilian functions separate.

R&D tax credits: One element of government policy that has been used to support R&D is the R&D tax credit, which allows companies that invest in R&D to reduce their taxes. However, efforts by some to make the R&D tax credit permanent have failed; credit in recent years has been repeatedly renewed but for limited periods of time (of the order of 1 year) by Congress.

State science and technology policy: With less emphasis on big government, states are assuming more of a role in a variety of areas, and science and technology policy may be no exception. Some states, such as California and North Carolina, have had significant science and technology policy offices and staffs since the 1970s. States often assume responsibility for federal functions as provided by law, such as environmental protection. There is a wide variation in the resources that states can bring to such an effort. At least one federal program, the Experimental Program to Stimulate Competitive Research (EPSCoR), has been established to encourage and support research in states that traditionally have received less federal funding.

University-industry cooperation: Closer cooperation between industry and universities in R&D activity is believed by some to be an essential element in keeping U.S. scientific and technological innovation healthy. In recent years, such cooperation seems to be increasing, and universities appear to be devoting more of their R&D portfolio to *strategic research* directed toward application areas of national importance. Government-funded programs, such as the National Science Foundation's Industry-University Research Centers and their Engineering Research Centers have contributed to this shift. University-industry cooperation, although growing and increasingly desired by both parties, presents some problems, including disputes over intellectual property and patent rights, and conflicts over the time frames in which the two parties are accustomed to working.

Careers for research scientists and engineers. Women and minorities: Doctoral graduates in the sciences have been experiencing difficulty in finding employment in universities in traditional faculty positions. In the biological sciences, growing numbers of postdoctoral research associates are populating medical schools for longer and longer period of time, waiting for permanent positions to open. There is even a growing class of temporary employees with substantial scientific and engineering credentials. Although some gains have been made by women and underrepresented minorities in certain branches of science, these groups are still substantially underrepresented compared with their percentage of the overall population in some fields, for example, in engineering. Efforts to increase women and minority need renewed attention and support in the face of financial pressures and attacks on affirmative action.

How Is Government Policy on Science and Technology Made?

Government policy is made by the democratic constitutional process characteristic of the United States. Policies may be established through passage of legislation by Congress followed by presidential approval. Where the proper authority exists, policies may also be put in place by executive order of the president.

A focal point for policy activity in the executive branch is the Office of Science and Technology Policy in the White House. The director of that office from 1993 to 1998, Dr. John Gibbons, is also Special

Assistant to the President for Science and Technology. Two high-level bodies that attempt to coordinate science and technology policy are the National Science and Technology Council (NSTC), which is chaired by the president and consists of cabinet officers and agency heads who have responsibility for activity that involves science and technology, and PCAST, the President's Council of Advisers on Science and Technology, consisting of leading individuals from universities and industry. Also of importance in the setting of government policy for science and technology are executive branch agencies, such as the Department of Energy, NASA, the National Institutes of Health, and the Office of Management and Budget (OMB); all of the president's budget proposals must pass through OMB and be coordinated.

Congress is also a vital player in the formulation of science and technology policy. The president's budget and program proposals must be examined and disposed of by the appropriate authorizing and appropriation committees. Sometimes committee jurisdictions do not correspond to cabinet department or agency jurisdictions; for example, in the Senate, budgets for the National Science Foundation and NASA must be approved by the same committee that also has responsibility for Housing and the Veteran's Administration. Conflict between the executive and legislative branches became particularly acute in 1995 when significant portions of the U.S. government were forced to shut down due to differences over the budget.

Congress has some support agencies that help with science and technology policy matters. They include the Congressional Budget Office, the Science, Technology and Medicine Division of the Congressional Research Service of the Library of Congress, and the General Accounting Office. In 1995, the Congress eliminated their Office of Technology Assessment, an agency that had performed substantive studies of science and technology policy issues and questions for 25 years.

In conclusion, technology managers need to be aware of the significant role that government plays in the formulation and implementation of science and technology policy. That policy has significant effects on the development of technology in private sector organizations; furthermore, it is changing in the face of a changing environment and conditions. Part of understanding science and technology policy is in understanding how this government of ours works, how policy is made. Furthermore, it involves understanding how to become a part of the policy-making process — how to make your views known [Wells, 1996]. There are numerous ways to do this: through lobbying, letter writing, visits to representatives, involvement in public interest organizations, and by running and being elected to office. Very few of our elected representatives have science and engineering backgrounds — they are mostly lawyers and business people. More citizen participation by scientists and engineers may very well be a key element of better science and technology policy.

Defining Terms

Applied research: Seeks to gain knowledge or understanding necessary for determining the means by which a recognized and specific need may be met.

Basic research: Seeks a fuller knowledge or understanding of the subject under study, without specific applications in mind.

Development: The systematic use of the knowledge or understanding gained from research directed toward the production of useful materials, devices, systems, or methods, including design and development of prototypes and processes.

Generic or precompetitive research: Research that falls within a gray area between basic research on one hand and research of proprietary interest for commercial products or processes on the other.

Policies for science: Includes research and development budgets, science advice to the President, government support for science education, other topics.

Science and technology policy: Government plans, programs and initiatives in support of science and technology, sometimes broadened to include all policy issues with significant scientific and technological components.

Technology policy: As opposed to science policy, tends to focus more on policies for getting research results into the private and public sectors, and on innovations and developments in industry.

References

Teich, A. H., Nelson, S. D., and McEnaney, C. *AAAS Science and Technology Policy Yearbook: 1996–1997*, American Association for the Advancement of Science, Washington, DC, 1997.

Bush, V. Science: The Endless Frontier, Report No. 90-8, National Science Foundation, Washington, DC, 1945.

Dupre, J. S. and Lakoff, S. A. 1962. *Science and the Nation*, Prentice-Hall, Englewood-Cliffs, NJ, 1962.

Branscomb, L., Ed. *Empowering Technology*, MIT Press, Cambridge, MA, 1993.

Morin, A. J. *Science Policy and Politics*, Prentice-Hall, Englewood Cliffs, NJ, 1993.

Smith, B. L. R., *American Science Policy Since World War II*, The Brookings Institution, Washington, DC, 1990.

Wells, W. G. Jr., *Working with Congress: A Practical Guide for Scientists and Engineers*, 2nd ed., American Association for the Advancement of Science, Washington, D.C., 1996.

Further Information

The American Association For the Advancement of Science publishes a Science and Technology Policy Handbook and an analysis of R&D in the U.S. budget annually. Their web site is www.aaas.org.

The National Science Foundation publishes *Science and Engineering Indicators* biannually. Their web site is www.nsf.gov.

The Office of Science and Technology Policy in the Executive Office of the President issues statements and reports on various aspects of science policy and technology policy periodically. Their web site is www.white house.gov/OSTP.html.

A major science and technology policy study was conducted by the Carnegie Commission on Science, Technology and Government. The concluding summary volume of the study is *Science, Technology and Government for a Changing World*, April 1993. Carnegie Commission, 10 Waverly Place, New York, NY 10003.

2.2   U.S. Tax Policy toward the Adoption and Creation of New Technology

Joseph J. Cordes

Although receipts from innovative activities are generally taxed in the same way as other business receipts, a number of specific tax provisions apply to the expenses incurred to develop innovations. These tax rules are especially favorable to investments that businesses make to develop their own technologies.

Costs of Acquiring New Technology

One way in which businesses can obtain the benefits of new technologies is to acquire such technologies from other companies. This can be done indirectly by buying new capital goods that embody new technologies or directly by buying patents.

Purchases of New Capital Goods

U.S. tax law subjects all new capital goods to the same general rules that govern the tax depreciation of capital assets, without regard to whether the capital asset embodies particular "new" technologies. Thus, aside from normal tax incentives for capital spending generally, there are no provisions that provide financial incentives for businesses to purchase capital goods that embody innovative technologies.

Tax Treatment of the Costs of Acquiring Patents

New technologies can also be acquired by purchasing patents from other businesses. When a company makes such a purchase it acquires an intangible asset. For example, in the case of tangible capital, such as machines or buildings, the commercial value of this asset will depreciate over time as the new technical knowledge protected by the patent declines in commercial value.

This fact is recognized in the U.S. tax code in that firms are allowed to claim depreciation deductions for the decline in the value of the patents they have acquired. As described in Section 2.8, these deductions are taken on a straight-line basis over a period of 15 years, depending on the legal life of the patent.

By comparison, tangible investments in machinery can be depreciated over recovery periods ranging from 3 to 10 years, using accelerated rather than straight-line recovery methods. Thus, the costs of acquiring patents are treated less favorably than those of purchasing equipment. The tax treatment of obtaining new technology through patents is also less favorable than that of developing the new technology internally through R&D, which is discussed in the next section.

Costs of Internally Developed Technologies

In contrast to the tax treatment of investments made by a business to acquire technologies from other companies, the tax treatment of the costs of developing technologies internally provide financial incentives for companies to make such investments. Special rules, set forth in Section 174 of the U.S. Internal Revenue Code, govern the deduction of costs that arise in connection with technological know-how developed by the firm; many businesses are eligible to claim a tax credit for certain expenses incurred in connection with R&D.

Broadly speaking, whether a particular item of expense is subject to these rules depends on whether it is considered to be incurred in connection with R&D, as defined by the tax code. If the particular item is not classified as R&D for tax purposes, it is subject to the relevant general tax rules that govern deductions of that type of business expense. The definition of R&D in the U.S. tax code encompasses basic and applied research, development of project requirements and specifications, construction of a pilot plant or production of a prototype, and obtaining a patent. Thus, operating expenses and capital costs incurred in connection with any of these activities are subject to certain special tax rules. By contrast, operating and capital costs incurred in connection with the tooling and manufacturing, manufacturing start-up, and marketing start-up phases of the innovation process are grouped together with "normal" product development and are not considered to be incurred in connection with R&D for tax purposes. Such expenses are subject to the same rules as expenses incurred in connection with noninnovative activities.

Expensing of Current Costs of R&D

To the extent a business activity is classified as R&D for tax purposes, current expenditures incurred in connection with such R&D may be deducted *immediately* or deferred and deducted over a period of at least 60 months, beginning with the month in which the taxpayer first realizes benefits from the R&D. Thus, the costs of creating new technical knowledge may be deducted more rapidly than the costs of acquiring plant and equipment or patents.

The option to expense rather than to capitalize R&D costs has been a long-standing feature of the U.S. corporate tax system. It appears that the provision was originally enacted for reasons of administration, rather than to provide incentives per se to R&D. However, notwithstanding the original intent, it is widely acknowledged that the effect allowing R&D to be expensed is to reduce the after-tax cost of investing in R&D as compared to other business investments (see Section 2.B).

The conceptual justification for this view among tax scholars is that industrial research is properly viewed as an activity whereby the firm creates for itself commercially valuable technical knowledge that would otherwise have to be acquired from other sources. By this logic, a firm that invests in industrial research in order to create the intangible asset, technological know-how, is in a position analogous to that of a firm that invests labor, materials, and capital to construct its own equipment or buildings rather than buys those tangible assets from other producers.

The uniform capitalization rule of the U.S. Internal Revenue Code normally requires that the costs of labor and materials and the annualized costs of capital used to "self-construct" an asset with a useful life that extends beyond the taxable year be capitalized into the value of the asset that is self-constructed and then deducted over several years once the asset is used in production. If this rule were applied to the current costs of R&D, "normal tax treatment" would require that R&D expenses be capitalized rather than expensed. Therefore, the option to expense, which is taken by virtually all producers, is viewed as a tax preference for R&D.

The position that expensing of R&D is something other than normal tax treatment is reflected in official publications of the U.S. government that treat expensing of R&D as a "tax expenditure". According to the Joint Committee on Taxation of the U.S. Congress (JCT), a tax expenditure is defined as "a revenue loss attributable to provision of the Federal tax laws which allow a special exclusion, exemption or deduction from gross income, or which provide a special credit, a preferential rate of tax or a deferral of tax liability." Because expensing of R&D is specifically identified in the Internal Revenue Code as an exception to the normal tax treatment of self-constructed assets as set forth in the uniform capitalization rule, it is considered to be a tax expenditure. The JCT has estimated that the value of R&D expensing, relative to normal tax treatment, is on the order of $2.0 billion per year.

Costs of Plant and Equipment Used in R&D

The after-tax cost of creating new technical knowledge can also be affected by tax depreciation rules that apply to the equipment and buildings used in the R&D process. The tax treatment of R&D capital appears comparable to that given to other forms of capital with durability similar to that of capital used in R&D activities.

R&D Tax Credit

Since 1981, eligible businesses have been able to claim a tax credit for research and experimental expenses excluding outlays for R&D capital and overhead. The current amount of the credit equals 20% of the amount by which eligible R&D exceeds a base amount. The base equals the product of a fixed ratio of R&D spending to sales (referred to as the fixed base percentage) times the average gross sales of the business in the preceding 4 years, with the proviso that the base amount cannot be less than 50% of the taxpayer's qualified research expenditures for the taxable year.

The fixed base percentage is defined as the lesser of .16 or the actual ratio of R&D spending to sales for at least 3 years between 1984 and 1988. Businesses without the required number of taxable years between 1984 and 1988 are assigned a fixed ratio of .03. Since 1994, businesses without the required number of taxable years have been allowed to retain the fixed base percentage of .03 for only the first 5 years, with additional calculations of the fixed base percentage required in the 6th through 10th years.

Since 1989, businesses have also been required to reduce the amount of R&D outlays that are expensed by the amount of the tax credit that is claimed. This has the effect of making the R&D tax credit taxable. For example, if the business' tax rate is .34 and it claims an R&D tax credit of $100, the amount of R&D that can be expensed under IRS Section 174 is reduced by $100, which costs the business $34 (= .34 × $100). This has the effect of reducing the net value of the R&D credit from $100 to $100 − $34 = $66. This is exactly equivalent to taxing the credit at a rate of 34%.

To illustrate how the credit would be calculated, assume that High Tech Corporation had R&D spending during the period from 1984 to 1988 that averaged 10% of sales and that sales in the 4-year period from 1993 to 1996 averaged $100,000,000. Its R&D base spending for 1997 would equal the fixed base percentage × average of 4 prior years sales = .10 × $100,000,000 = $10,000,000.

Once the base has been calculated, High Tech Corporation would be eligible to claim a tax credit equal to 20% of the amount of R&D spending that exceeded the $10,000,000 base. Thus, if R&D

spending in 1997 equaled $50,000,000, the initial amount of the credit that could be claimed would equal $.20 \times (\$50,000,000 - \$10,000,000) = \$8,000,000$. This amount would reduce the amount of R&D expenses that could be expensed by $8,000,000, which raises the business' tax liability by $.34 \times \$8,000,000 = \$2,720,000$. Thus, the overall effect of the credit would be to reduce the business's net tax liability by $5,280,000 = \$8,000,000 - \$2,720,000$. Looked at somewhat differently, for each $1 of R&D spending that exceeds the base, the business receives a (net) subsidy equal to $\$.20 - (\$.20 \times \$.34) = \$.20(1 - .34) = \$.20(.66) = \$.132$.

As discussed in Section 2.8, because a tax credit reduces a taxpayer's tax bill directly, the tax credit effectively provides a cash subsidy, delivered through the tax code, for eligible R&D expenses to companies that are able to use the credit to reduce their tax bills. Two broad features of the R&D credit have been intended to target the subsidy on certain forms of R&D spending.

Making the Credit Incremental

Since it was first enacted in 1981, attempts have been made to target the credit as much as possible on R&D spending that the business would not otherwise have made. This is the purpose of granting a credit only for R&D spending that exceeds the base amount described above.

Defining Eligible R&D Spending

A series of administrative rules has also been enacted in order to limit eligibility for the credit to R&D spending that is likely to lead to significant technological innovations. When it was first enacted in 1981, businesses were allowed to claim the R&D credit for costs incurred in connection with the development of an experimental or pilot model, a plant process, a product, a formula, an invention or similar property, and the improvement of already existing property of the type mentioned. In addition, to qualify for the credit the research or development activity had to be conducted in the United States, could not be funded by private or government grant, and could not involve the social sciences or the humanities.

These basic eligibility rules were tightened significantly in the Tax Reform Act of 1986 in response to two criticisms. One was that the initial regulations allowed businesses to claim R&D credits for relatively "routine" product development activities. The other was that the R&D credit in 1981 created incentives for firms to redefine activities as R&D in order to qualify for the credit.

Thus, since 1986, in order to qualify for the R&D credit, the company's R&D spending must satisfy the criteria established in 1981 and in addition be undertaken for the purpose of discovering information (1) that is technological in nature and also (2) the application of which is intended to be useful in the development of a new or improved business component of the taxpayer. In addition, such research is eligible for the credit only if substantially all the activities of the research constitute elements of a process of experimentation for a functional purpose.

The intended effect of these additional criteria is to limit eligibility for the credit to research activities that have the following characteristics:

1. The research activity must rely fundamentally on the principles of the physical or biological sciences, engineering, or computer science, rather than on other principles, such as those of economics. Thus, for example, research related to the delivery of new financial services, to advertising, or in some cases to the development of new consumer products, such as fast foods, does not qualify for the R&D tax credit.
2. The research activity must involve the "evaluation of more than one alternative designed to achieve a result where the means of achieving the result is uncertain at the outset." Thus, for example, expenditures for developing a new product or process does not qualify as research for purposes of the credit if the method for developing the new product or process is readily discernible at the outset.
3. The research activity must "relate to a new or improved function, performance, reliability, or quality." Thus, expenditures to assure the quality of a product or process that is already in commercial production or use or to implement cosmetic or style changes does not qualify as research for purposes of claiming R&D credit.

Policy Issues

The R&D credit has never been made a permanent feature of the U.S. tax code, but instead has been included as a temporary provision, requiring that the credit be reenacted every few years.

Most recently, the R&D credit expired on June 30, 1995, and was subsequently extended by the Small Business Job Protection Act of 1996 to apply to expenditures incurred between July 1, 1996, and May 31, 1997. The credit has since been extended once again in the most recent tax bill from June 1, 1997, through May 31, 1998.

There has been considerable policy debate about whether the R&D credit should be extended, and perhaps made permanent. Since 1981, The R&D tax credit has provide tax subsidies to industrial R&D spending in excess of $1 billion per year. The rationale for providing this subsidy is that investments in R&D often have social returns above the private return captured by the business that makes the investment. The existence of such benefits from R&D spending means that private investments in R&D may be less than would be socially optimal.

Economists generally agree that certain types of spending on industrial research can generate sizable external benefits and that, for this reason, government subsidies in some form for R&D may be justified. However, there is a range of views about whether, despite the efforts to target the credit, it encourages a significant amount of R&D spending that would not otherwise have been undertaken.

Supporters of the credit argue that it has been a cost-effective way of stimulating additional industrial R&D and should therefore be retained and made permanent. Those who hold this view maintain that R&D spending responds sufficiently to changes in its after-tax price so that each $1 of revenue lost in tax credits stimulates close to an additional $1 of industrial R&D. Moreover, supporters argue that the additional R&D stimulated by the credit provides significant benefits to society as a whole (external benefits) in addition to those captured by the firm financing the R&D.

Critics of the R&D credit disagree with this assessment of the credit's cost effectiveness. These critics argue that the weight of empirical research supports the view that each $1 of revenue lost in tax credits stimulates no more than perhaps $.40 of additional R&D. Moreover, while acknowledging that industrial R&D can generate significant external benefits, critics of the credit have pointed out that there is nothing in the way in which the credit is structured to ensure that the additional R&D that is encouraged necessarily has significant external as well as purely private benefits.

Despite the policy debate, the R&D credit has enjoyed broad bipartisan support. One indication of such support is that since 1981 the Reagan, Bush, and Clinton Administrations have all supported renewing the R&D tax credit in years in which it was due to expire. For example, in a web site posted during the 1997 debate about tax legislation, the Clinton Administration states that it ". . . recognizes the importance of technology to our national ability to compete in the global marketplace, and the research credit is one tool that is useful in supporting and fostering technology."

References

Cordes, J. J. Tax incentives and R&D spending: a review of the evidence, *Res. Policy*, 18. 119–133, 1989.

Cordes, J. J., Watson, H. S., and Hauger, S. The effects of tax reform on high technology firms, *Natl. Tax J.*, 40(3): 373–391, 1987.

Hall, B. R&D tax policy during the 1980s: success or failure, In *Tax Policy and the Economy, Vol. 7*, J. Poterba, ed., pp. 1–35, MIT Press, Cambridge, MA, 1993.

Moumeneas, T. and Nadiri, M. I. Public R&D policies and cost behavior of the U.S. manufacturing industries, *J. Public Econ.*, 63(1): 57–83, 1997.

Watson, H. S. The 1990 tax credit: a uniform tax on inputs and subsidy for R&D, *Natl. Tax J.*, 49(1): 93–103, 1996.

2.3   Policies for Industrial Innovation

Christopher T. Hill

While private industry is the major engine of commercial technological change, what governments do matters enormously to the technological change process. Every industrialized country has public policies intended to influence the pace and direction of technological change [Nelson, 1993], including incentives for private investment, government provision of key resources, and manipulation of the many "rules of the game" that influence private activity.

Public policy may be focused on encouraging the *first* commercially successful development and use of a new technology (i.e., on technological innovation) or it may be focused on applying existing technology in new places and/or new ways (i.e., on technology transfer and diffusion).

This section reviews the essential features of public policies for industrial innovation in the United States. The focus is on federal policy and on policies intended to stimulate the pace of technological innovation and change rather than on policies intended to influence the nature of that change.

Innovation Models, Technology Politics, and Technology Policy

The first significant attempts to establish explicit public policies intended to encourage technological innovation were made during the Kennedy and Johnson administrations in the mid-1960s. These early efforts did not survive, and the disagreements over them set the tone for debates about technology policy that have continued into the second Clinton administration.

On one side, the debate has featured those who believe that the U.S. economy can and should be more effective at developing and commercializing new technology and that government should help. On the other side are those who oppose government involvement on ideological grounds (the government has no business intervening in the private market place) or on the grounds that government intervention, however well intentioned, can never be as effective as the private market left to its own devices.

One of the key features of the political debate over technology policy is that most of the participants have limited understanding of how technological change happens and of the relationship of scientific understanding to such change. This has led to overdependence on the linear model of technological change.

Experiences during World War II lent credence to the linear model, in which technological change arises from the application of new scientific discoveries. For example, the discovery of nuclear fission in 1939 resulted in just 6 short years in the first successful atomic bomb. The Bush Report, which initiated public consideration of postwar science policy, emphasized the importance of national investments in basic scientific research as the key to a strong national defense, improved health, and a strong economy [Bush, 1945]. The message of the Bush Report was simple — valuable new technologies and a better life will flow almost automatically from government support of scientific research. The major policy problems within this framework were to develop new institutions to support fundamental research and the education of future scientists and engineers.

The linear model held sway in most public policy debates until the mid-1980s, and it continues to influence much contemporary consideration of technology policy. The same model motivated heavy investments by industry during the 1950s and 1960s in centralized, fundamental research laboratories far removed from the operations of the company. Like the industrial scientist encouraged to follow his or her own interests rather than address the express needs of the company, the large federally supported research and development (R&D) system in the United States that emerged after the war was designed to enable researchers in universities and federal laboratories to follow their own interests rather than to address the needs of society directly.

In critiquing the linear model, it is important to keep in mind that it was designed to replace the prewar model in which new technologies were thought to emerge *de novo* from the inspired insights of problem-solving men such as Whitney, Edison, Bell, McCormick, Ford, and the Wright brothers. It was widely believed that they were uniquely creative and not influenced by the work of scientists or of the inventors who preceded them. Clearly, the linear model preserved both the heroic aspect of the inventor model (for the hero, now an individual scientific genius, remained at the center of the linear model) as well as its essential disconnection from all that had come before (for the scientist depended on new discovery, just as the inventor proceeded without regard to the work of his predecessors). What was lost in the transition was the inventor's intense motivation to meet a societal need.

Today, owing to its failure to square with the facts, the linear model has become untenable. We understand that most new technologies are based on incremental improvements to existing technologies or on the "fusion" of two or more existing technologies rather than on scientific breakthroughs. Indeed, the first attempts at developing new technologies often occur before their scientific foundation is laid, not *after*. Information and understanding about the marketplace, societal expectations, and what can be manufactured, sold, and maintained successfully are as important as new scientific understanding and the preexisting technology base in determining which new technologies to pursue.

Supporters of limited government took heart in the linear model — it seemed to call for a policy based largely around government support of fundamental or basic research far removed from national need or economic purpose. To a substantial degree, it did not matter what basic research was done, since no one could say in advance where it might be useful. On the other hand, the more complex models call for a more diverse, nuanced, and purposeful government technology policy, and they require government agents to be more directly involved in choices that have traditionally been the prerogative of the private marketplace.

No good model has yet replaced the linear model in the policy debate. The most important notion is that the innovation process is complex and highly variable across time, place, technological fields, and firms; reality is considerably more complicated than the old inventor or linear models reflect. This complexity and variability militate, however, against the adoption of simple policies expected to have broad effect. Inevitably, the full richness of the innovation process argues for public policies that require public authorities to understand this diversity and that enable them to exercise considerable discretion. Furthermore, since innovation typically requires technical resources from many different firms and institutions, nurturing technical networks has emerged as an important aspect of how governments might encourage industrial innovation.

The Federal Government's Roles in Industrial Innovation

Broadly speaking, governments support industrial innovation by providing resources, establishing incentives for private investment, and setting the "rules of the game".

The U.S. government provides resources for industrial innovation by funding advanced training of scientists and engineers, R&D, and specialized facilities. For example, since the 1950s, the National Science Foundation (NSF) and other agencies have offered fellowships to outstanding students to help them pursue masters and doctoral degrees in science and engineering. Thousands of other graduate students are supported by research project grants to universities.

Federal financial support of R&D in industry has typically been intended to help pursue agency missions. For example, the Department of Defense, the Department of Energy (DOE) and its predecessors, NASA, and other agencies have contracted for new technology for missions such as national security, space exploration, and environmental protection.

More recently, modest federal support has been offered to firms and consortia of firms to develop new technologies, not for public missions, but for direct commercial purposes. The most prominent of these programs is the Advanced Technology Program (ATP) in the Department of Commerce (DOC), established in 1990 under the Omnibus Trade and Competitiveness Act of 1988 [Hill, 1997]. ATP pays part of the costs of industrial R&D projects selected in periodic competitions — some open to all technologies and others focused on specifics such as manufacturing, biotechnology, or materials. Both individual firms and consortia are eligible for ATP awards.

Another source of government funds for industrial technology is the Small Business Innovation Research (SBIR) Program [Wallsten, 1997]. Since 1982, each federal agency with a substantial R&D program has set aside part of its R&D funds to be awarded to small businesses. The set aside amount is now 2.5%, which yields more than 1 billion dollars annually for support of industrial R&D in firms with fewer than 500 employees by the 11 agencies with SBIR programs. So-called phase I SBIR projects typically involve awards of up to $100,000. Projects deemed likely to succeed commercially after phase I can compete for phase II awards, which today can range up to $750,000.

The federal government owns and operates, directly or through long-term contracts, more than 700 laboratories. Many have unique experimental facilities, most built to serve government missions, which industry can also use. The large NASA wind tunnels used by the aerospace and other industries were early examples of "user facilities". At several DOE laboratories, industry can use specialized and expensive instruments such as neutron sources, synchrotron light sources, high-power electron microscopes, combustion chambers, and hundreds of other devices [DOE, 1997].

In addition to funding aspects of the innovation process directly, the U.S. government creates indirect incentives to encourage firms to invest more of their own resources in innovative activity. For example, even though R&D is conceptually an investment that yields benefits over a sustained period of time, since 1954 firms have been able to deduct R&D-related expenses, except for fixed plant and equipment, from their gross income before payment of corporate income taxes for the year in which they are incurred. Beginning in 1981, firms have also been able to claim a credit against taxes owed for a portion of the increase in their R&D spending over the average of spending in the previous 3 years (see Section 2.2).

While direct funding and indirect incentives are both important to technology development, firms are also influenced by federal policies that affect the rules of the game under which they operate. For example, the Constitution gives Congress the power to establish a system of patents to help ensure that inventors can profit from their investments. In return, publication of patents facilitates dissemination of new knowledge to other inventors.

The antitrust laws help to preserve a competitive market environment that rewards innovation by limiting monopoly power and enabling new entrants and new ideas to enter the marketplace. At the same time, the National Cooperative Research Act (NCRA) of 1984 facilitates cooperative R&D among competitors. NCRA enables research consortia to disclose their membership and purpose and thereby obtain protection against being found guilty of antitrust law violations simply by the fact that they exist, as well as against having to pay triple damages if they are found guilty. None of the several hundred ventures registered under NCRA has been the 0bject of antitrust action.

A host of other policies establish rules for the conduct of R&D and the development and use of new technology. For example, environmental, health, and safety standards may limit the kinds of new technologies that may be used, while some of the governing legislation includes special provisions to encourage development and use of improved technologies to meet those goals. Government agencies such as the National Institute of Standards and Technology (NIST) and DOE participate in the establishment of performance and design standards for new products, some of which are voluntary and some of which have the force of law. The Food, Drug and Cosmetics Act requires premarket testing and approval of new drugs and food additives and imposes standards on the nature of the tests that may be done on such materials, including standards for the conduct of animal tests. Export controls limit the overseas sale of selected new technologies, such as data encryption, that may have important national security implications.

Recent Directions in Innovation Policy

Three key foci of recent federal policy and program development for technological innovation are cooperative R&D, dual-use technology, and technology transfer and diffusion.

Many recent government R&D programs, whether for public or private missions, have emphasized the value of R&D collaboration among firms, as well as among industry, universities, and government laboratories. Two rationales for encouraging collaboration are (1) R&D efficiency is improved when organizations with similar problems and capabilities cooperate to share costs and avoid duplicative efforts and (2) R&D effectiveness is enhanced when organizations bring complementary capabilities to projects. Several programs, most prominently NSF's Engineering Research Centers program, provide financial inducements to university-based cooperative research activities that include active participation by industry. The multiagency Program for a New Generation of Vehicles and the DOE Industries of the Future Program seek active participation by firms in defining a shared R&D agenda and in conducting a program of cooperative R&D projects, some of which enjoy federal cost sharing.

Increasingly, agencies are seeking to take advantage of the extensive technological capabilities of the private sector by supporting "dual-use" R&D programs that have as their objectives both the development of new technology for public purposes and the development, on the same base, of technology for commercial purposes [Cohen, 1997]. The Technology Reinvestment Project (TRP) of the early Clinton administration is the best known — and most controversial — illustration, but similar dual-use efforts are being mounted with less fanfare by DOE and NASA in pursuit of their objectives. Dual-use approaches are thought to be particularly useful in very fast-moving arenas such as microelectronics and software engineering, where the pace of change in the private sector is likely to be greater than the more cumbersome apparatus of government can facilitate in the public sector.

Finally, the federal government has become increasingly involved in helping to transfer new and improved technology from government laboratories and universities to industry, as well as among firms. The Stevenson-Wydler Technology Innovation Act of 1980 made technology transfer a mission of all the federal laboratories, while amendments to it, including the Federal Technology Transfer Act of 1986 and the National Competitiveness Technology Transfer Act of 1989, gave the federal laboratories authority to set up Cooperative Research and Development Agreements with industry to develop and extend new technologies. Under the Omnibus Trade and Competitiveness Act of 1988, NIST was given the authority to support a network of manufacturing technology transfer centers to help upgrade the technological activities and capabilities of private firms. Today, the resulting NIST Manufacturing Extension Partnerships program supports more than 100 such centers around the country, each of which is associated with a university or other nonprofit organization.

Issues in Innovation Policy

Three issues animate current debates over U.S. technology policy: assuring program accountability and measuring effectiveness, establishing workable federal/state relationships, and managing the participation of foreign firms in U.S. programs.

Despite the great ideological struggle over the proper federal role in industrial innovation and despite the substantial amount of public resources that supports the inherently uncertain industrial innovation process, the implementation of such programs has been remarkably free of allegations of inappropriate use of public funds. Instead, accountability has been concerned with identifying and measuring program results. Progress has been made in some areas, such as the ATP, which has been extensively although not conclusively evaluated [Hill, 1997]. However, because of the long time lags from federal R&D investment to measurable commercial outcomes and because the federal investment is rarely the sole contributor to project success, it has proven difficult to assess program impacts in a convincing manner. The Government Performance and Results Act of 1993 has stimulated attempts to assess government R&D programs, but the results have been limited to date.

Even as the federal role in technology development has grown over the past 2 decades, so have the individual states established programs and made investments in technology development and transfer to strengthen their industries. This has led, on occasion, to conflicting goals, priorities, and practices among the different levels of government. More recently, attempts have been made to improve coordination of state and federal technology programs [DOC, 1997].

Governments support technological innovation to achieve national purposes, whereas firms develop new technology to pursue corporate objectives, often without regard to national boundaries. U.S. firms have interests overseas and foreign firms have interests in the United States. This lack of congruence between public and private objectives has led to attempts by Congress to limit both the use of publicly supported technologies by U.S. firms overseas and the participation of foreign firms in U.S.-supported programs [NAE, 1996]. For the most part, industry opposes such restrictions. Workable compromises embedded in several programs require, for example, that foreign firms can participate in U.S. projects only to the extent that U.S. firms can participate in projects in their home countries on the same basis or that firms participating in U.S.-supported projects must promise to conduct operations based on the resulting technologies initially or principally in the United States.

References

Bush, V. *Science — The Endless Frontier: A Report to the President on a Program for Postwar Scientific Research*, reprint 1990, U.S. National Science Foundation, Washington, DC, 1945.

Cohen, L. R. Dual-use and the technology reinvestment program, In *Investing in Innovation: Creating a Research and Innovation Policy that Works*, L. M. Branscomb and J. H. Keller, eds., chap. 6, MIT Press, Cambridge, MA, 1997.

DOC. Technology Administration, U.S. Department of Commerce, United States Innovation Partnership, URL http://www.ta.doc.gov/Usip/default.htm, 1997.

DOE. U.S. Department of Energy, User Facilities, URL http://www.er.doe.gov/production/bes/dms/facilities/faclhome.html, 1997.

Hill, C. T. The advanced technology program: opportunities for enhancement, In *Investing in Innovation: Creating a Research and Innovation Policy that Works*, L. M. Branscomb and J. H. Keller, eds., chap. 7, MIT Press, Cambridge, MA, 1997.

NAE. *Foreign Participation in U.S. Research and Development: Asset or Liability?*, National Academy of Engineering, National Academy Press, Washington, DC, 1996.

Nelson, R. R. National Innovation Systems: A Comparative Analysis, Oxford University Press, 1993.

Wallsten, S. Rethinking the small business innovation research program, In *Investing in Innovation: Creating a Research and Innovation Policy that Works*, L. M. Branscomb and J. H. Keller, eds., chap. 8, MIT Press, Cambridge, MA, 1997.

2.4   Research and Development Laboratories

Hui-Yun Yu and Kuang S. Yeh

Due to increasing research and development (R&D) spending and similarly increasing complexity of technology, the majority of R&D activities are assumed by organizations funded primarily by government instead of private experimental labs. Since there is R&D spillover, the government is expected to have an active role in **R&D activities**. The role of R&D laboratories has changed significantly over the past few decades. Formerly, R&D laboratories were devoted to pure academic research or military mission-oriented research. This action avoided insufficient investment in R&D. However there was attendant losses of social welfare. More recently, however, enhancing the competitiveness of enterprises has become a significant goal for R&D laboratories. Currently R&D laboratories cooperatively engage in applied research and diffusion-oriented research to facilitate firms' adoption of technology [Noori, 1990].

According to Taiwan's R&D spending reports, in 1994, about 48% (excluding the defense R&D investment) of R&D funds came from the public sector. From this 48% of R&D resources, R&D laboratories became responsible for 52.6% of R&D spending [NSC, 1995]. R&D laboratories are expected to function as a vehicle of science and technology (S&T) policy, providing policy consultants, executing research projects, training research talent, and diffusing technological knowledge. The importance of industrial R&D laboratories and knowledge production are shown in Fig. 2.3. As the figure indicates, the emergence of industrial R&D laboratories stimulates the production of S&T knowledge. In order to take advantage of the abundant S&T knowledge base, the importance of firms' absorptive capacity and R&D investment can no longer be ignored. When firms emphasize R&D investments, the appearance of innovations is increased as well as the accumulation of knowledge. This driving force pushes firms and government to form R&D laboratories to jointly engage in R&D activities. Traditionally, public or nonprofit R&D laboratories have been treated as a means for government S&T policy without deliberate study of their organizational characteristics. This review focuses on public or not-for-profit R&D laboratories, excluding in-house R&D laboratories, and the existing conflict between the survival of the organization's economic goal and compliance with policy or a political goal [Bozeman and Crow, 1990].

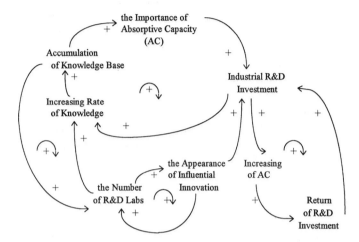

FIGURE 2.3 The mechanism of industrial R&D laboratories and knowledge production.

Classification of R&D Laboratories

Traditionally, R&D laboratories were classified into three categories: basic research, applied research, and technology development. Distinctions can be made between research-oriented (R-intensive) and development-oriented (D-intensive) organizations. R-intensive organizations are those that attempt to generate original inventions and expand the frontier of knowledge. D-intensive organizations focus their efforts on modifying inventions and existing innovations to meet their specific needs.

Another typical classification is according to the type of ownership: public, nonprofit, and private. Bozeman and Crow [1990] propose a new classification of R&D laboratories by funding for the appearance of a more hybrid organization, such as joint ventures, research consortia, and strategic alliances. They empirically suggest that R&D laboratories used for basic research should have stable and continuous funds to guarantee effective output. Alternatively, applied or development R&D laboratories should rigorously respond to industrial technological needs without extra public funds. R&D laboratories are formed according to professional fields with various combinations of ownership and research focus. In fact, there are no pure private or public laboratories nor are there pure R- or D-oriented laboratories.

Moreover, the expected roles of public R&D laboratories are diverse. Noori [1990] enumerates three specific roles that a government can play: an assessor role, an encouragement role, and a participative role. Tripsas et al. [1995] maintain that R&D laboratories have resource-providing mechanisms, institutional mechanisms, and administration mechanisms, that is, R&D laboratories play a role in providing resources, facilitating interfirm cooperation by improving R&D infrastructure, or directly joining cooperative projects.

The size of public R&D laboratories is also varied. For example, there are over 700 laboratories across the United States, employing one sixth of the nation's research scientists and engineers. Japan also has government laboratories, although the number is limited compared to the United States [Eto, 1984]. These differences notwithstanding, R&D laboratories face common management issues. A discussion of these issues follows.

Management Issues of R&D Laboratories

Professional Personnel

Since professional employees provide the bulk of labor for R&D laboratories, it is important to note some professional employee characteristics affecting organizational performance. They characteristically are not as motivated by economic rewards as are other workers, and they resist bureaucratic control. They tend to be more loyal to their profession's ethics than to organizational rules. They enjoy freedom and hate to be bound. They are more concerned about quality than cost control; more about technology research than technology diffusion; more about how to work independently than how to collaborate with others; and more about enhancing professional capability than improving management skill [Anthony and Young, 1994]. The R&D manager, therefore, must find a balance between individual freedom and organizational effectiveness.

There are some methods to aid managers in achieving this balance between individual freedom and organizational effectiveness. To motivate R&D personnel to achieve the organizational goals, managers are encouraged to adopt the following practices: use peer pressure to effectively finish routine tests or experiments for customers; hold conferences with senior experts to establish standards for research quality; connect output measures with reward systems, thereby establishing equity; survey customer opinions to keep staff aware of client needs; and frequently appear in work area to establish familiarity with research in progress [Lie and Yu, 1996].

Performance Assessment

The performance of R&D receives scrutiny from many points in the R&D process. Assessors include producers, mediators, technology adopters, and end customers. Output can be divided into two parts, direct and indirect output. For example, in the discovery of a new species of bananas, the new species is a direct output benefiting the R&D project. The indirect output is the increase of the existing knowledge base of bananas species. For public or not-for-profit R&D laboratories, the direct output of R&D is important for its economic revenue, as is the indirect output for building organizational legitimacy.

Uncertainty of output is the primary characteristic of an R&D organization. The well-known PARC, a laboratory resourcefully supported by Xerox, had little contributions to its funding since it was perceived as lacking clear goals and suffering from mismanagement [Uttal, 1996].

Performance evaluation of R&D laboratories is difficult for two additional reasons. One is a problem of measurement. As Mintzberg [1996] articulates: "Many activities are in the public sector precisely because of measurement: If everything was so crystal clear and every benefit so easily attributable, those activities would have been in the private sector long ago." The output of R&D laboratories is not easy to measure objectively; for example, technological capability, satisfaction, economic benefits of technology transfer, productivity improvement, employment creation, and customer responsiveness are performance criteria not amenable to precise quantification. The second reason performance evaluation is difficult is due to the selection of evaluation criteria. External and internal organizational members seldom agree on what should be evaluated. With diverse expectation, various actors in different stages of the R&D process prefer different criteria for R&D laboratories [Spann et al., 1995].

The assessment of R&D laboratories requires soft judgment on process that hard measurement on results cannot provide. Instead of examining what R&D laboratories' outputs are, managers should focus on influential factors of institutional effectiveness. As we know, the further the distance from R&D laboratories to customers, the more difficult for R&D laboratories to function effectively [Smilor and Gibson, 1991]. It is wise, therefore, to choose a location for the research facility near to ones' customers. Moreover, including R&D research personnel in laboratory customer relationship is critical to shorten the invisible distance and to promote interaction. The more diverse the R&D personnel, the easier is the effective boundary span with customers. Hence, R&D laboratory managers would not only better identify performance evaluation practices but would also create or improve the organizational infrastructure to facilitate members' job performance.

Resource Competition

Logically, the degree of competitiveness can be measured by the variance of one's annual revenue budget and growth rate. The factors relating to competitiveness include growth potential of industries, allocation of governmental research resources, and supply of research talent. The supply of research talent is dependent upon many factors. Governmental intervention is a major one. The policy of critical industries in most developing countries has profound impact on the flow of research talent. Critical industries have attracted abundant talent, but noncritical industries have a limited talent pool to draw from. Second, allocation of government funds has a profound impact on R&D laboratories. Funds coming from the public sector are characterized by long-time stable financial support, with little apparent evaluation criteria. This causes public or not-for-profit R&D laboratories to place more emphasis on the fund source rather than on their customers, thus preempting the public or nonprofit R&D laboratories from the pressure of competitive markets to a degree. Finally, the higher the potential for industrial growth, the more resources from industries become available. This causes R&D laboratories to pay more attention to industrial needs.

The work of scholars in the field provides some suggestions for R&D laboratories to face competition and to aid the management of resources. The first of these is to decrease the rivalry. R&D laboratories should increase efforts to collaborate with other R&D laboratories. Alternatively, R&D laboratories can increase size, thereby enlarging the available resource base. One method to increase the available resource base is for R&D laboratories to have a generalist strategy to cover various research fields to obtain funds and thereby lessen the uncertainty of external resources [Pfeffer and Salancik, 1978]. Third is to balance external demands. Meeting external needs results in having additional constraints on organizational autonomy. If R&D laboratories fail to obtain recognition, it limits their ability to attract financial support as well as to attract talented professionals. This then limits the ability of the laboratory to compete for other resources. The result is that the R&D laboratories must reposition themselves to find other funding sources or meet the specific demands of a fund source. For example, Taiwan Banana Research Institute (TBRI), internationally famous for banana research, was established in 1970 to support the banana-export industry. Owing to the shift of Taiwan industrial structure during the last 2 decades, banana export to Japan declined dramatically. The main source of funding, NT$1 per box of foreign banana sales from banana farmers, was decreasing and threatening TBRI survival. To counter, the TBRI tried to reposition itself to provide foreign technology support to obtain funding from sources other than the Council of Agriculture [Liu and Yu, 1996].

This study explores common characteristics and management issues of public or nonprofit R&D laboratories. These laboratories are characterized by a lack of clear-cut financial goals and direct market competition. With the changing role of public laboratories and high social expectation, they indeed do have certain R&D objectives to accomplish. The personnel in these laboratories are characterized by independence, not cooperation; they are loyal to their profession's ethics, not to organizational rules; and they are creative, not obedient. Managers in R&D laboratories face some common issues. They must identify the balance between personal freedom and organizational effectiveness. Concurrently, they have to concern themselves with performance assessment and obtaining survival resources. They must do this while simultaneously dealing with organizational autonomy and external effectiveness.

Defining Terms

R&D laboratory: The independent unit focusing on S&T R&D activities. The R&D laboratory in this review excludes domestic R&D laboratories within a specific company acting autonomously.

R&D activities: R&D comprises creative work undertaken on a systematic basis to increase the stock of knowledge. Criteria for R&D activities are (1) an innovative increase of knowledge and (2) updated application of knowledge. The ordinary or routine activities undertaken at S&T institutions or units are non-R&D activities.

References

Anthony, R. N. and Young, D. W. *Management Control in Nonprofit Organizations*, Irwin, Boston, 1994.

Bozeman, B. and Crow, M. The environment of US R&D laboratories: political and market influence, *Policy Sci.*, 3:25–56, 1990.

Eto, H. Behavior of Japanese R&D organizations — conflict and its resolution by informal mechanisms, In *R&D Management Systems in Japanese Industry*, H. Eto and K. Matsui, eds., North-Holland, New York, 1984.

Liu, C. Y. and Yu, H.-Y. *The Organizational Characteristics Study on Governmental Research Institutes*, NSC (NSC85-2416-H-110-015, Taipei, 1996.

Mintzberg, H. Managing government, government management, *HBR*, May–June: 75–83, 1996.

National Science Council (NSC). *Indicators of Science and Technology — Republic of China*, National Science Council, Taipei, 1995.

Noori, H. *Managing the Dynamics of New Technology*, Prentice Hall, Englewood Cliffs, NJ, 1990.

Pfeffer, J. and Salancik, G. R. *The External Control of Organizations — A Resource Dependence Perspective*, Harper & Row, New York, 1978.

Smilor, R. W. and Gibson, D. V., Technology transfer in multi-organizational environment: the case of R&D consortia, *IEEE Trans. Eng. Mgt.*, 38(1): 3–13, 1991.

Spann, M. S., Adams, M., and Souder, W. E. Measures of technology transfer effectiveness: key dimensions and differences in their use by sponsors, developers and adopters, *IEEE Trans. Eng. Mgt.*, 42(1): 19–29, 1995.

Tripsas, M., Schrader, S., and Sobrero, M., Discouraging opportunistic behavior in collaborative R&D: a new role for government, *Res. Policy*, 24: 367–389, 1995.

Further Information

A good discussion on performance assessment is presented in *Public Sector Efficiency Measurement* by J. A. Ganley and J. S. Cubbin (North-Holland, New York, 1992). "Technology Transfer in Taiwan's Information Industry: The Lessons" by Hui Yun Yu and Kuang S. Yeh (*Res. Technol. Mgt.*, 39(5): 26–30, 1996) proposes suggestions about building ethical norms among professionals to shorten the technology transfer gap between public laboratories and firms.

A detailed study of communication patterns by inter- and intralaboratories is introduced in by Thomas John Allen (MIT Press, Cambridge, MA, 1984).

2.5 Deregulation of Telecommunications: The Case of the Baby Bells

Yolanda Sarason

The evolution of telecommunication technology illustrates the upheaval technological changes can bring to an industry. This section discusses how technological change facilitated the break-up of the Bell System and includes a presentation of the divergent strategies of the Baby Bells since divestiture. A value chain analysis of the industry helps explain the rationale behind the strategic actions of the Baby Bells as well as actions of other telecommunication firms.

The Bell System

For almost 100 years, American Telephone and Telegraph (AT&T, also known as the Bell System) accompanied the United States into the industrial age, enabling access to the most reliable communication system in the world. As ubiquitous as the federal government, the company was seen as the definition of reliable and impenetrable. In the 1970s, AT&T was the world's largest corporation. The company's

$144 billion in assets was more than the gross national product of all but 20 countries [Kleinfield, 1981]. The company had evolved into a vertically integrated organization, providing what came to be known as "end to end service" [Coll, 1986]. AT&T Long Lines division provided long-distance service, the 24 Bell Operating Companies offered local telephone service, Western Electric manufactured telephone equipment, and Bell Labs conducted research and development.

In the 1970s several events began to threaten this vast corporation. The Justice Department initiated an antitrust lawsuit against AT&T, and legislation was pending in Congress that threatened to further constrain the Bell System's activities. Competition was entering every aspect of the telecommunication industry, from the manufacturing of telephones to long-distance service. AT&T Chairman Charlie Brown began to think of the telephone industry as a one-way hole, designed to let a myriad of competitors in, while it kept AT&T out [Coll, 1986]. The company was changing from denying competition in the telecommunication industry to an acceptance of the inevitability of a competitive environment.

In 1982, with the antitrust lawsuit that called for divestiture of all aspects of the Bell System coming to a conclusion, the senior management team agreed to a consent decree that would result in the full divestiture of the 22 operating companies that had been responsible for local telephone service. Key managers and the board of directors of AT&T agreed to this divestiture because they saw the operating companies as the most-regulated and least-profitable portions of their business. It was decided that these companies would be clustered into seven Regional Bell Operating Companies (**RBOCs**). These organization's would become known as the Baby Bells. Figure 2.4 shows the geographic boundaries of each of these organizations.

FIGURE 2.4 Baby Bell's geographic territories.

In January, 1982, the agreement to break-up the Bell System was announced. After 1984, each Baby Bell would be required to operate as a stand-alone organization, separate from AT&T, and would chart its future separate from the original Bell System. To date, after more than a decade since divestiture, these once-similar organizations have chosen different strategic paths. The following discussion will outline the Baby Bells strategic decisions since divestiture.

Strategic Differences among the Baby Bells

Even at divestiture, the Baby Bells could see pending threats to their business. Though they were cash rich, regulation prevented them from entering certain businesses, while changing technology was allowing competition into their core local telephony business. Since divestiture, the Baby Bells have raced to diversify their businesses into arenas in which they believe they can compete. The various strategies the

Baby Bells have chosen can be typified as: (1) focused telecommunication players, (2) diversified multi-media organization, or (3) organizations trying to compete both in telecommunication and multimedia. The following section outlines the choices each organization has made. The subsequent discussion will focus on the causes and rationale behind these strategic actions.

Focused Telecommunicators

The group of companies that have been typified as focused telecommunicators are NYNEX, Ameritech, and Pacific Telesis. Each of these organizations has chosen to stay relatively focused on the telecommunications industry, by moving into the wireless or the long-distance market or both.

Ameritech

Ameritech is the least diversified of the RBOCs with only 13.9% of its revenue generated from diversified activities. Ameritech kept the traditional organizational structure, which was organized by geography, throughout the 1980s. Just before William Weiss retired as CEO, he restructured the company from being organized by geographic regions to a company that was organized by major products and service groups. Ameritech has been the most vociferous Baby Bell in calling for open competition for local phone traffic and long-distance service. It has petitioned the FCC and the Department of Justice, overseers of the breakup decree, to let it "unbundle" its operations and allow any competitor to sign up customers for local telephone service and to hook up switches to Ameritech's network. In return, Ameritech wants permission to offer long-distance service in its region, especially between large cities such as Chicago and Milwaukee. Even though the company has not shown a desire to be an active player in the multimedia industry, it has joined a strategic alliance with Bell South, Southwestern Bell, and Disney for joint programming.

NYNEX

NYNEX diversified in an attempt to become a worldwide computer systems company before retrenching to address poor performance in the local exchange business. NYNEX helped fund Viacom's battle against QVC to take over Paramount, and this investment helped Viacom win its battle. NYNEX is also part of the alliance with Bell Atlantic, Pacific Telesis, and Michael Ovitz's Creative Artists Agency. NYNEX believes its capital is better invested abroad, as indicated by its ownership of 20% of Telecom Asia, which will install 2 million phone lines in Bangkok over the next 5 years. Subsequent to the passage of the Telecommunications Act, Bell Atlantic and NYNEX announced plans to merge their two organizations, which is scheduled for completion in 1997.

Pacific Telesis

Pacific Telesis set out to become the world leader in wireless networks. In June 1994, the company announced the planned spin-off of a new company called Air Touch, which provides wireless services worldwide. The spin-off allows the company to bid for personal communications services (**PCS**) licenses. Pacific Telesis has adopted the motto "California First" and plans to spend $16 billion in California through the year 2000. The company's goal is to make information superhighway-type services available to half of all homes in California by the century's end. As the construction gets up to speed, it is connecting 2000 homes a day or 700,000 homes a year. In a move that caught watchers of telecommunications industry by surprise, Southwestern Bell (SBC) announced plans to purchase Pacific Telesis, which was completed April 1, 1997.

Multimedia Players

The multimedia players have been more likely to enter the cable industry as their diversification strategy. This decision has required them to operate out of their original region because the divestiture agreement prevents them from operating competing services in their local service region. These companies are US West and Bell South. The following description outlines their strategic actions.

Bell South

Bell South is the largest regional phone company, measured by number of phone lines (19 million) and annual revenue ($16 billion). The growth in Bell South's territory has enabled it to roll out state-of-the-art fiber-optic technology without having to write off existing investments. Bell South paid $100 million for a 2.5% stake in Prime Cable of Austin, TX, which is part of SBC's turf. The company has indicated that the company does not have plans to buy any more large cable companies in the near future but instead will build fiber-optic and coaxial cable networks within its region. Bell South's plans are to develop high-tech networks as demand for **broadband** services increases. The company also joined the bidding for Paramount Communications with an investment in QVC contingent upon the latter's successful bid. Bell South joined the programming partnership with Disney, Ameritech, and SBC.

US West

US West has been the most active and widest diversifier, venturing into many unregulated activities. Initial efforts after divestiture were focused on diversifying into real estate and financial services, but subsequently the company exited these industries. Since 1989, the company has diversified into the cable and entertainment industries. In 1993 it paid $2.5 billion for 25.5% of Time-Warner Entertainment and its cable systems, cable channels, and movie studios as well as bought two Atlanta-area cable TV companies. In February 1996, US West announced the purchase of Continental Cablevision, which is the largest combination of a Baby Bell and a cable company. With its Time-Warner properties, US West manages 16.2 million domestic cable customers and has access to some 13.9 million homes abroad, making it the third largest cable company in the United States.

Multimedia and Telecommunication's Players

Two companies are trying to penetrate both the telecommunications market and the multimedia market. These companies are Bell Atlantic and SBC (formerly Southwestern Bell). Following is a synopsis of these companies' strategic actions.

Bell Atlantic

Bell Atlantic pursued scattershot diversification until Ray Smith became CEO in 1988. At this time, the company began focusing on offering integrated wireline and wireless service in its region. In 1993, Bell Atlantic proposed to buy TCI for around $33 billion. The much-publicized takeover of TCI was to be the largest merger in history, but was thwarted in the final stages of negotiations because of cable TV rate cuts ordered by the FCC. In May 1994, Bell Atlantic unveiled its plans to build a $11 billion flexible network architecture to service 20 markets by 1990. After passage of the Telecommunications Act, Bell Atlantic and NYNEX have pursued a merger of their two organizations, which is scheduled for completion in 1997.

SBC

Starting in the late 1980s the company bought Metromedia Cellular, which doubled its wireless business, and eventually SBC became known as the best cellular operator among the regional Bells. The company invested $1 billion to buy 10% of Mexico's telephone company, Telefonos de Mexico, which operates next door to SBC territory. The company was one of the first to get into cable television in the United Kingdom, jointly operating a cable company with Cox Cable and offering TV and telephone service over the same network. In October 1994, the company renamed itself SBC Communications. In 1997, the company purchased Pacific Telesis, bringing together two large RBOCs. In a move that surprised industry watchers, AT&T and SBC announced plans to merge organizations. This would bring together the largest long-distance company with the largest provider of local telephone service.

Value Chain Analysis

Why would these companies make these patterns of decisions? A value chain analysis is a useful tool for understanding the strategic actions of the Baby Bells since divestiture. This analysis traces an economic activity from inputs to outputs of goods or services. Most goods or services are produced by a series of vertical business activities such as acquiring supplies of raw material, manufacturing final products, and distribution [Porter, 1985]. If a company decides to enter a business along its value chain, it is deciding to vertically integrate.

The telecommunication's industry can be looked at as a series of value chain activities. The activities trace the generation, transmission, and delivery of information through devices, i.e., telephones (see Table 2.1). Traditionally information, generated by the consumers, is transferred via phone lines to other end users. After divestiture, this series of activities was divided. AT&T (and other long-distance carriers) provided the long-distance *transmission* of telephony services, and the Baby Bells provided primarily the *delivery* of services. The *devices* (telephones) were provided by a variety of companies. The Baby Bells were prevented from manufacturing devices because of the divestiture agreement.

TABLE 2.1   Value Chain of Converging Industries

Industry	Value Chain			
	Generation	⟷ Transmission ⟷	Delivery ⟷	Devices
Telecommunication	Public	Long distance and wireless (AT&T, MCI, Sprint)	Local access carriers (Baby Bells)	Telephones (GTE, AT&T, Sony)
Computer	Public	Long distance (AT&T, MCI, Sprint)	Local access carriers (Baby Bells)	Computers, software, operating systems (Microsoft, IBM)
Cable	Entertainment companies (Disney, Time-Warner)	Satellite (TCI, Continental Cablevision)	Cable (TCI, Continental Cablevision)	Cable box, TV (TCI, Continental Cablevision)

Vertical integration: telecommunications	Horizontal Integration: multimedia

Conceptualizing telephony as a series of value chain activities helps explain the behavior of the organizations that have chosen to focus on telecommunications. They are choosing to not only provide the *delivery* of service but also to provide the *transmission* of service through either wireless or long distance. These companies are making a decision to backward vertically integrate. Pacific Telesis and SBC are two that have invested heavily in the wireless industry. The companies are betting on the economies of scale and scope that can be captured by this backward vertical integration.

In order to understand the decision to diversify into multimedia, it is necessary to understand the technological changes of related industries. Changes in technology have meant that three once-distinct industries are converging. These industries are the telecommunications, computer, and cable industries. The principle reason that these industries are converging is that the same information can be transferred via these three mediums. The telecommunication's industry was defined by the two-way transmission of information, which was transmitted via twisted pair wire. The cable industry was defined as one-way transmission of information, which was transmitted via coaxial or fiber optic cable wire. Because of the availability of wireless communications and **broadband** networks, it is possible to transmit large amounts information with two-way communications. The computer industry is becoming more connected through the Internet, which for now is being transmitted primarily via phone lines. This convergence of technologies is becoming known as the multimedia industry [Maney, 1995]. True interconnectivity of technologies is technically, but not yet economically, feasible. The goal is to turn cable systems into mini-informational highways able to carry phone calls, movies on demand, and all types of Internet traffic.

When a company chooses to diversify by offering the same service in a similar industry, it is known as horizontal integration. Understanding technological changes in the three industries with a value chain analysis shows that the decision to move into multimedia is a decision to horizontally integrate (see Table 2.1), that is, traditionally the Baby Bells have offered the *delivery* of telecommunications services. Some of the Baby Bells have horizontally integrated by choosing to enter the *delivery* of cable services. These companies are choosing to diversify by buying or operating cable companies. These companies are betting that there will be synergy in the delivery of these services and that, when the cable, telecommunications, and computer industries do converge, they will be positioned to be viable players in the new competitive arena.

The decision to vertically integrate or to horizontally integrate depends on the size of the company's **footprint**. The footprint is the number of customers available to receive the company's services. For the wireless market, this is captured through PCS or cellular licenses. For the cable and telephone companies, this is represented by the number of homes wired. The mergers and alliances among Baby Bells, cable companies, and long-distance companies are intended, in part, to increase each companies' footprint.

If a company already has access to a large number of subscribers (a large footprint) as in the case of Pacific Telesis, they are more likely to exploit the economies of scale possible through long-distance or wireless services. If a Baby Bell is responsible for a sparsely populated area, such as US West, it is more likely to expand its footprint by moving into multimedia, because the territory by itself would not be a lucrative market to expand services. Therefore, a horizontal integration would be the more viable option.

The decision for the Baby Bells to vertically integrate into the telecommunication's transmission of information or to horizontally integrate into the delivery of broadband (cable) services also depends on when key decision makers believe the industries will converge. If a company's decision makers believe the convergence in industries is longer term, the decision to horizontally integrate becomes a riskier venture. Key managers at Ameritech, for example, express the opinion that convergence will happen in a 20-year time horizon. This is in contrast to US West managers, who speak of the convergence in a 5 to 10-year time frame [Sarason, 1997].

Since divestiture, the Baby Bell's variety of strategic activities has exploded. This rush to be a player in new arenas can be understood as diversifying through either vertical integration and/or through horizontal integration. Two factors are primarily driving their actions. The first factor is the attempt to capture synergy as companies try to lock in customers through their installed base or footprint. The second is the timing of technological changes in the telecommunications, cable, and computer industry and the convergence of these industries. Which strategies are successful will not be known until there is the newly merged set of industries consolidates. Each Baby Bell is betting billions that it has the correct answer for its company and each hopes it will be one of the players in this new competitive landscape.

Defining Terms:

Broadband: A broadband network refers to the network of coaxial cable and fiber-optic cables that can transfer large amounts of compressed data at high speeds.

Footprint: A company's footprint refers to its installed base of customers. For the cable and telephone companies, this is the number of homes wired either with cable or a telephone wire.

PCS: Personal communications services is a newly developed wireless technology that will allow the assignment of portable telephone numbers to users. Wireless communications is also less expensive with this technology than with cellular services.

RBOCs: After divestiture, the 22 operating companies were divided into 7 Regional Bell Operating Companies, which were to provide local telephone services for their region.

References

Coll, S. *The deal of the century: The Breakup of AT&T*, Antheneum, New York, 1986.

Kleinfield, S. *The Biggest Company on Earth: A Profile of AT&T*, Holt, Rinehart and Winston, New York, New York, 1981.

Maney, K. *Megamedia Shakeout: The Inside Story of the Leaders and the Losers in the Exploding Communications Industry*, John Wiley & Sons, New York, 1995.

Porter, M. *Competitive Advantage*, Free Press, New York, 1985.

Sarason, Y. Identity and the Baby Bells: Applying Structuration Theory to Strategic Management, unpublished dissertation, University of Colorado at Boulder, 1997.

Further Information

Megamedia Shakeout: The Inside Story of the Leaders and the Losers in the Exploding Communications Industry by Keven Maney provides a very readable book that takes a behind-the-scenes look at the major players in the multimedia industry.

A novel-like book that describes the events that lead up to the divestiture of the Bell System is *The Deal of the Century: The Breakup of AT&T* by S. Coll.

Telephony is a weekly publication for the telecommunications industry. For subscription information contact customer service at (800) 441-0294.

A popular business book that outlines contemporary thought on strategic management and that is appropriate across industries is *Competing for the Future* by Gary Hamel and C. K. Prahalad.

2.6   Commercialization of Public Sector Technology: Practices and Policies

Suleiman K. Kassicieh

This paper defines the importance of public sector technology commercialization outlining the issues faced by policy and decision makers in this area. It points out the important legislative accomplishments in the last 2 decades, suggesting that more is needed to get the full benefits of commercialization efforts. Issues that arise from questions posed by public debate and practical problems are explored with suggested solutions in the last two sections of the paper.

Importance of Technology Commercialization

There are several reasons why the commercialization of public sector technology is a beneficial idea contributing to the general welfare of the nation and to participants.

1. Returns required on the amount of money spent on public sector research. The United States spends a large amount of money (about $70 billion) to fund federal laboratories that employ about 200,000 scientists. These scientists work on projects that cover virtually every technological, engineering, and scientific subspecialty. In many cases, these scientists are well connected to their counterparts in private research and development (R&D) laboratories of major corporations and in universities around the world.

2. Increased competition in the global marketplace. As the technological content of processes and products has increased in the last 20 years, so has the competition between firms. In a global marketplace, the competition is now between firms located all over the globe. Several trends have affected this change:
   - A shorter life cycle of products and processes through an emphasis on fast design to market processes
   - More powerful global communications, allowing information about products and processes to travel fast from producer to customer
   - An emphasis on quality
   - A more sophisticated and demanding consumer

3. Dual use of technology. Traditionally, the major reason for public spending on technology has been for national security issues. As the Cold War changed the nature of national security from a dominantly military issue to a mixture of military and economic problems, the policy analysts have suggested the use of the federal laboratories in the economic arena.

4. Job creation. Another problem created by the change in the political situation in the world is the shift from a Cold War where military readiness was paramount to one where regional conflicts are the concern. This meant a reduction in the military research budget as well as reduction in the manpower required for military purposes. To maintain the economic vitality of the United States, high-paying jobs for skilled and semiskilled workers were needed. Firms created by new technologies were seen as a major contributor to the creation of jobs.

Congressional Mandate

Over the past 17 years, Congress has passed several bills aimed at establishing technology transfer as an integral part of public sector technology. A quick review of the important bills include

1. The Stevenson-Wydler Technology Innovation Act of 1980. This is the first of many acts that facilitated technology transfer. It mandated the federal laboratories to seek cooperative research with other interested parties in industry, academia, and other organizations as well as to establish offices to support research and technology applications at each federal laboratory. It provided for funding of technology transfer activities albeit at a comparatively low level compared to pure technical activities.

2. The Bayh-Dole Act of 1980. This allowed nonprofit organizations to have first right to title of inventions supported with federal monies. This did not include government-owned and contractor-operated laboratories (GOCOs). It also established the right of government-owned and -operated laboratories (GOGOs) to grant exclusive licenses for patents they own.

3. Cooperative Research Act of 1984. This act relieved companies from antitrust suits if they worked on joint precompetitive R&D.

4. Federal Technology Transfer Act of 1986. The Act makes a number of changes to existing laws. Scientists and engineers are now evaluated based on their ability to get technology transferred out of the laboratory. Inventors from GOGOs are required to receive a minimum 15% share of any royalties generated through patenting or licensing. Directors of GOGOs and not of GOCOs can enter into **cooperative research and development agreements** (CRADAs) to license inventions; they can waive rights to laboratory inventions and intellectual property under CRADAs. Federal employees can participate in commercial development if there is no **conflict of interest**.

5. Executive Order 12591 of 1987. This order established CRADAs between GOGOs and all other entities including other federal laboratories.

6. Competitiveness Technology Transfer Act of 1989. This Act protects CRADA arrangements from disclosure of inventions and innovations. It added nuclear weapons to the technology transfer mission and allowed GOCOs to undertake technology transfer activities.

7. The National Technology Transfer and Advancement Act of 1995. This act provides U.S. companies with exclusive license in a prenegotiated field of use for inventions resulting from a CRADA. It also allows the federal government to use invention for legitimate government needs. Agencies can use royalty revenues for research or other technology transfer costs. Inventors receive at least 15% of the royalties each year with a maximum royalty award of $150,000 per year. It also supports federal employees working on the commercialization of their inventions.

In the 1980s, as the Cold War continued, Congress supported technology commercialization at a lower level due to the importance of the national security mission. As the national security mission was redefined with the dismantling of the Soviet Union and the absence of a major nuclear threat, the importance of economic security based on the commercial success of peace-time products (as opposed to products of defense-related industries and companies) gained more support and Congress passed bills that reflected the collective thinking on the issue. As with any public policy question, there are issues that are of concern to policy makers. These issues result from discourse about advantages and disadvantages of different policies and are deeply connected to political beliefs. These issues are covered in the next section.

Issues

Most of the public debate about technology commercialization surrounds one of the following questions:

1. What is a reasonable measure of success in technology commercialization? While a majority of people support the commercialization of technology, they differ on this issue. It is a contentious point among policy makers, technology commercialization specialists, and the public in general. Measurement issues are directly related to declarations of who has succeeded and who has not. The degree of success determines the recipients of future public funds allocated to commercialization activities. This is a survival issue to many public sector organizations. Since the measurement of results is not an easy endeavor, we get arguments on the appropriateness of a measure, other intervening factors that affect the measurement process and/or the results, and the starting value of a measured variable, among many other issues. Carr [1994] indicated that measuring the numbers of CRADAs, licenses, and patents does not provide very useful information because the numbers frequently mask the quality and size of the counted event. He also laments the fact that good time-series data are not available and virtually no economic impact measures have been developed. He suggests that, for the policy makers and the public to support efforts in technology commercialization, long-term results should be analyzed in areas such as royalty income and economic impact on jobs. These concerns are easily corrected. What is needed is an independent organization that is charged with tracking and analyzing information about commercialization activities. The analysis should lead to conclusive results based on solid research principles. This effort can experiment with many options to answer questions regarding the measurement process and its parameters and, more importantly, how these results can help us in setting policies designed to achieve the stated goals.

2. Who benefits from it? The argument about benefits from technology commercialization arises mainly because of the "sour grapes" phenomenon. New technological innovations produced at public sector laboratories are designed to solve problems for a specific sponsor. The process of commercializing a product has many stages:
   a. Scientific discovery
   b. Needs assessment of potential users
   c. Handshake between source and recipient
   d. Maturation of the technology through prototyping and testing
   e. Legal steps of transfer
   f. Actual physical transfer of the technology
   g. Continued technical assistance from the source to the recipient

   As can be seen from the stages listed above, the person responsible for commercializing the product has to go through many steps before the commercialization effort is successful. In many situations, a number of intermediaries facilitate the technology transfer, adding layers of interaction between parties. This is indicative of the amount of time and resources that goes into the process of commercializing an innovation. To suggest that the benefits should accrue to the source alone (and by implication to the taxpayers who funded the research in the first place) diminished the worth of all of the activities needed to move the technology from an innovation to a product, process, and/or service that produces a stream of future income. Equitable sharing arrangements for the source and recipient are necessary to encourage this endeavor. The problem arises when one of the innovations is a "home run" and produces large income streams. Those who are not benefiting from it develop the sour grapes phenomenon. This is not suggesting that **fairness of**

opportunity issues are not important. Radosevich and Kassicieh [1994] indicated the importance of the issue but suggested that it should not be a "show stopper" if properly handled before these problems arise. Another issue related to the benefits question is the conflict of interest issue. This arises when a scientist funded by public funds to perform research produces an innovation. The scientist then acquires the license to commercialize the innovation. The scenario is the same as in the example above. In both situations, the assumption is that the technology is the important factor. This is usually the wrong assumption because venture capital managers usually point to business acumen as the necessary condition for success. The risk/reward structure also suggests that individuals who have taken the risk of commercializing an innovation by spending time and resources to make it a viable commercial product should be rewarded accordingly. This is not a difficult concept to explain but the reason that we still have problems with it is directly related to scientific organizations and individuals who place a much higher value on the scientific discovery *vis-à-vis* the other steps required in the commercialization process.

3. Who pays for it? The issue here is somewhat of an extension of the benefit question. Individuals and companies were unwilling to bear the total risk associated with commercialization of a technological innovation without spreading the risk across many projects (as is done by the venture capital community) or by asking for participation from a pool of public funds designated for such a purpose. CRADAs are an example (although the government has participated in funding other types of commercialization activities) where the government puts a portion of the money necessary for research geared toward commercialization activities. The question posed by opponents of government funding of these commercialization activities can be stated as follows: if there is a proven need for this product (process or service) and the company participating in the CRADA (or other commercialization activity) is going to make the profit from the results of the CRADA, why is the company not funding the project totally? The answer is not a simple one: the costs are shared but so are the benefits. The company will pay taxes on profits and hire more workers and the benefits will accrue to many in the economy. The real question is whether the costs and benefits are shared in equal proportions, and that is a very complex issue. The measurement of results as suggested earlier in the paper will play a large role in understanding the proper sharing of risk and reward among scientists, public sector laboratories, intermediaries, entrepreneurs, and business organizations that participate in the activities.

Practical Problems that Require Solution

A number of practical problems are needed to achieve the best results from technology commercialization efforts. They include

1. Investment in commercialization activities by public sector organizations is still small compared to the investment in technology: the problem here is that new technological discoveries require research funds to be expended by the government, which is under a lot of pressure to cut expenditures. Many research organizations are now looking at generating future research funds from intellectual property. The development of a major stream of income from licenses or spinning off new companies takes time and is subject to risk. This is, however, an activity that should be encouraged because it has the potential for us to "eat our cake and have it too". The Waste Management Education and Research Consortium is an example of a research organization that is following this strategy.

2. The risk capital available for commercialization of technological innovations is not large: This is a problem because venture capital funds are traditionally high-risk funds, some of which turn out to be high return. The solution to this problem is usually in applying the portfolio effect so that risk is spread among a large number of projects, and, if 5 out of the 100 turn out to be "home runs", then the fund returns an excellent dividend to the investors. Governmental agencies have

recently been asked to support this endeavor by supporting venture funds through the investment of a very small portion of the agency's pension fund monies in a matching fund for venture capital financing deals. New Mexico is one of the states that has recently passed legislation to this effect.

3. Strategic partnerships and alliances are very important for technology commercialization efforts: Carayannis et al. [1997] have shown that a large percentage of funds available for technology commercialization activities has been funded through strategic alliances between an agile entrepreneurial startup that has the new technology and a more established large organization that has the power of established market channels and funds. This is an important development that has not been noticed in the legislation designed to support commercialization. If this activity is encouraged through tax incentives, we might see more funding through this mechanism.

4. Predicting the success of new products and of entrepreneurs/companies due to the use of new innovations is an important consideration for research. Kassicieh et al. [1996] describe the situation in several federal laboratories in terms of support of commercialization activities. Kassicieh et al. [1997] use a model to predict which patent-producing scientists will become entrepreneurs that commercialize their technologies. The prediction of what will succeed and who will succeed can help us make better recruiting decisions.

Conclusion

A large number of positive steps have been achieved in the last 2 decades in supporting and nurturing the development of new products, processes, and services through the use of new innovations. The public sector has an incentive to continue in these endeavors so that we will have enough funds to perform research to solve new problems and have enough benefits such as new jobs created by the formation of new businesses accrued from the commercialization of these new technologies. We have many problems in this area but they are solvable given more analysis so that public policies are framed to achieve results.

Defining Terms

CRADA: An agreement between a commercial entity and a national laboratory for collaboration on research issues that are of benefit to the commercial entity and where the national lab oratory has expertise.

Fairness of opportunity: Fair access to the results of publicly funded research. The results are thought to be the property of the taxpayers and thus should be equally accessible to all.

Conflict of interest: A scientist files for a patent for which he or she later obtains the intellectual property right is thought to have a conflict of interest in divulging all information resulting from public funding.

Strategic alliances and partnerships: Collaboration between firms with different skills and resources is thought to provide the firms with a better chance for success in commercializing technology in today's complex markets.

References

Carr, R. A proposal for a framework for measuring and evaluating technology transfer from the federal laboratories to industry, In *From Lab to Market: Commercialization of Public-Sector Technology*, S. K. Kassicieh and H. R. Radosevich, eds., pp. 299–304, Plenum Publishing, New York, 1994.

Carayannis, E., Kassicieh, S., and Radosevich, R. The Use of Strategic Alliances in Early-Stage Seed Capital in Technology-Based, Entrepreneurial Firms, working paper, 1997.

Kassicieh, S. K., Radosevich, R., and Banbury, C. Predicting entrepreneurial behavior among national laboratory inventors, *IEEE Trans. Eng. Mgt.*, 1997.

Kassicieh, S. K., Radosevich, R., and Umbarger, J. A comparative study of entrepreneurship incidence among inventors in national laboratories, *Entrepreneur. Theor. Pract.*, 20(3): 33–49, 1996.

Radosevich, R. and Kassicieh, S. The King Solomon role in public-sector technology commercialization, In *From Lab to Market: Commercialization of Public-Sector Technology*, S. K. Kassicieh and H. R. Radosevich, eds., pp. 9–28, Plenum Publishing, New York, 1994.

2.7   Taxation

Keith E. Smith

Income taxation requires the measurement of income and expense. In the case of high-tech enterprises, some special questions arise concerning the character and timing of income and expense. The resolution of these issues involves balancing competing considerations: what is theoretically preferable as a matter of financial reporting?; what policy will most effectively encourage research?

Income from Patents

Income subject to tax may be either ordinary or capital gain. Ordinary income includes salaries, interest, dividends, and rent, while capital gain is a gain from the sale of a capital asset. For tax purposes, capital gain is preferable. If the asset sold has been held long term, i.e., for more than 1 year, then the gain is taxed at a rate no higher than 28%.[1] Ordinary income, on the other hand, may be taxed at rates as high as 39.6%.

Income may be realized from a patent either as a royalty or as a gain from the sale of the patent outright. The royalties received by the holder of a patent are ordinary income. Gain from the sale of the patent by a "holder", however, is treated as a long-term capital gain (even if the patent has been held for less than a year).[2] A "holder" for this purpose is the individual whose efforts created the patent or any other individual who bought it from him prior to actual reduction to practice of the invention covered by the patent, as long as the buyer is neither the creator's employer nor a person related to the creator.[3]

Example

> Inventor receives a patent for his invention, which he then sells the next day at a profit of $100,000. The gain is taxed as a long-term capital gain.

Deduction of Expenses for Research and Experimentation

Timing of Expenses in General

The federal income tax is based upon the net income of an enterprise for a particular period. Timing is critical. The amount of tax depends not only on how much revenue has been earned or expense incurred but also in what period. One of the pillars of accounting theory is the so-called "matching principle". The revenues of a particular undertaking should be matched with the related expenses, i.e., reported in the income statement of the same period. Therefore, the expense of acquiring property, plant, equipment, or any long-lived asset should be spread out over the period in which it is used by the enterprise to produce revenues (i.e., depreciated over its useful life), rather than allocated entirely to the period in which the asset is purchased or in which the cost is paid.

---

[1] A gain on the sale of a capital asset held for more than 18 months is, in general, taxed at a rate no higher than 20%.

[2] Internal Revenue Code, Section 1235(a).

[3] Internal Revenue Code, Section 1235(b).

Example

In year 1, taxpayer acquires an asset at a cost of $120. He expects the asset to last 3 years and be worthless at the end of that period, which proves to be the case. Using the asset in his business enables the taxpayer to earn revenues of $150 per year. There is no question that, over the 3 years, the taxpayer's profit is $330 (total revenues of $450 minus the expense of $120), but how much he has earned year by year depends on when the cost of the long-lived asset is recognized as an expense. If the acquisition cost is treated as an expense of the year of acquisition, then he has earned $30 in the first year and $150 in each of the next two:

	Year 1	Year 2	Year 3	Total
Revenues	150	150	150	450
Expenses	120	0	0	120
Net income	30	150	150	330

If, alternatively, the cost is spread out over the period in which the asset is used in the business, then he has earned $110 each year.

	Year 1	Year 2	Year 3	Total
Revenues	150	150	150	450
Expenses	40	40	40	120
Net income	110	110	110	330

The latter presentation is better financial reporting because it better matches the revenues of the 3-year period with the expenses that were incurred to earn them.

In general, firms are happy to defer the recognition of expense in their financial reports to shareholders and creditors. Doing so makes net income as high as possible as long as possible, although, as shown in the example above, it is "just" a matter of timing. For tax reporting, however, the firm's preference is usually the opposite. Deducting an expense makes taxes lower, and, the sooner, the better. In order to determine how much better early tax savings are than late ones, all tax savings must be discounted to their present value. Where

$$i = \text{interest rate}$$

$$n = \text{number of periods}$$

$$FV = \text{future value}$$

$$PV = \text{present value}$$

The present value of savings to occur in the future can be computed by the following formula:

$$PV(1 + i)^n = FV$$

or

$$PV = \frac{FV}{\left(1 + i\right)^n}$$

If the rate of interest is 10% per period, then $1 invested at present would grow to $1.21 by the end of the second period.

$$PV(1 + i)^n = \$1(1 + .1)^2 = \$1.21 = FV$$

or, putting it the other way around, $1.21 to be received two periods in the future has a present value of $1.

$$\$1 = PV = \frac{FV}{\left(1 + i\right)^n} = \frac{\$1.21}{\left(1 + .1\right)^2}$$

These formulas can be used to compute the amount of the tax benefit of deducting an expanse early for tax purposes.

Example

Assume that the tax rate is 40% and that 10% is the appropriate interest rate for computations of present value. An asset costing $120 will save $48 of taxes, whether it is depreciated over 1 year or 3, but the present value of the savings will be greater if the expense is recognized earlier

Depreciation over 1 year

Year	Depreciation	Tax saved	PV of tax saved
1	120	48	44
2	0	0	0
3	0	0	0
Totals	120	48	44

Depreciation over 3 years

Year	Depreciation	Tax saved	PV of tax saved
1	40	16	15
2	40	16	13
3	40	16	12
Totals	120	48	40

Timing of Expense of Research and Experimentation

The expense of research and experimentation often presents a more difficult question of timing than the cost of acquiring hard assets. If research is successful, it produces an intangible asset, knowledge (and perhaps a patent), which may benefit the firm far into the future. If it is unsuccessful, then there is no future benefit and the money spent on it is lost. In the latter case, the cost of the research should certainly be recognized as an expense of the period in which it is incurred, but, in the former, it would be theoretically preferable to spread the expense of the research over the future period to be benefited. As a practical matter, however, it may be difficult to estimate how long that period will be,[4] and, as a matter of policy, requiring firms to wait to realize the tax benefit of the cost of research may discourage them from doing it in the first place. Congress has, therefore, permitted firms, at their option, to deduct the expense of research and experimentation in the period in which it is incurred.

---

[4] Even a patent granted for a definite term may become worthless before it expires if a competitor patents a better product.

Research and Experimentation Expenditures Defined.

For tax purposes, the "research or experimental expenditures" means[5]

expenditures incurred in connection with the taxpayer's trade or business which represent research and development costs in the experimental or laboratory sense. The term generally includes all such costs incident to the development or improvement of a product. The term includes the costs of obtaining a patent, such as attorneys' fees expended in making and perfecting a patent application. Expenditures represent research and development costs in the experimental or laboratory sense if they are for activities intended to discover information that would eliminate uncertainty the development or improvement of a product. Uncertainty exists if the information available to the taxpayer does not establish the capability or method for developing or improving the product or the appropriate design of the product. Whether expenditures qualify as research or experimental expenditures depends on the nature of the activity to which the expenditures relate, not the nature of the product or improvement being developed or the level of technological advancement the product or improvement represents.

Specifically *excluded* from the category of research or experimental expenditures are ordinary quality control testing, efficiency surveys, management studies, consumer surveys, advertising or promotions, the acquisition of another's patent, model, production or process, and research in connection with literary, historical, or similar projects.[6]

Example

A firm that makes ladders undertakes a study to determine whether the design of the ladder is adequate to support a user who weighs 350 pounds. The expenses of the study are for "research" in the statutory sense of the term.

The firm also undertakes a study in which finished ladders are randomly sampled to determine whether they conform to the design specifications. The expenses of this study are not for research but for ordinary quality control.

The cost of land and buildings is also by definition not an expense of research or experimentation, although depreciation on the building is to the extent that it is actually used for research and experimentation.

Example

A firm buys land for $1 million and builds a building on it for $3.9 million. The land is not depreciable, but the building is depreciated over 39 years. For the first 10 years, the firm uses the building as a research facility, and depreciation of $100,000 per year on the building is treated as an expense of research, which may be either deducted immediately or amortized over a period of at least 60 months at the election of the firm.

In the 11th year, the firm converts the building to a manufacturing facility. In that year and thereafter, the depreciation of the building is treated as a cost of manufacturing the products and will be deducted as an expense of the year in which the products are sold.

The expenses of research and experimentation also include amounts paid by the taxpayer to others to conduct research on his behalf.

Immediate Expensing

At the election of the taxpayer, the expenses of research and experimentation may be deducted in the year incurred. The taxpayer may adopt this method of accounting without the consent of the Treasury if he does so in the 1st year in which he incurs expenses for research and experimentation. If, however, he begins by using some other method, then switching to immediate expensing in some subsequent year

---

[5] Treasury regulations 1.174-2(a)(1).
[6] Treasury regulations 1.174-2(a)(3).

requires the Treasury's consent.[7] This is consistent with the general rule that changing a method of accounting for taxable income requires the consent of the Treasury.[8]

Amortization over 60 Months

If the taxpayer does not elect to deduct the expenses of research and experimentation immediately, then they must be spread out (or "amortized") over "such period of not less than 60 months as may be selected by the taxpayer (beginning with the month in which the taxpayer first realizes benefits from such expenditures)."[9]

Example

Fast Corp and Slow Corp each secure a patent on September 30, year 1. The term of each patent is 17 years. Each firm incurred $60,000 of research and experimentation costs to obtain the patent. Fast Corp elects to deduct the expense immediately, white Slow Corp elects to spread it over 60 months. The deductions would, accordingly, be claimed by the two firms in the following years:

Year	Fast Corp	Slow Corp
1	60,000	3,000
2		12,000
3		12,000
4		12,000
5		12,000
6		9,000
Totals	60,000	60,000

Amortization of Patents Created by Others

As discussed above, the expenses of research and experimentation, including the cost of acquiring a patent, may be deducted immediately or, at the election of the taxpayer, amortized over a longer period. When a patent is acquired and used by one other than its creator, however, different rules apply. If the patent is acquired in connection with the purchase of the assets of a business or a substantial portion thereof, then the patent may be amortized over a period of 15 years, beginning with the month of acquisition.[10] The 15-year period applies whatever the actual legal life of the patent.

Example

As part of the purchase of another business, a firm acquires a patent on October 1, year 1, for $180,000. The cost may be claimed as an expense at the rate of $12,000 per year or $1000 per month. The deduction for the first year is $3000, then $12,000 per year for the next 14 years, then $9,000 in the final year.

Year	Deduction
1	3,000
2–15	168,000
16	9,000
Total	180,000

If the patent is acquired separately, rather than as a part of the purchase of a business, then its cost is amortized ratably over its remaining life.

---

[7] Internal Revenue Code, Sec. 174(a)(2)(B).
[8] Internal Revenue Code, Sec. 446(e).
[9] Internal Revenue Code, Sec. 174(b)(1).
[10] Internal Revenue Code, Section 197.

Tax Credit for Additional Research

A deduction reduces taxable income; how much the tax is reduced in consequence depends on the rate of tax to which the taxpayer is subject. A tax credit is a much greater benefit than a tax deduction because it reduces the tax directly, rather that merely reducing taxable income.

Example

Two taxpayers are both subject to tax at a rate of 30%. Both have gross income of $100,000. The first taxpayer also has a deduction of $10,000, while the second has a credit of $10,000.

	Taxpayer #1	Taxpayer #2
Revenue	100,000	100,000
Deductions	10,000	
Taxable income	90,000	100,000
Tentative tax	27,000	30,000
Tax credit		10,000
Tax net of credit	27,000	20,000

In the past, taxpayers have from time to time been permitted to claim a credit against their federal income tax equal to 20% of the excess of their research expenses for the year over the "base amount". The base amount was the percentage of their gross receipts spent on research from 1984 to 1988 (but not to exceed 16%) multiplied by the average annual gross receipts of the 4 years preceding the year in which the credit was claimed. The credit has never been made a permanent part of the Internal Revenue Code but instead has always been enacted with a fixed expiration date, making reenactment necessary in order for the credit to continue in force. As of this writing, the credit is scheduled to expire July 1, 1998. Further discussion of it is, therefore, omitted.

3

Innovation and Change

Clayton M. Christensen
Harvard Business School

Christopher M.
McDermott
Rensselaer Polytech Institute

Thomas H. Davenport
University of Texas at Austin

Andrew B. Hargadon
Stanford University

Timothy G. Kotnour
University of Central Florida

Richard Reeves
Cranfield University

Larry A. Mallak
Western Michigan University

Mario W. Cardullo
*Virginia Polytechnic Institute and
State University*

Philip Anderson
Dartmouth College

Marco Iansiti
Harvard University

0-8493-8577-6/99/$0.00+$.50
© 1999 by CRC Press LLC

3.1   The Evolution of Innovation

Clayton M. Christensen

Scholars have searched for patterns in the evolution of technological innovation for decades, hoping that regularities, frameworks, or theories might emerge to give guidance to managers charged with the complex and uncertain task of managing innovation. The character of these studies has varied. They range from deep historical studies of individual industries, to cross-industry surveys that search for frameworks that are robust enough to describe how innovation has evolved in very different competitive environments, to studies that focus on specific frameworks as being particularly useful.[1] These studies typically search not just for patterns in what types of innovation are likely to occur at different stages of a product or industry's maturity but for insights about what types of companies are most likely to succeed or fail at these innovations, in different circumstances.

Thomas Kuhn, the historian of science, observed that in the building of bodies of knowledge, *paradigms* eventually emerge [Kuhn, 1970]. In Kuhn's model, a paradigm can never explain everything — the world is too complex. Nor can a paradigm explain anything perfectly. However, it is a way of organizing and explaining enough about a class of phenomena that subsequent researchers find it to be a useful starting point for their own work.

No single paradigm has emerged in the study of patterns of innovation that would enable all researchers or managers to predict with certainty how technology is likely to evolve or what types of companies are likely to emerge victorious from innovative battles of various sorts. At this point in our understanding of the field, it seems unlikely that such a comprehensive, overarching theory will ever emerge, and Kuhn probably would not expect such a paradigm to develop. Indeed, the problems of managing technological innovation are so varied and complex that multiple bodies of knowledge are likely to be required to understand how to manage the evolution of innovation.

There now appear to be potential paradigms emerging in the study of how four particular dimensions or aspects of technological evolution occur. They are potential paradigms in the sense that each has helped make sense of what previously appeared to have been random or contradictory phenomena and because each has spawned streams of sustained, credible subsequent scholarship. These bodies of theory might be labeled as follows:

1. The **dominant design** theory, which asserts that the nature of innovation shifts markedly after a dominant design has emerged.
2. The technology s-curve theory, which states that the pace of performance improvement utilizing a particular technological approach will follow an **S-curve** pattern, flattening as certain natural limits to further improvement are reached. Theories of punctuated equilibrium are related to movement along a technology S-curve, intersected occasionally by a new S-curve.
3. The theory that patterns of innovation are determined by intersecting trajectories of performance demanded in the market, vs. performance supplied by technologists.
4. The study of how modularization of design can create options for the future, how it affects the optimal scope of the firm, and how it changes the nature of the competitive advantages that can and cannot be developed.

The following sections summarize each of these viewpoints, characterize the sorts of innovative problems to which the theories seem to have relevance, and describe the sequence of studies that have contributed most strongly to the building of each body of knowledge.

---

[1] Four studies exemplifying the tradition of deep, single-industry historical studies are Abernathy [1978], Constant [1980], Henderson [1989], and Christensen [1993]. Examples of cross-industry surveys are Sahal [1981], Tushman and Anderson [1986], Anderson and Tushman [1990], Rosenbloom and Christensen [1994], and Christensen [1997b]. Research that promotes the usefulness of specific frameworks includes Roussel [1984] and Foster [1986].

The Dominant Design Theory

The notion that a dominant design powerfully impacts the nature of innovation was articulated by Abernathy and Utterback [1978]. They proposed that in most industries there would be an initial period of *product* design ferment. In the early years of the automobile industry, for example, fundamental questions such as whether the power source should be a steam-, electric-, or gasoline-powered engine were not yet resolved. How the body would be supported over the drive train, what a transmission was, and how it would interface with the driver and with the engine were characteristic of the design issues that engineers in different firms approached differently from one product generation to the next.

Eventually, however, a *dominant design* emerged — a general acceptance of how the principal components would interface with others in the architecture of the automobile's design. The dominant design was not necessarily the *optimal* design. However, it became a standard architecture, with accepted metrics for determining the way in which components and modules would interact. This gave organizationally independent suppliers a well-defined technological context within which they could work to improve their pieces of the system. Abernathy and Utterback noted that the magnitude and rate of technological innovation directed at innovative *product* design declined markedly after the emergence of the dominant design.

The emergence of dominant designs enabled a surge of innovation in *process* technology, as suggested in Fig. 3.1. When designs were in flux, processes could not be standardized: volumes per process sequence were low, equipment had to remain flexible, and product costs were high. The dominant design enabled engineers to focus their innovative energies on process technology improvements. This enabled significant cost reductions in the product, allowing further volume growth and even further process refinement and cost reduction.

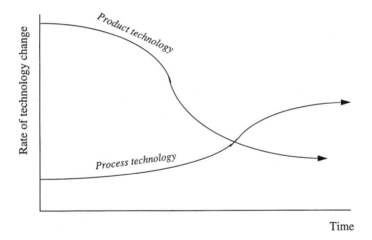

FIGURE 3.1 Impact of dominant design on the pace of technology change in products and processes. (*Source:* Adapted from Abernathy and Utterback [1978].)

Other scholars have subsequently built upon the dominant design paradigm, articulating how it operates and what its impact on patterns of innovation can be. For example, Suarez and Utterback [1995] and Christensen et al. [1997] found that, because the existence of dominant designs restricts engineers' freedom to differentiate their products through innovative design, there are fewer opportunities for small or entrant firms to find refuge in niche markets in the post–dominant design era. They found that firms entering a range of industries after dominant designs were defined faced lower posterior probabilities of survival.

They further concluded that dominant designs are *architectural* in character and that their elements coalesce one by one over time. They found that certain components came to be used in most models, but whether or not these components were adopted had little impact on survival. What mattered was adoption of the dominant elements of *architectural* design.

Finally, they noted that there appeared to have been a unique "window of opportunity" for entry into fast-moving industries. Not only did firms that entered after the dominant design emerged have low probabilities of survival, but firms that entered too *early* had low probabilities of survival as well — presumably because they had honed capabilities in that period of high design turbulence that were inappropriate for survival in the process-intensive post–dominant design era.

The Technology S-Curve Theory

The notion that the pace of technological progress follows an S-curve pattern has been featured in literature on technological innovation for years; Foster (1986) summarized the arguments most comprehensively. S-curve theory suggests that the rate of technological progress ultimately is subject to decreasing returns to engineering effort because trajectories of technological progress are eventually constrained by natural limits of some sort: they get too small, too large, too complex, or push the intrinsic properties of available materials to their theoretical maxima [Sahal, 1981]. As these limits are approached, it requires increasing amounts of effort to wring out additional performance improvement.

S-curve theorists argue that, when the rate of technological progress has begun to decline, the technology and its practitioners are vulnerable to being overtaken by a new technological approach, following its own S-curve pattern, as shown in Fig. 3.2. A key management task is therefore to monitor a company's position on its S-curve and, when it has passed its point of inflection, to find and develop the new technology that might overtake the present approach. S-curve patterns of technological progress and technology substitution have been shown to have occurred in foam rubber, aircraft engines, magnets, and disk drive components.[2]

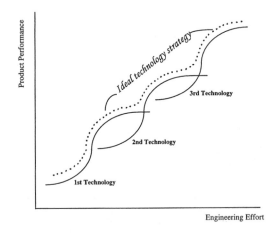

FIGURE 3.2 A view of technology strategy: switching S-curves. (*Source:* Adapted from Christensen [1992], p. 340.

Subsequent research has substantially refined the usefulness and limits of the S-curve paradigm. While it can be used to describe overall industry trajectories, it has more limited decision-making usefulness within companies, for two reasons. First, S-curves seem less relevant to performance of assembled products than to componentry. Christensen [1992] showed, for example, that the perception of the recording density at which disk drive makers felt they needed to jump to a new recording head S-curve differed across companies in the same industry by an order of magnitude. This was because, in the design of most assembled products, there is more than one route to achieving performance improvement. When the head technology reached its performance limit, some companies in the industry elected to jump to

---

[2]A listing of these publications can be found in Christensen, [1992].

the S-curve of the next-generation component technology, while other firms found design routes around the bottleneck by improving other components and architectural aspects of the design, which were not yet at the apex of their S-curves. Hence, some firms followed an S-curve switching strategy, component by component, as a means of relieving bottlenecks to performance improvement, while other firms followed an S-curve *extension* strategy — staying as long as possible with existing technologies, through clever innovations in system design. The result of both strategies was a steady improvement in the performance of the finished disk drive — a pace comprised of many incremental improvements and some radical S-curve leaps among the individual components comprising the drives. Iansiti [1995] measured the same phenomena in his study of the evolution of mainframe computer technology.

Another limit of S-curves' usefulness is that, with many types of innovations, the relevant attributes of performance offered by new technologies differ from those of the old. The new technology, while underperforming the established approach along accepted metrics in established markets, can become established in a new market segment, which values its different attributes. When its performance improves to the point that it satisfies the performance demanded in the original market as well, it then invades swiftly, knocking out the established firms in that market, as depicted in Fig. 3.3. S-curve theory, lacking a market dimension to its definition of performance, cannot account for this route of technological evolution, which subsequent scholars [Christensen, 1992, 1997; Levinthal, 1997] have shown to be quite common.

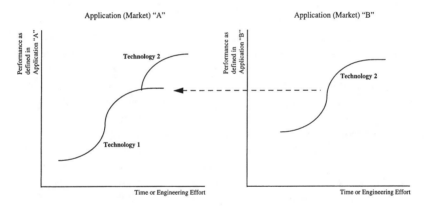

FIGURE 3.3 Route through which disruptive technologies penetrate established markets. (*Source:* Adapted from Christensen [1992], p. 361.)

A stream of research closely related to the S-curve concept might be characterized as a **punctuated equilibrium** theory. Tushman and Anderson [1986], who initially articulated this point of view, noted that in a range of industries they studied, technologies improved at a relatively measured pace across most of their histories. This incremental pace of progress was sporadically interrupted, however, by bursts of radical change that created discontinuities in the otherwise smooth trajectories of improvement. Levinthal [1997] has described the similarity of this pattern in technological evolution to the patterns that characterize biological evolution.

Organizations' technological capabilities are developed through the problems their engineers solve. The nature of the problems they confront or do not confront is determined to a significant degree by the prior choices of technology made earlier in the history of the company and industry: hence, technological understanding builds cumulatively [Clark, 1984]. Some scholars have observed that the leading companies in an industry are most likely to lead in developing and adopting new technologies when those technologies build upon the accumulated technological competencies they have developed. When new technologies render established firms' historically accumulated competencies irrelevant, however, entrants to the industry have the advantage [Tushman and Anderson, 1986; Henderson and Clark, 1990; Tripsas, 1997].[3]

The Technology and Market Trajectories Theory

The third stream of research around which a significant body of insight is accumulating combines theories about how and why technological progress is achieved, with insights about rates of technological progress customers are able to absorb. Its fundamental premise is that patterns of innovation are influenced heavily by the interaction of trajectories (rates of improvement in product performance) in what customers *need*, compared to trajectories of improvement that innovating companies *provide*.

The notion that technological progress of a class of products can be mapped as a trajectory of improvement over time was articulated by Dosi [1982]. A stream of research initiated by Christensen and colleagues has extended Dosi's notion of technology trajectories through a range of empirical studies [Christensen, 1993, 1997b; Rosenbloom & Christensen, 1994; Bower and Christensen, 1995; Jones, 1996]. They observe that the trajectory of technological progress frequently proceeds at a steeper pace than the trajectory of performance improvement that is demanded in the market. This means that a technology that squarely addresses the needs of customers in a tier in a market today may improve to overserve those needs tomorrow; a technological approach that cannot meet the demands of a market today may improve at such a rate that it competitively addresses those needs tomorrow, in the fashion described in Fig. 3.4.

Most technological innovations have a *sustaining* impact: they drive an industry upward along an established technology trajectory. Occasionally, however, *disruptive* technologies emerge — smaller, cheaper, simpler products that cannot be used in established markets because they perform poorly according to the attributes valued there. These disruptive products may enable the emergence of new market segments in which customers have a different rank ordering of product attributes than those of established markets. Practitioners of the disruptive technology can take root in this new segment, even while manufacturers and customers in mainstream markets ignore it.

Once commercially established in this new "low" end of their larger market, these disruptive technologists have very strong incentives to improve their products' performance at such a rapid rate that they can attack market tiers above them. This is because those market tiers typically are larger, and profit margins are more attractive in the higher-performance product models purchased there.

These scholars have found that the established companies in each industry they studied generally led their industries in developing and adopting innovations that were *sustaining* in character — even radical, discontinuous, competence-destroying technologies. It appears that, when the customers of established companies have demanded an innovation, the leading firms seemed somehow to find a way to get it. However, entrant firms consistently led in introducing disruptive technologies that could be used only in new or commercially unimportant markets. Strong, established companies that listened attentively to their customers and were skilled at directing their innovative investments to projects that promised the greatest profits typically found it nearly impossible to introduce disruptive technologies in a timely way, despite the fact that they typically are technologically simple. This pattern of innovation affecting the

---

[3]In her study of the semiconductor photolithographic aligner industry, Henderson [Henderson and Clark, 1990] noted that, after stable architectural designs had emerged, established firms' competence in architectural reconfiguration atrophied because they simply stopped tackling problems of architectural design. They continued to hone their component-level technological capabilities, however, because improvements in componentry from one generation to the next continued to be the vehicle for product performance improvement. Henderson then observed the same result that Tushman and Anderson saw: when a technological innovation occurred in the industry that did not build upon the established firms' practiced competencies in architectural technology development, they lost their positions of leadership to entrant firms. Tripsas [1997] observed that whether a firm failed at such a point of technological change depended on the existence of complementary assets, or, put another way, on whether there were multiple ways to compete in an industry. Where product performance constitutes the dominant basis of competition, as Henderson found, companies are likely to fail under these circumstances. When reputation, distribution or product line strength, field service capabilities, or customer relationships are important elements of customers' purchase decisions, then the difficulties encountered by established firms are less likely to prove fatal.

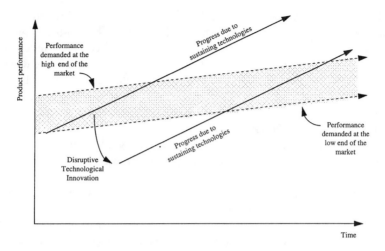

FIGURE 3.4 Intersecting trajectories of performance needed in the market vs. performance provided by technologists.

fortunes of established and entrant firms has been observed in retailing [McNair, 1958], telecommunications switching equipment [Jones, 1996], semiconductor testing equipment [Bower, 1997], commercial printing [Garvin, 1996], and the steel, mechanical excavator, computer, motorcycle, photocopier, accounting software, and executive education industries [Christensen, 1997a, 1997b].

The mechanism discovered by these scholars, through which disruptive technologies invade established markets and precipitate the decline of leading firms in those markets, is very similar to the way in which biologists have modeled the historical evolution of new species [Levinthal, 1997]. Rarely does speciation occur *within* established populations, in habitats to which they are well adapted. Rather, it occurs in peripheral or remote environments where the characteristics of a new (mutated) species give it an advantage vs. established species in the competition for survival. Established species can then be wiped out when members of the new species return to the home range from the remote environment in which their species initially took root.

The way that performance improvement is defined seems to be altered significantly in each market segment after a disruptive technology trajectory intersects the performance needed in the mainstream market, as depicted in Fig. 3.4. Once the older and the disruptive technology *both* provide adequate performance, how do customers choose between the alternative products? Researchers [Christensen, 1997a; Trent, 1997] have observed that, when two alternatives both overshoot the functionality that customers in a tier of the market actually need, the *basis of competition,* or the criteria by which customers choose one product over another, tends to change in that market tier, in a relatively consistent pattern. No longer can product functionality be a competitively relevant dimension of innovation: instead, a new trajectory of performance improvement is defined — often it centers on reliability — and the successful technologists innovate along that dimension instead. Ultimately, the trajectories of improvement in reliability overshoot the market need, stimulating *another* trajectory to coalesce — typically along *convenience* dimensions. It is the interaction of trajectories of technology improvement and market need that trigger changes in the basis of competition or in the definition and trajectory of competitively meaningful innovation.

Theories about the Modularization of Design

The fourth stream of ideas around which a substantial body of research is beginning to build relates the **modularization** of product designs. These concepts are substantially less developed than the other three — yet they appear to have the potential to enhance substantially what is understood about the evolution of innovation.

Most complex products consist of subassemblies and components, which themselves are often assemblies of yet smaller components, in a nested fashion [Alexander, 1964; Christensen and Rosenbloom, 1995]. In the earliest phases of the development of a product category, complex products often must be designed in an integrated way, meaning that each element of the product must be designed, simultaneously or iteratively, by the same organization. This is because the ways that the subassemblies, components, and materials interact with each other are not well understood: changes in the design of one component may change the way other components in the system perform, in ways that initially cannot be anticipated [Ulrich and Eppinger, 1995].

Such integrated design problems — essentially having to redesign the entire product in every generation — can be very expensive. For this reason, designers often work to understand the interactions among components and materials, to establish standards for how they should interface, and to define what attributes of each component must be measured and how to measure them. When these things are achieved, the design has become modularized.

Once modularization has been achieved, engineers are able to change or upgrade individual modules in the product without unexpected changes occurring in the performance or reliability of other pieces of the system. New and improved modules either can be "plugged and played", as can occur in most home stereo systems, or engineers at least can understand what changes they'll need to make in other parts of the product's design to account for what they've changed in one module, or to account for variation in the attributes of a component. Modularization makes improvements to product functionality and additions to features relatively fast and inexpensive. Modularity expands the economically viable options available to innovators and can be valued in the same way as financial options [Baldwin and Clark, 1997].

Industries can pass through cycles of modularization and demodularization as certain new technologies enter an industry. If a new technology changes the way components interact in the system design, engineers may no longer be able to predict how changes in the component or material (in which the new technology is embodied) might affect performance elsewhere in the system. They consequently will be unable to specify completely what critical attributes of the component must be met, meaning that at least some elements of the design have become integral again. Integrated design increases the cost and time required to create new product designs, however, — so engineers will likely work to understand how the new technology component interacts with other elements of the system in an effort to remodularize the product. Hence, industries can go through repeated cycles of modularization and demodularization [Christensen, 1994].

Modularity in product design also creates the potential for modularity in *organization* design. It is very difficult for a firm to source an integral component from an outside supplier. Because it cannot know how the component might interact with other elements of the system's design, the company cannot clearly specify what it needs its supplier to deliver. Once a modular design has been achieved in one or more components, however, then outsourcing is indeed a very viable option; standard interfaces between modules create a clear, "standard" way for the two organizations to interface [Sanchez, 1996]. There is substantial evidence, in fact, that, when product designs become modular, integrated organizations that produce components and design and assemble the final product lose market position to firms that flexibly can mix and match the most cost-effective components from the best independent sources. When new technology creates problems of integrality again, advantage shifts back to integrated companies that not only design and assemble the final products but manufacture the integral components as well [Christensen, 1994]. Indeed, the concept of the "virtual corporation" which moves alternately in and out of vogue in the business press, is viable only if clear, well-defined interfaces are established amongst all components and materials in the product, and across all elements of the value-added chain. A virtual organization is not a viable organization in an integral, non-modular world [Chesbrough and Teece, 1996].

Frequently, a core competence of a company lies in its processes for achieving exceptional product performance through the design of certain subsystems in its products whose components and materials have an integral character. If the components in this critical subsystem become modularized, this competence in integration essentially becomes embodied in the standard interfaces of the components. By this mechanism, proprietary competencies of early technology leaders can become diffused throughout an industry in the form of component interface standards. Hence, if a situation of complete product modularity were to occur in an industry, no firm could possess proprietary competence in product design: all competitors could mix and match components equally.[4]

We might expect, therefore, that, if an industry moved toward total modularity in its products, innovators would seek to establish competitive advantage by creating and maintaining new, nonstandard ways of integrating components in their products. As products became more highly integral in character, we would expect cost- and time-pressured engineers and marketers to always be searching for ways to modularize their products.

Defining Terms:

Dominant design: An explicit or *de facto* industry-wide standard architectural configuration of the components in an assembled product, in which the ways in which components interface with others in the product's architecture is well understood and established.

Modularization: A process by which the way that components and subsystems within an assembled product interact with each other becomes so well understood that standards emerge, defining how each component must interface with others in the system. When these standard interfaces exist, components and subsystems from multiple suppliers can be mixed and matched in designing and assembling a product, with predictable results for final system performance.

Punctuated equilibrium: A model of progress in which most of an industry's history is characterized by relatively steady, incremental, predictable improvement. This predictability is occasionally interrupted, or "punctuated", by brief, tumultuous periods of radical, transformational change.

S-curves: An empirical relationship between engineering effort and the degree of performance improvement achieved in a product or process. The improvement produced by an incremental unit of engineering effort typically follows an S-curve pattern.

References

Abernathy, W. *Productivity Dilemma*, Johns Hopkins University Press, Baltimore, 1978.

Abernathy, W. and Utterback, J. Patterns of industrial innovation, *Technol. Rev.*, June–July, 40–47, 1978.

Alexander, C. *Notes on the Synthesis of Form*, Harvard University Press, Cambridge, MA, 1964.

Anderson, P. and Tushman, M. Managing through cycles of technological change, *Res. Technol. Mgt.*, May/June: 1991, 26–31, 1991.

Baldwin, C. Y. and Clark, K. B. Sun wars: competition within a modular cluster, 1985–1990, In *Competing in the Age of Digital Convergence*, chap. 3, D. B. Yoffie, ed., Harvard Business School Press, Boston, 1997.

Bower, J. L. Teradyne: The Aurora Project, Harvard Business School, Case # 9-397-114, 1997.

Bower, J. L. and Christensen, C. M. Disruptive technologies: catching the wave, *Harv. Bus. Rev.*, January–February, 43–53, 1995.

Chesbrough, H. W. and Teece, D. J. When is virtual virtuous? Organizing for innovation, *Harv. Bus. Rev.*, January–February, 65–73, 1996.

Christensen, C. M. Exploring the limits of the technology S-curve, parts I and II, *Prod. Operat. Mgt.*, 1: 334–357, 1992.

---

[4]Although I know of no studies that measure this phenomenon directly, I suspect that the industry of designing and assembling personal computers was very nearly in this situation in the early and mid-1990s. The components from which they were built interfaced with each other according to such well-established standards that it was difficult for any manufacturer to sustainably assert that they offered proprietary cost-performance advantages in their products.

Christensen, C. M. The rigid disk drive industry: a history of commercial and technological turbulence, *Bus. Hist. Rev.* 67: 531–588, 1993.

Christensen, C. M. The Drivers of Vertical Disintegration, Harvard Business School, Division of Research, working paper, 1994.

Christensen, C. M. and Rosenbloom, R. S. Explaining the attacker's advantage: technological paradigms, organizational dynamics, and the value network, *Res. Policy,* 24: 233–257, 1995.

Christensen, C. M. Patterns in the evolution of product competition, *Eur. Mgt. J.,* 15: 2, 117–127, 1997a.

Christensen, C. M. *The Innovator's Dilemma: When New Technologies Cause Great Firms to Fail,* Harvard Business School Press, Boston, 1997b.

Christensen, C. M., Suarez F. F. and Utterback, J. M. Strategies for survival in fast-changing industries, *Mgt. Sci.,* 1998.

Clark, K. B. The interaction of design hierarchies and market concepts in technological evolution, *Res. Policy,* 14:235–251, 1985.

Constant, E. W. *The Origins of the Turbojet Revolution,* The Johns Hopkins University Press, Baltimore, 1980.

Dosi, G. Technological paradigms and technological trajectories, *Res. Policy,* 11: 147–162, 1982.

Foster, R. N. *Innovation: The Attacker's Advantage,* Summit Books, New York, 1986.

Garvin, D. R. R. Donnelly & Sons: The Digital Division, Harvard Business School, Case No. 9-396-154, 1996.

Henderson, R. M. *The Failure of Established Firms in the Face of Technological Change,* Ph.D. dissertation, Harvard University, 1989.

Henderson, R. M. and Clark, K. B. Architectural innovation: the reconfiguration of existing systems and the failure of established firms, *Admin. Sci. Q.,* March: 9–32, 1990.

Iansiti, M. Science-based product development: an empirical study of the mainframe computer industry, *Prod. Operat. Mgt.* 4: 4, 335–359, 1995.

Jones, N. When Incumbents Succeed: A Study of Radical Technological Change in the Private Branch Exchange Industry, Harvard Business School, DBA thesis, 1996.

Kuhn, T. S. *The Structure of Scientific Revolutions,* University of Chicago Press, Chicago, 1970.

Levinthal, D. The Slow Pace of Rapid Technological Change: Gradualism and Punctuation in Technological Change, working paper, Wharton School, University of Pennsylvania, 1997.

McNair, M. Significant trends and developments in the post-war period, In *Competitive Distribution in a Free High-Level Economy and Its Implications for the University,* A. B. Smith, ed., pp. 17–18 University of Pittsburgh Press, Pittsburgh, PA, 1958.

Rosenbloom, R. S. and Christensen, C. M. Technological discontinuities, organizational capabilities, and strategic commitments, *Indust. Corp. Change,* 3(3): 655–685, 1994.

Roussel, P. Technological maturity proves a valid and important concept, *Res. Mgt.* 27: January–February, 29–34, 1984.

Sahal, D. *Patterns of Technological Innovation,* Addison-Wesley, London, 1981.

Sanchez, R. Strategic flexibility in product competition, *Strat. Mgt. J.* 16: 135–159, 1995.

Sanchez, R. and Mahoney, J. T. Modularity, flexibility, and knowledge management in product and organization design, *Strat. Mgt. J.,* 17: Winter Special Issue, 63–76, 1996.

Suarez, F. F. and Utterback, J. M. Dominant designs and the survival of firms, *Strat. Mgt. J.* 16: 415–430, 1995.

Trent, T. R. Changing the Rules on Market Leaders: Strategies for Survival in the High-Performance Workstation Industry, unpublished Masters thesis, Management of Technology Program, Massachusetts Institute of Technology, 1997.

Tripsas, M. Unraveling the process of creative destruction: complementary assets and incumbent survival in the typesetter industry, *Strat. Mgt. J.,* forthcoming.

Tushman, M. and Anderson, P. Technological discontinuities and organizational environments, *Admin. Sci. Q.* 31: 439–465, 1986.

Ulrich, K. T. and Eppinger, S. D. Product architecture, In *Product Design and Development,* chap. 7, McGraw Hill, New York, 1995.

Further Information

References for each of the concepts noted in this chapter are listed below, including the articles and/or books that summarize the most important aspects within each body of scholarship.

> *Dominant design theory:* Abernathy, W. and Utterback, J. Patterns of industrial innovation, *Technol. Rev.*, June–July: 40–47, 1978.
>
> *S-curve theory:* Foster, R. N. *Innovation: The Attacker's Advantage,* Summit Books, New York, 1986; Christensen, C. M. Exploring the limits of the technology S-curve, *Prod. Operat. Mgt.* 1: 334–357, 1992; Tushman, M. and Anderson, P. Technological discontinuities and organizational environments, *Admin. Sci. Q.*, 31: 439–465, 1986.
>
> *The intersecting trajectories model:* Christensen, C. M. *The Innovator's Dilemma: When New Technologies Cause Great Firms to Fail,* Harvard Business School Press, Boston, 1997.
>
> *Modularization research:* Ulrich, K. T. and Eppinger, S. D. Product architecture, in *Product Design and Development,* chap. 7, McGraw Hill, New York, 1995; Sanchez, R. and Mahoney, J. T. Modularity, flexibility, and knowledge management in product and organization design, *Strat. Mgt. J.* 17: Winter Special Issue, 63–76, 1996.

The Technology and Innovation Management (TIM) section of the Academy of Management is an association of academics and managers whose research and practice focuses on issues of managing innovation. The activities of this organization, as well as the names of leading members of it, can be found on their web page, at http://www.aom.pace.edu/tim/.

3.2 Discontinuous Innovation

Christopher M. McDermott

When managing projects in an innovative environment, it is critical to be aware that projects differ in the extent to which they depart from existing knowledge. From a managerial perspective, these differences in the uncertainty of projects affect the way it needs to be directed and managed. As will be discussed in this section, the approaches used in managing an incremental project are often quite inappropriate when dealing with a project at the more discontinuous end of the spectrum. However, the relative infrequency of these discontinuous innovations (leading to the lack of experience most managers have dealing with these projects), coupled with the critical importance of such "leaps" to a firm's long-term viability, makes this a very important topic. This section provides an overview of research and practice in the management of projects at this more discontinuous end of the spectrum. It begins with a discussion of the measurement of innovativeness and the importance of discontinuous innovation within a firm's portfolio of projects. The section then moves on to discuss the impact of discontinuous innovation at a (macro) industry level and (micro) firm level, including a discussion of managerial approaches that appear to be appropriate in more discontinuous environments.

Fundamentals of Discontinuous Innovation

Traditionally, the bulk of academic research and practitioner attention in the management of technological innovation has focused on the development and implementation of practices that are appropriate and effective in directing incremental projects. This makes sense for a number of reasons:

- The majority of projects that are ongoing at any point in time in a firm *are* incremental.
- The short-term return on investment (ROI) mindset prevalent in many firms discourages discontinuous "leaps".
- Recent focus on lean production and reengineering absorbs many of the prerequisites for considering such projects.

However, the argument has been made [Morone, 1993; Tushman et al., 1997] that only through a balanced portfolio approach to innovation (with firms pursuing both incremental and discontinuous projects) can a firm continue to prosper in the long term. While incremental innovation can maintain industry leadership in the sort term, it is the projects at the discontinuous end of the spectrum that put firms in the lead in the first place.

There is no easy answer to the question: What makes a project discontinuous? Certainly, one common theme tied to **discontinuous innovations** is the high degree of risk and uncertainty throughout the process. The technologies involved are often unproven, and the projects are typically longer in duration and require a significantly greater financial (and managerial) commitment than their incremental counterparts. In addition, the end products themselves are often targeted at markets that are currently nonexistent or at least have the potential of being so significantly "shaken-up" by the introduction of the final product that existing, traditional market research is of questionable value. This discussion suggests one common way to conceptualize such projects which is shown in Fig. 3.5.

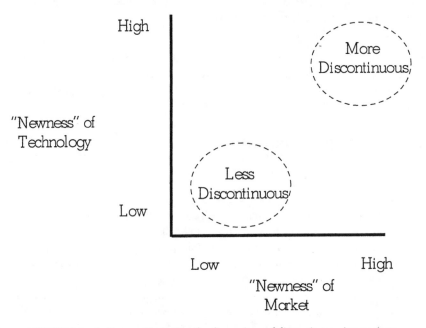

FIGURE 3.5   A diagram illustrating the dimensions of discontinuous innovations

In this figure, the vertical axis relates to the technological uncertainty associated with the project. At the low end of this axis, a firm is dealing with technology that is essentially an extension of current capabilities. The question here is more of *when* as opposed to *if* the technological advance embodied in the project will work. At the high end of this axis, however, firms manage projects where the functionality of the technology itself is in doubt or at least yet unproven. This may mean that a technology is being transferred to an entirely new set of applications (and production environment) where it has yet to be proven, or it might be a technology that emerged from research and development (R&D) as a promising opportunity but has yet to have all the kinks worked out.

The horizontal axis describes market uncertainty, again going from low to high. At the low end, the project is targeted at a market that is existing, well defined, and familiar to the firm. At the high end, the firm is moving into uncharted territory — a market that may not yet perceive the need for the product or understand the value it will add. The critical infrastructure required to make the innovation a reality (e.g., charging stations for the electric car) may also not exist.

Looking more closely within the matrix, projects positioned in the upper right face high degrees of uncertainty in both dimensions. However, if successful, such projects have the potential to be home runs (to create whole new markets even industries) while at the same time positioning the developers to be technological leaders in the future. Projects in the upper left quadrant of this scheme (low market uncertainty, high technological uncertainty) are going after existing (perhaps their own) markets with a new way delivering the product. A completely new, unproven technology that offers a major (30 to 40%?) decrease in product costs or enables significant improvement in product features to an existing market would fit here (e.g., Gillette's development of the Sensor line of razors). Projects in the lower right quadrant would be discontinuous in terms of the market while being more incremental in terms of the technology in the product. Projects here might be providing a new combination of existing technologies to a nonexistent market (e.g., Sony's Walkman).

Recent research by Green et al. [1995] divides characteristics of innovation into categories, indicating that it can be thought of as being comprised of at least four distinct, independent items:

- *Technological uncertainty* — Will the technology itself work as planned?
- *Technological inexperience* — Does the *firm* know enough about the innovation?
- *Business inexperience* — Is the firm familiar with the market and industry?
- *Technology cost* — What is the overall cost of the project?

Effects on Industry Structure

Tushman et al. [1997] and Ettlie, Bridges and O'Keefe [1984] cite the difficulties of large firms in creating an environment conducive to innovation. To summarize, a central problem in the management of more discontinuous innovation is the phenomenon that large, complex, decentralized, formal organizations tend to create a form of inertia that inhibits them from pushing toward the more discontinuous end of the spectrum. Their size and past successes push them toward patterns that strove to further enhance existing, dominant industry designs and practices. As a result, there is often a tendency for smaller firms to be innovation leaders. The larger firm's existing positions of leadership act to limit their desire to change current market paradigms, and their large, formal, structures make it difficult to do so even if they chose to. Companies develop core competencies [Prahalad and Hamel, 1990] and work to extend them incrementally.

Tushman et al. [1997] suggest that a form of technology cycles exists and "are made up of alternating periods of technological variation, competitive selection, and retention (or convergence)." These cycles are initiated by discontinuous innovations, which either enhance existing competencies (i.e., build off current expertise) or destroy existing competencies (i.e., render existing technological capabilities non-competitive). Thus, when a discontinuous innovation occurs, there is first a period of rapid change and multiple, competing designs vying for industry leadership. Finally, there emerges a winner — the dominant design — after which there is a prolonged era of incremental changes and elaborations on this industry accepted model. This period will continue until the next discontinuous innovation occurs. This cycle has been shown to exist across industries — from chemicals to computers.

Managing Discontinuous Innovations at the Firm Level

Challenges to the New Product Development Process

As was noted earlier, many of the approaches to the management of incremental innovation may not be appropriate as a given innovation becomes more discontinuous in nature. For example, the combination of speed to market pressures and the need for product innovation in the 1980s forced many firms to experiment with new ways of bringing new products to market. It was recognized early on that Japanese auto makers were consistently able to design and build a new automobile in less than 30 months. Until very recently, the Big Three automotive manufacturers required from 48 to 60 months to accomplish the same set of tasks. One common strategy that emerged during this time was to view the development of innovative new products as team based, rather than a sequential process, stressing the importance of getting the functional areas together early and frequently in bringing the product to market.

However, recent research on more discontinuous innovation [McDermott and Handfield, 1997] indicates that this parallel approach may be less than effective in discontinuous projects. Findings of this study show that much of the richness of this interaction is lost when the technological leap associated with the project is large. Often, team members find that the uncertainty levels are so high that there is little benefit to the frequent interactions encouraged by the more concurrent approach. Quipped one senior designer, "All we could prove to a typical manufacturing engineer is that we don't know enough to be wasting his time." This suggests that a return to some of the elements of the more sequential approach to innovative product development might be more appropriate in the face of such high uncertainty.

Evaluating Progress

Another interesting challenge associated with the management of discontinuous innovation is how a manager gauges success — both at the personal and project level. At the personal level, it quickly becomes clear that traditional evaluation methods based on meeting a set of articulated and predetermined goals fall apart in environments of discontinuous innovation. Unforeseen obstacles are the norm rather than the exception in this realm. If discontinuous project managers are measured by traditional, short-term methods such as payback, meeting technical milestones, etc., one will quickly find that it is difficult to find a person who will be willing to fill this leadership position. Further, technical team members need reassurance that their involvement in such projects, which typically have "failure" rates approaching 90%, will not limit their career. If a firm wants their best and brightest working on their leading-edge discontinuous projects, it becomes absolutely necessary to measure them on substantial progress toward technical goals rather than on performance against some external yardstick. The focus becomes "What have we learned?" and "Can we now do things better?" as opposed to "Did we meet our goals?".

At the project level, it is important to remember that these discontinuous projects are typically longer in duration and more costly than most of the other projects in a firm's portfolio. If a firm's traditional new product development metrics are used on discontinuous projects, they would most certainly be flushed out of the system. They are simply too high a risk for conventional management metrics. This is why many discontinuous projects in large firms are nurtured in corporate R&D, as opposed to the division's R&D. The short-term pressures on divisions (such as ROI performance) are frequently at odds with the very nature of discontinuous innovation.

Creating an Environment that Supports Change

The discussion above regarding short-term pressures is a key reason why upper-management support is critical for these projects. Clearly, without the protection and support of individuals in power, these more discontinuous projects could not continue to push on year after year. Dougherty and Hardy [1996] and Nadler and Tushman [1990] stress the importance of both upper-management and *institutionalized* organizational support for discontinuous change. Leonard-Barton [1992] argues that core competencies and capabilities too often become core "rigidities" in firms — limiting future moves and creating blinders that deny access to and interest in new ways of doing things. Dougherty and Hardy [1996] concur, arguing that it is the very nature of the organization systems in large firms that serves to limit their ability to innovate effectively. Their study of innovation in 15 large firms found that those searching for improvements need a new way of thinking, including

- More strategic discussions about innovation between senior and middle management.
- An overall increase in the relevance and meaning of innovation throughout the organization, through such means such as reconciling conflicts between cost cutting and innovation performance metrics within business units.

Thus, in large organizations, it is often not enough to have one senior manager or champion behind a project. This is a necessary but not sufficient requirement. The whole context of the firm and the way in which it views discontinuous activities is a critical part of building an environment that supports such projects. If the importance of such innovation is communicated and understood at all levels of the firm and throughout the management system, there is a greater chance of the survival. In short, senior management is very helpful, but a corporate environment that understands the importance of discontinuous innovation and nurtures promising projects is better.

One way in which a firm can nurture discontinuous projects is by providing **slack** resources. The creation of slack resources for discontinuous projects shows recognition that they *will* run into unforeseen difficulties and roadblocks along the way. There are simply more hurdles to go over and unknown entities in these projects than in other, more traditional ones. To fail to give them slack resources (in terms of extra time, people, or financial support) would essentially kill them off. Nohria and Gulati [1996] found that there is an inverse U-shaped relationship between slack and innovation. Firms that focus too much on lean operation may be hurting their long-term opportunities for innovation. Too much slack, on the other hand, was often seen to lead to a lack of focus or direction in innovating firms. The challenge for managers in environments of discontinuous innovation is then to provide enough slack so as not to stifle good projects, while at the same time provide enough direction and guidance to keep projects from floundering.

Core Competence and Discontinuous Innovation

The idea of building from core competencies and capabilities is also relevant to the management of discontinuous innovation. Prahalad and Hamel [1990] define core competencies as the "collective learning in the organization, especially how to coordinate diverse production skills and integrate multiple streams of technologies within an organization. " It makes a disproportionate contribution to the perceived customer value or to the efficiency with which that value is delivered. A core competence is an organization's hidden capability of coordination and learning that competitors cannot easily imitate, often providing a basis for access to new markets and dominance in existing ones. Prahalad and Hamel [1990] assert that it is necessary to seek competitive advantage from a capability that lies behind the products that serve the market. Clearly, working on projects that have some relevance to the firm's existing strengths increases the probability of success.

For example, Morone [1993] documented the effectiveness of Corning in their development of industry-leading, *discontinuous* process innovation. Beginning in the late 1970s with their development of an industry-leading "inside chemical vapor deposition process", Corning produced generation after generation of new-to-the-world, leading-edge manufacturing processes — each of which yielded both *substantially* improved fiber optics (with better attenuation, bandwidth, etc.) *and* lower costs. Often, these successive innovations were making obsolete the very processes that Corning had developed (and patented) just a few years earlier. Just as the competition began to benchmark their industry-leading practice, Corning would change the rules. As their capacity doubled and redoubled many times during the 1980s, fiber optics grew to be a central part of Corning's business. Due in no small part to these industry-leading advances in their manufacturing processes, Corning continues to be the dominant figure in fiber optics in the 1990s, with annual sales at an estimated $600 million.

Marketing and Discontinuous Innovation

One difficulty in working on discontinuous projects is in the development of and learning about new markets for products. Attempts to be customer driven are often missing the big picture — the customer may not yet realize the need for the product. Imagine trying to map the future of today's PC market 20 years ago. A recent study of this phenomenon [Lynn et al., 1996] reached the conclusion that conventional marketing techniques proved to be of limited value at best, and were often wrong in these environments. What proved to be more useful, they found, was what they termed the "**probe and learn**" process. In this process, the companies "developed their products by probing potential markets with early versions of the products, learning from the probes, and probing again. In effect, they ran a series of market experiments — introducing prototypes into a variety of market segments." This process, as described, understands the nature of emerging markets and uncertain technology. Probes are simply "feelers" to get a better sense of the market. From the outset, it is viewed as an iterative process, with the goal of the first several attempts to learn as much as possible about the market so that later attempts might be more on target.

Concluding Comments

Discontinuous innovation provides both unique challenges and opportunities to the firm. When working on a project that is a significant leap from existing technologies or markets, it is important to bear in mind that the techniques required to effectively manage the project are significantly different too. This section provides a brief overview of some of the critical issues relating to this topic at both the micro and macro level.

Defining Terms

Discontinuous innovation: Product development characterized by technical and/or market change ("newness") on a grand scale. Often creates new markets or destroys existing ones. Typically long term, with high risk and high potential return.

Slack: Excess resources (time and money) within a business unit. Often deliberately created to enhance discontinuous innovation.

Probe and learn: Running market "experiments" with new products, with the goal of gaining a greater understanding of end user needs and wants. Typically not viewed as product launch but rather just a means of further refining the product.

References

Dougherty, D. and Hardy, C. Sustained product innovation in large, mature organizations: overcoming innovation-to-organization problems, *Acad. Mgt. J.*, 39(5): 1120–1153, 1996.

Ettlie, J., Bridges, W. P., and O'Keefe, R. D. Organization strategy and structural differences for radical vs. incremental innovation, *Mgt. Sci.*, 30(6): 682–695, 1984.

Green, S., Gavin, M., and Aiman-Smith, L. Assessing a multidimensional measure of radical technological innovation, *IEEE Trans. Eng. Mgt.*, 42(3): 203–214, 1985.

Leonard-Barton, D. Core capabilities and core rigidities: a paradox in managing new product development, *Strat. Mgt. J.*, 13: 111–125, 1990.

Lynn, G. S., Morone, J. G., and Paulson, A. S. Marketing and discontinuous innovation: the probe and learn process, *Cal. Mgt. Rev.*, 38(3): 8–37, 1996.

McDermott, C. M and Handfield, R. The Parallel Approach to New Product Development and Discontinuous Innovation, working paper, The Lally School of Management and Technology, Rensselaer Polytechnic Institute, Troy, NY, 1997.

Morone, J. G. *Winning in High-Tech Markets*, Harvard Business School Press, Boston, 1993.

Nadler, D. and Tushman, M. L. Beyond the charismatic leader: leadership and organizational change, *Cal. Mgmt. Rev.*, 32(2): 77–97, 1990.

Nohria, N. and Gulati, R. Is slack good or bad for innovation?, *Acad. Mgt. J.*, 39(5): 1245–1264, 1996.

Prahalad, C. K. and Hamel, G. The core competence of the corporation, *Harv. Bus. Rev.*, May–June: 79–91, 1990.

Tushman, M. L., Anderson, P. C. and O'Reilly, C. Technology cycles, innovation streams, and ambidextrous organizations: organization renewal through innovation streams and strategic change. In *Managing Strategic Innovation and Change*, M. L. Tushman and P.C. Anderson, eds., Oxford University Press, 1997.

Further Information

A good, detailed example of three firm's successful development of discontinuous products can be found in *Winning in High-Tech Markets* by Joseph Morone.

The concept of the portfolio approach to managing new products is covered in great depth in *Portfolio Management for New Products* by Cooper, Edgett and Kleinschmidt.

Mastering the Dynamics of Innovation by James Utterback provides a good overview of the methods appropriate for the management in many different innovation environments.

Faculty from a variety of disciplines at The Lally School of Management and Technology at Rensselaer Polytechnic Institute in Troy, NY, are currently examining these and other issues relating to the management of discontinuous innovation in their longitudinal study supported by the Sloan Foundation and the Industrial Research Institute. Copies of their most recent findings can be acquired by calling (518) 276-8398.

3.3 Business Process Reengineering

Thomas H. Davenport

The 1990s will long be remembered as the decade of **reengineering**. Described in over 30 books that were translated into many languages, featured in articles in every major business publication, and discussed at hundreds of conferences, reengineering has penetrated into every continent except Antarctica. Thousands of companies and public sector organizations have implemented reengineering initiatives. Internal and external consultants on the topic have proliferated dramatically. Many universities have created courses on the topic for business school students.

However, reengineering means different things to different people. In the early writings on reengineering — and in more recent academic work — there was considerable consensus on its meaning (Davenport and Short, 1990; Hammer, 1990). In this "classical" view, reengineering means the radical redesign of broad, **cross-functional business processes** with the objective of order-of-magnitude performance gains, often with the aid of information technology.

Among the general business audience, however, this usage has become only one point in a broad spectrum of uses. At one end of the spectrum, reengineering means any attempt to change how work is done — even incremental change of small, subfunctional processes. At the most ambitious extreme, reengineering is synonymous with organizational transformation or major simultaneous change in strategy, processes, culture, information systems, and other organizational domains. However, reengineering is perhaps most commonly described today as a means of removing workers from corporations. Because of this association, which was not intended by the creators of the concept, the term reengineering has become less popular in recent years. Firms continue to do what they did under the name of reengineering, but they may no longer use that term to describe their efforts.

Many of the early thinkers on reengineering had their roots in the information technology (IT) field, specifically in IT consulting. IT began to be applied widely in organizations in the 1980s, but its increased use has not been associated with increased productivity — particularly in the service sector. One key driver of reengineering, then, was to make better use of the IT resource — to use it to change how work is done in an organization. The reengineering movement, however, quickly took on broader business objectives than those involving IT.

Key Components of Reengineering

Some observers, such as Peter Drucker on the cover of the best-seller *Reengineering the Corporation* (Hammer and Champy, 1993), argue that reengineering is new; others claim that there is nothing new about it. Both can be viewed as partially correct. The "components" of reengineering all existed prior to 1990, when the first articles on the topic were published. However, these components had not previously been assembled in one management concept. Reengineering is new, therefore, only as a new synthesis of previously existing ideas.

The components came from multiple sources that cut across academic disciplines and business functions. The idea of managing and improving business processes comes primarily from the quality or continuous-process-improvement literature, which itself modified the process concept from industrial engineering concepts extending back to the turn of the century and Frederick Taylor. This earlier work viewed processes as being small enough for single work groups to address. The notion of working on broad, cross-functional processes is somewhat more recent, but is certainly at least implicit in the value-chain concept popularized by Michael Porter and in the concurrent engineering and design-for-manu-facturing concepts employed in the manufacturing industry.

Another key aspect of reengineering is the "clean sheet of paper" design of processes. Although firms often disregard existing constraints in designing a new business process, the constraints must be taken into account during implementation, unless a new organizational unit is created. In any event, this idea is not new to reengineering. It was used prior to the reengineering concept at the Topeka pet food factory of General Foods in the early 1970s, at General Motors in the 1980s [in the Saturn project], and in the First Direct subsidiary of Midland Bank in the early 1990s. The idea of such a greenfield site has been explored in the work-design literature [Lawler, 1978].

Reengineering typically involves the use of IT to enable new ways of working. This idea has been discussed since the initial application of IT to business, though often not executed successfully. It is present in the concept of the systems analyst function and was also frequently expressed in the context of the strategic or competitive system.

Each of the component concepts of reengineering, however, had some "flaw"—at least from the perspective of someone desiring reengineering-oriented change. Quality-oriented business processes were too narrow and relied too heavily on bottom-up change to yield radical new work designs. The idea of broader, cross-functional processes was limited to certain domains and to manufacturing industries. Greenfield change before reengineering often involved an entire organization at a single site, with no smaller, exportable unit such as a process. Finally, while the notion that IT could change work was powerful, the systems analysts or other purveyors of IT could not by themselves create a context in which radical organizational change could occur. Reengineering appeared to "solve" all of these shortcomings.

The State of Reengineering

Reengineering has been under way for at least 7 years now — or more, if one defines the term somewhat loosely. It is possible at this point to give an overview of what is happening with reeengineering — how prevalent it is, what kinds of processes get reengineered, and the fate of reengineering projects.

A 1994 survey of large North American and European companies found that more than two thirds had at least one reengineering project under way [CSC Index, 1994]. Of those not then doing reengi-neering, half were planning or discussing projects. In both Europe and North America, companies had more than three reengineering initiatives completed or under way on average. As reasons, they cited more demanding customers, increased competition, and rising costs as the primary drivers. The most popular processes addressed by reengineering were customer service, manufacturing, order fulfillment, manufac-turing, distribution, and customer acquisition. Few firms in the survey, however, were taking on research or product development processes. Virtually none cited management processes as their target. These latter processes have historically been characterized by high levels of autonomy and very low structure and thus were difficult to address in process reengineering terms.

How did firms actually proceed to reengineer their processes? In practice, many companies did take a clean-sheet approach to *designing* a process. Design teams attempted to identify "the best of all processes", without regard to constraints of existing information systems or organizational factors. Not surprisingly, the new process designs that many firms created were quite radical, with ambitious plans for new technologies, new skills, and new organizational structures.

An unknown fraction of firms did achieve their process-change objectives. Table 3.1 lists several reengineering projects or corporate-wide initiatives that have been reported to achieve substantial ben-efits. These projects cut across many different processes and industries. There is thus little doubt that success in reengineering is eminently possible.

TABLE 3.1   Reported Reengineering Successes

Company	Process	Reported Benefit
Siemens Rolm	Order fulfillment	Order to installation completion time improved by 60%; field inventory cut by 69%
CIGNA Reinsurance	Many	Operating costs reduced by 40%; application systems reduced 17 to 5
CIGNA	Many	Savings of more than $100 million
British Nuclear Fuels	Fuel processing	Doubled the fuel processing rate
Eastman Chemical	Maintenance	80% reduction in order cycle time; savings of $1 million/year
American Express	Many	Cost reductions of more than $1.2 billion
Amoco	Capital allocation	33% staffing reductions
Xerox	Supply chain	30% inventory reduction
Rank Xerox	Billing	Cycle time reduction of 112 days to 3; annual cash flow increase of $5 million
Aetna	Many	Annual savings of $300–$350 million
Federal Mogul	New product development	Cycle time reductions from 20 weeks to 20 days
AT&T Business Communication Systems	Many	Product quality improved tenfold; 75% reduction in key cycle times

Firms that did succeed with reengineering typically cited senior management sponsorship, high-energy reengineering teams, customer feedback, and willingness to invest time and money in the effort as critical success factors. The broad survey cited above also found higher levels of success among companies that had particularly ambitious "stretch" goals for their projects.

In fact, most companies succeeded at creating much better designs for doing work. However, some fraction of companies did not attain their change goals. The problems typically arise at *implementation* time. Few firms can afford to "obliterate" their existing business environments and start from scratch. Assuming that the firm is not disillusioned by this situation, it proceeds to implement the new process incrementally. Thus, the general rule for reengineering is "revolutionary design, evolutionary implementation". This observation has been confirmed in a study of 35 reengineering projects by Jarvenpaa and Stoddard, (1996).

Perhaps as a result of the problems firms have faced in reengineering, there have been articles in the business press suggesting that reengineering projects have a high "failure" rate. However, it is not clear what failure means in the reengineering context. Reengineering can be viewed as a planning-oriented activity, in which plans for how the organization will do its work are drafted years ahead of actual implementation. It would not be surprising if the implementation of process designs did not exactly mirror the plans on paper. Furthermore, should success and failure be assessed relative to objectives or to actual accomplishment, that is, if an organization sets out to achieve tenfold improvement and then only betters its processes by threefold, is it success or failure? In general, large reengineering projects in large organizations are complex and multifaceted, and they defy easy characterization of success or failure.

Summary

Reengineering may have aspects of faddishness or management myth, but there are timeless aspects to it as well. It will always be important for firms to improve how they do work, sometimes incrementally and sometimes radically. It will always be necessary to use IT, not to automate existing modes of work, but to enable new work processes. Even when attention inevitably shifts to the next management nostrum, the factors that made reengineering popular will still be present, and its approaches will still be useful.

Reengineering has many facets and success factors. Successful redesign and implementation of a cross-functional business process amounts to major organizational change, not unlike massive restructurings, major product line changes, or a merger or acquisition. It will therefore always be complex and difficult.

Defining Terms

Business processes: An ordered set of activities by which an organization produces a specified output for a specified customer.

Cross-functional: Without regard to the boundaries of traditional business functions, e.g., engineering, manufacturing, and marketing.

Reengineering: The radical redesign of broad, cross-functional business processes with the objective of order-of-magnitude performance gains, often with the aid of information technology.

References

CSC Index. State of Reengineering Report Executive Summary, Cambridge, MA, 1994.

Davenport, T. H. and Short, J. E. The new industrial engineering: information technology and business process redesign, *Sloan Mgt. Rev.*, Summer: 11–27, 1990.

Davenport, T. H. *Process Innovation*, Harvard Business School Press, Boston, 1993.

Hackman, J. R. and Oldham, G. R. *Work Redesign*, Addison-Wesley, Reading, MA, 1980.

Hammer, M. Reengineering work: don't automate, obliterate, *Harv. Bus. Rev.*, Summer: 104–112, 1990.

Hammer, M. and Champy, J. A. *Reengineering the Corporation*, Harper Business, New York, 1993.

Jarvenpaa, S. L. and Stoddard, D. B. Business process reengineering: radical and evolutionary change, *J. Bus. Res.*, 1996.

Lawler, E. E. The new plant revolution, *Organ. Dyn.*, Winter: 2–12, 1978.

Venkatraman, N. IT-enabled business transformation: from automation to business scope redefinition, *Sloan Mgt. Rev.*, Winter: 73–87, 1994.

Further Information

For a broad collection of case studies on reengineering, see *Reengineering the Organization* (Harvard Business School Press) by Nolan, Stoddard, Davenport, and Jarvenpaa.

For a discussion of the factors involved in creating a process-oriented organization, see Michael Hammer's *Beyond Reengineering*.

A collection of academic papers on reengineering is presented in Grover and Kettinger, eds., *Business Process Change*.

An analysis of where reengineering went astray is in Thomas H. Davenport, "The Fad That Forgot People", Fast Company. November 1995.

3.4   Diffusion of Innovations

Andrew B. Hargadon

Turning an innovative idea into a new product or process and into a successful business venture requires getting others to adopt and use your innovation. Similarly, turning ideas into reality and getting them to market in time often requires adopting the latest in development and manufacturing technologies. These challenges are two perspectives of the same underlying process, the **diffusion of innovations**, and are central to engineering work.

Diffusion of innovations describes how new technologies spread through a population of potential adopters. On the one hand, it describes the process of getting new ideas adopted and, on the other, the process of adopting new ideas. A great deal of research has investigated how innovations diffuse in fields such as agriculture, healthcare, education, business, and manufacturing. The primary focus of diffusion research is the innovation. An innovation can be an object, a practice, or an idea that is perceived as new by those who might adopt it. Innovations present the potential adopter with a new alternative for solving a problem, but also present more uncertainty about whether that alternative is better or worse than the old way of doing things. The primary objectives of diffusion research are to understand and predict the rate of diffusion of an innovation and the pattern of that diffusion.

Innovations do not always spread in a straightforward way. For instance, the best ideas are not always readily and quickly adopted. The British Navy first learned in 1601 that scurvy, a disease that killed more sailors than warfare, accidents, and all other types of death, could be avoided. The solution was simple (incorporating citrus fruits in a sailor's diet) and the benefits were clear (scurvy onboard was eradicated), yet the British Navy did not adopt this innovation until 1795, almost 200 years later! The best ideas are not always the ones to achieve widespread adoption. The story of the QWERTY keyboard provides the example here. The QWERTY keyboard, the current pattern of letters on a typewriter keyboard, was created in 1867. The design was a conscious effort to slow a typist's keystrokes and came from the need, in the early days of the typewriter, to avoid jamming the relatively delicate machines. As machine designs evolved, this problem diminished and competing keyboard designs as early as 1888 achieved higher typing speeds (these alternatives peaked in 1932 with August Dvorak's now-famous alternative). Yet, despite clearly more efficient alternatives, the QWERTY design holds out to this day.

These examples show that diffusion of innovations is a complicated process and that it is difficult to predict what innovations will gain widespread acceptance or when. Yet recent research has gained considerable insights into the factors that influence the rate and pattern of diffusion. This section presents the findings of this current research, drawing especially from the integrating work of Everett Rogers [1995] within a framework of the three predominant research trajectories:

1. Characteristics of the innovation
2. Characteristics of the adopters
3. Characteristics of the social environment

This section then presents several important limitations of the existing diffusion research, providing insights into the strengths and weaknesses of the current understanding of how innovations diffuse.

Characteristics of the Innovation

Much of the early attention in diffusion research went to understanding and explaining the differences among adopters and differences in the environment in which the innovation was diffusing. More recently, attention has focused on the impact that characteristics of the innovation itself can have on the subsequent rate and pattern of diffusion [Rogers, 1995]. Research has centered around five main perceived attributes of innovations: relative advantage, compatibility, complexity, trialability, and observability. These characteristics are not purely objective measures but also represent the *perceptions* held by potential adopters about an innovation.

- **Relative advantage** represents the perceived superiority of an innovation over the current practice or solution it would replace. This advantage can take the form of economic benefits to the adopter, but also benefits in social prestige, convenience, and satisfaction. Innovations may also have advantages that cannot be compared to the existing solutions, meaning they offer advantages that were not previously possible or even expected.
- **Compatibility** represents the perceived fit of an innovation with a potential adopter's existing values, experiences, and practices. The more compatible an innovation is with a cultural values surrounding the new technology and with the preexisting skills that can be transferred from the current technology, the more rapid and widespread its diffusion.
- **Complexity** describes the extent to which an innovation is perceived to be difficult to understand or use. The higher the degree of perceived complexity, the slower the rate of adoption.
- **Trialability** represents the extent to which a potential adopter can experiment with the innovation before adopting it. The greater the trialability, the higher the rate of adoption.

- **Observability** represents the extent to which the adoption and benefits of an innovation are visible to others within the population of potential adopters. The greater the visibility, the higher the rate of adoption by those that follow.

Characteristics of the Adopters

Three central characteristics of potential adopters play a large role in the diffusion of innovations. First, the rate at which a population adopts an innovation often follows a typical pattern, known as the **S-shaped curve of adoption**. Second, individual adopters can be categorized based on their **innovativeness**, or willingness to adopt new innovations. Third, individual adopters follow an **innovation-decision process** when deciding to adopt a new technology.

S-Shaped Curve of Adoption

Individuals in a population do not all adopt a new technology at the same time nor do they adopt at a constant rate. Instead, diffusion studies have found a pattern that follows an S-shaped curve of cumulative adoptions, shown in Fig. 3.6. This adoption curve begins slowly but soon the rate of adoptions increases as more individuals in each successive time period make the decision to adopt. Ultimately, the rate declines again as fewer individuals in the population are left to adopt the innovation. There is variation in the slope of the curve, some innovations are adopted rapidly while others take longer, and in the length of the "tails" of the curve. While this curve decribes the diffusion rates of many innovations, there are circumstances under which it is not appropriate, for instance, when an innovation is difficult to communicate and the number of existing adopters does not affect a potential adopter's awareness or opinions of an innovation.

Adopter Categories

Because all individuals in a population do not adopt at the same time, diffusion researchers have characterized adopters by the degree to which they are more or less likely to adopt an innovation relative

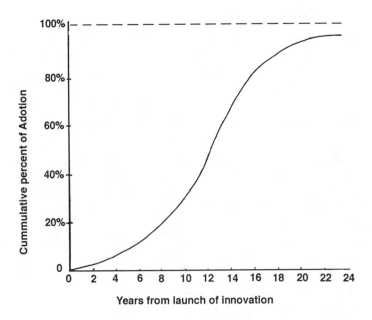

FIGURE 3.6 The S-shaped curve of cumulative adoption demonstrated by number of new adopters each year of hybrid corn seed in two Iowa communities. (*Source:* based on Ryan and Gross, 1943) in Rogers, E. M. [1995]. *Diffusion of Innovations*, 4th ed., Free Press, New York.

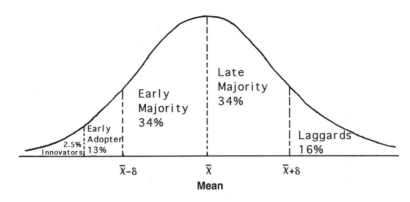

FIGURE 3.7   Adopter categories on the basis of innovativeness. (*Source:* Rogers, E. M. 1995. *Diffusion of Innovations*, 4th ed., Free Press, New York.

to others in the population, their **innovativeness** [Rogers, 1995]. When the cumulative adoption curve follows an S-shaped pattern, the distribution curve of adopters over time follows a normal distribution, as shown in Fig. 3.7. Diffusion research divides this curve into five categories of adopters and has found each category to have distinctive traits.

- **Innovators** typically make up the most innovative 2.5% of the population. They have been decribed as rash, risky, and daring. An ability to work with complex and often underdeveloped technology as well as substantial financial resources help them absorb the uncertainties and potential losses from innovations. Innovators are not usually opinion leaders when it comes to new innovations; their skills and risk-taking behavior often set them apart from the rest of the social system, but they often play a large role in importing innovations from outside and adapting them for broader adoption [see Von Hippel, 1988].

- **Early adopters** are more integrated with the existing social system than innovators, and often have the greatest degree of opinion leadership, providing other potential adopters with information and advice about a new technology. Change efforts surrounding new innovations often target this population first as they represent the successful, discrete, and objective adopters that hold the respect of the larger social system.

- **The early majority** adopts just ahead of the average of the population. They have much interaction within the social system but are not often opinion leaders; instead they typically undertake deliberate and, at times, lengthy decision making [Rogers, 1995]. Because of their size and connectedness with the rest of the social system, they link the early adopters with the bulk of the population, and their adoption signals the phase of rapid diffusion through the population.

- **The late majority** is described as adopting innovations because of economic necessity and pressure from peers. While they make up as large a portion of the overall population as the early majority, they tend to have fewer resources and be more skeptical, requiring more evidence of the value of an innovation before adopting it.

- **Laggards** are the last in a social system to adopt a new innovation. They tend to be relatively isolated from the rest of the social system, have little or no opinion leadership, and focus on past experiences and traditions. They are the most cautious when it comes to risking their limited resources on a new technology.

A number of generalizations have emerged from diffusions studies to differentiate between earlier and later adopters in a social system. In terms of socioeconomic characteristics, there tends to be no age difference between early and late adopters, but earlier adopters are more likely to have more years of formal education, to be more literate, and to have higher social status. In terms of personality traits, earlier adopters tend to have greater empathy, to be less dogmatic, to be more able to deal with abstractions and uncertainty, to have greater intelligence and rationality, and to have more favorable attitudes toward science. In terms of communication behavior, earlier adopters tend to have more social participation, to be more connected through interpersonal networks in their social system, to have traveled more widely and be aware of matters outside of their local system (to be more cosmopolite), and to have greater interaction with communication channels that provide information about possible innovations.

The Innovation Decision

The innovation decision process describes how an individual moves from knowledge of an innovation to adoption; research has focused on five stages [Rogers, 1995]:

- During the **knowledge** stage, an individual gains awareness of an innovation and some understanding of how it functions. This awareness is an interactive process; the existence of an innovation must be communicated to a population, but the potential adopters choose where to focus their attention and whether to value the problems the innovation is promising to solve.

- In the **persuasion** stage, a potential adopter attempts to gather more information and forms a positive or negative opinion about an innovation.

- The **decision** stage refers to the activities an individual adopter engages in that lead to a decision to adopt or reject an innovation. This includes experimentation on a limited basis, "trial-by-others" in which the experiences of other adopters are studied, and the decision to wait.

- The **implementation** stage describes when an adopter puts the innovation into use. During this phase much of the uncertainty surrounding the innovation is resolved, and the consequences, intended and otherwise, become apparent.

- The **confirmation** stage occurs when an adopter considers the consequences of adoption and may or may not continue with its use. In some cases the confirmation stage may never occur, as individuals neglect or avoid seeking information about an innovation's consequences.

Diffusion research suggests these stages exist in the innovation decision process. However, most social processes contain exceptions to the linearity implied by the stage models that attempt to describe them (e.g., stage C does not necessarily follow stages A and B) [Wolfe, 1994; Weick, 1979], and innovation decision models are no different. Individuals often form opinions before adequate awareness of an innovation, they often make decisions and then gather supporting evidence, and they often confirm (or reject) an innovation before completing its implementation. Stage models such as the innovation-decision model help to outline a general process but cannot adequately reflect all of its complexities.

Characteristics of the Environment

Potential adopters learn about, adopt, and implement new technologies within a larger social and technological environment. The characteristics of this environment have as much impact on the nature of the diffusion process as do the characteristics of the innovation itself or the adopters. Two aspects of the environment, the communication channels through which information about new innovations spreads and the social system within which potential adopters make their decisions, are discussed in this section.

Communication Channels

Communication channels refer to the means by which information about an innovation reaches the population of potential adopters. There are two primary distinctions between communication channels: mass media vs. interpersonal channels and cosmopolite vs. localite channels. The differences between these channels are best understood in terms of their effects at different stages of the innovation decision.

- **Mass media vs. interpersonal channels.** Mass media channels move information across a mass medium such as television, radio, or newspapers and typically involve one-way communication of information from a single or few sources to a large audience. Interpersonal channels are face-to-face exchanges of information between individuals. Mass media is more effective at quickly spreading knowledge of an innovation to a large audience; interpersonal channels are more effective at providing and clarifying the detailed information that will affect the persuasion of individual adopters.

- **Cosmopolite vs. localite channels.** Cosmopolite channels link potential adopters to others outside of the social system of interest. Localite channels, in contrast, represent communication between individuals within the social system. Cosmopolite channels bring new information into a social system and may be more influential in the early stages of the diffusion process, when new technologies are introduced into a population [see Attewell, 1992]. Cosmopolite channels are more valued at the knowledge stage of the innovation decision, while localite channels offer more value during the persuasion stage.

Social System

The social system within which innovations diffuse contains more than just the communication channels linking information and potential adopters. The social structure, norms and values, opinion leaders and change agents, decision environments, and consequences also influence how innovations diffuse.

- The **social structure**, or diffusion network, describes how individuals are connected to one another within a given social system. The more densely connected individuals are with one another, the more likely innovations will diffuse rapidly. The more connected some individuals are with outside sources of information, the more likely they will act as early adopters, importing new technologies into the social system.

- A social system's prevailing **norms and values** have profound effects on the rate and pattern of diffusion. Accepted behavior patterns within a society may inhibit, if not openly persecute, adopters of a new technology. The taboo nature of many subjects may prevent communication of a potential innovation to those who would adopt it. However, not all norms and values run contrary to innovation. A pro-innovation bias in, for example, fashion and business management often encourages adoption of innovations before their benefits are completely understood [see Abrahamson, 1991].

- **Opinion leaders**, as described earlier, are those members of a social system that other potential adopters respect and seek out when making their own adoption decision. **Change agents** represent outside interests that have interest in the diffusion of an innovation within a particular social system. Change agents often play an important role in introducing a new technology into the social system, while opinion leaders influence the rate at which that technology is adopted.

- Social systems will often have different **decision environments**, depending on the political stucture of the system under study. These decision environments vary according to whether the individual decision to adopt an innovation is optional, collective (consensus), authority based, or contingent. Optional means that each individual within the population makes his own choice to adopt or reject an innovation; collective means that the adoption decision is made as a group; authority based means that adoption decisions are made for the group by one or a few. Contingent refers to some combination of the former as, for example, when the individual adoption decision is optional only after an authority has approved of that option.

- The **consequences of an innovation** are not always clear before adoption and implementation, but must be considered for their implications to subsequent adoptions and for sustained use of the innovation within a social system. Consequences have been described in such terms as desirable vs. undesirable, direct vs. indirect, and anticipated vs. unanticipated.

Limitations of the Diffusion Model

The quantity and rigor of research in diffusion of innovations can be misleading. Any theoretical model of social processes oversimplifies a complex reality, and it is important to consider several of the underlying assumptions of diffusion research that limit the general applicability of its findings. This section discusses three related assumptions: that the primary focus of diffusion research, the innovation, remains unchanged; that the process of adoption is distinct from the process of invention; and that the adoption of one innovation is independent of the adoption of others.

- **The evolution of innovations.** The current diffusion research model and methodology originated in the field of rural sociology with the Ryan and Gross [1943] study of hybrid corn seed adoption but soon spread to a wide variety of fields, including education, anthropology, public health, marketing, geography, and communications. The methodology focused on measuring the rates and patterns of adoption of an innovation within a well-defined population. To do this, the research assumes that an innovation remains unchanged as it diffuses through a population. This is a more viable assumption for hybrid corn seeds than for computers; often early adopters play an active role in transforming a new technology to meet the needs of the market [Von Hippel, 1988]. Many modern innovations evolve more rapidly than they diffuse. For example, the personal computer has undergone significant changes and has yet to saturate the market of potential adopters. This evolution of innovations during the diffusion process makes it more difficult to compare the adoption decisions of the individuals of the population.

- **Invention and innovation adoption.** Diffusion research focuses on the adoption of innovations subsequent to their invention and development. Recent studies, however, suggest that the invention process itself involves significant amounts of adoption of component innovations, which are then combined to form larger systems as new products or processes [Hargadon and Sutton, 1997; Basalla, 1988]. When an innovation is combined with other technologies, new uses may emerge that greatly extend the population of potential adopters. This occurred, for example, during the diffusion of the microprocessor. First used in computers, this technology has now been incorporated as an innovative component in such products as automobile engines, toys, and medical devices. As innovations evolve through the diffusion process and are combined with other components, the population of potential adopters may change dramatically.

- **The interdependence of innovations.** The rate of adoption of a particular innovation is often dependent on the adoption of other innovations, forming what Rogers [1995] describes as "technology clusters". Synergies emerge, for instance, between innovations and create positive feedback loops in their adoption. The steam engine (as pump) increased the productivity of coal mines which, in turn, helped increase steel production through lower costs and higher quality inputs. Better and cheaper steel led to better steam engines and to railroads. Steam-driven trains led to a greater demand for coal, steel, and better steam engines. A similar feedback loop has emerged between computers, microprocessor designs, and the Internet, as computers both enable and drive the demand for more complex microprocessors and the Internet creates broad uses for the computer and more demands on the microprocessor.

Current diffusion research models have difficulty quantifying the evolution, adoption through invention, and interdependence of innovations. As a result, the impacts of such phenomenon have not been adequately studied and are less understood. Diffusion researchers must account for these overlooked complexities, and engineers and managers adopting or marketing innovations must also be aware of them, for their implications on the rate and pattern of diffusion are undeniable.

Summary

This section has described the research on diffusion of innovations. Diffusion research studies the process by which new innovations spread through a population of potential adopters. Its focus is on the innovation itself, and its objective is to understand and predict the rate and pattern of diffusion across the population. The section described the influence of three major aspects of the diffusion process: the characteristics of the innovation, the characteristics of the adopter and the adoption decision, and the characteristics of the environment surrounding the population of potential adopters.

The characteristics of an innovation that influence the rate of its adoption through a population include **relative advantage, compatibility, complexity, trialibility,** and **observability.** Adopter characteristics explain the **S-shaped curve of cumulative adoption** by separating individual adopters into categories distinguishing between **innovators, early adopters, early majority, late majority,** and **laggards.** Adoption involves an **innovation-decision process** characterized by five stages: **knowledge, persuasion, decision, implementation,** and **confirmation.** Environmental characteristics that influence the diffusion process include the **communication channels** along which information concerning an innovation spreads and the **social system** surrounding the population. These channels differ from **mass media** to **interpersonal** and from **cosmopolite** to **localite.** The social system influences adoption through its **social (or network) structure,** its **norms and values, opinion leaders and change agents, decision environment,** and the **consequences** of the innovation. Three limitations of the current diffusion model are noted: that innovations continue to evolve throughout the diffusion process, that invention and adoption are often indistinguishable, and that innovations often depend on (or drive) the adoption of other innovations.

Defining Terms

Communication channels: The means by which information concerning a new technology is spread through the population, for example, through mass media or interpersonal channels.

Diffusion of innovations: The term describing the process by which new technologies spread through a population of potential adopters.

Innovativeness: The likelihoood that an individual will adopt a new technology relative to other potential adopters in the population.

References

Abrahamson, E. Managerial fads and fashions, *Acad. Mgt. Rev.,* 16(3): 1991.

Attewell, P. Technology diffusion and organizational learning: the case of business computing, *Organ. Sci.,* 3(1), 1992.

Basalla, G. *The Evolution of Technology,* Cambridge University Press, New York, 1988.

Hargadon, A. B. and Sutton, R. I. Technology brokering and innovation in a product development firm, *Admin. Sci. Q.* 1997.

Rogers, E. M. *Diffusion of Innovations,* 4th ed., Free Press, New York, 1995.

Ryan, B. and Gross, N. C. The diffusion of hybrid corn seed in two Iowa communities, *Rural Sociol.,* 8: 15–24, 1943.

Von Hippel, E. *The Sources of Innovation,* Oxford University Press, New York, 1988.

Weick, K. E. *The Social Psychology of Organizing,* Addison-Wesley, Reading, MA, 1979.

Wolfe, R. A. Organizational innovation: review, critique, and suggested research directions, *J. Mgt. Stud.,* 31(3), 1994.

Further Information

An in-depth review of the research in the diffusion of innovations can be found in *Diffusion of Innovations,* 4th edition, by Everett Rogers. Rogers' research and earlier editions of this book set the direction for much of the current work in this field.

3.5   Knowledge Management

Timothy G. Kotnour

Knowledge management is a process for helping an organization continuously build its capabilities to maintain and improve organizational performance. Knowledge management supports an organization being a learning organization or "an organization skilled at creating, acquiring, and transferring knowledge, and at modifying its behavior to reflect new knowledge and insights" [Garvin, 1993]. The learning organization concept has been offered as a paradigm to use in managing an organization in the knowledge society in which knowledge is "*the* resource" [Drucker, 1993] for competitive advantage.

This section provides insights to help managers address fundamental issues such as improving performance through the creation and use of knowledge [Drucker, 1993]; ensuring the organization has accurate, relevant, and timely knowledge; and sharing what was learned from one part of an organization to another. First, knowledge and knowledge improvement are defined. Second, the aim and steps of knowledge management are discussed. Third, examples of knowledge management strategies are given.

Knowledge and Knowledge Improvement

Bohn [1994] describes a framework for defining and measuring the stages of technological knowledge. In his framework, Bohn defines knowledge as the item that "allows the making of predictions, causal associations, or prescriptive decisions about what to do" and provides examples:

When the control chart looks like that, it usually means machine A needs to be recalibrated' (causal association and prescriptive decision) or 'When the control chart is in control for the first hour of a new batch, it usually remains that way for the rest of the shift' (prediction).

Knowledge can be viewed as more than one piece of information in a pattern from which explicit inferences and predictions can be made to support decision making and action taking. Organizations must continuously improve their knowledge by increasing the certainty of the prediction, association, or decision. An organization increases the certainty of the knowledge through the organizational learning process.

Organizational learning is the process of creating, assimilating, disseminating, and applying knowledge in an organization [Huber, 1991]. Knowledge creation is the improvement of and/or increasing the certainty of a piece of knowledge and occurs during a learning experience. The creating and updating occurs through the detection and correction of errors [Argyris and Schon, 1978]. A lesson learned is an example of an output from knowledge creation. Knowledge assimilation is the collection, storage, and refinement of the created knowledge with existing knowledge in the organization's memory.

Organizational memory is the "stored information from an organization's history that can be brought to bear on present decisions" [Walsh and Ungson, 1991] and is contained in organizational members, files/records, culture, processes, procedures, organizational structure, and physical structure. Through knowledge assimilation support is given to best practices and is decreased for less-effective practices. Knowledge dissemination is the retrieval and distribution of the knowledge to use on another project. Knowledge application is the use of past knowledge to help solve the current problem. In applying the past knowledge, the decision maker must adapt the knowledge to the current situation. The relationship between a learning experience, organizational memory, and organizational learning is given in Figure 3.8. Knowledge is created in a learning experience (e.g., problem-solving experience, project, or task). The newly created knowledge is assimilated into memory. The contents of memory are disseminated and applied to another learning experience.

Knowledge Management

The aim of knowledge management is to continuously improve an organization's performance through the improvement and sharing of organizational knowledge throughout the organization (i.e., the aim is to ensure the organization has the right knowledge at the right time and place). Knowledge management is the set of proactive activities to support an organization in creating, assimilating, disseminating, and applying its knowledge. Knowledge management is a continuous process to understand the organization's knowledge needs, the location of the knowledge, and how to improve the knowledge [Bohn, 1994].

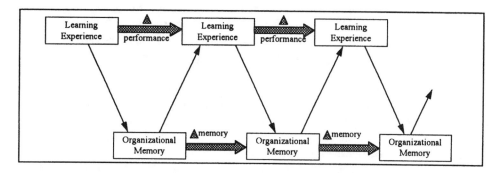

FIGURE 3.8 Learning and memory for organizational learning.

Determining the organization's knowledge needs. The aim of this step is to determine the core competencies or focused knowledge needs of the organization [Drucker, 1993]. The knowledge needs are driven by the nature of the business the organization is in and desires to be in. At an organizational level, the knowledge needs are a function of the organization's products and services and the processes by which the products are produced. At an individual level, the knowledge needs are a function of the things a worker is responsible and accountable for and the decisions to be made and actions to be taken. These knowledge needs must reflect both the current and future directions of the organization.

Determining the current state of organizational knowledge. The aim of this step to determine where and how the organization's current knowledge is being stored. Using the previously identified knowledge needs, the existing sources of knowledge or organizational memory are identified. The sources should be evaluated based on their ease of use.

Determining the gaps in knowledge and barriers to organizational learning. The aim of this step is to determine why the organization is not creating and applying knowledge that is accurate, timely, and relevant. The output of this step is a list of the improvement areas focused on the organization's knowledge and learning process. The improvement areas are based on the difference between the needed and current knowledge and barriers to creating, assimilating, disseminating, and applying the needed knowledge. Each of the organizational learning steps is a potential barrier to organizational learning. Purser et al. [1992] identified barriers to the organizational learning such as the failure to utilize knowledge (i.e., application), lack of sharing (i.e., assimilating and disseminating), or wrong parties involved in the project (i.e., needed organizational memory is not present). The lack of creating knowledge from a problem-solving or work activity is also a barrier. Knowledge management strategies provide the mechanisms by which the organizational learning process is supported and the barriers are overcome.

Developing, implementing, and improving proactive knowledge management strategies to support organizational learning. The aim of this step is to develop proactive strategies to support the creation, assimilation, dissemination, and application of the organization's knowledge. The knowledge needs and gaps provide the domain content areas for knowledge management strategies or what knowledge needs to be managed. The barriers to organizational learning provide the human, process, and tool issues for knowledge management or how to manage the knowledge. The next section provides examples of knowledge management strategies.

Knowledge Management Strategies

Knowledge management strategies are the proactive actions to support the creation, assimilation, dissemination, and application of organizational knowledge. These strategies must address both the human and information technology side of knowledge management. Knowledge management is as much about fostering/supporting personal relationships as it is about using electronic tools to share knowledge. Human interaction is vital throughout the organizational learning process. Knowledge creation is supported when individuals directly share through discussion (i.e., assimilate and disseminate) their experiences to other individuals internal and external to the organization [Nonaka and Takeuchi, 1995].

Knowledge creation strategies address the learning experience or how knowledge is created in the organization. For example, organizations can learn from their own experiences internal to the organization or the experiences and best practices of those external to the organization such as clients or competitors. Organizations create knowledge through processes such as a systematic problem solving process (e.g., TQM approach), experimentation with new processes, or benchmarking [Garvin, 1993]. Knowledge assimilation and dissemination strategies focus on the sharing of knowledge throughout the organization.

Two knowledge management strategies are described. The first strategy emphasizes the human or social side of knowledge management. The aim of this strategy is to provide a face-to-face and personal mechanism for individuals to share their knowledge. Van Aken et al. [1994] provide an example of knowledge management in describing the concept of affinity groups. An affinity group is a "collegial association of peers that meets on a regular basis to share information, capture opportunities, and solve problems that affect the group and the overall organization. A second example involves a government organization conducting information exchange workshops to provide an interactive forum to share information about existing problems, technology needs, and potential near-term technology solutions. Invited participants include the people conducting the work across the government organization as well as external entities such regulators and technology developers. Each workshop features a variety of interactive formats (e.g., panel, poster, and interactive poster sessions; spontaneous and scheduled breakout sessions; concurrent sessions; and site tours) designed to help peers share information about their successes and failures. The affinity group structure and the information exchange workshops allow for an organization's individuals to support the creation, assimilation, and dissemination of knowledge by providing a forum to share what they have learned internal and external to the organization.

The second set of strategies focuses on the use of information technology such as knowledge sharing or expert systems [Niwa, 1990]. The purpose of these tools is to provide a mechanism for placing (i.e., assimilating) and later using (i.e., disseminating) knowledge from a central source. Information technology supports assimilation by providing a set of tools for an individual to use to describe the results (e.g., a lesson learned) of a learning experience. Both the assimilation and dissemination are supported by having a centralized knowledge or database of the learning results from across the organization. Dissemination is supported by providing a set of tools for an individual to search and retrieve relevant lessons learned to use as an aid to the current problem. The use of information technology provides easy access to the organization's knowledge by providing mechanisms that facilitate the inputting, storage, search, and display of knowledge relevant to the user.

Conclusions

The goal of any organization is to improve performance, which is a function of the knowledge the organization uses or applies. Therefore, organizational knowledge must be accurate, relevant, and timely, that is, knowledge must be continuously improved. Knowledge is continuously improved through the organizational learning process, which creates and shares (assimilates and disseminates) knowledge from one part of the organization to another. Knowledge management provides the mechanism to improve both the knowledge and learning process by determining the organization's knowledge needs, current state of organizational knowledge, and the gaps in knowledge and barriers to organizational learning and then developing, implementing, and improving proactive knowledge management strategies to support organizational learning. These strategies need to focus on the process of organizational learning and balance the social and information technology needs.

Defining Terms

Knowledge: More than one piece of information in a pattern from which explicit inferences and predictions can be made.

Knowledge management: The set of proactive activities to support an organization in creating, assimilating, disseminating, and applying its knowledge.

Organizational learning: The process of creating, assimilating, disseminating, and applying knowledge in an organization.

Organizational memory: "Stored information from an organization's history that can be brought to bear on present decisions" [Walsh and Ungson, 1991].

References

Argyris, C. and Schon, D. A. *Organizational Learning: A Theory of Action Perspective*, Addison-Wesley, Reading, MA, 1978.

Bohn, R. E. Measuring and managing technological knowledge, *Sloan Mgt. Rev.*, Fall: 61–73, 1994.

Drucker, P. F. *Post-Capitalist Society*, HarperCollins, New York, 1993.

Garvin, D. A. Building a learning organization, *Harv. Bus. Rev.*, July–August: 78–91, 1993.

Huber, G. P. Organizational learning: the contributing processes and the literatures, *Organ. Sci.*, 2(1): 88–115, 1991.

Niwa, K. Toward successful implementation of knowledge-based systems: expert systems vs. knowledge sharing systems, *IEEE Trans. Eng. Mgt.*, 37(4): 277–283, 1990.

Nonaka, I. and Takeuchi, H. *The Knowledge-Creating Company*, Oxford University Press, New York, 1995.

Purser, R. E., Pasmore, W. A., and Tenkasi, R. V. The influence of deliberations on learning in new product development, *J. Eng. Technol. Mgt.*, 9: 1–28, 1992.

Van Aken, E. M., Monetta, D. J., and Sink, D. S. Affinity groups: the missing link in employee involvement, *Organ. Dyn.*, Spring: 1994.

Walsh, J. P. and Ungson, G. R. Organizational memory, *Acad. Mgt. Rev.*, 16(1): 57–91, 1991.

Further Information

An in-depth discussion of the learning organization can be found in *The Fifth Discipline: The Art and Practice of the Learning Organization* by Peter Senge.

Alvin Toffler in *Powershift* discusses the important role of knowledge in our society.

The Rise of the Expert Company by Feigenbaum, McCorduck, and Nii provides numerous examples of the use of expert systems in organizations.

Tom Peters in *Liberation Management* contains numerous case studies on the use of knowledge management processes and tools.

For further information on organizational learning see the journal *Organization Science* (Issue 1 of Volume 2, 1991) for a special issue on organizational learning.

3.6 Research and Development

Richard Reeves

Ideas concerning the management of research and development (R&D) have undergone rapid development in the last few years, and several new management methods have come into use in companies. For 80 years or more, R&D activity expanded in medium and large companies, but it has perhaps now reached saturation level. Scientists and engineers in companies often pursued research of their own devising while having only slight awareness of the commercial requirements. At the same time, corporate managers tended to leave R&D to its own devices, both from a feeling that this was something they did not understand and because R&D issues are never so urgent as the problems of production start-ups and market launches. This was often unwise because the products being launched determined the futures of the companies yet were sometimes selected for development by R&D without due attention by senior management.

Company spending on R&D in varies from 0.1% of turnover to over 20%, with an average of perhaps 4%. Despite its importance, R&D has received scant attention in management teaching and textbooks. R&D management almost never features in business school syllabuses or faculty lists or in the indexes of standard management texts.

Pressure for change has come from the deep worldwide recession of the early 1990s, causing companies to look closely at all classes of expenditure, and also from the rather belated recognition of technology and innovation as competitive elements. It has been recognized that although competition does indeed take place on the classic grounds of efficiency, price, promotion, and marketing, ownership of a technology can also give a profound advantage. Mastery of the difficult art of making silicon chips restricts that business to a few, while legal ownership of drugs that they have formulated enables pharmaceutical companies to make monopoly profits. Both of these advantages originate in the R&D laboratories of the companies concerned. **Innovation** is broader than technology development by R&D and can occur, for example, when business advantage is gained simply by buying something new from a catalog. Innovation is a concern of all the company, and the role of R&D in this is best defined simply as that part of innovation that happens to be done by the people in the R&D department.

Major contributions to new thinking about the management of R&D are the concepts of **third-generation R&D** brought together by the consultancy company Arthur D Little [Roussel et al. 1991], the **stage-gate** system of managing new product development (NPD) [Cooper 1993], portfolio management theory of R&D projects [Roussel et al., 1991; Cooper et al., 1997], and the methodology of **technology foresight**. These provide difficult but convincing tools for managing an activity that has caused much anguish in the past and enable constructive dialogue to take place between R&D and the rest of the company. There is one reservation to be made, which is that these methods all apply to corporate R&D, just at a time when R&D can be seen to be undergoing fundamental change: much R&D is now being contracted out. Independent laboratories supply an increasing share of R&D, and it may be that this trend will continue and that companies will in the future expect to buy-in most of their new development rather than source it in-house.

The Nature of R&D

R&D work is intrinsically risky. Almost by definition R&D tries out new ideas to see how well they work, seeks information that is as-yet unknown, and hopes for ideas as-yet unthought of. This means that success cannot be commanded, and most R&D is managed in terms of probabilities. The laboratory is gambler territory. It is useful to contrast R&D with design, where the assumption is that all the knowledge required exists and if the design is competent the product will work. Products that are designed tend to be of modest profitability because the knowledge required is likely to be fairly widely available. By moving into the unknown, R&D accepts risk in return for the chance of greater reward.

R&D staff are technical professionals. Many join a company because it provides an environment in which they can practice their congenial vocation. They are likely to adhere to standards of conduct set by their profession rather than by the company. They may aspire to recognition in their technical field rather than by their company directors. Industrial scientists can publish in the scientific literature, appear at conferences, and win Nobel prizes. They can also determine the futures of their companies. R&D staff tend to be fairly uniform in qualifications but to vary markedly in R&D ability. A very small proportion are highly creative and greatly outperform their colleagues in discoveries or inventions made. These people are management's greatest challenge in a technically based company.

Creating Intellectual Capital

New technology creation is nowadays usually more important than capital asset creation. The old idea that a company exists to exploit a capital asset does not so often apply now, as can be seen by the fact that stock markets now often value companies at far more than the value of their capital plus assets. Software and biotechnology are clear examples of fields in which capital assets are trivial and copyright, patents, and know-how are much more important. A company can more usefully be viewed as existing to trade in these intellectual assets [Budworth, 1996]. Just as a traditional company had to make provision for renewing its capital assets as they wore out, so a modern company must make provision for continuously renewing its intellectual assets, and R&D often plays the major role in this. R&D spending depresses current profits, and the question of how much should be spent on R&D has often been debated. Budworth proposes that, to maintain the current level of turnover in a mature company, the required ratio of R&D spend to turnover is equal to the ratio of the average expenditure on R&D per new product to the contribution per new product to revenues for the particular company. This can be found from past performance.

The Generations of R&D

In the 1950s and 1960s, it was felt that creativity must not be fettered, and the results of research were unpredictable, so a fixed percentage of turnover was given to the R&D department to spend on projects of its own devising. This was first-generation R&D. The problem was that the company could go out of business while waiting for a new nylon or transistor to emerge. In the 1970s and 1980s, second-generation R&D decreed that no R&D would take place unless there was a customer in an operating division who was prepared to pay for it. The problem was that only short-term work was funded, and work that might start a new division or protect against long-term problems was not undertaken. In the early 1990s, third-generation R&D emerged, in which no simple formula is applied: instead considerable work has to be undertaken by both R&D and the company to develop a common language that enables them to work together to consider their best policy for committing work on the company's technical future.

Types of Work Undertaken by R&D

Traditionally R&D has been described as passing through the phases of pure research, applied research, development, testing, engineering, and so on through to production. This model has been widely adhered to even though it is widely agreed to be wrong. Under the tenets of third-generation R&D, work is classified according to its strategic significance. **Incremental R&D** is work intended to bring about a definable modest improvement in a particular product or a process. **Radical R&D** is work that promises to produce a new product or process. **Fundamental R&D** is work that is not intended to produce a specific business benefit but is intended to produce scientific and technical information; this information might, for example, lead to a decision to try to develop a new line of business. In addition to these, there is **compliance** work, which is undertaken to meet environmental or other legislative requirements, and **customer support** work in which R&D staff work on the problems of existing products.

A common pattern of progress is that a product line or process is improved by many incremental developments over the years, until improvements become more difficult to find. It often then happens that a radical development replaces that product or process by a new one, which then enters into its own series of incremental improvements. From time to time a fundamental change renders the whole product line or process type obsolete.

For incremental work, the probability of success is fairly high, e.g., 70%, and the rewards are modest but reasonably easy to estimate. For radical work the chances of success are low, e.g., 20%, and the possible rewards high but are less predictable. Many companies have difficulty justifying fundamental work; its saving grace is that often it is not very expensive. Incremental work enables companies to compete on grounds of product performance, price, and profit. When radical change occurs companies can win or lose a dominant position in an industry. When fundamental change occurs, whole industries are likely to disappear or arise.

Financial Appraisal of R&D

In the recession years, R&D directors were asked to justify the cost of their departments in financial terms such as return on investment. Many were not able to do this, and the result was closed or reduced R&D departments. The classification described above makes it easier to see how R&D can be assessed, by breaking the work up into its components. Compliance work is a necessary cost of continuing in business; therefore, it is an overhead, and the only requirement is that compliance is achieved at as low a cost as possible. Customer support work assists production or marketing. It is a cost of sales, even though the work is performed by people employed in the R&D department, and their time should be booked to the appropriate function. Incremental work can be assessed in terms of the return on investment, modified by the risk involved. Since many incremental projects are undertaken, costs and benefits can be estimated in the light of past experience so long as the costs and outcomes of past projects have been recorded. Incremental work should be paid for by the product division that benefits, who should be a willing purchaser. Radical work may be radical to a product division or radical for the company as a whole and should be commissioned and paid for by the division or the company as appropriate. In most companies, fundamental work is paid for by the center: the cost of gaining new scientific information that can be used in strategic appraisal and that may lead to new business project ideas is an overhead of the business. It can be seen that the total spend on R&D is not globally determined but is the sum of the spends by the different types of customer for the different types of work.

A weakness of attempts to justify R&D financially is that the whole of the process of exploiting R&D output is not under the control of the R&D department. A perfectly good development may not proceed to manufacture because of a company decision to manufacture something else. An R&D investment has the character that, if it is successful, 10 or 100 times as much money is likely to be needed for investment in production start-up and launch as was spent on the R&D. The R&D project will either fail, so that the cost must be written off, or present the company with the option of spending a great deal more. The parallel with share options has been noticed, and it is argued that what the company buys with its R&D spend is the option to be able to invest more later if it chooses. Newton and Pearson, [1994] have put **option pricing of R&D** on a mathematical basis.

Stage-Gate Management of New Product Development

About 70% of R&D work can be classified as NPD. The term is also used in marketing, and indeed sometimes a new product is entirely a marketing concept, with trivial technical input, such as a new flavor or color. Marketing has an essential input even when most of the expenditure on a project is for technical work, as will be seen.

As R&D projects strive to proceed from mere ideas through stages to success as revenue-earning products or processes in the business, most fall by the way. Typically, when 20 ideas start as projects with budgets, only 1 will survive through to implementation. Most fail as accumulated information makes it clear that there are technical, marketing, financial, or strategic problems that render them unviable or at any rate less attractive than other projects. High infant mortality among projects is inevitable because of the nature of R&D, and, if very high completion rates are achieved, management should investigate to see whether the R&D work is too unadventurous. Even so, much failure can be due to weaknesses in managing the process. Sometimes money is spent to develop a product only to discover that manufacturing cannot make it or marketing cannot sell it or the company does not wish to move in that direction. Projects can be difficult to kill because of the commitment of their adherents long after facts are available that indicate that the money should be committed elsewhere. Sometimes managers attempt to pick winners at too early a stage, when it would be better to spend a little on a number of projects at first to find out more about them.

The stage-gate system of NPD project management provides go/kill decisions at about five gates, at each of which a project must apply for further funding. At the first gate, an idea seeks funding for an initial few days of work and has to meet only the simple criteria that it appears to have technical and market feasibility and that it is in line with company strategy. A financial appraisal is not required at this gate. The most critical gate is that allowing a project through to development because development is the most expensive of the stages but has a high failure rate. What is needed at this gate is a full business plan to which marketing, production, finance, and strategy have all contributed. Here formal discounted cash-flow estimates of project benefit are required, and plans must be available in appropriate detail for testing, production, and marketing if the project should pass through development successfully. The requirements to be met at each gate and the criteria for passing are set out in advance. Gate committees include representatives of all company functions. It is at the gates that managers of other functions are enabled to take part in R&D decision making, and the most senior managers should attend major gates such as the one leading to development.

It can be seen that the gate system is intended to ensure that the right homework has been done before commitments are made and introduces an appropriate element of parallel working among the company functions. It should also achieve buy-in for the projects selected. An important function of gate committees is to check that not only has the required work been done but that it has been done with sufficient quality.

Management of the R&D Portfolio

In the stage-gate system, projects are considered in isolation or at best in comparison with each other. A view of the whole group of R&D investments is also required to appraise whether there is a high probability of generating enough successful new products in the short to medium term, whether the long term is being appropriately catered for as well, and whether the total activity is fostering a coherent and synergistic technical capability. Senior managers are well used to handling commercial risks and should give considerable thought to their R&D portfolio.

R&D portfolio management is a subject that is still under development. Some of the factors to assess are the maturity of the technology the company uses, the maturity of the industry, technological competitiveness in each area in which the company operates, and the strategic needs of each part of the business. Project risks need to be estimated — it is usually sufficient to classify projects as small, medium, and high risk rather as company shares are classified. One useful tool is to classify the technologies that the company applies into **pacing technologies,** which have the potential to change the basis of competition, **key technologies,** which are critical to present success and may be proprietary, and **base technologies,** which may be essential but offer no competitive edge because they are widely available.

Management Structure and Location of R&D

There is much variation in how R&D is organized and funded according to the type of company. Typically nowadays the R&D laboratory operates as a profit center that gains 70% of its funding from the operating divisions and 30% from the company. Operating divisions tend to fund incremental work. The company tends to fund radical, fundamental, and compliance work of interest to the company as a whole. There is often a rule that R&D may add a surcharge of 10% to its contracts in order to be able to fund work of its own devising, which may include building new capability or generating ideas that will turn into proposals to the company and the divisions.

It is often argued that R&D should be located physically close to operations in order to prevent an ivory tower mentality from developing. The converse is that if R&D becomes drawn into too much firefighting work, its proper long-term work suffers.

Independent R&D

A company can prosper without carrying out R&D. A retail chain has argued that if it developed its own products then it would have to try to sell to the public whatever it happened to have developed, whereas it is better to concentrate on its specialty of understanding customer needs and then source widely to meet those needs. R&D is then the concern of the suppliers. Companies around the world that mine tin or use tin in their products are content to buy technical information as and when needed from the International Tin Research Institute, which is sustained by contracts and service provision for a worldwide clientele. Many companies are content for their R&D departments to work even for direct competitors, and this is sometimes a stage toward floating these departments off as independent R&D companies. The pharmaceutical industry has for some time recognized that compliance testing of new drugs is a routine operation that they can contract out to specialized testing companies. The industry is now going further and arguing that all stages of the R&D process can be contracted out, even the initial drug discovery programs, so long as ownership is secured by contract.

Technology Foresight

Foresight exercises are carried out within large companies, within industrial sectors, and on a national scale. Foresight is a systematic process for collating expert views on likely technical, economic, and social developments in the long term. Benefits are found to arise not just from the view of the future that results but from the bringing together of researchers and research users well before program ideas are committed or even formulated. Corporate managements give their time to such extramural exercises in the hope that the information accruing will help their strategy processes.

The Problem of Basic Research

There is widespread complaint by companies that they are unable to justifying doing long-term research and that universities are ceasing to do it as well. Results of fundamental research may turn out not to be applicable to the company that pays for the work; benefits may accrue in the far future, beyond present management's tenure, and results are usually made public and therefore available to competitors. An argument in favor of a company employing its own researchers is that this is necessary in order to have access to the research output of the whole world in the company's fields of interest. It is argued that research publications and conferences need to be assessed and interpreted by appropriate scientific specialists in the same way that a company needs a legal specialist to interpret law books and cases.

Defining Terms

Base technologies: Important to present success but offer no competitive edge because they are widely available.

Compliance: Meeting environmental or other legislative requirements.

Customer support: R&D staff work on the problems of existing products.

Fundamental R&D: Work that is intended to produce scientific and technical information.

Innovation: Introduction of improved products, processes, or operations by acquiring new technology or methods, by training, by improved management processes, and by R&D.

Incremental R&D: Work to bring about a modest improvement in a particular product or a process.

Intellectual capital: Knowledge that is a trading asset for a company. Includes patented inventions, copyrighted software, and confidential know-how.

Key technologies: Critical to present success and may be proprietary.

Option pricing of R&D: Valuing R&D output in terms of the further investment opportunities opened up for the company.

Pacing technologies: Have the potential to change the basis of competition.

Radical R&D: Work that attempts to produce a new product or process.
R&D portfolio: A company's collection of R&D projects.
Stage-gate: A system for managing the new product development process.
Technology foresight: A process of collating expert views to identify future technical and business opportunities.
Third-generation R&D: A philosophy of R&D management in which technology issues are couched in terms suitable for senior corporate management.

References

Budworth, D. *Finance and Innovation*, Thomson Business Press, 1996.

Cooper, R. G. *Winning at New Products*, Addison Wesley, Reading, MA, 1993.

Cooper, R. G., Edgett, S. J., and Kleischmidt, E. J., *Portfolio Management of New Products*, McMaster University, Hamilton, Ontario, Canada.

McNulty, T. and Whittington, R. Putting the marketing into R&D, *Market. Intell. Plan.*, 10(9): 10–16, 1992.

Newton, D. P. and Pearson, A. W. Application of option pricing theory to R&D, *R&D Mgt.*, 24(1): 83–90, 1993.

Roussel, P. A., Saad, K. N., and Erickson, T. J. *Third Generation R&D*, Harvard Business School Press, Boston, 1991.

Further Information

Older books on R&D management have been rendered out of date by the developments outlined here, and there are widely different situations ranging from corporate laboratories operating securely in the traditional manner to well-established independent laboratories. A description of the effect of the changes occurring in R&D is given by McNulty and Whittington [1992].

3.7 The Elusive Measures of R&D Productivity and Performance

Larry A. Mallak

Research and development (**R&D**) **units** face unique challenges concerning the measurement of productivity and **performance**. R&D outputs carry a high degree of **uncertainty,** and their impacts vary widely across many different organizational boundaries. This section frames the problem of measuring R&D productivity and extends the measurement to a larger set of performance criteria. Key problems and issues of measuring R&D performance are discussed, and relevant examples highlight how organizations have addressed these issues. Specific examples from industry and government R&D organizations show how R&D can be validly assessed in those environments. Advice and lessons learned are offered for those charged with measuring R&D performance and providing accountability for R&D activities.

R&D Productivity: What's the Problem?[5]

In laboratories across the world, scientists and engineers endeavor to find cures for diseases, develop faster and more powerful computers, and develop unimaginable new products that will both improve our quality of life and provide a strategic niche for the organization. However, how do organizations decide how well the estimated $98.5 billion [Whiteley et al., 1997] spent by U.S. firms in 1995 was put

---

[5]For the most current data on R&D organizations, the reader may wish to seek the latest survey of R&D organizations, a summary of which is published annually in the January/February issue of *Research Technology Management* or consult *Business Week's* R&D Scoreboard, published annually.

to work? Historically, corporations threw money at their R&D departments and let them loose. Now, with tighter reins on spending and the need for greater accountability for R&D budgets, technology managers must show how their R&D departments contribute to the bottom line — how the costs of supporting expensive laboratory space and Ph.D.-trained staff are value added.

In Brown and Svenson's [1988] classic paper "Measuring R&D Productivity", the authors clearly surface the problem by introducing "Peter" and "Andrea", two senior-level scientists working in the R&D laboratory of a major manufacturer. Peter works long hours, gets along with everybody, never misses a meeting, is considered a good project manager, meets his deadlines, and actively works with others to review research projects and proposals. His work has never resulted in any great value for the organization nor has he made any significant contributions to the discipline.

Andrea is just the opposite. She rarely works evenings or weekends, keeps a messy work space, frequently misses deadlines, must be coerced into attending meetings, and is considered abrasive in her interactions with others. However, she holds several patents, has designed processes saving the company millions of dollars, and her peers in the field formally recognize her expertise as "world class". She has publications in prestigious journals and has been invited to speak at many important conferences.

The crux of the R&D measurement problem can be exemplified with Peter and Andrea. If we measure behavior, then Peter is the superior performer — he's a hard worker, he's always at the meetings, and he's a team player. Andrea's behavior is rebellious — she's not a team player, and she doesn't have Peter's commitment to the organization. However, she has some enviable professional accomplishments.

R&D measurement plans that focus on behavior would punish Andrea and reward Peter. However, what have each of them contributed to the organization in terms of increased sales, competitive positioning from new products, time advantages (reduced cycle time or faster time to market), capital avoidance, and other metrics? The old adage, "You get what you measure" must be kept in mind to avoid motivating unwanted research behaviors such as generating many minor publications at the expense of new product innovations.

If R&D is one of the keys to an organization's ability to compete successfully, then technology managers ought to have decisive measures to assess an R&D unit's productivity and effectiveness. Technology managers can then make informed decisions for managing the R&D units most effectively. Unfortunately, even after more than 30 years of developing, testing, and using measures for R&D, no definitive benchmarks have been agreed upon. However, some common themes emerging from the literature suggest how to approach this difficult measurement issue.

Before presenting some highlights from the literature, the issue of R&D productivity needs to be reframed. Classically defined, productivity is the ratio of outputs to inputs. Therefore, it is no wonder most articles on R&D productivity look at outputs of the R&D unit — patents, publications, new products, and innovations — and compare those to input variables such as labor hours, R&D budget, etc. An article by Szakonyi [1994] is linked to R&D *effectiveness*, but the terms and concepts of measurement (i.e., productivity, effectiveness, and performance) are not carefully defined and differentiated from each other.

Therefore, this section treats the broader issue of R&D *performance* and addresses relevant performance measures including productivity, quality, effectiveness, and impacts on profitability and innovation. Table 3.2 has definitions of several performance terms. When measuring R&D performance, one must decide which performance measures are to be used. Several of the more "bean-counting" measures are easier to implement (e.g., efficiency and productivity) but not likely to produce desired results from the R&D unit. The measures more likely to produce desired outcomes (e.g., effectiveness and innovation) are more difficult to implement. However, these are the measures many R&D managers and researchers are trying to define and use.

Key Problems and Issues of Measuring R&D Performance

Organizations face many problems and issues in measuring R&D performance. Three key items presenting the greatest obstacles and confusion in R&D performance measurement are discussed.

TABLE 3.2 Performance Criteria

Criterion	Definition
Efficiency	Resources expected to be consumed/resources actually consumed
Effectiveness	Accomplish what you planned to accomplish; ratio of actual output to expected output
Quality	Quality checkpoints throughout the system, from supplier to customer
Innovation	Applied creativity
Quality of work life	Job satisfaction, working conditions, climate
Productivity	Ratio of output to input
Profitability (for-profit)	Ratio of revenues to costs, other financial ratios
Stewardship of funds	Responsible management of entrusted funds

Source: Based on Sink [1985], Kurstedt [1995], and Patzak [1995].

Unit of Analysis

The unit of analysis for measuring R&D should be the entire organization. When the R&D unit is singled out for analysis, the obvious outputs become the victims of measurement — patents, publications, and new products. When the organization is viewed as the unit of analysis, outputs include profits, sales, costs, cost avoidance, market share, and other strategic outcomes seen as desirable and necessary for successfully competing in the marketplace. In fact, this argument can be extended to assess the impact of R&D efforts upon its industry and society as a whole.

Bean Counting

Just because something is easy to measure doesn't mean it's a good measure. Scores of trivial patents and obscure journal articles result when beans are counted, but what is the impact on the firm's profitability and competitive position? Academic departments frequently fall into this trap when evaluating faculty members for promotion and tenure decisions. Refereed journal articles are counted, research dollars are toted up, student evaluations are compared against averages, and service to the department, college, university, and community are noted. However, where did their students get jobs, how much do those students earn, what have they contributed to their employer, can they write, do they have good problem solving skills, and how does the faculty member's research contribute to quality of life? The list of questions goes on and on. The busy behaviors are being counted and rewarded. The outcomes are relatively unknown.

In engineering education, for example, a new set of criteria — ABET 2000 — seeks to make outcomes more up front in evaluating engineering academic programs. ABET, the Accreditation Board for Engineering and Technology, recently announced a new method of assessing engineering education. Rather than rate each program against a predetermined set of criteria (e.g., counting courses, hours, assignments — bean counting), ABET requires these programs to define their own goals and objectives for their engineering programs, address those goals and objectives in their curricula, and perform valid measurement of how well those goals and objectives are being met.

Level of Uncertainty

R&D units face much higher levels of uncertainty than do production units. Staff in an a pharmaceutical laboratory never knows when a drug it developed will pass all the hurdles and become a significant source of revenue and profits for the firm. Such a laboratory "bets the company", meaning that a large amount of risk is undertaken in terms of funding research to launch a new drug that will take several years to show any eventual return. Many specific measures of performance and productivity have been defined, implemented, and benchmarked in production arenas. These measures include defective parts per million (ppm), unit cost, cycle time, etc. In Kurstedt's framework, these are processes and therefore have the least uncertainty — the same known end is repeatedly achieved (Fig. 3.9).

```
┌─────────────────────────────────────────────────────────────────┐
│ Uncertainty                                                       │
│                                                                   │
│     ▲    Perplexity - Can specify neither the start nor the end.  │
│     │                                                             │
│     │    Problem - Can specify the start but not the end.         │
│     │                                                             │
│     │    Program - Know the start and have qualititative fix on the end. │
│     │                                                             │
│     │    Project - Know the start and have specifications for the end. │
│     │                                                             │
│     │    Process - Repeatedly achieve the same known end.         │
└─────────────────────────────────────────────────────────────────┘
```

FIGURE 3.9   The pursuits framework explains part of the difficulty in measuring the more uncertain activities of an R&D unit. (*Source:* Kurstedt, 1995.)

These five pursuits range from the highest level of uncertainty (perplexity) to the least (process). Uncertainty, based on Galbraith [1973], refers to the comparison of the amount of information one has compared to the amount one needs to either make decisions or take action. Production units are primarily concerned with processes, although all pursuits will likely be represented over time. R&D units commonly have a mix of programs and problems with some projects on the development side. Therefore, an R&D unit inherently faces higher levels of uncertainty and correspondingly fewer readily applied measures than a production facility. The production facility may very well have its share of uncertainty but it knows what its goals are — to produce a specified number of products each shift each day.

We use different tools to monitor projects than those to monitor processes. For example, PERT and CPM tools help project managers track activities and sequencing, whereas meters, gauges, and production schedules help monitor production processes. R&D units often engage in research programs, where the starting point has been defined ("we have an electric car with limited range") and only a qualitative statement for the end is known ("we need a low- or no- emissions vehicle with a reasonable range"). The measurement of the program's effectiveness is much different from that for the processes or projects. The effectiveness of the vehicle is not the end result for the overall organization. R&D could develop a vehicle that meets all technical criteria and has very low emissions, but it has a body design that would not appeal to the typical consumer. For an R&D unit, as mentioned earlier, the unit of analysis must extend beyond the unit's boundaries and into the impact on the overall organization. In fact, the impacts of an R&D unit on its primary industry are also a measure of its performance as well as the impact on society. An R&D laboratory in a pharmaceutical firm that finds a cure for bone cancer benefits the firm, develops a reputation in the industry, and benefits society. However, as the drug came out of the R&D laboratory, it could have been just another also-ran.

Rarely does R&D occur in a crisis atmosphere. The high end of the pursuits framework, problems and perplexities are not events considered as expected activities in an everyday context (unless one is involved in emergency management). Certainly, all organizations need contingency plans, but that is outside the scope of this section.

Measuring R&D Performance in Organizations

Many arguments abound in the literature about why one measure is inadequate or why another measure "doesn't give the whole picture". Certainly, R&D activities themselves can be readily measured if we just focus on quantifiable outputs from the laboratories — patents, papers, drawings, etc. These alone won't help in characterizing the effectiveness of R&D efforts. We need to look at external measures from downstream systems — the organization, the industry, and society — to make better judgments as to the value of R&D outputs. Measuring activities and behaviors alone won't provide the information we need either, lest we have a laboratory full of "Peters". Better still, we should examine the more fundamental question of how to structure R&D activities to maximize the benefits of their capabilities.

Brown and Svenson's [1988] approach to R&D measurement provides advice on how to measure R&D accomplishments:

1. Focus on external vs. internal measurement.
2. Focus on measuring outcomes and outputs, not behavior.
3. Measure only valuable accomplishments/outputs.
4. Make the measurement system simple.
5. Make the measurement system objective.
6. Separate R&D evaluation.

Additionally, R&D plays a key role in getting the product to market more quickly. The first with the product to market has an advantage in locking up market share. Microsoft is a good example with its Windows operating system.

Internal vs. External Measures;

The argument for establishing measures that transcend the R&D unit boundaries points the way for external measures to be used, that is, external to the R&D organizational boundary. Alcoa [Patterson, 1983] used several external measures of R&D effectiveness (shown in Table 3.3).

TABLE 3.3 Alcoa's R&D Measurement Criteria

Criterion	Description
Cost reduction	Technical innovations that reduce the cost of existing operations
Sales advantage	Technical innovations that increase the sale or profitability of existing products or contribute to emergence of a new product
Sustain business	Technical activity that maintains existing business levels (e.g., averting customer rejections, upgrading products)
Technology sale/contracts	Profits realized from sale of Alcoa R&D technology or R&D contracts negotiated with government, industry, and universities
Capital avoidance	Technical innovations that obviate the need for planned capital expenditures
Capacity expansion	R&D contributions to planned expansion of capacity
Energy/raw materials	Technical innovations that ensure an adequate supply of energy or raw materials for operations
Knowledge feasibility	Significant new understanding or demonstrated feasibility of a new concept, process, or design of imminent value to Alcoa
Environmental/safety	Technical innovations or expertise that diminish adverse environmental effects, solve safety problems, help meet government regulations, etc.

Source: Patterson, 1983.

In contrast to the more general performance criteria in Table 3.2, these criteria are specific to Alcoa and to other R&D units. To evaluate R&D, one must know how the R&D output benefits the organization. Alcoa's criteria provide the basis for an organization to define in detail their R&D unit's contribution.

Measuring Behaviors vs. Measuring Results

The quality management movement left an early legacy: focus on the process and the results will follow. For many organizations, an exclusive focus on process left tangible results withering on the vine. Most TQM practitioners now see the value of directing processes toward key business results. Customer focus has played a central role in providing a target for processes.

Two recent efforts seek to provide greater outcome accountability in (1) engineering education and (2) the federal government. The engineering education example, ABET 2000, was discussed earlier. However, the U.S. government passed the Government Performance and Results Act (**GPRA**) in 1993. GPRA promotes goal setting, performance measurement, and public reporting and accountability. GPRA seeks to improve the provision of government services to improve program effectiveness and customer satisfaction.

A government R&D organization, the U.S. Army Tank-Automotive Research, Development, and Engineering Center (TARDEC), relies on several forms of measurement in its quest to meet the GPRA requirements. See Table 3.4 for the details. This is a sampling of the measures used; more operational level measures concerning protection of the soldier in combat are used on the team level. The bottom line of TARDEC's work is often relayed back from the front line. Recently, a soldier was driving a Humvee outfitted with TARDEC-designed ballistic protective blankets (BPBs). The BPB helps protect soldiers in non–combat-ready vehicles from enemy fire. The soldier drove over a land mine, which detonated, heavily damaging his vehicle. He and three other soldiers only sustained minor injuries — the BPB protected them from the mine's explosive force. This is the ultimate form of customer satisfaction TARDEC seeks from its R&D.

TABLE 3.4   Examples from a Government R&D Organization Measurement System

What TARDEC Measures	Criteria Addressed
Associate (employee) satisfaction and climate on an annual basis	Quality of work life
Survey their customers both at the team level and the corporate level to obtain feedback on their performance	Effectiveness, quality
Weapon lethality, friendly fire avoidance	Effectiveness, quality
Track progress against their strategic plan using Executive Assessment Report Card	Effectiveness
Number of patents awarded	Innovation
Papers published	Innovation, productivity
Educational profile (e.g., advanced degrees, graduate program enrollment)	Quality
Agreements with industry and academia	Effectiveness, innovation
Team performance via team self-ratings	Varies

Source: U.S. Army TARDEC, 1997.

Structure your Organization and its R&D Unit Differently

The more fundamental issue in R&D measurement is whether the R&D unit is properly structured to accomplish the organization's desired objectives. The stand-alone laboratory has attraction among its investigators who seek to replicate an academic environment, sufficiently removed from the day-to-day issues to allow serious study of new and basic research ideas. A stand-alone laboratory is also shielded from business cycles by receiving its entire budget from corporate. This can create resentment from R&D's internal customers, who are on the front line of the organization, making and selling products and services and whose workforces rise and fall with the business cycle.

Organization theory advises us to design an organizational structure based on the strategy of the unit being organized. Several organizations have looked at their R&D units critically and redesigned them to better meet their objectives. Industrial behemoth Asea Brown Boveri (ABB) decentralized its R&D function into business units rather than fund an isolated R&D laboratory [Coy, 1993].

A free-market approach to structuring R&D units has them seeking work from operating divisions rather than receive 100% funding from corporate. Several large firms have restructured their R&D budgeting to provide 25 to 30% of their budgets from corporate, with the balance being obtained through contracts with business units. For example, General Electric (GE) funds only 25% of its R&D costs directly from the corporate budget; Philips Electronics funds 30%. Corporate funds allow for basic research and exploratory work to continue but not as the exclusive focus. The need to generate work with business units forces interaction between R&D and business units to identify and define specific business needs. This links R&D to business unit measures such as new sales, capital avoidance, and lower manufacturing costs (see GE's measures in Table 3.5). Critics claim this approach turns R&D into technical service centers having a sole focus on the short-term. Properly managed, however, this strategy holds promise for R&D, not only to learn its customers' needs more intimately but to improve the ability to track R&D activities to definable outcomes. A basic research component is maintained through the corporate funding.

TABLE 3.5   GE Measures its R&D Units Using Four Methods

The "Jimmy Stewart" Test	Ask "What would our organization look like today if a R&D unit had not existed?"; this form of thinking helps identify key contributions of R&D to the organization, such as key patents and resultant products, entire manufacturing, or service divisions; GE attributes their R&D unit with hundreds of millions of dollars in revenue for the development of computed tomography (CT) and magnetic resonance imaging (MRI)
Counting Outputs	GE primarily counts patents and costs per patent (in patents per million dollars of R&D spending); other measures include how often a patent is cited, number of foreign filings for the patent, and licensing income; Also, GE tracks how much royalty costs are saved because they hold the patent
Analyzing Technology Transitions	To GE, a transition is a technology originating in the R&D center and moved to a business operation; GE wants dollars and cents — what is the discounted rate of return on each transition?; they called upon Booz Allen consultants to provide an outside opinion of the worth of each transition; the end result was a percentage for each transition that gave R&D credit for a portion of the benefits; the percentage was converted into a rate of return and present value; for the entire sample of 190 transitions, the present value was 2, meaning that the value of R&D's activities were twice the R&D costs
Free Market Measurement	This is how GE refers to funding R&D with 25% corporate dollars and having them contract with business units for the other 75%

Source: Robb, 1991.

Lessons Learned in Measuring R&D Unit Performance

Several lessons learned concerning the measurement of R&D unit performance are shared in the following table.

Lesson	Details
Expand the definition of productivity and effectiveness to performance	Productivity isn't the ultimate goal of R&D; more output isn't always good; you want output that makes a positive difference in your organization
Make the unit of analysis the entire organization, not just the R&D unit	R&D units exist to help the rest of the organization, so the measures should reflect that; use external as well as internal measures
You get what you measure — don't just count beans	If you measure publications, that's what you'll get
Measure outcomes, not just behaviors	Behaviors alone will not produce results; measure what those behaviors lead to and what impacts they have elsewhere in the organization and in the marketplace
Make the measurement system simple	Just because the staff has Ph.D.'s doesn't mean they should use them to decipher complex rating formulas; develop simple measures that capture R&D performance and that are understood at various levels in the organization

The nature of R&D activities inherently makes measurement difficult. However, by learning from other organizations, focusing on appropriate performance measures, defining a broader unit of analysis, and measuring outcomes linked to organizational goals, we can get a better handle on the elusive measures of R&D productivity and performance.

Defining Terms

GPRA: Government Performance and Results Act of 1993, designed promote goal setting, performance measurement, and public reporting and accountability in U.S. government organizations.

Performance: An assessment of system functions based on a set of criteria typically including productivity, effectiveness, quality, and others.

R&D unit: The organizational unit within a larger organization charged with carrying out R&D activities.

Uncertainty: Information gap between the amount of information one has compared to the amount of information one needs.

References

Brown, M. G. and Svenson, R. A. Measuring R&D productivity, *Res. Technol. Mgt.*, July/August: 11–15, 1988.

Coy, P. In the labs, the fight to spend less, get more, *Bus. Week*, June 28: 102–104, 1993.

Galbraith, J. *Designing Complex Organizations*, Addison-Wesley, Reading, MA, 1973.

Kurstedt, H. A., Jr. Management Systems Theory, Applications, and Design, unpublished working draft, 1995.

Patterson, W. C. *Res. Technol. Mgt.*, 26(2): 23–27, 1983.

Patzak, G. personal communication cited in Kurstedt [1995].

Robb, W. L. How good is our research?, *Res. Technol. Mgt.*, March/April: 16–21, 1991.

Sink, D. S. *Productivity Management: Planning, Measurement, Control, and Improvement*, John Wiley & Sons, New York, 1985.

Szakonyi, R. Measuring R&D effectiveness, I, *Res. Technol. Mgt.*, March/April: 27–32, 1994.

U.S. Army Tank Automotive Research, Development, and Engineering Center. U.S. Army Research and Development Organization of the Year, award application document, 1997.

Whiteley, R. L., Bean, A. S., and Russo, M. J. Meet your competition: results from the IRI/CIMS annual R&D survey for FY '95, *Res. Technol. Mgt.*, January/February: 16–23, 1997.

Further Information

Center for Innovative Management Studies (conducts the annual survey of R&D organizations), Rauch Business Center, Lehigh University, Bethlehem, PA 18015.

Industrial Research Institute, Inc., 1550 M Street, NW, Suite 1100, Washington, DC 20005-1712. Office: 202-296-8811. Fax 202-776-0756. Internet home page: http://www.iriinc.org/.

National Performance Review and Government Performance and Results Act, Office of the Vice President, Washington, DC. Internet home page: http://www.npr.gov/.

3.8 Technology Life Cycles

Mario W. Cardullo

No element within the existing universe exists eternally. All aspects of these elements follow parallel development. Each element emerges, grows to some level of maturity, and then declines and eventually expires. This cycle is nothing more than the results of the **Second Law of Thermodynamics**.

The life cycles of technology and technological developments are not removed from the Second Law of Thermodynamics. Each technology and associated developments emerges, grows to some level of utilization, and then declines and disappears. Some technologies have relatively long lives; others exist for an instant in relationship to others and never reach a maturity stage. However, each has an impact, even if it is to catalyze another development.

Emergence of new technologies is a never-ending phenomenon. Paraphrasing Kauffman's [1995] comment, "…life unrolls in an unending procession of change, with small and large bursts of speciations, and small and large bursts of extinctions, ringing out the old, ringing in the new."

The rules of life are the same for technology as they are for any species or system within this universe. The emergence of new technological developments may appear as planned, but, in reality, we are dealing with a complex process. This **nonlinear process** follows the laws of emergence and complexity with all their unforeseen consequences.

It is vital that managers of technology be aware of the emergent and complex nature of technological developments. Understanding the cross-impacts of technology life cycles such as how these life cycles interact with other technologies, systems, cultures, enterprise activities, and other interfaces is vital to technology management. A manager of technology must understand the life cycles of the technology being managed, as a parent understands the life cycle of a new family member.

Managers of technology must be fully aware of how these life cycles interact. Technology life cycles can have significant impacts on society. These impacts arise from interaction of a myriad of forces.

General Technological Life Cycles

Emergence or Birth

Kauffman [1995] has presented a theory of how emergence would account for the creation of order as a natural expression of the underlying laws of the universe. Accordingly, life is an emergent phenomenon. All technologies evolve from simple precursors. These technologies interact with each other and other elements such as culture to form complex technological ecologies.

The birth or emergence of a new technology development has numerous precursors or "parents" and "ancestors". Each of these precursors adds their technological "DNA", which interacts to form the new technological development. A new technological development must be nurtured similar to a new life. It must be fed and cared for if it is to achieve the hopes of its developers or parents. Similarly, many technologies perish before they are **embedded** in their intended environments. In fact, there is evidential material to indicate that as few as 1 technological development reaches maturity for each 60 to 100 newly conceived technological products [Pearce and Robinson, 1991].

Growth

If a technology survives its early phases, it begins to make inroads in its intended environment. This growth phase can also see the emergence of competitors for the same environment. The laws of natural selection play a role in the life cycle of technologies.

If the Darwinian metaphor is accepted, then differences in technologies will lead to differential success, culling out the fitter, leaving behind the less fit. However, according to Arthur [1989], technological lockout can occur, resulting in the decline of a competing technology, which may have more robust features. However, this is nothing more than a corollary of the Darwinian metaphor where a technology may by chance or strategy gain an early lead in adoption and thus lockouts competing technologies. This is similar to a competing species that corners a resource, such as food; little resources are left for its competition. This can be seen in the introduction of sheep to an area being grazed by cattle. The sheep crop the grass closer and make it difficult if impossible for cattle to survive in the same pasture.

The Arthurian concept of increasing returns does affect the life cycle of technologies. This concept of increasing returns causes some technologies to become standards and others to become also-ran, i.e., Microsoft's Windows operating system vs. IBM's OS/2 offering.

Maturity

The technological development that reaches a stable maturity state may have a very short stay at this level. The stable maturity stay may be measured in months. This has not always been the case. In fact, the time from invention to innovation or successful use took many decades before the 20th century. This time has been decreasing exponentially [Cardullo, 1996a]. The lifetime for some information technologies has decreased to as low as 6 months. However, it must be understood these short times usually represent times from one incremental change to another of the same "species" of the technology. This rapidity of change in the short term can have serious economic impact for enterprises [von Braun, 1996].

At some point, all technologies reach a point of unstable maturity where competing technologies or the environment in which they are embedded is changing significantly. This has been characterized by

Grove [1996] as a strategic inflection point. If technological enterprises experiencing these significant environmental changes are to survive, they must take significant technological steps to move into new technology regions. This was the case with Intel Corporation, which successfully survived a major strategic inflection point when the enterprise changed from a producer of memory to microprocessors [Grove, 1996]. The ability to survive is enhanced by the enterprise having a technological environmental monitoring system. The formal system captures the activities in both the internal and external technological environment of the enterprise.

Decline and Phase-out/Expire

However, all technologies eventually decline and are phased out or expire. Managers of technology must also consider this life cycle phase. An example is the impact of this consideration on the cost of retirement and disposal of nuclear power plants. The cost associated with this phase can be substantial and should be considered in determining the total life cycle cost of the technological development. Figure 3.10 shows a general technological life cycle.

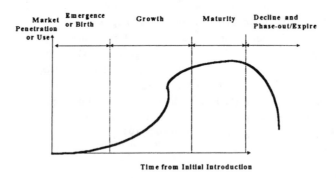

FIGURE 3.10   General technological life cycle.

Summary of Technology Life Cycles

Technology life cycles have been viewed in various ways [Cardullo, 1996b]. The manner of these viewpoints depends on from where the viewer is located, similar to looking out windows of a building. Each viewer is seeing the same world, but from a different perspective. Summed together, these views represent a clearer understanding of the world being viewed.

Technology must be translated into a useful product, process, or service, otherwise it will just remain technological. The development of new technology is an acquisition process, either internally or externally. The acquisition process of new technology, either internally or externally, is a dynamic process, with a well-defined sequence of actions. These are not random actions but proceed in expected patterns. All technologies progress through a series of life stages or life cycle. The acquisition process is the first stage of the life cycle of any technology.

Life cycles express the dynamic nature of various systems [Cleland and King, 1983]. The technology that is proposed as the solution to a particular problem moves from state to state as it evolves from a creative solution to final implementation and then to phase out. A manager of technology must consider the following life cycle models:

- General life cycles of technology
- Martino's [1993] view of the technological process divides the technology life cycle into the following phases: scientific findings, laboratory feasibility, operating prototype, commercial introduction or operational use, widespread adoption, diffusion to other areas, and social and economic impact.

- The life cycle technological product/process/service according to Cleland and King [1983] starts with the establishment phase where the product is introduced into its intended market. This phase is followed by growth, maturation, and finally declining use.
- The life cycle of a technological system according to Blanchard and Fabrycky [1990] is characterized as a consumer-to-consumer process with four major phases: acquisition, utilization, and finally phase-out and disposal.
- A technological project life cycle according to Cleland and King [1983] can be characterized in terms of the resource allocation, which has a very different level of requirements for human and fiscal resources and is defined by the information requirements. The three basic phases include preparation or initiation, implementation, and finally operation.
- Technical organizational life cycles exhibit life cycles as they evolve and can be divided into six phases: organization, directive, delegation, coordinate, collaborative, and finally dissolution or acquisition.
- A technological enterprise life cycle, according to Howard et al., [1992] can be divided into emergence, diffusion, development, and maturity.

The manager of technology should consider all these types of technological life cycle processes. This consideration should be given due to the nature of technology products/processes or services. Each system is embedded in a system, a project, and an organization within an enterprise. Many times, managers of technology do not understand the impact of the various interactions of these cycles.

Technology Life Cycle Interactions

Technology life cycles can have a significant impact on society. According to the Chairman of the Federal Reserve, Alan Greenspan, "Human skills are subject to obsolescence at a rate perhaps unprecedented in American history…the fallout from rapidly changing technology" [WSJ, 1996]. The introduction of any technology has various consequences. These impacts occur in the environment, to existing and future technological developments, and in the financial and social systems in which the new technology is embedded. The development and diffusion of new technologies has been widely studied. One of the results of the introduction of new technologies is usually increased industrial productivity. This has been one of the impetuses for reducing the scale of various enterprises. The development and introduction of information technology is a recent example of the impact of new technology and processes on employment.

The simplified model used by Cardullo [1996b] to describe the impact of technology on employment patterns was based on the assumption that new technology introduction followed the **Pearl equation**. The assumption was also made that the productivity was directly related to the number of new technological units, which were introduced. Thus, the resulting impact on employment is given by

$$ W = \frac{r}{p} = \left(\frac{R}{AU} \right) = \left(\frac{1 + a_u e^{-b_u t}}{1 + a_r e^{-b_r t}} \right) $$

where

W = number of workers required
U = upper limit of units that can be placed in use
R = upper limit of the requirement for new technology
A = productivity factor
p = productivity at time t
r = requirement for units in an industry or enterprise using the new technology at time t
a_u, b_u = coefficients for units of new technology placed within enterprise or industry
ar, br = coefficients for worker requirements

According to this model, the initial result of technological introduction is the reduction of workers to some extent. At some point, the productivity from the new technology reaches a plateau. If the enterprise requirements are increasing, it is possible that the firm or industry must start to hire new workers to meet the increased industry demands. The introduction of technological improvements starts the cycle anew. As new technology is introduced, the cyclic nature of technology becomes very evident. This cyclic nature influences the need for development of newer technology that will increase productivity. It is very possible to finally arrive at a situation that the time required to develop a technology has been significantly reduced. This seems to be the situation in the information technology sector.

The rapidity of the introduction of new technology may cause more employees to become displaced than in prior eras. It is quite possible that rapid changes in technology with shorter and shorter life cycles can lead to a highly unstable system. The complex nature of technological development coupled with short life cycles can cause serious economic consequences for industries and societies. As *The Wall Street Journal* [1996] stated, "*The anticipated boom in productivity, however, could well spawn more labor upheavals.*" It appears that 18 million U.S. workers may be at risk due to technology. It is important that we develop a better scientific understanding of the impact of technological life cycles, not only on the management of technology, but also on the underlying socioeconomic system upon which it is based.

Defining Terms

DNA: Deoxyribonucleic acid, a complex polymer, and the genetic instructions controlling development of organisms.

Embedded: Contained within an environment, such as a market or a larger system.

Emergence: Arising from a diversity of elements with increasing complexity. Structures that are more complex can evolve from simpler structures.

Nonlinear process: Small changes can have significant consequences; conversely, major changes in the process might yield very little impact.

Pearl equation: After U.S. demographer Raymond Pearl; also known as the logistic curve.

Second Law of Thermodynamics: The measure of disorder of system, i.e., entropy, will increase in all irreversible processes that isolated systems undergo.

References

Arthur, W. B. Competing technologies, increasing returns, and lock-in by historical events, *Econ. J.*, 99(March): 116–131, 1989.

Blanchard, B. S., and Fabrycky, W. J. *Systems Engineering and Analysis*, 2nd ed., W. J. Fabrycky and J. H. Mize, eds., Prentice Hall International Series in Industrial and Systems Engineering, Prentice-Hall, Englewood Cliffs, NJ, 1990.

Cardullo, M. W. *Introduction to Managing Technology*, Vol. 4, J. A. Brandon, ed., Engineering Management Series, Research Studies Press, Taunton, England, 1996a.

Cardullo, M. W. Technological Lifecycles: Causes and Effects, paper read at International Engineering Management Conference, August 18–20, Vancouver, Canada, 1996.

Cleland, D. I. and King, W. R. *Systems Analysis and Project Management*, McGraw-Hill, New York, 1983.

Grove, A. S. *Only the Paranoid Survive*, Doubleday, New York, 1996.

Howard, Jr., W. G. and Guile, B. R. *Profiting from Innovation: The Report of the Three-Year Study from the National Academy of Engineering*, W. G. Howard, Jr. and B. R. Guile, eds., The Free Press, New York, 1992.

Kauffman, S. *At Home in the Universe: The Search for Laws of Self-Organization and Complexity*, Oxford University Press, New York, 1995.

Martino, J. P. *Technological Forecasting for Decision Making*, 3rd ed., M. K. Badawy, ed., McGraw Hill Engineering and Technology Management Series, McGraw-Hill, New York, 1993.

Pearce, John A., II, and Richard B. Robinson, Jr. *Strategic Management: Formulation, Implementation, and Control*, 4th ed., Richard D. Irwin, Homewood, IL, 1991.

von Braun, C. F. *The Innovation War: Industrial R&D... The Arms Race of the 90's*, W. J. Fabrycky and J. H. Mize, eds., Prentice Hall International Series in Industrial and Systems Engineering, Prentice Hall, Upper Saddle River, NJ, 1996.

WSJ. Outlook, *Wall Street J.*, April 22: 1, 1996.

Further Information

A detail review of the concept of increasing returns by Arthur can be found in "Increasing Returns and the New World of Business" in the July-August 1996 issue of *Harvard Business Review*. Arthur is the principal exponent of this new economic concept and is a member of the Santa Fe Institute. The *Bulletin of the Santa Fe Institute* gives summaries of the newest work of the Institute dealing with complexity and emergence. These concepts are related to viewing technological developments within a wider framework than normal.

Methodologies for computing technology life cycles can be found in *Forecasting and Management of Technology* by Porter et al. The authors present various analytical methods for extrapolating technological trends. The text presents an introduction to the development of technological environmental monitoring systems and some sources of information that can be used.

Goodman and Lawless present the concept of technological antecedents in *Technology and Strategy*. The authors present ways of mapping technology's antecedents.

Betz presents the forecasting of directions of technological change in *Strategic Technology Management*. Also presented in this text is product life cycle analysis from both a technological and market viewpoint.

The *IEEE Transactions on Engineering Management* contains detail empirical studies and other studies that will give the reader an excellent starting point to developing analytical and qualitative information on technology life cycles.

3.9 Dominant Design

Philip Anderson

A significant technological innovation is generally introduced into a product class by means of a pioneering product. Once this initial product demonstrates the feasibility of an innovative concept, rival products appear (unless very strong barriers to imitation, such as patents, protect the pioneer). These competing products often vary in design features and choices. Therefore, significant innovations typically result in a period of ferment, characterized by two forms of competition: products that incorporate the new technology begin to displace products that do not and variants of the new technology compete with one another.

In principle, once the new technology supplants its predecessor, more and more variants could arise, each occupying a different niche. In practice, a single variant usually emerges as a **dominant design**, a general architecture or pattern that comes to account for a greater share of the market than all other variants combined. Sometimes, a dominant design is embodied in a single product, such as the IBM personal computer, the Ford Model T automobile, or the Douglas DC-3 aircraft, all of which accounted for 70 to 80% of their entire markets at the height of their popularity. Alternatively, a basic configuration of design choices may characterize a number of products, none of which individually dominates the market. Examples include the closed-steel-body automobile, the 4-hour VHS videocassette recorder, and the Winchester-type magnetic disk drive for data storage.

Dominant designs often incorporate **technological standards**, and standards often become accepted because they are incorporated in dominant designs, but the two concepts are distinct (see Section 14.14). A standard defines a common interface between two subsystems that must work together. The VHS videocassette standard, for example, defines a format that ensures machines and tapes are compatible. Standard thread sizes ensure that nuts and bolts fit each other, while various television signal transmission standards (e.g., NTSC, PAL, and Secam) specify how television sets receive and interpret broadcast signals. A dominant design specifies a broader range of design choices and parameters. For example, the Underwood Model 5 typewriter emerged as a dominant design in the early 1900s [Utterback, 1994], not only incorporating the QWERTY keyboard standard, but also establishing a pattern of features that quickly became conventional. Such features included visible typing (the ability to see the printed letter immediately after it was struck), a tab function, a cylinder carriage, and shift key operation.

In its essence, a dominant design is a set of engineering conventions, not a particular machine. It represents an accepted pattern of design choices. A body of knowledge and a set of taken-for-granted assumptions grow up around this set of choices, channeling the direction of technological progress and the dimensions of merit that engineers use to evaluate alternative designs. In the case of complex products, the dominant design constitutes an accepted configuration or architecture. It renders conventional a particular division of labor between a set of subsystems and a general understanding of how they interact to produce system performance. It therefore becomes a guidepost that often shapes the basic configuration of a type of product for decades [Sahal, 1981].

Why and How Dominant Designs Emerge

Why does one variant typically dominate all others? Why don't most innovations lead to a mosaic of different designs that incorporate the new technology via different sets of basic design choices? A consensus of modern scholarly opinion favors both economic and sociological explanations over technical explanations.

It is unlikely that one variant becomes dominant because it is clearly superior on technical grounds. In fact, dominant designs usually lie behind the state-of-the art technological frontier because designs that push the technical envelope are often too unstable or costly to meet the needs of a mass market [Anderson and Tushman, 1990]. Dominant designs typically synthesize a number of innovations that have already been introduced [Abernathy and Utterback, 1975]. Sometimes, this synthesis results in measurably superior cost/performance; for example, the DC-3 is often cited as the first plane capable of economic commercial operation. However, it is difficult to argue that VHS videocassette recorders were better than Beta VCRs or that the IBM personal computer was the best-performing microcomputer of its era.

Two economic explanations have more theoretical and empirical support than do arguments based on technical superiority. On the demand side, conventions of all types, including dominant designs, emerge because markets penalize uncertainty. When many variants exist, it is difficult for potential customers to compare them, and the risk of picking a dead-end design seems high. Customers, suppliers, channels of distribution, and producers of complementary goods all prefer greater certainty, so strong incentives contribute to the emergence of conventional solutions.

On the supply side, once a particular design begins to gain wide acceptance, powerful forces propel it ahead of rival designs. Process innovation, driving down costs, is more feasible when a design becomes stabilized [Abernathy and Utterback, 1975]. A network of firms and individuals supplying complementary goods and services emerges [Tushman and Rosenkopf, 1992]. Positive feedback loops often arise, driven by "network externalities"; the value of the product to each user increases as the customer base grows. Network externalities are especially salient when customers can interconnect with one another, but, even for stand-alone products, the larger the customer base, the greater the incentive for third parties to develop useful enhancements, spurring demand.

One consequence of these forces is that dominant designs often appear *in hindsight* to embody the best technology available, lending spurious force to the idea that superior technical performance accounts for the predominance of a particular design. For example, suppose a particular type of computer networking software captures a large share of its market, although it does not initially lie on the frontier of technical performance. Its subsequent rate of improvement should be steeper as a consequence. Eventually, the dominant technology is likely to become the best available. However, the design achieved technical superiority because it gained support, not the other way around. In just this way, Novell became the dominant operating system for local area networks in the 1980s, although rival designs often won "performance shootouts" during the infancy and adolescence of this industry.

If economic factors make the emergence of a single dominant design the likely outcome of an era of ferment, what determines *which* design prevails? Community-level social and political processes adjudicate among a number of feasible options [Tushman and Rosenkopf, 1992]. Different actors who share an interest in seeing a particular design prevail form a community through both formal and informal linkages. Competition among the interest groups favoring different designs is a battle for share of mind. Momentum matters; once a critical mass of customers becomes convinced that a particular design will prevail, their expectation becomes self-fulfilling. Hence, design competition is largely a struggle over meaning. Interest groups define and give meaning to the types of problems a design ought to solve, the dimensions of merit that should be emphasized when comparing alternative products, and the attributes that should be valued most highly in choosing a supplier of those products. Consequently, the victory of a particular design is shaped by the actions of people, typically key executives in organizations [McGrath, et al., 1992], not only by the technical characteristics of rival products.

The Economic and Social Impact of Dominant Designs

The emergence of a dominant design is a watershed event in a technology life cycle. It closes the era of ferment and inaugurates an era of incremental change [Tushman and Anderson, 1986]. The basis of competition shifts dramatically, as the trajectory of technical progress centers on refining and improving an accepted architecture.

Following a dominant design, the focus of innovative efforts frequently shifts from product improvement to process improvement [Abernathy and Utterback, 1975]. Before one design becomes accepted, it is difficult to maximize the efficiency of the production process because design parameters are fluid. Once a single stable configuration achieves sufficient scale, flexibility can be traded off against efficiency and process specialization.

Demand typically rises sharply following the emergence of a dominant design [Tushman and Anderson, 1990], and the minimum efficient scale of production tends to increase. Learning by doing and learning by using drive costs down as the scale of production increases. Incremental innovation begins to take place within an established framework, and performance dimensions become better understood, making it easier for customers to compare rival products. These changes lead to more cost competition, which is often accompanied by a shakeout, as firms that were able to survive during the turbulent era of ferment find themselves unable to adjust to a regime of continuous improvement and cost cutting [Suarez and Utterback, 1995].

Eventually, a dominant design typically becomes more deeply embedded in a web of linkages to other systems. Complementary inputs and peripherals are adapted to fit the prevailing technical architecture. Third-party services, such as specialized magazines, training centers, and maintenance operations, are organized to fit the logic of the dominant design. Standard linkages, interchangeable parts, and a used-equipment market appear, creating demand for newer product models to remain backward-compatible with older ones. All of these developments make it difficult for producers to alter an entrenched design significantly, slowing the pace of technical advance. As Thomas Hughes has noted, in the context of the General Electric incandescent lighting system that emerged from Edison's inventions, when a system becomes larger and more complex, it increasingly comes to shape its environment more than it is shaped by its environment.

A dominant design also has a profound effect on both the organization of industry and the organization of individual firms. A dominant design reinforces a technological trajectory [Nelson and Winter, 1982] and the emergence of a technological paradigm [Dosi, 1984]. A technological trajectory is a pattern of normal problem-solving activity. It singles out the direction of progress that most engineers in a field pursue. A technological paradigm is a pattern of selecting technological problems and principles for solving them. An established design has its own set of bottlenecks that become the focus of improvement efforts. The way knowledge in a field is organized and transmitted depends on the critical technical problems posed by an entrenched design. A set of legitimate, tried-and-true ways of coping with these problems constitutes the shared, taken-for-granted knowledge structure that professionals working in a particular industry must absorb.

This knowledge structure is one of the strongest influences on the organization of firms in the community that forms around a technology. The way in which labor is divided and coordinated, for instance, usually reflects the way that a product's subsystems are divided and coordinated. Product architectures are frequently reflected in organizational architectures. Social relations between organizations also tend to mimic the boundaries and linkages that become established features of a dominant design.

Dominant designs lead to inertia and industry maturity. A dominant design begins as a set of conventions and evolves into a world view, a framework for organizing the way that industry experts define what problems are interesting and what solutions are most promising. In many ways, the development of a coherent, consensual world view is a boon to an industry, greatly easing problems of coordination, training, and gaining customer acceptance for innovations. It is also dangerous, for, when a new set of principles emerges to challenge an entrenched technology, incumbent firms and their executives often respond defensively, unable to break free from deeply held assumptions. As a consequence, the displacement of a dominant design by a newer, superior technology is often accompanied by wrenching organizational changes and a turnover in industry leadership.

Continuing Controversies

Since the term "dominant design" was introduced by Utterback and Anthony in 1975, building on the insights of previous scholars such as James Bright [Utterback, 1994], it has achieved widespread recognition and usage. However, there remains no methodology that permits scholars or practitioners to identify a dominant design unambiguously. Differences in the level of analysis or the definition of an innovation can lead to different perspectives on what constitutes a dominant design.

A dominant design has usually been identified at the level of a system, or complete end item. Examples include automobiles, watches, and computers. However, dominant designs may also emerge for subsystems, components of a larger system. For example, automobile engines, watch movements, and computer communications buses also have evolved through a series of standard configurations. Additionally, dominant designs are sometimes identified with specific equipment models (e.g., the Boeing 747 jet) but are sometimes defined according to a set of design parameters broader than any single model (e.g., metal-bodied monocoque aircraft). With a sufficiently broad set of parameters (e.g., aircraft with wings), any industry can appear to have a dominant design.

Resolving these dilemmas requires penetrating to the essence of the concept. A dominant design is at the root a set of conventions and institutions, a collection of agreed-upon problems, principles, and parameters. Such conventions can characterize both systems and subsystems, both individual models and broader architectures. The topic of interest is how a set of conventions comes to predominate and reduce the number of variants of a technology following a significant innovation.

Defining Terms

Dominant design: A set of conventions specifying the basic pattern or configuration of a technical system or subsystem.

Technological standard: A defined common interface between two subsystems that must work together.

References

Anderson, P. and Tushman, M. L. Technological discontinuities and dominant designs, *Admin. Sci. Q.*, 35: 604–633, 1990.

Dosi, G. *Technical Change and Industrial Transformation*, St. Martin's Press, New York, 1984.

McGrath, R. I., MacMillan, I., and Tushman, M. L. The role of executive team actions in shaping dominant designs: towards shaping technological progress, *Strat. Mgt. J.*, 13: 137–161, 1992.

Nelson, R. R. and Winter, S. *An Evolutionary Theory of Economic Change*, Belknap Press, Cambridge, MA, 1982.

Sahal, D. *Patterns of Technological Innovation*, Addison-Wesley, Reading, MA, 1981.

Suarez, F. F. and Utterback, J. M. Dominant designs and the survival of firms, *Strat. Mgt. J.*, 16: 415–430, 1995.

Tushman, M. L. and Anderson, P. Technological discontinuities and organizational environments, *Admin. Sci. Q.*, 35: 604–633, 1986.

Tushman, M. L. and Rosenkopf, L. On the organizational determinants of technological change: toward a sociology of technological evolution, In *Research in Organizational Behavior, Vol. 14*, B. Staw and L. L. Cummings, eds., pp. 311–347, JAI Press, Greenwich, CT, 1992.

Utterback, J. M. *Mastering the Dynamics of Innovation*, Harvard Business School Press, Boston, MA, 1994.

Utterback, J. M. and Anthony, W. J. A dynamic model of product and process innovation, *Omega*, 3(6): 639–656, 1975.

Further Information

The quarterly journal *Research Policy* often contains studies describing technology life cycles, dominant designs, and the emergence of technology standards. For subscription information, contact the customer sales office, Elsevier Science, P.O. Box 945, New York, NY 10159-0945.

3.10 Technology Integration: Matching Technology and Context

Marco Iansiti

Product development at its core involves selecting and integrating technologies and matching them to their application context — the product's customer and manufacturing/service delivery environment. Choices must be made among a range of technological options, and this selection and integration process depends on knowledge — of technological options and of their context; in turn, this knowledge must be generated, retained, and applied. When technologies are novel and the context is complex, as is increasingly the case today, an effective process for technology integration is critical to research and development (R&D) performance (see [Iansiti, 1997] for more details.)

Unfortunately, technology choices are too often made in scattershot and reactive fashion, with commitments variously made in research, development, and supplier organizations. New technological possibilities are chosen for their individual potential rather than for their system-level integration, and insufficient attention is paid to the timing of technology integration activities. These activities represent a window of opportunity, which if closed too early can lead to inertia: the technological path has been fixed and the options left available to engineers and managers are constricted. When inevitable inconsistencies arise between the technology and existing system characteristics, the result will be delays, the need for additional resources, longer lead times, and poor R&D performance — even industry exit. When there is a proactive process of technology integration, however, one comprising a dedicated, authorized group of people armed with appropriate knowledge, experience, tools, and structure, performance improves dramatically.

This section begins by stressing the importance of matching technology and context, then introduces the building blocks of an effective technology integration process, showing how these are linked to performance.

Matching Technology and Context

A product that "fits" its context is one whose multiple technologies have been integrated into the requirements of that context (see Fig. 3.11). The most critical components of a product's context are the customer and manufacturing/service delivery environments. These environments are becoming increasingly complex, however, while novel technologies are proliferating.

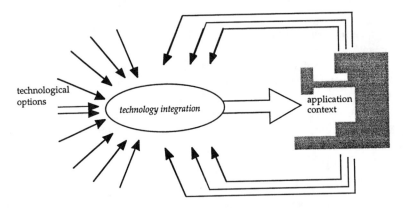

technological options

technology integration

application context

FIGURE 3.11 Matching technology and context.

In software, for instance, product design challenges greatly sharpen when customers demand reliability, user-friendliness, speed, compatibility, *and* "cool" features — in a constant stream of new versions. In consumer products, meanwhile, what is desirable in one part of the globe may be shunned in another — insulin pens, for example, are far more popular in Europe than in the United States, where syringe-type products are preferred. Rarely, in fact, is it possible to introduce the same product, unaltered, everywhere. A product's user environment, therefore, can be simultaneously broad, volatile, and subtle.

Meanwhile, the manufacturing/service delivery environment has grown equally complex. Intel's latest production facility, for example, cost more than $3.5 billion. Most of this was dedicated to equipment for a 600-step manufacturing process whose "pieces" had to work together in perfect coherence so that extremely high production yields were possible. At Yahoo Inc., the challenge is "scaleability": ensuring that a new Internet service can be scaled up overnight to be available to millions of customers.

At the same time, fluctuating market demand and/or novelty in the technology base exacerbate the challenge of matching technology to context. At Intel, typically at least one third of new process equipment has never been previously used. Internet search engines such as Yahoo!, while fitting the needs of a rapidly evolving — and fickle — customer base, are predicated on highly sophisticated mathematics and linguistics.

To master such challenges — to match technology and context — an effective technology integration process is needed.

The Foundations of Technology Integration

"Knowledge" is fundamental to technology integration. It consists of knowledge about different scientific and technological domains, knowledge about production systems, and knowledge about user environments. An effective technology integration process merges all this knowledge and has three broad mechanisms for doing so: mechanisms for knowledge generation, its retention, and its application/integration.

The ability to discover and learn — to generate knowledge — is a hallmark of an effective product development process; that ability, in turn, is founded on cycles of experimentation that enable technology and context to "fit". Such experimentation can range from scientific investigations of new materials to beta testing a product prototype with potential customers and from mathematically simulating product performance to studying product aesthetics via physical models. Likewise, experimentation can range from broad investigations of many concept possibilities to focused iterations aimed at refining them.

These experiments must be both designed and performed; their results must be interpreted and turned into decisions that guide the product's development. This need for interpretation and guidance underscores the importance of retaining individual expertise and experience. Experience can also come in many forms, from the breadth that arises from participating in many projects to the depth that comes from intense specialization.

Knowledge generated through experiments and interpreted by individuals must then be integrated to define a product concept that matches technology to context. This means having organizational mechanisms for applying/integrating knowledge across experiments, disciplinary domains, and organizational levels.

Knowledge Generation

The quality of technology choices is profoundly influenced by the quantity of knowledge is generated. Understanding in advance how technological possibilities will influence a product's application context greatly increases the chance that these technologies will work together in practice. In other words, the greater the knowledge of the interactions between technologies and system, the better the match between the two.

The idea of experimentation as a knowledge-building mechanism is not new nor is experimentation's critical role in helping organizational processes (particularly in production) evolve during periods of rapid technological change (see, for example, Arrow [1962], Leonard-Barton [1988, 1992], Adler and Clark [1991], Rosenberg [1982], and Tyre and Hauptman [1992]. The impact of experimentation on R&D processes has likewise been studied (e.g., Wheelwright and Clark [1992] and Bowen et al., [1994]). Thomke [1995], for instance, found that experimentation can substantially reduce development cost.

Experimentation's role in R&D is especially important when complexity and novelty both exist. On-site experimentation, for example, is essential to problem solving when information characterizing a production environment is complex and "sticky" [Von Hippel, 1994; Von Hippel and Tyre, 1995]. Pisano [1996], who also studied experimentation in production process optimization, found that the less explicit the knowledge base in the production environment, the greater the role experimentation plays in the actual production context. This, moreover, is not limited to production environments: Von Hippel has persuasively argued that user environments are challenged by sticky information.

All this and much more research lead us to conclude that effective technology integration demands an experimentation capability that links technological possibilities both to each other and to the application context. This capability comprises three dimensions, the first being capacity: how many different experiments an organization can perform in a given period. Experimental iteration time is the second dimension: in a single experiment, the minimum time that elapses between its design and execution. As such, experimental capacity drives the number of parallel experiments possible in an organization while iteration times drives the number of sequential experimental cycles possible in a project. These two dimensions, in turn, must be related both to understanding the fit between technical options and context and to project performance [Thomke et al., 1997].

The representativeness of the experimentation setup is the third dimension of experimentation capability: how similar the experimental context is to the actual context. No experimental setup, of course, perfectly replicates the context in which a product will function [Von Hippel and Tyre, 1995]; a real manufacturing plant or user environment is usually too complex to be completely reproduced in a controlled setting. Moreover, the product's evolution may influence its future context such that the user may develop new applications. Hence, an experimentation facility for technology integration should be designed not only in light of the previous context but also according to aggressive estimates of the future context. That means it allows for a diverse set of experiments that test a wide variety of technology–context combinations so that unanticipated interactions and their potential can be explored.

Knowledge Retention

Technology integration depends on more than knowledge generation and on more than the knowledge building that comes from effective experimentation. It requires, as well, individual skills and capabilities: both to guide experimentation and to interpret its results. Complementing knowledge generation, therefore, is knowledge retention, the capability that enables different projects to be linked and the R&D organization's knowledge base to evolve. In a nutshell, a solid foundation of system knowledge is crucial for technology integration.

System-level knowledge, however, as many researchers (e.g., Leonard-Barton [1988], Clark and Fujim-oto [1991], Von Hippel [1994], Tyre and Von Hippel [1997] have shown, is difficult to transfer from one person to another, making individual experience extremely valuable. Experienced people need to be retained, therefore, so that the impact of individual design decisions can be assessed. Such people become "architects", integrators of the future product, driving its specification. At the same time, organizations must stress the development of this experience.

However, "too much" experience can be detrimental, leading to rigidity and inertia and stifling team and career progress (see Katz and Allen [1985] and Leonard-Barton [1995]). New skills and fresh perspectives, therefore, are desirable along with experience, so that novel technical possibilities can be explored and existing solutions questioned. Thus, while some experience may be linked to effectiveness, too much experience might lead to rigidity, restricting the breadth of technical options considered in the decision-making process.

Knowledge Integration/Application

To be effective, knowledge retained through experience and generated through individual experiments must be integrated and applied. Hence, a project must have an organizational process that allows knowledge-generation activities to be coordinated and used coherently so that technology and context can be matched. The timing and sequencing of activities, particularly in experimentation, is crucial, both to avoid inertia and to ensure system-level "fit". With such a process in place, the consequences of individual decisions can be balanced against each other to make sure that the systemic impact of individual choices is assessed.

My empirical work in the computer industry (Iansiti, 1997) has *consistently linked these three foundations to product development performance*. Organizations that invested in experimentation capability, managed their experience base in a proactive way, and created an integrative process for matching technological possibilities to their application context consistently outperformed others in environments as different as Internet services and semiconductor manufacturing.

Integration and Performance

The performance of a product development project is defined on several dimensions. The first and most critical dimension of project performance is given by the *quality* of its outcome. Quality, here, is defined as the fit of the project's outcome with its context, which is embodied largely by its customer base and its manufacturing/service delivery environment, as described above.[6]

---

[6] Each of these four measures can be expressed in absolute terms or as a lead or lag relative to competitors. The product quality lead, for example expresses quality in relative terms in relative terms—a quality lead value of one implies that the product is 1 year ahead of its competitors when introduced. The productivity, elapsed time, and flexibility measures can also naturally be expressed in quality-adjusted terms, as long as quality measures can be defined analytically.

The second dimension of performance is the *productivity* of the project: the resources (human, financial, or other) used in its completion. In a complex environment, characterized by a variety of customers with fragmented needs, the productivity of a project (or of an R&D organization) is a crucial variable, particularly when adjusted for quality. The higher the productivity, the more products an organization can introduce per period of time.

The third dimension of performance is the *speed* of the project: the time elapsed between its formal start and its market introduction. This measure represents how quickly a new product generation can be brought to market.

The fourth dimension of performance is the *flexibility* of a development project: the time elapsed between concept freeze and market introduction. The later the concept freeze, the more the project is able to incorporate late-breaking ideas and respond to technology and market changes and the higher the flexibility of the project.

These measures of product development performance are not independent. Research provides ample evidence that some projects appeared consistently superior in a number of performance dimensions. Table 3.6 shows correlation coefficients between speed and productivity in mainframe projects, for example, showing that they are significant [Iansiti, 1997]. This is consistent with the work of a number of other academics [Clark and Fujimoto, 1991; Pisano, 1996], who found that some projects appeared to be both more productive and faster than others.

TABLE 3.6   Observed Relationships between Product and Project Performance

	Technological Yield	Technological Potential (GHz)	Product Performance (GHz)	R&D Resources (person years)
Technological yield	1			
Technological potential (GHz)	−0.527*	1		
Product performance quality (GHz)	0.257	0.614**	1	
Project productivity (person years)	−0.495*	0.365	−0.186	1
Project speed (years)	−0.172	−0.152	−0.465*	0.540*

The table displays correlation coefficients between different product and project performance dimensions, technological yield, technological potential, overall product performance quality, productivity, and speed. It shows that the most productive projects are also the fastest. Moreover, it shows that the yield variable is correlated with productivity. * indicates significance at the 5% level; ** indicates significance at the 1% level. (*Source:* Iansiti, 1997.)

Many authors have asserted that the structure of organizations and of the products they develop converge over time. Henderson and Clark [1990], for example, have argued that functional departments come to reflect product modules, such as optics and mechanical systems. Drawing from such research, it is possible to hypothesize that projects with the highest performance lead to products of the highest performance. Table 3.6 suggests this: the fastest projects also achieved the highest product performance.

The relationship between product performance and project productivity is more subtle, however, as the table implies, because product performance can be achieved in different ways. A project can improve existing technology, for instance, or it can aggressively pursue new technology. While the former approach should correlate with productivity (the more that is known about a technology, the more efficient the project will be), the latter should not. Since considerable new knowledge must be generated in order to introduce new technology, the project should not necessarily be efficient. My research has looked at this issue carefully, developing a methodology for doing so [Iansiti, 1997].

The methodology divides system-level product performance into two quantities: *technological potential* and *technological yield*. The technological potential estimates the maximum product performance given the technological base of a product and can be based on an actual physical model of the product. Technological potential should therefore be closely linked to the depth in fundamental research that a project draws from. The achievement of high technological potential, however, should not necessarily correlate with productivity, since changing the technological base of a product will require considerable new knowledge to be generated. Technological yield estimates how much of the product's potential is realized in actual product performance. It should therefore correlate with the capability to integrate the technologies with their context, and this should be correlated with productivity.

Table 3.6 shows that yield and productivity are correlated, while technological potential and productivity are not.[7] Table 3.7 shows a more direct test of the above hypotheses and indicates that technological potential is indeed correlated with the research process while yield is correlated with the process for technology integration. This result shows *explicitly* that the outcome of an effective technology integration process is a product whose technologies truly match their context in a superior way. More generally, it provides support that the structure of products reflects the nature of the process that developed them.

TABLE 3.7   Statistical Correlations between Basic R&D Process and Product Characteristics

Characteristics of the R&D Process	Technological Yield	Technological Potential
Tradition of research in related areas	0.202	0.661**
Targeted integration process, with dedicated group of individuals	0.753**	0.178

The double asterisk indicates significance at the 1% level.

Technology Integration and Product Development Process Design

Effective technology integration depends on having mechanisms for generating, retaining, and integrating/applying knowledge in place. These foundations, however, must be translated into an effective product development process, and such processes are not all alike. What works at Yahoo! differs considerably from what works at Intel — or at McDonald's — even though the foundations are shared in all three cases. Each has a different approach to experimentation, to retaining seasoned decision makers, and to the optimal sequencing of integrative tasks in a project. Each, as well, has a different strategic and competitive environment. Hence the "right" process — one that consistently achieves high performance — must reflect the characteristics of that environment (see Fig. 3.12).

Novelty and complexity are critical dimensions for product development process design (see Fig. 3.13). If the environment is characterized by high complexity but low novelty, the process must be optimized for task coordination. Integrative project management capabilities are needed, but tasks can be arranged sequentially (as in the "stage-gate" model) and only limited experimentation is required for effectiveness. If the environment is characterized by high novelty but low complexity, the process needs to maximize experimentation and responsiveness but does not require stressing coordination or integration. When both high novelty and high complexity are present, the challenge is the sharpest. In this case, the product development process must be flexible: it must be driven by rapid iteratives, a strategy of broad experimentation, and an organizational process that maximizes quick and frequent integration. Such product

---

[7]This argument does not work in the same way with project speed because when the change is revolutionary (high potential), projects tend to run experiments in parallel so as to examine different options. This increases project speed with respect to situations characterized by evolutionary change (high yield), which is driven by serial experimentation.

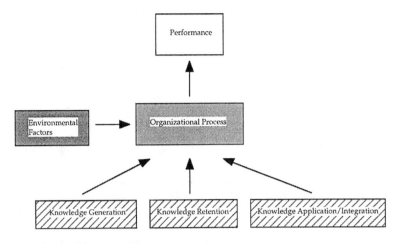

FIGURE 3.12   Linkages between foundations, process, and performance.

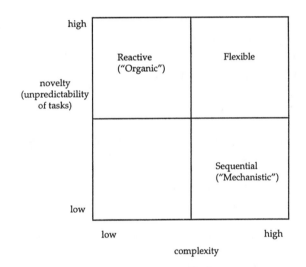

FIGURE 3.13   Matching organizational process to environment.

development processes are found in Internet software development firms, for example, which exist in an environment characterized by extreme novelty and complexity.

Clearly, designing the "right" product development process is not, by itself, sufficient for guaranteeing project performance. Projects function within a larger organization, and a multitude of factors, from resource allocation policies to investments in experimentation tools and facilities, come into play. As such, projects must be part of a proactive development strategy that includes higher-level processes for project selection [Wheelwright and Clark, 1992, 1995] and for defining product line architecture. Likewise, the role of R&D is likewise critical. The existence of a good process for technical decision making enables R&D activities to work more effectively, each dedicated to what it does best. With research, this means creating individual technological options. With development, this means executing a well-defined product concept. Technology integration leverages both capabilities by managing their interaction so that technology can truly match the product's application context.

References

Adler, P. S. and Clark, K. B. Innovation: mapping the winds of creative destruction, *Res. Policy,* 14(1): 3–22, 1991.

Arrow, K. Economic welfare and the allocation of resources of invention, In *The Rate and Direction of Inventive Activity: Economic and Social Factors,* R. Nelson, ed., Princeton University Press, Princeton, NJ, 1962.

Bowen, H. K., Clark, K. B., Holloway, C. A., and Wheelwright, S. C. *The Perpetual Enterprise Machine,* Oxford University Press, New York, 1994.

Clark, K. B. and Fujimoto, T. *Product Development Performance,* Harvard Business School Press, Boston, 1991.

Henderson, R. and Clark, K. B. Architectural innovation: the reconfiguration of existing product technologies and the failure of established firms, *Admin. Sci. Q.,* 35(1): 9–30, 1990.

Iansiti, M. *Technoloy Integration: Making Critical Choices in a Dynamic World,* Harvard Business School Press, Boston, MA, 1997.

Katz, R. and Allen, T. J. Project performance and the locus of influence in the R&D matrix, *Acad. Mgt. J.,* 28(1): 67–87, 1985.

Leonard-Barton, D. Implementation as mutual adaptation of technology and organization, *Res. Policy,* 17(5): 251–267, 1988.

Leonard-Barton, D. Core capabilities and core rigidities: a paradox in managing new product development, *Strat. Mgt. J.,* 13: 111–125, 1992.

Leonard-Barton, D. *Wellsprings of Knowledge,* Harvard Business School Press, Boston, 1995.

Pisano, G. Learning before doing in the development of new process technology, *Res. Policy,* 25(7): 1097–1119, 1996.

Rosenberg, N. *Inside the Black Box: Technology and Economics,* Cambridge University Press, New York, 1982.

Thomke, S. H. The Economics of Experimentation in the Design of New Products and Processes, Ph.D. dissertation, MIT Sloan School of Management, Cambridge, MA, 1995.

Thomke, S. H., Von Hippel, E. S., and Franke, R. Modes of Experimentation: An Innovation Process Variable, working paper 96-037, Harvard Business School, Boston, 1997.

Tyre, M. J. and Hauptman, O. Effectiveness of organizational response mechanisms to technological change in the production process, *Organ. Sci.,* 3(3): 301–320, 1992.

Tyre, M. J. and Von Hippel, E. The situated nature of adaptive learning in organizations, *Organ. Sci.,* 8(1): 71–83, 1997.

Von Hippel, E. The impact of 'sticky information' on innovation and problem solving, *Mgt. Sci.,* 40(4): 429–439, 1994.

Von Hippel, E. and Tyre, M. J. How learning is done: problem identification in novel process equipment, *Res. Policy,* 24(1): 1–12, 1995.

Wheelwright, S. C. and Clark, K. B. *Revolutionizing Product Development,* Free Press, New York, 1992.

Wheelwright, S. C. and Clark, K. B. *Leading Product Development,* Free Press, New York, 1995.

Minnesota Mining and Manufacturing Company (3M) was founded in 1902 at the Lake Superior town of Two Harbors, Minnesota. Five businessmen agreed to mine a mineral deposit for grinding-wheel abrasives. Eventually, they marketed sandpaper and related products.

<u>Growth in revenue</u>

1917	$1	million
1972	$2	billion
1979	$5	billion
1996	$14.2	billion

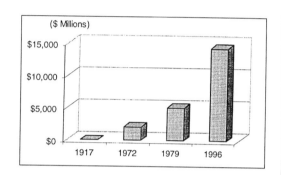

<u>Innovations:</u>

3M's success has come from producing innovative solutions. This list is just a sample of the wide range of innovations.

1920 — The world's first waterproof sandpaper

1925 — Masking tape

1940s — Reflective sheeting for highway markings, magnetic sound recording tape, filament adhesive tape

1950s — Thermo-Fax(TM) copying process, fabric protector, videotape

1960s — Dry-silver microfilm, carbonless papers

1970s — Fastened diapers, backup security for computers, new filling materials for dentists

1980s — Microreplication, highly reflective highway signs, lap-top computer screens first product, abrasives

1990s — Water-based version of Scotchgard

<u>Strategic Issues:</u>

- Delegate responsibility for innovation and encourage men and women to exercise their initiative.
- Provides 15% of an engineer's time for self-initiated products.

II

Knowledge for the Technology Manager

4

Economics

David R. Henderson
Hoover Institution and Naval Postgraduate School

Sharon M. Oster
Yale University School of Management

Paul Heyne
University of Washington

Richard B. McKenzie
University of California

Ivan Png
National University of Singapore

David M. Levy
George Mason University

Richard Reimer
The College of Wooster

William M. Frix
John Brown University

William Samuelson
Boston University

David Flath
North Carolina State University

Alok K. Chakrabarti
New Jersey Institute of Technology

Arnold Reisman
Reisman and Associates

Marc S. Gerstein
Massachusetts Institute of Technology

Lynne F. Baxter
Heriot-Watt University

Thomas W. Gilligan
University of Southern California

Peter S. Rose
Texas A&M University

4.1   Markets[1]

David R. Henderson

In his famous book, *The Wealth of Nations*, published in 1776, Adam Smith explained to his countrymen why some economies grew and others stagnated. Thus, the book's complete title: *An Inquiry into the Nature and the Causes of the Wealth of Nations*. Many of Smith's keen insights have weathered the test of time and evidence and thus are still relevant today. Particularly relevant, to the world and to this article, are Smith's insights about markets. Economists since Smith, particularly Friedrich Hayek and Ludwig von Mises, have added their own insights about the power of markets in creating wealth and opportunity. In this article, I will not always identify the particular economist who first had a given insight. Rather, my focus in this article is on the insights themselves no matter what their origin.

Exchange Makes Markets

"Market" is a summary term for all the exchanges that take place in a society. To qualify as an **exchange**, an action must have two characteristics. First, each party gives up something in return for what the other party gives up; second, the switching of items must be voluntary. Therefore, for example, if I sell a copy of my encyclopedia to you for $50, that is an exchange. You and I switch the $50 and the book, and we do so voluntarily. On the other hand, if I grab $50 out of your wallet and leave the book on your doorstep, that is not an exchange because you did not consent.

Because every exchange is part of the market, the market encompasses a wide range of exchanges, from the simple to the highly complex. The $50-for-book exchange is a simple one. A complicated exchange would be the "stripping" from a bond of its stream of interest payments. Stripping the bond divides one asset into two: the stream of interest payments and the principal. The sale of either asset is a complex exchange. An even more complex exchange, in which a large fraction of U.S. society has engaged at some point, is the purchase or sale of a house. The sale agreement for a house often specifies what must be repaired or replaced before the sale is complete, and many such contracts also reference covenants, conditions, and restrictions that, by a previous agreement, restrict the use of the property.

Gains from Exchange

One of the most powerful insights from economics is that both parties benefit from an exchange. **Trade** is never an exchange of equal value for equal value. If what I received in exchange equaled the value of what I gave up, I would not bother making the exchange. To make an exchange, I must receive something that I value more than what I give up.

The modern jargon often used to express this insight is that trade is a positive-sum game. Unfortunately, this statement is deficient. That trade is positive sum is necessary but not sufficient for it to benefit both sides. If one party to trade benefited while the other lost, but the first party gained more than the other party lost, the transaction would still be positive sum. However, trade benefits *both* sides, meaning that there is no loser. The statement that both sides gain from exchange is thus stronger than the positive-sum formulation.

When economists say that both parties to an exchange benefit, they are judging benefit by the values of those engaged in the exchange, not by their own values. Therefore, for example, someone who buys and smokes a cigarette gains from the exchange, judged by his values, even though I might think that he is unwise to buy or smoke cigarettes. Economists are also assuming, of course, that one of the traders did not defraud the other. When fraud is present, the conclusion that both sides benefit is no longer certain.

The simple insight that people gain from exchange has far-reaching and often radical implications. It means, for example, that actual cases of exploitation are far less common than is ordinarily believed.

Take an extreme hypothetical case in which Joe is out in the desert, perishing for lack of water, when he comes across Rita's Friendly Oasis. He knows that, if he drinks a quart of water, he will survive and be able to use his American Express card to get on the next flight home. If he gets no water, on the other hand, he will die. Rita's cost of pulling up an extra quart of water is, let us say, 10¢. Rita offers to sell Joe the quart of water — for $50,000. We can predict that, if Joe has the income and wealth of a typical American, he will commit to pay the $50,000, and will take out a loan, if necessary, to do so. Does Joe gain from this exchange? Yes. Does Rita gain? Yes. So the insight that both sides gain from exchange stands up even under extreme circumstances. Is Rita exploiting Joe? If by that term is meant that Rita is taking advantage of Joe's vulnerability by charging a high price, then Rita is exploiting Joe. However, if exploitation means that one side sets the exchange rate so that the other side is worse off, then Rita is not exploiting Joe. Both Joe and Rita are better off. The statement that both sides gain from exchange said nothing about the relative gains. Even if one side gains much and the other a little, both sides gain. In fact, Rita's and Joe's split of the gains from exchange is unequal, but not in the way that one might think: Joe, by preserving his life, gains far more than Rita, whose gain is only $50,000 minus 10¢.

The desert example, while useful for testing the limits of the gains-from-exchange idea, is fictitious. However, in the real world every day there are cases that come close. Take "sweatshops" for example. In many developing countries, people close to starvation work for very low wages, sometimes for wealthy multinational companies. Is this sweatshop work an example of gains from trade? Actually, yes. The multinational company, of course, gains by paying low wages in return for work. However, the workers gain also. Commentators in developed countries, noting the long hours, low pay, and absence of amenities on the job, have assumed that the workers in those jobs are losing in the exchange. In fact, closer investigation has found that factory jobs for low pay by U.S. standards are among the higher-paying and most desirable jobs in those countries. Workers who get factory jobs often move to those jobs from farms, where they were used to working even longer hours, at lower pay, and in more miserable conditions than in their new factory jobs. One apparel worker at a Honduras sweat shop told the *New York Times*, "This is an enormous advance and I give thanks to the maquila [factory] for it. My monthly income is seven times what I made in the countryside, and I've gained 30 pounds since I started working here." Whenever he returns to his home town, this worker explained, his friends and relatives want him to help them find jobs. Another wealthier Honduran noted that the good factory jobs were making it difficult to find a nanny or a maid. Sweatshops, in short, are a path from poverty to greater wealth.

Many commentators in the richer countries have seemed unable to detach themselves from their own circumstances and mentally put themselves in the situation of workers in poor countries. In denouncing jobs in poor countries in which workers earn less than $1 an hour, commentators seem to be comparing such jobs to their own economic situations, which are clearly better. However, this is simply an irrelevant alternative to workers in sweatshops. These workers don't have the option of moving to United States and earning the $30,000- to $250,000-a-year jobs that the commentators have or even of earning the U.S. minimum wage. The commentators' failure to get outside their own context and understand gains from exchange has had tragic consequences. Oxfam, the British charity, reported that, when factory owners in Bangladesh were pressured to fire child laborers, thousands of the children became prostitutes or starved.

That people gain from exchange also explains why the "drug war" is so difficult for the government to fight. Neither a drug buyer nor a drug seller is likely to report to police the crime committed by the other party to the transaction, because both the buyer and seller of illegal drugs see themselves as gaining from the exchange. Enforcing drug laws, therefore, is very different from enforcing laws against murder, rape, or burglary. In those cases, there is clearly a victim, or a victim's friend or relative, who objects to the crime and therefore has an incentive to report the crime to the police. However, when illegal drugs are bought or sold, there is, from the viewpoint of the participants, no victim. That is why, to enforce drug laws, the government has put in place an intrusive surveillance system at airports, has required that people paying $10,000 or more in cash fill out a special federal form that flags their case to federal law enforcement officials, has seized property that police suspect has been used or earned in the sale of drugs, and has carved out an exemption to the U.S. Constitution's prohibition on illegal search. The simple principle of gains from exchange implies that a government serious about preventing exchange would have to take extreme measures to do so, whether it is trying to restrict trade in cocaine or Coca-Cola.

Markets Facilitate the Division of Labor and Knowledge

No one knows how to make a pencil. This statement is literally true. No one single person has all the know-how to create all the elements that go into a pencil and to put those elements together. In his justly famous essay, "I, Pencil", Leonard Read told the dramatic story of how an international **division of labor** has created the simple pencil. The wood of the famed pencil was from northern California, the graphite was mined in Ceylon (then Sri Lanka), and the eraser was made by reacting rape seed oil from the Dutch East Indies with sulfur chloride. Each of these items, in turn, was produced by a complex division of labor, and no one person in the world understands all these complex production processes.

Markets facilitate the division of labor, as Adam Smith understood. The person trying to decide whether to become a butcher, baker, or brewer makes some estimate of his own productivity in each occupation, checks out the prices he can charge for meat, bread, and beer, and the amounts he must pay for materials and other factors of production, and then uses this information to decide which occupation to enter. Actually, the person trying to decide among occupations need not know much about prices of inputs or outputs in various industries: all he or she need do is check the wages that are offered in each. That simplifying step itself is made possible by the division of labor: some people in society have chosen to specialize as employers, making the employee's decision even easier. The division of labor allows each person to specialize in doing what he or she does best. The division of labor, noted Adam Smith, is limited by the extent of the market. The larger the market, the more specialized people can become. This specialization enhances productivity.

Markets allow people to coordinate their plans. If, for example, there is a sudden increase in the number of people who want to become butchers, the wages that a given butcher can earn will fall, pushing out those butchers who, at the lower wage they can now earn, do not wish to be butchers. Although recent news stories on layoffs in the *New York Times* and elsewhere have lamented this process, it is a necessary and even healthy aspect of markets. Markets let workers choose whether to accept lower wages or to find other work. The alternative to markets is government coercion. In the Soviet Union and China, where markets were ruthlessly suppressed for decades, workers were not free to choose. They were, in essence, long-term slaves.

Of course, people do not have to specialize in the occupation that pays them the highest financial returns. People can and do make tradeoffs between money and other aspects of jobs, including the pleasure they get from particular jobs, the hours they work, the people they work with, and the places they work. The division of labor helps them make such choices.

As Nobel laureate Friedrich Hayek emphasized in much of his writing, one of the underappreciated and least understood aspects of the market, even among economists, is its facilitation of the division of knowledge. In elaborating the important role of information in economies, Hayek, following his mentor, Ludwig von Mises, put the final intellectual nail in the coffin of socialism. Hayek showed that the standard formulation of the problem of socialist planning assumed the answer before the problem was solved. Hayek conceded socialist economist Oskar Lange's point that, if the central planner had all the relevant information about the economy, he could plan the economy efficiently. However, the problem in any complex economy, wrote Hayek, is that "the relevant knowledge" does not exist in any one mind or small group of minds. Rather it exists in the millions of minds of the millions of people who participate in the economy. Therefore, no central planner, no matter how brilliant and informed, could have the information that a government would need to run an economy efficiently. Well-known socialist economist Robert Heilbroner admits, about 50 years after Ludwig von Mises and Friedrich Hayek made the point about information, that von Mises and Hayek were right. "Socialism," writes Heilbroner, "has been the tragic failure of the twentieth century." The only institution that can solve the problem of knowledge, that is, the problem of how to let people's knowledge be used, is the market. The market leaves people free to make their own decisions and to bet their own resources on those decisions.

Somerset Maugham's story, "The Verger", illustrates beautifully how markets allow people to use their information. In that story, the verger of a church quits his job rather than learn how to read and write. The day he decides to quit, he wanders around London, pondering his predicament and his future. He starts longing for a cigarette and notices that he can't find a tobacco store. A fellow could make a lot of money setting up a tobacco and candy store in this area, he tells himself. That is what he proceeds to do. Later, he uses the walking-around method to choose where to build additional stores. The verger has information that is specific to him and he acts on it. No government planner, even in a society whose government was not hostile to smoking, would have had the knowledge about where to place a tobacco store. Only in a free market could someone be free to use that information and to bet his own capital on it.

Although "The Verger" is fiction, there are literally millions of examples of people in the real world betting their capital on their specific information. The business magazine, *Forbes*, in fact, regularly carries articles about someone who had an idea, invested in the idea, and made a few million dollars on it. Some investors, for example, had noticed the increase in the number of working wives and speculated that these wives would want for dinner something other than McDonald's fare or other fast foods. The fast-food chains were not catering to this segment, nor were the supermarkets. Thus began Boston Market, whose specialty is "home-meal replacement". Boston Market is a chain that sells wholesome, home-style meals with plenty of home-style side dishes. The typical mean of chicken feeds five and sells, in 1997, for about $18. Another *Forbes* story tells of Gregory Brophy, who, noticing that many large firms were outsourcing some of their more mundane tasks, decided to buy a 2000-pound, industrial-grade paper-shredding machine. His new idea was to collect and shred each customer's unwanted paper on the customer's premises. Until Brophy had come along with his business, no one had done that. In 1996, Shred-It, Brophy's company, earned about $800,000 after taxes.

Markets Happen

Markets are one of the most natural and spontaneous things in the human world. Start with a few people and no regulations preventing exchange, and, unless one group decides to attack the other, they will likely end up trading. In a fascinating article published just after the end of World War II, R. A. Radford described the intricacies of the markets that developed among British and American occupants of German prisoner-of-war camps. The currency, which evolved naturally without anyone dictating it, was cigarettes. Just as in larger markets outside the prison camps, the prices of goods fluctuated according to their relative supplies.

Indeed, the development of a market in a Japanese P.O.W. camp was a major part of the plot of James Clavell's novel, *King Rat*. Corporal King, an American in a prison camp full of Brits and Australians, was an entrepreneur who made money by arbitraging against price differentials within the camp. King also began a division of labor, hiring people, often superior to him in rank, to do various forms of labor for him. Although King's actions generated an enormous amount of hatred and envy, his effect was to enhance not only his own well-being, but also the well-being of everyone who traded with him. Clavell was a careful enough novelist to bring out this fact, even though the reader senses at times that Clavell was ambivalent about King Rat's morality.

Markets and Motives

Self-interest is the main human characteristic that causes people to enter markets whether as buyer or seller. Adam Smith put it best when he wrote, "It is not from the benevolence of the butcher, the brewer, or the baker, that we can expect our dinner, but from their regard to their own interest." Far from lamenting that fact, Smith celebrated it. Life would be tough, argued Smith, if our "affections, which, by the very nature of our being, ought frequently to influence our conduct, could upon no occasion appear virtuous, or deserve esteem and commendation from anybody." Translation: how can you attack self-love when it is an inherent part of being human and, indeed, a tool for human survival? Samuel Johnson, the author of the first dictionary, put it more succinctly when he wrote, "Man is never so innocently employed as when he is honestly making money."

Is there an inherent contradiction between pursuing one's interest in markets and caring about others? Not at all. Indeed, the connection is the opposite. Throughout history, the development of free markets has led to increases in wealth and income at all levels of society. This is what happened in the United States over the last 200 years, in Hong Kong in the last 40 years, and in many other parts of the developing world. This increase in wealth has led to a tremendous outpouring of generosity. Many of the major charities in the United States, such as the Red Cross, began during the 19th century, when government intervention was almost nonexistent and wealth was increasing. In economists' jargon, charity is a normal good; as our wealth increases, we give more to charity.

Moreover, consider the alternative. The only way to ban markets is to beat them down with force. Since markets are abstractions, the force is used against people. Therefore, the alternative to a market-oriented society in which everyone is required to respect everyone else's rights is a society in which those in power use force on whomever they can get away with using it on. Thus the irony to the charge that a free-market society follows the law of the jungle. The reality is that a society based on force — the antithesis of the market — is the real jungle.

Defining Terms

Division of labor: The dividing up of production into various component tasks, each of which is done by some person or firm that specializes in that and at most a few other tasks. The term can be used to describe tasks within a firm or within the whole economy.

Exchange: An action in which each of two parties voluntarily gives up something in return for what the other party gives up.

Normal good: A good that people buy more of as their wealth or income increases.

Trade: The same as exchange.

References

Clavell, J. *King Rat*, Dell, New York, 1962.

Hayek, F. *Individualism and Economic Order*, University of Chicago Press, Chicago, 1948.

Heilbroner, R. Socialism, In *The Fortune Encyclopedia of Economics*, D. R. Henderson, ed., pp. 161–165, Warner Books, New York, 1993.

Kroll, L. Fear of failing, *Forbes*, March 24, 1997.

Maugham, W. S. The Verger.

Radford, R. A. The Economic organisation of a P.O.W. camp, *Economica*, 12(4): 189–201, 1945.

Read, L. I, Pencil, *The Freeman*, reprinted in *Imprimis*, 21(6): 1992.

Schifrin, M. and Upbin, B. Crab Rangoon to go, *Forbes*, March 24, 1997.

Smith, A. *An Inquiry into the Nature and Causes of the Wealth of Nations*, E. Cannan, ed., University of Chicago Press, Chicago, 1776.

Further Information

Rothbard, M. N. Free market, In *The Fortune Encyclopedia of Economics*, D. R. Henderson, ed., pp. 636–639, Warner Books, New York, 1993.

Rothbard, M. N. *Power and Market: Government and the Economy*, 2nd ed., 1977.

Shane, S. *Dismantling Utopia: How Information Ended the Soviet Union*, Ivan R. Dee, Chicago, 1994.

Sowell, T. *Knowledge and Decisions*, Basic Books, New York, 1980.

4.2   Consumers

Sharon M. Oster

In developing new products, organizations typically confront considerable technical uncertainty. Once technical design hurdles have been met, however, new uncertainties emerge as firms face the test of the marketplace. If the product is produced, will anyone want to buy it, and how much will they be willing to pay for it? It is understanding these market uncertainties that forms the focus of this section.

The section begins with a general discussion of the economic model of the consumer. We outline the basic principles that govern consumer demand, with a focus on the relationship between prices and quantity demanded and on the role of substitute products in constraining the prices that firms can charge for new products. More advanced material on consumer demand in more high-technology products is also discussed.

2.1   Basic Principles of Demand

The central device that economists use to summarize consumer preferences is the **demand function**. A demand function is an equation relating the quantity demanded of a product to a set of characteristics of the product itself, the potential consumers of that product, and the overall market. In general, the quantity demanded of a product depends on the price of that product, the prices of its closest substitutes, and the income and tastes of consumers in the market. A demand function that represents the behavior of a particular consumer is known as an *individual demand function;* demand functions that represent an aggregation of all consumers in the market are known as *market demand functions.* In this Section, we focus our analysis on market demand.

In analyzing demand, economists often focus on the relationship between the quantity demanded of a good and its own price. For a given product, how many units will consumers buy at a given price? The answer to this, obviously, very important question is captured in a **demand curve**, which is a graphical representation of the relationship between the price of a good and the quantity demanded of that good, holding all other determinants of demand constant. A representative demand curve is given in Fig. 4.1 below.

In the figure, the demand curve slopes down. The downward slope of the demand curve is one of the fundamental

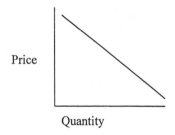

FIGURE 4.1   A typical demand curve.

theorems of economics: as the price of a good falls, the quantity demanded of that good rises. The demand response comes from two sources: first, as price falls, the number of consumers who wish to buy that good typically rises; second, the amount that each of them typically buys also goes up as the price falls. In pricing a product, the manager faces a trade-off: the lower the price, the less profit on any given transaction, since profit is the difference between price and costs, although there will be more transactions. In some cases, the increased volume will more than offset the loss in per-unit profit; in other cases not. The optimal price clearly depends on the shape of the demand curve, in particular, how rapidly the quantity increases with a price decrease or how fast the demand falls as price is raised.

The index that is used to summarize the responsiveness of quantity demanded to price as revealed by the demand curve is the **demand elasticity**. The demand elasticity is the percentage change in quantity demanded divided by the percentage change in the price of the good, all else equal. The demand elasticity is typically expressed as an absolute value. For example, an elasticity of 2 indicates that, for a 10% increase in price, the quantity demanded would fall by 20%; an elasticity of .5 similarly indicates that the same price increase would elicit a quantity fall of only 5%. A high elasticity indicates that the good is highly sensitive to price changes; a low elasticity indicates little responsiveness. On any given demand curve, the elasticity typically varies.

There is an important relationship between demand elasticity and revenues. Suppose we are on a segment of the demand curve that is inelastic, that is, the absolute value of elasticity is less than one. Here demand responds little to price changes. Thus, a price decrease will generate little new demand, and the firm will find that total revenues decline with price decreases. At the same time, a price increase will do little to choke off demand, and thus, in the inelastic portion, revenues are increased with a price increase. In the elastic portion of the demand curve, the opposite effects hold. Revenues are increased for price decreases and decreased for price increases.

As we can see, elasticity is a very important tool for understanding pricing, for it is a measure of the extent to which customers move in and out of the market when prices change. One of the most important determinants of elasticity is the number of substitutes in a market. The more good substitutes there are for a product, the more elastic in general will be the demand curve for that product. For example, the demand curve for a given type of ball-point pen is likely to be considerably more elastic than a demand curve for Microsoft's operating system, given the differences in the number of viable substitutes in the two markets. In this way, the existence of substitutes constrains a firm's ability to raise prices, for the resulting elastic demand tells us that a price increase will cause consumers to move into other markets. For truly revolutionary products, initial demand is often relatively inelastic, as the number of substitutes in the early part of the product's life cycle is generally small. Under these conditions, it is often possible to charge a relatively high price without sacrificing much demand. As the product ages, however, as similar products begin to appear in the marketplace, we expect the demand to become more elastic and the optimal price to go down. In most markets, even if there is little cost-reducing technical progress, prices fall as the market matures for the reasons just outlined.

In most markets, prices are set in a market mechanism that leads to the same prices being charged to everyone. In the typical grocery store, for example, the grocer cannot charge different prices to different customers for an identical box of cereal. In some markets, however, transactions are such that some **price discrimination** among customers is possible. Price discrimination occurs whenever exactly equivalent products are sold to different people at different prices. In using price discrimination, firms try to identify those customers located high up on the demand curve in Fig. 4.1 and charge them a high price, while charging lower prices to the customers on the low end of the curve. In general, price discrimination leads to revenue increases, sometimes quite substantial ones. Indeed, revenue would be maximized if I could charge each customer his or her *reservation* price and in this way expropriate the full value represented under the demand curve.

The key requirement for practicing price discrimination is the ability to separate markets. Absent such an ability, resale of the product among customers will thwart any attempts at price discrimination. For this reason, price discrimination is most common in service markets; there is no way I can resell you my dental visit or my college education, even if the prices I was charged are relatively low. Price discrimination also occurs in markets in which there is one-on-one bidding, particularly if there is product tailoring for the customer that makes product comparisons more difficult. In the market for data storage systems for mainframe computers, for example, a market dominated by EMC and IBM, all sales are negotiated and prices vary considerably among customers.

2.2   Pricing in More Complex Markets

While some new products stand on their own, others gain value only insofar as they are part of a broader *network*. Telephone systems are the classic example here. For any individual, the benefit of having a telephone increases with the number of phones in service. Economists refer to this phenomenon as a **network externality**. [Katz and Shapiro, 1985]. Network externalities exist whenever new users joining the network increase the value of that network to other users. A more recent example of a network externality is the automated teller machine (ATM) in the retail banking industry. For depositors, the value of an ATM system depends on the number of locations the network includes. [Saloner and Shepherd, 1995]. In these markets, firms that can produce quickly, in large volumes, and price relatively low are at an advantage since they can build a network more rapidly. In high-technology markets, network externalities are relatively common.

In some cases, network externalities can arise indirectly, through subsidiary markets. Consider, for example, a product that is part of as system. For example, a computer can be thought as a part of a system encompassing hardware and software. A music system is comprised of a compact disc (CD) player and compact discs. In these examples, the value of the machine depends critically on the supply of complementary products to use on the machine. Particularly when there are economies of scale in the production of the complementary good, as the network grows, more of the complementary product can be supplied at a lower price. For example, as the number of CD players increased, it became more profitable to produce CD's. This, in turn, increased the value of the CD player. The same effect is observed in the VCR market. The value of one product depends on the availability of a second product, and the production of the second product is stimulated by growth in the first.

Markets in which there are network externalities pose special challenges for the manager. The central concern in these markets is to build a network. However, because the value of the product depends on how many other systems are in place, consumers may be reluctant to be early adopters. Consumer inertia is especially likely if there are competing systems in the early market, as, for example, in the case of the early developments of the video recorder. Innovators in these markets face the challenge of creating a market.

There is a variety of pricing and product strategies used in these circumstances. First, it may be necessary to price the product very low initially in order to stimulate demand to begin to build the network. Licensing other producers to make the product may be necessary as a way to get the product out; here firms will earn their revenues through licensing fees. In some cases, it may be important to push for product standardization to ensure that the network has compatible parts. We see standardization in markets such as keyboards (the QWERTY keyboard), video recorders, operating systems, telephones, and numerous other areas. Some firms have found it necessary to enter multiple parts of the supply chain in order to build a network, for example, producing both CD players and CDs as a way of insuring demand for the product. The central economic fact in these markets is that the demand curve representing consumer value depends on the overall state of the networks, and thus each firm in the system has a stake in what is going on elsewhere in the system. Coordination problems are common in network markets.

Increasingly, in a range of different markets, firms have begun to sell their products in more complex combinations in a process known as **bundling**. In bundling, firms package together two or more products for sale. For example, theaters often sell series of concert tickets; airlines bundle tickets with hotel reservations; computers are generally sold with a variety of software. All of these are bundled products.

The advantages of bundling are several. When products are connected, bundling can help reap some of the rewards of price discrimination [Adams and Yellin, 1976]. In particular, bundling helps to meter demand and to distinguish consumers by their product valuation. The classic example here comes from the period in which computers used cards to feed in information. At that time, IBM sold both the mainframe computers and the punch cards; indeed, users were prohibited from using non-IBM cards on IBM machines. Now consider the pricing of the machines vs. the cards. Customers who most value the system are those who use it the most; high users are thus those customers on the upper end of the demand curve, with the highest reservation prices. One way to identify these users is by their card usage; here cards act as a metering device. Once we have identified these high-use/high-value customers, we cannot charge them higher prices for the machines. However, by changing high prices for the cards, while keeping machine prices closer to costs, we can effectively price discriminate against the high-use/high-value customers. In this way higher profits could be extracted from the high-use customers who are forced to pay for a large volume of relatively expensive cards than from the lower-use customers.

Metering product use in order to facilitate price discrimination is one reason bundling may be pursued. There are others as well. In the last few years, there has been a considerable increase in the use of bundling by firms in marketing their products. In some instances, the bundling cuts across company lines. An airline offer of frequent flier mileage as a reward for staying in particular hotels is an example of cross-company bundling. In general, bundling can be seen as a way of tailoring products more to individual consumer tastes and in this way increasing the profitability of serving particular customers.

Defining Terms

Bundling: Tying the sales of one product to a second product or offering combinations of the two products at a discount from the full individual price of the combination.

Demand curve: A graph illustrating how much consumers will buy of a good at a set of prices.

Demand elasticity: The percentage change in quantity demanded divided by the percentage change in price; a measure of market responsiveness.

Demand function: An equation relating the quantity demanded of a good to a series of explanatory variables, most prominently the price of the good.

Network externalities: Goods are more valuable to a user the more users adopt the same good or compatible ones.

Price discrimination: Charging different prices to different customers for identical products.

Reservation price: The maximum price a consumer is willing to pay for a good rather than do without it.

References

Adams, W. J. and Yellin, J. Commodity bundling and the burden of monopoly, *Q. J. Econ.*, 475–498, 1976,

Katz, M. and Shapiro, C. Network externalities, competition and compatibility," Am. *Econ. Rev.*, June: 424–440, 1985.

Saloner, G. and Shepherd, A. Adoption of technologies with network effects: an empirical examination of the adoption of automated teller machines, *Rand J. Econ.*, Autumn: 479–501, 1995.

Further Information

Katz, M. and Shapiro, C. Systems competition and network effects, J. Econ. Perspect., Spring: 93–116, 1994.

Oster, S. Modern Competitive Analysis, 2nd ed., chap. 16, Oxford University Press, 1994.

4.3   Profit

Paul Heyne

The word **profit** is regularly used in everyday speech and writing with different and even contradictory meanings. Since usage determines the meaning of words, there is no single correct definition of profit. The meaning most commonly employed today by economists was developed largely by Frank H. Knight in the 2nd decade of the 20th century as part of his effort to refine the theory of a "free-enterprise" economy [Knight, 1921].

Profit is a residual: what is left over from the undertaking of an activity. The profit from an activity or enterprise is what remains out of total revenue after all costs have been paid. Costs are payments promised by the undertaker of an activity to the owners or effective controllers of whatever productive resources the undertaker deems necessary for the successful completion of that activity. Costs include wages promised to obtain labor services, rent promised for the use of physical facilities, interest promised for the use of borrowed funds, payments promised to purchase raw materials and other inputs, as well as taxes paid to secure the cooperation or noninterference of government agencies. The undertaker of the enterprise is usually known today by the French term **entrepreneur** because we have surrendered the English word to the undertakers of funerals. When entrepreneurs themselves own some of the productive resources employed, the cost of using them is what those resources could have earned in their best alternative opportunity, or what economists call their **opportunity cost**. Thus, what the entrepreneur could have earned by working for someone else is a cost of production, not a part of the entrepreneur's profit.

In the absence of uncertainty, there would be no profits. Suppose it is generally known that a particular activity will certainly generate more revenue than the total cost of undertaking that activity. Then more of that activity will be undertaken. This will either reduce the price obtainable for the product or raise the cost of obtaining the requisite resources or do both, until expected total revenue exactly equals expected total cost. Profit, consequently, cannot exist in the absence of uncertainty. It is a residual that accrues to those who make the appropriate decisions in an uncertain world, either because they are lucky or because they know more than others. Losses accrue to those who make inappropriate decisions.

The extraordinary productivity of free-enterprise economies has evolved as entrepreneurs have sought out and discovered procedures for producing goods for which people are willing to pay more than their cost of production. This evolution has been largely a process of ever finer specialization, both in the productive activities undertaken and in the goods produced.

Critics of the free-enterprise system have generally failed to recognize what an extraordinary quantity of information is required to coordinate a highly specialized economic system and have vastly underestimated the uncertainties that permeate a modern economy. They have viewed profit for the most part as a simple surplus obtained by those who were in the fortunate position of being able to purchase the services of resources, especially labor, at prices less than their value or worth as measured by the revenue obtainable from the sale of the products of those resources. From this perspective, private profit and the private hiring of labor have no useful social function. The state or some other representative of society can simply take over the control of productive activity and apply the ensuing surplus to public rather than private purposes, thus ending exploitation and special privileges. The organization of production is a technical administrative task in a world without uncertainty.

However, in the actual and highly uncertain world, no one knows exactly what combination of productive actitvities will generate the largest "surplus" under current circumstances. What precise mix of goods should be produced and how should they be produced? No one knows! In a free enterprise system, however, entrepreneurs decide. Entrepreneurs are persons who act on their belief that a particular rearrangement or reallocation will extract a larger surplus or profit from available resources and that they themselves will be able to appropriate a significant share of this increased surplus or profit. To implement their beliefs, they must obtain command of the requisite resources. Entrepreneurs do this by promising to meet the terms of those who own the resources they want to use and then claiming for themselves the residual, that is, the profit.

The entrepreneurs' promises must be credible, of course. People will not work for someone else unless they confidently expect to be paid. Owners of physical resources will not let others employ the services of their resources without assurance that they will obtain the promised rent. Lenders of money want guarantees of repayment. Insofar as entrepreneurs cannot provide satisfactory assurances, they either cannot undertake the projects they have in mind or must surrender a measure of control over their contemplated projects. Lenders who have been given adequate security don't care what the borrower does because they know that their terms for cooperating will be met. However, lenders who cannot be certain they will be paid in a timely manner the amounts contracted for by the entrepreneur will want some kind of voice in the business, to "protect their investment", and will want to share in any profit that results, as a reward for accepting the uncertainty associated with their participation.

Thus, entrepreneurs obtain control over the projects they want to undertake, or become "the boss", and claim for themselves (or shoulder by themselves) the entire residual (profit or loss) by providing credible guarantees to all those whose cooperation they require.

How can one provide credible guarantees? The most obvious way is to grant the cooperator a "mortgage" on assets of well-established value. Entrepreneurs who don't themselves own such assets must employ other arts of persuasion and will often be compelled to accept some kind of co-entrepreneurship by agreeing to share both control and the residual with other members of the producing "team".

Marxists and others who wanted to do away with "capitalism" effectively outlawed the free-enterprise system when they obtained political authority by prohibiting the private hiring of labor and private profit. However, they never found an effective alternative way of coordinating productive activity in a world characterized by uncertainty. The central planning organizations they established reduced some uncertainties by curtailing the ability of consumers to influence what would be produced and by removing from the hands of individuals much of the power to decide how resources would be employed. However, this turned out to entail not only a radical reduction in individual liberty but also a drastic crippling of productivity. Central planners simply could not bring together under their oversight and control the information necessary for the coordination of a modern, highly specialized economic system [Hayek, 1945].

No one designed the free-enterprise or profit system *in order to* solve the information problems of modern economies. The system, including the social institutions that provide its framework and support, evolved gradually in response to the initiatives of entrepreneurs of many kinds who, in their efforts to profit from the projects in which they were interested, "were led by an invisible hand to promote an end which was no part of [their] intention" [Smith, 1776].

Defining Terms

Entrepreneur: The person who acquires control of a project by assuming responsibility for the project through guaranteeing to meet the terms of all those who own the resources required to complete the project; the **residual claimant**.

Opportunity cost: The value of all opportunities forgone in order to complete a project.

Profit: Total revenue minus total opportunity cost.

Residual claimant: The person who is entitled to everything that will be left over, whether positive (profit) or negative (loss), after a project has been completed and who thereby acquires the incentive to take account of everything that might affect the project's success; the entrepreneur.

References

Hayek, F. A. The use of knowledge in society, *Am. Econ. Rev.* 35(4): 519–530, 1945.

Knight, F. H. *Risk, Uncertainty and Profit*, Harper and Row, New York, 1921.

Smith, A. *An Inquiry into the Nature and Causes of the Wealth of Nations*, Liberty Classics, Indianapolis, IN, 1776.

4.4   Marginal Analysis in Economics

Richard B. McKenzie

Marginal analysis addresses the issue of how much should be done of anything by comparing the additional (or **marginal**) **cost** (MC) with the additional (or marginal) benefits. An important implication of marginal analysis is that any activity — production of computers or the painting of pictures — should be extended so long as the **marginal value** of an additional unit exceeds its MC. The expansion process should cease when the marginal value of the last unit equals the unit's MC.

James Buchanan, 1986 Nobel Laureate in Economics, aptly captured a central theme of the way economists approach analytical topics when he mused,

> The economists' stock-in-trade — their tools — lies in their ability and proclivity to think about all questions in terms of alternatives. The truth judgment of the moralist, which says that something is either wholly right or wholly wrong, is foreign to them. The win-lose, yes-no discussion of politics is not within their purview. They do not recognize the either-or, the all-or-nothing, situations as their own. Theirs is not the world of the mutually exclusive. Instead, it is the world of adjustment, of coordinated conflict, of mutual gain [1996].

No doubt, Buchanan exaggerated the extent to which economists shy away from "truth judgments" (given their strong, and often conflicting, opinions on policies) and avoid discussions of, for example, "all-or-nothing" deals, given their importance to the economics of politics and public choice literature to which Buchanan himself has contributed enormously. Nonetheless, Buchanan's comments still capture the profession's *proclivity* to consider all the possible adjustments people can and must make *on the margin* of any course of action.

Marginalism is a study of decision making, when any activity is either increased or decreased in any way, for example, in volume, scope, or quality.

Economics as a mode of discourse starts with the proposition that people are capable of purposeful action, which generally is taken to mean that people can imagine improved states for themselves and others and can work to make their stations in life better, as they define "better". The fact that people have alternative ways of improving their circumstances implies that there is always some cost of doing anything, whether building a bridge or reading a book. If any course of action, A, is taken, then some other course, B, is not taken. The value of B is what is garnered, i.e., the cost, of doing A.

Given that they cannot do everything, purposefully acting people will tend to do the best they can, which necessarily means maximizing their well-being, given the constraints of their physical surroundings and their evaluations of alternative courses of action. Maximizing people will do those things for which the value of the course taken exceeds the value of the courses not taken, and they will tend to shun those things for which the costs are greater than their value.

Granted, few things in life are certain, and the costs and benefits of actions are often realized over the course of time, for example, months and years. However, the conclusion remains essentially the same with the relevant values and costs being those that are *expected* (or those calculated using appropriate discount rates). The more risky or delayed the consequences of actions, the lower the present value of the discounted costs and benefits.

In more concrete terms, this means that, before an action A will be taken by a purposefully acting person, the *expected* value of A must exceed its *expected* cost (which is the *expected* value of what is not done, B). If the expected cost of doing A (which is the expected value of B) were greater than the expected value of A, then welfare would not be maximized; the acting person could expect to gain more by doing B instead of A. A book should be read only if its benefits exceed the cost of the time spent reading it; the same rule applies for the construction of bridges or, for that matter, most everything else.

Economics has long been called the "dismal science" because it emerged as a recognizable discipline in the early 18th century, when Thomas Robert Malthus was writing about how the world's population growth would ultimately be checked not by restraint on procreation but by war and pestilence. More recently, it has retained the dismal science tag because of the extent to which economists point out the hidden cost of supposedly "free" goods. Public education has never been free. The costs have simply been imposed on taxpayers, not consumers of the educational services. Similarly, when urban bridges are built with federal funds, the tax bill on the city citizenry may be zero. However, the bridge still has a cost, measured in terms of the public and private goods that could have been produced and supplied instead. People may eat out simply because they *want* to do so; but they may also eat out because restaurant meals are less costly when the price of the restaurant meal is compared with the total cost of the home-cooked meal, including the value of the cook's time.

Pollution is often treated in economics as a "market failure" because production costs are incurred by the general population in the form of a degraded environment or public and private clean-up efforts. Such "externalized" costs are not incurred by the firms that produce the good or by the consumers who buy the good that is produced with resulting pollution, thus implying that the value of the good might not cover *all* costs. Similarly, monopolies are viewed as a form of "market failure" because of the curbs they impose on output, given their goal to raise prices and profits. Monopolies are considered "inefficient", but only to the extent that the value of some unspecified number of the monopolized good not produced would, were they produced, be greater than their production costs. (Another way of saying the same thing is that the value of what *could* be produced by the monopolist is greater than what is produced, given that the **monopoly** curb on output releases resources to produce other goods.)

Marginalism is important in economics because it is a means of addressing an obvious question maximizing people and firms must inevitably face: How much of anything should be done? How much of any given book or how many books should be read in a given time period? How strong should a bridge be? How many stories should a building have? How fast should a computer run? How long should a meeting last? How well written should a chapter be?

The answers to those questions emerge not from consideration of *average* values and **average costs** (ACs), figures readily available from accounting records, but from a comparison of the *marginal* values and MC of each additional unit (i.e., their additional values and additional costs). Indeed, as will be shown, average calculations can be misleading.

Economists generally assume that, if the marginal values of successive units are not initially declining, they will decline eventually (or beyond the point of so-called "diminishing **marginal utility**", MU). Additional units of any good simply are not wanted with the same intensity. As more of any good is bought and consumed, the additional units are progressively used in changing proportion to other things consumed. Declining marginal value leads to the concept of **demand** (or **law of demand**), which predicts that people will buy more of any good when the price falls (and all other factors remain constant). A maximizing person will not only spend all of his or her income on selected goods and services but will also "equate at the margin". This means that the maximizing person will try to get the same satisfaction from the last penny (or dime or dollar) spent on the last unit of each good. More technically, it means the person will seek equality between the ratios of the MU to the prices of the goods consumed, that is, with just two goods, A and B, the maximizing person will distribute his or her expenditures so that

$$MUa/Pa = MUb/Pb$$

If that were not the case — if $MUa/Pa > MUb/Pb$ — then the person could consume less of B and more of A, gaining more utility. It follows that, if the maximizing person has established equality and the price of A falls, more of A will be bought (because then $MUa/Pa > MUb/Pb$). Hence, the law of demand follows: if the price of a good falls (*ceteris paribus*), more will be bought. What is significant for this discussion is that the law of demand is deduction of behavior-grounded marginal analysis.

Economists also assume that as production of anything is expanded, the MC of successive units will eventually increase (beyond the point of diminishing MUs). One explanation for rising MUs is physical: diminishing returns are inevitable. However, MC can also be expected to rise for another reason that is grounded in the behavior of maximizing people: in order to produce successive units, progressively more valuable units of other things will be given up, which necessarily implies increasing MC for the good in question.

The problem of how much should be produced can be decided conceptually with reference to the standard downward sloping demand curve (D in Fig. 4.2), which captures the declining marginal value of successive units of X, and the upward sloping supply curve (S), which captures the assumption of increasing MC of X. Given these constraints, quantity X1 should be produced to maximize the gain from X. At a lower output level, e.g., X2, the marginal value, MV1, is greater than the marginal cost, MC1. There is a gain to be had from producing the next unit of X equal to the difference, or MV1-MC1. There is a similar gap between the marginal values and MC for all other units between X2 and X1. The gain to be had from expanding production from X2 to X1 is the triangular area bounded by MC1MV1E. Production should not be extended beyond X1 because the MC of each unit, represented by the supply curve, would exceed the marginal value, represented by the demand curve, that is, in that region below the supply curve.

How many books should a person read? The general answer is that books should be read until the marginal value of the last unit just equals its MC. How strong should a bridge be? Its strength should be increased so long as the additional value of making the bridge sturdier (measured, for example, in the value of the time saved from increasing the number of cars that can cross during a given period) exceeds the additional cost (including the value of what could have been done with the labor and materials

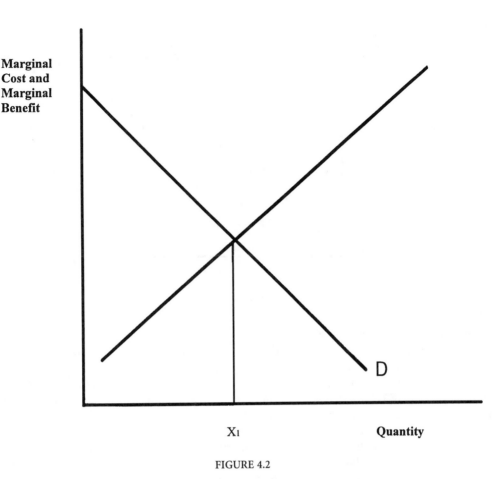

FIGURE 4.2

instead). Floors should be added to a building so long as the additional rental payments over the course of the building's life from additional floors exceed their additional cost.

Marginal analysis can help explain why bridges and buildings sometimes collapse: the cost of adding greater durability was, at the time the construction decision was made, greater than the expected value. More generally, marginal analysis helps explain why the old adage "Anything worth doing is always worth doing well" is so often violated. The value of doing something well, beyond some point (E in Fig. 4.2), is simply not worth it; the cost of extending how well something is done is greater than the value of the extension.[2]

Not considering marginal values and cost, in the case of business, can lead to serious errors, less-than-maximum profits and deflated stock prices for the firms. Consider Fig. 4.3. It depicts the typical structure of a firm's likely AC and MC over a wide range of output levels identified on the horizontal axis. The figure describes a typical production environment in which economies of scale (identified by the falling AC curve) are initially encountered, only to be followed by diseconomies of scale (identified by the rising AC curve). The MC curve must then initially fall and then rise (given that the two curves are mathematically tied together). The MC curve must intersect the AC curve at the low-point of the AC curve (given that the AC curve must fall when the MC is below the average and that the AC curve must rise when MC exceeds the AC).

---

[2]For more discussions of how marginalism has been applied by economists to a wide array of human endeavors, see McKenzie and Tullock [1994].

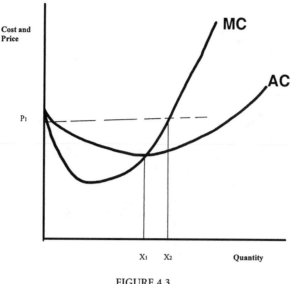

FIGURE 4.3

If the firm can sell as much as it wishes at P1, how much should be produced by a profit-maximizing firm? Business people, who are used to working with readily available AC figures and who are used to thinking about minimizing AC, usually point to Q1 as the profit-maximizing output level. Q1 is the output level at which the AC curve is at a minimum. No doubt, firms should do what they can to minimize the location of the AC curve in the figure. However, that mode of thinking does not extend to choosing the minimum on the AC curve, given its location. If output were limited to Q1, profitable opportunities would be missed. This is because, for each unit of output between Q1 and Q2, the price (or value of the unit to the firm) is greater than the additional cost, represented by the MC curve, that is, the firm can increase its profits by extending output to Q2, even if that means that the AC of production is raised. Of course, if output is extended to Q2, the firm's stock price will also rise.[2]

[2] For an expanded explanation of this cost model, see Browning [1996].

Again, this line of analysis does not preclude firms from seeking to push the whole AC curve as far down as possible. Indeed, a firm in a competitive market has strong motivation to seek the lowest possible AC curve. Otherwise, its competitors can lower their AC curves, expand output, and force the market price for the good down, leaving firms with unnecessarily high AC curves to contract or fold. In seeking to minimize the AC curve, firms should be alert to utilizing marginal thinking. Firm employees are not always motivated to minimize firm costs for the benefit of the firm owners. Accordingly, firms often must tie pay to employee performance. How strong should the tie be? A marginal perspective is helpful in addressing that question. The tie between performance and reward should be increased so long as the reduction in non–pay costs resulting from the greater incentive is greater than the added worker pay. Similarly, it follows that managerial oversight of workers should be extended so long as the added managerial costs are lower than the cost savings from added production.

Marginal analysis has been used to explain many forms of behavior. Pollution may be the product of "socially irresponsible" behavior. People just don't care about the damage they inflict by the trash they discard. Then again, they might pollute because they reason quite rationally that *on the margin* their trash doesn't materially affect the environment. If they withhold discarding their trash, the environment will not be significantly improved, given that everyone else may not be motivated to withhold their discarding of trash. Similarly, close to half of the electorate does not vote in presidential elections. The nonvoters might not vote because they do not care about the outcome of the election. However, they may also decide not to vote because their vote does not *on the margin* count for much in terms of changing the outcome of the election. Precautions to prevent accidents and crimes are often limited, leading to more accidents and crimes than could have been prevented. Nonetheless, extending the prevention could cause costs to rise by more than the value of the accident and crime prevention.

While instructive and useful, marginal analysis has its limitations. In most lines of analysis, marginalism assumes a given structure of benefits and constraints. For example, the legal structure of the firm can have a significant impact on how costs and benefits change with the expansion and contraction of the output level. For that matter, the economic system within which firms operate can powerfully influence the nature of firms. Analysts must be mindful of the fact that structural changes are possible, leading to different outcomes based on marginal analysis that is undertaken within the revised structure of the economic system.

If the marginal value of an addition unit exceeds its MC, it stands to reason that some gain is to be had by the production of the unit. Typically, as production is extended, the marginal value of additional units can be expected to decline while the MC of producing each successive unit will rise. Hence, all gains from production will be realized once production is extended to the point that MC of the last unit produced equals its marginal value.

Defining Terms

Average cost: Total cost divided by the quantity produced.

Demand (or law of demand): The assumed inverse relationship between the price of a good or service and the quantity consumers are willing and able to buy during a given period of time, assuming all other things held constant.

Marginal analysis: The process of determining how much of any activity should be undertaken by comparing the MC of each unit with its marginal value.

Marginal cost: The additional cost of one additional unit of a good or service.

Marginal utility: The additional satisfaction acquired from consuming one additional unit.

Marginal value: The additional benefit of one additional unit of a good or service.

Monopoly: The sole seller of a good or service.

References

Browning, E. K. *Microeconomics Theory and Applications*, HarperCollins, New York, 1996.

Buchanan, J. M. Economics and its scientific neighbors, In *The Structure of Economic Science: Essays on Methodology*, S. R. Krupp, ed., Prentice-Hall, Englewood Cliffs, NJ, 1966.

McKenzie, R. B. and Tullock, G. *The New World of Economics*, McGraw-Hill, New York, 1994.

4.5   Opportunity Cost[3]

Ivan Png

Managers must have accurate information about costs for many important business decisions. For instance, in deciding on the price for a new product, it is necessary to know the cost of the item. To decide whether to outsource the procurement of some input, a business must compare the cost of internal production with external supply. In assessing the performance of a business unit, it is necessary to know its costs as well as revenues. To allocate human, physical, and financial resources, a business must know the cost of the resources.

---

[3]©July 1997, Ivan Png
This chapter draws extensively from Png [1998].

The starting point for an analysis of the costs of a business is the accounting statements. In their conventional form, however, these statements do not always provide the information appropriate for effective business decisions. It is often necessary to look beyond accounting statements.

A key principle that managers must follow in measuring costs is **relevance**: only relevant costs should be considered, and all others should be ignored [Horngren et al., 1994].[2] There is no simple definition of relevant costs. Which costs are relevant depend on the nature of prior commitments and the alternative courses of action for the decision at hand.

Alterantive Courses of Action

The following example shows how an analysis of the alternative courses of action will uncover relevant costs that conventional accounting statements leave out. Suppose that JoyCo is a small hi-tech manufacturer, which was started by a group of scientists from a Fortune 500 company. The company's strength is in R&D. Recently, it spent $1.5 million to develop and patent its first product. JoyCo has successfully brought the product to market.

Table 4.1 presents the most recent income statement for the new product. JoyCo's annual sales revenue is $800,000 and its cash outlays are $500,000; hence, the contribution is $300,000. JoyCo is making a return on investment of 300/1500 = 20%. The new product seems to be doing very well.

TABLE 4.1   Conventional Income Statement

Revenue	$800,000
Cost	$500,000
Profit contribution	$300,000

This assessment of performance, however, overlooks a significant cost of JoyCo's business. To properly evaluate the business, we should investigate the company's alternative courses of action. For many hi-tech businesses whose strength is in R&D, an obvious alternative is to license the technology to larger manufacturers that can exploit **economies of scale** in production and marketing. (A business has economies of scale if the unit cost of production declines with the scale of operations.)

Suppose that Giant, Inc. is willing to pay $400,000 a year for an exclusive license to the technology. If JoyCo licenses the technology, it could receive a royalty of $400,000 a year. Therefore, if JoyCo considers the alternative of licensing the technology, it will see that this is better than continuing to produce the item itself.

The JoyCo example highlights a major deficiency of the conventional income statement, it does not present the revenues and costs of the alternative courses of action. In Table 4.2, we present an expanded income statement that explicitly shows the revenues and costs of JoyCo's two alternatives. It is then very clear that JoyCo should license the technology.

TABLE 4.2   Income Statement Showing Alternatives

	In-House Production	License
Revenue	$800,000	$400,000
Cost	$500,000	$0
Profit contribution	$300,000	$400,000

Opportunity Cost

By continuing in-house production, JoyCo foregoes the opportunity to earn $400,000 a year. The **opportunity cost** of the current course of action is the net revenue that would be generated by the best alternative course of action. In JoyCo's case, the opportunity cost of continuing in-house production is $400,000 a year.

The concept of opportunity cost can be applied to present the revenues and costs of continuing in-house production in another way. This includes opportunity costs among the costs of the business.

Table 4.3 presents a single income statement, in which costs include both the cash outlays as well as the opportunity costs. The cost of $900,000 consists of $500,000 in outlays plus $400,000 in opportunity cost.

TABLE 4.3   Income Statement Reporting
Opportunity Costs

Revenue	$800,000
Cost	$500,000
Opportunity cost	$400,000
Profit contribution	–$100,000

Using the opportunity cost approach, we find that JoyCo is incurring a loss of $100,000 a year; hence, it should license the technology. This approach leads to the same decision as with Table 4.2, which explicitly shows the two alternative courses of action.

Uncovering Relevant Costs

Generally, there are two ways to uncover relevant costs — one is to explicitly consider the alternative courses of action, while the other uses the concept of opportunity cost. When applied correctly, both approaches lead to the same business decision.

In JoyCo's case, there was one alternative to the existing course of action. Where there is more than one alternative, the explicit approach still works well. The opportunity cost approach, however, becomes more complicated: the procedure is to first identify the best of the alternatives and then charge the net revenues from that alternative as an opportunity cost of the existing course of action.

Conventional methods of cost accounting focus on the cash outlays associated with the course of action that management has adopted. Conventional methods ignore costs that are relevant but do not involve cash outlays. Hence, they do not consider the revenues and costs of alternative courses of action.

One reason for these omissions is that alternative courses of action and opportunity costs change with the circumstances and therefore are more difficult to measure and verify. Conventional methods of cost accounting focus on easily verifiable costs. Accordingly, they overlook opportunity costs.

Applications

The economic concept of opportunity cost is widely applied in management. For instance, it has been used in such diverse functions as electricity pricing [Marritz, 1995], environmental protection [Porter and van der Linde, 1995], hospital management [Strong et al., 1995], and financial portfolio management [Ellis, 1997].

The concept of opportunity cost is particularly important for effective management of resources. The introduction showed how the concept can be applied to the management of intellectual property. The following two examples show how the concept can be applied to the management of real estate and financial resources, respectively.

Example 1: Real Estate

Suppose that Giant, Inc. owns a building in the central business district. Giant's warehouse-style grocery superstore occupies 25,000 square feet on the ground floor of the building. The tenants of street-level floors of other nearby buildings include banks, restaurants, and gift shops.

In evaluating the performance of its superstore, Giant must consider the opportunity cost of the ground floor space. This opportunity cost is the net revenue that would be generated by leasing the space to third parties.

Suppose that the market rent for retail space around the Giant Building is $10 per square foot per month. Then the opportunity cost of the premises used by the superstore is $10 × 25,000 = $250,000 per month. A proper evaluation of the performance of the superstore should take account of this opportunity cost.

Conventional accounting statements do not show the opportunity cost of real estate. All real estate has alternative uses; therefore, it has an opportunity cost. A proper evaluation of the performance of a business should take account of the opportunity cost of the real estate.

Example 2: Cost of Capital

Suppose that Giant, Inc. has just sold its property portfolio for $1 billion in cash. Management is now considering how to invest the cash. The general manager of the semiconductor division is proposing to invest $500 million in a new manufacturing facility. He predicts that the facility will yield a profit of $3 million a year. Should Giant make this investment?

To decide on this investment, management must consider the alternative courses of action, which include other investments as well as increasing the cash dividend to shareholders and buying back shares. The proposed semiconductor investment is predicted to yield a return of 3/500 = 6%.

The opportunity cost of investing $500 million in the new semiconductor facility is the yield from the best alternative course of action. A proper evaluation of the proposed investment should take account of this opportunity cost.

Conventional accounting statements do not show the opportunity cost of capital. All funds have alternative uses; therefore, they have an opportunity cost. A proper evaluation of an investment should take account of the opportunity cost of the required funds.

Defining Terms

Economies of scale: The unit cost declines with the scale of operations.
Opportunity cost: Net revenue that would be generated by the best alternative course of action.
Principle of relevance: Only relevant costs should be considered and all others should be ignored.

References

Ellis, C. D. Small slam!, *Fin. Anal. J.*, 53(1): 6–8, 1997.
Horngren, C., Foster, G., and Datar, S. *Cost Accounting: A Managerial Emphasis*, 8th ed., Prentice Hall, Englewood Cliffs, NJ, 1994.
Marritz, R. It's all in the structure, *Elect. J.*, 8(9): 50–60, 1995.
Porter, M. E. and van der Linde, C. Green and competitive: ending the stalemate, *Harv. Bus. Rev.*, 73(5): 120–123, 1995.
Strong, J., Ricker, D., and Popiolek, L. Opportunity costs associated with long decision making, *Hosp. Mater. Mgt. Q.*, 17(1): 7–10, 1995.

Further Information

Horngren, C., Foster, G., and Datar, S. *Cost Accounting: A Managerial Emphasis*, 8th ed., chap. 2, 3, 10, 11, and 14, Prentice Hall, Englewood Cliffs, NJ, 1994.
Png, I. *Managerial Economics*, chap. 7, Blackwell, Malden, MA, 1998.

4.6   Government

David M. Levy

The economic analysis of the government is contested ground if only because economists have long supposed that our analysis has (or ought to have) an impact on policy. Economists have long renounced any superior access to individuals' goals. That aspect of economics that traditionally has been most concerned with government policy, welfare economics, supposes that individuals' goals/preferences can be taken as given. Much of welfare economics has implicitly assumed that economists give advice to a benevolent despot. When the despot was lacking, the device constructed as surrogate was a "social welfare function". Traditionally, the economic analysis of government has been to ask whether the social welfare function is maximized by a voluntary market process or by a coercive government policy. The modern economic analysis of the government has come about as economists recognize just how deeply democratic policy differs from despotic policy.

Three basic questions will be addressed. First, are traditional social welfare functions consistent with democratic politics? Second, there is the question associated with Arrow [1963]: does a democracy have preferences in the same sense that we might suppose a despot would? Third, supposing a democracy, Buchanan proposed that a policy proposal that an economist claimed would increase social well-being could be viewed as a prediction about the political process. If the proposal would in fact increase the well-being of citizens, then the citizens — who can be supposed to be interested in increasing their well-being — will adopt it [Buchanan, 1959].

Social Welfare Functions and Democracy

Social welfare functions come in many varieties; however, the most influential have their basis in reconstructed utilitarianism where the goal of policy is to maximize the total (or average) well-being of the members of society. If the population is fixed, then, of course, the two utilitarian criteria are the same. Perhaps, the single most widely used example of a social welfare function is that of "cost-benefit" analysis. It is testimony to the intuitive power of utilitarianism that cost-benefit analysis proceeds undisturbed by the theoretical economist's qualms about comparing using monetary measures of gains from trade instead of utility measures and making social comparisons of well-being by adding up gains and loss of different people.

Utilitarianism is indeed the hardiest of all methods to generate a social welfare function. The utilitarian slogan popularized by Bentham in the 19th century, to seek the "greatest happiness for the greatest number" — a definition that persists to this day in the *Oxford English Dictionary* — has allowed philosophers to slide between policies that maximize the well-being of the majority — which the "greatest number" suggests — and policies that maximize the well-being of the average — which "greatest happiness" suggests. It was not until late in the 19th century that it was pointed out that the slogan was mathematical nonsense: there are too many "maximizing" operations for the slogan to be well defined. The fact that both a mean and a median seem consistent with the definition ought to suggest that there is something seriously deficient with it.

It is an interesting and important fact that many distributions that interest economists, e.g., the distribution of income, wealth, and scientific citations, are heavy tailed and skew. Consequently, the policies aimed to maximize the well-being of the mean individual and policies aimed to maximize the well-being of the median individual may not be the same. It is necessary to consider this because one of the earliest results in the modern economic theory of government is that, when, in a two-party election, voters confront a single dimension, the median voter prevails [Downs, 1957]. Thus, if the goal of policies is to maximize the average well-being, there is no reason believe such will occur under a democracy. A utilitarianism based on median well-being is rather more in line with at least idealized democratic policies [Levy, 1995].

Fixed and Random Models of Government

The first well-attested debate over models of government occur between the Athenian philosopher Socrates and his disciples on one side and his fellow citizens on the other: is the government to be the province of experts or that of all the citizens? While this debate is understood in the history of political theory as one between aristocrats and democrats, it germane for our purposes because Athenian democratic practice was avowedly random. Many important offices were filled by a formal lottery. It is hardly surprising that nonquantitative histories of the debates gloss over the randomness of Athenian democracy — the requisite probability theory was 2 millennia in the future [Levy, 1989].

One of the most obvious characteristics of a random process is the possibility of policy cycles. A policy cycle has the technical name of an intransitive relation: that is, if *a* is chosen over *b*, and *b* over *c*, a policy cycle will allow *c* to be chosen over *a*! Suppose, the members of the society have transitive preferences. It is obvious that, except in the interesting case of unanimity, random policy selection will exhibit intransitivity. What about democracy by vote?

Democracy: Fixed or Random?

Median voter models present a fixed version of democracy. Given fixed desires of the voters and a fixed number of participants in the voting process, the policy is fixed by the preferences of the majority, the median voter. In one dimension the requirement for median voter model is convex preferences — moderation is preferred to extremes. Convexity is usually an assumption economists are willing to accept. However, without appeal to special preference configurations, the median voter approach does not generalize very well to more than one dimension. A line of argument starting with Arrow in 1951 (and Duncan Black somewhat earlier), continuing in Plott [1967], and culminating in McKelvey [1979] demonstrates that policy outcomes can be controlled by setting the agenda. Even with fixed preferences of the voters, the outcome is dependent upon the path selected. The random aspect is, of course, that, from the point of view of the outside observer, the preferences of the agenda controller is a random variable. The signature of this literature is that outcomes are intransitive even in the presence of transitive preferences of the voters.

The example that Black and Arrow employed to illustrate the issue is as following. Consider three polices: *a, b, c*. There are three voters each with the following preferences:

	Policy Preferences		
Voter 1	*a*	*b*	*c*
Voter 2	*b*	*c*	*a*
Voter 3	*c*	*a*	*b*

If an election is held between policies *a* and *b* then *a* wins 2 to 1: it is preferred by voters 1 and 3. If an election is held between policies *b* and *c* then b wins 2 to 1: it is preferred by voters 1 and 2. However, if an election is held between policies *a* and *c* then *c* wins 2 to 1: it is preferred by voters 2 and 3.

If we define "rationality" as transitivity, then there will be something deeply disturbing about such policy intransitivity. The Black-Arrow results depends upon nonconvex preference. However, the Plott-McKelvey results have convinced the economics profession that such intransitivity will not go away even with convexity when policies are generalized to multiple dimensions. If we view intransitivity as a signature of randomness, then this violation of transitivity tells us that there will be the same sort of randomness in democracy by vote as there is in democracy by lot.

Public Good Provision

The central problem in the modern economic analysis of the government is that posed by fact that what is optimal for individuals considered separately is not optimal for the group considered jointly. The most elegant example is provided by Samuelson's celebrated pure public good: a good that once provided can be used without marginal cost by all in the community [Samuelson, 1954]. Scientific knowledge is, of course, a paradigm of such a pure public good. The question is, "can the good be produced privately?"

The problem can be formulated as a two-agent game in which only one agent's contribution is required to produce the public good. The game has a name, the symmetric prisoner's dilemma. Each individual can contribute or not. The unproblematic elements of the game matrix are those on the diagonal where the agents' actions are the same. By stipulation R > P so both individuals prefer the Contribute cell to the Don't Contribute cell. Thus, considering individuals jointly, Contribute dominates Don't Contribute because (R,R) > (P,P).

Now, the question is whether this outcome is compatible with individual's incentives. Consider what results if the problem good is financed without an individual making the contribution: T> R, the best outcome possible. However, in the case that only one agent contributes, he regards himself as worse off than if he did not contribute at all (P > S). Now, what will the individual do? If the individual supposes that his action will not influence the action of the other individual then the answer is easy: he will not contribute. Don't Contribute dominates Contribute. Suppose that the other individual contributes; then not contributing results in a higher well-being than contributing since T > S. Suppose that the other individual does not contribute, then not contributing results (again) in a higher level of well-being since P > S. It is therefore "rational" for each individual to withhold contributions, and therefore the public good does not get financed by "rational" economic agents.

	Prisoner's Dilemma	
	Contribute	Don't contribute
Contribute	R,R	S,T
Don't contribute	T,S	P,P

One overhasty conclusion from this line of argument is that there is something "wrong" with private market choices — the market has "failed" — and therefore the government ought to produce the public good. The conclusion does not follow from the analysis because it is assumed that the government does not suffer from any such problem. However, consider the problem of voters in a democracy obtaining information about sensible government policy. Change the labels on the game from Contribute to Obtain Information and note that nothing else changes. It is to each individual's interest not to be well informed even though it is the interest of the individuals considered jointly to be so informed. Although this argument was known to Robert Filmer in the 17th century, it was widely discussed as a result of Down's work in 1957. The consequences of the argument of "rational ignorance" is that there will be "government failures" to match "market failures". Much of the "Chicago theory" of regulation in which the regulators are captured by the industry can be seen as a consequence of rational ignorance [Stigler, 1988]. Consequently, the theory of public goods does not provide an easy answer for those seeking policy prescriptions.

The Buchanan Test

Suppose we look at Samuelson's analysis of the provision of the pure public good of knowledge in a democratic context. If Samuelson is correct that knowledge is provided at an inefficient level, then we ought to observe government policy attempting to do something about it. The technical problem is that an optimal provision of a pure public good — no user charges — are something that complicates estimation techniques which are generally predicated on the supposition that there is no such thing as an input free for the taking.

Denoting output by Q, labor input by L, physical capital input by K, the stock of privately financed R&D capital by R, and the stock of government-financed R&D capital by G and then assuming a log-additive production function, the relation between inputs and outputs is

$$Q = L^\alpha \ K^\beta \ R^\gamma \ G^\delta$$

where the exponents are constrained to lie between 0 and 1. Suppose we attempt to estimate the production function and find that δ cannot be distinguished from 0. What do we make of this? One answer is that the government-financed R&D was wasted. Another answer is that we have asked the wrong question. Consider the decision for firms to employ inputs. The critical variable is the value of marginal product — the marginal product

$$\frac{\partial Q}{\partial G} = \delta L^\alpha K^\beta R^\gamma G^{\delta-1}$$

times the price of the product. If G is optimally provided at zero user fee, then G will be employed until the value of the marginal product is zero!

Perhaps a more fruitful method of evaluating the impact of G is to ask what impact an increasing amount of G has on other factors of production. Here the impact of an increasing G on the marginal product of R:

$$\frac{\partial Q^2}{\partial R \partial G} = \gamma \delta L^{\alpha} K^{\beta} R^{\gamma-1} G^{\delta-1}$$

If the supply of R is an increasing function of its marginal product — more R is produced because it is more valuable — then we should look for the impact of an optimally provided G on the responsiveness of R to G [Levy, 1990].

Conclusion

Perhaps the most fruitful modern method of viewing the government is an exchange between people. Government provides services in return for services. This point of view helps us bear in mind that government is not something outside the economy; it is not a benevolent despot eager to ask for our advice.

References

Arrow, K. J. *Social Choice and Individual Value,* 2nd ed., New Haven, CT, 1963.

Buchanan, J. M. Positive economics, welfare economics, and political economy, *J. Law Econ.,* 2: 124–139, 1959.

Downs, A. *The Economic Theory of Democracy,* New York, 1957.

Levy, D. M. The statistical basis of Athenian American constitutional theory, *J. Legal Stud.,* 18: 79–103, 1989.

Levy, D. M. Estimating the impact of government R&D, *Econ. Lett.,* 32: 169–173, 1990.

Levy, D. M. The Partial Spectator in the *Wealth of Nations*: a robust utilitarianism, *Eur. J. Hist. Econ. Thought,* 2: 299–326, 1995.

McKelvey, R. D. General conditions for global intransitivies in formal voting models, *Econometrica,* 47: 1085–1113, 1979.

Plott, C. R. A notion of equilibrium and its possibility under majority rule, *Am. Econ. Rev.,* 57: 787–806, 1967.

Samuelson, P. J. The pure theory of public expenditure, *Rev. Econ.Stat.,* 36: 387–389, 1954.

Stigler, G. J., Ed., *Chicago Studies in Political Economy,* Chicago, 1988.

4.7   Business Cycles

Richard Reimer

Fluctuations in economic activity have been an integral part of the U.S. economy since the very beginning of this country's founding. However, because these fluctuations vary considerably in terms of their length or frequency and the severity of the downturn, most economists no longer use the term business cycles. The word cycles implies a more or less regular rhythm, which is not true for economic fluctuations. In this century recessions have been as short as a quarter and as long as a decade as in the 1930s. Alternatively, in 1980 income dropped during the second quarter, and the subsequent recovery, although very weak, began the next quarter.

It should also be pointed out that some sectors of the economy fluctuate more than others. For instance, during the recession of the mid-1970s, private investment dropped by almost 29% while consumer expenditures dropped by only 0.6%.

The reasons for these changes in expenditures are numerous and may well vary over time.

An Economic Model

In order to analyze economic fluctuations or business cycles, a relatively simple **aggregate demand** and **aggregate supply** model is useful. The concept of aggregate demand is in some ways similar to the demand for a single product, but different in that the quantity variable is real gross domestic product (**GDP**), and the price variable is some measure of the general level of prices such as the **GDP deflator**. The aggregate demand (AD) can be written as follows:

$$AD = AD\ (C,I,G) \tag{1}$$

where C is consumption, I is private investment, and G is government expenditures on goods and services.

Of the three components of aggregate demand, consumption is the largest. To a large degree consumption is a function of disposable income and personal real wealth which may include expected future income. Other factors that may have some effect on consumption include interest rates and expectations about the future as well as other factors. Changes in interest rates are particularly likely to affect purchases of consumer durables, which are frequently purchased on the installment. Thus an increase in interest rates could mean a substantial increase in the monthly payments.

Changes in consumer taxes also influence consumption in that a tax increase will lower disposable income and vice versa.

Private investment as already mentioned fluctuates considerably and is more difficult to analyze. For the purposes of this model, it can be assumed that investment is largely a function of income and interest rates. Increased incomes mean higher sales and probably higher profits, both of which will stimulate firms to increase productive capacity. Thus, the firms will add capital equipment and possibly even increase the size of the plant.

Interest rates are also an important aspect of the investment decision because, as interest rates rise, the cost of borrowing increases. Even if internal funds are used to finance the investment, the opportunity cost of using internal funds needs to be considered.

Because investment projects last a number of years, expectations about future economic conditions, especially interest rates and potential profits, are important. Technological innovations may also stimulate private investment. Sometimes new products or new ways of producing old products will require new equipment.

Changes in business taxes may also impact on the investment decision in that, unless an increase in taxes is entirely passed on to the consumer in the form of higher prices, the increased taxes will decrease profits and the internal funds available to finance the investment project.

Government expenditures include all expenditures by the various levels of government on goods and services. **Transfer payments**, such as interest on government securities and social security benefits, are not included as government expenditures since no income is generated. They are instead classed as negative taxes, so that a decrease in welfare benefits can be treated the same as a tax increase. Government expenditures are determined by government officials at all levels of government and therefore are considered as autonomous to the model.

The aggregate demand function can now be plotted as shown in Fig. 4.4 where GDP is the real gross domestic product (**real GDP**) and P is the GDP price deflator. The position of the curve is determined by the level of the various types of expenditures that in turn are influenced by additional variables. Thus, an increase in business confidence would increase investment shifting of the AD curve to the right. An increase in consumer and business taxes would decrease consumption and investment, thus shifting the AD curve to the left.

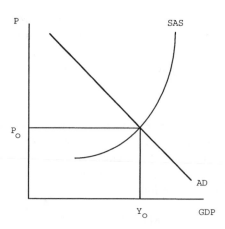

FIGURE 4.4   Short-run equilibrium.

Monetary policy is also assumed to have important effects on the position of the AD curve. An expansionary policy that increases the stock of money would lower interest rates, thus increasing investment and to some extent consumer spending particularly on consumer durables. Consumer expenditures might also be stimulated directly since an increase in the money stock increases real wealth.

The aggregate demand curve has a negative slope for three reasons. (1) A decrease in prices will increase the real **money supply**, which in turn will mean lower interest rates that will stimulate investment and consumer spending. (2) A more direct effect of the increase in the real supply of money is that real wealth goes up, which also stimulates spending. (3) Lower prices also mean that domestic prices relative to foreign prices have dropped, thus causing buyers to substitute domestic goods for foreign goods or decreasing imports. At the same time the relatively lower domestic prices will stimulate exports. This improvement in the balance of trade may be partially or even entirely offset by an appreciation of the domestic currency, thus eliminating the gain from lower relative prices.

Turning now to the short-run aggregate supply curve (SAS), it can be defined as the quantity of goods and services that firms are willing to supply at various price levels. The SAS curve is based on the cost of production. In order to keep the model relatively simple, assume that the cost of production is a function of the amount of materials used times the price of materials plus the wage rate divided by the marginal productivity of labor.

$$SAS = SAS\ (Pm \cdot m,\ w/n) \qquad\qquad (2)$$

where Pm is the price of materials, m the amount of materials used per unit of output, w is the wage is the wage date, and n is the marginal productivity of labor.

The SAS curve is illustrated in Fig. 4.4.

The slope of the curve is dependent on the degree to which resource prices, in this case, the price of materials and wages, adapt to changes in demand. The curve has been drawn relatively flat at low levels of GDP and much steeper at higher levels to indicate that, because unemployment will likely be high at low levels of GDP, wages and prices of materials are not likely to rise as much as when output expands when unemployment is low.

Given this SAS curve, a technological innovation that would increase the marginal productivity of labor or reduce the materials requirement per unit of output would shift the SAS curve down or to the right.

The short-run equilibrium is illustrated in Fig. 4.4 with P_0 and Y_0 being the equilibrium price level and output.

In the short run the equilibrium output and price level are determined by aggregate demand and the short-run aggregate supply. Fluctuations in output may result from various factors on the demand side such as changes in investment or consumption. Government policy (fiscal policy) may also influence aggregate demand through changes in either government expenditures or taxes.

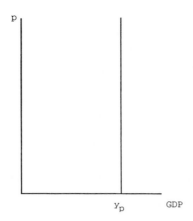

FIGURE 4.5   Long-run aggregate supply curve.

Most economists believe that in the long run the economy will move to a full employment or potential GDP level. In this case we have a long-run aggregate supply curve (LAS) that is vertical at full employment or potential GDP as illustrated in Fig. 4.5. Here Y_p is full employment or potential GDP.

In order to complete the model, it is necessary to reconcile the differences between the SAS and LAS positions. In Fig. 4.6 the economy is initially in a short-run equilibrium at Y_1, which is beyond potential output. Production at this level puts considerable pressure on wages and the prices of materials, and these prices will rise, causing the SAS_1 schedule to shift left to SAS_3, where potential output is reached. Here the upward pressure on wages and materials prices stops and a stable long-run equilibrium results. During a recession the SAS might be drawn as SAS_2, with the equilibrium output at Y_2. Here the decreased demand for materials would put downward pressure on materials prices, and the high unemployment would eventually cause real wages to drop, which would cause the SAS schedule to shift to the right until the long-run equilibrium is reached at Y_p.

The model that is presented here is self-correcting, that is, if the economy enters a recession, market forces in the resource market will tend to put pressure on wages and other resources prices, which in time will drop, thus lowering the cost of production and prices that would stimulate output. During a

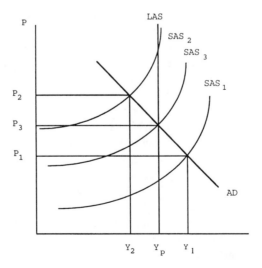

FIGURE 4.6   Short- and long-run equilibrium.

high employment inflationary period, the opposite would be expected to occur. The important question is, "How long will it take before these market forces can be expected to correct any imbalances?" The historical experience would suggest that, depending on the particular circumstances, an economic downturn may last a few months to as long as a decade as in the 1930s.

Based on this model, economic fluctuation can be caused by numerous factors. In the private sector autonomus changes in consumer expenditures or private investment would result in changes in income. These changes might result from changes in economic expectations or some other causes such as over investment in some areas resulting in excess capacity and a revision of investment plans. New technological innovations might also cause economic fluctuation either by stimulating consumer demand or by increasing private investment in certain sectors.

On the supply side changes in worldwide materials prices would shift the supply curve. In the 1970s, the large increase in petroleum prices shifted the aggregate supply curve left. This resulted in not only higher prices but also increased unemployment.

Policy Implications

If the economy is self-correcting as has been suggested and if fluctuations are likely to be relatively mild and of short duration, then an activist policy isn't needed to correct fluctuations in income. If, on the other hand, economic fluctuations are apt to be quite severe or to persist over a period of time, such as a few years, then policy actions may have considerable merit.

Traditionally, economists have discussed two types of policy that might be used to change aggregate demand. The first of these is fiscal policy, whereby changes in government expenditures or changes in taxes could be used either to restrain expenditures during a high-inflationary, low-unemployment period by decreasing government expenditures or increasing taxes or by stimulating aggregate demand by cutting taxes or increasing government expenditures during recessionary periods. The current fetish with a balanced budget at the federal level would appear to preclude any activist fiscal policy during a recession, and the current "no tax increase" mentality probably would rule out any major anticyclical policy during periods of rapid inflation.

It should be pointed out that there are those who argue that consumption is a function of permanent income or income over a long period of time. If this is the case, then a tax cut during a recession would be interpreted as an increase in current disposable income but the recognition that a tax increase will be needed in the future to pay off the debt resulting from the initial tax cut. In this case the argument continues that no increase in consumer expenditures will occur because households will make provisions to pay higher taxes in the future. To the extent that this is true, fiscal policy is much less effective than would be indicated by our model.

The second aggregate demand policy is that of monetary policy. The Federal Reserve System is also able to influence the level of economic activity; thus, during a recession they might increase the growth in the money stock. This would lead to lower interest rates and increased borrowing and spending both by households for consumer durables and also by businesses who now find conditions more favorable for investment. There may also be a direct increase in spending as households and businesses find themselves with more funds available with which to make purchases.

A contraction in the money stock would lead to higher interest rates and a decrease in aggregate spending and would be used during inflationary periods.

One of the problems of using monetary policy in a countercyclical manner is the long lags that may occur before the full effects of the policy change are felt throughout the economy. While the Open Market Committee of the Federal Reserve can change policy directives very quickly, many investment plans are made years in advance of the actual investment, and it is unlikely that relatively small changes in interest rates will have a major effect either in the timing or the extent of the investment, although some changes in spending are likely to occur much sooner.

Policies to affect the aggregate supply curve are much more difficult to implement. Those measures that would increase investments or labor productivity such as new technologies would over time cause the aggregate supply curve to shift to the right, thus increasing output and possibly lowering prices. However, the extent of the shift in a particular year will be quite small, and these policies might be used to affect the long-term growth rate.

Summary

Economic fluctuations or business cycles have been part of the American economic system for a long time and are likely to continue. The model that has been presented indicates that the causes of these fluctuations are many and varied. They may be the result of changes in aggregate demand either in the private or public sector or they may be the result of changes in aggregate supply.

The model also indicates that the economic fluctuations that occur are self-correcting, but what isn't known is the period of time required for these self-correcting forces to bring the economy back to a full-employment equilibrium. If this period of time is longer than a few months as appears likely in some cases, then activist policy measures may be called for. During recent years, decreasing emphasis has been placed on fiscal policy. This is due in part because of political problems as well as questions about the effectiveness of fiscal policy. Monetary policy has been receiving considerable emphasis. Monetary policy, although reasonably effective, also has problems, the most important one being that of long lags.

Economic fluctuations caused by aggregate supply changes are even more difficult to counteract through policy measures because the policies that change aggregate supply tend to be effective only over a long period of time.

Defining Terms

Aggregate demand: The quantity of goods and services that households, business firms, and governments would like to purchase at various price levels during a particular time period.

Aggregate supply: The quantity of goods and services that producers are willing to supply at various price levels during a particular time period.

GDP: Total amount of goods and services produced domestically during a particular year.

GDP deflator: A price index that measures the cost of purchasing the items included in GDP.

Real GDP: GDP deflated by some measure of the price level such the GDP deflator.

Money supply: The total of all currency in circulation and all checkable deposits at financial institutions.

Transfer payments: Payments made to persons or institutions without their providing a service or product in exchange, i.e., social security benefits or unemployment benefits.

References

Dornbusch, R. and Fischer, S., *Macroeconomics*, 6th ed., McGraw Hill, New York, 1994.

Economic Report of the President, Council of Economic Advisors, U.S. Government Printing Office, Washington, DC, various issues, 1964–1997.

Federal Reserve Bulletin, Board of Governors of the Federal Reserve System, Washington, DC, various issues, 1996–1997.

Gwartney, J. D. and Stroup, R. L. *Economics, Private and Public Choice*, 8th ed., The Dryden Press, Fort Worth, TX. 1997.

Further Information

Additional information can be obtained from any one of a number of intermediate textbooks. Two outstanding sources are *Macroeconomics*, by Dornbusch and Fischer, and *Macroeconomics*, by Mankiw.

4.8   Inflation

William M. Frix

The definition of **inflation** is a subject of debate, as is the cause of it. In theory, inflation is the increase of money (commodity, credit, and fiat monies) in an economic system. In practice and colloquially speaking, inflation is the increase in prices due to the increase of money in an economic system. There are 3 major theories about what causes inflation [Jones, 1996]:

1. The Monetary Theory holds that inflation results when the supply of money, usually defined as currency and credit, increases faster than the economy's real output capacity. The increase in money supply is partly attributed to both consumer and governmental borrowing.
2. The Keynesian (Fiscal/Quasi-Fiscal) Theory sees the cause of inflation as the result of large public sector deficits that need to be financed through money creation.
3. The Supply-Side Theory sees inflation as the result of a significant erosion of productivity due to trade problems, sporadic disasters, government regulations, changes in investment development, and other causes.

No matter which theory one supports in the cause of inflation, there is a consensus of opinion about the effect of inflation, which will next be explored.

Influence of Inflation on an Economic System

There is a lake in southwest Colorado that is popular with vacationers. The clear cold waters are filled with trout and a thriving tourist industry supports the few full-time residents there. There are several streams that feed the lake; the output of the lake passes through a spillway on its way to irrigating the farms below the dam that creates the lake. Optimally, the total of the waters flowing into the lake from the streams equals the water exiting the lake through the spillway. When this happens, the water within the lake remains fresh, the skiing and fishing is great, and the farmers below the dam are happy with the irrigation water (although they always want more).

During periods of excess rainfall, the streams feeding the lake swell, carrying excess water into the lake. The water leaving through the spillway initially remains constant as the lake expands to accommodate the increases in water volume. The tourist industry increases since the skiing becomes better. Initially, the fish within the lake are better fed, the result of insects being swept into the lake, resulting in an increase in the fish population. As the rains continue, soon the erosion from the land above the lake begins to cloud and poison the lake water, weakening the fish. Slowly, the water leaving the lake increases as the water level in the lake rises above the weir. The farmers have more water available for irrigation (which they may not need if the rains are also falling on them). If the output of the spillway does not match the excess flow from the streams, the lake water level continues to rise until it tops the dam. To prevent this, the managers attempt to either increase the flow through the spillway or to restrict the water flowing into the lake or divert it around the lake. If the lake waters continue to rise, the flow over the top of the natural dam weakens the levee with the potential of a catastrophic collapse, resulting in a flood of the farm land below the lake. The lake is destroyed, the fish are gone, as are the tourist industry and the farms. Nothing remains but to rebuild the dam and start over, hopefully with better insight as to how to avoid a similar tragedy in the future.

The discussion of the lake is illustrative of a country's economic system. The streams represent the money flowing into the system; the spillway output to the farms below represents the system output of useful, tangible goods and services. The fish within the lake represent the assets of the economic system (workers, capital equipment, facilities, natural resources, etc.), while the tourist industry represents the interests outside the economic system (e.g., foreign economic systems). The lake managers are the regulators of the economic system — the government usually.

When an economic system is stable, there is a balance between the money flowing into the system and the stream of goods leaving the system. The total economic assets neither increase nor decrease (although there may be isolated increases and decreases — just like there may be an increase in rainbow trout with a decrease in brook trout). There is a healthy interchange between the economic system and outside systems, represented by the tourist industry contributing to the system (e.g., "fishing" licenses) and removing some of the goods ("fish") or services ("skiing") from the system.

Inflation results when the flow of money into an economic system exceeds the output flow of useful, tangible goods and services. Initially, as occurs in **creeping inflation**, beneficiaries of the economic system enjoy the fruits of an expanding economy: the excess money (more water carrying more food for the fish) allows for increasing the system assets (fish populations) and in greater productivity (increased water output for the farmers). Outside interests are high and there is a general sense of growth and optimism. As long as the increased input supply of money is matched by an increased output in productivity and investment, the system remains healthy.

Problems begin when either the increased supply continues unabated (**chronic inflation**) or when the supply greatly exceeds the output (**hyperinflation**). This results in the popular concept of inflation — the supply of money being greater than the output stream of goods and services. As the excess water runoff carries erosion into the lake, with a resulting poisoning of the water and weakening of the fish, so a continual expansion begins to threaten the economic system. Assets begin to degrade in value, workers feel a sense of anxiety, and outside interests become concerned. Chronic inflation tends to disrupt normal economic activities [Jones, 1996]. Since prices typically increase, savings are sacrificed so that goods and services can be purchased before prices rise even higher. Therefore, the assets of an economic system degrade during inflation. Worse, chronic inflation tends to feed upon itself — present spending increases to avoid increasing prices, often using credit. This increases the amount of money within the economic system, thereby increasing inflation. Thus, chronic inflation tends to become permanent and ratchets upward to even higher levels as negative expectations accumulate [Jones, 1996]. If the money supply exceeds the capacity of the economic system by too large a margin, the entire economic system is threatened with collapse and the fleeing of outside interests. Finally, the economic system can no longer withstand the pressure of excess supply, and a catastrophic collapse of the system (usually with a destruction of society and industry) results.

The Influence of Inflation on Wages and Prices

We live in a world of limited resources: quantities of raw materials, time, and people are all finite. It is their limitation that gives assets their intrinsic value. For example, silicon (being plentiful in sand) is less valuable than gold, being rarer, although silicon has more uses than gold (such as glass and semiconductors). The value of a commodity is also dependent on its circumstances: fresh water is more valued in a desert than in a tropical rain forest.

Money also has a value that depends on circumstances and availability. When money is easily available, its value decreases with respect to other objects of value; when money is in short supply and difficult to attain, its value rises with respect to other objects of value. Consequently, the prices of goods and services (being a measure of the value of money compared to the goods and services) rise during inflationary periods since the value of money, being plentiful and readily available, decreases in value with respect to other objects (which are relatively more scarce).

Incorporating the Influence of Inflation in Economic Analyses

To evaluate the influence of inflation on economic analyses, it is necessary to distinguish between two types of value: actual and real.

The **actual value** of a commodity is the value of the commodity in terms of the prevailing monetary exchange used at a given point in time. If the commodity is to be purchased, this corresponds to the *actual cost* of the commodity — the total amount of current money (i.e., "out-of-pocket" dollars) required to purchase the commodity. If the commodity is to be sold, this corresponds to its *actual price* — the amount of current, out-of-pocket money that must be paid to secure the transaction. The key issue is that the actual value of a commodity at a given point in time is defined in terms of the amount and of the type of money exchanged as legal tender at the same point in time. The amount of legal tender (assuming dollars) that is involved in an actual value transaction is often referred to as the *actual dollars, then-current dollars, current dollars, future dollars, escalated dollars,* or *inflated dollars* [Thuesen and Fabrycky, 1984]. The actual value (cost or price) of a commodity is the customary viewpoint of value (costs or price).

The **real value** of a commodity is the value of the commodity, at a given point in time called the *evaluation time period,* in terms of the prevailing monetary exchange in effect at a defined point in time (called the *reference time period*), that is, in a real value evaluation, the value of a commodity is typically defined in terms of the amount and type of money exchanged as legal tender at another (usually earlier) point in time. If the commodity is to be purchased, this corresponds to the *real cost* of the commodity — the total amount of money (e.g., dollars), the value of which was defined during the reference period required to purchase the commodity. If the commodity is to be sold, this corresponds to its *real price* — the amount of money, the value of which was defined during the reference period, that must be paid to secure the transaction. The key distinction between the real value of a commodity and its actual value is that the evaluation time period and the reference time period for an actual value evaluation are the same (hence, the evaluation is made in terms of the then-current dollars), whereas in the real value evaluation the evaluation time period often is different from the reference time period. The amount of legal tender that is involved in a real value transaction is often referred to as the *real dollars, constant dollars, deflated dollars, today's dollars,* or *zero-date dollars* [Thuesen and Fabrycky, 1984].

Assume that a certain quantity of a commodity costs $X today. Assuming the true value of the commodity is unchanged but that the price of the commodity rises due to an **inflation rate** of $I\%$ per year. One year from now the actual cost of the same amount of the same commodity will be $X(1 + I/100)$. Assuming the inflationary effects continue, 2 years from now the actual cost of the same amount of the same commodity will be $X(1 + I/100)^2$. The pattern continues such that, if the inflationary rate, I, does not change, the actual cost of the same amount of the same commodity N years from now will be $XP(F/P, I, N)$ where $P(F/P, I, N)$ is the *single-payment, compound amount factor* $= (1 + I/100)^N$ [DeGarmo et al., 1979].[4]

We see that the cost of the commodity typically rises due to the influence of inflation. Alternatively, we can say that the buying power of actual (current) dollars decreases relative to the buying power of real (constant) dollars during inflationary periods. From the above analysis, it can be seen that $I/100$ more actual dollars are required per year to purchase the same amount of goods. Thus $X(1 + I/100)$ actual dollars per year have the same value as $X real dollars or that the value of each actual dollar decreases by $(1 + I/100)$ relative to a real dollar per year. Mathematically, this states that

$$\$A_N = \frac{\$1}{\left(1 + I/100\right)^N} = \$1\left(P/F, I, N\right) \tag{1}$$

---

[4] Note that this analysis explicitly assumes that the true value of the commodity (given by $X) is unchanged. This implies that the only effect of time on the commodity price was inflationary pressure due to the money supply. In some sectors of the economy this is not true. For example, the microcomputer industry has seen actual prices decreasing while inflation has been driving consumer prices upward. Predicting the influence of inflation on these situations is more difficult since the true value is no longer a constant but is a function of time and other factors, requiring techniques such as linear regression to predict trends. Such techniques are beyond the scope of this article.

where $\$A_N$ is the value of an actual dollar N periods after the reference period, and the quantity in the parentheses on the right is the *single-payment present worth factor* [DeGarmo et al., 1979].

The above analyses determine the value of money and the predicted cost of a commodity during inflationary periods. Often engineers and managers must consider how inflation affects investment decisions. The cardinal rule for evaluating the influence of inflation on investments is *inflation works to the benefit of a debtor but to the disadvantage of a lender*. That this is so can be easily seen by recognizing that a debtor receives money from a lender in today's (deflated) dollars but will repay the lender using future (inflated) dollars. A specific example should make this clear.

Able borrows $\$10,000$ from Baker. As a condition of the loan, Baker charges 5% interest, compounded annually, and stipulates that the loan is to be repaid (principle plus interest) in 10 years. Payments are to be made annually, at the 1-year anniversary of the loan. Inflation remains a constant 3% per year.

Ignoring inflation and using the *capitol recovery factor* at 5% interest over 10 years, Able pays Baker $\$1295.00$ in actual dollars per year in annuity for a total of $\$12,950$ in actual dollars. The loan thus costs Able $\$2950$ in actual dollars. However, due to the influence of inflation, each actual dollar is worth 0.9709 times its previous year's worth. Thus, Able pays Baker $\$1257.28$ real dollars the 1st year, $\$1220.66$ real dollars the 2nd year, $\$1185.11$ real dollars the 3rd year, and so forth. The total real dollars Able pays Baker is

$$T_R = \$1,295 \sum_{i=1}^{10} \frac{1}{\left(1+0.03\right)^i} = \$1,295 \left[\frac{\left(1+0.03\right)^{10} - 1}{0.03\left(1+0.03\right)^{10}} \right] = \$11,046.61 \qquad (2)$$

Able's real cost for the loan is $\$1046.61$, which is Baker's return on his investment. Thus, inflation worked for Able's benefit but to Baker's disadvantage. Note that the quantity inside the brackets is the *uniform series present worth factor* for 3% interest and ten periods, usually abbreviated $(P/A, I, N)$ for a uniform inflation rate I over N years. In general, the real cost of borrowing money is thus

$$C_R = P\left(A/P, i\%, N\right)\left(P/A, I\%, N\right) - P = P\frac{i}{I}\left(\frac{1+i/100}{1+I/100}\right)^N \frac{\left(1+I/100\right)^N - 1}{\left(1+i/100\right)^N - 1} - P \qquad (3)$$

where N is the period of the loan, i is the (percent) interest rate of the loan, I is the (uniform percent) inflation rate over the period of the loan, and P is the principle value of the loan.

When investments are made, there is a desirable minimum attractive rate of return (MARR) for the investment capital. When inflation is neglected, the MARR is normally equivalent to the desired interest rate to yield a given return on the investment. Mathematically, this means that the MARR is found from the actual dollars returned F from the present investment P according to the formula $F = P(1 + i)^N$ where N is the investment period and i is the MARR (without inflation adjustment). This formula gives the actual return; as shown previously, when inflation is present, the actual dollars decrease in value over the investment period.

When an inflationary rate of I% is present, an I% increase in actual dollars per dollar of desired return must be added to the desired return to compensate for the decrease in value of the actual dollars. We have

$$F = P\left(1+i\right)^N \left(1+I\right)^N = P\left[\left(1+i\right)\left(1+I\right)\right]^N = P\left(1+i+I+iI\right)^N = P\left(1+d\right)^N \qquad (4)$$

Thus, to adjust for inflation, a new, inflation-adjusted MARR d is determined according to the relation $d = i + I + iI$ where i is the unadjusted MARR and I is the rate of inflation [Potter, 1996].

Finally, the analyses have assumed a uniform inflation rate I (i.e., the inflation rate is assumed to be constant). Inflation rates are not typically uniform but are dynamic over time. The equations given earlier in this article are still useful when nonuniform inflation rates are encountered. In this case, the inflation rate I is replaced by the geometric inflation rate overline $\bar{I}$ given by Thuesen and Fabrycky [1984]:

$$\bar{I} = \sqrt[N]{\prod_{n=1}^{N} \left(1 + I_n\right)} - 1 \tag{5}$$

where N is the number of inflationary periods to be considered and I_n is the inflation rate for the nth period.

Defining Terms

Actual value of a commodity: The value of the commodity at a given point in time in terms of the prevailing monetary exchange in effect at the same point in time.

Chronic inflation: An economic condition characterized by a sustained drop in the value of money with a resulting sustained rise in prices for goods and services.

Hyperinflation: A severe economic condition characterized by rapid drops in the value of money with a resulting rapid rise in prices for goods and services.

Inflation: In theory, the increase of money or credit in an economic system. In practice, the rise in prices due to an increase of money or credit in an economic system.

Inflation rate: The percent change in the value of money over a period of time. Alternatively, the percent change in prices over a period of time.

Real value of a commodity: The value of the commodity at a given point in time (the evaluation time period) in terms of the prevailing monetary exchange in effect at a defined point in time (the reference time period).

References

DeGarmo, E. P., Canada, J. R., and Sullivan, W. G. *Engineering Economy*, 6th ed., Macmillan, New York, 1979.

Jones, S. L. Inflation and deflation, In *Microsoft® Encarta® 97 Encyclopedia*, Microsoft Corporation, Redmond, WA, 1996.

Potter, M. C., ed. *Fundamentals of Engineering*, 6th ed., Great Lakes Press, Okemos, MI, 1996.

Thuesen, G. J. and Fabrycky, W. J. *Engineering Economy*, 6th ed., Prentice-Hall, Englewood Cliffs, NJ, 1984.

Further Information

Consumer price indicies and inflation rates for the years since 1913 are available from the Woodrow Federal Reserve Bank of Minneapolis located at the World-Wide-Web site, http://woodrow. mpls.frb.fed.us/economy/calc/hist1913.html. In addition, the Woodrow Federal Reserve Bank of Minneapolis also provides computer programs to determine the value in current dollars of a good or service purchased in a previous year. The programs can be found at http://woodrow.mpls.frb.fed.us/economy/calc/cpihome.html. The Federal Reserve Bank of St. Louis' Federal Reserve Economic Data (FRED) database (http://www.stls.frb.org/fred/) provides up-to-date financial and economic data, including the monthly Consumer Price Indexes and monthly Producer Price Indexes, historical U.S. economic and financial data, including daily U.S. interest rates, monetary and business indicators, exchange rates, and regional economic data for Arkansas, Illinois, Indiana, Kentucky, Mississippi, Missouri, and Tennessee. Other Federal Reserve banks provide similar data for their regions. Finally, current price indicies (consumer and producer) as well as other economic indicators can be downloaded from the U.S. Department of Commerce's STAT-USA/Internet databases, located at http://www.stst-usa.gov/. Note, however, that this is a subscription service and requires the purchase of an authorization user name and password to access the databases.

4.9   Cost-Benefit Analysis

William Samuelson

"My way is to divide a half sheet of paper by a line into two columns; writing over the one Pro, over the other Con... When I have got them altogether in one view, I endeavor to estimate their respective weights; and where I find two, one on each side, that seem equal, I strike them both out. If I find a reason Pro equal to two reasons Con, I strike out the three, .. and thus proceeding, I find where the balance lies.... I have found great advantage from this kind of equation, in what might be called moral or prudential algebra." Benjamin Franklin, September 19, 1792.

Cost-benefit analysis is a method of evaluating public projects and programs. Accordingly, cost-benefit analysis is used in planning budgets, in building dams and airports, devising safety and environmental programs, and in spending for education and research [Musgrave, 1969]. It also finds a place in evaluating the costs and benefits of regulation: when and how government should intervene in private markets. In short, almost any government program is fair game for the application of cost-benefit analysis. The logic of cost-benefit analysis is simple. The fundamental rule is undertake a given action if and only if its total benefits exceed its total cost. (The rule applies equally well when taking one action forecloses pursuing another. Here, the action's cost is an opportunity cost: the foregone benefits of this next best alternative.) Applying the cost-benefit rule involves three steps: (1) identifying all impacts (pro or con) on all affected members of society, (2) valuing these various benefits and costs in dollar terms, and (3) undertaking the program in question if and only if it produces a positive **total net benefit** to society.

The aim of cost-benefit analysis is to promote economic efficiency [Gramlich, 1992]. While there is little controversy concerning the need to use resources wisely, there is criticism of the way cost-benefit analysis carries out this goal. Some critics point out the difficulty (perhaps impossibility) of estimating dollar values for many impacts. How does one value clean air, greater national security, unspoiled wilderness, or additional lives saved? The most difficult valuation problems arise when benefits and costs are highly uncertain, nonmarketed, intangible, or occur in the far future. Proponents of cost-benefit analysis do not deny these difficulties; rather they point out that any decision depends, explicitly or implicitly, on some kind of valuations. For instance, suppose a government agency refuses to authorize an $80 million increase in annual spending on highway safety programs, which is projected to result in 100 fewer highway deaths per year. The implication is that these lives saved are not worth the dollar cost, that is, the agency reckons the value of such a life saved to be less than $800,000. All economic decisions involve trade-offs between benefits and costs. The virtue of the cost-benefit approach is in highlighting these trade-offs.

A second point of criticism surrounds step three, whereby only total benefits and costs matter, not their distribution. A program should be undertaken if it is beneficial in aggregate, i.e., if its total dollar benefits exceed total costs. However, what if these benefits and costs are unequally distributed across the affected population? After all, for almost any public program, there are gainers and losers. (Indeed, citizens who obtain no benefit from the program are implicitly harmed. They pay part of the program's cost either directly via higher taxes or indirectly via reduced spending on programs they would value.) Shouldn't decisions concerning public programs reflect distributional or equity considerations? Cost-benefit analysis justifies its focus on efficiency rather than equity on several grounds. The first and strongest ground is that the goals of efficiency and equity need not be in conflict, provided appropriate compensation is paid between the affected parties. As an example, consider a public program that generates different benefits and costs to two distinct groups, A and B. Group A's total benefit is $5 million. Group B suffers a loss of $3 million. The immediate impact of the project is clearly inequitable. Nonetheless, if compensation is paid by the gainers to the losers, then all parties can profit from the program. A second argument for ignoring equity is based on a kind of division of labor. Economic inequality is best addressed via the progressive tax system and by transfer programs that direct resources to low income and other targeted groups. According to this argument, it is much more efficient to use the tax and transfer system directly than to pursue distributional goals via specific public investments. Blocking the aforementioned project on equity grounds has a net cost: foregoing a $5 million dollar gain while saving only $3 million in cost. Finally, though it is not common practice, cost-benefit analysis can be used to highlight equity concerns. As step one indicates, the method identifies, untangles, and disaggregates the various benefits and costs of all affected groups. This in itself is an essential part of making distributional judgments. The method's key judgment about equity comes when these benefits and costs are reaggregated: all groups' benefits or costs carry equal dollar weight.

Applying Cost-Benefit Analysis: A Simple Example

A city planning board is considering the construction of a harbor bridge to connect downtown and a northern peninsula. Currently, residents of the peninsula commute to the city via ferry (and a smaller number commute by car taking a slow, "great circle" route). Preliminary studies have shown that there is considerable demand for the bridge. The question is whether the benefit to these commuters is worth the cost. The planning board has the following information. Currently, the ferry provides an estimated 5 million commuting trips annually at a price of $2.00 per trip; the ferry's average cost per trip is $1.00, leaving it $1.00 in profit. The immediate construction cost of the bridge is $85 million dollars. With proper maintenance, the bridge will last indefinitely. Annual operating and maintenance costs are estimated at $5 million. Plans are for the bridge to be toll-free. Since the bridge will be a perfect substitute for the ferry, the ferry will be priced out of business. The planners estimate that the bridge will furnish 10 million commuting trips per year. The discount rate (in real terms) appropriate for this project is 4%.

Figure 4.7 shows the demand curve for commuter trips from the peninsula. The demand curve shows that at the ferry's current $2.00 price, 5 million trips are taken. Should a toll-free bridge be built, 10 million trips will be taken. Furthermore, the planning board believes that demand is linear.

Currently, the ferry delivers benefits to two groups: the ferry itself (its shareholders) and commuters. The ferry's annual profit is $(2.00 - 1.00)(5) = 5 million. In turn, the commuters' collective benefit takes the form of **consumer surplus** — the difference between what consumers are willing to pay and the

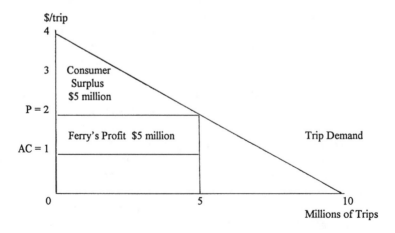

FIGURE 4.7   Benefit-cost analysis of building a bridge.

actual price charged. The triangular area between the demand curve and the $2.00 price line (up to their point of intersection at 5 million trips) measures the total consumer surplus enjoyed by ferry commuters. The area of this triangle is given by (.5) (4.00 − 2.00) (5) = $5 million. Thus, the sum of profit plus consumer surplus is $10 million per year. Supposing that this benefit flow is expected to continue indefinitely at this level, the resulting net present value is 10/.04 = $250 million. Here, the annual net benefit (in real terms in perpetuity) is capitalized by dividing by the appropriate (real) interest rate.

Now consider the cost-benefit calculation for the bridge. First note the adverse effect on the ferry: it is put out of business, so its profit is zero. Second, there is a burden on taxpayers. They must foot the bill for the construction and maintenance costs of the bridge. Since the bridge charges no toll, it generates no revenue. Third, the entire benefit of the bridge takes the form of consumer surplus (the triangle inscribed under the demand curve and above the zero price line). The dollar value is (.5) (4.00) (10) = $20 million per year. In present value terms, this benefit comes to $500 million against a total cost (also in present value terms) of 85 + 125 = $210 million. Thus, the bridge's net benefit is $290 million. Since this is greater than that of the ferry, the bridge should be built.

The decision to build the bridge depends crucially on charging the "right" toll. In the present example, no toll is charged. Here, the right price is zero because there is a negligible cost (no wear and tear or congestion) associated with additional cars crossing the bridge. The general principle behind **optimal pricing** is simple: the optimal price should just equal the marginal cost associated with extra usage [Layard, 1977]. For instance, since large tractor trailer trucks cause significant road damage to highways, they should be charged a commensurate toll. Here, a zero price insures maximum usage of the bridge and maximum benefit (with no associated cost). Setting any positive price would exclude some commuters and reduce the net benefit.

Regulating the Ferry

Before concluding that public provision is warranted, government decision makers should consider another option: regulating the private ferry market. Here, regulation means limiting the price the ferry operator can charge. From a cost-benefit point of view, what is the optimal regulated price? Basic economic principles provide a direct answer: the price that would prevail in a perfectly competitive market. If free entry of competitors were feasible, the ferry's price would be driven down to the zero-profit point: P = AC = $1.00. Thus, this is the price that the government should set for the (natural monopolist) ferry operator.

At a $1.00 price, the ferry makes 7.5 million in trips and makes a zero profit. Commuters realize total consumer surplus that comes to (.5) (4.00 − 1.00) (7.5) = $11.25 million per year. (As always, consumer surplus is given by the area under the demand curve and above the price line.) The present value of the net benefit from ferry regulation is 11.25/.04 = $281.25 million. Building the bridge, with a discounted net benefit of $290 million has a slight edge over the regulatory alternative and continues to be the best course of action.

Valuing Benefits and Costs

The main issues with respect to valuing benefits and costs concern (1) the role of market prices, (2) ways of valuing nonmarketed items, and (3) the choice of appropriate discount rate. In most cases, market prices provide the correct values for benefits and costs. In perfectly competitive markets, the price of the good or service is an exact measure of its marginal benefit to consumers and its marginal cost to producers, P = MB = MC. In some circumstances, however, correct valuation of benefits and costs based on market prices may require modifications. One instance occurs when "inframarginal" effects are important. We know that the current market price reflects the valuation of marginal units (i.e., the last ones consumed). However, if the output impact is large, a significant amount of consumer surplus will be created above and beyond revenue generated. In the previous example of the bridge, the benefit consisted entirely of consumer surplus. With a zero toll, the bridge generated no revenue.

A second problem concerns price distortions. Taxes are one source of distortions. For instance, suppose the government sets a tariff on the import of an agricultural good with the intent of protecting domestic farmers from foreign competition. The result is that the domestic price for the good is $1.00 above the world price. How should one value the new crops grown using water from a federal dam? The difficulty is that, instead of a single price reflecting marginal benefit and marginal cost, there are two prices. If all of the crop is sold on the world market, then that is the appropriate price, and similarly for the home market. If the crop is sold on both markets, then the appropriate value is a weighted average of the separate prices. A similar difficulty stems from the presence of monopoly, where the monopoly price is greater than marginal cost. Depending on circumstances, output could be valued at P or MC. A final example involves the employment of labor. Consider the cost of labor used to build the bridge. If undertaking the project has no effect on overall employment (construction workers are bid away from other jobs), the labor cost is measured by the going market wage. Alternatively, suppose that there is widespread unemployment so that new jobs are created. Then the analysis should include the wage paid (net of the cost of the worker's lost leisure time) as an additional benefit for the newly hired worker.

Nonmarketed Benefits and Costs

Several difficulties occur when valuing nonmarketed items. For instance, how might one measure the dollar benefits provided by public elementary and secondary schools? This is a tough question. The difficulty is that public education is provided collectively (i.e., financed out of local tax revenues); there is no "market" value for this essential service. Parents do not pay market prices for their children's education, nor are they free to choose among public schools. By contrast, valuing education provided by private schools is far easier to pin down. It is at least as much as these parents are willing to pay in tuition. If a private school fails to deliver a quality education, parents will stop paying the high market price.

This same point about valuation applies to all nonmarketed goods: national security, pollution, health risks, traffic congestion, even the value of a life. In the absence of market prices, other valuation methods are necessary. Roughly speaking, there are three approaches to valuing nonmarketed goods and so called "intangibles". One method is to elicit values directly via survey, that is, ask people what they really want. Surveys have been used to help ascertain the benefits of air quality improvement (including improved visibility in Los Angeles from smog reduction), the benefits of public transport, the cost of increased travel time due to traffic, and the value of local public goods. A second approach seeks to infer values from individual behavior in related markets. For instance, the benefit of a public secondary school education might be estimated as the expected difference in labor earnings (in present-value terms) between a high school graduate and a ninth-grade dropout. Appealing to labor markets provides a ready measure of the economic value of these years of schooling (though not necessarily the full personal value). One way to measure the harm associated with air pollution is to compare property values in high-pollution areas vs. (otherwise comparable) low-pollution areas. Finally, society via its norms and laws places monetary values on many nonmarketed items. Workman's compensation laws determine monetary payments in the event of industrial injuries. Judges and juries determine the extent of damages and appropriate compensation in contract and tort proceedings.

The Discount Rate

Evaluating any decision in which the benefits and costs occur over time necessarily involves discounting. The discount rate denotes the trade-off between present and future dollars, that is, one dollar payable a year form now is worth $1/(1+r)$ of today's dollar. In general, the net present value (NPV) of any future pattern of benefits and costs is computed as

$$\text{NPV} = [B_0 - C_0] + [B_1 - C_1]/(1 + r) + [B_2 - C_2]/(1 + r)^2 + \dots [B_T - C_T]/(1 + r)^T \quad (1)$$

As always, future benefits and costs are discounted relative to current ones. Moreover, the public-sector manager employs the present value criterion in exactly the same way as the private manager. The decision rule is the same in each case: undertake the investment if and only if its NPV is positive. For most public investments, costs are incurred in the present or near-term, while benefits are generated over extended periods in the future. Consequently, lower discount rates lead to higher project NPVs, implying that a greater number of public investments will be undertaken.

What is the appropriate discount rate to use in evaluating public investments? The leading point of view is that they should be held to the same discount rate as private investments of comparable risk. For instance, consider a municipal government that is considering building a downtown parking garage. Presumably, this project's risks are similar to those of a private, for-profit garage, so its NPV should be evaluated using a comparable discount rate. A fundamental principle of modern finance is that individuals and firms demand higher rates of return to hold riskier investments. Thus, setting the appropriate discount rate means assessing the riskiness of the investment.

A second point of view holds that the choice of the discount rate should be a matter of public policy. Proponents of this view argue that the overall rate of investment as determined by private financial markets need not be optimal. They contend that left to their own devices private markets will lead to underinvestment. For instance, there are many investments that generate benefits in the distant future — benefits that would be enjoyed by future generations. However, unborn generations have no "voice" in current investment decisions, so these investments (beneficial though they may be) would not be undertaken by private markets. The upshot is that private markets are too near sighted and private interest rates are too high, thereby discouraging potentially beneficial investment. According to this view, public investments should be promoted by being held to lower discount rates than comparable private investments.

Concluding Remarks

When used appropriately, cost-benefit analysis is a useful guide to decisions concerning public programs and government regulation. Inevitably, some aspects of the analysis — specific benefits, costs, or the discount rate — may be uncertain or subject to error. The appropriate response is to utilize sensitivity analysis. For instance, if the discount rate is uncertain, the project's present value should be computed for a range of rates. If the NPV does not change sign over the range of plausible rates, the optimal investment decision will be unaffected.

Defining Terms

Consumer surplus: The benefit enjoyed by consumers, equal to the difference between what they are willing to pay and the price they actually pay for a good or service.

Optimal pricing: The price set for a public project should equal the marginal cost associated with additional usage.

Total net benefits: The objective of cost-benefit analysis is economic efficiency, that is, maximizing the sum of net benefits measured in dollar terms.

References

Layard, R., ed., *Cost-Benefit Analysis,* Penguin Books, New York, 1977.
Gramlich, E. M. *A Guide to Benefit-Cost Analysis,* Prentice Hall, Englewood Cliffs, NJ, 1992.
Musgrave, R. A. Cost-benefit analysis and the theory of public finance, *J. Econ. Lit.,* 7: 797–806, 1969.

Further Information

The two following articles discuss the potential uses and abuses of cost-benefit analysis. The textbook reference offers an extensive treatment from which parts of this condensed survey are adapted.

Arrow, K. et al. Is there a role for benefit-cost Analysis in environmental, health, and safety regulation?, *Science*, 272(12): 221–222, 1996.

Dorfman, R. Why benefit-cost analysis is widely disregarded and what to do about it, *Interfaces*, 26: 1–6, 1996.

Samuelson, W. and Marks, S. *Managerial Economics*, chap. 14, Dryden Press, Fort Worth, TX, 1995.

4.10   Interest Rates

David Flath

An understanding of the meanings and implications of interest rates, crucial to correct evaluation of business decisions involving future consequences, distinguishes economic sophisticates from others. Everyone is familiar with the interest rates on bank loans or savings accounts, but how many know why those interest rates are greater than zero or what accounts for their movements over time? The answers are simple, but, to many otherwise well-educated persons, elusive. Engineers and business managers in particular should hold a firm grasp of these principles of economics.

The Meanings of Interest Rates

Interest rates represent the premia one must pay for actual delivery of goods as opposed to the mere promise of future delivery, and are usually expressed as a percentage of the price of the latter. In other words, an interest rate is the percentage by which a good's **spot price** exceeds one of its **forward prices**. Notice that the everyday notion that an interest rate corresponds to the terms of a money loan fits this definition. The spot price of a unit of currency equals one, by definition. Lending of money means forward purchase of currency, paying principal for a promised future delivery of principal plus interest. The interest rate corresponding to this transaction is the percentage by which principal plus interest (spot value of that amount of currency) exceeds principal (forward value of the currency). Though we usually think of interest rates as connoting the terms of loans or rates of accumulation of bank deposits, the phenomenon of interest is far more pervasive than simple usury.

There are as many different interest rates as there are goods and delivery periods, and because the supply and demand of each good reflects some considerations unique to it, the many interest rates need not coincide. However, the interest rates corresponding to differing goods do reflect some common forces. For example, the interest rates of all the goods denominated in units of currency reflect the market's anticipation of money price inflation in a similar way, for anticipation of price inflation erodes the present value of future claims on all such goods. The interest rates corresponding to real commodities do not reflect anticipated inflation in this same manner, for anticipated inflation does not erode the present value of real goods' future deliveries, as it does for monetary assets. Macroeconomic analysis attaches importance to movements in both the **nominal rate of interest** and **real rate of interest**, in actuality, indices of the economy's many monetary interest rates (those corresponding to assets denominated in currency units) and commodity interest rates, respectively.

The interest rates corresponding to real commodities are generally, but not always, greater than zero because spot delivery usually entails costs that future delivery avoids and because spot delivery is usually more valued than the mere promise of future delivery. In other words, as depicted in Fig. 4.8, because the supply of spot delivery is less and the demand for spot delivery greater than that of forward delivery, the equilibrium spot price exceeds the equilibrium forward price. This observation, however obvious once properly understood, eluded even the best minds in economics until cogently framed by Irving Fisher of Yale University early in the 20th century. Fisher's masterwork, *The Theory of Interest*, remains a paragon of clarity on the subject.

The nature of the cost of spot delivery, avoidable by future delivery, varies from good to good. In each instance, however, it represents an opportunity sacrificed if delivery is immediate, but exploitable

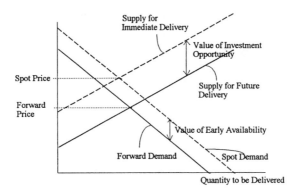

FIGURE 4.8 Equilibrium interest rate. An interest rate is the percentage by which the spot price of a good (a price paid now for immediate delivery) exceeds one of its forward prices (a price paid now for a promise of future delivery). Supply and demand determine the equilibria of both kinds of price. Most commodity interest rates are positive because, as depicted here, immediate delivery entails a cost that deferred delivery avoids, the sacrifice of a valued investment opportunity, and because most persons are willing to pay more for early availability.

otherwise, an *opportunity to invest*. For example, seeds can be planted in the ground and a greater quantity harvested if delivery is delayed a season but not otherwise. The lumber of a tree becomes more abundant only with the passage of time. A road or a building can be constructed without diverting resources from as high valued alternatives if the period to completion is extended, and so on. The preference for early enjoyment of goods may be a rational response to human mortality and the vagaries of life or an inbred trait of human nature. Whatever the case, evidence of the near universality of *impatience to spend income* is ubiquitous. Children must be educated in the rewards of delaying gratification of their wishes. Adults often need reminding of the same fact.

To illustrate the centrality of the two forces just identified in effecting positive interest rates, Fisher contrived a fanciful example in which interest rates would be negative. In Fisher's "figs" example [Fisher, 1930], shipwrecked sailors floating on a raft trade their only remaining provisions among themselves, a store of figs. The figs are spoiling in the hot sun, a cost of delayed delivery, avoided by early delivery, an inversion of the usual situation. Also, the sailors well anticipate the desperation of their future requirements of sustenance and so attach a greater value to promises of future delivery of a fig than they do to spot delivery, a further inversion of the usual case. On the raft, the figs' interest rate will be less than, not greater than, zero.

The relations among the multitude of interest rates are potentially subtle and complex, but in a fanciful economy can assume simplicity. For instance, in a stationary economy in which the supply and demand for each claim, including both forward claims and spot claims, continually and identically repeated itself as time passed, arbitrage would assure that all the interest rates of a given maturity equaled one another. A moment's reflection reveals why. If the interest rates differed, an astute trader could obtain a sure profit by combined forward purchase of any good with a relatively high interest rate and forward sale of any good with a relatively low interest rate. At the delivery date, the forward purchased good could be exchanged on the spot market for the forward sold good, and, because of the assumed stationarity of spot prices, the forward sale more than covered. Such considerations suggest, at the least, strong inter-relations among the many different commodity interest rates.

As already noted, anticipation of money price inflation erodes the present value of future claims on monetary assets and pushes up their corresponding interest rates but has no such influence on commodity interest rates. We can make this a bit more precise. The stationary economy of the previous paragraph is one in which inflation is absent; the anticipated rate of inflation is zero and rates of interest on monetary assets are all equal to one another and equal also to the commodity rate of interest. If, instead, inflation at a particular rate is anticipated by all, then lenders of money will insist that their principal be compounded not only at the real rate of interest but also at that particular anticipated rate of inflation;

borrowers, too, will acquiesce in this. For only then will the promised principal plus interest have the same expected purchasing power as under the regime in which no inflation was anticipated. Algebraically, the implied relationship is known as Fisher's equation:

$$(1 + \text{money rate of interest}) = (1 + \text{real rate of interest})(1 + \text{anticipated rate of inflation})$$

If the rates of inflation differ across goods and the aggregate inflation rate is measured by the growth of a price index, then the real rate of interest in Fisher's equation is an index of the various commodity rates of interest.

Historically, the real rate of interest in the sense just defined has held about the same order of magnitude as annual rates of economic growth, 3% or less. If anticipated inflation rates are similarly small, then Fisher's equation is well approximated as

$$\text{money rate of interest} \doteq \text{real rate of interest} + \text{anticipated rate of inflation}$$

With continuous compounding, that is, in which the various rates stand for exponential rates of increase rather than discrete rates of increase, the relation just stated is exact. One immediate implication of Fisher's equation is that, if the market anticipates deflation, the money rate of interest can become less than zero. The Nobel laureate Milton Friedman once advocated contraction of the nation's money supply to effect price deflation just sufficient to assure a money rate of interest equal to zero on the grounds that doing so would induce optimally enlarged holding of wealth as money instead of interest-bearing financial assets [Friedman, 1969].

The Fisher equation just discussed has an analog relating the money rates of interest of different nations to one another in the presence of international arbitrage. International arbitrage will assure that the compounding factor pertaining to domestic loans equals the compounding factor applicable to a loan of foreign currency by a lender planning to trade principal plus interest for home currency upon maturity:

$$(1 + \text{home country money rate of interest}) =$$
$$(1 + \text{foreign money rate of interest}) \times$$
$$(1 + \text{anticipated rate of depreciation of foreign currency in terms of domestic currency})$$

Where the anticipated rate of depreciation of foreign currency in terms of domestic currency exactly equals the differences in the anticipated rates of inflation in the two nations, then this Fisherian interest parity relation reduces to the statement that arbitrage assures equality of real rates of interest in each nation.

Understanding Interest Rate Movements

In thinking about interest rate movements, it is first useful to recognize that the real rate of interest, besides its other meanings, is an index for the relative price of durable goods. Relative price is the price of one good in terms of another, the ratio of the two goods' money prices. For instance, if in a particular bar, beer is priced $.50 per glass and wine is priced $1.00 per glass, then the relative price of wine equals two glasses of beer per glass of wine, that is, $1.00/glass of wine ÷ $.50/glass of beer. If buying a nondurable good represents a spot purchase, then buying a durable good represents a combined spot purchase and forward purchase. For instance, the price of durable redwood relative to that of less-durable pine represents the following:

price of redwood/price of pine =

$$(\text{spot price of lumber} + \text{forward price of lumber}) / \text{spot price of lumber}$$

$$= 1 + \text{forward price of lumber} / \text{spot price of lumber}$$

$$= 1 + 1 / (1 + \text{lumber rate of interest})$$

A rise in the price of durable redwood relative to that of less-durable pine means a fall in the lumber rate of interest and vice versa.

Now in macroeconomic analysis, it is customary to differentiate between durable goods and others, for the accumulation of durable goods represents investment while the production of other nondurable goods and services represents consumption. For this reason, the real rate of interest occupies an important niche in macroeconomic analysis; it equilibrates saving and investment. To keep matters simple, consider this process in a closed economy, one without international trade. In such an economy, saving, which is wealth accumulation, represents effective demand for durable goods, and investment represents additions to the stock of durable goods. The real rate of interest, inversely related as it is to the price of durable goods relative to that of nondurable ones, adjusts so that the actual supply of added durables matches the effective demand. As an empirical matter, national saving is insensitive to movements in the real rate of interest, but investment is sensitive. A lower real rate of interest, which represents a higher relative price of durables, induces expanded production of durables, expanded investment. A higher real rate of interest reduces investment. Government policies (such as reduced taxes or increased government spending) that contract national saving push up the real rate of interest to reduce investment by the necessary amount, as depicted in Figure 4.9. In an open economy, a contraction of domestic saving attracts foreign saving, mitigating the upward pressure on the domestic interest rate, but the other implications just described are qualitatively unchanged.

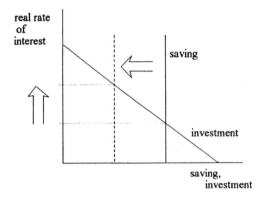

FIGURE 4.9 A contraction of national saving raises the real rate of interest.

The investment and saving paradigm also frames other important sources of variation in the real rate of interest. For instance, natural disasters such as earthquakes, hurricanes, or droughts generally reduce current production of goods and services far more than they impair anticipated future production. A sensible response in such extremities is to draw on wealth (that is, to reduce additions to the stock of wealth) to maintain consumption in spite of temporarily low income. As the saving of the entire community temporarily contracts, the real interest rate temporarily rises. Technological advance also influences the real rate of interest in an oddly similar way. At the earliest point that a technological advance is recognized, the community finds itself in the same position as an heir with great expectations. Its income, for the moment, unchanged, is nevertheless temporarily low in relation to its anticipated future level. As in the case of natural disaster, temporarily reduced saving enlarges the real rate of interest. As the technological advance finds application — raising the national income — saving returns to its normal level as does the real rate of interest. Historically, decades of remarkably rapid technological advance, such as that seen in the 1920s in the United States, have witnessed high real rates of interest. An intellectually respectable but still controversial hypothesis is that business cycles are nothing more than random accretions in the process of technological advance [Plosser, 1989]. Most macroeconomists have judged that monetary policy is also an important contributor to business cycles. Contraction of the rate of growth of the nation's money stock temporarily impedes the intermediation of funds between savers and investors, contracting the effective saving and temporarily enlarging the real rate of interest. Expansionary monetary policy has the opposite effect.

Defining Terms

Forward price: A price paid now for a promise of future delivery of a good.
Nominal rate of interest (or money rate of interest): An index of the rates of interest pertaining to money and other assets that are denominated in units of currency (for example, corporate debentures, but not stocks that are claims on real assets).
Real rate of interest: An index of the percentages by which the spot prices of commodities and services produced in the economy exceed their corresponding 1-year forward prices.
Spot price: A price paid now for immediate delivery of a good or service.

References

Fisher, I. *The Theory of Interest*, Macmillan, New York, 1930.
Friedman, M. The optimal quantity of money, In *The Optimum Quantity of Money and Other Essays*, chap. 1, Aldine Publishing, Chicago, 1970.
Plosser, C. Understanding real business cycles, J. Econ. Perspect., 3: 51–77, 1989.

Further Information

Hirshleifer, J. *Investment, Interest and Capital*, Prentice-Hall, Englewood Cliffs, NJ, 1970.

4.11   Productivity

Alok K. Chakrabarti

Productivity is the measure of efficiency of using factors of production. **Labor productivity**, denoting the productive efficiency of labor, is the most important factor in any policy analysis. However, we are increasingly becoming concerned with **total factor productivity**, measuring the efficiency of using all other factors of production.

Productivity has been the most important economic indicator for understanding and predicting economic prosperity. The economic decline in the United States and elsewhere during the mid-1970s through early 1980s had much to do with decline in both labor and total factor productivity. Consequently many speculations have been proffered to explain the slowdown of productivity.

Western countries such as the United Sates, Germany, the United Kingdom, and Japan have followed the trajectory of growth primarily by a growth of productivity, rather than the increase in quantity of labor. Edward Denison [1985] has estimated that 68% of economic growth of the United States from 1929 to 1982 may be attributed to increase in labor productivity. Increase in labor productivity has helped nations such as Japan and Germany to catch up with the United States and the United Kingdom during the postwar period. From 1950 to 1981, Japan experienced more than 6% annual growth of labor productivity, Germany experienced 4.5% growth, while the United States, Canada, and the United Kingdom experienced approximately 2% growth [McConnell, 1987]. However, from 1973 to 1990, the productivity of these nations plunged to a very low level, causing a worldwide recession [Denison, 1994].

Changes in productivity have three major consequences. First, the change in standard of living is related in growth in productivity. Real wage increase is only possible when there is a growth in productivity. Second, a decline in productivity is likely to induce a cost-push inflation. Third, a decline in productivity reduces the competitive strength in the international trade. This eventually sets up a vicious circle for further decline in productivity.

There are several factors that contribute to changes in productivity: **technological progress,** quantity of capital, education and training, economy of scale, improved resource allocation, and legal- human

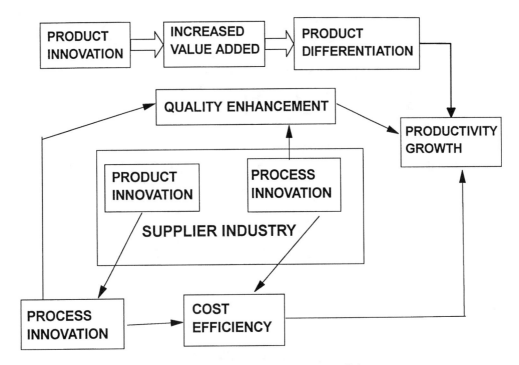

FIGURE 4.10   Productivity and innovation linkage.

environment. Denison [1985] estimated that the legal-human environment had a negative impact on the growth of productivity. Of course, technological advances have the largest impact on the productivity growth. The pioneering work in this area was done by Robert M. Solow [1957], who first established the quantitative linkage between technology and productivity.

Baily and Chakrabarti [1988] examined the linkage between technological innovation and productivity changes in several industries — chemical, textiles, and machine tools — in the United States during the 1970s. Their conclusion was that a slow down in innovation in these industries was linked with a slow down in the productivity. Figure 4.10 depicts the productivity and innovation linkage. The product innovations lead to increased **value-added** product that may help the firm command a higher price in the market, particularly if it creates a product differentiation. Innovation in process also leads to cost efficiency as well as quality enhancement. However, one should recognize the interindustry linkage in changes in productivity. For example, much of the productivity improvement in the textile industry was due to the textile machinery developed in the machinery industry. Scherer [1984] has shown that innovations in the supplier industry have a multiplier effect in productivity increase in the economy.

The complex linkage between the textile industry and various other industries is shown in Fig. 4.11. Production of textile products may simply consist of several steps:

- Fiber preparation: cotton is ginned and separated from foreign particles.
- Spinning: yarn is drawn from the cotton (or fiber material).
- Weaving (or knitting) of the yarn into cloth.
- Finishing: this step involves bleaching, dyeing, and finishing the cloth.

Several industries contribute to the productivity enhancement of these steps in textile production. The innovations in the equipment industry help the carding and ginning of fibers in fiber processing. New looms have substantially improved the speed and quality of the finished products. New knitting machines have made new products possible at a much higher efficiency. Innovations in the chemical industry have provided new materials for dyeing and finishing. Finally, the advancements in computer technology have

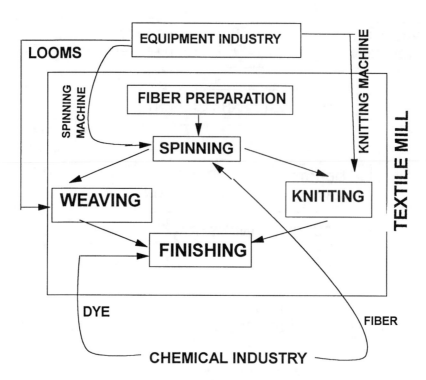

FIGURE 4.11   The textile industry.

contributed to increasing the productivity of various aspects of manufacturing. For example, computerized color control instruments have helped improve the quality control in finishing cloth. Other computer control systems have facilitated the productivity of weaving, spinning, knitting, etc.

In a cross-country comparison of labor productivity in manufacturing, McKinsey & Co. [1993] has observed that the United States has higher labor productivity than Japan and Germany. However, industries in these countries do differ in productivity for various reasons. For example, the steel industry in Japan is 45% more productive than its counterpart in the United States. However, the United States shows a better productivity in industries such as computer, beer, food processing, and soaps and detergents. The following factors explain the productivity differences in different industries in these countries:

- Output: mix, variety, and quality
- Production factors

    Technology intensity and age of plant and machinery
    Scale of operations
    Design for manufacturing
    Basic labor skills and motivation
    Raw materials and parts

- Capacity utilization
- Organization of functions and tasks.

In processed food and soaps and detergents industries, the output, in terms of mix, quality, and variety, is important. Machinery and plant and equipment as well as scope of operations are important in several industries, such as steel, metalworking, beer, and processed foods. Design for manufacturing is important in industries such as metalworking, automobiles, parts, and consumer electronics. Organization of functions and tasks is important in industries such as steel, metalworking, automobiles, parts, and consumer electronics.

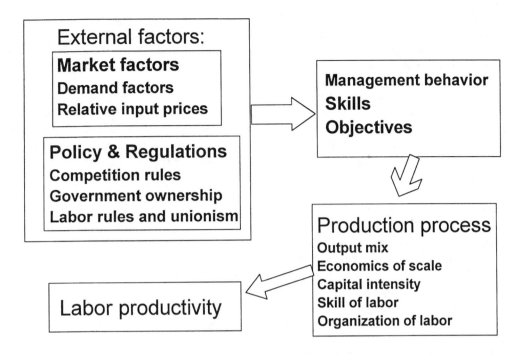

FIGURE 4.12   Determinants of labor productivity service sectors.

The McKinsey Study pointed out that "foreign direct investment (transplants) has been a more powerful way of improving productivity than trade has been, especially in Germany and in the U.S. Transplants from leading edge producers: (1) directly contribute to higher levels of domestic productivity, (2) prove that leading edge productivity can be achieved with local labor and many local inputs, (3) put competitive pressure on other domestic producers, and (4) transfer knowledge of best practice to other domestic producers through natural movement of personnel."

McKinsey & Co. also made a comparative study of several service industries, such as the United States, United Kingdom, airlines, telecommunications, retail banking, general merchandising, and restaurants in the United States, the United Kingdom, France, Germany, and Japan. Except for the restaurants, the United States has higher productivity than any other country. France has the highest productivity level in the restaurant industry. Figure 4.12 provides a schematic diagram for the determinants of labor productivity in service sector. External factors such as government policy and regulations related to competition rules affect labor productivity in all the industries. Government ownership also contributed to the differences in productivity in the airlines and the telecommunication industries. Among the factors related to the production process, the most important one is the organization of labor. Capital intensity and vintage contributed to the productivity difference in retail banking. The banks in the United States have invested heavily in information technology in processing many of the tasks. This investment has yielded difference in labor productivity compared with the banks in other countries. The most important factor explaining the productivity difference in service sector is management behavior in terms of the innovative skills and objectives for improving productivity. The study concluded that "managers in the U.S. make different trade-offs when they formulate and prioritize their objectives and that the choices are primarily determined by external factors. Managers in Europe and Japan pursue explicitly more social objectives, whereas management goals in the U.S. are heavily focused on economic objectives. Thus, in the U.S. managers focus more on pursuing productivity, while in Europe and Japan, managers trade productivity against other possibly social goals."

In a discussion of productivity, one should understand the implications of the policies aimed at increasing productivity. While productivity is the key variable in economic growth, but there are other sociopolitical goals that may not be sacrificed. Public health and safety are critical variables that cannot be ignored when one is trying to maximize productivity. Any policy that boosts labor productivity resulting in mass unemployment, homelessness, and increased crime should be troublesome in any country. Governmental regulations and laws are enacted to prevent such actions.

Increasingly environmental concerns are becoming important. This leads to the need for trading off actions for short-term increase in productivity with protection of the environmental quality. Conservation of nonrenewable resources is also gaining widespread importance. Various public interest groups are helping formulate the policies regarding usage of nonrenewable resources. This may temporarily hinder productivity growth. Baily and Chakrabarti [1988] observed, that during the 1970s, firms in the chemical industry were more involved in innovations dealing with environmental protection than in innovations that would lead to productivity growth.

Changes in productivity have a long-term impact on future jobs and development of human capital. As the productivity increases in certain industries, jobs may become fewer. This would necessitate people to adjust by seeking jobs in different industries. Government policies ought be formulated to facilitate such transitions in vocations and jobs. Regional imbalances also develop due to industrial changes. For example, the massive changes in employment in the steel industry have adversely affected the so-called rust belts in the midwest in the United States, and the Ruhr districts in West Germany. The government has a role to play in helping these regions to cope with these problems.

The dynamic relationship between government and the business firms in influencing productivity changes has been best described in Fig. 4.13. As the figure shows, the relationship is indeed a circular one. The McKinsey Study has termed it "productivity cycle". External factors create a competitive environment that changes management behavior. Innovations in production processes are created through changes in management behavior. Innovation changes the productivity, which in turn changes the economic and social well-being of the society affecting the external factors. The cycle then repeats itself.

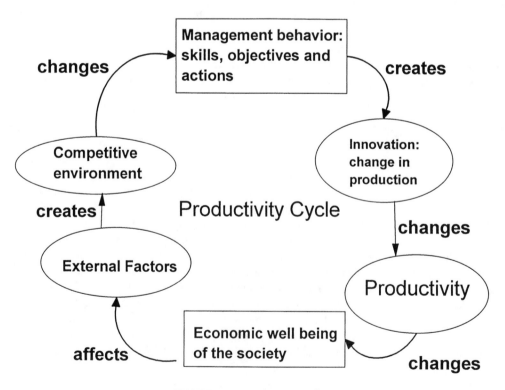

FIGURE 4.13   Productivity cycle.

Defining Terms

Labor productivity: Labor productivity is the efficiency with which labor is utilized in producing goods and services.

Productivity: Productivity is the efficiency with which resources are used to produce goods (and services) for the market. It is measured by computing the ratio of an index of the output to the index of the input.

Technological progress: Technological progress is said to occur when the same combination of factors of production yields a higher level of output.

Total factor productivity: Total factor productivity is the efficiency of using all the factors such as labor capital, raw materials, and energy in producing goods and services.

Value added: It is an index of output for an industry or a firm. It is computed as the market value of the goods and services produced minus the purchases of goods and services used in production.

References

Baily, M. N. and Chakrabarti, A. K. *Innovation and the Productivity Crisis* , Brookings Institution, Washington, DC, 1988.

Denison, E. F. *Trends in American Economic Growth* 1929–1982, Brookings Institution, Washington, DC, 1985.

Denision, E. F. Productivity, In *The Encyclopedia of the Environment*, R. A. Eblen and W. R. Eblen, ed., pp. 579–580, Houghton Mifflin, New York, 1994.

McConnell, C. R. *Economics: Principles, Problems and Policies*, McGraw Hill, New York, 1987.

McKinsey Global Institute. *Service Sector Productivity*, McKinsey & Co., Washington, DC, 1992.

McKinsey Global Institute. *Manufacturing Productivity*, McKinsey & Co., Washington, DC, 1993.

Scherer, F. M. *The World Productivity Slump: Discussion Paper IIM/P 84-25*, International Institute of Management, Berlin, 1984.

Solow, R. M. Technical change and aggregate production function, *Rev. Econ. Stat.*, 39: 1957.

Further Information

The Bureau of Labor Statistics at the Department of Labor, Washington, DC, collects and publishes data on labor productivity in different industries in the United States.

The Brookings Institution in Washington, DC, conducts studies on productivity and economic growth and their impact on the government policy. The journal *Brookings Papers on Economic Activities* publishes papers based on the research conducted at Brookings.

4.12 Trade: Import/Export

Arnold Reisman

Trade — the exchange of goods or services — can be subdivided into two basic categories: foreign and domestic. In domestic trade the exchange takes place within the confines of a single nation. On the other hand, in **foreign trade**, sometimes referred to as world or international trade, the objects of trade cross international boundaries. Foreign trade can be further subdivided into export or import. In the case of exports, goods and services flow out of a country. The reverse is true for imports.

International trade amounts to several trillion dollars a year, and most of it involves private individuals or firms. Government-to-government trade accounts for a small portion of the volume as measured in any of the "hard currencies" (HCs). A **hard currency** istypically the money of the major trading countries. These are the United States, Germany, Japan, France, and the United Kingdom. However, any currency that is freely convertible on the open/established foreign exchange markets into currencies of the above nations can also be considered as being "hard". The foreign exchange rate, e.g., the number of yen required to buy or sell a U.S. dollar is widely known on any given day. Minor variations from this rate can be found in any given locale. Trade with firms in the less-developed nations, including those of the former Soviet Union and its satellites, is usually done in the HCs of Western nations. The U.S. dollar dominates that market. However, not all international exchanges of goods and services are fully paid for in cash. Much trade involves some form of reciprocity in lieu of cash — the essence of **barter** or countertrade (B&C); it will be discussed later. Also, much domestic trade within nations facing high inflation rates and/or having weak currencies is often transacted in U.S. dollars, deutchmarks, etc. For example, in the late 1990s, such is the case in all of the countries arising from the Soviet Union.

Highly industrialized nations typically export goods having a high value-added content such as aircraft, computer and telecommunications equipment, machinery, etc. and import basic commodities such as oil, metals, coffee, etc. as well as goods that require a high, yet cheap, labor content in its manufacture, e.g., clothing, shoes, toys, etc. Nevertheless, the United States is still a major exporter of agricultural commodities such as grain, tobacco, cotton, etc.

Given a certain level of quality, delivery, service, and/or style, most goods are bought and sold on the basis of price. In a relatively free market economy, these goods will be made in those countries where they can be made at the lowest total cost. Thus, it is cheaper to assemble clothing in developing economies than it is to do the same in the United States, Germany, or the United Kingdom. At the other end of the spectrum, as we approach the turn of the century, some Western firms find that it is cheaper to contract out sophisticated computer software writing to firms in developing countries such as India than it is to do it domestically.

According to the theory of economics, each country should concentrate on producing the goods in which it has a comparative advantage, that is, the goods that can be made most efficiently within that country as compared to all others and import goods that can be made more efficiently elsewhere. However, in reality, market economies are not always perfect nor are they completely independent from government interventions. For instance, in 1997, the government of Israel contributed around U.S. $600,000, so that a company called Intel would expand its chip manufacturing facilities over there and by doing so create jobs requiring technical knowledge and skills, in other words, better-paying jobs. Such subsidies and various tax abatement formats are not atypical. They are practiced in highly developed countries. In the United States, this is typically done by local community or state governments. There are many other ways that governments impact international trade across its borders. The **tariff** or duty imposed on imports is a favorite mechanism. Some governments impose quotas on imports. This often obliges the importing firm to secure a government permit. The word protectionism is used to generically describe such government policies. The various formats of protectionism are used to encourage growth of local industry, protect existing industries, develop geographic regions of the country, counter the dependence on agriculture, and to improve the balance (exports vs. import) of trade. Also, for reasons that are strategic in nature, some governments prohibit the export of any equipment that can change the military capabilities of unfriendly nations. In the late 1990s, the United States placed a ban on such exports to Iran, Iraq, etc. and required licensing for the exporting of supercomputers to even friendly nations such as Israel. Trade barriers have a negative impact on the volume of business transacted, they increase prices paid by the consumer, and they deprive nations of the benefits that can be accrued from specialization.

On the other hand, for economic or strategic reasons, developed countries often eliminate most or all barriers to trade with certain other countries. Most such agreements involve only two countries and are known as bilateral or reciprocal trade agreements. The United States gave China the most favorite nation status in spite of misgivings in Congress about China's human rights record, exploitation of its prison and other populations, its shipment of strategic weaponry to nations unfriendly to the United States, such as Iran, its widespread pirating of U.S.-owned intellectual property rights, e.g., computer software, books, various electronic formats of entertainment (discs, cassettes, etc.) as well as its threatening posture vis-a-vis Taiwan — a friendly nation that has survived, for decades, due to U.S. military presence in the area.

In recent years however, agreements involving several countries have been recorded. Such is the case with the European Union, originally involving countries of Western Europe but gradually bringing in others as full or partial members. NAFTA was signed by the United States, Mexico, and Canada.

Governments typically keep track of the known transactions with one another's country. Thus, all transactions involving imports and exports of goods and services, investments abroad, the income from such investments, foreign aid, money spent by tourists, etc. are recorded. Some of these, such as tourist spending, are, in part, estimates. The value of exports in any given year compared to that of imports between the two countries is known as the balance of payments.

Importers who offer to pay for the goods or services received typically are asked to provide a **letter of credit** along with the purchase agreement. A letter of credit is issued by a third party, usually a well-established bank. It assures payment to the exporter once the goods have passed a certain milestone such as having been shipped. A bill of lading — the principal document between carrier/transporting company and the shipper usually suffices to have the money released according to the letter of credit. It acts as a receipt or evidence of the ownership of the goods. However, some foreign trade is executed on no greater formality than a letter of intent or just a handshake. In the former the signer declares an intention to buy, manufacture, or deliver, prior to executing a formal contract. Surprisingly, a number of airliners have been delivered by Boeing Aircraft on the basis of no more than a letter of intent to purchase. Much international trade in the wholesale diamond industry is handled on a handshake basis. However, any person who reads this article would do better with a formal contract accompanied by a letter of credit from a recognized bank located in a Western country.

As indicated earlier, the best-known and most-efficient form of trade wherein the goods and services are fully paid for in HC is not always an option to an exporter. Many firms and governments, for lack of HC or other reasons, demand full or partial payment, in some form of reciprocity. This is true not only in developing but also in developed nations. For instance, in concluding a multimillion dollar contract for AWAC aircraft to be supplied to the Royal Air Force. Boeing agreed to buy back the same dollar volume of British-produced goods. Thus, **barter**, the most ancient form of commerce, and several modern variants on the theme are indeed invoked in many cross-national transactions

Palia and Liesch [1991] provide statistics showing the extent of B&C involving Digital, Apple, Hewlett Packard, Honeywell, Unisys, IBM, and other well-known corporations. Moreover, they provide data of contract values by country. In addition to the United States, United Kingdom, Germany, and Japan, they show Canada, Israel, and Sweden. Lastly, they break the contract values down by type of "commodity", e.g., aircraft, spacecraft, and software systems, among others. The number of countries involved in countertrade now exceeds 100 and is expected to grow [Verzariou, 1992]. This phenomenal growth in its popularity suggests that any business that is currently engaged in or is planning to engage in international marketing cannot avoid B&C.

Verzariu [1992] summarizes the position of the U.S. government on all aspects of B&C. All of the relevant statutes and executive orders are discussed in a user-friendly manner for business decision makers. A companion volume by Versariu and Mitchell [1992] provides a summary of country-by-country practices. B&C accounting policies and practices are discussed in Bost and Yeakel [1992] while Schaffer [1990], among others, addresses the business strategy issues. However, developing prescriptive decision-aiding models for B&C has drawn little attention in the literature [Reisman, 1996].

The Taxonomies of Barter and Countertrade

The literature of B&C is disjoint for a number of reasons. This field of knowledge and practice is broad in scope and is addressed by disciplines that do not often communicate one with the other. Following is a taxonomical approach to B&C. The objective is to describe the field in a manner that is both efficient in the use of language and effective in communicating the coverage of the breadth of the field's scope.

The variety of B&C forms can be classified based on four major factors: (1) contract time duration. (2) requirements of financing, if any, (3) the number of parties involved, and (4) the extent of technology transfer [Reisman et al., 1988] More specifically,

1. Classification by the time factor
   a. Short-term, one time transaction (denoted by t)
   b. Long-term, multiple transaction (denoted by T)
2. Classification by financing requirement
   a. No financing is needed (denoted by f)
   b. Some form of financing (or cash payment) is required (denoted by F)
3. Classification by the number of parties involved
   a. Only two parties are involved (denoted by n)
   b. More than two parties are involved (denoted by N)
4. Classification by technology transfer
   a. No technology transfer is involved in the transaction (denoted by k)
   b. Technology transfer is involved in the transaction (denoted by K)

It can be seen from the above classification that the most complicated (or general) case of B&C is described by T, F, N, and K. At the other extreme the simplest case is described by t, f, n, and k. Figure 1 graphically represents all the major conceptual subgroups of B&C formats based on the above taxonomy. Some of the most popularly used types of B&C are defined and classified next.

Summary

One can find many examples of cooperation, based on real reciprocity, between large and small firms home based in large and small countries. Such reciprocity often involves R&D as well as marketing [Reisman, 1997]. Moreover, the marketing is not limited to a partner's national market. Because of the existence of trading blocks, political or other barriers, blocked currencies, unfavorable exchange rates, etc. as well as simple logistics, such reciprocity can transcend national boundaries of either partner. Often and at first glance, such partners look more like natural competitors. Yet, experience has shown that they offer one another synergy that could not have been otherwise attained for any one of a multitude of reasons. To be competitive in today's global markets, one must be creative in structuring deals and open to some form of nonmonetary compensation, e.g., some form of B&C.

Defining Terms

Barter: A barter is a short-term transaction between two parties calling for the direct exchange of goods and/or services. In terms of the above taxonomy, it corresponds to "tfnk".

Blocked currency: This is a method of getting products out of a country to compensate for the currency restrictions and the difficulties of repatriating holdings ("Tfnk", "TFnk").

Clearing: This is a bilateral agreement between two countries to exchange a specified amount of each other's products over a period of time ("Tfnk").

Compensation: Two separate and parallel exchanges; in one contract a trader agrees to build a plant or provide equipment or technology with a down payment usually made by the opposite party upon delivery. In the second contract the trader agrees to buy a portion of the plant's products. Compensation may take on the "tfnK", "TfnK", and the "TFnK", or the "TfNK" variety and joint ventures are represented most often by "TFnK".

Cooperation agreement: Three separate exchanges among a party in a developing country and two Western suppliers with one specializing as a seller and the other as a buyer; these three-party arrangements may be handled through a triangle barter deal or counterpurchase ("tfNk", "TfNK")

Counterpurchase: Two separate contracts: one in which a trader "sells" products for cash and/or credit and in the other the supplier agrees to "purchase" and market products from the opposite party ("tFnk").

Joint venture: There are two different versions of joint ventures, *equity* and *contractual*. In an equity joint venture, whatever the form of investment, the equity ratio must be calculated in money terms; the risks and profits are shared according to this ratio. In a contractual joint venture, on the other hand, investment may be made in different forms and the equity ratio is not necessarily in money terms, but is stipulated in the contract. Moreover, in an equity joint venture a partner's investment can only be recovered from his share of profits. In contractual joint ventures, the investment can be recovered in a variety of ways. Most joint ventures can be represented by "TFnK".

Offset: Two separate exchanges operating similarly to a counterpurchase, with the distinction that the reciprocally acquired goods/services can be used by the trader in his business ("tFnk").

Swap: This is an arrangement where in goods destined to different locations are redirected in the most economical way. In other words, they find their way to markets physically closer to the point of origin ("tfNk", "TFNK").

Switch: Switch is an arrangement in which the party to a bilateral agreement transfers some or all of its goods to a third party or nation ("tfNk", "TFNk").

References

Bost, P. J. and Yeakel, J. A. Are we ignoring countertrade?, *Mgt. Account.*, 74(6): 43–47, 1992.

Neale, C. W., Shipley, D. D., and Dodds, J. C. The countertrading experience of British and Canadian firms, *Mgt. Int. Rev.*, 31(1): 19–35, 1991.

Palia, A. P. and Liesch, P. W. Recent trends in Australian countertrade: a cross-national analysis, *Asia Pac. J. Mgt.*, 8(1): 85–103, 1991.

Reisman, A. *Management of Science Knowledge: Its Creation, Generalization, and Consolidation.* Quarum Books, Westport, CT, 1992.

Reisman, A. Barter and countertrade the well established form of trade: a new frontier for OR/MS, *Int. J. Operat. Quant. Mgt.*, 2(1): 35–47, 1996.

Reisman, A. Impact of technology transfer, barter and countertrade on the competitiveness of firms in international markets, In *Globalisation and Competitiveness Implications for Policy and Strategy Formulation*, pp. 53–73, M. Oral and O. Kettani, eds., Bilkent University Press, Ankara, Turkey, 1997.

Reisman, A., Aggarwal, R., and Fuh, D. C. Seeking out profitable countertrade opportunities, *Indust. Market. Mgt.*, 18(1): 1–7, 1989.

Reisman, A., Fuh, D. C., and Li, G. Achieving an advantage with countertrade, *Indust. Market. Mgt.*, 17(1): 55–63, 1988.

Schaffer, M. Countertrade as an export strategy, *J. Bus. Strat.*, 11(3): 33–38, 1990.

Verzariou, P. *International Countertrade: A Guide for Managers and Executives*, U.S. Department of Commerce, Washington, DC, 1992.

Verzariou, P. and Mitchell, P. *International Countertrade: Individual Country Practices*, U.S. Department of Commerce, Washington, DC, 1992.

4.13   Understanding the Value Chain

Marc S. Gerstein

What Is a Value Chain and Why Is it Important?

A value chain is a sequence of activities describing the movement of products or services from their basic *inputs* or "raw materials" through a series of enhancement or "value-adding" steps such as manufacture, distribution, and after-sales service to become a series of *outputs* [Porter, 1990].

In general, the relationships between value-adding activities are best described as a system, web, or network rather than a sequential chain. However, for semantic simplicity, I will employ the term value chain because of its widespread use even though the processes it describes constitute a complex dynamic system rather than a simple, linearly sequential one.

Value Proposition, Competitive Advantage, and Risk

The actual or projected costs at each stage in the value chain provide information about the economic performance and potential of the enterprise. Market prices at various value-added stages permit first-order profit margin ewstimates by subtracting the total cost accumulated for a set of activities from the total market value of the goods and services at each stage.

In order to create profit, any business must be based on an economically valid **value proposition**. While this may seem self-evident, there are many cases in which businesses created during one time period face changes in circumstances such as new technology, new regulations, or new market conditions that undermined their original value proposition in a dramatic manner. For example, the prestige of Swiss watches was based on their accuracy, which was in turn a product of precision engineering and manufacturing. The invention of the quartz movement, less expensive and more precise than a physical mechanism could ever be, changed all that, shifting the locus of world watch production from Europe to the Far East in the process and driving Swiss market share from 50 to 12% [Hirasuna, 1997].

Beyond the basic economic foundation, in order to prosper over the long run from one's value-adding activities, it is necessary to create **competitive advantage**. Such advantage may arise from unique access to raw materials or markets, specialized technology, licenses or know-how, or unusually effective ways to promote, market, sell, and distribute one's goods and services.

Regrettably, all advantages are only temporary and must be reinforced through innovation or they inevitably become vulnerable to erosion or direct attack. However, in addition to the need for a break-through of some kind, most upsets require that market leaders be unwilling or incapable of responding to the innovative threat [Moore, 1991].

For example, in the 1950s, the major U.S. vacuum tube manufacturers entered the semiconductor industry, but none could master the new core tasks well enough to succeed. For different reasons but with similar results, the major U.S. commercial aviation players of the 1950s left the door open to Boeing when they deliberately shelved their jet aircraft development projects when the British-made Comet proved unstable in flight. The message is that there may be risks when one innovates but there are also risks when one does not.

Thus, as the context for formulating strategy or designing a new organization, one must recognize that the value chain upon which the business is based is not likely to remain stable. Its underlying economics and its current pattern of competitive advantages, revenues, costs, and profits are likely to be altered by changing technology, customer needs, and societal conditions.

While such changes were sufficiently problematic when everything was under one corporate roof, today's abrupt shifts in external players' value-adding potential makes the design of "virtual organizations" (in which a business' value chain is put together from elements that span multiple organizations) profoundly more difficult and risky than designs in which everybody draws the same paycheck and ultimately has the same goals. If we are to construct value chains in the dynamic world of virtual organizations, we need a more effective way to analyze them.

Analyzing the Value Chain Using Coordination Theory

An analytically precise language to describe value chain's interdependencies has recently been developed by Tom Malone and his colleagues at the Center for Coordination Science at MIT [Malone et al. 1997], building on the work of Thompson [1967] and others (see Mintzberg [1979]).

Dependencies describe the relationship between *activities*, which delineate some work action, and resources, which constitute the inputs and outputs of that work. There are three primary dependency types:

Flow dependencies	Activities produce resources that are used by other activities in typical "flow chart" fashion; effectiveness requires that the "right things" be in the "right places" at the "right times" (this can become highly complex when delays and feedback loops exist)
Fit dependencies	Multiple activities contribute to the creation of a single resource; achieving needed fit involves time, iteration, and mutual adjustment to resolve interdependencies, make trade-offs, and debug complex interaction effects
Sharing dependencies	The mirror image of fit: multiple activities share the same resource, such as a person, machine, or budget; sharing dependencies are subject to capacity constraints and scheduling conflicts.

TABLE 4.4 Characteristics of Dependency Types

Dependency Type		Definition	Example Coordination Mechanism
Activity	Resource		
Flow		One activity produces a resource that is used by another activity	
Usability ("right thing")		Resources must have certain characteristics in order to be usable by the next activity	Use standards (e.g., sized clothing)
			Assess individual user needs and produce custom versions
Accessibility ("right place")		Resources must be transferred from point of creation to point of use	Ship by various transportation modes
			Make at point of use
			Access "remotely" (if feasible)
Prerequisite ("right time")		The resource must be produced in time for it to be used; often implies production scheduling	Make to order
			Make to inventory
Fit		Multiple activities collectively produce a simgle resource, such as the design of parts of the same car (engine, body, transmission)	Total simulations (Boeing)
			Daily build (Microsoft)
			"Debugging" (computer software)
Sharing		Multiple activities use the same resource, such as a common person, piece of equipment/software, or common budget	First come, first served
			Market-like bidding systems
			Allocation or budgeting schemes
			Managerial decision

Source: Malone et al. [1997].

Coordination is the process of managing these three types of dependencies. Over the years, a wide variety of techniques and methods has arisen to accomplish this, some so commonplace that we hardly notice them. For example, a parent adjudicating a dispute among his children over access to the family's only personal computer is managing a sharing dependency not dissimilar to management allocating access to an organization's scarce high-tech equipment. The relationship between dependencies and their respective coordination mechanisms is described in Table 4.4.

Managing Flow Dependencies

Today, we typically manage flow dependencies by seeking to orchestrate the movement of a large number of "standardized" items. For instance, by managing the logistics of "ready-to-wear" clothing, we allow a wide variety of people to obtain articles that meet their requirements for style, size, and price.

However, to satisfy a wide range of needs using standard sizes and styles, retailers have to undertake complex planning and distribution activities directed toward placing the right merchandise in the stores at the right time, which includes forecasting demand by style, color, and size, then moving the goods from each manufacturer through the distribution system and into the stores. Moreover, one must also replenish the stock based upon what"s selling.

With potentially hundreds of suppliers, thousands of stock-keeping units, multiple warehouses and distribution points, and tens (or hundreds) of stores in the typical chain, this is a daunting logistical undertaking. It is no wonder that we've all experienced the problem that something we need happens to be somewhere other than where we are. The 32-inch waist, 32-inch length stone-washed jeans are in stock somewhere — just not in this store.

Many industrial value chains (often called "supply chains") are analogous to that of the retailer and, over the years, management tools, including a variety of production scheduling, distribution system modeling, and demand forecasting systems, have been developed to coordinate the many interdependencies of this generic value chain.

In every field, however, some businesses inevitably take a different approach to coordination from their industry norm. Catalog sales companies, for instance, have eliminated the need to get items to the "right place", i.e., the store, by eliminating it altogether. Using a combination of detailed catalog descriptions and knowledgeable telephone sales personnel, merchandise in carefully selected standard units is shipped directly from the warehouse or factory to the customer's premises.

For those organizations with critical flow dependencies, the role of organization design is to create a framework in which these dependencies can be most effectively managed. Almost without exception, this requires highly effective horizontal information and logistics systems which, ever increasingly, extends beyond the boundaries of the organization to its suppliers, distribution partners, and customers.

Managing Fit Dependencies

Fit dependencies involve interactions between components that have to be created with other components in mind. The engine, transmission, and body of a racing car have to be designed to work together, as do parts of a complex digital system such as a computer. Thus, fit dependencies often characterize creative rather than traditional production tasks. In many cases, such as new product design, the work is done one time only.

In such undertakings, design of the separate components often occurs in parallel, and methods are employed to test whether the whole can be created from the sum of the parts. As in flow dependencies, complexity in the interaction of fit dependencies is reduced through standardization. In computer software, for example, "application programming interfaces" or APIs, standardize the manner in which program components "talk to" one another, thus reducing the complexity of the overall fit problem and facilitating more parallel work. (Such shifts from fit-dependent work to a flow-based work is a key strategy in a number of process redesign approaches.)

Standardization notwithstanding, fit-dependent efforts are characterized by the need for mutual adjustment and trade-offs throughout the work process. Consequently, diagnosing problems arising from unforeseen interactions is often fiendishly complex. Things may work separately, but not together. They may work under certain circumstances but not others. In this excerpt from Tracy Kidder's [1981] Pulitzer Prize-winning story about the design of a new minicomputer, two engineers struggle with a "fit dependency bug":

> Holberger and Veres hook the probes of two logic analyzers to various parts of Gollum [a prototype machine], and they set the analyzers so that they will snap their pictures when the machine fails. They call this 'putting on the trace.' They back up the program just a little ways from the point of failure; they run it, and it doesn't fail. Another clue. It suggests that they may be facing a "cache interaction problem." In a machine with accelerators, history is important; often it's some complex combination of previous operations that leads to failure later on. So now Holberger and Veres start the diagnostic program all the way back at its beginning and go out to the cafeteria for a cup of coffee. About fifteen minutes later, when they have returned to their chairs in front of Gollum, there is a quick flash on the screens of the analyzers. The machine has failed. They have their pictures. They pull up their chairs and start studying snapshots of signals.

Managing Sharing Dependencies

If the coordination challenge of flow dependencies is that of logistics, and that of fit dependencies the dilemmas of creation, then the challenge of sharing dependencies is that of scarcity. Since it is seldom possible to have enough of every resource, managerial solutions have arisen to address this problem. Whatever the approach, managing a sharing dependency must contend with two issues:

- *Allocation*: who will get the resource and how much. Decision tools may employ decision making by authorities, bidding, and other "market-based" mechanisms or the applica;tion of rules based on overall goal-related criteria.
- *Scheduling*: the timing of usage. Scheduling seeks to maximize the effectiveness of the resource by effectively utilizing slack time, resequencing activities to minimize contention, and so on.

The organization design issues of sharing dependencies arise because the context for deciding the optimum use of a scarce resources extends beyond the boundaries of the direct supervision of the resource. In both allocation and scheduling, *prioritization* determines who will get how much and when they will get it. Prioritization always involves judgment. Consequently, the organization design must develop (1) measures of overall benefit that form the basis for prioritization, (2) mechanisms to include various organizational perspectives in decisions, and (3) conflict resolution mechanisms to resolve the disputes that inevitably arive whenever there is not enough to go around.

Improving Quality and Cycle Time in the Value Chain

Starting in the 1980s, manufacturing organizations in the United States undertook a fundamental rethinking of their production management philosophy as a result of a serious deterioration of their competitive position at the hands of foreign (mostly Japanese) competitors. The views of David Kearns [Kearns and Nadler, 1992], former CEO of Xerox corporation, were typical:

A lot of our problems boiled down to the Japanese. They were really eating us for lunch. I had made something like two dozen trips to Japan… to understand how far ahead of us the Japanese were. With each trip, the revelations grew more sobering. … they were selling products for what it cost us to make them.

A series of initiatives — total quality, just-in-time and reengineering, among others — transformed both the techniques for managing production systems as well as the basic assumptions upon which traditional approaches rested. In contrast with many earlier initiatives that were based on creating formal rules, procedures, and supervisory practices, the process improvement initiatives of the 1980s focused on *how* work was done rather than the manner in which it was administratively organized. Some of the most significant problems unearthed were:

- Accounting-based cost, not time, was the principle focus of the performance measurement system and the focus of improvement programs. In the absence of measurement and focus, only between .05 and 5% of the time that products were in the value-creating system were they actually receiving value. The rest of the time, they were waiting [Stalk and Hout, 1990].
- Defects tended to be "inspected out" rather than quality "engineered in". This reflected a simple linear view of the production process in which defects were valued at their direct costs and the dynamic complexities of rework, delays, customer dissatisfaction, etc. were neither understood nor accurately valued by traditional cost accounting systems.
- Raw material and work-in-process inventories were *assumed* to be essential to the smooth flow of the supply chain. It was not considered technically feasible to create high reliability supply chains without inventory buffers between steps; thus, their creation was not attempted.
- Most performance improvements were typically undertaken within the boundaries of individualized budgetary departments and tended to focus on local cost reductions. In many cases, this either simply reduced the value created, exported the same cost elsewhere, or magnified the system-wide cost to obtain a small local benefit. In general, there was no concept of an overall "cost of quality".

The movement to rethink manufacturing during the 1980s was marked by its questioning of *basic assumptions* about the design of productive systems. The result was a transformation of both the *production paradigm* and the dominant *organizational architecture*. Change was revealed in seven important ways:

1. Rather than an internal, cost-oriented focus, the new initiatives tended to focus on *external customers* and those aspects of the product or service that customers, rather than management, considered of value. The slogan became "the customer is king".

2. A *process focus* rather than a structural chain-of-command orientation came to preoccupy management consciousness. Management learned to look horizontally, not just vertically.

3. *Time reduction* emerged as a cornerstone of virtually all improvement efforts. Starting with production, but eventually moving to all other process steps, reducing the "concept-to-cash" cycle time became a key yardstick for charting improvement.

4. A *combination* of improvements was undertaken rather than a single "magic bullet" approach. In addition, *organization development*-related approaches, participative methods, and the widespread use of teams would come to be considered inseparable from production innovations in the planning and implementation of improvement programs as well as in the operating organization's basic design.

5. The *boundaries* of the improvement effort did not stop at the boundaries of the organization. Following the early lead of the Japanese, suppliers, business partners, and customers were included in the scope of the effort from the planning phases through redesign and implementation.

6. The aggressive use of *information technology* to create "end-to-end process capability" changed the historical pattern of creating "islands of automation" to one of creating "enterprise systems" that focused on upgrading the entire supply chain and its key processes, including those activities that involved suppliers, partners, and customers.

7. *Radical breakthroughs* followed by *continuous improvement* characterized the most successful initiatives. This parallels the adoption of similar philosophies in Japan, but stands in sharp contrast to the static benchmarks and modest improvement targets that tended to characterize goal setting in earlier eras.

By the mid-1990s, these initiatives had evolved into efforts to transform one's entire supply chain to become *lean* and *agile*. Lean production systems are flexible, tightly coupled systems requiring that the rapid, frequent flow of goods and ease of manufacture, high reliability, and service efficiency be built in at the design stage [Levy, 1997]. Agility reflects the organization's capacity to anticipate and embrace changing conditions, whether a new customer requirement, a competitive threat, or a techn ological breakthrough. To paraphrase Tom Peters, agile firms "thrive at the edge of chaos."

Conclusion and Implications

The value chain is a dynamic, nonlinear system whose components and relationships evolve in response to technological innovations, market evolution, and social changes. Today, with the advent of virtual companies and networked organizations, the value chain is more complex ot manage than ever before.

Viewing the relationship between value chain elements as a series of dependencies (flow, fit, and sharing) enables us to analyze them effectively. Whereas earlier flow-chart representations of the value chain described the links between its "nouns", charting dependencies provides invaluable additional information by describing its "verbs". This perspective is essential if we are to construct real world organizations capable of world-class performance in speed, flexibility, and cost, all of which depend on effective coordination.

Defining Terms

Competitive advantage: The relative strengths of a company (or other entity) in a competitive marketplace. May arise from unique access to raw materials or markets, specialized technology, licenses, or know-how or unusually effective ways to promote, market, sell, and distribute one's goods and services.

Coordination theory: A set of concepts that can be used to describe business processes within the value chain.

Dependencies: The specific relationships between *activities*, which delineate some work action, and *resources*, which constitute the inputs and outputs of that work. Three generic dependency types have been identified: flow, fit, and sharing.

Value proposition: The basis for the exchange between a buyer and a seller, or a company and its customers, in which goods and services are exchanged for money or "services in kind".

References

Hirasuna, D. Design time, *J. Bus. Design*, 2(2): 1997.

Kearns, D. T. and Nadler, D. N. *Prophets in the Dark*, Harper Business, New York, 1992.

Kidder, T. *The Soul of a New Machine*, Avon, New York, 1981.

Levy, D. L. Lean production in an international supply chain, Sloan Mgt. Rev., 38: 94–102, 1997.

Malone, T. W. et al. Tools for Inventing Organizations: Toward a Handbook of Organizational Processes, Center for Coordination Science, working paper, Sloan School of Management, MIT, Cambridge, MA, 1997.

Mintzberg, H. *The Structuring of Organizations*, Prentice-Hall, Englewood Cliffs, NJ, 1979.

Porter, M. E. *The Competitive Advanbtage of Nations*, Free Press, New York, 1990.

Stalk, G. and Hout, T. M. *Competing against Time*, Free Press, New York, 1990.

Thompson, J. D. *Organization in Action: Social Science Bases of Administrative Theory*, McGraw-Hill, New York, 1967.

Further Information

The Center for Coordination Science, MIT Sloan School of Management, believes that a powerful source of intellectual leverage on questions about how groups of people can use computers will result from a better understanding of the nature of coordination. Work in the Center focuses on how coordination can occur, both with and without technology, including projects in the following areas: coordination technology (designing and studying innovative computer systems that help people work together in small or large groups); organizational structures and information technology (observing, analyzing, and predicting how information technology affects organizational structures and processes); and coordination theory (developing and testing theories about how activities can be coordinated among people and in distributed or parallel processing computer systems. For further information use the World Wide Web at http://ccs.mit.edu/ccsmain.html.

4.14 Supply Chain Management

Lynne F. Baxter

A **supply chain** is a series of customer-supplier links through which goods and services pass to a final customer. The concept is similar to the value chain; the operation and management are related, but they differ in emphasis. The primary focus is on the materials flow as distinct from the processes carried out on them. **Supply chain management** is concerned with the coordination of all the processes that enable products and services to be brought to the final customer at the right time, cost, and quality.

Recent influences in the supply chain center around five areas: time, configuration, location, the environment, and technology. Competitive pressures are encouraging the chain to operate much faster; this impacts the length and complexity of a chain; the distances between links are shortening in some instances and extending in others; and what used to be a highly face to face management activity is increasingly being influenced by communications and simulation technologies.

Fundamentals of Supply Chain Management

Supply chains are usually mapped out in terms of links between organizations.

Figure 4.14 shows a typical chain of organizations that would link together to bring, for example, an electric screwdriver to an end customer. A toolmaker fashions a tool that is sold to a printed circuit board (PCB) manufacturer, which supplies an original equipment manufacturer (OEM), which assembles a product which is sent via a distribution company to a retail chain, where a person buys it. It is common to call suppliers further away from a recognizable final product as **upstream** and those nearer the customer as **downstream**. The diagram implies that the links of the chain are close together, although this has not always been the case. There have always been management choices as to how many or few or how distant or close one needs to be from another chain member.

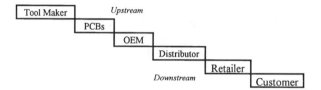

FIGURE 4.14 A diagram of a Consumer Goods Supply Chain.

Schonberger [1982] popularized the perception that Japanese organizations achieve faster cycle times and higher quality at lower cost through having closer relations with their suppliers. Their chains are demonstrably more innovative and flexible. Western organizations were more "distant" from their suppliers in terms of their business relationship, introducing the market mechanism to punctuate shorter relationships than Japanese organizations who **partner**ed their suppliers in a way that sometimes makes the boundary between the two companies untenable. The trend towards focusing on core competences leading to **outsourcing** has increased pressure for organizations to increase their supply base and elaborate the chain, which has been tempered by the Japanese influences that call for simplification and cooperation.

Various authors, including Sako [1992] and Hakansson [1989], have argued that this distinction is too finely drawn, with Western organizations having closer, longer relationships than most commentators would allow. For example, Hakansson carried out a survey of supply relationships involving technical developments and found that the weighted average length was 13 years. Mapping organization relationships this way enables a **pull** from the customer demand logic, which is consistent with **just in time** (JIT) philosophy (see below). As early as 1961, Forester demonstrated that demand information is corrupted through the chain operation, which results in high levels of distortion upstream. Unfortunately for the optimization of the whole chain, this is where the weaker and smaller organizations usually exist, which are less likely to be able to cope with such distortions.

The diagram describes the chain at the level of the organization, but many people will have encountered the term within organizations, referring to relations between individuals carrying out processes or departments. This is where the supply chain concept overlaps with that of the value chain. The diagram also highlights by omission that most organizations are actually involved in supply networks, having several options for both customers and suppliers, which does affect any link in the chain's ability to perform its task. The perspective adopted in the literature is usually that of the OEM, and it is common to talk about "first-tier" and "second-tier" suppliers, indicating that OEMs are in a position to pick and choose at levels in the chain from their position as the most privileged customer [Lamming, 1996].

This has led to supply chain management being an activity carried out by relatively powerful organizations and being reduced to selecting, performance monitoring, and frequently berating smaller, less-powerful organizations.

Applications and Examples

Ideas from the theory and Japan have been taken up in organizations in the West, but it would be fair to say that they have lost something in the translation and been diluted. For example, the upstream-downstream concept is analogous to one used in JIT manufacturing to suggest that materials should flow through the chain like a river. Most of the impetus has come from OEMs in the automotive and electronics sectors who face direct competition from organizations who are used as models of excellence, the best example being Toyota.

Early studies showed that the total cycle time for the typical Japanese supply chain to operate was far shorter, enabling a capital turnover ratio up to ten times better than their U.S. counterparts [Imai, 1986]. The process simplification, elimination of waste and employee involvement required of a JIT system yielded considerable improvements in all areas of operations. For example, a press tool machine set-up change for a front fender took 6 hours in GM, 4 hours in Volvo, and 12 minutes in Toyota. This was achieved through coupling JIT with a smaller supply base. Gadde and Grant [1984] identified that Toyota only managed 168 suppliers as compared to 3500 for General Motors and 800 for Volvo (a much smaller organisation than the other two). The relationships between the organizations differed as did the tasks performed.

Although at the time of writing the term has been around for over a decade, very few studies have addressed whole chains or networks, and most studies refer to pairs of organizations. Lamming [1996] draws a questionable distinction between **Lean supply** and supply chain management, pointing out that supply chain management practices that are overly skewed in favor of a more powerful link will not be as effective as ones that have taken a more even, whole-chain perspective. The distinction is questionable in that lean supply appears to be good supply chain management, except that good supply chain management does not happen in practice. He makes some excellent points about the translation of taking a whole-chain perspective into day-to-day management; for example, he argues that costs incurred in the chain impact the whole chain, not just the local organization. For performance measurement to mean anything, a whole-chain perspective needs to be adopted, with the OEM being examined along similar criteria as the supplier. Last, the tendency to blame another member of the chain for an error or fall in performance is the result of drawing a false barrier.

Innovations in technology such as electronic data interchange (EDI) and CALS (initially computer-aided acquisition and logistic support, now continuous acquisition and logistics support) are supportive of cooperation and waste minimization. EDI supports communications between organizations; CALS provides a framework for sharing data and information. Success stories are not that numerous, with the benefits projected for the technologies not being realized. A question that could be asked is how far has the theory influenced or been influenced by practice?

In order to be part of fewer chains, the Japanese OEMs were perceived to operate "partnering" relations with their reduced supply bases [Schonberger, 1982] instead of the market-based competitive tendering ones preferred by Western organizations. Japanese OEMs frequently had a stock holding in the supplier and cooperated with first-tier suppliers on scheduling, quality, and new product development in a way that was theoretically atypical in the West. The first-tier supplier was of higher status and capability than their Western counterpart.

Although the term partnership implies mutual benefits, it is not evident that these benefits are shared equally; the customer usually comes off best but is least willing to enter into the spirit of the relationship with formal ties [Helper, 1991]. Indeed, most cases written for public consumption about successful partnerships do not cover repetitive manufacturing environment contexts but rather new product development or major capital purchases. An example of this, which is in keeping with the theme of the example, would be Frey and Schlosser's [1993] description of the ABB-Ford collaboration to bring a $300 million paint facility to fruition. The process started off along traditional adversarial lines, but soon fell into more cooperative behavior once the companies realized their mutual dependence. Frey and Schlosser maintain that partnerships are more achievable when the organizations are equally powerful, but power need not be correlated with size.

OEMs are reluctant to relinquish control; the reverse is true, despite the "core competence" rhetoric. Increasingly larger manufacturing organizations are extending their supply chain to become involved in the disposal of products (e.g., German automotive companies are involved in the recycling and disposal of their products once the customer as a driver is finished with them). Technology might reinforce this negative trend for cooperation:

Future Developments

Globally applied virtual technologies involving simulating scenarios could result in changing the way new products are made as well as who is involved in the process. For example, through operating virtual supplier networks, OEMs can capture more information on the ultimate customer's requirements and not base design or production on mere perceptions. In theory, supply chains will operate more quickly and smoothly as a result. It should not matter where an organization is located from a communications point of view. Globalized companies may even have different product development disciplines situated in different parts of the world, suppliers and customers linking into the chain as they are required in the process. The technology will make it appear as if the chain is one organization but in reality it will be composed of a more-complicated version of the current position. This technology is not likely to be used to its full advantage if its application is not thought out properly and implemented sensitively.

OEMs will be able to use this technology to test scenarios and even run the manufacturing operations in supplier companies anywhere in the supply chain. This implies another barrier to becoming a supplier, in that the technological infrastructure would need to be in place. Whereas previously the OEM negotiated with a supplier as to whether they could make a part to certain performance criteria, they will be able to test whether this is the case and, in theory, link into the supplier's information and manufacturing systems, directly affecting and influencing the making of supplied parts. Whereas some supply chains that were optimal in terms of cost were not chosen because of control issues, the technology should reassure the customer that they retain control. The consequences of such a situation for suppliers are enormous and could interfere considerably with their profitability and their ability to serve other customers. A more positive note could be that customers have not implemented technology quickly or particularly well in the past (e.g., EDI).

Defining Terms

Downstream: The part of the supply chain nearest the end customer.

Just in time: Approach to operations that incorporates the pull system initiation with waste minimization, employee involvement, process simplification, and total quality.

Lean Supply: Applying the "lean" concept from Womack et al. to supply chain management. A holistic, almost ecological-philosophical approach to managing supply chains focusing on minimizing reduplication of effort and resource utilization.

Outsourcing: Divesting of activities where an organization is not perceived to add sufficient value to other organizations on the assumption that the price offered by the other organization coupled with the cost of managing the new relationship is less than the current one.

Partner: A term applied to a supplier by a customer in a supply chain that implies a close relationship, with cooperation leading to improved performance resulting in shared rewards. Sometimes achieved in the short term.

Pull system: Production initiated from actual (or as close to actual as possible) customer demand. Can be contrasted with production initiated by desire to maintain machines or people working at full capacity.

Supply chain: A series of people, groups, or organizations that transform raw materials into finished goods, which reach the end user.

Supply chain management: Is concerned with the coordination and operation of whole chains so that goods and services progress at improved speed and quality with reduced cost to the end user.

Upstream: The part of the supply chain farthest from the end customer where the product is in its component or raw material state.

References

Forrester, J. *Industrial Dynamics*, MIT Press, Cambridge, MA, 1961.

Frey, S. C. and Schlosser, M. M. ABB and Ford: creating value through co-operation, *Sloan Mgt. Rev.*, 35(1): 65–73, 1993.

Gadde, L.-E., and Grant, B. Quasi-Integration, Supplier Networks and Technological Co-operation in the Automotive Industry, Proceedings from the International Research Seminar on Industrial Marketing, 1984.

Gadde, L.-E., and Hakansson, H. *Professional Purchasing*, Routledge, London, 1993.

Hakansson, H. *Corporate Technological Behaviour — Co-operation and Networks*, Routledge, London, 1989.

Helper, S. How much has really changed between U.S. automakers and their suppliers?, *Sloan Mgt. Rev.*, 32(4): 15–28, 1991.

Imai, M., *Kaizen*, Random House, New York, 1986.

Lamming, R. Squaring lean supply with supply chain management, *Int. J. Oper. Prod. Mgt.*, 16(2): 183, 1996.

Sako, M. *Prices, Quality and Trust: Inter-Firm Relationships in Britain and Japan*, Cambridge University Press, Cambridge, 1992.

Schonberger, R. J. *Japanese Manufacturing Techniques: Nine Hidden Lessons in Simplicity*, Free Press, New York, 1982.

Further Information

Visit http://www.manufacturing.net/magazine/purchasing/ on the Web for an electronic journal covering many relevant topics.

The U.S. National Association of Purchasing New York's site has many useful links: http://www.solcon.com/napm-ny/links.htm.

Another good starting point is http://www.cyberbuyer.com/frameset.htm, the cyberbuyer's page, which has a variety of links, including links to several organizations' supplier management pages.

Finally, http://www.supply-chain.com/ contains good supply chain definitions and a group pooling resources on this topic.

4.15   Monopoly and Antitrust

Thomas W. Gilligan

One nominal goal of antitrust regulation is to encourage the **allocative efficiency** of economic resources. Antitrust regulation may do this by attacking industrial practices that promote **monopoly** and retard **competition**. Antitrust regulation is animated by noneconomic rationales as well and, paradoxically, is often used by firms to stifle competition. A thorough understanding of the motives and practice of antitrust regulation is essential for modern technology managers.

Monopoly and Allocative Efficiency

A monopoly exists when a single firm produces a good for which there are no close substitutes. In contrast to competition in which many firms produce a relatively homogenous good, consumers must transact with the monopolist or forgo consumption of the good.

A market or industry is said to be allocatively efficient when society's willingness to pay for the last unit of the good produced equals the **opportunity cost** of its production. The opportunity cost of producing a good is the value of all necessary resources in their best alternative use.

A profit-maximizing monopolist charging the same price for each unit of its good does not promote allocative efficiency. For a monopolist, the opportunity cost of producing the last unit of its good is strictly less than society's willingness to purchase that good. Society would be more than willing to compensate the monopolist for increasing its production, but the price reduction caused by additional sales is not in the monopolist's best interest.

In contrast, a competitive industry in which the same price is charge for each unit of the good does promote allocative efficiency. A profit-maximizing competitor continues producing so long as the society's willingness to purchase the good, as expressed by the good's price, exceeds the opportunity costs of production.

Antitrust Regulation

Antitrust regulation results from the behavior of governmental agencies and private individuals acting under the authority of state and federal antitrust statutes. The major federal statutes governing antitrust regulation are the Sherman Act, the Clayton Act, and the Federal Trade Commission Act. Broadly speaking, these statutes outlaw behavior that lessens competition or is deemed unfair or deceptive. Two federal agencies — the Antitrust Division of the Justice Department and the Federal Trade Commission — administer antitrust regulations. Private individuals injured by antitrust violations can also sue in Federal Court and, if successful, recover treble damages plus court costs.

Antitrust regulation targets price-fixing conspiracies or other agreements among competitors that may limit or eliminate price competition. Explicit pricing or production agreements among firms are obvious violations of antitrust regulation. Some early examples of antitrust regulation were against firms with explicit agreements to set prices or production quotas. For example, railroads and oil refiners were some of the earliest targets of antitrust regulation against explicit price agreements. Recent examples include airlines and construction companies.

Antitrust violations may also be leveled against firms or individuals even in the absence of explicit price agreements. In recent years, tacit collusion has been alleged to exist in industries where the pattern of prices over time was inconsistent with a truly competitive market. For example, the Justice Department recently attacked the stock-quoting practices of 24 major NASDAQ securities dealers based chiefly on the magnitude of dealer margins (i.e., the difference between the bid and ask price). Practices that may facilitate tacit collusion are also often the targets of antitrust regulation. Previously, the Justice Department had attacked the price-quoting practices of U.S. major airlines on the basis that these practices might facilitate noncompetitive pricing.

Agreements among competitors regarding non–price factors may also be subject to antitrust regulation. For example, the actions of industry or trade associations can be circumscribed by antitrust regulation. Agreements designed to establish standards of performance may also be subject to scrutiny. Such agreements are more likely to be challenged when pricing issues are involved.

Agreements among firms in the same value chain — among firms that do not compete — can also be subject to antitrust challenge. For example, establishing exclusive territories for independent retailers or requiring that retailers sell all or exclusively a single manufacturer's products can trigger antitrust regulation. Antitrust regulation can also emerge when a manufacturer attempts to control retail prices.

Antitrust regulation also targets structural changes — mergers and acquisitions — that might limit competition. Broadly speaking, a proposed industrial combination is likely to be challenged if it greatly reduces the number of major or potential competitors in an industry. Since 1982, the standards used to judge proposed combinations have been fairly explicit and consistent. These standards are codified in the Horizontal Merger Guidelines jointly issued by the Department of Justice and the Federal Trade Commission.

Antitrust regulation also targets unilateral actions by dominant firms that might be argued to lessen competition. For example, a jury recently found that Kodak's policy of not selling photocopier spare parts to independent service organizations (chiefly, ex-Kodak employees) violated the relevant antitrust statutes, even though Kodak has less than a 25% market share. Many of Microsoft's competitive strategies, including the addition of a Web browser to the introduction of its operating system, Windows 95, have been subject to intense scrutiny by antitrust regulators. Actions designed by a firm to increase the value of an innovation or enhance the functionality of a product to consumers can often be interpreted as anticompetitive and a violation of antitrust regulation.

Alternative Goals of Antitrust Regulation

In the spirit of the times in which it was conceived, antitrust regulation has always been and continues to be a reflection of the populist sentiment of distrust toward big business. Indeed, the trusts the Sherman Act were intended to combat were increasing output and reducing prices faster than other segments in the American economy. Far from generating the allocative inefficiency suggested by the static model of monopoly, trusts seemed to promote the types of productive and innovative efficiencies facilitated by large-scale production and distribution. Indeed, the populist "muckraker" Ida Tarbell and "trustbuster" Teddy Roosevelt conceded the productive efficiencies of the trusts. However, the fear of big business nevertheless gave rise to early antitrust regulation.

The fear of big business at the turn of the century, no doubt, can be attributed to the effects of large-scale enterprise on small, local businesses. National enterprises attempting to exploit scale economies in production and distribution tended to displace small, local, and less-efficient companies. This displacement gave rise to the political sentiments reflected in early antitrust regulation. These same sentiments exist in present-day antitrust regulation. Wal-Mart, an international consumer products distribution company, is often the target of state antitrust litigation initiated by small, local retailers concerned about their survivability in the face of a more efficient, lower-priced competitor. Clearly, the use of antitrust regulation in this manner has little to do with promoting allocative efficiency.

By far, most antitrust regulation is done by private, as opposed to government, litigants. For every case brought by the government, private plaintiffs bring 20. The most frequent private antitrust case involves a dealer or franchisee suing a supplier or franchiser over a simple contract dispute (e.g., whether conditions sufficient for dealer termination were present). In all but the smallest number of cases, there is little possibility for the disputed actions to harm competition. Troublingly, the second most frequent private antitrust case is brought by competitors. Since competitors should welcome actions that lessen competition and abhor those that heighten rivalry, many scholars have argued that the existence of such cases is *prima facie* evidence of the misuse of antitrust regulation.

Conclusions

There are two ways that managers should view antitrust regulation. First, antitrust regulation embodies principles designed to promote allocative efficiency. As such, antitrust regulation prohibits actions, such as overt price fixing, that are thought to lessen competition. Antitrust regulation, then, constitutes a set of prohibitions to which managerial decisions must conform. Understanding these prohibitions is important for effective management.

Second, antitrust regulation is often a threat to highly innovative and efficient managerial strategies. Since private litigants can administer antitrust regulation, and antitrust prohibitions are admittedly vague and dynamic, the antitrust statutes are often used to stifle economic leaders. Indeed, antitrust regulation is often the final option employed by losers in the economic race. Understanding these threats is essential, particularly for the management of firms in dynamic markets.

Defining Terms

Allocative efficiency: An industry is said to be allocatively efficient when the value of the last unit of the good or service produced by that industry to the consumer equals the opportunity costs of producing that good.

Competition: Competition exists when many a given good or service is produced by a large number of firms.

Monopoly: Monopoly exists when a firm produces a good or service for which there are no close substitutes, that is, a consumer must either purchase the good or service from the monopolist or forgo any utility associated with its consumption.

Opportunity costs: The value of a resource in its best alternative use.

References

Baumol, W. J. and Ordover, J. A. Use of antitrust to subvert competition, *J. Law Econ.*, 28: 247–265, 1985.

Bork, R. H. *The Antitrust Paradox: A Policy at War with Itself*, Basic Books, New York, 1978.

DiLorenzo, T. J. The origins of antitrust: an interest-group perspective, *Int. Rev. Law Econ.*, 5: 73–90, 1985.

Kaserman, D. L. and Mayo, J. W. *The Economics of Antitrust and Regulation*, Dryden Press, Fort Worth, TX, 1995.

4.16   Money and Banking

Peter S. Rose

The field of *money and banking* is changing in both content and focus along with the other branches of economics. The term "money and banking" is an old and, in many ways, an outdated label — a relic of prior centuries when "money" — the only asset with immediate spending power — and credit granted by banks dominated the global financial system. Today money and banking is a far broader field that encompasses the whole financial-services sector — bank and nonbank financial institutions supplying an ever-widening array of monetary and nonmonetary assets, including stocks, bonds, mutual funds, loan-backed securities, futures, options, and other derivative financial instruments. Money and banking is also becoming increasingly *internationalized* as governments and private decision makers increasingly recognize the interdependence of all national banking systems and financial service markets today. Moreover, there is widespread recognition that no one nation's economy, particularly those economic systems having significant export or import sectors, can operate under a money and credit policy that is completely isolated from the policies of other nations.

Banks and other Financial Institutions within the Financial System

Banks, pension and mutual funds, insurance companies, credit unions, and other financial service providers dominate the creation of credit and the investing of the public's saving in global financial markets today. The modern banking and financial marketplace fulfills multiple roles within the global economy — allocating savings into investment in plant and equipment that spurs the economy's growth and development, providing credit and liquidity to facilitate the public's purchases of goods and services, and supplying risk protection for life, property, and income as well as offsetting possible losses due to movements in market prices and interest rates.

As Table 4.5 reveals, banks continue to be the leading financial services firm in the United States and in most other nations as well. However, selected nonbank financial institutions — particularly mutual funds and pensions — are growing at a much faster rate than commercial banks, reflecting the shifting demands of the public away from traditional banking products centered around deposits toward nonbank-provided services that seem to promise greater long-run inflation-adjusted returns. These changing public demands for different financial service products within the financial system reflect two

TABLE 4.5   Financial Assets Held by Banks and other Financial Service Institutions within the U.S. Financial System

Financial Institutions Operating within the U.S. Financial System	Volume of Financial Assets Held as of Year End 1996	Percent of All Financial Asset Held
Commercial banking organizations (including U.S.-chartered commercial banks, foreign banking offices in the U.S., bank holding companies, and banks in U.S.-affiliated areas)	$4,710.0	22.8
Savings and loan associations, mutual savings banks, and federal savings banks	1,034.9	5.0
Credit unions	327.3	1.6
Trust companies (including personal trust and estates administered by nondeposit trust companies)	808.2	3.9
Life insurance companies	2,239.4	10.8
Other insurance companies	803.7	3.9
Private pension funds	2,020.8	14.7
State and local government employee retirement funds	1,734.8	8.4
Money market mutual funds	891.1	4.3
Open-end investment companies (mutual funds)	2,348.8	11.4
Closed-end funds	142.9	0.7
Government-sponsored enterprises (including the Federal National Mortgage Association and other government-sponsored mortgage agencies, the Farm Credit System, and the Student Loan Marketing Association)	981.0	4.7
Finance companies	896.7	4.3
Mortgage companies	47.8	0.3
Real estate investment trusts	31.2	0.2
Security brokers and dealers	629.0	3.0
Total of all financial assets held	$20,657.6	100.0

Note: Column figures may not add to totals due to rounding.

Source: Board of Governors of the Federal Reserve System, Flow of Funds Accounts of the United States, Statistical Release, Second Quarter 1996.

powerful external forces — changes in the structure and dynamics of the global economy and changes in the demographic makeup of populations all over the world. An aging population, particularly in the United States, Western Europe, and Japan, have led to a greater emphasis today upon planning for the long run — controlling inflation, minimizing tax exposure, managing interest rate risk, and preparing for higher educational and health care costs. The result is a decisive shift by millions of investors away from fixed-value claims issued by banks and other depository institutions toward more price-volatile stocks and bonds, sold primarily through security dealer houses and mutual funds, with many of these investments flowing into pension plans on behalf of individuals who today have much longer average life expectancies.

Trends such as these have led to a substantial decline in banking's share of all the assets held by the entire financial institutions' sector. In the United States during the 19th century, for example, banks represented well over half of the financial system's assets. Today they hold little more than a quarter of the U. S. financial sector's total financial assets. This long-range decline in banking's share of the financial marketplace has resulted in some shift of emphasis among researchers in the money and banking field toward more detailed studies of the behavior and services supplied by both bank and nonbank financial service firms.

One of the most pressing research issues in this field today focuses upon the *production economies* of financial service firms. Is there an optional size financial firm that yields the lowest cost per unit of services produced and delivered to the public? If so, and if there are no artificial or natural barriers to prevent financial firms from reaching their optimal minimum-cost size, the public should benefit in the long run in terms of lower resource costs and better quality services where effective competition prevails.

In the absence of regulation or other impediments, market forces will drive bank and nonbank financial institutions toward their optimal production point and eliminate those financial firms unable or unwilling to make cost-saving adjustments in their scale of operations to achieve the lowest-cost output.

Research findings on this so-called **economies of scale** issue among banks and nonbank financial institutions today are both voluminous and very mixed in their outcomes and implications (see, for example, Akhaveinet al. [1996], Berger and Humphrey [1991], and Bernstein [1996]). Most commercial bank cost studies, for example, suggest that banks probably reach their lowest-cost production point when they achieve a size of $100 million to perhaps $500 million or $1 billion in total assets — a relatively small banking firm by world standards. Further increases in bank profitability must, then, come mainly from the *revenue* side of the business.

These scale studies often find, however, that once optimal size is reached, a bank's production cost curve stays fairly flat until its assets reach perhaps $10 to $20 billion at which point its cost per unit of service tends to rise at a faster rate. Larger banks do appear to enjoy greater economies in their cost of capital, which allows them to continue to grow assets at somewhat lower long-run borrowing costs. There is, however, no compelling evidence that any financial service firms are natural monopolies that possess sufficient economic power to eliminate all of their competitors. Competitive forces can and should operate with full force in the financial services sector to benefit the consumer just as they should in the rest of the economy.

We must note, however, that the *structure* of financial service markets is changing rapidly today, generally moving toward greater and greater market concentration in the largest firms. For virtually all types of financial service companies, industry populations are falling as the leading firms merge with smaller companies and grow through both external and internal means. For example, the 100 largest U.S. banks held about 51% of all U.S. banking assets in 1980, but that proportion had risen sharply to account for more than 70% of industry assets by the late 1990s.

While, in theory at least, this shift toward greater concentration of industry resources in the largest financial services suppliers might suggest declining competition and potential damage to the public in the form of excessive service prices and a suboptimal allocation of scarce resources, the changing product mix of most financial firms toward broader and broader service menus — invading each other's service markets — suggests that we can no longer consider just one type of financial institution in assessing competition in the financial services marketplace. For example, in assessing competition for supplying home mortgage loans to the public in a major city such as New York or Chicago, we must now consider banks, credit unions, finance companies, mortgage companies, pension funds, and savings and loans as significant providers to the public of this important credit service. More realistic measures of industry and market concentration must be developed over time in order to derive a truer picture of the supply alternatives that consumers of financial services really face today.

Then, too, governments at all levels need to become more proactive in encouraging the emergence of a highly competitive climate for the production and sale of financial services by permitting *new* financial service firms to be formed more easily, through vigorous enforcement of the antitrust laws and by expanding the permissible range of financial services that each type of financial firm is allowed to offer the public. Beginning in the 1970s and 1980s several leading nations around the world, including Australia, Canada, Great Britain, Japan, and the United States, began to roll back some of the regulations that had held frozen for decades the service options, prices, and market territories that financial institutions were allowed to offer or enter.

For example, the United States began an aggressive financial deregulation movement in 1980, gradually lifting government-imposed interest rate ceilings on the deposits offered by banks and such thrift institutions as savings and loans, savings banks, and credit unions and significantly expanding the service options thrifts could adopt in an effort to bring them into direct competition with banks for credit cards, non–mortgage-related consumer loans, and checking accounts. Later in the 1990s, the first U. S. banking law at the national level to permit acquisitions of banks across the boundaries of all 50 states by bank holding companies — the so-called Riegle/Neal Interstate Banking and Branching Efficiency Act — was passed. This sweeping piece of banking legislation also authorized interstate branching activity via merger, allowing bank-holding companies to convert their banks acquired across state lines into full-service branch offices, provided the states involved did not forbid such conversions before June 1, 1997. Thus, for the first time in its history, American banking today is gradually opening up to the prospect of nationwide branch-office banking, with the largest and soundest banks capable of entering hundreds of communities in every region of the nation. Given sufficient time, the structure of American banking is likely to approach more closely (though not completely) the structure of banking in other leading countries, such as Canada and Great Britain, where a few large banking firms tend to dominate the industry.

Central Banking and Monetary Policy

Money and banking has always been one of the most fascinating parts of the field of economics because it encompasses **monetary policy** — the control of money and credit conditions by a **central bank** (or other agency of government) in order to achieve a nation's economic goals. Beginning in the years immediately following World War II, many nations came to the conclusion that they could effectively shape and control the performance of their economies by pursuing well-designed economic policies. While most economists at that time argued that **fiscal policy** — taxing and spending decisions by government — was likely to be the most powerful policy tool that governments could use to shape the performance and growth of their economies, many nations soon found that achieving a consensus of public opinion on what taxes to impose and on appropriate levels of government spending and borrowing was usually very difficult to achieve. In contrast, monetary policy typically is formulated and carried out by central banks that usually can reach a consensus and execute appropriate policy actions very quickly, often with relatively small announcement (i.e., psychological) effects on the public. The result is that today, in Europe, the United States, and in many other countries as well, government economic policies are largely pursued through central bank control and influence over interest rates and the growth of money and credit.

One of the sharpest controversies in the field of money and credit policy today centers on whether the *degree of independence* from government enjoyed by central banks affects their success or lack of success in achieving each nation's broad economic goals (such as full employment or low inflation). Certainly the degree of independence enjoyed by central banks today varies significantly around the world. For example, the Federal Reserve System is considered to be one of the more independent central banks because it generates its own revenue from service fees and earnings from the securities it holds and does not depend upon the U. S. government for its operating funds. Moreover, each member of the governing body of the Federal Reserve System, the Board of Governors, is appointed for a nominal term of 14 years, which limits the ability of the president or Congress to exercise significant control over central bank decisions.

Another central bank that, historically, has also exercised great independent decision-making power is Germany's Bundesbank. In fact, the Bundesbank Act of 1957 explicitly declares that the German central bank is "independent of instructions from the federal government". The primary goal of the Bundesbank is *price stability* — control over inflation — which that central bank is allowed to pursue vigorously even if the government objects. Both the Bundesbank and the Federal Reserve in recent years have reasonably good track records in controlling domestic inflation. However, by a strange twist of fate, the historical independence of the Bundesbank has proven to be something of a barrier to the final welding together of the European Economic Community into one true economic and political union because several other European nations exercise tighter control over their own central banks and have been less successful in controlling inflation and avoiding substantial government budget deficits.

Research on the relationship between the degree of central bank independence and the economic performance of nations began in the 1980s, and the issue continues to be explored in new research studies. While the outcomes are often mixed, there is evidence — for example, in the studies by Alesina [1988] and Bode and Parkin [1987] — that more independent central banks generally have a superior record in keeping domestic inflation in check. However, with respect to other important economic goals, such as promoting economic growth, there appears to be *no* consistently verifiable relationship between central bank autonomy and economic performance. These research findings regarding inflation control have been persuasive enough to lead several central banks, such as the Bank of France and the Bank of Japan, to seek greater independence from their nation's central government in recent years. Moreover, the central banks of Canada, Great Britain, New Zealand, and Sweden have recently adopted *inflation targeting* — adopting specific numerical inflation goals encompassed by an acceptable band of variation around the targeted inflation rate and a timetable for goal achievement — as their principal policy objective. Recent research provides at least some evidence that interest rates in these countries carry smaller inflation premiums, suggesting that public opinion has begun to accept inflation targeting as a viable central bank objective.

How do central bank policy tools — such as reserve requirements, discount rates, and trading securities in the open market — promote the achievement of a nation's economic goals? One route relies upon the level and growth of reserves in the hands of the banking system in order to influence the rate of growth of a nation's money stock, while the other consists of changing market interest rates, particularly the interest rate attached to short-term loans of reserves between banks (such as the interest rates we observe in the federal funds and Eurocurrency markets each day).

Central banks influence the money stock in the public's hands primarily by controlling the growth of the *monetary base* of the banking system. The key relationship involved may be illustrated by the following equation:

$$\text{money multiplier} \times \text{monetary base} = \text{money stock}$$

or

$$\frac{1 + \text{CASH}}{\text{RR}_\text{D} + \text{CASH} + \text{EXR} + (\text{RR}_\text{T}\text{XTIME})} XB = M \tag{1}$$

where M is a measure of the nation's money stock; B represents the monetary base (or sum of reserves available to the banking system plus currency and coin held by the public); CASH stands for the amount of pocket money (currency and coin) the public wishes to hold per dollar of their transactions (checkable) deposits; TIME reflects the volume of time and savings accounts the public wishes to hold per dollar of its transactions deposits; EXR represents the volume of excess reserves the banking system wishes to hold per dollar of outstanding transactions deposits; and RR_D and RR_T stand for the legal reserve requirement ratios for demand deposits and time deposits normally set by the central bank. Each reserve requirement ratio spells out how much in liquid reserves each depository institution must hold per dollar of demand deposits or other reservable liabilities. For example, suppose that CASH equaled 0.25, TIME was 0.50; EXR was 0.05; and RR_D and RR_T were both 0.10. This would yield a money multiplier of 1.25/0.45 or 2.78. If the nation's monetary base currently stands at $750 billion, its money stock in the form of checkable deposits would be: 2.78 × $750 billion = $2085 billion.

The above formula reminds us that central banks can influence a nation's stock of money through several different routes: (1) through changes in the monetary base (B) or (2) through changes in reserve requirements (RR_D and RR_T). The most popular approach today is to use **open-market operations** (buying and selling securities or other instruments) to change the size and growth of the monetary base. Open-market trading is often preferred because of the great flexibility and quick impact of this policy tool and the relatively small effect that central bank security trading usually has upon public psychology.

As an alternative to money-stock control a growing list of central banks (including the Federal Reserve System) has adopted a *money market indicator approach*, focusing on a key short-term market interest rate (for example, in the United States, the Federal funds rate; in Britain and Canada, the interest rates on short-term government and commercial bills). The central bank can use open market operations to put the desired amount of pressure on interest rates. For example, it could *sell* shorter-term securities to push the targeted money market interest rate higher or *purchase* shorter-term securities to nudge the targeted interest rate toward lower levels. The central bank can set the money market indicator rate within a desired range by using its policy tools to change the supply of reserves so as to offset any changes in the demand for reserves by the banking system. In the most recent period, the Federal Reserve in the United States has been announcing immediately when it changes its target money market interest rate in an effort to reduce public uncertainty regarding the objectives and direction of monetary policy.

An Overview of the Money and Banking System

As we have seen, the field of money and banking is undergoing great changes today. The rapidly changing technology of information has impacted this field because the services provided by banks and other financial institutions are information based. A loan of money, for example, is nothing more than a bundle of information bytes, indicating who owes what volume of funds to whom and under what terms of exchange. The high cost of new information technology and the broadening of markets as information technology leaps over geographic barriers is driving many smaller financial service firms from the market or causing them to be absorbed by larger companies, leading to consolidation and a more concentrated money and banking system, offering more services to the public but often at higher prices.

Central banks — the agents of money and credit policy in most countries — have come to play a critical role within the rapidly changing money and banking system. The Federal Reserve System, Germany's Bundesbank, the Bank of England, and the Bank of Japan have come to exert great influence over the well-being of national and international economies as they employ powerful tools to influence the stock of money, interest rates, public borrowing and spending, and the availability of jobs. Indeed, central banks around the world today are among the most powerful of all governmental institutions in modern society in shaping the public's economic welfare.

TABLE 4.6 Measures and Volume of the U.S. Money Stock

Money Stock Measure	Volume in Billions of Dollars (as of December 1996)
M1 (= Currency outside the U.S. Treasury, the Federal Reserve Banks, and the vaults of depository institutions + travelers checks issued by nonbank institutions + demand deposits at banks other than interbank and government deposits + other checkable deposit accounts	1,081.0
M2 (= M1 + savings and small-denomination [under $100,000] time deposits + balances in retail money market funds with minimum initial investments of less than $50,000)	3,833.0
M3 (= M2 + time deposits in denominations of $100,000 or more issued by all depository institutions + institutional money market fund balances (with minimal initial investments of $50,000 or more) + repurchase liabilities issued by all depository institutions + overnight and term Eurodollars held by U.S. banks worldwide and at all banking offices in the United Kingdom and Canada)	4,927.7
L (= M3 + nonbank public holdings of U.S. savings bonds, short-term Treasury securities, commercial paper, and bankers' acceptances [net of money market fund holdings of these assets])	6,058.1
D (= outstanding credit market debt of domestic nonfinancial sectors, including U.S. government and federal agency securities, mortgages, consumer debt, commercial paper, and bank and nonbank loans)	14,622.0

Source: Board of Governors of the Federal Reserve System, *Federal Reserve Bulletin*, June 1997.

Defining Terms

Central bank: An agency of government created to affect money and credit conditions, supervise the banking system, and help a nation achieve its broad economic goals.

Economies of scale: The relationship between individual firm or plant size and the firm's cost per unit of product or service produced.

Fiscal policy: Tax and spending decisions by governments in an effort to affect the level and growth of economic activity.

Monetary policy: The activities and policies employed by governments to influence money and credit conditions in order to achieve their economic goals.

Open market operations: The trading of securities or other financial assets in the marketplace by a central bank in an effort to achieve a nation's economic goals by influencing money and credit conditions and the growth of reserves in the banking system.

References:

Akhavein, J. D., Berger, A. N., and Humphrey, D. B. The effects of bank megamergers on efficiency and prices: evidence from the profit function, *Rev. Indust. Organ.*, 11, 1996.

Alesina, A. Macroeconomics and politics, In *National Bureau of Economic Research Macroeconomic Annual*, MIT Press, Cambridge, MA, 1988.

Anderson Consulting. *Vision 2000: The Transformation of Banking*, Bank Administration Institute, Chicago, 1994.

Berger, A. N. and Humphrey, D. B. The dominence of inefficiencies over scale and product mix economies in banking, *J. Mon. Econ.*, 28: 117–148, 1991.

Berstein, D. Asset quality and scale economies in banking, *J. Econ. Bus.*, 48: 157–166, 1996.

Bode, R. and Parkin, M. Central Bank Laws and Monetary Policy, unpublished paper, Department of Economics, University of Western Ontario, 1987.

Kasman, B. A comparison of monetary policy operating procedures in six industrial countries, *Q. Rev.*, Summer: 5–24, 1992.

Further Information

Dietrich, J. K. *Financial Services and Financial Institutions: Value Creation in Theory and Practice*, Prentice Hall, Upper Saddle River, NJ, 1996.

Rose, P. S. *Money and Capital Markets: Financial Institutions and Instruments in a Global Marketplace*, 6th ed., Richard D. Irwin, Burr Ridge, IL, 1996.

Sundaresan, S. *Fixed Income Markets and their Derivatives*, South-Western College Publishing, Cincinnati, 1997.

5

Statistics

Amy E. Herrmann
Georgia Institute of Technology

Janet K. Allen
Georgia Institute of Technology

Kenneth E. Case
Oklahoma State University

Jerry Dechert
University of Oklahoma

Terry E. Dielman
Texas Christian University

Frederick A. Rossini

Alan L. Porter
Georgia Institute of Technology

Marlene A. Smith
University of Colorado at Denver

Stephen R. Schmidt
Air Academy Associates

5.1 Data Collection

Amy E. Herrmann and Janet K. Allen

A typical experimentation strategy is given in Fig. 5.1. The experimenter carefully formulates the problem, forms a testable hypothesis, and identifies suitable statistical methods for evaluating the truth of the hypothesis. An appropriate sample is identified (see Section 5.2), and the required data are collected, as discussed in this section. Subsequently, these data are analyzed, and the validity of the hypothesis is determined. The truth or falsehood of this hypothesis then forms the foundation for another hypothesis, which proceeds through the validation cycle. There is a very close relationship between data collection and sampling (Section 5.2).

Types of Data

There are several types of data that can be collected; each has distinct characteristics and appropriate methods for analyzing them.

- Random variables both continuous and discrete. Occasionally it is appropriate to model discrete random variables as continuous, e.g, thousands of units on a production line.

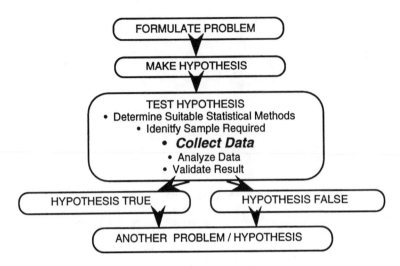

FIGURE 5.1   A typical strategy for experimentation.

- Ordinal (ranked) variables. Often it is difficult to judge how much difference there is between several items, but it is possible to determine the first choice, the second choice, etc. (e.g., a typical customer will choose this product first, that next, etc.)
- Measurements using an interval (or rating scale) scale.

For example,

It is important to maintain consistency among those who are doing the rating, so it is usually advisable to decide exactly what each interval means before attempting to do any rating.
- Attributes (blue, male/female, beautiful, etc.) In this case, it is not possible to quantify the difference between blue and red, for example; numerical values may be assigned to attributes for convenience, however. Attributes can be used to identify classes of subjects to be studied. Attributes can be modeled as sets, each bounded by a region of uncertainty represented as fuzzy numbers [Dubois and Prade, 1978]. For example consider the attribute "blue". Many colors would be generally agreed upon as blue; however, there is also a continuum of colors between blue and green that might or might not be members of the set blue and could conveniently be represented by fuzzy numbers.

Each of these types of data is useful in determining an answer, yet each type of variable must be dealt with in specific ways. Obviously, it is desirable to collect data that are directly relevant to the question being asked. In an experiment, there are two types of variables — the independent and the dependent. The experimenter controls the independent variable(s) and observes the results on the dependent variable(s).

Describing Continuous Random Variables with a Gaussian Distribution

Since many of the statistical relationships that have been developed are based on the assumption that the data are distributed according to a known distribution (usually a normal or **Gaussian distribution**), it is important to demonstrate that the collected data in fact do follow the **normal distribution**. If n data points have been collected and they follow a normal distribution, then, for the continuous random variable, x, the normal distribution of the **relative frequency**, f(x) [Roussas, 1997],

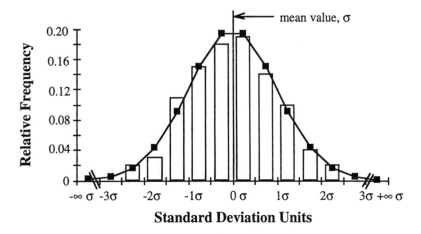

FIGURE 5.2 A normal distribution superimposed on an experimental frequency distribution.

$$f\left(x\right)=\frac{1}{\sigma\sqrt{2\pi}}\,e^{-\frac{1}{2}\left[\frac{\left(y-\bar{y}\right)}{\sigma}\right]^{2}}\qquad x\ \in\ \mathbb{R}\qquad\qquad(5.1)$$

where $\bar{y}$ is the **mean** value:

$$\bar{y}=\frac{\displaystyle\sum_{i=1}^{n}y_{i}}{n}\qquad\qquad(5.2)$$

and σ is the **standard deviation**:

$$\sigma=\sqrt{\frac{\displaystyle\sum_{i=1}^{n}\left(y_{i}-\bar{y}\right)^{2}}{n-1}}\qquad\qquad(5.3)$$

To determine whether or not collected data are in fact normally distributed, compute the sample mean and standard deviation, sequence the data, and create a relative frequency distribution by grouping the data into classes and then dividing by the total number of data points, Fig. 5.2. Next, using Eq. (5.1), compute a theoretical distribution with the mean and standard deviation of the sample data.

To compare frequency data to a theoretical distribution, compute

$$X^{2}=\sum_{i=1}^{a}\frac{\left(f_{i}-\hat{f}_{i}\right)^{2}}{\hat{f}_{i}}\qquad\qquad(5.4)$$

where f_{i} is experimental frequency and $\hat{f}_{i}$ is the expected frequency at the midpoint of a class for the theoretical distribution. In this case, a normal distribution with the mean and standard deviation of the experimental data is superimposed on the data, a is the number of classes into which the data are grouped. To determine whether the experimental distribution differs significantly from a normal distribution, compare the computed value of X^{2} with a value of chi-squared, χ^{2}, with (a-1) degrees of freedom. χ^{2} is available in tables [Beyer, 1991].

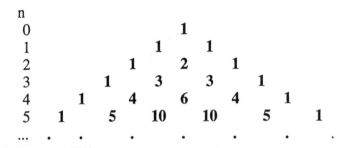

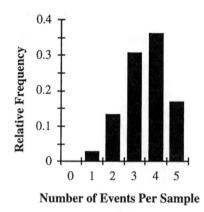

FIGURE 5.3   Pascal's triangle for evaluating the term $\binom{n}{x}$ for the coefficients of the binomial distribution.

FIGURE 5.4   A typical binomial distribution, p = 0.3, q = 0.7, n = 5 computed from Eq. (5.5).

A Discrete Distribution for Discrete Variables with Two Possible Outcomes: The Binomial Distribution

If the data are discrete instead of continuous, they may follow a binomial distribution. The binomial distribution is useful when there are two possible outcomes: success, S, and failure, F. These outcomes occur with probabilities p and q. In reality, there can be any number of outcomes as long as at least one is termed success and the others are grouped as failure. Relative frequencies for the binomial distribution are computed from Eq. (5.5).

$$f\left(x\right)=\binom{n}{x}p^{x}q^{n-x} \quad \text{where } 0<p<1,\ q=1-p \text{ and } x=0,1,2,\ldots n \tag{5.5}$$

It is convenient to determine the binomial coefficients, from the term $\binom{n}{x}$ in Eq. (5.5) using Pascal's triangle, Fig. 5.3.

The mean value and standard deviation for the binomial distribution are

$$\mu=np \qquad \sigma=\sqrt{npq} \tag{5.6}$$

A typical binomial distribution is shown in Fig. 5.4.

To determine whether experimental data conform to a binomial distribution, compute X^2 from Eq. (5.4). If the data do not follow a binomial distribution exactly, it may be because the events are not

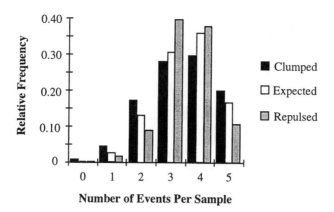

FIGURE 5.5   Expected frequencies of the binomial distribution shown in Fig. 5.4, compared with frequencies from a sample in which the data are clumped or repulsed.

truly random. If one occurrence of the event makes other occurrences of that event more likely, then the distribution is clumped, Fig. 5.5. If it makes other occurrences of that event less likely, then the distribution is said to be repulsed, Fig. 5.5.

A Discrete Distribution for Rare Events: The Poisson Distribution

The Poisson distribution is also a discrete frequency distribution that predicts the number of times a rare event occurs. A Poisson distribution can be used either to study the behavior of a system over time or to study the behavior of a number of similar systems spatially distributed. For example, it would be suitable for determining the number of telephone calls arriving at a particular telephone within a certain period of time or for studying the frequency of calls arriving at several telephones at any given time (assuming that the overall frequency of calls is low). More formally, the requirements for a variable to be distributed as a Poisson distribution are

- Its mean must be small relative to the maximum number of possible events within the sample.
- The occurrence of the event must be independent of prior occurrences of similar events within the sample.

Usually data on the occurrence of rare events are compared to a Poisson distribution to determine whether these events occur independently and are thus a test for randomness. Given the number of events per sample x = 0, 1, 2, ... and the parameter of the distribution $\lambda > 0$ is the mean value of the number of events over the entire sample, then the distribution of probabilities of the even occurring a specific number of times is

$$f\left(x\right) = e^{-\lambda}\frac{\lambda^x}{x!} \tag{5.7}$$

Notice that λ is the mean of the Poisson distribution and it completely determines its shape, Fig. 5.6. The standard deviation of the Poisson distribution is

$$\sigma = \sqrt{\lambda} \tag{5.8}$$

Nonprobability-Based Descriptions of Data

If the data does not follow any of the above distributions, they can still be described by a **median** value and a range and analyzed using **nonparametric statistics** [Lehman, 1995; Roussas, 1997]. In a data

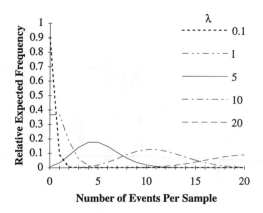

FIGURE 5.6 Poisson distributions for several values of λ. The Poisson distribution is a discrete distribution; however, for easy visualization, discrete values are connected.

distribution, one half of the values are equal to or less than the median; one half are larger. For an odd number of values, the median is the middle value:

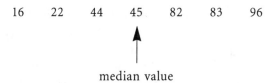

For an even number of values, the median is the midpoint between the two center values.

$$22 \quad 44 \quad 45 \quad 82 \quad 83 \quad 96$$

median value = 63.5

The sample range

$$\text{Sample range} = \text{largest value} - \text{smallest value} \tag{5.9}$$

Since the range is very sensitive to extreme values, the interquartile range is used as a measure of spread of the data. This represents the middle 50% of the data and is computed

$$\text{Interquartile range} = \text{value at the 75th percentile} - \text{value at the 25th percentile} \tag{5.10}$$

The semiinterquartile range, one half of the interquartile range, is often used as a measure of spread.

Data Collection over Time and Space

The study of any process, especially a process that is to be continuously improved, requires the accumulation of large amounts of data over a very large period of time. It is often assumed that the variability of sampled data evens out in the end, that is, given a large enough sample, the positive and negative errors will cancel each other out. However, over a long period of time, the measurement procedure itself may change, and there may be a tendency for the measured values to deviate in one direction or the other from actual values. This drift in the measurement procedure is called **bias**. Methodologically, it is important to insure that, when collecting data, both the **accuracy** and the **precision** (reproducibility) of the measurements remain constant.

The current trend is toward automated sampling or monitoring. Typical automated sampling systems can be divided into three parts: (1) sensors, (2) a data relay system, and (3) a data processing system. Advantages of automated data collection systems include cost effectiveness, flexibility, ease of implementation and lack of maintenance. Disadvantages include difficulties in dealing with a large accumulation of data and high initial expenditures. More information about statistical quality control of automated sampling is available in Section 5.2.

Closure

There are several types of variables that can be collected. Different kinds of data need to be treated differently. It is useful to determine whether the data conform to a known distribution, providing the foundation for further data analysis.

Acknowledgment

We gratefully acknowledge support from NSF grant DMI-96-12327.

Defining Terms

Accuracy: The extent to which a measurement reflects the actual value being measured.
Bias: The tendency of a measuring system to give values that consistently drift away from the actual value.
Gaussian distribution: See Normal distribution.
Mean: The average value of a variable, see Eq. (5.2).
Median: One half of the values in a sample are equal to or less than the median; one half are larger.
Nonparametric statistics: Inferential statistical methods that are not based on the assumption that data are distributed normally.
Normal distribution: A symmetrical, bell-shaped distribution that describes the behavior of many variables, Eq. (5.1). Many inferential statistical tests are based on the assumption that the variables obey a normal distribution.
Precision: The reproducibility of a measurement.
Relative frequency: The frequency of occurrence of each value relative to the total number of values possible.
Standard deviation: A measure of the spread of a distribution, see Eq. (5.3).

References

Beyer, W. H., Ed. *CRC Standard Probability and Statistics Tables and Formulae,* CRC Press, Boca Raton, FL, 1991.
Dubois, D. and Prade, H. Operations on fuzzy numbers, *Int. J. Syst. Sci.,* 9: 357–360, 1978.
Lehman, R. S. *Statistics in Behavioral Sciences,* Brooks/Cole, Pacific Grove, CA, 1995.
Roussas, G. G. *A Course in Mathematical Statistics, 2nd ed.,* Academic Press, San Diego, CA, 1997.

Further Information

For processes, there may be several inputs to dependent variables. This problem will be discussed Section 5.5.

5.2   Sampling

Janet K. Allen

Sampling involves identifying the amount and types of information required to prove a hypothesis. Thus it is essential for the experimenter to carefully formulate the hypothesis under investigation. Typically an investigation proceeds as shown in Fig. 5.7. Note that the material in this section on sampling is closely related to the data collection information given in Section 5.1.

Choosing Data for a Sample

The first important sampling decision is to determine exactly which **population** the sample is to represent. Any answers obtained are very sensitive to the group being queried. There are several ways of determining the best population to consider [Gibbons et al., 1977]. For example, the first and most obvious choice would be those who are currently using the product or service. However, it might also be appropriate to consider those who considered buying the product and those who didn't. Other populations that could be considered include those who currently use competing products, those who inquired about the product, or those who are demographically similar to current product owners, etc. Each choice has its limitations and its advantages. In some situations, it is not clear *a priori* whether certain parameters will make a difference in product use, e.g., the customer's geographic location, and it may be necessary to do preliminary surveys to determine which variables are important in the decision being made. However, a mistake in identifying the correct population can yield misleading results. It is often advisable to compare results from several populations; however, this can be an expensive, time-consuming process.

In some situations it may not be necessary to use a sample; all of the members of the population can be assessed directly. Although this can be costly, it removes a layer of uncertainty about the parameters being assessed. For example, it would be fairly easy to determine directly the different numbers and types of repairs done at a particular automobile service center. It is likely that, as information becomes more and more accessible, studies involving complete populations will become more and more common. In determining whether to use a sample of a population and the size of that sample, the value of the expected results must be balanced against the cost of obtaining the results.

In other situations, for example, when generating ideas or gaining insight, it may be appropriate to obtain information from individuals or small groups. One method for achieving a better understanding of a target market is by using a **focus group** to determine the voice of the customer. The focus group is an interview conducted by a trained moderator in a nonstructured and natural manner with a small

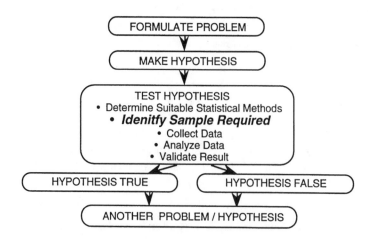

FIGURE 5.7   The role of sampling in experimentation.

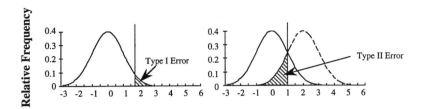

FIGURE 5.8 Type I and type II errors. In type I error, members of a population are determined not to be part of that population by the statistical test being used. In type II error, members of another population are recognized as being part of the population under investigation.

number of respondents [Malhotra, 1996]. Although the members of the group must be from the appropriate population, the size of the group is determined more by social dynamics rather than statistics; 8 to 12 has usually been found to be a convenient size. Alternatively, individual interviews can be an effective method of generating information [Griffin and Hauser, 1993]. Although there is substantial literature dealing with these methods, neither is the focus of this section. We focus on sampling as a foundation for statistical analysis and experimentation.

A statistically significant sample is a **random sample** if it is chosen from a population in such a way that every other possible sample of the same size has an equivalent chance of being selected [Lehman, 1995]. Notice that the procedure for random sampling only refers to the method by which the sample is selected; it does not guarantee that the sample is identical to the population. The most commonly used method of choosing a random sample of size N from a population of size P is by assigning a number to each member of the population and using a random number generator or a table of random numbers to select the required number of members of the sample.

Testing Hypotheses

Typically, an experimenter attempts to disprove the null hypothesis, H_0, that there is no difference between the sample and the population from which it is drawn; H_0 is compared with a specific hypothesis, H_1. There are two possible situations. In the first, suppose St is a sample statistic and St_p is the statistic for the entire population; then the experimenter merely asks whether the statistic is different from that of the entire population, thus H_0: St = St_p and H_1: St ≠ St_p. In the second situation, the experimenter asks whether the statistic is higher (or lower) than that of the entire population and H_0: St = St_p and either H_1: St > St_p or H_1: St < St_p. In the first situation, a two-tailed test of significance is required (because there are two possibilities that may be proven); in the later case, a one-tailed test of significance is used.

Once the null hypothesis has been identified, there are two ways in which errors can happen. The first is that, if the null hypothesis is actually true, the inference from the statistical experiment is that it is false; this is called a **type I error**. The second type of error occurs when the null hypothesis is actually false, but the statistical inference is that the null hypothesis is true; this is a **type II error**, Fig. 5.8 and Table 5.1.

To test hypotheses, suppose we have a distribution and want to determine whether it is drawn from a population. (The null hypothesis is that they are the same.) In the most general case, for any statistic, e.g., mean, standard deviation, etc., a value of t_s is computed:

$$t_s = \frac{St - St_p}{\sigma_{st}} \tag{5.11}$$

where σ_{st} is the estimated standard error of the statistic, i.e., the standard deviation of the sampling distribution of that statistic, and n measurements have been included in the sample.

TABLE 5.1   Type I and Type II Errors

		Actual Situation	
		The Null Hypothesis Is True	The Null Hypothesis Is False
Statistical Inference from Experiment	The Null Hypothesis Is True	Correct inference	Type II error
	The Null Hypothesis is False	Type I error	Correct inference

Using Eq. (5.11), to determine whether a particular mean value is drawn from a population, test the null hypothesis that the sample is drawn from the population with a mean value of $\overline{Y}$. If the mean value and the standard deviation of the sample are $\overline{y}$ and σ, respectively, for n-1 degrees of freedom, the estimate of the standard error of the mean is, $\sigma_{\overline{y}}$:

$$\sigma_{\overline{y}} = \frac{\sigma}{\sqrt{n}} \tag{5.12}$$

Then, Eq. (5.11) becomes:

$$t_s = \frac{\overline{y} - \overline{Y}}{\dfrac{\sigma}{\sqrt{n}}}$$

and t_s can be computed, and these computed values of t_s are compared with critical values of Student's t [Beyer, 1991]. If α is the level of type I error that is acceptable, and the number of degrees of freedom, $v = n-1$, then, for a two-tailed test, compare the computed t_s with the critical value of $t_{\alpha[v]}$; for a one-tailed test, compare the computed t_s, with the critical value of $t_{2\alpha[v]}$.

For samples from a Gaussian population, with n-1 degrees of freedom ($n > 15$), the estimate of the standard error of the standard deviation is

$$\sigma_\sigma = \left(0.7071068\right)\frac{\sigma}{\sqrt{n}} \tag{5.13}$$

As we will see in the next section, the size of samples required is dependent on the type of test being used and the degree of the difference that the experimenter accepts as being significant.

Sample Sizes

For samples with normal distributions, Eq. (5.11) is used to determine the significance between the difference of a sample statistic and the same statistic for the population. If it is important to reduce the probability of type II error in which the null hypothesis is false but is recognized as true by the statistical test being used, four variables must be considered: (1) the level of probability of which determines the likelihood of a difference in populations being missed, i.e., the significance level, α, (2) the sample size, n, (3) the variability of the population, usually measured by the population standard deviation, S, and (4) the actual difference between the true value and the sample value. In order to compute the sample size required, n, the other three parameters must be specified. The significance level is determined by the experimenter. Estimates of the population standard deviation can either be obtained by a pilot survey or by guesswork.

Given this information, the minimum sample size required to test a hypothesis, the value of n, can be computed by selecting an acceptable α and solving Eq. (5.11) for n. However, this is complicated by the fact that t_s is a function of n [Cochran, 1977]. Although it is possible to solve this problem iteratively, it is sometimes more convenient to use tables to determine the sample size, n, [Beyer, 1991]. The sample size, n, can also be obtained from operating characteristic curves, which show sample size requirements for various levels of type I and type II error in comparisons between means or standard deviations [Beyer, 1991].

If two or more groups of samples are being compared, it is preferable to replace a series of *t* tests by analysis of variance (ANOVA). Suppose δ is the smallest true difference that it is desirable to detect, a is the number of groups with n samples per group, P is the desired probability that the difference will be found to be significant, v is the number of degrees of freedom of the sample standard deviation, and $t_{\alpha[v]}$ and $t_{2(1-P)[v]}$ are the values of Student's *t* for a two-tailed test. Then the appropriate formula for the sample size required is [Sokal and Rohlf, 1981]

$$n \geq 2\left(\frac{s}{\delta}\right)^2 \left\{t_{\alpha[v]} + t_{2(1-P)[v]}\right\}^2 \qquad (5.14)$$

Eq. (5.14) is solved iteratively for n.

If data do not come from a normal population, binomial tests are used for the population median, and the number of samples required can be calculated similarly to those for the normal population [Lehman, 1995].

If data are measured on interval or ratio scales and are roughly linear, for example, Pearson product-moment correlation coefficients, tables are available to test the null hypothesis H_0: $r = 0$ [Lehman, 1995]. The numbers of samples required are determined from these tables by working backward from the desired significance levels and degree of difference. To test for values other than 0, the data must first be transformed using the Fisher r-to-z transformation [Lehman, 1995].

If data are not from normal distributions, but instead are ranked, a *t* test is used to test the hypothesis that Spearman's rank-order correlation (sometimes called Spearman's rho) is equal to 0 [Lehman, 1995]. To compute Spearman's rho, first rank the data and then compute the Pearson product-moment correlation coefficient using the ranks instead of the raw data. Again, the procedure for determining the sample size required is similar to that discussed for normal distributions.

Design of Experiments

For large experiments in which there are several variables, traditionally one variable is varied at a time and the effects are observed. Alternatively, different combinations of variable settings can be assigned randomly. However, a random approach requires large numbers of experiments to obtain statistically significant results. To increase the efficiency of large-scale experiments (and to reduce the number of samples needed), experimental design techniques have been developed using the power of statistics to logically reduce the numbers of experiments needed. Formally, an experimental design represents a sequence of experiments to be performed. Each is determined by a series of **factors** (variables) set at specified **levels** (predefined values of the factors). An extended discussion of experimental designs is presented by Montgomery [1997].

The most basic experimental design is the full factorial design in which each factor is tested at every level, full factorial designs are represented as (levels)$^{(\text{factors})}$, and the number of experiments required is determined by multiplying the levels and factors. The most common full factorial designs are 2^k (for evaluating main effects and interactions) and 3^k (for evaluating main and quadratic effects and interactions). The number of experiments is the product of the number of factors and their levels. Unfortunately, the size of a full factorial experiment increases exponentially with the number of factors, and this can

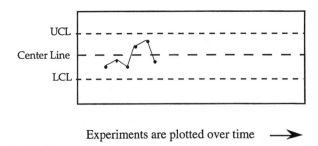

Experiments are plotted over time $\longrightarrow$

FIGURE 5.9 A statistical quality control chart.

lead to an unmanageable number of experiments. Therefore, especially when experiments are costly, fractional factorial experiments can be used. A fractional factorial design is a fraction of a full factorial design; the most common are $2^{(k-p)}$ designs where the fraction is $1/2^{(p)}$. The reduction in the number of experiments comes with a price, however. In fractional factorial designs, some of the variables cannot be determined independently unless the interactions are known (or assumed) not to exist.

Often 2^k or $2^{(k-p)}$ designs are used to identify or *screen* for important factors when the number of factors is large. In these situations, the sparsity of effects principle [Montgomery, 1997] is invoked, in which the system is assumed to be dominated by main effects, and their low-order interactions and two-level fractional factorial designs are used to screen factors to determine which have the largest effects and eliminate those that are unimportant. One specific family of fractional factorial designs often used for screening is the two-level Plackett-Burman (P-B) designs. These designs are constructed to study $k = N-1$ factors of $N = 4m$ design points. The class of Plackett-Burman designs in which N is a power of 2 are called geometric designs and are identical to 2^{k-p} designs. When N is a multiple of 4, the Plackett-Burman designs allow an unbiased estimation of all main effects using only one more design point than the number of factors. Orthogonal arrays are the experimental designs used for Taguchi's robust design. They are most often simply fractional factorial designs in two or three levels ($2^{(k-p)}$ and $3^{(k-p)}$ designs). Using any of these experimental designs substantially reduces the numbers of samples required.

Statistical Quality Control

For automated experiments, it has proven valuable to develop methods for statistical process control [Montgomery, 1996]. If a process is operating with only chance causes of variation, it is said to be in statistical control; otherwise, it is out of control.

Two important issues in successful quality control are the size and frequency of samples that must be checked in order to ensure the process remains in control. Small shifts in the process usually can be detected more easily with larger samples, and, of course, infrequent sampling means that the process can be out of control for longer periods of time. The problem then becomes one of allocating sampling efforts. Industry practice today tends to favor smaller, more frequent samples, particularly in high-volume manufacturing processes [Montogmery, 1996]; however, automatic sensing and measurement techniques will eventually make it possible to measure every unit as it is produced and thus reduce the required sampling frequency.

Suppose w is a sample statistic that measures some quality of a product, Fig. 5.9. If the mean of that statistic is μ_w and the standard deviation of that statistic is σ_w, a quality control chart may be developed in which

$$\text{Upper control limit (UCL)} = \mu_w + L\sigma_w$$

$$\text{Center line} = \mu_w$$

$$\text{Lower control limit (LCL)} = \mu_w - L\sigma_w$$

One way of selecting sample size and frequency is by using the average number of experiments required and length of time before the process is determined to be out of control. Usually, the expense of experiments is balanced against the cost of producing unacceptable products.

Closure

Although there is a variety of situations to be considered, determining sample size requirements is usually a process of identifying the hypothesis, determining the method to test the hypothesis, specifying the results that would be satisfactory to prove/disprove the hypothesis, and reasoning backward to obtain the sample size required to identify these results. In large-scale experiments, the numbers of samples required can be reduced by introducing various experimental designs.

Acknowledgments

This work has been supported by NSF grants DMI 96-12327 and DMI-96-12365.

Defining Terms

Factors: Variables used in experiments that have been designed.

Focus group: An interview conducted by a trained moderator in a nonstructured and natural manner and with a small group of respondents [Malhotra, 1996].

Levels: Predetermined values of factors that model the spread of values of a factor.

Population: A group being studied.

Random sample: Data that have been drawn from a population in such a way that each grouping of data has an equally likely chance of being drawn from the population.

Type I error: Occurs when the null hypothesis is actually true, but the statistical tests performed indicate the null hypothesis is false.

Type II error: Occurs when the null hypothesis is actually false but it is identified as being true.

References

Beyer, W. H., Ed. *CRC Standard Probability and Statistics: Tables and Formulae,* CRC Press, Boca Raton, FL, 1991.

Corchran, W. G. *Sampling Techniques, 3rd ed.,* John Wiley & Sons, New York, 1977.

Gibbons, J. D., Olkin, I., and Sobel, M. *Selecting and Ordering Populations,* John Wiley & Sons, New York, 1977.

Griffin, A. and Hauser, J. R. The voice of the customer, *Market. Sci.,* 12(1), 1–27, 1993.

Lehman, R. S. *Statistics in Behavioral Sciences,* Brooks/Cole, Pacific Grove, CA, 1995.

Malhotra, N. K. *Marketing Research: An Applied Orientation, 2nd ed.,* Prentice Hall, Upper Saddle River, NJ, 1996.

Montgomery, D. C. *Introduction to Statistical Quality Control, 3rd ed.,* John Wiley & Sons, New York, 1996.

Montogmery, D. C. *Design and Analysis of Experiments, 4th ed.,* John Wiley & Sons, New York, 1997.

Sokal, R. R. and Rohlf, F. J. *Biometry: The Principles and Practice of Statistics in Biological Research, 2nd ed.,* pp. 262–263, 1981.

Further Information

The texts listed above by Lehman, Malhotra, Sokal and Rohlf, and Montgomery are espeically readable. Statistical tables are available from many sources; the book edited by Beyer is especially complete, however, and also offers suggestions on using these tables.

5.3   Quality Control

Kenneth E. Case and Jerry Dechert

To be competitive on a worldwide basis requires that the needs of the customer be known and consistently met or exceeded. Products, including services, that satisfy and even delight the customer require processes that are stable and capable of meeting specifications. Even meeting specifications, however, is not enough. Leading-edge customers today expect critical manufacturing, service delivery, and laboratory measurement process characteristics to be "on target with minimum variance" [Wheeler, 1992].

The primary area of statistical quality control (SQC) is known as statistical process control (SPC). SPC is used proactively to help prevent or correct problems due to centering and variation. Acceptance sampling is a lesser area of SQC used reactively for lot disposition.

The major statistical tool employed in SPC is the **control chart**, which is widely used to help understand variation, increase knowledge of a process, and provide guidance for process improvement. There are many varieties of control charts, and the correct chart to be used is a function of the data being collected and evaluated. The control chart must then be designed correctly so that it defines **common cause** and **special cause** variation and provides insight needed for improvement actions to be taken. Control charts are most effective when they are used to study key product or process data on characteristics that are either (1) of direct importance to the customer, or (2) in-process factors that affect a characteristic of direct importance to the customer.

The flowchart, cause and effect diagram, data collection form, Pareto chart, and histogram are particularly helpful SPC tools used in conjunction with control charts for identifying key factors affecting a process and increasing knowledge about a process [Case, 1994]. **Natural process limits, process capability indices**, or **process performance indices** may be used to provide a measure of process performance relative to specification limits. Also, designed experimentation, analysis of variance, and regression analysis techniques are other statistical tools often used to identify key causal factors and interactions that affect the ability to hit target and reduce variation [Schmidt, 1994].

Fundamentals of Control Charting

Control charts help determine whether a process is stable. If a process is stable, it is also said to be consistent, predictable, in control, and affected by common cause variation only. An unstable process is said to be inconsistent, unpredictable, out of control, and affected by both common and special cause variation. Common causes are those that are inherent in the process. Special causes are not a part of the process all the time or do not affect everyone working in the process [Nolan and Provost, 1990]. Eliminating special causes returns a process to the in-control state.

Control charts have a center line, upper and lower control limits, and data (a statistic) to plot. The average of the statistic is usually taken as the center line. Control limits are actually calculated from the process data and represent the voice of the process. Ideally, 30 subgroups of data are available for calculation; about 12 subgroups are a practical minimum to establish a control chart. Specifications are entirely different from control limits and represent the voice of the customer.

Data are plotted on a control chart, usually in time order of occurrence. Either variables or attributes data may be plotted, but the principles underlying control chart use remain the same. The keys to successful use are (1) correct selection of the specific chart and chart formulas, (2) rational subgrouping during data collection, and (3) use of proper calculation procedures. Unfortunately, each of these is sometimes a problem. Table 5.2 provides a guideline for chart selection. Tables 5.3 and 5.4, respectively, provide the correct formulas and a limited set of factor constants for common charts.

Rational Subgrouping

Variables control charts will not work well, if at all, without rational subgrouping. Subgroups should be chosen in such a way that will give maximum chance for the measurements within a subgroup to be alike and the maximum chance for data from subgroup to subgroup to be different [Grant and Leavenworth,

TABLE 5.2 Guide for Control Chart Selection

Type of Data	Subgrouping Practice	Recommended Charts
Variables data	Subgroups of size n > 1 taken	$\bar{X}$ and R (2 < n < 10) $\bar{X}$ and s (n ≥ 10)
Variables data	Individual samples (n = 1) taken; artificial subgroups used	X and MR MA and MR
Attributes data: classification of items as good or defective	Often 50 < n < 500; n may be constant or variable	p chart np chart only if n constant
Attributes data: count of defects on surface or other "area of opportunity"	Constant "area of opportunity" Variable "area of opportunity"	c chart u chart

Note! Many other kinds of charts exist, but the above cover the vast majority used in practice. Worthy of note are cumulative sum (CUSUM) and exponentially weighted moving average (EWMA) charts when sometimes used in place of $\bar{X}$ or X charts.

TABLE 5.3 Common Control Chart Formulas

Chart	Center Line	UCL	LCL
$\bar{X}$	$\bar{\bar{X}}$	$\bar{\bar{X}} + A_2\bar{R}$	$\bar{\bar{X}} - A_2\bar{R}$
R	$\bar{R}$	$D_4\bar{R}$	$D_3\bar{R}$
$\bar{X}$	$\bar{\bar{X}}$	$\bar{\bar{X}} + A_3\bar{s}$	$\bar{\bar{X}} - A_3\bar{s}$
s	$\bar{s}$	$B_4\bar{s}$	$B_3\bar{s}$
MA	$\bar{X}$	$\bar{X} + A_2\overline{MR}$	$\bar{X} - A_2\overline{MR}$
MR	$\overline{MR}$	$D_4\overline{MR}$	$D_3\overline{MR}$
X	$\bar{X}$	$\bar{X} + E_2\overline{MR}$	$\bar{X} - E_2\overline{MR}$
MR	$\overline{MR}$	$D_4\,\overline{MR}$	$D_3\,\overline{MR}$
p	$\bar{p}$ ($\bar{p}$ IN PERCENT)	$\bar{p} + 3\sqrt{\bar{p}(100-\bar{p})/n}$	$\bar{p} - 3\sqrt{\bar{p}(100-\bar{p})/n}$; none if <0
np	$n\bar{p}$ ($\bar{p}$ IN FRACT DEF)	$n\bar{p} + 3\sqrt{n\bar{p}(1-\bar{p})}$	$n\bar{p} - 3\sqrt{n\bar{p}(1-\bar{p})}$; none if <0
c	$\bar{c}$	$\bar{c} + 3\sqrt{\bar{c}}$	$\bar{c} - 3\sqrt{\bar{c}}$; none if <0
u	$\bar{u}$	$\bar{u} + 3\sqrt{\bar{u}/n}$	$\bar{u} - 3\sqrt{\bar{u}/n}$; none if <0

TABLE 5.4 Abbreviated Table of Control Chart Factors (Shaded Area not Recommended)

n	A_2	E_2	d_2	D_3	D_4	A_3	c_4	B_3	B_4
2	1.880	2.660	1.128	NONE	3.267	2.659	0.7979	NONE	3.267
3	1.023	1.772	1.693	NONE	2.575	1.954	0.8862	NONE	2.568
4	0.729	1.457	2.059	NONE	2.282	1.628	0.9213	NONE	2.266
5	0.577	1.290	2.326	NONE	2.114	1.427	0.9400	NONE	2.089
6	0.483	1.184	2.534	NONE	2.004	1.287	0.9515	0.030	1.970
10	0.308	0.975	3.078	0.223	1.777	0.975	0.9727	0.284	1.716
15	0.223	0.864	3.472	0.347	1.653	0.789	0.9823	0.428	1.572

1996]. Put differently, subgroups should be chosen in a way that only common causes of variation are within a subgroup while special causes of variation, if any, occur between subgroups.

Charts for spread or dispersion such as the range (R), standard deviation (s), or moving range (MR) are used to determine if there are significant differences in common cause (within subgroup) variation over time. They ask, "Is the variation within subgroups consistent from subgroup to subgroup?" Since MR charts inherently have time between samples, they should ideally be of size n = 2 to minimize the

TABLE 5.5   Formulas for Estimating Process Standard Deviation, Capability Indices, and Performance Indices

What	1	2	3
$\hat{\sigma}$ (stable $\bar{X},R$; $\bar{X},s$; or X,MR)	$\hat{\sigma} = \bar{R}/d_2$	$\hat{\sigma} = \bar{s}/c_4$	$\hat{\sigma} = \overline{MR}/d_2$
C_P, C_{PK}, C_{PM} (stable process)	$C_p = \dfrac{U-L}{6\hat{\sigma}}$	$C_{PK} = \min\left(\dfrac{U-\bar{X}}{3\hat{\sigma}}, \dfrac{\bar{X}-L}{3\hat{\sigma}}\right)$	$C_{PM} = \dfrac{U-L}{6\sqrt{\hat{\sigma}^2 + \left(\bar{X}-TARG\right)^2}}$
P_P, P_{PK}, P_{PM} (unstable/unknown)	$P_p = \dfrac{U-L}{6s}$	$P_{PK} = \min\left(\dfrac{U-\bar{X}}{3s}, \dfrac{\bar{X}-L}{3s}\right)$	$P_{PM} = \dfrac{U-L}{6\sqrt{s^2 + \left(\bar{X}-TARG\right)^2}}$

opportunity for special causes to infiltrate the MR. Charts for centering such as the Xbar ($\bar{X}$), individual (X), and moving average (MA) are used to determine if there are significant differences due to special causes (between subgroup variation) occurring over time. They ask, "Are there any detectable differences from subgroup to subgroup?"

Proper Calculation Procedures

Each characteristic described by variables data requires two charts, one for spread (R, s, or MR) and one for centering ($\bar{X}$, X, or MA). Attributes data requires only one chart (p, np, c, or u). In the case of variables data, begin with the chart for spread (R, s, MR) first. The steps for variables charts (e.g., $\bar{X}$ and R) include

1. Determine the center line (CL) of the spread chart: $CL_R = \bar{R}$.
2. Calculate a "trial" upper control limit (UCL) for the spread chart: $UCL_R = D_4\bar{R}$.
3. Temporarily eliminate every value of the spread statistic that is above the UCL and recalculate the CL and UCL. Do this only once.
4. Proceed to calculate the lower control limit (LCL) of the spread statistic and the CL and control limits of the chart for centering

$$LCL_R = D_3\bar{R}; \quad CL_{\bar{X}} = \bar{\bar{X}}; \quad UCL_{\bar{X}} = \bar{\bar{X}} + A_2\bar{R}; \quad LCL_{\bar{X}} = \bar{\bar{X}} - A_2\bar{R}$$

5. Plot both charts, including those spread chart points that were temporarily eliminated.

Attributes charts follow a similar procedure, except there is only one chart. The trial UCL procedure is used in a similar manner.

Once a process is determined to be stable, its $\pm 3\sigma$ natural process limits and *capability* index may be calculated. If the process is not known to be stable, a process *performance* index is often calculated. Table 5.5 presents the correct formulas for variables data. Such indices are rarely calculated for attributes data.

Applications of Control Charting

Control charting is widely used in many industries, particularly those having to do with manufacturing and laboratories such as electronics, automotive, primary metals, chemicals, refining, mining, foods, health care, etc. More progressive organizations are using control charts to better understand their service operations, too.

Consider a health care manufacturer that prepares lots (batches) of reagent. These reagents, virtually homogeneous within a lot, are then used to fill bottles, which are then sold to hospitals and medical laboratories worldwide for use in conjunction with automated blood analyzers. Control charts are used

TABLE 5.6 Coded Data for Reagent Concentration (Coded Target is 40)

Lot	1	2	3	4	5	6	7	8	ooo	23	24	25	26	27	28	29	30	AVGS
Data	43	40	43	44	42	45	34	43	ooo	37	52	50	47	54	47	50	43	42.5
MR(n = 2)		3	3	1	2	3	11	9	ooo	0	15	2	3	7	7	3	7	3.9

in the reagent preparation process as well as the bottle fill process to (1) assess their stability and (2) estimate the capability to meet specifications.

Concentration Example — X and MR Charts

A new lot of reagent is prepared periodically and analyzed for concentrations as measured in RLUs (relative light units). Concentrations are actually measured in different levels in a laboratory setting, but only one level is illustrated below. Coded RLU data on the past 30 batches of reagent are presented in Table 5.6. Desired specification are 40.0 ± 13.0. Following the steps presented in the previous section: (1) $CL_R = \overline{MR} = 3.9$; (2) $UCL_{MR} = 3.267 * 3.9 = 12.6$; (3) eliminating the MR of 15, $CL_R = \overline{MR} = 3.5$ and $UCL_{MR} = 3.267 * 3.5 = 11.4$; (4) $LCL_{MR} = $ none, $CL_X = \overline{\overline{X}} = 42.5$, $UCL_X = 42.5 + 2.66 * 3.5 = 51.8$, $LCL_X = 42.5 - 2.66 * 3.5 = 33.2$; and (5) the X and MR charts are plotted in Figs. 5.10 and 5.11.

It is clear from the X chart that a special cause influenced the process to change abruptly beginning with lot 24. Investigation showed that the microparticle component of the reagent had itself been from a new lot beginning at that time. The MR chart also clearly shows that the process had shifted dramatically from lot 23 to lot 24. Both the X and MR charts show that the new level of concentration caused by the microparticle change remained at the higher level. Since the process is unstable, natural process limits and a process capability index are inappropriate, but common process performance indices can be calculated using the equations in Table 5.5 as follows:

$$P_p = \frac{U-L}{6s} = \frac{53-27}{6*4.8} = 0.90 \text{ and}$$

$$P_{pk} = \min\left(\frac{U-\overline{X}}{3s}, \frac{\overline{X}-L}{3s}\right) = \min\left(\frac{53-42.5}{3*4.8}, \frac{42.5-27}{3*4.8}\right) = 0.73$$

Neither of these indices is desirable, both being less than a commonly desired value of 1.33 to 2.00. Process capability indices can be calculated only to determine the *potential* of the process to achieve specifications, but not as a reporting value. These capability indices are calculated as follows from Table 5.5:

$$C_p = \frac{U-L}{6\hat{\sigma}} = \frac{53-27}{6\overline{MR}/d_2} = \frac{26}{6*3.5/1.128} = 1.40$$

$$C_{pk} = \min\left(\frac{U-\overline{X}}{3\hat{\sigma}}, \frac{\overline{X}-L}{3\hat{\sigma}}\right) = \min\left(\frac{53-42.5}{3\overline{MR}/d_2}, \frac{42.5-27}{3\overline{MR}/d_2}\right) = \min\left(\frac{10.5}{3*3.5/1.128}, \frac{15.5}{3*3.5/1.128}\right) = 1.13$$

The index C_p shows the potential value of C_{pk} if the process were centered on 40.0. An index of 1.40 would then result. The value of C_{pk} is less than that of C_p because it explicitly accounts for the fact that the process average is high due to the shift. To actually realize and report these capability indices, the process must be brought under control.

Bottle Filling Example — $\overline{X}$ and R

The bottle filling process has a minimum specification of 200 ml. Outsourced sterile bottles are used and filled from a four-in-line machine with each of the four heads individually set. Only one of the heads is

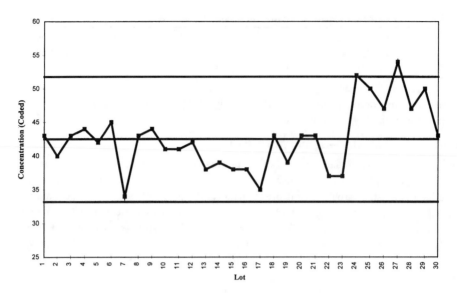

FIGURE 5.10   X chart of reagent lot concentrations.

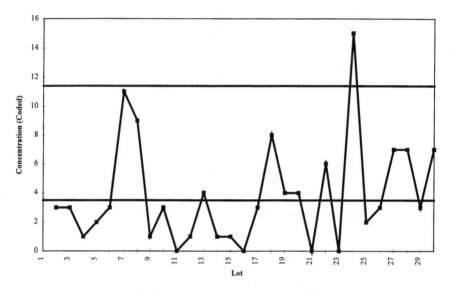

FIGURE 5.11   MR chart of reagent lot concentrations.

examined in the fill data presented in Table 5.7. Here, subgroups of $n = 4$ consecutive bottles are taken from a single head of the fill line every half hour and measured for fill volume. The $\bar{X}$ and R charts are now set up following the steps of the previous section: (1) $\bar{R} = 3.3$; (2) $UCL_R = 2.282 * 3.3 = 7.5$; (3) no ranges exceed the UCL_R, so LCL_R = none, $CL_{\bar{X}} = 203.05$, $UCL_{\bar{X}} = 203.05 + 0.729 * 3.3 = 205.46$, and $LCL_{\bar{X}} = 203.05 - 0.729 * 3.3 = 200.64$. The $\bar{X}$ and R charts are plotted in Figs. 5.12 and 5.13.

It is clear that the process is stable (in control). Process capability indices are now calculated to assess performance. Since there is a single lower specification of 200 ml, only C_{pk} is calculated.

$$C_{pk} = \min\left(\frac{U - \bar{X}}{3\hat{\sigma}}, \frac{\bar{X} - L}{3\hat{\sigma}}\right) = \frac{\bar{X} - L}{3\hat{\sigma}} = \frac{203.05 - 200}{3 * \bar{R}/d_2} = \frac{3.05}{3 * 3.3/2.059} = 0.63$$

TABLE 5.7 Reagent Bottle Fill Volumes

Subgroup	1	2	3	4	5	6	7	ooo	24	25	26	27	28	29	30
X1	200	203	200	202	202	203	203	ooo	203	204	202	202	205	202	202
X2	203	203	203	201	203	201	205	ooo	203	204	205	205	202	205	202
X3	205	202	202	202	204	203	205	ooo	206	204	204	202	204	203	202
X4	205	205	201	203	205	204	201	ooo	204	203	204	201	203	203	204
AVG	203.3	203.3	201.5	202.0	203.5	202.8	203.5	ooo	204.0	203.8	203.8	202.5	203.5	203.3	202.5
R	5	3	3	2	3	3	4	ooo	3	1	3	4	3	3	2

Grand average = 203.05   Average range = 3.3

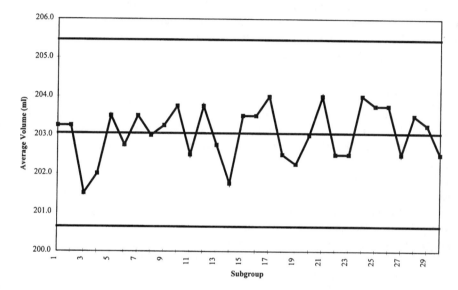

FIGURE 5.12   Xbar chart of reagent bottle fill volumes.

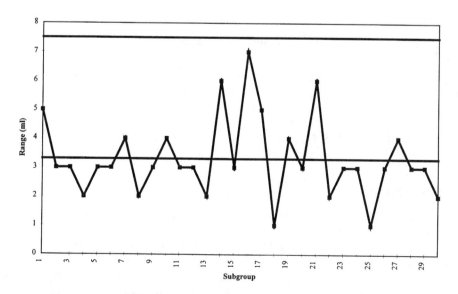

FIGURE 5.13   R chart of reagent bottle fill volumes.

The natural process limits are at $[\text{LNPL}_X, \text{UNPL}_X] = \bar{X} \pm 3\hat{\sigma} = \bar{X} \pm 3\bar{R}/d_2 = 203.05 \pm 3 * 3.3/2.059 = [198.24, 207.86]$. Even though the bottle fill process is stable, it is not capable of meeting the lower specification of 200 ml. Either the average fill volume must be increased or the variation must be decreased. Also, note that the measurement resolution of 1 ml should likely be improved, e.g., to 0.5 or 0.2 ml, if the process is to run near the lower specification in the future.

Summary

Control charting has proven to be a powerful tool for those who use it to display and understand variation, as presented in the examples. It is less effective when used only to decide when process adjustments are to be made, even though that is helpful. When used for learning, with the other tools of SPC, control charts can lead to (1) increased knowledge of factors affecting variation, (2) stable processes through elimination of special causes, (3) improved processes through reduction in common causes, (4) improved natural process limits, (5) higher process capability indices, (6) improved customer satisfaction and delight, and (7) worldwide competitiveness.

Defining Terms

Common causes and special causes: Common causes are those that are inherently part of the process, hour after hour, day after day, and affect everyone working in the process. Special causes are those that are *not* part of the process (or system) all the time or do not affect everyone but arise because of specific circumstances.

Control chart: A statistical tool that operationally defines the concept of a stable process. It seeks to determine if a sequence of data may be used for predictions of what will occur in the future.

Natural process limits: Limits within which the individual measurements of a stable process may be expected to fall. These are set at $\pm 3\sigma$ from the process average.

Process capability index: A simple way of expressing the relationship between the voice of the customer (specifications) and the voice of the process (natural process limits), expressed as a dimensionless ratio. It is used only when the process is known to be stable.

Process performance index: This is similar to the process capability index, but used when the process is either not known to be stable or is known to be out of control.

References

Case, K. E. *Statistical Process Control, 3rd ed.*, SPC Seminar Manual, Oklahoma State University, Stillwater, OK, 1994.

Grant, E. L. and Leavenworth, R. S. *Statistical Quality Control, 7th ed.*, McGraw-Hill, New York, 1996.

Nolan, T. W. and Provost, L. P. Understanding variation, *Qual. Prog.*, 23(5): 70–78, 1990.

Schmidt, S. R. and Launsby, R. G. *Understanding Industrial Designed Experiments, 4th ed.*, Air Academy Press, Colorado Springs, CO, 1994.

Wheeler, D. J. and Chambers, D. S. *Understanding Statistical Process Control, 2nd ed.*, SPC Press, Knoxville, KY, 1992.

Further Information

Good introductions and more advanced materials are available in the vast list of books, standards, monographs, and so on available from the American Society for Quality Control, 611 East Wisconsin Avenue, P. O. Box 3005, Milwaukee, WI 53201-3005. Call 800-248-1946 and request a free copy of their resource catalog.

Excellent examples and other materials are available from the ASQC's Annual Quality Congress held each May, the ASQC Statistics Division and its Fall Technical Conference, and the ASQC's *Journal of Quality Technology* and *Quality Engineering*.

Excellent elementary treatments and examples appear in the books referenced above by Grant and Leavenworth and Wheeler and Chambers.

Many software packages are available. Some of the most frequently encountered packages in client organizations include JMP, Statgraphics, Minitab, and SAS. Once per year, a software review appears in *Quality Progress*. Review that issue and request a demonstration copy of those packages of interest.

5.4   Regression

Terry E. Dielman

Regression analysis is one of the most useful of statistical procedures. In business, economics, engineering, medicine, and both the natural and social sciences, regression is used extensively to solve important problems. Regression analysis can be thought of as a way of expressing the relationship between a variable called the **dependent variable** and one or more **independent variables**. For example, consider a manufacturing firm that installs communications nodes at its various manufacturing sites. The company would like accurate predictions of the installation cost of future nodes. They have data available on the following variables: y = installation cost at 14 previously installed nodes, x_1 = number of communications ports at each node, and x_2 = bandwidth at each node. The data are shown in Table 5.8. The installation cost, y, is the dependent variable. The number of ports and the bandwidth are the independent variables. The question is, "How can these data be used to provide accurate predictions of the installation cost of future nodes?" We could ignore the information provided by the two independent variables and just use the average cost from the 14 nodes as our prediction. However, if cost is related to the number of communications ports and the bandwidth, we should be able to use this information to find a more accurate prediction of cost at future nodes. We will refer to this example later to illustrate certain aspects of regression analysis.

The Linear Regression Model

Regression considers the relationship of a dependent variable to one or more independent variables using an equation of the form

$$y_i = \beta_0 + \beta_1 x_{1i} + \beta_2 x_{2i} + \ldots + \beta_K x_{Ki} + e_i \tag{5.15}$$

for $i = 1,\ldots,n$, where y_i is the ith value of the dependent variable, $x_{1i},\ldots, x_{Ki}$ are the ith values of each of the K independent variables, β_0 is called the **constant**, $\beta_1,\ldots,\beta_K$ are the **coefficients** of the independent

TABLE 5.8   Communications Nodes Data

Cost of Node	Number of Ports	Bandwidth
52,388	68	58
51,761	52	179
50,221	44	123
36,095	32	38
27,500	16	29
57,088	56	141
54,575	56	141
33,969	28	48
31,309	24	29
23,444	24	10
24,269	12	56
53,479	52	131
33,543	20	38
33,056	24	29

variables, and ϵ_i is a random **disturbance**. To choose the "best" coefficient values for the regression equation, some criterion for best must be established. The standard used is to minimize the sum of the squares of the prediction errors for the sample data. This procedure is known as **least squares**. For example, in Eq. (5.15), we would choose estimates $b_0, b_1,..., b_K$ of $\beta_0, \beta_1,..., \beta_K$ to minimize $\sum(y_i - \hat{y}_i)^2$. The predicted values of the y variable, $\hat{y}_i$, are created using the coefficient estimates: $\hat{y}_i = b_0 + b_1 x_{1i} + ... + b_K x_{Ki}$. We are choosing the coefficient estimates to place the predicted values "close" to the actual y values in the sense of minimizing the sum of squares of the prediction errors. There are other criteria for choosing the coefficient estimates. For example, the least absolute value criterion chooses $b_0, b_1,..., b_K$ to minimize the sum of the absolute values of the prediction errors: $\sum |y_i - \hat{y}_i|$. Although there are pros and cons to each of these criteria, the least squares criterion is well established as the standard in most fields and is found in statistical packages and spreadsheets. For these reasons, least squares will be the procedure discussed here.

As with any statistical technique applied to a sample of data, it is desirable to have measures of the accuracy of the statistics produced. For example, do the coefficients obtained suggest that the independent variables are related to the dependent variable, or could the results have occurred by chance? The standard error of each coefficient, s_{b_k}, can be viewed as a measure of the accuracy of b_k as an estimate of the true coefficient, β_k. A t statistic can be created to test whether the true coefficient is different from zero: $t = b_k/s_{b_k}$. This statistic should be compared to a t value with n-K-1 degrees of freedom. Using t statistics can help to insure that only important variables are included in the equation. Another useful number is the **R-square** of the regression (sometimes written R^2). The R-square is used as a measure of the quality of the fit of the regression. The R-square will vary between 0 and 100%, with 0% indicating no linear relationship and 100% indicating a perfect linear relationship. The R-square can be used to determine how much of the variation in the y variable has been explained by the regression. For most regression applications, the goal should be to explain a reasonable amount of the variation in the dependent variable with a relatively simple model. This dictum has been found to provide useful results in practice. Include in the regression only those variables that are shown to be important by a criteria such as the t statistic. Including unimportant variables complicates the equation needlessly and also decreases the accuracy of predictions of the dependent variable.

Once a regression model has been developed, it should be validated. The regression **residuals** are the primary tool used to examine validity of the model. The residuals represent the difference between the true y values in the sample and the y values predicted by the regression equation: $e_i = y_i - \hat{y}_i$. Since the residuals represent the difference between the actual data values and the predictions from our fitted model, we can view the residuals as that part of the data that has not been explained. We can then examine the residuals to determine whether there are any unexplained patterns that might be useful in obtaining better predictions of the data. Examination of the residuals is usually done by looking at **residual plots**. Plots of the residuals vs. the x variables or vs. the predicted y values are often useful in validating a regression. In examining these plots, any systematic patterns in the residuals would suggest that the model developed is not yet complete. Also, residual plots can be used to detect outliers that are the result of data coding errors or unusual circumstances not incorporated into the model. Typically, the **standardized residuals** will be used in these plots rather than the raw residuals. The standardized residuals are the residuals, e_i, divided by their standard deviation. They provide essentially the same information as the residuals, but they do so in standardized units. The only difference between the residuals and standardized residuals is the associated scale.

When examining residual plots to validate the regression, we watch for systematic patterns or outliers. For example, a curvilinear pattern in a residual plot would indicate that we have fit a linear model when, in fact, the data are related in a curvilinear manner. A "cone-shaped" pattern in a residual plot indicates that the variance of the y values around the regression line is not constant. **Nonconstant variance** can result in poor estimates of the regression coefficients and incorrect inferences when judging the importance of the x variables included in the regression. Both curvilinearity and nonconstant variance can typically be corrected by data transformations. Regression can still be used in these cases, but modifications must be made to insure a valid model. For example, a curvilinear model can be fitted to data by using **polynomial regression** (powers of the x variable are included). Nonconstant variance can often be corrected by using the natural logarithm of the y variable as the dependent variable.

When data are collected over time (time-series data), the disturbances in Eq. (5.15) will often be correlated over time. This **autocorrelation** of the disturbances may be due to an important but omitted variable that is correlated over time. In business and economic data, the patterns produced by autocorrelation are often artifacts of business cycles. Rather than counting on residual plots, autocorrelation is typically detected using a test called the Durbin-Watson test. Autocorrelation can also often be corrected by data transformations, thus producing models that will generate better predictions of the y variable.

Another problem that sometimes adversely affects the quality of a regression is the presence of outliers, or extreme data points. Outliers can also be detected by examining plots of the standardized residuals. When standardized residuals are plotted, it is easy to see when a residual is extreme. Standardized residuals larger than ±2 or ±3 are often "flagged" as possible outliers that deserve further investigation. Outliers may be caused by miscoded data. These coding errors need to be corrected before the regression results are valid. Outliers may also be the result of unusual circumstances affecting a small part of the data. Corrections in these cases are less clear.

Cost Estimation Example

Figure 5.14 shows a typical regression output from a statistical package called Minitab. The regression equation shows the relationship between cost of communication node installation and the two independent variables, number of communications ports and bandwidth:

$$\text{cost} = 17086 + 469 \text{ numports} + 81.1 \text{ bandwidth}$$

The constant in the equation is 17086. The practical significance of the constant needs to be judged according to the application. In this example, the constant represents the cost when the number of ports and the bandwidth are both zero. Since there are no cases when both number of ports and bandwidth will be zero, it is somewhat questionable what practical significance this number may have. (One possible interpretation might be as the fixed cost for installing a new communications node. The remainder of the cost would be dependent on the number of ports and the bandwidth.) The coefficients of the independent variables show how much the dependent variable will change, on average, for each one unit change in the associated independent variable. Thus, the addition of each new port at the communications node will cost, on average, $469 (assuming that bandwidth is not changed). Unit increases in the bandwidth will result in $81.10 increases in the cost of the communications node (assuming that the number of ports is not changed).

The standard errors for the coefficients of numports and bandwidth are, respectively, 66.98 and 21.65. On average, we expect cost to increase by $469.00 for each new port installed, but our measure of uncertainty in this estimate is $66.98. Dividing 469.00 by 66.98 gives the t statistic to test whether we should view the true coefficient of numports as being different from zero. The t statistic is 7.00 (also given on the output). This number should be compared to a value from a t table using 11 degrees of freedom ($n\text{-}K\text{-}1 = 14\text{-}2\text{-}1 = 11$). Since the absolute value of the t statistic exceeds the tabled value ($t_{0.025,11} = 2.201$ if a 5% level of significance is used), we conclude that there is a relationship between cost and numports that can be used to obtain more accurate predictions of cost. As an alternative to the t statistic, a p value is provided on the output. The p value given for numports is 0.000. P values also measure the usefulness of the variables included in a regression. A standard rule used with p values is the variable is useful if p value < 0.05. In this case the variable numports is judged to be useful. Note that bandwidth is also judged useful since the p value for bandwidth is 0.003. Therefore, both of the these variables add to the ability of the regression to accurately predict cost.

Note that the R-square for this regression is 95% (R-sq). This means that 95% of the variation in cost has been explained by the regression.

The standard error of the regression, s = 2983, is a measure of the standard deviation around the regression surface and can also be used as a measure of the quality of the fit of the regression. The smaller

```
The regression equation is
cost = 17086 + 469 numports + 81.1 bandwidth

Predictor        Coef        StDev           T        P
Constant        17086        1865         9.16    0.000
numports       469.03       66.98         7.00    0.000
bandwidth       81.07       21.65         3.74    0.003

S = 2983         R-Sq = 95.0%      R-Sq(adj) = 94.1%

Analysis of Variance

Source       DF           SS          MS          F        P
Regression    2   1876012662   938006331     105.45    0.000
Error        11     97849860     8895442
Total        13   1973862521

Source       DF      Seq SS
numports      1   1751268376
bandwidth     1    124744286

Unusual Observations
Obs   numports      cost        Fit   StDev Fit   Residual    St Resid
  1      68.0      52388      53682       2532      -1294       -0.82 X
 10      24.0      23444      29153       1273      -5709       -2.12R

R denotes an observation with a large standardized residual
X denotes an observation whose X value gives it large influence.

  Fit   StDev Fit        95.0% CI              95.0% PI
57389        1439   (   54221,    60556)  (   50099,    64679)
```

FIGURE 5.14   Regression output from Minitab for the communications nodes data.

the value of the standard error, the better the fit. The standard error is in the original units of the y variable (dollars in this case), but not scaled as is the R-square, so it is sometimes more difficult to interpret.

Suppose we want to predict cost for a new communications node that will have 60 ports and a bandwidth of 150. Using the equation developed we have

$$\text{cost} = 17086 + 469.0(60) + 81.1\,(150) = \$57{,}391$$

We might also want a measure of the accuracy of this prediction. Some statistical packages such as Minitab will provide such measures. Note the following at the bottom of the Minitab output:

```
  Fit   StDev Fit        95.0% CI                95.0%  PI
57389        1439   (   54221,   60556)  (    50099,   64679)
```

This part of the output resulted from a request for a prediction of cost when 60 ports and a 150 bandwidth are required. The fit is the predicted value. The value computed previously as $57,391 is slightly different due to rounding error. Note that a 95% confidence interval (95.0% CI) is provided: ($54221, $60556). This is an estimate of the average cost of all communications nodes with 60 ports and a bandwidth of 150. A 95% prediction interval (95.0% PI) is also provided. The PI is used to express the uncertainty in a prediction for a single communications node with 60 ports and a 150 bandwidth. We can be 95% sure that the cost of an individual node with 60 ports and a 150 bandwidth will be between $50099 and $64679. The StDev Fit shown on the output is the standard deviation used to construct the CI. The prediction standard deviation is not shown (but is equal to $\sqrt{\text{StDev Fit}^2 + \text{standard error}^2}$).

The plot of the standardized residuals from the regression in Fig. 5.14 vs. the x variable numports is shown in Fig. 5.15. If the estimated equation is valid, there should be no explainable patterns in this plot, that is, the residuals should be random. Looking at the plot in Fig. 5.15, no patterns are observed. The same is true for the plots of the residuals vs. the variable bandwidth and vs. the fitted y values for this regression, although these plots are not shown. The equation has been validated.

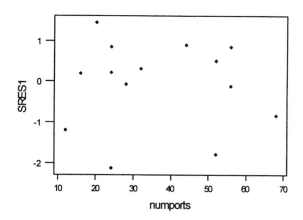

FIGURE 5.15 Plot of standardized residuals vs. numports variable.

Defining Terms

Autocorrelation: A common problem in time-series regression where disturbances in adjacent time periods are correlated.

Coefficients: The multipliers of the independent variables in a regression equation. The regression coefficients show the change in the dependent variable for a one-unit change in the associated independent variable.

Constant: The intercept term in a regression equation.

Dependent variable: The variable in a regression analysis that is to be explained or predicted.

Disturbance: The random error component in the theoretical regression equation.

Independent variable: The variable (or variables) in a regression analysis believed to be related to the dependent variable.

Least squares: The typical criterion used in choosing the best regression coefficient estimates.

Nonconstant variance: A problem encountered in some regression applications where the variance of the y values around the regression is not constant.

Polynomial regression: An equation using powers of the x variable(s) to model curvilinear relationships.

Regression analysis: A statistical method used to relate a dependent variable to one or more independent variables in the form of a linear equation.

Residual plots: Plots used to validate a regression model. Typically, the standardized residuals are plotted vs. each x variable in the equation and vs. the fitted y values.

Residuals: The difference between the true values of the dependent variable and the values of the dependent variable predicted by the regression equation.

R-square: A measure of the proportion of variation of the dependent variable that has been explained or accounted for by the regression.

Standardized residuals: Regression residuals divided by their standard deviation.

References

Dielman, T. E. *Regression Analysis for Business and Economics, 2nd ed.,* Duxbury Press, Belmont, CA, 1997.

Graybill, F. A. and Hariharan, I. *Regression Analysis: Concepts and Applications,* Duxbury Press, Belmont, CA, 1994.

Minitab Reference Manual, Release 11, Minitab, Inc., 1996.

Neter, J., Wasserman, W., and Kutner, M. H. *Applied Linear Regression Models, 2nd ed.,* Irwin, Homewood, IL, 1989.

Further Information

For an applied introduction to regression analysis, see *Regression Analysis for Business and Economics* by Terry E. Dielman.

For a more advanced discussion of regression, see *Applied Linear Regression Models* by John Neter, William Wasserman, and Michael Kutner or *Regression Analysis* by Franklin Graybill and Hariharan Iyer.

5.5   Technology Forecasting

Frederick A. Rossini and Alan L. Porter

A forecast predicts likely future states by a systematic and reproducible method. A **technology forecast** (TF) predicts the nature, extent, and timing of changes in technological developments or parameters of concern.

TF is based on the understanding that the processes involved in technological innovation are sufficiently orderly at some level of aggregation to allow a meaningful forecast. Innovation involves socioeconomic processes as well as technological development. Therefore, both contexts must be dealt with for effective TF.

Effective management of technology, especially in contexts where the technology is rapidly changing, depends on the ability to forecast technological developments as well as to anticipate their significant impacts (a process known as technology assessment). This "look before you leap" approach has been dubbed "managing the present from the future".

The track record of TF parallels that of management science in one important respect: highly mathematical, sophisticated analyses that may arouse admiration from academics have to date, in our knowledge, largely failed to deliver significant, usable results in the contexts that most managers encounter daily. Invariably, there is a lack of time, resources, and/or data for such approaches. Our approach in this article is to follow the KISS principle (keep it simple, stupid) and present relatively simple approaches that have often proved to be effective.

Forecasting Principles

There are a number of useful guidelines for conducting TFs. We state and comment on them in this section.

Get the subject right. First and most important is to be clear about the purpose of the forecast, its expected outcomes and their uses, and the users of the TF. The subject of the forecast, critical aspects of the methodology, and the content and form of the output are determined by this basic understanding. Whatever the purpose, the TF is constrained by available time, resources, and data. Bounding the subject of study, the technology, its context, the time frame of the forecast, the appropriate techniques, and the output form is the first critical activity of the TF. Involving the target users, from bounding onward, can bolster TF utility.

Get the technology right. It is necessary to be able to clearly describe the current state of the art of the technology being forecast. This implies a clear understanding of the systems and subsystems involved, their range of implementations, and the functions the technology can serve. It is also helpful to denote the most important parameters that define the technology and its performance. Getting the technology right also involves understanding supporting technologies, those required in order that the technology being forecast functions effectively, and competing technologies — alternatives to the technology being forecast, including unanticipated alternatives. An effective TF also attends to the technological barriers to developing the technology further — the "bottlenecks".

Get the context right. The context of a technology refers to the socioeconomic institutions that are involved in its development and implementation. Institutions include developers, regulators, marketers, competitors, and their complex interactions. The context includes critical decision points in the development of the technology as well as the changes in institutions and their interrelationships over the period covered by the TF. One concept that has been of significant help to us in performing TFs is the technology delivery system or TDS [Wenk and Kuehn, 1977]. The TDS is a simple boxes and arrows model that identifies the critical institutions involved with a technology and their linkages. The institutions and organizations in a TDS include both those intrinsic to the development of the technology (the "technological enterprise") and broader societal institutions that interact with the technology or are impacted by it. The TDS for a specific technology evolves over time. In the context of a TF, its boundaries are determined by the boundaries of the forecast as described in the paragraph on getting the subject right.

Get the assumptions right. After bounding, this is the most important consideration in a TF. Assumptions underpinning a TF are both technological and contextual. Such assumptions relate to the dynamics of technological innovation in the case of the particular technology being forecast. In effect they, along with the TDS, present a model for the development of a technology. These models are idiosyncratic as there is no general theory of innovation that works. Thus, the quality of judgment captured in the core assumptions underlying the TF is a major determinant of the quality of the TF.

Multiple approaches are best. The history of forecasting has been plagued by forecasts by extremely knowledgeable individuals that were dead wrong. In the same manner, using a single approach in a TF carries grave risk. Looking at the same subject from a variety of perspectives and techniques allows the forecaster to see the areas of convergence and to build confidence in the accuracy of the common outcome of the forecasts. Using multiple approaches in forecasting is analogous to triangulating when trying to locate a position. The next section of this paper deals with forecasting techniques and how different techniques can be combined to strengthen a TF.

Do it simply. The KISS principle works well in TF. All evidence known to the authors covering decades of forecasting indicates that large additional commitments of resources and time to a forecast produce few, if any, benefits to the forecasters. Using the five principles articulated above with sound judgment and clarity as to subject, context, and assumptions does not guarantee a successful TF, but they have proven to be more useful than costly and sophisticated techniques of modeling and time series analyses.

Forecasting Techniques

The three intellectual pillars of forecasting are theory, data, and technique. We lack effective predictive theory of sociotechnical change. Social change theories tend to be extremely "macro" in aggregation, such as Marxism, and empirically suspect. There are a wealth of conceptual models of technological innovation, technology substitution, technology diffusion, and technology transfer. Their net value in predicting technological changes tends to be little more than the observation that S-shape growth curves are common in many technological (and natural) processes. However, consideration of these models can aid a broader form of TF, more contextually sensitive, with clues as to potentially fruitful indicators of prospects for successful technological innovation in a given case [Watts and Porter, 1997].

Gathering data on technological processes often proves difficult. Rapid technological change usually means short time series, often guarded as having proprietary value. Data for societal changes are usually not gathered with reference to sociotechnical change. Social indicators and indicators of technological change lack a systematic framework for selection and organization.

Forecasting techniques must compensate for the deficiencies of theory and data. These techniques allow the use and integration of information available from disparate sources. At their best they facilitate the use of human judgment in forecasting. Quality judgment is the rock on which the pillars of forecasting rest.

Many forecasting techniques have been developed. Remember that sophistication contributes little, if anything, to forecasting quality. Here we present five families of techniques that have proven useful. We describe them briefly, noting their strengths and weaknesses.

Monitoring. This is the simplest forecasting technique and the most fruitful. Monitoring involves scanning the environment for information relevant to the topic of the forecast. Information to be monitored resides in computerized databases, the Internet, publications, experts, the physical environment, etc. In the past few years, clipping articles from newspapers has largely been replaced by searching electronic databases using software designed especially for this task [Porter and Detampel, 1995]. While vast amounts of information can be collected by monitoring, unless it is filtered of useless material and structured to assist in the TF, monitoring is not helpful. Insight into implications for successful innovation of the target technology is the most important component in filtering and structuring information.

One effective strategy is to tailor an area to be monitored to needs, resources, and level of familiarity using the "temperature" approach in which you "heat up" as you progress [Porter et al., 1991]. There are three stages: cold or unfamiliar with the area being monitored, warm or somewhat familiar, and hot or very familiar.

In starting cold, take a shotgun approach and grab what is convenient. Emphasize recent materials, especially state-of-the-art reviews. Find a mentor who is accessible and familiar with the technology. Typical questions in this stage include, "What are the state of the art and functions of this technology?" "What are the vital related technologies, contextual issues, and major players?" "What are possible future development paths?"

To warm up, focus on key issues and track development over time. Look for alternatives to the technology and seek forecasts that identify pivotal issues and offer comparisons. Identify and use experts with diverse perspectives. Typical questions include, "What influences are driving/blocking this technology's development?" "Can we map interdependencies with other technological and contextual factors?" "What are the key uncertainties ahead?"

When red hot, make the search as complete as needed. Develop a model of the innovation emphasizing drivers and obstacles. Seek confirmation of your model for this technology's further development from experts and periodically update it. Build toward more comprehensive forecasts and begin to consider impacts of the technology. Questions include, "What are the key factors to watch?" "What are the most likely alternatives for the near future? The longer term?" "What technology management recommendations will you make to your organization?"

Monitoring provides the basis for TF and for assessing the impacts of technologies. It is the first and most essential technique to master. Establishment of an ongoing monitoring system may be often desirable.

Expert opinion. Expert opinion assumes that experts exist whose knowledge can be tapped by the forecaster. A most basic caveat is not to rely on the forecast of a single individual as these have proven to be outstandingly wrong. In tapping expertise, there are three major activities: talking (T), estimating (E), and feedback (F). Techniques consist in various combinations of these activities. Problems arise in talking within an expert group as some individuals tend to dominate independently of the merit of their views. Feedback seems to help. However, it is better presented anonymously. Estimation too is best done anonymously. For quick generation of possibilities, brainstorming (T without criticism) works well. Survey (E) can get at the views of a group of experts but without feedback. Delphi (EFE until E converges) avoids dominance effects and provides anonymous, but limited, feedback. Nominal group process and EFTE use two or more rounds of EFTE with live discussion but anonymous voting and feedback. The EFTE pattern is perhaps the best all around, but requires the presence of the experts at a single site for the exercise [Nelms and Porter, 1985].

Trend analysis. Trend analysis involves the use of statistical and/or graphical techniques to extend time series data into the future. Some equate trend analysis with forecasting and develop precise techniques that require long time series of accurate data. It is our view that trend analysis is a useful technique but not the only one and that in a typical TF the number and quality of data points are typically low so that sophisticated techniques are largely useless. Trend analysis is especially vulnerable to discontinuities. Potential major sources of discontinuity should be identified as part of a trend analysis.

To conduct trend analysis, it is first necessary to determine the parameter (simple or compound) whose trend is being analyzed and why. Since any data set can be fit to any equation for a trend, it is essential to have a clear view of the dynamics underlying the trend so that the data can be fitted to a specific equation. There are an enormous number of forms that are candidates for being fitted to data. However, in our judgment only three are really significant for TF. These are the line, exponential, and S-curves.

Why the straight line? It is the ultimate fallback position for changes over time. Other curves can be transformed to lines. It is often a good approximation of complex processes, the details of which cannot easily be sorted out. It is a good model for many processes for limited time periods.

Exponential growth describes some change processes effectively, usually for limited periods. Continued exponential growth for many years is rare — the notable exception has been "Moore's Law" of exponential computer performance gains holding generally true for some 4 decades (e.g., microprocessor capabilities doubling every 18 months).

Why the S-curves (more formally, sigmoidal)? These are various mathematical formulations that represent slow initial growth, followed by a period of rapid (essentially exponential) growth, then slowing as they asymptotically approach a limit. The most prominent in TF are probably the Pearl, or Fisher-Pry, and the Gompertz (see Porter et al. [1991]). Amazingly enough, the S-curves fit many processes associated with technological innovation. This is in spite of the lack of specific theory that says why this is so. Among other things, they fit superbly the adoption of an innovation (the TV set, the pocket calculator) by a population over time. S-curves can also describe the change of a technical parameter central to an innovation over the course of the innovation's lifetime as well as the changes in that parameter due to successor technologies — a succession of S-curves is frequently observed. They can be extremely useful in identifying phases in the technological progress of an innovation as well as subpopulations of adopters. S-curves are the friend of the analyst of technological trends.

Models. This label covers a multitude of techniques, from global modeling using complex systems of equations to judgmental modeling for decision making, to games, to simple boxes and arrows models of institutions involved in the development of a technology. The strength of models lies in simplification of complex systems. Their weakness is poor choice of parameters and weak and/or unsound assumptions. The most important factors in modeling are simplicity and good judgment.

Scenarios. Scenarios are imaginative representations of future situations that can combine many forecasts and views of the future. They may be textual or visual (e.g., videos). They can take the form of snapshots incorporating information from a diversity of sources into a view at a point in time. Alternatively, they can be future histories that track the development of a technology over time using a wide range of sources. They can effectively integrate TF information. They also excel as vehicles for communicating forecast information, especially to nontechnical audiences. Their strengths are their richness, diversity, and comprehensiveness. Their weaknesses include poor or unconsidered assumptions, weak forecasts, or poor data underlying their rich presentation. They often come in threes — baseline, high, and low. The baseline or "muddle through" scenario is the most likely in almost all situations. We do not have to tell you why this is so. Scenarios, more so than other techniques, can be either extremely brilliant or terribly inane.

Whatever techniques are applied, sensitivity analysis is crucial — determining how forecasts change in response to specified variations in assumptions, initial conditions, variable values, and so forth. A key value offered by modeling is explicit sensitivity analysis. Expanding this notion of ascertaining a range of potential futures points to the highly desirable notion of forecasting "alternative futures" rather than making singular projections. These provide decision makers a far more valuable form of information.

"Scenario management" is an attractive vehicle to present alternative futures in a form that helps management plan by taking into account opportunities and uncertainties [Gausemeier et al., 1995].

A survey of technology foresight practitioners found that most studies entail use of multiple techniques, on average three [Lemons and Porter, 1992]. Expert opinion and monitoring are most often used, followed by trend analyses and scenarios, with qualitative and quantitative modeling following. The practitioners report that the gravest difficulty in performing technology foresight analyses is the shortage of quality data.

Using TF

TF used in conjunction with purposeful strategic analysis, incorporating sound assumptions and good judgment, can be of great benefit. TF, or cast more broadly — technology foresight — is used in many governmental and private sector organizations. Joe Coates has reviewed many TFs in his recent "Project 2025", finding fruitful, but mixed results. National Delphi (an iterative survey, with feedback, approach to expert opinion) has generated informative projections on likely technological advances. The Japanese Delphi conducted about every 5 years, addressing over 1000 potential technological advances, has set the pace, now emulated by German, Korean, and British efforts. Monitoring has provided the backbone for the increasingly popular "competitive technological intelligence" that seeks to profile emerging technologies and competitor interests in them.

A companion in technology foresight analyses is technology assessment (TA). TA shares many techniques with TF in dealing with future possibilities. TA, or impact assessment, addresses the potential consequences of technological innovation. The demise of the Congressional Office of Technology Assessment has hurt, but impact assessment flourishes worldwide. Of particular note is the emergence of "strategic environmental assessment" (SEA) in recent years (see Canter and Sadler [1997] and Porter and Fittipaldi [1998]).

We believe in TF as a valid, useful, and flexible approach to managing the present from the future. As needed, it can play a significant role in developing technological strategy for an organization or as a component in a broad societal analysis. With the rapid sociotechnical changes occurring today, we cannot neglect it.

References

Canter, L. and Sadler, B. *A Tool Kit for Effective EIA Practice* — *Review of Methods and Perspectives on Their Application*, Environmental and Ground Water Institute, University of Oklahoma, Norman, OK, 1997.

Gausemeier, J., Fink, A., and Schlake, O. *Szenario-Management* — *Planen und Fuhren mit Szenarien*, Carl Hanser Verlag, Munich, 1995.

Lemons, K. E. and Porter, A. L. A comparative study of impact assessment methods in developed and developing countries, *Impact Assess. Bull.*, 7(4): 5–15, 1992.

Nelms, K. R. and Porter, A. L. EFTE: an interactive delphi method, *Technol. Forecast. Soc. Change*, 28: 43–61, 1985.

Porter, A. L. and Detampel, M. J. Technology opportunities analysis, *Technol. Forecast. Soc. Change*, 49: 237–255, 1995.

Porter, A. L. and Fittipaldi, J. J. *Environmental Methods Review: Retooling Impact Assessment for the New Century*, U.S. Army Environmental Policy Institute, Atlanta, GA, 1998.

Porter, A. L., Roper, A. T., Mason, T. W., Rossini, F. A., and Banks, J. *Forecasting and Management of Technology*, John Wiley & Sons, New York, 1991.

Watts, R. J. and Porter, A. L. Innovation forecasting, *Technol. Forecast. Soc. Change*, 56: 25–47, 1997.

Wenk, E., Jr. and Kuehn, T. J. Interinstitutional Networks in Technological Delivery Systems, In *Science and Technology Policy*, J. Haberer, ed., pp. 153–175. Lexington Books, Lexington, MA, 1977.

Further Information

Technological Forecasting and Social Change is the leading TF journal.

The International Institute of Forecasters produces the *Journal of Forecasting* and holds annual meetings called the International Symposium on Forecasting.

Impact assessment is promoted by the International Association for Impact Assessment, which published *Impact Assessment* and holds annual conferences. Its Web site is <http://IAIA.ext.nodak.edu/IAIA/>.

The authors can be reached through the Technology Policy and Assessment Center, c/o ISyE, Georgia Tech, Atlanta GA 30332-0205. Its Web site is <http://tpac.gcatt.gatech.edu>. Our e-mail addresses are alan.porter@isye.gatech.edu and frossini@mindspring.com.

5.6 Data Analysis

Marlene A. Smith

Formal statistical inference (also called **confirmatory data analysis**) makes probabilistic statements about our ability to generalize sample results to a larger population. Exploratory data analysis (EDA) does not. Unlike inferential methods, EDA shies away from stochastic assumptions, significance tests, and interval estimation. Instead, it relies on graphical and algebraic tools to uncover patterns in data.

EDA searches might be conducted for various reasons. Commonly, exploratory searches precede confirmatory data analysis, since EDA can help us discover whether various parametric assumptions are plausible or whether certain observations could have undue influence on subsequent inferential results. Alternatively, EDA is useful when we don't have specific *a priori* hypotheses; advocates of EDA are quite comfortable allowing the *data* to suggest interesting questions. Finally, EDA might be the preferred method of statistical analysis when mathematically optimal estimators are known to produce poor estimates under certain circumstances such as small sample size.

Velleman and Hoaglin [1992] link EDA to continual process improvement practiced by many of today's businesses: "EDA is particularly well-suited to analyses of economic and business data intended to help in business decisions because it recognizes these analyses as part of an on-going process." This observation may be especially pertinent to high-technology companies, since these firms (typically producing innovative products) have limited historical data to rely on and need to engage in continual data analysis to understand their new and rapidly changing markets.

Tukey [1977], a pioneer in EDA, advocated pencil-and-paper explorations of data. Modern-day data analysis has benefited from advances in computing technologies. Statistical packages such as JMP permit three-dimensional graphical explorations of data (including axis rotation and dynamic data displays) and interactive statistical computing. High-speed processors make manageable some of the calculation-intensive smoothing, graphical, and estimation methods needed for EDA.

General Principles of Data Analysis: The Four Rs

The philosophical foundations of EDA are illuminated by its four basic principles: **revelation, residuals, reexpression**, and **resistance**.

Although statistical analysis involves the "reduction of data", EDA warns against too much reduction too early in the process. **Revelation** entails careful scrutiny of data using visual tools. By exploring graphical displays, we might discover miscoded data, interesting patterns and relationships, or unusual data points. Such discoveries guide further analysis: discarding influential observations or straightening curvilinear relationships, for example. Stem-and-leaf displays and **median traces** [Goodall, 1990] are two of many graphical methods used in exploratory stages of data analysis.

Residuals, the differences between observed and predicted values, provide evidence about statistical models. Although both confirmatory data analysis and EDA involve construction of statistical models,

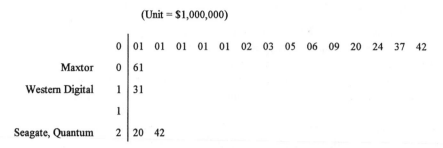

FIGURE 5.16   Stem-and-leaf display of R&D expenditures.

residual analysis plays different roles. Confirmatory analysis uses residuals to examine the relevance of an *assumed* stochastic model; EDA uses residuals to discover what that model might be.

EDA involves an iterative process that transfers patterns in the residuals to the statistical model. The process is complete when the residuals become uninformative, i.e., displaying no evidence of nonlinearities, outliers, unequal variances, or autocorrelation. The methodology is illustrated by Hoaglin and Velleman [1992] in an application to modeling major league baseball salaries.

Reexpression means transforming the original data. This might involve changing the units of measurement of the data by rescaling (altering the minimum and maximum values, but retaining relative distances between data points), performing a nonlinear transformation (e.g., logarithms, reciprocals, roots, exponents, and powers), or some combination such as the **froot** ("folded root"). Typically, the goal of reexpression is to convert unconventional patterns to ones more familiar, and useful, to the analyst. Appropriate reexpression can equalize nonconstant variances, induce symmetry to skewed distributions, and straighten curvilinear patterns.

Resistance advocates letting the body of the data have the largest impact on statistical results and not the outliers. This can be accomplished in two ways. One is to simply remove unusual observations. The other is to use resistant estimation methods: the median is used in place of the mean, or resistant lines [Johnstone and Velleman, 1985] replace the least squares regression line.

Illustration: Research and Development in High-Technology Firms

A sample of 18 firms whose primary product is computer storage devices (Standard Industrial Classification 3572) was acquired for fiscal year 1995. For each firm, data were collected on income before extraordinary items, net sales, and research and development (R&D) expense. Suppose our purpose is to understand how R&D activity relates to firm profitability.

Data analysis projects typically begin with revelation. A stem-and-leaf diagram, like a histogram, reveals information about location, spread, and shape of a univariate distribution, but, unlike the histogram, the stem-and-leaf shows the numerical values of each of the original data points. For instance, the stem-and-leaf display of R&D expenses (Fig. 5.16) shows the maximum value in our sample is $242M (242 x 10^6, the last entry in the display). The first stem (top line) of the diagram reveals other patterns — gaps in the data occur after $9M and $24M, suggesting multiple subgroups among the lower-ranked R&D firms. The data are skewed rightward; firms with the highest net sales in our sample (Maxtor Corp., Western Digital Corp., Seagate Technology Inc., and Quantum Corp.) have the highest level of R&D activity. These four firms also appear as large outliers on the stem-and-leaf display of income (not shown).

Two things have been learned from the stem-and-leaf diagrams of R&D and income. Since both variables display asymmetric distributions, the analysis might benefit from reexpression. We've also learned that our original research question leads to an uninteresting discovery: that large firms have larger incomes and more R&D activity.

Reexpression typically involves an algebraic transformation of the data. In our case, because firm size appears to be a lurking variable behind the relationship between R&D and income, we might transform the data by dividing both variables by some measure of firm size.

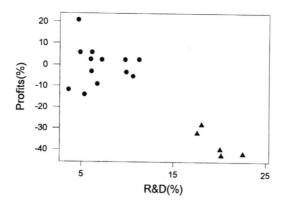

FIGURE 5.17   Scatterplot of profits(%) against R&D(%).

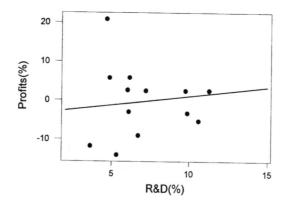

FIGURE 5.18   Detailed scatterplot of Profits(%) against R&D(%) with resistant line.

R&D and income are reexpressed by dividing both variables by net sales of the firm: R&D(%) is the percentage of net sales going to R&D, and Profits(%) is return on net sales (100 × income before extraordinary items divided by net sales). These transformed variables are somewhat more symmetric and smoother than the original variables as indicated by their stem-and-leaf displays and normal probability plots.

Figure 5.17, a scatterplot of Profits(%) and R&D(%), reveals two distinct clusters of data. For the five firms spending more that 15% of net sales on R&D (those marked with a triangle in Fig. 5.17), it appears that Profits(%) and R&D(%) are negatively related. However, there appears to be little relationship between Profits(%) and R&D(%) when R&D(%) is lower than 15% (refer to the circular points in Fig. 5.17). We might conclude that the two groups are behaving quite differently and should be subjected to individual analyses.

Figure 5.18 displays only those firms with R&D(%) less than 15%. The three-group Tukey line is used to fit the data. To find this line, the data in Fig. 5.18 are split into three groups: the four "low" R&D firms [R&D(%) < 6], the five midrange R&D firms [6 < R&D(%) < 8], and the four "high" R&D firms [R&D(%) > 8]. The Tukey line results from using the medians of Profits(%) and R&D(%) from the "low" group and the medians of Profits(%) and R&D(%) from the "high" group [Hoaglin et al., 1983]. This line is Profits(%) = −3.6 + 0.5 R&D(%), suggesting a slight, positive relationship between Profits(%) and R&D(%) for these firms.

The Tukey line shown in Fig. 5.18 demonstrates the data analytic principle of resistance. Because the Tukey method uses medians, outlying observations have a lesser influence on the placement of the line. To contrast, the least squares fit of the data in Fig. 5.18 gives Profits(%) = 0.7 − 0.2 R&D(%). The negative slope of the least squares regression line reflects the influence of the observation for Komag, Inc., that point hovering above the rest at the 20% profit level.

The next step of the analysis is the examination of the residuals. In Fig. 5.18, the residuals are the vertical distance between the observations and the Tukey line. Because this line is nearly flat, we can easily see the patterns in the residuals by slightly rotating Fig. 5.18 clockwise until the Tukey line is horizontal. The residuals display a pattern of nonconstant variance, with data points at low values of R&D(%) exhibiting more scatter around the Tukey line than those at high R&D(%) levels. Indeed, we might even ask ourselves, as we did in Fig. 5.17, if there isn't yet another subgrouping within the data separated at the R&D(%) = 8% fence.

At this stage, the data analyst might attempt to further understand the information being conveyed by the inhomogeneous residuals. Sometimes nonconstant variance is addressed by another reexpression of the data. Alternatively, the mixed variability of the residuals might be related to a third variable: are profits more variable at low R&D levels because the firm has more discretionary funds for other (uncertain) activities such as advertising? To study this proposition, the *residuals* from Fig. 5.18 would be plotted on the vertical axis against advertising expenses. Further exploration continues until the residual plot displays randomness.

Different insights arise from the data, depending on whether we take a classical modeling and inference approach or an EDA perspective. Referring again to Fig. 5.17, the least squares fit gives Profits(%) = 17 − 2.6 R&D(%). The coefficient on R&D(%) is statistically significant at the 1% level; the R-squared value is 74.5%. We conclude that there is a statistically significant, negative relationship between Profits(%) and R&D(%), and that the least-squares line is a good fit for cross-sectional data. Further, White's test of homoskedasticity fails to reject homoskedasticity at the 5% significance level.

EDA provides a different picture. We find that, if a negative relationship exists, it is only for those firms whose R&D(%) exceeds 15%. Among those firms spending less than 15% of their net sales on R&D, there is a positive, if any, relationship between Profits(%) and R&D(%), and lower levels of R&D(%) display more variability in Profits(%) around the statistical model.

Summary

At least in practice, confirmatory data analysis often rushes toward reduction of data in the quest for statistical significance. EDA relies on unstructured exploration of patterns in data to more fully understand the complexities of the underlying relationships. Our simple example in the last section provides an illustration:

- Because EDA is less concerned with degrees of freedom, individually analyzing subgroups of data is acceptable practice.
- EDA lets the *data* suggest interesting hypotheses. In our example, the data beg the questions of why there appears to be a gap at R&D(%) = 15% and why there is more variability in return on sales when R&D(%) is low. Who would have thought to ask these questions prior to examining the data?
- Resistant methods provide estimates that more closely reflect the majority of the data.
- Reexpression of data can reveal interesting patterns or help us to more carefully understand what questions can and should be asked of the data.

Practicing statisticians find value in both confirmatory and exploratory data analysis. Even so, exploratory techniques are often underutilized. Proponents of EDA argue that rigorous application of EDA tools uncovers information otherwise hidden in the data.

Defining Terms

Confirmatory data analysis: A term used in the EDA literature referring to inferential methods based on the theory of repeated sampling.

Froot: The folded root spreads out tails of univariate distributions. See Tukey [1977].

Median trace: Smoothing techniques that replace data with medians of surrounding data.

Reexpression: Transforming the original data, typically by rescaling or applying nonlinear transformations.

Residual: The difference between the observed (sample) value and the value obtained from the statistical model.

Resistance: Resistant statistical estimators are less affected by outliers than nonresistant methods, e.g., the median is a more resistant estimator of location than the arithmetic average.

Revelation: Guided examination of data using graphical methods.

References

Goodall, C. A survey of smoothing techniques, In *Modern Methods of Data Analysis*, J. Fox and J. S. Long, eds., pp. 126–176, Sage Publications, Newbury Park, CA, 1990.

Hoaglin, D. C., Mosteller, F., and Tukey, J. W. *Understanding Robust and Exploratory Data Analysis*, John Wiley & Sons, New York, 1983.

Hoaglin, D. C. and Velleman, P. F. A critical look at some analyses of major league baseball salaries. In *1991 Proc. of the Am. Stat. Assoc., Section on Statistical Graphics*. American Statistical Association, Alexandria, VA, 1992.

Johnstone, I. M. and Velleman, P. F. The resistant line and related regression methods, *J. Am. Stat. Assoc.*, 80: 1041–1054, 1985.

Tukey, J. W. *Exploratory Data Analysis*, Addison-Wesley, Reading, MA, 1977.

Velleman, P. F. and Hoaglin, D. C. Data analysis, In *Perspectives on Contemporary Statistics*, D. C. Hoaglin and D. S. Moore, eds., pp. 19–39, Mathematical Association of America, Washington, DC, 1992.

Further Information

Beginners will want to peruse Velleman and Hoaglin [1992] and Tukey [1977] cited in the reference list and Hoaglin, D. C., Mosteller, F., and Tukey, J. W. *Exploratory Data Tables, Trends, and Shapes*, John Wiley & Sons, New York, 1985.

More advanced discussions are found in *Modern Methods of Data Analysis* (see Goodall [1990]) and Hoaglin et al. [1983].

Multivariate data analysis techniques are covered in Gnanadesikan, R. *Methods for Statistical Data Analysis of Multivariate Observations*, John Wiley & Sons, New York, 1977.

Several books on regression and analysis of variance techniques are Hoaglin, D. C., Mosteller, F., and Tukey, J. W. *Fundamentals of Exploratory Analysis of Variance*, John Wiley & Sons, New York, 1991; Mosteller, F. and Tukey, J. W. *Data Analysis and Regression: A Second Course in Statistics*, Addison-Wesley, Reading, MA, 1977; and Weisberg, S. *Applied Linear Regression*, John Wiley & Sons, New York, 1985.

5.7 Design of Experiments

Stephen R. Schmidt and Kenneth E. Case

Design of experiments (DOE) has become one of the most popular statistical techniques of the 1990s. DOE was originated in 1920 by a British scientist, Sir R. A. Fisher, as a method to maximize the knowledge gained from experimental data, and it has evolved over the last 70 years. Unfortunately, most of the development of DOE was mathematically complex and thus, its use has been restricted to those well versed in mathematics. Recent widespread popularity of DOE is associated with the works of Taguchi [1987], a Japanese engineer, who focused on the practical use vs. mathematical perfection of the technique. In short, Taguchi's work began a revolution in the presentation of DOE material where mathematical theory is downplayed in order to enhance the clarity and practicality of the subject. Thus, scientists, engineers, technicians, and managers who are not mathematical experts are now becoming experimental design

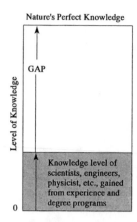

FIGURE 5.19   The gap between the current level of knowledge and nature's perfect knowledge tends to widen as industry today seeks to understand more complex processes.

practitioners. However, close scrutiny of the Taguchi method [Pignatiello, 1988] revealed limitations, which has led to the most recent evolution in DOE, a blended approach of Taguchi and classical techniques.

To understand DOE, consider a specific area of science, engineering, physics, etc. as a process or activity within nature. Complete or perfect knowledge of any aspect of nature is known only by nature, while scientists, engineers, etc. typically have only some subset of that knowledge (Fig. 5.19).

The typical engineering, scientific, or physics approach is to spend years of research trying to narrow the gap utilizing theoretical knowledge. DOE, on the other hand, will allow one to quickly narrow the gap through proper planning, designing, data collection, analysis, and confirmation.

Consider again nature's perfect knowledge of any process (activity or specified area of science, etc.) where there is a true mathematical relationship that can describe process outputs as a function of all the process inputs, e.g., Newton's law of motion $F = MA$. If nature knows this relationship but we don't, how can be obtain the relationship quickly without getting tied up in years of theoretical research as Newton did?

A simple DOE for two inputs such as mass (M) and acceleration (A), each tested at two levels, can be set up as the four experimental runs (or trials) of low (Lo) and High (Hi) settings shown in Table 5.9.

The DOE set of test conditions is used to interrogate nature about how things perform under certain conditions. The specific set of conditions (or **design matrix**) is constructed in order to analyze the resulting data for the purpose of building a model to approximate the real model contained in nature (i.e., approximate nature's perfect knowledge). The example in Table 5.9 is a full factorial of two factors each at two levels. Several other experimental design types are available to meet your specific objective, i.e., linear modeling, nonlinear modeling, screening, etc. [Schmidt and Launsby, 1994]. Assume our experiment generated an average set of force results for each experimental run shown in Table 5.10. The low and high settings of M (5, 10) and A (100, 200) are chosen based on ranges of interest to the experimenter.

TABLE 5.9

Run	M	A
1	Lo	Lo
2	Lo	Hi
3	Hi	Lo
4	Hi	Hi

To maximize the amount of knowledge gained, the modeling analysis of a DOE is conducted on standardized input values where low input settings are coded (−1) and high input settings are coded (+1). These coded values will standardize the scale and the units of the input variables. The result is a new matrix of coded inputs shown in Table 5.11.

Notice the addition of an M · A column in Table 5.11, which is generated by the product of the coded M and A columns and will be used to analyze an interactive or combined effect of M and A. The columns M, A, and M · A represent the three effects to be evaluated, i.e., the linear effects of M and A and the interaction effect of M with A (M · A). The analysis is conducted as shown in Table 5.12.

To obtain the numbers in the shaded section, you have to do the following for each of the M, A, and M · A columns: find the average of the output F̄ when the effect column values are −1, find the average of the output F̄ when the effect column values are +1, and then find the difference of the two effect

TABLE 5.10

	Inputs		Average Output
Run	M	A	$\bar{F}$
1	5	100	500
2	5	200	1000
3	10	100	1000
4	10	200	2000

TABLE 5.11

Run	M	A	$M \cdot A$	$\bar{F}$
1	−1	−1	1	500
2	−1	1	−1	1000
3	1	−1	−1	1000
4	1	1	1	2000

TABLE 5.12

	Effects			
Run	M	A	$M \cdot A$	Avg. Output $\bar{F}$
1	−1	−1	1	500
2	−1	1	−1	1000
3	1	−1	−1	1000
4	1	1	1	2000
Avg (+1)	1500	1500	1250	$\bar{\bar{F}}$ 1125
Avg (−1)	750	750	1000	
Avg (+1) − Avg (−1) = Δ	750	750	250	

averages (Δ). For example, when the M effect column is at −1 (runs 1 and 2) the output is 500 and 1000, which averages to 750.

The model generated from the DOE is built using least-squares regression, which can be simplified for two level designs as shown below:

1. $\hat{F} = \bar{\bar{F}} + \dfrac{\Delta_M}{2} M_c + \dfrac{\Delta_A}{2} A_c + \dfrac{\Delta_{MA}}{2} M_c A_c$

   where $\hat{F}$ = predicted average for force; M_c = coded variable for mass; A_c = coded variable for acceleration; $\bar{\bar{F}}$ = grand experimental mean; and Δ_M is the size of the linear effect for M, etc.

2. Thus, our actual model becomes

$$\hat{F} = 1125 + \frac{750}{2} M_c + \frac{750}{2} A_c + \frac{250}{2} M_c A_c$$

3. Remembering that our prediction model is for coded values (i.e., −1, +1 scale) of M and A, we can transform it into a model of actual M_a and A_a values using the following relationship.

$$M_a = \left(\frac{M_H + M_L}{2} \right) + \left(\frac{M_H - M_L}{2} \right) M_c$$

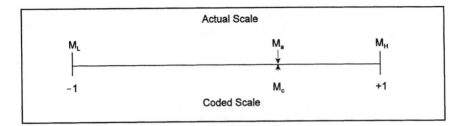

FIGURE 5.20   Actual vs. coded scale.

where M_a = actual setting value of mass; M_c = coded setting value of mass; M_H = experimental high setting of actual values; and M_L = experimental low setting of actual values. For our example,

$$M_a = \left(\frac{10+5}{2}\right) + \left(\frac{10-5}{2}\right)M \quad \text{or} \quad M_a = 7.5 + 2.5M_c$$

Thus,

$$M_c = \left(\frac{M_a - 7.5}{2.5}\right)$$

The previously described calculations are a simple way to go back and forth from coded to actual setting values, as shown on the combined scale in Fig. 5.20.
A similar transformation occurs for acceleration (A), as shown below.

$$A_a = \left(\frac{200+100}{2}\right) + \left(\frac{200-100}{2}\right)A_c$$

or

$$A_a = 150 + 50\,A_c$$

Thus,

$$A_c = \left(\frac{A_a - 150}{50}\right)$$

4. Substituting the actual transformed variables into the $\hat{F}$ equation in (2) will produce the following new $\hat{F}$ model:

$$\hat{F} = 1125 + \frac{750}{2}\left(\frac{M_a - 7.5}{2.5}\right) + \frac{750}{2}\left(\frac{A_a - 150}{50}\right) + \frac{250}{2}\left(\frac{M_a - 7.5}{2.5}\right)\left(\frac{A_a - 150}{50}\right)$$

$$\hat{F} = M_a A_a$$

In our case, we obtained the exact same model as Newton because our F data did not contain noise. In reality, there exists noise or error in experimental data and thus the DOE models will only be approximations to nature's true relationship. However, imagine how quickly Newton could have closed the gap if he would have had DOE in his box of scientific tools!

TABLE 5.13

Factor	How Factor Will Be Addressed in DOE
Current (I)	An experimental variable — use two levels of 0.005 and 0.020 amperes
Resistance (R)	An experimental variable — use two levels of 100 and 470 ohms
Wattage of resistor (W)	An experimental variable — use two levels of 0.25 and 1.0 watts
Room temperature and humidity	Hold constant by running the entire experiment in one sitting; randomize experimental runs to spread any unknown variation of these over the entire experiment
Measurement equipment	Hold constant by using the same calibrated laboratory-grade instruments, a voltmeter and an ammeter, throughout the experiment
Technician	Hold constant by employing a trained electronics technician to run the entire experiment
Experimental technique	Hold constant by closely following standard operating procedures agreed upon in advance.

Since DOE was not developed until the 1900s, Newton never had access to the tool. Newton, however, was not opposed to the use of new tools. For example, when he ran into a roadblock in his research, he developed calculus as a new tool to assist him. It is, therefore, logical to speculate that Newton, if he was living today, would eagerly embrace new tools such as DOE to assist him in gaining knowledge from experimental data.

The previous example used data generated from an F = MA model instead of a physical experiment. In our next example, we will tackle a similar problem using actual experimental data. Suppose we go back in time to Germany during the early 1800s. In 1826, German scientist George Simon Ohm determined the formula now known as Ohm's law. Ohm's law establishes the mathematical relationship between voltage (V), current (I), and resistance (R) as V = IR. Today, we know that this law applies to both direct and alternating currents, but the formula, as given in its original simple form, applied only to steady DC situations. It is known that Ohm conducted much careful experimental work. Unfortunately, he did not have access to DOE, or his work might have been shortened considerably.

Suppose we pretend to be ignorant of Ohm's law and see how Ohm might have proceeded with even the most basic of today's DOE tools. Of course, nature knows those factors affecting voltage, so we will interrogate nature in a methodical way. First, however, we have to establish those **experimental variables** to include in a test using DOE. Specifically, we need to identify any factors we believe may affect the response of interest — in this case "voltage". Suppose the factors in Table 5.13 are felt to be important; each is addressed as to how it will be incorporated into the DOE.

We all know that wattage of the resistor (W) should have no effect, but remember that we are pretending to be ignorant of Ohm's law; we also know that room temperature and humidity should have a negligible or imperceptible effect under normal room conditions. The other factors are "real" considerations worthy of treatment as described above.

A full factorial is chosen for this experiment. In a full factorial, all possible combinations of experimental levels of all variables are run. Since there are only three experimental variables of interest, the number of runs in the full factorial equals $2^3 = 8$, which is certainly feasible. This will permit us to evaluate all possible effects and interactions of the three experimental variables (I, R, and W) and to write a mathematical model including any and all contributing terms. The experimental array is shown in Table 5.14. Note that availability of resistors at particular wattages, safety through low voltages and currents, the need to be within the wattage ratings of all resistors, and instrumentation ranges and precision were major factors used in the selection of experimental variable settings.

Table 5.15 presents the array of coded input values, actual responses, and averages used in the modeling analysis of the DOE. The coded values of I, R, and W are set at −1 for low settings of each experimental variable and at +1 for high settings. The −1 and +1 values in the interaction columns do not reflect experimental settings, but are determined from the product of the coded values in their component columns and are used to analyze the strength of the interactive effect of the variables.

TABLE 5.14

Run	Current (I)	Resistance (R)	Wattage (W)
1	0.005	100	0.25
2	0.005	100	1.00
3	0.005	470	0.25
4	0.005	470	1.00
5	0.020	100	0.25
6	0.020	100	1.00
7	0.020	470	0.25
8	0.020	470	1.00

TABLE 5.15

Run	Effects I	R	IR	W	IW	RW	IRW	Reps (of Voltage V) Y1	Y2	Y3	Y4	Y5	Ybar
1	−1	−1	1	−1	1	1	−1	0.489	0.503	0.496	0.497	0.520	0.501
2	−1	−1	1	1	−1	−1	1	0.485	0.487	0.518	0.518	0.486	0.499
3	−1	1	−1	−1	1	−1	1	2.328	2.340	2.349	2.441	2.270	2.346
4	−1	1	−1	1	−1	1	−1	2.379	2.327	2.425	2.419	2.375	2.385
5	1	−1	−1	−1	−1	1	1	1.995	2.018	2.012	2.002	1.961	1.998
6	1	−1	−1	1	1	−1	−1	2.204	2.058	2.015	1.971	2.129	2.075
7	1	1	1	−1	−1	−1	−1	9.464	9.103	9.352	9.481	9.582	9.396
8	1	1	1	1	1	1	1	9.506	9.255	9.486	9.382	9.567	9.439
Avg (+1)	5.727	5.892	4.959	3.600	3.590	3.581	3.570						Avg =
Avg (−1)	1.433	1.268	2.201	3.560	3.569	3.579	3.589						3.580
Δ	4.294	4.624	2.758	0.040	0.021	0.002	−0.019						

Multiple replications (reps) of the experiment provide five different voltage measurements. These voltages vary somewhat due to the "noise" always involved in experimental research. The "noise" is due to the continued reestablishment of the operating conditions at each of the experimental run settings (different resistors were used for each rep, and current was reestablished for each rep). In this experiment, each of the runs was performed in a random order for the first rep. Then, each was performed again in a different random order for the second rep. This was continued throughout the five reps. The term $\bar{y}$ represents the average of the five reps for each of the eight experimental runs. Appended to the bottom of Table 5.15 (in the shaded area) are the relevant average calculations, similar to those illustrated in Table 5.12.

We can now proceed with the analysis and modeling. First, note the values of Δ in the last row of Table 5.15. The first three are considerably larger than the rest, indicating that their columns (I, R, and IR) are of substantial importance in modeling voltage, while the other column effects are due to experimental noise and are of negligible importance for the model.[1] Returning to the four-step modeling illustration presented earlier for F = MA, the following equations result:

1. $\hat{V} = \bar{\bar{V}} + \dfrac{\Delta_I}{2} I_c + \dfrac{\Delta_R}{2} R_c + \dfrac{\Delta_{IR}}{2} I_c R_c$

where $\hat{V}$ = predicted average for V; I_c = coded variable for current; R_c = coded variable for resistance; $\bar{\bar{V}}$ = grand experimental mean; and Δ_I = is the size of the linear effect for I, etc.

---

[1]The significance of effects is customarily determined using statistical criteria. In this experiment, the software package Q-Edge [Bishop] was used for analysis and decisions regarding significant effects.

2. The actual model becomes

$$\hat{V} = 3.580 + \frac{4.294}{2}I_c + \frac{4.624}{2}R_c + \frac{2.758}{2}I_cR_c$$

3. Remembering that our prediction model is for coded values (i.e., −1, +1 scale) of I and R, we can transform it into a model using actual I_a and R_a values as follows:

$$I_a = \left(\frac{0.020+0.005}{2}\right) + \left(\frac{0.020-0.005}{2}\right)I_c \quad \text{or} \quad I_a = 0.0125 + 0.0075I_c$$

Thus,

$$I_c = \left(\frac{I_a - 0.0125}{0.0075}\right)$$

Likewise,

$$R_a = \left(\frac{470+100}{2}\right) + \left(\frac{470-100}{2}\right)R_c$$

or

$$R_a = 285 + 185R_c$$

Thus,

$$R_c = \left(\frac{R_a - 285}{185}\right)$$

4. Substituting the actual transformed variables into the $\hat{V}$ equation in (2) will produce the new $\hat{V}$ model.

$$\hat{V} = 3.580 + \frac{4.294}{2}\left(\frac{I_a - 0.0125}{0.0075}\right) + \frac{4.624}{2}\left(\frac{R_a - 285}{185}\right) + \frac{2.758}{2}\left(\frac{I_a - 0.0125}{0.0075}\right)\left(\frac{R_a - 285}{185}\right)$$

$$\hat{V} = 3.580 + 286.267I_a - 3.578 + 0.012R_a - 3.562 - 0.994I_aR_a$$

$$- 0.012R_a - 283.254I_a + 3.541$$

$$\hat{V} = -0.019 + 3.013I_a + 0.000R_a + 0.994I_aR_a$$

Under actual experimental conditions, we got a model very close to nature's V = IR. Note that the wattage of the resistor (W) dropped out as insignificant during the analysis and does not appear in the model. Of course, we knew this should happen, but it is always comforting to know that if we had not known that W is not an affecting factor, DOE would have told us! The coefficient for I_a of 3.013 appears large relative to the coefficient of I_aR_a. However, one must remember that the actual range of I_a is 0.005 to 0.020. Thus, the prediction of V is only changing from about 0.015 to 0.060 over the range of I_a. Therefore, the most likely reason for a non-zero coefficient for I_a is the low test values (0.005 and 0.020) used in the DOE.

If we were to now use the model to predict the voltage across a resistor of 220 ohms when the current is 0.012 amps, the model would predict $\hat{V} = 2.641$ volts. Nature (using Ohm's law) knows that the voltage would be $V = IR = 2.640$ volts. Even though the experimental model did not yield a perfect $V = IR$, knowing that the voltage is in large part determined by the interaction or combined effect of I and R might have reduced Ohm's research time significantly.

Obviously, not all models in nature are as simple as $F = MA$ and $V = IR$; however, the purpose here is to show how DOE can enable today's scientist, engineers, and physicists to close the gap quickly with a model from experimental data. For more information on evaluating nonlinear relationships and/or large numbers of input variables, see a good text on DOE!

Defining Terms

Design of experiments: Purposeful changes to the inputs (factors) to a process (or activity) in order to observe corresponding changes in the outputs (responses). A means of extracting knowledge from a process (or activity).

Design matrix: A set of experimental runs that lists the settings for each factor in the experiment.

Experimental variable: Variable (e.g., x, y, or z) used to designate the settings of specific factors (e.g., time, pressure, or temperature) in a designed experiment.

Further Information

Our Web site address is www.airacad.com.

References

Bishop, M., *The Quality Edge*, Statistical Software, Air Academy Press, Colorado Springs, CO.

Pignatiello, J. J., An overview of the strategy and tactics of Taguchi, *IIE Trans.*, 20: 247–254, 1988.

Schmidt, S. R. and Launsby, R. G., *Understanding Industrial Designed Experiments*, 4th ed., Air Academy Press, Colorado Springs, CO, 1994.

Taguchi, G., *System of Experimental Design*, Kraus International, White Plains, NY, 1987.

6

Accounting

Paul A. Griffin
University of California

Henry E. Riggs
Keck Graduate Institute

David M. Cottrell
Brigham Young University

Daniel L. Jensen
The Ohio State University

Michael W. Maher
University of California

John D. Lyon
University of California

Tung Au
Carnegie Mellon University

Masako N. Darrough
University of California

6.1   Financial Statements

Paul A. Griffin

Key Issues for the Technology Manager

This section develops a broad economic framework for financial statements and reporting. Its goal is to help the technology manager understand better the following four key issues: why investors and analysts use financial statements, why managers are often very concerned about what goes in financial statements, why governmental bodies and private-sector agencies regulate the information that companies produce, and why capital markets can impose major economic consequences on companies that publish financial statements. For a technology manager interested in maximizing **shareholder value**, understanding these accounting issues is vital.

For example, how we think about the responsiveness of the stock market to financial statements can greatly affect a decision about the accounting method we choose, where to disclose information in the financial report, and how investors might perceive and react to that information. Technology company stock prices have traditionally demonstrated a great sensitivity to changes in reported earnings, especially when fourth quarter and hence annual earnings fall short of analysts' expectations. Poorly managed financial statements can have a disastrous impact on a technology company.

This first section discusses what should be the role of accounting in a competitive economy and what factors influence the use of accounting information by investors, creditors, and managers. The second section focuses on the supply of accounting information and how it is influenced by regulation and management incentives. The third section briefly discusses the impact of the Securities and Exchange Commission's (SEC) new electronic filing system. The fourth section focuses on financial information in an **efficient market**. A final section summarizes the key points.

A Technology Manager Should Understand the Role of Financial Accounting

How should a technology manager view financial accounting? Financial accounting is the process whereby the managers of business enterprises supply information to investors and others who have a use for that information. Such users demand financial and other information because they have some kind of interest — present or future — in the company. For example, investors usually want information that will enable them to assess the future cash flow, earnings, or distribution potential of the company in order to decide on a market price at which they would be willing to buy or sell some or all of their shareholdings (or options regarding those shareholdings).

Managers and users often disagree about how much information is enough. Users generally want more than managers are willing to reveal, and mechanisms are needed to reconcile these differences. Investors, for example, often want more information about future company plans and products. Some conflicts are resolved privately. For instance, the information needs of lenders are normally specified in the bank loan agreement, and major trade creditors may be privy to key information about the production process. However, when private-sector mechanisms cannot the reconcile these conflicts, a government or government-designated agency ususally steps in.

The primary body in the United States that oversees the regulation of company financial information is the SEC.[1] However, the SEC has delegated the Financial Accounting Standards Board (FASB) to establish **generally accepted accounting principles** (GAAP) for all public companies. The SEC also requires that public companies engage an independent accountant (external auditor) to attest to whether their financial statements are in accordance with GAAP.

---

[1]The details are contained in the Securities Act of 1933 and the Securities and Exchange Act of 1934. These acts give the SEC authority to prescribe all accounting and reporting standards for all public companies, though it has delegated much of this to the FASB.

The objective of financial reporting has been defined by the FASB in Concepts Statement No. 1. This states: "Financial reporting should provide information that is useful to present and potential investors and creditors and other users in making rational investment, credit, and similar decisions." The characteristics of accounting information that make it useful to investors, creditors, and others have also been defined. FASB Concepts Statement No. 2 presents a hierarchy of qualities that should be inherent in useful financial information. Relevance and reliability are of primary importance. However, one key factor applicable to all interested parties constrains those qualities. All standards should pass a cost-benefit test that weighs the impacts of all parties involved — managers, users, auditors, government, and others.

Though many would agree with the FASB's focus on investors and creditors as the primary users of financial reporting information, others offer a different interpretation. For instance, company managers often state that financial reporting should provide information about how effectively and efficiently managers have performed and how properly they have taken care of investors' funds. This is the stewardship role of financial reporting. The stewardship role can also be viewed from a principal-agent perspective. In accounting, managers are the agents and shareholders the principals. According to this view, shareholders need information to monitor compliance with the terms, obligations, and responsibilities of the various contracts with their managers. Also, contracts may be written to encourage senior managers and employees to act in the best interests of the shareholders (e.g., stock option compensation plans, performance-based compensation plans). These incentive contracts are designed reduce the costs and conflicts that would occur without such contractual arrangements and should benefit all parties.

So what is the precise role of financial accounting in a competitive economy? How should a technology manager think about these issues? As the preceding suggests, it depends critically a manager's views about (1) the appropriate balance among the different reporting objectives and (2) the accounting and reporting environment, particularly the groups that have vital interests in the accounting and reporting process and the appropriate mechanisms that enable those groups reconcile their differences. In thinking about the role of accounting, it is also worthwhile to distinguish between financial statements and external financial reporting. This distinction is important as it is only the financial statements that are subject to a detailed audit. In the United States, financial statements means (1) a statement of financial position,[2] (2) a statement of income and retained earnings,[3] (3) a statement of cash flows,[4] (4) a statement of changes in owners' or shareholders' equity,[5] and (5) all footnotes to those statements. In addition to the financial statements, a company's financial report would include supplementary disclosures, management discussion and analysis, reports on products and employees, lists of officers and directors, and other special reports. In this section, the predominant focus is financial statements and footnotes.

Who Creates the Demand for Financial Statements?

Let us now proceed according to the FASB assumption — that the primary users of financial reports are investors and creditors.[6] Let us further assume that the ultimate object of accounting analysis is the assessment of **return** on a security, namely, the sum of the expected dividends and distributions or interest payments plus any capital gain. These two assumptions enable the benefits of financial information to be

---

[2] $Assets_t - liabilities_t = shareholders' equity_t$.

[3] $Net\ income_t = revenue_t - expenses_t + gains_t - losses_t$; retained earnings$_t$ = retained earnings$_{t-1}$ + net income$_t$ − distributions$_t$.

[4] Cash flows from operations$_t$ ± cash flows from investing$_t$ ± cash flow from financing$_t$ = change in cash$_t$.

[5] Change in shareholders' equity$_t$ = change in contributed capital$_t$ + change in retained earnings$_t$.

[6] A technology company manager may disagree with this assumption. For example, the FASB recently proposed that the value of certain stock options granted to employees be viewed as an expense of doing business, presumably, because this would be useful information for investors and creditors. Most technology companies opposed this treatment and "persuaded" the FASB to modify its position by giving the company the option of either recording an expense or showing the impact on profit as a footnote. Most companies have chosen not to record an expense but to disclose as a footnote only.

defined as follows. To have benefit, accounting information must be capable of changing an individual's assessment of return by changing some aspect of the distribution such as expected value (future returns are likely to be higher or lower due to the new information) or variance (future returns are likely to be more or less risky than before). Understanding an individual's use of financial report information, consequently, implies understanding how return assessments are made and how they might change in response to information. Presently, our knowledge about how investors form assessments of returns is limited, though there is some evidence that investors hold underperforming stocks for too long (and thus would do better if they would cut their losses early).

Though we have identified investors and creditors as the primary users of financial report information, experience suggests that most investors hold diversified portfolios and, as such, they use accounting information only indirectly. They seldom have the time, inclination, or expertise to pore over stacks of annual reports. Instead, the job is left to people who are highly trained and paid to perform that service for them, namely, equity and bond analysts. These people have strong incentives to understand financial reports and accounting measurement issues. However, the detailed work of equity and bond analysts is skewed toward larger companies of institutional interest. This suggests that financial report information can be more critical for companies with limited analysts' scrutiny, for example, newer and smaller companies listed on the NASDAQ or a regional market. The stock market, also, looks very closely at the first set of annual financial statements following an initial public offering. This is the first full disclosure of how management uses accounting and financial reporting to describe company performance and position and whether the company is able to live up to initial market expectations.

Investment Decisions Determine What Goes into Financial Statements

Distinctions between stock investments and bond investments have been blurred in recent years as companies increasingly offer securities that have the characteristics of both (e.g., convertible debt, debt with attached common stock warrants) and as interest rates have become more unpredictable. While both emphasize the need to assess future enterprise cash flows, equity analysts usually spotlight the assessment of return for purposes of assessing shareholder value, whereas bond analysts price company borrowings on the basis of **yield**.

The need for information, however, in both cases derives from the particular way in which the valuations are made. Shareholder value, for example, can be assessed on the basis of the present value of future cash flow (before interest) from the invested capital, with an appropriate deduction for debt. Shareholder value can also be assessed on the basis of the present value of **abnormal profit** (profit in excess of what an equity investor considers normal). This is sometimes referred to as an economic-value-added approach. However, when questioned, equity analysts traditionally state that they assess common stock returns most often on the basis of the ratio of price per share to earnings per share (P-E ratio) or the market value of the company's net assets in relation to the book value of those net assets (market-to-book ratio). Thus, for the majority of equity analysts, earnings per share and net assets (book value) are the lifeblood of the assessing returns and shareholder value. However, irrespective of whether the analyst uses cash flow or earnings-based approaches, the technology manager is well advised to understand the valuation processes used by the financial analysts that follow the company. Much of this can be learned during the **conference call** and other interactions between managers and the investment community, since the latter will be prodding managers for key inputs into their valuation models.

Management Contracting also Determines What Goes in Financial Statements

So far we have emphasized investors' and creditors' needs for and uses of accounting information and have only mentioned briefly the needs of company managers as people who have a demand for such information. Because parties external to the firm judge management's performance and determine the market price of the company based on financial report information, most managers are very concerned about the content and format of the company's statements. First, managers who own shares or options to purchase shares in their company have a natural interest in maximizing shareholder value. They use their own firm's statements in the capacity of investors, though their use is not identical to that of an outside shareholder. For example, managers have a lower propensity for risk as failure can hurt them more than outside shareholders (e.g., they can lose their job). Second, managers are more likely than the average investor to take a position in another company for purposes of control.

However, managers are also concerned about factors beyond shareholder value. This has led to the use of management performance and bonus plans that contractually link compensation to *long-term* financial statement variables such as net income and return on equity averaged over several periods and/or benchmarked to competitors. Such financial statement variables may also be grouped with nonshareholder measures so as to achieve a **balanced scorecard** of the company's goals. Some companies state and discuss their balanced scorecard objectives in the annual financial report.

Company Financial Statements Are the Responsibility of Management

The primary responsibility for the preparation of financial statements resides with management. Management is expected to establish controls and systems such that the resources entrusted to them are utilized in accordance with approved plans and strategies. However, the interests and incentives of managers and shareholders are not necessarily the same, and thus arrangements must be made to deal with any conflicts. This section discusses ways in which management might influence how much information is given to persons outside the firm. We first identify key elements of the company's internal environment.

Normally, the process begins in the accounting or controller's department and proceeds to the chief financial officer and then to the chief executive officer. Company's lawyers scrutinize the financial report so that it complies with the requirements of regulatory agencies, state laws, and stock exchanges. The internal audit department may also be involved, primarily in reviewing control and accountability systems and, more generally, insuring that the statements are in accordance with **generally accepted auditing standards** (GAAS) and GAAP. The company's audit committee reviews key issues of internal control and special questions regarding GAAP and assists in reconciling material accounting and reporting differences between management and the external auditor. GAAS requires that the most serious differences be communicated to the board of directors; otherwise, the auditor must resign.

While these elements exist in most public companies, financial reports vary from company to company in shape, size, format, quality, and indeed just about as many ways as there are companies. Some make available a videotape or multimedia CD-ROM; others present their data in comic-strip fashion. An increasing number now provide detailed financial report information on the company's Web site. Collectively, these differences reflect variations in the **financial reporting culture**, which in turn often arise from either the personalities of top managers or the image that the company wants to project to outsiders (e.g., conservative, growth oriented). Much can be gained from understanding the reporting culture. It can provide subtle signals about the style of managers and their intentions and strategic thinking.

For example, in the 1980s H. J. Heinz was known to have an aggressive top management team. Pressure on lower-level divisional managers to achieve consistently tight profit goals (that determined their bonuses) caused them to shift earnings in periods of good performance to periods of poor performance. However, such income smoothing was in violation of GAAP and the firm was required by the SEC and its own audit committee to restate its income for those years. A similar situation struck Bausch & Lomb in the 1990s. Fierce pressure from top management to maintain double-digit sales and earnings growth yielded dire unintended consequences. Unit managers inflated revenues with fictitious sales, shipped items that were never ordered, and forced distributors to take unwanted inventories. Eventually, after several years and an SEC investigation, the accounting fiction was discovered and corrected. Interestingly, at that point, top management ordered the company to follow the most conservative accounting practices possible.

Five Factors that Influence a Manager's Supply of Financial Information

What general factors influence a managers' supply of financial information? This section discusses five: (1) who pays for the information, (2) capital market effects, (3) compensation and corporate control, (4) external auditing, and (5) institutional and legal framework.

First, with respect to the "who pays" issue, economists suggest that, when users of a good do not have to pay for it, the quantity supplied is less than if all who used the good had to pay full price. Economists use the term "public good" to describe this notion. Applied to accounting information, this suggests that, absent any outside factors (e.g., regulation), managers will supply less rather than more information to outsiders. Managers also have a natural tendency to want to operate in relative privacy, claiming that can be disadvantaged by supplying more outside information than necessary. For instance, privacy can be critical for activities such as research and development since it may give competitors an unfair advantage.

Second, regarding capital markets, a competitive and efficient market constitutes a powerful, disciplinary force on the behavior of company managers, and this, many would argue, is the best protection for shareholders against unsavory acts by managers. Some managers are driven more by the capital market than others. For example, managers with a "Wall Street" perspective would always want to consider whether the market would approve or disapprove of their intended actions. Their choices about accounting measurement rules and disclosures would be similarly affected by Wall Street's possible reactions. Some financial analysts believe that managers preoccupy themselves with reported "bottom line" net income not because it is a most valid measure of company performance but because that is what Wall Street wants.

How capital markets respond to accounting information is a subject of great discussion and debate, though we will not go into that here. At a minimum, a manager should appreciate that both income-increasing and income-decreasing accounting events can increase shareholder value. For instance, statements of intention by companies to buy back shares or recognize unrealized gains have been shown to have positive effects on prices. However, other income-increasing accounting actions (e.g., anticipation of revenues prior to sale) have been interpreted negatively, and critics often lambaste companies involved for such questionable practices with obvious adverse repercussions. Paradoxically, accounting changes that decrease net income can have beneficial stock price effects: First, those changes may mean lower tax payments; second, they may signal a stronger financial position than the competitors' in that the company's competitors are not willing to risk taking a similar move; third, they may be favorably perceived as signaling a conservative management style.[7]

Managers often cannot ignore market signals because their own personal wealth and compensation is linked to them. Because market prices and other means of monitoring management are imperfect indicators, managers have incentives to choose accounting rules and report numbers to benefit themselves, potentially at the expense of the shareholders. One study, for example, reports that managers act as if they use **income accruals** to maximize their bonuses by making negative income accrual entries when no bonus (or no additional bonus) was able to be earned and positive income accruals when such actions would increase their bonus. The Heinz and Bausch & Lomb examples, though more extreme, also demonstrate this notion.

In the area of corporate control, common sense suggests that an incumbent manager would not want to produce information that increases the likelihood that others would take control of the company. For example, no manager would want reporting policies that obscured economic value, disclosed excessive perquisites, or reflected questionable activities. Such policies increase the threat of takeover. However, some managers apparently do act in this way, thought they try to disguise those accounting actions by taking them jointly with another financial or investment decision (e.g., the timely sale of an asset at a gain that offset the losses or abnormal expenses taken elsewhere).

---

[7]Research also indicates that managers are more likely to engage in income-decreasing actions in periods of bullish market activity. For example, companies typically record restructuring losses in periods of stock market growth.

The external auditor exerts a fourth influence on the supply of financial information. The auditor's role, ideally, is to make the financial representations of managers more credible by expressing an opinion that the financial statements show fairly the company's results of operations and financial position. The auditor bases the examination on GAAS and also makes sure that the financial statements are in accordance with GAAP. In general, the interests of both managers and shareholders or owners are served by the external audit function. However, because the external audit function brings regulation into the picture, a company that hires the auditor may pay for services beyond what would be optimal absent the regulation. Company managers are often perplexed by the high cost of an audit relative to the benefits they receive. However, from a longer-term perspective, it is not the firm but the shareholders who pay for those services. Shareholders accept a lower return as a payment for an increase in the quality of information due to the audit.

A fifth factor is the regulatory environment. In Western countries, the accounting and reporting rules are wedded to the legal and institutional environment. For instance, countries such as the United Kingdom, Australia, and New Zealand enforce much of their accounting and disclosure through company law. On the other hand, members of the European Economic Community (EEC) countries are subject to EEC directives (on financial reporting policies) as well as local legislation (e.g., tax and company law). In the United States, the SEC and FASB determine the company financial reporting process.

The details of the SEC filing requirements are contained in the Securities Act of 1933 and the Securities and Exchange Act of 1934. For financial reporting, the most important legal concept is that the various SEC filings shall contain all **material** information about the company. However, materiality is normally left to the courts to define and interpret, which in recent years has been defined in terms of whether the information significantly affected the company's stock price at the time it became known publicly. Thus, in complying with the U.S. securities acts, public companies must always make sure that their reports have not misrepresented or omitted material financial information.

When shareholders claim that the compay has not given them all material information, they may bring a court action against the company asserting they were damaged as result of a violation of the materiality provisions of the securities laws. Such actions may be brought against the company, by the SEC (e.g., as an enforcement action) or as a class action by a group of similarly affected shareholders. Technology companies have been frequent targets of class actions in recent years. This may be due in part to the volatile nature of their sales and earnings growth and their choice of accounting policies often designed to accentuate that growth.[8]

Another regulatory factor that can concern managers is whether they must change their decision making when forced to change their financial statements. To the extent that this happens, decisions to supply accounting information and decisions to manage a business are inextricably connected. For example, when the FASB mandated in the 1980s that promised pension and health benefit obligations be shown on the balance sheet as liabilities, companies began switching to defined contribution plans for pensions and managed care plans for health benefits. These new plans eliminated many of the liabilities required under the old plans. Many claimed that the FASB standards rather than economic common sense drove these changes. In short, imposed accounting alternatives can significantly change business decision making. The effects of such changes need not always be adverse, however. Some accounting changes may better sensitize managers to the underlying economic issues as well as give them improved data. If managers may make better decisions than before despite the costs of changing to a new measurement or disclosure rule, such changes can enhance shareholder value.

Electronic Filing Achieves an Important SEC Objective

A third key issue for a technology manager involves how financial reporting might change as corporate America shifts to the SEC new electronic medium for the delivery of financial information to investors

---

[8]A good repository of information about shareholder class actions is the Securities Class Action Clearing House (http://securities.stanford.edu/). This Web site contains details of all federal securities class action lawsuits since December 1995.

and creditors. Certainly, this new medium serves well the SEC's doctrine of equal access. This doctrine is best illustrated by the SEC's **EDGAR** database (http://www.sec.gov/cgi-bin/srch-edgar) that collects and makes available all SEC company documents via the Internet. As soon as a company document is filed, EDGAR makes it freely available for all to read and analyze. Sites such as FreeEdgar (http://www.FreeEDGAR.com) "push" the latest filings to the user, and software such as hOTeDGAR (http://www.hottools.com) assists in the analysis by converting the SEC documents to analysis-friendly spreadsheets. This kind of electronic approach to the dissemination and analysis of financial information, ultimately, will allow the user to construct financial statements according to any alternative set of accounting assumptions, almost costlessly. It also allows for more resources to go to analysis rather than data collection, which should be to everyone's benefit.

Accounting Choices Cannot be Viewed Apart from the Markets in which They Occur

So far, we have made only limited references to the markets in which investors and creditors trade stocks and bonds. Investors' and creditors' activities, however, cannot be viewed apart from these markets, and so a complete understanding of financial reporting requires knowledge of how capital markets operate. Business enterprises sell securities to shareholders and bondholders, and those securities are subsequently traded on the national and regional exchanges. Managers who make decisions about what goes in a financial report want to know what affect their decisions might have on the securities markets, since the anticipated market reaction constitutes an expected cost or benefit of their decision. Users too need to know whether and how to respond to company information. For instance, an announcement that confirms what investors have already forecasted would normally require a different kind of response than an announcement that investors view as new and surprising.

The relationship between security price changes and information is central to the concept of an efficient market. A capital market is perfectly efficient when all available information is instantaneously and unbiasedly impounded in prices. Thus, a market can be characterized as inefficient when the competitive actions of buyers and sellers of securities are such that the prices do *not* quickly impound the information or that buyers and sellers in aggregate react in a biased fashion (i.e., over- or underreact to that information). A perfectly efficient market, however, is a notion that exists in the minds of theoreticians.

A more practical way to think about an efficient market is in terms of degrees, where that degree depends on the speed of reaction and the duration of over- or underreaction to the new information. For example, when information is very costly to obtain or analyze, the market is less efficient with respect to that information than when that same information is available at low cost. In a related manner, when information is of low quality, not well understood, and perhaps presented in isolation without corroborating data, market reaction may be temporarily delayed or biased as it works out the correct implications.

A vast literature now exists describing the market's response to announcements, events, and other news. These studies support the proposition that U.S. capital markets are highly efficient processors of financial information, as are the markets in Australia, most of Europe, Japan, New Zealand, and Scandanavia. For companies on the U.S. NYSE and AMEX exchanges, prices have been observed to respond well within a few seconds to announcements of earnings and dividends. Even NASDAQ stocks appear to anticipate and respond promptly and without over- or underreacting to accounting news announcements. Any limited attention by analysts and institutions is often counterbalanced by the dealers who hold inventories and "make" the market in the stock.[9]

---

[9]However, a small but growing set of studies has drawn attention to numerous puzzles that efficient market theory does not adequately explain. Some argue that excessive stock market volatility cannot be satisfactorily explained by changes in investors' assessments about future cash flows from dividends and that social phenomena (e.g., fads and fashions, irrational exuberance) may also be driving market prices.

Managers Perceptions about Market Efficiency Affect Accounting Decisions

Because market prices ultimately reflect the aggregate consequences of their actions, managers who make accounting choices implicitly make assumptions about the degree to which the capital market is efficient. Managers who believe that reported profit numbers based on accrual accounting are the principal signals that guide market behavior are likely to take changes in accounting measurement rules very seriously. For them, any change in a measurement rule could cause the market to react, regardless of the economic situation. On the other hand, managers who do not believe markets respond naively to reported numbers but look behind those numbers to assess changes in the firm's prospective cash flows would be much less concerned. The latter focuses much more on the quality and quantity of information disclosed relative to what they believe is presently available and reflected in prices. Finally, in providing unbiased signals about company decision making, an efficient market protects investors generally against inappropriate actions by managers, since it is assumed that an efficient market unbiasedly anticipates those actions. Price protection, however, is neither absolute nor universal. For example, managers and others may engage in policies that misrepresent or omit material company facts — either intentionally or unintentionally — in ways that would be almost impossible to anticipate. The only protection here is the threat of litigation when such misrepresentation or omission is eventually discovered.

Good Financial Statement Choices and Increased Shareholder Value Go Together

This section concludes by summarizing the key points. In doing so, the underlying challenge for the technology manager is underscored: good financial accounting and reporting means understanding the proper role of financial statements — especially in terms of financial statement users and how accounting rules affect managers' incentives and contracts. Good financial reporting means good decision making, which ultimately should increase company or shareholder value. Technology managers should recognize the following:

- Subject to cost-benefit considerations and regulatory constraints, financial reporting should provide all material information that is useful for making rational investment and credit decisions.
- While the principal users of financial statements are investors and creditors, financial accounting should also serve managers' needs for information that is useful for measuring performance, compensating managers and employees, and contracting with outside parties.
- Managers have the primary responsibility for the financial reporting and have incentives to choose the form and content of financial statements on the basis of shareholder value and their own self interest. Managers should understand the incentive effects of acounting and reporting well since they could have a profound effect on company and shareholder value.
- The SEC and the FASB regulate accounting and reporting primarily in the interests of outside investors and creditors. Fast electronic dissemination of that information now enables equal access at very low cost. This should improve the overall quality of outsiders' financial analyses of the company.
- Efficient capital markets process financial report information very quickly and without over- or underreaction. Recent evidence suggests that stock markets are not as efficient as once thought, and it is best to think about market efficiency as a matter of degree. Market efficiency depends crucially on the cost of obtaining information about the company.
- A manager's perception about whether the stock market is efficient or inefficient can influence greatly the form and content of financial statements.
- By unbiasedly anticipating the actions of a company manager, an efficient market helps protect investors and creditors from managers' actions not in their best interests. The threat of litigation also protects investors and creditors from inappropriate actions by a company manager.

Defining Terms

Abnormal profit: The difference between actual or expected profit and normal profit. Normal profit is calculated as the normal rate of return times the beginning-of-period book value of shareholders' equity.

Balanced scorecard: A formal approach to evaluating company performance based on an integrated set of financial statement and nonfinancial statement measures. These measures are normally integrated with the company's strategic goals.

Conference call: A mechanism whereby companies announce and discuss their financial results simultaneously with a large number of financial analysts, all of whom are telephonically linked to the company executives.

EDGAR: An electronic document gathering and retrieval service maintained by the SEC covering almost all SEC registrants' filings, including companies' annual (form 10-K) and quarterly (form 10-Q) financial statements. SEC documents may be retrieved free of charge via the Internet.

Efficient market: A market in which prices reflect quickly and unbiasedly all available information, which is assumed to be costlessly available. When information and market transactions are costly, not all available information is reflected in prices. Such a market, though not perfectly efficient, is efficient to the degreee that it is not possible to earn an abnormal profit by exploiting any possible inefficiency.

Financial reporting culture: The values and customs inherent in a company's decisions regarding its financial measurement and disclosure practices. These values and customs often mirror the views and attitudes of senior management, and they can also reveal important insights into a company's plans and strategies.

Generally accepted accounting principles: A body of accounting standards, principles, and practices that has "substantial authorative support". FASB standards currently have official authority as generally accepted accounting principles under AICPA Rules of Conduct, Section 203, and SEC Financial Reporting Release No. 1, Section 101.

Generally accepted auditing standards: A body of standards and rules issued by the U. S. Auditing Standards Board that defines the scope and quality of audit services to be performed to provide the necessary assurances that a company's financial statements are free of material misstatement.

Income accruals: Non-cash accounting entries that adjust cash revenues received to reflect revenue earned for a period and cash expenses paid to reflect expenses incurred for a period. Positive accruals increase net income; negative accruals decrease net income.

Material: A legal threshold defining what is significant for investors. According to the U.S. Supreme Court, an omitted fact is material if there is a substantial likelihood that a reasonable shareholder would consider it important in deciding how to vote, that is, there must be a substantial likelihood that the disclosure of the omitted fact would have been viewed by the reasonable investor as having significantly altered the "total mix" of information made available.

Return: A measure of actual or future investment performance expressed on a periodic basis as the sum of the (actual or expected) share dividends and distributions or interest payments on a security plus the (actual or expected) price change or capital gain. For a single period, that sum is the security's *return* (often stated as a percentage of the beginning of period investment.

Shareholder value: The net present value of the future operating earnings or cash flows available to all contributors of capital after investment, discounted at the weighted average cost of capital, plus the fair value of the net financial assets less the fair value of all nonshareholder invested capital.

Yield: A single rate of *return* that is used to produce the same market price as that based on the individual or spot rates for several periods. Investors usually price bonds, for example, in terms of an expected *yield to maturity*, namely, that single rate of return that "averages" the individual or spot rates to maturity.

Further Information

Books

Mulford, C. W. and Comiskey, E. E. *Financial Warnings*, John Wiley & Sons, New York, 1996. How to manage financial statements to avoid unnecessary stock market surprises.

Murray, D., Neumann, B. V. R., and Elgers, P. *Using Financial Accounting: An Introduction*, South-Western College Publishing, Cincinnati, OH, 1997. General introduction to financial accounting and reporting, references to financial accounting Web sites.

White, G. I., Sondhi, A. C., and Fried, D. *The Analysis and Use of Financial Statements, 2nd ed.*, John Wiley & Sons, New York, 1998. Comprehensive coverage of how investors and creditors use financial statements for shareholder value and credit analysis.

Internet Sites

Financial Accounting Standards Board: http://www.rutgers.edu/Accounting/raw/fasb/; FASB standards, interpretations, exposure drafts, latest decisions, and more.

Financial Times: http://www.ft.com/hippocampus/; company news and financial information, worldwide.

Paul A. Griffin: http://www.gsm.ucdavis.edu/~griffin/; learn more about the author, his publications, and current research and teaching interests.

Securities and Exchange Commission: SEC filings: http://www.sec.gov/cgi-bin/srch-edgar; company financial statements and other SEC filings, U.S. focus.

Wall Street Journal: http://www.wsj.com/; company news and financial information, U.S. focus.

6.2 Valuation in Accounting[10]

Henry E. Riggs

How does an accountant value what the enterprise owns, as well as its obligations? How is a monetary value assigned to the many and varied transactions to which the enterprise is a party? Some valuations are straightforward and relatively indisputable. Others present very real dilemmas.

Valuation Methods

The accounting framework requires that we value both assets (what the entity owns) and liabilities (what it owes). Valuing assets and liabilities is the fundamental accounting task and the one that draws most of the arguments among accountants and most of the controversies between accountants and their audiences.

These arguments and controversies typically center on which one of the following three valuation methods is most appropriate in a particular circumstance:

1. The *time-adjusted method*, viewing values as a function of future benefits and costs arising from the item owned or from the obligation.
2. The *market value method*, equating value to the price at which the item owned could now be bought or sold.
3. The *cost value method*, equating value to the price paid for the item when it was originally acquired.

Keep in mind a few blunt facts about these alternative methods. They do not typically result in the same values. No single method is inherently more correct than the other two. The cost value method is the predominant method used but is in some sense the least appealing intellectually. To better understand

---

[10]Adapted from *Financial and Cost Analysis for Engineering and Technology Management*, Riggs, H. E., Wiley Interscience, New York, 1994, chap. 2.

the advantages and shortcomings of each method, you need to understand all three. We begin with an explanation of the time-adjusted value method.

Calculating Time-Adjusted Values

The truism on which this valuation method is based is that money has a time value. You would rather receive a certain amount of money today than the same amount at a later date, say, a year from today, since you can earn a return[11] on money during the intervening year. Similarly, you would prefer to delay the payment of a specified amount as long as possible so that you can earn a return on the money during the period of the delay.

If you know the rate of the available return, you can calculate the advantage or disadvantage associated with accelerating or delaying receipts and payments. Thus, values of cash flows (receipts or payments) are a function of their timing; timing differences have monetary value consequences. The time-adjusted valuation method values, as of today, future cash flows that will arise because of assets owned or obligations undertaken, utilizing an appropriate equivalency (interest) rate.

Suppose, for example, that you are the manager of a retail store. When the store has a small amount of excess cash, you invest it in a short-term deposit or security that earns interest at a specified rate, $r\%$ per year; when the store runs short of cash, you withdraw (disinvest) from this deposit or security the amount necessary to cover the cash deficiency. Now suppose that the owner of your building offers to reduce (or discount) the store's rent if you agree to pay the rent a year in advance. You need to determine the rent discount that would be attractive, that is, you need to know how much to pay in rent today in lieu of a $20,000 rental payment otherwise due a year from today.

If you do not make the advance rental payment, the money will remain in the deposit or security that earns at the rate r. Your equivalency rate is r. Therefore, the maximum amount that you would be willing to pay today is that amount that would otherwise grow at the rate of r per year to $20,000 at the end of the year, that is, you will leave the money in the interest-bearing deposit or security if the advance rental payment is not made. Therefore, an equation for the advance payment P is

$$P\left(1+r\right) = \$20,000$$

$$P = \frac{20,000}{\left(1+r\right)}$$

If r is 8%:

$$P = \frac{20,000}{\left(1.08\right)} = \$18,520$$

Your decision? If the building owner will accept today $18,520 or less in lieu of $20,000 a year from today, you should pay the rent in advance. P is referred to as the present worth or time-adjusted value of the future rental payment at rate r. The time-adjusted value today of $20,000 a year from now is $18,520, assuming an interest rate of 8%.

Now suppose the building owner asks you to consider paying in advance the rentals due at the end of each of the next 3 years. Certainly, you will demand a greater discount on the 2nd year's rent than on the 1st and a still greater discount on the rent otherwise due at the end of year 3. Therefore, you need to know the time-adjusted value (or present value) P of a stream of three payments, each of $20,000,

---

[11]The return may be in the form of interest earned or in the form of wants satisfied by spending the money.

occurring at 1-year intervals, assuming an interest rate of r. P is the sum of the individual time-adjusted values for each of the 3 years:

$$P = \frac{20,000}{(1+r)} + \frac{20,000}{(1+r)^2} + \frac{20,000}{(1+r)^3}$$

If r is once again 8%, this equation produces a value for P of $51,540. You are willing to pay an amount up to $51,540 — but not more — to be relieved of the obligation to make the three annual rental payments of $20,000 each.

Assume now that the building owner, badly needing cash, will have to borrow money at very high interest rates if you don't prepay the rent. Under these conditions the owner might be willing to accept substantially less than $51,540, that is, the owner undoubtedly has an equivalency rate well above 8% and thus a time-adjusted value of these future cash flow that is less than yours. Assuming an r of 12%, P for the building owner — the time-adjusted value of the next three annual rental payments — is $48,040.

Note that the higher the interest, or equivalency rate, r used in the calculation, the lower the time-adjusted value of future cash flows. Standard present value tables allow you to calculate time-adjusted values.

To apply the time-adjusted method, an asset's owner (or the accountant for the owner) must forecast the future benefits derived from ownership. For certain assets, such as loans, investments in common stock, paid-up insurance policies, or customers' promises to pay, estimating future benefits is relatively easy. These benefits are either future cash inflows or the elimination of future cash outflows that would be required in the absence of the asset. Payment in advance of 1 year's rent on a building creates an asset — the right to use the building for 1 year or, equivalently, the elimination of the requirement to make rental payments during the year.

This valuation method is particularly useful for assets that have a long life and a predictable benefit flow. What is the value of a loan requiring the borrower to make level payments, incorporating both principal and interest, at the rate of $1000 per year for 10 years? The value of the loan depends upon the equivalency rate that the lender assigns to loans (investments) of this type. Its value is $10,000 only under the very unlikely assumption that the lender has a zero interest equivalency rate, implying that the lender has nothing else productive to do with the money. The higher the equivalency interest rate assumed, the lower the value of the loan:

Equivalency Interest Rate(%)	Time-Adjusted Value ($)
6	7360
8	6710
10	6144
12	5650
15	5019

Consider this valuation method for other assets, for example, a machine tool owned by a manufacturing company and a cellar of fine wines owned by an individual. In both cases, asset ownership promises benefits stretched out into the future, but measuring these benefits is a real challenge.

The benefits of the machine tool depend upon a host of factors, including

1. The rate of obsolescence of the tool, which in turn depends upon the rate of technological development by machine tool builders.
2. The future demand for the product or products produced on the machine tool (this demand in turn depends on the rate of change in the company's product lines, changes in the competitive climate, the strength of the economy, and changing consumer preferences).

The cellar of fine wines may be even more troublesome to value. The benefits are the personal pleasure of consuming the wine and serving it to friends; but here value is also dependent upon factors such as how well the wine ages, the volume and quality of wines to be produced in upcoming years, and future supply-demand imbalances that may affect the cost of wines.

Therefore, this method of valuation, while appealing in concept, may be less useful in practice, at least for valuing such assets as machine tools and wine cellars.

Can obligations (liabilities) also be similarly valued by the time-adjusted method? Consider a company's obligation to perform warranty service on the products it sells. If the company has some experience with the warranted product, it can estimate the timing and extent of the warranty service that will have to be provided. Once the pattern of warranty expenditures — future cash flows — has been estimated, the time-adjusted valuation method can be applied to arrive at a present-day equivalent value.

Similarly, a company's obligation to make payments to employees who have retired (i.e., pensions) or to others on vacation or sick leave can be valued on a time-adjusted basis. Typically, the timing and amounts of these flows are fairly predictable if large numbers of employees are involved.

Market Value

Another indicator of value of both assets and obligations is market value, or the price at which similar assets and obligations currently trade in the marketplace, that is, if you can determine the price at which the asset can be purchased or sold, that price is a basis for valuing the asset.

For example, a used automobile has a ready second market. Because used automobiles are regularly traded in the secondhand market in this country, information is collected, compiled, and published as to typical, or market, prices for a variety of makes and models.

Of course, actively traded common stocks (referred to as being publicly held) have market prices. Loans are also bought and sold among financial institutions and thus have a market price. On the other hand, common shares in a company predominantly owned by a single family, or loans of a specialized nature, or loans between individuals, are not actively traded in any organized market and thus do not have an easily determinable market price. Nevertheless, even these securities do have some market price, that is, a price at which they could be bought or sold.

What about valuing obligations (liabilities)? Can the market value method also value these obligations? For example, is there a market for warranty obligations? A manufacturer of devices subject to warranty could contract with another organization, perhaps one with an extensive network of service centers, to undertake the warranty work. Such a contract fixes a market price for this obligation: the price the manufacturer will pay to be relieved of the obligation to repair during the warranty period. In fact, contracts of this nature are entered into regularly in certain industries, such as the telecommunications industry.

How about employee pension obligations? Contracting with outside firms, generally insurance companies, to fulfill employee pension obligations is widespread — another example of paying a market price to be relieved of a future obligation. Many organizations also contract with insurance companies to pay wages to employees in the event of extended sickness.

Therefore, valuing in terms of current market price for both assets and liabilities is feasible; in many instances undue effort or research is not required, and the resulting values are quite defensible. This evaluation method has some strong appeal: it does not require predictions about the future, as the time-adjusted value method does; it is rooted in the reality of today's marketplace; and it is understandable, explainable, and not complex.

However, the method does have its shortcomings. Consider, for example, a specialized machine tool for which the market is very limited. It might command quite a low price in a sale. Does this imply that the specialized tool has little value to the owning company, which, in fact, has no intention of selling it? Indeed, just how relevant are current market prices to the task of valuing an asset if its owner does not intend to sell it? The value of the specialized machine tool resides in its use, not in its resale. The fact that there may exist no other potential users of the specialized tool does not denigrate or destroy its usefulness to the company that now owns it.

Thus, the market value method of valuation, while appealing, has limitations, as did the time- adjusted value method.

Cost Value

A more complete name for this method is the historical cost method, since it calls for assets to be valued at the prices paid at the time they were acquired.

A decided advantage of the cost value method is that typically you can determine with both ease and accuracy what was actually paid for the asset. The acquisition prices of such assets as securities, an automobile, household furniture, and machine tools (whether standard or custom) are known.

Note, however, that these historical cost values bear no necessary relationship to the values arrived at by the time-adjusted value or market value methods. If the automobile or household furniture was purchased long ago, its cost value may be considerably higher than its current market value. On the other hand, if the automobile is now considered a classic or if the owner of a particular asset made an unusually astute purchase, current market prices may be considerably above the acquisition or historical cost value. Consider valuing a parcel of land, a building site. If the land was acquired some time ago, its original cost is probably much below today's current market values, unless, of course, the development of the community or roadway system has left the parcel in less-desirable surroundings, in which case the reverse may be true.

If the decline in an asset's value is caused simply by the passage of time, for example, wear, tear, and obsolescence on buildings or machine tools or vehicles, we could value the asset at a declining percentage of its original cost as the asset ages. You probably already know that this procedure is in fact widely followed: cost values in the accounting records are reduced each year by an amount called *depreciation*.

Even if we undertake a systematic write-down of an asset's value over its life, this procedure takes care of only one set of causes of value decline. A number of other factors create wide differences between market value and historical cost value. The most important is inflation.[12] Inflation can cause an old asset recorded at historical cost to appear in accounting records to have substantially less value to the company than its replacement cost. This condition frequently affects public utilities (electric, gas distribution, telephone). The facilities built by these companies many years ago may continue to be very productive, but inflation in land and construction costs causes them to be valued very much below the cost value of newly constructed, equivalent facilities. This dilemma has been widely discussed by the accounting profession, but with essentially no resolution.

Can the cost method be used to value liabilities? By their nature, obligations are settled in the future and have no historical cost; history has not yet caught up with the obligations. However, it is not unduly difficult to measure the future cost of meeting these obligations.

Choosing among the Valuation Methods

What are the advantages and disadvantages of the three valuation methods just discussed? Which one does the best job of valuing the assets and liabilities of the company? By now you realize that different valuation methods fit different situations. The time-adjusted value method well suits the valuation of common stocks, notes, and other investment securities. The market value method seems workable in valuing automobiles, standard machine tools, a pantry full of canned goods, and perhaps even a cellar of fine wines. The cost value method seems widely applicable, but, the longer the time between acquisition date and evaluation date and the higher the rate of inflation, the less comfort we can have in this method of valuation.

We need some criteria to compare these three valuation methods. Here are six useful ones.

Currently Relevant

The accounting information resulting from the valuation of assets and liabilities is used to make decisions, such as operating decisions by management, investment decisions by shareholders, and credit decisions by

---

[12]Or, more generally, changes in purchasing power, both inflation and deflation. The history of virtually all organized economies is one of persistent inflation at various rates, punctuated by relatively few periods of deflation.

lenders. These decisions require currently relevant data, data that reflect today's situation and expectations. Almost by definition, the historical cost method suffers in fulfilling this criterion, and the more ancient the history, the more it is likely to suffer.

Feasible

One has to be able to develop the data required by the valuation method. If the future flow of an asset's benefits simply cannot be quantified, the time-adjusted value method is infeasible for that asset. Market values are readily ascertainable for some assets, but not for others. Some liabilities (obligations) can be transferred for a (market) price; others cannot. The cost value method is typically the most feasible — historical cost data are readily available.

Effective

While it may be feasible to develop certain data for time-adjusted valuation or market valuation, it is not always practical to do so. The expense incurred to engage expertise, computing power, or the service of outside appraisers may be prohibitive.

One cannot afford to spend more to develop accounting information than the information is worth. As a result, the time-adjusted and market value methods are ineffective for valuing very many assets and liabilities that do not have a predictable flow of future benefits and readily available market prices.

Timely

The usefulness of accounting information declines rapidly with time. Management needs information to make today's operating decisions, to correct problems, and to seize opportunities. Investors need information to decide to buy, sell, or hold securities, and they want it as soon as possible. A month-end accounting report that takes 6 months to prepare is much less useful than one that is available in 1 week following month end. To get rapid valuation, therefore, accountants are willing to sacrifice some accuracy.

Free from Bias

By now, one undoubtedly sees that an accountant has a good deal of latitude in deciding what should be recorded and at what value. The ideal valuation method is objective and little affected by an accountant's conscious or unconscious bias.

Verifiable

Financial records and statements of major enterprises are audited by independent, professional accountants. For these auditors to fulfill their role of confirming that the financial reports fairly represent the company's position, the valuations of assets and liabilities by the company's own accountants must be verifiable by the outside accountants, that is, tangible evidence that is subject to independent verification must be used. Again, if the method used is objective, it will probably satisfy the final two criteria — it will be free from bias and verifiable.

The Challenge of Alternative Valuation Methods

Clearly, none of the three valuation methods — time-adjusted value, market value, and cost value — satisfies all six criteria. The first criterion, that data be currently relevant, seems best satisfied by the time-adjusted value and market value methods and least satisfied by the cost value method. However, the remaining criteria are well satisfied by the cost value method.

The challenge that faces the accountant is to develop accounting information and reports that combine the relevancy and usefulness inherent in time-adjusted values with the efficiency and reliability inherent in cost values.

We must deal with reality, and reality is that current accounting practices and rules are built to a very large degree on the cost value method. In the vast majority of cases today, accountants use historical cost data as the best evidence of what the company owns — the value of its assets — and what the company owes — the value of its liabilities. While expedient and pragmatic, the method is not wholly adequate. It is precise but, in the view of many, not sufficiently accurate.

Today accepted valuation rules and techniques adhere closely to the cost value methodology; but, increasingly, both accountants and financial audiences are pressuring for a correction of some shortcomings of the cost value method. The time-adjusted value and market value methods are now accepted in more and more circumstances.

6.3 Ratio Analysis[13]

Henry E. Riggs

The fundamental method of analyzing financial statements is by ratio analysis. Analysts of financial statements learn to think in ratios. Current assets are instinctively compared to current liabilities. Cost of goods sold is automatically compared to total sales to measure gross profitability and to total inventory to judge the rate of inventory usage.

Calculation of ratios is straightforward. Their interpretation, however, requires judgment, and judgment is sharpened by experience. An analyst who has studied a variety of financial statements — statements of companies in different industries and in both prosperous and recession periods — is able to glean more reliable conclusions from a particular set of statements than is an inexperienced analyst.

Categorization of Ratios

There are about as many financial ratios as there are analysts calculating them. Here we deal with only the most commonly employed ratios. The four primary categories of information that they provide are

1. Liquidity: how able is the company to meet its near-term obligations?
2. Working capital utilization: how efficiently is the company using the various components of its current assets and current liabilities?
3. Capital structure: what are the company's sources of capital?
4. Profitability: how profitable is the company in light of both its sales and its invested capital?

The ratios discussed here are particularly applicable to manufacturing enterprises. Companies engaged in different industries may find some of these ratios irrelevant; other ratios, not discussed here, may be highly relevant to them. For example, a commercial bank is concerned with the ratio between loans outstanding and customer deposits — the loan-to-deposit ratio — since customer deposits provide the funds that the bank lends to its borrowers.

Liquidity

A company unable to meet its obligations as they come due runs the risk of bankruptcy. Trade suppliers and employees must be paid on time; interest and principal payments on borrowed money must be made when due. A company that has substantial liquid assets in relationship to its near-term obligations has strong **liquidity**.

Current Ratio
The most widely quoted financial ratio is the **current ratio**:

$$\text{Current ratio} = \text{current assets} \div \text{current liabilities}$$

Current assets are those assets that are either presently the equivalent of cash or within the next 12 months will be turned into cash. Cash, accounts receivable, and inventory are the primary current assets, listed in order of decreasing liquidity. Current liabilities are obligations that must be met within the following

---

[13]Adapted from *Financial and Cost Analysis for Engineering and Technology Management*, Riggs, H. E., Wiley Interscience, New York, 1994, chap. 10.

12 months, including primarily accounts payable, wages and salaries payable, short-term bank borrowing, current portion of long-term debt, and miscellaneous accruals.

The higher the current ratio, the greater the margin of safety, that is, the more likely the company is to have sufficient liquid assets to meet its obligations as they come due.

Acid-Test (or Quick) Ratio

Another liquidity measure is the **quick ratio**, often referred to as the **acid-test ratio**:

$$\text{Quick ratio} = (\text{cash} + \text{marketable securities} + \text{accounts receivable}) \div \text{current liabilities}$$

Working Capital Utilization

Current assets, particularly accounts receivable and inventory, are major investments for many companies. The more efficiently the company uses its current assets, that is, the faster it collects from customers and the less inventory it requires to accomplish its sales, the less capital the company will require.

Working capital is defined as the difference between current assets and current liabilities. The following ratios indicate how efficiently the company is using the primary elements of working capital.

Accounts Receivable Collection Period

Most manufacturing companies extend customer credit. The longer customers take to pay their bills, the more the manufacturer must invest in accounts receivable. A comparison of sales volume with outstanding accounts receivable indicates how promptly customers are paying, the **accounts receivable collection period** ratio:

$$\text{Collection period (in days)} = \frac{\text{accounts receivable}}{\text{average sales per day}}$$

where

$$\text{Average sales per day} = \text{annual sales} \div 365$$

The collection period is the number of days of sales remaining uncollected (and therefore in accounts receivable) at the end of the accounting period. Stated another way, the ratio equals the average number of days between the customer invoice date and the date that payment is received. Actual time between invoice date and collection date varies widely by customer; some pay very promptly and others take a distressingly long time to pay. The mean of a frequency distribution of days from invoice date to payment date approximates the collection period expressed in days.

Inventory Turnover

Most manufacturing and merchandising companies invest substantial amounts in inventory in order to serve customers and assure the uninterrupted flow of manufacturing processes. Carrying inventory is expensive when one includes the costs of storage, insurance, risk of obsolescence, and the tied-up capital. Thus, in deciding on inventory levels, every company trades off (explicitly or implicitly) the costs with the benefits of carrying inventory.

When inventory is sold, inventory values are reduced and a charge is made to "cost of goods sold". A comparison of inventory and cost of goods sold, then, indicates the rate at which inventory is used, that is, the speed with which inventory is moving from receipt to final sale.

This **inventory turnover** ratio parallels the accounts receivable collection period: the collection period ratio compares sales and accounts receivable, both valued at sales prices; while the inventory turnover ratio compares cost of goods sold and inventory, both valued at cost values.

The inventory turnover ratio is typically expressed in times per year, but an alternate form of the ratio is the **inventory flow period** ratio, expressed in number of days. The ratio in times per year is

$$\text{Inventory turnover} = \frac{\text{cost of goods sold}}{\text{inventory}}$$

and in number of days, the ratio is

$$\text{Inventory flow period} = \frac{\text{inventory}}{\text{cost of goods sold} \div 365} \quad \text{or} \quad \frac{365}{\text{inventory turnover}}$$

Accounts Payable Payment Period

A major determinant of working capital is the amount of trade or vendor credit utilized. Just as a comparison of sales and accounts receivable reveals the collection period, a comparison of credit purchases and accounts payable yields the **accounts payable payment period** — the average time a company takes to pay its suppliers. This payment period compared to the normal terms of purchase assesses how well the company is meeting its obligations to suppliers.

The payment period in days is

$$\text{A/P payment period} = \frac{\text{accounts payable}}{\text{credit purchases} \div 365}$$

Capital Structure Ratios

Capital structure ratios help to evaluate the liabilities + owners' equity side of the balance sheet, that is, how the company is financed. These ratios assess the financial riskiness of the business as well as the potential for improved returns through the judicious use of debt.

When a company borrows money, it undertakes a firm, ironclad obligation to pay interest and principal repayments on schedule. Failure to make these various payments when due (i.e., default on the provisions of the loan agreement) subjects the company to the risk of bankruptcy. For any particular company, the higher the debt, the greater the risk. If the company's operating performance is erratic or encounters difficulties, the company may not be able to comply with its borrowing agreements. By contrast, when a corporation obtains additional funds by the sale of new capital stock, it undertakes no such firm obligation to its new shareholders; these shareholders will receive dividends only if and when declared by the corporation's board of directors. Failure to pay dividends does not constitute default and does not subject the company to the risk of failure, although it may subject the management to other pressures. Thus, from the point of view of the corporation, common stock financing is much less risky than borrowing.

However, the judicious use of borrowing benefits shareholders. If a company can borrow funds at an interest rate equal to $x\%$ and invest those funds to earn consistently at a rate greater than $x\%$, then this incremental return results in a higher return on shareholders' invested capital. This phenomenon is known as **debt leverage**.

Bear in mind that the use of borrowed funds is inherently neither good nor bad. The greater the use of borrowed funds, that is, the greater the use of debt leverage, the greater the financial risk to which the company is exposed but also the greater the potential return to shareholders.

Total Debt to Owners' Equity Ratio

The debt of most corporations is composed of both current and long-term liabilities; owners' equity consists of both capital invested by the owners and earnings retained in the business. The ratio between the two sums indicates the relative contribution of creditors and owners to the company's financing:

$$\text{Total debt to owner's equity} = \frac{\text{current liabilities} + \text{long-term liabilities}}{\text{total shareholders' equity}}$$

Long-Term Debt to Total Capitalization Ratio

This ratio requires careful definition of both the numerator and denominator. Long-term debt is that portion of total borrowings having a maturity longer than 1 year (and therefore not included in current liabilities).

Total capitalization is defined as the total permanent capital. Current liabilities are not a permanent source of capital since they arise spontaneously from operations; amounts owing to trade creditors and employees and miscellaneous accruals would not be present if the company were not actively engaged in trade. Total capitalization, then, is defined as long-term debt plus owners' equity. This ratio indicates the percentage that permanent, or long-term, borrowed funds are of total permanently invested capital:

$$\text{Long-term debt to total capitalization} = \frac{\text{long-term debt}}{\text{long-term debt} + \text{owners' equity}}$$

Profitability

The ratios dealing with liquidity, working capital utilization, and capital structure focus primarily on the company's financial position. By contrast, the profitability ratios measure the company's performance, the rate at which it is earning financial returns.

Two types of profitability ratios are useful. The first measures profit in relation to sales levels and is obtained by comparing data solely within the income statement; the second measures profit in relation to investment and involves comparisons of income statement and balance sheet data.

Percentage Relationships on Income Statement

The ratio of net income to total sales is a useful indicator of the company's profitability. In addition, the percentage that each line item on the income statement is of total revenue, or sales, also provides useful insights. For example, the gross margin percentage shows the relationship between sales revenue and product cost. Sales expense as a percentage of sales revenue shows what percent of the sales dollar the company spends on selling and marketing activities.

Return on Equity

Profitability can also be related to investment. The most fundamental ratio is net income to total owners' equity, the so-called **return on equity** (ROE). To compensate for growth (or decline) in equity during the year, the ratio compares earnings for the year to average equity or

$$\text{Return on equity} = \frac{\text{net income}}{\left(\text{opening equity} + \text{ending equity}\right) \div 2}$$

The ROE ratio compares net income after payment of all expenses including interest and taxes with the total book value of the shareholders' investment including both invested capital and earnings retained by the business. Since all shareholders invest for a return, ROE is one ratio that can be compared across different industries.

ROE does not, however, indicate the degree of risk inherent in the return. Shareholders are typically willing to accept a lower ROE in exchange for a lower risk. Note also that this ratio compares net income to book shareholders' equity. An investor typically has to pay more or less than equivalent book value for a share of stock, as market values of securities bear no necessary relationship to book values. Thus, the ROE generally does not indicate for a particular investor the rate of return on an actual securities investment.

Return on Assets

The ROE for a company is, of course, influenced by its capital structure. A company employing high debt leverage is typically subject to wider swings in ROE than another company obtaining its permanent capital primarily through shareholders' equity. To factor out the influence of capital structure when appraising the company's earnings on investment, total assets can be used as the measure of investment — the total amount owned by the company, whether financed through debt or owners' equity.

The **return on assets** (ROA) ratio compares earnings to total assets. However, net income is inappropriate as the numerator since net income is profit after payment of interest, and the amount of interest is a function of the capital structure. Thus, we use earnings before interest and taxes (EBIT) in calculating ROA:

$$\text{Return on assets} = \frac{\text{earnings before interest and taxes}}{\left(\text{opening assets} + \text{ending assets}\right) \div 2}$$

Bear in mind that the ROA percentage and the ROE percentage are not comparable, since ROA is a before-tax percentage and ROE is an after-tax percentage.

Interpreting Ratios

By now, one undoubtedly wonders, "What should be the value of these various ratios?" "What represents an appropriate current ratio, or collection period, or ROE?"

Unfortunately, these questions cannot be answered. Adequate liquidity or appropriate debt leverage is a function of industry characteristics, the company's stage of development, philosophy of the management and owners, and many other factors. While conventional wisdoms do exist, such as a manufacturing company should have a current ratio of 2.0, these wisdoms are dangerous. Some companies enjoy very adequate liquidity with a current ratio less than 2.0, for example, service companies with low inventories or companies that sell for cash and thus have no accounts receivable. Other companies need a current ratio well in excess of 2.0 to assure adequate liquidity, for example, companies whose manufacturing processes require large inventories.

In short, judging the appropriateness of a particular ratio is not easy. An analyst's judgment is aided by two techniques, however: reviewing trends over time and comparing the ratios with those of similar companies in the same industry.

Suppose that companies A and B now both have current ratios of 2.0; they apparently have equivalent liquidity. Now, suppose that the current ratios for these two companies over the past 3 years have been

	2 Years Ago	1 Year Ago	This Year
Company A	1.4	1.8	2.0
Company B	2.6	2.2	2.0

Company A seems to be building somewhat more comfortable liquidity (its current ratio has strengthened over recent years), while company B's liquidity has deteriorated. Trend information shapes our conclusion.

A company's ratios must be interpreted in light of its particular industry. A comparison of a high-technology manufacturing company to an electric power utility is irrelevant. Utilities typically have high fixed assets and little working capital; their fixed asset and total asset turnovers are low, but they employ substantial debt leverage to earn competitive ROE. Manufacturing companies have substantially greater investment in current assets than in fixed assets and many avoid debt financing.

6.4   Cost Allocation

David M. Cottrell and Daniel L. Jensen

Cost is fundamentally a function that associates monetary flows with objects. Usually we understand cost to mean **acquisition cost** (the monetary outlay to acquire an object) — and that will be the meaning used here — but other meanings are possible, including current cost (the money outlay required to replace an object acquired earlier) or opportunity cost (the profit sacrificed by using an object in one way rather than another). Once acquired, objects or assets, as accountants call them, are used, transferred, combined, and otherwise altered in ways that result in new objects or assets for which cost must be determined. When two or more acquired inputs are merely combined to form a new object, the cost of the new object is simply the sum of the input costs, and no allocation is required. However, when an input, such as a machine, is used to produce several products, the cost of the machine must be allocated to determine the full cost of the products. Thus, **cost allocation** may be defined as the partitioning of one object's cost among two or more related objects. As the relationship among the inputs and interme-diate products (the production function) becomes more complex, allocation of cost to the outputs becomes more difficult to rationalize. These complexities of allocation must be resolved in virtually every decision or policy determination dependent on cost data.

Allocations in Income Determination

Perhaps the most commonly used cost allocations are those incorporated in business net income. **Depre-ciation**, which represents a significant part of expense for most businesses, is the cost of long-lived assets that has been allocated to the period in which use of the asset is judged to have provided a benefit to business. For example, if a building is used by the central administration of a large business, its cost is allocated to annual depreciation charges, which the company recognizes as expenses year by year over the life of the building. Accounting literature emphasizes the cost allocation nature of depreciation by defining depreciation as "a system of accounting which aims to distribute the cost or other basic value of intangible capital assets, less salvage (if any), over the estimated useful life of the unit (which may be a group of assets) in a systematic and rational manner. It is a process of allocation, not of valuation" [AICPA, 1953].

Depreciation is frequently used to allocate costs related to individual assets and groups of assets. As described by the International Accounting Standards Committee, these allocations can also be applied to component parts of a single asset. "In certain circumstances, the accounting for property, plant and equipment may be improved if the total expenditure is allocated to its component parts, provided they are in practice separable, and estimates are made of the useful lives of these components. For example, rather than treat an aircraft and its engines as one unit, it may be better to treat the engines as a separate unit if it is likely that their useful life is shorter than that of the aircraft as a whole [IASC, 1982].

Several depreciation methods are widely used to compute the amount of cost allocated to any one operating period. These methods include straight-line depreciation, which assigns an equal amount of an asset's cost to each period of its expected life, and various "accelerated" depreciation methods, which assign higher amounts to early periods of an asset's life and lower amounts to the later periods of an asset's life. A third method, called the units-of-production method, allocates the cost of an asset over its life in proportion to the units of output or production generated by the asset in each period. The table below shows that the straight-line method of depreciation is the most popular for public companies.

Depreciation Methods

	Number of Companies			
	1995	1994	1993	1992
Straight line	572	573	570	564
Accelerated methods	88	85	91	100
Units of production	49	60	55	52

Source: AICPA, *Accounting Trends and Techniques*, 50th ed., G. L. Yarnall and R. Rikert, eds., Jersey City, NJ, 1996, p. 357. These data were obtained by a survey of 600 annual reports to stockholders undertaken for the purpose of analyzing the accounting information disclosed in such reports. The annual reports surveyed were those of selected industrial, merchandising, and service companies. Several companies report using multiple depreciation methods.

In production or service settings, allocations frequently involve multiple steps or levels. For example, costs may first be allocated to time periods over the life of an asset, then within these time periods costs may be further allocated to individual products or services. This situation would occur where a factory building is used to produce several products. An annual depreciation charge would be computed for the factory and then allocated within the year to the products actually produced. Thus, the cost of manu-facturing facilities is allocated twice, once to the time period in which it is used and then to the products manufactured during that time period. The business would account for these allocated costs as part of the expense item called "cost of goods sold", which requires a further allocation of the cost of goods manufactured in this or prior periods to the goods sold during this period. This latter allocation is usually accomplished by using one of several standardized inventory cost methods, for example, the first-in first-out (FIFO) method, the average cost method, or the last-in first-out (LIFO) method.

Income determination also requires other types of cost allocation, including, for example, allocations of interest expense accrued in one period but paid in the next; the **accrual accounting system** controls many of these short-term allocations in the interests of measuring expenses when they occur rather than when they are paid.

Given the complexity and interrelatedness of most production systems, allocations of input cost to outputs tend to be rather arbitrary and difficult to rationalize in terms of the relationships among the inputs, intermediate cost objects, and outputs. This fact has led some scholars to conclude that allocations are at best unnecessary and at worst misleading [Thomas, 1974]. One well-known consequence of allocating fixed costs to inventories is the potential for temporarily inflating income by increasing the rate of production and thereby increasing the amount of fixed cost held back from income and assigned to the inventory. An income determination system called "variable (or direct) costing" would avoid this problem by simply taking fixed cost as expense in the period in which the related inputs are used and assign only variable costs to inventory. However, variable costing never "caught on" — probably owing to difficulties associated with variable cost measurement in complex production settings — and "full (or absorption) costing" remains the generally accepted approach to income determination. Other accounting scholars, e.g., Chambers [1966] and Sterling [1970], have argued for market-value approaches to income determination that would minimize the role of allocation; the problem is that reliable market values are not readily available for all assets.

Allocations and Economic Behavior

Others scholars, e.g., Jensen [1977] and Zimmerman [1979], note the pervasiveness of cost allocations in practice and argue that cost allocations perform functions in organizational and decision context that cannot be fully understood in terms of the underlying relationships of production. Instead, cost allocations may be used to influence behavior. In an ideal world, allocations could be used to proxy for difficult-to-observe opportunity costs and ensure optimal economic behavior in an organization.

For example, when several departments use an internal computer and telecommunication system, allocations of the system's cost among the departments can be used to make the departments accountable for the system's cost and to give the department an incentive to use an efficient central system rather than to use outside service providers. Of course, if not calculated with good information about outside service opportunities, such cost-sharing allocations can discourage use of the internal system and result in excessive expenditures by the organization.

Allocations in Practice

Traditional accounting systems usually allocate costs in fairly standardized ways that reflect the general structure of the allocation problem. This "cookbook" approach to cost allocation has received increasing criticism in recent years, resulting in new approaches to cost allocation incorporated in activity-based costing systems and strategic cost management approaches.

Service Department Allocations

The service department allocation problem is a cost-sharing problem that arises in most organizations. Central departments engaged in human resource functions, computer and telecommunication support, maintenance, and general administration — to name just a few examples — all give rise to costs that are frequently allocated to the departments directly engaged in producing the products and services of the organization. The problem in complicated by the fact that the central service departments may perform services for one another. This problem is easily solved by formulating the allocation as a system of linear equations and solving for the allocation parameters following a procedure known to cost accountants as the "reciprocal method". Simpler methods, called the "step-down" and "direct" methods, are also found in practice, but these methods do not take full account of the interactions among service departments.

Service department costs may be either fixed or variable with the level of service activity. A refinement of service department allocations uses a two-part allocation to recognize the long-term character of the responsibility for fixed costs and the short-term character of the responsibility for variable costs by allocating the two cost types on different bases. Variable costs are allocated on the basis of actual service usage, making the variable cost allocation responsive to variation of actual from budgeted usage. Fixed costs are allocated on the basis of budgeted usage (sometimes long-term budgeted usage), making the fixed cost allocation independent of short-term variation in actual usage. If fixed costs were allocated on the basis of actual usage, then variation in service usage by one department would affect the allocation to other departments; the two-part allocation avoids the dysfunctional consequences of such fixed cost allocations.

Joint Product Cost

When products are processed together, it may be difficult or impossible to measure the joint production activity as a basis for allocating its cost to the products, e.g., Jensen [1974]. Further, if the proportions in which the products emerge from the joint process are fixed or difficult to vary, it may be impossible to attribute even the variable costs of the joint process on any nonarbitrary basis. In the face of this difficulty in justifying such joint production cost allocations, accounting systems employ a number of standard joint cost allocation methods. The "net realizable value method", which allocates the joint cost in proportion to the total sales value net of separate processing costs, is one such method; it and other methods are described in cost accounting texts.

Activity-Based Costing

Activity-based costing (ABC) is a comprehensive approach to cost allocations that requires a detailed analysis of the activities that drive the costs of production or service. ABC systems endeavor to provide a flexible approach to cost allocation that can respond to contemplated changes in service and product lines and that reflects the interrelatedness of the underlying activities. ABC is described in Section 6.5.

Cost Justification and Reimbursement

Cost allocation is frequently necessary in order to justify claims for reimbursement under contracts with agencies of public and private organizations. All contracts with the U.S. government must comply with the cost accounting standards issued by the Cost Accounting Standards Board (CASB), an independent board within the Office of Federal Procurement Policy. These standards, which are incorporated in the Federal Acquisition Regulations (FARs), define terms and prescribe procedures related to the allocation of costs. See Alston [1993] for additional discussion.

Defining Terms

Accrual accounting: A system of accounting that measures expenses and revenues when they occur rather than when the related cash is paid or received.

Acquisition cost: The monetary outlay to acquire an object.

Cost: A function that associates monetary flows with objects.

Cost allocation: The partitioning of one object's cost among two or more related objects.

Depreciation: The cost of long-lived assets that is first allocated to time periods over the life the asset and then allocated to the period in which use of the asset produces revenue.

References

American Institute of Certified Public Accountants. Accounting Research Bulletin No. 43, Chapter 9C, Paragraph 5, 1953.

Chambers, R. J. *Accounting Evaluation and Economic Behavior*, Prentice-Hall, Englewood Cliffs, NJ, 1966.

International Accounting Standards Committee. International Accounting Standard No. 16, Paragraph 10, 1982.

Jensen, D. L. The role of cost in pricing joint products: a case of production in fixed proportions, *Acocunt. Rev.*, 49(3): 465–476, 1974.

Jensen, D. L. A class of mutually satisfactory allocations, *Account. Rev.*, 52(4): 842–856, 1977.

Thomas, A. L. *The Allocation Problem: Part Two*, Study in Accounting Research No. 9, American Accounting Association, Sarasota, FL, 1974.

Sterling, R. R. *Theory of the Measurement of Enterprise Income*, Lawrence, KA, 1970.

Zimmerman, J. L. The costs and benefits of cost allocations, *Account. Rev.*, 54(3): 504–521, 1979.

Further Information

Alston, F., Worthington, M., and Goldsman, L. *Contracting with the Federal Government*, John Wiley & Sons, New York, 1993.

Biddle, G. C. and Steinberg, R. Allocations of joint and common cost, *J. Account. Lit.*, 3: 1–42, 1984.

Horngren, C. T., Foster, G., and Datar, S. *Cost Accounting: A Managerial Emphasis, 8th ed.*, Prentice-Hall, Englewood Cliffs, NJ, 1994.

Young, H. P. *Equity in Theory and Practice*, Princeton, NJ, 1995.

Zimmerman, J. L. *Accounting for Decision Making and Control*, Irwin, Chicago, 1997.

Internet References (current as of 5/1/1998)

Cataloging and catalog maintenance: functional cost allocation system, http://www.lib.uwaterloo.ca/~wroldfie/function.html

How to treat the costs of shared voice and video networks in a postregulatory age, http://www.cato.org/pubs/pas/pa-264.html

Minimum cost spanning extension problems: the proportional rule and the decentralized rule, http://greywww.kub.nl:2080/greyfiles/center/1994/96.html

Order-based cost allocation rules, http://greywww.kub.nl:2080/greyfiles/center/1996/56.html

6.5 Activity-Based Costing and Management

Michael W. Maher

Starting with innovations at Hewlett-Packard and John Deere in the mid-1980s, activity-based costing and management has emerged as one of the leading tools for managing a company's resources. This section gives a brief overview of activity-based costing and discusses how it can be used in managing resources.

Activity-based costing is a costing method that assigns costs first to activities and then to the products based on each product's use of activities. An **activity** is any discrete task that an organization undertakes to produce a product. For example, quality inspection is an activity used to produce a product. Using activity-based costing, accountants or other staff ascertain the cost of the activity "quality inspection". Then they assign that cost to products that use quality inspection. In principle, activity-based costing is a straightforward and logical method of assigning costs to activities and products. In practice, it is a difficult and time-consuming process, as discussed in the following sections.

Fundamentals

Activity-based costing is based on the concept that products consume activities and activities consume resources. If managers want their products to be competitive, they must know (1) the activities that go into making the good or providing the service and (2) the cost of those activities. To reduce a product's costs, managers will likely have to change the activities consumed by the product.

Activity-based costing involves the following four steps:

1. Identify the activities — such as purchasing materials — that consume resources.
2. Identify the cost driver(s) associated with each activity. A **cost driver** causes, or "drives", an activity's costs. An example of a cost driver for purchasing materials is "number of orders placed". Each activity might have only one cost driver or it might have multiple cost drivers.
3. Compute a cost rate per cost driver unit or transaction. If the cost driver is "number of purchase orders", then the cost driver rate would be the cost per purchase order.
4. Assign costs to products by multiplying the cost driver rate times the volume of cost driver transactions consumed by the product. For example, to compute the cost of purchasing materials for product X for the month of January, multiply the cost driver rate by the number of purchase orders placed during January.

Often, the most interesting and challenging part of the exercise is identifying the activities that use resources. A Deere & Company plant identified six major activities required to produce its products, for example, and used one cost driver for each activity. Then it developed two cost rates for each cost driver, one for variable costs and one for fixed costs. A Hewlett-Packard plant, on the other hand, developed more than 100 cost drivers.

Strategic Benefits

Many experts view activity-based costing as offering companies strategic opportunities. Porter [1985], among others, has pointed out that certain companies have learned to use the information gained from their cost systems to cut prices substantially to increase market share. One key way that companies develop competitive advantages is to become a low-cost producer or seller. Companies such as Wal-Mart in retailing, United Parcel Service in delivery services, and Southwest Airlines in the airline industry have created a competitive advantage by reducing costs.

Activity-based costing and activity-based management play an important role in companies' attempts to develop a competitive cost advantage. Whereas activity-based costing focuses on determining the cost of activities, activity-based management focuses on managing costs by managing activities.

Activity-based management can be used to identify and eliminate activities that add costs but not value to the product. Nonvalue-added costs are costs of activities that could be eliminated without reducing product quality, performance, or value to the customer. For example, storing product components until needed for production does not add to the finished product's value to the customer.

The following types of activities are candidates for elimination because they do not add value to the product:

- *Storage.* Storage of materials, WIP, and finished goods inventories are obvious **nonvalue-added activities**.
- *Moving items.* Moving parts, materials, and other items around the factory floor does not add value to the finished product.
- *Waiting for work.* Idle time does not add value to products.

Managers should investigate the entire production process, from purchasing, to production, to inspection, to shipping in the search for activities that do not add value to the customers. Managers should ascertain whether the company needs as many setups, whether the cost of higher-quality materials and labor could be justified by a reduction in inspection time, whether the cost of ordering could be reduced, and so forth.

Complexity as a Resource-Consuming Activity

One lesson of activity-based costing is that costs are a function of both volume and complexity. It might be obvious that a higher volume of production consumes resources, but why does complexity consume resources?

To understand the answer to that question, imagine that one company produces 10,000 units per month of one single-size cardboard box designed for shipping one particular type of software. Another

TABLE 6.1 Examples of Cost Drivers and Related Costs

Cost	Cost Driver
Cost of power to run machines	Machine time
Wages of assembly line workers	Labor time
Materials handling costs	Pounds of materials handled, number of times materials are handled
Costs of having documents printed	Pages printed
Cost of machine setups	Number of setups, setup time
Cost of purchasing	Number of purchase orders
Cost of quality inspection	Quality inspection time
Cost of servicing customers	Number of customers serviced, number of different customers served

company also produces 10,000 cardboard boxes per month for shipping software, but makes 100 different types of boxes with a variety of styles. Although both companies produce the same total volume of boxes per month, it is easy to imagine how the overhead costs in the single product company will be less than those in the company that makes 100 different types of boxes.

After installing activity-based costing, managers have often found that the low-volume products should be allocated more **overhead**. Low-volume products may be more specialized, requiring, for example, more drawings, specifications, and inspections. Further, low-volume products often require more machine setups for a given level of production output because they are produced in smaller batches. In addition, a low-volume product adds complexity to the operation by disrupting the production flow of the high-volume items. (One appreciates this fact every time one stands in line behind someone having a special and complex transaction.)

Choosing Cost Drivers

Table 6.1 presents several examples of cost drivers used by companies. How do managers choose a cost driver? The best cost drivers are those that cause the costs. If a machine consumes energy, then the time of the machine's operation is a good cost driver for energy costs. Any product that uses the machine would then be charged for energy costs based on the time the machine was used to make the product.

Applications and Illustrations

The following example illustrates how unit costs are computed when companies use activity-based costing. Assume that ABC Company makes two products, Alpha and Beta. The Alpha product line is a high-volume product line, whereas the Beta line is a low-volume, specialized product. In this example, ABC's staff uses activity-based costing to assign overhead costs to each of the two products: Alpha and Beta.

Step 1: Identify activities that consume resources. ABC's staff identified the following four activities that were the cause of overhead costs: (1) purchasing materials, (2) setting up machines when a new product was started, (3) inspecting products, and (4) operating machines.

Step 2: Identify the cost drivers associated with each activity. ABC's staff identified the following four cost drivers for the above activities: (1) number of pounds of material, (2) number of machine setups, (3) hours of inspection, and (4) number of machine hours.

Step 3: Compute a cost rate per cost driver unit or transaction. To compute the cost driver rate, ABC's staff estimated the amount of overhead and the volume of transactions for each cost driver. For example, the staff estimated that the company would purchase 100,000 pounds of materials requiring overhead costs of $200,000 for the year. Examples of these overhead costs are the salaries of people who purchase, inspect, and store materials. Consequently, each pound of materials used to make a unit of product was assigned an overhead cost of $2 ($200,000/100,000 pounds). Table 6.2 shows the cost driver rates computed for each of the four activities. The estimated overhead cost is divided by the cost driver volume to derive the rate.

Step 4: Assign costs to products by multiplying the cost driver rate times the volume of cost driver transactions consumed by the product. For example, to compute the overhead cost associated

TABLE 6.2 Computations of Cost Driver Rates

Activity	Cost Driver	Estimated Overhead Cost for the Activity	Estimated Cost Driver Volume	Cost Driver Rate
1. Purchasing materials	Number of pounds of materials	$200,000	100,000 pounds	$2 per pound
2. Machine setups	Number of machine setups	800,000	400 setups	$2000 per setup
3. Inspections	Hours of inspection	400,000	4000 hours	$100 per hour
4. Running machines	Number of machine hours	600,000	20,000 hours	$30 per hour

with purchasing materials for the Alpha product for the month of January, multiply the cost driver rate by the number of pounds of material purchased during January to obtain materials for the Alpha product. The following display indicates the volume of cost driver transactions for Alpha and Beta for the month of January.

	Alpha	Beta
1. Purchasing materials	6000 pounds	4000 pounds
2. Machine setups	10 setups	30 setups
3. Inspections	200 hours	200 hours
4. Running machines	1500 machine hours	500 machine hours

Given this information, ABC's staff computed the overhead cost to be assigned to each of the two products for the month of January as shown in Table 6.3.

Unit Costs Compared

Assume that ABC Company produced 1000 units of Alpha and 500 units of Beta in January. Using activity-based costing, ABC's staff would have allocated $97 ($97,000/1000 units) of overhead per unit to the higher-volume Alpha and $206 ($103,000/500 units) to Beta. Using activity-based costing, more overhead is allocated per unit to the more specialized, lower-volume Beta product. Although Beta was a lower-volume product, it consumed as many resources as the higher-volume Alpha product did in January. Further, Beta consumed more than double the resources that Alpha did on a per-unit basis.

Many companies have found their situation to be similar to this example. For example, a Hewlett-Packard division in Boise, Idaho, found that too much overhead had been assigned to many of its high-volume products by conventional costing because conventional costing allocated costs proportional to volume. Activity-based costing revealed that low-volume, specialized products were the cause of higher costs than managers had realized.

Activity-Based Costing of Services

Activity-based costing also can be applied to services, including marketing and administrative activities. The principles and methods are the same as those discussed in manufacturing. Instead of computing the cost of a product, however, accountants compute a cost of performing a service. The following is an example.

SU Software Company has an order-filling service. Customers can call an 800 number and order either the Standard or Unique product. Management is concerned about the cost for this service and is considering outsourcing it to another company. As a result, SU Software sought bids from outside companies to perform the order-filling service, the lowest of which was $30 per unit. (A unit is receiving a telephone call, completing the order, shipping the product, and dealing with customer returns.) Managers of SU Software wanted to know the internal costs of this service so they can decide whether to continue filling orders internally and, if so, to identify ways to improve efficiency. The team appointed to the task proceeded as follows:

- Identified the activities that cause costs. The team identified order taking, order filling, shipping, and customer returns.
- Identified cost drivers, such as number of orders for order taking.
- Computed cost driver rates.

TABLE 6.3   Assigning Overhead Costs to Products Using Activity-Based Costing

Product Alpha

1. Purchasing materials	$2 per pound × 6000 pounds =	$12,000
2. Machine setups	$2000 per setup × 10 setups =	$20,000
3. Inspections	$100 per inspection hour × 200 hours of inspection =	$20,000
4. Running machines	$30 per machine hour × 1500 machine hours =	$45,000
	Total overhead costs assigned to product Alpha	$97,000

Product Beta

1. Purchasing materials	$2 per pound × 4000 pounds =	$ 8,000
2. Machine setups	$2000 per setup × 30 setups =	$ 60,000
3. Inspections	$100 per inspection hour × 200 hours of inspection =	$ 20,000
4. Running machines	$30 per machine hour × 500 machine hours =	$ 15,000
	Total overhead costs assigned to product Beta	$103,000

Based on this analysis, management determined that order-filling costs averaged $21 per unit, much lower than the best outside bid of $30 per unit. Management decided to reject the idea of outsourcing this activity. Further, managers learned that customer returns averaged $8 per order, which seemed large. They found that, by improving their descriptions of the products in advertisements, the company was able to reduce the number of customer returns by nearly 50%.

Summary

Activity-based costing is a costing method that assigns costs first to activities and then to the products based on each product's use of activities. Activity-based costing is based on the premise that products consume activities and activities consume resources. Activity-based costing involves the following four steps:

- Identify the activities that consume resources and assign costs to those activities.
- Identify the cost driver(s) associated with each activity.
- Compute a cost rate per cost driver unit or transaction.
- Assign costs to products by multiplying the cost driver rate times the volume of cost driver units consumed by the product.

Experience with activity-based costing and activity-based management over the past 10 to 15 years indicates that they add value to companies in two ways.

1. *Better information about product costs.* Activity-based costing uses more data than conventional costing and provides more informed estimates of product costs. Better product cost information helps managers make decisions about pricing and whether to keep or drop products.
2. *Better information about the cost of activities and processes.* By identifying the cost of various activities, managers gain useful information that the accounting system previously buried.

So far, the discussion has implied that implementing activity-based costing will add value to the organization. That is likely to be true in varying degrees. Companies that have complex production processes producing many different products and that operate in highly competitive markets probably stand to benefit the most. That's why companies such as Hewlett-Packard, Chrysler, and IBM have implemented activity-based costing. Companies such as Starbucks and Nike would probably benefit also but less than more complex companies.

In considering how much value activity-based costing adds to a company, remember that implementing activity-based costing is costly. These costs include the costs of the accountants and other people who develop and implement activity-based costing, additional recordkeeping costs, software costs, and, possibly, consulting costs. It also shakes up the organization by changing the accounting rules. This can be a good thing, but many companies also have found it to be painful.

Defining Terms

Activity: Any discrete task that an organization undertakes to produce a product.

Activity-based costing: A costing method that assigns costs first to activities and then to the products based on each product's use of activities.

Activity-based management: The management of activities with the idea of reducing costs, particularly seeking to eliminate nonvalue-added activities.

Cost driver: A cause or "driver" of costs.

Nonvalue-added activities: Activities that do not add value to the customer.

Overhead: Costs that are not directly traced to the product. Examples are maintenance on machines that make products, salaries of supervisors, and miscellaneous supplies used in production.

Value-added activities: Activities that add value to the customer.

Reference

Porter, M. *Competitive Advantage*, Free Press, New York, 1985.

Further Information

See J. Shank and V. Govindarajan, *Strategic Cost Analysis*, McGraw/Irwin, Burr Ridge, IL, 1989, for an extensive discussion of the strategic use of cost analysis.

The following two practical articles present excellent discussions of the application of activity-based costing: R. Cooper and R. S., Profit priorities from activity-based costing, *Harv. Bus. Rev.*, May–June: 130–135, 1991; and J. A. Ness and T. G. Cucuzza, Tapping the full potential of ABC, *Harv. Bus. Rev.*, July–August: 130–138, 1995.

See Chapters 7 and 8 of M. Maher, *Cost Accounting: Adding Value to Management*, McGraw/Irwin, Burr Ridge, IL, 1997, for more a more detailed discussion about activity-based costing and management.

6.6 Auditing

John D. Lyon

Auditing is the process of assessing the accuracy, fairness, and general acceptability of accounting records, then rendering a conclusion about this assessment. This section focuses on external auditing, where the assessment process takes place with respect to management's published financial statements, the conclusion is embodied in the audit report, and the primary beneficiaries of this report are stockholders, banks, creditors, and any potential stakeholders in the company.

The key feature of the external audit process is its performance by a licensed party, who is totally independent of the company management. External auditors are members of the American Institute of Certified Public Accountants (**AICPA**) and as such must adhere to **generally accepted auditing standards**. These standards set the level of professional competence in the performance of an audit. Drawing a conclusion about the fairness of presentation of financial statements is based on **generally accepted accounting principles** to which all AICPA members must adhere. These documented principles provide the general guidelines on "good accounting". However, it should be stressed that, even though auditors have both accounting and performance standards, a good deal of judgment is still necessary in drawing conclusions about whether a set of financial statements accurately depicts the company's operating results. This is because of the many choices available to management in preparing the financial statements. The professional skill of the external auditor rests on assessing whether management's disclosures are likely to mislead a reasonable user of the statements about the company's operating results.

An auditor's professional opinion about the fairness of presentation of the financial statements is formalized in the audit report. Statement users rely heavily on the audit report since they are unable to verify management's assertions about the company's operating results for themselves. Consequently, it is important that statement users know what an audit report does and does not say as well as have a rudimentary understanding of the procedures undertaken in forming the opinion. An experienced statement user always recognizes that, in rare circumstances, bad audits can be performed or management has been able to cleverly disguise the operating results. To guard against these situations, an experienced user seldom accepts the auditor's opinion as complete assurance without first making his own assessment of a company's financial reputation. In addition, astute users are aware of corporate conditions that might indicate the possibility of financial fraud. Encountering these conditions suggests a cautious reliance on the auditor's opinion.

The remainder of this discussion considers the contents of the audit report and its meaning, the strategy and procedures generally followed in arriving at a financial statement opinion, and corporate conditions that history has shown to be indicative of financial fraud.

The Audit Report

The audit report is addressed to the directors and stockholders of the client corporation. The following is the example of the audit report illustrated in the generally accepted auditing standard AU508.08. It is referred to as the standard audit report.

Independent Auditor's Report

Penney and Nichols, CPAs
45789 Beachwood Drive
Centerville, New Jersey 08000

Board of Directors and Stockholders
X Company

> We have audited the accompanying balance sheet of X Company as of December 31, 19X5, and the related statements of income, retained earnings, and cash flows for the year then ended. These financial statements are the responsibility of the Company's management. Our responsibility is to express as opinion on these financial statements based on our audit.

> We conducted our audit in accordance with generally accepted auditing standards. Those standards require that we plan and perform the audit to obtain reasonable assurance about whether the financial statements are free of material misstatement. An audit includes examining, on a test basis, evidence supporting the amounts and disclosures in the financial statements. An audit also includes assessing the accounting principles used and significant estimates made by management, as well as evaluating the overall financial statement presentation. We believe that our audit provides a reasonable basis for our opinion.

> In our opinion, the financial statements referred to above present fairly, in all material respects, the financial position of X Company as of December 31, 19X5, and the results of its operations and its cash flows for the year the ended in conformity with generally accepted accounting principles.

[Report Date]
[Signature]

The first paragraph of the standard audit report is called the "introductory paragraph". It indicates that the financial statements were audited, management is responsible for the financial statements, and the auditor's responsibility is to provide an opinion only on the financial statements identified in the audit report.

The second paragraph is the "scope" paragraph. It states that the audit was conducted in accordance with generally accepted auditing standards and outlines the auditor's basis for forming an opinion on the financial statements. It should be noted that the auditor's objective as stated in this paragraph is to provide reasonable assurance, not a guarantee, that the financial statements taken as a whole are free of material misstatement.[14]

The third or "opinion" paragraph presents the auditor's conclusions as a result of the audit. The phrase "in our opinion" implies that the auditor is reasonably sure of his or her conclusions. The phrase "present fairly" means that the opinion applies to the financial statements taken as a whole and does not imply that any single item in the statements is exact or precisely correct. Instead, the auditor is attesting that the statements are a complete disclosure and free from material bias or misstatement.

The audit report above expresses an unqualified opinion. This means that the auditor has no reservations about the fairness of management's assertions in these financial statements. Any disagreements that arose in the course of the audit have been resolved to both the management's and the auditor's satisfaction, and as a result an unqualified opinion has been issued. In cases where a dispute has arisen, and resolution has not been possible, the auditor can issue what is termed a qualified opinion. This results in an extra paragraph being introduced into the report explaining the effect on the financial statements of the disputed item. An opinion is given on the financial statements "except for" the disputed item. In certain rare circumstances the auditor may issue an "adverse" opinion, that is, the financial statements taken as a whole do not fairly present the results of operations. Finally, in rare cases where the auditor has been unable to obtain sufficient information to form an opinion, the report documents that no opinion is given. This is referred to as a disclaimer opinion.

Audit Procedures

At the basis of any audit is the evaluation of the company's system of internal control. Internal control includes those procedures instituted by management to (1) guard against errors or irregularities in the financial data, (2) protect and safeguard the company assets from theft or misappropriation, and (3) control and evaluate operations. It is impossible for an auditor in today's modern corporation to check the accuracy and validity of all transactions undertaken by the corporation. The auditor's preoccupation with internal control stems from a desire to know how prone the accounting system is to making mistakes in recording all transactions that affect the company under audit. A strong system of internal control instituted by an honest control-conscious management goes a long way to providing reasonable assurance that the resulting financial statements fairly present the results of operations. In the absence of a strong system of internal control, considerably more detail checking must be done. Consequently, providing a good audit at reasonable cost to the client very much hinges on the system of internal control.

Based on an analysis of internal control, the auditor determines the quantity of testing he or she must perform on the financial statements to form an opinion. The testing takes a variety of forms but in all cases revolves around gathering evidence that supports the assertions made in the financial statements. These forms of evidence can be categorized as (1) internal evidence, (2) physical observation, (3) external third-party evidence, and (4) analytical review.

Internal Evidence

A major part of the auditor's time is devoted to an analysis of the internal financial records. Many of the accounts are analyzed for an independent verification of changes and balances. For example, receivable and payable accounts may be analyzed and listed for subsequent investigation. Major movements in account balances are usually analyzed for supporting business documents such as purchase invoices, checks issued, and cash remittance receipts.

---

[14]Misstatements are material if they are significant enough to make a difference to the decision making of a reasonable financial statement user.

Physical Observation

Auditors make extensive visual inspections of their client's properties to satisfy themselves that the assets disclosed in the financial statements do exist in reality. For example, cash on hand is counted and securities are inspected for reconciliation with accounting records. Physical inventories taken by the client's employees are observed by the auditor. New additions to plant assets are inspected to ensure conformity with accounting records.

External Third-Party Evidence

Auditors routinely communicate with other businesses and individuals who do business with the client under audit. These external communications and confirmations aid the auditor in verifying many of the business dealings recorded in the accounting records independently of the company's own documentation. Examples include enquiries of (1) banks, to verify balances on deposit and outstanding loans, (2) trade creditors, (3) customers, to verify debts owed to the company, and (4) attorneys, to investigate any pending litigation against the company. In addition, the auditor may use specialists to confirm management assertions of the value of items beyond the auditor's competence. Typically, these external communications with third parties produce highly credible evidence for the auditor in providing an independent opinion.

Analytical Review

An analytical review of the relationships between data shown by the financial records and revealed during the audit examination adds significantly to the auditor's degree of satisfaction with the resulting financial statements. For example, comparisons of the client's current bad debt losses with those of prior periods and with those of other businesses in the industry provide insights into the adequacy of the company's current bad debt provisions. Analysis of changes in departmental gross profit percentages and inventory turnover may help substantiate the recorded income and inventory levels. Property tax payments should corroborate property ownership. This general area of analysis usually requires a good deal of ingenuity and imagination on the part of the auditor.

Detecting Financial Statement Fraud

Auditors cannot be relied upon to detect every case of financial statement fraud. The best protection for statement users is always to be alert to the possibility of statement fraud and to be aware of the conditions likely to encourage it. In most cases of revealed financial statement fraud one or more of the following conditions existed:

- The chief executive officer managed by setting ambitious simple financial objectives, such as an earnings growth of 25% per year.
- The chief financial officer had a low regard for the financial and accounting functions.
- The nonmanagement members of the board of directors played a passive role in the board and company affairs.
- The management believed that aggressive accounting choices and actions are a legitimate means to achieve corporate goals.
- The company had a rapid growth in earnings that was becoming more difficult to sustain because of internal weaknesses, market changes, or competitive development.
- Management repeatedly issued optimistic statements about the company's future in the face of declining industry results.
- Management was in the hands of a single dominant person.

While these conditions do not always indicate financial fraud, statement users need to be aware that the company management is not always honest and that the conventional auditing has been unable to detect most cases. However, it should also be noted that the incidence of financial fraud is relatively low and that the vast majority of audits do provide statement users with useful information about the financial statements.

Defining Terms

AICPA: Stands for the American Institute of Certified Public Accountants. It is the professional body to which most certified public accountants belong. As such, members are bound by the professional pronouncements on how an audit should be performed and how they should conduct themselves in the performance of an audit. Its main office is located at Harborside Financial Center, 201 Plaza Three, Jersey City, NJ 07311-3881.

Auditing: Is a process performed by an auditor that provides an independent assessment of a set of financial statements produced by a company's management. The purpose of this assessment is to report to the users of the financial statements whether these statements accurately depict the results of operations for the period under review.

Generally accepted accounting principles: Often abbreviated GAAP, are the accounting methods approved by the Financial Accounting Standards Board for use in preparing financial statements.

Generally accepted audited standards: Often abbreviated GAAS, are the professional pronouncements promulgated by the AICPA. They detail for members how to conduct an audit to ensure that sufficient evidence is gathered to provide a financial statement opinion as well as provide a guide to handling various ethical issues that might arise during an audit.

References

Arens, A. A. and Loebbecke, J. K. *Auditing an Integrated Approach, 7th ed.*, Prentice Hall, Englewood Cliffs, NJ, 1997.

Bailey, L. P. *Miller GAAS Guide*, Harcourt Brace, New York, 1997.

Bolgna, J. G. and Lindquist, R. J. *Fraud Auditing and Forensic Accounting — New Tools and Techniques, 2nd ed.*, John Wiley & Sons, New York, 1995.

Deflisse, P. L., Jaenicke, H. R., O'Reilly, V. M., and Hirsch, M. B. *Montgomery's Auditing*, John Wiley & Sons, New York, 1996.

Knapp, M. C. *Contemporary Auditing — Issues and Cases, 2nd ed.*, West, St Paul, MN, 1996.

6.7   Depreciation and Corporate Taxes[15]

Tung Au

Depreciation refers to the decline in value of physical assets over their estimated useful lives. In the context of tax liability, **depreciation allowance** refers to the amount allowed as a deduction in computing taxable income, and **depreciable life** refers to the estimated useful life over which depreciation allowances are computed. Historically, an asset could not be depreciated below a reasonable salvage value. Thus, depreciation allowance is a systematic allocation of the cost of a physical asset between the time it is acquired and the time it is disposed of.

The methods of computing depreciation and the estimated useful lives for various classes of physical assets are specified by government regulations as a part of the tax code, which is subject to periodic revisions. Different methods of computing depreciation lead to different annual depreciation allowances and hence have different effects on taxable income and the taxes paid.

Let P be the historical cost of an asset, S its estimated salvage value, and N the depreciable life in years, Let D_t denote the depreciable allowance in year t, and T_t denote the accumulated depreciation up to and including year t. Then for $t = 1, 2,..., N$,

$$T_t = D_1 + D_2 + \mathrm{L} + D_t \tag{6.1}$$

---

[15]This article is based on material in chapters 10, 11, 12, and 16 of Au and Au [1992]. The permission of Prentice Hall, Inc. is gratefully acknowledged.

An asset's book value B_t is simply its historical cost less any accumulated depreciation. Then

$$B_t = P - T_t \tag{6.2}$$

or

$$B_t = B_{t-1} - D_t \tag{6.3}$$

Among the depreciation methods acceptable under the tax regulations, the straight-line method is the simplest. Using this method, the uniform annual allowance in each year is

$$D_t = \left(P - S\right)/N \tag{6.4}$$

Other acceptable methods, known as *accelerated depreciation methods*, yield higher depreciation allowances in the earlier years of an asset and less in the later years than those obtained by the straight-line method. Examples of such methods are sum-of-the-years'-digits depreciation and double-declining-balance depreciation, which are treated extensively elsewhere [Au and Au, 1992].

Under the current IRS regulations on depreciation, known as the Modified Accelerated Cost Reduction System, the estimated useful life of an asset is determined by its characteristics that fit one of the eight arbitrarily specified categories. Furthermore, the salvage value S for all categories is assumed to be zero, whereas all equipment with life of 10 years or less is assumed to be purchased at midyear.

Tax Laws and Tax Planning

Capital projects are long-lived physical assets for which the promulgation and revisions of tax laws may affect the tax liability. For the purpose of planning and evaluating capital projects, it is important to understand the underlying principles, including the adjustments for the transition period after each revision and for multiyear "carry-back" or "carry-forward" of profits and losses.

The federal income tax is important to business operations because profits are taxed annually at substantial rates on a graduated basis. Except for small businesses, the corporate taxes on ordinary income may be estimated with sufficient accuracy by using the marginal tax rate. **Capital gain**, which represents the difference between the sale price and the book value of an asset, is taxed at a rate lower than on ordinary income if it is held longer than a period specified by tax laws.

Some state and/or local governments also levy income taxes on corporations. Generally, such taxes are deductible for federal income tax to avoid double taxation. The computation of income taxes can be simplified by using a combined tax rate to cover the federal, state, and local income taxes.

Tax planning is an important element of private capital investment analysis because the economic feasibility of a project is affected by the taxation of corporate profits. In making estimates of tax liability, several factors deserve attention: (1) number of years for retaining the asset, (2) depreciation method used, (3) method of financing, including purchase vs. lease, (4) capital gain upon the sale of the asset, and (5) effects of inflation. Appropriate assumptions should be made to reflect these factors realistically.

Decision Criteria for Project Selection

The economic evaluation of an investment project is based on the merit of the **net present value (NPV)**, which is the algebraic sum of the discounted net cash flows over the life of the project to the present. The discount rate is the minimum attractive rate of return specified by the corporation.

The evaluation of proposed investment projects is based on the net present value criteria, which specify the following: (1) an independent project should be accepted if the NPV is positive and rejected otherwise and (2) among all acceptable projects that are mutually exclusive, the one with the highest positive NPV should be selected.

A more general treatment of the net present value decision criteria for economic evaluation of investment projects may include effects of different reinvestment assumptions [Beaves, 1993] and different scales of investment [Shull, 1992].

Inflation Consideration

The consideration of the effects of inflation on economic evaluation of a capital project is necessary because taxes are based on then-current dollars in future years. The year in which the useful life of a project begins is usually used as the baseline of price measurement and is referred to as the **base year**. A **price index** is the ratio of the price of a predefined package of goods and service at a given year to the price of the same package in the base year. The common price indices used to measure inflation include the consumer price index, published by the Department of Labor, and the gross domestic product price deflator, compiled by the Department of Commerce.

For the purpose of economic evaluation, it is generally sufficient to project the future inflation trend by using an average annual inflation rate j. Let A_t be the cash flow in year t, expressed in terms of base-year (year 0) dollars, and A'_t be the cash flow in year t, expressed in terms of then-current dollars. Then

$$A'_t = A_t \left(1 + j\right)^t \tag{6.5}$$

$$A_t = A'_t \left(1 + j\right)^{-t} \tag{6.6}$$

In the economic evaluation of investment proposals in an inflationary environment, two approaches may be used to offset the effects of inflation. Each approach leads to the same result if the discount rate i, excluding inflation, and the rate i', including inflation, are related as follows:

$$i' = \left(1 + i\right)\left(1 + j\right) - 1 = i + j + ij \tag{6.7}$$

$$i = \left(i' - j\right) / \left(1 + j\right) \tag{6.8}$$

The NPV of an investment project over a planning horizon of n years can be obtained by using the constant price approach as follows:

$$NPV = \sum_{t=0}^{n} A_t \left(1 + i\right)^{-t} \tag{6.9}$$

Similarly, the NPV obtained by using the then-current price approach is

$$NPV = \sum_{t=0}^{n} A'_t \left(1 + i'\right)^{-t} \tag{6.10}$$

In some situations the prices of certain key items affecting the estimates of future incomes and/or costs are expected to escalate faster than the general inflation. For such cases the differential inflation for those items can be included in the estimation of the cash flows for the project.

After-Tax Cash Flows

The economic performance of a corporation over time is measured by the net cash flows after tax. Consequently, after-tax cash flows are needed for economic evaluation of an investment project. Since interests on debts are tax deductible according to the federal tax laws, the method of financing an investment project could affect the net profits. Although the projected net cash flows over the years must be based on then-current dollars for computing taxes, the depreciation allowances over those years are not indexed for inflation under the current tax laws.

It is possible to separate the cash flows of a project into an operating component and a financing component for the purpose of evaluation. Such a separation will provide a better insight to the tax advantage of borrowing to finance a project, and the combined effect of the two is consistent with the computation based on a single combined net cash flow. The following notations are introduced to denote various items in year t over a planning horizon of n years:

A_t = net cash flow of operation (excluding financing cost) before tax
A_t = net cash flow of financing before tax
$\mathbf{A_t}$ = $A_t + A_t$ = combined net cash flow before tax
Y_t = net cash flow of operation (excluding financing cost) after tax
Y_t = net cash flow of financing after tax
$\mathbf{Y_t}$ = $Y_t + Y_t$ = combined net cash flow before tax
D_t = annual depreciation allowance
I_t = annual interest on the unpaid balance of a loan
Q_t = annual payment to reduce the unpaid balance of a loan
W_t = annual taxable income
X_t = annual marginal income tax rate
K_t = annual income tax

Thus, for operation in year $t = 0, 1, 2, \ldots, n$,

$$W_t = A_t - D_t \tag{6.11}$$

$$K_t = X_t W_t \tag{6.12}$$

$$Y_t = A_t - X_t\left(A_t - D_t\right) \tag{6.13}$$

For financing in year $t = 0, 1, 2, \ldots, n$,

$$I_t = Q_t - A_t \tag{6.14}$$

$$Y_t = A_t + X_t I_t \tag{6.15}$$

where the term $X_t I_t$ is referred to as the **tax shield** because it represents a gain from debt financing due to the deductibility of interests in computing the income tax.

Alternately, the combined net cash flows after tax may be obtained directly by noting that both depreciation allowance and interest are tax deductible, Then,

$$W_t = A_t - D_t - I_t \tag{6.16}$$

$$Y_t = A_t - X_t\left(A_t - D_t - I_t\right) \tag{6.17}$$

It can be verified that Eq. (6.17) can also be obtained by adding Eqs. (6.13) and (6.15), while noting $A_t = A_t + A_t$ and $Y_t = Y_t + Y_t$.

Evaluation of After-Tax Cash Flows

For private corporations the decision to invest in a capital project may have side effects on the financial decisions of the firm, such as taking out loans or issuing new stocks. These financial decisions will influence the overall equity-debt mix of the entire corporation, depending on the size of the project and the risk involved.

Traditionally, many firms have used an adjusted cost of capital, which reflects the opportunity cost of capital and the financing side effects, including tax shields. Thus only the net cash from operation Y_t obtained by Eq. (6.13) is used when the net present value is computed. The after-tax net cash flows of a proposed project are discounted by substituting Y_t for A_t in Eq. (6.9), using after-tax adjusted cost of capital of the corporation as the discount rate. If inflation is anticipated, Y_t' can first be obtained in then-current dollars and then substituted into Eq. (6.10). The selection of the project will be based on the NPV thus obtained without further consideration of tax shields, even if debt financing is involved. This approach, which is based on the adjusted cost of capital for discounting, is adequate for small projects such as equipment purchase.

In recent years another approach, which separates the investment and financial decisions of a firm, is sometimes used for evaluation of large capital projects. In this approach the net cash flows of operation are discounted at a risked adjusted rate reflecting the risk for the class of assets representing the proposed project, whereas tax shields and other financial side effects are discounted at a risk-free rate corresponding to the yield of government bonds. An adjusted net present value reflecting the combined effects of both decisions is then used as the basis for project selection. The detailed discussion of this approach may be found elsewhere [Brealey and Myers, 1988].

Effects of Various Factors

Various depreciation methods will produce different effects on the after-tax cash flows of an investment. Since the accelerated depreciation methods generate larger depreciation allowances during the early years, the NPV of the after-tax cash flows using one of the accelerated depreciation methods is expected to be more favorable than that obtained by using the straight-line method.

If a firm lacks the necessary funds to acquire a physical asset that is deemed desirable for operation, it can lease the asset by entering into a contract with another party, which will legally obligate the firm to make payments for a well-defined period of time. The payments for leasing are expenses that can be deducted in full from the gross revenue in computing taxable income. The purchase-or-lease options can be compared after their respective NPVs are computed.

When an asset is held for more than a required holding period under tax laws, the capital gain is regarded as long-term gain and is taxed at a lower rate. In a period of inflation, the sale price of an asset in then-current dollars increases, but the book value is not allowed to be indexed to reflect the inflation. Consequently, capital gain tax increases with the surge in sale price resulting from inflation.

Example 1. A heavy-duty truck is purchased at $25,000 in February. This truck is expected to generate a before-tax uniform annual revenue of $7000 over the next 6 years, with a salvage value of $3000 at the end of 6 years. According to the current IRS regulations, this truck is assigned an estimated useful life of 5 years with no salvage value. The straight-line depreciation method is used to compute the annual depreciation allowance. The combined federal and state income tax rate is 38%. Assuming no inflation, the after-tax discount rate of 8%, based on the adjusted cost of capital of the corporation, is used. Determine whether this investment proposal should be accepted.

Solution. Using Eq. (6.4), the annual depreciation allowance D_t is found to be $5000 per year, since a useful life of 5 years is specified and the salvage value is zero, according to the current IRS regulations.

Following the midyear purchase assumption, the actual depreciation allowances for years 1 through 6 are as shown in the following table. The actual revenues over 6 years are used in the analysis.

t	A_t	D_t	$A_t - D_t$	K_t	Y_t
0	-25,000	—	—	—	-25,000
1	7000	2500	2500	1710	5290
2-5	7000	5000	2000	760	6240
6	7000	2500	4500	1710	5290

Using the adjusted cost of capital approach, the net present value of the after-tax net cash flows discounted at 8% is obtained by substituting Y_t for A_t in Eq. (6.9),

$$\text{NPV} = -25,000 + (6240)\left(P \mid U, 8\%, 5\right) - (6240 - 5290)\left(P \mid F, 8\%, 1\right)$$
$$+ (5290)\left(P \mid F, 8\%, 5\right) = 2369$$

In which $(P \mid U, 8\%, 5)$ is the discount factor to present at 8% for a uniform series over 5 years, and $(P \mid F, 8\%, 1)$ and $(P \mid F, 8\%, 5)$ are discount factors to present at 8% for a future sum at the end of 1 year and 5 years, respectively. Since NPV = \$2369 is positive, the proposed investment should be accepted.

Example 2. Consider a proposal for the purchase of a computer workstation that costs \$20,000 and has no salvage value at disposal after 4 years. This investment is expected to generate a before-tax uniform annual revenue of \$7000 in base-year dollars over the next 4 years. An average annual inflation rate of 5% is assumed. The straight-line depreciation method is used to compute the annual depreciation allowance. The combined federal and state income tax rate is 38%. Based on the adjusted cost of capital of the corporation, the after-tax discount rate, including inflation, is 10%. Determine whether this investment proposal should be accepted.

Solution. To simplify the calculation, the assumption of midyear purchase is ignored. Using Eq. (6.4), the annual depreciation allowance D_t is found to be \$5000 for $t = 1$ to 4. This annual depreciation allowance of \$5000 will not be indexed for inflation, according to the IRS regulations.

The annual before-tax revenue of \$7000 in base-year dollars must be expressed in then-current dollars before computing the income taxes. From Eq. (6.5),

$$A'_t = (7000)(1 + 0.05)^t$$

where $t = 1$ to 4 refers to each of the next 4 years. The after-tax cash flow Y'_t for each year can be computed by Eq. (6.13). The step-by-step tabulation of the computation for each year is shown in the following table.

t	A_t	A'_t	D_t	$A'_t - D_t$	K_t	Y'_t
0	-20,000	-20,000	—	—	—	-20,000
1	7000	7350	5000	2350	893	6457
2	7000	7718	5000	2718	1033	6685
3	7000	8013	5000	3103	1179	6924
4	7000	8509	5000	3509	1333	7176

Using the adjusted cost of capital approach, the net present value of the after-tax cash flows discounted at $i' = 10\%$, including inflation, can be obtained by substituting the value of Y'_t for A'_t in Eq. (6.10) as follows:

$$NPV = -20,000 + (6457)(1.1)^{-1} + (6685)(1.1)^{-2} + (6924)(1.1)^{-3} + (7176)(1.1)^{-4}$$

$$= 1498$$

Since NPV = $1498 is positive, the investment proposal should be accepted.

Example 3. A developer bought a plot of land for $100,000 and spent $1.6 million to construct an apartment building on the site for a total price of $1.7 million. The before-tax annual rental income after the deduction of maintenance expenses is expected to be $300,000 in the next 6 years, assuming no inflation. The developer plans to sell this building at the end of 6 years when the property is expected to appreciate to $2.1 million, including land. The entire cost of construction can be depreciated over 32 years based on the straight-line depreciation method, whereas the original cost of land may be treated as the salvage value at the end. The tax rates are 34% for ordinary income and 28% for capital gain, respectively. Based on the adjusted cost of capital, the developer specified an after-tax discount rate of 10%. *Solution.* Using Eq. (6.4), the annual depreciation allowance D_t over 32 years is found to be $50,000. Noting that $P = 1,700,000$ and $T_t = (6)(50,000) = 300,000$, the book value of the property after 6 years is found from Eq. (6.2) to be $1.4 million.

Ignoring the assumption of midyear purchase to simplify the calculation and assuming no inflation, the after-tax annual net income in the next 6 years is given by Eq. (6.13):

$$Y_t = 300,000 - (34\%)(300,000 - 50,000) = 215,000$$

The capital gain tax for the property at the end of 6 years is

$$(28\%)(2,100,000 - 1,400,000) = 196,000$$

Using the adjusted cost of capital approach, the net present value of after-tax net cash flows in the next 6 years, including the capital gain tax paid at the end of 6 years discounted at 10% is

$$NPV = -1,700,000 + (215,000)\left(P \,|\, U, 10\%, 6\right)$$

$$+ (2,100,000 - 196,000)\left(P \,|\, F, 10\%, 6\right)$$

$$= 311,198$$

in which $(P \,|\, U, 10\%, 6)$ is the discount factor to present at 10% for a uniform series over 6 years and $(P \,|\, F, 10\%, 6)$ is the discount factor to present at 10% for a future sum at the end of 6 years. Since NPV = $311,198 is positive, the proposed investment should be accepted.

Defining Terms

Base year: The year used as the baseline of price measurement of an investment project.
Capital gain: Difference between the sale price and the book value of an asset.
Depreciable life: Estimated useful life over which depreciation allowances are computed.
Depreciated: Decline in value of physical assets over their estimated useful lives.
Depreciation allowance: Amount of depreciation allowed in a systematic allocation of the cost of a physical asset between the time it is acquired and the time it is disposed of.

Net present value: Algebraic sum of the discounted cash flows over the life of an investment project to the present.

Price index: Ratio of the price of a predefined package of goods and service at a given year to the price of the same package in the base year.

Tax shield: Gain from debt financing due to deductibility of interests in computing the income tax.

References

Au, T. and Au, T. P. *Engineering Economics for Capital Investment Analysis, 2nd ed.*, Prentice Hall, Englewood Cliffs, NJ, 1992.

Beaves, R. G. The case for a generalized net present value formula, *Eng. Economist,* 38(2): 119–133, 1993.

Brealey, R. and Myers, S. *Principles of Corporate Finance, 3rd ed.*, McGraw-Hill, New York, 1988.

Shull, D. N. Efficient capital project selection through a yield-based capital budgeting technique, *Eng. Economist,* 38(1): 1–18, 1992.

Further Information

Up-to-date information of federal tax code may be found in the following annual publications:

Federal Tax Guide, Prentice Hall, Englewood Cliffs, NJ.
Standard Federal Tax Reporter, Commerce Clearing House, Chicago, IL.

6.8   Performance Evaluation

Masako N. Darrough

Performance evaluation is the process by which performance of a company, a subunit, a project, a manager, or an employee is measured, evaluated, and rewarded. This process is one of the most important tools in managing any organization. Performance evaluation is by definition an *ex post* (after the fact) activity: to assess how well an organization has achieved its goals, how well a project has turned out, and how good a job a manager did. This information is indispensable in making future managerial decisions, since the past is often indicative of the future. Performance evaluation is also useful *ex ante* (before the fact) in providing appropriate incentives for decision makers to achieve better performance. To do so, however, performance evaluation should be tied to incentive or pay-for-performance schemes. What gets measured then gets managed.

Many issues must be considered to evaluate performance such as what to measure, how, on what basis, and how often. Clearly, answers to these questions depend heavily on what aspect of performance (the "object" of evaluation) we are evaluating and the purpose for which such evaluation is used. For example, the method for evaluating an entire organization should differ from that used in evaluating a divisional manager. The performance of a project might be evaluated differently from that of the manager in charge.

In this section, we first examine the issues relevant to measuring performance and then examine how various compensation schemes are used for promoting better performance.

Performance Measures

Objects of Performance Evaluation

Most organizations of any size are structured as a hierarchy of some sort. Various subunits such as strategic business units, divisions, or departments are engaged in various activities to achieve the goals of and enhance the value of the organization. At each level of the organization, managers and workers make strategic decisions and exert effort to implement the decisions. In order to promote better decisions and induce appropriate behavior, performance evaluation is carried out for different objects, ranging in scope from the entire organization to an individual worker. However, depending on the object of evaluation, the performance measures available might differ. For example, market-based measures such as stock prices are available for publicly traded companies, but not for privately held companies or governmental or nonprofit organizations. Even for a publicly traded company, its stock price measures the public assessment of the company as a whole, but not of any specific subunit. Therefore, the stock price may not be an appropriate performance measure for subunits. In addition, to the extent that measures are used to motivate behavior and decisions, appropriate measures would likely differ from different objects. Subunits and individual employees all contribute toward the entire company's performance, but the nature and the degree of contributions are not the same.

Most of the non–market-based financial-performance measures (as opposed to nonfinancial measures such as customer satisfaction and new product development) are accounting based. Accounting-based measures are useful because they tend to be more objective, more understandable, and less controversial. Thus, measures such as operating income, net income, residual income, earnings per share (EPS), return on investment (ROI), and return on equity (ROE) are widely used as performance measures. Economic-value added (EVA), a variation of residual income, has been introduced only recently, but is gaining wide acceptance.

In selecting appropriate performance measures, a useful framework is the concept of **responsibility accounting**. A hierarchical structure is organized with clear-cut responsibilities and the span of control assigned to each subunit. The four major categories of responsibility are investment, profit, revenue, and cost centers. These responsibility centers are then evaluated based on how well they have carried out their responsibilities. Thus, an investment center would be measured based on "returns" in relation to the assets that generated those returns, while profit centers would be measured based solely on returns (or profits). Those measures that involve returns are clearly relevant for organizational units and their managers who are responsible for and have control over their profitability. However, the responsibilities of some units might not be directly linked to profits. For example, service departments such as accounting, legal, or information technology provide essential services but cannot directly control the organization's profits. They are responsible for providing necessary services in the most efficient manner. They are cost centers: they should be evaluated according to how well they managed the costs of providing the services. Likewise, revenue centers would be accountable for revenues only but not necessarily for profits.

Another important underlying notion is that of controllability: a manager (subunit) should be evaluated based on those variables over which the manager (subunit) has control. The **controllability principle** is intuitive and well accepted in managerial accounting. It seems only fair to make a manager accountable for only what he can influence and control. However, a mechanical application of the principle can result in suboptimal results. Although some variables are not under the direct control of a manager, he can take actions that reduce the adverse effects of or take advantage of the uncontrollable events. For example, fluctuations in interest rates, exchange rates, or business cycles are not controllable by a manager. Hurricanes, flood, and earthquakes are acts of God. However, the manager can take appropriate actions to minimize the risk from these events. Simply making a manager accountable for only controllable variables does not provide correct incentives to the manager.

Another example that illustrates the problem of the controllability principle is the use of relative performance. Clearly, the performance of a comparison group (such as other managers in the same organization or in other similar organizations) is not under the control of the manager in question. A relative performance measure, however, could be a better measure than an absolute measure, since it reduces risk in the manager's compensation. It does so by removing stochastic elements that are common to all managers but are not under the control of any manager.

To induce a manager to exert effort and make optimal decisions, we need to focus on the variables that are "informative" about the manager's decisions and actions (managerial inputs). The very idea of decentralization or delegation through responsibility centers implies that the optimality of managerial inputs are not easily observable. Hence, we use outcome-based measures. The best variables for incentives, therefore, are those outcome variables that are informative of the manager's input. What is important is not whether the manager can control the variables, but rather whether he can control their information content. Granted, those variables tend to be under a manager's control, but as mechanical application of controllability can miss the point as illustrated in the above examples.

In the planning process, strategies that have been developed are summarized into a master **budget**. Actual results can then be evaluated against targeted results. The sources of the discrepancy between the budgeted and the actual results are identified by **variance analysis**. For example, a shortfall in the actual profit level might be attributed to various factors such as smaller sales, lower prices, different sales mix, higher prices of raw materials, higher wage rates, higher usage of raw materials, longer labor hours, higher overhead costs, etc. For analyzing cost variances, actual costs are compared to **standard costs** after adjusting for the actual volume of production through the use of a **flexible budget**. Managerial performance could then be based on the variances (**management by exception**).

Two important issues arise with respect to the choice of performance measures: (1) how to define and calculate the measures and (2) how to choose the appropriate measures for performance evaluation and compensation. These two issues are not independent of each other. Since performance measures are used for making decisions and influencing behavior, how performance is measured affects outcome. To induce desirable outcome, in turn, an appropriate measure must be chosen.

Measurement

Even though the basic definitions of various accounting measures are straightforward, actual implementation of these measures requires a careful selection of appropriate variables. For example, measures that evaluate returns could be operating income, operating income after taxes, net income, or other accounting measures. In some situations, cash flows might be appropriate. Accounting income figures also depend on the accounting methods adopted (e.g., inventory valuation and depreciation). Measures that evaluate asset bases can be gross or net based. The measurement issue is further complicated for subunits of an organization. For example, "income" for a subunit would depend upon how joint or common costs are allocated (**cost allocation**) as well as how **transfer prices** are set when there are intracompany transactions. Similarly, asset bases for subunits are subject to the same allocation problems of commonly shared assets. Similar issues arise in calculating EVA in terms of how "income", "cost of capital", and "asset base" are defined. The basic definition of EVA is

$$EVA = income - assets \cdot cost\ of\ capital$$

A common definition of income is operating income after tax with possible adjustments such as (1) adding back goodwill amortization, (2) capitalizing R&D, and (3) capitalizing advertising and promotion. Such adjustments may be made to induce desirable behavior (such as taking a long-term perspective). In defining performance measures, the most important criterion is, of course, whether the measures are effective in promoting desirable behavior of decision makers.

Impact on Behavior

Current earnings, ROI, ROE, residual income, and EVA are calculated for a specific time period, which tends to be short run (SR), such as a quarter or a year. An excessive focus on these SR measures would induce myopic behavior: for example, obsession on quarterly earnings is often criticized as the cause of the short-term orientation of corporate America. Long-term or deferred compensation schemes can be combined to induce a more balance view. Other dysfunctional behavior might be induced by these measures. For example, ROI and ROE appear to be appropriate measures for investment centers, since they calculate ROIs. However, excessive focus on ROI or ROE could induce managers to maximize these measures. As a result, they might forego opportunities that are value enhancing (to the investment center and the entire organization) because these measures lower average ROI or ROE. Hence, goal incongruence emerges between the subunit and the corporate level.

Because of this potentially dysfunctional impact, managerial accountants have advocated that residual income (RI) is superior to ROI or ROE. The basic notion of RI is closest to that of "profit" in economics.

In the last decade or so, a more refined notion of RI has been adopted by a number of companies under the new title of EVA. Major corporations such as Bank of America, Coca Cola, Eli Lilly, Quaker Oats, and AT&T are using EVA as a tool for strategic decision-making, performance evaluation, and as a basis for compensation.

Since EVA is accounting based and not market based, it is also used by governmental and nonprofit organizations including the U.S. Postal Service. The advantage of EVA is that it forces managers to identify value-increasing opportunities and to account explicitly for the scarcity of capital. Although EVA is an SR measure, the basic approach is more consistent with capital budgeting. Since EVA does not require a stock price, it is especially suitable as an internal measure of operating performance. When stock prices are available, EVA can be extended to market value-added (MVA), which focuses on the total market value of a company in excess of the cost of assets that generate its value.

Judicious use of EVA will induce managers to engage in activities that increase the EVA of the organizations by taking on new investments that have a positive EVA, by improving efficiency of existing businesses, or by eliminating negative EVA segments of the existing business. All these actions are value-enhancing strategies that an organization would wish to induce mangers to take.

Compensation

The employment relationship is not a zero-sum game. By providing appropriate incentives to employees (agents of the owners), everybody can be better off (**Pareto improvement**). It is often difficult, however, to measure the contributions made by each individual, since individual ability, effort, or output is not easily observable. This information asymmetry leaves room for opportunistic behavior, resulting in goal incongruence. Thus, an appropriate compensation scheme should be designed as a tool for motivating employees and managers to achieve organizational goals by aligning incentives with those of the organization (**goal congruence**).

One way to achieve better goal congruence is by making manages and employees behave as if they are owners of the organization. Stock ownership is the most direct way of making managers and employees owners. However, the inside ownership is usually a fraction of the entire organization: the free-rider problem remains. Moreover, increasing ownership levels implies increased risk. Since managers and employees have all of their human capital tied up in the organization, they are underdiversified. Thus, imposing risk on risk-averse managers and employees through increasing ownership has limitations.

"Pay-for-performance" is another approach to better align the interests of employees and the organization (shareholders). Many pay-for-performance methods are based on accounting performance measures. In governmental or nonprofit organizations, since "bottom-line" accounting measures are not relevant, a heavier reliance can be put on nonaccounting, but outcome-based, performance measures. For example, the police department in the City of Sunnyvale uses a host of service measures, one example being the percentage of 911 calls that have been serviced within a targeted interval of time. Determining exactly how much of the total compensation of a manager should be "at risk" and how much should be fixed requires the weighing of risk and incentives. The proportion of at-risk compensation typically increases with one's position in the hierarchy and the level of total (expected) compensation.

Types of Incentive Schemes for Top Management

There are many accounting and market-based compensation schemes. Widely used pay-for-performance schemes are (1) a lump-sum bonus (if a target is achieved), (2) profit sharing (a predetermined percentage share up to some maximum), (3) long-term performance plans (based on performance over a time period, for example, over 5 years), (4) gainsharing (based on measures such as productivity increases), and (5) stock options.

In addition to a fixed base salary, top manages are paid via various incentive schemes. For most of them, the base salary is a relatively small portion of their total compensation. For example, the salary of Mr. Bill Gates of Microsoft was a mere $275,000 in 1995. Even with the bonus of $140,580, he would not have been the richest person in the United States without stock ownership that amounts to 23.9%

of the company. A survey conducted by Towers Perrin found that 50.3% of total compensation of American executives in 1995 was performance based.[16]

Stock Options

Especially in high-tech and startup companies, stock options are increasingly utilized as a means of employee compensation. An article in *Fortune* states flatly that "Silicon Valley wouldn't be what it is today without stock options…"[17] Typically, stock options are granted to mid- to top-level managers, but more companies are adopting stock options to lower-level employees. For example, both Intel and Microsoft offer stock option plans to all employees. 3Com in the Silicon Valley offers a stock plan to all new hires.[18] Stock options are obvious methods of aligning the interests of employees and owners. In addition, stock options offer cost advantages for fast-growing companies. First, at the time of granting, stock options are free to the granting company. No cash payment is required and no charge is taken on the income statement. Only a footnote disclosure is required.[19] Clearly for the grantee, however, options have value. When stock options are exercised, the resulting dilution of stock, on the other hand, imposes a cost to the other shareholders. Second, current tax codes allow companies to deduct option values at the time of exercise, yet GAAP do not require a deduction of the values on the income statement. Third, in cases where companies already have shares to give out, companies can reduce tax bills without having to lay out any cash. *The Wall Street Journal,* May 13, 1997, reports that the tax saving for Microsoft due to the deductibility of the difference between the market value and exercise price amounted to $352 million in fiscal 1996 on revenues of $8671 million.

Designing of Incentive Systems

Since there are many ways to motivate and compensate managers, the designing of incentive schemes requires careful consideration. First, incentive schemes should be understandable and not overly complex so that individuals can see how to improve their performance and be rewarded by good performance. Second, incentive schemes should not impose excessive risk. Third, performance measures must measure goals and achievements in a systematic and consistent manner. Finally, the incentive system need not be individually based. In fact, it can be explicitly based on the performance of a team. Alternatively, an incentive scheme might have two components, one based on individual performance and another based on company (or divisional) performance.

American corporate executives are highly paid by any standard. Their compensation might be justified by the excellent performance of American corporations in the recent years as measured by stock prices, earnings, or EVA. However, some critics, such as Graef [1991], are concerned with the "excessive" level of executive compensation. Two issues are relevant: (1) the absolute level and (2) the link between performance and compensation. The gap in compensation between the CEO and factory workers has been widening over time. *Business Week* reports that, in 1996, CEOs' average total compensation increased by 54%, while that of factor workers increased by 3%.[20] One might argue that top management needs and deserves to be paid well to take risks and to make right decisions. It is not uncommon for a CEO of a major corporation to receive close to $100 million in a 1-year period in combined compensation. Such compensation might have been handed out while aggressive cost cutting, reengineering, and downsizing of the organization is taking place. A growing income gap between top and bottom corporate ranks might be a matter of social concern, but, for each individual company, competition for top talent dictates the level of compensation.

---

[16] *The Economist,* May 4, 1996.

[17] The Next Best Thing to Free Money", July 7, 1997.

[18] Radford and Kove [1991].

[19] Statement of Accounting Standards No. 123, issued in October 1995 by the Financial Accounting Standards Board, encourages but does not require a recognition of stock option expense at the time of granting.

[20] *Business Week,* April 21, 1997, p. 59.

The designing of link between pay and performance is another matter. There is anecdotal evidence that, although upside potential in compensation is enormous, downside risk is often limited. For example, exercise prices for stock options are usually set at market level. Sometimes these prices are lowered if stock prices decline. Nevertheless, when stock prices appreciate, many top managers can realize enormous capital gains on a large number of stock options. Even worse, many CEOs are paid extremely well despite the poor performance of their companies. If returns to shareholders and executive compensation are not aligned, it is difficult to argue that the incentives of the top management are aligned with the interests of the owners of the company.

There are many studies that compare the performance of firms with some incentive plans to that of firms without such plans. The evidence, though not definitive, seems to show that firms with plans do better on the average (see, for example, Banker et al. [1996] and Mehran [1995]). The important question is no longer whether to have an incentive plan, but rather what kind of plan to have and how to structure it. After all, what gets measured and rewarded gets managed.

Defining Terms

Budgeting: The process of formalizing plans into estimated financial results.
Cost allocation: The process of allocating costs to different objects.
Goal congruence: The alignment of an employee's personal goals with those of the organization.
Management by exception: An approach to management by setting targets and focusing on the deviations.
Pareto improvement: An improvement in which at least one party is made better off without making anybody else worse off.
Responsibility accounting:
Standard costs: Benchmark costs for the production of goods and services that are efficient and attainable.
Transfer price: The price charged for goods and services transferred within the same organization.
Variance analysis: Decomposition of differences between actual and budgeted amounts into specific factors.

References

Banker, R., Lee, S.-Y., and Potter, G., A field study of the impact of a performance-based incentive plan, *J. Account. Econ.*, 21(3): 1996.
Graef, C., *In Search of Excess: The Overcompensation of American Executives*, W. W. Norton & Company, New York, 1991.
Holmström, B., Moral hazard and observability, *Bell J. Econ.*, Spring: 1979.
Mehran, H., Executive compensation structure, ownership, and firm performance, *J. Fin. Econ.*, 38(2): 1995.
Radford, J. and Kove, S., Lessons for the Silicon Valley, *Personnel J.*, 70(2): 1991.
Stewart, B., *The Quest for Value*, Harper Business, New York, 1991.

Further Information

Boschen, J. F. and Smith, K. J., You can pay me now and you can pay me later: the dynamic response of executive compensation to firm performance, *J. Bus.*, 68(4): l1995.
Demski, J., *Managerial Use of Accounting Information*, Kluwer Academic, 1994.
Merchant, K., *Rewarding Results: Motivating Profit Center Managers*, Harvard Business School Press, 1989.

7
Organizations

Ralph Katz
Northeastern University and Massachusetts Institute of Technology

Janice A. Klein
Massachusetts Institute of Technology

Marc S. Gerstein
Massachusetts Institute of Technology

Jesse Peplinski
Georgia Institute of Technology

Patrick N. Koch
Georgia Institute of Technology

Farrokh Mistree
Georgia Institute of Technology

Urs E. Gattiker
Aalborg University

John P. Ulhøi
The Aarhus School of Business

Karen Stephenson
University of California

Donald Palmer
University of California

Andrew Ward
Emory University

Loren Falkenberg
University of Calgary

Ron Franklin
University of Calgary

Dick Campion
Rational Solutions

Kimberly D. Elsbach
University of California

Francis T. Hartman
University of Calgary

Sim B. Sitkin
Duke University

Kathleen M. Sutcliffe
University of Michigan Business School

Karl E. Weick
University of Michigan Business School

Charles A. O'Reilly III
Stanford University

Jahangir Karimi
University of Colorado

Yash P. Gupta
University of Colorado

Jay A. Conger
University of Southern California

Michael D. Kull
George Washington University

7.1   The Motivation of Technical Professionals

Professor Ralph Katz

Motivation is a critical ingredient for high-performing individuals and groups and yet it is often the most difficult to understand. While the presence of motivation does not guarantee high performance, its absence seems to guarantee long-term problems. Highly motivated professionals and project teams usually push themselves to overachieve, often stretching themselves to accomplish considerably more than their brighter and more-capable associates. In fact, as managers gain experience, they soon realize that, in order to get new ideas and innovative advances commercialized more successfully, they are probably better off having individuals with A-rated motivations and B-rated capabilities than having individuals with A-rated capabilities but with B-rated motivations [Pinchot, 1985]. In over 25 years of teaching and consulting on issues related to the management of innovation and technology, I have found that research and development (R&D) managers usually list the motivation of engineers and scientists as one of the most difficult and perplexing aspects of their leadership roles.

A generalized **model** of the motivation process, according to Steers and Porter [1995], can be characterized by three basic common denominators. Motivation is primarily concerned with (1) what energizes particular behaviors, (2) what directs or channels these behaviors, and (3) how these behaviors are sustained or altered. Each of these three components represents an important aspect of human motivation. The first component is concentrating on those needs, drives, or expectations within individuals or their work settings that trigger certain behaviors, while the second component emphasizes the goals and visions of the individuals and groups toward which the energized behaviors are directed. The last component of any motivational model has to deal with feedback, focusing on those forces within individuals or their organizational environments that can either reinforce and intensify the energy and direction of the behaviors or can dissuade them from their prevailing course of action, thereby redirecting their efforts.

Cognitive Models of Motivation

Although a wide range of models of motivation has been put forth, many of them are psychological in nature, focusing on the willingness of the professional to undertake action in order to satisfy some **need**. An unsatisfied need creates tension, which stimulates a drive within the individual to identify particular goals that, if attained, will satisfy the need and lead to a reduction of tension. As a result, motivated employees are in a state of tension and engage in activity to relieve this tension. The greater the tension, the greater the drive to bring about relief. Cognitive models of this type require managers to understand the psychological needs of their R&D workforce, for, when technical professionals are motivated and working hard at some set of activities, they are driven by a desire to achieve goals they value. Consequently, if organizations want to have motivated engineers and scientists, they must create the kinds of job assignments, careers, and work-related conditions that allow these professionals to satisfy their individual needs.

Maslow's Hierarchy of Needs

One of the best-known approaches in this psychological arena is Maslow's (1954) **hierarchy** of needs model. According to Maslow, within every human being there exists a hierarchy of five classes of needs as follows:

1. Physiological: These involve bodily needs such as hunger, shelter, and sex.
2. Safety: These include one's needs for security and protection from physical and emotional harm.
3. Social: These include one's needs for affection and a sense of belonging and acceptance.
4. Self-esteem: These include one's internal needs for self-respect, autonomy, and achievement as well as one's external needs for status and recognition.
5. Self-actualization: This involves the need for self-fulfillment, to grow and achieve one's full potential.

Based on this hierarchical model, as each class of needs becomes substantially satisfied, the next class of needs in the hierarchy becomes dominant. Maslow separated the five classes of needs into lower (physiological and safety) and higher (social, self-esteem, and self-actualization) orders. He then postulated that lower-order needs are satisfied externally through wages, bonuses, job security, and the like, while higher-order needs are satisfied internally through the individual's own sense of personal growth and development. Since needs that are essentially satisfied no longer motivate, individuals tend to move up the hierarchy as their lower-order needs are met. (Of course, they can also move down the hierarchy when, for example, their job security is threatened.) The managerial implications of this motivational model are rather straightforward, that is, R&D settings should be organized and designed to satisfy the higher-order needs of their technical professionals. To be strongly motivated, technologists need to feel that their jobs are both important and meaningful and that their contributions are truly valued by their organizations, their professions, and even by society. Nevertheless, research studies continue to show that, even though engineers and scientists significantly value achievement, recognition, responsible involvement, high levels of autonomy, and task assignments that provide opportunities for growth , most organizations have trouble providing these kinds of job experiences on a consistent basis to their knowledge workers [Badawy, 1993].

Herzberg's Two-Factor Theory

In the belief that an individual's attitude toward work can greatly affect success or failure on the job, Herzberg et al. [1959] asked professional employees to write two separate paragraphs; one to describe a situation in which they felt exceptionally satisfied about their jobs and one describing a situation in which they felt especially dissatisfied about their jobs. Based on these paired comparisons of critical incident descriptions from many hundreds of professionals, Herzberg discovered that the kinds of replies professionals gave when they felt good about their jobs were significantly different from the replies given when they felt bad. After categorizing the responses from these paired comparisons, Herzberg argued that certain job characteristics, labeled motivators, are more strongly related to job satisfaction, while other characteristics, labeled hygiene factors, are more consistently connected with job dissatisfaction. Motivating or **intrinsic**-type factors included items involving achievement, recognition, responsibility, challenging work, and opportunities for growth and advancement, while the hygiene or **extrinsic**-type factors associated with job dissatisfaction were comprised by items involving company policies, administrators, supervisory relationships, working conditions, salary, and peer relationships.

Based on this differentiation, Herzberg postulated that it may be more useful to think of those factors that can lead to greater levels of job satisfaction, and therefore more motivated individuals in the workplace, as separate and distinct from those hygiene factors that seem to only affect one's level of job dissatisfaction. As a result, R&D managers and supervisors who seek to eliminate factors that lead to job dissatisfaction may bring about more harmony but not necessarily greater motivation. It is more likely that they are placating their workforce rather than motivating them. When hygiene factors are adequate, employees will *not* be *dis*satisfied but neither will they feel motivated. In similarity with Maslow's theory, if organizations want to enhance the motivations of technical professionals on their jobs, they need to emphasize the set of motivating factors that engineers and scientists find intrinsically rewarding. In short, R&D managers should understand whether they are leading and creating change in ways that motivate and excite their technical professionals or whether they are merely behaving and pursuing only the kinds of changes that end up reducing levels of dissatisfaction.

McClelland's Theory of Needs

Other psychological or needs-based models have been developed to look at motivation as a function of the fit between the individual and the organizational job setting. McClelland and colleagues, for example, contend that individuals' needs and **motives** can be measured along three critical dimensions [McClelland and Boyatzis, 1982]. The *need for affiliation* describes an individual's desire for friendly and close interpersonal relationships. Employees with a high need for affiliation want to be well liked and accepted by their colleagues. They prefer job situations that are cooperative rather than competitive, environments in which relationships are built on high levels of mutual trust and understanding. A second dimension depicts an individual's *need for power*, that is, the drive to influence others and have an impact. People with a high need for power strive for control, prefer competitive situations, and seem to enjoy being in charge. The third dimension captures a person's *need for achievement*, the desire to excel or succeed at some challenging activity or project. Individuals with a high need for achievement seek to overcome obstacles in order to do things better or more efficiently than they were done in the past. They work to accomplish difficult goals that have intermediate levels of risk, and they are willing to accept the personal responsibility for a project's success or failure rather than leaving the outcome to chance or to the actions of others.

Studies in R&D settings have consistently found that technical professionals with a high need for achievement are more motivated and successful in entrepreneurial activities [Roberts, 1991], although a high need to achieve does not necessarily lead to being a good technical manager. In fact, McClelland reports from his most recent research that the best managers have a relatively low need for affiliation but have a relatively high need for power — a need for power, however, that is carefully kept under restraint by the individual. Since innovation often requires entrepreneurial risk-taking behavior, R&D managers can either select professionals with high achievement needs to lead such efforts or they can establish appropriate training programs to stimulate the achievement needs of those technologists undertaking entrepreneurial kinds of projects and activities.

Motivation through the Design of Work

All of these cognitive theories of motivation would agree that, when professional employees are well matched with their jobs, it is rarely necessary to force or manipulate them into working hard and trying to perform their jobs well. When there is a good *fit* between the individual and the job, the person typically describes a very high level of internal work motivation, feeling good both about himself and what he is accomplishing. Good performance becomes self-rewarding, which, in turn, serves as an incentive for continuing to do well. In similar fashion, poor performance creates unhappy feelings, which prompts the person to try harder in the future to avoid unpleasant outcomes and regain the intrinsic rewards that good performance can bring. The result is a self-perpetuating cycle of positive work motivation powered through the intrinsically rewarding nature of the work itself. The critical issue is for organizations to structure and design requisite project activities in a way that professionals will also find the work personally rewarding and satisfying.

Several theories have been developed to examine job-related conditions that would affect an employee's level of work motivation. *Equity theory*, for example, asserts that employees' motivations center around *relative* rewards and what they believe are comparatively equitable. Basically, individuals weigh what they put into a job situation, i.e., their inputs, against what they get from their jobs, i.e., their outcomes, and then compare their input-outcome ratios against the input-outcome ratios of relative others. If employees perceive their ratios to be equal to that of relevant others with whom they choose to compare themselves, the motivational system is in **equilibrium** since it is viewed as being fair. However, if the ratios are unequal, then individuals will see themselves as either under- or overrewarded and will be motivated to reestablish equity either by changing the levels of effort they put into a job, by changing the kinds of outcomes they seek from their jobs, or by altering their comparative others.

Expectancy theory argues that the strength of a tendency to act in a certain way depends on the strength of an expectation that the act will be followed by a given outcome and on the attractiveness of that outcome to the individual. According to this model, the strength of a person's motivation is conditioned on three sets of variables: (1) the *effort-performance linkage*, that is, the probablility perceived by the individual that exerting a given amount of effort will lead to high performance, (2) the *performance-reward linkage*, that is, the degree to which the individual believes that this level of performance will result in the attainment of a certain reward or outcome, and (3) the *importance* or *attractiveness* that the individual places on that particular reward or outcome. This motivational theory not only requires R&D managers to understand what outcomes and rewards professionals value from the workplace, but it also requires managers to establish the kind of organizational practices and processes that will help professionals build more reassuring reward linkages and expectations.

While equity and expectancy theories emphasize relationships between work motivation and extrinsic rewards, the job design model developed by Hackman and Oldham [1980] is the one most known for showing how *internal* work motivation is linked to the way job tasks are organized and structured. First, an individual must have *knowledge of results,* for, if the employee who does the work never finds out whether it is being performed well or poorly, then he or she has no basis for feeling good about having done well or unhappy about doing poorly. Second, the individual must *experience responsibility* for the results of the work, believing that she or he is personally responsible for the work outcomes. This allows the individual to feel proud if one does well and sad if one doesn't. Third, the individual must *experience the work as meaningful,* something that counts in the person's own system of values. According to Hackman and Oldham, all three of these factors, labeled psychological states, must be present for strong internal work motivation to develop and persist. Even individuals with important jobs, for example, can exhibit low levels of internal work motivation if they feel they are being micromanaged or have to follow too many bureaucratic procedures and, consequently, experience little personal responsibility for the outcomes of the work.

Most individuals display motivational problems at work when their tasks are designed so that they have little meaning, when they experience little responsibility for the work outcomes, or when they are separated from information about how well they are performing. On the other hand, if tasks are arranged so that the individuals who perform them richly experience the three psychological states, then even engineers, who often view themselves as basically lazy, will find themselves exerting more effort to do their jobs well. Motivation at work, then, may have more to do with how project tasks are designed and managed than with the personal dispositions of the professionals who are doing them.

However, what are the task characteristics that create conditions for high levels of internal work motivation? Hackman and Oldham suggest that three task dimensions are especially powerful for shaping the degree of meaningfulness experienced in one's job:

1. Skill variety: The degree to which a job requires a variety of different activities and/or involves the use of a number of different skills and talents.
2. Task identity: The degree to which a job requires completion of a whole and identifiable piece of work, that is, doing a job from beginning to end with a visible outcome.
3. Task significance: The degree to which the job has a substantial impact on the lives of other people either within the organization or in society at large.

Each of these three task characteristics contributes to the overall experienced meaningfulness of the work. A fourth task characteristic and the one that fosters increased feelings of personal responsibility for the work outcomes is *autonomy*, defined by the degree to which the job provides substantial freedom, independence, and discretion in scheduling the work and in determining the procedures to be used in carrying out requirements. As autonomy increases, the individual becomes more reliant on his or her *own* efforts, initiatives, and decisions. As a result, he or she begins to feel more personal responsibility and is willing to accept more personal accountability for the outcomes of the work.

Knowledge of the results of one's work is affected most directly by the amount of feedback one receives from doing the work itself. *Job feedback*, a fifth task characteristic, is the degree to which carrying out the requisite work activities provides the individual with direct and clear information about the effectiveness of his or her performance. Because a given job can be high or low on one or more of these five task characteristics simultaneously, Hackman and Oldham argue that the overall *motivating potential* of a job for fostering internal work motivation must consider the combined effects of experienced meaningfulness, responsibility, and knowledge of results in a *multiplicative* rather than in an *additive* manner. This kind of interaction among the task characteristics implies that a deficiency in the task characteristics associated with any one of the three psychological states cannot be compensated for by higher levels in the task characteristics connected to the other psychological states. The researchers of this job design model also emphasize that jobs with high levels of task characteristics do not *cause* individuals who work on those jobs to do well or experience job satisfaction. Instead, a job that is high in motivating potential simply creates the conditions for high levels of internal work motivation such that, if the person performs well, he or she is likely to experience a reinforcing state of affairs.

Based on this model of motivation, R&D managers can diagnose the jobs of their technical professionals to determine whether any of them are relatively weak with respect to each of the five task dimensions, namely, skill variety, task identity, task significance, autonomy, and job feedback. Jobs and assignments could then be restructured and redesigned to *enrich* them along any of the deficient dimensions. Fractionalized project tasks, for example, could be combined to form new jobs that would have more skill variety and task identity; engineers could be given more discretion or control over certain decisions in order to enhance their sense of autonomy; or technologists could join their marketing colleagues in project discussions with customers in order to increase their sense of task significance. Research by Katz [1982] has shown that the task characteristic that is generally most deficient in a technical environment, that is, the task dimension that has the lowest mean value is job feedback. There are probably several reasons for this: (1) in R&D settings, we often don't know what good performance is or what the best ideas are for quite some time, (2) technical managers and leaders are not particularly comfortable or well trained at giving feedback, and (3) technical professionals are also not particularly receptive to feedback that is not extremely positive. Although engineers and scientists may complain the most about the lack of feedback surrounding job performances and career paths, it is not the dimension that is most powerful for augmenting internal work motivation. Task significance is by far the most critical dimension for creating conditions for high levels of motivation. Technologists have to feel that they are working on something that is not trivial or low priority but something that is important — something that will make a difference! Interestingly enough, even though it is task significance that really enhances the motivational potential of a job assignment, most survey responses show that engineers and scientists say they want autonomy. Technical professionals may say they want a lot of autonomy; this does not mean, however, that they will thrive or feel very motivated in highly autonmous work situations.

Tenets of Motivation

After many years of carefully observing and studying successful motivational practices in R&D laboratories, Manners et al. [1997] have put forward a number of interesting **tenets** of motivation for the practicing R&D manager. These researchers also assert that too many managers confuse the concept of motivation with performance and behavior. Just because a group of professionals is excited about work and is exhibiting lots of activity, it does not necessarily mean that the group is productively or effectively active. Technologists could be highly motivated and end up running amuck perhaps because their motivational spirit is not well focused or well managed.

The first tenet of motivation is simply that generating incremental excitement about work is very difficult. On the other hand, destroying excitement and morale is relatively easy. The inexperienced R&D manager needs to understand this difficulty so that he or she does not get discouraged too quickly and give up trying to create a highly motivated work group. Another important observation offered by Manners et al. is that "fat, happy rats never run mazes." This does not mean that R&D managers should keep their technical professionals deprived. What is implied is that a *positive tension* needs to be present in order to have productive motivation. Rather than focusing only on the credentials of technical professionals, managers should also consider recruiting and selecting professionals who are capable of generating excitement or who have high achievement needs. This is similar to Pelz and Andrew's [1977] findings that technical professionals perform best and are most innovative when forces for stability and change are both present at the same time, that is, when technologists experience *creative tension*. A third interesting facet of motivation is the notion that emotion has almost no intellectual content. This creates problems for R&D because it is staffed by people of high intellect who believe you can intellectualize all motivational problems away. People, however, will do things simply because it *feels* good and so professionals need to have the celebrations and fun times, that is, the emotional opportunity to feel good not only about what they are doing but also with and for whom they are doing it.

Although the concept of *hedonism* is the fundamental principle that people seek pleasure and avoid pain, individuals are usually very different in what they like and dislike. One of the most critical errors a manager can commit is to make broad generalizations about what motivates all of his or her people. The more managers can recognize the individual differences among their personnel, the more they can tailor the use of informal rewards, including travel, equipment, assignments, etc., to motivate their employees. Manners et al. also discuss how managers need to communicate clearly how they intend to protect their technical professionals if they truly want their people to take more risks. They also advocate from their experience that incremental rewards and status symbols should only be associated with success and not the risk taking act itself. A more controversial tenet of the researchers is that managers should learn to distinguish between time spent on supervision and time spent on motivation. Although low performers generally require disproportionate amounts of supervisory time, managers should not spend more motivational time on them. Instead they should *invest* this kind of time on the higher performers. The argument is that one moves a group's mean performance to a new plateau by motivating the high performers not by rescuing the low performers.

Finally, the researchers conclude that the more-effective managers do not rely solely on the organization's formal reward system; instead, they employ a continual stream of informal rewards that they can deliver on a timely basis to generate employee excitement. These managers do not reward incremental performance every time nor do they use or rely on the same rewards every time. By keeping the system from getting too stable, they are able to prevent complacency and the expectations that rewards are entitlements. Ultimately, the capacity to motivate is dependent upon the manager's credibility. If the subordinate does not believe in the manager, he or she cannot motivate him or her. It is in this ability of the manager to control or get access to rewards and not let them get lost in a bureaucratic system that allows the employees to trust and build continued confidence in their manager.

Defining Terms

Motivation: The energy or willingness to do something and is reflected by the degree of excitement or arousal.

Model: An abstract representation of the relationships among a set of variables and concepts.

Need: A physiological or psychological deficiency that makes certain outcomes appear attractive.

Hierarchy: A governing body of individuals in job positions organized into orders or ranks each subordinate to the one above it.

Motive: An emotion, desire, or impulse acting as an incitement to action.

Intrinsic: Items that essentially reside within the individual.

Extrinsic: Items that essentially originate from outside the individual.

Equilibrium: A system that is in a state of balance.

Tenet: An opinion that is held to be true.

References

Badawy, M. *Management As A New Technology*, McGraw-Hill, New York, 1993.

Hackman, J. R., and Oldham, G. R. *Work Redesign*, Addison-Wesley, Reading, MA, 1980.

Herzberg, F. B., Mausner, B., and Snyderman, B. *The Motivation to Work*, John Wiley & Sons, New York, 1959.

Katz, R. The effects of group longevity on project communication and performance, *Admin. Sci. Q.*, 27(1): 81–104, 1982.

Manners, G., Steger, J., and Zimmerer, T. Motivating your R&D staff, In *The Human Side of Managing Technological Innovation*, R. Katz, ed., pp. 3–10, Oxford University Press, New York, 1997.

Maslow, A. *Motivation and Personality*, Harper & Row, New York, 1954.

McClelland, D. C. and Boyatzis, R. E. Leadership motive pattern and long-term success in management, *J. Appl. Psychol.*, 67:737-743, 1982.

Pelz, D. and Andrews, F. *Scientists in Organizations, 2nd ed.*, John Wiley & Sons, New York, 1977.

Pinchot, G. *Intrapreneuring*, Harper & Row, New York, 1985.

Roberts, E. *Entrepreneurs and High Technology: Lessons From MIT and Beyond*, Oxford University Press, New York, 1991.

Steers, R. M. and Porter, L. W. *Motivation and Work Behavior, 6th ed.*, McGraw-Hill, New York, 1995.

Further Information

Humphrey, W. *Managing for Innovation: Leading Technical People*, Prentice-Hall, Englewood Cliffs, NJ, 1987.

Katz, R. *The Human Side of Managing Technological Innovation*, Oxford University Press, New York, 1997.

Shapero, A. *Managing Professional People*, Free Press, New York, 1985.

7.2   Job Design

Janice A. Klein

Ever since the Industrial Revolution, people have been searching for the ideal job design, which is productive and cost effective, while also providing meaningfulness and motivation for the incumbent. There are several different schools of thought as to what is the "ideal", i.e., most efficient, job design. Although each was initially developed in response to organizational or competitive pressures during a specific time in industrial development, variations of each of the models continue to be in use today.

A Brief History of Job Design Models

Probably the oldest of the job design models, often referred to as the **craft** or **apprentice** model, dates back to the craft guilds of the Industrial Revolution. The primary goal in craft jobs is to develop complete knowledge and depth of expertise in a particular skill or discipline, such as machining, tool and die, welding or pipefitting. This expertise is typically acquired through multiple years of apprentice training, including both formal training and on-the-job instruction by a skilled craftsperson. Although this section focuses primarily on the design of traditional manufacturing jobs, this model is also the basis for many professional and quasiprofessional jobs, e.g., engineering, finance, marketing, etc.

The craft model is typically quite rewarding for incumbents in that they are viewed as the experts in a field that they often have chosen. From an organizational point of view, employees who are experts in a particular area require minimal supervision and can be looked to solve complex problems provided those problems do not transcend the boundaries of their craft. Unfortunately, as products and technologies become increasingly integrated and more complex, many problems tend to overlap various crafts. As a result, there is a need to find ways to better integrate skills either through the use of multidisciplinary teams or multiskilling (a subject we will return to in the discussion of **sociotechnical theory** and **lean manufacturing**).

All jobs do not require the extensive training inherent in the craft model. With the advent of the production line, many jobs were deskilled based on the principles of **scientific management**. Frederick Winslow Taylor, the father of scientific management, recognized that jobs could be broken down to minimize the amount of training required and, provided there was sufficient volume, the repetition of those narrowly defined jobs could move one down the **learning curve** to become quite proficient in a minimum amount of time. The critical element here is the need for high-volume production typically associated with assembly lines.

In contrast to craft jobs, which can be quite motivating, jobs designed based on scientific management principles (often referred to as Taylorized jobs) are typically viewed as boring and demeaning. Furthermore, the process of deskilling the work elements and knowledge required to perform those tasks takes away any understanding of how the individual jobs fit together as a whole. Problem solving and any improvements to the process must fall outside the job definition due to lack of depth of knowledge concerning the individual task and the integration of various individual jobs. As a result, Taylorized jobs require close supervision and high overhead support.

In response to negative social consequences of scientific management, social scientists developed an alternative model that aims to jointly optimize the social and technical systems. Sociotechnical theory was born in a British coal mine, where it had become apparent that the advent of new technology allowed workers to work in teams with minimal direct supervision [Emery and Trist, 1960]. This lead to the development of sociotechnical design principles, which include minimizing critical specifications to allow for individual worker autonomy and designing whole work tasks where there is minimal interdependence between work groups [Pasmore and Sherwood, 1978]. In an effort to meet these objectives, a structured methodology was developed to jointly analyze both the technical and social systems [Taylor, 1975]. Derivatives of this theory include high-commitment work organizations, high performance work systems, and self-directed work teams.

Sociotechnical work design attempts to create jobs that are motivating and enhance employee commitment and ownership toward the task at hand. In this regard, five job characteristics have been found to create individual commitment to work: task variety, task identity, task significance, autonomy, and feedback [Hackman and Oldham, 1980]. In order to accomplish these objectives, early sociotechnical researchers suggested that jobs of individuals and/or teams be decoupled in an effort to provide maximum independence, i.e., autonomy. In recent efforts to streamline value chains, there has been a greater emphasis placed on meeting internal customer needs as opposed to individual autonomy [Klein, 1991].

International competition within the automobile industry forced American manufacturers to investigate how Japanese manufactures achieved superlative levels of productivity. Researchers uncovered the Toyota Production System, which has since been labeled lean manufacturing [Monden, 1983; Womack et al., 1990]. To counter the inflexibility of traditional scientific management, production-line workers under lean manufacturing perform multiple tasks (typically eight to ten) within their team. In addition, there is a recognition that production-line workers are in the best position to identify quality defects on the line and find incremental continuous improvements. This is in contrast to Taylor's belief that the

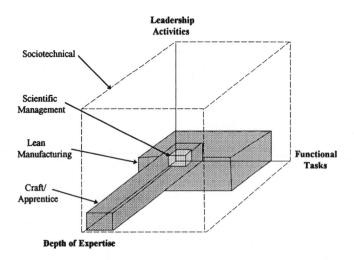

FIGURE 7.1   Comparison of job design models. (*Source:* Adapted from Klein, J. A. *Advance in Interdisciplinary Studies of Work Teams, Vol. 1*, M. M. Beyerlein and D. A. Johnson, eds., p. 155. JAI Press, Greenwich, CT, 1994.)

industrial engineer could identify the one, most-efficient work method. Another critical element of lean manufacturing is standardized work procedures, often developed by the workers themselves, to minimize process variation.

A Comparison of the Four Models

Figure 7.1 provides a graphical comparison of the four job design models along three dimensions:

1. The breadth of tasks that workers are expected to perform.
2. The degree to which workers assume their own leadership and administrative activities.
3. The overall depth of expertise that is expected of workers performing jobs.

The breadth of tasks performed can be readily identified and compared. The scientific management model restricts workers to one task to minimize training costs and speed learning, while the craft model expands the number of task to encompass all activities within a functional group. Although the lean model broadens the number of functions within the teams purview, the boundary is still within production. Hence, the sociotechnical model is the broadest since it includes functional support activities such as materials, quality, and maintenance.

The degree to which workers assume their own leadership and administrative activities follows a similar pattern. In the scientific model, workers are not expected to assume any leadership or administrative activities. Although the same could be said for the craft model, the depth of expertise associated with knowing the entire function typically leads to a small number of administrative tasks, such as making entries into a maintenance journal, etc. The horizontal axis becomes a conscious factor in both the lean and sociotechnical models. In the lean model, workers are expected to "**kaizen**" or continuously improve their work tasks. In addition, they are given the authority to stop the line if a defect occurs. Self-direction does not come into the picture, however, until one moves to the sociotechnical model, which makes self-management of daily tasks a conscious objective. In this model, many of the activities of the traditional first-line supervisor are assumed by the team members.

The depth of expertise axis is the most complex because it includes the operational knowledge of how to perform a task, an understanding of the scientific or analytic principles that underlie why a task should be performed a particular way, and the ability to integrate all the individual tasks into a whole. It is a combination of these three that leads to a high level of expertise in both the craft and sociotechnical models. However, the craft model is probably greater in the first two elements, while the sociotechnical model will have more integrative depth since the tasks typically cross multiple functions. While not

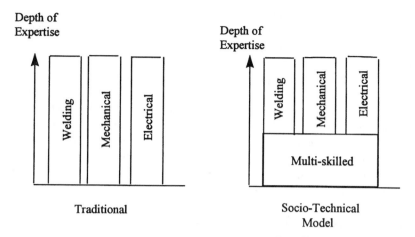

FIGURE 7.2 Balancing breadth and depth within the socio-technical model. (*Source:* Adapted from Klein, J. A. Advance in Interdisciplinary Studies of Work Teams, Vol. 1, M. M. Beyerlein and D. A. Johnson, p. 155. JAI Press, Greenwich, CT, 1994.)

having as deep an expertise as the craft and sociotechnical models, the lean model requires greater depth of expertise than the scientific management model, primarily due to the requirement to kaizen the work.

Balancing Breadth and Depth

Another way to compare the models is to look at the extent to which each worker is to fill the entire volume of the space identified in Fig. 7.1. While job incumbents in the scientific management, craft, and lean models are expected to acquire the skills and competencies of the entire cube, the breadth, depth, and height of the sociotechnical model is typically beyond the capabilities of any one team member. Hence, it is the team, rather than any single individual team member, that covers the three dimensions of the sociotechnical space.

Within a sociotechnical team, there is a need to create the appropriate balance of breadth and depth of skills. Figure 7.2 illustrates the difference between traditional functional job design and the multiskilling that can occur under the sociotechnical model. The level of multiskilling will vary based on the degree to which cross-training is necessary for daily task execution and team decision making as well as a cost-benefit analysis of the training costs. In highly technical fields, such as engineering, the focus on multi-skilling is in the area of integrative knowledge as opposed to specific functional tasks.

The idea that individual team members may have different task assignments or levels of expertise is a relatively new idea for production work even though it is commonplace for professional and quasiprofessional work. In the past, a supervisory work group would typically be comprised of individuals with similar skill levels in the same job classifications. This assured that the supervisor had sufficient knowledge of all the tasks to monitor the work. Since workers in the sociotechnical model manage their own daily tasks, the supervisor becomes more of a coach or facilitator for the team and a boundary manager, assuring the team has proper resources to meet their goals.

Infrastructure Needed to Support Job Design

The choice of job design model should be based on the organizational and process technology needs, coupled with the management philosophy of how to organize work. In addition, managers should recognize the importance of aligning the support infrastructure with the job design model. The following list identifies several areas that should be considered:

Redesigned managerial and support roles — In addition to changes in the role of first-line supervisors, there are also significant changes in functional support roles, especially under the sociotechnical model where production workers assume many of the more routine functional support tasks. This frees up functional staffs to apply their expertise to more in-depth problem solving or long-term planning. There is often, however, fear of letting go of the daily tasks by both supervisors and support staff due to a concern over job security.

Human resource systems — Job designs directly impact job classifications. In unionized settings where job classifications are part of the labor contract, any changes will require negotiations with the local union bargaining committee. Since most job classifications within the United States have been based upon either the craft or scientific management models, there is often some apprehension on the part of unions to move toward either the lean or sociotechnical models. The lean model appears to be most problematic since it has often been equated with job reductions, i.e., the job flexibility provided by multiskilling reduces the number of workers provided there is no increase in production levels. The sociotechnical model, on the other hand, provides an opportunity to broaden the jurisdiction of traditional labor contracts into staff areas, which alleviates the impact of employee layoffs due to increased flexibility.

Measurement systems and rewards/recognition — The craft and scientific management models have traditionally been associated with individual performance measures. Since the individual jobs revolved around a specific task or set of tasks, there was no need to consider broader team or organizational metrics. Both the lean and sociotechnical models place more emphasis on team activities and the contribution of each team member to the overall team performance. Hence, new measurement systems are needed. On the surface, this appears to be a rather straightforward task, but there needs to be a fine balance among organization, team, and individual contribution. If only individual or team performance is measured, there is a strong possibility that a recommendation or action may optimize the local situation while suboptimizing the total system. On the other hand, measuring only the total organization masks the contribution of individuals or teams, and it is also difficult for individuals and/or teams to act on overall system measures. There is also the potential of losing individual accountability and demotivating individuals within a team. The obvious solution is a combination of measures with the situation dictating the amount of emphasis given to any one in particular.

Communication/information sharing — Although the amount of information shared is a philosophical decision based on management's willingness to share information, it is also a critical component concerning the viability of the different job design models. Under scientific management, the assumption is that production workers do not need any information other than to be instructed in how to perform their immediate task; managers and engineers are responsible for coordinating activities and problem solving. In the craft model, workers may need a bit more information to handle the full range of functional activities, but there is typically little need to share information beyond one's functional group. Again, it is management's responsibility to coordinate between individual craft workers. However, as one moves to the lean or sociotechnical models, there is an increased need for information sharing as workers assume greater leadership and administrative activities.

In summary, organizations that are considering utilizing either the lean or sociotechnical job design models must undertake a redesign of their entire infrastructure to reap the maximum gains of the increased flexibility.

Defining Terms

Apprentice: Historically a training period of several years where a master craftsperson instructs a trainee.

Craft: A trade (e.g., machinist, tool and die, welding), discipline (e.g., electrical engineering, mechanical engineering), or function (e.g., marketing, finance).

Kaizen: A process for continuous improvement.

Lean manufacturing: Typically synonymous with the Toyota Production System.

Learning curve: The reduction in hours per unit that is associated with increased cumulative units produced.

Scientific management: The breaking down of work into individual discreet units.

Sociotechnical theory: The joint optimization of the technical and social systems.

References

Emery, F. E. and Trist, E. L. Socio-technical systems, In *Management Science, Models, and Techniques,* C. W. Churchman and M. Verhurst, eds. pp. 83–97, Pergamon Press, London, 1960.

Hackman, J. R. and Oldham, G. *Work Redesign,* Addison-Wesley, Reading, MA, 1980.

Klein, J. A. A reexamination of autonomy in light of new manufacturing practices, *Hum. Relat.,* 44(1): 21–38, 1991.

Klein, J. A. Maintaining expertise in multi-skilled teams, In *Advance in Interdisciplinary Studies of Work Teams, Vol. 1,* M. M. Beyerlein and D. A. Johnson, eds., pp. 155. JAI Press, Greenwich, CT, 1994.

Monden, Y. *Toyota Production System: Practical Approach to Production Management,* Industrial Engineering and Management Press, Norcross, GA, 1983.

Pasmore, W. A. and Sherwood, J. J. *Sociotechnical Systems: A Sourcebook,* University Associates, San Diego, CA, 1978.

Taylor, F. W. *The Principles of Scientific Management,* Harper & Brothers, New York, 1911.

Taylor, J. C. The human side of work: the socio-technical approach to work system design, *Personnel Rev.,* 4(3): 17–22, 1975.

Womack, J. P., Jones, D. T., and Roos, D. *The Machine that Changed the World,* Rawson Associates, New York, 1990.

Further Information

Baugh, R. *Changing Work: A Union Guide to Workplace Change,* AFL-CIO Human Resources Development Institute, Washington, DC, 1994.

Majchrzak, A. *The Human Side of Factory Automation: Managerial and Human Resource Strategies for Making Automation Succeed,* Jossey-Bass, San Francisco, 1988.

Weisbord, M. R. *Productive Workplaces: Organizing and Managing for Dignity, Meaning, and Community,* Jossey-Bass, San Francisco, 1987.

7.3   The Logic of Organization Design

Marc S. Gerstein

Organization design consists of the configuration of owned and "virtual" business capabilities; the structuring of reporting relationships and administrative units; the engineering of operating, management and information systems; and the fostering of the organization's different "cultures" to enable its various constituencies to accomplish their respective objectives.

In contrast to previous approaches that focused primarily on organization *structure,* this section encompasses a wider perspective. The approach is based on the simple premise that organizations are fundamentally a means to accomplish the work and purposes of society.

The Nature of Design

Except when computers do the job, the process of design, be it architectural, graphic, engineering, or philosophical, is always a messy, iterative affair. Organization design is one of the most problematic areas because there are so many component parts to the design problem, the relationships are so complex, and, after all is said and done, the end product must be explained to the large number of people who have a significant stake in the outcome.

Thus, the content in this section can only provide a rough guide to a process that inevitably will be different in practice than it is in theory. It is also likely to be different for each instance of design, even when practiced by experienced designers.

Design involves three sets of related issues: functional problem solving, engineering, and aesthetics.

- Functional problem solving deals with the *capability* of a design to fulfill its objectives. In other words, it addresses the essential viability of the design as a solution to the stated problem.
- Engineering deals with the *realization* of the design utilizing the people and technology available and within the conditions under which it must operate. This addresses issues of practicality, including economic considerations.
- Aesthetics deals with design characteristics other than function, performance, and cost that might cause us to prefer one solution over another. Surprisingly, perhaps, in organization design, intangibles such as aesthetics matter as much as they do in any other arena. Let us now consider each of the three sets of issues [Mitchell, 1990].

Functional Problem Solving

A functionally effective design is one that satisfies the purposes for which it was created. Generalized statements such as "satisfying customers" or "making money" are of some use in this analysis, but a more productive approach may be to articulate the organization's specific functional requirements such as "facilitates rapid new product development based on market needs." We call such statements **design criteria** because they specify the basis upon which a design can be evaluated and against which alternative designs can be compared.

With design criteria in mind, it is possible to develop a range of *alternatives* to perform the desired functions. This creative phase is what most people think of when they use the word "design". In fact, design has been described as "an iterative conversation with materials" (see Hirasuna [1997]), which suggests that at any point in time there will be alternative ways to create analogous outcomes, although some are likely to be better at some things than are others. Consequently, clearly understanding one's purposes and carefully evaluating one's options using design criteria is at the center of the functional aspect of the design process. Alternative designs produce similar results but not identical ones, and there are often differences in the details that have a significant impact on the outcome in a competitive setting.

Organizational Engineering

The engineering elements of design address whether functional solutions can be built in the real world, what they will cost, and what they require for ongoing operations. Just as the design of a bridge has a set of engineering requirements having to do with stresses and loads, so too does an organization. It is perhaps surprising that organizations often fail because of simple "engineering mistakes" such as conflicting lines of authority, ambiguous boundaries, or incentive systems that undermine the structure.

It might be useful to note at this point that information technology has become a *structural material* for organization design. Just as the telegraph and typewriter "enabled" the large-scale functionally specialized firm at the turn of the century, and structural steel enabled the skyscraper during the same time period, modern information technology enables organizational architectures such as the multi-organizational network that were simply not realizable in earlier eras. Note, however, that employing new materials does not *create* a new architecture, but, in the hands of a gifted designer, it can make one possible [Gerstein, 1992].

The Human Factor

Although one might not automatically include it, the role of people as well as technology figures prominently in an organization's "engineering". The defining case was the introduction of mechanized mining equipment into the British coal mines in the 1940s. Work system changes encouraged by the new technology inadvertently destroyed the teamwork and safety procedures that had long evolved to deal with the materials flow requirements and hazards of working underground. The lesson from this case is that one cannot design one aspect of a work system without consideration of the others, and one cannot, for that matter, change one part without having a significant impact upon the rest.

This means that organizations are "sociotechnical systems", a term coined half a century ago (by researchers studying those same coal mines) to capture the interconnections between people and technology. As an obvious example, if two groups must closely collaborate due to the requirements of the functional design, it is essential that they speak the same language (both literally and figuratively) and that the technical infrastructure and physical logistics make it easy for them to communicate.

In this context, people are not just a "change management" issue to be added to the equation at the implementation phase; rather, their needs, aspirations, and values are cornerstones in the initial design of the enterprise and just as basic to shaping its eventual form as its economic and technical foundations.

When one considers the modern organization's need to achieve coherent effort by bringing together people from different companies, ethnic groups, and national cultures, the human aspects of organization design — including establishing the very purposes for which modern organizations exist — are certainly among the most challenging aspects of enterprise creation. Furthermore, as intellectual contribution continues to become ever more important, harnessing human "capital" will become increasingly central to organization design just as vertically integrated structures harnessed financial capital in the last century.

Design "Aesthetics"

When most people think of good design in their everyday lives, they inevitably consider aesthetics as well as functional performance. Although we often are forced into trade-offs, most of us would rather have the best of both worlds. In organization design, aesthetic considerations include clarity and simplicity, recognizable repeating patterns, and graceful harmony among design elements.

Organizations that look like diabolical Rube Goldberg machines — even if the engineering is technically viable — are often incomprehensible to the people who have to make them function and to their customers. We have all been victims of impenetrable corporate or governmental bureaucracies with arbitrary rules and nonsensical procedures. Like a good appliance, an organization should operate the way we expect it to. If it requires a thick instruction manual or the need to engage in counterintuitive behavior in order to operate successfully, it will probably waste a great deal of time, since people tend to follow their instincts rather than the "book". Harnessing people's intuition and common sense is not essential to making things work, but more often than not it is good design.

Graceful Failure

Finally, as one formulates the overall design solution, it is important to keep in mind that sociotechnical systems interact in complex, often counterintuitive ways. For example, if any system is stressed beyond its design limits it is likely to fail. Good design requires that it "fail gracefully". Chaotic failure, in which small changes in conditions lead to dramatic and irreversible consequences, makes it almost impossible for the organization to learn and "do better next time". When things collapse totally, the damage is too great, the human cost of failure too high, and the road back destroyed.

Unfortunately, in their quest for better risk management, many organizations have inadvertently made it difficult for people to learn from mistakes. Ironically, while "failure proofing" has often made mistakes less frequent, they now often occur with striking suddenness, are bigger when they hit, and are more complex to diagnose and remedy. Since most organizations can only afford to "underwrite failure" when mistakes are modest and noncataclysmic, graceful failure is essential to learning and must therefore be a key design objective [Sullivan and Harper, 1996].

Eras in the Evolution of Organization Design

From a design perspective, at their most basic level, organizations are tools — albeit complex ones — and, like all of mankind's tools, they must be understood in context. To explore the choices in organization design that are available to a manager today, it is therefore useful to have a historical perspective on the evolution of organizations and to place their development in the framework of larger environmental circumstances.

TABLE 7.1   Eras in Organization Design

Individual businesses	Before 1870	Modest, owner operated enterprises
		Narrow range of economic activities; e.g., manufacturing
		Small scale, local markets
		Market-based coordination and pricing
Functional organization	1870–1920	National markets
		Range of economic activities; vertical integration
		Internal coordination
		Consolidated manufacturing and distribution
		Increasing range of products in later years
Multibusiness divisionalized form	1920–1960	Business groups with corporate support staff organizations
		Specialized senior management roles focusing on strategy, resource allocation, and performance review
Matrix	1960s	Dual chain of command, typically functional and either market, product, or project
		Matrix label sometimes applied to project management systems of a variety of forms although they are not technically matrix structures
Strategic business units (SBUs)	1970s	Businesses expected to act largely like separate companies, despite their corporate membership
		Formal profit and loss responsibility
Outsourcing	1980s–present	Initially applied to data processing and nonessential services
		Incentives were operational improvements, cost benefits arising from market efficiencies, and financial benefits arising from structure of outsourcing contracts
Virtual organizations	1980s–present	Logical expansion of outsourcing to full range of business activities, including strategically critical functions such as manufacturing
Asymmetrical networks	1980s–present	Relaxing of "one size fits all" organization design rigidity
		Businesses played a variety of strategic roles based on capabilities and marketplace characteristics
"Deconglomerization"	1980s–present	Break-up of multibusiness firms into smaller groups of businesses with similar characteristics and clear synergies
		Often motivated by desire to improve stock market valuations of corporations containing businesses with different price/earnings characteristics

New Organizational Arrangements

Table 7.1 reveals a historical evolution from market-based coordination to centralized companies and back to markets again. It suggests that organization design is moving to where it once was — a clear case of "the wheel has come full circle". This is not, however, the whole story. In response to new conditions, two new organizational forms have come into being that did not exist to anywhere the same degree in earlier times. The first is the **decentralized, connected network**, and the second is the **radically decentralized organization** or "meta-organization".

Decentralized, connected networks are comprised of entities acting separately yet in a coordinated fashion. They obtain the benefits of overall coordination without the need for a central authority by taking advantage of their ability to communicate directly. The global trading of currency, for example, operates in this fashion. Despite adding value that would, under other circumstances, be provided by traditional organizations, networks are typically not considered to be "designed" organizations as such, especially if they appear to arise spontaneously as a result of a rich set of direct connections.

A somewhat different set of circumstances has arisen to support radically decentralized meta-organizations, such as Visa International and the Internet. These organizations are extremely complex to set up and maintain because they must integrate the needs of a large number of separate entities, many of whom are competitors in their respective products and markets.

Despite the effort required, the benefits are clearly worth it. Visa has grown to $650 billion in annual transaction volume and a position as one of the world's most recognized brands, results that could never have been achieved without considerable coordinated effort. Yet Visa has no traditional hierarchy as such, and does not make profit in the conventional sense of the term. It is run by a series of regional organizations, each receiving a mandate from those entities *below* it (rather than from those above in a conventional organizational sense) [Malone, 1997].

The Internet is similar in the sense that it has no conventional center. There is no one to shut it down or to give permission about what business can and cannot be done on it. As long as the technical "rules of the road" are obeyed, the Internet belongs to everyone.

While such unbridled freedom may not be everyone's cup of tea, the Internet has been doubling in size every year since 1988. Its uniquely egalitarian character has facilitated knitting together of a large number of separate and previously incompatible e-mail and networking systems. Such a feat could never have been accomplished by a single vendor (or consortium, for that matter), given the practical impossibility of creating a central organization analogous to AT&T in 1900, which leveraged its dominance of "long lines" traffic into a monopoly position that could dictate standards [Chandler, 1977]. In the absence of the Internet, it is unclear what shape the world's electronic networking would be in.

Phases in the Organization Design Process

In practice, the organization design sequence is less tidy than is outlined below, and will likely loop back on itself. Furthermore, since a new design always implies change, experience suggests that it is useful to consider overall change issues, such as ensuring participation and the need for communications, from the beginning rather than leaving them to the end, as is common practice.

1. Clarify the purpose of the organizational redesign, articulating the design criteria to be used to contrast and compare design alternatives. Consider overall change management requirements in view of the nature and scale of the redesign being undertaken.
2. Analyze the industry and company value chains to develop an understanding of the business' fundamental processes, basic economics and risks, and potential alternative foundations for competitive advantage.
3. Identify and evaluate alternative configurations of "strategic components", focusing on those parts of the value chain that should be conventionally owned vs. those that should be provided through partnership, supply, and outsourcing arrangements. Decide on the nature of the macrostructure and other arrangements required to achieve the necessary coordination between separate organizations.
4. Generate specific "design options" for one's own organization, including the overall organizational architecture, major processes, fundamental information technology infrastructure, structural grouping and linking mechanisms, etc. Include external organizations as well as internal ones within the scope of the design to ensure necessary horizontal integration.
5. Identify additional requirements to achieve overall objectives in areas such as performance measurement and reward, staffing and selection, knowledge leverage, organizational culture, and leadership.
6. Identify implementation issues, such as managing the approval process, planning the communications/roll out logistics, overcoming specific sources of resistance, and so forth.

This process is elaborated upon in Section 7.5.

Defining Terms

Design criteria: The basis upon which a design can be evaluated and against which alternative designs can be compared.

Decentralized, connected network: A group of entities acting separately, yet in a coordinated fashion. The benefits of overall coordination are obtained without the need for a central authority by taking advantage of the entities' ability to communicate directly.

Radically decentralized organization (meta-organization): An organization consisting of other organizations, most typically linked on a voluntary basis and deriving its power from the "bottom up" rather than the "top down".

References

Chandler, A. D. *The Visible Hand,* Harvard University Press, Cambridge, MA, 1977.

Chesbrough, H. W. and Teece, D. J. When is virtual virtuous? Organizing for innovation, *Harv. Bus. Rev.,* Jan–Feb, 1996.

Gerstein, M. S. From Machine Bureaucracies to networked organizations: an architectural journey, In *Organizational Architecture,* D. N. Nadler, M. S. Gerstein, and Shaw, eds., chap. 1, Jossey-Bass, San Francisco, 1992.

Gillet, S. E. and Kapor, M. The Self-Governing Internet: Coordination by Design, Center for Coordination Science working paper 197, Sloan School of Management, MIT, 1997.

Hirasuna, D., quote of Don Schon of MIT in Interview with James Moore, *J. Bus. Design,* 3(1), 1997.

Malone, T. W. Is empowerment just a fad? Control, decision-making, and IT, *Sloan Mgt. Rev.,* 38(2): 23–36, 1997.

Mitchell, W. J. *The Logic of Architecture: Design, Computation, and Cognition,* MIT Press, Cambridge, MA, 1990.

Sullivan, G. R. and Harper, M. W. *Hope Is Not a Method,* Random House, New York, 1996.

Further Information

The Center for Coordination Science, MIT Sloan School of Management, believes that a powerful source of intellectual leverage on questions about how groups of people can use computers will result from a better understanding of the nature of coordination. Work in the Center focuses on how coordination can occur, both with and without technology, including projects in the following areas: coordination technology (designing and studying innovative computer systems that help people work together in small or large groups), organizational structures and information technology (observing, analyzing, and predicting how information technology affects organizational structures and processes), and coordination theory (developing and testing theories about how activities can be coordinated among people and in distributed or parallel processing computer systems. For further information use the World Wide Web at http://ccs.mit.edu/ccsmain.html.

The MIT Organizational Learning Network. MIT Sloan School of Management, is a community of scholars and practitioners interested in ideas, research, and practice related to organizational learning. For more information, use the World Wide Web at http://learning.mit.edu/index.html.

7.4   Hierarchy and Integration in Organizations and their Products

Jesse Peplinski, Patrick N. Koch, and Farrokh Mistree

Vertical and horizontal integration are terms that have received considerable attention in the research communities of **organization design** and strategic management. Interestingly, however, neither of these terms has single commonly accepted definitions; instead, separate definitions exist, depending on the context of the example. In this section, a broader perspective on vertical and horizontal integration is taken by recognizing the fundamental underlying concept of *hierarchy*. This concept is used in the first section to unite and explore these separate definitions of integration; the advantages and disadvantages are explored for integration both within a single firm and between separate firms. In the second section, the concept of hierarchy is explored in more detail in the context of designing the *products*, or technical systems, of manufacturing organizations. The activities of decomposition, analysis and recomposition are introduced. Finally, the potential for integration and alignment between an organization and its products themselves is proposed.

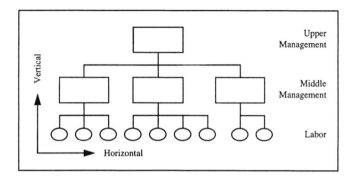

FIGURE 7.3 Hierarchy within an organization.

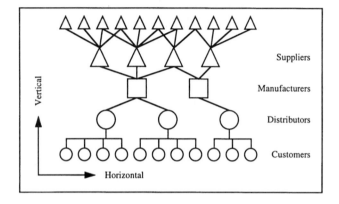

FIGURE 7.4 Hierarchy in a supply chain.

Hierarchy in Organizations: Vertical and Horizontal Integration

Hierarchy arises as a way for dealing with complexity and interdependence. It is perhaps an *imperative*, evident in the structure of natural systems [Simon, 1981] as well as in the information-processing structure of organizations [Galbraith, 1994]. Instead of requiring every employee in an organization to communicate and coordinate with every other employee, a hierarchical structure allows localized groups of people to form around common tasks, functions, or projects, thus reducing the span of interaction to reasonable proportions. The leaders of these groups then communicate and coordinate with each other, and if the number of these leaders is again overwhelming then further levels in a hierarchy can be formed. The notion of a hierarchy within an organization is illustrated in Fig. 7.3. On a larger scale, hierarchies also exist in the marketplace in the form of supplier chains. In this case each element of the hierarchy may be an entire organization in itself. This is illustrated in Fig. 7.4 in which multiple supply chains are represented, starting from component and subsystem suppliers to manufacturers, distributors, and customers.

Although hierarchies help filter out complexity to a manageable level, there are several disadvantages of hierarchical structures that sometimes become significant. These disadvantages provide the impetus for *integration* efforts, both vertically and horizontally within a single organization and across entire supply chains. The motivations and potential risks for each of these integration efforts are discussed next.

Integration within Organizations

Vertical integration within an organization is equivalent to "flattening out" the hierarchy. These efforts are evident in firms that reduce the size of middle management or remove measures of status or rank between employees. However, these efforts are not truly aimed at doing away with the hierarchy but instead are focused on "eliminating the dysfunctional effects" [Galbraith, 1994] of hierarchy — the barriers to communication that grow from differences in rank and the managerial behaviors that grow from a command and control model. Trends in vertical integration may continue as information technology advances allow larger and larger numbers of people to communicate and coordinate with one another, but hierarchical structures will persevere.

Galbraith [1994] defines horizontal (or lateral) integration in terms of coordinating different functions (or business units or divisions) without communicating through the hierarchy. In other words, people in different functions communicate directly with each other instead of through their managers. This integration within an organization is akin to the development of cross-functional teams across an organization devoted to the resolution of a particular issue. This form of integration is evident in matrix organizations [Ulrich and Eppinger, 1995] in which a functional hierarchy exists but product development teams are also formed to guide the development and production of individual products, each guided by project managers.

The advantages of horizontal integration are that decisions can be made more quickly and by people with access to better (more current and more local) information. Such integration also may give the organization more flexibility, and it frees upper management time to focus on longer run or external issues. However, there are several potential disadvantages: the quality of decisions may decrease because local decision makers do not see the big picture, each decision maker is likely to spend much more time communicating across the hierarchy, thus taking time away from dealing with customer and supplier issues, and, as with any increase in distributed and group decision making, the potential level of conflict within the organization rises.

Integration within Supply Chains

Vertical integration within a supply chain refers to the linkage of a firm to "upstream" or "downstream" firms in the complete production process, whereas horizontal integration describes the merger or alignment of competitors at the same level of similar supply chains. If a potential for integration across both dimensions is possible, then vertical integration appears to be dominant; a move for vertical integration can be employed to preempt the threat of a detrimental horizontal integration. Vertical integration has received the majority of attention in the strategic management literature and thus is the focus in the remainder of this section.

A strong motive for vertical integration, most noted in the semiconductor industry, is the drive to capture more of the value added of a product along the supply chain. By acquiring its suppliers a firm is able to reap a higher percentage of the rewards of bringing a product to market. Similarly from the perspective of an innovator integration is a strategic option if the innovation needs complementary assets for commercial success [Teece, 1986]. If the assets are specialized and critical, if the innovator firm's cash position is solid, and if its competitors are not better positioned, then an integration strategy can be pursued to bring the innovation to market. (Otherwise the firm can contract for access.) However, it is important to note that vertical integration is a risky proposition; there is a history of firms that vertically integrate and fail to improve their growth or control over the industry. Stuckey and White [1993] examine four common reasons to integrate and warn managers against a number of other spurious reasons.

Hierarchy in Products: Theory, Definitions, and Implications

Integration efforts within organizations and supply chains are initiated to deal with interdependence, and especially in manufacturing organizations this interdependence is due to the complexity of the products they produce. Complex products are often represented through a hierarchical structure of information, and these structures drive the structure of the organizations themselves. In this section hierarchy is defined in the context of **product design**, and the notions of system decomposition, or partitioning, and recomposition are introduced. These concepts are then applied to explore the correlations between product and organization hierarchies.

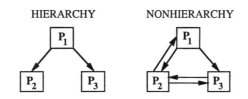

FIGURE 7.5   Hierarchy vs. nonhierarchy: strict definitions.

Hierarchy vs. Nonhierarchy — Definitions

Strictly defined, hierarchy refers to two or more levels of parent-child relationships. In this definition information in a hierarchy only flows from parent to child; a parent can have more than one child, but each child is independent and can have only one parent. This definition of hierarchy is depicted in Fig. 7.5 and contrasted with a representation of nonhierarchy, where each boxed "P" refers to a product or product design problem or a part thereof. On the left side of Fig. 7.5, P_2 and P_3 are subsystems or components of product P_1 in a product hierarchy. On the right side of Fig. 7.5 a nonhierarchy is characterized by information flowing between all boxes; a single parent cannot be identified. A benefit of hierarchy is that a sequential order for solving the product design problems can be determined; in a nonhierarchy the subsystems are interdependent and thus iteration with feedback is necessary.

Product Decomposition and Partitioning

In contrast to the usually preexisting hierarchies in organizations, hierarchies in products are *created* through a process of *decomposition* (formal methods) or a process of *partitioning* (intuitive methods). General guidelines for system decomposition [Sobieszczanski-Sobieski et al., 1984] are

1. Breaking the overall large task into a number of smaller, self-contained subtasks along interdisciplinary lines or the physical divisions among the subsystems.
2. Preserving the couplings between subtasks.
3. Carrying out concurrently as many subtasks as possible to develop a broad workfront of people and computers.
4. Keeping the volume of coupling information small relative to the volume of information that needs to be processed internally in each task.

The goal in decomposition is to break the system or problem along lines (disciplinary or physical) that make the subsystems or subtasks as *independent* and *self-contained* as possible and to minimize the number of couplings that complicate the system level problem (integration). Many studies have been devoted to the decomposition of large systems and optimization problems, and many approaches for performing decomposition exist; an excellent review of hierarchical decomposition is presented by Renaud [1992]. The development of nonhierarchical decomposition schemes is relatively new; a review of the early work in nonhierarchical decomposition is also presented by Renaud [1992].

Two basic classes of decomposition methods exist [Sobieszczanski-Sobieski et al., 1984]: formal methods and intuitive, or heuristic, methods. With formal methods a mathematical structure of the problem is used to derive a decomposition scheme. In these approaches the strict definitions of hierarchy and nonhierarchy, as depicted in Fig. 7.5, are observed. Alternatively, intuitive or heuristic approaches, often referred to as *partitioning* rather than decomposition, are based on the natural functional or physical partitioning of a system into its subsystems. In these less-formal partitioning methods, the definitions of hierarchy and nonhierarchy given at the beginning of this section begin to break down.

In Fig. 7.6 a system is partitioned into three levels (two levels of subsystems). Clearly this representation is a hierarchy. For a technical system, however, subsystems must interact with each other and are thus not independent, creating a paradox in hierarchical system partitioning. Subsystem behavior and performance drives system behavior and performance, and thus information travels up the hierarchy as well as down, thus conflicting with the strict definition of hierarchy.

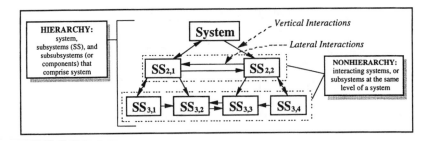

FIGURE 7.6   Hierarchy and nonhierarchy represented through system/subsystems (SS) partitioning.

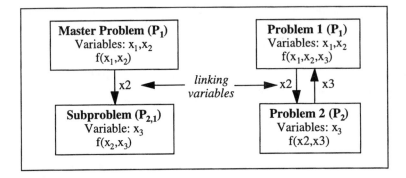

FIGURE 7.7   Hierarchy in the analysis and design of decomposed problems.

At any given level of a product or system hierarchy, subsystem level 2 or 3 in Fig. 7.6 for example, a nonhierarchy exists: multiple interdependent subsystems interact with each other with no apparent ordering. Applications such as in Fig. 7.6 create a need for modified definitions [Shupe et al., 1987] of hierarchy. In hierarchical design

- There are interactions between the various levels of subsystems (vertical interactions). These interactions may go one way or both ways.
- There are interactions between the subsystems at the same level of the same parent or of different parents (lateral interactions). These interactions can go one way or both ways.

The many interactions between the system, subsystems, and components of a complex product design, as shown in Fig. 7.6, define the need for integration, both lateral and vertical. The solution techniques for solving these multidimensional problems seek to facilitate the (subsystem analysis) integration necessary to effectively evaluate a system design alternative.

Analysis of Hierarchically Decomposed/Partitioned Systems and Recomposition

The decomposition of large technical systems has long been viewed as being beneficial to the efficient design of that system. Although breaking a system up into smaller, less-complex subsystems may allow for the effective solution at the subsystem level, decomposition complicates the system design problem by requiring the coordination of compatible subsystem solutions into a feasible system solution. The problem of analyzing and synthesizing various subsystems poses a difficult problem. Given the strict definition of hierarchy and the formal (mathematical) decomposition methods, the analysis of a hierarchical problem proceeds sequentially from a master (parent) problem to one or more subproblems as depicted on the left side of Fig. 7.7. In this figure, the master problem mathematically is a function of two variables: x_1 and x_2. Once solved, x_2 is passed to the subproblem, which is a function of x_2 and x_3.

The system representation of Fig. 7.6 and the associated analysis are more complicated as both lateral and vertical interactions, occurring in both directions, must be modeled and analyzed. The right side of

Fig. 7.7 represents two interdependent problems, strictly defined as a nonhierarchy, which could be viewed as two subsystems of a system. In this case the problems cannot be solved sequentially; each problem requires information from the other, and iteration must occur. This second case is unfortunately more common, as complex systems are multidisciplinary and multidimensional in nature. Both single-level and multilevel approaches for solving the interdependent problems of a decomposed system are being developed. A review and comparison of multidisciplinary design optimization approaches is given by Balling and Wilkinson [1996].

Integration across Organizations and their Products

In the previous two sections the concepts of hierarchy and integration have been explored in the separate domains of organization design and product design. Although the discussions have gone in substantially different directions, it is important to note the underlying relationship that brings the two of them together. It is that the hierarchical decomposition, concurrent design, and integration of the subsystems of a product often parallel the hierarchical structure of an organization. Often an organization is structured to facilitate the effective development of a product that has been hierarchically decomposed, with the interactions between product subsystems creating the need for integration among the design groups within the organization. In an indirect sense, then, the hierarchical decomposition of an organization may ultimately be traced to disciplinary lines or natural physical partitioning lines of the products that it produces.

Therefore, an interesting relationship appears: if a horizontal or vertical integration effort is initiated in an organization, it appears advantageous to consider the hierarchies embodied by the organization's products so that compatibility may be maintained. A change in organization structure may impede its capabilities to integrate the design of a product's interdependent subsystems. Similarly, if a major restructuring of a product design is initiated, then it becomes important to consider the potential effects on the organization's hierarchy itself. A given organizational hierarchy may become less efficient if the architectures of its products are altered. It appears worthwhile to coordinate efforts in both product design and organization design so that proper alignment between organization structure and product structure can be ensured.

Along the same vein, it is important to recognize the relationship between the hierarchical structure of a product and the structure of its supporting chain of suppliers. An intriguing option to vertical integration is therefore to alter the basic hierarchy of the product, thereby reconfiguring the supplier chain to the organization's advantage.

Finally, perhaps the mathematical formulation and solution procedures developed for designing product hierarchies will find potential applications in the design of organization hierarchies. A cross-pollination of these research fields may ultimately become a source of competitive advantage.

Defining Terms

Organization design: Shaping organizations to meet common goals by setting variables such as organization structure, technologies, strategies and policies, tasks, reward systems, information systems, and decision-making processes.

Product design: The process of defining and realizing a physical artifact, aided by the technical and scientific factors of our culture, to fulfill human needs.

References

Balling, R. J. and Wilkinson, C. A. Execution of Multidisciplinary Design Optimization Approaches on Common Test Problems, 6th AIAA/NASA/ISSMO Symposium on Multidisciplinary Analysis and Optimization, AIAA-96-4033-CP, pp. 421–437, Bellevue, WA, 1996.

Galbraith, J. R. *Competing with Flexible Lateral Organizations*, Addison-Wesley, Reading, MA, 1994.

Renaud, J. E. Sequential Approximation in Non-Heirarchic System Decomposition and Optimization: A Multidisciplinary Design Tool, Doctoral dissertation, Rensselaer Polytechnic Institute, 1992.

Shupe, J. A., Mistree, F., and Sobieski, J. S. Compromise: an effective approach for design of hierarchical structural systems, *Comput. Struct.*, 26(6): 1027–1037, 1987.

Simon, H. A. *The Sciences of the Artificial*, MIT Press, Cambridge, MA, 1981.

Sobieszczanski-Sobieski, Barthelemy, J.-F. and Giles, G. L. Aerospace Engineering Design by Systematic Decomposition and Multilevel Optimization, 14th Congress of the International Council of the Aeronautical Sciences (ICAS), Toulouse, France, pp. 828–840, 1984.

Stuckey, J. and White, D. When and when not to vertically integrate, *Sloan Mgt. Rev.*, 34: 71–83, 1993.

Teece, D. J. Profiting from technological innovation: implications for integration, collaboration, licensing and public policy, *Res. Policy*, 15: 285–305, 1986.

Ulrich, K. T. and Eppinger, S. D. *Product Design and Development*, McGraw-Hill, New York, 1995.

Further Information

Allen, J. K., Krishnamachari, R. S., Masetta, J., Pearce, D., Rigby, D., and Mistree, F. Fuzzy Compromise: An Effective Way to Solve Hierarchical Design Problems, Proceedings Third Air Force/NASA Symposium on Recent Advances in MDO, San Francisco, pp. 141–147, NASA, 1990.

Koch, P. N. Hierarchical Modeling and Robust Synthesis for the Preliminary Design of Large Scale Complex Systems, Doctoral dissertation, Georgia Institute of Technology, 1997.

Mistree, F., Smith, W. F., Bras, B., Allen, J. K., and Muster, D. Decision-based design: a contemporary paradigm for ship design, *Trans. Soc. Naval Architects Mar. Eng.*, 98: 565–597, 1990.

Peplinski, J. Enterprise Design: Extending Product Design to Include Manufacturing Process Design and Organization Design, Doctoral dissertation, Georgia Institute of Technology, 1997.

7.5 Organization Structure and Design Decisions

Marc S. Gerstein

The organization designer's job is to select the least-managerially demanding organization that best fits the "design criteria" appropriate to the situation and strategy. In the past, this task principally focused on creating an appropriate organization structure.

Today, however, organizational work is increasingly undertaken across multiple boundaries and is heavily supported by advanced information systems. Specifying the broader strategic, process, and technological context is essential to working out reporting relationships and to designing horizontal people-to-people coordination mechanisms. Furthermore, in contrast to focusing exclusively on traditional "owned organizations", it is essential to include value-chain participants from partnership organizations within the scope of the design.

Table 7.2 places traditional structural grouping and linking mechanisms in the middle of the design process. Thus, structural decisions are made *after* selecting the configuration of **strategic component**s, identifying major processes, and conceptualizing information technology infrastructures but before completely specifying measurement/reward systems, staffing and selection, and organizational culture requirements.

This approach departs from the traditional design paradigm in which process design and technology infrastructure were subordinated to organization structure. Processes were often optimized *within* organizational units using administrative boundaries as the context for improvement. The approach detailed in Table 7.2 marks an important evolutionary shift in the nature of the organization design process.

Strategic Component Decisions: "Plug 'n' Play" Organization Design

From the late 19th century until the 1980s, the design of most business organizations in the West involved making decisions about people who worked for you. Today, in contrast, many organizations are critically dependent on outsiders for the provision of critical functions that in earlier times would have been

TABLE 7.2 Summary of Phases of Organization Design

Design Phase	Comments
1. Clarify the purpose of organizational redesign, and articulate "design criteria"	Establish the purposes for making organization changes Articulate specific criteria against which alternative design options can be evaluated
2. Analyze the industry and company value chains, develop an understanding of the business' fundamental processes, basic economics and risks, and basis for competitive advantage	Clarifies specific requirements between steps in the value chain as the basis for detailed organization design decisions; may have implications for partnership agreements (see phase 3)
3. Identify and evaluate alternative configuration of "strategic components"	Develop alternative arrangements of conventionally owned organizational units and "virtual" arrangements with partners Identify strategic and logistical dependencies; identify coordination requirements and options Assesses alternative configurations against overall objectives using design criteria Develop requirements for partnership agreements and contracts
4. Generate alternative organizational options, including the basic organizational architecture, major processes, information technology infrastructure, structural grouping and linking mechanisms, etc.	Generate organizational architectural options and assess against design criteria Clarify major processes Identify needed IT infrastructure and systems to support process requirements and needed coordination Generate alternative structural grouping and linking options, including external linking requirements, and assess them against design criteria. Narrow design choices to a few "best alternatives"
5. Identify additional requirements to achieve overall objectives. Include areas such as measurement/reward, human resource management, knowledge leverage, organizational culture, etc.	Identify each design alternative's implications for various aspects of the organization such as measurement/reward, staffing, culture, systems, etc. (the use of an organizational model such as the McKinsey 7S, Tushman-Nadler Congruence Model, etc. is recommended)
6. In the context of an overall change management strategy, identify implementation issues and formulate appropriate plans	Making staffing decisions Communications/roll out logistics Managing the approval process; overcoming specific sources of resistance, etc.

provided in-house. While this "outsourcing" was originally applied to noncore activities, such distinctions no longer apply. Producers such as Nike, most computer and electronics firms, and many others operate with significant portions of their value chains provided by companies that are not owned by them.

Unfortunately, in comparison to internally focused organization design, crafting relationships between voluntarily associated firms is significantly more complicated. This added difficulty stems from three facts: (1) separate firms have legitimately differing objectives as well as varied capabilities to add value to relationships, (2) these goals and capabilities change over time, and not necessarily in synch with those of other firms, and (3) legal contracts are an inefficient means, at best, to structure long-term relationships (although such contracts are often the primary form of expression for such arrangements).

The critical questions one must ask when considering **virtual organization** design are

1. Which portions of the entire industry value chain should we participate in? In other words, which *strategic components* do we need to include in our scope of business?

2. Of these components, which should be within our organization and which provided by others? In other words, which should be conventionally owned entities and which "virtual"?

3. How exclusive or market involved should each organizational component be? In other words, should the component be completely captive in the traditional model, should it have modest noncompetitive relationships with others, or should it be a full-market participant able to sell to all comers including direct competitors?

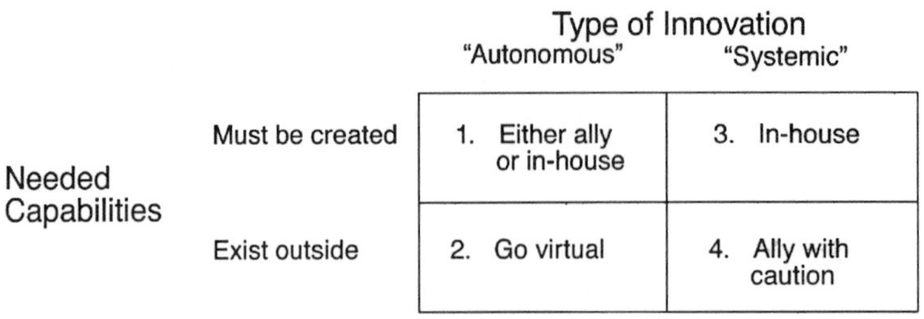

Type of Innovation
"Autonomous"   "Systemic"

	Must be created	1. Either ally or in-house	3. In-house
Needed Capabilities	Exist outside	2. Go virtual	4. Ally with caution

FIGURE 7.8 Framework for matching organization to innovation type. (Adapted from Chesbrough and Teece, 1996.)

Since even the simplest alliances posses significant complexities and risks, it is tremendously helpful to have a conceptual framework to guide one's decision making. Such a model has been developed by Chesbrough and Teece [1996] to outline the appropriate organizational choices based upon strategic conditions (Fig. 7.8).

Autonomous innovations are those that occur in relatively isolated "pockets" in the value chain and thus are not dependent on fundamental changes elsewhere in order to be successful. For instance, a new specialized accounting software package can be implemented without redesigning either the computer system it runs on, the books and records it documents, or the organization that employs the new system. *Systemic innovations,* in contrast, typically involve the redesign of multiple aspects of the value chain to create fundamental benefits. Conversion to the music CD required changes by the consumer electronics companies, record producers, record manufacturing equipment companies, music retailers, retail display manufacturers, etc.

The matrix's "main diagonal", cells 2 and 3, represent the most clear-cut cases in terms of strategy formulation. In the first instance, when autonomous innovations already exist, one is advised to take advantage of marketplace efficiencies by allying with others. Pharmaceutical companies acquiring new conventional drugs, computer manufacturers acquiring the latest hard disk drives, and the purchase of outside telemarketing and customer fulfillment services are examples of such alliances.

On the other hand, if new systemic innovations must be created (cell 3), one is advised to maintain control by undertaking such development in-house. The historical development of computer operating systems by hardware manufacturers illustrates the benefits of this approach.

The creation of new autonomous innovations (cell 1) provides organizations with a greater strategic choice, since the risks associated with systemic innovations are not present. In such cases, the individual business case may be the best guide, although time-to-market and specific functional requirements may encourage the in-house route.

Finally, we come to the complexities of cell 4. As previously stated, although separate organizations may have objectives in common, they rarely have completely aligned interests. In fact, their objectives may be diametrically opposite along some dimensions. This makes achieving the necessary strategic fit between the players difficult to create, and even more difficult to sustain over time.

The Structural Spectrum of Organizational Forms

The "Anchor Points"

Within any given organization, the three pure structural alternatives — *functional, product/market,* and *matrix* — are described in Table 7.3. The table reveals that each of these pure forms is far from perfect: each has tremendous strengths but glaring weaknesses. Nevertheless, except for adopting a truly radical design, such as the one employed by Visa International, the Internet [Gellet and Kapor, 1997], or other radical organizations, one of these three structural forms will likely be the cornerstone for most organization designs for the near-term future.

Except for the matrix, the structural options described in Table 7.3 are rarely implemented in their pure form. Rather, one or more horizontal *integrating mechanisms* are used to coordinate the work of

TABLE 7.3   Characteristics of Principle Structural Options

Functional	Matrix	Product/Market
Description		
Grouping by similar skills/activities	Subgrouping by project, specific product, or product/market *in addition* to functional and product/market groupings	Grouping by customer group (e.g., geography, segment) or output (e.g., product)
Easiest to understand	Difficult to understand	Generally easy to understand
Single chain of command	Two (or more) chains of command	Single chain of command
Performance measurement		
Principally cost center, with additional measures for technical excellence	Cost, profit and/or project accounting including time-to-market and customer-related measures	Typically profit center, with additional measures of customer satisfaction and competitive performance
Authority and control		
Clear lines of authority (e.g., functional chain of command)	Ambiguous and conflicting authorities are commonplace	Clear lines of authority (e.g., product chain of command)
Clear decision making	High-conflict decision making	Clear decision making
Flexibility and responsiveness to change		
Slow to sense and react to marketplace change; rapid response to technological developments	Responsiveness rapid, but sensitive to team membership	Highly responsive to customer issues; less to technology
Generally slow to reallocate resources	Quickly reallocates resources but may be slow to decide if conflict exists	Quick to request additional resources; slow to give up own resources to others
Lack of customer focal point	Clearest customer focal point, although large customers may have multiple points of contact	Clear customer focal point
Organizational effectiveness and efficiency		
Lowest managerial overhead, but tends to require staff groups elsewhere to achieve integration	Very high management overhead, high levels of conflict, and stress	Moderate overhead; significant levels of effort required for global optimization
Most efficient use of technological resources; R&D can reach world class levels	Dedicated technical resources often generalists vs. specialists	Tends to duplicate technical resources; R&D technical excellence tends to be spotty
Under emphasizes role of business resources	Balances business and technical resources	Maximizes use of business resources

interdependent groups. These familiar mechanisms are shown in Table 7.4, which also includes the matrix structure as a reference.

The table shows that each of the structural alternatives differs along a set of organizational dimensions. In addition, as will be seen in Fig. 7.9, the power held by the poles of the organization (functional or product/market) ranges from 100% for the pure structures to 50% for the dual-hierarchy matrix form. (The numbers along the x-axis correspond to the columns in Table 7.4 and are symmetrical about the center point.)

As increasingly powerful integration mechanisms are employed (moving from either pole to the center), the power of the dominant pole falls and that of the group representing the opposite perspective increases, reflecting its correspondingly greater influence in decision making and resource allocation.

As the power relationship approaches equilibrium, the organization's overall *management complexity* rises sharply. Figure 7.10 portrays this shift in management complexity as a function of changing structural form. The graph is based on a subjective assessment of a large number of cases, and it reflects managers' perceptions of what it is like to get work done in each of the various configurations.

TABLE 7.4 Structural Integrating Mechanisms

Structural Integrating Mechanisms	Liaison Roles	Project Manager + Cross-unit Groups	Project Managers + Teams	Autonomous Teams	Matrix Structure
Reference No.	1	2	3	4	5
Example	Coordinator	Task forces and committees: "lightweight teams"	Project Teams: "heavyweight teams"	"Tiger team"	
Principal responsibilities	Communications conduit; little direct contact between groups	Coordination through direct contact	Team-based project management	End-to-end responsibility for deliverable	Project results
Responsibility for overall results	No	No	Usually	Yes	Yes
Formal leadership	No	No	Yes	Yes	Yes
Source of authority	Informal, expertise based	Informal, expertise based	Budgetary/task approval authority	Line control	Dual command
Coordinator organizational position	Junior-middle	Middle	Senior	Senior	Full chain of command
Team staffing	None	Part time	Part time or full time	Full time	Full time
Cross-unit colocation	No	No	Often	Yes	Generally
Management complexity	Low	Low	Moderate–High	High	Very High
Results measures	Informal, with emphasis on traditional measures	Informal recognition of project results	Frequent recognition of project results + traditional measures	Formal recognition of project results; traditional results may suffer	Multiple, formalized measures
Staff evaluation	Line boss only	Line boss only	Line with informal input	Project only	Formal dual evaluation

The figure shows that, from a baseline of the pure functional or product/market form, managerial complexity rises slowly until "heavyweight teams" are employed (scale point 3). Such teams, often colocated and led by a senior individual, represent an effective counterbalance to the prevailing functional or product/market power structure. It is, of course, precisely such teams' ability to do things differently from the traditional organization that is a key source of their effectiveness. However, these benefits come at the expense of a greater communications load (meetings, e-mail, etc.), higher levels of task conflict, greater stress levels, etc.

Management complexity rises still further in the case of autonomous "tiger teams" (scale point 4), which are often employed for projects of sufficient priority to justify abandoning existing practices. Such teams are often very disruptive because they play by different rules than the rest of the organization, including shortcutting traditional approval processes, disrupting existing priorities, employing "non-standard" technologies, etc.

Even after their work is done, autonomous teams create management problems. Since "writing a new rule book" is exciting, team members are often reluctant to give up their newfound freedom. In addition, since it is common for tiger teams to step on a number of management toes as they bulldoze through obstacles on their way to success, such teams often engender feelings of resentment and envy among both managers and co-workers. Consequently, team members frequently encounter problems reintegrating themselves into the organization when their assignments are completed. While the company may get

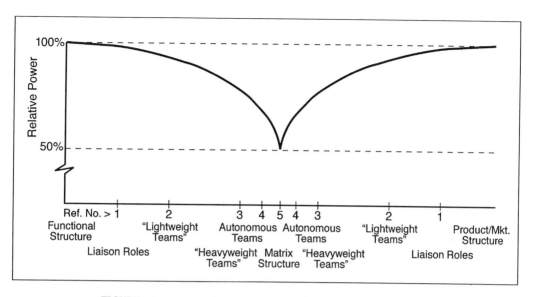

FIGURE 7.9   Concentration of power in various organization structures.

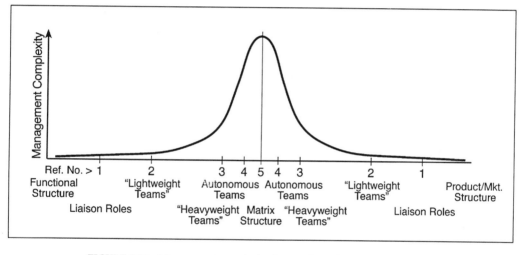

FIGURE 7.10   Management complexity for a variety of organizational forms.

the results it needs, autonomous team solutions come with a price tag: their members continue to pay long after the work is completed.

Special Requirements of Matrix Organizations

The "small shift" in the power relationships between team-based designs and a matrix structure grossly understates a much greater change in the character of the organization. The requirements of a matrix organization fall into four categories:

1. *Clear matrix structure and decision-making process.* Specifically clarifying what and who is matrixed, clarifying decision-making power and authority, especially for matrixed groups, and developing decision-making processes capable of dealing with high complexity and rapid change.
2. *Appropriate measures and rewards.* Developing relevant, appropriately differentiated measures of organizational performance to be used by the poles of the matrix and by matrixed groups; developing specific criteria for performance evaluation, and creating appropriate mechanisms to allocate recognition and rewards to matrixed and nonmatrixed staff.

3. *Selecting appropriate matrix managers.* Choosing managers with the proper mix of technical know-how and market savvy to balance conflicting demands; ensuring that matrix managers possess adequate conflict management, problem-solving and influence skills to deal with the requirements of the job; and selecting managers with the maturity, credibility, and interpersonal strength to deal with the stress, workload, and complex politics associated with large-scale, high-stakes program management and the dual chain of command.

4. *Supportive organizational climate and culture.* Tolerance of conflict and role, task, and decision-making ambiguity; ensuring senior management support of matrixed managers; facilitating high levels of undistorted communications to facilitate decision making and coordination, managerial willingness to relentlessly "push down" decisions vs. allowing them to escalate; and ensuring that the power of the matrix remains balanced in the face of constant attempts by each of the poles to increase its influence.

Conclusion: Selecting an Organization Design

By this point, it should be clear that there is no one best-organization design. The designs that are relatively simple, clear, and easy to operate are relatively weak at integration, and the designs that are effective integrators are highly complex and difficult to staff and manage. Of course, some managers remain ever hopeful that a "magic bullet" design will be invented that provides the benefits without the costs, but so far such a solution has not been forthcoming, nor are there any candidates on the horizon. While newer forms of organization such as the networked virtual form have unparalleled benefits, they also possess demanding managerial requirements and unique risks, as discussed earlier in this section.

Fortunately, as revealed in Table 7.1, organizations have evolved to be "asymmetrical" rather than rigidly consistent in their design. This frees us to build the overall organization as if it were a set of "modules". Such components can be individually optimized, which significantly reduces overall design complexity by dividing the problem into pieces. However, an effective overall design demands that the process and inter-unit flow, fit, and sharing dependencies be effectively managed and that the overall design possess coherence and harmony (see Section 7.3). Thus, while the design problem is less complex than it could be, it is far from simple.

The essence of creating a design, as detailed in Table 7.2, is therefore to develop a series of "design alternatives" for each of the relevant stages in the value chain and then to evaluate each of these alternatives against design criteria formulated on the basis of one's competitive situation, overall goals, and strategy. Some parts of the design problem may best be approached by one of the conventional design solutions presented in Table 7.4 while others may require more unconventional solutions based upon various partnership arrangements, market models, or radically decentralized designs.

For each decision, choosing the best alternative is similar to making any complex choice involving multiple trade-offs, such as buying a house or car, or selecting a person for a job. In each case, one creates a profile of the "ideal" solution based upon one's design criteria and then evaluates the "candidates" against the criteria in order to find the one that best fits one's requirements.

In this process it is important to be sensitive to the manner in which trade-offs are made. For example, if one employs an approach based upon a simple "weighted average" to evaluate alternatives, one is implicitly saying that having more of one attribute "makes up for" having less of another. This may not always be what is required. A car that has more performance than one needs may not make up for its lack of trunk space.

To avoid such errors, in evaluating alternatives, one may wish to rely on careful judgment or use more-sophisticated mathematical models, such as those that set thresholds or score overachievement and underachievement of requirements in an asymmetrical manner. (For an example of an implementation of this approach for executive selection, see Nadler and Gerstein, [1992]).

Of course, common sense and experience suggest that the choice of a design alternative can never be a completely rational decision. Many factors will play themselves out in the process, and they will be based on both performance and politics. However, by structuring the process and making the trade-offs between alternative designs explicitly rather than implicitly, one can improve the fit between the final design and contextual requirements. Despite what may appear as the additional work of rationalizing the process, the rewards will be well worth the effort.

Defining Terms

Autonomous innovations: Innovations that are narrowly constrained to a single portion of the value chain and do not require simultaneous innovations elsewhere in order to be implemented.

Strategic component: A portion of the value chain with identifiable inputs and outputs. It may be a separate company or a portion of a larger enterprise.

Structural integrating mechanisms: A variety of organizational means, including project teams, to achieve needed coordination.

Systemic innovations: Innovations that require simultaneous changes to a number of parts of the value chain.

Virtual organizations: A value chain created by connecting together strategic components. Has given rise to the notion of "plug 'n' play" organization design.

Reference

Nadler, D. A. and Gerstein, M. S. Strategic selection: staffing the executive team, In *Organizational Architecture*, Nadler, D. N., Gerstein, M. S., and Shaw, eds., ch. 9, Jossey Bass, San Francisco, 1992.

Chesbrough, H. W. and Teece, D. J. When is virtual virtuous? Organizing for innovation, *Harv. Bus. Rev.*, Jan.–Feb., 1996.

Further Information

Inventing Organizations for the 21st Century, MIT Sloan School of Management. The overall goal of this landmark interdisciplinary effort is to work with business leaders not just to understand, anticipate, and exploit novel ways of working but also to build on a broad-based knowledge of organizations, economics, and emerging information technologies to actually invent entirely new approaches that can be put into practice. The initiative is a thoughtful and creative approach that actively synthesizes the best insights from the worlds of theory and practice to lay the groundwork for the future. For further information, use the World Wide Web at http://ccs.mit.edu/21c/index.html.

7.6 The Matrix Organization Revisited

Urs E. Gattiker and John P. Ulhøi[1]

In a literature review, Ford and Randolph [1992] stated that "matrix organization continues to elude definition even after more than 30 years of use in work settings (p. 268)." Often forms differ across organizations and industry but here, following Mintzberg [1983], we tentatively start out by defining matrix as

> An organization where multiple command structures are used whereby employees experience dual lines of authority, responsibility and accountability (e.g., employee reports to project manager and manager of electrical engineering) with associated support mechanisms, cultural and behavioural patterns.

A **matrix structure** still includes the normal vertical hierarchy within the individual functional department; however, this arrangement is "overlaid" by a form of lateral influence. Often in a matrix-type

---

[1]Both authors contributed equally to this paper. We would like to thank Anne-Mette Hjalager and Steen Hildebrandt for their comments made on an earlier draft, usual disclaimers apply. Comments should be addressed to Urs E. Gattiker, Dept. of Production, Aalborg University, Fibigerstraede 16, DK-9220 Aalborg, Denmark. Telephone:+45 9635-8990 (GMT + 1); Fax: +45 9815-3030; Electronic Mail via Internet: Urs_The_Bear@bigfoot.com, or the Web: http://www.TIM-Research.com

structure the project or divisional manager has primary control over the resources of the division. In turn, the functional manager (e.g., product development) provides support and advice while being responsible for his or her area (e.g., the following of accepted engineering principles, codes of conduct by team members, etc.). Accordingly, matrix structures are also team-oriented arrangements that promote coordinated and multidisciplinary activity across functional areas, while the sharing of knowledge is implied. These structures were originally designed to foster greater technological innovation while helping rapidly growing organizations (e.g., Burns and Wholey [1993]).

In this section, we give a short overview of matrix structure and technology management. We outline some of the characteristics and also point out that many organizations may actually be **hybrid**s (i.e., mix several ways of organizing to allocate resources effectively). Finally, culture and new technologies may further complicate matters and make organizing, and the matrix structure in particular, an elusive concept. The section concludes with some suggestions on how some matrix-type organization efforts may be put into place in a firm and with some implications for decision makers.

Matrix Structure and Technology Management

The literature usually distinguishes between the unitary organization (U-form) and the multidivisional firm (M-form). For the U-form, functional authority and structure guide decision-making processes. In the M-form, the product influences decision making. Of interest here is that organizations may use neither "pure" U- or M-form nor matrix structure. Instead, a firm employing matrix management may also have some structural features usually associated with a U- or M-form organization, thereby representing a hybrid.

Concerning pros and cons of the matrix structure, Denis [1986] found a rather mixed picture in large Quebec engineering firms. On the one hand, the matrix structure was found cumbersome in that it imposes dual authority, permits power struggles and conflicts to develop, and decreases motivation. On the other hand, matrix structure was found advantageous since it is generally well accepted, offers organizational flexibility, and tends to improve technical quality and individual development of capacities [Denis, 1986].

Literature also suggests that organizations introducing a matrix structure tend to keep the structure over time [Burns and Wholey, 1993]. In turn, this requires managers and engineers to carefully prepare the organization for a matrix structure, before it is introduced. Moving a firm from a U-form to a hybrid, which also involves matrix elements, increases employees' job complexity (e.g., working in two teams plus doing one's own "regular" task at acceptable performance levels), and workers are unlikely to appreciate that some of their past skills have now become largely obsolete (e.g., being a lone ranger vs. an effective team member). Also, job ambiguity, as well as performance demands upon employees, is likely to rise with matrix management [Gattiker, 1988]. Moreover, in a matrix structure, various stakeholders (e.g., team members) are now scrutinizing one's performance (i.e., whether one carries his or her weight), thereby increasing one's accountability.

As this section concentrates on matrix organization specifically, issues such as size of organization, type (e.g., manufacturing vs. service), market (e.g., oligopoly), type of technology and much more are beyond the scope of this section. However, in some cases these issues are discussed elsewhere in this book and the reader is referred to those chapters for a more in-depth assessment of these matters.

Hybrid and Matrix-Type Structures of Organizations

Often, organizational researchers describe a hybrid as an organizational arrangement that uses resources from more than one existing organization [Borys and Jemison, 1989]. Here economic and other pressures result in mergers and licence agreements and result in a broad range of organizational combinations and forms (e.g., organization networks). In the context addressed in this paper, *hybrid organizational forms* focus on

> the *governance structure*, whereby the firm makes an attempt to improve coordination in complex and interdependent activities; thereby reducing its transaction costs and/or finding the most effective structure for itself and various relationships with its stakeholders (e.g., suppliers, in order to manage just-in-time delivery).

TABLE 7.5 Project Organization and Matrix

In practice, project management organization structure can be distinguished by using the following *three categories* [Chuah et al., 1995]:

Functional organization of project management whereby sub-tasks are being divided and assigned to the appropriate functional areas, while project members' primary responsibility remains with their functional group.

Matrix organization of project management involves a project manager who either jointly, with or without support from functional managers, supervises the project.

Team organization of project management may encompass functional experts on a part-time or full-time basis reporting to the project manager; functional managers are involved in advisory capacity (e.g., as far as supporting team members regarding functional issues), while authority and responsibility rests with the team manager(s).

The above illustrates that in a U-form of organization a functional approach to project management is likely, while in the case of a matrix structure a team organization for project management seems obvious; as the sections in this section show (see moderating variables), however, various factors may influence the above and be reflected in different ways for organizing project management across firms and/or countries/industries.

Hybrid approaches may also be utilized in the research and development (R&D) domain. For instance, collaboration and various types of alignments with other firms may be pursued to lower transaction costs, such as outsourcing certain R&D activities or parts of projects to a research-intensive firm. Alternatively, the firm may take over the smaller partner in order to reduce transaction costs, e.g., interunit coordination (e.g., Grandstrand and Sjölander [1990]) and get access to new technology and technology competence not available in-house (Gattiker and Willoughby [1993]). These hybrid organizational arrangements and strategic alliances are discussed in detail elsewhere and are often linked to matrix management (e.g., Borys and Jemison [1989]).

Organizational Structures and Research and Development Projects

Ford and Randolph [1992] pointed out that terms such as matrix and project management are often used interchangeably, despite the fact that there are some distinct differences between the two. Project structure allows for efficient coordination among specialities and it provides clear responsibilities for all activities related to a specific project. However, in its pure form, lack of responsibility for the long-range technical development of specialities exists. Both matrix and project imply that a functional structure (U- form) is "overlaid" by a horizontal structure consisting of projects, products, and business areas/subsidiaries (more M-form). The key characteristic of a matrix organization is its dual lines of authority, responsibility and accountability that contradict the traditional one-boss principle in management.

Reviewing the literature Chuah et al. [1995] pointed out that project management can involve several forms and structures as outlined in Table 7.5. In their study of Hong Kong industries, the authors reported that there was a clear preference for matrix project management compared to functional and project teams (see also Table 7.5). McCollum and Sherman [1993] remarked that empirical literature on this subject is limited and primarily based on anecdotal evidence. Based on their findings in the Southeastern United States from a sample of high-technology organizations in the public and private sector, McCollum and Sherman [1993] reported that assigning R&D personnel to two projects at the same time was most effective, such as growth in sales, return on investment (ROI). However, assigning personnel to either one or three at any one time was not as beneficial as far as performance measures are concerned. In another study, McCollum and Sherman [1991] found that if an organization assigned a large percentage of employees to two projects at the same time (e.g., within their division or in projects, cross-functional teams, and/or work groups), there was a positive effect upon ROI as well as the 5-year rate of growth (sales) for the firm.

Managing projects and being involved in more than one project team requires good interpersonal skills for managers as well as employees. Barker et al. [1988] reported that managers using a combination of cooperative and confirming approaches to conflict were judged, by team members, to have a positive impact upon the management of a project in contrast to other mediation styles. Additionally, this requires that employees are able to manage job diversity and complexity while keeping on top of both projects, that is, not neglecting one for the other (planning and coordinating, see Gattiker [1990a] p. 220). However, in today's fast moving-business environment, such requirements (e.g., handling complexity and regular skill upgrading) are no longer limited to matrix organizations. Most importantly, increasing group work in various forms (e.g., project teams, working teams such as information system end-user council) requires employees to manage job diversity and complexity. Moreover, as outlined below, various factors may influence how a firm may best organize itself, while organization diagrams may not tell us the whole story.

Is Matrix Management Dead? Some Moderating Variables

The above sections outline that, while firms may use matrix structures in some forms or another, various hybrid forms of organizing resource management may be used. For engineers it is interesting to note that project management may sometimes be synonymous with matrix organization, but as Table 7.5 suggests, differences do exist. The section below will give the reader a short overview of some moderating variables that may relate to structural and technology management issues. Burns and Wholey [1993] pointed out that although much has been written about matrix-type organizations, empirical research investigating what spurs the use of this type of structure and what may affect its success is scarce. In the following section we provide a short overview of some of the important moderator variables affecting matrix management of organizations and projects. Due to space limitations, the variables included are not complete. Nevertheless, for in-depth discussion of other moderating variables (e.g., rewards and motivation, organization size) the reader is referred to the appropriate chapters in this book.

Technological Competence and Training

A great challenge for managing today's rapid pace of change in the technology domain is **technological competence**. Technological competence builds upon the firm's distinctive capabilities and draws upon its strengths in various areas. We define technological competence as "the ability to *retrieve, process,* and *use* information for solving technology and economic-related problems and for making appropriate decisions." Employees have a certain degree of technological competence (micro). The firm's challenge is to secure the most appropriate levels of *technological competence* in its workforce (macro) to promote success in the marketplace.

In a matrix-type organization, dual lines of authority, reporting, and responsibility already require the employee to keep abreast of functional developments (e.g., ethical and moral concerns about genetically manipulated vegetables) and disciplinary developments (e.g., new codes in electrical engineering). To succeed, the employee must be willing and able to regularly update job and technology skills. For this, both internal (e.g., on-the-job or in-house training) and external (e.g., seminars and courses) training may be applied. Moreover, continuous education may also occur through informal (e.g., talking to others) and formal (e.g., attending continuous education seminars) means, thereby keeping one abreast of new developments while, most importantly, ensuring one's skills remain valuable to the firm as well as trying to maintain a certain degree of overlap between individual and corporate values. However, the firm must also be willing to invest in its employees to reap benefits as (e.g., Gattiker [1995]).

Effective technology management requires appropriate technological competence by the employee and the appropriate mix for the firm. This ensures that the employee offers the organization the level of competence needed. One might assume that in a matrix-type organization the required competence levels are likely to be higher for both the employee and the firm than in a U- or M-form type of organization (e.g., Burns [1989]). Today's rapidly changing work environment (e.g., deregulation and new technology), however, results in firms seeing their workers' competence levels rise continuously (e.g., through working in various project teams over a couple of years) or, possibly, threaten the firm's competitiveness. Hence, these factors influence organizations regardless of their structure. Moreover, as the section below shows, organization diagrams outlining the structure may not tell the whole story about the inner workings of a firm.

Cultural Similarities and Differences

Empirical work by Maurice et al. [1980] shows that organizational processes develop within an institutional logic that is unique to a society. For instance, German manufacturing plants place decision making with crafts workers (e.g., tool and die makers), while in France and the United Kingdom such decisions may be made by technicians and an engineer or manager, respectively. Therefore, while structure, specialization, and technology may appear similar across countries, their interpretation and application may differ according to the national context.

Gattiker and Willoughby [1993] pointed out that a firm's effective management of new technology may be greatly influenced by the moderating effect of cultural factors. These factors even effect training and acquiring of technological competence. For instance, in North America, workers are often keen on improving their skills in order to secure jobs and interesting assignments. In Germany, however, workers may more likely expect their employer to take the initiative (in Denmark it is often the unions that do). Moreover, in Germany, as in Denmark, the firm is expected to pick up the tab for additional training. Internal labor markets (i.e., rules and regulations pertaining to employees — such as hiring first from within) may partially explain this; further, labor contracts may stipulate that the employer must provide the employee with a certain amount of days for continuous education during a year [Gattiker, 1990b].

Heller et al. [1988] found in a three-country study (United Kingdom, Netherlands, and the former Yugoslavia) about decision making that power equality for lower-level workers is greater in the former Yugoslavia than in the United Kingdom or Netherlands. Also, the participating U.K. firms showed the lowest overall participativeness for lower-level workers in the decision-making process. Here, hierarchical structure and power levels or distance affect who participates in mid- and long-range decisions of a more strategic type. Democratization of the decision-making process may erode the hierarchical structure of the firm. Moreover, using a Danish sample Ulhøi et al. [1997] reported a high degree of participation in the decision-making process. In particular, the authors found that, in the case of implementing integrated environmental management systems, high participation in decision making was apparent, in turn, giving room for various ad hoc structures within organizations.

The above suggests that organizational structures and systems (e.g., internal labor markets and their influence upon training of employees) may give one a limited picture about a firm's way of functioning. For instance, in relatively egalitarian societies where decisions are made on the shop floor or by groups, even a functional organization chart may fail to portray the involvement in decision making, and thus job complexity, experienced by lower-level workers [Gattiker and Willoughby, 1993]. Such complexity is more likely to be attributed to a matrix-type work environment in a more hierarchical society, where workers are less likely to participate in decision making than elsewhere, such as the United States. Accordingly, the structure of the organization may tell one little about the inner workings of the organization. Moreover, while using M-form or profit centers in some firms may be indicated by the organization chart, internal workings may suggest that various "matrix" forms are used for project groups and between divisions trying to develop new competencies and/or markets. Finally, looking at the literature, matrix organization may be a North American phenomenon transferred to, or adopted by, firms abroad with some success; however, cultural intricacies may question the fruitfulness of such efforts. A matrix structure tells little about the institutional logic of organizing unique to a society.

Adoption and Abandonment of Matrix Management

While much has been written about adoption and abandonment of matrix organization and management, empirical research about moderating variables, and cultural effects in particular, is primarily anecdotal and lacks empirical assessment of issues of importance to managers and researchers [Ford and Randolph, 1992]. Moreover, sometimes matrix issues are addressed, but methodological and sample issues about the firm participating in a particular study are not explained, and acceptable levels of research rigor required to reduce threats to validity are not being met (e.g., Abo [1995]).

Burns and Wholey [1993] suggested that organizations may adopt matrix structures for reasons unrelated to information-processing demands or task diversity. Burns [1989] found, in his sample of hospitals, that matrix management is not a transitional form but instead is quite stable. Neither did he find confirmation that matrix complexity increases over time. Burns and Wholey [1993] reported that hospitals with high diversity are more likely than their counterparts to adopt matrix management but, most importantly, regional and local hospital networks, professional media, and decisions by neighboring hospitals affect the organization's decision about using a matrix structure. Again, it needs to be pointed out that the work of Burns et al. deals with hospitals in the United States where medical markets are becoming increasingly competitive but still work under rigid regulations (e.g., health care insurers' need for documentation of services). Unfortunately, these factors may moderate the findings of Burns et al. [1989; 1993], thus limiting their applicability to other industries and countries.

The Virtual Organization and Groups/Teams

Technology has enabled us to create virtual organizations, which could be defined as "a firm which make take on different forms…" and has a physical/spatial, time, and social/organizational network dimension [Gattiker and WG1, 1997]. With teleworking and satellite offices, workers can perform their tasks in various geographical locations (e.g., Dürrenberger et al. [1995], Soares [1992]). Moreover, new opportunities arise and geographical boundaries are removed, which helps smaller firms especially (e.g., Rice and Gattiker [in press]). Effective governance of organizations without geographical boundaries (e.g., cyberspace shopping mall), while keeping transaction costs as low as possible, may increase complexity and skill demands upon workers in order to secure continuous employment. However, at this stage, we are uncertain how much project management, matrix organization, and other forms will either flounder or bloom under such a new regime of realities.

What seems certain is that cross-functional work in groups and teams is becoming ever more important. In addition to organizational teams, however, an employee's membership in interorganizational teams (e.g.,working groups in professional organizations) using listservers or other electronic means for discussions and solving challenges are on the rise. Accordingly, the employee may be involved as a group member in various teams and organizations (formal and informal, within and outside the firm). Moreover, the virtual firm may consist of various organizations, or people, associating themselves with others to perform certain tasks and services, while dissolving after completion of a contract. Additionally, the increase in contract work by professionals working out of their homes may further raise the complexity and result in myriad ways used by firms to organize tasks and processes. This may also suggest that the formal structure of the firm may be of little consequence in comparison to its actual inner workings, which may be affected by cultural differences and similarities, interorganizational projects, teleworking, and other variables as discussed in this section.

Conclusion and Implications

This section discussed the matrix organization/structure as used in organizations. As our discussions showed, while we have failed to come up with a clear definition, even after nearly 4 decades of its use, matrix-type structures can be found in most organizations around the globe. For instance, matrix approaches for technology and R&D projects (both in people and participating organizations) seem advantageous. For starters, both project and functional concerns can be addressed while, most importantly, the team can draw upon the expertise of members from various backgrounds (e.g., engineering and accounting). Managing technology requires both technical specialists (e.g., engineers) as well as social issues experts (e.g., consumer and environmental management specialists) to assure that both technical aspects and social/political/marketing issues are understood before giving the go ahead for a particular new invention to be readied for the market.

Implications for Technologists and other Decision Makers

While we have learned a lot about matrix organizations, it appears that empirical evidence is not as great as one might hope. Nevertheless, the evidence suggests that managers and engineers take the factors outlined in Table 7.6 into consideration when trying to implement and/or work in a matrix structure.

Table 7.6 points out that issues in various areas must be considered before we can be sure that matrix organizations benefit the firm and facilitate its governance, while reducing its transaction costs in order to help improve shareholder value and job security.

A matrix structure may help by providing an interdisciplinary work environment, thus expanding employees' horizons, skills mix, and understanding of issues (e.g., technology) beyond their immediate

TABLE 7.6   Strategic Issues: Managing "Matrix-Type" Organizations Successfully

The points listed below are not all inclusive or in order of particular importance

General

(1)   If a matrix structure is to be implemented, it is recommended to proceed quickly, i.e., make the necessary changes and avoid a slow and more costly process of change.

(2)   Greater task diversity is more likely to fit a matrix structure.

Leadership and Reengineering

(3)   Matrix organization requires leaders who are able to foster a conciliatory work environment where team spirit and participation is being supported and encouraged.

(4)   Conflict resolutions should be accomplished in a peaceful and conciliatory manner, which may increase time requirements but, more importantly, is likely to be supported and carried by all stakeholders thereafter.

(5)   Some reengineering is required to accommodate matrix management while, as importantly, making best use of information technology to support decentralized work such as telework.

Information Technology

(6)   A matrix structure suggests an information technology structure that is decentralized and cooperative, thereby providing an increased capacity for resource sharing and communication between various stakeholders (e.g., suppliers and employees).

(7)   Distributive cooperative computing is likely to thrive in a matrix environment.

Project and Technology Management

(8)   Increased technological complexity in R&D and project management may suggest a matrix-type organization structure, whereby employees are members of more than one team.

(9)   Increasing complexity of technology issues (e.g., technical and social) necessitates interdisciplinary project and work teams, which is facilitated, if not amplified, by matrix management.

Technology Competence

(10)   Striving to reduce costs and take better advantage of market and technological opportunities requires the organization to assure that technological competence is available; hence, continuous on-the-job as well as other means for skills upgrading must be used and, most importantly, mutual investment (e.g., employee invests time during off-working hours) motivates all stakeholders (e.g., company pays for tuition).

(11)   Technological competence and work in a matrix structure requires that people comprehend issues beyond their immediate job and acquire the skills required to take advantage of technology (e.g., Windows 95) to the fullest; otherwise, new releases of software or technology are introduced while employees have still barely mastered the old version/technology.[a]

[a] When an old skill becomes obsolete one's technological competence is also lowered. Resistance to the change may, therefore, increase, since the lowering of one's technological competence may decrease one's job security as well

job; in turn, development of human resources becomes a paramount issue for assuring the appropriate mix of skills to attain the desirable level of technology competence (e.g., Gattiker [1988; 1990a]). A matrix structure that inherently exhibits a culture of continuous change should make employees more receptive to change efforts which tend to affect their jobs and technological competence and thus necessitate skills upgrading. Such support and acceptance from employees requires that bread and butter issues (e.g., job security and improving of skills) are taken into careful consideration when proceeding with change.

While a matrix structure may be the way to go for a firm (see Table 7.6), an organization may already be infected by the matrix bug in various parts of its structure (e.g., interdepartmental task force to cut costs or reengineer production). Employees who care about their work and use their skills and competence levels actively while being on the job together with a system that encourages and rewards initiative may still ultimately result in the greatest rewards for all stakeholders. Accordingly, for decision makers it may be of little consequence if such a construct is called a matrix or an M-form organization, as long as it delivers what it is supposed to and satisfies all important stakeholders.

Conclusion

The material presented herein indicates that firms may use various forms of organizing of which one might be the matrix form. This section suggests that cultural factors may be of greater importance than the apparent structure of the firm as exhibited by the organization chart. In conjunction with various project teams and work teams (within the firm and outside as well as having employees be members of such teams in professional associations), more and more employees may experience dual lines of authority even though the organization chart may suggest a functional type of structure.

In addition to the above, and addressed only partially in this section, is the challenge for firms with new information technology to remove geographical and time-zone boundaries, i.e., people are more and more expecting 24-hour service, 365 days a year. Unfortunately, employees' possible unwillingness to be accessible during long breaks (e.g., Denmark during July — employees take 3 or more weeks of vacation) makes managing the "borderless" organization difficult. Managing long employee absences when cutting costs is a must, and coordinating teams in various time zones and countries naturally increases demands upon employee flexibility and accessibility (e.g., e-mail on weekends) as well as skills. Increasing use of interdisciplinary teams involving employees from various departments and/or organizations (e.g., suppliers and key customers) are here to stay, with or without matrix structures. While the option for having an overall matrix structure for a firm may be dead, organizing certain activities along matrix lines is alive and kicking and will certainly grow considering such new forms of organizing as the virtual firm.

Defining Terms

Hybrid: Various organizational arrangements are being used to improve governance structure within an organization and between its various units; hence, while one unit may be using a matrix form, another may have a functional structure.

Matrix structure: Multiple command structures are used whereby employees experience dual lines of authority, responsibility and accountability.

Technological competence: The ability to retrieve, process, and use information and data for solving technology and economic-related challenges to make the most appropriate decisions.

References

Abo, T. A comparison of Japanese "hybrid factories" in U.S., Europe, and Asia. *Mgt. Int. Rev.*, 35: 79–93, 1995.

Barker, J., Tjosvold, D., and Andrews, R. Conflict approaches of effective and ineffective project managers: a field studying a matrix organization, *J. Mgt. Studies*, 25(2): 167–178, 1988.

Borys, B. and Jemison, D. B. Hybrid arrangements as strategic alliances: theoretical issues in organisational combinations, *Acad. Mgt. Rev.*, 14(2): 234–249, 1989.

Burns, L. R. Matrix management in hospitals: testing theories of matrix structure and development, *Admin. Sci. Q.*, 34: 349–368, 1989.

Burns, L. R. and Wholey, D. R. Adoption and abandonment of matrix management programs: effects of organisational characteristics and interorganizational networks, *Acad. Mgt. J.*, 36(1): 106–138, 1993.

Chuah, K. B., Tummala, V. M. R., and Nkasu, M. M. Project management structures in Hong Kong industries, *Int. J. Project Mgt.*, 13(4): 253–257, 1995.

Denis, H. Is the matrix organization a cumbersome structure for engineering projects?, *Project Mgt. J.*, 17: 49–55, 1986.

Dürrenberger, G., Jaeger, C., Bieri, L. and Dahinden, U. Telework and vocational contact, *Technol. Studies*, 2(1): 104—131, 1995.

Ford, R. C. and Randolph, W. A. Cross-functional structure: a review and integration of matrix organization and project management, *J. Mgt.*, 18(2): 267–294, 1992.

Gattiker, U. E. Technology adaptation: a typology for strategic human resource management, *Behav. Inform. Technol.*, 7: 345–359, 1988.

Gattiker, U. E. Where do we go from here? Directions for future research and managers, In *Studies in Technological Innovation and Human Resources, Vol. 2*, U. E. Gattiker and L. Larwood, eds., pp. 287–303, Walter de Gruyter, Berlin, 1990a.

Gattiker, U. E. *Technology Management in Organizations*, Sage Publications, Newbury Park, CA, 1990b.

Gattiker, U. E. Firm and taxpayer returns from training of semiskilled employees, *Acad. Mgt. J.*, 38: 1152–1173, 1995.

Gattiker, U. E., and WG 1. Internet security: strategic and social issues, Proc. eicar Security Workshop 97, Hamburg, Germany, pp. 173–208, 1997.

Gattiker, U. E. and Willoughby, K. Technological competence, ethics, and the global village. Cross-national comparisons for organization research, In *Handbook of Organizational Behaviour*, R. Golembiewski, ed., pp. 457–485, Marcel Dekker, New York, 1993.

Grandstrand, O. and Sjölander, S. The acquisition of technology and small firms by large firms, *J. Econ. Behav. Org.*, 13: 367–386, 1990.

Heller, F., Drenth, P., Kooman, P., and Rus, V. *Decisions in Organizations. A Three-Country Comparative a Study*, Sage, Newbury Park , CA, 1988.

Maurice, M., Sorge, A., and Warner, M. Societal differences in organizing manufacturing units: a comparison of France, West Germany, and Great Britain, *Org. Studies*, 1: 59–86, 1980.

McCollum, J. K. and Sherman, J. D. The effects of matrix organization size and number of project assignments on performance, *IEE Trans. Eng. Mgt.*, 38: 75–78, 1991.

McCollum, J. K. and Sherman, J. D. The matrix structure: bane or benefit to high tech organizations? *Project Mgt. J.*, 24(2): 23–26, 1993.

Mintzberg, H. *Structure in fives. Designing effective organizations*. Englewood Cliffs, NJ: Prentice-Hall.

Rice, R. and Gattiker, U. E. Computer-mediated organizational communication and structure, In *New Handbook of Organization Communication*, F. Jablin and L. Putnam, eds., Sage, Newbury Park, CA, in press.

Soares, A. Telework and communication in data processing centres in Brazil, In *Technology-Mediated Communication, Studies in Technological Innovation and Human Resources, 3*, U. E. Gattiker, ed., pp. 117–145, Walter de Gruyter, Berlin, 1992.

Ulhøi, J. P., Madsen, H., and Villadsen, S. *Training in Environmental Management — Industry and Sustainability: The Role and Requirements of Categories of Lower Managers and Workers*, European Foundation for the Improvement of Living and Working conditions, Dublin, 1997.

Further Information

http://www.aom.pace.edu/tim/   Academy of Management, Technology and Innovation Management (TIM) Division's home page

The above page provides interesting links to other sites and information about TIM's electronic newsletter as well as copies of the printed newsletter which can be found here for downloading

http://www.brint.com/   Brint Research Initiative

A nice research resource for management, business, and technology matters. "Council Partner of the U.S. Federal Government Inter-Agency Benchmarking & Best Practices Council"

http://www.TIM-Research.com   Technology and Innovation Management, Denmark

Not-for-profit virtual research organization involving various universities and organizations. Provides information about new developments on TIM issues, especially new research on environmental, R&D, Internet, e-commerce, privacy, and other technology issues. If you so desire, you can also participate on-line in ongoing research. Download aggregate results (descriptive statistics) of the latest studies' findings, or obtain research as well as technical reports, press releases, and much more (e.g., interesting links to other sites).

http://www.uti.ca/   University Technologies International Inc. Calgary, Canada

Provides information about inventions, patents, new products, interesting links, and reports for downloading and much more.

7.7 Networks

Karen Stephenson

The term **network** has three meanings in social life. The verb *to network* refers to the activities of social discourse or *socializing*, hence, the endless jabberwocky about networking in popular psychology, business press, and tabloids. In organizational behavior, *network analysis* refers to an analytical tool desperately seeking a theory. This comes as no surprise to veterans who know all too well that practice precedes theory by way of methodology. In management science, things are more mercurial: the noun *network* exists at the level of metaphor with minimal regard for rigorous empiricism. It is network as a noun that is in need of explanation and the objective of this précis.

We find networks on either side of the atomic divide: from the macromolecular in chemistry to the subatomic in physics. From Feynman and physics, the source and course of subatomic travel can be traced via networks as in Fig. 7.11. From organisms and organic chemistry, rates of reaction are differentially distributed along a marcomolecular chain of benzene building blocks as in Fig. 7.12. Can we look at extinct and extant human organizations in the same way? Can we "trace" the ancient trade networks and deduce the dynamics of dynastic expansion? Instead of tracking family lineages, can we track "financial and organizational lineages" and follow the money to unravel contemporary management and moral mazes? What if these ideas are not metaphorical conveniences or artifacts of the past but rather, scientific organizing principles?

Elementary Structures

Valuable insights stem from the desire to unearth buried structure. For example, look at the anthropologist Levi-Strauss's own failure to find a real "atom" of kinship in families [Levi-Strauss, 1969]. His suggestion that we should be less concerned with the theoretical consequences of a 10% increase in the population in a country having 50 million inhabitants than with the changes in structure occurring when a "two-person household" becomes a "three-person household" [Levi-Strauss, 1955] is a calculated statement about the relative importance of structure over scale. It is true that scale, i.e., going from 1 million to 10 million to 50 million, can and does make a difference. Just ask Donald Trump. However, what if there was an iterative pattern to the structuring of the five deals that produced Trump's 50 million? What if management mazes on the ground form recognizable patterns when viewed from afar? In other words, what if there was a structural principle at work in work? For Levi-Strauss, the original idea of an elementary structure or atom

FIGURE 7.11   Feynman diagram: scatter/transformation map of subatomic fission/collision.

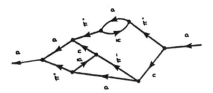

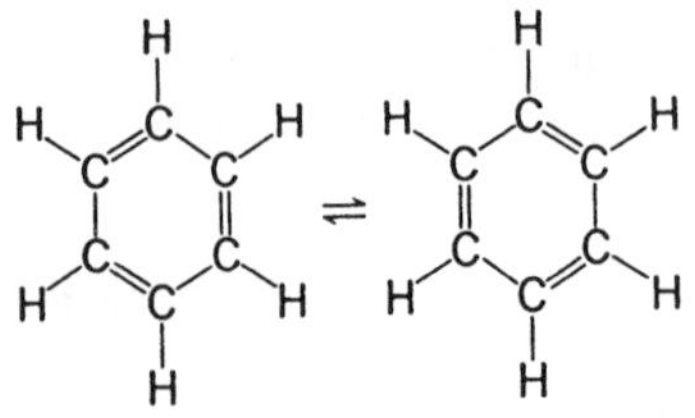

FIGURE 7.12   Benzene ring: six-carbon building block of organic compounds.

of kinship was more provocative than practical. It turned out there was no reliably repeating structure in the biological or fictive family. However, what if there was an atom of organization, a recurring structure of how people organize? Is there a network at work in work and, if so, can it be measured?

If one closely examines patterns in ancient trade networks and early settlements, we don't find a carbon copy of benzene (Fig. 7.12), but we do find a constellation of three prototypical patterns. These patterns emerge from the remnants of past civilizations to tell us why certain people traded with certain partners and avoided others. From the remains of daily routines, these same patterns unfold to tell us how people organize at work. To share identical patterns with our ancestors is a hint that humans are responding to scientific principles of organization.

Let's examine this notion more closely by reviewing the patterns. The first repeating pattern is to be **central** [Freeman, 1979], like the **hub** in a "hub and spoke" system on a bicycle wheel. This pattern represents an optimal distribution system for trading, settling in the flat lands and for centralizing work processes. The second pattern is the **gatekeeper** on the critical pathway between hubs and thereby connecting hubs to each other. These gatekeepers serve as important links or bridges along waterways, connecting one part of a society to another, or one part of an organization to another. The third pattern is the **pulsetaker**, someone who is maximally connected to everyone via the shortest routes. Pulsetakers have their finger on the pulse of the organization and know what everyone is thinking and feeling. Machiavelli was a pulsetaker, someone behind the scenes and arguably all seeing if not all knowing. These three **culture carriers** are pivotal and operate as change agents in a general sense, that is, they can resist change if they want or, by the same token, can rapidly catalyze change if they choose. Once identified, culture carriers can be used to retard or speed the rate of a restructuring or an acquisition or divestiture of another company. The speed of light pales in comparison to the speed and synchronicity of messages coursing through human networks.

A simplified example of tactically leveraging a communication network would go something like this: if a message needs to be sent to 500 employees, enlist the help of the hubs to broadcast the message effectively and efficiently. While the hubs are so deployed, do due diligence with the gatekeepers. Be sure they are working with you, not against you. Once 3 months or so have elapsed from the time the first message was sent, check with the pulsetakers regarding the veracity of the message. If "apples" represent the original message and the pulsetakers report back to you "apples", then the message was accurately and uniformly received among the 500 employees. You don't need to ask all 500 employees, just the pulsetakers — if they know, rest assured everybody else knows. If the message reported back to you was not "apples", but "oranges", then you know you have a problem and that problem is most likely located with the gatekeepers. Damage control consists of resending the message through them.

If one adds these three structures together, a "benzene ring" of network structure is approximated and is illustrated in Fig. 7.13. It may not be the atom of kinship Levi-Strauss envisioned, but it is a highly charged molecule or *network* of social interaction. This molecule of interaction appears in both extinct and extant records and is a highly structured form of social capital.

In summary, a network is a structured pattern of relationships typified by reciprocal patterns of communication and exchange. A seamless and often invisible web of differential and deferential reciprocity achieved largely through face-to-face and/or frequent interactions holds these trust-based relationships in place [Mauss, 1990]. Trust, typically conceived as a "warm and fuzzy" form of social capital, can be highly coercive and used to groom and maintain network contacts for monopolizing resources.

A network *is* the structure of culture: HGP (an acronym for the pattern of Hubs — Gatekeepers — Pulsetakers) is the structure of a network just as DNA is the structure of biological identity. Networks are the reason why culture is difficult to change and why rearranging the hierarchical or organizational chart is an impotent attempt at change. How do networks work? Trust is the glue that holds human networks together and is not unlike (1) the shared electrons that bind benzene or (2) field theory that prevails upon protons to produce subnuclear cohesion. Cursory calculations reveal that "matter" matters little: rather it is the field of energy that makes brick walls, steel plates, and diamonds impenetrable. If networks are the structure of culture, then the hardness of culture gives lie to the notion that organizational science is a "soft science".

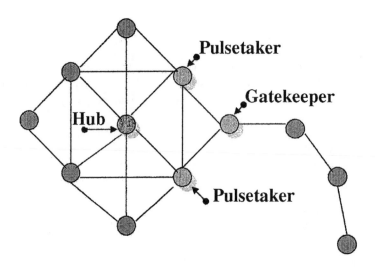

FIGURE 7.13   Culture carriers: the hub, gatekeeper, and pulsetaker of social interaction.

Of Myth, Metaphor, and Mystification

Assume a hierarchy. It was long thought that business organizations are islands of planned coordination in a sea of market relations. This hubbub over hierarchy eclipses the network or shadow organization and instead evokes images of a pristine paradise inhabited by vertically integrated tribes of happy and naive natives. However, the myth of the noble savage in primitive society was debunked by a generation of anthropologists who produced map after map of kinship diagrams charting the intricate webs of relationships among tribal society; theories about the *elementary structures* of kinship resulted.

In like fashion, postmodern explorers debunked the myth of the firm by discovering tangles of communication channels as intricate as kinship. General theories, such as the ABCs of kinship [Rivers, 1900] or the XYZs of organizational theory resulted [McGregor, 1957; Ouchi, 1981]. Map after map of hidden networks underlie notions of an organization in much the same way that generation after generation of biological or fictive kin define tribal or clanlike behavior. Networks bolster the hierarchy when needed or unravel it when it becomes burdensome. They nimbly collide and collude to produce asynchronous and asymmetric exchanges and thereby elude the visible hand of direct control [Chandler, 1977] and even the invisible hand of the market [Smith, 1978]. Bureaucracies live in infamy because of the capability to leverage networks against its own hierarchy; they are nothing more than knots of networks hanging on hierarchies of convenience.

The Deep Structure of Organizations

There is more truth than fiction to the phrase, "In the beginning there was a network…and from networks, all organizational life derives." The network is the embryonic beginning of culture and the initial stage of structure. It absorbs a certain amount of information or energy until it reaches a threshold and then subdivides into layers of networks. This incremental layering of networks is the birth of hierarchy. After continuous replication, hierarchical growth will slow to a glacial pace and ossify. Networks will then reemerge to nimbly dance like a whirling dervish around **tradition**, tradition being no more or less than fossilized hierarchy. The network contains the seeds of innovation, made visible against the backdrop of hierarchical tradition, and so it goes, organizations learn, grow, and perpetuate themselves through the networks producing hierarchies and hierarchies producing networks in a chain of leapfrog. Together, networks and hierarchies are pressed into organizational service and form a natural tension in the dance to discovery. Let's examine how this happens.

TABLE 7.7 Comparison Chart of Features of Hierarchies and Networks

Features	Pure Types	
	Hierarchy	Network
Organizing metaphor	Highways	Surface streets
Organizing principles	Depth, breadth, inverse relationship	Centrality: hub, gatekeeper, and pulsetaker
Protocol	Formal	Informal (casual) /nonformal
Elasticity	Rigid	Flexible
Visibility	Explicit	Implicit
Relationship	Authority	Trust
Diversity	Heterogeneous	Homogeneous
Knowledge stored in	Procedures	Relationships
Half-life	Centuries	Generations
Power	Direct: command and control	Indirect: influence and persuade
Resource investment	Money: financial capital	Time: social capital

Networks and Hierarchies: The Yin and Yang of Organizational Structure

When viewed from afar, organizational structure is not one structure but a *set* of distinct subsets of structure, that of hierarchies and that of networks that exist in a symbiotic relationship with one another, each having its own inherent set of organizing principles summarized in Table 7.7. Many of the features have already been discussed or alluded to in the text. A few additional features are explained below.

Organizing metaphor: Charting the formal and mapping the nonformal networks produces two kinds of representations. To give an apt analogy, there are maps of a city's highways system and separate maps of the complex web of surface streets. Each kind of map has its own rules that produce different sets of travel itineraries. However, most organizational journeys require both kinds of maps. More importantly, the behavior of the transportation system depends fundamentally on the interaction of traffic on the freeways and the surface streets. Not only does one system feed the other but the overload or failure of one system spills over into the other as drivers innovate to solve unexpected congestion problems.

This is a simplified view of organizations. Two kinds of organizational structures exist side by side and interact in important and predictable ways to determine the behavior of the firm. The analog of the highway system is the corporate hierarchy; it is apparent and obvious and changes design infrequently, tightly controlling behavior. The analog of the flow of traffic on the surface streets is the network. It is apparent, but not obvious; the network is not the system of surface streets but rather their pattern of use. The network changes organically and frequently as drivers constantly explore new routes, changing overall traffic patterns daily in ways that are beyond the direct control of traffic engineers. The network imposes a *different* set of constraints on behavior; if maneuvering becomes difficult, people will drive through parking lots, across lawns, around corners, and even over pedestrians!

Organizing principles: The three organizing principles of networks have been previously discussed. Three principles of hierarchical organization are: depth, breadth, and the inverse relationship between the two. Depth or layers of hierarchical levels are found in every organization and range from 2 in entrepreneurial organizations (the president and everybody else) up to and over 200 in government. Most organizational or management literature deals with the dark side of depth, that is, the multiple hierarchical layers that serve as filtering mechanisms — making information disappear all together or adding the patina of personal bias to information as it is handed off to those above. With so many hierarchical levels handling information flow, it can be difficult to track where and when information may get off track. Thus, deep hierarchies are often perceived as a *black hole* into which accountability is drawn and disappears.

Breadth (or span of control) is the number of people directly reporting to the person above them. Span of control is a way of segmenting or compartmentalizing information into buckets that can be sorted. These direct reports in turn have multiple direct reports to them and so on, as this self-replicating pattern cascades down the hierarchy.

A critical but subtle connection is the inverse relationship between these two organizing principles of depth and breadth, that is, a change in one (e.g., depth) will have the inverse or opposite effect in the other (e.g., breadth). For example, when one decreases the depth (flattens the organization), the breadth or span of control will increase. If one increases the depth (adds more hierarchical levels), the breadth will automatically decrease, that is, different departments representing the elements of span of control will merge or be eliminated. This relationship consistently holds if one does not substantially change the population of the organization through a divestiture or acquisition. The organizing principles of hierarchies have inherent constraints, allowing organizations to change only incrementally. Thus, hierarchies are beautifully designed structures for slow and incremental change. Networks, on the other hand, are exquisitely designed structures for rapid and radical change. For significant and substantive change, we must alter the networks.

Diversity: Networks are based on trust and trust takes time. Because trust is determined through face-to-face interactions, one needs to appreciate the profound truth that the face of culture is still a human face. What's coded in a face: gender and race for starters. After that, dress, height, accent, and a host of other personal attributes, which when aggregated with network formation reveal the stark truth about networks: you don't look like me, you don't dress like me, you don't think like me, therefore, I don't want to know or understand you. Such an opposition to diversity comes from a fetish for the familiar and is fundamentally tribal and resistant to the more-heterogeneous qualities of a hierarchy. Therefore, the last and perhaps the most important point to make about networks is that, contrary to popular opinion, networks have a dark side: they form exclusionary groups based on *like seeking like* and mask a fundamental fear of differences [Stephenson and Lewin, 1996]. A network is the most natural (and most ancient) form of grouping; its cultural complement to be found in hierarchical organization. That is why it is so important that the two organizational forms of hierarchy and network are forever yoked together to assure balance and accountability.

Defining Terms

Central (centrality):   A mathematical (combinatorial or statistical) measure to indicate maximum connection as algorithmically defined.

Culture carriers: A collective term including hubs, pulsetakers, and gatekeepers as key to controlling the rate and substance of cultural change.

Network: Pattern of reciprocal communications or exchanges in a human group.

Hub: To be central in a "hub and spoke" pattern.

Pulsetaker: To be central by being maximally connected to everyone in a network over all possible paths.

Gatekeeper: To be central by being a bridge between highly connected groups or individuals.

Tradition: Fossilized hierarchy: a codified record of past organizational failures.

References

Chandler, A. *The Visible Hand*, Cambridge, MA, Harvard University Press, 1977.
Freeman, L. Centrality in social networks, *Social Networks*, 1: 215–239, 1979.
Levi-Strauss, C. The mathematics of man, *Int. Social Sci. Bull.*, 6: pp. 581–590, 1955.
Levi-Strauss, C. *The Elementary Structures of Kinship*, Beacon Press, Boston, 1969.
McGregor, D. The human side of enterprise, In *Management Review*, The American Management Association, 1957.
Mauss, M. *The Gift*, W.W. Norton, New York, 1990.
Ouchi, W. *Theory Z*, Addison-Wesley, Reading, MA, 1981.

Rivers, W. A genealogical method of collecting social and vital statistics, *J. R. Anthropol. Inst. GB Ir.,* 30: 74–82, 1900.

Smith, A. *The Wealth of Nations,* 1978.

Stephenson, K. and Lewin, D. Managing workforce diversity, *Int. J. Manpower,* 17(4/5): 168–196, 1996.

Further Information

Allen, T. *Managing the Flow of Technology,* MIT Press, Cambridge, MA, 1977.

Axelrod, R. *The Evolution of Cooperation,* Basic Books, New York, 1984.

Baglivo, J. and Graver, J. *Incidence and Symmetry in Design and Architecture,* Cambridge University Press, New York, 1983.

Barrow, G. *Physical Chemistry,* McGraw-Hill, New York, 1973.

Bourdieu, P. *Outline of a Theory of Practice,* Cambridge University Press, New York, 1977.

Bull, G., Ed. [Machiavelli, 1514]. *The Prince,* Penguin Books, New York, 1981.

Burt, R. *Structural Holes,* Harvard University Press, Cambridge, MA, 1992.

Feynman, R., Leighton, R., and Sands, M. *The Feynman Lectures on Physics, Vol. 1, 2, and 3,* Addison-Wesley, Menlo Park, CA, 1963.

Hage, P. and Harary, F. *Structural Models in Anthropology,* Cambridge University Press, New York, 1984.

Lamberg-Karlovsky, C. and Sabloff, J. *Ancient Civilizations,* Benjamin/Cumings, Menlo Park, CA, 1979.

Mizruchi, M. and Schwartz, M., Eds. *Intercorporate Relations,* Cambridge University Press, New York, 1987.

O'Neill, R., DeAngelis, D., Waide, J., and Allen T. *A Hierarchical Concept of Ecosystems,* Princeton University Press, Princeton, NJ, 1986.

Rogers, E. *Diffusion of Innovations,* The Free Press, New York, 1983.

Rogers, E. and Kincaid, L. *Communication Networks,* The Free Press, New York, 1981.

Salthe, S. *Evolving Hierarchical Systems,* Columbia University Press, New York, 1985.

Stephenson, K. Diversity: a managerial paradox, *Clin. Sociol. Rev.,* 189–205, 1994.

Stephenson, K. and Zelen, M. Rethinking centrality, *Social Networks,* 11: 1–37, 1989.

Williamson, O. *Markets and Hierarchies,* The Free Press, New York, 1975.

7.8 Power in Organizations

Donald Palmer

Power is the capacity to attain one's goals over the resistance of others. Thus power is a social relationship — between more-powerful people and less-powerful ones. Power, as the term is used here, should be distinguished from **formal authority**. Formal authority is rooted in an organization's chain of command. Those at the top of the hierarchy possess authority over those below them insofar as the latter recognize their legitimacy. Legitimacy is based partly on evidence of a superior's ability to successfully fulfill the obligations of his/her office and partly on his/her ability to meter organizationally sanctioned rewards and punishments to subordinates conditional on their conformity to organizational rules and directives. Power, as we will discuss in greater detail below, is rooted in the control of **resources**. Power is analytically separable from the formal authority structure. Subordinates can possess power over their superiors just as surely as superiors can possess power over their subordinates. However, insofar as power may improve a superior's record of success, it can help boost his/her authority over subordinates.

Power, as the term is used here, should also be distinguished from **influence**. While power is a "capacity" to attain one's goals against the resistance of others, influence is "action". Influence may be related to power, as when it translates power into action. However, it may also be unrelated to power, as when it entails the altering of others' perceptions of what they want. Importantly, powerful people can obtain the compliance of others without exerting perceptible influence. The powerless sometimes comply with the inferred interests of the powerful because they perceive the costs of resistance to be high and the

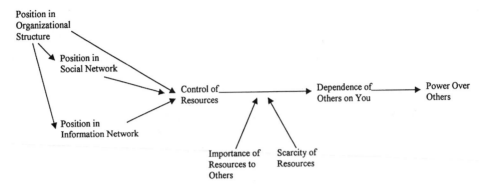

FIGURE 7.14   A diagram of the resource-dependence model of power in organizations.

probability of successful opposition to be low. For the most part, powerful people prefer to avoid engaging in overt influence to get their way. Overt influence typically requires the expenditure of valued time and resources and may engender increased resistance. The tendency of powerful people to avoid overt acts of influence is important to keep in mind when diagnosing the power structure of an organization, which we discuss below.

The Bases of Power in Organizations

People or subunits possess informal power, hereafter used interchangeably with power, to the extent that others are dependent upon them for valued resources that cannot be obtained elsewhere. Resources consist of an unlimited variety of "things" that others believe they need — either as ends in themselves or as means to desired ends. Resources include tangible items such as money, equipment, and offices as well as intangible quantities such as time and expertise. Access to resources is largely a function of a person's position in one of three social structures: the organizational structure, the social network, and the information network. People who have the right job assignments, friends, and contacts find it easier to obtain valued resources such as money, equipment, offices, time, and advice. Not all resources, however, build dependence. Resources must be important to others and scarce. If others do not need a resource, or if others require a resource but it is easily available from a variety of sources, then they will not become dependent on those people who possess it [Mechanic, 1962]. This simple resource-dependence model of informal power in organizations is depicted in Fig. 7.14.

The Functions of Power-Oriented Behavior

It is easy to identify the drawbacks of violating the chain of command in pursuit of informal power. Power-oriented behavior, activity geared to the accumulation and use of power, may waste an organization's time and resources insofar it is not directly focused on (indeed, may divert attention from) the attainment of organizational goals. Power-oriented behavior may also generate interpersonal animosities that inhibit communication and coordination, impeding the performance of necessary organizational tasks. Finally, power-oriented behavior may result in an allocation of resources that does not facilitate the attainment of organizational goals. Powerful persons may obtain higher salaries and powerful subunits may win larger budgets, even though neither is justified by organizational imperatives.

However, power-oriented behavior may have benefits. Power can be used to overcome organizational inertia, the tendency to resist change, which sometimes inhibits the pursuit of valued organizational goals. Organizational inertia may stem from sunk costs in plant and equipment or from taken-for-granted assumptions about the way things should be done. The limits of the chain of command, and by inference the possible benefits of power-oriented behavior, are suggested by the experience of Roger Boisjoly at Morton Thiokol [Boisjoly, 1987]. When confronted with information suggesting inadequacies in the field joints of the solid rocket boosters used to propel the Challenger space shuttle into orbit, Boisjoli used every available formal channel to bring this to the attention of his superiors and to correct the problem. However, his attempts were thwarted, and the result was a catastrophic failure of the booster and the loss of ten lives. Had Boisjoli been able to develop and used informal power to draw attention to and correct the defects in the field joint, and in so doing avert the Challenger disaster, his power-oriented behavior would have been laudable.

Indeed, some believe that power-oriented behavior tends to realign organizations with their environments in times of change [Pfeffer and Salancik 1978]. While organizational participants differ widely in the interests they hold, most share the desire to sustain and advance the organization. Thus, organizational participants will tend to value resources that are needed to meet immanent threats or take advantage of timely opportunities — referred to as **critical contingencies** — and will become dependent on those other participants who control those resources, especially if these resources are scarce. In this way, those people and subunits in the best position to help their organization cope with its critical contingencies tend to become powerful.

Of course, this system is not foolproof. Organizational participants can misdiagnose the critical contingencies confronting their organization or can misidentify the resources needed to cope with these contingencies. Further, powerful persons and subunits have an interest in preserving existing perceptions of their organization's critical contingencies and the resources they control because these existing perceptions are the foundation for their position of power, and their power provides them with an advantage in conflicts involving these determinations. Put another way, entrenched power structures may be a third source of structural inertia impeding organizational change. In sum, power-oriented behavior may have positive or negative consequences for an organization, depending upon the purpose to which it is directed. While there is a tendency for power-oriented behavior to help organizations realign themselves to changing environments, this realignment process is likely to contain frictions.

How to Develop Power in Organizations

Insofar as power can help organizations attain their goals, it is appropriate to analyze how people can develop power. The resource-dependence model of power implies the following five steps to developing power in organizations.

Determine who can help or hinder you in your attempt to achieve your objectives. As noted above, power is a relationship between you and others. Thus, the first step in building power is to identify over whom you need power. If you do not obtain power over those who can help or hinder you, it may be more difficult to advance your position. Conversely, if you expend time, effort, and resources developing power over others who are not in a position to help or hinder you in the pursuit of your goals, you have wasted time, effort, and resources. Xerox's Palo Alto Research Center (PARC) invented the first personal computer, the first hand-held mouse inputting device, the first icon-driven windows operating system, and other major new products. However, other companies — most notably, Apple — were the first to bring these innovations to market. PARC scientists underestimated the extent to which other Xerox subunits in Rochester, NY and Stamford, CT, might influence their ability to develop and market their innovations. Further, they assumed that other subunits would automatically champion their inventions, given — what to them were — their inventions' obvious merits. Thus, PARC scientists did not cultivate company-wide support for their projects [Smith and Alexander, 1988].

Determine what others want and what resources they need to satisfy these wants. Knowing what others want may help you to identify natural allies, people who are predisposed to help you in your cause. Knowing what resources others need to obtain their goals helps you to determine what resources you need to obtain control over in order to make others dependent on you. Peter Peterson was an exceptionally successful CEO of Lehman Brothers, a major New York investment bank. However, Lew Glucksman was able to depose him partly because Peterson out of touch with the interests of the firm's partners, the most senior of which dominated Lehman's board of directors. Peterson preferred to focus on external relations (spending the bulk of each work day outside the office) and was interpersonally aloof. Thus, he did not realize how much the partners worried about the possibility that he would sell the firm and that the firm's sale would end their careers and provide them with little monetary consolation. Glucksman focused on internal affairs (spending each day in the office from early in the morning until late at night) and was interpersonally accessible. Thus, he was attuned to the partners' concerns and could present himself as a champion of their interests and a challenger to Peterson's rule who was worthy of their support. [Auletta, 1985].

Determine and develop the resources at your command. Often we overlook resources to which we have access by virtue of our position in the organization. Even more frequently, we forgo opportunities to obtain access to resources not under our direct control. Lyndon Johnson was a master at identifying and developing his control of resources. Like many other congressional secretaries, when Johnson arrived in Washington, DC, he was confronted by large sacks of mail containing letters from constituents requesting favors from his boss. Unlike the vast majority of other secretaries, who viewed such mail as a nuisance, Johnson opened each letter and satisfied each request with zeal. With each fulfilled request, another dependence was built, and, if the mail was light for a time, Johnson stimulated it — by sending letters of congratulations to every high school graduate in his district, reminding them that he was at their service. Soon he was known in his congressional district (and beyond) as the person to go to in the nation's capital if one wanted something accomplished [Caro, 1982].

Improve your position in the organization with respect to the control of resources. Insofar as access to resources is contingent on one's position in the organizational hierarchy, information network, and social network, building one's power requires improving one's position in these three structures. Shortly after World War II, Henry Ford II became convinced that cost cutting was the only way to return the Ford Motor Company to profitability, and the best way to cut costs was to turn over the company's reigns to a new breed of managers who were trained in finance and facile with highly sophisticated quantitative decision-making tools. Once these individuals, among them Robert McNamara, Tex Thornton, and Lee Iacocca, attained positions of formal power, they assiduously positioned their proteges in key functions throughout the organization. In this way, they gained access to information about and obtained influence over key decisions in production, marketing, and research and development. Before long, every decision, whether it concerned a model change or an investment in new plant and equipment, was a "financial" one [Halberstam, 1986].

Analyze the power structure of your organization. Of course, you may not be the only one interested in developing and using power in your organization. Others engaged in power-oriented behavior may have objectives that conflict with your own. In this case, which we consider typical, it is important to understand the power structure of your organization. Krackhardt [1990] has conducted research showing that, the more accurate employees' perceptions of their organization's power structure, the more power they possessed. There are three ways to diagnose an organization's power structure, each of which can be useful — sometimes in combination. The decisional method entails identifying key decisions in the recent past and determining who participated in those decisions and as a result realized their objectives. The representational method entails identifying key positions in the chain of command (particularly important offices, task force, team assignments, etc.) and determining who occupies them. The reputational method entails identifying knowledgeable organizational participants and obtaining their assessments regarding who is powerful. The representational and reputational methods are the least-objective means of diagnosing an organization's power structure, insofar as they focus on the likely results of power (as sedimented in the distribution of formal authority in the organization or the perceptions of informants) rather than the actual process of power. However, these methods take into account that power may shape organizational outcomes in the absence of overt influence.

How to Manage Power Dynamics

As managers move up the organizational hierarchy, they increasingly find themselves not only in a position to accumulate and use power but also in the position of supervising and coordinating others engaged in power-oriented behavior. In such cases, managers have to manage power dynamics that unfold around and beneath them in the chain of command. There are two things to keep in mind when managing power dynamics. First, as a leader, one must diagnose and herald his/her organization's critical contingencies. These are the imperatives that power dynamics should be steered toward resolving. Second, one can manipulate the conditions that underpin the power of people and subunits so that those best able to cope with the organization's critical contingencies acquire the most power. These conditions pertain to the distribution of valued and scare resources in the organization. However, one must remember that, like all organizational processes, power processes are emergent, that is, they have a life of their own. Managing an organization's power dynamics may feel more like riding a bucking bronco in a rodeo that taking a Sunday drive in the family car.

Defining Terms

Critical contingencies: The most consequential threats and opportunities facing an organization.

Formal authority: The right to demand compliance to organizational rules and directives from subordinates.

Influence: Action geared toward changing the attitudes (in particular, the perceived interests) and behaviors of others.

Power: The capacity to achieve one's objectives over the resistance of others, whether they reside below, above, or at the same level in the organizational hierarchy.

Resources: Anything other people need, tangible (e.g., money) or intangible (advice), as an end in itself or as a means to an end.

References

Auletta, K. Power, Greed, and glory on Wall Street: The fall of Lehman Brothers, *New York Times Magazine*, February 17, 1985.

Boisjoly, R. Ethical Decisions — Morton Thiokol and the Space Shuttle Challenger Disaster, paper presented at the America Society of Mechanical Engineers Annual Meeting, Boston, December, 13–18, 1987.

Caro, R. *The Path to Power: The Years of Lyndon Johnson,* Alfred A. Knopf, New York, 1982.

Halberstam, D. *The Reckoning,* William Morrow, New York, 1986.

Krackhardt, D. Assessing the political landscape: structure, cognition and power in organizations, *Admin. Sci. Q.,* 35: 342–369, 1990.

Mechanic, D. The sources of power of lower participants in complex organizations, *Admin. Sci. Q.,* 7: 349–364, 1962.

Smith, D. and Alexander, R. *Fumbling the Future: How Xerox Invented, Then Ignored, the First Personal Computer,* William Morrow, New York, 1986.

Further Information

A more thorough elaboration of the ideas presented in this chapter that is well suited for practitioners can be found in *Managing with Power: Politics and Influence in Organizations,* by Jeffrey Pfeffer [1992]. A more academic treatment of these ideas can be found in *Power in Organizations,* also by Jeffrey Pfeffer [1981].

7.9   Influence

Andrew Ward

Power was defined in the previous section as a social relationship, rooted in the control of resources — by controlling resources others want, a position of power is established. Power is usually defined in terms of a relationship where one person or unit has power over another.

Influence, however, does not require a social relationship. While, as stated in the previous section, influence may be related to power to the extent that power is used to exert influence over another, we are most commonly influenced by our own perceptions, not by the force of others. Therefore, if you can alter the perceptions of another person, you can exert influence over them.

Oftentimes in today's organizations, people are expected to exert influence over others without any formal power to do so. Indeed, many technical and professional organizations are deliberately structured in such a manner, where projects thrive or die by the ability of individuals to influence and gain the cooperation of others whose input is essential to the survival of their project but over whom they have no organizational power or control over needed resources. The ability to influence others to do what you want them to do without direct power over them in order to force compliance is then an essential skill in today's technical and professional organizations.

We will first examine how we are influenced by our own minds — those deep-rooted, ingrained psychological forces at work within us that produce certain behaviors. Once we have examined how we are influenced, we will then look at the corresponding tactics that can be used to exert influence based on the awareness of these characteristics.

How We Influence Ourselves

While we might chuckle at the thought of Ivan Pavlov's dogs salivating at the sound of a bell ringing, because they have learned to associate the sound with the imminent arrival of food, we ourselves have learned many automatic responses that save us from undue cognitive thought. In order to save ourselves time, we guide our actions by many automatic responses, from buying the same brands at the supermarket without reconsidering the options at each opportunity to brushing our teeth with the same motions every morning.

Reciprocation

Our society has set up norms that guide our behavior to the general good. These norms are taught to us as children and are reinforced by society and our interactions with it. These conventions include everything from saying an automatic "Fine, thanks. And you?" when asked how we are, regardless of how we actually feel, to lining up in an orderly fashion at the supermarket checkout. These automatic reactions that we perform without thinking, for the most part, serve both for the benefit of society and ourselves as functioning members of that society. One of the strongest norms in our society is the norm of **reciprocation**, requiring repayment, in kind, for what has been given by another person. How many times have you received in the mail a free gift of address labels or notecards or picture stamps from a charitable organization? How many times have you been approached at an airport by a member of the Hare Krishna thrusting a flower into your hand or by a mute giving you a bookmark or pen as a gift? In all these circumstances, whether personally given or received through the mail, it is always emphasized that the token is a gift for you, and it is yours to take whether you reciprocate or not. However, the giver knows the norm of reciprocation is strong in our society, and most people will feel obliged to give something in return — in these cases, money. The emphasis on the token being a gift and not something to be paid for also disassociates the value of the token from the value of the return gift. This usually results in a return gift of much more than the cost of the token gift.

The norm of reciprocity pervades a variety of different circumstances in business. From calendars sent out by printers during the holiday season to free samples of cheese along the supermarket aisles to hospitality at prime sporting events, reciprocity abounds. Not only does the norm of reciprocation apply to gifts, it also applies to concessions. If one person makes a concession to another, the other usually feels obliged to make a return concession to that person. In business, budgeting often is as a result of concession reciprocation. People put in inflated requests for project or department funding, but concede certain points in order to bring funding down to the original hoped-for level, where the person (or group) dispersing the budget feels that he or she has gained a concession that must be reciprocated.

Commitment and Consistency

Consistency is one of the most valued traits in society. Not only is it a societal value, but most people have a strong need to be consistent within themselves; their view of their world and themselves needs to be consistent. Consistency is also of much practical value. Being consistent with past actions and behaviors relieves the burden of thinking through all the information used previously in making past decisions. Once you have gone through the decision process to purchase one brand of computer disks vs. another, the next time it comes to purchase disks, it is much easier to continue to purchase the same brand without reevaluating the decision.

However, not only do we want to be consistent because it is easier for ourselves and because it is a strongly held value in society, we want to also appear to be consistent to others. This is where commitment comes in. If we have previously said or done something that we think would lead others to expect one form of behavior, we are likely to act in that way in order to appear consistent to others. Making initial small commitments may appear harmless at first, but it often leads to increasingly larger commitments later. This is especially true if the commitments are active, public, effortful, and are seen as freely made. Many a venture capitalist has continued to sink good money after bad in an awry fledgling enterprise where he would not if it were a newly considered venture because of his prior investment and his need to show consistency and commitment to the failing enterprise. This pattern of escalating commitment to a previous course of action even when the action is obviously doomed is more common than we might think, and can happen on a massive scale. A business example of large-scale escalating commitment occurred when Fred Smith at Federal Express continued commitment to his Zap Mail project to the point where it cost FedEx many millions of dollars. This pattern of commitment and consistency is difficult to break. Robert Stansky, head of Fidelity Investments' massive Magellan Fund, tries to break this potentially devastating escalation of commitment to the stocks in his portfolio by consciously not considering his past behaviors on the stock, "It is more important to me to start every day by asking, 'What should I do with these stocks?' I don't care what I paid for them. I don't care where they've been. I only care about where they're going."[2]

This need to appear consistent to the point where we will escalate commitment to past decisions is taken advantage of by salesmen in a number of ways. Sometimes they may ask for a small commitment first such as a car salesman asking, "What color do you like the car in?" Other times they may ask for a commitment that is seemingly innocuous or vague, "If I can get you the right price, are you willing to drive it away today?" Frequently, they will follow up a purchase with a further sale, "Surely you want to protect your investment with an extended warranty?" A good salesman can use our need for consistency to lead us to drive out with a car and a greatly extended warranty, when we were only there for an initial test drive!

Social Proof

Just as we like to be internally consistent and to be seen to be consistent by others, we also desire to be consistent with others. We want to immitate what others are doing in order to be seen to behave in a

---

[2]*Time*, May 26, 1997, p. 60.

socially accepted manner. While, in most situations, we have had past experience of appropriate behavior, in ambiguous situations, or in situations that are new to us, we do not always know how to behave by ourselves and look to others for cues for appropriate behavior. This is known as **social proof** — because everyone else is behaving in a certain way, that must be the appropriate way to behave in this situation. We also look to "similar others". What people who are similar to us are doing is an important influence on what we ourselves do. If we can identify with a person, we are likely to imitate their behavior, especially in ambiguous situations.

A good example of the power of social proof is the experience of joining a new company. Because the norms of behavior in the organization are unknown to us, we tend to imitate what others, particularly what others at the same level or similar to us are doing, whether it be in style of dress, ways of addressing others, or other behavioral norms in the workplace.

Scarcity

Entire industries thrive around the principle of scarcity. From the art world to makers of exotic cars, vendors rely on our increased desire for something that is scarce. A wonderful demonstration of this principle was devised by Stephen Worchel and his colleagues [Worchel et al., 1975], who ran an experiment where participants in a consumer preferences study were asked to taste and rate the quality of chocolate chip cookies. For half of the participants, the jar they were given contained ten cookies, and, for the other half, the jar contained two cookies. Even though the cookies were identical, those who ate their cookie from a jar with only two cookies rated it more favorably in terms of being more desirable to eat in the future, more attractive, and more expensive than the people who took an identical cookie from a jar containing ten cookies. Moreover, if the participant was originally given a jar of ten cookies and then this was replaced by a jar of two cookies when the experimenter told them that they needed the cookies for other participants, the participant rated the cookies as even more desirable than even the cookies rated by those who had been given a jar with two cookies in the first place.

Thus, the scarcity principle is based on three phenomena. First, things that are rare and difficult to obtain are typically more valuable, and, therefore, scarcity serves as a proxy for value to us. Second, scarcity of something that is normally or was previously in abundance implies its popularity with others and so gains the force of social proof. Third, as things become less accessible, it constricts our freedom of choice that we strive to preserve. Once again, good salesmen of all varieties play on our increased desire for scarce and popular items by creating false scarcities, or "last-chance" opportunities, for purchase.

Authority

One of the most famous, yet most hard-to-believe, experiments in social science demonstrates the power of authority. This series of experiments, conducted by Stanley Milgram in the 1970s [Milgram, 1974], was disguised as a group of experiments on the effects of punishment on learning but was really about the power of authority. Professor Milgram set up the experiment where two people drew straws to select a teacher and a learner. Unbeknownst to the participants, the draw was rigged so that the real participant became the teacher, and an assistant of Milgram's became the learner. The learner (Milgram's assistant) was then strapped in a chair and connected to electrodes. Then the teacher (participant) and Milgram then went to the next room, where the teacher proceeded to ask the learner questions over an intercom between the rooms. The teacher was instructed by Milgram to administer an electric shock to the learner whenever the learner gave a wrong answer. The shock level was to start at 15 volts and was to increase by 15 volts for each mistake. To do this, the teacher had a dial that showed voltage ranges from 15 to 450 volts. The dial was also marked with words ranging from "slight shock" at 15 volts to "intense shock" at 255 volts to "DANGER: Severe shock" at 375 volts to finally "XXX" at the 450-volt level. Milgram gave the teacher a sample shock of 45 volts (still in the "slight shock" range), so that he could appreciate the learner's experience.

Unbeknown to the participant, the learner (Milgram's assistant) was not really being shocked. However, he started making grunts of pain at the 75-volt level. At 120 volts, the learner started shouting that the shocks were causing pain. At 150 volts, the learner screamed, "Experimenter, get me out of here! I won't be in the experiment anymore! I refuse to go on!" The teacher, if he complained to the experimenter of the learner's pain, was told simply to continue the experiment. Despite continued agonized screams from the learner, and even when the learner stopped responding and the teacher did not know whether the learner was even conscious, the majority of "teachers" continued to administer the shocks up to the maximum 450-volt level, stopping only when the experimenter told them to do so.

In an equally frightening experiment [Hofling et al., 1966], one of the researchers made a phone call to a nurse's station posing as a hospital physician. This "physician" was not known to the nurses, but directed the nurse to give an obviously dangerous amount of an unauthorized drug to a particular patient. In 95% of the cases, the nurse unquestioningly went straight to the ward's medicine cabinet and obtained the drug and was headed for the patient's room before she was stopped by another researcher.

This remarkable evidence of willingness to obey authority, despite the obvious harm it was doing to an apparently innocent victim, in the case of the Milgram experiments, demonstrates the power of authority and shows how we are indoctrinated from an early age to succumb unquestioningly to the influence of authority. This willingness to obey authority is the backbone of many more-traditional organizations that are run according to a strict hierarchy of power.

Liking

Put simply, we are much more likely to agree with people that we like. Liking has to do with a number of factors, but principal among these are how well we know a person, how closely we identify with the person, and how attractive that person is to us.

It is little wonder that advertisers try to associate attractive people with all sorts of products. Research suggests that we are more influenced by attractiveness than we think, and that attractive people have a "halo effect" that extends to traits unrelated to attractiveness, such as intelligence, talent, and kindness.

We are also likely to be influenced by people who are, or appear to be, similar to us. Another advertising trick refrains from employing attractiveness *per se* by using a model to endorse a product but rather shows a "Mr. or Mrs. Average" for the segment they are targeting. Even though the person is someone we don't know, because they are similar to us and appear to have similar wants and needs, we are prone to listen to what they have to say.

Other things being equal, we tend to prefer people that we know well, people that we spend a lot of time with or see frequently. It makes our lives considerably more pleasant if we enjoy the company of those people that we see frequently, and so we tend to like them more. In organizations, this tends to be reflected in people liking people in their own "in group", be it a division or a project group. While this has positive effects for cohesion of the group, it can also result in intense rivalries and can cause resentment to build between divisions or groups, to the detriment of the organization (particularly if a high degree of interdependence between divisions is necessary for success).

Influence Tactics

Now that we have examined the ways in which we are influenced by our predispositions and cultural norms, let us now turn to the tactics that can be used to exert influence based on these norms and dispositions. Influence tactics, although widely varied in their manifestation, broadly fit into three categories: tactics that draw upon **legitimacy**, tactics that use persuasion, and tactics that rely on **ingratiation**.

Legitimacy

The use of influence tactics involving legitimacy calls on our natural obedience to authority. In organizations, one often hears phrases such as, "The boss wouldn't like it" when someone is trying to stifle a new idea. Even though the proposal has not yet been put to the boss, the boss is being used as adding legitimacy to the argument to stifle the innovation. Similarly, outside consultants are often used in organizations not so much for providing fresh insight but more for providing legitimacy and "independent" evidence that a proposed plan of action is the most effective.

Often, when you receive an invitation to a conference in the mail, there are either endorsements of previous participants or names of well known attendees to the conference prominently featured in the literature. In this way, the organizers gain legitimacy to their conference through coalition and social proof. Similarly, restaurants will often display photographs of famous people who have dined in the restaurant as symbols of legitimacy.

While these previous forms of legitimacy use apparent or actual endorsements from other legitimate people to gain legitimacy for the idea or conference, legitimacy can also be gained from the use of symbols. Frequently, when trust is important and not immediately assumed between a service provider and a client, the service provider will often try to gain legitimacy and trust through the display of credentials, such as degrees or by proof of membership in professional associations, which are framed on the walls.

Persuasion

While legitimacy involves an indirect association with legitimate others in order to prove the worth of association in the hope that you will be influenced to join those that have gone before, tactics using persuasion rely more on the direct benefit for complying or on the cost of not complying to the will of the influencer. Thus, tactics of persuasion encompass the most overt forms of influence, such as the use of sanctions and the direct control over rewards or punishments for behaving in a certain manner. However, these types of influence tactics also make use of the more-subtle forms of reciprocity discussed previously, as well as bargaining, argument through reasoning, and the use of assertiveness.

Ingratiation

Ingratiation relies on the predisposition discussed above to show preference to those people we like. While in the previous discussion, we showed that liking depends upon how well we know a person, how closely we identify with the person, and how attractive the person is to us, there are a number of ways that these factors can be influenced.

We tend to like people who like us; thus, friendliness itself is an effective influence tactic. Friendliness also contributes to increased knowledge and familiarity with a person. The more friendly someone is to us, the more time we will spend with them, and the more closely we will identify with them.

Flattery is also deceptively effective as an influence tactic. While we think that we are not more likely to like someone who obviously flatters us, we are surprisingly prone to such a tactic. If someone flatters us, we have basically two possible reactions. We can think that the person is lying and just trying to influence us, which makes us feel bad that the flattering remark is not true and makes us feel bad about the other person. Alternatively, we can believe what the person is saying about us and feel good about ourselves and the other person. Generally, we prefer to feel good rather than bad and would prefer to ·think that the flattering remark is true, so we tend to react the more positive way and have positive feelings toward the flatterer.

The outward display of emotion is another means of exerting influence. The use of positive emotions reinforces friendliness and makes us feel closer to the person displaying the emotion, provoking similar emotional responses in us. However, the use of other emotions can also be effective in influencing behavior, such as sadness provoking sympathetic reactions or anger reinforcing assertive tactics.

Conclusion

Many of today's professional and technical companies are organized in project-based teams or in matrix structures that require organizational members to exert influence over others in order to get the job done without the organizational power to force the compliance of others. Thus, the ability to exert influence on others and to recognize when influence is being exerted upon oneself is an essential skill in today's technical and professional organizations.

In this section, we have examined the nature of influence, being that we are often influenced by ingrained norms and predispositions to behave in a particular manner when faced with a particular circumstance, and have also discussed how various tactics can be used to trigger such response, thereby gaining influence over others.

Defining Terms

Ingratiation: The gaining of favor of another.
Legitimacy: To gain credence by association with another person or institution.
Reciprocation: A societal norm that promotes the repaying in kind of a gift or concession received.
Social proof: Validation of one's behavior or the learning of appropriate behavior by observing the similar behavior of others.

References

Cialdini, R. B. *Influence: Science and Practice, 3rd ed.,* HarperCollins, New York, 1993
Hofling, C. K., Brotzman, E., Dalrymple, S., Graves, N., and Pierce, C. M. An experimental study of nurse-physician relationships, *J. Nerv. Ment. Dis.,* 143: 171–180, 1966
Milgram, S. *Obedience to Authority,* Harper & Row, New York, 1974.
Worchel, S., Lee, J., and Adewole, A. Effects of supply and demand on ratings of object value, *J. Personality Soc. Psychol.,* 32: 906–914, 1975.

Further Information

An excellent and highly readable book on influence and its sources and uses is Robert Cialdini's *Influence: Science and Practice* [3rd ed., 1993].

7.10 Negotiating Effective Business Agreements

Loren Falkenberg, Ron Franklin, and Dick Campion

Business negotiations range from very simple to unbelievably complex processes, making it very difficult to describe a "right" approach or to outline specific steps that should be followed. However, an overriding framework can be developed that guides negotiators through the planning stage, increases the exchange of information during negotiations, and leads to agreements that create value for both parties.

Effective Planning

A critical element of the planning stage is identifying the appropriate strategies for the negotiation. Negotiating strategies range from a traditional competitive approach to a more integrative, interest-based process [Lewicki et al., 1997]. In general, interest-based bargaining is a more effective strategy; however, under certain situational factors and for certain types of issues, a competitive approach is more appropriate. The ability to recognize when each strategy should be employed, and when each strategy is being used by the other party, is one of the differentiating factors between skilled and unskilled negotiators.

In a competitive strategy, the focus is on claiming as much of the resource(s) as possible by opening with extreme positions and progressing through minor concessions to a compromise solution. Relatively little information is exchanged between the parties, and this information is implicitly released through concessions. This strategy is appropriate if one of the parties

- Has a very short time frame
- Has little, if any, room to move off a position
- Has limited resources or limited flexibility to consider different alternatives
- Must meet the expectations of constituents or superiors on a particular issue
- Has a reputation for competitive behaviours

There are two serious limitations with this negotiating strategy. The first is that it leads to a compromise where the final agreement is somewhere in between the opening positions. Compromise is rarely the best solution; it "leaves money on the table". Another potential negative outcome of this strategy is a deterioration in the relationship between the parties; given the competitive nature of the process, there is a tendency to interpret the other party's actions as negative or threatening. A poor relationship at the close of the formal negotiations can lead to difficulties when implementing the terms of the agreement.

An integrative strategy involves the exchange of information in the belief that the parties share common areas of interest, and the negotiation process can lead to the creation of value for the parties. In contrast to the focus on "claiming" in competitive negotiations, the initial focus is on "creating", eventually moving to claiming. Information is exchanged through questions designed to explore the needs or interests of each party and to identify potential alternatives. The parties may open the negotiations with positions, but the discussion then shifts to the situational factors, needs, and/or priorities of each party. This approach to negotiations involves a more comprehensive exchange of information, often taking significantly longer than a competitive approach. An integrative strategy is appropriate if the parties

- Have the time and resources available to explore underlying interests.
- Have the flexibility to consider other alternatives (i.e., their constituencies do not expect a specific outcome).
- Will continue to interact after an agreement is reached.
- Have the power, authority, and/or resources to follow up on the commitments made.

A complication with this approach is that it requires trust. Trust is fragile and can be unknowingly broken. Part of the fragility of trust arises from the difficulty in differentiating between the emotional reactions of the parties and the issues being negotiated. A strong emotional reaction to another negotiator or to an issue can limit the rational assessment of the issue and/or the alternatives being discussed. Another potential problem arises from the constituencies rather than the negotiators themselves. A negotiator's constituencies are often not involved in the lengthy exploratory discussions of underlying interests and alternatives and, consequently, fail to comprehend why unexpected alternatives are better than desired positions. Rather, they assume the negotiator was not skilled or forceful enough to win their position.

Other elements of planning are the thorough exploration of the underlying interests to your initial positions, anticipating the other parties' needs and positions, and reviewing the expectations of constituents or outside parties [Lewicki et al., 1997]. Initial information-gathering steps may include informal meetings with the other parties and consulting with other groups who have negotiated or worked with an identified party. Opening positions are often based on some type of public quantitative information that may be obtained from sales information, financial databases, industrial standards, etc. In contrast to positions, underlying interests are much more difficult to identify, with parties often entering negotiations without a full understanding of their own needs. Negotiators should extensively review situational factors, time frames, scarce/surplus resources, risk preferences, and/or priorities that could be addressed by an agreement with the other party(ies). Some questions that may help negotiators identify their own interests and positions, as well as those of the other parties, are highlighted in Table 7.8.

Exchanging Information

Negotiations generally open with each party stating its positions. Positions are a narrow description of a party's needs/desires and can lead to errors in interpretations of a party's needs and/or flexibility to consider alternatives. Consequently, negotiating from positions tends to limit the exchange of information. Communication involves repeating or rewording statements that support a desired position while finding ways to reduce the value of the other party's position. In order to move from positions to the exploration of needs and interests, a negotiator must be skilled at structuring questions so that the other party can accurately interpret and respond to the questions. There are some simple rules, based on the broad classification of open and closed questions, for asking effective questions in negotiations.

Open questions begin with what, where, who, why, or how and should be used to elicit general information about the other party's situational factors, available resources, depth of knowledge, and/or

TABLE 7.8 Identifying Underlying Interests and Opening Positions

Questions that should be asked during the planning phase	How the information can be used in the negotiation
What are the available resources you have when entering the negotiations? What are the other party's available resources?	The answers to these questions should identify resources that may be traded or included in the agreement (e.g., financial, time, in-kind support, expertise)
What specific needs must be met by an agreement? Why must these needs be met?	These questions should help identify the underlying interests behind the positions (e.g., Does one party have a specific deadline? Must the demands of a constituency be met? Is the reputation of a negotiator important? Is there a need for a scarce resource? Is there a surplus resource that needs to be reduced?)
What are the costs to each side of the different alternatives considered? Can these costs be compensated?	The answers to these questions will guide negotiators to understanding the trade-offs that can create value (e.g., absorbing a short-term risk for the other party if they are willing to reduce your surplus)
What facts/logic support your position? What information will the other party use to support their position?	By answering these questions, negotiators can better understand the differing views or frameworks of each party as well as identify where differences in perception or interpretations could limit the exchange of information
What are your best alternatives to a negotiated agreement (BATNAs)? What are the BATNAs of the other parties?	Identifying BATNAs provides a benchmark as to what is a satisfactory agreement and limits the trap of focusing on the sunk costs of negotiating rather than walking away from a poor agreement

priorities. They are most appropriate when opening negotiations, introducing a new issue, or when wanting to shift the focus of the discussion from justification of positions to a broader exchange of information [Ury, 1993]. They should not be used when a negotiator wants to maintain control of the discussion, when time/deadlines are impending, and/or when closure on an issue is needed.

A rule of thumb is that closed questions should follow open questions. They should be used to probe for specific pieces of information and/or to check assumptions and interpretations about the other party's needs. When used too early, they can limit the discussion and lead to a focus on positions. Closed questions are useful as negotiations close on an issue and to ensure that the parties share similar interpretations and understanding of an agreement. Used at the wrong time or worded in the wrong way, closed questions can be perceived by the other party as threatening and prompt a defensive reaction.

An ironical twist to negotiations is that it is the differences in preferences or priorities between the parties, not the common areas of interest, that lead to the creation of value [Fisher and Ertel, 1995]. Each party enters the negotiation with different situational factors, flexibility, availability of resources, and interpretations of specific outcomes. Examples of how differences can lead to the creation of value are when one party can absorb risk in the short term while the other party wants to shift its risk to a longer time frame or when resources can be traded to ensure that one party is able to reach a looming deadline in return for the provision of resources at a later date. Value may also be created by reframing a loss into a gain [Bazerman, 1994]. A classic example of reframing a loss is in the selling of real estate. A house is for sale with a list price of $155,000, the currently assessed market value. The seller paid $100,000 for it 5 years ago. A final offer is made for $150,000; however, the seller does not want to accept it as he believes it represents a $5000 loss. In contrast, if the benchmark is the original price, a $50,000 gain has been made and the sale may be made.

Finally, it is sometimes difficult for negotiators to recognize that what is valuable to one party may be of little value to the other party (i.e., that each party has a different utility for a given resource) [Fisher and Ertel, 1995]. In the heat of negotiations, there can be a tendency to perceive that what is valuable to one party must be valuable to the other party(ies). One way of determining the value of a specific resource or outcome is to ask "is it nice, but not necessary?" or "is it necessary rather than nice?". Often value is created when one party views an outcome as nice while the other party views it as necessary. A trading of priorities can then occur, particularly if there is another nice-vs.-necessary tradeoff that can be included.

Closing with Value

Closing negotiations can be one of the most difficult steps, partially because it is not always apparent whether sufficient information has been exchanged to create an optimal agreement. It is unlikely that an answer to this issue is available during the negotiations. One step to prevent inefficient and dragged-out negotiations is to set time lines during the initial discussions. Reasonable time lines can prevent inefficient and lengthy discussions and create a joint need to close the negotiations.

In order to determine if sufficient value has been created, the parties need to continually review their needs and priorities and assess whether the alternatives being negotiated address these needs. If the negotiations are not creating any value for a party, that party should close the negotiations without an agreement. Sometimes parties are reluctant to close the negotiations because they do not have a sense of having won, of having achieved their opening position, or even of having taken a tough stance. Application of the following criteria may help the parties recognize a potentially good agreement:

- It is better than known alternatives
- It well satisfies our interests, it acceptably satisfies the other party's interests, and it is tolerable enough for outside parties to be durable
- It includes commitments that are well planned, realistic, and operational
- The process has helped build the desired long-term relationship

Unfortunately, integrative negotiations can break down during the closing phase because perceptions of fairness have been violated. As negotiations close, there is a shift from listening and exploring to ensuring that self-interest is maximized. Given this shift from "creating to claiming value", managing perceptions of cooperation and fairness are critical. Trust can easily be broken when the other party perceives a sudden lack of interest in their needs. Three steps or actions that can facilitate the closing process are

- To develop terms of agreement that incorporate each party's interests, not just those of one party
- To introduce potential terms of agreement by describing how they satisfy your needs followed by asking how they address the other party's needs
- To review the specific details and interpretations of each clause or item in the agreement and address all "what ifs" before final agreements are concluded

Although the last step can be frustrating as deadlines loom, it will determine whether a workable agreement has been achieved and will influence the amount of conflict that will occur as the agreement is implemented.

In closing, this short article is only the "tip of the iceberg" in terms of developing negotiation skills. The intent of this section was not to provide a quick answer to negotiating problems but to provide a framework from which to more effectively plan, exhange information, and produce an effective outcome.

Defining Terms

Competitive bargaining: Occurs when negotiators open the bargaining process by stating positions on the different issues involved and move to a compromise agreement through a series of concessions.

Integrative bargaining: Occurs when negotiators exchange information through discussion and questions designed to identify interests with the goal of the negotiations to find an agreement that creates value for both parties.

References

Bazerman, M. H. *Judgement in Managerial Decision Making, 3rd ed.,* John Wiley & Sons, New York, 1994.

Fisher, R. and Ertel, D. *Getting Ready to Negotiate: The Getting to Yes Workbook,* Penguin Books, New York, 1995.

Lewicki, R. J., Saunders, D. M., and Minton, J. W. *Essentials of Negotiation,* Irwin, Chicago, 1997.

Ury, W. *Getting Past No: Negotiating Your Way from Confrontation to Cooperation,* Bantam Books, New York, 1993.

For Further Information

A good overview of the negotiation process is provided in the book *Essentials of Negotiation* noted in the references.

An interesting review of the different types of negotiating situations, the relevant factors to consider in these situations, and the perceptual traps that should be avoided is given in *The Dynamics of Bargaining Games* by J. Keith Murnighan.

A good book for developing communication skills is *What to Ask When You Don't Know What to Say* by Sam Deep and Lyle Sussman.

7.11 Rewards for Professionals: A Social Identity Perspective

Kimberly D. Elsbach

According to a recent poll, the 1993 MBA class at the Harvard Business School ranked salary seventh among the reasons for the career choices they made. Job content and level of responsibility were at the top of the list, followed by company culture and the caliber of colleagues [Labich, 1995]. Over the past 5 years, my own in-class polls of undergraduate and graduate business and engineering students have also shown pay to consistently rank third, behind "interesting work" and "the chance to use your mind" as job motivators. In response to this growing desire among professionals for interesting work, corporations have begun to move away from traditional jobs and reward systems (i.e., well-defined responsibilities, clear task boundaries, clear hierarchies of power, and **tiered pay systems**) into what is called a "jobless" environment. Today's large corporations are increasingly organized around work that is boundaryless, providing professionals with responsibilities and opportunities that are more typical of a family than of a bureaucracy. In essence, there is a "New Deal" afoot in corporate America that trades the rights of professionals for interesting and important work, the freedom and resources to perform it well, and the pay that reflects their contribution for a workplace with no job security and the personal responsibility of continually finding ways to add value to the company.

In line with this new deal, the identity of the new professional appears to be more closely aligned with that of a leader or an entrepreneur rather than that of a foot soldier. For many of today's professionals, the rewards of work are tied to freedom and to the risks of managing one's own career rather than to the safety and stability of working in a highly structured environment. As Ross Webber, chairman of the Management department at the Wharton School of the University of Pennsylvania puts it,

> There's been a change in the myths that talented people in this new generation guide their lives by, and an entrepreneurial, rather than corporate connection is a strong part of that mythology [Labich, 1995].

It is this new entrepreneurial identity that has changed the nature of work and of rewards for today's professionals. Theory and research on *social identity* (i.e., individuals' self-concepts based on their ties to social groups, such as their work organizations and professional associations) suggests that, to the extent that being associated with an organization enhances and sustains a positive self-concept for an employee, he or she will be more loyal, committed, and supportive of the organization [Dutton et al., 1994]. An employee who sees herself as a creative, individualistic entrepreneur may be more likely to identify with (and, thus, be committed to) an organization that rewards and supports worker autonomy and individual initiative than a company that rewards conformity and obedience. The evolution toward more "entrepreneurial" identities in today's professionals suggests a need for a similar evolution in how organizations sustain and support those identities.

In this section, I propose that corporate reward systems (i.e., the means by which corporations purposefully remunerate employees for their work) are a primary means by which organizations can support and enhance the evolving social identities of their employees. It seems likely that reward systems are correlated with organizational identification among employees because they are salient and because they say something about the people earn them (i.e., they indicate what kinds of compensation people are willing to work hard for). Thus, although issues of fairness and equity have been most commonly linked to successful reward systems, it may be the case that only those workers whose social identities are enhanced by a given compensation system will thrive in an organization, independent of the equity of the rewards provided. From this perspective, I will argue that a social identity perspective may be a useful tool for assessing the appropriateness of reward systems for today's professionals.

In the remainder of the section, I will discuss how corporate reward systems involving (1) pay and (2) work opportunities can affect professional employees' self-concepts in ways that affirm positive social identities.

Pay and Social Identity

Although it consistently ranks low as a job motivator for employees, pay in its various forms (i.e., salary, bonuses, vacation-time, and other material perks) remains the primary means by which organizations reward and motivate employees for their work. In most modern corporations, pay is commonly regarded as the most recognizable and easily understood indicator of status and self-worth. As a consequence, a person's self-esteem and self-concept may be affected both by the quantity of his or her pay and by the manner in which it is earned. Pay systems indicate, publicly, the goals people strive for at work as well as their effectiveness in achieving those goals.

Traditional Tiered Pay

Many large corporations continue to rely on a version of the tiered pay system devised by Edward Hay at General Foods nearly 50 years ago. The Hay system ties pay to an employee's ranking in a hierarchy of job difficulty and importance. The system measures the degree to which three factors (i.e., know-how, problem solving, and accountability) are required in a job. Within each of these factors there are several specific job requirements (i.e., know-how includes expertise, education, and individual ingenuity) along with a hierarchy of expectations for each job requirement at different pay levels. The Hay system is popular because it is flexible enough to apply to almost any job and can be updated as jobs change.

In terms of social identity, working for a company that uses a traditional tiered pay system may promote a sense of distinctiveness or uniqueness by placing employees into large numbers of highly specific job categories. Distinctiveness has been defined as a central component of social identity [Ashforth and Mael, 1989]. In the same vein, social psychologists have found that most people have a need for uniqueness (i.e., a need to be distinct from others on important identity dimensions) as well as a "uniqueness bias" (i.e., a tendency to see ourselves as different from others). By providing employees with positively defined social categories to which pay is tied (e.g., employees that get an MBA through night school will be automatically moved to a higher grade with higher pay), organizations may satisfy the needs of employees for distinctiveness and uniqueness.

Reward systems that are linked to numerous categorizations of employees also increase the likelihood that employees may make favorable social comparisons with their peers. Recently, researchers have found that by emphasizing its favorable identity traits, while downplaying its unfavorable ones, organizational members could place their organization and themselves into categories that most positively compare to others. Thus, given enough different possible categorizations, almost anyone can be a "big fish" in the pond.

On the downside, tiered pay systems may promote the notion that jobs are merely a sum of their parts and that the value of a worker can be mathematically determined by traditional scientific management principles. Such hierarchy-based systems tend to conflict with the trend toward flatter organizational structures with fewer pay levels (referred to as "broadbanding" by many corporations) and with the notion that today's most valuable workers cannot be so easily categorized. Marilynn Brewer's [1991] theory of "optimal distinctiveness" suggests that people may desire identities that place them into social categories that are exclusive enough to satisfy their needs for uniqueness but are also inclusive enough to provide them with legitimacy of the larger group and to allow them to claim the traits that are widely sought and accepted as normative. Thus, working for rewards that are highly differentiating may not appeal to entrepreneurial workers who see themselves as more broadly skilled.

Incentive Pay

In contrast to tiered pay systems that tie pay to rank or job description, incentive pay systems tie large amounts of compensation (e.g., year-end bonuses, pay raises, stock options, or other perks) to predetermined performance goals, such as meeting a companywide or divisional productivity improvement goal. For example, year-end bonuses may be tied to the ability of managers to achieve earnings better than their cost of capital (i.e., what is commonly called economic value added, or EVA) as an incentive for managers to think more about shareholder value. Incentive pay systems became popular in the 1980s as a means of encouraging creativity, innovation, and "healthy" competition among workers as they attempted to improve the profitability of their organizations.

This competitive feature of incentive pay may have important consequences for the social identity of employees. When companies reward only those employees that meet specific performance goals, they promote competition by implicitly (or explicitly) ranking employees against each other. Such ranking may provide employees with a sense of "positional status", which is an important aspect of social identity for many people [Frank, 1985]. As Frank [1985] notes, many of the rewards for which individuals compete are positional goods, "sought after less because off any absolute property they possess than because they allow people to compare favorably with others in their own class." Similarly, social psychologists have found that people often choose to make downward comparisons (i.e., comparisons to those worse off than themselves) in order to enhance their self-esteem and self-concept. Incentive pay systems provide a ready means for making such downward comparisons.

Yet, after little more than a decade of use, corporate reviews of incentive pay systems have become increasingly critical. Recent surveys report that incentive pay often promotes greed, short-sightedness, and decidedly unhealthy competition among employees [Nulty, 1995]. An article about its failure at a small manufacturing organization reported,

> In the real world, pay for performance can also release passions that turn workers into rival gangs, so greedy for extra dollars they will make another gang's numbers look bad to make their own look good... some employees even argued over who would have to pay for the toilet paper in the common restrooms. One aspiring bean counter suggested that toilet paper costs should reflect the sexual makeup of the division, on the shaky theory that one gender uses more tissue than the other [Nulty, 1995].

These problems underscore the double edge of positional status as an identity attribute; while a few may claim the rank of "top-tier" employee, many more must suffer the rank of "average" or "second rate". The fact that only a few employees will acquire positive social identities from most incentive pay systems while many more will actually acquire negative social identities may help explain why incentive pay systems fail at many organizations. When only extreme performance is rewarded, many moderately high-level performers become dissatisfied and leave the organization. This effect may be explained by the fact that the relevant comparison group for moderately high-performing employees is the extremely high-performing group that received the maximum bonus (i.e., an unfavorable social comparison). Thus, by disappointing high-performing employees who may expect a large bonus, incentive pay systems dole out what is perceived as "punishment" to many employees [Kohn, 1993].

Work Opportunities and Social Identity

While the above discussion suggests that pay systems can be tailored to fit the social identities of workers, it also suggests that most pay systems have several inherent weaknesses in their ability to enhance and support positive social identities. First, pay tends to focus on quantitative aspects of identity (e.g., your status or comparative worth as a human being may be related to your salary — and pay is the only way to determine the value of your job). Today's more entrepreneurial professionals may view these traits as less central to their self-concepts than are *qualitative* identity traits, such as skills, abilities, or values, because they are less transient across time and context (e.g., a person ranked highest in salary at one company may be lowest at another). Second, because pay is viewed as manipulative (i.e., do this and you'll get that), identifying with an organization that rewards primarily with pay may suggest that one is the type of person who is willing to be manipulated for money [Kohn, 1993]. A large amount of psychological research has shown that, if people focus on extrinsic motivators (e.g., pay), they are less likely to be motivated to work hard than if they are motivated by intrinsic motivators (i.e., identity-enhancing effects of expressing a valued skill). As a consequence, pay may be motivating as a reward if status is important to one's identity, and working for pay is not incongruent with that identity.

Yet, as noted in the beginning of this section, these conditions do not appear to be consistent with the identities of a great deal of today's professionals. As an alternative to pay, organizations that reward employees with work opportunities that allow them to enhance skills and abilities, as well as to sustain their ideals and values, may be most effective in supporting positive social identities in today's professionals [Dutton. et al., 1994]. These types of opportunities affirm valued and distinctive identity attributes that are likely to be more enduring than status and thus are likely to be more intrinsically motivating than pay [Kohn, 1993].

Skill-Affirming Opportunities

One of the most important antecedents of organizational identification is employee successes in meeting their professional goals [Ashforth and Mael, 1989]. By extension, organizations that allow employees to build and affirm important and valued skills and abilities through work may increase those employees' identification with the organization and enhance their social identities. In this vein, a study of U.S. Forest Service employees showed that employees who felt that their jobs were challenging, that they allowed them to do the things they liked best and to properly use their talents, and whose work was important in helping the Forest Service achieve its goals were most likely to identify with the Forest Service and to acquire a positive self-concept from working there [Schneider et al., 1971]. Demonstrating and building valued work skills may be even more important to today's entrepreneurial professionals who define themselves as "jacks-of-all-trades" managers — able to step across functional boundaries with ease [Labich, 1995].

Organizations may use skill-affirming opportunities as rewards by giving employees the option of working on special assignments or projects (especially in areas that interest the employee) following satisfactory performance on a previous assignment. For example, in a recent study, I found several of the corporate sponsors for the Olympic Games in Atlanta rewarded high-performing managers who also had an interest in "crisis management" or "special projects management" by assigning them as special Olympic projects managers in marketing, operations, and community relations for the 2 or 3 years just preceding the Games. These highly visible assignments may also support status dimensions of employees' social identity.

It is important to note that these opportunities are not promotions but are selective assignments tailored to meet the identity needs of particular employees. In this way, they may also be more than mere job enrichment (i.e., making jobs more intrinsically interesting by increasing things such as skill variety and task identity) — they may be a means of identity affirmation. The opportunity to affirm an important identity dimension through one action (i.e., exhibiting a valued skill) may actually compensate for taking other actions that appear to damage the identity (working for less pay than is perceived adequate).

Value-Affirming Opportunities

In addition to rewarding employees with the opportunity to affirm identity-relevant skills, organizations may reward employees with the opportunity to affirm identity-relevant values. Research on person-organization fit suggests that employees prefer, seek out, and remain at work organizations whose values are perceived to be congruent with their own. By contrast, a mismatch between a person's values and the environmental attributes that allow that person to fulfill those values has been shown to be a strong predictor of employee dissatisfaction with work.

This research suggests that employees may be motivated by opportunities that go beyond merely affirming their skills and abilities to affirming their character and moral worth. Increasingly, when asked "why do you work?", employees answer with remarks about affirming important life values and fulfilling individualistic needs rather than achieving material ends. In a recent study on reputation building and social identity, for example, Elsbach and Glynn [1996] found that United Parcel Service employees who were able to affirm their values for community service by participating in community outreach programs for literacy improvement, drug abuse intervention, and sheltering the homeless found their own identities affirmed at the same time. Similarly, Tom Chappell, CEO of the environmentally conscious personal products corporation Tom's of Maine, rewards his employees by allowing them to volunteer one workday a month at the nonprofit organization of their choice. As Chappell suggests, this kind of work is rewarding because it affirms who you are, "It's not winning at all costs. It's challenging yourself to win according to who you are" (cited in Dumaine [1994]).

Conclusion

In this section, I have argued that corporate reward systems might be most efficiently aligned with the needs and desires of today's professionals if viewed from a social identity perspective. This perspective suggests that the central, distinctive, and enduring dimensions of employees' identities should be supported and enhanced by pay systems and work opportunities that comprise the bulk of corporate rewards. Because professional workers are increasingly adopting more entrepreneurial identities, reward systems that signal the importance of innovation, autonomy, and self-direction may best support and enhance the self-concepts of today's workforce. Thus, rewarding employees with opportunities to affirm their individual creativity or to hone new skills, rather than with pay incentives that focus on their positional status, may be more effective in the 1990s. The most important benefit of a social identity perspective to reward systems, however, may be that it stresses the importance of continual adaptation of reward systems to be congruent with the ever-changing identities of today's professionals.

Defining Terms

Self-affirmation: Behaviors or cognitions a person carries out to reinforce established beliefs about him or herself.

Social identity: A person's self-concept based on his or her association with social groups.

Tiered pay systems: Compensation programs that reward people based on their level of expertise and rank within an organization.

References

Ashforth, B. E. and Mael, F. Social identity theory and the organization, *Acad. Mgt. Rev.*, 14: 20–39, 1989.

Brewer, M. B. The social self: on being the same and different at the same time, *Personality Psychol. Bull.*, 17: 475–482, 1991.

Dumaine, B. Why do we work?, *Fortune*, December 26, 196–204, 1994.

Dutton, J. E., Dukerich, J. M., and Harquail, C. V. Organizational images and member identification. *Admin. Sci. Q.*, 39: 239–263, 1994.

Elsbach, K. D. and Glynn, M. A. Believing your own PR: embedding identification in strategic reputation, In *Advances in Strategic Management, Vol. 13*, J. A. C. Baum and J. E. Dutton, Eds., pp. 65–90, 1996.

Frank, R. *Choosing the Right Pond: Human Behavior and the Quest for Status*, Oxford, New York, 1985.

Kohn, A. Why incentive plans cannot work, *Harv. Bus. Rev.*, Sept-Oct: 54–63, 1993.

Labich, K. Kissing off Corporate America, *Fortune*, 131(3): 44–52, 1995.

Nulty, P. Incentive pay can be crippling, *Fortune*, November 13, 235, 1995.

Schneider, B., Hall, D. T., and Nygren, H. T. Self image and job characteristics as correlates of changing organizational identification, *Hum. Relat.*, 24: 397–416, 1971.

7.12   Teams and Team Building

Francis T. Hartman

Teams are more than a buzzword. They are a necessity for survival. As technology continues to double every few years, we are forced to respond by addressing the need to manage the people who work with an increasing body of knowledge. There are three obvious options as to how this may be achieved:

- Double our technology staff every few years and quickly go out of business.
- Double our intellectual capacity: no evidence that we have tried or succeeded.
- Do something else: we already are.

The "something else" is happening in several ways in business strategy and tactics:

- Focus on core business: we divest ourselves of such elements as IT, payroll, recruiting, and training, items considered part of core business a scant few years ago.
- Enter into alliances, channel relationships, and partnering agreements, even with our competitors.
- Joint development of strategic products with other organizations.
- Boundaries between buyers and suppliers are blurred.
- Concurrent engineering is increasingly required for businesses to stay competitive and for products to get to market on time.

These changes reflect an underlying need to share scarce resources. We can no longer rely on people as commodities, a common mindset as recently as the early 1990s. Individual skill profiles and expert knowledge in specific technologies and their application are two of the key drivers in selecting members of a team at corporate and individual levels.

Today's teams are invariably cross-functional interdisciplinary groups of specialists and generalists from different parts of organizations or from different organizations. Often they may not even meet. Today's teams face a whole set of new challenges. Some of the emerging issues in assembling and managing teams in this environment are addressed in this section.

Fundamentals of Teams and Team Building

Pinto and Kharbanda [1995] summarize the classical team-building cycle of forming, storming, norming, performing, and adjourning. The implication is that, once we have formed a team, we need to rage (or storm) to establish the norms of behavior for the group. We can then move on to performing the work we set out to do. Conventional processes for team building have led us to this conclusion through extensive observation on many projects. Most of us can relate directly to this too. There are a number of other features that Pinto and Kharbanda mention, drawn from earlier work by Pinto and Slevin [1987] and Wilemon and Thamhain [1983]. These are important as they reflect some of the key issues in effective team formation. These works and many others repeatedly identify certain factors that are present in successful teams (Table 7.9).

TABLE 7.9   Factors leading to Team Success

Clear sense of mission	The purpose around which a team will focus its creativity; a clear set of measurable goals shared by team members
Trust	Difficult item to achieve, measure, and define; best characterized as the level of mutual respect and confidence team members have with one another
Job ownership	Also interdependency, the understanding of individual team members of their own responsibilities and contribution to the team task, as well as appreciation of the contribution of the other team members
Alignment	The degree to which a team sees the same mission and the team members buy into this goal; also a measure of how consistent the view of success of the project is between team members
Cohesiveness	An effective team *wants* to work together
Enthusiasm	Elements include energy and fun

TABLE 7.10 Factors in a Regenerative Team

Have fun	Fun needs to be planned; it will not happen by itself; it will generate energy and commitment, and help foster the right work climate for the other items in this table to grow.
Open communication[a]	This is about avoiding hidden agendas and making sure that everyone on the team is kept well informed
Job ownership[a]	See Table 7.9
Propensity to take risk[a]	To be effective and competitive we need to step "outside the box" when considering possible solutions to a problem; conventional or standard corporate solutions will not normally generate sufficient innovation; in many organizations, this type of approach is potentially career limiting!
Trust[a]	See Table 7.9
Unique culture[b]	Each team needs to develop its own culture; this culture must not be unduly influenced by any individual or member company if it is to be effective for the team

[a]These items were corroborated by an independent study by Cahoon and Rowney [1996].
[b]This item has not been specifically tested in the described research but has been identified through observation.

Teams inherently are new organizations. They need time to develop as effective working units. Graham [1989] quotes the following, "We trained hard … But it seemed that every time we were beginning to form up into teams we would be reorganized. I was to learn later in life that we tend to meet any new situation by reorganizing: and what a wonderful method it can be for creating the illusion of progress while producing confusion, inefficiency and demoralization." The source of this quote is Petronius Arbiter, Greek Navy, 210 B.C. Clearly, team building issues are not new! However, the necessity for the constant reorganization of teams and for individual team members to be active on more than one team at a time is being driven by the requirements of technology projects that need to integrate many skills from the various technologies involved to the operational and business elements of the product.

Randolph and Pozner [1992] developed a set of ten rules for successful projects. The relevant ones to team development supplement the list provided by Pinto and Kharabanda:

- Develop people as well as teams.
- Keep everyone informed.
- Empower the team and its members.
- Be willing to take risks by approaching problems creatively.

These factors are consistent with observations made in a best-practices review undertaken at The University of Calgary in which a survey of over 150 projects identified potential areas for improvement or learning. The study led to five firm conclusions on the elements required to make teams effective and a sixth element likely to influence the success of a team (Table 7.10). These observations formed part of a new model for project management (SMART project management) subsequently tested and validated through success on some 25 live projects.

The concept of a regenerative team came from observation in action research to test SMART project management. There was no observed storming phase in the development of the teams we worked with. To validate this observation, other researchers were consulted. Cahoon and Rowney [1996] had recently completed a study on stress in the workplace and had observed the absence of the storming activity where four of the six items listed in Table 7.10 were observed. The two remaining items were not specifically measured in their study.

SMART project management is based on four guiding principles:

- **S**trategically **M**anage your project and its team.
- Ensure that we **A**lign the team and the project objectives.
- Work at developing a **R**egenerative organization.
- Recognize that we work in a **T**ransitional environment.

Application of Teams and Team Building

The principal issues influencing effective teams and team building are summarized below. How these issues affect processes and tools are discussed with examples. The transient and continuously changing nature of team organizations are discussed, and one solution for mapping the organization is presented. One process for effective and fast team building is outlined. Tools to consider using are listed, and the section concludes with items leading to the success and failure of project teams in order to reinforce key points.

Summary of Issues

A review of the common issues identified in the literature and through observation of effective teams may be summarized as follows:

- A clear team mission. "The project plans will reflect a top-down approach using the total wisdom of the project team which sets the stage for more effective, detailed, bottom-up validation of the plans" [Archibald and Russell, 1992].
- Communication is open, effective, two-way, and with no hidden agendas, keeping the team informed on all essential points.
- The team is regenerative, with trust, energy, focus, and commitment.
- There is buy-in to the plan of work, mission, and culture of the team.
- There is consistency in team management and alignment of team members.
- The culture of the team is developed independently of any parent organizations, departments, or groups from which the members are drawn and is protected from undue influence from these other cultures.

Where the Issues Fit in the Process of Team Building

From the above, we can see that a clear mission is critical to success. We can use this to our advantage by developing a mission as part of the initial planning and team-building process.

Communication needs to be open and as complete as possible. An organization chart is usually used to establish the formal communication links of a team. However, since teams are dynamic, another tool is needed (RACI+ charts are discussed below). To start the communication process, we need to develop a climate that promotes openness. This needs to happen at the outset and, as with all such management issues, needs to stem from the top. Open communication requires trust. As trust needs to be built first, this is often difficult to achieve.

A regenerative team will help to build trust and is developed at the team formation and planning stage. Part of the process uses team involvement to develop the direction and culture of the team from the outset, generating buy-in to the mission, plan, and culture of the team.

Finally, we need to obtain management consistency. This is best done through the development of a unique culture that will encourage fun and nurture a regenerative spirit.

Organizations

The structure of most technology teams can best be described as "fluid". Membership will change over time, as will size. Individuals will rarely be involved throughout the process of delivering on the mission. In other words, change within the team is inevitable. This means that any organization chart will be somewhat like a broken watch — in any 12-hour period it will be right exactly once, but you have to know precisely when to look! If we challenge the effectiveness of an organization chart, we should consider its function. There are two practical reasons for such a chart. We want to understand the pecking order of the team, and we need to know who needs to formally communicate with whom. If we can replace this functionality with another tool that reflects the changing world of teams, we will be better off. One such tool has been around for a long time in various forms, the RACI+ Chart.

Process

Involvement of the team in developing the way in which it will achieve its mission at the greatest practical level is the key to success in team building. The major stakeholders should be active in the development of the team's mission and the criteria upon which success will be measured. The team needs to identify common ground as the basis for building trust and openness in communication. One useful tip is to announce that any changes to the plan after it has been set by the core team will be to the account of the person, department, or other entity that requires the change. The onus is then on individuals in the team to ensure their agenda items are included in the mission and definition of success for the team. If they are not, or the definition of success is unclear, there will be some direct accountability for the consequences. On one project where this was announced, the result was that someone in the room who had not been active until that moment felt forced to join the discussion. The first comment he made was that his directive from his department head was to see what was going on and then to try to scuttle the project. This announcement, handled well by the team leader, led to discussion of the reasons why this department was against the project. Issues that would otherwise have quietly festered and remained unknown could be effectively dealt with at the outset.

To maintain open communication, there should be no bad ideas. This means that all ideas should be accepted and discussed by the team. Ones that are inappropriate to the mission of the team, or do not materially add to what is being discussed, should be put to one side but not discarded or negatively commented on in an open forum. All ideas need to be acknowledged and either used, discussed and modified, or parked. Parked ideas need to be addressed and their owners satisfied that they are not relevant to the team's work. The buy-in of the persons who generated "parked" ideas is important for several reasons. They need to be encouraged to continue to come up with ideas and to participate in the process. The idea and associated expectations need to be managed and aligned with the team's goals. If someone expects something that is totally disconnected from the team's mission, this needs to be identified as quickly as possible so that it will not later interfere with the success of the team.

One way to build the culture of a team to include the elements above is to plan to have them present. This can be done in a number of ways. One is in planning to have fun. In one blame-culture organization, the team decided to have fun by introducing a project tie. They picked one with pink pigs all over it. The original version of the plan was based on awarding the tie (by team consensus) to the person who contributed the least to the team mission each week. This was quickly found to be both offensive and ineffective! Therefore, they ceremoniously discarded the original tie and someone brought an old tie in to work to replace the disgraced one. It had horizontal stripes. One team member suggested that it be awarded to the person that the team felt had contributed the most to the project and they got to wear it for the next week. Another team member suggested that each deliverable the team needed to produce be written on one stripe. Any surplus stripes would then be cut off. Another person suggested that if they were going to cut off stripes, then cut off the ones the team had delivered, and present the stripe to the person who had contributed the most to the project. Silly as this all may seem, it worked. Within no time the tie was known throughout the company's head office. Whoever wore it carried some significant kudos. Written and e-mail project communications carried the by-line "may the tie be with you". They had a great team and a very successful project.

In summary, the process for planning a team's tasks should be used to help build the team. The six elements of a regenerative team need to be put in place as the plan is developed. The planning should only be done to the level of detail that allows bottom-up involvement and validation. Wherever possible team members should decide on how their own work will be done, working within the limitations imposed by the overall time frame and resource constraints, and recognizing the involvement and contribution of other team members.

Tools

Three tools are included here. There are many others, but these three have been tested as part of the SMART process with consistent success. The tools are RACI+ charts to lay out the organization and to provide a communications road map, Three Key Questions that help test and develop team alignment, and "2×4 planning". This last tool is a procedure that validates the involvement of members and ensures the team's mission is in line with corporate strategy.

RACI+ Charts

The acronym RACI represents four generic roles team members play:

- Responsibility: the role of management or coordination and the person ultimately responsible for the delivery of an item or component of the team's mandate.
- Action: the role of doing something that delivers or contributes to delivery of all or part of the team's mandate.
- Coordination: the role of ensuring that team members and others affected by or affecting the work of the team are coordinated with, to avoid problems or conflicts.
- Informing: the role of keeping others informed of what is going on or of being kept informed of what is happening with all or part of the team's activities and success.

Without the "+" on the end of the acronym, a RACI chart is simply a matrix chart that lists activities or deliverables for the team on one axis and the team members on the other. The intersecting points are used to identify the role of each team member in each particular activity or deliverable. This then serves as a road map for the communication for the team. If you add other information (e.g., timelines for key budgets and work hours), we then add the "+" to the acronym. Now, if we get buy-in from each of the team members who appear on the RACI chart, we can use this as a tool for planning and for building the team and its shared expectations.

Three Key Questions

Alignment is a big issue for most teams, particularly if they are cross-functional. Asking the following three questions will test and improve alignment:

1. How do we know when this team has finished its' work? What have we delivered to mark this milestone?
2. When we are finished, how do we know that the team was successful? What was it that the team achieved that led everyone to say that it was hugely effective?
3. Who gets to vote on the first two questions?

The third question should really be asked first since it is the one that identifies the real stakeholders. Generally, the closer to the team all the stakeholders are, the more successful teams seems to be. The first two questions require discussion as they rarely generate the same answer from any two team members. Obtaining consistency may change the direction of the team and will certainly improve alignment. Here are two examples:

1. Team 1: System implementation project: Definitions of completion ranged from "we have finished beta test" to "the system has been through one complete annual cycle". A time span of about 18 months. Success was variously defined from "the operators are using the system" to "we have reduced head count in the department by 30 people".
2. Team 2: Expansion to a petrochemical plant: Completion was defined as "the plant is mechanically complete" to "we have been operating successfully for a year". Again, an 18 month difference. Success ranged from "the operators like the plant" to "we are producing our product at $x below current costs".

Clearly, it would be difficult to assess success at the end of the project consistently unless these issues were known and managed from the outset.

2×4 Planning

This approach to planning requires that the team plans its work twice. The first step is to develop a mission. This is then developed into a set of key result areas (KRAs), which reflect the perspectives of the major stakeholders. On a software system, these may be as follows:

- Operators: The system is easy to use and helps save time.
- Management: The cost of the system is recovered in 12 months.

- Customer: It is easier to do business, as measured in terms of the cost of business.
- IT department: The system can be implemented without disrupting service to others.

Each KRA is described in terms of what needs to be delivered to achieve the needs of the stakeholder or stakeholder group. This part of the process is designed to define expectations as completely as practical. It also serves, if handled carefully, to bring out hidden agenda items.

This planning process uses left-brain type thinking and moves forward from the mission to specific detail. The team's work is then planned again, this time backward, and starting with the shared definition of completion and success discussed earlier. From this end point, the work is described that was needed to achieve final completion. This task will require other elements to have been completed (for example, you cannot finish fixing bugs until software testing is complete). The preceding deliverables that enable the final task to be performed can then be identified. The task required to produce these deliverables is then identified, and the preceding deliverables for this task are found. The process is repeated until the preceding deliverables needed for a task are the ones we have in hand now. This process requires all team members to consider interactions and their respective roles in the process. Simply by using the process, the team is able to work out interrelationships. By working backward, we force some right-brain thinking into the process, opening up discussions that would probably not occur in more-conventional planning procedures. Two important by-products are alignment of the team's mission with corporate objectives and a second check on the team's plan that builds confidence.

Conclusions: Success and Failure

The success of projects may be defined as having all stakeholders happy with the result. This self-fulfilling definition is delightful in its simplicity. It is also a signpost for effective team building. It tells us to involve our stakeholders and make them feel part of our team. If our stakeholders are happy with the result, and we have a team that is having fun and being productive, we will invariably have an effective team. Conversely, those that break down or do not form effectively ultimately suffer from just one thing: a breakdown in communication. This can be avoided if effective communication is actively planned. This requires developing a regenerative work environment as well as sensitivity to, and respect for, the different cultures in the groups and individuals that make up our teams. Finally, if we develop effective teams, we will nurture trust and probably achieve a better quality end product.

References

Archibald and Russell D. *Managing High-Technology Programs and Projects, 2nd ed.,* John Wiley & Sons, New York, 1992.

Cahoon, A. and Rowney, J. *Stress in the Workplace,* unpublished study, The University of Calgary, Alberta, Canada, 1996.

Graham, R. *Project Management, As If People Mattered,* Primavera Press, Bala Cynwyd, PA, 1989.

Pinto, J. K. and Kharabanda, O. P. *Successful Project Managers,* Van Nostrand Reinhold, New York, 1995.

Pinto, J. K. and Slevin, D. P. Critical factors in successful project implementation, *IEEE Trans. Mgt.,* EM-34: 22–27, 1987.

Randolph, W. A. and Posner, B. Z. *Getting the Job Done!,* Prentice-Hall, Englewood Cliffs, NJ, 1992.

Wilemon, D. L. and Thamhain, H. J. Team building in project management, *Project Mgt. Q.,* 14: 21–33, 1983.

Further Information

Adams, J. R. and Adams, L. L. The virtual project: managing tomorrow's teams today, *PM Network,* 11(1): 1997.

Cleland, D. Team building: the new strategic weapon, *PM Network,* 11(1): 1997.

De Bono, E. *Handbook for the Positive Revolution,* Penguin Books, London, 1992.

Fisher, R. and Ury, W. *Getting to Yes,* Penguin Books, New York, 1981.

Frame, J. D. *The New Project Management*, Jossey-Bass, San Francisco, 1994.

Quinn, R. E., Faerman, S. R., Thompson, M. P., and McGarth, M. R. *Becoming a Master Project Manager*, John Wiley & Sons, New York, 1990.

Thamhain, H. Engineering management, In *Managing Effectively in Technology Based Organizations*, John Wiley & Sons, New York, 1992.

Whetten, D. A. and Cameron, K. S. *Developing Management Skills*, 2nd ed., Harper Collins, New York, 1991.

7.13   Organizational Learning

Sim B. Sitkin, Kathleen M. Sutcliffe, and Karl E. Weick

In this section, we pursue three goals. First, we provide an overview of research on **organizational learning**, including key ideas and research results. Second, we highlight key lessons that can help technology managers understand and manage **learning** processes in their organizations. Finally, we discuss new developments and directions in research on organizational learning and identify several promising areas of future inquiry and practice.

Defining Organizational Learning

Organizational learning has been defined in many ways in the sociological and organizational literatures. Most definitions have focused on learning as a conscious improvement in organizational knowledge, actions, and performance in response to a change in stimulus. However, some recent work has highlighted limitations inherent in these definitions by pointing out that learning can occur without conscious choice, need not always be in the direction of improvements, and need not result in observable changes in beliefs or behavior.

Far more complex or enumerative definitions have been offered in the past, but we believe that these have not reflected the range of learning issues either as comprehensively or as well as the following simpler definition: *Organizational learning is a change in an organization's response repertoire.*

Three points about our definition should be highlighted. First, this definition is framed intentionally in general terms to capture changes in beliefs, practices, relationships, or formal structures and processes. Second, our emphasis on repertoires captures the notion that it is not so much that a particular thing must be learned but that some part of the organization's range of available skills and knowledge can be elaborated or modified, whether or not those new capabilities are used. Thus, this definition includes strengthening existing skills or routines as well as pruning outdated or inappropriate capabilities. Third, we highlight that learning enables the potential for action but that action need not be manifest at the time learning has already occurred. In addition, whereas some have focused on changes in beliefs, actions, or performance as indicators of learning, we focus on **response repertoires** to reflect the idea that not all learning results in observably changed actions or articulated changes in beliefs. For example, Weick (in Cohen and Sproull [1996]) points out that learning has occurred when an organization improves existing response sets such that they are now effective for problems that previously required unique responses. The learning here is in making simpler repertoires more robust and general by understanding more deeply how to apply and adapt them.

Key Themes in the Research on Organizational Learning

There are several major schools of thought concerning organizational learning, each built around the work of Argyris (e.g., Argyris [1992] and Argyris and Schon [1978]), March (e.g., see Cohen and Sproull [1996]), or Senge (e.g., Senge [1990]) and their students and colleagues. It is striking in reviewing these streams of research that they share a number of similarities yet have remained almost entirely insular, rarely referring to or building on the work in the "other" streams.

Although the history of research on organizational learning has remained isolationist, often representing the same basic ideas with unique terminology, these discrete streams of work are both similar and complementary; for in-depth reviews of the learning literature see Duncan and Weiss [1979], Fiol and Lyles [1985], Hedberg [1981], Levitt and March (reprinted in Cohen and Sproull [1996]), Shrivastava [1983], Huber (reprinted in Cohen and Sproull [1996]), Miller [1996], and Weick and Westley [1996]. Thus, the learning literatures can be clarified in terms of three cross-cutting themes: what is being learned (learning content), the process by which learning takes place (learning process), and the level at which learning occurs (learning level).

What Is Being Learned

Learning can involve changes in **action repertoires** (what an entity does) or changes in **knowledge repertoires** (what it knows). A large body of work has examined how learning is manifest in what is referred to as "adaptive behavior", "procedural learning", or "knowing" in which an organization develops new rules, routines, or patterns of behavior that respond to environmental or technological changes. Other research has focused on learning in terms of cognitive development rather than action or structural change. In this second group are studies of increased "cognitive complexity", more subtle associations or "causal maps", "knowledge", and increased volumes of useful information being processed and stored in the organization.

Learning Processes

A second organizing theme in the learning literature concerns the degree to which learning processes are aimed at improving existing routines vs. those processes associated with exploring new routines [Argyris, 1996; Cohen and Sproull, 1996; Miller, 1996; Sitkin et al., 1994]. The first type of process emphasizes the pursuit of reliability through reduced variation (e.g., "zero defects") against preset standards and is characterized by incremental learning around already-known problems and solution sets. The second type of learning process stresses enhancing organizational adaptability by avoiding premature clarity and closure when facing ambiguous problems. This is done by experimenting with new strategies, paradigms, practices, and ideas — and by rewarding those who take reasoned risks, even if those risks do not work out.

At What Level Is Learning Occurring

Individual learning has been viewed traditionally as a process of knowledge transfer. Much organizational learning research has reflected this view, merely aggregating individual learning to represent what the organization has learned. More recent work on organizational learning has taken a much more systemic view of how organizations learn [Levitt and March in Cohen and Sproull, 1996; Miller, 1996; Senge, 1990], focusing on changes in organizational structures, processes, and norms vs. individual beliefs or actions.

This recent shift reflects an important change in learning research. Using terms such as "situated learning", "communities of practice", "transactive memory" [e.g., Brown and Duguid in Cohen and Sproull, 1996; Lave and Wenger, 1991; Wegner, 1986], recent studies have found that learning is embedded in cultural norms, work routines, and shared practices rather than in individual's minds. This research has helped to explain the widespread failure of training focused on individuals or transfer efforts based on individual agents of change. This research exemplifies this shift and highlights that beliefs and skill sets are not lodged in the individuals, but reside in the interactive practices and understandings of organizational subgroups. Even expert members of effective organizations and groups can lose what they have apparently learned when they are removed from the social or physical setting in which they learned it. That is what led Lave and Wenger to refer to "situated learning" and has led to the rise of work that has begun to explain why it can be so difficult to transfer effectively what was thought of as simple knowledge or skill sets from one part of an organization to another. This work suggests that such "learning" is embedded in the group and its practices rather than individual skills or knowledge.

In summary, research has moved beyond the overly simplistic view that organizational learning can be seen in terms of accumulating and transferring information by atomistic individuals. Newer work in this area reflects a more accurate, nuanced, and practical set of insights because it is now recognized that groups and organizational units can learn in ways distinct from their individual members and in ways that are grounded in action and interaction patterns.

Key Lessons for Technology Managers

Our definition of learning involves enhancing an organization's range of possible responses to threats and opportunities. This suggests that learning is important for managers because of its strategic and tactical implications. Thus, the study of organizational learning may provide significant insights to practicing managers and it is toward this end that we direct our attention. In this section, we distill six key insights for the practitioner about managing learning in organizations.

Learning too Quickly Can be a Problem

It is commonly presumed that accelerating an organization's response rate and speed of learning is desirable. Speed-accuracy trade-offs can be a real curb on learning quickly. Research suggests that, when decision makers are under pressure for speed, they are likely to search for evidence that confirms prior expectations, which leads to mindless, superstitious, and superficial learning. In contrast, when decision makers are under pressure to develop a more accurate understanding, they are more likely to be more thorough in their search and analysis activities and to be more reflective, which may lead to better learning. Encouraging a tradeoff of speed for accuracy may undermine an organization's efforts to achieve the very goals they seek.

Researchers associated with the approach of March (e.g., see Cohen and Sproull, 1996) have built an impressive array of simulations and studies that identify the conditions under which slower learning is superior to fast learning. There are two fundamental insights from this body of work. The first is that actions that are smaller in scale and quicker in cycle time are more amenable to learning than are actions that are large and slow. This lesson runs counter to most planned organizational efforts at learning, which are characterized by larger or more radical change, rather than the smaller steps more typical of retro-spectively labeled learning. Yet, the research suggests that organizations that break large projects into smaller pieces can enhance their learning because smaller actions are more easily understood and inter-preted, can be implemented more quickly, and thus more fine-grained feedback can be acquired in shorter periods of time. This also leads to the second insight, which is that, while action cycles should be fast and small, learning should be slower to incorporate the larger action set results. If early results are adopted too quickly (i.e., fast learning), then overgeneralization or erroneous conclusions can lead organizations to adopt changes that are then very difficult to reverse when subsequent, more-reliable results begin to come in. The key problem is learning too quickly from unreliable or unrepresentative early results — whether positive or negative. For today's managers who are under increasing pressure for quick adjust-ments and quick results, this important lesson is often overlooked with terrible consequences.

Learning too Well Can be a Problem

This theme appears in a slightly different form ("competency traps", "success traps", or "skilled incompe-tence") in many of the key writings on learning. The counterintuitive research finding is that learning to do anything well begins to foreclose other learning options because high levels of proficiency uninten-tionally make it very inefficient to try out new skills or approaches. This problem is exacerbated when high levels of learning are associated with success, since the combination can lead to lower levels of attention to alternatives and more resistance to trying new things. What the learning literature highlights is that it is important to incorporate secondary or peripheral knowledge into organizational (and even individual) response repertoires to minimize the potential detrimental effects of high levels of specific competency.

Developing an Organizational Learning Capacity Involves More than Just Acquiring Specific Technical Information or Practices

The previous point highlights why cutting-edge knowledge is not a protection against competitive erosion — since specific knowledge can change so quickly, the only protection is to develop a more general capacity to build on existing strengths in developing new ones. This represents another insight from the learning literature: that critical organizational learning does not only involve the acquisition of specific technical knowledge but involves the general capacity to create new knowledge through proactively exploring new problems or solution sets as well as the capacity to link ideas and people in new ways. Thus, an organization's learning capacity involves two complementary aspects: efficiently absorbing what is already known inside or outside the organization and effectively pushing the boundaries of that knowledge. General learning capabilities are reflected in research that stresses double-loop learning, resilience, improvization, self-design, and exploration skills — or under the more overarching label, the learning organization [Argyris, 1992; March in Cohen and Sproull, 1996; Senge, 1990; Sitkin et al., 1994; Weick and Westley, 1996].

Evaluation of Learning Achievements Drives Out Learning Opportunities

The desire to evaluate learning is a natural one for managers who take seriously the importance of organizational learning. However, this desire often leads to the development of systems that integrate information used to evaluate personnel with information used to learn. For example, Tamuz [1987] found the integration of such systems was a critical problem for air traffic safety reporting systems that were charged with improving understanding of near misses and accidents, while also being used to assess individual responsibility for reported problems. Such integrated systems unintentionally undermine learning in two ways. First, when information is used to evaluate, that which is ambiguous or critical tends to be omitted or modified (e.g., to protect individual careers). Second, learning is best achieved when problems are not prematurely or artificially defined in clear and precise ways — yet, when evaluations are involved, ambiguity tends to make people uncomfortable and, thus, the information that may be best suited to promote learning is lost.

Avoiding Mistakes Is an Inefficient Way to Learn

Catch phrases such as TQM's "zero defects" suggest that failure avoidance is desirable; however, recent research has highlighted the perils of success and the dangers of such extreme risk aversion. Researchers examining organizational innovation processes, high-risk technologies (e.g. nuclear power plants or airlines), and public health settings have found that it is extremely difficult for organizations to sustain high levels of reliability in dealing with highly uncertain, complex problems unless they are able to incorporate mechanisms for sensing and learning from errors and mistakes.

Despite the truism that "we learn more from our mistakes than our successes," most organizations today are not truly open to the honest reporting of errors, and as a result they circumscribe their opportunities for learning. Today's managers are held to a tight standard of efficiency and continuous success. The downside of this is that information that could foster learning is inadvertently disregarded, distorted, or suppressed for a variety of understandable but destructive reasons. For example, in one set of studies it was found that data about errors was used to punish as well as to learn — and this made the goal of learning impossible to achieve.

Unlearning as a Precursor to Learning

Sometimes an organization has to discard outdated understandings and routines before further learning can occur. Pruning of outdated knowledge or routines is learning because it involves new understandings of old problems and new response repertoires. Rigidly holding onto techniques that have worked effectively in the past can hamper the recognition of new opportunities, the pursuit of learning and effective responses. This important finding manifest in the individual learning literature emphasizes that, when in stressful or complex situations that may require creative adaptation, individuals often rely on over-learned responses.

Weick [1996] uses the metaphor of "dropping your tools" to capture the idea that it can be essential to let go of the old (even if it is usually effective) in order to be prepared to grasp the new. For example, he notes [Weick, 1996] that "Karl Wallenda, the world-renowned high-wire artist, fell to his death still clutching his balance pole, when his hands could have grabbed the wire below him." One of the lessons of social science research as applied to organizational learning is that organizations and their members tend to keep clutching tools that helped them remain balanced through past crises rather than dropping these tools when the situation requires grasping the new. One way of fostering organizational capabilities in "dropping" is to create practice situations in which old tool sets are clearly inapplicable. For example, some firms that try to use acquisitions to further their adaptability have avoided the temptation to simply graft their traditional procedures or structures by purchasing firms with clear and unmistakably unique needs and then insulating them from such grafting efforts [Haspeslagh and Jemison, 1991].

Future Directions in Organizational Learning

We have suggested that research on organizational learning has traditionally focused on individual learners and their improved ability to act efficiently over time. In the preceding sections, we have suggested how learning research has also begun to move toward recognizing the importance of organizational learning that resides in the group or the community. In this final section we will highlight promising new directions in future research.

First, an important body of work in psychology has just recently begun to enter the organizational literature and holds much promise for understanding tacit learning at the group and organizational level. Some of this work focuses on what is referred to as "transactive memory" (e.g., Wegner [1986]) and involves the distribution of skills and knowledge to different parts of a group (i.e., not everyone knows everything) but also in which members share information about core competencies and where critical information or skills are located within the group. Members are adept at different things but also know where the strengths and weaknesses of the group lie. Not everyone is good at or knows about everything, but the group as a whole can take advantage of its distributed and complementary resources. What is perhaps most interesting in this work is that the members are often unaware of this knowledge and skill in the sense that they cannot communicate it, despite using it effectively. Furthermore, as is shown in related research on "communities of practice" [Brown and Duguid in Cohen and Sproull, 1996; Lave and Wenger, 1991], the knowledge and skills each individual member appears to have while they are part of the group are ineffective or may even entirely disappear once they are removed from the group context. That is, the individuals appear to be carriers of the group's learning but to not "have" that learning in any transportable sense. This research, which is already beginning to be used in corporate change programs, suggests when learning efforts must be centered around enhancing group or unit-level response repertoires rather than focusing on individual training or formal systems. The implication of this line of research is that traditional actions aimed at fostering learning (e.g., training more managers or instituting more procedures) are not effective. While it is clear that some managerial support is helpful and standardization is harmful, current research is just beginning to identify the specific ways that these learning groups function in organizational settings.

Second, we anticipate increased attention will be paid to the linkage between action-based and cognition-based views of learning. For example, one promising idea is that a sense of control and efficacy is associated with the capacity to grasp fine-grained variations in the environment and increase the likelihood that organizations will see opportunities for effective intervention. In other words, the beliefs that are brought into a learning situation interact with the capabilities for action and either lead to enhanced learning or obstruct it.

Third, there is an opportunity for those who study learning to contribute to the burgeoning body of work on alliances, joint ventures, and other forms of cooperative partnerships in organizational settings. The rapid rise in interest in issues of trust and relationships in organizations is part of a general trend that recognizes that learning is often done within relationships rather than by isolated entities. Although specialists in the areas of cooperative relationships and learning have not yet joined forces, we believe that this is one of the most-promising areas for future work that can benefit theorists and practitioners alike.

Fourth, recent work on how integrated evaluation and learning-oriented information systems can foster rather than impede developing a capability for exploratory learning holds some promise for influencing the use of formal control systems (e.g., TQM programs) in organizations in fast-changing industries. Historically, learning research presumed that highly formalized systems could be implemented to foster learning in a wide variety of settings. More recently, work has begun to emphasize the relevance of less-structured, more-emergent approaches to learning when problems are ambiguous or potential solutions are less well understood. One promising direction for work in the future is to examine how highly structured approaches can under certain conditions provide a more predictable and powerful springboard for exploratory activity. Evidence of this process has been found in a series of studies of TQM programs in which high competence in traditional control-oriented TQM served as a foundation for more exploratory and learning-oriented project achievements. Future work may be able to generalize to other domains.

In contrast with where we believe the field is going, we should note that there is one area that we believe merits more attention but is unlikely to receive it: emotional learning. As a number of recent studies have suggested, individuals and organizations vary greatly in their capacity to process emotional material constructively. There is little doubt, from our perspective, that enhancing this aspect of an organization's response repertoire could only be helpful in facilitating more-effective relations with employees, customers, partner organizations, regulators, and other stakeholder groups.

In conclusion, research on organizational learning has generated a great deal of attention and has made progress in identifying frameworks and insights of potential value to practitioners. However, there remain several independent streams of work that have not been effectively integrated in the literature or in practice. Thus, it difficult for the practicing manager to capture the range of insight generated through just a few key books, articles, or training sources. Notwithstanding these difficulties, we have highlighted how the area of organizational learning does in fact have a few common crosscutting themes and insights. These will, we believe, provide the basis for future work.

Acknowledgments

The first and second authors gratefully acknowledge support from the National Science Foundation (Grant No. SBR-94-96229 and SBR-94-20461) for the program of research on which this paper is partially based.

Defining Terms

Learning: A conscious improvement in knowledge, actions, and performance in response to a change in stimulus.

Organizational learning: A change in an organization's response repertoire.

Response repertoires: A range of available skills and knowledge.

Action repertoires: What an entity does (or can do).

Knowledge repertoires: What an entity knows.

References

Argyris, C. *On Organizational Learning,* Blackwell Scientific, Cambridge, MA, 1992.

Argyris, C. and Schon, D. *Organizational Learning: A Theory of Action Perspective,* Addison-Wesley, Reading, MA, 1978.

Cohen, M. D. and Sproull , L. S., eds., *Organizational Learning,* Sage, Thousand Oaks, CA, 1996.

Duncan, R. and Weiss, A. Organizational learning: Implications for organizational design, In *Research in Organizational Behavior, Vol. 1,* B. Staw and L. Cummings, eds., pp. 75–123, JAI Press, Greenwich, CT, 1979.

Fiol, C. M. and Lyles, M. A. Organizational learning, *Acad. Mgt. Rev.,* 10: 803–813, 1985.

Hedberg, B. L. T. How organizations learn and unlearn, In *Handbook of Organizational Design,* P. C. Nystrom and W. H. Starbuck, eds., pp. 8–27, Oxford University Press, London, 1981.

Miller, D. A preliminary typology of organizational learning: synthesizing the literature, *J. Mgt.,* 22: 485–505, 1996.

Senge, P M. *The Fifth Discipline: The Art and Practice of the Learning Organization*, Doubleday/Currency, New York, 1990.

Shrivastava, P. A typology of organizational learning systems, *J. Mgt. Studies*, 20: 7–28, 1983.

Sitkin, S. B, Sutcliffe, K. M., and Schroeder, R. G. Distinguishing control from learning in total quality management: a contingency perspective, *Acad. Mgt. Rev.*, 19: 537–564, 1994.

Tamuz, M. The impact of computer surveillance on air traffic safety reporting, *Columbia J. World Bus.*, 22(1): 69–77, 1987.

Wegner, D. M. Transactive memory: a contemporary analysis of the group mind, In *Theories of group behavior*, G. Mullen and G. Goethels, eds., pp. 185–208, Springer-Verlag, New York, 1986.

Weick, K. E. Drop your tools: an allegory for organizational studies, *Admin. Sci. Q.*, 41: 301–313, 1996.

Weick, K. E. and Westley, F. Organizational learning: affirming an oxymoron, In *Handbook of Organization Studies*, S. R. Clegg, C. Hardy, and W. R. Nord, eds., pp. 440–458, Sage, London, 1996.

Further Information

Center for Creative Leadership, Greensboro, North Carolina
Institute for Research on Learning, Menlo Park, California
Society for Organizational Learning, Cambridge, Massachusetts

7.14   Organizational Culture

Charles A. O'Reilly III

Few concepts of the past decade have so captured the attention of scholars and practitioners as that of *organizational culture*. What is *culture*, and why should managers worry about it? The answer to this question can be illustrated in Table 7.11, which shows the leading firms in the semiconductor industry from 1955 to 1995. During this period there was an almost complete turnover in industry leadership. Great firms such as RCA, General Electric, and Hughes were forced out, and new firms such as Intel and Motorola emerged. Why should this happen? Firms such as RCA had significant first-mover advantage, great technology, and the financial strength to dominate the industry — but they failed. Foster [1986], in an industry study, attributed this failure in part to the inability of these companies to manage the cultural tensions of new and mature technologies. Tushman and O'Reilly [1997] extend this observation and demonstrate the importance of culture in achieving long-term success, to generate streams of innovation. It is organizational culture that lies at the roots of an organization's ability to be innovative. Michael Porter [1985], Harvard's competitiveness guru, says speed and innovation have become crucial for success, "It's gone from a game of resources to a game of rate-of-progress. Competition today is a race to improve."

TABLE 7.11   Semiconductor Industry 1955–1995

1955 (Vacuum Tubes)	1955 (Transistors)	1965 (Semiconductors)	1975 (Integrated Circuits)	1982 (VLSI)	1995 (Submicron)
1. RCA	Hughes	TI	TI	Motorola	Intel
2. Sylvania	Transition	Fairchild	Fairchild	TI	NEC
3. General Electric	Philco	Motorola	National	NEC	Toshiba
4. Raytheon	Sylvania	GI	Intel	Hitachi	Hitchi
5. Westinghouse	Texas Instrument	GE	Motorola	National	Motorola
6. Amperex	GE	RCA	Rockwell	Toshiba	Samsung
7. National Video	RCA	Sprague	GI	Intel	TI
8. Rawland	Westinghouse	Philco	RCA	Philips	Fujitsu
9. Eimac	Motorola	Transitron	Philips	Fujitsu	Mitsubishi
10. Landale	Clevite	Raytheon	AMD	Fairchild	Philips

In this section, we consider "culture" not as a diffuse or abstract concept but as a potential control system that managers can use to implement business unit strategy. In this sense, culture can be thought of as a control system that can help or hinder the execution of strategy and promote or retard organizational innovation. Much of management is really about control, about getting people to do what is necessary to get the work done, preferably in a way that uses their full potential and leaves them feeling motivated and engaged. Culture is a *social control system* that can help coordinate and control collective action.

This section has four parts. First we elaborate on the notion of culture as a social control system, defining culture in concrete terms that can be used by managers. Second, we briefly describe the psychology underlying social control and illustrate how culture as a control system operates. Third, we show how culture is a critical determinant of organizational innovation. Finally, we briefly discuss how culture can be managed.

Culture as a Social Control System

Where does control come from? Consider the following generic definition: *Control comes from the knowledge that someone who matters to us is paying close attention to what we are doing and will tell us when we are doing a good or bad job.* From this perspective, organizational control systems, whether they are financial planning systems, budgets, inventory control processes, or safety programs, are effective when those being monitored are aware that others who matter to them, such as a boss or staff department, know how they're doing and will provide rewards or punishments for compliance or noncompliance.

Formal control systems typically rely on direct supervision to monitor performance. Yet, direct supervision is one of the most-expensive methods by which information on work activities can be acquired due to the large time expenditures required by evaluators. Further, direct observation of some aspects of performance may not even be possible in jobs that require autonomy, flexibility, and initiative. Further, even if possible, the personal scrutiny required to directly observe others may be difficult for evaluators to manage given the potential negative effects on those being supervised. In addition, even if such rewards could be calibrated, it is not clear that people are as motivated by extrinsic rewards as they are by feedback that highlights the intrinsic value of a task. Research has shown that relying solely on extrinsic rewards can reduce performance and intrinsic motivation. This is especially true for performance on tasks that individuals engage in volitionally and from which they derive intrinsic satisfaction. However, as tasks become more unpredictable and uncertain and the need for flexibility and adaptability increases, formal control systems can become less effective and more costly [Caldwell and O'Reilly, 1995]. This creates a dilemma: as uncertainty and the need for change increase, traditional control systems become less useful and the specter or loss of control rises.

If control comes from the knowledge that someone who matters to us is paying close attention to what we are doing and will tell us when we have behaving appropriately or inappropriately, then it can emanate from a formal system such as rules, procedures, and organizational hierarchies or from personal relationships. In this sense, to the extent that we care about others and have some agreement about what constitutes appropriate behavior, then, whenever we are in their presence, we are also potentially under their control. In this way, social control targets values, attitudes, and behaviors that may be relevant to desirable organizational outcomes such as service, safety, and respect for others. Of course, social control can also increase undesirable outcomes if the norms and values to which members attend to are not strategically appropriate. Either way, rather than being based on legitimate or formal authority, social control is based informational and normative influence (e.g., O'Reilly and Chatman [1996]). Further, a reliance on the opinions of valued others implies that social control may be far more extensive, and less expensive, than formal systems. The paradox is that strong social control systems often result in positive feelings of solidarity and a greater sense of autonomy among people, rather than a feeling of being a cog in a machine that sometimes results from close formal control systems.

Culture as a form of social control operates when members of a group or organization share expectations about values, or what is important, and how these values are to be manifest in norms, that is, in words and actions. Norms refer to the expected behaviors sanctioned by the system. For group norms and values to exist, there must be beliefs about appropriate and required behavior for group members as group members, that is, there must be a commonality of such beliefs such that, while not every member of the group must hold the same idea, a majority of active members are in agreement.

If we define organizational values as the beliefs shared by organizational members and norms as the expectations about appropriate attitudes and behaviors derived from these organizational values, organizational culture can be viewed as *a system of shared values defining what is important and norms, defining appropriate attitudes and behaviors that guide members' attitudes and behavior's*. A "strong culture" can be said to exist when there are a set of norms and values that are widely shared and strongly held throughout the organization [O'Reilly, 1989]. It is important to note that the operative norms that characterize a group or organization may not necessarily be those espoused by senior management or articulated in the company mission or vision. Similarly, norms may exist in one part of the organization but not be widely shared in other parts. For example, the marketing department may value meeting customer's needs through new products while the manufacturing department values stable product designs and long production runs. Variations of this sort may result in strong subcultures. However, we use the term strong culture to refer to organizational norms that are widely shared and strongly held across the units that comprise an organization.

Under these circumstances, it makes sense to talk about an *organizational* culture and to consider its implications as a control system. The question is whether these norms are intensely held, this is, whether they enhance commitment or not, and whether they are aligned with strategic demands, that is, whether they enhance organizational performance and permit adaptation to changing circumstances. In this manner, behavior is adapted to and controlled by the situation. This, of course, is not an argument not to use formal control systems. Clearly, formal controls have critical ingredients for managers. However, a close examination of successful firms and managers suggests the need for both *formal* as well as *social* control systems, with the latter often being an overlooked but powerful alternative to the former.

The Psychology of Social Control

Organizational culture operates through informational and normative influence, that is, in organizational settings in order to understand what is important, where to focus our efforts, and what attitudes and behaviors are appropriate, we rely on the signals we receive from others [O'Reilly and Chatman, 1996]. Indeed, most of what we "know" in organizations results from these social agreements or norms. What is the appropriate dress? How formal or informal should we be? How do we treat customers? Is initiative expected? How about working long hours? The answers to these questions comes from a social consensus of others who are important to us, such as senior managers and our colleagues.

Although the effects of normative and social influence are well documented and powerful, they are often not intuitive [Cialdini, 1993]. Because we as individuals tend to "explain" other's beliefs and actions in terms of personality or individual dispositions, we often miss how much of behavior in organizations is driven by situational pressures. Yet, although we are all familiar with the idea of norms, we are less appreciative of is how powerfully norms can be in shaping our own behavior, especially in organizational settings. Shared agreements and social expectations can constitute a powerful and pervasive social control system within groups and organizations.

In order to think about using culture as a social control system, we need to make two important distinctions. First, culture should not be equated with the vision or values espoused by senior management. Simply having a formal statement of the company values or laminated cards that employees carry in their wallets is no guarantee that these values are shared. A second important distinction in diagnosing and managing culture is to appreciate that norms vary in two ways. First, norms can vary in their *consensus* or in how widely shared they are. Before a strong culture or social control system can operate there needs to be agreement that certain values, attitudes, and behaviors are important. However, simple consensus is not sufficient. With sufficient publicity, espoused values such as "quality" or "customer service" may become widely known but not necessarily practiced. In order to have a strong culture, the norm must also be characterized by *intensity*, that is, people who share the norm must also be willing to tell others when the norm is violated.

Using Culture for Competitive Advantage

With this framework, it is possible to consider how managers can diagnose and align cultures with strategic objectives. Conceptually, this requires a simple six-step process: (1) clarity about the business unit strategy (How will we compete? Why should any customer, internal or external, prefer our product or service to a competitor's?); (2) clarity about the half-dozen critical tasks necessary to execute the strategy (Concretely, what are the tasks that must be accomplished if the strategy is to be implemented?); (3) Decide on the three or four norms that are crucial to help accomplish the critical tasks (What norms and values, if they were widely shared by people in the organization, would facilitate the critical tasks?); (4) diagnose the current culture (What are the central norms that currently characterize the organization, that is, what are the expectations currently shared?); (5) identify the important cultural gaps (What current norms may hinder the accomplishment of the critical tasks? What norms are needed that are not currently present?; and (6) design interventions to attenuate undesired norms and to promote those that are needed for future success.

Consider the dilemma managers face in designing control systems to foster organizational innovation and how social control may be a potential solution. It is obvious that organizations must be innovative to survive. At the same time, designing formal control systems to ensure innovation is difficult, since by its very nature innovation involves unpredictability, risk taking, and nonstandard solutions. If there were a guaranteed way to promote innovation, such as there is to optimally schedule inventory or manage cash flow, most organizations would adopt it and there would be little competitive advantage gained. It is precisely because of this uncertainly that a competitive advantage is possible for innovative firms.

How can this be done? A piece of the puzzle is in the use of social control. What are the norms, if they were widely shared and strongly held, that would help promote innovation? What are the shared expectations about attitudes and behavior that would result in higher levels of innovation? To answer this, we must first recognize that innovation is an outcome. We know innovation has occurred only because something has changed. This means there are two component processes that underlie all innovation: (1) *creativity* or the generation of a new idea and (2) *implementation* or the actual introduction of the change. Innovation occurs only when both components are present. Thus, to promote innovation in organizations requires that managers both stimulate new ideas and put these ideas into practice. In a study of teams in a set of high-technology firms, Caldwell and O'Reilly [1995] found four norms that were significantly related to innovation. They reported two main ingredients for stimulating creativity: (1) *support for risk taking and change* and (2) *a tolerance of mistakes*. They also found two primary norms associated with implementation or execution: (1) *effective teamwork and group functioning* and (2) an *emphasis on speed and urgency*. O'Reilly et al., [1998] have recently replicated some of these findings.

In sum, it is important for managers to align the culture of an organization with the business unit strategy and the critical tasks. Doing this ensure that the social control system promotes the execution of strategy. A failure to do this may result in a culture that works against the strategic objectives of the business. Further, when the critical tasks include innovation and change, managers can seldom rely on formal control systems to help drive the process. Instead, social systems are required. How can a manager shape or change the culture in his or her unit?

Managing Culture

Four mechanisms are commonly used by strong culture organizations to generate commitment and manage through social control: (1) systems of participation that promote choice and lead people to feel committed, (2) management actions that set goals, focus attention, and help people interpret events in ways that emphasize their intrinsic importance. (3) consistent information from valued others signaling what is and is not important, and (4) comprehensive reward systems that are seen as fair and emphasize recognize, approval, and individual and collective contributions.

1. Participation. The literature demonstrating the power of participation to produce commitment is substantial (e.g., Cialdini [1993]). Getting people incrementally involved has been shown to be a powerful play to induce subjects in an experiment to voluntarily eat an earthworm, increase bone marrow donors, conserve energy, or even to secure religious converts. When people make choices, they usually feel responsible and hold more positive attitudes toward the activity.

2. Management as symbolic action. A second mechanism for developing and managing through social control comes from management, in the form of signals about what is important and the intrinsic significance of the work. Managers are sensitive to the influence of language, symbols, and consistency of action as a means for cuing organizational members about what is important. Managers act as signal generators sending messages about what is important through their own behavior, often in mundane ways such as consistently asking certain questions or following up on desired activities. Although particular symbols by themselves are not likely to be effective, when they reflect an important and widely shared value, they may shape interpretations and enhance the intrinsic importance attached to specific attitudes and behaviors. In this sense, managers who influence others' interpretation of events and see the intrinsic value of their efforts shape the social control system.

3. Information from others. Clear, consistent messages from co-workers also shape an individual's beliefs and behaviors. A large body of social psychological research provides dramatic examples of the power of informational influence. For instance, studies have shown that face-to-face requests for blood donations were successful 25% of the time, but, when requests were made in the presence of a model who complied, the rate more than doubled to 67%. Organizations capitalize on the impact of others' behaviors on us in a number of ways. Some emphasize equality among members by reducing distinctions between management and workers (e.g., no special perks such as parking spaces, common titles, open office space, informality, etc.). Other emphasize close relations among members through social activities and family involvement.

4. Comprehensive reward systems. A final important lever for shaping culture involves the comprehensive use of rewards and recognition for exemplary compliance with the core norms and values. For instance, direct sales organizations routinely use continual recognition and reinforcement to motivate employees. These may take the form of small gifts, recognition from peers, or even awarding vacations and automobiles. For example, providing people with verbal reinforcement and positive feedback, compared with external rewards, increases their intrinsic motivation in tasks. Further, providing people with small rewards may be more effective in shaping behavior than offering large rewards, especially when the rewards are framed in terms of "appreciation" rather than "control".

These four mechanisms (participation, management as symbolic action, information from others, and informal reward and recognition systems) are the primary levers organizations use to develop culture as a social control system. Each capitalizes on the importance of strong informational and normative influence as a potential determinant of attitudes and behavior. Each acts to provide organizational members with consistent signals about which attitudes and behaviors are important, either from one's own previous behavior or from information provided by values others.

Conclusion

Culture is a prevalent social control system operating in organizations. Based on the psychological mechanisms of participation, management as symbolic action, information from others, and comprehensive reward and recognition systems, managers can create strong situations and shape collective action. Culture as social control can, under certain circumstances, be an important means for accomplishing strategic objectives and motivating the work force. The dynamic capabilities or core competencies of successful organizations may rest, in part, on norms that promote organizational norms that encourage autonomy, innovation and constant change are at the heart of the long-term success of many organizations.

Strong cultures that embody norms of creativity, innovation, and change may be the most effective mechanisms for promoting organizational adaptability. The good news is that managers can diagnose and manage the cultures in their organizations. To do this requires an emphasis on understanding the norms and values that operate within the organization and an appreciation for the psychology of social control. The bad news is that, for those managers who choose not to pay close attention to the organization's culture, the same social control processes may result in an inability to change and adapt to new technologies and circumstances.

References

Caldwell, D. and O'Reilly, C. Promoting Team-Based Innovation in Organizations: The Role of Normative Influence, paper presented at the Fifty-Fourth Annual Meetings of the Academy of Management, 1995.

Cialdini, R. *Influence: Science and Practice*, Harper Collins College, New York, 1993.

Foster, R. *Innovation: The Attacker's Advantage*, Summit Books, New York, 1986.

O'Reilly, C. Corporations, culture and commitment: Motivation and social control in organizations, *Calif. Mgt. Rev.*, 31: 9–25, 1989.

O'Reilly, C. and Chatman, J. Culture as social control: corporations, cults, and commitment. In *Research in Organizational Behavior, Vol. 18*, B. Staw and L. Cummings, eds., pp. 157–200, JAI Press, Greenwich, CT, 1996.

Porter, M. *Competitive Advantage*, Free Press, New York, 1985.

Tushman, M. and O'Reilly, C. *Winning through Innovation: A Practical Guide to Leading Organizational Change and Renewal*, Harvard Business School Press, Cambridge, MA, 1997.

7.15   Telecommuting Technologies

Jahangir Karimi and Yash P. Gupta

Telecommuting has the capacity of redrawing the geographical and organizational boundaries of the traditional, centralized enterprise. It enhances individual autonomy, control, flexibility, convenience, and family togetherness and improves flexibility in work arrangements that, in turn, results in higher productivity. US West Communications, for example, reported that the productivity of telecommuters increased, some by as much as 40%. Further, the company reported savings of $4000 to $21,000 annually per telecommuter in terms of space, reduced absenteeism, and retention of workers who might otherwise have left the company. A recent survey of Fortune 1000 executives, released by the General Services Administration, reported that 92% of the executives with telecommuting experiences said it produced some advantages for their companies. Fifty-eight percent cited increased productivity, 61% reduced absenteeism, 63% improved employee retention, 64% savings on office space costs, 63% reduced employee stress, and 79% improved employee morale.

A recent survey of 49 Fortune 1000 companies by Forrester Research states that 11% of the companies classify their telecommuters as daytime, home-based telecommuters, while 89% of companies classify their telecommuters as casual, after-hours workers. **Hoteling** practice clearly support telecommuting, with home and/or a center serving as alternate work locations when appropriate. Early adopters of the hoteling concept, including Xerox, AT&T, Ernst and Young, Dun and Bradstreet, and Kodak, report real expected savings in the millions of dollars. Telecenters, in particular, may find a niche, as hoteling filters down to smaller organizations for which reserving hoteling space in their own facilities may not be practical but which can use multiemployer telecenters to reduce their permanently owned or leased space needs.

Telecommuting on the Rise

Today, an increasing number of firms are providing telecommuting opportunities to their employees. A 3-year government-backed initiative, aimed at studying the impact of telecommuting on corporate America, suggests that 64% of Fortune 1000 companies in the United States have implemented telecommuting programs, and 60% of the companies currently without such programs expect to institute one within the next 3 years [Moore, 1995]. According to a recent survey of top executives reported, 62% of 305 senior executives surveyed report that their companies are encouraging telecommuting arrangements for their employees. This is a significant increase over the last few years: 49% in 1995 and 39% in 1994. Forty-two percent of companies have telecommuting programs in place, while 70% predict that telecommuting will increase in their companies throughout 1996. Oddly though, overall employee participation in telecommuting remains extremely low — with only 7% of workers taking advantage of the available programs. Insurance companies reported the highest use of telecommuting at 52%, followed by high-tech firms at 50%, retail/wholesale businesses at 46%, and utilities and transportation companies at 40%.

Data on current levels of telecommuting are typically based on small and often nonrepresentative samples, and a lack of consensus on how telecommuting is defined can lead to widely disparate estimates and forecasts. For example, a recent report suggested that, by the end of the year 1995, 9.2 million Americans will be working from home or telework centers; in the next 15 years, the number of telecommuters is expected to triple, representing 20% of the total U.S. workforce [Wird, 1995]. The U.S. Department of Transportation (DOT) [1993], however, forecasts that 5.2% to 10.4% of the labor force would be telecommuting in 2002 — a 100% difference between the low forecast and the high forecast. Currently, workers who telecommute do so an average of 1 to 2 days per week, on a part-time basis; the U.S. DOT report forecasts an increase to 3 to 4 days per week approaching a full-time basis. Together the numbers produce forecasts ranging from 1.0% (if telecommuting continues on a part-time basis) to 8.3% of workers telecommuting on a given day in 2002 — a wide range for a relatively short-term forecast.

The U.S. Bureau of Labor Statistics estimates that approximately 30% of the U.S. workforce spends an average of 6 to 8 hours per week telecommuting. In information-intensive industries such as insurance, banking, and transportation, however, this average ranges from 30 to 40 hours per week [Snizek, 1995]. Approximately 20 million nonfarm employees were engaged in some work at home as part of their primary job in May 1991, representing 18.3% of those at work. More than 60% of those who worked at home were simply taking some work home from the office and were not paid specifically for that. Of those who were paid, or were self-employed, only about half worked at home for 8 hours or more per week. The projected total number (in millions) and percentage of people in several key occupational categories who spend 20% or more of their time away from their desk or immediate work areas are 27.3 (74%); the distribution in the categories are technical professionals 7.1 (75%), nontechnical professionals 14.4 (74%), owners 2.4 (75%), managers 3.4 (72%) [The Wall Street Journal, 1994]. **Information workers** are more likely to have tasks that could be performed away from the office. Currently, information workers are generally estimated as being 50% or more of the labor force.

Telecommuting Technologies

For technology managers and providers, there are a number of pressing issues: what services and technology to offer, where, when, to whom, and what is the best way to promote it? With the changing marketplace and escalating availability of advanced technologies, and increasing user demands, this task is immense. Concomitantly, the characteristics and needs of telecommuters are evolving, due to the changing regulatory and work environments, and the growing use of technology. This evolution has also led to a different set of technology needs at the person's residence than is traditionally required by the residential market.

There are a variety of computer and communication technologies that support telecommuting, such as telephones, computer modems, fax machines, electronic mail, and computer information networks. The standard equipment for most telecommuters is a PC and/or laptop computer, a modem, and a fax — and these are often available in one integrated hardware package. It may be necessary to have an additional telephone line installed, and it is now possible to link the worker's own telephone into the company telephone network during working times. Some people also find it useful to have a small photocopier, a pager, or a mobile phone. More sophisticated communication services, such as call waiting, call forwarding, three-way calling, speed calling, and caller ID will also make it easier for telecommuters to be away

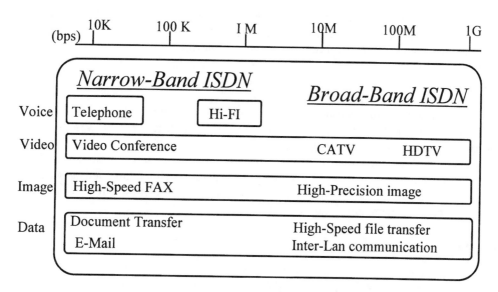

FIGURE 7.15   New communications technologies.

from the office. In 1992, 30% of telecommuters were linked to the office by modem — a significant share of the 52% of workers with computers at home. A modem hook-up provides another communication link, giving telecommuters electronic mail and data-transfer capability.

ISDN for Telecommuting

New communication technologies that facilitate telecommuting include narrow-band and broadband **Integrated Service Digital Networks (ISDN)** (see Fig. 7.15). These technologies can link together a variety of media, such as facsimile, audio and video conferencing, and multimedia electronic mail. ISDN services in the United States have grown from a $5.7 million industry in 1989 to $26.4 million in 1993. Overall, service revenues for 1994 are estimated to be $47.6 million and, within the next 4 years, ISDN services will blossom into a $2.25 billion industry [Riggs, 1995].

According to one industry observer, today the majority of medium- and large size businesses in the United States has access to high-capacity networks that carry two-way video. Interactive two-way entertainment, information, and education services for business will be available in the near future. In addition, since 1988 AT&T has deployed wideband ISDN, which can deliver high-quality color images simultaneously with voice and data, at more than 300 locations in United States and in a dozen countries abroad. The seven regional Bell operating companies and the largest independent telephone companies have filed more than 200 ISDN tariffs in 46 states, the majority already in effect. By the end of 1994, there were 66 million ISDN-capable local access lines that could support simultaneous voice, data, and image services. In a little more than 3 years, 70% of all access lines in the United States are ISDN capable. These technologies will have the potential to be increasingly used for telecommuting work. According to 1993 Gartner Group Market projections, by 1997 the ISDN lines for personal conferencing units is expected to grow to close to a six million level.

As recently as in 1995, ISDN offered the best balance of low cost, connection availability, higher speeds, and industrial-scale manageability, and 15 years after it was first rolled out, ISDN is finally making it big as a tool for U.S. corporate information system (IS) managers who need affordable, high-speed access for remote users. In fact, as much as 80% of all ISDN being put in today comes from IS managers plugging in their remote and home users.

New routers also replace the terminal adapters and other ISDN access equipment with a single integrated unit with both standard Ethernet connections and links to analog phone lines. These new routers take the maximum bandwidth of an ISDN basic rate interface, 128k bps, and mix in compression techniques to stretch that bandwidth even further. These features enable remote workers to appear to have a constant connection to an office LAN that can be miles away.

Other equipment high on the new functionality list is the ability to put the ISDN line to use only when it's needed, while giving the telecommuters the appearance of a constant connection. This is made possible by the ability of ISDN to quickly set up a call, using the intelligent signaling that is inherent in the service, and a newer functionality called "**spoofing**". The spoofing and fast reconnect combine to make telecommuters feel like they are in constant touch. This is necessary because with ISDN, which is a switched service, charges are on a usage basis. With spoofing behind ISDN, the LAN and the user think they are both on-line, but the user is not tying up the line. Other new services offered by Bell Atlantic in support of telecommuting are service packages that include everything from desktop video, remote LAN access, and multipoint group video-conferencing to professional services for reengineering business processes to match telecommuting model.

Other Information Technologies

Computer and peripheral manufacturers and software developers are now developing products specially targeted to telecommuters. As early as 1992, Apple Computer and Dell Computer introduced easy to use, low-cost computers aimed at the small office, home office, or "SoHo" market. Not surprisingly, the industry is pushing a high-technology definition of telecommuting. One industry journal describes the "simple needs" of an employee who works at home 1 or 2 days per week: "a PC and fax, a remote E-mail package, and a second phone line for faxes and E-mail".

New technologies in support of telecommuting include a **mini satellite dish**, an integrated printer, a fax modem and copier, **document-conferencing software**, **security ID card**, and an **E-mail filter**. Beside transferring E-mail and sending files, recent services offered in conjunction with local phone companies allow home users to transfer their home voicemail messages to other staff members (whether at the office or at their own home offices). Better yet, they won't have to directly call the person they want to transfer the message to. They simply dial an extension just like they would in the office, and the voicemail is transferred.

Security for Remote Work

As more employees routinely tap into networks and databases through remote access, companies are concerned, more than ever, about protecting valuable information that their competitors would love to see. Recently, vendors have offered several options to secure laptops, such as lock-down devices useful primarily for securing the laptop while at the office, security-slot options, and the option of locking a device inside the slot to a disk drive. In addition, special security precautions are required to protect corporate data. These include **dial-in protection**, **virus protection**, **data privacy**, and **data integrity**. The dial-back system approach fell apart for mobile telecommuters such as the traveling sales force. The alternative for both at-home telecommuters and the mobile workforce is often a package that uses a multiplatform authentication server along with a credit-card-sized automatic password generator. The generator ensures that users won't keep using the same password.

The most common methods for securing data are password access control, hard-drive data encryption, automatic screen blanking and lock-out for unattended portable computers, and virus protection software. Products, such as a remote network server, facilitates telecommuting by allowing multiple users to dial in and connect to a central network over standard phone lines and then work exactly as if they were in the office. This technology also improves network security. Other security technologies are also being developed, including "**tokens**". To achieve high-quality, front-end authentication, most vendors use what's known as a **two-phase approach**. However, cost can be a major consideration when it comes to two-phase authentication.

Finding the right remote security options depends on planning. Companies need to make sure their site has a well-thought-out security policy in place for all users and resources. It is also important to know what data are most sensitive and put a value on those data, thinking about how well the business could continue if the worst happened.

Telecommuting Economics

According to a survey of 1000 large companies conducted by Forester Research, Cambridge, MA, every telecommuter costs his employer an average of $4000 for set up, remote installation, and network support and more than $2000 a year for maintenance. The money is spent for new computers and software, high-speed modems, hefty phone bills, and technical support back at corporate headquarters. The director of marketing for AT&T Virtual Office Solutions says, "for every dollar spent, we saved $2," on their tele-commuting project. With approximately 8000 employees functioning in the virtual world, managers report productivity up 45% and office space saving up 50% [San Francisco Examiner, 1994]. In 1995, AT&T's 35,000 telecommuters saved the company $80 million in real estate costs alone.

Companies should assess the cost of data transmission and advise the employee of the most cost-effective methods of communication. Telephone line charges should be met by the employer, as should the cost of insuring the equipment. However, telecommuters would be well advised to check their own home building and contents insurance as well. Companies may also consider installing high- bandwidth lines, such as ISDN.

Three factors are making it easier for companies to adopt ISDN: (1) equipment prices have dropped, (2) local ISDN access charges seem to be leveling off at between $25 to $35 per month, and (3) there is a growing number of wide-area network integrators that better understand ISDN installs. In fact, for less than $1000 per user, companies can hook up a high-capacity ISDN line that provides up to 128 kbps of data access to remote computers or the Internet. This cost will include the access device, the software configuration, the RJ-45 jack, and the labor. This simple ISDN access arrangement can get the telecom-muters into the corporate system, across the local area network (LAN) to the Internet router, and out into cyberspace. In addition, ISDN access devices now allow the companies to run data and phone off the same line, cutting the expense incurred from giving their telecommuters two separate lines.

Legal Implications

The Fair Labor Standards Act (FLSA) poses several challenges for companies with nonexempt employees interested in telecommuting. Keeping track of work hours for off-site nonexempt employees will be crucial. If the number of hours a telecommuting employee works is overreported, the employer will pay for unworked hours. On the other hand, if a telecommuter underreports work hours, the company could be liable for an FLSA violation. As a result, employers considering telecommuting programs need to have a good idea of how long it should take an employee to accomplish specific tasks. Requiring employees to log in and log out via a telephone or computer system also can increase the accuracy of an employee's report of off-site work hours. Formal log-in/log-out procedures for telecommuters will be critical for FLSA compliance, but they may also be important for worker's compensation cases. The courts have traditionally held employers responsible for employees injuries that occur while performing a job-related task, even if the accident occurs off site. Health and safety regulations apply to telecommuters, and this can affect furniture and the position of computer screens. Experts have advised that it is not enough simply to issue employees with a package of equipment and leave it to them to site it and provide their own chair and desk. Each home workplace should be inspected and passed as suitable and in line with health and safety standards.

Defining Terms

Data privacy: The company may have gone to great lengths to protect its information from the outside world, forgetting that the telecommuter may be saving the same information on unsecured home PC. Data privacy is enforced by making user aware of problems associated with protecting company data on unsecured home PC.

Data integrity: Employees often lose data at work and it is more possible they lose data at home where support is not readily available. Some losses can be prevented by training and data protection and recovery utilities.

Dial-in protection: Naturally most telecommuters need to connect to the office network. A dial-back system can be an effective security measure, since telecommuters' modem can be reached at predictable telephone numbers.

Document-conferencing software: Allows telecommuters to share information and have a conversation at the same time.

E-mail filter: Helps prevent information overload by filtering E-mail downloads by sender or subject and can even stop large incoming file attachments.

Hoteling: Hoteling is a telecommuting concept under which workers share desks and office space through a system of reservations that assigns facilities on days when a worker is not on the road.

Information workers: The U.S. DOT report [U.S. DOT, 1993] defined information workers as "individuals whose primary economic activity involves the creation, processing, manipulation or distribution of information" and predicted that they will increase from 56% of the workforce in 1992 to 59% of the workforce in 2002. However, labor force data are usually broken down into relatively crude occupational classifications. Some occupations may be primarily information oriented and other primarily not, but few are entirely one or the other. This limits the analysis to rather rough estimates of the number or share of information workers.

Integrated Service Digital Network (ISDN):   ISDN replaces 28.8 modem with a high-speed digital connection that is over four times faster with digital reliability and accuracy. The ISDN basic rate interface comprises two B-channels and one D-channel, and the ISDN primary rate interface comprises 23 B-channels and one D-channel. A B-channel is a 64 kbps (called DS-0) ISDN user-to-network channel that carries a voice, a data, or an image call, but not signaling for the call. N-ISDN (narrowband ISDN) includes a basic interface (2B + D) and primary rate interface (23B + D). It is copper based with rates at or below 1.5 Mbps. B-ISDN (broadband ISDN) extends N-ISDN with new services and provides voice, data, and video in the same network. It is fiber based with rates at 150 and 600 Mbps.

Mini satellite dish: Designed to speed Internet document downloads with 400k-bps data transfer rate.

Security ID card: A credit-card-sized ID card, which provides users with a unique six-digit access code every 60 seconds: this code as well as user ID and password must be repeated upon log-in to a server with Access Control Module software.

Spoofing: New products are designed in a way that, after 60 seconds have gone by with no data appearing over an ISDN link, the connection is automatically dropped. The equipment then spoofs the traditional "keep alive" messages that would be transmitted between a LAN and a remote access point, so that each behaves as if the connection were live. When the remote user does something that requires data to be sent or downloaded from the LAN, the network connection is automatically recreated.

Telecommuting: As telecommuters become larger percentage of U.S. workforce, lawmakers, employers, and employees will need to reexamine traditional definitions of work and the workplace. The definitions used in current laws and regulations may be outmoded, given the anticipated changes in where, when, and how employees perform work assignments. Telecommuting is defined here as work carried out at home or at an office close to home (remote from central offices or production facilities) where the worker has no personal contact with co-workers but is able to communicate with them and perform work-related activities using computer and communication technologies. This definition intents to include daytime home-based telecommuters, part-time workers, full-time workers, and the casual after-hours workers who work in an office or at home during regular hours or after normal hours.

Tokens: A small circuit card that plugs into the telecommuter's computer and that must be read by the central computer before access is allowed.

Two-phase approach: Requires two components: a one-time use password, usually tied to a code generated by a separate smart card that the users carry with them, and a dedicated authentication server that holds a database of the users and their password.

Virus protection: Most viruses are introduced into office systems on disks transported from homes. Security managers suggest having a policy that prohibits employees from bringing disks from home to use on office PC or setting up a dedicated PC that can be used to screen disks for viruses before they are used at work

References

Moore, M. Telecommuting is on the rise in U.S. firms, says survey, *PC Week*, October 10, 1995, 1995.

Riggs, B. US ISDN renaissance, *IEEE Computer*, 28(1): 11, 1995.

Snizek, W. E. Virtual offices: some neglected considerations, *Commun. ACM*, 38(9): 15–17, 1995.

San Francisco Examiner, May 29, 1994.

Telecommunications, *The Wall Street Journal Reports*, February 11, 1994.

U.S. Department of Transportation. Transportation Implications of Telecommuting Washington, DC, April, 1993.

The future of telecommuting, *Wired*, October, 1995, p. 68.

Further Information

Handy, S. L. and Mokhtarian, P. L. The future of telecommuting, *Futures*, 28(3): 227–240, 1996.

7.16 The Nature of Technological Paradigms: A Conceptual Framework

John P. Ulhøi and Urs E. Gattiker

The development of new technology, chaotic as it may seem, tends to follow certain procedures and preconceptions, or paradigms; these "dictate" the kind of problems the technology community deals with. In this section, a conceptual framework based on a survey of current literature in the field is outlined, describing the nature and mechanisms of technological development at the macro level. This may be of help to technology managers trying to "discern an overall picture" from the ongoing renewal of the technological landscape. Identifying and understanding the ruling **technological paradigm**(s) can thus prove very useful prior to deciding the direction of further technological development [Green, et al., 1994]. The framework is designed to identify the key actors and elements that influence the direction of technological development in general and the design of technological strategies in particular. Understanding the process of technological change requires studying the firms involved, since these are major actors in the process of innovation. By viewing industrial innovation processes in terms of company-specific configurations, the existing macro-oriented technological paradigm theory can be greatly improved.

As recently pointed out by Freeman [1994], one of the ongoing paradoxes of economic theory is the increasing recognition that technological change is both the most important source of drivers in capitalist economies and relatively neglected in mainstream economic theory. While various explanations have been offered for this, most have in common an explicit or implicit assumption that technological change is outside the domain of economists and is best left to technologists and scientists. This "black-box" approach has led to the dangerous assumption that science and technology are best treated as exogenous variables.

The public perception of new technology is characterized by two extremes: a positivistic and a pessimistic view. *Techno-optimists* argue that technology can be controlled (to the extent that this is necessary), whereas *techno-pessimists* are more defeatist, believing that there can be little or no strategic influence on the direction of technological development. Such an "either/or dichotomy" is too rigid, however.

In this section, the overall process of technological change is seen not as the result of a deliberate "plan", but of a dynamic and chaotic web of local actions, decisions, and increases in knowledge, which spread and interact in ways that no one can predict or control [Ulhøi, 1996]. This does not, *a priori*, exclude the possibility for exerting deliberate influence on the process of technological change, however. We therefore propose an "in-between" view, one that offers some latitude for strategic actions that may influence the overall process of technological development. The importance of the individual firm in this connection is beyond question. The business community has most of the resources and know-how necessary to sustain the process of technological change. Notwithstanding, the actual level of firms' research and development (R&D) is heavily influenced by local government business policy, infrastructure, etc. This will not be discussed further here, however.

The framework outlined in this section can be used by decision makers in their attempts to identify certain characteristics of and trends in the overall process of technological change. By interpreting technological "patterns" within the framework of technological paradigms, strategic potentials for influencing the outcome or direction of development may be substantially improved. We provide an outline of a normative conceptual framework for use in the assessment of the developmental directions of new generic technologies through the concept of technological paradigms.

Current literature on technological paradigms tends to overlook the importance of the individual firm, thus inadequately reflecting the multidimensionality and complexity of the technological innovation process. This is a serious flaw that may invalidate the overall value of the technological paradigm framework. We discuss the theory of technological paradigms in relation to the individual firm and present conclusions and implications.

Technological Paradigms and Trajectories

The concepts of technological paradigms and technological trajectories are closely intertwined, both focusing on strategic and normative prescriptions for the direction technological change must take and the technological options that must be rejected. According to Dosi [1982], a *technological paradigm* can be defined as a set of procedures, a definition of "relevant" (i.e. relative to the group of technologists involved) technological problems and the specific knowledge needed for their solution. It follows from this that, based on specific technological, economic, and social trade-offs, each technological paradigm strongly influences its own concept of progress. Thus, a **technological trajectory** indicates the direction of progress within a certain technological paradigm. History provides many examples of how technologies have followed specific directions or trajectories [Rosenberg, 1976; Hughes, 1989]. Rosenberg [1976] described this phenomenon as technological development directed by "focusing devices".

A more recent example may illustrate how the technological paradigm framework works. When Steve Wozniak, Apple Computers, first introduced the idea of the personal computer, it was based on user friendliness. However, the existing technological paradigm at that time was based on the philosophy that the computer was a specialist product not able to be used by "ordinary people" (see TP_1 in Fig. 7.16). Nonetheless, subsequent software research no longer focused solely on technological performance but devoted increasing attention to the user-interface problem. Two fundamentally new design issues were at stake: (1) the idea that computing could be done on a personal computer (as opposed to the mainframe) and (2) that computing could be done by nonspecialists. For a number of years, this new user-friendly software paradigm existed side by side with the predominant "specialist" paradigm (see TT_1 in Fig. 7.16). Then, during the 1980s, as it became increasingly institutionalized through its widespread use in business, education, and administration, the new paradigm gradually began to take over. Today, no one in the computer industry would question the importance of user friendliness.

According to the theory of technological paradigms, technological innovations are supposed to follow general prohibitive and/or permissive rules, leading to accumulative and continuous improvements. The paradigm metaphor is taken from the Kuhnian framework of scientific development [1962]. This in turn

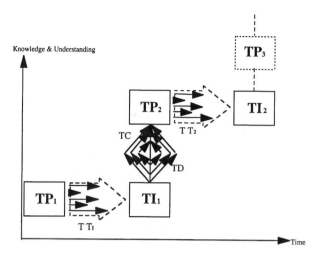

FIGURE 7.16 Increase of knowledge over time through the development of technological paradigms. A technological paradigm results in a technological trajectory being followed by industry and, for example, scientists and technologists in the field concerned, which may ultimately result in radical or derivative innovations [Gattiker, 1990]. This leads to divergence in developments (e.g., Betamax vs. VHS, Unix vs. Windows, Apple vs. PCs), but ultimately, the convergence of these approaches results in the establishment of a new technological paradigm. Once more and more people share the new paradigm it becomes TP_2 and is again shared by a larger group, ultimately becoming the dominant paradigm. The process of change then starts a new with TT_2. The process described in this figure may sometimes be gradual and other times consist of rapid leaps by certain groups, countries, and/or firms. During the process, a firm or country can increase its knowledge and understanding while at the same time improving its technological competency (e.g., Gattiker and Willoughby [1993] and Gattiker and Ulhøi [1998]). TP_n: Technological paradigm, which can be defined as a set of procedures with relevant technological problems and specific knowledge related to their solution; well-developed paradigms result in shared meanings, values, and beliefs by technologists and/or scientists involved. TT_n: Technological trajectory, which can be defined as the direction of advance within a technological paradigm being followed by scientists and developers/technologists. TI_n: Technological innovations may be radical or derivative. TD: Technological divergence, by which is meant that the paradigm is being constantly bombarded by new potential paradigms and ideas, i.e., shared values and beliefs are being frequently challenged, ultimately resulting in the paradigm's demise. TC: Technological convergence, develops from TD by bringing one or more ideas and approaches to the forefront. Alternatively, new ideas and approaches shared by various groups of scientists and/or developers/technologists gradually take over, thus leading to a new paradigm (see TP_2).

generates technological sequences according to preferential paths or trajectories [Biondi and Galli, 1992]. *Radical* technological innovation, which occasionally occurs, involves some change in the organization of production and markets. In other words, organizational and institutional innovations are inextricably associated with the technological innovation process. The fact that technological paradigms become institutionalized suggests that they can be seen as social as well as technical resources. However, due to lack of space, we are unable to discuss the importance of the institutional aspect further.

The concept of technological trajectories was first introduced by Nelson and Winter [1977] and later supplemented by alternatives, e.g., the notion of *technological guideposts and avenues* [Sahel, 1985]. The idea that "something" is guiding the direction of the innovative search process dates back to the late 1960s, however, when the notion of "technological imperatives" was first coined [Rosenberg, 1969].

Figure 7.16 shows that, for a certain period, the development of new technology tends to follow a dominant paradigm, which in turn defines relevant research questions and provides guidelines for how to deal with technological problems. Gradually, a stream of more- or less-focused technological developments leaves new technological "tracks", which indicates a temporary direction of technological development within the overall trajectory (see TT_1 in Fig. 7.16). During this period, during which the established paradigm is becoming more and more worn down, i.e., the scope for new innovations and improvements of existing innovations is drastically decreasing, the need for a more fundamental revision and/or substitution increases. At this point, continuity in R&D is time and again replaced by discontinuity and the existing paradigm is gradually dissolved, leaving room for maximum "freedom" for the technologists. This period can best be described as a period of technological turbulence, in which new technological development is rooted in rival technological variants, each aiming at dominance.

At the beginning of the "intermediate" phase between two paradigms, i.e., the turbulent period, new technological developments tend to diverge to the point where the contours of a "winner" can be identified and convergence (see TC in Fig. 7.16) sets in, leading to a new technological paradigm (see TP$_2$ in Fig. 7.16). During such a period of "post-paradigm" or "preparadigm" development, there tends to be a mismatch between the new technology and the old framework (markets and capital, educational, managerial, and regulative frameworks), which adds to the level of turbulence during the overall process of technological change. Sooner or later, radical innovations (i.e., technological revolutions) succeed, to be followed by other firms, thus leading to new dominant designs and a new technological paradigm. In particular, during such periods, the potential role of the individual firm (i.e., its possibilities for introducing a new dominant design) reaches a peak. As pointed out by Saviotti [1995], this implies a relatively short period of *technological discontinuities,* followed by relatively long periods of incremental development within an established paradigm, as suggested by the time axis in Fig. 7.16. As one paradigm is gradually replaced by another, there will be a corresponding increase in knowledge and understanding.

A general framework has been outlined to account for the way in which major technological innovations tend to follow certain rules and paths. We expand the framework to include the role of the firm in the technological innovation process.

Technological Paradigms and Firms

The role of the individual firm during the life and death of technological paradigms has not been very well understood. Firms inject a significant amount of "fuel" into the overall process of technological innovation. It is therefore necessary to expand the macro-oriented theory of technological paradigms and trajectories to include the firm perspective.

It is somewhat paradoxical that, in the general theory of technological paradigms and trajectories, the firm does not seem to play a particularly important role, despite the fact that firms are central actors in the shaping of technological trajectories. However, some authors have explicitly pointed out the importance of firm-specific technological accumulation (e.g. Dosi [1984], Pavitt [1986], Granstrand and Sjölander [1992], and Christensen [1995]).

Technological development can be seen as social construction capable of being institutionalized. Once institutionalized, they tend to become paradigms, dominating development for a period of time (see Fig. 7.16). During this process of social institutionalization, existing technological paradigms become embedded in designers', engineers', and managers' frameworks of perceptions, calculation, and routines. This directs attention toward the way in which technological paradigms change, including how socially constructed beliefs and expectations are created and maintained. This may be of interest from the point of view of governance, since it takes account of how such social mechanisms can be influenced.

When a technological paradigm becomes institutionalized, the identification of its trajectory may enable designers, engineers, managers and entrepreneurs to visualize likely future paths of development [Nelson and Winter, 1977]. The individual firms in the industry in which the paradigm rules will typically pursue incremental technological innovations, i.e., carry out continuous, but typically *incremental,* changes in the product and/or process [Gattiker, 1990]. Over time, such changes will enable the individual manufacturer to reduce the costs of machinery, labor and material inputs. However, the specific trajectory pursued by a specific firm within a particular paradigm is mainly governed by decisions and actions in the world outside.

It is only natural to focus on the individual technological actor (the firm level), since most R&D activities either spring from, or are directly linked to, industry (which, incidentally, also has most of the available financial and HRM resources, see Gattiker [1988]). Furthermore, focusing on the individual firm allows for both a less-aggregated understanding of the nature and mechanisms of technological paradigms and a focus on the related social processes underlying technological development. The ***technological potential* of the individual firm** can be defined by its accumulated financial and know-how resources, explicit as well as implicit [Gattiker and Willoughby, 1993] and its relative technological performance compared with its competitors [Bye and Chanaron, 1995]. The individual firm's potential for influencing technological development depends on its own technological fit with the dominant technological paradigm and its trajectory of products and/or services.

Adopting a resource-based perspective (e.g., Wernerfelt [1984], Prahalad and Hamel [1990], and Grant [1991]) of the firm may facilitate a less-aggregated and rigid understanding of the technological innovation process, since this tends to understand the firm in terms of its present configuration of resources and capabilities. By perceiving industrial innovation processes in terms of firm-specific configurations of technological innovation assets associated with specific market opportunities, the framework of technological paradigm and trajectories can be further expanded. A change in the firm's actual configuration of technological assets can thus be seen as a strategic response to market and/or technological opportunities or threats. The trajectory of a technological paradigm allows the individual firm to identify the opportunities for further incremental improvements in innovation, which in turn serves as an "early warning" of increasing instability and new market and technological "breakthroughs".

Conclusion and Implications

Green et al. [1994] suggest that the mechanisms by which one paradigm is overthrown and replaced by another are not satisfactorily explained by the technological paradigm framework. We agree with Anderson and Tushmore [1990] that the domination of technological paradigms is due not only to technological forces but also to social, political, and organizational forces. This calls for a stronger focus on the social embeddedness of the technological innovation process. Thus, the importance of the institutionalization of technology is directly linked to the "selection environment", which, as we have indicated, is not given but is the result of an active and social construction. This in turn implies that more research is needed on the formation, maintenance, and change of the social, political, and organizational forces underlying technological development. A deeper insight into these processes may significantly improve the options for managing (i.e., influencing the direction and speed of) technology.

An important practical implication of the framework outlined above is that major technological investments during periods of technological divergence (TD), e.g., Sony's investment in Betamax technology, is very risky because a rival technology (VHS technology in this example) can emerge as the winning paradigm. However, as soon as a converging trend sets in (see TC in Fig. 7.16), the risk decreases. Put another way, this conceptual framework may support the corporate technology management function in making the difficult choice between alternative technologies [Ulhøi, 1996] and related assessments [Madsen and Ulhøi, 1992].

Future research ought to pay more attention to the social, economic, and political embeddedness of technological paradigms in general and to mechanisms related to the technological selection environment in particular. This in turn may improve the understanding of the key inhibiting and catalysing forces influencing technological development; these cannot be explained solely by technology itself but must be ascribed to social, economic, and political aspects.

Defining Terms

Technological paradigm: A set of procedures, a definition of relevant, i.e., relative to the group of technologists involved, technological problems and specific knowledge related to their solution.

Technological potential of the individual firm: The firm's accumulated financial and know-how resources (explicit as well as implicit) and its relative technological performance compared with competitors.

Technological trajectory: Indicates the direction of advance within a certain technological paradigm.

References

Andersen, P. and Tushman, M. L. Technological discontinuities and dominant designs: a cyclical model of technological change, *Admin. Sci. Q.*, 35: 604–633, 1990.

Biondi, L. and Galli, R. Technological trajectories, *Futures*, 24(6): 580–592, 1992.

Bye, P. and Chanaron, J.-J. Technological trajectories and strategies, *Int. J. Technol. Mgt.*, 19(1): 45–66, 1995.

Christensen, J. F. Asset profiles for technological innovation, *Res. Policy*, 24: 727–745, 1995.

Dosi, G. Technological paradigms and technological trajectories. A suggested interpretation of the determinants and directions of technical change, *Res. Policy*, 11: 147–162, 1982.

Dosi, G. *Technical Change and Industrial Transformation*, Macmillan, London, 1982.

Freeman, C. The economics of technical change, *Cambridge J. Econ.*, 18: 463–514, 1994.

Gattiker, U. E. Technology adaptation: a typology for strategic human resource management, *Behav. Inform. Technol.*, 7: 345–359, 1988.

Gattiker, U. E. *Technology Management in Organization*, Sage, Newbury Park, CA, 1990.

Gattiker, U. E. and Willoughby. Technological competence, ethics, and the global village. Cross-national comparisons for organization research, In *Handbook of Organizational Behavior*, R. Golembiewski, ed., pp. 457–485, Marcel Dekker, New York, 1993.

Gattiker, U. E. and Ulhøi, J. P. The matrix organization revisited, In *The Handbook of Technology Management*, R. C. Dorf, CRC Press, Boca Raton, FL, 1998.

Granstrand, O. and Sjölander, S. Managing innovation in multi-technology corporations, *Res. Policy*, 19(1): 35–61, 1992.

Grant, R. M. The resource-based theory of competitive advantage: implications for strategy formulation, *Calif. Mgt. Rev.*, 33: 114–135, 1991.

Green, K., McMeekin, A., and Irwin, A. Technological trajectories and R&D for environmental innovations in UK firms, *Futures*, 26(10): 1047–1059, 1994.

Hughes, T. P. The evolution of large technological systems, In *The Social Construction of Technological Systems: New Directions in the Sociology and History of Technology*, W. E. Bijker, T. P. Hughes, and T. J. Pinch, eds., pp. 51–82. MIT Press, Cambridge, MA, 1989.

Kuhn, T. S. *The Structure of Scientific Revolutions*, Chicago University Press, Chicago, 1962.

Madsen, H. and Ulhøi, J. P. Strategic considerations in technology management: some theoretical and methodological perspectives, *Technol. Anal. Strat. Mgt.*, 4(3): 311–18, 1992.

Nelson, R. R. and Winter, S. G. In search of useful theory of innovation, *Res. Policy*, 6(1): 36–76, 1977.

Pavitt, K. 'Chips' and 'trajectories': how does the semiconductor influence the sources and directions of technical change?, In *Technology and the Human Prospect*, R. Macleod, ed., pp. 31–55, Frances Pinter, London, 1986.

Prahalad, C. K. and Hamel, G. The core competence of the corporation, *Harvard Bus. Rev.*, May-June: 79–91, 1990.

Rosenberg, N. The direction of technological change: inducement mechanisms and focusing devices, *Econ. Dev. Cultur. Change*, 18: 1–24, 1969.

Rosenberg, N. *Perspectives on Technology*, Cambridge University Press, Cambridge, 1976.

Sahel, D. Technological guideposts and innovation avenues, *Res. Policy*, 14: 61–82, 1985.

Saviotti, P. P. Technology mapping and the evaluation of technical change, *Int. J. Technol. Mgt.*, 10(4/5/6): 407–424, 1995.

Ulhøi, J. P. The technological innovation management — strategic management relationship, In *Innovation Strategies. Theoretical Approaches — Experiences — Improvements. An International Perspective*, H. Geschka and H. Hübner, eds. pp. 15–26, Elsevier, Amsterdam, 1992.

Ulhøi, J. P. Greener technologies for a cleaner environment. Hope or hot air?, In Proceedings of "The IASTED International Conference on Advanced Technology in the Environmental Field", Gold Coast, Australia, 1996.

Wernerfeldt, B. A resource-based view of the firm, *Strat. Mgt. J.*, 5: 171–180, 1984.

Further Information

http://www.aom.pace.edu/tim/   Academy of Management, Technology and Innovation Management (TIM) Division's home page

The above page provides interesting links to other sites and information about TIM's electronic newsletter. Copies of the printed newsletter can be found here for downloading.

http://www.brint.com/   Brint Research Initiative

A useful research source for management, business, and technology matters. "Council Partner of the U.S. Federal Government Inter-Agency Benchmarking & Best Practices Council"

http://www.TIM-Research.com   Technology and Innovation Management

Not-for-profit virtual research organization, involving various universities and organizations. Provides information about new developments on TIM issues, especially new research on environmental, R&D, Internet, e-commerce, privacy and other technology matters. You can also participate in ongoing research on-line. Download aggregate results (descriptive statistics) of the latest studies' findings, or obtain both research and technical reports, press releases and much more (e.g. interesting links to other sites).

http://www.uti.ca/   University Technologies International, Inc., Calgary, Canada

Provides information about inventions, patents, new products, interesting links, and reports for downloading and much more.

7.17 Leadership

Jay A. Conger

Leadership has been a subject of great interest for several millennia. Early Greek and Roman thinkers such as Plato and Aristotle pondered the sources of leadership many centuries ago, and to date thousands of studies have been conducted on the topic. It is a subject of such enduring interest because leadership plays a vital role in the well-being of both societies and organizations. In this section, we will explore what we know today about the subject of leadership. We will discuss primarily leadership in organizational contexts since we are concerned about the **management** of technology and the role of leadership in that process.

Until recently, it was assumed that leadership and management were essentially interchangeable activities. Today, however, we now consider the two roles as distinct. For example, leadership is defined around activities that facilitate significant levels of change and innovation while management is defined around activities that ensure the efficient and productive functioning of the status quo . As it relates to technology management, leadership plays an especially critical role in the championing of technological innovations.

Leadership vs. Management

With the dramatic rise of global competition over the last half of the 20th century, organizations in the corporate world began to face the enormous task of transforming themselves to be far more competitive and responsive. As they attempted to change, they found that the process of reinvention was extremely difficult. Rarely did company insiders possess the courage and change management skills necessary to orchestrate large-scale organizational transformations. The leadership talent needed for such undertakings was essentially in short supply. At the same time, companies faced problems with employee commitment. In the midst of their change efforts, firms had resorted to extensive downsizing and new organizational arrangements such as flatter hierarchies and strategic business units. While improving corporate performance, these initiatives took their toll on employee satisfaction and **empowerment**. In the process, an old social contract of long-term employment in return for employee loyalty was broken.

The net result was a deep sense of disenfranchisement for the workforce. This occurred just at the moment when companies were demanding ever-greater commitment from employees. The challenge for organizations became a question of how to orchestrate transformation while simultaneously building employee morale and commitment. These twin events of transformation and disenfranchisement would lead both organizations and academics to realize that the roles of leading and managing were quite different from one another. After all, there was a healthy supply of managerial talent within the corporate world, yet it was insufficient to bring about the radical changes desired. As a result of this discovery, we today draw several critical distinctions between managing and leading.

At its most basic, we can think of the managerial role as one requiring individuals to focus on administrative, day-to-day functions. In essence, a manager's responsibility is to ensure that the current state of their organization is operating efficiently. The focus is short term and operational. To achieve its goals, management relies principally on the successful implementation of formal systems and procedures that effectively monitor, control, and motivate the performance of an organization. Therefore, for example, effective management involves the designing of appropriate plans and budgets to match the operating demands of a business unit. It involves identifying and putting in place salary and bonus systems that contain the proper incentives to ensure the achievement of operating goals.

In contrast, the leadership role requires an individual to focus on the longer-term, adaptive challenges facing his organization. The leader's central role is as an agent of change. Given the dynamism of marketplaces and the rapid pace of technological change, the managerial routines and procedures developed by an organization often become outdated. For example, competitors may introduce innovations or tactics that overnight alter radically "the rules of the game" in an industry. As a result, the formal plans, measurement systems, and organizational structures that once proved so effective for one company may no longer match the reconfigured requirements of an industry. The dilemma is that organizations become invested in the status quo that brought them their original success. In turn, they lose their adaptability. At this point, leadership must play a pivotal role. The leader's ability to continually and effectively challenge the status quo of his firm ensures that the organization remains adaptive to ever-changing markets.

A critical activity of the leader is therefore to develop new strategic directions or **visions** of the future, to actively communicate and gain commitment to these future goals, and to energize individuals to overcome internal and external obstacles to adapting the organization to the future vision. In essence, there are three categories of leadership activities: (1) visioning, (2) aligning, and (3) mobilizing.

The Leadership Skill of Vision

Leaders actively search out existing or potential shortcomings in the status quo of an organization [Conger and Kanungo, 1988]. For example, the failure of the organization to exploit new technologies or new markets becomes a strategic or tactical opportunity for the leader. To be effective in this dimension, leaders must also be able to make realistic assessments of the environmental constraints and resources needed to bring about change in their organizations. For example, instead of launching an initiative as soon as the sense of direction is formulated, a leader's environmental assessment may dictate that he or she prepare the groundwork and wait for an appropriate time and for the availability of resources. In addition, leaders must also be sensitive to the abilities and emotional needs of their subordinates, who, as we shall see, are critical resources for attaining organizational goals.

After assessing the environment, the leader must then formulate future goals for the organization, which we today refer to as strategic visions. Here the term vision refers to idealized goals that the leader wishes to achieve. While the content of visions varies between leaders and companies, they share a simplicity built around a positive image of the company's role in the future [Conger, 1989]. Unlike managerial or tactical goals, which often aim at standard business performance measures such as return on assets or shareholder value or increased market share, visions involve more abstract and personally satisfying goals. For example, Steven Jobs in the early days of Apple Computer described his company's vision as revolutionizing the educational system of America through computing power for young people. The founder of Silicon Graphics, James Clark, set out as his company's vision that they would transform computer screens into windows onto visual worlds.

Visions play two important roles for organizations by providing, first, a strategic focus, and, second, a motivational stimulus [Conger, 1989]. In their first role, they are a statement of purpose for the company. Their simplicity encourages a clear focus unlike company strategic plans that are typically too lengthy and complicated for most employees to follow. For example, Jan Carlzon, the former CEO of Scandinavian Airlines (SAS), was one of the earliest proponents of business class for the airline passengers. His vision was for SAS to build the best business class experience for international travelers. This clarity of focus shaped most of the company's major decisions. For example, opportunities or investments unrelated to this vision would never be considered — they simply distracted from the core focus.

Equally important as clarity of focus is the ability of an effective vision to generate motivational commitment. When an organization has a clear sense of purpose that is widely shared and perceived as highly meaningful, employees inevitably find their mission and work more rewarding. Participating in such a highly worthwhile enterprise sparks greater commitment and enthusiasm as well as the motivation to work hard. It creates a sense of being at a special place where real change and innovation are taking place. For instance, imagine being on the ground floor of the new Boeing 777 or the turnaround of IBM or the startup of Netscape.

The Leadership Skill of Alignment

Aligning an organization behind a new sense of direction requires several abilities on the part of leaders [Conger, 1989; Kotter, 1990]. First, they must be able to articulate their future vision in such a manner that it is widely understood and appealing. Second, their own deeds must reinforce the central message of their vision.

To effectively articulate a vision, its message must be keep relatively simple and straightforward. For example, Southwest Airlines' vision is often described as being "a friendly, reliable flying car". In their communications about the organization's vision, leaders may also describe why the status quo is no longer acceptable with the aim of creating discontent with it and a stronger identification with future goals. Leaders also articulate simultaneously their own motivation for leading and for the firm's goals. They will describe their convictions, self-confidence, and dedication to goals.

In their deeds, leaders role model the important values and actions necessary to achieve the future vision. Returning to Carlzon for a moment, we can see how deeds and role modeling send a powerful message. On his first day as President of SAS, he was scheduled to take a flight on his own airline. While waiting at the boarding gate, he realized that it was 5 minutes to departure time and none of the passengers had been boarded. Concerned about a possible delay, he approached an employee at the gate and asked if something was wrong. She explained that they were simply waiting for Carlzon to board as it was customary for company executives to get on board before other passengers. Amazed at this policy, Carlzon explained that customers were now the company's number-one priority and that he would board last. The story about this deed circulated throughout the entire company within a matter of days. As a result, the organization's values shifted to customers as the number-one priority over company officials. In this case, the leader's own behaviors powerfully aligned with the vision he was communicating. It heightened the credibility of both the leader and his own personal commitment to the vision.

The Leadership Skill of Mobilization

The third dimension of leadership is the ability to motivate subordinates and the organization to accomplish future goals. Under the management role, motivation is achieved through formal systems such as pay and promotions for performance. Under leadership, motivation is more personal and built around actions that empower — empowerment being the act of strengthening an individual's beliefs in his or her sense of effectiveness [Conger and Kanungo, 1988]. Any leadership practice that increases an individual's sense of self-determination will tend to make that person more effective and powerful. From research in psychology [Bandura, 1986], we know there are at least four empowering practices: (1) helping individuals actually experience the mastering of a task with success, (2) presenting models of success with whom people can identify, (3) giving positive emotional feedback during experiences of stress and anxiety, and (4) offering words of encouragement and positive persuasion. Effective leaders employ such practices to promote greater performance motivation as well as to ensure an enduring commitment to the vision's goals. The latter is particularly crucial since visions tend to have lofty aims, which imply considerable obstacles that must be overcome for the vision to be realized.

Leaders as Champions of Innovation

Related to the three activities just described is the notion of championing. This is one of the most important activities of a leader when it comes to the management of technology and technological innovation. The term champion describes individuals who very actively and enthusiastically promote an innovation through all the organizational stages of decision making to transform it into a reality [Achilladelis et al., 1971]. Schon [1963] argued that, to overcome the obstacles associated with innovation, it was necessary to have an individual who could champion the idea. Such individuals identified personally with the innovation and vigorously promoted it through informal organizational channels. They were willing to risk their own status and credibility to ensure the realization of the innovation. Numerous field studies have shown that indeed the success of an innovation is strongly correlated with the presence of these champion leaders (e.g., Burgleman [1983], Howell and Higgins [1990], and Roberts [1968]). In one of the most recent studies, Howell and Higgins [1990] found that leaders in these champion roles shared certain characteristics. Specifically, they were effective at assessing both the resources and constraints for bringing about the innovation and could then articulate a compelling vision of the innovation's potential for the organization. They employed unconventional, innovative tactics to align critical constituencies. They also continually expressed confidence in others who participated in the initiative, in essence, the qualities of vision, alignment, and mobilization that we have been describing.

One additional important finding emerged. The champions in the Howell and Higgins study were matched with nonchampions on the variables of age, salary, job level, functional background, and education. Despite these shared characteristics, the researchers found that the champions' career experiences in their companies involved many different jobs spanning multiple functional areas, operating groups, and geographic locations in contrast to nonchampions who had less diversity in their backgrounds. This suggests that the broad exposure developed during the champions' careers enabled them to have far greater opportunities to build information networks for locating new ideas, for developing a broader vantage point on emerging opportunities and trends in their industries, and for assembling a large number of personal contacts for coalition building. These findings have particularly important career implications for those wishing to effectively manage technological innovations.

Defining Terms

Empowerment: The act of strengthening an individual's belief in his or her sense of effectiveness and self-determination.

Leadership: The act of establishing future direction or vision for a working group or organization, aligning the necessary constituencies and resources to achieve that future direction, and motivating these same constituencies to overcome internal and external obstacles to the achievement of the future goals.

Management: The act of identifying and implementing formal organizational systems and procedures that effectively monitor, control, and motivate the performance of organizational members toward operational and strategic goals.

Vision: Idealized strategic and organizational goals that convey a positive and highly meaningful image of the company's role in the future.

References

Achilladelis, B., Jervis, P., and Robertson, A. *A Study of Success and Failure in Industrial Innovation,* University of Sussex Press, Sussex, England, 1971.

Bandura, A. *Social Foundations of Thought and Action. A Social Cognitive Theory,* Prentice-Hall, Englewood Cliffs, NJ, 1986.

Burgelman, R. A. A process model of internal corporate venturing in an organization, *Admin. Sci. Q.,* 8: 223–244, 1983.

Conger, J. A. *The Charismatic Leader,* Jossey Bass, San Francisco, 1989.

Conger, J. A. and Kanungo, R. N. Toward a behavioral theory of charismatic leadership, *Acad. Mgt. Rev.,* 12: 637–647, 1987.

Conger, J. A. and Kanungo, R. N. The empowerment process: integrating theory and practice, *Acad. Mgt. Rev.,* 13: 471–482, 1988.

Howell, J. M. and Higgins, C. A. Leadership behaviors, influence tactics, and career experiences of champions of technological innovation, *Leader. Q.,* 1(4): 249–264, 1990.

Kotter, J. *The Leadership Factor,* The Free Press, New York, 1988.

Roberts, E. B. A basic study of innovators, *Res. Mgt.,* 11: 249–266, 1968.

Schon, D. A. Champions for radical new inventions, *Harv. Bus. Rev.,* 41: 77–86, 1963.

Further Information

Conger, J. A. Leadership for the year 2000, In *The New Portable MBA,* E. Collins and M. A. Davenna, eds., pp. 388–412, John Wiley & Sons, New York, 1994.

Kotter, J. *The Leadership Factor,* Free Press, New York, 1988.

7.18   The Golden Rules of Power and Influence

Michael D. Kull

Familiar to most managers is a contemporary rendering of the Golden Rule phrased as "he who has the gold makes the rules." This version recognizes the need for managers to control resources in order to get things done. However, as pressure grows for organizations to become more project oriented, team based, flat, and virtual, managers are faced increasingly with doing more with fewer resources, or none at all. In this way influence differs from power in that one's ability to get things done does not necessarily rely solely on one's control over organizational resources.

Power, influence, persuasion, inducement, and related concepts are not neatly segmented into conceptual boxes [Abell, 1975]. Ordinarily, influence is distinguished from power in that influence focuses on changing the perceptions and behavior of others regardless of an influencer's basis of power. Those bases of power can take several forms [French and Raven, 1959] that are often reduced to two: **position power** and **personal power**. Position power, or institutional power, is usually vested in formal managerial authority and involves the use of tangible rewards and punishments to command and control people and processes. Influence also can be based on personal power, which includes expert knowledge as well as respect reflected in the interpersonal skills of the individual [Gibson et al., 1991]. A person may have position power but not be influential. Similarly, an influential person may not have position power but may effectively leverage personal power to shape perceptions and values.

Position power is conventionally understood to be the source of "real" influence in most organizations. A manager can use position power to alter the accumulation and flow of resources through an organization, to coordinate activities, to fashion rewards and punishments, and to define vertical reporting relationships. An understanding of position power is important if one is trying to understand how a boss influences an employee in a hierarchical bureaucracy but is increasingly insufficient for understanding how people gain influence in networked organizations staffed by members who share power. When influence is understood in terms of position power, disputes arise as members struggle for bureaucratic position. Solutions to these political conflicts are seen usually as mutually exclusive, or win/lose. When influence is understood in terms of personal power, however, then members of networked organizations can benefit mutually, and win/win solutions emerge.

Fewer levels of management as well as lateral, informal networks define the postbureaucratic organization, despite the many organizational analysts and managers who all too frequently still think of the organization in terms of a descending pyramid [Salancik and Pfeffer, 1988]. Flat organizations decrease power distance so that individual managers have more influence as formal structures dissolve and lines of authority are shortened, strained, or eliminated. For example, in the area of project management, where control of organizational resources or material incentives are frequently lacking, the importance of building personal influence has gained recognition. Project managers are often not superiors but peers with the project staff. In such a climate cooperation is obtained through personal influence alone.

Influence and Innovation

Almost all industries have become more dynamic in recent years requiring organizations to become more innovative. Institutional power bases are being eroded through downsizing, outsourcing, and project-based enterprises. Many scholars have noted the shift from autocratic to democratic organizations [Ackoff, 1994], and others have pointed out the uses of power to align organizational members during turbulent times [Salancik and Pfeffer, 1988]. For organizations to adapt successfully to changes in the environment, managers must create information and knowledge and encourage all members to become active agents of innovation [Nonaka and Takeuchi, 1995]. Active participants of innovation do not respond well to a climate of command and control. Today's manager is more likely to deal with professionals who resent being treated as subordinates and who expect to be consulted in the process of setting organizational goals and objectives [Tannenbaum and Schmidt, 1973]. At the heart is a shift to a management philosophy that sees workers as self-actualizing and engaged collectively. The concept of **empowerment** has redefined power in corporations as the ability to have objectives accomplished through people rather than the ability to control individuals through authority or resource dependency. Empowerment is the foundation of self-managing teams and creative groups [Rogers, 1995].

It has often been said that invention is only one half of innovation. The other half is salesmanship. Creative individuals and groups utilize influence strategies to "sell" ideas in order to build cooperation with leaders and groups throughout the organization. Whether the challenge is inventing new technology, launching a new product, providing a new service, or designing a new system, innovation is the result of creative groups and individuals influencing others. The role of the manager of innovation is to create the objectives, norms, and a climate for innovation. In this endeavor an effective manager must allow for the creation and the free flow of ideas. The manager must deputize or be themselves innovation **champions**. Champions are change agents who may or may not hold position power but who influence the organization to recognize and implement or sell entrepreneurial initiatives.

With this in mind, the daily task of a manager is to align people with organizational objectives and foster innovation. Innovation requires trust, mutual respect, due consideration of different perspectives, communication, and confident risk-taking. Innovation leaders are thus better understood as facilitators of the creative process. They outline the desired results, provide guidelines and standards of performance, and then allow teams to make the tactical and often unorthodox decisions.

Rules for Building Influence

The ability to influence others is a craft requiring the talent, training, and the proper perspective for proficiency. The shift in management philosophy from influence based on position power to influence based on personal power leads to the following heuristics or rules of thumb for building influence.

Golden Rule #1: Do unto Others as You Would Have Others Do unto You

When groups of students entering MBA programs are asked how they would treat subordinates, they routinely opt for the command and control model. Yet, when the same students are asked how they would like to be treated as subordinates, they prefer the empowerment model. This realization helps the students learn to treat subordinates as well as peers the way they themselves would like to be treated. The lesson reveals that intelligent, creative professionals prefer to work in environments that are participative, communicative, and performance based, where they are allowed to make their own decisions but are clearly aware for what they are being held accountable.

Golden Rule #2: Do unto Others as Others Would Have You Do unto Them

This adage modifies the original by recognizing that an individual does not want to be treated the way their manager would like to be treated, but rather the way they themselves would like to be treated. Influence tactics should be those that the individual would respond to, not those that the manager would respond to were he or she in the other person's shoes. One must seek to understand the motivations of others to be influential, not simply project one's own motivations onto other people. This skill requires the ability to listen and value diverse perspectives especially when those perspectives or the people holding them are not well accepted in the organization. Influence is built when an individual feels that a manager genuinely understands his or her perspective, even if the manager does not agree with that perspective. Such a demonstration of understanding creates favorable perceptions that lead to trust, admiration, and confidence in the manager's abilities.

Golden Rule #3: Do unto Others as You Would Have Others Do unto Others

Influence tactics require time and effort and are relied upon most often by front-line managers who may not hold much position power but who must create networks of support. In order to align members with an organization's goals, a manager influences organizational behavior recursively, that is, a manager leads by example. By encouraging identification and emulation, a manager leverages his or her personal power. Under this model, managers empower others through coaching, mentoring, and delegation. One of the best ways to learn model behavior is to see it in others and follow their example; those examples may come from real or fictional people, living or dead.

Influence Strategies and Tactics

Given that innovation managers must build lateral relationships, align innovation teams with organizational goals, and sell innovations to the larger organization, the choice of influence strategy is a function of the personal characteristics and interpersonal skills of the individual in addition to the prevailing organizational culture. Synergy requires high levels of trust and cooperation that cannot be created through the exercise of position power alone, but rather through strategies of **networking**.

Build Social Networks

Leaders can create interdependencies by influencing perceptions of the leader's actual power and influence [Kotter, 1996]. Social expectations are learned and tend to become crystallized over time. Thus, personal power is usually established early in a project's initiation or an individual's arrival. However, given time an individual can shape or reshape perceptions of power and influence and build a positive reputation through delegation, accessibility, consistent and exemplary conduct, participate planning, communication channels and technology, showing appreciation, doing favors, and increasing the social status of others. Through social networks, coalitions can be formed to sway the larger organization to try a new idea, adopt a new technology, or transform the value-creating systems of an organization.

Build Knowledge Networks

Knowledge, information, and expert power are intangible sources of influence, unlike traditional resources. Middle managers who controlled information and considered it a resource to shore up their position power were the first casualties of downsizing. For innovative organizations, all participants must have access to expert knowledge and strategic intelligence. The job of the manager is not to control knowledge and information but to see that it is generated, transferred, diffused, and re-used in managing projects and for shaping the shared mental models of the network. This includes mediating knowledge within the organization from its periphery to the center and back again. Networks often extend beyond the visible organization to suppliers, distributors, strategic allies, and other partners. A manager can increase his or her influence through education and experience gained through the accumulation of special knowledge and special access to information, experts, and stakeholders.

Build Interpersonal Skills

Managers are most often faced with the need to influence individuals within social or knowledge networks whether those individuals hold power or are linked with the organization's goals. Paramount to becoming an effective influencer of others, one must first know oneself. First, one should become aware of his or her underlying assumptions, prejudices, and mental models. Second, one should elicit ideas and advice from others. Good solutions spring from considering competing viewpoints and learning the premises that make these positions seem reasonable. Furthermore, one should recognize that another person may hold a deep understanding of a situation or problem and that everyone has a unique set of learning experiences, regardless of position. Third, one should restate the other person's perspective to demonstrate understanding and attention. Fourth, one should address concerns by emphasizing areas of mutual benefit and common interest; these usually far exceed the differences. Fifth, one should discuss alternatives for meeting mutual goals.

Three basic influence approaches are used in management: hard, soft, and rational [Kipnis and Schmidt, 1985]. Hard tactics are generally demanding and assertive and used when the influencer has a position power advantage or a transgression of social norms has occurred. The greater the position power distance, the greater the likelihood that hard tactics will be used. Soft tactics are used when the influencer has a position power disadvange, when there exists the potential for mutual benefit, or when a favor is sought. The third or rational approach is the most popular and is used when there is a clear advantage to cooperation, when there is no clear power advantage, and when resistance is not expected. Kipnis breaks these three basic strategies into seven [Kipnis et al., 1980]:

Assertiveness: Using a tough, forceful approach; pointing out formal roles.
Bargaining: Using exchange of benefits or favors as the basis for negotiation.
Coalition: Using alliances with other people or groups to build support.
Upward appeals: Gaining sanctions from higher authorities to demonstrate support.
Ingratiation: Using flattery, good will, and making positive impressions.
Reason: Using facts, data, and details to support a logical argument.
Sanctions: Using organizationally derived rewards and punishments.

Project or innovation managers frequently wonder how they can build influence for themselves and their teams when the team members are often borrowed from other functional areas of the organization. There are several ways this can be accomplished without formal control. For example, a manager can hold productive meetings, create a team space, create team "signs", publicize team efforts, develop recognition rewards for good behavior such as letters of commendation, give first choice on scheduling assignments, give first choice for new technology, allow flex time and comp time, let team members give briefings to senior management, and organize informal events celebrating individual and team achievements.

Which influence tactic to use depends on the objective, the context, and the ability of a professional manager to leverage effectively his or her personal power to strengthen and extend interdependent relationships. To do this over time, one must genuinely understand and care for others. If others sense they are simply the target of an influence technique, they will feel manipulated and suspect the motives of the manager. This, of course, reduces the manager's influence. By following the Golden Rules, one is more likely to learn trust, respect, and consideration for others, and more often than not these attitudes are reciprocated fully. In networked organizations, an individual's ability to influence others is one of the most elementary but arguably the most personally challenging steps in creating communities of innovation and for leading change.

Defining Terms

Champions: Individuals who exercise influence to turn invention into innovation.

Empowerment: The process by which managers help others acquire and use sources of power needed to affect work decisions.

Networking: The application of personal power within a social or knowledge network.

Personal power: Interpersonal abilities and the extent of an individual's knowledge, expertise, and social relationships.

Position power: The control over resources and formal authority that legitimizes the extent to which a manager can offer or withhold rewards and punishments.

References

Abell, P. Measuring intra-organizational power and influence, In *Organizations as Bargaining and Influence Systems*, P. Abell, ed. pp. 10–40, John Wiley & Sons, New York, 1975.

Ackoff, R. L. *The Democratic Corporation: A Radical Prescription for Recreating Corporate America and Rediscovering Success*, Oxford University Press, New York, 1994.

French, J. and Raven, B. The bases of social power, In *Studies in Social Power*, D. Cartwright, ed., pp. 150–167, University of Michigan Institute for Social Research, Ann Arbor, MI, 1959.

Gibson, J., Ivancevich, J., and Donnelly, J. *Organizations*, Irwin, Homewood, IL, 1991.

Kipnis, D., Schmidt, S. M., and Wilkinson, I. Intraorganizational influence tactics: explorations in getting one's way, *J. Appl. Psychol.*, 65: 440–452, 1980.

Kipnis, D. and Schmidt, S. The language of persuasion, *Psychol. Today*, April: 40–46, 1985.

Kotter, J. P. *Leading Change*, Harvard Business School Press, Boston, 1996.

Nonaka, I. and Takeuchi, H. *The Knowledge-Creating Company*, Oxford University Press, New York, 1995.

Rogers, E. M. *Diffusion of Innovations, 4th ed.*, Free Press, New York, 1995.

Salancik, G. R. and Pfeffer, J. Who gets power — and how they hold on to it: a strategic-contingency model of power, In *Readings in the Management of Innovation*, M. L. Tushman and W. L. Moore, eds., pp. 179–195, HarperBusiness, New York, 1988.

Tannenbaum, R. and Schmidt, W. H. How to choose a leadership pattern, *Harv. Bus. Rev.*, May-June, 1973.

Futher Information:

In addition to the references listed above, popular guides to influence are available:

Blanchard, K. and Johnson, S. *The One Minute Manager*, Blanchard-Johnson, La Jolla, CA, 1981.

Carnegie, D. *How to Win Friends and Influence People*, Simon & Schuster, New York, 1936.

Covey, S. R. *The Seven Habits of Highly Effective People*, Fireside, New York, 1989.

Steven Jobs and Steven Wozniak designed the first Apple computer (Apple I) on April 1, 1976 when they were members of the Homebrew Computer Club, Palo Alto. It was only a year after the Apple II debuted in 1977 at a local computer trade show that orders for the first computer with friendly graphic interface started.

- Steven Jobs becomes Apple's chairman in March 1981.
- John Sculley becomes Apple's CEO in Feb. 1983.
- Gilbert Amelio becomes Apple's CEO in 1996.
- Steven Jobs becomes Apple's CEO in 1997.

Growth in sales

1979	$47.9 million
1981	$334.8 million
1984	$1,516 million
1987	$2,661 million
1990	$5,558 million
1993	$7,977 million
1996	$9,833 million

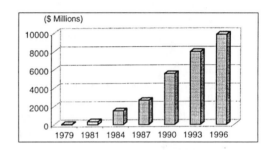

Innovations:

Apple's history has been a volatile one, and there have been many ups and downs for the company. But there is an underlying theme in the history of Apple: innovation. Apple has had more industry firsts than any other personal computer company in the world, and it is this innovation that has always set Apple apart from the norm.

1976 — Apple I
1977 — First computer with friendly graphic interface (Apple II)
1978 — The first inexpensive floppy drive (Apple Disk II)
1986 — Affordable postscript (laser printer)
1987 — Mac II
1991 — PowerBook
1994 — PowerMac

Strategic and Management Issues:

- On January 22, 1984, during the third quarter of the Super Bowl, Apple aired its widely seen 60-second commercial introducing the Macintosh, which depicted the IBM world being shattered by a new machine. Initially, the Mac sold very well.
- Did not allow any company but Apple to manufacture Apple computers until 1996.

Tools for the Technology Manager

8

Finance

Kenneth J. Purfey
Executive Financial Consultant

Robert G. Beaves
Robert Morris College

S. Bhimjee
San Francisco State University

James E. Hodder
University of Wisconsin

Robert Keeley
University of Colorado — Colorado Springs

Donald E. Vaughn
Southern Illinois University

Jules J. Schwartz
Boston University

Peter K. Clark
University of California

Gordon Brown
Digital Image Technology

Kevin P. Prykull
Senstar Capital Corporation

Harold E. Marshall
National Institute of Standards and Technology

Wolter J. Fabrycky
Virginia Polytechnic Institute & State University

Benjamin S. Blanchard
Virginia Polytechnic Institute & State University

Walter D. Short
National Renewable Energy Laboratory

Chris Hendrickson
Carnegie Mellon University

Sue McNeil
Carnegie Mellon University

0-8493-8577-6/99/$0.00+$.50
© 1999 by CRC Press LLC

8.1   Cash Flow and Financial Management

Kenneth J. Purfey

The terms "**cash**" and "**cash flow**" mean many things to many people. Their interpretations depend on both the user's requirements and the audience for which the discussion is intended. There are perhaps over a dozen terms and management tools describing the various types and uses of cash flow. Most of these are discussed or defined from both the accounting and the financial viewpoint. They are

Cash	Cash conversion cycle
Cash equivalent	Cash flow estimate
Cash flow	Cash management
Cash flow before tax	Cash budget
Cash flow after tax	Operating cash flow
Free cash flow	Incremental cash flow
Statement of cash flow	

Accountants have strict definitions for cash and cash flow as the Financial Accounting Standards Board (FASB) has promulgated several rules and regulations for the definition and measurement of these items. Financial managers sometimes carry the nomenclature further by developing specific ways of measuring cash flow that are more useful to their needs.

By cash flows we mean the actual amount of cash to be received or paid. This is not the same as Net Income or earnings in the accounting sense. Thus, there is a fundamental difference between accounting uses for cash flow and financial uses.

"Cash" does not mean "cash flow". The accounting system is not primarily designed to report the inflow and outflow of cash, whereas the financial system is designed to measure and evaluate these type of cash flows. Cash flow is the financial lifeblood of the firm, flowing to all parts of the corpus, ensuring the firm's overall health and well-being. Cash flow is critically important to the firm; more so than cash, because cash flow is both theoretically correct and unambiguous. Also, one cannot spend Net Income because Net Income is not cash. Net Income is a function of accounting rules involving depreciation, deferred taxes, inventory methods, and other noncash charges. Thus we will attempt to define and differentiate among the dozen or so meanings and interpretations of cash and cash flow with emphasis on the financial tools.

Cash and Cash Equivalents

Cash is the most liquid of all assets. It is used as the medium of exchange and the basis for measuring and accounting for all other items within the firm. **Cash** includes coins and currency, balances on deposit with financial institutions, petty cash, and certain negotiable instruments accepted by financial institutions for immediate deposit and withdrawal. These negotiable instruments include ordinary checks, cashier's checks, certified checks, and money orders. On the other hand, **cash equivalents** are items similar to cash but not accounted for or classified as cash. They include treasury bills, commercial paper, and money market funds. Cash equivalents are very near cash but are not in negotiable form. Cash also excludes postage stamps, travel advances to employees, receivables from employees, and cash advances to employees or outside parties [Dyckman et al., 1995]. From an accounting aspect, cash and cash equivalents are reported on the Balance Sheet, but they are not normally used in financial analysis. Financial managers use other tools such as cash flow, **operating cash flow**, and **free cash flow** to serve their requirements.

Cash Flow

Cash flow is the actual amount of cash coming into a firm (cash inflow) or paid out by a firm (cash outflow). Cash flow is measured over some regular time period, often 1 year. Cash, on the other hand, is the amount of liquid assets on hand at one particular moment in time, i.e., a balance sheet date.

Cash flow is not the same as earnings or Net Income and is quite different. While accounting statements prepared in accordance with Generally Accepted Accounting Principles (GAAP) are oriented toward recording what happened in the past (Net Income), the financial manager is concerned with what will happen in the future (cash flow), and thus it is cash inflows and cash outflows that are more important.

The simplest definition and calculation for cash flow is

$$\text{Cash flow} = \text{net income} + \text{depreciation} + \text{depletion} + \text{amortization}$$

Within the terms depreciation, depletion, and amortization (DD&A), depreciation refers to the allowance for normal wear and tear on fixed assets; depletion refers to the allowance for using up of natural resources and minerals, and amortization refers to the obsolescence of leasehold improvements. While the purchase of fixed assets, natural resources, or leasehold improvements all use up cash at the time of original purchase (time 0), the accounting DD&A costs related to them are all considered to be *noncash* items beyond time 0 because no additional cash actually leaves the company. These are sometimes referred to as noncash charges. Thus though DD&A charges may continue for many years for accounting purposes, they are irrelevant for financial management purposes and are therefore added back to any accounting results or Net Income. Thus, for both accountants and financial managers, the traditional term cash flow refers to Net Income plus DD&A.

The bedrock upon which all financial management is based is the management of cash flows so as to maximize the value of the firm. Cash flow is what makes firms more valuable or less valuable. Theoretically, the value of the firm is determined by the (1) magnitude, (2) timing, and (3) riskiness of future cash flows. In other words, in analyzing new projects or new businesses, the financial manager must address the critical cash flow questions of how much, how soon, and with what risk are project cash flows acceptable. Traditional net present value (NPV) techniques address the answers to these questions. However, traditional NPV techniques require additional descriptions and definitions of the numerous types of cash flows required by a financial analyst.

Cash Flow in Financial Analysis

While adding back DD&A to Net Income gives the financial manager a better picture of the cash flow and financial health of the firm, it is not particularly useful when dealing with project evaluations, new business evaluations, or mergers and acquisitions. Takeover artists, LBO (leveraged buyout) specialists, merger and acquisitions professionals, and business/project appraisers often use operating cash flow instead. **Operating cash flow** is cash flow before interest and taxes. It can be calculated as follows:

$$\text{Operating cash flow} = \text{cash flow} + \text{interest expense} + \text{income tax expense}$$

By adding back interest expense and taxes, the financial professional can see how much cash flow is available to service the firm's debt — which is often increased substantially in a takeover or LBO. The business valuation appraiser sometimes uses various multiples of operating cash flow to estimate the value of the business. This helps a potential buyer ascertain what is, or what is not, a fair purchase price.

Some merger and acquisition professionals go one step further to determine free cash flow. **Free cash flow** is the cash flow remaining after paying all capital expenditures and debt principal. It can be defined and calculated as follows:

Free cash flow = cash flow − capital expenditures − debt principal repayment

By focusing on the discretionary cash flow left over after maintaining the firm's capital plant and servicing debt, the financial manager has a better picture of the firm's financial flexibility. Flexibility is important because free cash flow surpluses can be used to repurchase stock, pay dividends, expand, acquire other businesses, pay debts, or invest in securities. If free cash flow is negative, the deficiency must be made up with additional debt or equity financing [Dyckman, Dukes, and Davis 1995]. Therefore, free cash flow is perhaps one of the most useful management tools available, especially for the project evaluator or the chief financial officer.

However, typically the project evaluator needs three additional tools to properly evaluate a project: **incremental cash flow, cash flow before tax** (CFBT), and **cash flow after tax** (CFAT). Without them the project's financial attractiveness cannot be determined, especially when using traditional NPV techniques. **Incremental cash flow** for a project is defined as the difference in the firm's cash flows for each period if the project is undertaken vs. if it is not undertaken. It is the change in the value of the firm attributable to the project, holding everything else constant [Brigham and Gapenski 1994]. All other costs are irrelevant for project evaluation. CFBT is the cash flow that occurs while the project or asset is in operation over its economic life. CFBT is calculated as the project's revenues (inflows) less the project's operating costs (outflows), usually forecasted on an annual basis. CFAT removes the associated taxes, if any. CFAT is important because only after-tax cash flows are available to the firm. A simple example may help to clarify this.

Assume a firm wishes to financially justify a new printing press costing $7500 that is intended to penetrate new markets not currently served by the firm. Thus all cash flows will be incremental to current operations. It is estimated that the press will last 10 years and will generate new (incremental) revenues of $5000 annually. However the press will require operating and maintenance (O&M) expenses of $2000 annually. Thus, the CFBT will be $3000 ($5000 revenues − $2000 O&M) annually. Since the $3000 CFBT is a net addition (increase) to the firm's taxable income, it is subject to income taxes. Fortunately, the IRS allows taxpayers a tax deduction for depreciation. Assuming tax depreciation was $750 per year, then the taxable cash flow would be $2250 ($3000 − $750). If the firm's tax rate was 40%, taxes on $2250 would be $900 ($2250 × 40%). The resulting Net Income from the project would be $1350 ($2250 − $900), and the CFAT would be $2100 ($1350 + $750). Depreciation of $750 is added back because we need CFAT, not net income (remember depreciation is not cash). Then using traditional NPV analysis, the financial analyst must calculate whether the present value of $2,100 annually taken over the printing press' useful life of 10 years is greater than the $7500 original cost. Thus, if the NPV is positive (benefits greater than costs) the firm would accept the project. Otherwise, the firm would reject the project.

One of the most critical responsibilities of the project analyst is to obtain forecasts of all the relevant incremental annual cash flows so as to arrive at reasonable estimates for CFBT, CFAT, and DD&A. Therefore, other tools, definitions, and cash flow management concepts are required.

Other Cash Flow Management Concepts

The most important step in analyzing a project is performing the **cash flow estimation**, which involves forecasting the cash flows (CFBT and CFAT) after the project goes into operation. The financial manager must obtain many variables from a variety of sources and thus must work closely with all departments within the firm. For example, forecasts of prices and sales are normally made by the marketing department. Similarly, the original capital expenditure estimate is generally obtained from the engineering department, while operating costs are estimated by cost accountants, production experts, purchasing agents, and others.

Cost accountants, production experts, engineers, and others who make cash forecasts should also be knowledgeable about the firm's cash conversion cycle. The **cash conversion cycle** is the net elapsed time interval of the flow of cash through the firm. It measures how many days it takes the firm to purchase its raw materials, produce its goods, sell its products, and collect its cash [Pinches, 1996]. Generally, the shorter time the better, because it makes the firm more liquid. Many firms have cash conversion cycles of 45 to 75 days.

The cash conversion cycle information is useful to have when constructing a cash budget. A **cash budget** is a detailed forecast of all expected cash inflows and cash outflows by the firm for some period of time. Typically this is performed monthly, quarterly, and/or annually. Cash budgets are important for two reasons. First, they alert the firm to future cash needs and second, they provide a standard budget against which subsequent performance can be judged. Thus, the cash budget is a forward-looking document useful to the financial manager. A document that is similar but backward looking and therefore useful to the cost accountant and investment community is the **statement of cash flows**. The **statement of cash flows** is a financial document that categorizes and reports the actual cash receipts, cash payments, and net change in cash resulting from the ongoing operating, investing, and financing activities of the firm on a historical basis [Kieso and Weygandt, 1989].

Most of the cash and cash flow tools and techniques mentioned here make up a major portion of the management function called cash management. **Cash management** refers to the activities performed by management designed to optimize the risk/return trade-off for liquid assets [Pinches, 1992]. This activity involves the following:

1. Having enough cash and liquid reserves to meet all the firm's obligations as they come due and take advantage of growth opportunities.
2. Not holding excess liquid reserves because investment in long-term projects generally provides higher returns than short-term investments.
3. Maintaining a minimum cash balance while actively managing the firm's portfolio of marketable securities to ensure as high return as possible commensurate with the risk involved.

These trade-offs are critically important for the health of the firm and guide financial managers in the cash flow generation, measurement, estimating, forecasting, budgeting, evaluation, and shareholder value optimization decision-making process. Thus, cash management is a discipline involving many different management tools requiring substantial financial expertise to implement successfully.

Summary

The effective management of cash is an important function for the financial manager. There are at least a dozen ways of looking at cash and cash flow, and each depends on the function of the analyst and the audience for which the information is intended. Effective management of the firm's cash and cash flow requires understanding the techniques for cash gathering, disbursement, and measurement. How the firm's cash balance can be minimized while maximizing cash flow is a juggling act requiring an understanding of risks and returns.

The project analyst requires special tools and techniques involving sophisticated financial evaluation and projection methods. The tools are quite different for accountants who are normally only concerned with historical costs.

Nevertheless, whether an accountant or financial manager, cash flow is the lifeblood of the corporation, without which the firm would wither away and die. This is why most financial managers pay more attention to cash flow than to cash itself, thus correctly choosing to "go with the flow".

Defining Terms

Cash: Coins and currency, balances on deposit with financial institutions, petty cash, and certain negotiable instruments accepted by financial institutions for immediate deposit and withdrawal.

Cash budget: A detailed forecast of all the firm's expected cash inflows and cash outflows to obtain a net amount. Net shortfalls indicate a cash shortage, while net excesses indicate a surplus.

Cash conversion cycle: The time (usually measured in days) it takes for a dollar to flow through the firm, starting with the purchase of raw materials and ending when the firm collects its money on the sale of finished goods.

Cash equivalents: Items similar to cash, but not cash. They include treasury bills, commercial paper, and money market funds.

Cash flow: The actual amount of cash coming into or going out of a firm; not the same as Net Income. Cash flow is usually calculated as Net Income plus DD&A.

Cash flow after tax (CFAT):   Usually used in project economics, refers to the cash flow remaining after project revenues and expenses have been considered, taxes removed, and DD&A added back.

Cash flow before tax (CFBT):   Usually used in project economics, refers to the cash flow remaining after project revenues and expenses have been considered.

Cash flow estimation: The art and science of forecasting future cash flows, usually involving the input of many different skills and management personnel.

Cash management: The activities performed, usually by the treasurer or chief financial officer of a firm, designed to minimize cash, maximize cash flow, and optimize the risk/return trade-off for the firm's liquid assets.

Free cash flow: The discretionary cash flow left after paying for the firm's capital expenditures and servicing debt. This measure gives the financial manager an indication of the cash flows available to do other things, such as pay debts, repurchase stock, buy other businesses, etc.

Incremental cash flow: The difference in cash flows caused by a project, usually measured as the change in value of the firm before and after implementation of the project.

Operating cash flow: Cash flow before interest and taxes; useful when valuing businesses.

Statement of cash flow: A primary financial statement used by accountants to measure and categorize cash flows occurring within the operational, financial, and investing activities of the firm. It is a backward-looking historical document ascertaining where the firm's cash flow has been.

References

Brigham, E. F. and Gapenski, L. C. *Financial Management Theory and Practice, 7th ed.*, p. 423, Dryden Press, Orlando, FL, 1994.

Dyckman, T. R., Dukes, R. E., and Davis, C. J. *Intermediate Accounting, 3rd ed.*, pp. 319 and 1185, Irwin, Chicago, 1995.

Kieso, D. E. and Weygandt, J. J. *Intermediate Accounting, 6th ed.*, p. 1174, John Wiley & Sons, New York, 1989.

Pinches, G. E. *Essentials of Financial Management, 5th ed.*, p. 444, HarperCollins, New York, 1996.

Further Information

American Institute of Certified Public Accountants, New York.

Block, S. B. and Hirt, G. A. *Foundations of Financial Management, 7th ed.*, Irwin, Burr Ridge, IL, 1994.

Gitman, L. J. *Foundations of Managerial Finance, 4th ed.*, HarperCollins, New York, 1995.

Jones, C. P. *Investments Analysis and Management, 5th ed.*, John Wiley & Sons, New York, 1996.

Pinches, G. E. *Essentials of Financial Management, 4th ed.*, HarperCollins, New York, 1992.

Institute of Certified Management Accountants, Montvale, NJ.

8.2   Asset Valuation Methods

Robert G. Beaves

Like beauty, value is largely "in the eye of the beholder". An asset's value as a potential investment depends upon one's perception of (1) the future cash flows that it is expected to provide, (2) the risks involved in generating those cash flows, and (3) the appropriate rate of return to require in light of those risks. Although the concepts underlying the valuation process are relatively simple, practical application of those concepts is not.

The Basic Valuation Principle

This section presents the process generally used to determine the investment value of an asset. An asset is a resource purchased and owned with the expectation that it will generate positive net cash flows and, in turn, positive returns on investment. Assets may be tangible, such as land, buildings or equipment, or they may be intangible, such as trade secrets, licenses, patents, goodwill, a marketing program, or common stock. An asset's investment value (hereafter "value") is based solely on financial considerations, in particular, the effect that having the use of this asset will have on the investor's future cash flows. Psychic considerations such as "pride of ownership" or "image enhancement" are irrelevant beyond any contribution to expected cash flows.

As a basic principle, the value of an asset is equal to the sum of the present values of the future cash flows that that asset is expected to provide or

$$V_0 = \sum_{t=1}^{\infty} \left[\frac{C_t}{\left(1+k\right)^t} \right] \tag{8.1}$$

where V_0 = the value of the asset, C_t = the cash flow attributable to the asset during period t, and k = minimum acceptable rate of return required to compensate for the risk associated with this asset.

Because assets are usually evaluated over a finite investment horizon, Eq. (8.1) is often revised as follows:

$$V_0 = \sum_{t=1}^{n} \left[\frac{C_t}{\left(1+k\right)^t} \right] \tag{8.2}$$

where n = the number of periods in the finite investment horizon.

Relevant Cash Flows

The **relevant cash flows** for the purpose of determining the value of a particular asset have several key characteristics:

1. They are *net periodic* cash flows. All cash flows that occur during a period of time are netted by subtracting disbursements from receipts. By convention, these cash flows are assumed to occur at the end of the respective period.
2. They are *incremental* cash flows, which are calculated by comparing the investor's expected cash flows if the asset is purchased with the cash flows that would be expected were the asset not purchased. Increment cash flows represent the changes in the investor's expected cash flows that would occur if the asset where purchased and would not occur but for the purchase of that asset.

TABLE 8.1 Incremental Effects of New Factory — Year 1

Item Effected	W/O Factory	W/Factory	Δ in Item
Operating receipts	1,250,000	1,850,000	+600,000
Operating disbursements	500,000	725,000	+225,000
Depreciation	300,000	450,000	+150,000
Investment	50,000	50,000	0

TABLE 8.2 Incremental Effects and Relevant Cash Flows — New Factory

Year	ΔR	ΔE	ΔD	ΔI	C_t
1	+600,000	+225,000	+150,000	0	+285,000
2	+615,000	+230,000	+150,000	0	+291,000
3	+620,000	+235,000	+150,000	+100,000	+191,000
4	+635,000	+235,000	+165,000	0	+306,000
5	+645,000	+240,000	+165,000	-115,000	+424,000

3. They are *free* cash flows. Free cash flows include not only operating flows but also any future **investment flows** required to produce the operating cash flows over the asset's investment horizon.
4. They are *after-tax* cash flows. Like other disbursements, any tax liabilities associated with investment in the asset must be netted out in determining relevant cash flows.

The following equation can be used to determine the relevant cash flow associated with an asset for time period t:

$$C_t = \left(1-T\right)\left(\Delta R_t - \Delta E_t\right) + T\left(\Delta D_t\right) - \Delta I_t \tag{8.3}$$

where T = investor's marginal tax rate attributable to the asset, ΔR = incremental operating receipts attributable to the asset, ΔE = incremental operating disbursements attributable to the asset, ΔD = incremental depreciation expense attributable to the asset, and ΔI = incremental investment disbursements necessary to support operating flows.

As an example, consider Acme Manufacturing Corporation (AMC), which may purchase an existing factory for the purpose of increasing its production capacity. The projected incremental effects of this purchase during the first year subsequent to the purchase are calculated in Table 8.1. If the factory is purchased, Acme's operating receipts, disbursements, and depreciation will increase by amounts of $600,000, $225,000, and $150,000, respectively, while its investment disbursements would be unchanged. When given the fact that AMC's marginal tax rate (T) is 40%, we can calculate the period 1 relevant cash flow associated with buying the factory:

$$C_1 = 0.60(600,000 - 225,000) + 0.40(150,000) + 0 = \$285,000$$

In other words, AMC's after-tax cash flow will be $285,000 higher if it purchases the factory than it would have been had it forgone that purchase. The incremental effects of purchasing the factory over a 5-year investment horizon are presented in Table 8.2.

The relevant annual cash flows associated with purchasing the factory are also presented in Table 8.2. During year 3 there is an incremental investment disbursement representing expenditures necessary to produce the operating flows projected for subsequent years. The negative investment disbursement during year 5 represents a return of capital in the form of after- tax proceeds of liquidating the factory at that point in time. Note that the cash flows relevant for determining the asset's (factory's) value to the prospective investor (AMC) must not include a projected purchase price of the asset (factory) itself.

TABLE 8.3 Present Value of Relevant Cash Flows — New Factory

Year	C_t	$PV(C_t)$
1	+285,000	237,500
2	+291,000	202,083
3	+191,000	110,532
4	+306,000	147,569
5	+424,000	170,396
	Value	$868,080

The Minimum Acceptable Rate of Return

The minimum rate of return (MARR) required for investing in a particular asset can be viewed as having two component parts, a risk-free rate and a risk premium:

$$MARR = \left(risk - free\ rate\right) + \left(risk\ premium\right) \tag{8.4}$$

The risk-free rate is the rate that the investor can currently earn on investments that involve virtually no risk (a short-term, insured certificate of deposit or Treasury bill) and can be determined objectively by examining current financial market rates. The risk premium is the minimum additional return that this investor will accept as sufficient compensation for bearing the level of risk associated with the asset being evaluated. To determine the appropriate risk premium, one begins by assessing the amount of risk associated with investing in the asset under consideration. Risk exists if the actual cash flow that an asset will produce during a specific period may differ from the flow that the investor expects. Generally speaking, the wider the dispersion of possible cash flow values about the expected cash flow, the riskier the asset. Nevertheless, measuring risk is more an art than a science. Suffice it to say that reasonable investors may disagree as to the level of risk associated with a particular asset. Even if two investors were to agree as to the amount of risk, they may demand different risk premiums as compensation for bearing that risk. While all investors are assumed to be risk averse (i.e., they will bear additional risk, but only if compensated by higher expected returns), some investors are more risk averse than others and consequently demand higher risk premiums for bearing the same risk. Useful benchmarks for determining an appropriate MARR for a specific investment include the **opportunity cost** of that investment and the investor's **cost of capital**. In terms of benchmarking, the investor's MARR should equal its opportunity cost or its risk-adjusted cost of capital, whichever is greater.

Consider again the example of AMC and its potential investment in a factory. After careful analysis of alternative investments and of its cost of capital, AMC has decided that a risk premium of 13.0% is appropriate for investing in this factory. Adding this risk premium to the current risk-free rate, which AMC believes to be 7.0%, produces an MARR of 20.0% for investment in the factory. In Table 8.3, AMC's 20% MARR is applied to the relevant cash flows expected from the factory to determine their present values. By summing those present values it can be determined that the value of the new factory to AMC is $868,080. Although AMC will, of course, be willing to pay less than $868,080 for that factory, it should not pay more than $868,080.

Specialized Valuation Methods

Within specific industries "rules of thumb" may arise with respect to valuing assets. For example, the rule of thumb for valuing a commercial bank might be to multiply the bank's "book value" by a factor between 2.0 and 2.5. Real property, on the other hand, is often value on the basis of X dollars per square feet of space. P/E (price/earnings) multiples are often used in valuing common stock. To the extent that these rules of thumb provide valuations that are consistent with those provided by the basic valuation principle, they may be useful, time-saving devices. Nonetheless, as circumstances change, these rules are often slow to adjust and can provide significant "misinformation". Such valuation shortcuts should be employed only if one has determined their current relevance as compared to one's own subjective values based on the basic valuation principle.

Related Concepts

An asset can have many types of values, some of which are directly related to its investment value and some of which are not. Among these valuation concepts are

Market value — the price that a willing buyer has paid and a willing seller has accepted for an asset that is essentially the same as the one being evaluated. That price must be "timely" in the sense that the transaction has occurred recently enough that the price has current relevance. Market price represents the only "objective" value that an asset has. Current market price essentially sets a "floor" under an asset's investment value based on the assumption that one at least has the opportunity to sell the asset for current market value.

Purchase price — the price that the investor will have to pay to obtain the asset being evaluated.

Net present value — the difference between the investment value one assigns to an asset and the asset's purchase price. To the extent that an asset's investment value exceeds its purchase price, it increases the investor's value or wealth. Thus, assets with positive net present values should be purchased and those with negative net present values should not. For example, if AMC could purchase the existing factory for $750,000, that asset would offer a net present value of $118,080 (i.e., $868,080 − 750,000) and should be purchased.

Intrinsic value — the true value or the value "inherent" in the asset. Because value is subjective to the investor, an asset has no unique intrinsic value.

Book value — the value at which the asset is carried on the investor's balance sheet. Book value is generally calculated by subtracting the asset's accumulated depreciation from its historical cost. An asset's book value is important for tax purposes but is generally irrelevant for the purpose of determining the investment value or market value of an asset. For example, the fact that the factory that AMC may purchase is carried on its current owner's books at $450,000 should have no effect on AMC's determination of the factory's value.

Replacement value — the cost of creating the asset "from scratch". In the AMC example, replacement value of the factory would be the cost of building and equipping its own factory rather than buying an existing factory. Replacement cost should include any costs of the delay in putting the factory into operation as opposed to simply purchasing an existing facility. Replacement cost generally represents a "ceiling" on the purchase price the investor would be willing to pay.

Summary

The value of an asset to a particular investor depends on that investor's intended use of the asset, the cash flows that the investor expects the asset to provide in that use, the investor's perception of the risk associated with the asset, and the minimum rate of return required by the investor given that perceived level of risk. Thus, valuation is a subjective rather than an objective process. An asset's market value, when available, provides the only objective measure of that asset's worth. One should be wary of rules of thumb for valuing assets that may be popular within specific industries. Such valuation shortcuts should be used only when the investor is confident that the values that they generate are consistent with values obtained using the basic valuation principle. Although the basic valuation model may appear to be mechanical and straightforward, actual application of that model requires considerable use of judgment on the part of the investor. As is true with any quantitative model, the asset value generated is only as good as the information input into the valuation model. Valuation is as much an art as a science!

Defining Terms

Cost of capital: The investor's weighted average cost of long-term funds. This concept is sufficiently complex that it cannot be adequately developed within the confines of this section. Suffice it to say, the investor's cost of capital should be risk adjusted to conform with the level of risk involved in the asset being evaluated. See any financial management text for further information.

Investment flow: A cash flow resulting from the purchase or sale of long-term assets such as land, plant, and equipment or stock in a subsidiary corporation.

Opportunity cost: The highest rate of return offered by an alternative investment involving a similar level of risk. By selecting to invest in one asset, the investor forgoes the opportunity of investing in the best alternative and thereby incurs this cost.

Relevant cash flow: A cash flow that must be taken into account in determining an asset's value.

References

Damodarian, Aswath *Corporate Finance: Theory and Practice*, John Wiley & Sons, New York, 1997.

Gitman, Lawrence J. *Foundations of Managerial Finance*, 4th ed. Harper Collins College, New York, 1995.

Rao, R. K. S. *Financial Management: Concepts and Applications, 3rd ed.*, South-Western College Publishing, Cincinnati, OH, 1995.

Further Information

Most financial management and most investments textbooks address the issue of valuation.

8.3   Capital Budgeting

S. Bhimjee

Most businesses have a continuous planning process that lays down the missions, goals, and processes by which a business will achieve these goals. Different categories of budgets are developed for the implementation of these goals. The two main parts of a corporate budget are the operational budgets and the capital expenditure budget — each one representing the use of funds during the planning cycle. The capital budget represents a plan for implementing strategic decisions, such as expenditures for investments in buildings, equipment replacement, development of new products and services, research, new manufacturing plants, opening new outlets, and acquisitions of other businesses. All businesses have limited resources. Capital budgeting techniques are useful tools for making these decisions

Availability of Capital

A technology-based firm is constantly in need of funds to maintain a rapid pace of growth. Most start-up and high-growth companies pay little or no dividends. The income from operations is the source for internally generated investment capital. Investment capital may also be obtained from external sources by issuing additional stocks and bonds or through borrowings. The statement of cash flows, which forms a part of a company's financial statement, gives a historical view of the uses and sources of both internally and externally generated funds. However, the statement of cash flows does not reveal the future planning regarding the sources and uses of funds.

Projects

The need for investment funds is represented by a collection of funding requests for projects. These requests for funds may originate in departments or divisions or can be a part of strategic planning formulated by the board of directors. Projects include product development, such as a new model of a computer, development of new software, development of totally new innovative products such as Web TV or NetPC, introduction of on-line marketing through a Web site, opening of new stores, setting up new manufacturing plants, expansion of physical space, and acquisition of other companies through mergers or takeovers. Each project proposal includes a financial analysis giving a year-by-year cash inflows and outflows for the project.

Some project proposals list alternate ways of accomplishing a single objective. For example, a manufacturing facility may be located in the United States or off-shore. Only one of these options can be selected. These choices are "mutually exclusive" alternatives. In selecting the list of projects to fund, a company must not only make a decision about which projects to fund but also which option to select from the list of mutually exclusive alternatives.

The interrelationships between project proposals are also difficult to include. If two projects A and B are strongly interrelated then a new set of alternatives are shown: Choose project A only, choose project B only, or, choose both A and B. Such constraints are more difficult to include in simple selection models.

A technology-based company with a low return would have a low market capitalization value. If such a company has a few good products, they would be a target for takeover by another company. Takeovers and mergers are included in the set of strategic investment decisions.

Criteria for Selection of Projects

A variety of decision criteria are used for the selection of projects to be funded. Subjective criteria are often used. A project may be funded simply because it is a pet project of the board of directors. A business may wish to gain market share by sacrificing profits. However, from the viewpoint of long-term growth of a business, objective financial criteria are needed. Formal analysis techniques have a goal of maximizing the future wealth of the company and therefore the wealth of its stockholders. The stockholders' return is measured by a combination of the increase in the value of stock and the dividends received by stockholders.

Two common criteria used are maximizing present value and maximizing the weighted return on capital. When correctly applied, these two criteria will result in the selection of the same set of projects.

Project Cash Flows

Before an analysis for the selection of investment projects can begin, the cash flows for each project proposal must prepared using a uniform set of assumptions. An investment project has an outflow of cash for the first few years as capital expenditures are made. This is followed by a series of net cash inflows as revenues are earned from these investments. The common set of assumptions should include:

Are the cash flows based on constant dollars or do they include inflation?
Have the cash flows been adjusted for risk?
Are all cash flows based on the most likely outcomes, or do they represent an optimistic or pessimistic view of outcomes?
In case of multinational projects, are the assumptions of currency translation consistent?
Are all cash flows before income taxes or after income taxes?

A consistent set of assumptions is required for all proposals to provide a meaningful comparison.

Project Selection

Once the cash flows for each project have been consistently defined, we can begin the project selection process All unacceptable projects are first eliminated. Projects may be eliminated because of societal or political reasons. In eliminating such projects, the company is implicitly asserting that the quantitative cash flow does not represent total consequences that may result from the acceptance of such a project. If such adverse outcome were to be included, the resulting cash flow would be quite different. For example, the cash flows for a project may not include the environmental and political consequences that could impact the long-term future of the company. Thus, no pharmaceutical company in the United States was willing to undertake the manufacture or distribution of the birth control pill RU52 because of potential backlash and boycott from a group or customer. Projects are unacceptable if they do not yield a minimum acceptable return established by the company policy.

Minimum Acceptable Return

What is a minimum acceptable return? This varies from industry to industry and from one business to another. In a perfect capital market, the stock prices adjust themselves to account for differences in risks and returns. Therefore, if a company such as Microsoft has a consistently high annual growth rate of 50%, its price to earnings ratio is also high. Because an investor pays a very high earnings multiple, the return to the stockholder is evened out unless the company continues the high growth rate. However, to maintain an average annual growth rate of 50%, the projects undertaken by Microsoft would have to yield a return of more than 50%. Thus, Microsoft would be rejecting most proposals that have a return less that 50%. High return expectations are the norm for new technology-based companies. On the other hand, a mature company may be willing to accept projects with a return of only 15% and its price earnings ratio would be in the neighborhood of 10.

Cash Flows, Present Value, and Internal Rate of Return (IRR)

The most frequently used criterion for choosing investment projects is to maximize the present value of the set of projects selected. It is quite simple to compute the internal rate of return and present value based on cashflow for a project. A simple example, created using Microsoft Excel, is shown in Table 8.4. To construct this graph, the value of (cash flow at $t = 0$) + (NPV cash flow for $t = 1$ to 4) is plotted for different interest rates. The point at which the graph crosses the x-axis (45%) is the return on this cash flow (Fig. 8.1).

TABLE 8.4

t	Cash Flow ($)
0	3,000
1	1,750
2	1,750
3	1,750
4	1,750

The present value of the cash flow at any interest rate can be simply read off on the y-axis. To be acceptable, the present value of each individual project must be greater than zero at the minimum acceptable rate of return. The set of proposals that maximizes the total present value is then selected.

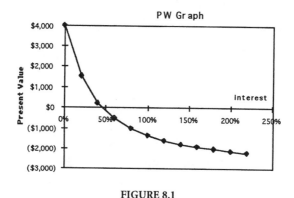

FIGURE 8.1

Ranking Proposals on IRR

Another method, which is very easy to apply, is to rank the proposals on their IRR. Once the proposals are ranked, we first choose the one with the highest return and work down the list until all available internal funds are used up. If projects with high return are still unfunded, then management needs to make a decision whether to use external funding for these remaining projects. The calculations of the

TABLE 8.5

Project	Investment ($)	Return (%)
A	100	55
B	200	48
C	80	40
D	120	30
E	300	25
F	250	20

Return

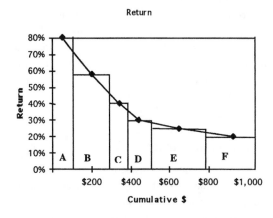

FIGURE 8.2

IRR from the cash flows are again very simple using a program such as Microsoft Excel, which provides a built-in IRR function. A list of proposal is shown in Table 8.5. Figure 8.2 shows the cumulative funds used up by these projects.

Thus, if $400 million were available, then projects A, B, and C would be funded and $20 million would be left over. The selection of a set of proposals may not exactly use up all the funds, and the "lumpiness" of investment proposals may not give a mathematically optimal solution. However, the selection of proposals obtained by using the above method will in most cases give reasonably good solutions. Management can also easily grasp which projects are being selected and the funds being used.

Capital Rationing and Mathematical Programming Solutions

The limited availability of capital is known as captial rationing. To obtain optimal solutions under capital rationing situation, many mathematical programming models are available. These models resolve the difficulty of choosing an optimal set of projects when many projects are being evaluated and we wish to choose the best set taking into account the lumpiness effect.

Decentralizing Capital Budgeting Decision Making

Decentralizing capital budgeting decisions may obviously lead to different criteria and cutoffs being used by each department and division. This may seem to reduce the overall growth rate. When a company becomes too large, decentralized budgets with each department or division making its own capital investment decision can effectively increase the growth rate of a company. In a technology-based company, the need to make quick decisions is very important. The flexibility of more rapid decision making in a decentralized environment may lead to timely selection of investment projects. A combination of centralized decision for projects exceeding a preset limit and decentralized decision for projects less than the limit is often a good compromise to maximize growth.

Postaudits

Many companies neglect the postaudit of the projects that were implemented. To gain experience in selection of projects and to assess the weaknesses in either cash flow forecast or the use of an acceptable growth rate, a postaudit comparison of the project's financial outcome with the original project proposal is necessary. The results of the postaudit should be given to all those who are involved in initiating, funding, implementing, and operating the project.

Defining Terms

Internal rate of return (IRR):   It is the return earned by an individual project. It is the value of the interest rate at which the present value of a cash flow is zero.

Minimum acceptable return: This is the cutoff set by management. Projects with returns lower than the minimum accpetable return are rejected without further analysis.

Net present value (NPV):   It gives the amount of cash today that would be equivalent to the series of cash inflows and outflows of a project. This equivalent amount is computed using a specific value of a discount rate or interest rate, usually the minimum acceptable return.

Price-earnings ratio: The current market value of one share divided by the sum of the previous four quarterly earnings per share.

References

Northcott, D. *Capital Investment Decision-Making,* Academic Press, Orlando, FL, 1992. This small book is excellent for giving a complete view of capital budgeting. The appendix has a well-designed sample project proposal form.

Johnson, H. *Strategic Capital Budgeting,* Irwin, 1994.

8.4   Risk Management and Assessment

James E. Hodder

The term "risk management" means different things to different people. For example, the insurance industry tends to refer to risk management as the use (or not) of insurance contracts to deal with potential physical or legal hazards such as fire, earthquake, workplace accidents, product liability, or environmental pollution. In this section, we will focus on risk management from a finance perspective. In that context, risk management (sometimes called financial risk management) refers to the assessment and management of a firm's exposure to financial risks such as movements in input or output prices, interest rates, or exchange rates (for example, see Smithson et al. [1995] or Jorion and Khoury [1996]).

Sometimes these risks are mitigated (managed) via supply contracts that fix the quantity and price of particular inputs or outputs over an extended time horizon. However, such contracts are frequently unavailable or have unattractive pricing. In such cases, it may be possible to hedge (manage) a particular risk by taking a position in a financial instrument (e.g., **forward contract**) such that a loss on the underlying risk exposure is offset by a gain on the financial instrument.

Helping companies assess and manage financial risks is a becoming a significant business for large banks as well as consulting firms. Although hedging in agricultural commodities and in foreign exchange has been around for a long time, major commercial banks and investment firms have dramatically extended and enhanced risk management techniques over the last decade. Initially, their motive was self-preservation. Subsequently, some financial institutions discovered that superior risk management capabilities could be used as a competitive advantage. More recently, regulators have mandated the use of relatively sophisticated risk management and reporting systems at large banks and investment firms. Recognizing the potential for this technology to be adapted for use by nonfinancial firms, several large banks as well as a number of consulting firms have gotten into the "risk advisory" business.

Risk Assessment

Fundamentally, risk assessment is an identification and measurement problem, i.e., to what risk(s) is a firm exposed and to what extent. Logically, the first step in risk assessment is to specify an objective or performance measure about which one is concerned. Hypothetically, someone from marketing might be concerned about total sales revenue. From a broader perspective, one might focus on profit or cash flow at the project, division, or firm level.

Suppose we focus on the profit from a particular transaction. We can then address the sensitivity of that measure to a change in the value of a particular risk factor. For example, suppose the firm has an account receivable of 500 million yen from a recent sale. The dollar value of the firm's profit on that sale is at risk due to a possible movement in the dollar/yen exchange rate prior to receiving the yen payment and converting it into dollars. If the value of the yen declines by 0.001 $/¥, then the value of that receivable declines by $500,000 — as does the firm's profit. Symmetrically, an increase in the value of the yen by 0.001 $/¥ results in $500,000 additional profit. Moreover, if we wrote down a mathematical expression for the profit in this example, we would see that it is a linear function of the dollar/yen exchange rate. Conceptually, it is useful to think of the sensitivity of profit (or an alternative performance measure) to a change in some risk factor (e.g., dollar/yen exchange rate) as the mathematical derivative of such an expression.

In the above example, profit was a linear function of the exchange rate; however, nonlinear functions can easily arise. Suppose, for example, a firm exports to a Japanese trading company using a dollar invoice price and with payment in dollars. The exchange rate is still a risk factor if the quantity sold depends on the effective yen price at the time of sale. Typically, a decrease in the yen's value results in a smaller quantity sold since the effective price in yen is higher (vice versa for an increase in the yen's value). In such situations, profit will generally be a nonlinear function of the exchange rate at the time of sale. We can still think of the sensitivity of profit to the exchange rate as a mathematical derivative; however, it is no longer a constant, which has important hedging implications as we shall see below.

The next step toward realism is to recognize that there are usually multiple risk factors. The firm may be selling in multiple markets. Prices for (potentially several) inputs may be random. Perhaps, some input or output prices are best represented as functions of multiple random variables (risk factors). Conceptually, we should now thing of the sensitivity to a particular risk factor as a partial derivative (of either a linear or nonlinear function).

Thus far, we have implicitly been talking about profit (or some other performance measure) at a particular point in time. However, profits or cash flows will typically occur at multiple times in the future. One approach is to estimate sensitivities for each relevant point in time. As a practical matter, a relatively small number of time points may be chosen (e.g., quarterly) with nearby profits lumped at the closest time point. Assuming multiple risk factors, we would have a matrix of sensitivities with time along one dimension and the risk factors along the other.

The alternative approach for dealing with multiperiod risks is to focus on a performance measure that already takes into account multiperiod profits or cash flows. For example, the **net present value** (NPV) of a project represents the current value of both present and expected future cash flows. One can then think in terms of that NPV's sensitivity to changes in next quarter's dollar/yen exchange rate (as well as other risk factors). This results in a single vector of sensitivities; however, it implicitly discards information about cash flow timing that may be important for risk management. Also, as time passes and some cash flows are realized, the sensitivities to various risk factors will usually change, perhaps dramatically (even switching sign). This means that the sensitivities will need to be adjusted as time passes.

Thus far, we have acted as if we can construct a precise mathematical function with which to describe the relevant performance measure (e.g., profit) as a function of the specified risk factors. It is this function that we would differentiate to obtain the sensitivities to the risk factors. The reality is usually much more complex; and with some exceptions (such as our initial account receivable example), it is typically not possible to specify a precisely accurate function. It is still useful to think of risk factor sensitivities as partial derivatives; however, we will need to obtain numerical values for those sensitivities by some method other than mathematical differentiation. A standard approach is to estimate the sensitivities using a statistical regression with historical data.

In this approach, the dependent variable is the relevant performance measure (e.g., profit), and the independent variables are the specified risk factors. Virtually always, the estimation technique is for a linear regression; and the estimated regression coefficients are the sensitivities for the risk factors. There are several caveats with this approach. First, the true underlying relationship may well be nonlinear; and the regression represents a linear approximation. Moreover, the sensitivities will generally be estimated with error even if the true underlying relationship is linear. Furthermore, it is necessary to have historic data to perform the estimation. This is not usually a problem for risk factors such as exchange rates, interest rates, or commodity prices. However, historical data will not be available for the price of a newly created part or the NPV and cash flows of a new investment project. One can sometimes use proxies (e.g., a similar previous project or a similar existing part) to obtain historical data, but this will tend to introduce additional error in the sensitivity estimates.

An essentially equivalent approach involves estimating the statistical variances and covariances (or correlations) between all of the risk factors and the performance measure using historical data. Not surprisingly, the caveats for the regression approach also apply here.

A rather different alternative for obtaining sensitivity estimates is to build a mathematical model for the project (or division or firm) which allows for changes over time in pricing decisions, production, etc. in response to changes in the risk factors. This may be a very simplified model, or it may be rather detailed and complex. The model builder will need to consider issues such as how product pricing will respond to a change in, for example, the dollar/yen exchange rate. In turn, the price response will influence sales volume and production, with consequent changes in input requirements, etc. With such a model in hand, one can then calculate net cash flow, profit, etc. at different exchange rates and estimate a sensitivity to that risk factor.

With multiple risk factors, this modeling approach becomes more complicated since one has to consider simultaneous changes and correlations across the risk factors. A typical procedure is to use a multivariate statistical simulation to generate a time series of risk factor values. At each relevant point in time, the pricing, production, etc. outcomes are determined based on the risk factor values for that time point. Then the net cash flow, profit, etc. are calculated. By using a large number of replications (e.g., 10,000), probabilistic estimates of the desired sensitivities can be obtained.

Although this simulation technique is rather different from the regression approach mentioned above, there still tends to be a dependence on historical data for estimating the correlations and other statistical distribution characteristics of the risk factors. Under these circumstances, there is an implicit assumption that the statistical characteristics of the risk factors will remain the same in the future as during the historical estimation period. On the other hand, the project is being explicitly modeled (albeit approximately) rather than using a previous project as a proxy; and nonlinear relationships are included. In this sense, the simulation approach has a benefit relative to the regression procedure for estimating risk sensitivities. The attendant cost is the effort required to build and simulate the model. In any case, one should recognize that the estimated sensitivities from all these approaches may be imprecise and should be used with some caution.

Risk Management

Let's assume the performance measure of concern (e.g., profit) has been determined, the relevant risk factors (exchange rates, prices, etc.) identified, and their respective sensitivities estimated. The issue now is altering those sensitivities in desirable ways — hopefully, keeping in mind that the sensitivity estimates may be imprecise.

One possibility is to change the firm's operating procedures in ways that mitigate the sensitivity to one or more risk factors. An example would be the use of a long-term contract to fix the price for a substantial portion of a key input. To maintain some output flexibility, one might precommit to only a portion of the (random) input requirement and plan to purchase the rest in the future. Alternatively, one might change a pricing policy to provide more production stability. The firm might decide that a particular product or market was too risky and withdraw from that business.

The firm might invest in more flexible manufacturing facilities or larger inventories. It might decide to invest in a foreign production facility in order to have more foreign currency production costs as an offset to foreign currency sales revenue. One might even structure a capability for shifting a portion of production to the lowest-cost plant in response to exchange rate or local price movements. This requires excess capacity on a global basis, but it might be a worthwhile investment. A benefit of the simulation approach in risk assessment is that it provides a ready mechanism for examining the implications and desirability of the sort of operational hedges just mentioned.

Alternatively, the firm can use financial hedges to alter the sensitivity to various risk factors. Most texts on options or international finance have chapters on hedging (for example, see Chapters 4 and 14 in Hull [1995]). Frequently, financial hedging is less costly and/or more flexible than operational hedges; however, both have their merits and should be considered. Before delving further into hedging techniques, it is important to consider whether the desired objective is to reduce total volatility or only to reduce downside risk.

To illustrate this issue, consider our earlier account receivable example. That yen receivable could be hedged by selling a ¥500 million forward contract with a maturity matching the receivable collection date. A decrease in the value of the yen results in a loss on the receivable, but this is now offset by a gain on the firm's forward position. Symmetrically, an increase in the yen results in a gain on the receivable, which is offset by a loss on the forward position. There is no explicit price for this hedge in the sense that no initial payment is required for the forward contract. However, there is an implicit cost in that the firm forgoes possible profits from an increase in the yen's value.

An alternative hedging approach would be to use a ¥500 million currency **option** instead of the forward contract. By using an option, the firm could structure a hedge that offsets any possible loss on the receivable while keeping the potential gain (if the yen increased in value). In other words, the firm can eliminate the downside risk without forgoing the upside potential. However, there is a explicit cost for this protection represented by the option's price; and this can be thought of as analogous to an insurance premium.

To summarize, one can create a hedge that trades upside potential for reduced downside risk. Alternatively, one can pay an explicit price for reducing downside risk while retaining the upside potential. Combinations of these fundamental approaches are clearly possible. It should also be clear that one can choose to "partially hedge" by reducing a risk sensitivity without eliminating it entirely. This might be done to minimize hedging costs or to preserve more upside potential. Recalling the likely imprecision of our risk assessment process, the notion of completely hedging a risk should typically be viewed as an abstraction.

There are a host of financial instruments used for hedging (risk management) purposes, and investment banks frequently create new ones. Fundamentally, these instruments fall into two broad categories: those with linear payoffs and those with nonlinear payoffs. Frequently, these categories are often referred to as linear and nonlinear securities. A forward contract is an example of a linear security, whereas an option is in the nonlinear category.

To illustrate this, consider a forward contract to purchase yen at a future date. Let S_T denote the value of the "underlying asset" (yen) at time T (the contract maturity date). The yen's forward price (F) is specified now, although the purchase will not take place until time T. The size of the contract (number of yen to be purchased) is also specified initially. At the maturity date, the payoff for the forward contract will be $S_T - F$ times the contract size. Hence at maturity, the payoff increases linearly with S_T (the value of the underlying security).

Now consider a "call option" on yen, which gives the right (but not obligation) to purchase yen for a prespecified "exercise price" at time T. Letting X denote the exercise price, the payoff at maturity for this option is Max $(0, S_T - X)$. Since the option's owner is not obliged to purchase yen at the exercise price, it is only rational to do so if the yen exchange rate is above X. Otherwise the option is worthless, and hence the nonlinear payoff. It is such a nonlinearity that is the key to creating a hedge that reduces only downside risk. Numerous other types of options exist, including ones with rather complex payoff structures. However, some nonlinearity in the payoff is a fundamental characteristic of all options. Consequently, one can view options as tools for making nonlinear adjustments in a firm's risk exposure.

Considering multiple risk factors suggests using multiple hedge positions (linear or nonlinear). The basic approach is to use a hedging instrument that matches the risk factor, e.g., a yen forward contract to hedge yen currency risk. If you think that a hedge instrument to match your risk factor doesn't exist, consider calling a major investment bank. They will almost certainly be willing to create the instrument you want — for a fee. This can, however, be a reasonable alternative if your hedging demand is large. Effectively, the investment bank is being paid to manage your risk; and they have developed considerable expertise in risk management.

Alternatively, a "cross hedge" will sometimes be employed where the hedging instrument does not perfectly match the risk factor. For example, a Deutsche mark contract might be used to hedge a Dutch guilder risk exposure. Since the values of these two currencies are very highly correlated, the cross hedge may work well; however, there is the additional risk of an unusual circumstance that causes the hedge to perform poorly.

In a similar vein, multiperiod risks can be addressed using hedge instruments whose maturities match the risk exposures. For example, a sensitivity to yen at the 6-month horizon is adjusted with a 6-month yen forward, while a 2-year French franc forward is used for hedging an anticipated cash flow in francs 2 years from now. Recall that, when discussing risk assessment, we talked about developing a matrix of sensitivities designated by time and risk factor. In an idealized world, one would adjust each of those sensitivities using a hedging instrument that matched both the risk factor and the timing (maturity) of that sensitivity; however, this may be impractical.

We have already mentioned the possibility of using a cross hedge where the hedge instrument is highly correlated with the risk factor, but not a perfect match. There is an analogy in the time dimension, where an instrument with one maturity is used to hedge a risk sensitivity with a different maturity. This will result in a need to adjust the hedge over time. For example, a 6-month forward would need to be "rolled over" once in order to match a risk exposure at the 1-year horizon. This brings us to a discussion of dynamic hedging.

Dynamic hedging refers to a procedure where hedges are adjusted over time. This may be necessary or desirable for several reasons. We just mentioned rolling a shorter-term hedge instrument to match a longer-term exposure. To be more precise, there should also be an adjustment in the size of the position at each roll. When discussing risk assessment, we mentioned that risk sensitivities for a measure such as project NPV will tend to change over time. Again hedge positions will need to be adjusted. This is related to the rolling hedge concept, in that project cash flows at multiple dates were being hedged with instruments of a single maturity. If the relationship to the underlying risk factor (e.g., yen) is nonlinear, the estimated sensitivity will typically change over time as the value of that variable (yen) changes. Again, dynamic hedging will be appropriate. Indeed, one can use dynamic adjustments of positions in linear securities to approximate nonlinear securities and, hence, to hedge nonlinear risk exposures.

Some Concluding Comments

Risk management can be a complex undertaking. The starting point is risk assessment, and that requires determining the performance measure of concern. One also needs to decide whether to reduce total volatility or to focus only on downside risk. The costs and hedging techniques can be rather different. Then one needs to identify relevant risk factors and estimate their respective sensitivities. To simplify the process, it may be appropriate to restrict the analysis to a relatively small number of risk factors. Where several risk factors are highly correlated, it is reasonable to focus on one (implicitly cross hedging the others). Also, some risks will be small enough to ignore. Remember, there is going to be much imprecision in this process. One should probably plan on using a dynamic hedging approach unless the risk assessment reveals a rather simple structure of risk factors and sensitivities.

As a final word, there should be an oversight mechanism for monitoring actions taken by risk managers. There have been too many instances where supposed hedge positions were, in effect, speculations. Sometimes these were attempts to cover up previous errors. Sometimes people were trying to make money for the firm (and be heros). Sometimes there was a miscommunication between operations and risk management. There is also considerable potential for modeling or estimation errors. Such problems need to be deterred and/or identified before they become a major risk. The best mechanism for that seems to be oversight.

Defining Terms

Forward contract: An agreement that specifies the quantity and price for a transaction that will take place at a specified future date. Such agreements are widely used with currencies and commodities (e.g., crude oil). Specialized agreements can be created for virtually any asset. Futures contracts are very similar but trade on exchanges (e.g., Chicago Mercantile Exchange) and are available only for very standardized items.

Net present value: The mathematical sum of present and discounted future cash flows for a project or investment using a specified discount rate. The terms present value and present worth refer to the same concept.

Option: An agreement that specifies the price and quantity at which the option's owner may choose to transact at a future date. A call option gives its owner the right to buy a specified asset. A put option gives its owner the right to sell. In contrast with forward contracts, option holders are not required to exercise (use) their options.

References

Hull, J. C. *Introduction to Futures and Options Markets, 2nd ed.,* Prentice-Hall, Englewood Cliffs, NJ, 1995.
Jorion, P. and Khoury, S. P. *Financial Risk Management,* Basil Blackwell, Cambridge, MA, 1996.
Smithson, C. W., Smith, C. W., and Wilford, D. S. *Managing Financial Risk,* Irwin, Chicago, IL, 1995.

Further Information

Risk is a monthly magazine from Risk Publications, 104–112 Marylebone Lane, London W1M 5FU, England. It tends to focus on risk management at financial institutions but does contain useful material for industrial firms.

Judy Lewent and John Kearney's 1990 article "Identifying, Measuring, and Hedging Currency Risk at Merck" in the *Journal of Applied Corporate Finance,* 2(4), provides an interesting discussion of some early risk management efforts at a major pharmaceutical firm.

A 1997 monograph with a variety of useful information is *The J.P. Morgan/Arthur Andersen Guide to Corporate Risk Management,* Risk Publications, London, England.

8.5   Equity Finance

Robert Keeley

Equity is the money provided to a firm by its owners. As the owners the equity holders have several collective rights and powers:

- Control of the business is commonly exercised indirectly through a board of directors who in turn hire a company's management.
- The right to sell the assets and retain the proceeds, net of the costs of satisfying all debts.
- Ownership of the annual net profit, which may be retained in the business or distributed.
- The right to change, expand, or shrink the business and to raise money through borrowing or sale of additional equity.

They also have the individual right to sell their ownership to another party, as occurs daily for **public companies** (companies with many stockholders that provide information on their performance and operations to the general public) in the stock market. Equity is one of two principal sources of capital. The other is debt (see Section 8.6). Of the two, equity is fundamental. In mature companies, it serves as a "cushion" to allow lenders to provide debt with little risk of loss. In start-up firms, it is often the only source of funds. A business may exist without debt; it cannot exist without equity.

Conceptual Issues Associated with Equity

The term "equity" has two related meanings. One is based on history, the other on the future:

- The **book value** of a firm's equity summarizes its investment history. It is simply the cumulative amount invested in a firm by its owners (including cumulative net profits earned and subtracting any losses incurred) less the cumulative amount returned to the owners via dividends or repurchases of stock.
- The **market value** or **market capitalization** ("**market cap**") of a firm's equity is the value that the owners can obtain by selling their shares. Market value is measured as the price per share of common stock multiplied by the number of shares outstanding. It represents the collective judgment of investors, interacting through the stock market, about the value of the future cash flows associated with owning the stock. **Private companies** also have market values, although the values are seldom observable because the stocks do not trade often.

In principle, the market value of a share is given by Eq. (8.5), a **discounted cash flow (DCF)** expression stating that the market value of a firm's stock is simply the discounted value of its future dividends per share:

$$P_t = \sum_{\tau=t}^{\infty} \frac{D_\tau}{\left(1+r_e\right)^\tau} \tag{8.5}$$

where P_t = price per share at time $t \geq 0$, D_τ = dividends paid at time τ, $\tau \geq t$, and r_e = discount rate (**cost of equity**) appropriate for the company's level of risk. Future dividends and prices are inherently uncertain. Eq. (1) refers to their statistically expected values.

In practice, a shorter summation is used by capturing the value of all dividends beyond a time horizon (T) with the expected stock price at the horizon (P_T), resulting in Eq. (8.5'):

$$P_t = \sum_{\tau=t}^{T} \frac{D_\tau}{\left(1+r_e\right)^\tau} + \frac{P_T}{\left(1+r_e\right)^T} \tag{8.5'}$$

A firm's managers should strive to maximize the market value minus the book value of equity. Decisions that expand the gap, such as investing in profitable new products, benefit the owners. Decisions that reduce the gap or cause it to become negative destroy the wealth of the owners. Rather than reduce the gap, managers should simply pay the money back to the owners — a neutral action with respect to the difference between market and book values.

Cost of equity: Each type of capital has a cost. The cost of equity, expressed as r_e in Eq. (8.5), is analogous to the interest on debt. In contrast to debt, no contractual payments are demanded by, or promised to, stockholders. However, those stockholders clearly expect some reward, albeit an uncertain one. The key difference between the cost of equity and any other cost is that future payments to equity holders are statistical expectations — expected dividends and prices.

In practice the cost of equity can be difficult to estimate. The most widely used method is the **capital asset pricing model (CAPM)** which has three components:

1. The **cost of capital** for a riskfree invesment (**riskless rate**) — r_f in Eq. (8.6). The riskless rate represents the pure time value of money plus compensation for expected inflation. In 1997, the riskless rate was about 5.6%.[1]
2. The additional rate of return expected by investors from an average stock (the **equity risk premium**) — π in Eq. (8.6). The equity risk premium is around 8.3% — estimated as the average difference between 70 annual historic rates of return on the stock market and the riskless rates in the same years. Estimation error (standard deviation of the estimate) is about 2.3%.
3. The **relative risk** of the company in question vs. an average stock (**beta**) — β in Eq. (8.6). Relative risk is measured using a linear regression of the stock's return on the return of the overall stock market. The average value of β is 1.0, with most stocks falling between 0.8 and 1.2. Technology stocks tend toward the high end.[2]

The three components are combined in a linear equation:

$$r_e = r_f + \beta\,\pi \tag{8.6}$$

A typical technology-based company, with a β of 1.15, would have a cost of equity of 15.1% in 1997, as shown in the following calculation:

$$r_e = 5.6\% + 1.15 \cdot 8.3\% = 15.1\% \tag{8.6$'$}$$

As interest rates shift upward or downward, the cost of equity shifts by the same amount. Thus, in 1981, when the riskless rate was 15.6%, the cost of equity would have been 25.1% for the company in this example.

The cost of equity (and of other forms of capital) shifts with inflation, highlighting the fact that a company's forecasts of future profits, cash flows, dividends, and stock price should all incorporate a forecast of inflation. Alternatively, the forecasts can be in real dollars, in which case inflation should be subtracted from the cost of equity to obtain a real cost of equity. The real cost of equity in the example above is 15.1% − 2.0% = 13.1% because the inflation forecast built into the riskless rate is 2.0%.

In principle, a company's overall β is an average of the βs of its different businesses, weighted by their respective values. Thus, for each of its business units, a company should use a different cost of equity, one appropriate to the risk of the business unit. The principle can be extended to each individual project — a project cost of equity based on a project β. In practice, relative risk is usually estimated by a finding **public company** in the same business and using its β, a method that does not easily extend to individual projects. Thus, manager's must make subjective adjustments to r_e when a project's risk differs from the firm's.

So far this discussion has focused on the cost of equity. In fact, most firms use a combination of debt and equity to finance their operations. Debt has a lower cost than equity. Thus, the overall cost of capital is a weighted average of the cost of debt and the cost of equity and is known as the **weighted average cost of capital** (r_{WACC}). It is calculated as follows:

$$r_{WACC} = \frac{D}{D+S}\,r_D\left(1-T_C\right) + \frac{S}{D+S}\,r_e$$

---

[1]The time value of money can be estimated as the interest rate on inflation-indexed treasury securities (3.6% in September 1997). The difference in interest rates between a treasury note, say, of 1 year maturity, and inflation-indexed treasury securities is an estimate of inflation. In September 1997, 1-year treasury notes had a rate of 5.6%. The difference, 2.0%, is an estimate of inflation.

[2]Estimates of β based on statistical regressions are subject to estimation errors of 0.2 to 0.3 in most cases, which increases the estimation error in the cost of equity above the 2.3% error in the market risk premium — commonly to about 2.8%.

where $\frac{D}{D+S}$ = proportion of debt, T_C = corporate income tax rate, $\frac{D}{D+S}$ = proportion of equity, r_D = cost of debt, and r_e = cost of equity.

The cost of debt includes an adjustment for taxes because debt is tax deductible and payments to stockholders are not. The weighted average cost of capital is used in investment decisions. Any project whose rate of return on capital exceeds r_{WACC} will also provide a return to the equity component of capital above r_e.

Decisions on Equity Financing

The principal decisions for company managers regarding equity are two: (1) how much to raise from (or return to) investors and (2) from whom to raise it.

How much equity to raise: A company begins by estimating its total capital needs using pro forma projections of the income statement and balance sheet, and through its capital budgeting process (see Sections 8.3, 8.15, 8.16, and 8.17). The total requirements are divided into debt (long and short term[3]) and equity capital according to a company's preferences. The optimum mix is a subject of much debate and is beyond the scope of this section. Once the proportions are set, the amount of equity needed is easily derived from the total capital needs.

$$\text{Equity needed} = \left(1 - \frac{\text{debt}}{\text{capital}}\right)\left(\text{total capital needed}\right) \tag{8.7}$$

The amount raised in a given year may vary from the optimum calculated from Eq. (8.7). Over time a firm adjusts the proportions of debt and equity raised to keep debt/capital near its target.

Where to raise equity funds: Companies have three broad sources of equity capital:

1. Retained earnings
2. Private sources including (1) founders, family and friends, (2) informal venture capital ("business angels"), (3) institutional venture capital, (4) private placements, (5) industry partners.
3. Public sources

The sources vary in the amount of capital available, the process for obtaining it, and the total cost (including the cost of equity as defined earlier and the costs of finding, negotiating, and closing the investment — often called **issue** or **flotation costs**).

Retained earnings are the cheapest source of equity because they involve no flotation costs. All other sources involve flotation costs, costs that raise the cost of equity as described in Eq. (8.8):

$$r_{e,e} = \frac{r_{e,i}}{1-f} \quad \text{where } r_{e,e} = \text{the cost of equity from external sources} \tag{8.8}$$

where $r_{e,e}$ = the cost of equity from external sources, $r_{e,i}$ = the cost of equity from internal sources, (i.e., r_e from Eq. (8.6), and f = flotation costs as a fraction of the gross amount raised.
For example, if $r_{e,i}$ = 15.1% and f = 10%, $r_{e,e}$ = 16.8%. Because of the added costs of external financing, firms prefer retained earnings. However, retained earnings may provide less funding than a rapidly growing firm needs.

---

[3]The formal rules of accounting do not treat short-term debt as "capital," because it is not "permanent" financing. However, most technology-based companies that have short-term debt use it as a lower cost substitute for long-term debt. It is de facto a source of permanent funding and should therefore be lumped together with long-term debt as a component of capital.

The distinction between the two external sources, private and public, stems from the number of investors participating, their degree of sophistication, and the securities regulations that apply. Public offerings of equity must follow a set of securities laws and regulations. Those laws prescribe a process for registering the sale of stock with the Securities and Exchange Commission, for disclosing information about the company, for soliciting investors, and for selling the stock. Private sales are exempt from those rules to varying degrees on the assumptions that they do not involve many investors and that the investors are sophisticated enough to know and get the information needed in order to make an informed choice. Stiff penalties apply for failure to make an honest disclosure about a company in the process of selling stock, whether by public or private sale. Private sales are discussed in Chapter 1, so only public offerings will be discussed here.

Public offerings of equity are rare events for most companies. Only 6300 companies are traded on national stock markets (the New York Stock Exchange, the American Exchange, and the NASDAQ), far less than 1% of the total number of businesses in the United States. Several thousand others trade on smaller markets, and a handful of companies have recently raised equity through small offerings over the Internet. Public offerings usually raise $10 million or more (often much more) in equity at a time. Thus, they are for larger companies, those with at least $20 million in sales. Smaller companies generally rely on private sources of funds (described in Part I), although small, local public offerings are possible.

A company's **initial public offering (IPO)** may not always imply a need for additional equity. The company's investors from earlier private offerings may wish to sell their stock, and employees may wish to cash in their stock options. In order to do so at the highest possible price they need a public market. Later offerings by a company (**secondary offerings**; note that secondary offerings can also refer to sales by large shareholders, someone other than the primary issuer — the company) occur only if the company truly needs additional equity.

Investment bankers act as intermediaries in almost all public offerings, selling the stock on behalf of an issuer. Their services may be a "best-efforts" attempt to sell the stock or a firm commitment underwriting in which the investment banking syndicate guarantees the price received by the company. Offerings of less than $5,000,000 may frequently be best efforts, but larger offerings are usually a firm commitment underwriting. Each investment banking firm has criteria regarding the types of firms it will take public. Most require, at a minimum, that a company be profitable, growing, and well managed, with a need for $10 million or more in new equity.

The process of selecting investment banker, filing a registration statement, and contacting potential buyers takes at least 3 months. When selecting an investment banker a company should balance reputation, recent performance on public offerings, follow-on support given other public companies, aggressiveness on pricing, and personal relationships with the investment bank's officers.

Flotation costs of public offerings often reach 10 to 20% of gross proceeds:

1. Legal, accounting, and filing fees, printing, and other direct expenses: often about 2% of gross proceeds.
2. Underwriter's discount: difference between price paid by public and price received by company. Often about 7% of gross proceeds.
3. Underpricing: The extent to which the price rises shortly after the offering. Although not a cash expense, it is as much a flotation cost as the underwriter's discount. Averages 10% of gross proceeds on IPOs, but only 2% on later offerings.

Example: Cirrus Logic in 1996

Cirrus Logic, Inc., a semiconductor company with sales of $1.1 billion in 1996, illustrates the points discussed above. Table 8.6 summarizes shareholders' equity. Cirrus Logic has obtained $429 million from shareholders (including $99 million in retained earnings) and has a market value of $1390 million — a gain of $962 million over the shareholders' investment.

Table 8.7 shows that Cirrus Logic has not relied on equity alone. Thirty percent of its capital (treating short-term debt as capital) is debt — relatively high for a technology-based company.

TABLE 8.6 Book and Market Values ($ in thousands) of Equity for Cirrus Logic

Book value	
Paid in (from stockholders to company)	329,574
Retained earnings	99,092
Total shareholders' equity	428,666
Market value	
Number of shares outstanding (thousands)	63,951
Market price per share (1996 average)	$21.75
Total market value	$1,390,934
Market value – book value (goal is to maximize)	$962,268

TABLE 8.7 Cirrus Logic's Capital ($ in thousands)

Debt	
Short term	$106,575
Long term	76,727
Total (interest bearing) debt	$183,302
Shareholders' equity (book value)	$428,666
Capital = debt + equity	$611,968
Debt/capital	0.30

TABLE 8.8 Cirrus Logic's Cost of Capital, 1996

Cost of equity	
Riskless interest rate (r_f)	5.4%
Market risk premium (π) (average 1926–1995)	8.3%
Relative risk (β)	1.90
Cost of equity from Eq. (8.6) (r_e)	21.2%
Cost of debt	
Average (estimated) interest rate on debt (r_D)	10.0%
Tax rate (T_C)	31.0%
After-tax cost of debt ($r_D \cdot [1 - T_C] = 10\% \cdot [1 - 0.31]$)	6.9%
Weighted average cost of capital (note weights are 30% debt, 70% equity)	
$r_{WACC} = \dfrac{D}{D+S} r_D \left(1 = T_C\right) + \dfrac{S}{D+S} r_e = 0.30 \cdot 6.9\% + 0.70 \cdot 21.2\%$	16.9%

Table 8.8 estimates the costs of equity (r_e) and the weighted average cost of capital (r_{WACC}). High relative risk (β) of 1.9 gives Cirrus Logic an unusually high cost of equity — 21.2%. Its weighted average cost of capital — used for making investment decisions — is 16.9%.

Table 8.9 summarizes how Cirrus Logic has financed its operations. It has grown rapidly, about 50% per year, well beyond its ability to grow on retained earnings alone. Consequently, it held a public stock offering in 1993, netting $136 million. Flotation costs were 5.1% of proceeds. Employees invested $75 million through exercise of stock options and through a stock purchase plan. Cirrus Logic has also increased its proportionate use of debt, raising 35% of new capital with debt.

Conclusion

As managers strive to maximize the excess of market value over the book value of shareholder equity, they must translate the stockholder's view into guidelines for decisions. The cost of equity (r_e) is such a guideline, telling managers whether their investment decisions will create or destroy value. In a growing company, managers will need to obtain additional equity in order to finance the growth in operations.

TABLE 8.9 Cirrus Logic: Historic Growth and Financing Activities

Average Growth 1992–1996 (% per year)	
Sales	51.5%
Assets	51.9%
Increase in debt 1993–1996 ($ in thousands)	$155,391
Increase in equity 1993–1996 ($ in thousands)	$282,250
Internal source (retained earnings)	$70,587
External source (public offering)	136,025
External source (employee options)	75,638
Total increase in capital 1993–1996 ($ in thousands)	$437,641
Percentage of capital provided by debt	35%
Percentage of equity provided by external sources	75%

The cheapest source of equity, with a cost of $r_{e,i}$, is retained earnings. If growth exceeds the rate sustainable with retained earnings alone, the company must raise external equity through private or public sources. In either case the time delays associated with raising the funds are substantial, and the flotation costs raise the cost of external equity above that of retained earnings.

Defining Terms

Beta (β):   A measure of **relative risk**, usually obtained from a linear regression of a time series of returns on a company's stock against the return on the overall stock market. A stock with average relative risk has a beta of 1.0.

Book value of equity: The value of stockholder equity or net worth shown in a company's balance sheet. It is the cumulative amount invested in a firm by its owners (including cumulative net profits earned and subtracting any losses incurred) less the cumulative amount returned to the owners as dividends or repurchases of stock

Capital asset pricing model (CAPM):   A widely used method for estimating the **cost of equity (r_e)**, that adds a risk premium appropriate to each company to the interest rate on a riskless investment.

Cost of capital: The (statistically) expected rate of return to investors who provide capital. Different forms of capital have differing risks and therefore different costs of capital. The cost of capital depends on the risk of the activity receiving the investment.

Cost of equity: The cost of equity capital. It is (statistically) the expected rate of return to investors who buy a company's stock. In general, riskier companies have higher costs of equity. In 1997 the cost of equity for an average company was estimated as 13.9 % per year.

Discounted cash flow (DCF):   A method based on compound interest mathematics for assigning a current value to future cash flows. The current value is less than the expected future value, hence the term "discount". The cost of capital is used as the discount rate.

Equity: Used in two ways. See **Book value of equity** and **Market value**.

Equity risk premium: The expected return (above the interest rate on a riskless investment) earned by investors on a stock of average relative risk.

Flotation costs: See **Issue costs**.

Initial public offering (IPO):   The first sale of stock by a company to the general public. Must be done in compliance with securities laws.

Issue costs: The costs of issuing equity to new stockholders. Issue costs often equal 10% or more of gross proceeds from the sale of stock.

Market value or **market capitalization (market cap):**   The value of a company's stock in the stock market. May be given as a price per share or as an aggregate value for all shares. **Private** companies do not have observable market values, although one may estimate the market value based on similar public companies or from **DCF** methods.

Private company: A company that has not sold stock to the general public. Usually has only a few stockholders. Over 99% of companies are private.

Public company: A company that has sold stock to the general public. About 6300 U.S. companies are public and have their stock traded on a national stock market.

Relative risk: A ratio of the average percentage return on a company's stock to the percentage return on the market. See **Beta**. Relative risk is used to calculate the risk premium for a company's stock using the **capital asset pricing model**.

Riskless rate: The rate of return earned on an investment with no risk of default or variation in price. Often the interest rate on a short-term U.S. Treasury security is used.

Secondary offering: A sale of stock to the general public by a party other than the company or a sale to the general public by the company that is subsequent to an **IPO**.

Further Information

Every college-level textbook on financial management deals with equity financing and the decisions associated with equity. Many are available in introductory (for a first undergraduate course) and intermediate (for advanced undergraduate or MBA course) versions. The intermediate versions are more detailed and generally serve as better references. Two excellent texts are *Corporate Finance* by Ross, Westerfield, and Jaffe (Irwin McGraw Hill) and *Fundamentals of Financial Management* by Van Horne and Wachowicz (Prentice Hall). Additionally, *Analysis for Financial Management* by Higgins (Irwin McGraw Hill) is a shorter, practical treatment of financial management. Prices are between $50 and $100.

NASDAQ publishes a guide to IPOs (*Going Public*, NASDAQ, 1735 K Street, NW, Washington, DC 20006-1500) as do all of the large accounting firms. These are usually free.

CLE International (1541 Race Street, Denver, CO 80206) periodically offers short courses on public and private financing. The courses are specifically designed as continuing education for lawyers but are open to the public. Materials from a course can often be purchased for about $100.

8.6   Debt Finance

Donald E. Vaughn

The U.S. economy has been built on debt usage. Major components of debt include government issues (federal, federal agencies, and municipal debt), business debt, and consumer debt. Since over two thirds of business firms are single proprietorships, consumer debt and business debt are comingled for this group of mostly small firms, a section on consumer debt is included below. Moreover, partnerships and corporations must compete with governmental and consumer borrowers for funds. Over the past century, the debt in the United States has run approximately 1.6 times its Gross Domestic Product (GDP).

Debt is used to provide an immediate purchasing power rather than to save funds and pay cash for desired purchases of goods and services. The U.S. federal budget usually runs about 20 to 22% of GDP. In the late 1980s and early 1990s, the expansion of federal debt was running about 5 to 10% of federal government spending. Some federal tax increases occurred in early 1995, and the U.S. Congress (Senate and House) were reducing spending. By fiscal 1997, the federal annual deficit had dropped to about $150 billion a year, compared to federal outstanding debt of about $6000 billion.

Major components of tax collections include income taxes on corporations and individuals, **OASDHI** (Social Security) taxes, fuel taxes, so-called sin taxes on tobacco and alcoholic beverages, federal estate taxes, luxury taxes, etc. [Lang, 1995]. Federal deficit spending (spending exceeds tax collections) is financed with a number of debt instruments, covered below. States and political subdivisions each year issue municipal debt averaging about $100 billion for long term (over 1 year to maturity) and a similar amount for short term (1 year or less to maturity). Such municipal debt proceeds are largely used for constructing public facilities, such as prisons, schools, highways, parks, hospitals, etc. Consumers use debt not only to buy large-ticket items such as residences, furnishings, and vehicles, but they also use credit card debt and cash loans for purchases of consumer and semidurable goods and services. Business firms use a combination of short-term and long-term debt issues. Debt outstanding for an average businesses generally runs about 45% of total assets, with equity ownership accounting for the balance, but the mix varies widely by type and size of business firm, relative maturity of the industry, the levels of interest rates, and the strength of the national economy.

Major Types of Debt Issues

Federal Government Issues

The U.S. Treasury uses a mixture of debt issues, with roughly 75% in marketable government issues, meaning that they may be traded in the secondary markets, and the balance being in savings bonds and special issues (tax anticipation notes, etc.) that have no secondary trading. Of the marketable issues, roughly one third are in treasury bills (T-bills), which are sold on a discount basis from face amount and redeemed at par. The T-bills have original maturities of 3, 6, and 12 months to maturity. The 3- and 6-month T-bills are usually issued weekly through auctions held by the Federal Reserve System. Denominations are usually in $10,000 or $25,000 units, and major buyers include commercial banks, other savings institutions, pension accounts, foreign buyers, and affluent individual investors. Several large commercial banks and money-market institutions stand ready to buy or sell large volumes of T-bills in the secondary market. Spread (profit margin to the dealers) runs about 2/32nds to 3/32nds percent of value. Yields in the secondary and primary (new issue) markets are often quoted in whole percentages and fractions, with a basic point being 0.01% return. Issues of 12-month T-bills are made at less-frequent intervals. T-bills are held in vast amounts by commercial banks and other savings institutions, large domestic corporations, pension accounts, money-market mutual funds, affluent U.S. citizens, and foreign buyers.

U.S. Treasury notes and bonds are longer-term debt issues of the U.S. Treasury Department, with their distinction being in original maturities at the time of issue. U.S. Treasury notes usually have original maturities of 2 to 9 years, while treasury notes have original maturities of 10 years or longer. The most recently issued 30-year Treasury bond yield is often quoted so as to compare market returns on treasury bonds with that for municipal issues or corporate issues. Spreads between bid (i.e., what a bond dealer will pay for bonds) or ask (what they will sell them for) runs from about 4/32nds to 12/32nds percent of value, though yield spreads narrow and broaden due to buying and selling strength of the particular issue and the strengths and weaknesses in the national economy, Federal Reserve monetary action to raise or lower yields, and other factors.

Marketable treasury notes and bonds pay interest semiannually to registered holders. Pension accounts, life insurance companies, and bond mutual funds are the largest three types of U.S. holders of long-term government issues. The intermediate-term treasury notes are held by these and many types of savings institutions (e.g., commercial banks, savings banks, savings associations, and credit unions), mutual funds, U.S. individual investors, foreign holders, etc. [Hempel et al., 1994]. A breakdown of major U.S. Treasury holders is provided in each monthly issue of the *Federal Reserve Bulletin*.

The interest earned on municipal notes and bonds (munis) makes them particularly attractive investment vehicles to commercial banks, whose marginal tax rate might go as high as 39.6% (in 1997) or to individuals in a 28% or higher federal tax rate, as such interest is exempted from Federal income taxes. State taxes on municipal note and bond interest varies widely from state to state, being exempt by some and taxed at the regular state income tax rate by others [Gitman and Joehnk, 1996]. Capital gains on the trading of munis are often subjected to tax at the prevailing rates (which are likely to be changed in 1997 or 1998 by the U.S. Congress). A taxable, 10-year Treasury yield of 7% would be equivalent to $0.07 \times (1 - \text{tax rate})$ or $(1 - 0.40)$ or a rate of 4.2% annually on a tax-free municipal issue to a bank or U.S. holder in the 40% tax bracket. Prices are quoted as percentages of par and common fractions on municipals with yields quoted as decimal fractions.

Business Debt

Most business firms use a mixture of short-term and long-term debts [Maness, 1993]. The short-term debts may be subdivided into self-generating debt, such as trade payables on purchases of goods and services, and on accruals for wages and salaries, payroll and income taxes, accrued expenses for insurance, ad valorem (property) taxes, etc., and debt arrangements for short-term cash loans from commercial banks and other lending institutions [Van Horne and Wachowicz, 1995]. The rate of interest charged depends on the credit rating assessment by the lender, often running **prime rate** for low-risk customers to much higher rates on more risky clients. Smaller firms and very risky and cyclical firms (e.g., residential building contractors) often pay rates of prime plus 2 to 4% for bank, working-capital loans. Farms and ranches, some retail firms, and some manufacturers have heavy seasonal demand for funds, needed to help finance temporary buildups of inventories and receivables, and then repay the short-term loans after a few months when sales or production slacks off, inventories are sold off, and accounts receivable are collected.

Business intermediate-term debt, in the form of installment, equipment purchase notes, and cash loans of over a year from institutional lenders, the Small Business Administration (SBA), some state agencies created to meet such credit demand, small business investment companies, and other lenders are used by many business firms. Lease financing of business equipment, machinery and vehicles may be easier to obtain by firms than cash loans, so leases have somewhat replaced intermediate-term business loans during the last 20 years. Capitalization of lease debt sometimes appears on the balance sheet of a firm, but many firms elect to provide the main factors of lease amounts and terms of repayment in footnotes to financial statements. Periodic payments by lease users are often paid monthly while term-loan repayments are monthly, quarterly, or annually.

Long-term corporate debt, 10-years or longer to initial maturity, may be in term loans, shown on the borrower's balance sheet as notes payable or in the form of bonds payable. Such bonds are often issued for 20, 25, or 30 years years to maturity, but numerous 100-year bond issues were made in late 1996 and early 1997. Installment repayment of principal and interest is typically negotiated on the notes at origination and may be monthly, quarterly, or annually. The amounts owed within 12 months of the balance sheet date usually appear as short-term notes payable with the balance shown as long-term payables. Some accountants show categories for short-term, intermediate-term, and long-term debts while others merely use the two-division category as short term (a year or less to maturity) and long term (over a year to maturity). A financial analyst might wish to refer to the annual survey of 600 firms, *Accounting Trends and Techniques*, published annually by the American Institute of Certified Public Accountants (AICPA), for such comparisons.

Commercial banks, other savings institutions, life insurance companies, the SBA, and other lenders make intermediate- and long-term credit in vast amounts to business borrowers. A bank might participate with a life insurance company in making a term loan, with the bank taking the first half of the repayments and the life insurance company underwriting the balance. Bank-SBA immediate or deferred participation, guaranteed loans are also made by some interested bank lenders. A nearby small business development center or a Service Core of Retired Executives (SCORE) chapter can usually provide information and application forms on SBA loan types. Beginning in 1995, the paperwork on very small, SBA loans was reduced significantly. Interest rates on such variable-rate, bank-SBA participation loans often run 2.75% above the prime loan rate, with quarterly adjustments. In many cases, short-term loans may be signature loans while those with longer maturities usually carry a pledge of assets equal to 1.25 times the outstanding loan balance.

Corporate Long-Term Debt

Long-term debt with maturities over 10 years may not be easy to obtain by small and intermediate-sized business firms, unless it can be secured with long-term assets equal to about 1.25 times the amount of the loan. This type of credit is referred to as an asset-based loan. Large firms with adequate debt-servicing ability may borrow from a wide range of financial institution lenders, mentioned above. Small business investment corporations (SBICs) were first licensed by the SBA in 1958 to make long-term credit and equity funds available to small business concerns. Other lenders include development corporations and some state agencies that provide funds to small and intermediate-sized firms. Roughly 99% of the business firms in the United States meets the SBA's classification of small. Smallness might be based on a certain level of sales, for trade and service firms, or by work force size for others.

The largest 100,000 U.S. corporations generally have access to financial markets for selling mortgage bonds (i.e., bonds secured by real assets with long-term lives), chattel bonds (secured with a pledge of personal real assets), collateral trust certificates or bonds (secured with a pledge of financial assets), debenture bonds (no special asset pledge is made), or convertible bonds. These convertibles, at the option of the holders, may be exchanged for common shares in the issuing firm at some set conversion ratio. The annual updated volume of *Moody's Investors Service* provides the most complete description of financials and outstanding issue characteristics known to this writer.

Some bond issues are retired at maturity by corporate issuers from cash set aside in a sinking-fund deposit. Some bonds are callable, at the option of the issuing firm, at face amount for retirement purposes [Van Horne and Wachowicz, 1995]. Still others are issued with a deferred call of about 5 years but callable thereafter at any interest payment date at a reducing call premium. The call premium on a 15-year debenture issue might be 10% in year 6, 9% in year 7, 8% in year 8, etc., and callable at par during the last year or two of the life of the bonds. The annual report of the issuing corporation, the **offering prospectus,** or *Moody's Investors Service* should provide this information to the interested investor. In the early 1970s, the Securities and Exchange Commission (SEC) began to require that corporations under its reporting jurisdiction report earnings per share on a fully diluted basis. The SEC's jurisdiction is over exchange-listed firms and over-the-counter firms with 500 shareholders or more (except for portions of the insurance industry that are state regulated). Fully diluted assumes that the bonds would be converted, if to the advantage of the holders, and replaced with outstanding common shares. Dilution in earnings per share could also occur from exercising outstanding warrants so as to buy additional shares of common stock at the fiscal year's closing market price. The capital structure and debt structure portions of the balance sheet for such a firm would be reported on a proforma basis. Some firms report earnings per share on a regular basis and also on a fully diluted basis.

Utility companies — electric, gas distribution, gas transport, telecommunication firms, and water — often finance upward of 50 to 60% of assets with long-term debt, finance working capital needs with short-term debt, and finance the balance of about 30 to 40% of assets with equity. Some preferred stock and common stock often makes up this equity portion. The Federal Power Commission, a division of the Department of Energy, regulates the capital structure mix and wholesale utility rates charged by the firms under its jurisdiction (large electric and gas distribution and transmission companies that operates in more than one state). State commissions designed for that purpose regulate other utilities. A significant amount of deregulation of transport firms occurred during the Reagan years (i.e., 1980 to 1988). By 1997, financial institutions and transport firms had substantially been deregulated.

Consumer Debt

Consumers with earning potential, as well as those that earn a good credit rating by borrowing and repaying on time, have access to credit. A college student will usually receive several offers of credit cards whose balances, up to some preset line of credit, have been underwritten by banks or other savings institutions. While prudent usage of credit cards helps to build a good credit rating for its holder, overextension of credit card debt during early adult years can cause repayment difficulties for years to come. The monthly rate of carrying charge often runs from about 0.5% as an introductory, 6-month rate, to about 1.5% or higher beyond that point. Visa, MasterCard, and Discover are the most widely used credit cards. The user is expected to make a minimum monthly payment on the debt balance, with 2% being common. Some three fourths of this monthly payment might go to service the interest on the loan while only 0.5% is applied to principal reduction. The credit card holder usually continues to buy goods on credit, only to see his or her balance maxing out on the credit card. He or she uses another card and repeats the process. The total outstanding balance might exceed a year of income by the college graduation date. Ten or 20 years might be required for the student to dig himself or herself out of the *credit card hole.* Most chain retailers accept major-brand credit cards as well as their own if they issue them. In the mid-1990s, most chain supermarkets also began to accept major brand credit cards. Heavy users sometimes find themselves with excessive amounts of debt and seek federal personal bankruptcy debt elimination. When this happens, though, purchase money credit on vehicles or homes might become impossible to obtain. The stigma from the personal bankruptcy lingers for 10 years or longer. By 1997, some estimates of distressed credit card debt were set at about 4% of outstanding balance. A wise credit user pays off the balance in full each month and usually has no interest assessment for goods or services charged. Cash and credit card loans usually carry interest from the date of loan until paid. Business borrowers, unincorporated and incorporated, use credit cards for convenience of payment.

While in college, a student might qualify for financial aid in the form of guaranteed student loans. A quasifederal agency, often referred to as **Sally Mae**, underwrites a majority of these loans. Repayment of principal and interest is expected to begin about 6 months after graduation by the borrower. A 10-year retirement of such debt is thought to be about average. In recent years, the IRS has agreed to pay federal income tax refunds due to delinquent student loan borrowers to the holder of such federal guaranteed loans. Increasingly vigorous credit collection policies were being exerted by the federal agency to reduce such nonpaying borrowers in the mid-1990s.

A large ratio of adults either lease or buy a motor vehicle (e.g., auto, truck, van, cycle, etc.) with a consumer installment loan. The *big three* U.S. auto manufacturers (Chrysler, Ford, and General Motors) offer low down payment and low monthly payments but with a substantial balloon payoff amount on many of their models. Some auto import firms do the same. The notes are sometimes held by the subsidiary financing firm for the auto manufacturer, but they may be sold to other financial institutions. While the deal appears to be attractive to budget- strapped consumer, the payoff amount might exceed the value of the vehicle at the leases' expiration. Mileage above a certain annual average (12,000 annual miles are common) might assess a closeout of $.15 or $.20 per additional mile on the lease. Any scratches or mechanical malfunction to the vehicle might have to be repaired at the user's expense. No equity is being built by the lessee. Thus, a prospective lessee of a vehicle should weigh the advantages and disadvantages of leasing vs. owning and financing a vehicle.

The largest single purchase by a family unit is likely to be its primary residence. About 2.5 times annual family gross earnings is a rule of thumb assessed by many purchase money mortgage lenders (e.g., savings institutions, mortgage banking firms, and some life insurance companies). Down payments might be low, perhaps as low as zero plus closing costs on Veterans' Administration (VA) loans, a graduated scale of 3 to 25% on FHA-insured loans, up to some preset FHA maximum (that varies among states, depending on average cost of living figures), or 10% or more of the purchase price on conventional loans (non-FHA or -VA). The payout might be at a fixed rate for 15, 20, 25, or 30 years or at a variable rate tied to some benchmark rate (such as the five-year Treasury note rate plus 1%) with rates changing quarterly or annually. The borrower gets a bargain rate the first year, only to see the rate being charged begin to parallel market rates on the outstanding mortgage balance from then on. Some variable rate mortgages will permit the borrower to convert to a fixed rate after a certain time period, such as 2 or 3 years. Mortgage rates were running close to 16% in 1979 and 1980, and a possible future return to such a high charge on a variable rate mortgage would build the outstanding principal amount even with a reasonable monthly payment by the borrower. It might be possible for a borrower to have a variable rate mortgage with a 2% annual escalation limit but not more than 6% over the life of the loan. In order to assume this additional risk, the lender might insist on one or two points up front (a point is 1% of the loan amount and paid as a portion of the loan's closing costs).

Asset-based, consumer loans might also be obtained from an insurance company up to the cash surrender value of permanent life insurance or by an institutionally granted home equity loan (in most states). Home equity loans have increased in usage since the itemization for federal income tax purposes of personal interest was phased out in the 1986 Tax Revision Act. Interest (up to some maximum for wealthy filers) on one or two residences, either in the form of first mortgages, second mortgages, or home equity loans, may be used by taxpayers that itemize for federal income tax purposes. Information published annually by the Internal Revenue Service shows the maximum interest that may be itemized. Interest on personal credit, such as auto loans, consumer cash loans, credit card loans, cash surrender value loans, etc., has not been an itemizable item since about 1990. Interest on brokerage credit for margin accounts can be claimed when there has been annual trading activity in the account during the tax filing year.

Merchants of big-ticket items, such as furniture stores, may offer 90-days of credit the same as cash, requiring about a 25% down payment with monthly payments of one third of the balance. The notes are often discounted with regional or local finance companies that offer longer payout terms to the furniture buyer, but at higher interest rates. After the loan is repaid satisfactorily, the secondary lender offers direct lines of credit to the previous borrower.

Before a consumer or business owner borrows, he or she should review carefully the terms of credit and compare alternate credit sources [Lang, 1993]. While the prudent usage of credit is helpful to the consumer, excessive usage can lead to financial pressures and reduced standard of living as debt payments are made from limited family funds.

Timing of Debt Issuance

The U.S. Congress often uses fiscal spending to stimulate a weak economy, thus pumping more purchasing power into many benefitting industries. This stimuli might be aimed toward liberalizing the depreciable life or tax guidelines or granting an investment tax credit on investments into real business assets other than structures. Lower down payments might be required on mortgage loans. Highway building or repair might be funded to a larger degree with federal fuel tax collections. Exports of goods and services might be encouraged.

When signs of general weakness or overheating begins to appear in the U.S. economy, the Board of Governors of the Federal Reserve System, along with the Open Market Committee, might take several steps to create more (or curtail) loanable funds that are available from **member commercial banks**. Changing reserve requirements, lowering or raising rediscount rates of borrowing from Federal Reserve Banks, and following a buy-or-sell T-bills policy for the Federal Reserve Banks are the normal monetary tools used. Selective credit controls might be aimed at changing the minimum stock margin, minimum down payment ratios on loans, and shortening or increasing loan payouts, though these steps have not been taken by the Fed in recent years. As government yields fall or rise, so do costs of business and other classes of debt [Vaughn, 1997].

The U.S. Treasury Department manages the nation's debt. It attempts to keep a balance between marketable bills, notes, and bonds outstanding and consumer savings bonds and special issue notes and bonds. Demand for certain maturities might dictate types to issue, but the Treasury often reduces long-term maturities to intermediate range when interest rates are exceedingly high. Long-term treasury bonds are usually callable during the last 20% of the years of its maturity so as to provide the Treasury with some flexibility in refinancing debt when favorable market conditions exist. Foreign holders own about 10 to 12% of marketable treasuries, bought because of their prime credit rating

Agency issues are often short and intermediate in term and are offered to direct funds into certain industries. The agricultural sector, real estate construction industry, home financing industry, the Tennessee Valley Authority, and college student loan programs are large beneficiaries of such funding guarantees by the U.S. Treasury. The interested reader should consult current issues of *The Wall Street Journal* or the *Federal Reserve Bulletin* for daily trading and amounts outstanding of these issues, respectively.

The mix of corporate debt varies widely from industry group to industry group. A farm might be operated with 15% debt and 85% equity, while a commercial bank may have 6 to 8% equity and the balance in debt [Hempel, 1994]. Most other types of firms use debt/equity mixes somewhere between these extremes. Annual reviews by Dun & Bradstreet, Inc. (D&B) (the parent of Moody's) and published in *Industry Norms and Key Business Ratios* are compiled for about 800 Standard Industrial Classification (SIC) groups. The *Quarterly Report of Manufacturing* (QRM), a government document, is another excellent source of benchmark ratios. Most university libraries subscribe to these publications. A review of these common-sized financial statements with some key ratios are provided for 18 industries in a recently published text [Vaughn, 1997]. The D&B key ratio data are collected from credit report information compiled by D&B. The annual, *Statement Studies*, published yearly by Robert L. Morris Associates, with data gathered from association member bank credit files, is an alternate source. QRM-published data are largely from quarterly reports filed with the Department of Trade by the largest 500 firms in the United States. Breakdown of expense accounts for unincorporated and incorporated firms, including interest paid as a percentage of sales, for about 400 SIC groups, are found in *Almanac of Business and Industry Financial Ratios*, updated annually by Prentice-Hall, or found in IRS publications.

Top-level managers for expanding business firms usually have a target mix between debt and equity. Interest on contractual debt arrangements is a business expense while dividends on preferred or common shares are not. Thus, in keeping with a firm's management assessment of risk of possible debt servicing problems and the lower after-tax cost of debt than equity, debt will often be issued in reasonable amounts. Industry benchmarks, ± about 10%,will usually be accepted by capital markets. Greater debt usage than near average for industry groups may cause bond ratings assigned by Moody's or Standard & Poor's Corporation (a division of McGraw-Hill) to be lowered. Downgrading of the issue, perhaps from one of the top four grades of Aaa/AAA, Aa/AA, A/A, or Baa/BBB, respectively, by Moody's and S&P to a speculative level of Ba or BB (or even lower) that falls below the level of rated investments that qualify for buying by **fiduciary accounts** cuts out many holders. Speculative investors and high-income mutual funds then become the major buyers of speculative-grade corporate bonds. To keep their debt/equity in balance, growing business firms often retain 70 to 100% of their after-tax earnings to increase equity at about the pace of planned growth. Debt, then, might be issued so as to maintain the appropriately deemed balance between debt and equity. Investment banking firms, commonly referred to as underwriters, provide information on debt/equity mixes, market timing of sales, etc., for their clients and prospective clients.

Consumers should monitor their own usage of credit, though some merchants and fund lenders set standards before approving requested credit. Mortgage credit of 2.5 times gross family income is one benchmark. The payment on interest, mortgage principal, ad valorem taxes, and insurance are likely to take 25 to 30% of the disposable family income. Payments on installment debts should not exceed another 10 to 15% of disposable income.

Defining Terms

Fiduciary accounts: Accounts managed for the benefit of others, where the prudent man rule is often applied.

Member commercial banks: Nationally chartered commercial banks and electing state commercial banks.

OASDHI: The old age, survivors, dependents, and health insurance programs managed by the Social Security Administration.

Offering prospectus: Financial disclosure statement required by the SEC for most public sales of securities.

Prime bond rating: The highest, usually reserved for direct issues of the U.S. Treasury Department.

Prime interest rate: Interest rate on business loans to large, financially strong firms made by large banks.

Sally Mae: The acronym for the 3- to 10-year bonds issued by the Student Loan Marketing Association.

References

Gitman, L. and Joehnk, M. *Fundamentals of Investing*, Harper Collins College, New York, 1996.

Hempel, G. et al. *Bank Management*, John Wiley & Sons, New York, 1994.

Lang, L. *Strategies for Personal Finance*, McGraw-Hill, New York, 1995.

Manis, T. and Zietlow, J. *Short-Term Financial Management*, West, St. Paul, MN, 1993.

Van Horne, J. and J. Wachowicz, Jr. *Fundamentals of Financial Management*, Prentice-Hall, Englewood Cliffs, NJ, 1995.

Vaughn, D. *Financial Planning for the Entrepreneur*, Prentice-Hall, Upper Saddle River, NJ, 1997.

Further Information

Almanac of Business and Industry Financial Ratios (annual), Prentice-Hall, Englewood Cliffs, NJ.
Industry Norms and Key Business Ratios (annual), Dun & Bradstreet, New York.
Statement Studies (annual), Robert L. Morris Associates, Philadelphia.
U.S. Department of Commerce. Business Census (Retailing, Service, Wholesaling), Government Printing Office, Washington, DC, 1992, 1997.
U.S. Securities and Exchange Commission (annual), Annual Report, Government Printing Office, Washington, DC.

8.7   The Nature of Costs

Jules J. Schwartz

Years ago it was traditional for engineering deans to greet their new freshman class by lettering the word SCIENTISTS across the blackboard at the front of the hall and telling them that the difference between what they planned to be and these people is $¢IENTI$T$, dollars and cents! Engineers must concern themselves with the economics of their designs. With this in mind, the purpose of this section is help you examine the nature of costs and how they affect your ability to sell your proposals in-house and the ability of your firm to make a profit from your ideas. Estimating the initial cost of any design is still another issue we shall not deal with here. Our concern is how trade-off decisions regarding the nature of costs, scale, location, and duration of the manufacturing operation will affect your costs.

Fixed and Variable Costs

It is often useful to think about the costs that go into a product as either fixed or variable. Fixed cost are those that generally don't change with the volume of production. These include salaries, most rents, interest paid on the debt used to finance the enterprise, depreciation on equipment, and utilities used to light and heat the facility. Since many firms allocate such fixed costs on a formula basis to their various products, these costs are often called *overhead*. Variable costs, on the other hand, are those that do vary with number units produced. These include the raw materials that go into the product, the direct labor to produce it, and the power used to run the production equipment. It should be obvious that in the short run many variable costs are really fixed. If we have nothing for the production crew to do during the last 15 minutes of a given day, we don't lay them off. On the other hand, in the long run all costs become variable; if we exit a particular line of business, we lay off managers, sell equipment, and pay off debt, eliminating fixed costs. Incidentally, if your company does assign overhead, check to see that it's done on a rational basis, not just one that can be easily done by a bookkeeper!

So why bother to differentiate between the two kinds of cost? As we shall now learn, it enables us to see how profits will vary with volume and to determine how many units must be sold to break even. To do this, we need to write the formula for pretax profits using these costs.

$$\text{Pretax profit} = \text{sales} - \text{total variable costs} - \text{fixed costs}. \tag{8.9}$$

If we state both variable costs and sales price on a per-unit basis, we get

$$\text{Pretax profit} = \text{volume} \times (\text{unit price} - \text{unit variable cost}) - \text{fixed costs}. \tag{8.10}$$

Switching to algebraic notation:

$$P = N\,(p - v) - F \tag{8.11}$$

where lower case is a unit value and upper case is a total. Now let's put some numbers in the formula and draw some conclusions. Suppose a company makes and sells 10,000 widgets at $20 each, incurrring $12 dollars in variable cost for each, and that it's fixed costs are $40,000:

$$P = 10,000 \times (20 - 12) - \$40,000 = 40,000. \tag{8.12}$$

Let's next suppose that it has sufficient capacity so that in the next accounting period its able to make and sell 11,000 units. Assuming no inflation or increase in fixed costs, pretax profits would now be

$$P = 11,000 \times (20 - 12) - 40,000 = \$48,000! \tag{8.13}$$

Notice that a 10% increase in volume (and sales) resulted in a 20% increase in pretax profits. Why? First you may note that the total cost of producing one unit, with volume at 10,000, is the sum of the unit variable cost plus an allocated share of the fixed cost of 40,000/10,000, or 12 + 4, or $16 per unit. On the other hand, when volume increases to 11,000 units, the fixed cost borne by each unit decreases, so total unit cost drops to 12 + 40,000/11,000, or only $15.64 per unit. Not only has it sold more units at $20 each, but it produced each for $0.36 less. This fortuitous combination accounts for the disproportionate increase in profit, termed *leverage*. If the fixed costs are finanacial ones such as interest expense, we call this result *financial leverage*; if the fixed costs are from salaries or depreciation, the result is termed *operating leverage*. You can have both.

How many units must the company sell to break even? By definition, breakeven is where the firm neither makes a profit nor loses any money, that is, profit before tax is zero. If we set $P = 0$ and solve for volume, N, in Eq. (8.11), we see that

$$N_{b.e.} = F/(p-v) = 40,000/(20-12) = 5000 \text{ units}. \tag{8.14}$$

The term (p -v) is often called *unit contribution*, since each unit that is sold contributes $(20 - 12) = \$8$ toward covering fixed costs, and ultimately to profits. Notice that the breakeven volume goes up directly proportionately with fixed costs. Think about this when deciding whether a product should be produced by a labor-intensive (variable cost) process or an automated one, entailing the fixed costs associated with the depreciation of machines, and perhaps added interest expense, if debt financing is used to buy such equipment. When you have fixed costs, business is great as long as sales go up. Try exercising Eq. (8.11) for a decrease in volume of 10% from the base level of 10,000 units. You'll discover that leverage works in both directions.

Scale Affects

When economists speak of economies of scale, they are referring to the lower unit costs incurred when operating at near-full capacity and spreading fixed costs over the maximum number of units. Still another advantage of large-scale operations was discovered by Lang, a chemical engineer who collected empirical data on the cost of building chemical plants. He concluded that the cost of such facilities does not rise in direct proportion to capacity but rather in the following way:

$$Cost_2/cost_1 = (capacity_2/capacity_1)^L \tag{8.15}$$

where the Lang exponent, L, varied between 0.4 and 0.7. Later studies confirmed that an exponent of about 2/3 seemed to apply to all capital investments in buildings and capital equipment. This suggests that one can build a plant with eight times the capacity of an earlier one at only four times the cost; similarly, 27 times the capacity should only cost nine times as much. Why then don't we produce all of the autos in the United States in one location in Kansas City and realize the huge reduction in the fixed costs of depreciation and interest expense on a single, cheap, big plant compared to those attached to multiple smaller ones with the same total capacity? The answer, of course, lies in the transportation costs. As a plant gets bigger, it is necessary to reach farther to amass all the factors of production, including labor, power, and raw materials. Similarly, the cost to transport to the larger market it must serve also rises.

To understand this trade-off between the Lang Affect and transportation expenses, we must examine the cost of moving materials around. My own studies indicate that transportation costs are best related to *bulk value*, the value of an item compared to the volume it occupies. We ship things in ship loads and boxcar loads, mostly volume related; so, the higher an item's bulk value, the lower transportation costs are relative to its total value. A Chevrolet car, selling for $18,000, occupies perhaps 300 cubic feet and thus has a bulk value of about 18000/300, or $60 per cubic foot. Compare this car to Intel's microprocessor chip, selling for $500. You can pack about 4000 of them in a cubic foot, resulting in a bulk value of 4000X500/1, or $2,000,000 per cubic foot! Without debating the accuracy of this estimate, it should be clear that the cost of transporting chips is very low compared to their value and compared to the cost of shipping cars. Therefore, Intel can afford to produce chips in very large plants and realize the full advantage of the Lang Affect. All the diamonds in the world could be cut in a big plant in Antwerp. Concrete blocks, on the other hand, must be made in relatively small local facilities. Indeed, the affect of a manufacturing process on bulk value will also often tell you where the plant must be located. Copper, mined as low-quality ore, typically experiences a 200-fold increase in bulk value when its refined, so refining capacity is built near the source of the ore to reduce product transportation costs, even if it means building in a jungle in Indonesia. On the other hand, blow molding a plastic bottle for a soft drink may increase its value, but it increases the volume it occupies much faster, so these bottles are normally made in a relatively small plant right next to where they are to be filled.

The other factor that influences transportation decisions is the mode to be utilized. The following is a nonexhaustive list of potential modalities, roughly in order of how cheap they can move product:

Pipelines
Barges and ships
Railroads
Trucks
Parcel service
Air
Mail and courier

Notice that the cheapest means are a bargain only if you are willing to sacrifice schedule and flexibility in pickup and destination and ship large consignments. These modes also benefit from the Lang Affect, if they're big.

The trade-offs then are to build as big as transportation considerations will allow and locate so that you transport the product while it is in its highest bulk-value form. Another consideration that applies particularly to chemical process plants is the need to build on a "world scale", where you produce sufficient quantities of any related by-products to permit their efficient use in still other Lang-efficient plants. This explains why refinery complexes become so large and tend to congregate. Don't forget, however, that building a huge facility in one location increases the risks associated with process obsolescence, fire, natural diasters, and strikes and leaves one captive to local tax decisions.

Project Duration Costs

Certain costs are related to the length of time you expect to be in a particular business. Obviously, the depreciation schedule, established to recover the cost of the capital investment, depends on the length of the project. Less obvious, perhaps, the interest rate on any debt incurred is normally higher if the duration of the loan is longer.

Even more significant is the *experience* or *learning affect* on costs. As an operation gains experience, it generally is able to cut costs. Part of this is a self-fullfilling prophecy; it's expected! Still greater economies are realized as the process is optimized, the product design is made more robust, reject rates improve and yield increases, greater sales result in longer runs, less down time, and more volume to spread fixed costs over, specialization of labor permits the use of lower-skilled, cheaper labor, and larger volume results in volume discounts for goods and services purchased.

Empirical research confirms that there is a log-log relationship between the average unit cost of all units produced to date and the total number of units produced to date. This means that, each time the total volume to date doubles, you expect a constant percentage reduction in the average unit cost for all the units. The slope of this curve gives the experience curve its name. A 20% curve (very typical) exhibits a 20% drop for each doubling of total volume. You must find the slope empirically for your product. Some results for a 20% curve are summarized as follows:

Total Volume to Date	Average Unit Cost
1000	10.00
2000	8.00
4000	6.40
8000	5.12
16000	4.10

Think what advantages accrue to someone who enters a business first or manages to seize the dominant share of any market. Since this firm is further along its learning curve than others, it has the choice of maintaining a price umbrella and realizing higher profit margins than its competitors or lowering its prices to grab still greater market share, while perhaps preventing competition, with their higher costs from making any profit at all. Intel has used this strategy quite successfully to make it tough for others to play catch-up.

Further Information

We have seen how the nature of costs, fixed and variable, scale related, duration related, and transportation related, combine to suggest trade-offs on where, on what scale, and how capital or labor intensive the production process should be. We have also seen how the design of this process will affect the cost and profitablity of your product.

To learn as much as most managers know about these things, I suggest you read the paperback *Techniques of Financial Analysis* by E. A. Helfert. If you want to know as much as MBAs are taught (but more than they learn), try *Financial Management and Policy* by J. C. van Horne. If you learn with your eyes and ears, the Teaching Company in Arlington, VA offers my 12-hour videotape series, *Finance and Accounting for the Non-Financial Manager*, a part of their SuperStar Teacher Series®.

8.8   Mergers and Acquisitions

Kenneth J. Purfey

There is probably no topic within the world of finance that is more fascinating to the executive manager than mergers and acquisitions (M&A). M&A work combines the cool logic of a chess grandmaster with the inquisitiveness of a Sherlock Holmes, all supported by the endurance of a marathon runner. Successful acquisitions require a team whose members posses these type of rare talents and financial skills.

As with any problem or opportunity in finance, striving to maximize shareholder value is paramount, and this can only be accomplished when the benefits expected from the transaction exceed the costs. Thus, the lowest common denominator for justification of any acquisition is to approach the financial decision-making process as a typical net present value (NPV) problem.

There are over 12 million corporations in the United States. One of the most difficult and time-consuming processes can be the search for and identification of a suitable acquisition candidate. The process is not unlike the search for the traditional needle in a haystack. Thus, acquisition consummation rates of 1 to 2% of all candidates analyzed are not uncommon.

Typically, the bidder has only a vague notion of what kind of company would be a good strategic fit and thus make a good acquisition target. Therefore, the M&A professional must help management formulate criteria and financial parameters so as to narrow the field of potential targets. Only then can the M&A professional focus on the very best candidates. This process of searching for targets, screening out undesirable candidates, and ranking desirable candidates is an area where analytical and deductive logic skills are important, not unlike those of a Sherlock Holmes.

Once a target candidate is selected, the owners are contacted to ascertain their interest in selling. This can be a contentious issue and can result in the deal being pursued via either a friendly or unfriendly, i.e., hostile, route. Assuming the owners wish to sell, the M&A professional then attempts to calculate a preliminary value of the target company, and price negotiations are begun. This is one area where good old-fashioned horse trading takes place and much depends upon the negotiating skills of both parties. In large deals, sometimes a person with the skills of a chess grandmaster are necessary, especially if a bidding war breaks out with other interested parties and/or litigation ensues.

Once a preliminary price agreement is reached, a process called **due diligence** is begun. A due diligence is an in-depth analysis of the company's financial and operational condition. It may be as detailed as an accounting audit but is much broader in scope because the operational condition and efficiency of the target's assets are investigated as well. The mission of the due diligence team is to ascertain the true economic values and results of operations and express them in financial terms. The objective is to find, identify, and estimate the impact on the preliminary purchase price, if any.

There are many ways to acquire the target company, taking into account a multitude of accounting rules and tax regulations. There is no one best way to acquire a company, and all the alternatives must be examined to find the optimum method where all parties are satisfied with the ultimate structure. Financial models are constructed to assist with both the target company valuation and pricing issues as well as the optimum financing structure.

There is no substitute for good legal advice along the way. Some attorneys have reputations as "deal killers" as just about any deal can be killed if enough time is expended and problems are found. The trick is to separate the truly important issues from the mundane. Therefore, it is equally important to employ attorneys who are "deal makers" and who find ways to matriculate through the vast forest of laws and regulations without hitting a lot of trees.

Depending on the size of the acquisition, the entire process can take anywhere from a few weeks to 1 or 2 years or more. Typically, the larger the companies the longer the process. This is one area where the acquisition team of professionals must have the endurance and staying power of a marathon runner. It is not uncommon for companies in regulated industries, i.e., utilities that report to multiple agencies to have the regulatory approval process itself add upward of 1 year to the overall timetable.

Nevertheless, the business professional who is a part of the acquisition team can normally take an immense amount of pride in building a successful new entity and in accomplishing the ultimate goal of enhancing shareholder value.

Financial Analysis

For any acquisition to increase shareholder value, the benefits must exceed the costs. While measurement of the costs involved is usually straightforward, determining the benefits derived from the acquisition is usually more difficult. From the bidder's standpoint, an acquisition is another capital budgeting problem with financial estimates required of the after-tax benefits, costs, and risks involved.

The basic concept for NPV is

$$\text{NPV} = \text{benefits} - \text{costs}$$

where

$$\text{Benefits} = \text{value of the target} + \text{incremental value gained}$$

and

$$\text{Costs} = \text{cash and/or other consideration (stock, assets, etc.) given up}$$

The incremental value gained is the financial benefit expected to be gained from the acquisition. The value of the target is the current or preoffer market value of the target firm. In an efficient market, the

NPV would be zero because no prudent investor would pay more for the target than it is worth. Therefore, for the acquisition to make sense, and for a positive NPV to result, the bidding firm must be able to find and realize incremental economic values or tax benefits not available to the target firm.

There are perhaps dozens of economic, financial, and strategic reasons to merge with or acquire another company. Most reasons have to do with synergies, where the whole is greater than the sum of the parts $(2 + 2 = 5)$, or with tax advantages. Some of these include

Cost reductions	Better management
New markets	Defensive protection
Tax savings	Lower costs of capital
Lower risk	Better credit rating
Economies of scale	Better access to financial markets
New products	Better access to suppliers and customers

Screening and Analysis of Target Companies

Sometimes executive management only has a vague idea of what kind of company it would like to acquire. On those occasions, the M&A professional must usually guide management in determining what type and size of target company it should attempt. Thus, the M&A professional should first develop a target company profile. A target company profile is a matrix of key items, elements, dimensions, and financial parameters outlining management's "preferences" and "consideration" characteristics that describe a potential target company. Preference items are those characteristics management would prefer to have, while consideration items are those characteristics broader in scope that management would at least consider if it cannot have the preferred characteristic. These key characteristics include size, revenue, earnings, geography, etc. For example, management might prefer to acquire a company with a market value of $20 to $50 million, but would also consider a company in the $5 to $100 million range.

Once the target company profile is developed, the M&A professional can use a variety of techniques to locate target companies fitting most or all of the preferred parameters. With the advent of computer on-line systems such as Dow Jones Newservice, CompuServe, Dialog, America On-Line, Value Line, etc., it is a fairly straightforward matter to quickly screen through tens of thousands of companies. Progressively tighter "screens" (tighter or more restrictive ranges of parameters) can be applied to the total population of companies to screen out undesirable candidates and finally arrive at a short list of desirable targets.

Software such as Dow Jones Newservice or Value Line can be used to calculate traditional financial ratios, obtain operational information, or publicly reported data such as that found in an annual report or a 10-K report.

Assuming the companies on the short list of targets (probably numbering 5 to 15) still meet management's requirements, the financial analyst might begin a strategic assessment and in-depth financial analysis of the companies. This would include a brief outline of the strengths and weaknesses of each company, advantages and disadvantages of acquisition, a preliminary valuation, and discussion of possible synergies. From this analysis, a recommendation to management would be helpful, advising whether to proceed, i.e., make a "go"/"no go" decision.

When management has identified a primary target and is ready to make "first contact", it is a good time for the legal team to get involved. Once the principals are introduced, there is generally a get-acquainted period, followed by a series of strategic meetings and an in-depth examination period that ultimately leads to the acquisition. This is very similar to the dating, engagement, and marriage process in society. For larger companies, there are generally three legal steps or milestones that lead to the eventual acquisition. These steps include development of the (1) **term sheet**, (2) **letter of intent** (LOI), and (3) **definitive agreements**. These documents reflect the progressive nature of the understanding and terms and conditions of the purchase of the target. A term sheet is usually a one-page document outlining only the most critical aspects of the bidder's offer, including the preliminary purchase price and conditions of closing. It is usually wise to create a term sheet after a few meetings between the principals to indicate a certain degree of interest on the part of the bidder. Then the ball is in the target's court.

Generally, after continued negotiations and agreement in principle on all major points, but not necessarily on small items, the parties are ready to sign an LOI. The LOI is a nonbinding document, usually 5 to 20 pages in length, spelling out in greater detail the terms and conditions precedent to closing. Though the LOI is nonbinding, it represents a significant milestone in the progression because it indicates the parties intend to consummate the transaction. It is very similar to an "engagement" precedent to marriage, in that the LOI is usually publicly announced. This serves two purposes. First, it makes it more difficult for either party to back out, and, second, it indicates agreement on most of the basic terms and conditions for closing. Sometimes the LOI includes a break-up fee, which acts like a "prenuptial agreement". This is necessary because, by the time the LOI is entered into, both parties have likely expended considerable time, effort, and money, and therefore damages would result if one party unilaterally walked away. This action would then trigger a cash payment to the damaged party.

Nevertheless, when all legal documents and every major element of the transaction are agreed upon, the parties are ready to consummate the union by signing the definitive agreements, which are the actual purchase and sale agreements, and the deal is concluded.

Due Diligence

About midway through the process, sometime after the term sheet is developed but before the definitive agreements are signed, the bidder performs a due diligence. A due diligence is an in-depth intensive examination of the most critical aspects of the target's operations and financial status. A thorough due diligence can be (and sometimes is) compared to the traditional blood test given prior to a marriage where the body is examined for diseases. If the findings are material, it is generally not too late to make appropriate adjustments.

The due diligence process is broader than an audit because it includes an examination of the operational aspects of the business as well as the financial status. For example, a typical due diligence might include the review of

Accounting policies	Analysis of assets, liabilities, and earnings
Debt status and equity	Form of incorporation
Board of directors minutes	Status of patents, trademarks, copyrights, etc.
Operational policies	Examination of tax filings, records, and audits
Long- and short-term contracts	Sales, marketing, and pricing analysis
Lines of business, condition of production assets	Personnel and payroll policies
Union agreements, etc.	Evaluation of environmental liabilities
Existing or potential lawsuits outstanding	Management efficiency and control

Forms of an Acquisition

We normally tend to think of the term "acquisition" as the purchase of a smaller company by a larger one. We also tend to think of a "merger" as the joining of equals. Though this is probably acceptable for common terminology, it is not quite accurate for the financial technician.

More specifically, in an acquisition involving a purchase of the target, the bidder absorbs the target's operations, and the bidder's original identity is maintained, whereas in a merger, a new identity is generally created for both of the companies. Unfortunately, there are no clear lines of demarcation between the two as, for example, when a consolidation results. A consolidation occurs when two or more firms combine to form a completely new firm. A new legal entity is formed, and there is no bidder and no target [Pinches, 1992].

Nevertheless, in everyday parlance there are four basic forms an acquisition can take:

1. Stock for stock
2. Stock for assets
3. Assets for stock
4. Assets for assets

Selection of any these forms normally depends on the desires of the parties involved, and therefore combinations and variations of the above are nearly infinite. We will concentrate on the four basic forms.

1. *Stock for stock* — Common stock of the bidder is exchanged for the common stock of the target according to an agreed-upon exchange ratio. This method results in the bidder taking on ownership of both the assets and liabilities of the target. Taking on liabilities of the target is usually a disadvantage for the bidder because of the additional unknown risks and possibly unrecorded liabilities. This situation is therefore usually favorable to the target.
2. *Stock for assets* — Common stock of the bidder is exchanged for the assets of the target according to some preagreed exchange of values. This is usually a safer route for the bidder than stock for stock because the bidder is acquiring only the assets it really wants and absorbing no liabilities of the target. Consequently, this is probably a disadvantage to the target. The target must then normally take the cash value and pay off its liabilities to liquidate the company.
3. *Assets for stock* — In this form, the asset being offered by the bidder is almost always cash or predominantly cash, which is exchanged for common stock of the target. In addition to cash, the bidder may offer assets such as property, plant, equipment, or other special items in combination. Of the four basic forms of acquisition, this form is probably the least-utilized due to the complexity and unilateral disadvantage of acquiring stock of the target. However, it is the fastest.
4. *Assets for assets* — Again the asset being offered by the bidder is almost always cash but could include other assets. Acquisition of assets of the target is probably not as risky as acquiring stock due to the bidder not taking on the unknown liabilities and risks of the target. However, valuation of assets is a critical concern.

Forms 2, 3, and 4 are usually taxable events to the target, who will have to pay capital gains tax, ordinary gains tax, or both, depending on the tax basis in the stock or assets. A major advantage of form 1 — stock for stock — is that it usually can be constructed to be virtually *tax free* because it represents an exchange of like-kind investments. According to IRS regulations, taxes on this type of like-kind exchange can be deferred for long periods of time, until the stock is eventually sold or perhaps when the taxpayer is in a much-lower tax bracket. This advantage sometimes results in the target being more amenable to sharing this valuable tax-saving benefit with the bidder through a lower price, and therefore the stock-for-stock form is one of the most-popular methods.

Accounting for Acquisitions

There are two ways a firm can account for an acquisition: the **pooling method** and the **purchase method** [APB, 1970]. There are 12 requirements an acquisition must meet to qualify for a pooling. If an acquisition does not meet all 12 requirements, the acquisition *must* be accounted for as a purchase. Pooling is generally more beneficial to use than purchase accounting because of the favorable restatements of the balance sheet and income statement allowed in addition to the favorable tax treatment.

While the 12 requirements are too detailed to discuss here, one of the requirements for pooling is that the bidder must exchange its common stock for at least 90% of the target's common stock. Thus, this method generally coincides with the stock-for-stock form of the acquisition, allowing the transaction to be tax free. None of the other forms of acquisition qualify for pooling because they are not stock for stock, and therefore purchase accounting must be used.

Some of the major differences between the two methods, and benefits to pooling, are as follows:

1. In a *pooling*, net income of the target company for the entire year is added to the bidder's consolidated net income regardless of the date of pooling. Thus, a pooling consummated on the last day of the year will result in additional consolidated net income for the bidder for the entire year. In a *purchase*, the net income of the target company can only be included in the bidder's consolidated net income from the date of the acquisition.
2. In a *pooling*, all of the retained earnings of the target company are added to the bidder's consolidated retained earnings. It is simply "added across" on a consolidating worksheet. In a *purchase*,

only those retained earnings from the date of the acquisition forward are included in the bidder's consolidated retained earnings.

3. In a *pooling*, net book values of the newly pooled companies remain the same. In a *purchase*, net book values of the target company are adjusted — generally upward — to reflect the fair market values paid on the date of acquisition. This is referred to as a "step-up" in book and tax basis.

4. In a *pooling*, any acquisition **goodwill** (an intangible asset) is credited or debited to "paid in capital" of the bidder, or, more simply put, they are blended together and no new goodwill is created. Thus, no goodwill appears on the balance sheet of the bidder. In a *purchase*, any difference between the purchase price and the fair market value of the assets results in positive or negative goodwill. Goodwill is normally amortized for financial accounting purposes over a period not to exceed 40 years but can be deducted for tax purposes over a 15-year period.

More specifically, goodwill occurs when the bidder pays more than fair market value for the target company. There are several reasons why a bidder would do this. One of the most important is that the bidder expects to realize significant synergies and perhaps excess future earning power as a result of the acquisition. Nevertheless, shareholders and investors sometimes question the appearance of goodwill on the bidder's balance sheet as an admission by management that perhaps they "paid too much" for the target. For these reasons, goodwill is generally an item to be avoided if possible.

Thus, there are some significant accounting benefits to pooling in the form of higher recorded earnings and the avoidance of goodwill. This advantage is not lost on the marketplace because since 1992, companies have used pooling accounting nearly *ten times* as often as purchase accounting [Securities Data Corp., 1997].

Summary

For any acquisition to be truly successful, it must add value to the company over the long run. Thus, traditional NPV analysis is the preferred method to use. Companies acquire other companies for a multitude of reasons, most having to do with synergies and tax savings. Acquisition professionals must carefully interpret management's requirements to find viable targets that fit most of the optimum criteria. Thus, the screening and ranking process is extremely important as is the due diligence examination process. Large acquisitions can take as long as a few years to complete.

There are at least five ways to accomplish an acquisition when considering both the tax and accounting criteria as shown by an (X) in the table below:

Form of the Acquisition	Purchase	Pooling
1. Stock for Stock	X	X
2. Stock for Assets	X	
3. Assets for Stock	X	
4. Assets for Assets	X	

Only the stock-for-stock form of the acquisition normally results in a tax-free transaction, while pooling generally has more-favorable accounting and financial benefits. Companies clearly prefer using stock for stock in combination with pooling, as in 1996 alone there were $140 billion of poolings vs. only $2 billion of purchases. However, this advantage may be coming to an end as the Financial Accounting Standards Board is contemplating ending or restricting the pooling method [MacDonald, 1997].

M&A activities are important to the corporation because they can result in dramatic change to the corporate structure, employees, and shareholders alike. If chosen carefully and financially synergistic, an acquisition can be an exciting and challenging opportunity for the M&A professional, ultimately resulting in a significant enhancement to shareholder value and increase in corporate wealth.

Defining Terms

Definitive agreements: The full and final purchase and sale agreements concluding the deal. Including all important details of the deal, this document could be as long as 100 pages or more.

Due diligence: An in-depth examination of the target company's financial and operational health, but broader than an audit because it includes the company's operations as well.

Goodwill: The amount in excess of fair market value that a bidder may pay for the target. Bidders may be willing to pay more for a target if there is an abundance of financial, operational, or strategic synergies or tax advantages.

Letter of intent: A nonbinding document that indicates the intent of the two parties to enter into a merger or acquisition. This is usually the first public announcement of the transaction. The document itself is an outline usually covering 10 to 30 pages of all the major points and legal conditions precedent to closing.

Pooling method: One of the methods allowed by the Financial Accounting Standards Board (FASB) to account for a merger or acquisition. There are 12 requirements both companies must meet to qualify for pooling; otherwise, the purchase method must be used. There are several advantages to pooling.

Purchase method: The other method allowed by FASB to account for a merger or acquisition. Companies who cannot pool must use this method.

Term sheet: A summarized outline of the broad terms and conditions envisioned in making the acquisition, including a preliminary price. This document is usually only one or two pages in length.

References

Accounting Principles Board, *Business Combinations*, Opinion No. 16, 1970.

MacDonald, E., *The Wall Street Journal*, April 15, 1997, p. 4.

Pinches, G. E. *Essentials of Financial Management* — *4th ed.*, HarperCollins, New York, pg. 723, 1992.

Securities Data Corporation, New York, 1997.

Further Information

American Institute of Certified Public Accountants, New York.

Block, S. B. and Hirt, G. A. *Foundations of Financial Management, 7th ed.*, Irwin, Burr Ridge, IL, 1994.

Brigham, E. F. and Gapenski, L. C. *Financial Management Theory and Practice, 7th ed.*, Dryden Press, Orlando, FL, 1994.

Gitman, L. J. *Foundations of Managerial Finance, 4th ed.*, HarperCollins, New York, NY, 1995.

Institute of Certified Management Accountants, Montvale, NJ.

8.9 Pension Funds

Peter K. Clark

Many employers provide for employee income after retirement by establishing a pension plan, which states how much the retired employee is to be paid as a function of his or her income at retirement and number of years of service or as a function of the amount the employee and business have set aside for this purpose over the years. In the United States, businesses are required by law to set up a trust account, or **pension fund**, to make these retirement payments. In some cases (typically larger, older firms) there is one fund that is professionally managed and pays all retirement annuities; in other cases, individual funds are assigned to each employee. In total, these pension funds hold a large fraction of U.S. financial assets amounting to trillions of dollars and are becoming a dominant force in the U.S. capital market.

Pension Plans

A pension plan is an arrangement for providing and administering retirement annuities for a company's future and current retired employees. For older, larger businesses, and many governmental entities, this plan is typically a **defined-benefit pension plan**, which specifies the retirement benefits payable to retirees as a function of their number of years of employment with the employer, their age at retirement, and their salary history. For example, an employee for the state of California retiring at age 65 with 35 years of employment and a salary averaging $50,000 over the last three preretirement years of work might be entitled to 2% × 35 × $50,000 = $35,000 per year as a retirement annuity. This annuity would be paid out of a large pension fund (totaling billions of dollars) established by the state and funded by tax revenues. Note that with this arrangement, the sponsoring company or government agency bears all the investment risk associated with the fund. If the prices of the assets in the fund go down, the plan sponsor is obligated to make up the difference; if assets in the fund appreciate rapidly, the plan sponsor reaps the benefit by being able to make smaller payments into the fund. One advantage that at least partially offsets this investment risk is that funding of a defined-benefit plan is to some extent discretionary; funding can be diminished to enhance reported earnings relative to cash flow when cash flow is weak (and conversely), producing a less-variable earnings stream.

For smaller and newer businesses, the retirement annuity arrangement is typically a **defined-contribution plan**. In it, the plan sponsor (employer) and the employee make periodic payments into an account (trust fund) that belongs to the employee, and the employee is usually allowed to make investment decisions for the fund. One of the standard problems with these individual, employee-managed funds is that employee investment choices tend to be more risk averse than larger, professionally managed funds find appropriate. On the other hand, some defined-contribution plans require that some or all of the investment be made in the employer's stock; this requirement exposes employees to large (and unnecessary) levels of risk.

When the employee retires, he or she receives annuity payments based on the value of the fund. Thus, an individual's IRA (individual retirement arrangement) balance is a pension fund, as are the balances in accounts established under Section 401(K) of the U.S. Internal Revenue Code. With a defined-contribution plan, investment risk is borne entirely by the individual pension fund owner; if asset prices go up, retirement payments go up, and the converse is true. As the concept of lifetime employment fades in the United States, defined contribution plans are becoming more prevalent, both because they are less costly for new companies to set up and because they are portable.

In either case (defined benefit or defined contribution), these plans are regulated in the United States by the federal government under a complicated set of laws and rules known as ERISA (Employees' Retirement Income and Security Act) enacted in the mid-1970s. ERISA requires that pension plans that are underfunded (i.e., the present value of promised pensions is less than current assets), receive a minimum level of funding, and that low-paid workers are included along with higher-paid ones. Pension plans in other developed countries are regulated under different, sometimes equally complex arrangements.

Pension Funds

Usually, the term pension fund refers to large trust funds, sometimes totaling many millions or even billions of dollars, owned by large corporations, state governments, or large unions as the asset backing a defined benefit pension plan, as discussed above. The money in these large funds is usually managed by institutional money managers, who buy and sell assets on behalf of the plan sponsor. For example, AT&T has a multi-billion-dollar pension fund for its defined-contribution plan. The pension staff and the pension plan board of trustees at AT&T meet regularly to decide how to allocate these billions among a wide variety of money managers. The financial returns on each investment are closely monitored, sometimes with the help of outside consultants. Part of the fund might be invested in a Standard and Poor's 500 index fund whose return performance is designed to mirror the movement of the Standard and Poor's U.S. large stock index. Another part of the pension fund might be put into a corporate bond fund, run by a different manager that specializes in bond portfolios. Additional investments might be made in international equities or bonds, commercial or residential real estate, funds specializing in new ventures, or funds dedicated to commodity futures.

Periodically (e.g., four times a year), the pension fund's board of trustees meets to discuss the performance of the fund. Typically, the questions to be decided at such a meeting fall into two broad categories. (1) Are the fund's assets correctly allocated among various broad asset classes? (2) Within each asset class, is the current set of money managers achieving an acceptable level of returns? The answer to the the the first of these questions is called the **asset allocation** decision. Possible investments are divided into broad asset classes (e.g., domestic stocks, domestic bonds, international stocks, international bonds, real estate, and less-traditional investments, such as venture capital, hedge funds, or commodities). The investment committee or investment officer decides what fraction of the pension fund should be invested in each of these categories, usually based on the long-term historical performance of assets in each class. The asset allocation decision is typically the major determining factor in the risk and return characteristics of the overall pension fund. Asset allocation is reviewed periodically, because returns vary across asset classes, causing the asset mix to change over time. If the original asset allocation is deemed to be still appropriate, the pension fund's asset mix is **rebalanced** to maintain the desired proportion of the fund in each category. Asset allocation decisions are complicated by the fact that expected returns within each asset class may move over time; random variations in returns are so large that it is difficult to detect changes in asset-class expected returns and covariances. Still, reliance on long historical averages for guidance on asset allocation decisions may involve the use of data that are totally irrelevant under current circumstances.

The answer to the second question, "How well are the fund's managers doing?" is called **performance evaluation**. Performance evaluation is notoriously difficult because the return on any financial asset has a huge random component relative to its mean. For example, an individual stock may have an expected return of 2/3% in a month; the standard deviation of return might be 8 or 10% over the same period. A portfolio manager picking assets within a particular asset class might have substandard results over lengthy periods of time even though her investment strategy had an above-average expected value. Conversely, a portfolio manager with a superior track record may have been following an inferior strategy and just been lucky. By the time a particular money manager can show statistically that his performance is superior, he may be past retirement age. Thus, the choice of portfolio managers for a pension fund's money is necessarily heuristic, more art than a science. Despite the known unreliability of short-term return statistics, money managers posting high returns tend to receive new money from pension funds, and then tend to have their returns retreat to average or even below. The realization that this is true has greatly increased the popularity of **index fund** managers, who seek to match asset-class return averages at minimum expense rather than pay transaction fees trying to beat the average.

Conclusion

Due to both the tax advantage (retirement funds can be accumulated on a before-tax basis) and the need to encourage employees to save for their retirement, firms in the United States should seriously consider establishing a pension plan even if they have a small number of employees. Creating such a plan can be simple and easy to administer (in the case of a defined-contribution plan) or more complicated (in the case of a defined-benefit plan), but, either way, the pension fund or funds created will help fund income security for retired employees in the future.

Defining Terms

Asset allocation: The division of a pension fund's total into the amount to be invested in different general classes of asset (e.g., stocks, bonds, cash, international, real estate, etc.).

Defined-benefit pension plan: A legal document specifying retirement benefits to be paid by an employer to its retired workers as a function of their number of years of employment, age at retirement, and preretirement salary history.

Defined-contribution pension plan: A legal document specifying employer and employee payments to an individual trust fund intended to fund the employee's retirement payments.

Pension fund: A trust fund dedicated to the payment of employee retirement benefits. In the case of a defined-benefit plan, this large fund is owned by the employer, who is responsible for any shortfall in the fund due to inadequate investment returns. In the case of a defined-contribution plan, each smaller fund is owned by the employee.

Performance evaluation: Quantitative and qualitative analysis of a money manager's return performance, typically comparing it to asset-class averages and to the performance of other money managers.

Rebalancing: Buying and selling securities in a portfolio to reattain a previously determined asset mix.

References

Arnott, R. D. and Fabozzi, F. J. 1992. *Active Asset Allocation*, Probus, Chicago, 1992.

Logue, D. E. *Managing Corporate Pension Plans*, HarperCollins, New York, 1991.

Maginn, J. L., and Tuttle, D. L. *Managing Investment Portfolios*, 2nd ed., Warren, Gorham, and Lamont, New York, 1990.

8.10   Real Estate : Rights, Development and Value

Gordon Brown

Real estate is land and all things attached to land in such a manner as to make them permanent. Fixtures are nonpermanent improvements to real property that may be removed at any time. There is often argument between buyers and sellers regarding the delineation of permanent real estate and fixtures.

Ownership of real estate (also called real property) is contained in a series of rights. The ownership of rights to all or part of a real property may, and often is, divided among several parties. Rights to the surface, subsurface, and air space may be divided. The "**Bundle of Rights**" associated with a particular real property include four specific rights, one or more of which may be conditioned by private or public limitations: (1) the "right of possession", which is the right to occupy or use and usually to keep other out, (2) the "right of control", which is the right to make changes in the property, physically or economically, (3) the "right of quiet enjoyment", which protects the current owner from interference by past owners or others, including neighbors, and (4) the "right of disposition or transfer", which enables the current titled owner to transfer all or part of the real property or rights. Be absolutely certain of the rights you hold or those you intend to purchase.

The "bundle of rights" owned is legally termed an estate. A "freehold estate" contains the highest quality of real property rights because they generally include all rights categories. The terms "fee" and "fee simple" are equivalent terms meaning the unconditional use, control, enjoyment, and disposition of real property. A defeasible fee estate contains some condition to the rights granted either permanent or for a specified period of time. "Life estates" are granted only for a specified period of time, usually a person's lifetime, after which ownership passes to another.

"Leasehold estates" grant only certain rights, usually the right of use or possession for a time. A landlord retains the other rights plus the right of reversion of possession after the stipulated time period. Leaseholds should be written but may be oral under certain circumstances. Lease payments called rent may be of a fixed or flat amount, graduated according to some agreed upon increase or decrease over time, escalated along with a predetermined index such as the Consumer Price Index, or calculated as some percentage of the gross or net revenues generated by the use of the real property.

Many people have rights to real estate but may not own it. For example, an "easement" is the right to use a property but not the right to possess. Power and telephone companies and other utilities own many rights-of-way, which give them use but not possession. An "express easement" is one explicitly stated in writing. An implied easement is created by necessity or years of unrestricted use.

Other nonpossessionary interests in real estate include a "license", which is the right to go on the land of another, including the use of certain improvements, for a specified purpose. A ticket to a sporting event is a license. "Profit" is the right to take a portion of that which is owned by another. Examples are timber, soil, gas, and oil rights.

Valid Deeds

Real estate deeds specify the quantity and qualtity of rights owned or titled to the current owner of record. A history of title called an Abstract is in the public record on file in the courthouse of the county in which the property is situated. Changes in ownership rights should be recorded on the public record whenever any portion of that title has been transferred.

To be valid, deeds to real property must be in writing. The deed must identify the person or persons to whom title is being conveyed. The deed must accurately describe the property. Finally, the deed must be signed by the grantor or conveyor. Valid deeds should be immediately recorded at the courthouse to give the world "notice" or "constructive notice" of the grantee's interest in the title. All claims to title are recorded on a first-come, first-served basis, with the exception of taxes, which supersede other claims and interests in title.

The type of deed you possess specifies the quality of rights you are granting or receiving. Deeds may be one of three types. (1) General Warranty deeds warrant or guarantee title. The grantor of a deed promises that the bundle of rights associated with a specifically described real property are owned by that grantor and can be transferred to the grantee. The warranty is, in effect, a pledge that the grantor has liability if the bundle of rights as identified in the deed are not whole or complete. A title insurance company often is paid a fee to research the title and to stand liable for clear and complete title as known at the time of the search. (2) Special Warranty deeds warrant only against defects to title while the grantor owned the property and against defects to title which the grantor created. Theoretically, if clear and complete title passed to the grantor and the grantor did not create any title defects while owning the property, clear and complete title rights can be passed to the grantee. Nonetheless, Special Warranty deeds offer less guarantee of title to the grantee than general warranty deeds. (3) Quitclaim deeds transfer only the rights owned by the grantor. No warranty is given regarding those rights or title. No explicit description of the rights or title is proffered. A statement by the grantor giving "any and all interest owned by the grantor" is generally all that is provided to the grantee. When ownership rights are unclear and many owners are involved, quitclaim deeds signed by all prospective interested parties may be the simplest way of uniting title under one grantee.

Once again, be certain of the rights you own or those that are available for purchase. A real estate attorney is able to identify the quantity and quality of the rights associated with a specific real property.

Public Limitations to Title

Regardless of the privately deeded rights to real estate that you possess, public policies and administration may affect both the quality and quantity of your rights. Federal, state, and local governments use the planning process to establish land-use controls through zoning regulations. Subdivision ordinances and building codes control the location, building footprint, and materials used in construction.

Public agencies use the power of "eminent domain" to actually take all or a portion of a vital property for the public good. The government initiates legal action, "condemning" the property and proposing "just and fair compensation" to the owner for the property or the portion of the property taken. Each owner of condemned property has the right to seek an independent definition of just and fair compensation. That owner may also seek "consequential damages" or "severance damages" resulting from damages arising from the taking of only a portion of the property.

Common law restrictions prevent the uncontrolled use of real property for purposes that are not in the interests of the general public. The laws of nuisance restrict uses that may cause excessive noise, pollution, traffic, moral decay, or other potentially harmful by-products.

The Land Development Process

Once you own rights in real property and possess a valid deed or contract verifying those rights, the land development process can begin. Actually, you will want to protect yourself by ascertaining your potential rights prior to acquisition. You will also want to determine any and all limitations to use (which includes land and building construction) as part of a feasibility analysis. Feasibility analysis explores legal, physical, market, and financial opportunities and limitations, determining at its end whether an investment should be undertaken.

Be certain that all necessary utilities and services are available and deliverable to the site. Commitment letters from service suppliers reduce uncertainty and speed the land development and construction process. Physical site issues should be analyzed in depth by architects, engineers, and construction specialists. Cost estimates for both building and land preparation should be obtained for every phase of development. Add 10% or more to the bids for unknown site costs that may arise in the land development process. Trade associations and local building organizations can provide guidelines for expense estimation. (See the list of essential trade associations on the final page of this section.)

Market feasibility analysis attempts to predict the demand for the proposed space and associated enterprises. Potential consumers for the specific type of developed real estate should be listed and qualified. Rental rates or sales prices of competing properties, complete with amenity packages, advantages or disadvantages of the location, and marketing strategies to attract consumers should be presented in detail.

Financial feasibility analysis projects whether the real estate investment will be profitable. Once costs of acquisition and construction are calculated, seek mortgage financing (try the local financial institutions first) and estimate those costs. Then all of the legal, physical, financial, and market analysis numbers can be fed into a financial model to help determine the overall feasibility of the proposed development. The Proforma statement appearing in Fig. 8.3 illustrates one sequence of income and expense categories to be analyzed.

GROSS POTENTIAL INCOME FROM THE PROPERTY   _____

      Less: Vacancy Allowance & Credit Loss   _____

EFFECTIVE GROSS RENT   _____

      Plus: Other Income   _____

EFFECTIVE GROSS INCOME   _____

      Less: Total Operating Expenses   _____

             Reserves for Replacement   _____

NET OPERATING INCOME   _____

      Less: Interest Expense   _____

NET INCOME BEFORE TAXES   _____

      Less: Income Taxes   _____

             Principal Repayment   _____

      Plus: Depreciation   _____

NET INCOME AFTER TAX (CASH FLOW)   _____

FIGURE 8.3   Proforma income statement.

Construction

Each construction cost estimate needs ultimately to be finalized in contract form. Cost estimation forms are usually available through real estate or construction associations. Labor rates and materials vary from market to market. Bids, generally three or more in number for each component, are requested from contractors and subcontractors for each critical job. One general contractor may be asked to assume the risk of the total construction for a fee. Alternatively, several subcontractors may manage and construct their own specific area. The choice of one general contractor or many subcontractors depends on the ability of the owner or investor to supervise the construction. Each contractor should be expected to coordinate the flow of materials, labor, necessary permits and inspections, bill payments, and end-user scheduling as part of job completion. As an owner or investor, do not underestimate the complexity of construction.

The Economic Value of Real Property

Real estate derives its value from location and from the improvements made to that location that are measured by the future benefit stream anticipated from that property in use. The quantity of and the probability of receiving those anticipated future benefit streams is a vital part of real estate appraisal methodology.

Real estate is different from other economic commodities in that each property is unique and fixed in place. Since it cannot be moved, the local market dominates demand/supply interactions for the land and any improvements thereto. The linkage between the subject property's use and that of other properties around it establishes demand and supply preferences in the local market. The availability of essential utilities, zoning, transportation, and financing are key factors in determining locational preferences within the local market. New construction generally requires one or more years to be put in place, which adds value to existing properties with "situs" value. The complexity of constructing new space serves as a barrier to new competition.

Since real estate is a long-lasting and expensive commodity, long-term financing is generally available in both public and private financial markets. The cost of capital is very important to a consumer or a business desiring additional or improved space because that cost is usually fixed over 15 to 30 years. On the positive side, tax incentives exist to spur new construction and to refurbish historic structures.

The local nature of real estate means that the land development process is controlled by the local municipality. The local township, borough, or city building or the county courthouse maintains records of ownership, development, taxation, utility services, highway construction, building code compliance, and environmental issues. The local municipal administrator should be able to direct you to the appropriate sources of information and control officers.

Real Estate Valuation Terminology

The term "**market value**" is but one of several valuation concepts that are often confused one with another. Market value is that expected price at which a ready, willing, and able buyer would pay and a ready, willing, and able seller would sell, each under no duress and with reasonable market knowledge. The term "cost" is historical. It is accounting in nature. It represents the price paid to purchase or to construct. It may equal market value or it may not. "**Investment value**" differs from both market value and cost in that it defines what something is worth to you, the investor. A good or service may be worth more or less to one person or another.

"Highest and best use" is a term used mainly by appraisers. It is that "reasonable and probable use that will generate the highest present value as of the effective date of an appraisal." It is the end use that all real property owners seek. However, due to changes in local economies, highest and best use is ever changing. Land and the improvements thereto often function at a less-than-highest return level. Highest and best use does not necessarily mean that more-extensive improvements are necessarily better. Those investments that produce the highest return are considered "best".

Be careful to note what value is being discussed. Tax value, condemnation value, loan value, insurance value, and estate value may all differ from general market value because each is calculated for a different purpose. Appraisers generally qualify their appraisals according to the type of value sought and the uncertainties that exist in the local and national marketplace.

Appraisal Methods

Three different approaches are used to estimate value. The relative importance of the cost approach, the market comparison approach, and the income approach in estimating market value depends upon use of the property and the availability of meaningful data for analysis. For example, a single-family home that is not rented to others has no rental history or reason for income analysis; hence, the focus is on the cost and market approaches to value. An office for rent is primarily income producing, which makes the income approach most applicable.

The *cost approach* is concerned primarily with the physical cost of the land and producing or reproducing the land and the improvements of the subject property. A fundamental assumption is that a reasonably informed purchaser would not pay more than the cost of producing a substitute property possessing the same utility. The site value plus the cost of the improvements serves as a good measure of physical and locational value.

The appraisal should determine whether deductions to that measure of value should be made for physical depreciation and for functional and economic obsolescence, which represent diminished utility of the property. Each of the three types of obsolescence should be calculated separately. Physical depreciation measures wear and tear of the structure due to time and use, including the absence of orderly and systematic maintenance. Functional depreciation or obsolescence represents the decreased capacity of the structure to serve the function for which it was intended particularly with reference to current standards of performance. Economic or locational obsolescence results from external changes in the locational desirability of the property.

All three types of depreciation may be either curable or incurable. If physical depreciation can be cured by improving the structure, that amount is the cost to cure. If the physical structure, all or in part, is not able to be cured with reasonable economic investment, then the depreciation is said to be incurable. Functional deficiencies such as style, size, or old technologies may be curable or incurable as well. Locational obsolescence may require off-site investment to cure. Generally speaking, physical depreciation is more apt to be cured than functional or locational obsolescence.

The *market comparison approach* uses the actual sales of similar properties in the local real estate market to predict the market value of a subject property. Market-comparable properties assume that ready, willing, and able buyers have interacted with ready, willing, and able sellers, each under no duress and with reasonable knowledge. Market researchers seek comparable sales that have occurred at arms' length. Properties with unusual financing terms are either rejected for comparison purposes or adjusted to reflect current cash transactions. Only recent sales are considered. Sales of properties in close proximity to the subject property are considered primarily because they are influenced by the same neighborhood market conditions.

Adjustments to comparable property sales must accurately reflect differences between those properties and the subject property. Location, lot size, materials, square footage of the structure, age, materials locational differences, physical condition, neighborhood, traffic patterns, zoning regulations, deed restrictions, taxes, assessed values, etc. are just a few of the variables to be considered and/or adjusted when comparing sales prices of recently sold properties with the subject property.

The *income approach* to value translates income produced by a real property in use into market value using the following formula:

$$\text{Market value} = \text{net income/capitalization rate}$$

Net income is really net operating income, which is that net revenue produced by the operation of the property in its present use. When looking at the proforma income statement (see Fig. 8.3) for a property, know that low vacancies result from locational advantage, favorable rental rates, amenity differences, and good management. Income other than property rent can be earned from parking, vending machines, laundry facilities, or other on-site activities. Operating expenses are generally 25 to 40% of effective gross rents for a mature property. Properties operating outside that range may have either rental problems or expense control problems. Reserves for replacement (depreciation) may include more than the straight-line method used for income tax purposes. Long-term financing fluctuates according to market conditions and managements' ability to maximize cash flow to cover debt. In summary, the income statement tells both the operating and the financing story in brief. An appraiser may find it necessary to adjust the numbers for abnormal results in either category.

Income Capitalization

The purpose of the capitalization process is to convert anticipated income streams into present market value. Net operating income (NOI) reflects the income-producing ability of the property. The capitalization rate reflects the financing structure and investor preferences for rates of return and risk within current financial markets. Both NOI and capitalization rate numbers relate to a property's current use, not an alternative use.

A capitalization rate is a compilation of the rates of return required to attract and retire debt capital and equity capital. Those rates of return reflect investor (debt and equity) discounts for putting capital at risk for the return period. An appraiser selects a rate that approximates the risk and cost of capital in the local market. Historically, capitalization rates have ranged from 7.5 to 12.5%. That 5% rate variability can produce a significantly different market value vis-a-vis the income approach; therefore, appraisers are just as careful in selecting a capitalization rate as they are in accepting NOI statistics.

Choosing the Best Estimate of Value

Correlating the three market value estimates produced by three differing methods to arrive at a single value estimate for a specific moment in time is an art, and increasingly a science. Property type, age, and location and the availability of meaningful data favor one method's application over the others. All three approaches should be used if meaningful comparative data are available as of the date of the appraisal. Note that appraisers use the three approaches to merely estimate value. The best estimate of value is the one that correctly anticipates the future benefits to be derived from the subject property, those future benefits discounted into present value terms.

Reference Materials

Real estate manuals and books are available in a variety of places. The local board of realtors generally has its own library of reference materials as do public libraries. The Internet can provide information on real estate, construction, law, homes, appraisal, and property.

Feasibility data relating to population, traffic, building permits and other local construction data, retail sales, home prices, schools, shopping, mass transit, and other facts relating to real estate development are generally available in local libraries and in municipal buildings. Metropolitan statistics are available in university libraries, county buildings, and chambers of commerce. Public utilities also maintain files of data relating to urban and suburban population trends and macroeconomic statistics.

Defining Terms

Bundle of rights: The bundle of rights associated with real estate includes for specific rights: right of possession, right of control, right of quiet enjoyment, and right of disposition or transfer.

Income capitalization: The process of converting income into market value.

Investment value: Is neither market value nor cost but what it is worth to the investor.

Market value: The worth of the future benefit stream.

Further Information

Trade associations are potentially the best sources of information. Books and manuals on how to estimate costs and what you should look for when designing or renting or building real estate are a part of ongoing educational interests. Significant numbers of publications are available to the general public. The trade associations listed below can provide help to interested parties:

National Association of Realtors
430 North Michigan Avenue
Chicago, Illinois 60611
phone: 312-329-8200
fax: 312-329-8576

Institute of Real Estate Management
430 North Michigan Avenue
Chicago. Illinois 60611-4090
phone: 312-329-6000
toll-free: 1-800-837-0706
fax: 312-661-0217

American Institute of Architects
1735 New York Avenue NW
Washington. DC 20006
phone: 202-626-7300
fax: 202-626-7421

Construction Specifications Institute
601 Madison Street
Alexandria. VA 22314-1791
phone: 703-684-0300
toll-free: 1-800-689-2900
fax: 703-684-0465

Appraisal Institute
875 North Michigan Avenue. Suite 2400
Chicago, IL 60611-1980
phone: 312-335-4100
fax: 312-335-4400

References

The Appraisal of Real Estate, Special Technical Committee, American Institute of Real Estate Appraisers, Chicago.

Essentials of Real Estate Investment, David Sirota, Dearborn Financial Publishing, Chicago.

Modern Real Estate, Charles Wurtzebach and Mike Miles, John Wiley & Sons, New York.

Facilities Manager's Reference: Management, Planning, Building Audits, Estimating, Harvey Kaiser, R. S. Means Company, Kingston, MA.

Data Sources for Real Estate Market Analysis, C. F. Sirmans, University of Connecticut Center for Real Estate & Urban Economics Studies, Storrs, CT.

Real Estate Development Workbook and Manual, Howard Zuckerman and George Blevins, Prentice Hall, Englewood Cliffs, NJ.

Real Estate Accounting and Reporting Manual, Richard Carlson, Warren, Gorham & Lamont, Boston.

Real Estate Investments and How to Make Them, Milt Tanzer, Prentice Hall, Englewood Cliffs, NJ.

Real Estate, Jerome Dasso, Prentice Hall, Englewood Cliffs, NJ.

Real Estate Finance, John Wiedemer, Prentice Hall, Englewood Cliffs, NJ.

8.11 Understanding Equipment Leasing

Kevin P. Prykull

History

Leasing was first recorded by the Sumerians sometime before 2000 B.C. in the City of Ur. Sumerian lease documents were written in clay tablets and recorded transactions ranging from the lease of agricultural tools, land, and water rights to oxen and other animals. In the Middle Ages, many knights of old were known to have leased their armor. Modern leasing arose with the railroads during the late 19th century as a result of cash and capital shortages within that industry. Leasing provided a solution to the shortage of cash and capital, a major reason to this very day why companies engage in lease financing.

Industry

Annual leasing volume in the United States during 1996 was approximately $170 billion. The industry is a fairly mature one, at least in the United States, with approximately a 2 to 3 % growth rate per year. Interestingly, about one third of all equipment acquired in the United States is leased! Worldwide, annual volume is approximately $500 billion and is concentrated in North America (41.4%), Europe (26.4%), and Asia (25.6%).

Overview

A **lease** is a contractual arrangement by which the owner of an asset (**lessor**) grants the use of his/her property to another party (**lessee**) under certain conditions for a specified period of time. Leasing is a very practical and flexible financing vehicle for today and serves as an alternative form of long-term debt financing for many companies.

The key to understanding a lease transaction hinges on the following: "In a lease the opportunity exists to separate ownership of the asset from its usage." The lessee is paying to use the equipment, in many cases for a period of time that is less than the equipment's economic life. Thus, lease payments can be lower than would be available in a similar loan financing. Many lessors assume an expected value to be realized from the equipment itself at lease maturity. This salvage value is known as **residual**. Lessors hope to realize the minimum assumed residual in a lease by either having the lessee purchase the equipment at the end of the lease, extend or renew the lease for an additional period of time, or return the equipment to the lessor, who in turn either will sell the equipment or re-lease the equipment to another party.

TABLE 8.10 New Leasing Business by Type of Equipment in the United States (%)

Transportation	31.2
Aircraft	9.3
Rail	9.1
Trucks and trailers	12.2
Computers	21.9
Industrial/manufacturing	9.1
Office machines	7.9
Construction equipment	7.7
Telecommunications	4.1
Medical	3.1
FF&E	2.7
Material handling	2.6
Electric power	2.0

Source: Equipment Leasing Association of America, 1995. Automobile lease volume is excluded. Equipment categories of less than 2% are not shown.

A legal document says that the word "lease" or "master equipment lease agreement" at the top does *not* necessarily mean that we actually treat it that way from an accounting, tax, or legal perspective in the United States. In the United States, what matters is the substance of the transaction and not its form! The tax and legal dimensions of leasing require the

TABLE 8.11   The Dimensions of Leasing

Accounting[a]	Operating or capital
Tax	Tax or nontax
Legal	True or secured transaction

[a]Lessor and lessee are not required to agree on classification.

lessee and lessor to agree on the classification of the lease, thus a mirror image requirement. The accounting treatment however differs in that the lessor and lessee do not have to agree on the identical classifcation of the lease. Thus four possible combinations exist in the accounting for a lease.

Accounting

Both lessors and lessees are required to classify their lease transactions into one of two categories — operating or capital.

Operating Lease vs. Capital Lease

Operating lease	Behaves like a usage agreement and typifies the lease concept; the lessee uses the equipment, while the lessor is treated like the owner of the equipment.
Capital lease	Is analogous to a purchase of an asset by the lessee and a related financing, i.e., a loan, provided by the lessor.

Disclosure Requirements

Operating leases are disclosed only in the footnotes to the financial statements. On the other hand, **capital leases** are required to be shown in the actual financial statements themselves under fixed assets and under liabilities (with that principal portion of the lease coming due within the next 12 months shown as the current portion of term debt and the remainder shown as long-term debt). Given this accounting treatment, lessees generally prefer the off-balance sheet nature of operating leases.

FASB #13

A complete discussion of **Generally Accepted Accounting Principles (GAAP)** and the **Financial Accounting Standards Board Statement #13 (FASB #13)** is beyond the scope of this section; however, its mention is critical. FASB #13 is the guideline by which lessors and lessees use to determine if a lease transaction is to be classified as operating or capital. If any *one* of the four criteria is met, the lease will be treated as capital. The four capital lease criteria are:

1. Automatic ownership transfer
2. Bargain purchase option
3. Lease term is greater than or equal to 75% of the asset's economic life
4. The present value of the minimum lease payments is greater than or equal to 90% of the asset's fair market value

Both lessor and lessee consider the criteria and draw their own independent conclusion on the lease classification. The lease classification into the operating or capital category is made at the inception of the lease and is irrevocable, notwithstanding future changes in economic circumstances. Finally, the tax status of the lease does not impact the accounting determination.

Tax vs. Nontax Lease

Why does a **tax lease** work? Because of **MACRS** depreciation. MACRS — **Modified Accelerated Cost Recovery System** is one of the acceptable and by far the most-favorable methods of depreciating assets for tax purposes. MACRS provides for a more rapid write-off of the depreciation expense for tax compared to other alternative methods, such as straight line. The IRS recognizes depreciation expense as a legitimate tax deduction, thus reducing a firm's taxable income. This creates a benefit for the owner of equipment. In a

| Tax lease | lessor gets MACRS | this behaves like a lease |
| Nontax lease | lessee gets MACRS | this behaves like a loan |

There is no "sharing" of MACRS depreciation. Thus, either the lessor takes 100% of the available benefit (in a **tax lease**) or the lessee takes 100% of the available benefit (in a **nontax lease**). Only one party is entitled to take the MACRS depreciation. MACRS has "value" to the party who claims it. Thus, a lessee may be able to obtain a lower lease payment on a tax lease, all things being equal, than on a similar nontax lease or a loan.

True Lease vs. Secured Transaction

Under the **Uniform Commercial Code**, which governs certain business transactions in all 50 states, two specific articles apply to leases:

| Article #9 | secured transaction | this behaves like a loan |
| Article #2A | true lease | this behaves like a lease |

Article #9 for **secured transactions** is applied to certain leases that behave more like a loan with the equipment acting as collateral for the financing. In bankruptcy, the lessor is treated like any other secured creditor and subject to the normal bankruptcy procedures. Article #2A, on the other hand, was written recently (approved in all but three states) and gives a better frame of reference for leasing and lease transactions. In a Chapter 11 bankruptcy, the lessee through the bankruptcy court can petition to have the lease either "accepted" or "rejected". If the lease is accepted, the equipment will remain with the bankrupt entity and the lessee normally will resume paying the lease payments to the lessor. If the lease is rejected, the equipment will be returned to the lessor immediately.

Tax and Legal Dimensions are Similar

In the assessment of whether a lease transaction is a tax lease or a nontax lease for tax purposes and a secured transaction or **true lease** for legal purposes, two questions tend to be frequently asked:

1. What was the intent of the parties? That is, which party is to receive the benefits of MACRS and will Article #2A or #9 apply? Normally, these areas are addressed in the lease documentation and agreed to by both lessor and lessee in writing.
2. Who is "at risk" with respect to the equipment? That is, does the lessor or lessee assume the risks of ownership? Both the IRS and the courts look to things such as residual, end-of-lease term options, title transfer, purchase options, etc. in their final determination.

Dynamics of Leasing

Understanding the dynamics of leasing is essential in evaluating the lease decision. There are a number of reasons why people lease:

Lessee Reasons and Motives to Lease

Major categories	Benefits
Financial reporting considerations	Favorable financial statement impact
Income tax factors	Tax advantages may be available
Financial reasons	Less restrictive than loan financing
Ownership issues	Do not have to own equipment to use it
Cash management	Conservation of cash and 100% financing
Obsolescence avoidance	Hedge against technological obsolescence
Service, flexibility, and other factors	Convenience and service factors

Financial Reporting Reasons

The reported net profit for a lessee under an operating lease will be higher than that reported under a capital lease or similar loan arrangement during the early years of the lease term. This favorable earnings treatment results from lease expense being less than the sum of depreciation plus interest expense. On the balance sheet the lessee is *not* required to report any asset nor liability under an operating lease. Thus, financial ratios used by both creditors and by investors "appear" better. Specifically, an operating lease can give the financial "illusion" of improved liquidity, less leverage, higher profitability, and better returns on asset and equity. Further, this treatment is completely acceptable under GAAP. Accordingly, operating leases are very much sought as a preferred financing methodology by corporate treasurers and chief financial officers of corporations.

Income Tax Factors

Leasing can assist customers in avoiding certain tax penalties, such as the mid-quarter convention and alternative minimum tax and/or preserve certain tax benefits available through MACRS depreciation. A firm that has generated significant net operating losses for tax purposes does not pay taxes and cannot make use of MACRS. In these situations lessees wish to "exchange" MACRS to the lessor for lower periodic lease payments. In this way both lessor and lessee can readily benefit from a tax lease.

Financial Reasons

Leasing provides an additional source and diversification of capital that leaves other financing sources intact. Lines of credit provided by the customer's bank can remain available for working capital needs and not be improperly used to acquire equipment. Bank term loans can be kept available to finance the nonequipment needs of the firm.

Leasing is a less-restrictive form of financing compared to loan financing, particularly loans from commercial banks. There are no or few **financial covenants** contained within the equipment lease agreement. These restrictive covenants can hamper a firm's ability to operate independent of the loan source and can cause problems for the lessee when violated. Finally a lease may help a lessee in circumventing existing loan covenants, as leases sometimes are not included in the definition and calculation of debt and leverage ratios frequently used in covenants.

Cash Management

Clearly the main reason to lease for most lessees is for preservation of liquidity and the minimization of cash flow requirements over the term of the financing. Leases can offer the lowest possible payment to a lessee due to the residual position assumed by the lessor on the equipment, tax benefits exchanged to the lessor in a tax lease, and longer lease terms due to the equipment's economic life. Leasing generally does not require any down payment and provides for 100% financing of the asset's purchase price. Leases offer fixed rate, fixed payment financing, unlike commercial banks. Finally, with leasing one can acquire more equipment for the budget dollar compared to loan financings. This may provide great assistance to plant managers, engineers, and operations people. If a treasurer has a liquidity preference, leasing makes a great deal of sense.

Obsolescence and Ownership Issues

Leasing provides a hedge against technological obsolescence of the equipment. For example, the leasing of computers shifts much of the technology risk of the equipment to the lessor. Lessors normally permit selected upgrades and modifications to leased equipment, the swap of similar-type equipment, and the rollover of a current lease into a new one. Ownership of equipment is not always available or feasible. For example, the exclusive ownership of a satellite is not likely. Thus, with leasing one can obtain the use of the equipment without its actual ownership. Finally in some cases certain limited environmental risks may be shifted to the lessor under a lease arrangement as owner of the equipment.

Service, Flexibility, and other Motivations

Convenience is frequently cited as a reason for leasing. Some lessors provide "full service" or "bundled leases" where the lessee's payment includes such things as insurance, maintenance, taxes, etc. Leasing offers one-stop shopping for the acquisition and financing of equipment needs. Particularly with captive leasing companies (those owned by the manufacturer of the leased equipment), lessees perceive an element of control over the manufacturer in the event of an equipment problem. Leasing can also help avoid disposal/salvage hassles associated with the equipment at the end of the lease term. Unlike commercial banks and traditional loan financings, lessors are prepared to offer flexible payment structures to include selected skips payments, step payments, and contingent payments (based on usage) to better meet cash-flow requirements. Finally, leases can sometimes be paid from an operating rather than a capital budget, thus expediting the equipment acquisition process.

Lease Products

There are numerous lease products available in the marketplace. Each is designed to meet specific needs for both lessees and lessors. *Conditional sales agreements* are capital leases that behave much like a loan financing with the lessee taking MACRS and the courts treating the lease as a secured transaction. Normally these are structured with a $1 buyout or a low prestated fixed purchase option at lease maturity. A *single investor tax lease* is normally treated by the lessee for accounting purposes like an operating lease and for tax and legal purposes like a tax lease and true lease, respectively. A single investor tax lease can offer low periodic payments because the lessor assumes a residual position in the equipment and receives the MACRS tax depreciation. The lessee normally has the option to purchase the equipment at the end of the lease term for its then-fair-market value, renew the lease at some to-be-determined amount, or return the equipment to the lessor. *Leveraged leases* use a significant amount of nonrecourse debt in very large deals (e.g., $30 million or more) to provide a low-cost solution to the lessee by offering longer-lease terms and low payments. Leveraged leases work well with long-lived equipment and better-rated credits as a lessee. A *TRAC lease* (terminal rental adjustment clause) is a tax lease even though the lessee guarantees all (or a portion of) the residual value at the end of the lease term. TRAC leases are only permitted for truck, trailer, and autos for business use. A TRAC lease can provide very low payments to the lessee, due to the tax benefits and the assumed residual, while fixing today the end of lease term costs and consequences for the lessee. *Synthetic leases* are a relatively recent development in leasing. In a synthetic lease the transaction is structured to be treated by the lessee as an operating lease for accounting purposes and a nontax lease for tax purposes. Accordingly, the lessee gets off-balance-sheet treatment for accounting, yet claims tax depreciation under MACRS. This product has become quite popular for firms in cyclical industries that have turned tax positive due to the long economic recovery period that we have experienced during the 1990s. A *first-amendment lease* is a tax lease for tax purposes but is designed to fail the 90% test under FASB #13 and thus be treated as an operating lease by the lessee for accounting purposes. Most have both a purchase option at lease maturity for a predetermined amount (representing a cap on the equipment's fair market value) and/or a structured renewal in a predetermined amount and for a certain predetermined period of time. Many of these leases are structured with early buyout options as well. Lessees receive low periodic payments due to the exchange of tax benefits to the lessor and the assumed residual value at the end of the lease. Also, the lessee knows today what are the end-of-lease options and consequences. Finally, there are *municipal leases* and *tax-exempt government leases* to meet the needs of the local municipalities and school districts and the federal government, respectively.

Conclusion

From the knights of old in the Middle Ages to the chief financial officer of today's most noteworthy corporations, leasing has provided the right financing solution for the acquisition and use of equipment. With over $500 billion dollars in equipment leased last year alone worldwide, it is no wonder leasing is considered so frequently in the capital budgeting process at most firms today. By separating ownership of an asset from its usage, a lease can be crafted to meet the sophisticated needs of a lessee. Although the accounting, tax, and legal aspects of a lease seem burdensome, the astute lessee can realize the flexibility and opportunity created in a lease compared to a staid loan financing. Numerous lease products have been created, as discussed in the prior pages, to reflect the different dimensions of leasing. Lessees are also motivated to lease for qualitative reasons in addition to leasing's quantitative financial benefits. In conclusion, leasing is rapidly becoming a preferred choice of alternative long-term debt financing for treasurers and chief financial officers worldwide.

Defining Terms

Capital lease: Analogous to a purchase of an asset by the lessee and a related financing (i.e., loan) provided by the lessor.

FASB #13: The guideline used in the accounting treatment of a lease in determining whether the lease is operating or capital.

Financial covenants: Financial performance targets and restrictions placed on a borrower by a bank.

GAAP: Generally Accepted Accounting Principles. GAAP is a collection of accounting principles that have substantial authoritative support.

Lease: A contractual arrangement by which the owner of an asset (lessor) grants the use of his or her property to another party (lessee) under certain conditions for a specified period of time.

Lessor: The owner of the equipment and/or the party that provides the lease financing arrangement.

Lessee: The user of the equipment and/or the party that requires the lease financing arrangement.

MACRS: Modified Accelerated Cost Recovery System is an accelerated method of tax depreciation allowed by the Internal Revenue Service. The tax owner of an asset can use MACRS depreciation and obtain the related tax benefits.

Nontax lease: Lessee is entitled to take the MACRS depreciation for tax.

Operating lease: Behaves like a usage agreement and typifies the lease concept. The lessee uses the equipment, while the lessor is treated like an owner of the asset.

Residual: The expected value of a piece of equipment at the end of a lease term.

Secured transaction: The lessor will be treated as a secured creditor by the courts. Article #9 under the Uniform Commercial Code will apply.

Tax lease: Lessor is entitled to take the MACRS depreciation for tax.

True lease: The legal system will treat the lessor as the legal owner of the asset. Article #2A of the Uniform Commercial Code will apply. In bankruptcy, the lessee will petition the court to "accept" or "reject" the lease.

References

Brigham, E. F. and Gapenski, L. C. *Financial Management: Theory and Practice*, 7th ed., pp. 929–961, Dryden Press, Orlando, FL, 1994.

Contino, R. M. *Handbook of Equipment Leasing — A Deal Maker's Guide*, 2nd ed., AMACOM, New York, 1996.

Financial Accounting Standards Board. *Statement of Financial Accounting Standards Number 13*, FASB, Norwalk, CT, 1976.

Halladay, S. D. and Amembal, S. A. *A Guide to Accounting for Leases*, Publishers Press, Salt Lake City, UT, 1992.

Halladay, S. D. and Amembal, S. A. *The Handbook of Equipment Leasing*, 2nd ed., P.R.E.P. Institute of America, Salt Lake City, UT, 1995.

1996 Survey of Industry Activity and Business Operations, Equipment Leasing Association, Arlington, VA, 1997.

Further Information

A world-renown leader in equipment leasing education and consulting is Amembal, Deane and Associates (AD&A) located at Suite 1115, 50 Broad Street, New York, 10004-2307. AD&A offer numerous courses, seminars, workshops, and conferences dealing with all aspects of equipment leasing in the United States and abroad. Telephone number is (212) 224-3984.

The Equipment Leasing Association of America is a trade group representing the equipment leasing industry. The organization is based in Arlington, VA, and publishes ten times a year a magazine entitled *Equipment Leasing Today*. Telephone number is (703) 527-8655 or Web site is http://elaonline.com/.

The *Monitor* Leasing and Financial Services is published by Molloy Associates with a circulation exceeding 18,000. Telephone is (610) 649-7112 or Web site is http://monitordaily.com.

United Association of Equipment Leasing (UAEL) offers courses and conferences in leasing. The organization has an academy and awards a certified lease professional designation. For information call (510) 444-9235.

8.12   Sensitivity Analysis[4]

Harold E. Marshall

Sensitivity analysis measures the impact on project outcomes of changing one or more key input values about which there is uncertainty. For example, a pessimistic, expected, and optimistic value might be chosen for an uncertain variable. Then an analysis could be performed to see how the outcome changes as each of the three chosen values is considered in turn, with other things held the same.

In engineering economics, **sensitivity analysis** measures the economic impact resulting from alternative values of uncertain variables that affect the economics of the project. When computing **measures of project worth**, for example, sensitivity analysis shows just how sensitive the economic payoff is to uncertain values of a critical input, such as the **discount rate** or project maintenance costs expected to be incurred over the project's **study period**. Sensitivity analysis reveals how profitable or unprofitable the project might be if input values to the analysis turn out to be different from what it is assumed in a single-answer approach to measuring project worth.

Sensitivity analysis can also be performed on different combinations of input values, that is, several variables are altered at once and then a measure of worth is computed. For example, one scenario might include a combination of all pessimistic values, another all expected values, and a third all optimistic values. Note, however, that sensitivity analysis can in fact be misleading [Hillier, 1969] if all pessimistic assumptions or all optimistic assumptions are combined in calculating economic measures. Such combinations of inputs would be unlikely in the real world.

Sensitivity analysis can be performed for any measure of worth. Since it is easy to use and understand, it is widely used in the economic evaluation of government and private-sector projects. Office of Management and Budget [1992] Circular A-94 recommends sensitivity analysis to Federal agencies as one technique for treating uncertainty in input variables. The American Society for Testing and Materials (ASTM) [1994], in its *Standard Guide for Selecting Techniques for Treating Uncertainty and Risk in the Economic Evaluation of Buildings and Buildings Systems*, describes sensitivity analysis for use in government and private-sector applications.

Sensitivity Analysis Applications

How to use sensitivity analysis in engineering economics is best illustrated with examples of applications. Three applications are discussed. The first two focus on changes in project worth as a function of the change in one variable only. The third allows for changes in more than one uncertain world.

The results of sensitivity analysis can be presented in text, tables, or graphs. The following illustration of sensitivity analysis applied to a programmable control system uses text and a simple table. Subsequent illustrations use graphs. The advantage of using a graph comes from being able to show in one picture the outcome possibilities over a range of input variations for one or several input factors.

---

[4]Contribution of the National Institute of Standards and Technology. Not subject to copyright.

Sensitivity Table for Programmable Control System

Consider a decision on whether to install a programmable time clock to control heating, ventilating, and air conditioning (HVAC) equipment in a commercial building. The time clock would reduce electricity consumption by turning off that part of the HVAC equipment that is not needed during hours when the building in unoccupied.

Using **net savings** (NS) as the measure of project worth, the time clock is acceptable on economic grounds if its NS is positive, that is, if its **present value** savings exceed present value costs. The control system purchase and maintenance costs are felt to be relatively certain. The savings from energy reductions resulting from the time clock, however, are not certain. They are a function of three factors: the initial price of energy, the rate of change in energy prices over the life cycle of the time clock, and the number of kilowatt hours (kWh) saved. Two of these, the initial price of energy and the number of kWh saved, are relatively certain. However, future energy prices are not.

To test the sensitivity of NS to possible energy price changes, three values of energy price change are considered: a low rate of energy price escalation (slowly increasing benefits from energy savings), a moderate rate of escalation (moderately increasing benefits), and a high rate of escalation (rapidly increasing benefits).

Table 8.12 shows three NS estimates that result from repeating the NS computation for each of the three energy price escalation rates.

To appreciate the significance of these findings, it is helpful to consider what extra information is gained over the conventional single-answer approach, where, say, a single NS estimate of $20,000 was computed. Table 8.12 shows that the project could return up to $50,000 in NS if future energy prices escalated at a high rate. On the other hand, it is evident that the project could lose as much as $15,000. This is considerably less than **breakeven**, where the project would at least pay for itself. It is also $35,000 less than what was calculated with the single-answer approach. Thus, sensitivity analysis reveals that accepting the time clock could lead to an uneconomic outcome.

TABLE 8.12 Energy Price Escalation Rates

Energy Price Escalation Rate	Net Savings ($)
Low	−15,000
Moderate	20,000
High	50,000

There is no explicit measure of the likelihood that any one of the NS outcomes will happen. The analysis simply shows what the outcomes will be under alternative conditions. However, if there is reason to expect energy prices to rise, at least at a moderate rate, then the project very likely will make money, other factors remaining the same. This adds helpful information over the traditional single-answer approach to measures of project worth.

Sensitivity Graph for Gas Heating Systems

Figure 8.4 shows how sensitive NS is to the time over which two competing gas heating systems might be used in a building. The sensitivity graph helps you decide which system to choose on economic grounds.

Assume that you have an old electric heating system that you are considering replacing with a gas furnace. You have a choice between a high-efficiency or low-efficiency gas furnace. You expect either to

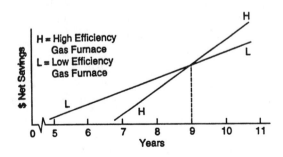

FIGURE 8.4 Sensitivity of NS to holding period.

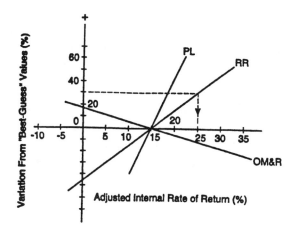

FIGURE 8.5   Spider diagram showing sensitivity of the adjusted internal rate of return to variations in uncertain variables. PL = project life, RR = reinvestment rate, and OM&R = operation, maintenance, and replacement costs.

last at least 15 to 20 years. You do not expect any significant difference in building resale value or salvage value from selecting one system over the other. So, you compute the NS of each gas furnace as compared to the old electric system. You will not be able to say which system is more economical until you decide how long you will hold the building before selling it. This is where the sensitivity graph is particularly helpful.

NS are measured on the vertical axis, and time on the horizontal axis. The longer you hold the building, the greater will be the present value of NS from installing either of the new systems, up to the estimated life of the systems. However, note what happens in the ninth year. One line crosses over another. This means that the low-efficiency system is more **cost-effective** than the high-efficiency system for any holding period up to 9 years. To the left of the crossover point, NS values are higher for the low-efficiency system than for the high-efficiency system. However, for longer holding periods, the high-efficiency system is more cost-effective than the low-efficiency system. This is shown to the right of the crossover point.

How does the sensitivity graph help you decide which system to install? First, it shows that neither system is more cost-effective than the other for all holding periods. Second, it shows that the economic choice between systems is sensitive to the uncertainty of how long you hold the building. You would be economically indifferent between the two systems only if you plan to hold the building 9 years. If you plan to hold the building longer than 9 years, for example, then install the high-efficiency unit. However, if you plan to hold it less than 9 years, then the low-efficiency unit is the better economic choice.

Spider Diagram for a Commercial Building Investment

Another useful graph for sensitivity analysis is the **spider diagram**. It presents a snapshot of the potential impact of uncertain input variables on project outcomes. Figure 8.5 shows — for a prospective commercial building investment — the sensitivity of the **adjusted internal rate of return (AIRR)** to three uncertain variables: project life (PL), the reinvestment rate (RR), and operation, maintenance, and replacement costs (OM&R). The spider diagram helps the investor decide if the building is likely to be a profitable investment.

Each of the three uncertain variables is represented by a labeled function that shows what AIRR value results from various values of the uncertain variable. (Although these functions are not necessarily linear, they are depicted as linear here to simplify exposition.) For example, the downward-sloping OM&R function indicates that the AIRR is inversely proportional to OM&R costs. By design, the OM&R function (as well as the other two functions) passes through the horizontal axis at the "best-guess" estimate of the AIRR (15% in this case), based on the best-guess estimates of the three uncertain variables. Other variables (e.g., occupancy rate) will impact the AIRR, but these are assumed to be known for the purpose of this analysis. Since each of the variables is measured by different units (years, percent, and money), the vertical axis is denominated in positive and negative percent changes from the best-guess values fixed at the horizontal axis. The AIRR value corresponding to any given percent variation indicated by a point on the function is found by extending a line perpendicular to the horizontal axis and directly reading the AIRR value. Thus, a 30% increase in the best-guess reinvestment rate would yield a 25% AIRR, assuming that other values remain unchanged. Note that, if the measure of AIRR were also given in percent differences, then the best-guess AIRR would be at the origin.

The spider diagram's contribution to decision making is its instant picture of the relative importance of several uncertain variables. In this case, the lesser the slope of a function, the more sensitive is the AIRR to that variable. For example, any given percent change in OM&R will have a greater impact on the AIRR than will an equal percent change in RR or PL, and a percentage change in RR will have a greater impact than an equal percentage change in PL. Thus, an investor will want to know as much as possible about likely OM&R costs for this project because a relatively small variation in estimated costs could make the project a loser.

Advantages and Disadvantages

There are several advantages of using sensitivity analysis in engineering economics. First, it shows how significant any given input variable is in determining a project's economic worth. It does this by displaying the range of possible project outcomes for a range of input values, which shows decision makers the input values that would make the project a loser or a winner. Sensitivity analysis also helps identify critical inputs in order to facilitate choosing where to spend extra resources in data collection and in improving data estimates.

Second sensitivity analysis is an excellent technique to help in anticipating and preparing for the "what if" questions that are asked in presenting and defending a project. For instance, when one is asked what the outcome will be if operating costs are 50% more expensive than expected, one will be ready with an answer. Generating answers to "what if" questions will help assess how well a proposal will stand up to scrutiny.

Third, sensitivity analysis does not require the use of probabilities, as do many techniques for treating uncertainty.

Fourth, sensitivity analysis can be used on any measure of project worth.

Finally, sensitivity can be used when there are little information, resources, and time for more-sophisticated techniques.

The major disadvantage of sensitivity analysis is that there is no explicit probabilistic measure of **risk exposure**. That is, although one might be sure that one of several outcomes might happen, the analysis contains no explicit measure of their respective likelihoods.

Defining Terms

Adjusted internal rate of return (AIRR):   The annual percentage yield from a project over the study period, taking into account the returns from reinvested receipts.

Breakeven: A combination of benefits (savings or revenues) that just offset costs, such that a project generates neither profits nor losses.

Cost-effective: The condition whereby the present value benefits (savings) of an investment alternative exceed its present value costs.

Discount rate: The minimum acceptable rate of return used in converting benefits and costs occurring at different times to their equivalent values at a common time. Discount rates reflect the investor's time value of money (or opportunity cost). "Real" discount rates reflect time value apart from changes in the purchasing power of the dollar (i.e., exclude inflation or deflation) and are used to discount constant dollar cash flows. "Nominal" or "market" discount rates include changes in the purchasing power of the dollar (i.e., include inflation or deflation) and are used to discount current dollar cash flows.

Measures of project worth: Economic methods that combine project benefits (savings) and costs in various ways to evaluate the economic value of a project. Examples are life-cycle costs, net benefits or net savings, benefit-to-cost ratio or savings-to-investment ratio, and adjusted internal rate of return.

Net savings: The difference between savings and costs, where both are discounted to present or annual values. The net savings method is used to measure project worth.

Present value: The time-equivalent value at a specified base time (the present) of past, present, and future cash flows.

Risk exposure: The probability that a project's economic outcome is different from what is desired (the target) or what is acceptable.

Sensitivity analysis: A technique for measuring the impact on project outcomes of changing one or more key input values about which there is uncertainty.

Spider diagram: A graph that compares the potential impact, taking one input at a time, of several uncertain input variables on project outcomes.

Study period: The length of time over which an investment is evaluated.

References

ASTM. Standard guide for selecting techniques for treating uncertainty and risk in the economic evaluation of buildings and building systems. E1369-93. *ASTM Standards on Buildings Economics, 3rd ed.*, American Society for Testing and Materials, Philadelphia, 1994.

Hillier, F. The derivation of probabilistic information for the evaluation of risky investments, *Manage. Sci.*, p. 444, April 1963.

Office of Management and Budget. 1963 *Guidelines and Discount Rates for Benefit-Cost Analysis of Federal Programs*, p. 12–13, Circular A-94, October 29, Washington, DC, 1992.

Further Information

Marshall, H. E. *Techniques for Treating Uncertainty and Risk in the Economic Evaluation of Building Investments*, Special Publication 757, National Institute of Standards and Technology, Gaithersburg, MD, 1988.

Ruegg, R. T. and Marshall, H. E. *Building Economics: Theory and Practice*, Chapman and Hall, New York, 1990.

Uncertainty and Risk, part II in a series on least-cost energy decisions for buildings, National Institute of Standards and Technology, 1992. VHS tape and companion workbook are available from Video Transfer, Inc., 5709-B Arundel Avenue, Rockville, MD 20852. Phone: (301)881-0270.

8.13   Life-Cycle Costing[5]

Wolter J. Fabrycky and Benjamin S. Blanchard

A major portion of the projected **life-cycle cost** (LLC) for a given product, system, or structure is traceable to decisions made during conceptual and preliminary design. These decisions pertain to operational requirements, performance and effectiveness factors, the design configuration, the maintenance concept, production quantity, utilization factors, logistic support, and disposal. Such decisions guide subsequent design and production activities, product distribution functions, and aspects of sustaining system support. Accordingly, if the final LCC is to be minimized, it is essential that a high degree of cost emphasis be applied during the early stages of system design and development.

---

[5]Material presented in this section adapted from chapter 6 in W. J. Fabrycky and B. S. Blanchard, *Life-Cycle Cost and Economic Analysis*, Prentice Hall, 1991.

The Life-Cycle Costing Situation

The combination of rising inflation, cost growth, reduction in purchasing power, budget limitations, increased competition, and so on has created an awareness and interest in the total cost of products, systems, and structures. Not only are the acquisition costs associated with new systems rising, but the costs of operating and maintaining systems already in use are also increasing rapidly. This is due primarily to a combination of inflation and cost growth factors traceable to the following:

1. Poor quality of products, systems, and structures in use
2. Engineering changes during design and development
3. Changing suppliers in the procurement of system components
4. System production and/or construction changes
5. Changes in logistic support capability
6. Estimating and forecasting inaccuracies
7. Unforeseen events and problems

Experience indicates that cost growth due to various causes has ranged from five to ten times the rate of inflation over the past several decades. At the same time, budget allocations for many programs are decreasing from year to year. The result is that fewer resources are available for acquiring and operating new systems or products and for maintaining and supporting existing systems. Available funds for projects, when inflation and cost growth are considered, are decreasing rapidly.

The current situation is further complicated by some additional problems related to the actual determination of system and/or product cost.

1. Total system cost is often not visible, particularly those costs associated with system operation and support. The cost visibility problem is due to an "iceberg" effect, as is illustrated in Fig. 8.6.
2. Individual cost factors are often improperly applied. Costs are identified and often included in the wrong category; variable costs are treated as fixed (and vice versa); indirect costs are treated as direct costs; and so on.

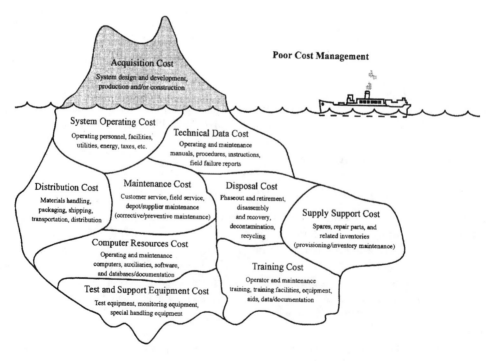

FIGURE 8.6　The problem of total cost visibility.

3. Existing accounting procedures do not always permit a realistic and timely assessment of total cost. In addition, it is often difficult (if not impossible) to determine costs on a functional basis.
4. Budgeting practices are often inflexible regarding the shift in funds from one category to another, or from year to year, to facilitate cost improvements in system acquisition and utilization.

The current trends of inflation and cost growth, combined with these additional problems, have led to inefficiencies in the utilization of valuable resources. Systems and products have been developed that are not cost-effective. It is anticipated that these conditions will become worse unless an increased degree of cost consciousness is assumed by engineers. Economic feasibility studies must address all aspects of life LCC, not just portion thereof.

LCC is determined by identifying the applicable functions in each phase of the life cycle, costing these functions, applying the appropriate costs by function on a year-to-year basis, and then accumulating the costs over the entire span of the life cycle. LCC must include all producer and consumer costs to be complete.

Cost Generated over the Life Cycle

LCC includes all costs associated with the product, system, or structure as applied over the defined life cycle. The life cycle and the major functions associated with each phase are illustrated in Fig. 8.7. LCC is employed in the evaluation of alternative system design configurations, alternative production schemes, alternative logistic support policies, alternative disposal concepts and so on. The life cycle, tailored to the specific system being addressed, forms the basis for LCC.

There are many technical and nontechnical decisions and actions required throughout the product or system life cycle. Most actions, particularly those in the earlier phases, have life-cycle implications and greatly affect life LCC. The analysis constitutes a step-by-step approach employing LCC figures of merit as criteria at a cost-effective solution. This analysis process is iterative in nature and can be applied to any phase of the life cycle of the product, system, or structure. Cost emphasis throughout the system/product life cycle is summarized in the following sections.

Conceptual System Design

In the early stages of system planning and conceptual design, when requirements are being defined, quantitative cost figures of merit should be established to which the system or product is to be designed, tested, produced (or constructed), and supported. A **design-to-cost** (**DTC**) goal may be adopted to establish cost as a system or product design constraint, along with performance, effectiveness, capacity, accuracy, size, weight, reliability, maintainability, supportability, and so on. Cost must be an active rather than a resultant factor throughout the system design process.

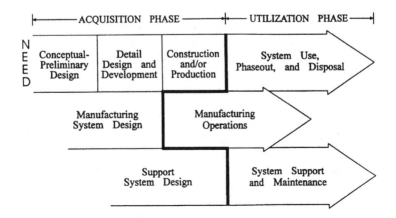

FIGURE 8.7 Product, process, and support life cycles.

Preliminary System Design

With quantitative cost requirements established, the next step includes an iterative process of synthesis, trade-off and optimization, and system/product definition. The criteria defined in the conceptual system design are initially allocated, or apportioned, to various segments of the system to establish guidelines for the design and/or the procurement of needed element(s). Allocation is accomplished from the system level down to the level necessary to provide an input to design and also to ensure adequate control. The factors projected reflect the target cost per individual unit (i.e., a single equipment unit or product in a deployed population) and are based on system operational requirements, the system maintenance concept, and the disposal concept.

As system development evolves, various approaches are considered that may lead to a preferred configuration, **Life-cycle cost analyses (LCCAs)** are accomplished in (1) evaluating each possible candidate, with the objective of ensuring that the candidate selected is compatible with the established cost targets, and (2) determining which of the various candidates being considered is preferred from an overall cost-effectiveness standpoint. Numerous trade-off studies are accomplished, using LCCA as an evaluation tool, until a preferred design configuration is chosen. Areas of compliance are justified, and noncompliant approaches are discarded. This is an iterative process with an active-feedback and corrective-action loop.

Detail Design and Development

As the system or product design is further refined and design data become available, the LCCA process involves the evaluation of specific design characteristics (as reflected by design documentation and engineering or prototype models), the prediction of cost-generating sources, the estimation of costs, and the projection of LCC as a **life-cycle cost profile (LCCP)**. The results are compared with the initial requirement, and corrective action is taken as necessary. Again, this is an iterative process, but at a lower level than what is accomplished during preliminary system design.

Production, Utilization, and Support

Cost concerns in the production, utilization, support, and disposal stages of the system or product life cycle are addressed through data collection, analysis, and an assessment function. High-cost contributors are identified, cause-and-effect relationships are defined, and valuable information is gained and utilized for the purposes of product improvement through redesign or reengineering.

The Cost Breakdown Structure

In general, costs over the life cycle fall into categories based on organizational activity needed to bring a system into being. These categories and their constituent elements constitute a **cost breakdown structure (CBS)**, as illustrated in Fig. 8.8. The main CBS categories are as follows:

1. *Research and development cost.* Initial planning, market analysis, feasibility studies, product research, requirements analysis, engineering design, design data and documentation, software, test and evaluation of engineering models, and associated management functions.
2. *Production and construction cost.* Industrial engineering and operations analysis, manufacturing (fabrication, assembly, and test), facility construction, process development, production operations, quality control, and initial logistic support requirements (e.g., initial consumer support, the manufacture of spare parts, the production of test and support equipment, etc.).
3. *Operation and support cost.* Consumer or user operations of the system or product in the field, product distribution (marketing and sales, transportation, and traffic management), and sustaining maintenance and logistic support throughout the system or product life cycle (e.g., customer service, maintenance activities, supply support, test and support equipment, transportation and handling, technical data, facilities, system modifications, etc.).
4. *Retirement and disposal cost.* Disposal of nonrepairable items throughout the life cycle, system/product retirement, material recycling, and applicable logistic support requirements.

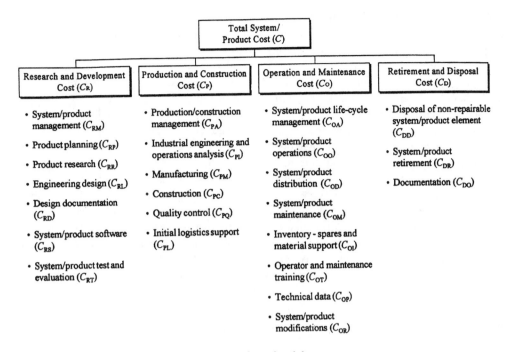

FIGURE 8.8 A general cost breakdown structure.

The CBS links objectives and activities with organizational resource requirements. It constitutes a logical subdivision of cost by functional activity area, major system elements, and/or one or more discrete classes of common or like items. The CBS provides a means for initial resource allocation, cost monitoring, and cost control.

Life-Cycle Cost Analysis

The application of LCC methods during product and system design and development is realized through the accomplishment of LCCA. LCCA may be defined as a systematic analytical process of evaluating various designs or alternative courses of action with the objective of choosing the best way to employ scarce resources.

Where feasible alternative solutions exist for a specific problem and a decision is required for the selection of a preferred approach, there is a formal analysis process that should be followed. Specifically, the analyst should define the need for analysis, establish the analysis approach, select a model to facilitate the evaluation process, generate the appropriate information for each alternative being considered, evaluate each alternative, and recommend a proposed solution that is responsive to the problem.

Cost Analysis Goals

There are many questions that the decision maker might wish to address. There may be a single overall analysis goal (e.g., design to minimum LCC) and any number of subgoals. The primary question should be as follows: What is the purpose of the analysis, and what is to be learned through the analysis effort?

In many cases the nature of the problem appears to be obvious, but its precise definition may be the most difficult part of the entire process. The design problem must be defined clearly and precisely and presented in such a manner as to be easily understood by all concerned. Otherwise, it is doubtful whether an analysis or any type will be meaningful. The analyst must be careful to ensure that realistic goals are established at the start of the analysis process and that these goals remain in sight as the process unfolds.

Analysis Guidelines and Constraints

Subsequent to definition of the problem and the goals, the cost analyst must define the guidelines and constraints (or bounds) within which the analysis is to be accomplished. Guidelines are composed of information concerning such factors as the resources available for conducting the analysis (e.g., necessary labor skills, availability of appropriate software, etc.), the time schedule allowed for completion of the analysis, and/or related management policy or direction that may affect the analysis.

In some instances a decision maker or manager may not completely understand the problem or the analysis process and may direct that certain tasks be accomplished in a prescribed manner or time frame that may not be compatible with the analysis objectives. On other occasions a manager may have a preconceived idea as to a given decision outcome and direct that the analysis support the decision. Also, there could be external inhibiting factors that may affect the validity of the analysis effort. In such cases the cost analyst should make every effort to alleviate the problem by educating the manager. Should any unresolved problems exist, the cot analyst should document them and relate their efforts to the analysis results.

Relative to the technical characteristics of a system or product, the analysis output may be constrained by bounds (or limits) that are established through the definition of system performance factors, operational requirements, the maintenance concept, and/or through advanced program planning. For example, there may be a maximum weight requirement for a given product, a minimum reliability requirement, a maximum allowable first cost per unit, a minimum rated capacity, and so on. These various bounds, or constraints, should provide for trade-offs in the evaluation of alternatives. Candidates that fall outside these bounds are not allowable.

Identification of Alternatives

Within the established bounds and constraints, there may be any number of approaches leading to a possible solution. All possible alternatives should be considered, with the most likely candidates selected for further evaluation. Alternatives are frequently proposed for analysis even though there seems to be little likelihood that they will prove feasible. This is done with the thought that it is better to consider many alternatives than to overlook one that may be very good. Alternatives not considered cannot be adopted, no matter how desirable they may actually prove to be.

Applying the Cost Breakdown Structure

Applying the CBS is one of the most significant steps in LCC. The CBS constitutes the framework for defining LCC categories and provides the communications link for cost reporting, analysis, and ultimate cost control.

In developing the CBS one needs to proceed to the depth required to provide the necessary information for a true and valid assessment of the system or product LCC, identify high-cost contributors and enable determination of the cause-and-effect relationships, and illustrate the various cost parameters and their application in the analysis. Traceability is required from the system-level LCC figure of merit to the specific input factor.

Cost Treatment over the Life Cycle

With the system/product CBS defined and cost-estimating approaches established, it is appropriate to apply the resultant data to the system life cycle. To accomplish this, the cost analyst needs to understand the steps required in developing cost profiles that include aspects of inflation, the effects of learning curves, the time value of money, and so on.

In developing a cost profile, there are different procedures that may be used. The following steps are suggested:

1. Identify all activities throughout the life cycle that will generate costs of one type or another. This includes functions associated with planning, research and development, test and evaluation. production/construction, product distribution, system/product operational use, maintenance and logistic support, and so on.

2. Relate each activity identified in step 1 to a specific cost category in the CBS. All program activities should all into one or more of the CBS categories.

3. Establish the appropriate cost factors in constant dollars for each activity in the CBS, where constant dollars reflect the general purchasing power of the dollar at the time of decision (i.e., today). Relating costs in terms of constant dollars will allow for a direct comparison of activity levels from year to year prior to the introduction of inflationary cost factors, changes in price levels, economic affects of contractual agreements with suppliers, and so on, which can often cause some confusion in the evaluation of alternatives.

4. Within each cost category in the CBS, the individual cost elements are projected into the future on a year-to-year basis over the life cycle as applicable. The result should be a cost stream in constant dollars for the activities that are included.

5. For each cost category in the CBS and for each applicable year in the life cycle, introduce the appropriate inflationary factors, economic effects of learning curves, changes in price levels, and so on. The modified values constitute a new cost stream and reflect realistic costs as they are anticipated for each year of the life-cycle (i.e., expected 1996 costs in 1996, 1997 costs in 1997, etc.). These costs may be used directly in the preparation of future budget requests, since they reflect the actual dollar needs anticipated for each year in the life cycle.

6. Summarize the individual cost streams by major categories in the CBS and develop a top-level cost profile.

Results from the foregoing sequence of steps are presented in Fig. 8.9. First, it is possible and often beneficial to evaluate the cost stream for individual activities of the life cycle such as research and development, production, operation and support, and so on. Second, these individual cost streams may be shown in the context of the total cost spectrum. Finally, the total cost profile may be viewed from the standpoint of the logical flow of activities and the proper level and timely expenditure of dollars. The profile in Fig. 8.9 represents a budgetary estimate of future resource needs.

When dealing with two or more alternative system configurations, each will include different levels of activity, different design approaches, different logistic support requirements, and so on. No two systems alternatives will be identical. Thus, individual profiles will be developed for each alternative and ultimately compared on an equivalent basis utilizing the economic analysis techniques found in earlier sections. Figure 8.10 illustrates LCCPs for several alternatives.

Summary

LCC is applicable in all phases of system design, development, production, construction, operational use, and logistic support. Cost emphasis is created early in the life cycle by establishing quantitative cost factors as "design to" requirements. As the life cycle progresses, cost is employed as a major parameter

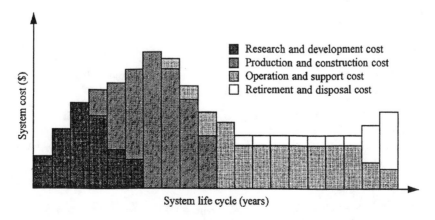

FIGURE 8.9 Development of LCCPs.

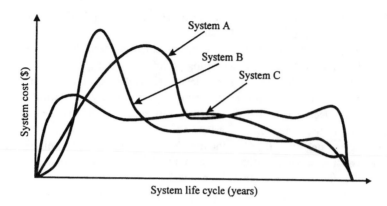

FIGURE 8.10   LCCPs of alternatives.

in the evaluation of alternative design configurations and in the selection of a preferred approach. Subsequently, cost data are generated based on established design and production characteristics and used in the development of life-cycle cost projections. These projections, in turn, are compared with the initial requirements to determine the degree of compliance and the necessity for corrective action. In essence, LCC evolves from a series of rough estimates to a relatively refined methodology and is employed as a management tool for decision-making purposes.

Defining Terms

Cost breakdown structure (CBS):   A framework for defining life-cycle costs; it provides the communications link for cost reporting, analysis, and ultimate cost control.

Design-to-cost (DTC):   A concept that may be adopted to establish cost as a system or product design constraint, along with performance, effectiveness, capacity, accuracy, size, weight, reliability, maintainability, supportability, and others.

Life-cycle cost (LCC):   All costs associated with the product or system as anticipated over the defined life cycle.

Life-cycle cost analysis (LCCA):   A systematic analytical process for evaluating various alternative courses of action with the objective of choosing the best way to employ scarce resources.

Life-cycle cost profile (LCCP):   A budgetary estimate of future resource needs over the life cycle.

References

Fabrycky, W. J. and Blanchard, B. S. *Life-Cycle Cost and Economic Analysis*, Prentice Hall, Englewood Cliffs, NJ, 1991.

Further Information

The reference above should be studied by readers who want a complete view of life-cycle cost and economic analysis. Further information may be obtained from the following:

Blanchard, B. S. and Fabrycky, W. J. *Systems Engineering and Analysis, 3rd ed.*, Prentice Hall, Upper Saddle River, NJ, 1990.
Canada, J. R. and Sullivan, W. G. *Economic and Multiattribute Evaluation of Advanced Manufacturing Systems*, Prentice Hall, Upper Saddle River, NJ, 1989.
Fabrycky, W. J., Thuesen, G. J., and Verma, D., *Economic Decision Analysis*, 3rd ed., Prentice Hall, Upper Saddle River, NJ, 1998.
Ostwald, P. F. *Engineering Cost Estimating, 3rd ed.*, Prentice Hall, Upper Saddle River, NJ, 1992.
Thuesen, G. J. and Fabrycky, W. J. *Engineering Economy, 8th ed.*, Prentice Hall, Upper Saddle River, NJ, 1993.

8.14 Financial Ratios

Donald E. Vaughn

The term **financial ratios** carries different connotations to different types of users. Utilizers include financial planners for a firm, short-term and long-term creditors, equity owners, and current and potential investors and speculators in the securities of the business.

Financial ratios to a corporate finance officer may refer to several things. The term may be applied to **common-sized financial statements,** or those shown as 100% of some base. For the balance sheet, the total of assets equals 100%. Assets, debts, and equity items are shown as percentages of this total. For an income statement, the total of net sales of a merchandising firm, revenue of a service, utility, or transport company, or income for a financial organization is set as 100%, and costs and expenses are shown as percentages of this total. The finance officer sets up several years of financial statements in parallel so as to make a **trend analysis. Benchmark ratios,** such as those shown for about 800 **standard industrial classification (SIC) codes,** as developed by several governmental agencies in 1972 and expanded in 1987, are published annually by Dun & Bradstreet (D&B) in *Industry Norms and Key Business Ratios.* D&B uses its database of roughly eight million firms on which it collects credit information in the United States and Canada for making the annual updates of four-digit SIC comparisons. A somewhat similar study is updated from member commercial bank credit file information by Robert L. Morris Associates and published in the annual, *Statement Studies.* U.S. federal agencies, such as the Internal Revenue Services, the Department of Labor, and the Department of Commerce, likewise collect and publish industry group data in financial statement form.

Historical financial statements may be used for constructing many financial ratios. The 14 key ratios, selected to show solvency, efficiency, and profitability of an industry group, are updated annually by D&B. An example is shown in Table 8.13. Short-term and long-term creditors are interested in these and other ratios.

Equity investors are interested in growing or waning financial strengths of their firm. They employ trend analysis for the firm over several years, perhaps 5 or 10, and compare its position with that of other

TABLE 8.13 SIC 5812, Eating Places (no breakdown, 1474 Establishments)

	$	%	RATIOS	UQ	MED	LQ
Cash	57,964	17.8	Solvency			
Accounts receivable	14,979	4.6	Quick Ratio (times)	1.5	0.6	0.3
Notes receivable	2,931	0.9	Current Ratio (times)	2.4	1.2	0.6
Inventory	25,074	7.7	Curr Liab to Nw (%)	19.0	48.7	120.4
Other current	19,538	6.0	Curr Liab to Inv (%)	153.0	335.0	571.8
Total current	**120,486**	**37.0**	Total Liab to Nw (%)	33.2	94.2	228.4
Fixed assets	152,724	46.9	Fixed Assets to Nw (%)	46.3	95.4	198.5
Other noncurrent	52,428	16.1	Efficiency			
Total assets	**325,638**	**100.0**	Coll Period (days)	1.5	4.0	11.3
Accounts payable	32,564	10.0	Sales to Inv (times)	124.3	75.0	39.3
Bank loans	316	0.1	Assets to Sales (%)	18.3	30.9	55.0
Notes payable	10,420	3.2	Sales to Nwc (times)	46.5	20.9	9.4
Other current	56,661	17.4	Acct Pay to Sales (%)	1.7	3.1	4.8
Total current	**99,971**	**30.7**	Profitability			
Other long term	81,084	24.9	Return on Sales (%)	7.7	3.7	0.7
Deferred credits	651	0.2	Return On Assets (%)	21.0	8.0	1.5
Net worth	143,932	44.2	Return On Nw (%)	52.7	21.9	5.8
Total liability and net worth	**325,638**	**100.0**				
Net sales	1,433,308	100.0				
Gross profit	746,753	52.1				
Net profit after tax	51,599	3.6				
Working capital	20,515	—				

Source: Dun & Bradstreet, Inc. *Industry Norms and Key Business Ratios,* New York, 1996. With permission.

competing firms or benchmark ratios for its industry group, such as those sited above. Would-be, long-term investors take about the same approach, perhaps using Standard & Poor's or Value Line's data, while speculators attempt to judge the short-term likelihood of price change for a financial issue. One of several types of charts is used by technicians.

Common-Sized Financial Statements

The financial analyst often compares historical financial statements for about 5 years ago, last year, and the most recently audited financial statements. Interim statements are sometimes shown for partial years. The balance sheet, of course, shows assets, debts, and equity items at a point in time, usually the end of a fiscal year or quarter, while the income statement is for a duration of time, such as a month, quarter, or year. Commercial banks and other institutional lenders may want audited financial statements for about 3 years and **proforma**, or forecasted, financial statements by months for about 6 months, and by quarters for another six quarters, as a part of a business plan/credit request. Annual income statements and balance sheets may be forecasted into the future, often for about 5 years or longer. Trends in growth rates of sales and assets for the firm under review and its comparative industry are important for proper development of such statements. The inflation rate on which forecasts are made is important for proper development of proformas. Lenders may also want cash budgets.

One of about 800 SIC industry designations, published by D&B, is shown in Table 8.13. The data are for SIC 5812, eating places, with no further breakdown, for 1995 credit report collections, and embraces some 1474 establishments. The dollar figures for the balance sheet are in current U.S. dollars, and the second column shows percentages. A skeleton statement for the income statement, showing only net sales, gross profits, and net profit after taxes, is shown immediately below the balance sheet data. Fourteen important ratios, six showing solvency, five reviewing efficiency of the industry, and three profitability measures, are shown below the financial statements. These profit measures include return on sales (%), return on assets (%), and return on Nw (net worth, %). Comparison ratios are for **UQ, MED**, and **LQ** meaning upper quartile (the 25th percentile in an array of the 1474 business establishments), the midpoint between the 737th and 738th firm, and the 75th percentile, respectively. Thus, in comparing the ratios for a given company by a credit analyst or some other interested reviewer, the firm's ratios can be grouped as falling into the upper quartile, the upper-median quartile, the lower-median quartile, or the lower quartile, respectively.

A potential business entrepreneur may also construct estimated statements by using these benchmark ratios. For example, in this 5812 SIC designation, the average size of net worth at year-end 1995 was $143,932. Assets to sales, using the median figure, were 30.9% and net worth was 44.2% of total assets. Thus, on average, $44,200 in equity should, on an ongoing basis, support about $55,800 in debts, with 30.7% being in short-term and 24.9% being in long-term. The total in assets would be $100,000; average supportable sales would be about 3.3 times assets; gross profit should run about 52.1% of the $330,000 annual sales; and operating expenses would reduce the net profit after taxes to about 3.6% of sales, or roughly the $11,000 to $12,000 range. Salaries and wages (including fringes), food costs, and housing and utility costs would be some of the largest expense categories for a restaurant. Rent or ownership would influence the breakdown of housing costs (depreciation, structure insurance, real estate taxes, etc.).

The solvency ratios, for the most part, merely show the relationship between two items. For example, the quick ratio is computed by relating current assets, minus inventories, to current debts. The exception is the collection period, which is expressed in days rather than as a ratio of sales. The year-end accounts receivables are compared to net sales, and the common fraction is multiplied by 360 days in a year. Banks sometimes use 360 rather than 365 days in a year.

The relationships are about as follows, for the 14 key D&B ratios. A brief explanation of each follows the ratios.

Quick Ratio (times): (current assets − inventory)/total current debts: and shows very liquid assets to short-term debts.
Current Ratio (times): current assets/total current debts: and is a standard measure of current debt-paying ability.

Curr Liab to Nw (%): total current debts/net worth (or equity): and shows relationship of short-term creditors and equity supplied funds. This ratio is another measure of the near-term, liquidity of the firm.

Total Liab to Nw (%): total debts/equity ownership (Nw): and shows each dollar of debt to each one of equity.

A bank or the Small Business Administration, for a startup firm, may insist on at least 33.3% equity before making a loan approval. A very small firm might have 25% equity, 75% debt, especially if it rents facilities.

Fixed Assets to Nw (%): total fixed assets/net worth: and shows nonliquid assets compared to the firm's equity.

Coll Period (days): (ending receivables/net sales) × 360 days: and shows the average age of receivables.

Sales to Inv (times): net sales/year-end inventory: and gives a rough estimate of inventory turnover for the year.

Assets To Sales (%): total assets/net sales: shows the asset turnover for the year.

Sales to Nwc (times): net sales/net working capital (or current assets – current debts): and shows the velocity of turnover of net working capital for the industry.

Acct Pay to Sales (%): accounts payable/net sales: and compares payables to annual net sales.

Return on Sales (%): net profit after taxes/net sales: and shows after-tax earnings per dollar of sales.

Return on Assets (%): net profit after taxes/total year-end assets: and shows after-tax return on resources.

Return on Nw (%): net profits after taxes/net worth (or equity): and shows after-tax return on year-end net worth.

The analyst computes ratios for his or her firm and contrasts them to the above benchmark ratios for comparison.

An accountant, financial analyst, or engineer should be familiar with the above balance sheet and income statement terms, so most of them will not be defined. A brief explanation is provided on certain items that contain several accounts that are grouped together. Accounts receivable to a restaurant might be sales on credit to schools, hospitals, and the like. Notes receivable could be charge slips that have not yet been processed. Other noncurrent assets are a mixture of leaseholds, leasehold improvements, trade names, trademarks, and prepaid items, such as insurance.

On the liability side of the balance sheet, certain of the items might need clarification. Accounts payable are for trade accounts. For a restaurant, it would be purchase of food items, paper and plastic products, packaging containers, etc. Bank loans are small at restaurants, while notes payable might be to finance vehicles or equipment from regular institutional installment lenders or notes owed to a manufacturing firm — provider of restaurant equipment. Other current debts contain a mixture of salaries and wages payable, fringes on these, accrued payroll and other taxes, accrued insurance, etc.

In interpreting a trend in financial ratios, where one or two appear to be out of line, the analyst should not be unduly alarmed. For example, the current ratio might have fallen from one year to the next from 4.0 to 3.0 times. However, suppose that profits had gone up by 300% and the largest current debt was for taxes payable. This really is a positive situation, if the firm has enough liquidity to service its current debts in a timely fashion. Ratios should be reviewed as a group.

Trade creditors and commercial lenders are likely to use different benchmark ratio sources. A large wholesale or manufacturing firm supplier probably subscribes to D&B's *Reference Book* and probably uses published D&B ratios. Such a credit grantor may also obtain D&B credit reports on new clients or for very large credit applicants. A commercial bank is likely to be a member of the banking association, Robert L. Morris Associates, and obtain the annual 1000-page publication for a nominal fee, referred to as *Statement Studies*. The latter provides about four pages of data for each SIC reviewed, with ranking by several sizes of firms, and ranking by sales and also by assets. The comparison also shows several years of trend data for each of about 250 SIC designations.

An insurance or pension fund making a term loan or bond investment in a large firm might use some of the above ratios, but such a financial analyst would more likely use data published in *Moody's Investment Services* (about six different volumes), which is also published by D&B, publications of Standard & Poor's (S&P) Corporation, and the Value Line *Value Line Investment Survey* with weekly updates. Value Line and some of the other services also have monthly updates of data banks on PC data disks. *Compustat* (S&P) and *Value Line Database* + are two widely used databases that contain a vast number of financial ratios on several thousand large firms.

Fundamental Market Ratios

Fundamentalists are number crunchers, whereas **technicians** attempt to interpret trends in the market, trends in stocks, volume of trading, relative market momentum of a stock or group of stocks, and so forth. Several chapters in a standard textbook on investments are devoted to each of these two approaches toward investing in stocks and bonds, hybrid issues, options, etc. The reader may also wish to refer to a recently published trade book on the topic [Aby and Vaughn, 1995].

A security analysts for either bonds or stocks begins with a review of the overall market. Is the climate likely over the next year or two to be a good place for the funds under question? What industries are probable to be leaders and which ones are likely to fall behind? What companies, selected from several or a few dozen industry groups, seem to be better bargains at their recently traded prices? Financial ratios are one of a few tools of the fundamentalist, so this section will concentrate on approaches usually followed by fundamental financial analysts.

Published financial statements are available in several sources, but the annual and interim reports published for the benefit of security holders provide a glossy advertisement and comparative financial statements for 2 or 3 years. The balance sheet, income statement, cash flow statement, footnotes to the statements, and sometimes other supporting statements are included. An audit statement from a CPA firm is almost always present. Many brokerage firms, public libraries, and social science libraries at colleges and universities maintain historical, vertical files of past annual reports. The annual and interim reports filed with the Securities and Exchange Commission may be preferred to the published ones by serious investors.

Some firms are more highly regulated than are others. Many financial institutions, especially if they are regulated by a federal or state regulatory agency, must file annual reports with the respective agency. Listed firms and OTC companies with over 500 stockholders must file annual and interim reports with the Securities and Exchange Commission in Washington, DC. Most large, publicly held electric and gas utilities and gas transmission firms must file annual reports to their regulatory agency, the Federal Power Commission (a part of the Department of Energy) in Washington, DC. *Moody's Investment Services* uses the financial statements filed with regulatory agencies, when available, in gathering information published in its many volumes of data. From 2 to about 7 years of comparative financial statements are provided on many thousands of companies by Moody's, a division of D&B.

Industry comparative balance sheets and income statements may be obtained on large firms from the federal document, *Quarterly Report of Manufacturing, Trade and Service Firms*. Roughly 40 industry groups are included in the quarterly publication, recently available on CD rom disks. IRS-published financial statements from information supplied on Schedule C of proprietorships, partnership returns, and corporate income tax returns are also published by the IRS. The five-year *Census of Business, Retailing, Wholesaling, and Trade* provides a vast amount of data on these three (of about ten) sectors of the economy. The annual *County Business Patterns* is published on each of the U.S. states and for the entirety of the United States and is available in published form and on data disks from the U.S. Government Printing Offices. Key financial ratios are provided on most SIC designations in the above federal publications. Annual reports from the appropriate federal or state agency also contain valuable financial comparisons on firms reporting to them. The *Annual Report of the Securities and Exchange Commission*, for example, provides a wealth of information on securities markets and securities laws. The monthly publication of the Federal Reserve System, *Federal Reserve Bulletin*, contains about 100 pages of information on money, banking, and credit. Each of the 12 Federal Reserve Banks usually publish their own monthly review of monetary articles and data of interest from business persons, educators, and state agencies in their geographic region of operation.

A small investor is likely to be a fundamentalist who attempt to discover worthwhile stock candidates from a review of Moody's *Handbook of Common Stocks,* S&P's *Corporate Reports,* or the *Value Line Investment Survey.* For the above sources, one or two pages of statistical information with brief interpretative comments, are provided on each firm covered, with 10 to 15 years of annual and/or quarterly data being common.

The stock fundamentalist is usually interested in the trend in sales, profit margins, earnings per share, cash flow per share, dividends per share, book value per share, etc. These and other fundamental ratios are provided in the above-referenced sources. Morningstar's publication on mutual funds is a quality service, and selected top-performing mutual funds from about a dozen investment objective categories are reviewed monthly by *Money Magazine* and *Kiplinger's.* Computer analysts may prefer to glean recently compiled data banks from S&P (*Compustat*), from *Value Line Database* +, or from other sources.

Some technical chart services provide up to about 15 fundamental stock measurements and about an equal number of technical strength comparisons. One such source is *Trendline's Current Market Perspective,* a midmonth publication of Trendline Corporation, a division of S&P Corporation (a division of McGraw-Hill).

This study provides weekly range charts and 30 important types of information on about 2370 large listed and OTC firms. Certain studies and business comparisons are made in the weekly publication by McGraw-Hill, *Business Week.* The list goes on and on. Some investment texts publish a list of six to eight pages of financial publications.

Traders in speculative stocks (high betas, volatile prices), options, futures, etc., may wish to follow daily quotes on delayed TV financial channels, the ticker tapes at brokerage firms, or daily quotes in many regional newspapers, *The Wall Street Journal,* and so on. *Barron's, Commercial and Financial Chronicle,* and other sources provide weekly trading information. Both Moody's and S&P Corporation publish monthly *Stock Guides* and *Bond Guides* with comparative data on about 5500 of the largest companies, several thousand mutual funds, and several hundred real estate investment companies.

The bond investor is likely to be interested in long-term growth in the company under review, its ability to service short- and long-term credit, and the mix between assets supplied by creditors and equity holders. The number of times that fixed charges (e.g., times interest earned, sinking fund payments, and lease payments) are earned, pretaxes, is one such measurement emphasized for bonds in some of the bond publications listed above. The debt/equity ratio, as well as the times charges earned, and trend and comparison to its industry, largely influences the bond grade assigned to the issue by Moody's or S&P Corporation. Many states limit bond purchases by fiduciary accounts to the top four grades of corporate bonds, investment grade of AAA/Aaa, AA/Aa, respectively, by S&P and by Moody's and to the intermediate grades of A/A or BBB/Baa, respectively, for S&P and Moody's. They sometimes add a + or – to these major categories. Direct issues of the Treasury (T-bills, notes, and bonds) are rated prime, while federal agency issues are rated triple A by both S&P and Moody's. Municipal notes and bonds are usually rated in six categories by these two bond rating agencies, and a prudent investor might be wise to consider only the top three municipal grades.

Bonds are usually bought for a given grade, with the desired maturity, and with yield to maturity slightly above average for the category grouping. The same might be true for the shares in a bond or note mutual fund, selected from a group of funds with similar investment objectives. Past performances over the last year, 3 years, or 5 years are considered to be importance by many security analysts.

Stocks are sometimes analyzed with certain evaluation approaches. A standard investment text usually provides coverage on the dividend discount model, the price earnings model, the cash flow model, the intrinsic value model, etc. The brevity of this writing precludes the detailed explanation of these, but growth rates in earnings or dividends, the actual level of dividends or earnings per share, and the book value per share of common stock are factored into some of the above models. Mutual fund managers probably compare price earnings levels for similar companies, choosing those that are trading at the lowest P-E multiples with acceptable growth trend in earnings and dividends per share. The Value Line model uses a 5-year, moving regression between cash flow per share and price that provides the line of best fit, with the last 2 years being forecasted cash flows per share. Their ranking of I, II, III, IV, and V for 12 months and 3- to 5-year stock growth considers these measures, earnings per share shock (i.e., above or below general market expectation of security analysts), market momentum in stock prices, and a few other factors.

Recent Trends in Security Analysis

There has been a general acceleration in the retrieval and analysis of financial data for investment vehicles. Programs are developed for mainframe or PC computers that will review a large data bank of candidates and extract a few or a few dozen that meet certain prescribed attributes. For example, the 1615 or so firms monitored by *Value Line Database +* are divided into about 98 industry groups, including some geographical distribution in the utility and banking industries. About 50 ratios are provided on each of the 1615 firms. The data bank of Compustat is even more gigantic, but so is the price — a few hundred dollars yearly vs. seven to ten thousand for the S&P database. A knowledgeable investor can extract a list of securities with preset characteristics in a few minutes from these data banks. Large trades by institutional buyers, and programmed buying and selling based on the spreads between prices of financial futures and weighted stock indicators (indexes) have brought about wide swings in stock market prices. The urge to invest in stocks by baby-boomers that wish to catch up on pension contributions appears to be fueling the market for corporate stocks and mutual fund shares (indirect investment into corporate issues). Dividend yields have (in early 1997) fallen to historical lows, with the market hovering around 6800 on the Dow Jones Industrial Average (DJIA, Dow) of 30 large firms. The S&P's 500 Index also made a recent historical high. In time, the market will reach a peak, decline by 10 to 25%, and begin an upward run again. Some stock forecasters believe the market will reach 10,000 on the Dow by the turn of the century. Some see only doom and gloom, a repeat of 1929 or 1987.

Over the past 70 or 80 years, stocks have returned about 10% yearly, corporate bonds about 7 to 8%, and treasury bonds a percentage less in yield. Intermediate notes and short-term issues usually return less, since they are less risky in the sense that they are less volatile in price. Fundamental investors usually diversify their security holdings to minimize risk of losses. Speculators assume higher levels of risks by seeking more volatile issues and attempting to improve on their timing of purchase and sale. Both use ratios, of one kind or another, but this brief discussion is intended as an appetite whetter. A serious investor, or speculator, should read, simulate investment strategies with play money until some mastery is achieved, and then commit only a portion of his or her savings to risky ventures. Both fundamental and speculative measurements for issues are important.

Defining Terms

Benchmark ratios: Key balance sheet and income statement ratios for an industry group or a business sector.

Common-sized statements: Financial statements shown as percentages of assets or of net sales.

Comparative analysis: Comparing a firm's common-sized financial statements to industry averages (benchmarks).

Fundamentalist: A long-term investor who bases decisions on trends in sales, growth in profits, P-E ratios, etc.

Proforma statements: Statements prepared before the fact; forecasted statements based on stated assumptions.

SIC codes: A Federal agency system of classifying businesses into two-, three-, or four-digit Standard Industrial Classifications.

Trend analysis: Comparing changes in financial statements of a firm over time.

References

Industry Norms and Key Business Ratios, Dun & Bradstreet, New York, 1996.

Statement Studies. Robert L. Morris Associates, Philadelphia, 1995.

Trendline's Current Market Perspectives, Trendline Corporation, New York, January 1997.

Van Horne, J. C. and Wachowicz, J. M., Jr. *Fundamentals of Financial Management,* Prentice-Hall, Englewood Cliffs, NJ, 1995.

Vaughn, D. E. *Financial Planning for the Entrepreneur,* Prentice-Hall, Englewood Cliffs, NJ, 1997.

Further Information

Aby, C. D., Jr. and Vaughn, D. E. *Asset Allocation Techniques and Financial Market Timing,* Quorum
Books, Westport, CT, 1995.

Gitman, L. J. and Joehnk, M. D. *Fundamentals of Investing,* Harper Collins College, New York, 1996.

U.S. Department of Commerce. *Business Census (Retailing, Service, Wholesaling),* Government Printing
Office, Washington, DC, 1992 and 1997.

U.S. Department of Commerce. *County Business Patterns,* Government Printing Office, Washington, DC,
1996 or latest.

U.S. Department of the Treasury. *IRS Statistics of Income (Proprietorship, Partnership, and Corporation),*
Government Printing Office, Washington, DC, latest annual.

8.15 Project Analysis Using Rate-of-Return Criteria

Robert G. Beaves

Many decision makers find rate-of-return investment criteria to be more intuitive and therefore easier
to understand than **net present value** (NPV). As a result the **internal rate of return** (IRR) continues to
be widely used despite its legendary quirks and the well-established superiority of NPV. A second rate-
of-return criterion, the **overall rate of return** (ORR), provides more NPV consistency than does the IRR
but is not widely known or used. This section compares NPV, IRR, and ORR and demonstrates some
shortcomings of rate criteria.

Net Present Value

A project's NPV represents *the change in the value of the firm* that occurs if and when the firm implements
that project. The NPV of a project is calculated by summing the present values of all cash flows associated
with it while preserving negative signs of flows to the project and positive signs of flows from the project.
NPV can be defined as follows:

$$\text{NPV} = \sum_{t=0}^{n} a_t (1+k)^{-t} \qquad (8.16)$$

where a_t = the cash flow at the time t (end of period t), k = the firm's **opportunity cost**, the discount or
"hurdle" rate, and n = the number of periods in the project's life.

Implementation of projects with positive NPVs increases firm value, whereas implementation of
projects with negative NPVs reduces it. Because management's goal is to maximize firm value, a project
should be rejected if its NPV is negative and accepted otherwise. When forced to choose among competing
positive-NPV projects, the project that offers the highest NPV should be preferred. It is well established
that the NPV criterion provides theoretically correct accept/reject and project-ranking decisions.

TABLE 8.14 Projects A, B, and C*

	Project A	Project B	Project B–A	Project C
Time				
0	−1000	−1000	0	−2000
1	700	0	−700	1400
2	300	0	−300	600
3	200	1423	1223	400
Method				
IRR	12.48%	12.48%	12.48%	12.48%
NPV	$34.56	$69.12		$69.12
IB	$1000.00	$1000.00		$2000.00
ORR	11.25%	12.48%		11.25%
ORR_{2000}	10.63%	11.25%		11.25%

*$k = 10\%$.

Internal Rate of Return

A project's IRR is that discount rate k^* for which that project's NPV would be zero. Note that a project's IRR is independent of the firm's actual hurdle rate k. A project's IRR represents *the average periodic rate of return earned on funds while they are invested in that project*. Because IRR is an average rather than a constant rate, the rate or return generated by a project in any one period need not equal its IRR.

Consider project A in Table 8.14, which has an IRR of 12.48%. Project A's rate of return during the first period (time 0 to time 1) is unknown. The "balance" in project A at time 1 need not be $1125 ($700 released to the firm and $525 remaining in the project) as would be the case if its 12.48% IRR was a constant rate. Although a project's IRR represents the average rate of return earned on funds invested in that project, the amount of funds so invested varies from period to period and is often unknown.

A project should be rejected if its IRR is less than the firm's **opportunity cost** k and accepted otherwise. NPV and IRR provide the same accept/reject decisions for projects that have unique IRRs. The use of IRR for accept/reject decisions is complicated, however, by the fact that some projects have multiple IRRs whereas others have none.

It is well established that ranking competing projects on the basis of their IRRs can provide an incorrect choice from among those projects. Consider projects A and B in Table 8.14. Both projects require initial investments of $1000 and both have the same IRR, 12.48%. No clear preference exists between projects A and B if ranked on the basis of their IRRs. Nonetheless, project A's NPV is only $34.56, whereas project B's is $62.12. Project B has a high NPV because its entire initial investment and all accumulated returns on that investment remain in the project, earning an average 12.48% rate for three periods. On the other hand, much of project A's initial investment and accumulated returns are released before the end of its three-period life.

Any two projects can be correctly ranked by examining the IRR of an incremental or "difference" project. Project B–A in Table 8.14 is an incremental project for projects A and B created by subtracting each cash flow of A from the respective cash flow of B. Because the IRR of incremental project B–A exceeds the firm's 10% opportunity cost, project B (the project from which a second project was subtracted) should be preferred over A. If the IRR of the incremental project was less than k, project A would have been preferred. Because incremental IRR analysis can only provide pairwise rankings, its use becomes tedious where more than two projects are being ranked. Further, an incremental project may have multiple IRRs or no IRR at all.

Overall Rate of Return (ORR)

A project's ORR represents the *average periodic rate earned on a fixed investment amount* over the expected life of the project. That fixed investment amount is known as the project's **investment base**. A project's ORR can be defined as

$$ORR = \left[\frac{\left(IB + NPV\right) \times \left(1+k\right)^{n}}{IB} \right]^{1/n} - 1.0 \qquad (8.17)$$

where NPV = the project's NPV, with k as discount rate and IB = the project's investment base, to be defined in greater detail later.

In calculating a project's ORR, that project's investment base plus any accumulated returns are assumed to earn the project's IRR while invested in it and to earn the firm's opportunity cost when not needed by the project. Thus, project B in Table 8.14 has an ORR of 12.48% because, for its entire three-period life, its $1000 investment base and all accumulated returns remain invested in the project earnings its 12.48% IRR. Project A's ORR is less because significant amounts of its investment base and accumulated returns are released at times 1 and 2 and earn the firm's 10% opportunity cost for the balance of that project's life.

Projects with ORR greater than or equal to the firm's opportunity cost k should be accepted, whereas those with lower ORRs should be rejected. As Eq. (8.17) suggests, the ORR always provides accept/reject decisions that are consistent with those provided by NPV. In contrast to the IRR, the ORR is uniquely defined for all projects and is generally a function of the firm's opportunity cost k. Further, the ORR provides an NPV-consistent ranking of competing projects that have the same scale or investment base IB.

When comparing projects having different investment bases, ORRs adjusted to some common scale will provide an NPV-consistent ranking. Scale adjustments are required because the ORR, like all rate criteria, does not preserve scale. Consider project C (Table 8.14), which was created by doubling project A's cash flows. Although a project's NPV doubles when its size is doubled, its IRR and ORR are unchanged. Scale is lost in the calculation of any rate of return, an important shortcoming in ranking competing projects.

Project Investment Base

A project's investment base (IB) represents the time 0 value of all external funds (i.e., funds not provided by the project itself) needed to finance the project. All projects considered thus far required a single cash inflow (i.e., negative flow), which occurred at time 0. The IB of such projects is simply the amount of that initial inflow, $-a_0$. Because some projects have more complex cash flow patterns, it is necessary to establish the following general rule for determining a project's IB:

A project's IB is determined by multiplying the minimum cumulative present value associated with its cash flows times − 1.0.

Consider project D in Table 8.15, for which the firm's opportunity cost is assumed to be 10%. Using 10% as the discount rate, present values and cumulative present values are calculated for each of project D's cash flows. The minimum cumulative present value associated with project D's cash flows is −$1057.85. Thus, project D's IB is $1057.85. Assuming a 10% opportunity cost for all funds not currently needed by project D, $1057.85 is the minimum time 0 amount that will fund project D to its termination at time 5.

Note that the final cumulative present value listed in Table 8.15 is necessarily project D's NPV of $57.94. Project D's ORR is calculated by substituting into Eq. (8.17) as follows:

$$\left[\frac{\left(57.94 + 1057.85\right) \times \left(1.10\right)^{5}}{1057.85} \right]^{1/5} - 1.0 = 0.1118$$

In other words, $1057.85 committed to project D at time 0 will earn an average return of 11.18% per period for five periods.

TABLE 8.15   Project D*

	Cash Flow (a_t)	Present Value	Cumulative PV
Time			
0	−1000	−1000.00	−1000.00
1	300	272.73	−727.27
2	−400	−330.58	−1057.85
3	700	525.92	−531.93
4	−500	−341.51	−873.44
5	1500	931.38	57.94
Method			
IRR	11.48%		
NPV	$57.94		
IB	$1057.85		
ORR	11.18%		

*$k = 10\%$.

Scale-Adjusted ORR

ORR as defined in Eq. (8.17) should only be used to compare projects of identical scale (i.e., the same IB). Assuming projects B and C in Table 8.14 are mutually exclusive, they cannot be directly compared, because project C's 11.25% ORR is earned on an IB of $2000, whereas project B's 12.48% ORR is earned on an IB of only $1000. A project's ORR can be adjusted to any scale by replacing its IB in Eq. (8.17) with the desired scale as follows:

$$\text{ORR}_S = \left[\frac{\left(S + \text{NPV}\right) \times \left(1 + k\right)^n}{S} \right]^{1/n} - 1.0 \tag{8.18}$$

where ORR_S is the project's ORR adjusted to scale S and S is the desired scale to which the project's ORR is being adjusted.

These scale adjustments are based on the assumption that differences in project scales can be invested at the firm's opportunity cost k. Investing at the firm's opportunity cost increases a project's scale but does not affect its NPV. Where two or more competing projects are being compared, the ORRs of all must be adjusted to a common scale (preferably the largest IB among the projects). ORRs adjusted to a common scale of $2000 were calculated in Table 8.14 for projects A, B, and C. Those ORRs reveal that the firm should be indifferent in choosing between projects B and C but should prefer either to project A, whose ORR_{2000} is 10.63%. These preferences are consistent with the NPV ranking of projects A, B, and C.

Project Life Differences

Comparing projects that have different project lives creates certain problems — no matter what decision criterion is being used. Such comparisons generally require an assumption as to what occurs at the end of each project's life, that is, whether the project is replaced with a similar project. Where all competing projects are one-time expenditures with no replacement, their NPVs can be directly compared even if they have different project lives. Such comparison assumes that funds released by any project earn the firm's opportunity cost k until the end of the longest-lived project. In contrast, the ORRs of competing one-time projects cannot be directly compared if those projects have different lives. Rate criteria such as ORR and IRR measure average performance per period over a project's life, whereas NPV measures cumulative performance over a project's life.

TABLE 8.16 Projects E and F*

	Project E	Project F
Time		
0	−5000	−4000
1	2500	−1100
2	2000	2000
3	2000	2000
4		2000
5		1500
Method		
IRR	15.02%	13.15%
NPV	$428.25	$452.93
IB	$5000.00	$5000.00
ORR	13.05%	11.92%
$ORR_{5000,5}$	11.82%	11.92%

*$k = 10\%$.

Consider projects E and F in Table 8.16. If we assume that both projects are one-time expenditures that will not be replaced at the end of their respective lives, the ORRs of projects E and F are not directly comparable. Project E earns an average rate of 13.05% per period for three periods, whereas project F earns an average rate of 11.92% per period for five periods. The ORR of any project can be calculated over a common life z periods longer than its project life as follows, by assuming that all funds can be invested at the firm's opportunity cost k during those z additional periods:

$$ORR_{S,n+z} = \left[\frac{(S+NPV) \times (1+k)^{(n+z)}}{S} \right]^{1/(n+z)} - 1.0 \tag{8.19}$$

where $ORR_{S,\,n+z}$ is the project's ORR adjusted to common scale S and common life $n + z$, and $n + z$ is the common life over which the project's ORR is being calculated. Adjusted to project F's five-period life, project E's ORR is 11.82% and is less than project F's ORR of 11.92%. Thus, ORRs calculated over a calculated over a common five-period life provide the same ranking of projects E and F as is provided by their NPVs.

Where competing projects are assumed to be replaced at the end of their initial project lives, project rankings for projects having different lives cannot simply be based on the NPVs of the respective projects. Instead, revised projects are generated for each competing project by extending their cash flow streams to some common terminal point. NPV's and ORRs calculated for such revised projects should provide consistent project rankings.

Example — Project Analyses. A firm has been approached by two different manufacturers who want to sell it machines that offer labor savings. The firm will buy one machine at most because both machines being considered perform virtually the same services and either machine has sufficient capacity to handle the firm's entire needs (i.e., mutually exclusive projects). The firm's cost accounting department has provided management with projected cash flow streams for each of these machines (see Table 8.17) and has suggested a 15% opportunity cost for these machines.

When one considers the IRRs and unadjusted ORRs of machines A and B in Table 8.17, it is clear that both are acceptable, since the IRR and the ORR of each exceeds the firm's 15% opportunity cost. It may appear that machine B should be favored because both its IRR and its unadjusted ORR are higher than those of machine A. Note, however, that machines A and B have different IBs and different lives. Adjusting machine B's ORR to A's larger scale of $25,000 and to project A's longer five-period life lowers it from

TABLE 8.17   Machines A and B*

	Machine A		Machine B	
	Cash Flow	Cumulative PV	Cash Flow	Cumulative PV
Time				
0	−25,000	−25,000.00	−12,000	−12,000.00
1	7,000	−18,913.04	−4,000	−15,478.26
2	7,000	−13,620.04	10,000	−7,916.82
3	9,000	−7,702.30	9,000	−1,999.18
4	10,000	−1,984.86	6,000	1,431.34
5	7,000	1,495.38		
Method				
IRR		17.41%		19.10%
NPV		$1,495.28		$1,431.34
IB		$25,000.00		$15,478.26
ORR		16.34%		17.57%
$ORR_{5000,5}$		16.34%		16.292%

*$k = 15\%$.

17.57 to 16.29%. When adjusted to a common scale and a common life, the ORRs of machines A and B provide the same ranking of those two projects as does NPV.

Conclusion

Rate-of-return criteria can provide correct accept or reject decisions for individual projects and can correctly rank competing if used properly. The concept of a rate of return seems easier to understand than NPV. After all, most investments (stocks, bonds, CDs, etc.) are ranked according to their rates of return. The NPV criterion, however, is the "gold standard", and its characteristics of preserving project scale and measuring cumulative performance make it more convenient to use when ranking projects.

Rate-of-return criteria should be viewed as complementing the NPV by providing project performance information in a slightly different form. The IRR provides the average rate earned per period on funds invested in the project itself while they remain to invested. In contrast, the ORR provides the average rate earned per period on the project's investment base over the entire life of the project, assuming that funds earn the firm's opportunity cost when not invested in the project itself.

Defining Terms

Internal rate of return: The average rate per period earned on funds invested in the project itself.
Investment base: The time 0 value of the funds that must be provided to (invested in) the project during its life.
Net present value: The expected increase in a firm's value if and when it implements the project.
Opportunity cost: The rate of return the firm believes it can earn on projects of similar risk. Also referred to as the *minimum required return, the discount rate,* or the project's *hurdle rate.*
Overall rate of return: The average rate earned per period on the project's investment base over the life of the project.

References

Bailey, M. J. Formal criteria for investment decisions, *J. Polit. Econ.,* 67(6): 476–488, 1959.
Beaves, R. G. The case for a generalized net present value formula, *Eng. Econ.,* 3l8(2): 119–133, 1993.
Bernhard, R. H. Base selection for modified rates of return and its irrelevance for optimal project choice, *Eng. Econ.,* 35(1): 55–65, 1989.
Hirshleifer, J. On the theory of optimal investment decision, *J. Polit. Econ.,* 66(4): 329–352, 1958.
Lin, S. A. Y. The modified rate of return and investment criterion, *Eng. Econ.,* 21(4): 237–247, 1976.
Mao, J. T. The internal rate of return as a ranking criterion, *Eng. Econ.,* 11(1): 1–13, 1966.

Shull, D. M. Efficient capital project selection through a yield-based capital budgeting technique, *Eng. Econ.*, 38(1): 1–18, 1992.

Shull, D. M. Interpreting rates of return: a modified rate-of-return approach, *Financial Pract. Educ.*, 3(2): 67–71, 1993.

Solomon, E. The arithmetic of capital budgeting decisions, *J. Bus.*, 29(12): 124–129, 1956.

Further Information

Au, T. and Au, T. P. *Engineering Economics for Capital Investment Analysis*, 2nd ed., Prentice Hall, Englewood Cliffs, NJ, 1992.

8.16   Present Worth Analysis[6]

Walter D. Short

Evaluation of any project or investment is complicated by the fact that there are usually costs and benefits (i.e., **cash flows**) associated with an investment that occur at different points in time. The typical sequence consists of an initial investment followed by operations and maintenance costs and returns in later years. **Present worth** analysis is one commonly used method that reduces all cash flows to a single equivalent cash flow or dollar value [Palm and Qayum, 1985]. If the investor did not care when the costs and returns occurred, present worth (also known as *net present value*) could be easily calculated by simply subtracting all costs from all income or returns. However, to most investors the timing of the cash flows is critical due to the time value of money, that is, cash flows in the future are not as valuable as the same cash flow today. Present worth analysis accounts for this difference in values over time by discounting future cash flows to the value in a base year, which is normally the present (hence the term "present" worth). If the single value that results is positive, the investment is worthwhile from an economic standpoint. If the present worth is negative, the investment will not yield the desired return as represented by the **discount rate** employed in the present worth calculation.

Calculation

The present worth (PW), or net present value, of an investment is

$$PW = \sum_{t=0}^{N} \frac{C_t}{\left(1+d\right)^t} \tag{8.20}$$

where N is the length of the **analysis period**, C_t is the net cash flow in year t, and d is the discount rate.

Analysis Period

The analysis period must be defined in terms of its first year and its length as well as the length of each time increment. The standard approach, as represented in Eq. (8.20), is to assume that the present is the beginning of the first year of the analysis period [$t = 0$ in Eq. (8.20)], that cash flows will be considered on an annual basis, and that the analysis period will end when there are no more cash flows that result from the investment (e.g., when the project is completed or the investment is retired) [Ruegg and Petersen, 1987]. For example, the present worth for a refinery that will require 6 years to construct and then last for 30 years (to 2031) can be represented as

---

[6]The material in this section was previously published in *The Engineering Handbook*, R. C. Dorf, Ed., CRC Press, Boca Raton, FL, 1996.

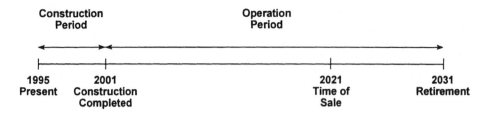

FIGURE 8.11   Analysis period for a refinery.

$$\text{PW}(1995) = \sum_{t=0}^{36} \frac{C_t}{(1+d)^t}$$

(8.21)

In this case the analysis period covers both the construction period and the full operation period, as shown in Fig. 8.11. In Eq. (8.21) the parenthetical (1995) explicitly presents the base year for which the present worth is calculated. This designation is generally omitted when the base year is the present.

However, any of these assumptions can be varied. For example, the present can be assumed to be the first year of operation of the investment. In the refinery example the present worth can be calculated as if the present were 6 years from now (2001, that is, the future worth in the year 2001, with all investment costs between now and then accounted for in a single turnkey investment cost (C_0) at the time ($t = 0$). In this case the base year is 2001.

$$\text{PW}(2001) = \sum_{t=0}^{30} \frac{C_t}{(1+d)^t}$$

(8.22)

The length of the analysis period is generally established by the point in time at which no further costs or returns can be expected to result from the investment. This is typically the point at which the useful life of the investment has expired. However, a shorter lifetime can be assumed with the cash flow in the last year accounting for all subsequent cash flows. For the refinery example, the analysis period could be shortened from the 36-year period to a 26-year period, with the last year capturing the salvage value of the plant after 20 years of operation.

$$\text{PW} = \sum_{t=0}^{26} \frac{C_t}{(1+d)^t}$$

(8.23)

where C_{26} is now the sum of the actual cash flows in C_{26} and the salvage value of the refinery after 20 years of operation. One common reason for using a shorter analysis period is to assess the present worth, assuming sale of the investment. In this case the salvage value would represent the price from the sale minus any taxes paid as a result of the sale.

Finally, the present worth is typically expressed in the dollars of the base year. However, it can be expressed in any year's dollars by adjusting for inflation. For example, in the refinery case, in which the base year is 2001 [Eq. (8.22)], the present worth value expressed in the dollars of the year 2001 can be converted to a value expressed in today's (1995) dollars, as follows:

$$\text{PW}_{1995\$}(2001) = \frac{1}{(1+i)^{2001-1995}} \text{PW}_{2001\$}(2001)$$

(8.24)

where i is the annual **inflation rate** between 1995 and 2001 and the subscripts indicate the year of the dollars in which the present worth is presented.

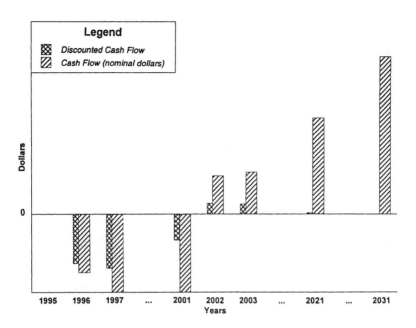

FIGURE 8.12 Cash flows.

Cash Flows

Cash flows include all costs and returns where costs are negative and returns or income are positive (see Fig. 8.12). The cash flow for each year of the analysis period must include all actual costs and income associated with the investment during that year. Thus, cash flows must include all initial capital costs, all taxes, debt payments, insurance, operations, maintenance, income, and so forth.

Cash flows are typically expressed in terms of the actual dollar bills paid or received (**nominal** or current **dollars**). Alternatively, cash flows can be expressed in terms of the dollars of a base year (**real** or constant **dollars**). Nominal dollar cash flows, C^n, can be converted to real dollar cash flows, C^r (and vice versa) by accounting for the effects of inflation.

$$C_0^r = \frac{C_t^n}{\left(1+i\right)^t} \tag{8.25}$$

where i is the annual inflation rate and t is the difference in time between the base year ($t = 0$) and the year of the nominal dollar cash flow.

A discounted cash flow is the present worth of the individual cash flow. Discounted cash flow in period s equals.

$$\frac{C_s}{\left(1+d\right)^s} \tag{8.26}$$

As shown in Fig. 8.12 (which used a nominal discount rate of 0.2 or 20%), discounting can significantly reduce the value of cash flows in later years.

Discount Rate

The discount rate is intended to capture the time value of money to the investor. The value used varies with the type of investor, the type of investment, and the opportunity cost of capital (i.e., the returns

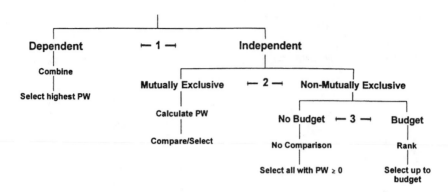

FIGURE 8.13   Comparison of investments.

that might be expected on other investments by the same investor). In some cases, such as a regulated electric utility investment, the discount rate may reflect the weighed average cost of capital (i.e., the after-tax average of the interest rate on debt and the return on common and preferred stock). Occasionally, the discount rate is adjusted to capture risk and uncertainty associated with the investment. However, this is not recommended; a direct treatment of uncertainty and risk is preferred (discussed later).

As with cash flows, the discount rate can be expressed in either nominal or real dollar terms. A nominal dollar discount rate must be used in discounting all nominal dollar cash flows, whereas a real discount rate must be used in discounting all real dollar cash flows. As with cash flows, a nominal dollar discount rate, d_n, can be converted to a real dollar discount rate, d_r, by accounting for inflation with either of the following:

$$1+d_n =\left(1+d_r\right)\left(1+i\right)$$

$$d_r =\frac{1+d_n}{1+i}-1 \tag{8.27}$$

Since cash flow should include the payment of taxes as a cost, the discount rate used in the present worth calculation should be an after-tax discount rate (e.g., the after-tax opportunity cost of capital).

Application

Present worth analysis can be used to evaluate a single investment or to compare investments. As stated earlier, a single investment is economic if its present worth is equal to or greater than zero, but is not economic if the present worth is less than zero.

PW > 0. The investment is economic.
PW < 0. The investment is not economic.

The use of present worth to compare multiple investments is more complex. There are several possible situations, as shown in Fig. 8.13, including investments that impact one another (dependence, point 1 on Fig. 8.13), investments that are mutually exclusive (only one investment is possible from a set of investments, point 2 on Fig. 8.13), and/or investments for which there is a budget limitation (point 3 on Fig. 8.13).

If the investments do have an impact on one another (i.e., they are dependent), then they should be considered both together and individually. Investments that impact one another are reevaluated as a package, with new cash flows representing the costs and returns of the combination of investments. The present worth of all individual investments and all combinations of dependent investments should then be compared. This comparison will, by definition, include some mutually exclusive investments (i.e., they cannot all be undertaken because some are subsumed in others). Each set of these mutually exclusive alternatives can be resolved to a single alternative by selecting the combination or individual investment with the higher present worth from each set. For example, if a refinery and a pipeline are being considered as possible investments and the pipeline would serve the refinery as well as other refineries, then the presence of the pipeline might change the value of the refinery product (i.e., the two investments are not independent). In this case the present worth of the refinery alone, the present worth of the pipeline alone, and the present worth of the combined pipeline and refinery should be calculated and the one with the highest present worth selected.

Once each set of dependent investments has been resolved to a single independent investment, then the comparison can proceed. If only one investment can be made (i.e., they are mutually exclusive), then the comparison is straightforward; the present worth of each investment is calculated and the investment with the highest present worth is the most economic. This is true even if the investments require significantly different initial investments, have significantly different times at which the returns occur, or have different useful lifetimes. Examples of mutually exclusive investments include different system sizes (e.g., three different refinery sizes are being considered for a single location), different system configurations (e.g., different refinery configurations are being considered for the same site), and so forth.

If the investments are not mutually exclusive, then one must consider whether there is an overall budget limitation that would restrict the number of economic investments that might be undertaken. If there is no budget (i.e., no limitation on the investment funds available) and the investments have no impact on one another, then there is really no comparison to be performed, and the investor simply undertakes those investments that have positive present worth and discards those that do not.

If the investments are independent but funds are not available to undertake all of them (i.e., there is a budget), then there are two approaches. The easiest approach is to rank the alternatives, with the best having the highest benefit-to-cost ratio or savings-to-investment ratio. (The investment with the highest present worth will not necessarily be the one with the highest rank, since present worth does not show return per unit investment.) Once ranked, those investments at the top of the priority list are selected until the budget is exhausted. Present worth can be used in the second, less-desirable approach by considering the total present worth of each combination of investments whose total initial investment cost is less than the budget. The total present worth of the investment package is simply the sum of the present worth values of all the independent investments in the package. That combination with the greatest present worth is then selected.

Other Considerations

Three commonly encountered complications to the calculation of present worth are savings vs. returns, uncertainty, and externalities.

Savings vs. Income/Returns

In many, if not most, investments, the positive cash flows or benefits are actual cash inflows that result from sales of products produced by the investment. However, there are also a large number of investments in which the positive cash flows or benefits are represented by savings. For example, the benefit of investing in additional insulation for steam pipes within a refinery is savings in fuel costs (less fuel will be required to produce steam). These savings can be represented in a present worth evaluation of the insulation as positive cash flows (i.e., as income or returns).

Uncertainty

In the present worth discussion up to this point, it has been assumed that all the input values to the present worth calculation are known with precision. In fact, for most investments there is considerable uncertainty in these values — especially the future cash flows. The preferred method for including such uncertainties in the calculation of present worth is to estimate the probability associated with each possible cash flow stream, calculate the present worth associated with that cash flow stream, and assign the probability of the cash flow to the present worth value. This will produce one present worth value and one probability for each possible cash flow stream (i.e., a probability distribution on the present worth of the investment). This distribution can then be used to find statistics such as the expected present worth value, the standard deviation of the present worth value, confidence intervals, and so forth. These statistics, especially the expected present worth value and confidence intervals, can then be used in the decision process.

Externalities

In many cases not all costs and benefits are included in the cash flows because they are not easily quantified in terms of dollars or because they do not benefit or cost the investor directly. Such costs and benefits are referred to as *externalities* because they are generally considered external to the direct economic evaluation (i.e., they are not included in the present worth calculation). For example, the cost of air emissions from a refinery are often not considered in a present worth calculation, even though they impact the local community as well as the employees of the refinery. Such emissions may result in lost work days and more sick pay as well as the loss of the local community's goodwill to the refinery, making future refinery expansion more difficult. Likewise, emissions may affect the health and quality of life of local residents.

Externalities can be considered qualitatively, along with measures such as the present worth, when evaluating an investment, or externalities can be explicitly considered within a present worth calculation by estimating their costs and benefits in terms of dollars and including these dollars in the present worth cash flows. In this case these costs and benefits are said to have been *internalized* and are no longer externalities.

Defining Terms

Analysis period: The period during which the investor will consider the costs and benefits associated with an investment. This period is usually determined by the period during which the investor will develop, operate, and own the investment.

Cash flow: The dollar value of all costs and benefits in a given period. Cash flows are normally expressed in nominal dollars but can be expressed in real dollars.

Discount rate: The interest rate that represents the time value of money to the investor. For most investors this is the opportunity cost of capital (i.e., the rate of return that might be expected from other opportunities to which the same capital could be applied). Discount rates can include inflation (nominal discount rate) or exclude inflation (real discount rate).

Inflation rate: The annual rate of increase in a general price level, frequently estimated as the gross domestic product deflator or the gross national product deflator. Estimates of future value are generally provided by macroeconomic forecasting services.

Nominal dollars: Current dollars; dollars in the year the cost of benefit is incurred (i.e., number of dollar bills).

Present worth: Net present value. The sum of all cash flows during the analysis period discounted to the present.

Real dollars: Constant dollars; value expressed in the dollars of the base year. Real dollars are the value excluding inflation after the base year.

References

Palm, T. and Qayum, A. *Private and Public Investment Analysis*, Southwestern, Cincinnati, OH, 1985.

Ruegg, R. and Marshall, H. *Building Economics: Theory and Practice*, Von Nostrand Reinhold, New York, 1990.

Ruegg, R. and Petersen, S. *Comprehensive Guide for Least-Cost Energy Decisions*, National Bureau of Standards, Gaithersburg, MD, 1987.

Further Information

Au, T. and Au. T. P. *Engineering Economics for Capital Investment Analysis*, Allyn & Bacon, Boston, 1983.

Brown R. J. and Yanuck, R. R. *Life Cycle Costing: A Practical Guide for Energy Managers*, Fairmont Press, Atlanta, 1980.

Short, W. et al. *A Manual for the Economic Evaluation of Energy Efficiency and Renewable Energy Technologies*, National Renewable Energy Laboratory, NREL/TP-462-5173, Golden, CO, 1995.

Samuelson, P. A. and Nordhaus, W. D. *Economics 12 ed.*, McGraw-Hill, New York, 1985.

Stermole, F. J. *Economic Evaluation and Investment Decision Methods, 5th ed.*, Investment Evaluations, Golden, CO, 1984.

Weston, J. F. and Brigham, E. F. *Managerial Finance, 7th ed.*, Dryden Press, Fort Worth, TX, 1981.

8.17 Project Selection from Alternatives

Chris Hendrickson and Sue McNeil

Practical engineering and management requires choices among competing **alternatives**. Which boiler should be used in a plant? Which computer should be purchased for a design office? Which financing scheme would be most desirable for a new facility? These are practical questions that arise in the ordinary course of engineering design, organizational management, and even personal finances. This section is intended to present methods for choosing the best among distinct alternatives.

Problem Statement for Project Selection

The economic project selection problem is to identify the best from a set of possible alternatives. Selection is made on the basis of a systematic analysis of expected revenues and costs over time for each project alternative.

Project selection falls into three general classes of problems. Accept-reject problems (also known as *determination of feasibility*) require an assessment of whether an investment is worthwhile. For example, the hiring of an additional engineer in a design office is an accept-reject decision. Selection of the best project from a set of mutually exclusive projects is required when there are several competing projects or options and only one project can be built or purchased. For example, a town building a new sewage treatment plant may consider three different configurations, but only one configuration will be built. Finally, capital budgeting problems are concerned with the selection of a set of projects when there is a budget constraint and many, not necessarily competing, options. For example, a state highway agency will consider many different highway rehabilitation projects for a particular year, but generally the budget is insufficient to allow all to be undertaken, although they may all be feasible.

Steps in Carrying Out Project Selection

A systematic approach to economic evaluation of projects includes the following major steps [Hendrickson and Au, 1989]:

1. Generate a set of project or purchase **alternatives** for considerations. Each alternative represents a distinct component or combination of components constituting a purchase or project decision. We shall denote project alternatives by the subscript x, where $x = 1, 2, \ldots$ refers to projects 1, 2, and so on.
2. Establish a **planning horizon** for economic analysis. The planning horizon is the set of future periods used in the economic analysis. It could be very short or long. The planning horizon may be set by organizational policy (e.g., 5 years for new computers or 50 years for new buildings), by the expected economic life of the alternatives, or by the period over which reasonable forecasts of operating conditions may be made. The planning horizon is divided into discrete periods — usually years, but sometimes shorter units. We shall denote the planning horizon as a set of $t = 0$,

1, 2, 3, ..., n, where t indicates different periods, with $t = 0$ being the present, $t = 1$ the first period, and $t = n$ representing the end of the planning horizon.

3. Estimate the **cash flow profile** for each alternative. The cash flow profile should include the revenues and costs for the alternative being considered during each period in the planning horizon. For public projects, revenues may be replaced by estimates of benefits for the public as a whole. In some cases revenues may be assumed to be constant for all alternatives, so only costs in each period are estimated. Cash flow profiles should be specific to each alternative, so the costs avoided by not selecting one alternative (e.g., $x = 5$) are not included in the cash flow profile of the alternatives ($x = 1, 2,$ and so on). Revenues for an alternative x in period t are denoted $B(t, x)$, and costs are denoted $C(t, x)$. Revenues and costs should initially be in **base year** or constant dollars. Base-year dollars do not change with inflation or deflation.

    For tax-exempt organizations and government agencies, there is no need to speculate on inflation if the cash flows are expressed in terms of base-year dollars and a MARR without an inflation component is used in computing the **net present value**. For private corporations that pay taxes on the basis of then-current dollars, some modification should be made to reflect the projected inflation rates when considering depreciation and corporate taxes.

4. Specify the **minimum attractive rate of return** (MARR) for discounting. Revenues and costs incurred at various times in the future are generally not valued equally to revenues and costs occurring in the present. After all, money received in the present can be invested to obtain interest income over time. The MARR represents the trade-off between monetary amounts in different periods and does not include inflation. The MARR is usually expressed as a percentage change per year, so that the MARR for many public projects may be stated as 10%. The value of MARR is usually set for an entire organization based upon the opportunity cost of investing funds internally rather than externally in the financial markets. For public projects, the value of MARR is a political decision, so MARR is often called the *social rate of discount* in such cases. The equivalent value of a dollar in a following period is calculated as (1 + MARR), and the equivalent value two periods in the future is (1 + MARR) · (1 + MARR) = (1 + MARR)². In general, if you have Y dollars in the present [denoted $Y(0)$], then the future value in time t [denoted $Y(t)$] is

$$Y(t) = Y(0)(1 + \text{MARR})^t \qquad (8.28)$$

   or the present value, $Y(0)$, of a future dollar amount $Y(t)$ is

$$Y(0) = Y(t) \big/ (1 + \text{MARR})^t \qquad (8.29)$$

5. Establish the criterion for accepting or rejecting an alternative and for selecting the best among a group of mutually exclusive alternatives. The most widely used and simplest criterion is the net present value criterion. Projects with a positive net present value are acceptable. Only one form of mutually exclusive alternatives can be chosen. For example, the alternatives might be alternative boilers for a building or alternative airport configurations. From a set of mutually exclusive alternatives, the alternative with the highest net present value is best. The next section details the calculation steps for the net present value and also some other criterion for selection.

6. Perform sensitivity and uncertainty analysis. Calculation of net present values assumes that cash flow profiles and the value of MARR are reasonably accurate. In many cases assumptions are made in developing cash flow profile forecasts. Sensitivity analysis can be performed by testing a variety of such assumptions, such as different values of MARR, to see how alternative selection might change. Formally treating cash flow profiles and MARR values as stochastic variables can be done with probabilistic and statistical methods.

Selection Criteria

Net Present Value

Calculation of net present values to select projects is commonly performed on electronic calculators, on commercial spreadsheet software, or by hand. The easiest calculation approach is to compute the net revenue in each period for each alternative, denoted $A(t, x)$:

$$A(t,x) = B(t,x) - C(t,x) \tag{8.30}$$

where $A(t, x)$ may be positive or negative in any period. Then, the net present value of the alternative, $NPV(x)$, is calculated as the sum over the entire planning horizon of the discounted values of $A(t, x)$:

$$NPV(x) = \sum_{t=0}^{n} A(t,x) / (1 + MARR)^t \tag{8.31}$$

Other Methods

Several other criteria may be used to select projects. Other discounted flow methods include **net future value** [denoted $NFV(x)$] and **equivalent uniform annual value** [denoted $EUAV(x)$]. It can be shown [Au and Au, 1992] that these criteria are equivalent where

$$NFV(x) = NPV(x)(1 + MARR)^n \tag{8.32}$$

$$EUAV(x) = \frac{NPV(x)(1 + MARR)^n}{\left[(1 + MARR)^n - 1 \right]} \tag{8.33}$$

The net future value is the equivalent value of the project at the end of the planning horizon. The equivalent uniform annual value is the equivalent series in each year of the planning horizon.

Alternatively, benefit-to-cost ratio (the ratio of the discounted benefits to discounted costs) and the internal rate of return [the equivalent MARR at which $NPV(x) = 0$] are merit measures, each of which may be used to formulate a decision. For accept-reject decisions, the benefit-to-cost ratio must be greater than one and the internal rate of return greater than the MARR. However, these measures must be used in connection with incremental analysis of alternatives to provide consistent results for selecting among mutually exclusive alternatives (e.g., see Au and Au [1992]).

Similarly, the payback period provides an indication of the time it takes to recoup an investment but does not indicate the best project in terms of expected net revenues.

Applications

To illustrate the application of these techniques and the calculations involved, two examples are presented.

Example 1—Alternative Bridge Designs. A state highway agency is planning to build a new bridge and is considering two distinct configurations. The initial costs and annual costs and benefits for each bridge are shown in the following table. The bridges are each expected to last 30 years.
Solution. The net present values for a MARR of 5% are given as follows:

	Alternative 1	Alternative 2
Initial cost	$15,000,000	$25,000,000
Annual maintenance and operating costs	$15,000	$10,000
Annual benefits	$1,200,000	$1,900,000
Annual benefits less costs	$1,185,000	$1,890,000

$$\text{NPV}(1) = (-15{,}000{,}000) + (1{,}185{,}000)\big/(1+0.05) + (1{,}185{,}000)\big/(1+0.05)^2$$

$$+ (1{,}185{,}000)\big/(1+0.05)^3 + L\ + (1{,}185{,}000)\big/(1+0.05)^{30}$$

$$= \$3{,}216{,}354$$

$$\text{NPV}(2) = (-15{,}000{,}000) + (1{,}890{,}000)\big/(1+0.05) + (1{,}890{,}000)\big/(1+0.05)^2$$

$$+ (1{,}890{,}000)\big/(1+0.05)^3 + L\ + (1{,}890{,}000)\big/(1+0.05)^{30}$$

$$= \$4{,}053{,}932$$

Therefore, the department of transportation should select the second alternative, which has the largest net present value. Both alternatives are acceptable since their net present values are positive, but the second alternative has a higher net benefit.

Example 2—Equipment Purchase. Consider two alternative methods for sealing pavement cracks [McNeil, 1992]. The first method is a manual method; the second is an automated method using a specialized equipment system. Which method should be used? We shall solve this problem by analyzing whether the new automated method has revenues and benefits in excess of the existing manual method. *Solution.* Following the steps outlined earlier, the problem is solved as follows:

1. The alternatives for consideration are (1) the existing manual method and (2) the automated method. The alternatives are mutually exclusive because cracks can only be sealed using either the existing method or the new method.
2. The planning horizon is assumed to be 6 years to coincide with the expected life of the automated equipment.
3. The cash flow profile for alternative 2 is given in the following table:

System acquisition costs	$100,000
Annual maintenance and operating costs	$10,000
Annual labor savings	$36,000
Annual savings over costs	$26,000

The values are estimated using engineering judgment and historical cost experience. We assume that the productivity and revenues for both alternatives are the same and treat labor savings as additional benefits for alternative 2. Therefore, only the net present value for alternative 2, which represents the result of introducing the automated method, need be computed.

4. The MARR is assumed to be 5%. The net present value is computed as follows:

$$\text{NPV}(2) = -100{,}000 + (26{,}000)\big/(1+0.05)$$

$$+ (26{,}000)\big/(1+0.05)^2 + L\ + (26{,}000)\big/(1+0.05)^5 \tag{8.34}$$

$$= \$12{,}566$$

TABLE 8.18 Relative Value of New Equipment with Different Assumptions

Acquisition Cost ($)	Labor Saving ($)	Maintenance and Operaiton ($)	MARR		
			0.05	0.1	0.15
50,000	36,000	10,000	$62,566	$48,560	$37,156
100,000	36,000	10,000	$12,566	($1,440)	($12,844)
150,000	36,000	10,000	($37,434)	($51,440)	($62,844)
50,000	45,000	10,000	$101,532	$82,678	$67,325
100,000	45,000	10,000	$51,532	$32,678	$17,325
150,000	45,000	10,000	$1,532	($17,322)	($32,675)

5. Using the criterion NPV(2) > 0, alternative 2 is selected.
6. To determine the sensitivity of the result to some of the assumptions, consider Table 8.18. The table indicates that additional investment in the automated method is justifiable at the MARR of 5% if the acquisition costs decrease or the labor savings increase. However, if the MARR increases to 10% or the acquisition costs increase, then the investment becomes uneconomical.

This example illustrates the use of the net present value criteria for an incremental analysis, which assumes that the benefits are constant for both alternatives and examines incremental costs for one project over another.

Conclusion

This section has presented the basic steps for assessing economic feasibility and selecting the best project from a set of mutually exclusive projects, with net present value as a criterion for making the selection.

Defining Terms

Alternatives: A distinct option for a purchase or project decision.
Base year: The year used as the baseline of price measurement of an investment project.
Cash flow profile: Revenues and costs for each period in the planning horizon.
Equivalent uniform annual value: Series of cash flows with a discounted value equivalent to the net present value.
Minimum attractive rate of return (MARR): Percentage change representing the time value of money.
Net future value: Algebraic sum of the computed cash flows at the end of the planning horizon.
Net present value: Algebraic sum of the discounted cash flows over the life of an investment project to the present.
Planning horizon: Set of time period from the beginning to the end of the project; used for economic analysis.

References

Au, T. and Au, T. P. *Engineering Economics for Capital Investment Analysis, 2nd ed.*, Prentice Hall, Englewood Cliffs, NJ, 1992.
Hendrickson, C. and Au, T. *Project Management for Construction*, Prentice Hall, Englewood Cliffs, NJ, 1989.
McNeil, S. An analysis of the costs and impacts of the automation of pavement crack sealing, *Proc. World Conf. on Transp. Res.*, Lyon, France, July 1992.
Park, C. S. *Contemporary Engineering Economics*, Addison Wesley, Reading, MA, 1993.

Further Information

A thorough treatment of project selection is found in *Engineering Economics for Capital Investment Analysis*. Many examples are presented in *Contemporary Engineering Economics*.

9

Decision and Modeling Methods

Ralph L. Keeney
University of Southern California

Robert T. Clemen
Duke Universtiy

Ralph F. Miles, Jr.
University of Southern California and California Institute of Technology

William Samuelson
Boston University

9.1 Value-Focused Thinking

Ralph L. Keeney

Decision making usually focuses on **alternatives**. Decision problems are thrust upon us by the actions of others — competitors, customers, government, and stakeholders — or by circumstances — recessions and natural disasters. Faced with a **decision problem**, the so-called solving begins. Typically, the decision maker concentrates first on alternatives and only afterward addresses the objectives or criteria to evaluate the alternatives. I refer to this standard problem-solving approach as **alternative-focused thinking**.

Focusing on alternatives is a limited way to think about decision situations. It is reactive, not proactive. Alternatives are relevant only because they are means to achieve your **values**. Thus, your thinking should focus first on values and later on alternatives that might achieve them. Naturally, there should be iteration between articulating values and creating alternatives, but the principle is "values first". I refer to this manner of thinking as **value-focused thinking** [Keeney, 1992].

Value-focused thinking is a way to channel a critical resource — hard thinking — to lead to better decisions. Values are used to create better alternatives and to identify better decision situations. These better decisions situations, which you create for yourself, should be thought of as **decision opportunities** rather than as decision problems.

Making Values Explicit

Strategic thinkers have long emphasized the need to clarify values as a key step in making informed decisions (e.g., Franklin [1956]). Peters and Waterman [1982] refer to their "one all-purpose bit of advice for management" in the pursuit of excellence as "figure out your value system". However, figuring out a value system requires more than simply listing values. To create a useful set of values requires the following steps:

TABLE 9.1   Techniques to Use in Identifying Values

A wish list   What do you want? What do you value? What should you want? If you had no limitations at all, what would you want?

Alternatives   What is a perfect alternative, a terrible alternative, some reasonable alternatives? What is good or bad about each?

Problems and shortcomings   What is wrong or right with your organization? What needs fixing?

Consequences   What has occurred that was good or bad? What might occur that you care about?

Goals, constraints, and guidelines   What are your aspirations? What limitations are placed upon you?

Different perspectives   What would your competition or your constituency be concerned about? At some time in the future, what would concern you?

Strategic values   What are your ultimate values? What are your values that are absolutely fundamental?

Generic values   What values do you have for your customers, your employees, your shareholders, yourself? What environmental, social, economic, or health and safety values are important?

Structuring values   Follow means-ends relationships: why is that value important, how can you achieve it? Use specification: what do you mean by this value?

Quantifying values   How would you measure achievement of this value? Why is value A three times as important as value B?

Source: Keeney, R. L. Creativity in decision making with value-focused thinking, *Sloan Mgt. Rev.,* 35(4): 33–41, 1994.

- Develop a list of values.
- Convert each of these values to an objective.
- Structure these objectives to clarify relationships among them.

Developing a List of Values

The foundation for any model of values is a comprehensive and logical list of what one cares about in the particular decision situation. Values are articulated by asking individuals what it is they value. A natural way to begin the discussion is to ask the person to write down everything that is important for valuing alternatives or opportunities in the given situation. To help stimulate the generation of values, the questions in Table 9.1 may be helpful. Although different questions may lead to the same values, this is not a shortcoming. It is easy to recognize redundant values but difficult to identify values that are not initially listed.

Converting Values to Objectives

The initial list of values will come in many forms. It may include constraints, criteria, measures, alternatives, targets, aspiration levels, or concerns. To develop consistency, convert each item into a corresponding **objective**, meaning something one wants to strive toward. An objective has three features: a decision context, an object, and a direction of preference. For example, one objective of a firm concerned with shipping hazardous material is to "minimize accidents". The decision context here is shipping hazardous material, the object is accidents, and the preference is to minimize.

Structuring Objectives

It is important to understand how different objectives are related. **Fundamental objectives** concern the ends that decision makers value in a specific decision context; **means objectives** are methods to achieve those ends. Fundamental objectives for strategic decisions, meaning the broadest class of decisions facing an organization, are defined as **strategic objectives**. The strategic objectives provide the common guidance for all decisions in an organization and form the logical foundation for the fundamental objectives appropriate for all specific decisions.

The list of objectives developed from the values will include both fundamental objectives and means objectives. For each objective, ask "why is this objective important in the decision context?" Two answers are possible. One is that the objective is an essential reason for interest in the situation. Then it is a fundamental objective. The other answer is that the objective is important because of its implications for achieving some other objective. In this case, it is a means objective. The response to the question

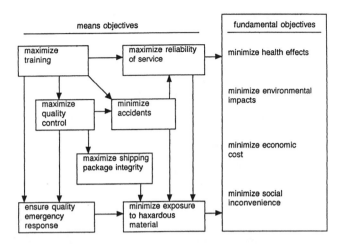

FIGURE 9.1   Illustrative means-ends objectives network for shipping hazardous material. (An arrow means "influences the achievement of".)

about why it is important identifies another objective that may not be on your original list. If it is not, add it. The "why is it important?" test must be given in turn to this objective to ascertain whether it is a means or a fundamental objective. Tracing means-ends relationships will lead to at least one fundamental objective.

Again consider shipping hazardous material. One stated objective may be to ensure quality emergency response. When asked why this is important, a response may be that this minimizes exposure to hazardous material. When asked why this is important, it may be that it minimizes health effects to humans. When asked why it is important to minimize health effects, that individual may simply say that health effects are one of the reasons for being interested in the problem. This would suggest that minimizing health effects is one of the fundamental objectives regarding the shipping of hazardous material.

The means-ends relationships among objectives and the identification of the fundamental objectives are indicated in a means-ends objectives network. An illustration of a means-ends objectives network for shipping hazardous material, which is not meant to be complete, is shown in Fig. 9.1.

To better understand the fundamental objectives, each is specified into logical parts that are easier to consider. For example, one of the fundamental objectives of shipping hazardous material is to minimize health effects. To be more specific, ask "what health effects should be minimized?" The response, which might be possible health effects to employees and to the public, would lead to a better understanding of that objective and better focus subsequent thinking and action. For each of these more-detailed components of the fundamental objective, one can specify further. For instance, regarding health implications to employees, one might specify the health implications of exposure due to normal operations and due to accidents.

The relationships of component fundamental objectives can be indicated in a fundamental objectives hierarchy. An illustration for shipping hazardous material is shown in Table 9.2. When evaluating alternatives, it is often useful to collect the information and evaluate the consequences at the lower levels of this hierarchy because the data and information relevant to those circumstances may be generated from different sources or from individuals with different expertise and because the different consequences may be differently valued. For instance, one may consider a given health effect to a member of the public to be more significant than that same health effect to a worker because the worker may knowingly accept that risk and be compensated for accepting that risk.

Creating Alternatives

The range of alternatives people identify for a given decision situation is often unnecessarily narrow. One often quickly identifies some viable alternatives and proceeds to evaluate them without making the effort to broaden the search for alternatives.

TABLE 9.2 Illustrative Fundamental
Objectives Hierarchy for Shipping
Hazardous Materials

Minimize health effects
    To employees
        Due to normal operations
        Due to accidents
    To the public
Minimize environmental impacts
Minimize economic costs
    To shipping company
    To government
        Local
        State
        Federal
    To other business
Minimize social inconvenience
    Due to disruption
    Due to property damage

The first alternatives that come to mind in a given situation are the obvious ones, those that have been used before in similar situations and those that are readily available. Once a few alternatives — or perhaps only one, such as the status quo — are in the table, they serve to anchor thinking about others. Assumptions implicit in the identified alternatives are accepted, and the generation on new alternatives, if it occurs at all, tends to be limited to a tweaking of the alternatives already identified. Focusing on the values that should be guiding the decision situation removes the anchor on narrowly defined alternatives and makes the search for new alternatives a creative and productive exercise.

Numerous guidelines facilitate the search for alternatives or, more precisely, the search for good alternatives. The principle is that alternatives should be created that best achieve the values specified for the decision situation. Hence, each of the fundamental objectives should be systematically probed to initiate creative thought. To begin, take one objective at a time and think of alternatives that might be very desirable if that were the only objective. Consider every objective regardless of its level in the hierarchy. This exercise should generate many alternatives, most of which evaluate rather poorly on some objectives other than the one for which they were invented. If this is not the case, you have not been very creative in generating the alternatives.

The next step is to consider objectives two at a time and try to generate alternatives that would be good for both. These alternatives are likely to be refinements or combinations of those created using single objectives. Then take three objectives at a time, and so on, until all objectives are considered together.

The means objectives are also fruitful grounds to stimulate thinking about alternatives. Any alternative that influences one of the means objectives should influence at least some of the associated fundamental objectives. More-complicated decision situations tend to have more means objectives, which should suggest more alternatives. For each of the means objectives concerning the shipping of hazardous material in Fig. 9.1, a number of alternatives could be created.

Descriptive research on decision making has made it clear that the processes typically used to arrive at decisions are often not as logical or systematic as we would like them to be (see Russo and Schoemaker [1989]). Thus, at the end of a decision process, when you are about to choose an alternative, you should once more think about new alternatives. At this stage, you know how good that about-to-be-chosen alternative is. You should also have a good idea of the time, money, and effort necessary to implement it. Thus, carefully search for some new alternative that can better achieve the fundamental objectives by using the same or fewer resources in a different way.

Decision Opportunities

Who should be making your decisions? The answer is obvious: you. Well then, who should be deciding what decision situations you face? The answer here is the same: you. At least you should control far more of your situations than many of us do. Controlling the decision situations that you face by identifying decision opportunities may have a greater influence on the achievement of your objectives than controlling the alternatives selected for the decisions. A decision opportunity may alleviate some decision problems or allow you to avoid many future problems.

There are two ways to create decision opportunities. One is to convert an existing decision problem into a decision opportunity. Often this involves broadening the context of the problem (see Nadler and Hibino [1990]). The other way to create decision opportunities is from scratch. Use your creative genius, which can be stimulated by value-focused thinking, to examine how you can better achieve your objectives.

Decision opportunities are identified in a manner similar to creating alternatives. Now, however, for each objective, you ask what is a decision opportunity that you could address that will lead to better achievement on that objective. Sometimes it is useful to use the strategic objectives of an organization and ask, "where are the opportunities?"

Decision opportunities can be very helpful when you do not have direct control over a decision that you care about. In an important class of such decisions, one stakeholder wishes to have a certain alternative selected, but a different stakeholder has the power to make the decision. A firm wants its proposal to supply a product to another firm accepted, a government wants another government to sign an agreement, a manager wants his proposal for a sabbatical accepted by the organization, and so on.

The key to "solving" a problem of this nature is to view it from the perspective of the decision maker. First, structure her values as much as possible. In this process, begin by identifying the negative impacts of your desired alternative relative to the status quo in terms of her values. These impacts probably affect her means objectives. Follow their implications through a means-ends objective hierarchy to the fundamental objectives of the decision maker. Now you should be able to create modified alternatives that can improve matters in terms of her fundamental objectives while maintaining the key consequences desired by you. In many cases, the modified alternatives should be better than the status quo to the decision maker. In this way, you create a win-win alternative that is better for the decision maker, so she chooses it and you get what you want.

Defining Terms

Alternative: One of several choices available to a decision maker facing a decision problem or decision opportunity.

Alternative-focused thinking: Thought organized to evaluate alternatives to select the best or a good one.

Decision opportunity: A decision situation that the decision maker creates and chooses to face.

Decision problem: A decision situation that a decision maker must face as a result of the actions of others or circumstances.

Fundamental objectives: Statement of an essential reason for a decision maker's interest in a given decision situation.

Means objective: Means to better achieve fundamental objectives in a given decision situation.

Objectives: What a decision maker strives toward consisting of an object and direction of preference for a given decision situation.

Strategic objectives: Broadest class of objectives of a decision maker that serves to guide all decisions.

Value-focused thinking: Thought organized first on what the decision maker wants and then how to get it.

Values: What a decision maker wants from facing a particular decision.

References

Franklin, B. Letter to Joseph Preistly, in *The Benjamin Franklin Sampler*, Fawcett, New York, 1956.

Keeney, R. L. *Value-Focused Thinking: A Path to Creative Decisionmaking*, Harvard University Press, Cambridge, MA, 1992.

Nadler, G. and Hibino, S. *Breakthrough Thinking*, Prima Publishing and Communications, Rocklin, CA, 1990.

Peters, T. J. and Waterman, R. H. *In Search of Excellence*, Harper & Row, New York, 1982.

Russo, J. E. and Schoemaker, P. J. H. *Decision Traps*, Doubleday, New York, 1989.

Further Information

Value-focused thinking has been used to structure objectives and create decision opportunities in several firms. The range of applications includes integrated resource management, providing professional negotiation and dispute resolution service, developing mineral resources, creating a fair compensation system, designing waste water facilities in a metropolitan area, managing medical care, and reducing seismic risks. The methodology and procedures of value-focused thinking are illustrated in applications discussed in the following articles:

Gregory, R. and Keeney, R. L. Creating policy alternatives using stakeholder values, *Mgt. Sci.*, 40: 1035–1048, 1994.

Keeney, R. L. Creativity in decision making with value-focused thinking, *Sloan Mgt. Rev.*, 35(4): 33–41, 1994.

Keeney, R. L. Value-focused thinking: identifying decision opportunities and creating alternatives, *Eur. J. Opera. Res.*, 92: 537–549, 1996.

Keeney, R. L., McDaniels, T. L., and Ridge-Cooney, V. L. Using values in planning wastewater facilities for metropolitan Seattle, *Water Resour. Bull.*, 32(2): 293–303, 1996.

9.2   System Models for Decision Making

Robert T. Clemen

The **system model** concept is fundamental to this section. Scientists often represent systems — physical, biological, economic, social, or psychological — using models. In many cases, the ultimate goal is to be able to exploit or control the system in some way. For example, a freight-hauling company may create a model of its transportation network to improve customer-service quality. In other cases, systems are modeled to gain deeper understanding; a model of star formation in nebulae can provide insight on stellar and galactic evolution.

Many system models are mathematical. Such models consist of mathematical expressions, equations, or inequalities that represent the relationships among the elements of the natural system. Analysis of a mathematical model can lead to insights and understanding about the workings of the natural system.

Probabilistic system models are mathematical models distinguished by the incorporation of uncertainty regarding aspects of the system. The uncertainty might relate to randomness in the process, such as the random arrivals of patients at a health clinic. In addition, uncertainty can relate to unknown system parameters, such as the average arrival rate of patients at the clinic or the average cost of a patient visit. Sometimes one can obtain information about these system parameters, but often one must work with less than full information; decisions may be required before information becomes available, or the information could be prohibitively expensive or impossible to obtain.

Decision models are system models used specifically for decision-making purposes. Such models explicitly incorporate decisions, the available alternatives, and a way to measure the value of possible outcomes. Many decision models incorporate probabilistic system models as a representation of the uncertainty faced by the decision maker.

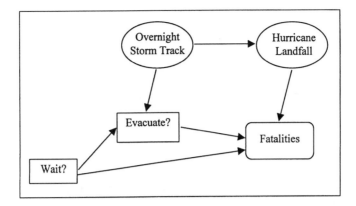

FIGURE 9.2   An influence diagram of an evacuation decision.

We introduce three model types that are widely used in decision and risk analysis: influence diagrams, decision trees, and spreadsheet risk-analysis models. In the final section, we discuss principles that underlie good modeling and analysis practice.

Influence Diagrams

An **influence diagram** is a directed, acyclic graph. Figure 9.2 shows a simple example of a decision model of an evacuation decision; it is directed because the arcs between nodes show direction, and it is acyclic because no cycles exist among the nodes and directed arcs. In this example, imagine an emergency-management agency that must make the evacuation decision for a coastal area. Tracking an approaching hurricane, the agency could issue the evacuation directive immediately or it could track the storm overnight before deciding. In the influence diagram, ovals represent uncertain variables or chance events, rectangles represent decisions, and the rounded rectangle (the sink in the directed graph) represents the value of the eventual outcome. For simplicity in this example we assume that the agency is interested in minimizing the number of fatalities. In reality, the agency's objectives would be more complex, including such things as injuries, human suffering, cost, and perhaps political influence; these objectives would be modeled according to principles described in other chapters in this book, and the resulting influence diagram might include a complex hierarchical representation of the value structure.

Details of the decision situation are suppressed in favor of a more compact display of the relationships among the elements. For example, the "Wait?" node consists of two alternatives: decide now (e.g., 6:00 p.m.) or track the storm overnight, delaying the decision until the following morning. The "Hurricane Landfall" node may include a wide variety of different locations along the coast where the storm could hit.

The arrows in an influence diagram indicate relationships among the variables, and the nature of the relationship depends on the downstream node. If the arrow points to a decision, it indicates time sequence and hence what information is known at the time of that decision. For example, the "Evacuate" decision is made after the "Wait" decision and also after the outcome of "Overnight Storm Track" is known. An arrow that points to a chance node indicates probabilistic dependence; the probability distribution associated with the downstream chance variable is conditioned on the outcome of the upstream node (whether chance or decision). In our model, the arrow from "Overnight Storm Track" to "Hurricane Landfall" indicates that the probability distribution for the landfall location would be assessed conditional on the overnight track. Finally, an arrow pointing to a value node indicates that the upstream variable is needed to determine the value of the outcome. In our example, whether the agency waits, the evacuation decision made, and the landfall location all have an impact on the number of fatalities. In contrast, the overnight storm track has no direct influence on the number of fatalities, although it is important in the model because of its relevance for forecasting the eventual landfall location.

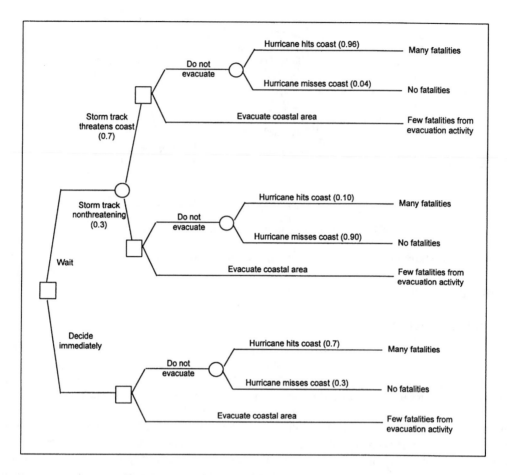

FIGURE 9.3   A decision tree for the evacuation decision.

Although this example is quite small, some applications involve very large diagrams. When a system is highly complex, the influence diagram can provide a compact representation useful for high-level discussions among decision makers and stakeholders. Some examples of such systems, along with more information about influence diagrams, are available in Clemen [1996].

Influence diagrams are often used for decision-making purposes. Doing so requires analyzing them to characterize the value of the different available alternatives and to determine an optimal decision strategy. Analysis requires a solution algorithm; although solution algorithms for influence diagrams are available, the most widely used software converts the influence diagram to a decision tree prior to solution.

Decision Trees

A **decision tree** is a hierarchical representation of a decision situation. The tree flows chronologically from left to right, and, as with influence diagrams, squares represent decisions and circles represent uncertainties. Figure 9.3 shows the decision-tree representation of the evacuation example. In contrast to the influence diagram, the details regarding the possible decisions and chance outcomes are explicitly represented. In addition, the parenthetical numbers on the chance branches are the probabilities (presumably estimated from data or assessed by experts) of the various chance outcomes.

Figure 9.3 is largely self-explanatory. The current decision is whether to wait overnight or decide immediately. It the choice is to decide immediately, then one must do so without knowing the storm track. If the choice is to wait, then the next event is whether the storm track threatens the coast. After following the storm overnight, the evacuation decision must be made. Regardless of the path taken to the evacuation decision, the agency must still decide whether to evacuate or not; if the choice is made not to evacuate, then the number of fatalities depends on whether the storm hits the coast. Note that in the upper part of the decision tree the probabilities of the storm hitting the coast are conditional; as in the influence diagram, the chance of the storm hitting the coast is conditioned on the storm track.

As noted, this decision model is highly simplified. The agency may have more choices than waiting overnight, and, when it does face the evacuation decision, more than just two options may be available. Naturally, the probabilistic system is much more complex than that pictured in the decision tree and influence diagram; the agency may follow many different meteorological variables, using them as information in making the evacuation decision.

Like influence diagrams, decision trees are used in decision situations, and, in order to be useful, they must be analyzed. The decision-tree solution algorithm, a version of dynamic programming, is often known as "averaging out and folding back". To do so requires that the tree be complete; all chance events must have probabilities, and the end points at the tree's right side must have numerical values. Details of analysis procedures are omitted here but are available in Clemen [1996]. From a decision tree, an analyst can calculate the expected value and **risk profile** (probability distribution over possible consequences) for each alternative or decision strategy. Other types of analysis include various types of sensitivity analysis, expected value of information, and calculation of expected utility and certainty equivalents when the decision maker's risk preferences are included in the model.

Spreadsheet Risk-Analysis Models

A **spreadsheet model** is a collection of algebraic relationships encoded into an electronic spreadsheet on a computer. The widespread availability of electronic spreadsheets has put this type of modeling capability in the hands of many individuals and managers at all levels within organizations.

Spreadsheet models are typically deterministic, that is, inputs are specified and used to make calculations of items of interest (net present value, market share, and so on). It is common to perform "what if" analysis by changing the inputs to determine the implied change in the quantities of interest. A primitive version of sensitivity analysis, "what if" analysis can be implemented more or less systematically.

With the advent of risk-analysis spreadsheet add-ins such as @RISK and Crystal Ball, spreadsheet users are now able to perform **Monte Carlo simulation** in the spreadsheet environment. These programs allow the user to specify probability distributions for the inputs and, via the simulation process, determine the expected values and risk profiles for possible decision alternatives. In addition, the programs provide some graphing and sensitivity-analysis features. Although there are some practical limitations on the type and size of simulation model that can be created and analyzed in this way, the ability to use these spreadsheet add-ins puts considerable analytical power in the hands of managers who deal with spreadsheet models for decision making.

Principles of Modeling and Analysis

What is the purpose of creating and analyzing system models for decision making? Many students and practitioners make the mistake of looking for the "correct answers" to the problems they face. Because models are no more than representations of reality, however, there is no "right answer". Indeed, some models are better; they more faithfully reflect reality, especially in a way that allows the analyst to find relevant insights. For example, a complex financial model of the value of a salvageable sunken ship may be of some use in deciding what to pay for salvage rights. However, if the salvage rights are to be auctioned, then a game-theoretic model that is less rich in financial detail may actually be more useful for choosing a bidding strategy.

A fundamental principle of modeling and analysis is that the ultimate purpose is to gain insight into the workings of the natural system and the impact of decisions with respect to that system. When analyzing a system model for decision-making purposes, the analyst may look for many different insights, all of which fundamentally revolve around how the various decision alternatives compare. For risky decision situations, how do the expected values compare? What about the risk profiles? On the basis of these, how do the different alternatives compare? Is it possible to screen out any alternatives using stochastic dominance? Is it necessary to make a trade-off between riskiness and expected value in order to choose? If so, can we use what we know about the decision maker's risk preferences to compare the risky alternatives?

For situations complicated by conflicting objectives, the insights we seek relate to trade-offs among the decision maker's objectives. For example, what makes each alternative attractive? Can any alternatives be screened out because they are dominated by others? How can we characterize each alternative in terms of the "package" of benefits and costs that it offers? How much do we need to know about the decision maker's preferences in order to select a preferred alternative?

Regardless of the type of decision model, analysts also seek insights from sensitivity analysis. Throughout the modeling and analysis process, questions arise regarding the robustness of the model. For example, how much can a given parameter change without substantially changing the results of the analysis? If a parameter changes, does the preferred alternative change? What are the relative impacts of changes in different inputs; to which inputs are the results sensitive or robust? What happens when multiple parameters change simultaneously?

The goal of modeling and analysis is to obtain insights by answering questions such as these. Ultimately, the decision maker should understand clearly what course of action is appropriate and why. This is not simply a matter of creating a model, analyzing it to find the "right answer", and then adopting that answer blindly. It may be that, in a complex model, highly detailed understanding is impossible; nevertheless, the analyst should still work to understand the essential risks and trade-offs inherent in a particular choice and to characterize the sensitivity of the analytical results.

In the iterative process of constructing, analyzing, and refining the model and its subsequent analysis, how does the analyst know when to stop? Phillips [1984] introduces the idea of a **requisite model**: "A model can be considered requisite only when no new intuitions emerge about the problem." Related ideas are "just enough modeling" and "just enough analysis". The analyst should create a model that just captures the essential elements for making the decision at hand, and the analysis should be just enough to gain clarity regarding the choice to take.

Defining Terms

Decision model: A system model that is used for decision-making purposes. Decision models typically include a system model (often probabilistic) describing the natural system that is the context for the decision(s) to be made, a model of the decision maker's objectives, and a description of decisions that may be made to control or otherwise exploit aspects of the system in order to achieve objectives.

Decision tree: A type of decision model: a hierarchical representation of the decision and chance events that make up a decision situation.

Influence diagram: A type of decision model: a directed, acyclic graph that uses nodes and directed arcs to represent the elements in a decision situation and the relationships among those elements.

Monte Carlo simulation: A procedure for analyzing probabilistic system models. Given specific probability distributions for some inputs in a model, the simulation chooses random values for the inputs and uses these particular values to calculate the value of the consequence from those random draws. After repeating this procedure many times and recording the consequence for each iteration, one can use the resulting simulated data to estimate the expected value and risk profile for the consequence.

Probabilistic system model: A system model that uses probability to describe aspects of uncertainty or randomness in a system.

Requisite model: A model that captures the essential elements of a decision situation for the purposes of the decisions to be made. In the iterative process of constructing the model, analysis, model refinement, and subsequent analysis, when no additional intuitions emerge from further analysis, the model is said to be requisite.

Risk profile: A probability distribution over possible consequences that result from a particular decision alternative (or set of alternatives chosen in a sequence of decisions).

Spreadsheet model: A representation of a system using an electronic spreadsheet.

System model: A representation of a natural system that contains several related components. The model describes the characteristics of the components, the nature of the relationships among them, and the overall workings of the system.

References

Clemen, R. T. *Making Hard Decisions, 2nd ed.*, Duxbury Press, Belmont, CA, 1996.
Phillips, L. D. A theory of requisite decision models, *Acta Psychol.*, 56: 29–48, 1984.

Further Information

For a thorough introduction to decision models and the practice of decision analysis, see Clemen [1996]. In particular, Clemen provides a brief annotated bibliography of related work in Chapter 17.

The **Decision Analysis Society** of INFORMS (Institute for Operations Research and Management Science) maintains a page on the World Wide Web at http://www.fuqua.duke.edu/faculty/daweb/. This web page includes background information on decision analysis, a directory of resources, books, course syllabi, software, membership list, consultants, conferences, and related links.

9.3   Uncertainty

Ralph F. Miles, Jr.

Uncertainty is ubiquitous [Morgan and Henrion, 1990]. It is with us at all times, in all scenarios, and is inescapable. This is particularly true of complex technological systems. Managers ask for conservative, fail-safe, and even failure-free systems. Engineers attempt to design to worst-case analyses. Operators are ordered to follow standard procedures proven by analysis, test, and past experience. Yet, the reality is that some uncertainty still exists, and within that uncertainty lies scenarios and failure modes either unpredicted or that reside outside even worst-case limits.

It is difficult to imagine a reality without uncertainty. We are uncertain of events that have occurred in the past, events for which we have received little or imperfect information. We are uncertain of events that are transpiring in the present beyond our span of information. We are uncertain of events that will occur in the future. Knowledge of any of these events may imply a reality that is not consistent with either our observations or our actions. Uncertainty can arise because the events are inherently random, for which further knowledge can only reduce the uncertainty to a certain limit. Uncertainty can arise because the underlying event-producing process is not well understood, for which further knowledge can in principle eliminate.

One of the most difficult problems facing a manager is that of predicting the **outcome** of future **events**. Knowledge of events transpiring in the past is in principle, at some cost, available. Knowledge of events transpiring in the present can be known in some cases. However, knowledge of events transpiring in the future can only be guessed or through analysis forecast, albeit with imperfection.

At the highest level of uncertainty, a technology manager desires to know if a proposed technology or system will function as specified, will function reliably, can be maintained, will comply with all programmatic constraints including legal, and can be implemented within budget and on schedule.

Given that uncertainty is inescapable, the manager must ask not how to eliminate the uncertainty but how to confront it. The manager's role is to recognize the uncertainties that exist and to incorporate the uncertainties into the decision-making process so as to minimize the effects of the uncertainties upon the technology or system. Assuming that the technology or system is sufficiently complex such as to require modeling, then, to adequately support decision making, this uncertainty must be explicitly incorporated into the value and system models as described in the preceding sections.

Probability

The **measure** of uncertainty that will be presented in this section is **probability**. There are other measures of uncertainty, but probability is predominately the measure of uncertainty. Probabilistic statements and analyses appear in business, engineering, finance, insurance, medicine, political surveys, and weather reporting.

Before the mid-17th century, the word "probable" meant "worthy of approbation". After that time, the word came into its present usage as a measure of uncertainty. Probabilistic calculations and statements at that time were made initially concerning the outcomes of games of chance but quickly spread to other enterprises [Hacking, 1975]. Probability, as a mathematical discipline, was not formally recognized until 1933, when Kolmogorov demonstrated that three simple axioms were all that were needed to develop the formalism of probability [Kolmogorov, 1956]. These three axioms are

1. Probabilities are numbers equal to or greater than zero.
2. The probability of something happening (the **universal event**) is one.
3. The probabilities of **mutually exclusive** events (they both can't happen) add.

Probabilities are associated with events selected from the set of all possible events concerning the outcome of experiments or observations. The probability of an event that cannot occur is 0. The probability of occurrence of one of the complete set of possible events (**collectively exhaustive**) is 1. If the set of possible events is partitioned down to the lowest level (necessarily mutually exclusive), the events are called **elementary events**. If the events under consideration consist of subsets of elementary events, the events are called **compound events**. If the probabilities of two (necessarily not mutually exclusive) events are not dependent upon whether the other has occurred, the events are said to be **independent events**. The probabilities of two mutually exclusive events can be added to obtain the probability of the occurrence of at least one of them. The probabilities of two independent events can be multiplied together to obtain the probability of both of them occurring. The probability of an event (event A) can be formulated as conditional upon the occurrence of a second event (event B), and that **conditional event** is written P(A|B), usually stated as the probability of event A *given* that event B has occurred. Events can be **discrete** (only a finite number constitute the universal event such as for a coin or dice) or **continuous** (the events are associated with a range such as the unknown cost of developing a new technology) or a mixture of both. The table listing the probabilities of a complete set of discrete events is called a **probability mass function**. The function describing the probability values for continuous events is called a **probability density function**. Each point of a probability density function has probability 0 of occurring, but a range over a probability density function has a probability, e.g., the probability that the cost of developing a new technology lies between $10 million and $20 million might be assessed as 0.7. The probability that an event takes on a value equal to or less than some stated value is called a **cumulative distribution function**. These definitions constitute the majority of terms used in probability analysis for technology management.

Probabilities are sometimes stated descriptively, such as "unlikely", "possibly", "probable", "almost certain", and many similar terms, but repeatedly these descriptive statements have been shown to have different meanings by persons, severely limiting their value in decision making.

Books on probability theory abound, and any technical bookstore will have at least a dozen different books on probability, each tailored to readers with different mathematical skills and interests.

The Interpretation of Probability

There is general agreement that Kolmogorov's axioms capture the mathematical nature of probability. It is one thing to define probability mathematically and quite another to interpret it in terms of contemplated or observable events in reality. There are many **interpretations** of probability [Fine, 1973; Weatherford, 1982]. The three most commonly encountered interpretations are the **classical interpretation**, the **frequentist interpretation**, and the **subjective interpretation**.

The classical interpretation states that if *n* possible events are possible for an experiment or observation, then the probability of any one of the events is *1/n*. This is a valid interpretation for "fair" coins, dice, or balanced roulette wheels. It also is valid where a "principle of indifference" applies and events can be shown by some line of argument to be equiprobable. In engineering and science, statistical mechanics with a large ensemble of identical and/or indistinguishable particles would be a valid example. For most technologies or systems of interest, the feasible states are not equiprobable, so the classical interpretation is not relevant. The definitive theory of the classical definition of probability is perhaps best expressed by Laplace [1951].

The frequentist interpretation states that the probability of an event is the ratio of the number of times the event is observed divided by the number of trials (occurrences when the event could have happened), i.e., the frequency with which the event was observed or could be expected to occur. In the repeated operation of a system, the probability of a successful event would be the ratio of the number of times the system operated successfully to the number of times the system attempted to operate. The frequentist interpretation can further be divided into "actual frequentism", the actual ratio observed, and "hypothetical frequentism", the ratio that would be expected to occur if the experiment or observation were carried out an infinite number of times. For the frequentist interpretation to be valid, the trials must be identical (except perhaps for time and space). For any sufficiently complex technology or system, this identity of trials will not hold rigorously, and the degree of identity will limit the validity of the **probability assessment**. For a small number of trials, the statistical variation to be expected will also limit the validity of the probability. Of course, it is impossible to actually determine a probability rigorously adhering to the infinite-trial requirement of hypothetical frequentism. Thus, all probability assessments based on frequentism must be qualified by statements of errors introduced by either the finiteness of the number of trials or the degree of identity of the trials. The frequentist interpretation is the most-common interpretation in engineering and science. It forms the basis of **classical statistics**. The definitive theories of the frequentist interpretation of probability are perhaps best expounded by Reichenbach [1949] and von Mises [1957].

The subjective interpretation of probability says that probability is a state of mind rather than a state of an object. A bent coin clearly does not have a probability of ½ of landing heads, but a person can have a subjective (personal) probability for it landing heads. If that subjective probability is expressed as a prior (before consideration of the data) probability density function, then **Bayes' Theorem** can be used to update the prior probability density function to form a posterior (updated after consideration of the data) probability density function based upon the data of the results of repeated tosses of the coin. The important use of Bayes' Theorem in technology management is in the combining of expert judgment or models with uncertain parameters with observed data to refine probabilistic assessments. An important school of statistics has developed around Bayes' Theorem, called **Bayesian statistics**. The definitive theory of the subjective interpretation of probability is associated with Ramsey [1964] and de Finetti [1964].

The Assessment of Probability

Probabilities can be assessed from symmetries, from observed data, from validated models, or from expert judgment. The classical school of statistics would only use symmetries or observed data. The Bayesian school of statistics claims that *all* probabilities are based at least in part on expert judgment. Reliability and risk analysis for the management of technology may use either the classical school of statistics [MIL-HDBK-217F, 1995] or the Bayesian school of statistics [Martz and Waller, 1982]. The nuclear power industry has the most well-developed approach to assessing probabilities using expert judgment [Office of Nuclear Regulatory Research, 1990]. Morgan and Henrion [1990] and von Winterfeldt and Edwards [1986] have extensive discussions on the methods for assessing probabilities and the heuristic biases that can enter into the assessments.

Defining Terms

Bayesian statistics: That school of statistics based on Bayes' Theorem and the subjective probability theories of Ramsey and de Finetti.

Bayes' Theorem: A theorem that relates a prior probability distribution (discrete or continuous) to a posterior probability distribution by combining the prior probability distribution with the probability of observing the data to form a posterior probability distribution.

Classical interpretation of probability: That school of probability interpretation that says that each event has a probability of 1/n.

Classical statistics: That school of statistics based on symmetries or observed data and developed from the not-completely-compatible theories of R. A. Fisher, J. Neyman, and E. S. Pearson.

Collectively exhaustive: A set of events that equals the universal event.

Compound event: An event consisting of more than one elementary event. The event that a die lands with an even number up (2, 4, 6) is a compound event.

Conditional event: An event whose probability is conditional on the outcome of another event, written P(A|B) and stated as the probability of the event A *given* event B.

Continuous event: An event that can have a range of values. A battery might produce anywhere from 0 to 1.6 volts, depending upon its state of charge.

Cumulative distribution function: The probability that an event takes on a value less than or equal to some specified value. It is the integral of the probability density function from the lowest possible value up to some specified value.

Discrete event: An event that can have only discrete values. The sum of two die takes on integer values between 2 and 12.

Elementary event: The smallest partition of the universal event. The event, "the coin lands heads", is an elementary event.

Event: One possible outcome (an elementary event) or one of several possible outcomes (a compound event) of an experiment or observation that are unknown in advance.

Frequentist interpretation of probability: That school of probability interpretation that says the probability of an event is the ratio of the number of times the event is observed to the number of trials.

Independent events: Two events are independent events if the occurrence of one of the events does not affect the probability of the occurrence of the other event.

Interpretation: A specific relationship between a mathematical structure and an empirical structure. The three schools of probability interpretation considered here are the classical interpretation, the frequentist interpretation, and the subjective interpretation.

Measure: A relationship between a mathematical structure and an empirical structure. For probability, the objects of the mathematical structure correspond to events, and the operations of the mathematical structure correspond to the acts of combining events. A measurement of an event occurs when a number is assigned to the event.

Mutually exclusive: Two or more events that cannot simultaneously occur. A coin cannot land both heads and tails.

Outcome: One possible result from an experiment or observation.

Probability: A measure of uncertainty.

Probability assessment: The act of assigning a probability to an event. The assignment may be based on symmetry, on actual data, on validated models, on expert judgment, or on any combination of these.

Probability density function: The probability measure of an event that can have a continuum of outcomes.

Probability mass function: The probability measure of an event that can have only discrete values.

Subjective interpretation of probability: That school of probability interpretation that says probabilities are states of mind and not properties of objects.

Universal event: That compound event containing all possible elementary events. The probability of the universal event is 1, as defined by the Kolmogorov axioms.

References

de Finetti, B. Foresight: its logical laws, its subjective sources (1937), In *Studies in Subjective Probability*, H. E. Kyburg, Jr. and H. E. Smokler, eds., pp. 93–158, John Wiley & Sons, New York, 1964.

Fine, T. *Theories of Probability*, Academic Press, New York, 1973.

Hacking, I. *The Emergence of Probability*, Cambridge University Press, London, 1975.

Kolmogorov, A. N. *Foundations of the Theory of Probability (1933)*, Chelsea, New York, 1956.

Laplace, P. *A Philosophical Essay on Probabilities (1812)*, Dover, New York, 1951.

Martz, H. F. and Waller, R. A. *Bayesian Reliability Analysis*, John Wiley & Sons, New York, 1982.

Morgan, M. G. and Henrion, M. *Uncertainty: A Guide to Dealing with Uncertainty in Quantitative Risk and Policy Analysis*, Cambridge University Press, Cambridge, England, 1990.

Office of Nuclear Regulatory Research. *Severe Accident Risks: An Assessment for Five U.S. Nuclear Power Plants*, NUREG-1150, Vol. 1, U.S. Regulatory Commission, Washington, DC, 1990

Ramsey, F. P. Truth and probability (1926), In *Studies in Subjective Probability*, H. E. Kyburg, Jr. and H. E. Smokler, eds., pp. 61–92. John Wiley & Sons, New York, 1964.

Reichenbach, H. *The Theory of Probability*, University of California Press, Berkeley, California, 1949.

U.S. Department of Defense, *Reliability Prediction of Electronic Equipment*, Military Handbook MIL-HDBK-217F, Notice 2, February 28, 1995.

von Mises, R. *Probability, Statistics and Truth (1951), 2nd ed.*, Dover, New York, 1957.

von Winterfeldt, D. and Edwards, W. *Decision Analysis and Behavioral Research*, Cambridge University Press, Cambridge, England, 1986.

Weatherford, R. *Philosophical Foundations of Probability Theory*, Routledge & Kegan Paul, London, 1982.

Further Information

For recent advanced books on probability, see A. Papoulis, *Probability, Random Variables, and Stochastic Processes, 3rd ed.*, McGraw-Hill Book Company, 1991 and P. Billingsley, *Probability and Measure, 3rd ed.*, John Wiley & Sons, New York, 1995. The American Statistical Association (1429 Duke Street, Alexandria, VA) publishes a monthly journal containing articles on both classical and Bayesian statistics.

9.4   Decision Analysis

William Samuelson

Decision analysis is a systematic approach to making choices under uncertainty. As such, it is a useful tool in a wide range of settings where risks are important, including business strategy, government policy, and scientific, medical, and legal decisions. Decision analysis is a normative method outlining how decisions should be made. In particular, it extends the logic of optimal decision making from deterministic to probabilistic settings. In addition, decision analysis is a practical method, guiding the user to "think straight" under uncertainty. Faced with complex problems, real-world decision makers can't consider everything. The decision analytic approach prods him to model and scrutinize the choices, risks, and values that matter.

The normative model underlying decision analysis is often referred to as the **subjective expected utility model** of choice [Bell et al., 1988]. The model presumes that the decision maker assesses subjective probabilities for uncertain outcomes and is ready to revise these probabilities in light of new information according to **Bayes rule**. Furthermore, he or she expresses preferences for different outcomes by assessing corresponding utility values. Finally, the decision maker chooses the preferred course of action that has the highest expected utility value. In a short survey, one cannot review all aspects of the decision analytic method. The approach taken here is to offer a practical introduction to the method by emphasizing the widely used decision tree methodology. This allows a careful consideration of probability assessments, sequential decisions, the expected utility rule, and the value of information. However, it is fair to

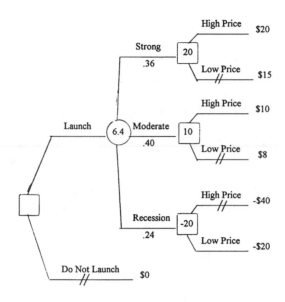

FIGURE 9.4   Launching a product.

acknowledge the omission of a number of important topics: statistical decision theory [Raiffa and Schlaifer, 1972], multiattribute utility theory [Keeney and Raiffa, 1976], and behavioral decision making [Kahneman et al., 1982]. The accompanying sections in this volume on uncertainty, simulation, and value-based decisions consider a number of these areas.

To apply decision analysis, the user should (1) Identify the available choices, not only for direct action but also for gathering information on which to base later action, (2) list the outcomes that can possibly occur (these will depend on chance events as well as on his actions), (3) evaluate the chances that any uncertain outcome will occur, and (4) assign values to the possible outcomes. All four of these steps can be incorporated in a **decision tree** diagram. The following managerial application, launching a new product, is perhaps the most direct way to motivate a discussion of the decision analytic approach.

A consumer products firm is contemplating launching a dramatically new version of an existing product in the face of an uncertain economy. Its economic forecasting unit envisions three possible macroeconomic scenarios playing out over the next 3 years: strong economic growth, moderate economic growth, or a prolonged recession. The product's midterm profit depends directly on the strength of the economy. The firm's dilemma is that it must commit to the enormous cost of the product launch now more than a year before it will come to learn the future course of the economy. Thus, in deciding whether or not to launch the product, it must rely on imperfect economic forecasts. There is one piece of good news: after the product is launched and the course of the economy (and the strength of the product's sales) becomes known, the firm can fine tune its strategy by choosing to sell the item at either a high price or a low price.

The highly simplified decision tree in Fig. 9.4 depicts the firm's basic decision. The tree starts with a point of decision, by convention represented by a square, from which emanate two branches: the decisions to launch or not to launch. If the firm decides not to launch, the story ends there. The firm's final profit outcome is $0. If the firm decides to launch, a chance event occurs (represented by a circle). From this chance event emanate the three possible macroeconomic scenarios mentioned above. Attached to each branch is a probability indicating the likelihood that the outcome is expected to occur. Which outcome occurs (which branch will be taken) will only be known in a year's time. Finally, once the course of the economy is known, the firm is free to establish its pricing choice, high or low. Note that, once a branch tip of the tree has been reached, all uncertainty has been resolved, that is, the profit consequence is known with certainty. Accordingly, the firm's economists have supplied profit projections for each of the possible ultimate outcomes.

The decision tree usefully summarizes the chronology of firm actions and chance events as time moves forward from left to right in the figure. Though the present example is kept intentionally simple, the decision tree framework is flexible enough to include multiple decisions and multiple uncertainties. For instance, if the cost of producing or promoting the product were uncertain, additional branches would be included to portray the possible cost outcomes. Clearly, expanding the number of uncertainties directly increases the number of final branch tips in the tree (one for each different profit outcome). Though it is still possible to solve very "bushy" trees by hand, the practical recourse is to turn to decision tree software that is widely available and easy to use [Buede, 1994].

The expected, utility criterion instructs the decision maker to choose the course of action that has the maximum expected utility. To apply the criterion to the decision tree, we first consider the "last" decision, what price to charge. As indicated in the tree, the company should set a high price if economic growth is strong or moderate but set a low price if a recession unfolds. (In each case, the inferior price branch has been crossed out, and the maximum attainable profit has been placed inside the decision square.) Anticipating these contingent pricing actions, should the company launch the product in the first place? The answer is yes. Launching the product implies an expected utility of $(.36)(20) + (.4)(10) + (.24)(-20) =$ $6.4 million. (Here, the utility of each outcome is measured simply in dollar profit. Therefore, expected utility coincides with expected profit.) Expected utility represents a weighted average of the possible value outcomes, where the weights are the probabilities of the respective outcomes. Launching the product is much more profitable on average than doing nothing. Indeed, even if the company were offered $5 million by a rival to transfer all rights to the product, the company would refuse the offer and launch the product itself. (Note that maximizing expected profit offers no place for risk aversion. For a so-called risk neutral decision maker an expected profit of $6.4 million, though highly risky, is preferred to a certain $5 million receipt.)

Decision trees are always solved backward, from right to left. In making a choice now, the decision maker must first look forward to the risks and decisions that lie ahead. To make the right decision now, he or she must anticipate making the right future decisions if and when they should occur. The philosopher Sören Kierkegaard expressed a similar sentiment, "Life can only be understood looking backwards but it must be lived forwards." Mathematicians recognize this as an application of the well-known principle of dynamic programming. As an example, suppose the company were to charge a high price (the wrong price) during a recession. (Perhaps, it insisted shortsightedly that its superior product should always command a premium price.) This mistake would generate a $40 million loss (instead of a $20 million loss) and would reduce the expected profit of launching to $1.6 million. Thus, making a suboptimal future decision clearly robs value from the launch opportunity. Indeed, it could lead to a suboptimal initial decision, such as selling out for $5 million when the potential launch value is really $6.4 million.

Plotting sequential decisions in a tree illustrates another general principle. "Never commit to a decision today if it can be costlessly postponed and made later in light of new information." For example, suppose that the firm (for whatever reason) were obligated to set its price at the same time it launched the product. This means redrawing the decision tree, interchanging the order of the "macro risk" circle and the pricing square. As one can readily compute, the expected profit of launching at a high price is $1.6 million, while the expected profit of launching at a low price is $3.8 million. The point is that committing to a price up front, even a low price, is inferior to postponing this decision and employing a contingent pricing strategy. Here is an additional observation. If growth is strong or moderate (a 76% chance), a high price is optimal, after the fact. Nonetheless, before the fact committing to a high price is far worse than committing to a low price. In short, tailoring a decision to the "most likely scenario" while ignoring the range of uncertainty can readily lead to suboptimal decisions.

The Value of Information

An important key to making better decisions lies in acquiring additional information about uncertain events. For example, superior economic forecasts command high prices in the market for information. As a dramatic example, suppose the firm, prior to its launch decision, could obtain a *perfect* prediction

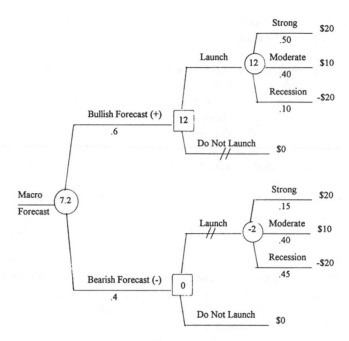

FIGURE 9.5   Using a macroeconomic forecast.

of the future course of the economy from an outside forecaster. After it pays the forecaster and obtains the perfect forecast, its actions are straightforward. If strong or moderate growth is forecast, it should launch the product at a high price (for profits of $20 million and $10 million, respectively). If a recession is predicted, it forgoes the launch (for a $0 profit). Before hearing the prediction, the firm judges its expected profit to be: (.36)(20) + (.4)(10) + (.24)(0) = $11.2 million. Note that, before learning the actual prediction, the firm uses its own probability assessments in computing the expected value. Recall that the company's original expected value was $6.4 million in Fig. 9.4. Therefore, the value of a perfect forecast is 11.2 − 6.4 = $4.8 million. In short, the company would be willing to pay up to this amount (but no more) for the forecast. Though this value is enormous, remember that perfect forecasts of important risks are not available in the real world.

Now consider an *imperfect* macroeconomic forecaster who predicts the course of the economy 18 months in advance and makes two kinds of forecasts: "bullish" (+) indicating a strong economy or "bearish" (−) indicating a weak economy. The forecaster boasts the following 40-year track record. Preceding six separate episodes of strong economic growth, the forecaster has made bullish predictions in five cases. Preceding ten episodes of moderate economic growth, the forecaster has made bullish predictions in six cases. Finally, prior to the four postwar recessions, the forecaster has made three bearish predictions (and one incorrect bullish forecast). With a record of the forecaster's past accuracy in hand, management's task is to determine whether to purchase a current forecast and to decide its launch strategy contingent on the forecaster's report.

Figure 9.5 shows the decision tree incorporating the macroeconomic forecast. (To conserve space, the firm's pricing decision is not shown; the firm's maximum profit from its optimal price choice is placed on the branch tips.) The tree begins with an uncertain event, whether the forecast will be bullish or bearish. Next comes the firm's contingent launch decision, actually two separate choices, whether to launch in light of a bullish forecast and in light of a bearish forecast. In either case, the uncertain macroeconomy then evolves. In short, at its point of decision, the company knows the forecast but not the eventual course of the economy.

Left to be explained are the new probabilities appearing in Fig. 9.5. Suppose the firm were to launch the product after receiving a bullish forecast. The revised probabilities shown in Fig. 9.5 are .5, .4, and .1 for strong growth, moderate growth, and recession. Given the forecaster's enviable track record, a bullish

TABLE 9.3 Joint Probabilities

	Strong	Moderate	Recession	Total
Bullish (+)	.30	.24	.06	.60
Bearish (−)	.06	.16	.18	.40
Total	.36	.40	.24	1.00

forecast increases the chance of strong growth and reduces the chance of a recession. Table 9.3 demonstrates how to compute these revised probabilities. The table lists joint probabilities. For instance, the company's assessed probability of receiving a bullish forecast and subsequently observing strong growth is Pr(+&S) = .30. The six joint probabilities are computed using the company's prior probability assessments and the forecaster's track record for accuracy. For instance, one invokes the probability rule:

$$\Pr\left(+\,\&\,S\right) = \Pr\left(+\,\middle|\,S\right)\Pr\left(S\right) \tag{9.1}$$

to compute Pr(+&S) = (5/6)(.36) = .30. Remember that the company's prior probability assessment for strong growth is .36 and that five of six periods of strong growth were preceded by bullish forecasts. The other joint probabilities are computed analogously. Summing the entries in each row, one finds the marginal probabilities Pr(+) = .6 and Pr(−) = .4. Finally, we are ready to compute revised probabilities. For example, the chance of strong growth, given a bullish forecast is

$$\Pr\left(S\,\middle|\,+\right) = \Pr\left(+\,\&\,S\right)\middle/\Pr\left(+\right) \tag{9.2}$$

or from Table 9.3, Pr(S|+) = .30/.60 = .5. In other words, of the 60% chance of a bullish forecast, 30% of the time economic growth will also be strong. Therefore, the revised probability of strong growth given a bullish forecast is 50%. What if the forecast is bearish instead? Turning to the second row of Table 9.3, one can compute: Pr(S|−) = .06/.4 = .15, Pr(M|−) = .16/.4 = .4, and Pr(R|−) = .18/.4 = .45. In light of a bearish forecast, the probability of strong growth is reduced (relative to the company's original assessment), and the chance of a recession increases. These probabilities are listed in the lower right-hand branches of Fig. 9.5. One other observation is in order. Substituting Eq. (9.1) into Eq. (9.2), we obtain

$$\Pr\left(S\,\middle|\,+\right) = \left[\Pr\left(+\,\middle|\,S\right)\Pr\left(S\right)\right]\middle/\Pr\left(+\right) \tag{9.3}$$

This is the most familiar form of Bayes rule for probability revision, that is, updating one's prior probability Pr(S) to a revised probability Pr(S|+) in light of new information. In fact, there are many equivalent ways of updating probabilities: using Bayes rule, constructing a table of joint probabilities, or "flipping" decision trees [Raiffa, 1968].

The final step in assessing the impact of new information requires analyzing the decision tree in Fig. 9.5. According to the tree, the company should set a *contingent* launch strategy: introducing the product given a bullish forecast but forgoing the launch given a bearish forecast. In the latter case, the high chance of a recession accounts for the expected $2 million loss from a launch. Moving from right to left in the tree, we can now compute the company's expected profit before learning the actual forecast: (.6)(12) + (.4)(0) = $7.2 million. Comparing this to the expected profit of $6.4 million of Fig. 9.4, the expected value of the additional information afforded by the macroeconomic forecast is $.8 million. Presuming that the cost of obtaining the forecast is considerably less than this amount, the company should purchase the forecast.

This simple application is useful in illustrating the general principles surrounding the value of new information. First, new information is valuable only if it affects actual decisions. In Fig. 9.5, the company profits by avoiding an unprofitable launch following a bearish forecast. Cognizant of the bearish forecast, the company saves the $2 million expected loss it would have incurred had it launched the product absent the information (as in Fig. 9.4). This $2 million "savings" occurs 40% of the time (the chance of a bearish forecast), implying an expected savings of $(.4)(2) = \$.8$ million. This exactly accounts for the $.8 million difference between Fig. 9.4 and 9.5. In general, we can say that the value of new information depends on (1) the revisions in probability assessments that cause changes in decisions, (2) the magnitude of expected savings from these changed decisions, and (3) the frequency with which decisions are affected. Thus, the large value associated with the macroforecast is owing to its (unrealistically) high accuracy and the great savings it affords the firm by avoiding product launches when a recession is likely.

Concluding Remarks

This brief overview of decision analysis has centered on the use of decision trees to implement the subjective expected utility model. For expositional purposes, the focus has been on a risk-neutral decision maker whose objective is to maximize expected profit. Of course, it is equally plausible that the decision maker is risk averse, that is, values a risky distribution of possible profits at less than its expected value. The task of a risk-averse individual is to assess a concave utility function that accurately reflects his attitude toward risk. Procedures for assessing risk preferences are discussed in Section 9.1. Finally, the validity of a decision analysis depends on the quality of the probability and utility assessments that go into it. Thus, it is imperative to use sensitivity analysis to test the decision effect (if any) of changing problem inputs.

Defining Terms

Subjective expected utility model: After assessing probabilities for uncertain events and utilities for possible outcomes, the decision maker should choose the course of action that has the highest expected utility.

Bayes rule: A principle of probability that calls for revising one's prior probability assessment in light of the predictive accuracy of new information.

Decision tree: A diagram capturing the sequence of actions and chance events over time. Following the principle of dynamic programming, an optimal decision is found by averaging and folding back the tree from right to left.

References

Bell, D. E., Raiffa, H., and Tversky, A. *Decision Making*, Cambridge University Press, Cambridge, England, 1988.

Raiffa, H. and Schlaifer, R. *Applied Statistical Decision Theory*, MIT Press, Cambridge, MA, 1972.

Keeney, R. L. and Raiffa, H. *Decisions with Multiple Objectives*, John Wiley & Sons, New York, 1976.

Further Information

The first two references provide guides to behavioral decisions and a variety of formal decision methods. Raiffa offers a classic, rigorous treatment of decision analysis, while Buede provides a survey of decision analysis software.

Kahneman, D., Slovic, P., and Tversky, A. *Judgment under Uncertainty: Heuristics and Biases*, Cambridge University Press, Cambridge, England, 1982.

Kleindorfer, P. R., Kunreuther, H. C., and Schoemaker, P. J. H. *Decision Sciences*, Cambridge University Press, Cambridge, England, 1993.

Raiffa, H. *Decision Analysis*, Addison-Wesley, Reading, MA, 1968.

Buede, D. Aiding Insight II, *OR/MS Today*, 12: 62–71, 1994.

10

Legal Issues

Constance E. Bagley
Stanford Business School

Paul A. Beck
Paul A. Beck & Associates

Carol I. Bordas
Thorp, Reed & Armstrong

L. James Ristas
Alix, Yale & Ristas, LLP

10.1 Contracts[1]

Constance E. Bagley

A **contract** is a legally enforceable promise. Contracts are essential to the conduct of business, both in the United States and internationally. Without contract law, a seller of goods would be reluctant to ship them because the seller would have no way of ensuring that it would be paid. Similarly, a manufacturer that outsourced production of a key component would have no way to ensure that the key component is delivered on time. A software developer would not be willing to license another to use her software without knowing whether the promise to pay royalties and the obligation to use the software on only one machine would be enforceable. Similarly, an entrepreneur would be reluctant to leave one company and go to another if he could not be sure that the new employer would honor its promise of stock options.

Governing Laws

In the United States, the common law of contracts, which is law developed by judges in court cases, governs contracts for the rendering of services, the purchase of real estate, and the lending of money. Contracts for the sale of moveable items — called *goods* — are governed by the Uniform Commercial Code (UCC). The UCC is a body of statutes, enacted in some form in every state in the United States, designed to codify certain aspects of the common law applicable to commercial transactions and to release those engaging in commercial transactions from some of the requirements of the common law.

    Many contracts for the sale of goods internationally are governed by the United Nations Convention on Contracts for the International Sale of Goods (CISG). As of September 1995, 44 countries had become parties to the CISG, including the United States, Mexico, Canada, and most of the European nations.

---

[1]The following is a general discussion of certain contract principles and is not, and should not be relied on as, legal advice. Readers are responsible for obtaining such advice from their own legal counsel.

The CISG applies only to commercial sales transactions between merchants, i.e., persons in the business of selling goods. The CISG automatically applies to international sales when both parties are from countries that are CISG-contracting states, unless the parties affirmatively opt out of its application.

Except when otherwise noted, the discussion that follows is based on the common law.

Elements

For a contract to exist, there must be (1) an offer, (2) acceptance, and (3) consideration. An *offer* is a statement by a person (the *offeror*) that indicates a willingness to enter into a bargain on the terms stated. *Acceptance* occurs when the person to whom the offer was addressed (the *offeree*) indicates a willingness to accept the offeror's proposed bargain. **Consideration** is anything of value that is exchanged by the parties. It can be property, money, a promise to do something a person is not otherwise legally to do, or a promise to refrain from doing something a person would otherwise be legally entitled to do.

Offer

For an offer to be effective, the offeror must objectively manifest a willingness to enter into a bargain that justifies another person to believe that his consent to that bargain is invited and will conclude it. An offer that is made in obvious jest or anger will not be effective. On the other hand, if a reasonable person would construe the offeror's words and conduct as indicating a genuine intent to enter into a bargain, then unstated, subjective intent not to enter into such a bargain will not prevent the offer from being effective.

Acceptance

Acceptance occurs when the person to whom the offer is addressed indicates a willingness to accept the terms proposed. If the offeree instead proposes substantially different terms, then that constitutes a *counteroffer*. A counteroffer extinguishes the original offer and itself becomes a proposal to enter into a contract on the new proposed terms. If the counteroffer is rejected, then the original offeree is not free to accept the original offer.

The UCC rules are less stringent. For example, the UCC presumes the existence of a contract if the parties act as if there is one, such as when a seller has shipped goods and the buyer has paid for them. Under the UCC, the offer and acceptance do not have to be mirror images of each other. If an offer to sell goods on certain terms is accepted by the buyer with a minor modification in the contract terms, a contract has been formed, even though the offer and acceptance are not mirror images of each other.

Consideration

Consideration is a legal concept that means a bargained-for exchange. For example, suppose that an employer asks a current employee to sign an agreement to be bound by a personnel manual that provides that the employee's employment is at-will and can be terminated by the employer or the employee at any time. If the employer had previously given assurances that the employee would not be fired absent just cause, then, even if the employer later persuaded the employee to sign and agree to be bound by the new personnel manual providing for at-will employment, the new agreement would generally not be enforceable unless the employer gave the employee something of value — other than just continued employment — as consideration. On the other hand, if the employment is at-will, with no implied contract not to terminate absent just cause, courts in many states will uphold a promise to abide by a new employment manual that is given in exchange for continued employment.

Generally, the relative value of the promises exchanged is irrelevant to the issue of whether a contract has been formed. However, if there is a great disparity between what one party is receiving under the contract compared to what the other party is giving up, then that may be evidence that the contract was procured by fraud, duress, or was otherwise *unconscionable* (i.e., so one-sided as to offend the conscience of the court). A court will not enforce an unconscionable contract. For example, in one case the court refused to enforce a contract entered into by a welfare recipient to buy a freezer worth $300 on credit for a total purchase price of $1234.80, including time credit charges, credit life insurance, credit property insurance, and sales tax.[2]

---

[2]*Jones v. Star Credit Corp.*, 298 N.Y.S. 2d 264, 6 UCC Rep. 76 (N.Y. Sup. 1969).

A purely gratuitous promise, i.e., one that is given without anything received in exchange, is not enforceable. For example, if an electrical engineer promised that she would share the technology she developed with a colleague in another company, then changed her mind and decided not to, the colleague in the other company could not enforce her promise to share. On the other hand, if she promised to share in exchange for the payment of money or the promise by the colleague to share anything that he developed, then her promise to share would be enforceable.

Promissory Estoppel

Under limited circumstances, a court will grant some relief when a gratuitous promise is relied upon by a person to whom a promise was made. For example, suppose that your grandmother promises to send you to Europe and, in reliance on that promise, you purchase a nonrefundable airline ticket. If your purchase of that ticket could have been reasonably anticipated by your grandmother, then a court may require her to reimburse you the out-of-pocket cost of the ticket. This is done under a doctrine called **promissory estoppel**. In order for promissory estoppel to apply, the person to whom the promise was made (the promisee) must show that the promisee's reliance on the promise was both justified and reasonably foreseeable to the promissor and that injustice would result if some relief were not given.

The doctrine of promissory estoppel can be used not only with respect to gratuitous promises but also with respect to negotiations that have not yet ripened into a contract because the terms being negotiated are too indefinite or unsettled. For example, in one case, a person named Hoffman negotiated with Red Owl Stores, Inc. to open a Red Owl grocery store. The negotiations went through several stages and continued for more than 2 years before they broke down. When Hoffman first approached Red Owl about buying a franchise, he said that he had only $18,000 to invest. Red Owl assured him that this amount would be sufficient. With Red Owl's encouragement, Hoffman bought a small grocery store to get more experience. Although the store was profitable, Red Owl advised Hoffman to sell it because he would have a larger Red Owl store within a few months. Hoffman also paid a $1000 deposit for a site for the new store and sold his bakery. Negotiations broke down when Red Owl increased the size of Hoffman's required investment from $18,000 to $24,100, and a few weeks later to $26,100. Hoffman successfully sued Red Owl to recover the amount that he had lost by reason of his reliance on Red Owl's promise. It is important to note that the court did not enforce the promise itself, that is, Hoffman was not given a Red Owl store for $18,000, but only permitted Hoffman to recover an amount equal to the out-of-pocket damage he suffered by reason of Red Owl's failure to honor its promise.[3]

Mistake and Misunderstanding

Sometimes misunderstandings may arise from ambiguous language in a contract or from a mistake as to the facts. If the terms of a contract are subject to different interpretations, then the party that would be adversely affected by a particular interpretation can void, or undo, a contract when (1) both interpretations are reasonable and (2) either both parties knew or both did not know of the different interpretations. If only one party knew or had reason to know of the other's interpretation, the court will find for the party who did not know or did not have reason to know of the difference.

A contract may also be voidable if there has been a mistake of fact that has a material effect on one of the parties. A classic case involved two parties who had signed a contract in which Wichelhaus agreed to buy 125 bales of cotton to be brought by Raffles from India on a ship named *Peerless*. Unbeknownst to the parties, there were two ships named *Peerless*, both sailing out of Bombay during the same year. Raffles meant the *Peerless* that was sailing in December; Wichelhaus meant the *Peerless* that was sailing in October. When the cotton arrived on the later ship, Wichelhaus refused to complete the purchase, and Raffles sued for breach of contract. The English court held that the contract was voidable due to a mutual mistake of fact.[4]

---

[3]*Hoffman v. Red Owl Stores,* Inc., 26 Wis. 2d 683, 133 N.W. 2d 267 (Wis. 1965).
[4]*Raffles v. Wichelhaus,* 159 Eng. Rep. 375 (Exch. 1864).

On the other hand, a mistake of judgment about the value or some other aspect of what is being bargained for will usually not result in a voidable contract. For example, if A buys a piece of land from B for $100,000, erroneously thinking that she can resell it for $150,000, but in fact is able to resell it for only $90,000, that is a mistake of judgment, not fact. Accordingly, A could not get out of what turned out to be a bad bargain.

Illegal Contracts

A contract to do something that is prohibited by law is illegal and, as such, is void from the outset and thus unenforceable. For example, a contract requiring the payment of interest in excess of the maximum rate permitted by the jurisdiction in which the loan is made is illegal, and the lender is guilty of *usury*. Similarly, many jurisdictions require that members of certain professions or callings obtain licenses allowing them to practice. For example, many states require contractors, real estate brokers, and architects to be licensed. In general, a contract to pay money to an unlicensed individual is not enforceable. Thus, suppose that Massachusetts requires all real estate brokers to be licensed. A homeowner agrees to pay an unlicensed broker 6% of the sale proceeds from the sale of her house. The unlicensed broker is successful in selling the house, but the homeowner refuses to pay the commission. In that case, the promise to pay the commission would not be enforceable.

Oral or Written Contracts

Most contracts are enforceable regardless of whether they are oral or in writing. However, almost every state has a statute, called the **statute of frauds**, that requires certain types of contracts to be in writing before they are enforceable. These include (1) contracts involving an interest in real property, such as a lease or contract of sale of land, (2) contracts that cannot by their terms be performed within 1 year from the date the contract is formed, (3) prenuptial agreements, and (4) under the UCC, contracts for the sale of goods priced at $500 or more. Even if a contract does not have to be written to be enforceable, it is usually best to put the agreement in writing, even if the writing is informal.

Damages and Specific Performance

If a party breaches its contract with another, the nonbreaching party is usually entitled to monetary damages. The nonbreaching party is usually entitled to the benefit of the bargain, also called **expectation damages**. Expectation damages compensate the plaintiff for the amount he lost as the result of the defendant's breach of contract; in other words, it puts the defendant in the position he would have been in if the contract had not been breached.

Even if a contract has been breached, however, the nonbreaching party is required to make reasonable efforts to minimize the damage suffered by reason of the breach. This is called *mitigation of damages*. For example, if a supplier fails to deliver goods in accordance with the contract, the buyer must try to procure them elsewhere. Similarly, if an employee is fired in violation of an employment agreement, the employee must try to find comparable work. If the employee fails to take a comparable job elsewhere, then she will be able to recover only the difference between what she would have been paid under the employment agreement and what she could have earned at the comparable job.

Occasionally, a court will issue an injunction ordering the breaching party to do what he promised to do. This usually is done only when a unique item is involved. For example, if A contracted to sell a piece of land to B then fails to execute and deliver a deed transferring title to B, then a court may grant **specific performance**, i.e., require A to execute and deliver the deed. Similarly, a court would order specific performance of a promise to sell a one-of-a-kind Morgan automobile.

Special Types of Contracts

Entrepreneurs are likely to encounter a variety of contracts, including leases, contracts for the purchase of real property, loan agreements, contracts for the sale or acquisition of a business, employment contracts, and license agreements. Each type presents its own special issues.

Real Property

For example, calculation of the rental charge and restrictions on subleasing the space or assigning the lease to a third party are often critical aspects of a lease negotiation. The acquisition of real property, such as a building or an empty lot, involves highly technical laws that vary markedly from state to state, depending on where the property is located. An entrepreneur should never enter into a contract to buy real property without first consulting an experienced real-property lawyer in the state in which the property is located.

Loan

In negotiating a loan agreement, the borrower needs to pay attention to a number of issues, including the logistical details of receiving the loan (e.g., whether the amount will be transferred in full or in installments), what conditions must be met by the borrower before the loan will be funded (*conditions precedent*), what promises the borrower must make either to do something or to refrain from doing something that, if breached, will result in an event of default and termination of the loan (*covenants*), and the repayment terms, including any rights to cure an event of default due to a late or missed payment.

Sale of Business

Contracts for the sale or acquisition of a business almost always involve important tax issues as well as concerns by the buying entity for liability it might have for acts that occurred or products that were sold prior to the acquisition. Most acquisition agreements will include representations and warranties by the target and its shareholders about the business (e.g., a representation or warranty that the target owns its intellectual property and that its activities do not violate the intellectual property rights of any third party); representations and warranties by the buyer about its business if the buyer's stock is to be the acquisition currency; covenants of the target, including promises not to take any action out of the ordinary course of business prior to closing; covenants of both the target and the buyer to take certain actions necessary to permit the transaction to occur; conditions to closing, including delivery of legal opinions and the absence of any material adverse change in the target's business; procedures for termination of the acquisition contract prior to closing; and indemnification provisions to cover the liability of the parties for breach of their representations, warranties, and covenants.

The buyer will normally request that representations and warranties be absolute so that the seller is liable to the buyer for any damages arising out of the misrepresentation even if the seller believed the representation to be true at the time that it was made. To protect against such liability, the seller will often want to modify the representation and warranty by a knowledge qualifier, that is, to state that to the best of the seller's knowledge the stated representation is true. Because knowledge qualifiers shift the risk that a representation and warranty may be innocently untrue, the sellers will seek to have as many knowledge qualifiers as possible, whereas the buyer will resist, arguing that the resulting harm to the buyer is the same, regardless of whether the seller knew of the problem.

Employment

Employers should memorialize the terms of the employment relationship in a written document that briefly describes the employee's duties, the compensation and employment benefits, the terms of any stock options, and the conditions under which the employment can be terminated. It is generally preferable for an employer to specify that employment is at-will, meaning that it can be terminated at any time by either the employer or the employee.

Intellectual Property

Intellectual property licensing agreements can range from a few pages to several hundred pages in length. Patent, trademark, and copyright licenses all have different provisions that are of particular importance to each type of intellectual property. However, key considerations that affect many intellectual property licenses include the specification of what is to be licensed, the scope of the license (e.g., exclusive or nonexclusive, geographic limits, duration), license fees (including calculation of royalties), representations and warranties by the licensor as to its ownership rights and notification by the licensor in the event that those representations prove incorrect, and promises to do (or not to do) certain things (e.g., in a trademark license, the promise by the licensee to use the trademarks in ways that maintain their values as symbols of goodwill for the business).

Defining Terms

Consideration: Something of value provided as part of a bargained-for exchange.

Contract: A legally enforceable promise. Requires offer, acceptance, and consideration.

Expectation damages: The amount of money necessary to put the nonbreaching party in the same position it would have been had the breaching party fulfilled its contractual obligations.

Promissory estoppel: An equitable doctrine that gives limited relief to a person who has relied to her detriment on the promise of another. Elements include (1) a promise, (2) justifiable reliance, (3) foreseeability, and (4) injustice.

Specific performance: A court order that requires the defendant to do what she contracted to do.

Statute of frauds: A legal requirement that certain types of contracts be in writing to be enforceable. Contracts involving real property, contracts that cannot be performed within 1 year, and prenuptial agreements must all be in writing.

Further Information

Bagley, C. E. *Managers and the Legal Environment: Strategies for the 21st Century, 2nd ed.*, West, St. Paul, MN, 1995.

Bagley, C. E. and Dauchy, C. E. *The Entrepreneur's Guide to Business Law,* West, Cincinnati, OH, 1998.

10.2 Patents, Copyrights, Trademarks, and Licenses

Paul A. Beck and Carol I. Bordas

The origins of U.S. intellectual property law — which includes patents, copyrights, and trademarks — can be traced to medieval England. Letters patent were issued by the king of England under the Statue of Monopolies in 1623 to confer the exclusive privilege of working or making a new manufacture for a 14-year term to the true and first inventor. In 1710 the Statute of Anne was enacted to both limit and create copyright monopolies in the publishing industry. Within the medieval guilds, individual artisans such as jewelers placed marks on their products in order that a purchaser could identify the source of their product.

Today a patent, copyright, trademark, or license can be used to protect intellectual property rights. For example, a computer program used in a process for manufacturing a new drug can be protected by a patent, a copyright, a trademark, and a license. A patent can be obtained for the process of manufacturing and to protect the use of the computer program in manufacturing the drug. A trademark can be used to identify the origin of the computer program. A copyright is created in the source code, which prevents the copying of the program. The owner of the copyrighted program can enter into a license agreement in which the licensee can distribute and reproduce the program in return for royalties payable to the licensor-owner.

Patents

A patent can be viewed as a contract between the inventor and the government in which the government grants the inventor the right to exclude others from making, using, or selling the invention in exchange for the inventor fully disclosing the invention to the public. In order for an inventor to obtain patent protection in the U.S., an inventor must apply for a patent and be granted a U.S. Letters Patent by the U.S. Patent and Trademark Office (USPTO). Three types of patents are granted by the USPTO. Utility patents are granted for processes, machines, manufactures, compositions of matter, and any improvements thereof that are useful, new, and unobvious to one of ordinary skill in the art. Design patents are granted for any new, original, and ornamental design of an article of manufacture. Plant patents are granted for any distinct and new variety of plant, including those asexually produced and those found in an uncultivated state.

A utility patent application includes (1) a description of the invention, (2) at least one **claim** the defines the boundaries of protection sought by the patent applicant, and (3) any drawings necessary to understand the invention. The description of the invention must disclose the **best mode** of practicing the invention known by the inventor at the time of filing the patent application.

In order to be granted a U.S. patent the utility patent application must describe an invention that is useful, novel, and unobvious. The invention is novel if the invention was conceived and reduced to practice before any **prior art**. The invention is unobvious if the invention would have been unobvious to one of the ordinary skill in the art at the time of the filing of the patent application in view of the prior art [Chisum, 1994]. If the invention was patented or described in any printed publication, in public use, or on sale in the United States more than 1 year before the U.S. patent application filing date, the inventor is barred by law from being granted a patent.

There is no one patent that can grant worldwide protection for an invention. Generally, if an inventor seeks to protect an invention in a foreign country, the inventor must apply for a patent in that specific country. The term of a patent in a foreign country usually extends for 20 years from the filing date of the patent application.

Patent infringement is the making, using, or selling of a patented invention during the term of the patent without the permission of the owner of the patent. The patented invention is defined by the claims of the patent. When every element recited in a claim of the patent or an equivalent of the element is present in an accused infringing device, direct infringement is established.

Copyrights

A copyright prohibits others from copying a work. Copyrights protect expressions of original work of authorship fixed in a tangible medium. Works that can be copyrighted include literary works, musical works, dramatic works, pantomimes and choreographic works, pictorial, graphic and sculptural works, motion pictures and other audiovisual works, sound recordings, and architectural works [Nimmer, 1993]. Copyrights only protect nonfunctional artistic features of a work.

The owner of a copyright has the exclusive right to reproduce the work, prepare **derivative works** based on the work, distribute the work, perform the work, and display the work. Generally, the owner of the copyright is the author of the work unless the work was specially commissioned or was produced in the scope of the employment of the author, in which case the employer is the author and owner of the copyright and the work is considered a work for hire.

Because a copyright is created at the moment the expression is fixed in a tangible medium, a copyright need not be registered with the U.S. Copyright Office to be valid. However, if the copyright is registered within 5 years of the publication of the work, the copyright is presumed to be valid. Further, registration is a prerequisite for a copyright infringement action in the courts.

A federal copyright registration application includes (1) an application form, (2) a filing fee, and (3) two **best editions** of the copyrighted work. The copyright application is examined for completeness, consistency of information, and appropriateness of subject matter. The term of a registered copyright is generally the life of the author plus 50 years. If the copyrighted work is a joint work, the term of the copyright is the length of the last surviving author plus 50 years. If the work is a work for hire, the copyright term is 75 years from the first publication or 100 years from the year of creation, whichever comes first.

To establish infringement one must prove both (1) ownership of a valid copyright and (2) copying of the original work. Copying is usually established if a substantial similarity between the accused work and the copyrighted work exists and if the accused infringer had access to the copyrighted work. Although copyright notice is no longer required, if notice is affixed to the infringed work, innocent infringement cannot be successfully claimed by the infringing party. Copyright notice consists of the word "Copyright" or an abbreviation thereof, the year of publication, and the name of the copyright owner. One can affix copyright notice to a work that has not been registered with the Copyright Office.

The U.S. is a signatory to the **Berne Convention** and as a result copyrighted works of U.S. citizens enjoy the same copyright protection as other signatory countries accord their nationals.

Trademarks

A trademark is a badge of origin placed on goods to signify the origin of the goods. A word or several words, a symbol, a color, or a scent can be utilized as a trademark. A mark that is distinguishable from other trademarks should be chosen in order that there is no confusion as to the origin of the goods [McCarthy, 1994].

Trademarks can be registered in the USPTO as well as with a state agency within each of the 50 state and in most foreign countries. In the case of federal trademark registration, the trademark must be based on actual use or based on a *bona fide intent to use* the trademark. A federal trademark registration will generally be granted unless the trademark (1) includes immoral, deceptive, or scandalous matter; (2) includes a flag or coat of arms or other insignia of a municipality or nation; (3) includes a name, portrait, or signature identifying a particular living individual without written consent of the individual or includes the name, signature, or portrait of a deceased president of the U.S. during the life of his widow without written consent; (4) resembles a trademark registered in the USPTO that when used can cause confusion; or (5) when used with the goods is merely descriptive, deceptively misdescriptive, geographically misdescriptive, or is merely a surname. However, if the applicant for registration can establish that the mark is distinctive such that the public identifies the mark with the goods — i.e., the mark gains **secondary meaning** — then the mark's being descriptive, misdescriptive, or a surname will not prevent the applicant from obtaining a trademark registration.

An application for federal trademark registration based on actual use consists of (1) a written application that includes a description of the goods to which the mark is affixed, the date of first use of the mark in commerce, and the manner in which the mark is used; (2) a drawing of the mark; (3) **specimens** of the mark as it is used on the goods; and (4) the filing fee. An application based on an intent to use contains all of these four items except that the date of the first use of the trademark and the description of the manner of use of the mark are both supplied to the USPTO within 6 months after a trademark has been allowed. A trademark registration has a term of 10 years form the data of registration, which can be renewed for 10-year periods indefinitely.

Trademark infringement is established if a likelihood of confusion between the accused mark and the protected trademark exists. The following are some factors that are considered when determining if there is a likelihood of confusion: similarity of marks, similarity of goods, or similarity and character of markets.

Licenses

A license can be granted by the owner of a patent, a copyright, or a trademark to permit another to use the owner's exclusive rights while retaining title in exchange for a **royalty**. A patent owner can license others to make, use, and sell the invention. The three rights can be licensed separately or in combination with one another. In other words, the patent owner can license the right to make the invention to one individual and license the rights to use and sell the invention to a second individual. The license may be exclusive or nonexclusive. An **exclusive license** prohibits the patent owner from licensing the same right to another [Lipscomb, 1993], whereas a nonexclusive license permits the patent owner to enter into the same nonexclusive licensing agreement with more than one individual. Other terms and conditions can be added to a patent licensing agreement, such as a territory restriction.

A copyright license must take the form of a nonexclusive license. An exclusive license of the copyright owner's rights, even if limited in time and place, is considered a transfer of copyright ownership. An owner of a copyright can enter into a nonexclusive license agreement that grants any of the exclusive rights of a copyright. For example, the license can be granted the right to reproduce the copyright subject matter and distribute the copyrighted work but not display the copyrighted work. One joint author does not need the permission of the other joint author when he enters into a nonexclusive license agreement of his copyright. No recordation requirement for nonexclusive copyright licenses exists.

For a trademark license to be valid, the licensor must control the nature and the quality of the goods or services sold under the mark. Controlling the use of the licensed trademark ensures that the licensee's goods are of equal quality to that of the licensor. If the quality of the goods and services designated by the trademark are changed, then the public could be deceived. If proper quality control is not practiced by the licensee, the trademark could be considered abandoned by the licensee.

Defining Terms

Berne Convention: The Convention for the Protection of Literary and Artistic Works, signed at Berne, Switzerland, on September 9, 1986.

Best edition: The edition of a copyright work, published in the United States before the deposit, that the Library of Congress determines suitable.

Best mode: The specification must set forth the best mode contemplated by the inventor in carrying out the invention. The requirement is violated when the inventor knew of a superior method of carrying out the invention at the time of filing the application and concealed it.

Claim: An applicant must include one or more claims in a patent application that set forth the parameters of the invention. A claim recites a number of elements or limitations and covers only those inventions that contain all such elements and limitations.

Derivative work: A work based on one or more preexisting copyrighted works.

Exclusive license: A license that permits the licensee to use the patent, trademark, or copyright and prohibits another from using the patent, trademark, or copyright in the same manner.

Prior art: Those references that may be used to determine the novelty and nonobviousness of claimed subject matter in a patent application or a patent. Prior art includes all patents and publications.

Royalty: A payment to the owner of a patent, trademark, or copyright that is payable in proportion to the use of the owner's rights in the patent, trademark, or copyright.

Secondary meaning: A trademark acquires secondary meaning when the trademark is distinctive such that the public identifies the trademark with the goods. A descriptive trademark is protected only when secondary meaning is established. A generic trademark can never acquire secondary meaning.

Specimen: Trademark specimens consist of samples of promotional material bearing the trademark used for labeling of applicant's goods.

References

Chisum, D. S. Nonobviousness, In *Patents*, §5.01, June, Matthew Bender, New York, 1994.

Lipscomb, E. B., III. Licenses, in *Lipscomb's Walker On Patents, 3rd ed.*, Lawyers Co-operative, Rochester, NY, 1993.

McCarthy, J. T. The fundamental principles of trademark protection, in *McCarthy on Trademarks and Unfair Competition, 3rd ed.*, §2.07, June, Clark Boardman Callaghan, Deerfield, IL, 1994.

Nimmer, M. B. and Nimmer, D. Subject matter of copyright, In *Nimmer on Copyright*, §§2.04–2.10, Matthew Bender, New York, 1993.

Further Information

The monthly *Journal of the Patent and Trademark Office Society* publishes legal article sin the fields of patents, trademarks and copyrights. For subscriptions, contact P.O. Box 2600, Subscription Manager, Arlington, VA.

Marketing Your Invention is a booklet prepared by the American Bar Association Section of Patent, Trademark and Copyright Law. To obtain a copy, contact American Bar Association Section of Patent, Trademark and Copyright Law, 750 North Lake Shore Drive, Chicago, IL.

The International Trademark Association produces and publishes a range of authoritative texts concerning trademark law. Among them are *Worldwide Trademark Transfer, U.S. Trademark Law,* and *State Trademark and Unfair Competition Law.* For more information regarding these publications, contact International Trademark Association, 1133 Avenue of the Americas, New York, NY 10036-6710. Phone: (212)768-9887. Fax: 212.768.7796.

10.3   Intellectual Property Rights in the Electronics Age

L. James Ristas

Intellectual property (IP) is an umbrella legal term that encompasses certain rights with respect to ideas, innovation, and creative expression. Of greatest relevance for engineers and the management of engineering-related activities in the electronics age are rights arising under the laws pertaining to patents, copyrights, trademarks, and trade secrets.

In general, the owner of an intellectual property right has the power to prevent another who is not licensed by the owner from engaging in particular activity that the law deems to be under that owner's exclusive control. In addition to stopping another from engaging in activities that violate (i.e., infringe on) the owner's intellectual property right, the owner can obtain a compensatory damages award and sometimes an enhanced award as well as attorney's fees.

With respect to computer-related innovation, it is often difficult even for an engineering manager who is familiar with basic intellectual property law concepts to recognize what subject matter is eligible for protection, what steps should be taken to secure that protection, and whether a product or service can be marketed without infringing the IP rights of another.

Fundamentals of Intellectual Property Rights

Patent rights in the United States arise only under federal law and are enforceable in federal court. The owner of a patent has the exclusive right to make, use, sell, and offer to sell the invention that is claimed in the patent. In a very crude way, a patent can be viewed as protecting how the invention accomplishes a useful technological function or result. Inventions eligible for patent protection include computer-related apparatus or methods. Of interest here are inventive methods having novel steps executed under the control of a **computer program** and inventive apparatus such as devices or systems that contain executable instructions in the form of stored computer programs.

Copyright also arises only under federal law. The owner of a copyright can prevent another from copying the creative expression embodied in the copyrighted work of authorship. Copyright protects the manner of expression of an idea but not the underlying idea itself. Thus, a copyright in a computer program can be enforced against another who copies the program or translates the program into a different program language but not against another who achieves the same program functionality using an independently developed program. Copyright in the program covers displays and "look and feel" to the extent these contain copyrightable expression.

Trademark rights can arise under both federal and state law. A trademark is a word, logo, design, symbol, or the like that is carried on a product and that serves to indicate the source of the product. A necessary condition for securing federal trademark protection is that the product carrying the trademark must have been shipped in interstate commerce. The owner of a trademark has the right to prevent another from using the same or similar trademark on the same or similar products. A service mark is analogous but for indicating the source of a service rather than a product. The essence of trademark or service mark infringement is whether the ordinary purchaser of the particular products or services at issue would likely be confused as to the source of the products or services.

Trade secret protection arises under state law. The owner of information covered by trade secret rights can prevent another who acquired secret information via such owner from using or disclosing the secret. This right protects secret information in any form, including printed, digitally stored, displayed, or transmitted. Information cannot be protected by trade secret law unless it is in fact substantially secret (i.e., not generally known) and is the subject of reasonable efforts by the owner to maintain such substantial secrecy.

Typical Product Development Scenario

Virtually any engineering undertaking can be characterized as a "project". The typical sequence of development of a product or service begins with the "concept development" phase, which includes (1) defining the objective, (2) understanding the current state of the art, (3) conceiving potential solutions that improve on the state of the art, (4) analyzing these to weed out the least desirable, and (5) demonstrating proof of principle of one or more surviving concepts. This is followed by the "commercial development" phase, which includes (6) developing a prototype that can be tested in situ, (7) testing the same or an improved prototype at a user site, and (8) completing the commercial version. The final phase concerns marketing and distributing the product or service.

Relevant IP issues will be addressed in the context of the typical engineering development of a product or service in which the innovation is embodied in a computer program or **software**.

IP Considerations during Concept Development

At the outset, the **sponsor** of the project should be sure to have written agreements with all participants (employees, consultants, subcontractors, vendors, investors, etc.) specifying the terms and conditions regarding ownership and obligations with respect to the handling and confidentiality of trade secrets, ownership of copyrights, ownership of inventions, freedom from restrictions as to former employers, noncompetition in future activities, and cooperation in perfecting and enforcing IP rights.

To the surprise of many sponsors, the law does not automatically confer full ownership of IP rights on the sponsor of the project. Especially if independent contractors are consulted or otherwise work on a project, they will retain ownership of the copyright on their work product unless they have agreed in writing to assign ownership of the copyright to the sponsor.

Without a written agreement in place, an independent programmer for the project could retain ownership of the copyright on the computer program and the right to secure a patent on the computer-implemented process or system. The sponsor would have only the right to use the program and process, whereas the independent programmer could sell or license the program and process to others.

At the concept development and future phases of development, documents such as system and performance specifications, flow charts and schematics, pseudocode and source code listings, screen display renditions or artwork, photographs, user manuals, maintenance manuals, instruction videos, and the like, whether in hard copy or digitally stored as software, are protectable by copyright and, to the extent substantial secrecy is maintained, as trade secrets.

If the product or service as conceived will provide a new functionality that is not merely an obvious or straightforward improvement over the state of the art known among the project participants, the possibility of a **patentable invention** should be explored with a patent attorney. The attorney can advise managers of the project on how to assure that good records are kept to identify the **inventor(s)** and the inventive concept.

If the new functionality relates to the control of machines or mechanical equipment that physically actuates or transforms a tangible thing, patent protection is strongly indicated. If the new functionality relates to the human senses, such as unique characteristics of a user display interface, patent protection may or may not be available. Incremental improvements for automating prescriptive and repetitive computations, such as payroll processing, are not likely to be patentable. Along this spectrum, a wide variety of computer-related innovations have been patented, including artificial intelligence or knowledge systems, financial or engineering database management systems, network communications protocols, analysis and display of system failure probability, and human-factors enhancements to process control panels.

One commentator [Laurenson, 1995] has suggested the following summary of the eligibility of software for patent protection: Software is patentable provided that the underlying patent application is drafted in such a way that the software is described as embodying, representing, or being intimately associated with a process, machine, or article of manufacture, while keeping in mind that abstract ideas, laws of nature, or natural phenomenon are not patentable. According to court decisions cited by Laurenson, "process" includes a series of steps that result in a transformation of data or signals, provided that the data or signals represent or constitute physical activity or objects [*In re Schrader*, 1994]. "Machine" includes a specific configuration of discrete hardware elements programmed in a certain way to perform data manipulation functions [*In re Alappat*, 1994]. The term "manufacture" includes a computer memory on which is stored functional computer software [*In re Beauregard*].

The strong possibility now exists that computer programs in the form of executable (digitized) data stored in tangible media apart from the computer could be eligible for patent protection as an article of manufacture. This would permit securing a patent for CD-ROMs and the like on which innovative software, games, and graphics are stored.

The patent attorney can advise not only on what aspects of the project development are likely to be patentable but also on whether the project as contemplated by the sponsor can be completed and the new product or service marketed, without infringing on the IP rights of another. The infringement issue can arise with respect to emulation or improvement of, or substitutability, compatibility, or interfacing with, existing products. Especially in the realm of multimedia products and services, permission may be required from the copyright owner of, for example, photographs, video clips, music, etc. In areas of young and rapidly growing technology, a leading competitor could have a dominating patent position that would preclude commercialization of your company's improvements.

IP Considerations during Commercial Development

During this phase, the favored concept may become ripe for the filing of a patent application. Typically, variations of a given inventive concept can be covered in a single patent application. A patent attorney can advise on what information is needed for the attorney to prepare and file the application in the U.S. Patent and Trademark Office. Failure of the patent application to be filed before the first publicly accessible disclosure of the invention, or before the first offer for sale, could jeopardize the right to obtain valid U.S. or foreign patents.

Any off-site testing should be made only according to terms and conditions of a written agreement, which includes restrictions analogous to those imposed on independent contractors or vendors involved in the project.

Copyright registrations should be secured for at least the computer program and associated documentation that is intended to be distributed commercially.

IP Considerations during Marketing and Distribution

Under some circumstances, marketing and/or distribution cannot be accomplished by the sponsor alone; some kind of formal arrangement must be made with a third party. This situation often necessitates entry into a licensing agreement covering IP rights. Trademark considerations are usually part of such agreement. Marketing and distribution of computer-related technology or services can be a fundamental design issue; the strategic planning for the project should include an early review of this issue to assure that negotiations with a necessary third party can be initiated on a timely schedule to avoid negotiating handicapped by the press of time.

Internet Issues

IP law has continually been required to adapt to fundamental changes in technology and the relationship to other fields of law. A current topic of interest concerns how certain IP rights are to be established and interpreted in regard to products and services promoted and distributed electronically, especially over a global communications network (e.g., the Internet).

Copyright and trademark law appear to raise the most interesting new issues. Under copyright, one can imagine innumerable fact patterns relating to the unauthorized posting of or providing access to copyrighted materials on the Internet. An access provider, bulletin board operator, or Web site administrator who, with the knowledge of the infringing activity, induces, causes, or materially contributes to the infringing activity of another may be held liable as a contributory copyright infringer. Under trademark law, if Company A registers and uses an Internet domain name that is identical to a trademark that has been used for many years by Company B, does A infringe on B's trademark rights? Can B require that A relinquish the domain name? Does B have any right to secure that domain name for its own use? These issues are currently in litigation.

Conclusion

One who is responsible for the management of a computer-related technology project should not be expected to make the strategic and tactical decisions regarding associated IP, without the advice of an IP attorney who is experienced in handling computer related technology. Important IP issues arise from the start of the project and continue through marketing and distribution. This brief overview does not address the enforcement of one's IP rights. The owner of IP rights who is contemplating enforcement should consult with an IP attorney and approach the enforcement decision in a manner analogous to the justification of other investments aimed at establishing and protecting one's market share.

Defining Terms

Computer program (a):   Under U.S. copyright law, "A computer program is a set of statements or instructions to be used directly or indirectly in a computer in order to bring about a certain result". (17 U.S.C. §101)

Computer program (b):   Under a functional definition more suitable for but not defined by patent law, a computer program means the sequence of operations performed by a computer in order to bring about a certain result.

Inventor: The inventive entity of a patentable invention means a natural person who alone or jointly with other inventors was the first to conceive a workable embodiment of the invention.

Patentable invention: A novel or improved process, machine, article of manufacture, or composition of matter that originates with or is discovered by a natural person and that is not a mere obvious variation of the state of the art at the time the invention was made. (35 U.S.C. §§101, 102, 103)

Software: A vague term that has no legal meaning except as may be defined in writing between parties, usually considered to include the copyright law definition of a computer program and sometimes meant to include documentation.

Sponsor: The person or entity that provides the funding for and/or who employs others to engage in a project.

References

Laurenson, R. C. Computer software "article of manufacture" patents, *J. Pat. Trademark Off. Soc.*, 77(10): 811–824, 1995.

In re Schrader, 22 F.3d 290 (Fed.Cir. 1994).

In re Alappat, 33 F.3d 1526 (Fed.Cir. 1994).

In re Beauregard, U.S. Patent and Trademark Office Appeal No. 95-1054.

Further Information

For more background on IP law and forms of agreements concerning computer technology, see Epstein, M. A. and Politano, F. L. *Drafting License Agreements, 3rd ed.,* Aspen Law & Business, New York, 1994, 1997. For a more general treatment of legal issues beyond IP, see Nimmer, R.T. *The Law of Computer Technology, 2nd ed.,* Warren Gorham Lamont, Boston, 1992.

Other helpful information can be found in Scott, M. D. *Scott on Multimedia Law, 2nd ed.,* Aspen Law & Business, New York, 1997, and Smith, G. V. and Parr, R. L. *Valuation of Intellectual Property and Intangible Assets,* John Wiley & Sons, New York.

11

Information Systems

Herman P. Hoplin
Syracuse University

Clyde W. Holsapple
University of Kentucky

John Leslie King
University of California

Timothy Morgan
The Automated Answer, Incorporated

Lynda M. Applegate
Harvard Business School

11.1 Database

Herman P. Hoplin

Most database management systems (**DBMS**s) and related development tools are based on **relational** technology. This technology has largely replaced the earlier classic (legacy) systems, which were state of the art in the 1960s and 1970s. However, relational systems are now being challenged by **object-oriented** developments, which may better serve managers in the 21st century. It is expected that the database model of the future may be a hybrid DBMS composed of the major strengths of relational and object-oriented technologies.

Developments in database are now sensitive to the needs of **users**, both managerial and functional. Technology has made it possible to implement faster and simpler systems to serve all levels of management. The advent of micro (personal) computers and faster processing storage devices and networks has brought about paradigm shifts in the management of data. It is now possible to maintain the integrity of corporate databases even under distributed conditions that can be done functionally, geographically, or networked using new technologies such as **client servers**.

DBMSs have integrated the data management functions to make information available to all organizational elements in a timely and effective manner. Transaction processing, decision making, answering ad hoc queries, and preserving the integrity of the data result from this alignment. Primary developments now include distributing the databases, more powerful computing power, desktop processing, telecommunications and networking, software tools, and artificial intelligence (expert systems). Recent shifts toward relational and object-oriented databases, distributed DBMS, and integration with multiple **management information systems (MIS)** are pushing the state of the art even more.

The Concepts and Principles of Database Management

Prior to thinking about managing data as a system (or series of systems) data were stored in *files* made up of *records* composed of *characters* (individual pieces of information) [Cohen, 1986]. The management of these file systems proved to be difficult, duplicative, and time consuming, which spawned the concept of DBMS. Researchers such as Richard L. Nolan had strong opinions that there must be a better way to manage data [Nolan, 1973] and came up with the idea in the late 1960s of the database approach as a replacement for the traditional file-management approach. The big advantage to the database concept is that DBMSs create independence between data and programs, reduce gross duplications of data, and provide quick ad hoc expedited decision making in organizations needing fast solutions to both structured and unstructured problems. This organization is usually headed by a **database administrator (DBA).** Tools for implementing the database concept soon followed in the form of languages, data dictionaries, repositories, and data warehouses.

Computer-oriented DBMSs are powerful vehicles for gathering, structuring, storing, manipulating, and retrieving massive amounts of data. The organization and manipulation of data should be defined to meet the company's or organization's standards and expectations. In addition, query processing, report generation, multiuser integration, and nonprocedural languages support are expected characteristics of effective DBMSs. The basic criteria for a good database include performance, data integrity, understandability, expandability, and efficiency. For the most part, conventional DBMS are conducive for processing of large amounts of data and applications with high transaction rates where the data path is predetermined. The methodology and theory for creating and maintaining these databases have become well established by the **hierarchical, network,** and in more recent years by the **relational** approaches to database management.

The first two approaches to database management are directed toward meeting the traditional programming methods for solving well-structured tasks. The third approach, relational, is excellent for decision support situations and answering "what if" types of questions. The key aspects of a traditional DBMSs are that voluminous data can be collected, verified, stored, manipulated, and reviewed quickly and efficiently to meet the operational, managerial, and strategic informational requirements of an organization. In addition, query capabilities to support ad hoc requests are well within the realm of most current DBMS.

Historically, while database technology focused on storage and retrieval of alpha- numerical data, demands for non-numerical databases such as decision rules, graphics, and text have been increasing. The capacity of conventional DBMSs to handle nonstandard data has caused MIS personnel to review new technologies such as expert systems to add a dimension to conventional DBMSs to meet the needs of end users.

For purposes of this section, a DBMS can be considered a software program dedicated to the better management of data. As we move into the future, the conventional two-part system of processing and data are no longer the optimal alignment of functions. Object-oriented processing and systems now permit the merging of data and process so that we arrive at the higher goal of data handling (optimization) with both data and processing in the same module or model.

Current Developments of Database Applications

What Management Needs to Know about New Developments in Database

Historically speaking, the functional user's need for a better way to manage data forced the entry of database technology. Gaining a competitive advantage through the use of database drove this technology forward despite vendor proprietary systems and multifaceted compatibility problems. The database advantage offered many benefits in what otherwise would have been an ever-worsening data management condition. Therefore, the database concept prevailed even though it was back-end loaded with long-term benefits and conversion costs, which do little to improve short-range annual budgets. After much turmoil with mounting data communications problems, some of the more serious-thinking vendors are reversing their self-serving policies so that a more "open systems" cooperative environment will be possible to permit systems from other vendors to be compatible in the user environment. This will do much to solve the growing compatibility problems that now permeate the industry.

Shift in Database Models from Hierarchical and Network to Relational

In the past decade, the database industry has shifted market share from the hierarchical and network models to the relational model as developed and fostered by E. F. Codd [McFadden and Hoffer, 1991]. Under this model, the user can relate mathematical probability with the relations carried in the data and not in mathematical sets or pointers as is the case with the earlier models.

User involvement has also been a factor in the trend toward relational systems where the DBMS software changes its orientation from the computer perspective to the user's perspective. The older classic DBMS before the advent of the relational system had a human/computer boundary that excluded user participation in the design and development of a DBMS; the relational model moved the human/computer boundary to the right to include the database users in the design and development of the DBMS. This proved to be a quantum jump in enlisting users in providing "know-how" to technicians who were not previously conscious of the realities of what was actually needed in the database. In short, this means that more resources must be committed to training end users.

Integration of Database Management in Organizations

Primary developments including distributed databases, use of more powerful computing power, and exploiting telecommunications through the "PC revolution" have all radically changed database administration.

In an article on "information technology and the management difference", Peter Keen points out that, when all firms have access to the same information technology, it is the management difference that counts [Keen, 1993]. Although many firms do not avail themselves of the option they now have, the situation of access availability now exists. This places a challenge on the database administration people and forces them to consider the effects on changing trends and issues in the industry. This results in greater responsibility and more work in database operations if the firm is to compete successfully with its competitors.

Specifically, this requires

- More team activities and monitoring of operations in the user areas.
- The use of database and software tools initially require training of both technical and user personnel.
- That distributed DBMS and networks (LANs) be coordinated, dispersal of personnel be considered, and the scope and span of control be expanded. Recent studies show that, as corporate data gets distributed, MIS will inherit backup control from communications managers and, with users growing and tying LANs to existing host systems, DBMS increasing scalability becomes a key requirement [Rinaldi, 1992]. The recent development of client/server architecture underscores the MIS involvement since, in effect, it splits the DBMS function between the database server and client or user workstations.
- Developing knowledge or SMART DBMS will require additional research, increased use of computer systems, and possible recruitment and training of a new kind of systems analyst — the knowledge engineer.
- Research into the field of object-oriented DBMS. This will require new development teams and the monitoring of applications and training to change the mind set of technicians and users from files to objects and from entities to semantic objects.

The above examples of actual or impending changes require a dramatic shift in database administration from technology to management.

Summary

1. The database thoughts and evolution reviewed in this section are intended to update managers' business views of developments and issues critical for creating effective database management in the environment of the 1990s.
2. In this rapidly changing global environment, research and development is ongoing. Most of the leading-edge concepts and developments mentioned are undergoing intensive research. Dramatic developments will continue in all phases of database including management and applications.
3. Software tools and utilities are rapidly becoming available to aid in the design and implementation of database systems. Newer ones include repositories (data dictionaries or data warehouses), entity relationship diagrams, object diagrams, and CASE tools.
4. As different data models develop, it is now desirable to use more than a single conceptual framework to interpret database requirements. Two of these frameworks are a hybrid consisting of the major strengths of relational and object-oriented models, which may provide the best of both database worlds in the immediate future and in the next millennium.

Defining Terms

Client/server system: A system of two or more computers in which at least one provides services for other computers. The services can be database, communication, printer, or other functional services.

Database administrator (DBA): An individual responsible for the management of one or more data bases who controls the design and structure of these databases.

Database (DBMS): A collection of shared data that is used for multiple purposes.

Hierarchical model: A data model that represents all relationships using hierarchies or trees.

Management information systems (MIS): A system that collects, condenses, and screens data until it becomes information, then makes it available in a timely and usable form for decision making at required levels of management in an organization.

Network model: A system used in databases to store data by combining records through various linked lists identified by pointers.

Object-oriented model: A database where data and instructions (process) are combined into objects that are modules that perform specific tasks.

Relational model: A database system consisting of relations having the capability to combine data elements to form relations that provide flexibility in the use of the data.

Users: Any manager, planner, professional, or ultimate utilizers of a database facility whose jobs will be changed by database systems. (Note: DBMSs should exist only to serve users.)

References

Cohen, E. B. *A Study Guide to Introduction to Computers and Information Systems,* H. C. Lucas, Jr., ed., p. 81. Macmillan, New York, 1986.

Keen, P. G. W., Information technology and the management difference: a fusion map, *IBM Syst. J.,* 32(1): 17, 1993.

McFadden, F. R. and Hoffer, J. A. Relational database implementation, In *Database Management, 4th ed.,* Benjamin Cummings, Redwood City, CA, 1994.

Nolan, R. L. Computer data bases: the future is now, reprinted from *Harvard Business Rev.,* Sept.–Oct. 1973, Reprint Service, HBR, Soldiers Field, Boston, MA 02163.

Rinaldi, D. Plotting recovery routes, *Software,* March spec. ed., pp. 21 and 27, 1992.

For Further Information

For a good introduction to the database concept read *Computer Data Bases: The Future is Now,* by Richard L. Nolan. (Reprinted from HBR — see references)

For further information on *Building the Data Blueprint* to gain a "computer edge", read *Stage by Stage* (letter to management from Richard L. Nolan.), Vol. 3, No. 4, Winter 1983, Nolan, Norton & Company.

Today's managers who view information as the primary tool or resource may be interested in reading *Blueprints for Continuous Improvement* (Lessons from the Baldridge Winners), An American Management Briefing (AMA) by Richard M. Hodgetts (AMA Publication Services, P.O. Box 319, Saranac Lake, NY 12983).

Current database texts that should be consulted by managers or database administrators (DBAs):

- Watson, R. T. *Data Management An Organizational Perspective,* John Wiley & Sons, New York, 1996.
- Kroenke, D. M. *Data Processing Fundamentals, Design, and Implementation, 5th ed.,* Prentice Hall, Englewood Cliffs, NJ, 1995.
- McFadden, F. R. and Hoffer, J. A. *Modern Database Management, 4th ed.,* Benjamin/Cummings, Redwood City, CA, 1994.

Recent texts such as those listed above are a must for information on current and evolving databases involving

- Client/Server and other databases on computer networks.
- The object-oriented data model
- The entity-relationship data model
- The query-by-example language
- Personal database systems
- The SQL relational database standard

Leading computer societies have timely publications on seminars, conferences, databases, and journals. An example is ACM (largest association for computing), which publishes numerous journals such as *Communications of the ACM,* a monthly publication containing technical and management articles, columns, and departments. Special interest groups of ACM also feature journals and newsletters such as *The Data Base for Advances in Information Systems.*

In view of the rapid changes in database, particularly in state-of-the-art developments in client/servers, networking, and object-oriented developments, it is highly desirable that both managers and technicians keep abreast of what is taking place in the field today. Although there are many journals and magazines currently published, it is essential to keep up with current developments. Suggestions are to regularly read monthly journals or commercial publications such as *Database Programming and Design, DBMS Tools and Strategies for IS Professionals, Client/Server Computing,* and *Object Magazine* (i.e., the August 1997 issue of *Object Magazine* has an article on "Object Relational Databases: Is the Combination Stronger?").

11.2 Decision Support Systems

Clyde W. Holsapple

The 65th anniversary issue of *Business Week* contained a special report rethinking the nature of work. It argued that technological and economic forces are fostering a radical redefinition of work: "Better technology, better processes, and fewer better workers. The ideal: technology that actually helps workers make decisions, in organizations that encourage them to do so. That's the promise of computers, just now starting to be realized" [Hammonds et al., 1994]. Such technology, known as **decision support systems** (DSSs), includes many variations, ranging from spreadsheet applications to large-scale computer-based modeling, from executive information systems to facilities for multiparticipant decision makers, from hypertext storage and search systems to artificially intelligent mechanisms such as expert systems.

The work of decision makers is knowledge work. DSSs facilitate that work by representing and processing various types of knowledge. The raw materials, work in process, by-products, and finished goods of decision making are all units of knowledge. A DSS is technology that supplements or amplifies a decision maker's knowledge handling abilities, allowing more or better decisions to be made in a given time frame. It relaxes the cognitive, economic, or temporal constraints under which an unaided decision maker operates [Holsapple and Whinston, 1996]. As such, DSSs not only enable improved decision-making performance but can be instrumental in business process reengineering and the achievement of competitive advantages.

The Nature of Decision Support Systems

DSSs differ in several important ways from their forerunners: **data processing systems** and **management information systems**. The earliest business computing systems were designed to process the large volumes of data involved in various transactions. These data processing systems focus on keeping records about the entities (e.g., customers, employees) engaging in the transactions, updating those records as new transactions (e.g., deliveries, payments) occur, and generating transactions (e.g., invoices, checks) from the records.

While recordkeeping is also important for a management information system (MIS), it is designed to produce periodic standard reports for managers. The information conveyed in these reports gives managers a characterization of current or past states of an organization's operations. To the extent that this information is factored into managers' decision-making activities, an MIS can be regarded as a limited kind of DSS. The limitations on an MIS's ability to support decisions stem from several factors: each report type is predefined before the system is implemented, reports are periodically generated by computer professionals, and the system is concerned with managing descriptive knowledge (i.e., information) only.

Ideally, a decision maker should have timely access to whatever knowledge is desired in the course of decision making plus assistance in manipulating that knowledge as desired. Pursuing this ideal is what distinguishes DSSs from their predecessors. A DSS includes a storehouse knowledge that describes salient aspects of the decision maker's world. In addition to descriptive knowledge, the storehouse may contain procedural knowledge comprised of step-by-step specifications indicating how to accomplish certain tasks and/or reasoning knowledge that tells what conclusions are valid in various situations.

Aside from its knowledge system (KS), a DSS has three essential abilities: an ability to acquire and maintain any of the types of knowledge permitted in its KS, an ability to use its KS by selecting or deriving knowledge in the course of problem solving or problem finding, and an ability to present selected/derived knowledge on an as-needed basis in ad hoc or standard reports. These abilities are possessed by a DSS's problem processing system with which a user can interact directly in a way that permits flexibility in the choice and sequencing of support requests. The set of all possible requests a user can make is called the DSS's language system. The set of all possible responses is called its presentation system.

The generic framework for DSSs portrayed in Fig. 11.1 is adapted from Bonczek et al. [1981], Dos Santos and Holsapple [1989], and Holsapple and Whinston [1996]. It suggests that a DSS's purpose is to help users manage knowledge used in decision making [Holsapple, 1995]. In so doing, a DSS provides one or more of the following benefits:

1. It augments a user's capacity for representing and processing knowledge during decision making.
2. It solves large or complex problems that a user would not or could not otherwise handle.
3. It solves relatively simple problems more reliably or faster than a user can.
4. It stimulates a user's thought process during decision making (e.g., by facilitating exploration of knowledge, by offering advice, or by recognizing problems).
5. It furnishes evidence to help a user check, confirm, or justify a decision.

Decision Support System

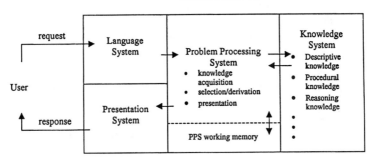

FIGURE 11.1 Generic DSS framework.

6. The act of developing a DSS can lead to new ways of viewing a decision domain or more formal approaches to some aspects of decision making.

7. A DSS can enhance an organization's competitive position by means of greater decision-making productivity.

Ultimately, the value of a particular DSS depends not only on its knowledge representation and knowledge processing characteristics but also on the nature of the user and traits of the decision context.

Decision Support System Classes

Decision support systems can be classified based on the **knowledge management techniques** they employ. These techniques include database, hypertext, spreadsheet, solver, and rule-oriented approaches to representing and managing knowledge [Holsapple and Whinston, 1996]. An alternative classification is based on the role a DSS plays in an organization, ranging from corporate planning system to local support system and from supporting an individual to supporting multiple participants [Philippakis and Green, 1988]. Appreciating the classes of DSSs is beneficial to their potential users, to developers who create DSSs, and to those who manage the deployment and operation of an organization's DSS technology.

Technique-Oriented Classification

The knowledge management technique used in developing a DSS strongly influences what that DSS can do: what user requests are allowed in the language system (LS), what representations are permitted in the KS, what processing can be performed by the problem processing system (PPS), and what responses are possible in the presentation system (PS). Each technique-oriented class of DSS can be examined in terms of the generic framework in Fig. 11.1, resulting in a specialized framework for that class. For example, a specialized framework characterizing the class of database-oriented DSSs is depicted in Fig. 11.2.

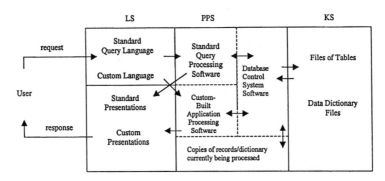

FIGURE 11.2 Database-oriented DSS framework.

A DSS built with a database technique of managing knowledge typically gives users a language for making requests that is customized to the users' preferences and the traits of the application problem domain. Users may also make requests in terms of a standard query language (SQL). Assuming that a relational database is being used, the KS contains descriptive knowledge represented as files of tables and a data dictionary holding assimilative knowledge (i.e., governing the system's assimilation of knowledge). The PPS is comprised of a database control system, which is software that manipulates (e.g., stores, retrieves, alters) KS contents subject to the KS's assimilative knowledge, a standard query processor that interprets query requests into retrieval instructions for the database control system and presents the retrieved data in responses to the user, and custom-built application programs that accept requests in a customized language to trigger appropriate database control system processing and produce customized responses with the results.

A DSS built with a spreadsheet technique represents knowledge in terms of objects other than tables, records, and fields. Its KS is comprised of spreadsheet files involving such objects as cells, expressions, and cell blocks. Whereas databases are oriented to holding large volumes of descriptive knowledge, spreadsheets typically hold a mix of descriptive and procedural knowledge. The PPS of a spreadsheet-oriented DSS is typically an off-the-shelf processor such as Excel or 1-2-3, which accepts requests (e.g., for what-if analysis) in terms of a mostly standard language (e.g., menu selections) and issues responses in terms of mostly standard presentations (e.g., a grid of cells).

A solver-oriented DSS is one that has a collection of algorithms (i.e., solvers), each of which can be invoked to solve any member of a class of problems pertinent to the decision domain. For instance, one **solver** may solve linear optimization problems, another may solve depreciation problems, and another may do portfolio analysis. There are two ways to incorporate solvers into a DSS: the fixed and flexible approaches. In the former, solvers are part of the PPS, and they can operate on data sets held in the KS. A data set is a grouping or sequence of numbers represented according to conventions that a solver requires for its input. With this approach, the DSS's collection of solvers tends to be relatively fixed, not readily amenable to addition, deletion, or modification. A user makes requests for the execution of solvers with particular data sets. In contrast, a flexible-solver DSS is one whose KS contains both solvers (procedural knowledge) and data sets (descriptive knowledge). The PPS is designed to add, delete, modify, combine, coordinate, as well as execute solver modules in response to user requests. It also has facilities for manipulating data set objects.

Another special case of the generic DSS framework is the class of DSS that focuses on representing and processing rules. The KS of a rule-oriented DSS holds one or more sets of rules, along with state variables that characterize a situation of interest. Rules are objects that represent reasoning knowledge by indicating what conclusions are valid when certain circumstances can be established as existing. The PPS of such a DSS is designed to use rules to draw inferences about a specific situation confronting a user. The result is a recommendation (and associated explanation of it) that can be presented to the user. Because the rules are often representations of reasoning knowledge obtained from an expert at making such recommendations, such DSSs are called expert systems. The PPS that infers the advice is often called an inference engine.

It can be beneficial to build a DSS that incorporates multiple knowledge management techniques when only one is inadequate for offering the desired support. The best-known (but by no means only) example of this is the specialized framework that involves both solver management and database management [Sprague and Carlson, 1982]. Figure 11.3 illustrates the usual way that this is portrayed. Here, the KS is comprised of a database and a model base. The latter refers to solver modules existing in the KS. The PPS has three parts: a database management system for manipulating records held in the KS's database, a model base management system for manipulating solver modules held in the model base, and a dialog generation and management system, which interprets requests and packages responses. The LS and PS are implicit in the recognition of dialog processing but are not explicitly indicated in the diagram.

Organizational Classification

A different way of classifying DSSs is based on the roles they play in an organization. Based on DSS practices in organizations, Philippakis and Green [1988] have identified four types of DSSs, which they

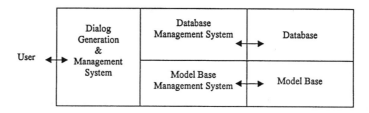

FIGURE 11.3   Example of a compound DSS architecture.

call corporate planning systems, functional DSSs, executive information systems, and local DSSs. All have the generic DSS traits noted in Fig. 11.1 and all can exist in the same organization. They differ in terms of the degree of system formality (high vs. low) and the organizational level/scope (operational/restricted vs. strategic/broad). Management of an organization's DSS resources depends on a recognition of the four types of DSSs, trade-offs that exist among them, and an appreciation of the benefits of coordinating their creation and operation.

Corporate planning systems are highly formal and have a broad/strategic nature. These DSSs derive and hold knowledge pertaining to overall planning decisions made by upper management. They cut across organizational divisions, departments, and functions. The KS of such a system tends to hold extensive data about the organization and its environment. Solvers that mathematically model an organization's processes and environment are also present in this kind of DSS. Nearly all large organizations use corporate planning systems and commit considerable effort to their development and upkeep.

A functional DSS is also relatively formal but is more restricted/operational in character. Such DSSs hold and derive knowledge used in making decisions about some organizational function (e.g., financial, marketing, operational). Unlike corporate planning systems, the decisions they support are restricted to a particular function, have a relatively shorter time horizon, and are not made by top management. An FDSS tends to be developed by computing professionals but usually does not require as many resources as a corporate planning system.

An executive information system is less formal than the previous two types. It holds knowledge pertinent to wide-ranging decisions made by top management, decisions that are not necessarily concerned with planning or any particular function and that are comparatively immediate and short term. This kind of DSS aims to satisfy an executive's ad hoc needs for information about an organization's current performance, current environmental conditions, and anticipated developments over the short run.

A local DSS supports decision makers below the upper-management level in the making of decisions localized within some organizational function. They hold and derive knowledge pertinent to a limited aspect of a function. Compared to the other kinds of DSSs, such systems typically require less time, effort, and expense to develop. They are often developed by end users rather than computer professionals.

Another distinction made by Philippakis and Green [1988] notes that some DSSs are designed to support an individual user and others are designed for multiparticipant use. The best-known example of the latter is a group decision support system (GDSS) which aims to enhance the gains that can be achieved by a group of persons that collectively makes a decision while minimizing the losses that can occur when the decision maker has multiple participants [Nunemaker et al., 1993]. Figure 11.4 expands on the generic architecture of Fig. 11.1 to portray a generic framework for multiparticipant DSSs. It is applicable to GDSSs as well as DSSs that support other configurations of multiple participants (e.g., hierarchical teams, organizations) [Holsapple and Whinston, 1996].

Notably, the PPS has an ability absent from Fig. 11.1: the coordination of the participants' actions, which can range from supporting their communications with each other to structuring the flow of their activities. The LS can be partitioned into public messages, which any individual user may submit, and private requests, which only a specific individual user can make. The PS can be similarly partitioned. The KS can hold system knowledge that characterizes the infrastructure of the multiparticipant decision

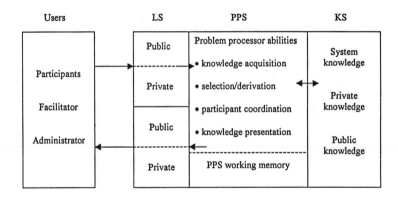

FIGURE 11.4   A generic multiparticipant DSS framework.

maker (e.g., in terms of roles, relationships, regulations), private knowledge that is accessible only to individual users, and public knowledge that is accessible to all users.

Emerging Developments in Decision Support

Although DSSs are widely used in today's organizations, the relentless march of technological innovations and the shift toward knowledge-based organizations continue to fuel developments in the decision support field. Increasingly, the fruits of artificial intelligence research are finding their way into DSS implementations yielding artificially intelligent DSSs [Dhar and Stein, 1997]. The rise of data warehousing [Devlin, 1997] is an important extension to database-oriented DSSs and is becoming an important backbone for coordinating the disparate DSSs existing in an organization. The explosion of electronic commerce [Kalakota and Whinston, 1996] is leading to DSSs with vast hyperlinked knowledge systems of which the World Wide Web's content is the most prominent example. We are rapidly entering an era dominated by knowledge-based organizations [Holsapple and Whinston, 1987] that treat knowledge as the key resource and will equip knowledge workers with increasingly powerful technological systems that amplify knowledge-handling capabilities in the interest of improved, competitive decision-making performance.

Defining Terms

Data processing system: A computer-based system that handles the processing of data that describe transactions and the updating of records based on those transactions.

Decision support system: A computer-based system comprised of a language system, problem processing system, knowledge system, and presentation system, whose collective purpose is the support of decision-making activities.

Knowledge management technique: A technique for representing knowledge in terms of certain kinds of objects and for processing those objects in various ways.

Management information system: A computer-based system that keeps current records about an organization or its environment and produces periodic, predefined kinds of reports from those records.

Solver: A program that solves problems of a particular type in response to corresponding problem statements.

References

Bonczek, R. H., Holsapple, C. W., and Whinston, A. B. *Foundations of Decision Support Systems*, Academic Press, New York, 1981.

Devlin, B. *Data Warehouse from Architecture to Implementation*, Addison-Wesley, Reading, MA, 1997.

Dhar, V. and Stein, R. *Intelligent Decision Support Methods: The Science of Knowledge Work*, Prentice Hall, Upper Saddle River, NJ, 1997.

Dos Santos, B. L. and Holsapple, C. W. A framework for designing adaptive DSS interfaces. *Decision Support Systems.* 5(1): 1–11, 1989.

Hammonds, K. H., Kelly, K., and Thurston, K. The new world of work. *Business Week.* October 17, 1994.

Holsapple, C. W. Knowledge management in decision making and decision support, *Knowledge and Policy,* 8(1): 5–22, 1995.

Holsapple, C. W. and Whinston, A. B. *Decision Support Systems: A Knowledge-Based Approach*, West, St. Paul, MN, 1996.

Kalakota, R. and Whinston, A. B. *Frontiers of Electronic Commerce*, Addison-Wesley, Reading, MA, 1996.

Nunamaker, J. F., Jr., Dennis, A. R., Valacich, J. S., Vogel, D. R., and George, J. F. Group support systems research: experience from the lab and field, In *Group Support Systems: New Perspectives*, L. Jessup and J. Valachich, eds., pp. 125–145, Macmillan, New York, 1993.

Philippakis, A. S. and Green, G. I. An architecture for organization-wide decision support systems, Proceedings of the Ninth International Conference on Information Systems, Minneapolis, December, 1988.

Sprague, R. H., Jr. and Carlson, E. D. *Building Effective Decision Support Systems*, Prentice Hall, Englewood Cliffs, NJ, 1982.

Further Information

Many special topics related to decision support systems can be found in the content and hundreds of links maintained at the World Wide Web site whose url is http://www.uky.edu/BusinessEconomics/dssakba/.

11.3   Network Management

John Leslie King and Timothy Morgan

The growth of data communication networks in the past 15 years has been astonishing. This is seen most easily in the world of the **Internet**. As Fig. 11.5 shows, the number of Internet host domains has grown from fewer than 1000 in the early 1980s to about 20 million today.

By most other measures as well, the growth of Internet use has been explosive. In mid-1980 there were approximately 15 Usenet sites; by mid-1995 there were over 130,000. In mid-1994 there were about 2,700 WWW pages; by mid-1997 there were more than 1.2 million. Today, over 170 countries have Internet nodes; over 150 have UUCP nodes, and more than 100 have FIDONET nodes.

Statistics about the Internet and other large, public network infrastructures reveal only part of the picture. Equally important, but seldom the subject of comprehensive statistical surveys, is the growth of local area networks (LANs) and wide area networks (WANs) that are used within and between organizations, often in proprietary configurations. Typical LANs of 10 years ago did little more than support file sharing and printer access; today they support electronic mail (e-mail), desktop video conferencing, shared databases, and other important applications. WANs have grown greatly as well, linking together globally distributed offices of multinational firms, tying together suppliers and customers, and providing the means of coordinating global enterprise. Some LANs and WANs use Internet technology, and are often called "intranets", but many use proprietary solutions provided by computer and networking vendors.

The growth attests to the popularity of computer networking and helps to explain the rapid increase in demand for qualified network administrators. In a more subtle way, it suggests the difficult challenges network managers face. The contemporary network manager must be technically astute, but, increasingly, the network manager must have skills in organization, economics, and politics to match. Networks are no longer merely useful in organizational life. Increasingly, they are vital to routine operations, customer and client relations, and management strategy.

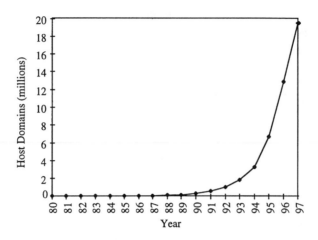

FIGURE 11.5   Internet Host Domain Growth 1980 to 1997.

The Importance of Network Management

Computer networks have become vital to organizations for three reasons. First, computer networks are communication networks, linking together various people in the larger social network of the organization. Every organization is made up of social networks that are vital to getting work done. Computer networks are complementary to other, long-established organizational communications modalities such as meetings, memos, and telephone conversations. Applications such as e-mail, the "killer application" of computer networking, and, increasingly, groupware systems allow for computer-supported cooperative work. The more tightly social networks are bound to computer network infrastructure, the more critical the computer networks are to the organization's welfare.

Second, computer networks exhibit an important economic characteristic called network externalities. An externality is an outcome of some action that results in benefits or costs for those who did not create the outcome. Positive network externalities arise when current members of a network benefit by having additional members join the network. A good example of this is seen in the fact that owning the first telephone in town is useless because there is no one to talk to, but each successive subscriber benefits the first subscriber by increasing the number of people available by telephone. At some point in a network's growth, the "critical mass" of participants is reached, and the cost of joining the network falls below the cost of not being part of the network. Then there is a stampede to join, as recent Internet growth rates suggest. On the downside, rapid increases in network users that outstrip network capacity generate congestion for all users, a negative network externality that also can be seen in the Internet. The dynamics of network externalities can be difficult to predict, while their effects can be profound.

Finally, the more networks are integrated into routine operations, the more dependent on them the organization becomes. Consider the case of the telephone system, which is now such a fundamental part of social and economic infrastructure that failures can have catastrophic consequences for everything from emergency services to airline reservations. In fact, a key purpose behind the development of the **ARPANet,** the precursor to the Internet, was the creation of a national communication system that would be more robust than the telephone system in time of war. Computer networks are remarkably reliable, but not because they are inherently reliable. They must be kept up and running through constant attention and care by network administrators. They are now vital infrastructure for organizations, for the economy, and for society. A key feature of infrastructure is that it is usually only visible on breakdown.

We see, then, that the network manager faces an extraordinary set of challenges. The network administrator is increasingly responsible for maintaining key aspects of the organization's social networks. This is a difficult and risky job because people dislike having their social networks disrupted. The network administrator also faces continuing uncertainty in capacity planning and legacy system management due to network externalities that affect the demand for services. The network manager's involvement with infrastructure creates the awkward situation where good performance is considered as "just doing the job", while poor performance causes outrage among those who depend on the infrastructure.

Technological Parameters

The contemporary network manager faces a bewildering array of technologies to deal with. Technological progress in the networking field is so rapid that new products and processes bring added capabilities every few months, but these changes make it difficult to keep the local network coherent and manageable as upgrades occur on a piecemeal basis. Also, while much progress has been made in network interoperability so different kinds of networks can share data, there is little hope that any universal solutions will be available in the near future. Most network managers must operate in environments of heterogeneous computer equipment, multiple network protocols, and multiple versions and releases of key operating system, networking, and application software. The network manager thus is required to be at least knowledgeable in several network technologies and must have on staff real experts in each technology. Also, the network manager must be constantly thinking about the upgrade pathways for the network infrastructure as demand grows and new technologies emerge.

LANs and WANs

One of the most important technical distinctions in the network world is between the infrastructure of LANs and WANs, in both hardware and communication protocols. Engineering realities have long dictated that higher-speed transmissions can be accomplished by trading off distance. For example, the early Internet backbone used 56-kbps links, whereas a typical LAN of the time operated at 4 to 16 Mbps — hundreds of times faster. Today, the Internet uses links of 45 Mbps, and experiments are under way with long-distance links operating in the range of 2 Gbps. Such high-speed links are economically feasible today because the number of users sharing the backbone has grown so dramatically, both increasing the traffic but also spreading the cost. Although the speeds of LANs and WANs are converging at high levels, there are important differences in software. The high-speed, short-distance environment of the LAN caused LAN protocols to develop very differently from WAN protocols.

Perhaps most important, LAN protocols benefit from the comparatively simple challenges of the LAN environment. They can be considerably simpler and often do not need centralized management because the demands on the network are relatively modest. In the simplest cases, such as the AppleTalk protocol, a computer user can simply plug into the network and start accessing shared resources, without the need for any centralized management. A large internet, in contrast, cannot operate without a central authority to govern network and host naming, topology, **gateways** machines, etc. Also, the rate of packet loss on a small token ring or ethernet segment is very small compared to that on a global network, and LAN-only protocols have been optimized to work under this assumption. The round-trip time for packets on a LAN is very small, so protocols need not optimize for delays. A WAN-oriented protocol, on the other hand, will usually have several packets "in the air" simultaneously and will automatically retransmit only those packets that are not acknowledged within some period of time. Acknowledgments can be "piggybacked" with data packets going in the other direction, minimizing the number of packets that must be sent, but the management of packet loss is still much more complicated in a WAN than in a LAN.

There is also a major difference in the way LANs and WANs handle broadcasts. LANs often have a mechanism by which a single computer can send a "broadcast" packet that will be received by all the other computers on the LAN. Many LAN-specific protocols were designed to take advantage of this ability, often as a substitute for centralized management. The hosts talk to each other to discover what services are available within the network, eliminating the need to consult a table of such information constructed by a network manager. The use of broadcasts also works against scalability, since the number of "housekeeping" broadcast packets being broadcast increases as a function of the number of hosts on the network. On large LANs, these packets can become a significant fraction of the total traffic. On a LAN such as ethernet, an originating machine will broadcast one packet but the "to" address on that packet is to everyone on the network. Each host on the network picks it up and decides what to do with it. In the WAN world, unlike in the simple LAN world, there are problems of gateways, multiple pathways, loops, and so on that would play havoc with broadcast messages. A packet could get into the network and loop forever. Few WANs can support broadcast packets, and exceptions require high-level solutions. For example, the protocol that drives Netnews assigns to each message a unique identifying number so that duplicates in the network can be suppressed. However, this truly is an exception and requires considerable overhead to manage.

In the early days of LANs, communication between heterogeneous computers and networks was not important, and the networks usually handled only simple problems such as file and print sharing. LAN network protocols were often proprietary, usually defined by the same company that provided server-side software in packaged, point solutions. Since LANs were seldom connected to each other or to outside networks, protocols such as IBM/Microsoft's NetBEUI were not designed to support multiple network segments or gateways. WAN protocols, in contrast, had to be designed to support a wide range of possible services across many proprietary platforms. They also had to enable additional features that were likely to be needed by the user community as use increased. The successful WANs proved to be highly extendable; for example, the World Wide Web, which operates on top of **TCP/IP**, was not invented until a decade after TCP/IP made its debut.

Today, it is very common to find TCP/IP, the Internet's basic protocol suite, in use on LANs alongside legacy proprietary protocols. Many network managers are pushing to move to a single protocol to simplify management of the network and procurement of hardware and software. Corporate LANs are now frequently connected to the Internet, or at least to a corporate-wide WAN. Since TCP/IP is not tied to any particular hardware implementation, and because network managers tend to favor nonproprietary solutions when they can find them, TCP/IP is fast becoming the worldwide standard in computer networking. This could be accelerated by the advent of IPv6, version 6, of the Internet Protocol, which provides 128-bit instead of 32-bit addresses and allows for specialized supports for certain functions such as giving isochronous higher priority than regular data packets.

One way that companies can create a WAN at a reduced expense is by connecting each office to the Internet and then using the Internet backbone as the WAN. Corporations are naturally worried about sending confidential information across the open Internet. The usual solution to this problem is to create a virtual private network (VPN) by encrypting the contents of all packets moving between the corporate LANs, while leaving other packets in plain text. Commercial **routers** used to connect LANs to the Internet often have this capability builtin and automatically encrypt/decript the packets as appropriate. Another feature provided by such routers is to "piggyback" non-Internet protocols by placing them inside of IP packets and then sending them, encrypted, through the VPN. This allows legacy proprietary protocols to work across the Internet even though they are based on incompatible protocols.

Another revolution that has been brought about by the Internet is connecting companies with their vendors and customers. Having real-time, shared-network connections between businesses was formerly nearly unheard of. Companies can now query databases of their suppliers and even submit purchase requests, facilitating just-in-time deliveries. Software products can be delivered almost instantly to customers, bypassing packaging and wherehousing costs. Consumers often need access to data stored in proprietary company databases: order status, airline schedules, etc. Companies want to share this information without putting their whole database or LAN completely on the Internet, and without having to worry about what type of computer or software the consumer might be using. This is often accomplished by creating an Extranet — a World Wide Web server that is a gateway to only the information that the company wishes to publish.

A great deal of networking discussion concerns TCP/IP. Despite TCP/IP's popularity, it is important to recognize that some network protocols, and particularly IBM's Systems Network Architecture (SNA) family, are likely to be important players in the network world for some time to come. SNA was developed in the early 1970s as a solution to problems of highly reliable commercial computer networking. Unlike TCP/IP, which routes individual packets of data based on an address number that identifies the destination machine, SNA requires a client and server to establish a "session" before the exchange messages begins.

The TCP/IP and SNA approaches are sometimes seen as competitors, but, in fact, the two architectures are complementary. An organization using both can probably deal with almost any network challenge. IP is highly flexible and can tie together heterogeneous devices, but errors in IP networks often go unreported and uncorrected because intermediate equipment reroutes subsequent messages through a different path. SNA is excellent for building reliable networks out of dedicated, centrally managed devices but requires a technically trained central staff ready and able to respond to problems as they are reported by the network equipment. The utility and resilience of SNA in the market has prompted many network product companies to create devices and software that allow ready exchange of data between TCP/IP and SNA architectures.

Aside from the issues of LAN and WAN configuration, network managers face a rapidly growing problem in managing the exponential growth of network name spaces. This problem arises because names in networks serve as addresses for interactive communications among people and systems. IP address, URLs for a web site, and user account names are "called" by services such as e-mail, network routing, and shared databases servers. Name spaces must be accurate over time, and, to achieve this end, network managers must make key decisions regarding the way names are selected (e.g., based on users, machines, and services or on values for user IDs or network addresses), on the tracking of name changes over time, on the enforcement of organizational naming standards, and on the allocation of privileges for changing or updating name information. The potential for trouble through careless name space management is high. Consider the case of a network manager on the night shift who innocently changes the name of a server without changing the pointers in the configuration files that point to particular servers by name. The next morning, key services such as e-mail and network routing can collapse because of the resulting name space infidelity. A good solution to such problems is to design a comprehensive database that not only maintains the names but also codifies the relationships between the names so that a server cannot be deleted while it still houses users' home directories, handles e-mail for a LAN, and so on. The name problem is not just a technical problem; it is political as well. Many users and departments are jealous of their system names and resent efforts of system administrators to control the name space. The ongoing controversy in the Internet world over who shall have authority to assign top-level domain names in the domain naming system (DNS) is an example of this.

Basic Configurations

LANs today typically use 10 or 100 Mbit ethernet, usually in a hub and spoke system, with offices wired to a central wiring closet along with telephone connections. This design is important to the network manager because it usually isolates the most common faults (shorted wiring, bad network adapter hardware) to an individual user. It also facilitates turning service on and off in individual offices since this merely involves plugging in or unplugging a cable. The hubs, or routers, in the wiring closets are connected to each other over a backbone network, which may in turn be routed to other corporate networks or the Internet. The cost of wiring a building for a network is dominated by labor costs, so it is important to pull "category 5" or better cabling even when 10-Mbps speeds are to be employed in the immediate future. Eventually, you will want to upgrade to 100-Mbps or faster speeds which will require the higher-capacity wiring. Gigabit-speed ethernet is currently beginning its way through the standardization process, so it may be in wide use in 5 to 10 years.

Routers fall into two categories. Simpler, cheaper routers merely repeat and amplify any signal they receive on one port out to all their other ports. This type of **repeater** essentially duplicates the shared cable of the original ethernet specification — each host on the network sees every signal. More sophisticated routers, called "switching hubs", know which ethernet addresses are connected to which port. When a packet arrives, it is routed only to the appropriate port. This feature greatly reduces the overall amount of traffic and hence the collision rate. It does nothing if there is a single host that is involved in most communications, as a central file server might be. In this case, and for backbone network connections, some switching hubs feature a one or two 100-Mbps ports while the remainder are at 10 Mbps. The high-speed ports are used to connect to resources where traffic will be aggregated and the higher capacity will be utilized.

Ethernet works by sharing a signal on (conceptually, at least) a single wire. When a computer wants to transmit, it simply goes ahead, and, if it turns out to have collided with another transmitter, both back off and try again. This approach works well and keeps implementation costs down as long as the collision rate remains low (no more than a few percent of transmitted packets). However, as the capacity of the medium is neared, this approach becomes increasingly inefficient.

Token ring networks, the other most-common LAN hardware approach, offer a solution to this problem by regulating which computer is allowed to transmit at any given time. Hosts are arranged in a ring, with an imaginary token circulating from node to node. When a particular node has the token, and the token isn't carrying a packet destined for another host, then the node may send its own packet out in the token. Token rings are, therefore, more efficient in heavy-traffic environments than ethernets. This advantage is offset by the higher cost and complexity of the hardware needed to connect a computer to the network and by the more difficult wiring topology (offices in buildings are rarely arranged in a circle).

Other technologies that are in varying levels of use today are the **fiber-distributed data interface (FDDI)** and the **asynchronous transfer mode (ATM)**. Because of their higher cost and special wiring requirements, these technologies are not often used all the way to the desktop level. Instead, they are more often employed as backbone networks, connecting floors of buildings, and/or buildings in a campus setting, together. FDDI can be configured in a dual concentric ring topology, which can allow the network as a whole to reconfigure itself and continue functioning if any one node fails. ATM's design was heavily influenced by the telephone companies, and therefore it is much more capable than ethernet or token ring in carrying isochronous traffic, such as digital voice and video. All ATM data are carried in "cells", which are 53 bytes in size. This uniformity allows gateways to route packets at extremely high speeds. Today, ATM networks run at speeds between 155 and 622 Mbps. ATM, in particular, has the further advantage of working at these high speeds over long distances.

Network Management Protocols

An important technological development in recent years has been the creation of network management protocols that allow the network manager to gather information from and, ideally, to control directly the many different devices in the network. The network manager must know about these protocols and their use with assorted network products.

At the simplest level, a network manager can use special protocols to determine network connectivity and routing. An example of this in the IP world is the internet control message protocol (ICMP), which allows "pinging" of other devices on the IP network. Higher level protocols, such as the **simple network management protocol (SNMP)** and the common management information services/**common management information protocol** (CMIS/CMIP) from ISO, define a mechanism whereby the network administrator can not only determine connectivity to a remote device but can query and even *control* that device in a standardized fashion. The ability to control many present and *future* network devices using a single software application and protocol is invaluable to the network administrator. These protocols work by defining a database architecture, called the **management information database (MIB),** and a mechanism for querying and changing this database remotely.

Typical information/parameters in the MIB include name of device, version of software in device, number of interfaces in device, number of packets per second at interface, and operational status. The MIB may also include many other entries that are specific to its function. A router, for example, would have a very different MIB from a network telephony server, yet the network administrator might use the same client software to control both devices.

Fundamental Challenges

Operational Reliability and Effectiveness

Like all complex technical systems, computer networks sometimes break down. A breakdown or fault can range from minor to catastrophic, but, in any case, the customary strategy is to find the fault, isolate it, and fix the problem. Faults generate symptoms recognized by users or by monitoring software in devices. A critical network event such as a dropped link usually shows up immediately; minor faults sometimes are identified first through polling or other behavior of network devices. The network manager collects the data on the symptoms and uses fault management techniques to execute the find, isolate, and fix sequence. The art of fault management includes figuring out what can and cannot be fixed by the local network managers (some problems require action from outside the local domain), initializing action on such faults, and prioritization of action on the faults within the manager's sphere of influence. Tools to facilitate fault detection and management, and in some cases even fault correction, are available after market and are increasingly bundled with network products. If and when network technologies stabilize, advances in fault tolerance might make fault management less of a concern for network managers. At this time, however, fault management is a major part of the network manager's routine.

Another key factor in reliability and effectiveness is configuration management. Unlike fault management, which deals with fixing problems that impede network performance, configuration management is aimed at preventing faults and performance degradation. Configuration management is the art of arranging the right devices in the right network architecture in the right sequence. Success depends on choosing the right equipment and the right architecture, setting the right parameters in the devices (e.g., addresses, routing instructions), and "tuning" the resulting system so it works well. Configuring a brand new network offers the chance to get key details right in advance, but, increasingly, managers must mix and match devices in an existing network as a result of upgrades and replacement. In such situations, network configuration tools can be very useful. These are basically data collection and analysis packages that watch the network in operation, store relevant performance data in a database, and allow manipulation of the data in ways that reveal problems. Advanced configuration management tools use expert systems or other techniques to provide guidance on steps to improve the configuration.

Security

Computer networks often contain and carry information that must be controlled carefully to ensure access by only a restricted set of individuals. Security concerns are particularly important in financial systems that can be abused for theft or fraud, systems containing personal data on individuals, databases of special competitive advantage to a firm, or systems holding data vital to institutional interests such as law enforcement or national security. In addition to those concerns, the network manager must also deal with threats to system integrity and performance that can be perpetrated through use of computer viruses and worms. The key challenge in computer network security is access control: allowing in only those network connections and those data streams that pose no risk. In short, the challenge is to identify the vulnerable data and devices in the system, identify the access points required to get to them, and secure those access points in a sustainable manner.

The only truly secure computer is locked in a room with no network connections to the outside, ideally with the power turned off. The whole idea of networking is to connect to other computers so information can be shared. Networks, by their nature, pose security problems. Experience has shown computer networks to be exceedingly difficult to secure.

There are several key strategies for securing networks. First are physical controls, such as restricting access to communications ports and host authentication that helps verify just what machine is making contact. Another method is callbacks. The calling computer connects to a properly equipped host computer, which accepts the call, notes the address of the calling computer, and immediately hangs up. The host computer then calls the calling computer back if its address (e.g., the calling computer modem's telephone number) is listed as an acceptable connection. The host computer also notes that it placed the call, which port was used, how long the connection was maintained, and so on. This is mainly used in dial-in applications at this point, but a simpler version of this strategy is embedded in a Unix utility called "tcp wrappers" that logs the source IP address for every connection to every service offered by the appropriately equipped server.

There are also software-based access controls, such as passwords and key authentication. Passwords are the most commonly used technique, and almost every operating system asks for some kind of password. More-sophisticated operating systems force users to select only "good" passwords that are not easily guessed or cracked. Key authentication uses a special "key server" that reduces transmission of password strings that might be intercepted and is found in MIT's widely known Kerberos scheme. Together with such software, a security-minded manager can be proactive by running the cracking software before the criminals do to find any weaknesses in software defenses. Organizations with high security requirements such as the military bring in expert "tiger teams" that attempt try to crack systems that must be secure. The software approach has the advantage of being relatively easy on the network manager — the manager need not be a security expert to operate such controls. However, downside is that the criminal does not have to be much of a security expert, either, to break in.

Security can also be enhanced by packet filtering (admitting or excluding packets based on access rules), "firewalls" (devices that prevent certain kinds of traffic from entering a host domain), and encryption (turning information into noise). These require a comparatively high degree of expertise to implement and manage and are often used in addition to physical and software controls to make networks particularly secure.

There are many pros and cons to the use of security strategies; preparation of a detailed security plan requires a high degree of expertise. As a general rule, the greater security, the greater the cost and difficulty imposed on the network manager and the users. Most network managers have little training in the details of system and network security, and more than a few implement the most basic software controls and hope for the best. This might be a good strategy if the network and the systems it supports are not vital for the organization's welfare, but, if that is not the case, it can pay in the long run to acquire the necessary expertise to develop and implement a strong security plan. In most cases is makes sense to address the security issue in the broader context of risk management. This allows consideration of trade-offs between various ways of reducing risk, such as increasing security vs. buying additional insurance.

Economic Efficiency

The whole purpose of a network is to facilitate communications and enhance organizational performance. This can usually be accomplished by throwing money at performance-degrading problems, but this is not an option for most organizations. Good performance management must serve the ends of economic efficiency, meaning good performance at reasonable cost. The typical measures of performance in networks are response time (how long it takes for a message and its response to make the round trip in the network), rejection rates (the percentage of failed transfers of packets due to resource constraints), and availability (measure of accessibility over time, often measured as mean time between failure). Many network software packages come with tools to generate such performance data, and after-market packages are available. Examination of historical data on the network is often useful to see whether changes in key aspects of the network (e.g., configuration or number of users) account for performance problems. The art of performance management is in effective capacity planning, meaning projection and acquisition of needed capability in advance of rising or falling demand so performance remains good while costs remain reasonable. As noted above, a serious challenge for capacity planning is the turbulence in demand resulting from network externalities.

Another important aspect of economically efficient network operations is accounting management. This consists of three aspects. The first is keeping track of resource use on the network according to use centers. Use centers could be individual users or, at a more aggregated level, departments or divisions. Use centers can also be defined by type, such as routine information broadcasts, routine correspondence, file exchange, or machine-to-machine address resolution. The object of such analysis is to determine where the network resources are being spent, from which it becomes possible to ascertain the benefits of network use within the organization. In addition, there is an important link between use metrics and economical capacity planning. Heavy uses ought to justify their demands on the resources, and, if costs outweigh benefits, action should be taken to remedy the problem. A common form of action is imposition of constraints on resource use, such as control of access times and limitations on storage space or packet traffic. Alternatively, a pricing mechanism can be installed to make users pay for the resources they use, with the expectation that higher prices will drive down demand. The worst "solution" to differential demands for network resources is to simply let everyone use whatever they wish. Heavy users, irrespective of the value of their uses for the organizations, will push the limited capacity into congestion, causing all users — even the most efficient and cost-effective — to suffer. Accounting management and imposition of effective controls on resource use can be expected to grow as high-demand applications using technologies such as streaming video and audio become more common.

General Management

The contemporary network manager faces daunting challenges on organizational, economic, and technical fronts. The organizational challenges arise from the growing organizational dependence on networks and the rising costs of network failure. It is no longer possible for a network manager to get away with explaining network failures as normal events. Network managers should take a lesson from the local telephone companies, for whom the key operational goal was availability of dial tone at all telephones at all times. This is what users expect from communication networks, and they will demand it. In addition, networks are increasingly interdependendent. Failures in one network can propagate, causing problems in other networks. Thus, trouble in one network manager's domain can cause trouble elsewhere. The liabilities inherent in network management are growing as society's dependence on networks grows.

Network management is fraught with technical complexity and turbulence as a result of the rapid improvements in network technology. Technological improvement is a two-edged sword. On one hand, new capabilities enable new benefits. On the other hand, new capabilities stretch the management capacity of most organizations. Management skill must be obtained through a process of organizational learning by doing. This usually takes longer than the cycle time of rapidly advancing technologies. It is now common in the world of information technology for new products to arrive before organizations have figured out how to exploit previous products. One might think that the solution to this problem is for the organization to lag behind the technology curve in order to master the earlier technology, but this can have serious costs as well. Older products are "orphaned" by their manufacturers, top-notch employees resist staying with the old technology while their marketable skills erode, and interoperability with other networks becomes a problem as the managers of the other networks upgrade. The successful network manager must strike a balance between using proven and manageable technologies and living on the cutting edge of technological progress, all while serving the needs of organizational infrastructure.

Many network managers hold out hope that their problems of technical complexity and turbulence will be solved by the widespread adoption of industry standards. This is a false hope. Standards always sound good in principle, and most manufacturers proudly announce their willingness to work together to achieve standardization. In practice, however, standards are extremely difficult to establish in highly competitive, rapidly changing technical fields. For one thing, standard setting takes time, and it is not uncommon for a new product or protocol to emerge before the standard-setting process for whatever it replaces to be completed. When this happens, customers often flock to the new and abandon the old, making the whole standards process moot. More important, standards have grave competitive implications. Manufacturers who favor standards usually are found to favor standards that will help them and hurt their competitors. Their competitors do not agree and propose alternatives that turn the tables. The best indicator of how long it will take to achieve a major standard is how long it took to achieve the last major standard. In the information technology field, the only standards that emerge quickly are those that are embedded in products that achieve a high degree of popularity rapidly. These de facto standards are sometimes proprietary in the sense that a manufacturer owns key aspects of the process or surrounding technology. More often, the de facto standards are enforced simply because of the advantages of large market share and resulting network externalities. Standards are important, and will continue to be so, but standard-setting processes are no solution to the problems of turbulence faced by network managers.

Conclusion

Network management poses real challenges. Data communications networks are increasinly vital to organizations and are highly complex and difficult to manage. When the networks work properly, they are invisible and the network manager is hardly noticed. When the networks break down, the network manager is blamed. In addition to the challenges of managing data networks, many organizations are tasking network managers with responsibility for telephone networks and other organizational communications as these become integrated into the infrastructure of organizational operations and communication.

Defining Terms

ARPANET: The progenitor of the modern Internet. A network set up in the 1970s by the Department of Defense.

Asynchronous transfer mode (ATM):   High-speed switching and multiplexing technology using 53-byte cells (5-byte header, 48-byte payload) to transmit voice, video, and data traffic simultaneously and independently without use of a common clock.

Abstract syntax notation one (ASN.1):   OSI language for describing abstract network entities using macros built on simpler entities but not dependent on underlying hardware.

Bridge: Network device operating at the data link layer connecting LANs together.

Customer network management (CNM):   Provides access to management information regarding available services in a public network management system.

Common management information protocol (CMIP):   Application-level OSI protocol for network management.

Element management system (EMS):   Manages specific portions of the network such as asynchronous lines, PABXs, multiplexers, systems, or applications.

Fiber-distributed data interface (FDDI):   High-speed fiber optic lines used in LAN backbones and some WANs.

Gateway: Device that can perform protocol conversion from one protocol to another.

Integrated services digital network (ISDN):   Switched digital transmission service from local telephone carriers that uses copper telephone lines for home or small office applications; available in BRI (two 64-kb data channels, one signaling channel) or PRI (23 data/voice channels and one signaling channel).

Internet: Colloquial term used to describe the largest collection of internetworked subnets around the world.

Internetworking: Connection of subnetworks using devices in such a manner that the collection of networks acts like a single network.

Management information base (MIB):   The objects that are available in a managed system. The information is represented in abstract syntax notation 1 (ASN.1).

Manager of managers system (MoM):   Integrates information from element management systems, correlating alarms between them.

Open Systems Interconnect, OSI: A multilayer hierarchical scheme developed by the International Standards Organization (ISO) as a blueprint for computer networking.

Repeater: Physical layer network device that propagates and regenerated bits between network segments.

Router: Network layer device that can decide how to forward packets through a network.

Simple Network Management Protocol (SNMP):   Application-layer protocol for remote management of networked devices.

Synchronous Optical Network (SONET):   High-speed fiber-optic network protocol for WANs.

TCP/IP: Transport control protocol/internet protocol. TCP is a transport-layer protocol documented in RFC 793 that provides reliable transmission on IP networks. IP is a network layer protocol described in RFC 791 that contains addressing and control information by which packets are routed.

Trap: Unsolicited (device-initiated) message often used to report real-time alarm information.

References

Baker, R. H. *Networking the Enterprise: How to Build Client/Server Systems that Work,* McGraw-Hill, New York, 1994.

Ball, L. L. *Cost-Efficient Network Management,* McGraw-Hill, New York, 1992.

Halsall, F. Network technologies, In *Network and Distributed Systems Management,* M. Sloman, ed., pp. 17–45, Addison-Wesley, Reading, MA, 1994.

Janson, P. Security for management and management of security. In *Network and Distributed Systems Management,* M. Sloman, ed., pp. 403–430, Addison-Wesley, Reading, MA, 1994.

Kimmins, J., Dinkel, C., and Walters, D. *Telecommunications Security Guidelines for Telecommunications Management Network,* National Institute of Standards and Technology, Gaithersburg, MD, 1995.

Krishnan, I. and Zimmer, W., Eds. *Integrated Network Management, II,* North-Holland, Amsterdam, 1991.

Leinwand, A. and Fang, K. *Network Management: A Practical Perspective,* Addison-Wesley, Reading, MA, 1993.

Meandzija, B. and Wescott, J. *Integrated Network Management, I,* North-Holland, Amsterdam, 1989.

Moffett, J. D. Specification of management policies and discretionary access control, In *Network and Distributed Systems Management,* M. Sloman, Ed., pp. 455-480, Addison-Wesley, Reading, MA, 1994.

Muller, J. *Network Planning, Procurement, and Management,* McGraw-Hill, New York, 1996.

Stallings, W. Simple network management protocol, In *Network and Distributed Systems Management,* M. Sloman, Ed., pp. 165-196, Addison-Wesley, Reading, MA, 1994.

Taylor, E. *Multiplatform Network Management,* McGraw-Hill, New York, 1997.

Udupa, D. K. *Network Management Systems Essentials,* McGraw-Hill, New York, 1996.

On-line sources of information on the growth of network services:

Matthew Gray (http://www.mit.edu/people/mkgray/net/web-growth-summary.html)

Hauben and Hauben (http://www.columbia.edu/~hauben/netbook/)

Larry Landweber (ftp://ftp.cs.wisc.edu/connectivity_table/)

Mark Lottor (ftp://ftp.nw.com/pub/zone/)

Netcraft (http://www.netcraft.com/survey/)

Network Wizards (http://www.nw.com).

Further Information

Copies of USENIX Association System Administration conference proceedings:

USENIX Association
2560 Ninth Street, Suite 215
Berkeley, CA 94710

Information on Standards:

American National Standards Institute (for ANSI and OSI standards)
1430 Broadway
New York, NY 10018

IEEE Computer Society Press
Customer Service
10662 Los Vaqueros Circle
Los Alamitos, CA 90720-1264

Sources on the World Wide Web:

Alphabetical listing of network management products:

> http://www.micromuse.co.uk:80/

University of Illinois, Urbana–Champaign network management resource site:

> http://tampico.cso.uiuc.edu/~gressley/netmgmt/

State University of New York at Buffalo network management sites:

> http://netman.cit.buffalo.edu/index.html
> http://netman.cit.buffalo.edu/MailingLists.html
> http://netman.cit.buffalo.edu/Archives.html
> http://netman.cit.buffalo.edu/FAQs.html

11.4 Electronic Commerce

Lynda M. Applegate

> The total value of goods and services traded between companies over the internet will reach $8 billion this year and $327 billion in the year 2002.
>
> Blane Erwin et al., Forrester Resarch Report, July 1997.

Daily managers hear claims about how the **Internet** will radically change the way companies do business in the future. To many, it appears that the Internet has burst on the scene — a new technology that offers new capabilities. One of the most visible is the potential of the Internet as a channel for marketing and selling products and doing business with partners, suppliers and customers. Many consider this a revolutionary new idea. However, like the Internet itself, **electronic commerce** is actually more than 3 decades old.

In fact, a *Harvard Business Review* article, published in 1966, first described how managers could use information technology (IT) to share information and do business across organizational boundaries [Kaufman, 1966]. Even though the computer industry was in its infancy at the time, several forward-thinking firms were making the article's "visionary" predictions a reality. An entrepreneurial sales manager at American Hospital Supply Corporation, for instance, had created a system that allowed his company to exchange order-processing information with hospital purchasing departments across telephone lines using punch cards and primitive card-reading computers. Enterprising sales managers at American Airlines were also paving new ground by giving large travel agencies computer terminals that allowed them to check airline schedules on American's computers. Indeed, from these entrepreneurial actions grew two legendary strategic IT applications that changed the face of their respective industries. In doing so, they also ushered in the era of electronic commerce.

Today, many of the most dramatic and potentially powerful uses of IT involve networks that transcend company boundaries. These **interorganizational systems (IOS)** enable firms to share information electronically — often called **electronic data interchange** or **EDI** — and to streamline supply chains (procurement processes) and "buy" chains (distribution processes), decreasing costs and cycle times and improving performance for all participants. In some industries, electronic commerce has altered the balance of power in buyer-supplier relationships, raised barriers to entry and exit, and shifted the competitive position of industry participants.

This section discusses the trends and opportunities for electronic commerce. It looks back at the evolution of electronic commerce and its impact on industries and markets and the organizations that participated within them. It then uses the lessons from the past to frame the potential of electronic

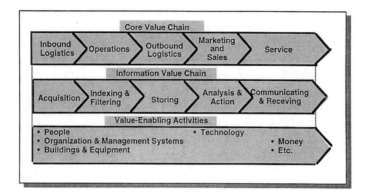

FIGURE 11.6 The value chain.

commerce on the Internet. The lessons discussed have been drawn from extensive study of the actions of over 100 firms within a broad range of industries — including financial services, retailing, manufacturing, consumer products, pharmaceuticals, oil and chemicals, and government. However, I draw on two well-known examples — American Hospital Supply and American Airlines.

Understanding Electronic Commerce: Looking Back

Most firms operate within a network of suppliers, producers, distributors, and retailers that work together to design, manufacture, market, and distribute products and services. These activities are often referred to as a **value chain** since they describe the steps through which the inputs from suppliers are transformed into outputs for customers that have an intrinsic market value [Porter, 1985]. Accompanying the physical value chain is a related information value chain through which involved parties coordinate and control activities (see Fig. 11.6).

As companies consider how to organize and manage the value chain, they are confronted with two fundamental questions:

1. Which activities should we keep *inside* the boundaries of our firm and which should we source from the *outside*?
2. How should we *interact* with our customers, suppliers, distributors, and others?

Where Should Activities Be Performed?

Value chain activities can be organized in three basic ways [Williamson, 1975, 1991; Powell, 1990], (see Table 11.1). First, activities can be incorporated within a single, "vertically integrated" firm. Second, one or more selective activities can be sourced from an external party. Finally, a "virtual corporation" can be created within which a firm retains only those activities that (1) are considered a "core competency" or (2) are required to manage, coordinate, and control value chain activities within and across organizational boundaries.

TABLE 11.1 Options for Structuring Market Activities

Options	Description
Vertical integration	Locate all but the most routine, transaction-oriented activities inside the firm
Selective sourcing	Source selected activities form the outside; traditionally, sourced activities were controlled through rigid contracts
Virtual corporation	Retain within firm boundaries only those activities that are (1) a core competency and (2) required to manage, coordinate, and control external value chain activities

Traditionally, managers have chosen to locate an activity within their organizational boundary (i.e., vertically integrate) when a significant cost (or risk) was involved in managing it on the outside. Costs and risks increase when (1) a firm is required to make a significant investment in physical facilities, people, or management systems to coordinate and control activities with outside suppliers, distributors, or other parties, (2) the services or activities are critical to the effective and efficient delivery of the firm's products and services, or (3) a high degree of uncertainty surrounds the ongoing nature of the relationship, which, in turn, makes it difficult to define the interorganizational relationship in a comprehensive, structured contract and/or to create an efficient set of interorganizational governance mechanisms.

Eastman Kodak, for example, was highly vertically integrated even into the 1980s. Founded in the late 1800s on the principle of "commitment to quality", the firm established its own laundry to ensure that cloth for wiping film was of the highest quality; it bought its own blacksmith and built its own foundry to make its machines; and it managed its own credit union to finance its products. Having defined their strategy around quality, Kodak managers believed that the costs and risks associated with managing these activities on the outside were greater than the benefits that would be achieved through sourcing. General Motors followed a similar strategy in the early and mid-1900s; many recall its familiar slogan — "Genuine GM Parts".

In the 1970s and early 1980s, visionary firms within a number of industries found that they could use IT to help manage the risks and cost of interorganizational coordination and control [Malone et al., 1987]. By establishing electronic linkages with suppliers, distributors, and even directly with customers, they were able to integrate and coordinate interorganizational activities much more efficiently, to monitor operations more closely, and to communicate with external organizations more interactively.

How Should We Relate to Internal Parties?

All firms also make choices about the nature of the relationships that they develop with customers, suppliers, and other external industry participants. These choices fall along a continuum from transactions to contracts to partnerships [Granovetter, 19985; Stinchcombe, 1985; Bradach and Eccles, 1989] (see Table 11.2).

Transactions involve the simple exchange of goods, services, and payments, usually during a specific time period and with limited interaction or information sharing between the parties involved.

In contractual relationships, the products or services to be provided by each party, and the length of the relationship, are well defined at the start of the relationship and are clearly documented. The formal "terms of the contract" become the basis for coordinating and controlling the exchange of goods, services, payments, and information throughout the length of the contract.

Partnerships are required when the activities to be jointly managed are complex, uncertain, and critical to the success of the firms involved. Partnerships require shared goals, complementary expertise and skills, and integration of activities across organizational boundaries. The exchange of goods and services is ongoing, and the interactions and relationships must adapt to the changing priorities of the parties involved. Partnerships often require significant investments in systems and people for carrying out, coordinating, and controlling shared activities.

As mentioned above, the cost and risk of coordinating and controlling all but the most routine transactions and well-structured contracts traditionally led managers to locate most activities within their organizational boundary [Williamson and Winter, 1993]. Exceptions were at times made to enable a firm to gain valuable expertise or resources that could not be developed inside the firm. In these cases, managers balanced the increased cost of interorganizational coordination and control with the benefits to be gained from sourcing the expertise from the outside.

Two Examples from History

With this in mind, it comes as no surprise that, in the 1960s, as firms began to install their first computers, they focused their efforts on improving coordination and control of activities inside the firm. This was

TABLE 11.2   Relationship Options

	Transaction	Contract	Partnership
Basis of interaction	Discrete exchange of goods, services, and payments (simple buyer/seller exchange)	Prior agreement governs exchange (e.g., service contract, lease, purchase agreement)	Shared goals and processes for achieving them (e.g., collaborative product development)
Duration of interaction	Immediate	Usually short-term; defined by contract	Usually long term; defined by relationship
Level of business integration	Low	Low to moderate	High
Corodination and control	Supply and demand (market)	Trems of contract define procedures, monitoring, and reporting	Interorganizational structures, processes, and systems; mutual adjustment
Information flow	Primarily one way; limited in scope and amount; low level of customization	One or two way; scope and amount are usually defined int he contract	Two-way (interactive); extensive exchange of rich, detailed information; dynamically changing; customizable

true for both American Hospital Supply, Inc. (AHSC) and American Airlines (AA). AHSC, for example, first installed computers to manage internal inventory and order-processing activities and AA used computers to manage their internal reservation process. In both cases, the value of these early systems came from the ability to structure, simplify, and coordinate internal operations. However, once they had simplified and structured these activities inside the firm, both AHSC and AA recognized that they could now outsource them, managing the relationships with outside parties as simple transactions and structured contracts. Thus, AHSC's internal inventory and order-processing systems served as the foundation for their legendary ASAP electronic commerce purchasing system, and AA's internal reservations systems became the foundation for SABRE.

Initially, AHSC gave hospitals the card readers that would be required to do business electronically and taught them how to use them. It even helped hospital purchasing clerks redesign their internal purchasing processes to fit with the new on-line process. AA did the same thing when they gave travel agents the SABRE system reservation terminals. Neither AHSC nor AA charged their customers for the computer equipment or the training. Why? The benefits to AHSC and AA from on-line purchasing, whether it was hospital supplies or seats on an airplane, more than offset the cost of giving away the terminals. For example, by 1985, AHSC saved over $11 million per year through on-line ordering and generated $4 to $5 million per year in additional revenue. More importantly, however, because AHSC and AA had created the system, they could coordinate and control both the transactions *and the information.*

Once AHSC and AA had succeeded in getting a large number of customers in their respective industries to do business electronically, other industry suppliers began to feel the pressure to participate. Sensing they were at risk from being excluded from the market and lacking the money, expertise, and time to respond, other industry participants in both the hospital supply and airline industries demanded that their catalogs and reservations be included in the AHSC and AA systems. Customers, having also achieved significant savings and benefits from doing business electronically, encouraged the channel consolidation; they recognized the value of a multivendor marketplace but were unwilling to put up with the problems of doing business using multiple different supplier systems. Within a short time, both AHSC and AA became powerful channel managers within their industries, controlling both the physical and information channels for conducting business.

Over the next few years, participation in these two electronic communities expanded; as participation increased so too did the range of services offered and the value to customers. This led to further consolidation of the industry — this time across traditional industry boundaries. For example, in 1987, AHSC, which by this time had been purchased by Baxter Healthcare, launched ASAP Express. This expanded ASAP to include suppliers of hospital equipment and office furniture. In 1994, with the launch of OnCall, additional suppliers, such as Eastman Kodak (a supplier of advanced imaging systems and

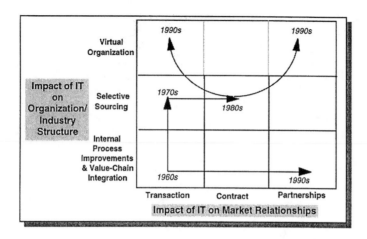

FIGURE 11.7   Impact of IT on the evolution of industry structure and market relationships.

supplies) and Boise Cascade (a supplier of paper products), joined the community. AA followed a similar strategy, negotiating agreements with others in the travel industry, including hotels and car rental agencies. Figure 11.7 provides an overview of the pattern of evolution of electronic commerce at AHSC and AA. This same pattern was identified in other industries, including financial services, package distribution, and pharmaceuticals.

The early pioneers in the history of electronic commerce blazed new trails. They created the networks, developed software to exchange data with outside parties, defined network standards, and set the rules. The systems were costly to develop but provided tremendous financial and technological barriers to entry. Ownership of these systems enabled firms such as AHSC and AA to assume the powerful role of channel manager, controlling both the information flows and business transactions among buyers and suppliers.

However, over time, the advantage these firms once gained from heavy investments in proprietary systems eroded. Even before the emergence of the Internet as a channel for electronic commerce, *reliable, low-cost, third-party on-line services and network facilitators had become viable alternatives to proprietary networks.* The interorganizational systems that once provided tremendous proprietary advantage were rapidly becoming viewed as commodities.

Learning from the Past and Predicting the Future

The low cost, flexibility, shared ownership, and global connectivity of the Internet dramatically expands the opportunities for electronic commerce. Yet, proprietary advantages from business ventures on the Internet appear fleeting. The old model for evaluating electronic commerce opportunities no longer applies, and new rules are yet to be written. As managers venture forward, the lessons from the history of electronic commerce can help guide the way.

Lesson # 1: Build Upon Internal Capabilities

Both AHSC and AA were successful in their electronic commerce ventures because they built their linkages to outside suppliers, distributors, and customers as extensions of systems that coordinated and managed activities inside their companies. The same technologies that were used to lower the cost and improve the ability to integrate, coordinate, and control operations inside the firm were expanded to enable integration, coordination, and control of activities across firm boundaries. This enabled AHSC and AA to oursource — and effectively manage — increasingly more complex activities and processes.

This also holds true when considering electronic commerce on the Internet. While the economic benefits from marketing and selling on the Internet remain elusive, there are sizable opportunities to be gained by creating intranets — internal systems developed and deployed on the Internet that enable efficient communication, coordination, and integration of global operations. These internal intranets can then become the foundation for electronic linkages with suppliers, partners, and customers in the future.

By targeting intranet projects to improve "high leverage-high leakage" information-intensive processes and/or to support key decisions at the interface with customers, suppliers, or business partners, managers can position their company for electronic commerce in the future. General Electric (GE), for example, laid the foundation for their successful internet-based Trading Process Network by redesigning their internal sourcing processes. In 1994, the year they started the intranet process redesign project, sourcing represented approximately $25 billion dollars per year in cost for GE. Within the first year of implementation of its redesigned internet-enabled sourcing process, GE had reduced the cost of sourced components by 25%, decreased cycle time by 2 weeks, and saved 100% of the costs of printing and mailing.

Cadence Design Systems developed their initial intranet systems to support redesign of their internal sales process. By providing Internet-based information and support tools for prospecting, selling, and customer service, the company was able to reduce the time it took for new salespeople to reach quota from 4 months to 2 months, generating over $600,000 per year in additional revenues.

Lesson #2: Penetrate Quickly

AHSC and AA achieved a dominant position within their industries through a strategy of rapid penetration and value creation for all members of the community. By giving away the terminals and software required to do business electronically, AHSC and AA created strong linkages to customers in the industry. They trained customers to use the technology and even helped them redesign their processes to ensure that all parties achieved maximum value. Once hospital customers were using AHSC systems, other hospital suppliers sought to join the community. The same pattern was evident in the travel industry. (Note: In the travel industry several large airlines — e.g., United — were fast followers. Thus, in the travel industry, we saw the emergence of several competing electronic markets.) As the new electronic markets evolved, membership in the community expanded, and the value to all participants increased. The strategic necessity of belonging to the community also increased.

Once more, we see history repeating itself on the Internet. Recall how Netscape and Microsoft gave away their browser software to ensure that they could control access to future Internet markets. Today, we see a number of companies — from traditional IT vendors (e.g., IBM, Hewlett-Packard, Microsoft, Oracle, and SAP) to communications vendors (e.g., Cisco, Lucent Technologies), to network services providers (e.g., AT&T, USWest, and MediaOne) to broadcasting companies (e.g., NBC and BBC) to content providers (e.g., Time Warner and Disney), to entertainment companies (e.g., Sega and Nintendo) — jockeying to set the standards for both the flow of information within Internet-based electronic markets and the software that will be used to conduct business over it. These players hope to be in a position to define the standards and rules that will govern Internet-based electronic commerce. If successful, they would be in a position to exert significant control over global electronic markets, which, in turn, could result in consolidation of power within the hands of several large players.

However, as these wars are being waged over the platform, managers are struggling to evaluate the opportunities for doing business on that channel. As they do, it quickly becomes clear that the advantages are elusive. Many believe that the low-cost, ease of entry, and shared control of the Internet will make it exceedingly difficult to establish a sustainable position in an Internet electronic market, which, in turn, would lead to rapid commoditization of any products and services offered on the Internet. However, another scenario also bears watching.

Because the Internet is built on "open standards" and no one person controls the platform, we are seeing a much earlier role for neutral, third-party electronic market facilitators. In fact, the emergence of any new market has been accompanied by an increased need for information brokers capable of simplifying and structuring the flow of information and business transactions. Recall the important role that travel agents played in the period of uncertainty and rapid change following the deregulation of the airlines in the late 1970s and early 1980s. In fact, the airlines themselves supported the development of these market facilitators to rapidly increase market demand for air travel.

The Internet is but one more example of a highly uncertain and rapidly changing market. As such, there is money to be made for those who are able to identify an opportunity to simplify and structure the flow of information and transactions between buyers and sellers on the Internet. On the Internet, however, these information brokering opportunities are fleeting. The low cost of participation and the ease with which others can respond dramatically shortens the window within which value can be extracted. To capture the "deal-making" flavor, I think of these brokering opportunities as a form of *information arbitrage*. But, the value that can be extracted from pure brokering is fleeting. Once positioned in the middle of the channel, a market facilitator must *create additional value by exploiting the economic value of information flowing through the channel* and by *leveraging the resources of the community itself.*

Lesson #3: Exploit the Economic Value of Information and Leverage the Community

The word "information" comes from a Latin word "infomare", which means to "put into form". Thus, at its most basic level, information can be used to structure and simplify what was previously complex. Both AHSC and AA recognized and capitalized on this inherent property of information to create tremendous value inside their organization and in the way they did business with customers, suppliers, and business partners. Within both the airlines and hospital supply industries, AHSC and AA began by using IT to automate and integrate value chain activities, but, once they had automated the processes, they also gained control of valuable information about each and every transaction that flowed down the information value chain — information that had tremendous economic value in its own right.

Since AHSC and AA owned the networks and the software that controlled the transactions, they also controlled the information. Both AHSC and AA exploited this information to increase internal expertise and to leverage that expertise in the marketplace. They used the information captured through direct links to their customers to improve the efficiency and effectiveness of internal operations, to develop a more detailed and timely understanding of market dynamics, and to fine tune their product line to continually drive customer expectations to higher and higher levels that competitors could not match. The information was also used to create new information-based products and services. By the late 1980s, both AHSC and AA had spun off new divisions charged with developing, marketing, and distributing information-based products and services.

As managers consider Internet-based electronic commerce opportunities, it is critical to understand how to exploit the full economic value of digital information and leverage the community (see Table 11.3). Once information is available in digital form, it can be packaged and delivered to increase organizational intelligence, to create new products and services, and to add value to existing ones. These information-based products and services possess some very interesting properties. First, they are reuseable. Unlike physical products, information can be "sold" without transferring ownership and "used" without being consumed. As one Internet information merchant observed: "I sell information to you, and now you own it. Yet I still own it, and we both can use it." Second, they are easily customized. The same information can be presented in different forms (e.g., text, graphics, video, and audio) and in varying levels of detail. It can be combined with information from other sources to communicate different messages and to create new products and services. Third, information-based products and services possess an inherent "time value". As the speed of business accelerates, the time value of information increases.

The ease with which information can be accessed, stored, packaged, and delivered on the Internet greatly expands its potential economic value. Companies that sell information-based products have been quick to reposition their traditional products and services over the Internet — an exceptionally low-cost channel for reaching new markets with a more-customizable product. Dow Jones, Inc., publisher of *The Wall Street Journal* (WSJ), for example, has exploited the power of digital information to transform its traditional newspaper for the Internet market. Not only is the cost of delivering the product over the Internet significantly lower than delivering a physical newspaper to a customer's front door each morning, but the value of the WSJ to its subscribers and advertisers has significantly increased. By combining the interactive, information-rich nature of the Internet with up-to-date news stories, full-text searchable business profiles, and customizable personal news features, Dow Jones has not only transformed its

TABLE 11.3 Options for Leveraging Information and Community

Connections
- Access to specialized experts
- Business partner identification and matching services
- Information searching and retrieval
- Network access, messaging, and routing services based on Internet protocols enable distributed information sharing and process coordination and control among buyers, sellers, and other market participants (24 hours per day/7 days per week operations)
- Internet e-mail services
- Security services (e.g., encryption, digital signatures, and authorizations)
- Directory services

Content
- Publishing services and systems
- Electronic catalogs
- On-line marketing programs
- On-line news and information services
- Analytical tools
- On-line training programs and certification programs
- Web server hosting and management
- Content hosting and presentation services

Commerce
- Supply chain management and services
- Distribution chain management and services
- Development of new channels
- Buying and selling products and services
- On-line auctions
- Contracting services
- Shared design services and systems (including computer-aided design/computer-aided manufacturing)
- Interorganizational control systems (including service-level agreements and performance measurement, transaction control systems, and workflow coordination)

---

existing product, it has paved the way for the company to assume a major role as an Internet market facilitator at the center of an expanded community of information suppliers and consumers. Its success will depend on the company's ability to penetrate quickly, leverage the community, and fully exploit the economic value of information throughout the community.

Summary

In approaching electronic commerce on the Internet, many managers want to position their firms as "leading edge", but not "bleeding edge". A thorough assessment of opportunities requires starting with the basics — a deep understanding of the value chain of activities that link your firm with suppliers, customers, and business partners. A deep understanding of your firm's and your competitors' current strategy and core capabilities is also required. What choices has your firm made about the activities and resources retained within organizational boundaries vs. those that are sourced from external parties? What types of relationship does your company cultivate with customers and business partners? Should the flexibility, low cost, and global reach of the Internet cause you to reexamine the activities that you perform and the relationships that you form? A deep understanding of today's business must be married with a deep understanding of the Internet can add value to the current business or enable the firm to create new businesses.

The following questions can be used as general managers contemplate opportunities for electronic commerce on the Internet.

1. Are you harnessing the power of the information embedded in your products to add value to on-line customers? Could new information-enabled goods or services enhance or replace your current offerings? If so, can — and should — these new offerings lead to a new basis of interaction with your customers?

2. Are your capturing the full potential of value chain integration? Could you eliminate intermediaries, simplify and streamline product or service delivery, dramatically improve quality, or decrease costs? Can your firm become a specialized expert, and, if so, can you effectively coordinate and control value chain activities with other market participants to ensure high quality and customer satisfaction?

3. Are you at risk for disintermediation? Can you preempt potential threats by differentiating your position as an Internet market facilitator? Are your products and services at risk for potential commoditization and/or obsolescence? What actions should you take to prevent erosion of market share and price?

4. Have you selected your partners wisely? Do you have a shared vision? Do you bring complementary power and resources (particularly each partner's specialized expertise) to the relationship? Is the relationship financially and competitively sustainable? Have you and your partners jointly redesigned business processes and put appropriate coordination and control systems in place?

5. Is your firm's technical infrastructure appropriate for the new types of electronic commerce you are considering? Do you maintain an appropriate balance between experimentation and control? Have you instituted appropriate levels of security and reliability? Are your systems flexible enough to respond to dramatic changes in capacity requirements and service offerings?

Many managers underestimate the Internet's far-reaching potential to rewrite the rules of electronic commerce. Whether it emerges as the foundation for global electronic commerce or is replaced by a more-powerful information and communication platform, there is much to be learned by keeping abreast of new developments, building organizational capabilities, and launching new business ventures. However, as you look to the future, don't abandon the past.

Defining Terms

Internet: Developed in the 1960s and 1970s by the U.S. Department of Defense, the Internet is a set of standards that define how computers package and distribute information over networks. By 1997, these Internet standards had been accepted by vendors, companies, public institutions, and government agencies around the world, thus enabling the development of a ubiquitous, global, open platform for information management and communication.

Electronic commerce: The use of computer networks and systems to exchange information, products, and services with buyers, sellers, and other industry participants and business partners.

Interorganizational systems (IOS):   A broad category of information system that refers to the use of networked computers to enable companies to share information and information processing across organizational boundaries.

Electronic data interchange (EDI):   A form of interorganizational system in which information is processed within a firm's boundaries and is then shared electronically. Although the two terms are often used interchangeably, in its pure form, EDI does not require process integration across firm boundaries.

Value chain: The series of interdependent activities that bring a product or service to customers. These include (1) "core" value activities that directly contribute to designing, building, marketing, selling, and maintaining a product or service and (2) "support" activities that enable core activities to take place (e.g., human resource management, facilities operations, finance, and activities related to management, coordination, and control).

References

Beniger, J. *The Control Revolution,* Harvard University Press, Cambridge, MA, 1986.

Bradach, J. and Eccles, R. Price, authority and trust, *Annu. Rev. Sociol.,* 15: 97–118, 1989.

Granovetter, M. Economic action and social structure: the problem of embeddedness, *Am. J. Sociol.,* 91: 481–510, 1985.

Kaufman, F. Data systems that cross company boundaries, *Harv. Bus. Rev.,* February, 1966.

Konsynski, B. Strategic control in the extended enterprise, *IBM Syst. J.,* 32(1), 112–130, 1993.

Malone, T., Yates, J., and Benjamin, R. Electronic markets and electronic hierarchies, *Commun. ACM,* June, 484–497, June 1987.

Porter, M. *Competitive Advantage: Creating and Sustaining Superior Performance,* Free Press, New York, 1985.

Powell, W. Neither market nor hierarchy: network forms of organization, *Res. Organ. Behav.,* 12: 295–336, 1990.

Stinchcombe, A. *Contracts as Hierarchical Documents, Organization Theory and Project Management,* Stinchcombe and Heimer, Eds., pp. 121–71, Norwegian University Press, Bergen, Norway.

Williamson, O. *Markets and Hierarchies,* Free Press, New York, 1975.

Williamson, O. Comparative economic organization: the analysis of discrete structural alternatives, *Admin. Sci. Q.,* 36: 269–296, 1991.

Williamson, O. E. and Winter, S. G. *The Nature of the Firm,* Oxford Press, New York, 1993.

Further Information

Hagel, J., III and Armstrong, A. *Net Gain, Expanding Markets through Virtual Communities,* Harvard Business School Press, Boston, 1997.

Kalakota, R. and Whinston, A. *Frontiers of Electronic Commerce,* Addison-Wesley, New York, 1996.

Tapscott, D. *The Digital Economy, Promise and Peril in the Age of Networked Intelligence,* McGraw-Hill, New York, 1996.

| Genentech | Headquarters — San Francisco, 94080 | Web Site — http://www.gene.com |

Genentech, Inc. was founded in 1976 by venture capitalist Robert A. Swanson and biochemist Dr. Herbert W. Boyer. In the early 1970s, Boyer and geneticist Stanley Cohen pioneered a new scientific field called recombinant DNA technology.

Growth in revenue

1980	$9.00 million
1985	$89.5 million
1990	$476 million
1995	$917 million
1996	$968 million

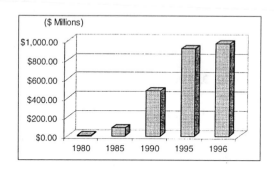

Innovations:

1977 — The first human protein

1978 — Human insulin cloned by Genentech scientists

1984 — First laboratory production of Factor VIII, a clotting factor for bleeding in hemo-philiacs

1993 — Approval to market Nutropin®

1994 — Approval to market Pulmozyme for treating CF

1995 — Clearance to market an accelerated infusion regimen of Activase

1996 — Regulatory clearance to market Nutropin AQ

Social improvements:

1986 — Genentech instituted the Uninsured Patients Program, providing free human growth hormone for financially needy, uninsured patients in the U.S.

1989 — Genentech opened its day-care center, Genentech's Second Generation, one of the largest corporate-sponsored day-care centers in the U.S.

Strategic and Management Issues:

- The company's strategy is to utilize biotechnology to produce unique and useful pharmaceutical products to provide treatments for currently untreatable diseases, or to improve upon existing, but less adequate, treatments. They have a clear four-point strategy in place which is based on a realistic assessment of the current competitive environment and real assets:

  1. Maximize sales of marketed products.
  2. Accelerate and expand product development.
  3. Increase the pace of forming strategic alliances.
  4. Improve financial returns.

IV

Managing the Business Function

12

Marketing of Technical Products

Russell S. Winer
University of California

Allen M. Weiss
University of Southern California

Robert J. Dolan
Harvard Business School

Mary Lou Roberts
University of Massachusetts

Richard C. Dorf
University of California, Davis

Eitan Gerstner
University of California

Derby A. Swanson
Applied Marketing Science, Inc.

John H. Mather
Carnegie Mellon University

Stephen M. Millett
Battelle

Timothy L. Smunt
Wake Forest University

Prasad Naik
University of California

Peter LaPlaca
The University of Connecticut

Kevin Lane Keller
Dartmouth College

Paul E. Green
University of Pennsylvania

Jerry Wind
University of Pennsylvania

Vithala R. Rao
Cornell University

0-8493-8577-6/99/$0.00+$.50
© 1999 by CRC Press LLC

Barbara E. Kahn
University of Pennsylvania

Cynthia Huffman
University of Pennsylvania

David P. Paul, III
Monmouth University

Earl D. Honeycutt, Jr.
Old Dominion University

Donald R. Lehmann
Columbia University

12.1   Situation Analysis

Russell S. Winer

A fundamental component of the marketing of technical products is the **marketing plan**. The marketing plan is a written document in which the manager analyzes the market situation and develops objectives, strategies, and programs for the product over a planning period. The marketing plan has a number of important benefits for the company (see Lehmann and Winer [1997] for more details).

An outline of a typical marketing plan is shown in Fig. 12.1. A key component of the plan is the background assessment, which provides a historical perspective on the product and the market, an evaluation of the current situation, and planning assumptions such as sales forecasts. Taken together, the three parts of the background assessment help the manager to determine what objectives and strategies are feasible and competitive at the point in time the plan is being prepared.

Perhaps the most important component of the background assessment is the situation analysis. The main parts of a situation analysis are (1) defining the product category in which the product competes, (2) analyzing the category's attractiveness, (3) a customer analysis, (4) a competitor analysis, and (5) a resource analysis. Descriptions of how to perform these analyses is the topic of this section.

I. Executive summary
II. Background assessment
  A. Historical appraisal
  B. Situation analysis
  C. Planning assumptions
III. Marketing objectives
IV. Marketing strategy
V. Marketing programs
VI. Financial documents
VII. Monitors and controls
VIII. Contingency plans

FIGURE 12.1 Outline of a typical marketing plan.

Defining the Product Category

It is critical that the manager have a good conceptual definition of the set of competitors against which the product is competing. We will refer to this set of products as the product category.

There are four different levels at which competition can be defined. Using plain paper fax machines as the focal product, competition can be defined at four levels of generality:

1. Product form competition: Other brands of plain paper fax machines. These are the products that are the most similar to the focal product in that they have the same features.
2. Product category competition: All fax machines, including plain paper and thermal paper machines would be included in this set of competitors. This would be the usual definition of an "industry" or "product category".
3. Generic competition: These competitors include all products and services that satisfy the same customer need or benefit; in reality, customers determine competitors, not managers. Thus, document-delivery services (e.g., Federal Express or DHL) and shipping files through the Internet would compete against brands of fax machines.
4. Budget competition: These competitors would include other products and services competing for the same part of the customer budget. Assuming that fax machines are purchased from a company's capital budget, many other products such as copying machines, personal computers, etc. might compete for the same customer dollar.

The output from this analysis is a set of competitors that we refer to as the product category. Note that this definition of the category may be broader than the normal conception as it includes all products and services that the manager feels competes against the focal product at a given point in time. In most circumstances, the manager will use either a product form or usual product category definition for operational purposes as the other definitions often result in a large number of competitors, which are difficult to process simultaneously in a plan. However, particularly for new products based on innovative technologies, the most important competitors are usually generic since the new product is substituting for an older technology satisfying a particular need.

Category Analysis

Given the definition of the category, the next step in the situation analysis is to assess its attractiveness. An outline of this analysis is shown in Fig. 12.2. The objective is to evaluate the category on the different dimensions shown in Fig. 12.2 and then draw implications of the analysis for the focal product.

Aggregate Factors

The factors include overall statistics describing the product category. Category size is a static measure of the sales revenue potential. Growth is a more dynamic indicator of the future potential. Taken together,

Aggregate category factors
    Category size
    Category growth
    Stage in the product life cycle
    Sales cyclicity
    Profits
Category factors
    Threat of new entrants
    Bargaining power of buyers
    Bargaining power of suppliers
    Current category rivalry
    Pressure from substitutes
    Category capacity
Environmental factors
    Technological
    Political
    Economic
    Regulatory
    Social

FIGURE 12.2   Category attractiveness analysis.

sales and growth provide managers with information about the stage of the **product life cycle**. Sales cyclicity is an indicator of the variability of category sales over time. Seasonality is an intrayear indicator of such variability. Profitability is an obvious indicator of attractiveness. Overall, size, growth, and profitability are positively related to category attractiveness while cyclicity and seasonality are negatively related.

Category Factors

The factors in this category are largely related to Porter's [1980] analysis of industry structure. Threat of new entrants is related to the category's barriers to entry and threat of retaliation by extant competitors. Examples of barriers to entry are marketing costs (advertising and distribution), economies of scale, technology, customer switching costs, etc. The likelihood of new entrants is negatively related to category attractiveness. The bargaining power of buyers refers to any leverage that any "downstream" organizations have in the category. Greater buyer power can result from undifferentiated products from sellers (i.e., the focal product category) when the product is a large proportion of the buyer's costs, when the buyer can backward integrate, and when substitutes exist for the seller's product. Bargaining power of suppliers is "upstream" leverage. This can result when suppliers are highly concentrated, when supply is limited, when there are no or few substitutes for the product being supplied, and when the suppliers have differentiated their products well. Both kinds of power, suppliers and buyers, are negatively related to category attractiveness. Category rivalry is also negatively related to attractiveness. Rivalry or intense competition between market participants normally results in reduced profits and, on occasion, aberrations in behavior such as personal attacks on managers. Rivalry is most likely to occur in categories where there are balanced competitors, i.e., several with equal resources, when the market is growing slowly, when fixed costs are high, and when there is little product differentiation. The threat of substitutes is directly related to the generic competition described in the section on defining competition. Finally, excess category capacity, particularly for manufactured goods, tends to lead to the need for increased sales volume and often lower prices producing lower profits.

Environmental Factors

A product category is also sensitive to changes in factors outside the control of the category, what are referred to as environmental factors. Technology can render a product category obsolete; a high threat

1. Who are they?
2. What do they buy?
3. How do they choose?
4. Where do they buy it?
5. When do they buy it?

FIGURE 12.3   Customer analysis.

of technological obsolescence makes a category unattractive. A category can also be affected by political factors, particularly if it has extensive sales in global markets. Economic factors such as interest or exchange rate fluctuations and regulations can also significantly affect a category's performance. Finally, some categories are particularly vulnerable to social changes such as the birth rate, aging of the population, health consciousness, etc.

In general, the manager must not only note which factors most affect the focal category but what can be learned and used from the analysis to improve the attractiveness from the product's perspective. For example, if the power of suppliers is strong, one way to reduce it is to write long-term supply contracts.

Customer Analysis

The objective of having a customer focus for both the organization and the marketing strategy has been a fundamental concept of marketing since the 1950s. The purpose of the customer analysis part of the situation analysis is to ensure that the marketing plan has the customer orientation necessary to develop strategies that account for customer behavior. This is particularly important in technology-based product categories where customer needs can change rapidly as the technology evolves and the products become more familiar.

As shown in Fig. 12.3, in the customer analysis, the manager attempts to answer a series of questions about the customers.

Who Are the Customers?

A basic question to ask is who are the customers buying (and not buying) the product? This is the job of finding ways to divide the marketing into homogenous **market segments** or subgroups. The manager's task is to collect marketing research information, either secondary (data that have already been collected for another or more general purpose) or primary (data that are collected specially for the analysis at hand, usually survey based), that describes the behavior of different groups of customers toward the focal product (the analysis can be done at either the brand or product category level).

For both consumer and industrial technology products, the kinds of variables used to segment markets divide into two major groups: descriptive and behavioral. As the label implies, the former variables basically characterize the customers using demographic/socioeconomic variables (consumer: age, income, occupation, gender, and geographic location; industrial: company size, location, and industry) or psychological variables (consumer: lifestyle and personality; industrial: degree of risk taking). Behavioral variables describe how the customer group has acted toward the brand or product category. For both consumer and industrial markets, these include purchase quantity (heavy, medium, light, or none), degree of loyalty, benefits obtained from the product, and others. The job the manager faces is developing a relationship between the descriptors and the behavioral variables. In other words, the manager has to establish how different groups as defined by descriptor variables behave toward the product. For fax machines, it is insufficient to divide the population of companies into groups by size; what has to be done is to link the groups to how many they buy, of what type, etc.

Besides developing segments by linking descriptor to behavioral variables, the "who?" question can also be addressed to the organization making the purchase. In any organization, whether a family decision-making unit or a company, there are usually a number of people involved with the purchase decision. Of primary importance is that the people in the different buying roles can have different needs and wants and act as "micro" segments within the buying unit. These people can be divided into the following types: the initiator(s) of the purchase, the influencer(s) of the purchase, the decision maker(s), the purchaser(s), and the user(s). In some cases, the same person or people may fill one role. However, in many situations, different people play the various roles.

What Do Customers Buy?

In this part of the analysis, it is critical to remember that customers purchase products to obtain the benefits and services from those products. Thus, the manager must understand, for each segment, the benefits that the customers hope to obtain from purchasing the product.

How Do Customers Buy?

While there are many models of both consumer and industrial buying behavior examining how brand choices are made within a product category, a commonly used framework for understanding how customers make purchasing decisions is the **multiattribute model**. This model has four components. First, the manager needs to determine the attributes/benefits that the customers are using to define and evaluate the product. Second, customers are asked to provide their perceptions of the different brands in the product categories on the attributes/benefits previously determined. This is important; perceptions drive purchase decisions, not reality or actual values of the attributes. Third, customers have **importance weights** for each of the attributes in terms of how critical that attribute is in the overall evaluation of a product. Fourth, by combining the attribute importance weights and the brand perceptions on each attribute, the manager can develop an overall preference measure for each customer for each brand.

Where Do They Buy?

In order to decide which channels of distribution to use, it is important to understand where customers buy the product or how they gain access to it. For example, a small business could purchase a fax machine at a retailer such as Home Depot, a consumer electronics store such as Circuit City, from an office supplies catalog, or directly from the manufacturer.

When Do They Buy?

When do they buy encompasses time of year, time of month, and even, potentially, time of day.

Competitor Analysis

While a customer focus is important, with the rapid change in the competitive environment due to joint ventures, increase globalization of business, and other trends, having a competitor focus is becoming equally important. A marketing strategy that ultimately satisfies customer needs may not be competitive, that is, may not be sufficiently different or better than competition to be successful. The competitor analysis part of the situation analysis helps to ensure that the strategy satisfies both a customer and competitor focus.

Competitor analysis involves two major activities: data collection and analysis. The former can be done relatively easily today using the extensive amount of information available from computer information services such as Compuserve and from the Internet. Figure 12.4 shows how the information is used to develop a picture of the competitors in the product category.

Assessing the Competitors' Current Objectives

An assessment of current objectives provides valuable information concerning the intended aggressiveness of the competitors in the market in the future. If a competitor is pursuing a market share growth objective, the manager of the product is likely to spend more money on activities that can improve sales (advertising, promotion, and sales force) and be competitive with pricing. Alternatively, a competitor pursuing profits would take an opposite approach to the market.

1. What are their objectives?
2. What are their current strategies?
3. What are their capabilities?
4. What are they likely to do in the future?

FIGURE 12.4   Competitor analysis.

Assessing the Competitors' Current Strategies

The manager needs to understand the current strategies employed by the competitor products including their target markets, how they are **positioning** the product against proposition, and what the key **value proposition** is being delivered to customers. In addition, the manager needs to know the **marketing mix** for each product, which includes the price(s) being charged, distribution channels, promotions and communications, and product features.

Assessing Competitors' Capabilities

A third step in the competitor analysis process is to evaluate the competitors' capabilities. This can be done by comparing the competitors on several dimensions: their abilities to conceive and design new products; finance, produce, or deliver the service; market; and manage.

Predicting Future Strategies

Finally, the manager must put all of the prior information together and attempt to forecast what the competitors' likely strategies are in the future. This is necessary to ensure that the strategy developed in the marketing plan is forward looking and accounts for where the competitors are headed rather than where they have been.

Resource Analysis

This part of the situation analysis compares the focal product and the company to the competitors. An overall assessment of comparison would focus on strengths and weaknesses, in particular, relative to those factors necessary for success in the market.

Defining Terms

Importance weights: How customers view the relative importance of product attributes.

Marketing mix: Specific marketing decisions about price, promotion/communications, channels of distribution, and product features.

Marketing plan: A written document detailing the product situation, objectives, and strategies for the planning period.

Market segments: Homogenous groups of customers who differ from other groups in terms of their behavior.

Multiattribute model: A model of customer decision making based on decomposing a product into its attributes.

Positioning: The image that a manager wants to project for the product relative to competition to a segment.

Product life cycle: How the sales and marketing strategies for a product category change from the introductory period to the decline period.

Value proposition: A statement that summarizes the target customer(s), the positioning, and key value/reason to buy to the customer relative to competition.

References

Lehmann, D. R. and Winer, R. S. *Product Management, 2nd ed.*, Richard D. Irwin, Inc., Chicago, 1997.
Porter, M. E. *Competitive Strategy*, The Free Press, New York. 1980.

Further Information

For further information on defining competition, see *Defining the Business* by Derek Abell.

Many good books on consumer/customer behavior exist. A notable example is William Wilkie's *Consumer Behavior.*

For those readers interested in using the Internet for performing competitor analyses, simply using some of current popular web search engines such as Yahoo (www.yahoo.com) or AltaVista (www.altavista.digital.com) provide a considerable amount of information quickly and at low cost. A leading consulting firm providing information about competitors is FIND/SVP based in New York.

12.2 The Nature of Technical Products

Allen M. Weiss

The marketing of technical products must begin with a sound understanding of what *is* a product and in what ways, if at all, do technical products differ from products that are nontechnical in nature. For if technical products are not fundamentally unique, managers can simply apply to these products the available strategies and tactics used in marketing other products and services. With a clear understanding in mind of the similarities and differences, managers of technical products can proceed to develop marketing strategies in a more focused manner.

Prior research in marketing and other disciplines has provided a rich framework for understanding the nature of products in general. Much of this research focuses on a customer-oriented definition of a product based on its perceived benefits and attributes. This work is indeed important and will provide our first perspective of technical products, and we will see that from this perspective technical and nontechnical products share several qualities and marketing implications. Recent academic research, however, puts technical products within the context of a possible ongoing stream of technological improvements and recognizes the level of scientific knowledge embedded in such products. Identifying these considerations helps to clarify our understanding of what is different about technical products and ultimately guides our attention to unique product-related decisions. We begin by examining the shared qualities of technical and nontechnical products.

Shared Qualities of Technical and Nontechnical Products

From a customer's perspective, all products can be described on several levels [Kotler, 1996]. The most primitive level is the core product. The **core product** indicates what benefits the customer receives from the product, typically stated in terms of what problem the product is helping a buyer solve. As an example, for a consumer the core product associated with a book might be knowledge about a certain subject or, in the case of a novel, the relaxation it provides. For an engineer designing a computer, the core product of a microprocessor is its ability to direct and manage digital information. Microprocessors are highly technical in nature, but at the core product level they still can (and should) be described in terms of the benefits they provide. Thinking of products at a core product level is central to marketing, for, if customers are unable to perceive the core benefits of a product and see how it solves a relevant problem, they have little reason to purchase it. In fact, engineering-oriented companies often produce highly technical products that fail in the marketplace precisely because nobody identifies the core product.

The next level of a product is the **actual product**, or the set of characteristics designed to deliver the product's core benefits. This set of characteristics is stated in terms of a product's **attributes**. A book can have several attributes, e.g., its physical dimensions, cover type (hard or soft), packaging (cover art), and price. The name of the book and its author are obviously important to customers, and these are product attributes as well. A microprocessor also has attributes, such as its speed, compatibility, price, and ability to efficiently process graphics algorithms. The attributes of technical products tend to be more objective than nontechnical products, but, as Intel has demonstrated with its Pentium chip, technical products can have a powerful brand name that prospective customers consider when choosing between purchase options. Subjective and psychologically based attributes, such as a brand name or even the appearance of the product (e.g., the package design of a testing device), are not typically associated with technical products. These attributes exist nonetheless, customers use them to make decisions, and thus they should be understood by the marketing manager of technical products.

All products can also be described by the extent to which they have **augmented product** characteristics. Warranties, after-sales service, installation, and delivery are examples of ways to provide additional benefits to customers beyond the core benefits embodied in the actual product. Firms that sell books can offer money-back guarantees and home delivery, while microprocessors can be sold with technical support and an array of compatible motherboard devices. Marketers are always trying to find ways to augment their products and provide competitively unique benefits to their customers, and this goes for both technical and nontechnical products.

Finally, products can also be described in terms of their **features** — a highly detailed product view — and technical products tend to be conceptualized and marketed this way. Thus, a microprocessor has a clock speed, thermal characteristics, floating-point capabilities, etc. Notice that nontechnical products also have features, such as the font and font size for a book. However, savvy marketers know that, while specific features must meet the needs of customers, this highly detailed view quickly gets lost in customer's minds and, regardless, these features must generate observable benefits before a prospective customer will buy.

Customers of all types of products tend to make purchase decisions based on the product levels described above, and in this way technical and nontechnical products are similar. Indeed many central product decisions, such as branding and packaging decisions, are based on describing products this way. Consequently, managers of technical products can profitably employ these ideas in their marketing efforts. However, if technical and nontechnical products are similar in these ways, are there ways that technical products are different, and are there any unique product decisions that follow from these differences? We now turn to these questions.

Unique Qualities of Technical Products and Implications for Product Decisions

Technical products can often be described by two distinguishing characteristics. First, technical products often evolve through a series of generational changes in the underlying technology. Therefore, customers tend to view a product in the context of an ongoing stream of technological advances. In addition, firms can advance the technology along different paths, creating product decisions based on the existence of industry standards. Second, technical products usually contain a significant degree of scientific knowledge. We now examine these two dimensions in turn and some of their implications for product decisions.

Consider first the series of generational changes in the underlying technology. If the technology advances at a rapid pace (in terms of the speed and/or size of technological improvements), as it tends to with "high-technology" products, two issues emerge. First, much of the information provided to customers about the current generation (or version) of a product becomes "**time sensitive**" [Glazer and Weiss, 1991]. Time-sensitive information is information that quickly loses its value. For example, computers and the microprocessors on which they are based have been rapidly improving in terms of speed, capabilities, etc. Consequently, knowledge associated with a given generation of computer quickly diminishes, and both customers and engineers are finding it difficult to maintain up-to-date knowledge. As it relates to their purchase behavior, Weiss and Heide [1993] found that customers who perceive a rapid pace of technological improvements in computer workstations do tend to recognize the short "shelf life" of the received information. A second result of rapid technological change is the expectations it generates in prospective customers that they may purchase a soon-to-be-obsolete technology. These expectations have been shown to induce prospective customers to leapfrog current generations of technical products [Weiss, 1994]. Presumably, these expectations reduce the perceived benefits of owning a current product generation.

When information is time sensitive and customers anticipate rapid improvements, marketers of technical products face several challenging product decisions. In particular, they must decide when to introduce new generations of a product and whether older generations should be sold concurrently. While the pressures of producing leading-edge products are high, managers who quickly introduce new generations may both cannibalize their existing products and increase buyer's perceptions that their technology is changing rapidly. This may reduce the benefits of owning a current generation, and ultimately encourage customers to leapfrog.

At the level of product features, managers of products that are technologically changing must also decide whether to make their products **backward compatible**. Backward compatibility refers to the technical compatibility between a current generation and previous generations of a product. If the generations are not backward compatible, prior customers may be stranded. In some cases, this will reduce the core benefit of a customer owning the latest generation. For example, Microsoft originally confronted this problem in its most recent version of its word-processing software (Word 7.0) by not allowing it to save documents in Word 6.0. This lack of backward compatibility lowered customers ability to communicate and distribute documents with users of older versions. This reduced the stated core benefit of the product (i.e., communicating and working together). Eventually, Microsoft corrected this problem. However, this example does point out the link between the generational changes in a technical product, a firm's product decisions, and the level of benefits ultimately provided to customers.

Another product decision that follows from an ongoing stream of technological advance is whether to introduce a product based on a substantially different technology, thus creating implications for industry standards. In these cases, customers must often choose between the competing standards and **network externalities**. When network externalities exist, the core benefits customers receive from a product based on a standard depend on the number of other customers who choose the same standard [Farrell and Saloner, 1992]. Research indicates that, since later adopters receive greater benefits than early adopters, the rate of new product adoption will be retarded. The infamous case of VHS vs. Beta versions of videotape devices demonstrates how the benefits of a videotape machine can depend on how many others adopt the same machine due to the availability of software (i.e., movies). Telephones, fax machines, and computer platforms are obvious other examples of how network externalities affect the extent to which a technical product can deliver a core benefit. When making product decisions regarding the introduction of a product based on a new technical standard, and in contexts where network externalities are present, managers must consider the extent to which the core benefit of their product is affected by these externalities.

When customers make purchase decisions at the **product system** level, rather than at the level of a single product, decisions about products and standards are highly interlinked. An example of a system purchase would be a telephone network, consisting of transmission and switching equipment, or a circuit board assembly system, consisting of separate products that mount, solder, and test devices. In these contexts, managers must make several product decisions, in particular the degree to which they want architectural control over the system. **Architectures** are the standards and rules of the system (e.g., interface protocols) that makes interconnections between components possible [Morris and Ferguson, 1993]. Architectural control is indicated by creating groups of component products with proprietary interfaces as compared to employing open standards and allowing other suppliers to provide the rest. The decision as to the degree of architectural control of a system is complex, and prior research (e.g., Wilson et al. [1990]) indicates it is based on several factors. These include the extent to which customers want to mix and match products from different suppliers and the added growth in the market from employing open standards.

The second distinguishing characteristic of technical products is their relatively high level of embedded scientific knowledge. A unique characteristic of scientific knowledge is that it includes a large "tacit" component. **Tacit knowledge** is defined as knowledge that is difficult to communicate via documentation [Polanyi, 1958]. The concept of tacit knowledge is easily understood by contrasting it with codifiable knowledge, which is amenable to the printed page and can be easily transmitted, such as in designs and specifications. Tacit knowledge, in contrast, is more easily exchanged via such methods as learning by doing. The tacit property of scientific knowledge has product decision implications for both customers and between firms involved in product development.

From the customer's perspective, the tacit quality makes it difficult to quickly understand the benefits of technical products. Simply put, technical products tend to be complex. Consequently, from a product perspective managers need to identify vehicles for exchanging that knowledge efficiently to customers aside from written product information. For example, consider the large amount of tacit knowledge required to use a computer effectively and the inability of inexperienced customers to gather relevant information from computer manuals. The products themselves can be designed to shield the customer from the need for tacit knowledge to understand the product's benefit, e.g., by using simpler product-user interfaces. Alternatively, managers of technical products can use video as a means of providing instructional information since it is more amenable to transmitting tacit knowledge. Finally, the difficulty of transferring tacit knowledge may put product-support services at a premium, and technical product managers may need to devise creative service strategies.

The tacit property of scientific knowledge also has significant implications for the manner by which firms organize product development. As is often the case with complex products, separate firms may join to develop a technical product, as through a joint venture. Here, a potential problem arises due to the likelihood that one firm's highly innovative tacit knowledge will leak to a partner firm, which can then use this knowledge opportunistically. For example, the partner firm may autonomously develop improved versions of the technical product based on leaked knowledge. Protecting this leakage by appropriate price contracts is difficult because tacit knowledge, by its nature, must be revealed before it is understood. However, once revealed it is essentially given away for free. Some research in marketing (e.g., Dutta and Weiss [1997]) and economics has shown that firms organize their joint product development activities in particular ways (e.g., via licensing) to respond to this problem. Managers must carefully think about the leakage of scientific knowledge that is necessary when joint product development efforts are required.

Summary

Are technical products uniquely different from products that are nontechnical? In one sense, no. All products must provide benefits to customers regardless of the level of technology involved. Without a core benefit that solves a customer's problem, products are likely to fail. Defining a technical product from this perspective helps to prevent this outcome. In addition, many of the techniques established in marketing to ensure that customers perceive and receive relevant product benefits can be directly applied to the technical product. However, managers of technical products must place their products in a wider context than that often considered in nontechnical markets. With ongoing technological change, for example, managers must recognize that the benefits customers perceive from current product offerings may be affected by multiple forces beyond the immediate product. Moreover, the scientific knowledge that forms the basis of technical products, the knowledge that firms that develop such products easily understand, may be quite tacit to customers and therefore require very keen marketing skills.

Defining Terms

Actual product: A set of product characteristics as perceived by a customer, which together delivers the core benefit.

Architecture: The complex of standards and rules that guides the interconnections of a system of components.

Attribute: A product characteristic as perceived by a customer, generally stated at a higher level than a feature. The specific characteristics depend on the product in question.

Augmented product: Additional benefits to customers beyond the core benefits embodied in the actual product.

Backward compatible: Technical compatibility between a current generation and previous generations of a product.

Core product: Indicates what benefits the customer receives from the product, stated in terms of what problem the product is helping a buyer solve. It is the most general and basic description of a product.

Features: A more detailed view of the product than attributes. Features are specific means of delivering attributes. For example, fast microprocessor clock speed is a feature that delivers performance (an attribute). Likewise, low defect level is a feature that delivers reliability.

Network externality: Exists when the core benefits customers receive from the product depend on the number of other customers who choose the same standard.

Product system: A group of products that, only when linked together, provides benefits to customers. An example would be a computer system made up of computer hardware, applications software, and possibly a network that allows several computers to be hooked together.

Tacit knowledge: Knowledge that is difficult to communicate via documentation. Unlike codifiable knowledge, which is amenable to the printed page and can be easily transmitted, tacit knowledge is far more difficult to codify and hence difficult to imitate.

Time-sensitive information: Information that quickly loses its value over time.

References

Dutta, S. and Weiss, A. M. "The relationship between a firm's level of technological innovativeness and its pattern of partnership agreements," *Mgt. Sci.,* 43: 343–356, 1997.

Farrell, J. and Saloner, G. Competition, compatibility and standards: the economic of horses, penguins and lemmings, In *Product Standardization and Competitive Strategy,* H. Gabe, ed., 1992.

Glazer, R. and Weiss, A. M. Marketing in turbulent environments: decision processes and the time-sensitivity of information, *J. Mktg. Res.,* 30: 509–521, 1993.

Kotler, P. *Marketing Management: Analysis, Planning, Implementation, and Control, 9th ed.,* Prentice Hall, Englewood Cliffs, NJ, 1996.

Polanyi, M. *Personal Knowledge: Towards a Postcritical Philosophy,* University of Chicago Press, Chicago, 1958.

Weiss, A. M. The effects of expectations on technology adoption: some empirical evidence, *J. Ind. Econ.,* 42: 341–360. 1994.

Weiss, A. M. and Heide, J. B. The nature of organizational search in high technology markets, *J. Mktg. Res.,* 30: 220–233, 1993.

Wilson, L. O., Weiss, A. M., and John, G. Unbundling of industrial systems, *J. Mktg. Res.* 28: 123–138, 1990.

Further Information

A good overview of basic product definitions can be found in several introductory marketing textbooks, including the Kotler book referenced above. In addition, consumer behavior textbooks in marketing give a rich description of products from several different perspectives. The following book is a good starting point: Hoyer, W. D. and MacInnis, D. J. *Consumer Behavior,* Houghton Mifflin, Boston, 1997.

In addition, the following articles are especially helpful to understanding the general effects of technology on the decision making of customers and firms.

Rosenberg, N. On technological expectations, *Econ. J.,* 86: 523–535, 1976.

Teece, D. Capturing value from technological innovation: integration, strategic partnering and licensing decisions, *Interfaces,* 18: 46–61, 1988.

12.3   Pricing of Technical Products

Robert J. Dolan

Pricing is an area of marketing about which managers feel least confident they are doing a good job. For example, in recent surveys of managers rating marketing tasks on a scale from a "low problem pressure" (= 1) to "high problem pressure" (= 5), industrial good marketers' average rating for pricing was 4.5, the highest of any marketing function [Dolan and Simon, 1996]. Yet, pricing is critically important to the firm. Given the cost structure of a typical large corporation, a 1% improvement in price realization yields a net income gain of 12% [Garda and Marn, 1993].

Generally, managers' concerns about how well such an important task is being done are justified. Two common errors are the attitude firms take toward pricing and a reliance on cost as the primary input to the pricing process. The firm's profit equation can be written as

$$\text{Profit} = \text{price} \times \text{unit volume} - \text{costs}$$

Firms proactively manage the last two terms in the profit equation. "Unit volume" is specified in the annual business plan and is everyone's concern. "Costs" have been the focus of the reengineering efforts, which have permeated the business landscape. Yet, many firms abdicate responsibility for managing price, e.g., "the market sets the price and there really isn't much we can do" or "our competitors set the price level and we pretty much have to go along."

The two keys to effective pricing are (1) adopting a proactive attitude toward pricing, i.e., regarding price as an element of the marketing mix to be managed as proactively as any other, and (2) adopting customer value as the prime input to pricing decision making. This section sets out the value perspective, describes how customer value can be estimated, and shows how to tailor pricing to value delivered to the customer.

The Value Perspective

Figure 12.5 presents a basic schematic for understanding the relationship of pricing to the rest of the marking mix and also the value perspective. Marketing planning typically begins with two analytical components. As shown at the top right, consumer analysis is done to determine important consumer segments, which may be chosen as target segments. In concert with this, competitive analysis identifies possible points of differentiation for the firm. This leads to a positioning, i.e., what the firm and its product/services are to be *to whom*. With the target market set, the value-creating elements of the marketing strategy are assembled, i.e., the product that delivers value to our chosen customers, the communication mechanism to inform target customers of the product and its attributes, and the distribution vehicles to deliver the product efficiently to the customer. All of this activity, set in a competitive context, determines the value that a given customer perceives in the product. As shown in the middle of Fig. 12.5, this perception is the key. Laboratory performance does not create pricing latitude in and of itself, but only through its impact on the customers' value perception. This perceived value is the maximum price that the customer is willing to pay. Typically, this varies across customers, making target market selection a critical means of focusing the value-creation efforts on those with the most potential. Since the perceived value is the maximum pricing the customer is willing to pay, it should be the primary input to pricing; since it is an outcome of the marketing activities of product, communication, and distribution, price has to be "in synch" with these activities.

A good example of the importance of the value perspective is Glaxo's introduction of Zantac, the ulcer medication, into the United States. Tagamet had been introduced about 5 years earlier as the pioneer in the category. Conventional wisdom on pricing a "second one in" to a category was to price 10% below the pioneer, given the reputation and distribution advantage held by the "first one in". However, Glaxo understood the superior value that Zantac delivered in terms of fewer side effects, drug interactions, and

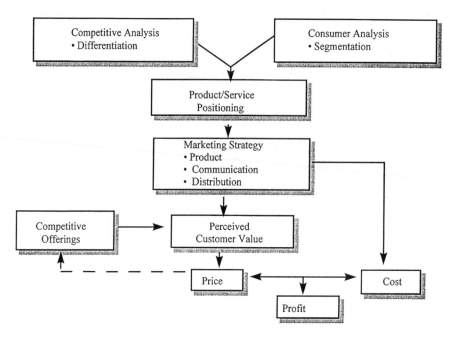

FIGURE 12.5 Schematic of value and pricing process.

dosing convenience. Market research had shown that, with proper investment in a communications effort, this value could be conveyed to consumers. Consequently, Glaxo developed a large-scale selling effort to create a perceived value in line with actual performance and accordingly priced Zantac at more than a 50% premium over Tagamet. Despite the price premium, Zantac quickly became the market leader, and prices increased as the performance of Zantac was proven in the marketplace and, consequently, the perceived value increased.

Implementing a Value Perspective

Glaxo was able to see the opportunity for and ultimately justify a significant price premium because it understood customer value. In a recent survey, however, while over 80% of respondents felt "well informed" about costs, only 21% felt that way about customers' perceptions of value and acceptance of prices [Dolan and Simon, 1996]. Simply put, in many firms, the knowledge base does not extend to value of the product to the customer. The initial step in implementing the value perspective is to estimate this value rigorously.

For technical products, the three methods that have proven most useful in estimating the value to customers are

1. Cost structure studies
2. Management judgment
3. Customer surveys

In a cost structure study, one seeks to understand the business operations of a potential customer to assess how costs would be impacted by adoption of the product. For example, when Nortel (then Northern Telecom) began development of its Meridian key systems (telephones and control units for small businesses), it considered its immediate customer set, resellers, such as the Regional Bell Operating Companies and GTE. Key managers interviewed these resellers as inputs to product/service design and pricing. Six major elements of reseller cost were identified and the underlying drivers of each element of cost understood. This enabled Northern to see key leverage points such as better MTBFs and delivery to the resellers' end user (to obviate the need for the reseller to hold inventory locally) and also to understand the true value delivered to resellers by such an offering (for details, see Dolan [1993a]).

Managerial judgment can also be used and is particularly appropriate for small markets, wherein the cost of an extensive value study may not be justified. If this method is used, it is important to follow a structured approach with experts from different functions (e.g., finance and marketing) and levels of the corporate hierarchy (e.g., VP-marketing and salesperson). Each expert should be asked individually to estimate the value to the customer of each particular type. This stage typically produces widely divergent estimates. The experts should then share their judgments and rationales for their estimates. Discussion usually leads to convergence of the experts or identification of a very specific area of uncertainty, which can be resolved through targeted inquiry. For example, the difference in value estimates may stem from different views on a factual issue such as whether the product can be used with existing systems. These structured process managerial judgment estimations are quite different from an isolated manager's "guesstimate".

The most common method for estimating value is the customer survey, of which there are two types:

1. Direct questioning about willingness to pay
2. Inference based on expressed preferences

In the direct questioning method, purchase intent (PI) scales are usually employed, e.g., questions are of the type:

What is the likelihood you would buy this product if priced at $1000?

1. Definitely would buy
2. Probably would buy
3. Not sure would buy
4. Probably would not buy
5. Definitely would not buy

The direct questioning method is simple to understand but does have some limitations in isolating the price variable and potential bias in response. However, given the number of studies research companies have done using PI scales, these biases are well understood and can be factored out.

With improvements in computer interviewing techniques, the second method of inferring product value based on expressed preference in now growing in favor. "Conjoint analysis" is the primary technique here, and it has been widely applied, e.g., see Dolan [1993b] for description and application to design and pricing of material requirements planning software. The conjoint technique is also described in Section 12.1.

While these value estimates are often not as precise as cost estimates, it is preferable to base pricing on the right variable (value) imprecisely estimated than on the wrong thing (cost) precisely estimated. As conjoint analysis does, it is important to consider the value at the level of the individual customer rather than the average "market" level. Valuations typically vary markedly across customers. If values vary across customers, astute pricing practice is to seek to vary price across customers as well to capture the value.

Price Customization: Value and Methods

If customers vary in their valuations of the good, a pricing policy constrained to charging a uniform price to all customers severely limits profit potential. For example, consider the situation shown in Fig. 12.6.

This shows a sales response curve running from A through B to C. This line shows the quantity that would be sold at different prices and depicts a situation in which willingness to pay declines linearly across customers. If point D is the firm's variable cost per unit, triangle DBC represents the excess of the group of customers' valuation of the good above its cost since each point on the sales response curve is the marginal customer's valuation of the good. A firm with perfect knowledge of these values, but

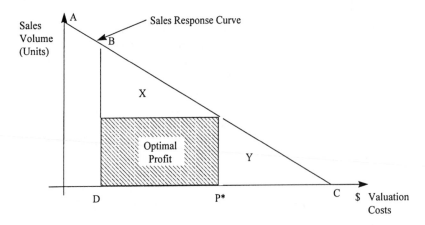

FIGURE 12.6 Sales response curve.

restricted to charging only one price, would maximize its profits by charging P* (the halfway point between D and C) since this sweeps out the largest profit rectangle from the triangle DBC. However, this profit rectangle is only 50% of the total excess value triangle DBC. Two triangles X and Y, each half as big as the optimal profit rectangle, are left behind. Triangle Y is "money left on the table" by the firm. It represents those customers who value the good at more than P* but were only asked to pay P* because the firm had to charge the same price to everybody. Triangle X represents a set of opportunities lost because these customers valued the product more than the firm's cost but did not buy it because the firm's asking price was P*, an amount above their valuations.

With linear sales response curves (which represent variation in the product value across customers according to a uniform distribution), the best one-price policy can capture only 50% of the value. This provides a strong impetus to find a way to break out of this one-price constraint when faced with significant variation in value across customers. Many firms have implemented pricing policies that entail prices "customized" to individual buyers based on their values as a key component.

For example, airlines use sophisticated yield management systems to vary prices, depending on when the reservation is made and whether or not a Saturday night stay is part of the trip. This is to differentiate business travelers and their high values for the service and pleasure travelers with low values. Software manufacturers charge different prices to upgraders and first-time buyers. Many firms offer quantity discounts as a way of effectively reducing price for large-quantity buyers. Effectively managing customized pricing is challenging but can boost profitability by 25% or more.

In general, there are four major ways to implement a policy of price customization.

1. Product line offering — A firm can offer a product line that allows customers to select among alternative offerings, and that selection itself identifies the customer as a high- or low-value customer. For example, consider Fig. 12.7 in which the valuation curves of a high-value, VH, and low-value, VL, customer are shown. The curve shows the valuation each customer places on the product as a function of its "performance", e.g., the speed of a computer, the accuracy of a measuring instrument, etc. The figure also shows the firm's cost of providing a unit of the good of a given performance level. If the firm offered a product of performance level, PH, at anything above its costs, the low-value customer would not be interested. An offering of two products of performance levels PL and PH with the price of PL near LMAX and a price of PH just slightly below HMAX would have the high-value person selecting PH, even though PL is freely available to him and low value selecting PL. Thus, the pricing and product line offering to the market generally succeeds in sorting the two types of customers.

2. Controlled availability — Especially in situations where the sale is made through a direct sales operation, price lists need not be published for the market, but the price can be tailored to the situation as perceived by the salesperson.

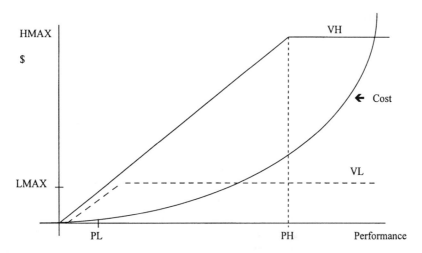

FIGURE 12.7 Valuation curves of customer high and low (VH and VL) ad firm cost curve.

3. Pricing based on buyer characteristics — In some situations, a characteristic of a buyer can be a good proxy for his value of the good. For example, an owner of a particular type of software program would typically value a new competitive offering less than if he did not own the product at all. Thus, Borland charged $325 for Quattro Pro to first-time buyers of spreadsheet programs but $99 to "upgraders" from Lotus 1-2-3. These upgraders were asked to identify themselves by sending in a piece of the 1-2-3 User's Guide.

4. Pricing based on characteristics of the transition — The most common customization of price to a transaction is through the offering of quantity discounts. The principle is that large-quantity buyers have more options available because they have the incentive to seek out information. By offering quantity discounts, firms can offer lower prices to these customers without having to offer similarly low prices to high-valuing, smaller buyers.

Conclusion

Pricing is a critical activity as added revenues to the top line through pricing initiatives drop right to the bottom line. The key to effective pricing is understanding customer value. Without this understanding, one is simply guessing and typically relying on often-used but inappropriate rules of thumb such as taking a historical mark-up on costs.

A variety of proven methods exists for estimating value. Which one is best depends on the specifics of the situation. A good pricing process meets the following tests.

1. It is integrated with the overall marketing strategy.
2. It drives off customer value and variation in value across segments.
3. It considers price customization opportunities.
4. It considers competitive reaction as perceived value is a function of competitive offerings.

References

Dolan, R. J. Northern Telecom (B): The Norstar Launch, Harvard Business School Case #593-104, Boston, 1993a.

Dolan, R. J. *Managing the New Product Development Process*, Addison-Wesley, Reading, MA, 1993b.

Dolan, R. J. and Simon, H. *Power Pricing: How Managing Price Transforms the Bottom Line*, Free Press, New York, 1996.

Garda, R. and Marn, M. Price wars, *McKinsey Q.*, 3, 87–100, 1993.

12.4   Distribution

Mary Lou Roberts

Distribution, as one of the four marketing mix variables, often represents a perplexing strategic decision for high-technology firms. It is also a decision that, once made, is difficult to modify. The difficulties arise for three primary reasons. The first is that some channels may require exclusive and/or long-term agreements. The second is that effective channels may be difficult to find, especially in highly specialized or international markets. The third is that channels involve relationships, which, once established, are difficult to sever.

The fundamental question, however, is always, "Why channels?" The basic answer is that channel intermediaries are specialists who can often perform essential channel tasks more efficiently and effectively than producers. Yet, the temptation is always to regard the margin earned by the intermediary as a cost that could be avoided if the intermediary were eliminated and the producer assumed the tasks. One of the objectives of this section is to specify the functions performed by various intermediaries so manufacturers can more accurately assess the value of channels. The second objective is to examine environmental factors that are changing the way manufacturers configure and manage their channels. Finally, we will consider the contemporary reality for most corporations — managing hybrid and multiple channels of distribution. In dealing with these objectives, we will concentrate on channels as a marketing strategy issue and will not deal in detail with the important, but well-understood, logistical aspect of distribution channels.

Channel Alternatives

The producer of high-technology products, and increasingly the purveyors of services in those markets, have several basic channels options from which to choose. These alternatives, shown in Fig. 12.8, include selling directly to the end customer and through an industrial distributor, with or without a dealer present in the channel, and through channels that include an agent or broker. Products flow from manufacturer, through channels, to the final customer. There should be a robust information flow in the opposite direction, with intermediaries supplying information about customer needs, activities, and satisfaction

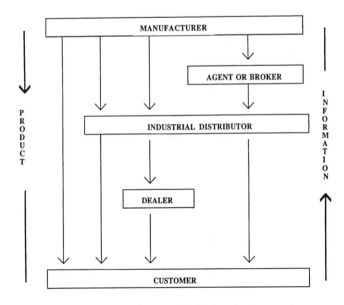

FIGURE 12.8   Channel options for high technology products.

back to the manufacturer. The firm that chooses to sell direct — whether through a field sales force, through inside salespersons, or via electronic means such as the Internet — must perform or outsource all these functions.

In high-technology channels, the industrial distributor may look very much like the traditional wholesaler. The distributor is an intermediary who accepts large shipments of product, warehouses them, performs a selling function to the next level of the channel, provides the entire set of fulfillment functions that include order processing, shipping, and inventory management, and acts as a liaison between the producer and the final customer. The value added of these activities is represented by the margin between the distributor's buying and selling prices.

Dealers mirror the retailer in consumer channels, selling directly to customers in small amounts, stocking inventory, usually at the point of sale, and providing customer service. In fact, in many high-technology situations these intermediaries often function both as retailers and as industrial dealers from the same physical location.

Finally, agents and brokers operate in high-technology channels just as they do in other channels of distribution. These intermediaries usually do not take title to or physical possession of goods. They facilitate contacts, negotiations, and physical exchange of merchandise between buyers and sellers. Agents and brokers are particularly important in international markets.

This enumeration of the functions of the basic types of intermediaries, while technically accurate, does not satisfactorily represent the activities of intermediaries or the multitude of distribution and customer-service functions they perform in high-technology channels. The reason has to do with the basic nature of high-technology products. For a product to genuinely represent the solution to a customer's needs, more than the physical product is required. Most high-tech products also require software of some type to be functional. Virtually all require installation and user training to perform satisfactorily in the purchaser's environment. Many hardware manufacturers do not produce their own software. They would also find it difficult to provide installation and support in the field, especially to small- to mid-sized customers. Specialized intermediaries have grown up to perform these vital functions.

The overall designation given to these intermediaries is "value-added resellers" (VARs). Within this general category, there can be specialized middlemen, with the systems integrator being the most common. These intermediaries perform a wide range of need assessment, product identification, installation, service, and support functions. To do this, they typically need to call on a wide range of products from a variety of, often directly competing, producers to satisfy customer requirements. The "value added" they provide comes from being able to identify, configure, install, and support an entire system for the customer. Because significant systems installations are infrequent for most customers, they ordinarily do not have internal personnel capable and/or available to do the time-consuming work required. Neither the hardware nor software producers, or even the suppliers of data or telecommunications systems, have field personnel with the wide range of expertise — much less the ability to draw upon the products of many manufacturers — to perform these functions. Therefore, VARs in high-technology applications provide an important service for all types of producers — hardware, software, and connectivity — alike.

This also provides insight into why high-technology channels are so complex in practice, even though they are configured from only a few basic options. At the distributor level, it is crucial for many VARs to have access to the product for it to be considered for as many systems as possible. At the dealer level, since this is essentially a retail business with customers choosing from nearby outlets, it is essential to have a sufficient number of dealers to provide market coverage and to match the activities of significant competitors. These requirements can lead to the use of many different middlemen, even though there are numerous multiunit operations at both levels that give national and even international coverage. While a diagram of actual channels may look like a cross between a spider web and a pile of spaghetti, this complexity may be necessary to meet marketplace needs. If the configuration of intermediaries and channels is providing the necessary services in a cost-effective manner, trying to streamline may do more harm than good.

This suggests an important question, "What are the necessary activities in any specific channel?"

Channel Requirements and Criteria

There is no "one size fits all" answer to the query about channel requirements. There is, however, an unequivocal response to the question, that is, "Ask the customers." However, manufacturers often make the mistake of translating "customers" in this statement as their own customers — the intermediaries. It is important to understand needs inside the channels, but that is the *second* step. The *indispensable first* step is to ask the end users of the product or service what it is that they really need. Only in this manner can channels be organized around real customer requirements. The manufacturer should not depend on regular information flows in the channel to provide intelligence about basic customer needs when channel design issues are considered. Intermediaries, and this often includes the manufacturer's own sales force, have an understandable tendency to focus on what is easy or what they do well at the expense of what customers really want and need. There is no substitute for primary research, conducted by the manufacturer, to assist in initial channel design and to monitor existing channels.

Some of the key issues that customers will be asked about include frequency of purchase, level of assistance required, desired lot size, type of contact and communications preferred, stocking intervals, system design and integration issues, and specific after-sale and support services expected from the supplier. While the manufacturer may well have to guide customer expectations toward reality, for instance, to rely more on 800 numbers and less on service calls, it is critical to understand these issues from the customer's point of view.

Intermediaries should also be the focus of research to understand their needs and issues. The day when even world-class manufacturers could wield absolute power in channels of distribution is gone, never to return. The reason should, by now, be evident. Manufacturers need the services that channels perform just as much as channels need their products. Adversarial relationships, characterized by squeezing every cent possible out of channel margins or by a "we win, you lose" attitude, do not work in an era of interdependency. Partnerships throughout the channel are the norm, not the exception. "Win/win" is the necessary perspective today when no one channel member can succeed, or even survive, without the others.

Channel members can no longer simply be viewed as separate entities, each of whom behaves independently and without necessarily taking the well-being of the entire channel into account. The needs of customers and the ever-changing market for high-technology products require a more comprehensive understanding of channel options.

The Changing Channel Environment

The unique nature of high-technology products creates a number of issues that must be dealt with by channels. These issues include the move to just-in-time (JIT) exchange relationships, more rapid product life cycles, shrinking margins and shortening channels, and customer needs for continued service after sale. First, however, we need to consider the two key enabling conditions — information technology and logistics innovations — that permit firms to service the new channel requirements.

The Enabling Conditions

Figure 12.8 indicates that information flows from the customer, back through channels, to the producer. Traditional information flows have typically been slow, incomplete, and, in fact, have been totally dependent on the good will of channel intermediaries for their existence. Because many manufacturers did not have direct contact with the end user, however, they have historically been dependent on intermediaries for customer intelligence.

Modern information technology has changed this situation. Information now flows directly between customer and supplier through a combination of modes including both telecommunications and computer networks. This has the potential to greatly improve the speed, accuracy, and comprehensiveness of the information flows and to greatly decrease the role of intermediaries as information conduits.

Developments in the physical movement of goods, primarily the development of cost-effective overnight delivery, have also helped to create a different channel environment. Combine almost-instantaneous information flows with quick delivery, and the stage is set for channel variants that meet the most pressing needs of high-technology customers while they maintain viable channels of distribution.

Just-In-Time Inventory Management

Many firms, especially original equipment manufacturers (OEMs), no longer maintain a warehoused inventory of components and systems to be used in producing their own products. They expect product to be delivered directly to the manufacturing plant where space is limited to an inventory of no more than a 24-hour supply. This causes a pervasive change in the nature of channel relationships, which begins by reducing the number of suppliers to a minimum and establishing close working relationships with the remaining supplier(s).

Since JIT is generally regarded as an operations management issue, marketers have paid little attention to its effects in general or on distribution channels in particular. An exception is a paper by Frazier, Spekman, and O'Neal, who point out that "the emphasis on per-unit price from the supplier is supplanted in a JIT exchange by the notion of 'total cost of ownership' and the array of value-adding services provided by the supplier. The OEM begins to recognize that per-unit price is less important than the costs associated with inspection and reinspection, handling, warehousing, inventory, scrap, and rework."[1] They emphasize that a "full partnership", which goes beyond the distribution relationship to include practices such as joint product design and sharing of cost data, is essential to make a JIT relationship work.

Shrinking Margins and Pressure to Sell Direct

While there are exceptional firms that have managed to withstand the trend, the majority of high-technology firms recognize the pressure on margins as a fact of life.

Intense competition, both domestic and international, makes it difficult to raise prices. If a firm makes a commodity product, there may even be downward pressure on prices. Without the luxury of regular price increases, companies turn to cost cutting. Manufacturing and operational costs are early targets, and many firms have pared those costs to a minimum. At that point, distribution costs often become a target.

Cutting distribution costs is usually more difficult than it sounds. A classic case occurs when a manufacturer considers the option of eliminating distributors from the channel and performing all the selling functions with the manufacturer's own sales force. A financial analysis will frequently suggest that the incremental costs of sales and physical distribution will be less than the margins earned by current channel intermediaries. However, it is extremely difficult to execute the change from channel interme-diaries to a direct sales force without seriously damaging sales for some period of time. First, putting the logistics together quickly may be difficult. Even more significant, building relationships with a myriad of customers who were accustomed to dealing with a distributor may be impossible within any reasonable time frame. It is also important to realize that eliminating intermediaries for direct distribution is a step that may be irreversible, since relationships with the intermediaries themselves may have been damaged beyond hope of repair.

Cutting distribution costs, therefore, often takes a different approach. The best solution may be for each member of the channel to specialize and to perform only the functions at which it clearly excels. The objective must be the lowest *total* channel cost, not the lowest cost for each individual channel member. It is not easy to balance lowest system cost with satisfactory results for each member of the channel while the service expectations of customers are fully achieved. Managing this type of channel relationship is challenging. We will return to this subject later in the discussion of hybrid channels of distribution.

On the other hand, the impetus to explore direct channels continues to expand in many areas of high technology. The pressure on profit margins is one reason. The advent of flexible manufacturing has also created a major impact. To react quickly to customer's specific product needs, the producer must be in

---

[1] Frazier, G. L., Spekman, R. E. and O'Neal, R. E. "Just-in-time exchange relationships in industrial markets," *J. Marketing*, 53: 53–54, 1988.

direct contact with the customer. A direct channel of distribution is a logical consequence of this type of relationship.

Direct channels have played an important role in high-technology marketing for many years. Catalog and telephone marketing have been used with great success by high-technology marketers of all types. More recently, the Internet has joined these media as a direct sales channel. Some companies use the direct media for consumable supplies and add-on products even as they employ a field sales force or channel intermediaries to sell and service the basic product. Others find that an entire product line may achieve satisfactory exposure in direct media. The addition of the direct sales channel to traditional channels is often a sign of market segments that can be best served by using a variety of distribution channels.

The Demand for Customer Service

The importance of customer service in the high-technology milieu has been stressed repeatedly. Complicating the issue is the fact that customers invariably prefer to obtain service from the manufacturer, even if they purchased from an intermediary. If manufacturers depend on the cost efficiency of channel intermediaries, how can they possibly provide direct service to a multitude of customers? The obvious answer is that, if a direct sales force is not cost effective, it is highly unlikely that a direct service force would be either. The alternatives become either forcing customers to obtain service from intermediaries or finding a way to provide high-quality, low-cost service directly.

Requiring customers to obtain support from distributors may be the only viable option, especially if physical adjustments and repair are necessary. If this is the case, manufacturers should provide training and support to intermediaries to ensure that they provide the best possible service.

On the other hand, technology provides many options by which customers can obtain service directly from the manufacturer without a physical visit by a service representative. Telephone service is one obvious way, and at lower levels of technological intensity teleservice is accepted practice today. High-technology firms have also found that it can be cost effective to assign highly skilled and expensive personnel to telephone duty to decrease the number of field service calls and increase the level of customer comfort with new and/or highly specialized products. If the caller has a high level of expertise, he or she may be more satisfied by telephone contact with an equally skilled specialist than by a service call by a less-skilled technician.

Technology provides other options for providing product service and customer support. These range from video training for customer personnel to on-line diagnostics of hardware malfunctions. There are two points to keep in mind when deciding how to provide customer service. The first is to determine what the customer needs and then to fulfill those needs. The second is that customers may need to be educated about the value of nonpersonal service. If the service can be immediate as well as being of consistently high quality, customers may come to prefer the nonpersonal approach.

Managing High Technology Channels

Multiple Channels

Considering the many complexities inherent in the production and use of high-technology products, it seems a foregone conclusion that most suppliers will use multiple channels to bring their products to market. The need for multiple channels stems from the existence of market segments that require different levels of price and service. To the greatest extent possible, it demands that the segments be kept cleanly separated so that strategies do not spill over from one into another. This is, of course, not entirely possible. The customer may request a full product demonstration from a sales representative or dealer and then obtain the lowest price by purchasing through a direct channel. The best way for manufacturers to keep channels separate is often to formulate a different product for each channel.

Hybrid Channels

The hybrid channel is an even more complex mechanism. In multiple channels, each level plays its own role in the system, acting with considerable independence from the upstream and downstream channel members. Each participant receives product from the preceding level, adds its own value, and transmits the product to the next level.

In hybrid channels, the demarcation lines between different levels of the channel become blurred. The producer may field a sales force that takes customers through the entire sales cycle. When the order is written, it is turned over to a local distributor or dealer who fulfills the order and provides customer support services. As long as the situation is understood by all parties to the transaction, it should proceed smoothly. If the customer does not expect the transition, or if the intermediary does not perform satisfactorily, there is likely be considerable customer dissonance.

Another variation is for the manufacturer to turn over sales leads that promise to produce revenue below a certain dollar amount to an intermediary. This assumes that the manufacturer promotes the product, generates sales leads, and qualifies the leads before allocating higher-value prospects to its own sales force and lower-value prospects to the channel. Two of the challenges in this model are to properly qualify the leads before they are sent to the field and to define higher- and lower-value leads in a way that satisfies the channel and maintains the profitability of the sales force effort.

Conclusion

To bring a product to market, the manufacturer must design and manage channels of distribution to meet customer needs for product and service quality while providing a satisfactory return to all levels of the channel. Shorter product life cycles and shrinking product margins are creating pressure toward direct channels of distribution. However, in high-technology channels, VARs perform functions and deal with customers in ways that often cannot be matched by manufacturers.

At the same time, manufacturers recognize the need to be in close touch with final customers who, in turn, like to receive information and support directly from the producer. Information technology, including the Internet, is enabling manufacturers to communicate directly with users even as products continue to move through conventional channels.

Conflicting requirements often result in a complex web of distribution channels and information flows, which require sophisticated management at all levels and will continue to present a challenge to all high-technology suppliers.

References

Anderson, E., Day, G. S., and Rangan, V. K. Strategic channel design, *Sloan Mgt. Rev.*, 38(4): 59–70, 1997.

Cespedes, F. V. Industrial marketing: managing new requirements, *Sloan Mgt. Rev.*, 35(3): 45–60, 1994.

Frazier, F. L., Spekman, R. E., and O'Neal, C. R., Just-in-time exchange relationships in industrial markets, *J. Marketing*, 53: 52–67, 1988.

Johnston, R. and Lawrence, P. R. Beyond vertical integration — the rise of the value-adding partnership, *Harv. Bus. Rev.*, 66(4): 94–100, 1988.

Moriarty, R. T. and Moran, U. Managing hybrid marketing systems, *Harv. Bus. Rev.*, 68(6): 2–11, 1990.

Rangan, V.K. Reorienting Channels of Distribution, Harvard Business School Teaching Note 9-594-118. 1–12, 1994.

Stern, L. W. and Sturdivant, F. D. Customer-driven distribution systems, *Harv. Bus. Rev.*, 65(4): 2–7, 1987.

Further Information

Finding information that is specific to high-technology marketing has always presented a challenge. Marketing material that focuses on "industrial" markets is often largely irrelevant. Publications that focus on "business-to-business" marketing tend to be more contemporary but frequently do not provide specific guidance for the high-technology marketer.

One approach is to continue to search for Web sites that serve the high-technology informational marketplace. As of this writing, http://www.techweb.com, http://cmpnet.com and http://www.idg.net all seem to hold promise.

There are also trade publications that serve specialized segments of high-technology marketers. *Reseller Management* and *VAR Business* from Cahners Publishing Company are two examples.

Finally, there are professional organizations that have local chapters that may have programming useful to high-technology organizations. The American Marketing Association, the Business Marketing Association, and the Sales and Marketing Executives Association are such organizations.

12.5 Promotion Activities for Technological Products

Richard C. Dorf and Eitan Gerstner

A main objective of marketing is to help a business obtain new customers and keep existing customers satisfied and loyal. Without customers, no amount of engineering know-how or operations expertise can make or keep a company successful. Marketing activities can be related to four major areas: the development and management of the product itself, the channels of distribution to be used, pricing decisions, and **promotion**. The purpose of this section is to review and discuss promotion activities that can be useful for technological products.

In general, promotion activities are intended to encourage customers to purchase a product or adopt a service [Tellis, 1998]. This can be done by using a variety of tools including **advertising, sales promotions, special events, personal selling,** public relations, **networking,** and **word of mouth.** The area of advertising is discussed in a separate section of this book and therefore will not be covered here.

Sales Promotions

Sales promotions consist of monetary or other valuable incentives to stimulate sales such as store coupons, rebates, or gifts such as mugs or posters [Blattberg and Neslin, 1990]. Some sales promotions such as manufacturers' coupons are typically targeted toward end users, while others such as trade deals or cooperative advertising are targeted toward distributors. Manufacturer's promotions targeted toward end users are called **pull promotions** because the intention is to pull demand by motivating the end users to seek the product from the retailer. Manufacturer's promotions targeted toward retailers are called **push promotions** because they are intended to motivate the retailer to push the product, that is, to convince end users to buy because of the larger profit margin or lower marketing costs that would result. Table 12.1 gives an example of push promotions that can be used for technological products.

A good strategy is to combine push with pull [Gerstner and Hess, 1991; 1995]. For example, Intel used a push-pull combination strategy when promoting its Pentium microprocessor. Pull was used by running ads on television and packing materials with the "Intel Inside" slogan. At the same time, Intel offered to share advertising with distributors who would use the Intel slogan on their own ads or packaging materials.

TABLE 12.1 Description of Pull Sales Promotions

Off invoice	A discount given to a distributor for a fixed period of time; the specific terms of the discount usually requires sales performance by the dealer
Cumulative volume rebates	A discount is given to the distributor after total purchases have reached a certain point
Floor planning or inventory financing	Financing of the distributor's inventory for a certain period, say 90 days; this gives the distributor incentives to sell the merchandise before payments are due
Free goods	Offering free merchandise along with the purchase of an item; the distributor can offer it to customers as an incentive
Cooperative advertising	Money given by a manufacturer to a distributor who agrees to advertise a manufacturer's product; it can be a fixed amount or a percentage of the expense
Contests	Free gifts such as appliances or vacations trips offered to distributors who win a contest; the criteria can be reaching a sales quota, exceeding the sales of the previous year, etc.

Pull promotions such as coupons, rebates, or special sales are effective tools for price discrimination between price-sensitive customers who go through the special effort of using these promotions, and less price-conscious customers, who would pay a higher price to avoid the hassle involved in using these promotions. Using pull promotions to price discriminate can lead to larger profits compared to a single price strategy [Gerstner et al., 1995].

A format of pull promotion that is widely used in recent years is **direct marketing**. Here, the promoter offers products or services via mail or television ads directly to customers, who are expected to respond with orders. The use of databases has made it easy to target customers via direct mail; however, customers have been overwhelmed with direct mail pieces, and, because of that, the term "junk mail" has been used to describe all the unsolicited mail. There are also growing concerns about privacy issues related to the use of databases because marketers have been selling their mailing lists to other companies without asking for the customers' permission.

Should sales promotions be targeted only to attract new customers, or should they also be offered to existing customers? Traditionally, more sales promotions were targeted to attract new customers, especially those who buy from the competition. When all competitors use this strategy, however, nonloyal customers switch from one competitor to the other, no competitor can gain market share, and profits are reduced.

To avoid this scenario, some companies started to offer sales promotions that reward customers for being loyal. The best example is the "frequent flier" type of promotion offered by airlines, hotels, restaurants, and automobile companies. These promotions are cumulative; customers can earn rewards only when they accumulate enough purchasing points. What is the advantage to the company? This is a defensive marketing strategy that makes it more difficult for competitors to switch customers because customers are caught in the process of earning points for obtaining the rewards. Therefore, this strategy helps enhance loyalty. It also slows down price competition because customers who accumulate points are not so eager to switch in response to a price cut by a competitor [Brandenburger and Nalebuff, 1996].

Special Events and Personal Selling

Special events are happenings organized for a selected target market with the objective of exchanging information, displaying products, and demonstrating their usefulness. Examples are news conferences, special sales, anniversary celebrations, or grand openings. Such events can help introduce new technological products, break out news, and get public relations coverage. Companies often use celebrity appearances, guest lectures, and good food and wine to attract the right people to these events.

Trade shows are special events often organized by trade associations. They typically attract potential customers with a high interest in the industry, and they are also useful for networking because the attendees come to see demonstrations and learn about new products and business opportunities. Trade shows also attract distributors, and therefore they can be used to open new channels of distribution.

Personal selling is communication efforts done in person for an individual or a group. This is the most expensive promotion method; a sales call can easily cost hundreds of dollars because of the travelling and time involved. However, personal selling is often vital to technological products because customers need demonstrations, installation, and fit with existing equipment. Before sending a sales person, however, it is desirable to use the telephone to qualify prospects and find out exactly what the customer wants. Doing this increases the efficiency and productivity of sales calls.

Public Relations and Networking

Public relations are activities designed to obtain coverage by reporters. Although this coverage seems to be free, it is not. To get good coverage, a company typically must pay public-relation specialists to write news releases and organize special events. The main advantage of articles and materials published by independent reporters is that they are perceived to be more credible than advertising because the latter is entirely controlled by the company.

Networking is informal communications with individuals or groups who can spread the word and provide leads to potential buyers. Networking should be done in an organized fashion with the objective of connecting to opinion leaders and other influential individuals who can help push the product and develop the business. Once the right group of people is identified, it is important to find events in which they can be approached. Examples of such events are meetings organized by trade associations and local chambers of commerce. Local business newspapers typically publish local business events.

A useful medium for networking is the Internet (Web). There are thousands of network communities organized around topics of interest. Contacts can be made by tuning in to relevant news groups, participating in their discussions, and offering help when needed. Again, the key is to hook up with the right group.

Word-of-Mouth Promotions

Word of mouth refers to recommendations made by acquaintances and experts for using products they like. It is a powerful promotional tool, especially for new technology products. A company can encourage word-of-mouth by appealing to crucial people who can influence other people to try the product. To target the right people, it is helpful to understand the process by which new technology products are adopted by customers.

The adoption of new technology products can be described in a diffusion process. The first group of people that tries the product is called innovators (2.5%), the second, early adopters (13.5%), the third, early majority (34%), the forth, late majority (34%), and the fifth, laggards (16%), where the percent of the total market is indicated in the parentheses [Rogers, 1983].

The innovators and early adopters are crucial groups for promoting the product. If they like the product, they would help push it to the other groups through word of mouth, and the company can use their names to convince other customers to buy. Because of their importance, companies can benefit by offering attractive terms to the innovators if they agree to try the product. For example, a common promotional technique used in the software and electronics industry is the placement of a new product with lead users who test the product for free. In return these provide user feedback and, if happy, good word-of-mouth promotions.

Conclusion

Promotion includes marketing tools used to motivate distributors and end users. It is important to use all these tools in harmony; all promotion activities should be consistent with each other and with the overall marketing strategy. It should also fit well with the target audience. Finally, it is important to realize that *promotion cannot sell the product, unless it is supported by good quality, excellent service, and a good value for the money spent.*

Defining Terms

Advertising: Paid communications via different media designed to raise interest and motivate purchases.

Direct marketing: Offering products or services via mail or television ads directly to customers who are expected to respond with orders.

Networking: Informal communications with individuals or groups who can spread the word and provide leads to potential buyers.

Personal selling: Communication efforts done in person for an individual or a group.

Promotions: Activities designed to raise interest and provide incentives to purchase a product or adopt a service.

Pull promotion: Promotional efforts directed to the end user within a channel of distribution.

Push promotion: Promotional efforts directed toward an intermediary within a channel of distribution.

Sales promotions: Special monetary or other valuable incentives to end user or distributors to help stimulate sales.

Special events: Happenings organized for a selected target market with the objective of exchanging information, displaying products, and demonstrating their usefulness.

Word-of-mouth: Recommendations made by acquaintances and experts for using a certain product or a service.

References

Blattberg, R. and Neslin, S., *Sales Promotion,* Prentice Hall, Englewood Cliffs, NJ, 1990.

Brandenburger, A. and Barry N., *Co-opetition,* Currency Doubleday, New York, 1996.

Gerstner, E. and Hess, J. D. A theory of channel price promotions, *Am. Econ. Rev.,* 81(4): 872–886, 1991.

Gerstner, E. and Hess, J. D. Pull promotions and channel coordination, *Market. Sci.,* 14(1): 43–60, 1995.

Gerstner, E., Hess, J. D., and Holthausen, D. Price discrimination through a distribution channel: theory and evidence, *Am. Econ. Rev.,* 81(4): 1437–1445, 1994.

Rogers, E. M. *Diffusion of Innovations,* Free Press, New York, 1983.

Tellis, G. J. *Advertising and Sales Promotion Strategy,* Addison-Wesley, Reading, MA, 1998.

12.6   The House of Quality

Derby A. Swanson

What Is The House of Quality?

The **House of Quality** is a management design tool that first experienced widespread use in Japan in the early 1970s. It originated in Mitsubishi's Kobe Shipyard in 1972, helped Toyota make tremendous improvements in their quality in the mid-1970s, and spread to the United States (via the auto industry) in the mid-1980s. A significant step in popularizing its use in the United States was the 1988 publication, in the *Harvard Business Review,* of "The House of Quality", by John Hauser and Don Clausing. Hauser and Clausing [1988] stated that "the foundation of the house of quality is the belief that products should be designed to reflect customers' desires and tastes — so marketing people, design engineers, and manufacturing staff must work closely together from the time a product is first conceived".

Now, in 1997, The House of Quality (and similar total quality management-related quality tools) have moved well beyond their initial application in the U.S. auto industry. These techniques are used in all types of settings for a wide variety of product and service improvement/development projects.

The House of Quality[2] refers to the first set of matrices used in **quality function deployment** (**QFD**). It acts as a building block for product development by focusing efforts on accurately defining customer needs and determining how the company will attempt to respond to them.

Building a House of Quality matrix starts by defining a project around product and/or service development/improvement and putting together a cross-functional team with responsibility for making it happen. The team begins by working to fully understand the customers' wants and needs for the particular product, service, or combination — known as the **voice of the customer** (**VOC**) or customer attributes (CAs). A determination of how important the needs are to customers and how well they are currently being met is used to help assess their criticality in the development effort.

The team then identifies the possible measures or product characteristics that could impact those customer needs. In product development customer needs are related to physical characteristics, commonly called engineering characteristics (ECs), of the product. For example, a customer's need for an *easy-to-carry* container will be impacted by the physical characteristic of its *weight*. For issues having to do with service development — for example, how to set up an order desk — the customer needs (e.g., *Quick and accurate from order to delivery*) would be impacted by performance measures (PMs) such as *time it takes to complete an order.*

---

[2]The name "House of Quality" comes from the shape of the matrix components.

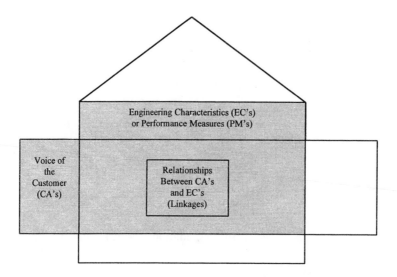

FIGURE 12.9   Diagram of House of Quality components.

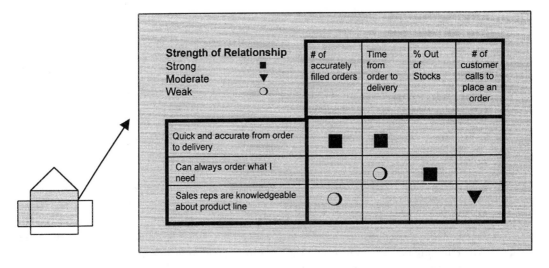

FIGURE 12.10   Example of CAs, PMs/ECs, and linkages.

The next step is to correlate the two — to assess the relationship (or impact) of the measures on the needs. This step fills in the matrix that makes up the center of the House (Fig. 12.9).

The strength of those relationships (example in Fig. 12.10) will be an important factor in telling the team where they should focus. After the links to customer needs are identified it will be possible to prioritize product characteristics or measures based on which have the biggest net impact on customers. The more impact a measure/specification has on important customer needs, the more attention it should get.

In addition to defining the customer needs, it is important to understand that all customer needs may not be created equal! Some will be more important to customers than others (e.g., *safety* vs. *easy to buy* for a medical product). Also, there may be areas that the team wants to target for improvement — where the company's performance is lacking vs. competition or where there are particular quality problems that need to be addressed. Both the importance to customers and any relative weight on the needs will also be addressed in the House of Quality matrix.

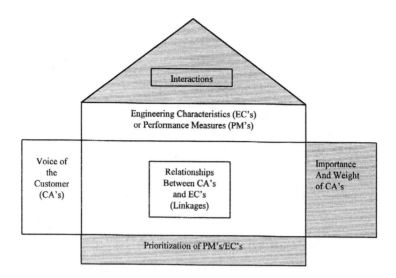

FIGURE 12.11 Diagram of House of Quality components.

Since both the importance/weighting of the customer needs and the relationships at the center of the House are expressed as numbers, the result of "crunching the numbers" in the matrix is a prioritization of the engineering characteristics/performance measures that tells the team where it should put its energy and resources.[3]

Finally, the roof of the House contains information regarding how the ECs and PMs interact with each other. They may be independent or there may be positive or negative correlations, for example, attempts to reduce a copier product's footprint will correlate positively with making it lighter in *weight* but there may be a negative effect on *the amount of paper it can hold* (Fig. 12.11).

Why Use The House of Quality?

The discipline of building a House of Quality leads to a design that is truly driven by the voice of the customer rather than the "voice of the loudest vice-president". Development time is invested at the beginning rather than in redesigning a product/service that doesn't work. A cross-functional team of people working together, focused on the customer instead of on their own departments, becomes a cohesive unit. Finally, a decrease in cycle time and increase in responsiveness from understanding the marketplace can lead to the kind of results that helped Japanese automakers to launch their successful foray into Detroit's marketplace.

Businesses succeed by offering products and services that meet customer requirements profitably. Businesses retain their customers by continuing to please them, which sounds very basic — but is, in fact, difficult. One of the causes has to do with the ways in which companies attempt to "hear" their customers. There are many stakeholders to be satisfied and customers speak different languages from one another and, most especially, from a company. Greising [1994] gives an example of what companies face as they pay homage to quality without understanding what it takes to be truly customer driven:

- A major computer manufacturer spent years using mean time between failures (MTBF) as their measure of reliability. However, along with their industry-leading MTBF went a customer reputation for unreliable systems. It turned out that to customers reliability had to do with availability — how much of the time did they have access to their equipment. The manufacturer needed to include diagnostic time as well as technician travel time, etc. to really understand how they were doing.

---

[3]Commercially available software such as QFD Capture helps manage this process.

Customers have different concerns and speak a different language than companies, and it is that language and thought process that must be understood. Commonly used "quality-oriented" market research words such as "reliable", "convenient", or "easy to use" will likely have very specific meaning to individuals on a project team, as in the MTBF example. However, they may have very different meanings for customers. Those differences, and similarities, must be understood in detail before a company can begin successful use of the House of Quality.

Another problem companies face is trying to aggregate data too soon, to look for the one number, the one answer on how to design a product. A rating of "85% satisfaction" tells you very little about the cause and effect of what the company is doing. Scoring on typical market research survey attributes doesn't give engineers and designers the level of detail they need to design. Plus, this type of data gives you information after the fact; it measures results based on history. The House of Quality is proactive — it enables you to build a model that relates the actions taken (or the things that you measure) directly to the impact that they will have on customers.

Beginning a House of Quality

There are many tools available for helping a team interested in using QFD and the House of Quality — including firms and individuals that can act as facilitators and books (e.g., Lou Cohen's 1995 *Quality Function Deployment — How to Make QFD Work for You*) that can act as manuals. The discussion that makes up the remainder of this section talks about some of the basic issues to be addressed while putting together a team and beginning a House of Quality. The emphasis on customer needs is intentional – if the VOC that begins the exercise is flawed, so is everything that follows.

Getting Started

As with any successful team effort the key here is assembling the team and defining the problem/issues. It is important the team be cross-functional — with representation from all interested (and affected) parties). This means marketing plus engineering plus manufacturing plus information technology, etc. It can be extremely helpful to have an experienced facilitator from outside. This is someone who has been through the process before and whose responsibility is taking the team through the exercise — not trying to promote his own agenda. An outsider can be objective, help resolve conflicts, and keep the team on track.

It is important to spend time up front assessing what the project is about and who the "customers" really are. First, there is likely more than one buyer for a product or service, e.g., medical instruments (medical technicians, purchasing agents, and doctors) or office furniture systems (facilities managers, architects, designers, MIS professionals, and end users). A university must consider students, faculty and industry recruiters. Utilities deal with everything from residential to industrial to agricultural customers — and in literally millions of locations.

There are also internal customers who may feel very differently from external ones. Employees are also customers and may well have significant impact on how external customers evaluate the service they receive.

Collect the Voice of the Customer

A focused and rigorous process for gathering, structuring and prioritizing the VOC will ensure that you have the customers' true voice and that the data are in a form that will be useable for a House of Quality (as well as other types of initiatives, e.g., developing a customer satisfaction survey).

Customers will not be able to tell you the solution to their problems, but they will be able to articulate how they interact with a product or service and the problems they encounter, e.g., wanting to get printer output that is "crisp and clear with no little jaggy lines." By focusing too early on a solution, e.g., resolution of *600 dots per inch*, you lose sight of other, even-better solutions. The goal is products that are "inspired by customers — not designed by them".

Beginning with one-on-one interviews (or focus groups) to gather the customer needs and using more quantitative customer input to structure and prioritize customer needs enables the customer's own words

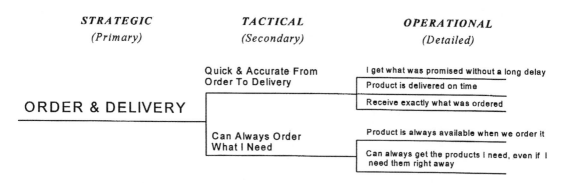

FIGURE 12.12   Section of attribute hierarchy.

to drive the development process forward. It maintains the purity of the customer data while allowing the developers to focus on the company voice. As Hauser and Clausing [1988] state,

CAs are generally reproduced in the customers' own words. Experienced users of the house of quality try to preserve customers' phrases and even clichés — knowing that they will be translated simultaneously by product planners, design engineers, manufacturing engineers, and salespeople. Of course, this raises the problem of interpretation: what does a customer really mean by "quiet" or "easy"?

The issue of interpretation is addressed by asking customers themselves to define these words. Twenty to 30 one-on-one interviews from a relatively homogeneous group of customers will ensure that 90% or more of the needs have been identified [Griffin and Hauser, 1993]. However, the topic and type of customer being interviewed should also be considered — more complex topics may require more interviews.

What you talk about in a VOC interview is critical to its success. Customers tend to talk in targets, features, and solutions — *"Why don't they just give me one rep?"* The goal of the interview is to understand the needs behind the solutions by asking "why". The key is to question everything — *"Why is that important?"; "Why is A better (worse) than B?"*. That way you hear things such as *"If I had a dedicated rep then I'd know that that person could **get my questions taken care of right away**".* By understanding underlying needs, the team can think about a number of different ways to meet them.

Team members can gain significant insight into customers by listening to them discuss their issues and concerns. However, this does not mean that team members should necessarily conduct the interviews. The keys are training and objectivity, which outside interviewers are more likely to have. For example, in one case an engineer spent an hour interviewing a customer; the engineer started out by being defensive, then tried to solve the customer's problem, and ended up learning nothing about customer's needs. However, with the right focus and training, company personnel **have** found it very useful to conduct interviews themselves.

A formal process (such as VOCALYST®[4]) will help efficiently reduce the thousands of words and phrases collected from the interviews to the underlying set of needs (e.g., 100 to 200) that can then be categorized into groups that provide a "primary", "secondary", and "tertiary" level of detail (Fig. 12.12).

The best way to do the necessary data reduction is to have customers develop their own hierarchy of needs (e.g., through a card-sorting process). Just as the language that customers use is different from a company's, so is the way in which they aggregate ideas. Members of a company team often organize ideas based on functional areas (things that are marketing's role or part of the information services department). Companies think about how they deliver or make a product, and customers think about a product based on how they use it or interact with it. No customer ever separated his needs into hardware and software — but many computer companies have organized themselves that way. For example, Griffin and Hauser [1993] show two hierarchies — team consensus and customer based — to a number of

---

[4]VOCALYST® stands for Voice of the Customer Analyst and is used by Applied Marketing Science, Inc.

professionals (including the team!) and find that the "customer-based (hierarchy) provided a clearer, more believable, easier-to-work-with representation of customer perceptions than the (team) consensus charts."

Develop Engineering Characteristics (ECs) or Performance Measures (PMs)

Next the team will develop lists of ECs/PMs, assess the relationship to customer needs, and set priorities. Each customer need should have at least one EC or PM that the team feels will strongly impact the customer need. One way to get a list is to brainstorm a list of possible contenders for each need and then have the team decide which ones are most likely to be predictive of meeting the need. As with the customer needs, it important that the "company voice" fit the criteria for a "good" measure:

- Strongly linked to need
- Measurable
- Controllable
- Predictive of satisfaction
- Independent of implementation
- Known direction of improvement or desired target value
- Impact of hitting the target value is known

Assign Relationships between ECs or PMs and Needs

The team votes on every cell in the matrix to determine the strength of relationship the measure will have with each need. For the most part, these relationships are positive (since that is the point of the House!) but there may be customer needs that end up negatively impacted by an EC/PM and that should be accounted for. There are many scales and labels, but the most common process is to assign the following types of values:

- Strong relationship "9"
- Moderate relationship "3"
- Weak relationship "1"
- No relationship

Calculating Priorities for the ECs/PMs

The "criticality" of each EC or PM comes from the strength of its relationship with the customer needs, weighted by both the importance and any weight assigned to the need. The importance comes directly from customers (either when they create the hierarchy of needs or via a survey). A weight can also be assigned to each need based on a target for improved performance (from customer data), a desire for improved quality in that area, etc. In the example here, the weight is expressed as an index. In the case of the example in Fig. 12.13, the PM with the highest priority is *number of accurately filled orders* — calculated by summing down the column for each linkage with that PM:

- (Importance of the need · the strength of the relationship) · weight of the need
- $[(100 \cdot 9) \cdot 1.0] + [(75 \cdot 1) \cdot 2.0] = 1050$

The ECs with the highest priority are the ones that the team absolutely needs to focus on to ensure that they are delivering on customer needs.

Finally

As companies make decisions about where to allocate resources for programs and development, they should be focused on the things that are important to customers but also play to their own strengths and core competencies. In "The House of Quality", Hauser and Clausing [1988] said "strategic quality management means more than avoiding repairs. It means that companies learn from customer experience and reconcile what they want with what engineers can reasonably build." Using the House of Quality can help make that so.

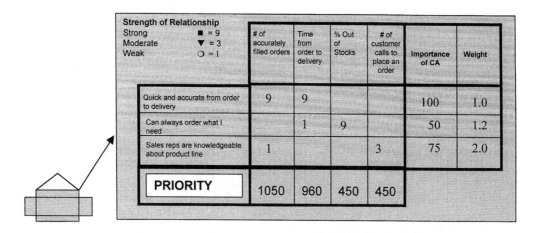

FIGURE 12.13 Example of CAs, PMs/ECs, and linkages plus weighting and prioritization.

Defining Terms

House of Quality: Basic design tool of management approach known as quality function deployment.
Quality function deployment (QFD): A series of planning matrices that allow planners to continually integrate customer needs with the technical responses to meeting those needs when designing a product or service
Voice of the Customer (VOC): Customer wants and needs for a particular product or service — developed and structured in such a way as to be used in a House of Quality or other initiative.

References

Greising, D. Quality: how to make it pay, *Business Week,* August 8, pp. 54–59, 1994.
Griffin, A. and Hauser, J. R. The voice of the customer, *Market. Sci.,* 12(1): 1–27m 1883.
Hauser, J. R. and Clausing, D. The house of quality, *Harv. Bus. Rev.,* 66(3): 63–73, 1988.

Further Information

Anyone interested in the basics should read the *Harvard Business Review* article "The House of Quality" by John Hauser and Don Clausing. A good overall introduction to the whole topic of new product design is the 2nd edition of Glen Urban and John Hauser's book, *Design and Marketing of New Products* (Prentice Hall, 1993).

One example of a "primer " for conducting a QFD/House of Quality program would be Lou Cohen's book *Quality Function Deployment — How to Make QFD Work for You,* published by Addison-Wesley (1995) in their Engineering Process Improvement Series.

12.7 Marketing Segments

John H. Mather

Market segments result from the utilization of the primary marketing strategy of **segmentation**. The very hard reality today for many companies is that mass markets do not exist anymore. This is especially true for technology-based organizations. This section describes the need, definition, processes, and results of segmentation. The fundamental focus of marketing is to achieve an exchange in the marketplace. Segmentation is the method by which the probability of any exchange increases for the product/service producer and the user. All the methods of segmentation described in the following material are at the strategic level in marketing planning. It is through diligent, persevering, and creative segmentation that technology-based ideas are accepted initially by the market, establish a market growth pattern, and then become market successes. The process of marketing the idea to the most likely end user is a critical role for marketing. The lack of a match, or the technology solution searching for a problem, too often leads to the unnecessary termination of another start-up.

Need and Definition of Segmentation

Marketing's primary role is achieving an exchange between a product and user. This seemingly easy task is very deceptive and challenging today in that over nine out of ten new products do not reach the commercialization stage of market development. The exchange process is based on mutual interest, willingness, and capacity to produce and to deliver a product from a producer to an end user. For both sides of the exchange, mutual satisfaction results from the exchange and should lead to a long-term relationship characterized by repeated purchases. Not every individual in the market is meant for every product exchange. The goal of segmentation is to generate market segments that are understood and ranked for an exchange potential. As a producer of a product, the seller enters the exchange with the segment that represents the highest level of success. The utilization of resources in the most efficient and productive manner is a requirement of survival. Segmentation can help ensure that survival by establishing a target user group for the product that exhibits the highest intent to purchase the product. This target segment is the result of many systematic activities, which are described in the ensuing material. Locating, understanding, and serving those key target market segments at a desired satisfaction level is the goal, process, and outcome of segmentation.

Segmentation has existed for many decades. The emergence of customer-driven marketing in the 1950s introduced the need for a better understanding of one's customers, and the consumer-based companies responded with products, promotion, and retailing. Customer driven is an accepted phrase in all industries today, yet the actual practice varies widely. Segmentation is the primary signal for being customer driven. The 1980s are often referred to as the marketing decade and the use of target marketing. A definition of segmentation that was short and poignant at that time was [Weinstein, 1987]

Segmentation is the process of partitioning markets into segments of potential customers with similar characteristics who are likely to exhibit similar purchase behavior.

The overall objective of using a market segmentation strategy is to improve your company's competitive position and better serve the needs of your customers.

Kotler [1994] states succinctly that market segmentation is "the act of identifying and profiling distinct groups of buyers who might require separate products and/or marketing mixes." The process of segmentation then yields market segments that are "large identifiable groups within a market." Combining these two general definitions into an expanded explanation that links the external process to the organization as well as indicates the desired results of segmentation produces the following definition:

Strategy of segmentation is the *subdividing* a *defined* market into smaller groups of potential product users that exhibit a *homogeneity* across one or more *meaningful* and *significant* variables that represent a *measurable* and *increased* opportunity for product *positioning* and *investing* the resources of the organization which raises the *probability* of *achieving* marketing objectives and also overall organization business goals.

Reviewing this definition again yields many key words. The critical elements of the definition include the following ideas:

- Subdividing — reducing a market into logical and mutually exclusive parts.
- Defined market — the intended market must be understood clearly and relevantly according to standard means of description.
- Meaningful and significant — the variables employed to generate the subsegments must provide an observable variance of exchange potential.

- Positioning and investing — the end result of segmentation is a segment(s) that represents a target user group for which the product can be positioned with a high likelihood of purchase (exchange) and therefore a realistic investment of resources.

Before ending this first section of defining segmentation, it is worthwhile to diffuse the naysayers of segmentation. The mental discipline and market sensitivity resulting for segmentation is often counter-intuitive to the technology entrepreneur of today — and many other marketers too. Numerous entre-preneurs see the technology-driven products as being desired and purchased by everyone (internal product-driven marketing). The total markets are perceived as enormous, and just a small portion will make them and their first investors (family and friends) wealthy. However, the probabilities are stacked against these capable and talented entrepreneurs because the start-up organization in no way can amass the resources required to enter a total market. Second, not everyone in the total market has the same need, interest, or capacity to purchase the product. Thus, segmentation minimizes the high chance of failure by defining, evaluating, and locating the segment of the total market that constitutes the best chance of entry success for the organization that can lead to eventual market growth and desired share.

Given the positive and probably required affirmative answer to the segmentation idea so far, the next step is to decide on the most useful ways to divide the market. Market segmentation approaches are explained and illustrated in the next section.

Segmentation Strategies

Segmentation strategies represent the approaches used to divide a defined market. Other terms mentioned for this process are **partitioning** or **disaggregating.** The terms are interchangeable — the associated process is the valuable operation. This is the process that makes market entrants smart and increases the success probability. The primary purpose of the partitioning activity is to stimulate thinking about the market and its exchange capacity with the organization and the product it offers. The correct mental state is to consider all the possible variables that would increase the chances of a possible customer entering the exchange mode for a product or service. In more formal terms, segmentation is constructing a sales function for your product based on the variables that can be used to define submarkets. The usual starting point is **demographics**. The market can be national, but a specific product may have a better chance in a region, local area, or even selected urban locations. As will be seen in the following material, basic as well as newer perspectives of segmenting markets are discussed. This is a very creative process of marketing, and those individuals that have developed market sensitivities or insights often discover that unique variable or two that divides a market into smaller groups of customers who exhibit a higher propensity to purchase the product. This is not a simple exercise but one that requires a keen under-standing of the market and its components as well as the basic strengths of the product and the corre-sponding benefits it can provide a user.

There are numerous segmentation strategies available to the partitioning of a market (Table 12.2). The following list is a traditional starting point of the partitioning alternatives. The list combines segmentation approaches for individuals as consumers or multilayered for business or high-technology customers.

The list is ordered as a funnel in that the first group of variables divides the market into the largest subsegments. Here the end result can be specific industries, different applications of the product, or simply current user vs. nonusers of the product or closely related products or technology. The funnel narrows now to examine the types of organizations within the previous segments. At this juncture, partitioning can be focused on the type of organizations, technology levels in each segment or the buying centers (committees or task forces charged with decision making), and budget development and amounts. Last, the segmentation can examine the individual decision maker in the organization.

An example will synthesize the process and use of the preceding segmentation approaches. Consider a company offering new scheduling software that is appropriate for emergency vehicle placement. The industry is the provision of emergency services in a community. The three largest segments (macro) are fire, police, and medical. All three segments need equipment tracking and positioning programs, but there are big competitors already in the police and fire segments. So the medical segment looks uncluttered

TABLE 12.2   Segmentation Strategies (Disaggregation)

Macro	Micro	Individual
Industry/SIC	Organization	Demographics
Competition	Descriptors	Title
End use/applications	Type	Motivation
User/nonuser	Culture	Needs/benefits
	Purchasing	Psychographics
	Budget	Value
	Policies	
	Culture	
	Technology levels	
	Buying center	
	Participants	
	Roles	
	Decision process	
	Situation	
	Location	
	Usage/experience	
	Criteria/benefits	

now. More importantly, budget pressures as well as an increased demand for services has placed the medical emergency organizations under extreme pressure for greater efficiencies. Moving now to the next micro level, the medical services can be divided by the descriptors of service organization such as employees, vehicles, number and type of calls, size of area, size of community served, age of service, or geographic location. Another way is to divide the target market by ownership or funding methods, budget size, equipment purchasing cutoff approval levels, and final-decision maker (director, purchasing agent, technology specialist, or outside consultant).

Continuing with the micro variables, medical emergency services vary by technology level, usage, present vehicle response capability, or next step for upgrade. By emphasizing these sequential steps like a funnel, the market is divided into numerous subcells representing different propensities to purchase the new software vehicle scheduling package. Each subsegment would be a different "fit" for the developed software and therefore a different probability of exchange success. As stated previously, the preceding is the standard approach today. It is systematic, reliable, and requires learning of the possible configurations of partitioning structures.

However, there are two additional methods that can increase market success by raising the chances of product acceptance. The first method is market sales and share growth based on product adoption rates. The second method is another model for product adoption based on the key variables of technology acceptance. Both of these approaches are very relevant to technology-based products.

Figure 12.14 contains the five growth periods of a **product life cycle** (PLC) with the corresponding buyer segments. The market growth cycle is descriptive of any product's acceptance in the U.S. market-place. The curve can vary by time from weeks to years. Today, the much used descriptive phrase is fast cycle time (12 to 18 months) for a high-technology growth period.

The PLC's shape is a result of the five consumer segments that comprise the sum total purchases for any product. These five segments are based on extensive research that measured attitudes and purchase patterns of new products. Therefore, in any defined market — using the emergency vehicle software example — there will be 2.5% organizations that will try the software first. These are the innovators or technology enthusiasts as Moore [1991] names these initial users of high-technology products. These persons are risk takers and enjoy the experience of being first to try something new. They can be helpful in suggesting final product improvements.

The next product acceptance group is the early adopters or visionaries. This segment adds another 13.5% to the sales curve. Both the innovators and early adopters equal 15% of a total market and generally are considered the market introduction phase. The mass market or commercialization phase for technology

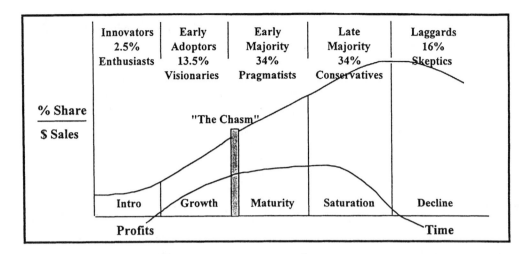

FIGURE 12.14 Product life cycle (technology). (*Source:* Moore, G. A. *Crossing the Chasm*, p. 17, Harbor Business, New York, 1991.)

products is composed of the early- and late-majority groups (pragmatists and conservatives). Moore maintains that, with high-technology products, there is significant difference between the introduction groups and the commercialization segment groups. Therefore, marketing efforts and product design must be altered to "cross the chasm". Further reading of Moore's book is suggested for the detailed background of each segment and strategy development.

Many questions can be directed at the adoption process as it progresses through the five segments of the total potential purchasers in a market. However, a primary query is the reason for the different rates of adoption. This is especially relevant for the reinterpretation of the PLC for technology-based products. One plausible explanation can be found in Roger's model [1983] concerning the diffusion of technology through the levels of U.S. society. The rate of acceptance is a function of the following five variables. The less hesitancy an individual or group of individuals can answer in the affirmative, the sooner the technology will be purchased. Roger's five variables are

- Relative advantage: an immediate recognition that a significant and positive organizational impact will occur with the new technology.
- Compatibility: the new technology will not require massive changes in present operations.
- Complexity: the new technology can be understood in less-complicated terms.
- Divisibility: testing is possible before full installation and commitment to the new technology.
- Communicability: the understanding and impact of the technology can be shared easily with internal employees and customers.

These five gatekeepers of new technology acceptance influence the decision process to purchase technology-based products. Thus, the speed of answering the five questions will discriminate the market into introductory groups and the commercialization market segments. Because the variables help explain adoption rates, the Rogers' variable set can be used to identify and to screen the early product triers from the later product users.

Returning to the example, the segmentation can proceed forward using the previously mentioned methods. Assuming a segment of emergency medical services has been defined preliminary as being located in communities of under-50,000 population, being privately funded, having two vehicles, and being directed by a local manager, the final trial segments can be identified. For this specific segment the number might be 5000 that are independent or part of a chain (national or regional). Not all of the organizations are interested, willing, or capable of purchasing the new software. Therefore, the next step is to determine the initial 15% that can be innovators or early adopters. The need is to determine those factors that will define a subset of these organizations as initial triers and provide the company that crucial market entry position from which to expand. Creativity and knowledge of the industry can provide the insight. It might be the community culture (university located there), current technology being used for vehicle allocation and positioning, personnel backgrounds, or local pressures for delivery of higher service levels that separate the previously defined market segment. There could be several factors that would make the emergency service organization more receptive to trying the new vehicle placement software.

Another approach to determining the initial market entry stage 15% would be to use Rogers' five variables as a multiple screening test. Depending on the initial sales number needed (beta sites) or a general market entry (local vs. regional vs. national), the emergency service organizations can be prioritized according to the five factors. Those organizations that answer a strong affirmative probably are willing to try the software. Further analysis of these firms can provide the needed profile for the remaining market introduction. Successive market screening according to the adoption cycle model will provide a systematic plan for market entry, growth, and share.

As indicated in the prior software example, the next step after the segments of a market are determined is the ranking of the segments. This is termed targeting. There are numerous methods for targeting, which could be the subject of another chapter; however, the key is to evaluate the segments according to the highest probability of the product being purchased. The question is always where the best fit exists between a segment and the product offering of the firm. Given the enhanced insight and knowledge gained from the segmentation process, the ordering of the segments is not a difficult task. The key is to remember the primary criterion: the best opportunity of a fit within the competitive context of the defined market segments.

The targeting process can be used also for selecting beta sites if that is the stage of product development. These then become learning sites for the next phase of the market entry, and the market entry phase becomes the added learning for the commercialization phase. At all stages the fit is confirmed for the product offering and target segment(s). Often, as the product moves into the targeted segment, the next-ranked segments become the growth increments. However, the discipline of segmentation permits the initial fit. Successive learning provides for the market expansion and eventual market share.

Conclusions

This section has defined, described, and illustrated the marketing strategy of segmentation. Segmentation is the process of dividing a market into smaller segments so that the most appropriate fit can be found among the product, producer, and the user. By evaluating potential segments within a defined market, the product producer targets the segment with the highest probability of exchange and then expands the purchase base from there. Even though several segmentation methods were presented, the key idea is that the best marketplace entry segment must be defined based on the use of several variables that could influence the potential product user. The goal is to determine what variables will separate the market into unique segments that represent differing levels of product impact or benefits. The highest impact of product benefits delivered is the segment that is targeted first.

It is by using the segmentation strategy to understand and to rank segments that the firm systematically finds the most likely entry point in the market. Correspondingly, the resources consumed on this entry endeavor are most effectively and efficiently used. Segments then are not smaller markets, but building blocks for market entry, growth, and desired share. Defining, prioritizing, and satisfying market segments is a critical and strengthening marketing responsibility in any organization that provides the basis of matching the firm's resources with a target market segment. Segments are the successive steps to marketplace success.

Defining Terms

Demographics: Physical descriptions of a market that are used to partition into segments such as age, name, gender, education, and occupation. Can be applied to organization also.

Disaggregating: A similar term as partitioning or segmenting that results in dividing a market into more-meaningful product purchase groups.

Partitioning: The method of dividing a market into segments based on a successive application of variables.

Product life cycle: The growth pattern of a product in the marketplace consisting of five stages based on the rate of acceptance of a new product.

Segmentation: The separation of a market into smaller parts (segments) based on variables that generate varying levels of purchase propensity for the product.

References

Kotler, P. *Marketing Management, 8th ed.,* Prentice Hall, Englewood Cliffs, NJ, 1994.

Moore, G. A. *Crossing the Chasm,* Harper Business, New York, 1991.

Rogers, E. M. *Diffusum of Innovations, 3rd ed.,* Free Press, New York, 1983.

Weinstein, A. *Market Segmentation,* Probus, Chicago, 1987.

Further Information

American Demographics magazine defines and describes a demographic segment of the U.S. population in every monthly issue. The magazine has published several books on the segmentation of the U.S. population also. American Demographics, Cawles Business Media, P.O. Box 68, Ithaca, NY, 14851; phone (607) 273-6343.

American Marketing Association offers numerous conferences annually on the subject of segmentation and targeting segments from research, description, and utilization perspectives. American Marketing Association, 250 South Wacker Drive, Chicago, IL, 60606-5819; phone (312) 648-9713.

The Burke Marketing Institute presents several multiday seminars at various U.S. city locations annually on segmentation strategies. These seminars are taught by experts in the field and provide an in-depth study for user understanding and application. The Burke Institute, 50 East River Center Blvd., Covington, KY, 41011; phone 1(800) 543-8635.

12.8 Technology Forecasting and Futuring

Stephen M. Millett

The term **technology forecasting** has both a narrow and an expanded definition. In the narrow sense, technology forecasting is the expectation for a technological parameter to be achieved by a date in the future. In a broader sense, technology forecasting is a group of methods to think about, plan, and implement technologies toward a desired goal. Because of the confusion in meanings, it is more useful to restrict technology forecasting to the narrow definition and to use technology **futuring** for the expanded meaning. Both definitions will be covered in this essay.

Technology forecasting, narrowly defined, is conducted much like economic forecasting. It is an exercise in two-dimensional analytical geometry. Indeed, the tools of technology forecasting came from mathematics and economics rather than engineering.

In a typical technology forecasting graph, time is the x-axis and a physical parameter is the y-axis. The x-axis is usually marked in years with the present year in the middle. Some graphs show as many years in the past as in the future, so that, to illustrate the point, the year 1997 would be in the middle with 1987 at the far left and 2007 at the far right. Some scholars have asserted that there should be twice as many years in the past shown as for years in the future, but there are too many practical reasons to make a firm rule.

The y-axis can be any physical parameter or performance measure. It might be the number of calculations per second made by a semiconducting computer chip. It might be the number of bytes contained in a computer's random access memory. It might be the 1-year charge of a battery. The y-axis can be virtually anything according to the needs of the forecast.

Trend Analysis

Visualized in graphic terms, technology forecasting is trend analysis. This fact explains why trend analysis is the most popular form of technology forecasting. It also illustrates the shortcomings of the approach.

Technology forecasting requires historical data to start the plot from the past and carry it to the present. When done well, historical data can be very educational in showing how rapidly technological performance has moved. All forecasters of whatever kind employ historical trend analysis to one extent or another because the past explains how one got to where one is and what are the bounds of practical reality based on experience. However, the temptation then is to extrapolate the past into the future. Extrapolations look compelling, but they may prove very misleading.

Historical data extrapolated into the future can result in accurate predictions when the data are accurate, the relationships among variables remain constant, and there are no disruptive events that break the continuity of the vector. Trend projections, therefore, work best when a great deal of historical data exists, when the topic is microscopic rather than macroscopic, and when the time horizon is short. This generalization holds for economic and financial as well as technological forecasts.

Of course, not all trend data projections are straight lines. The most famous trend line is the "S curve". Starting in the lower left corner of the graph, the line grows very little over the first few years. Then it accelerates rapidly before cresting and then declining. The plot looks very much like an "S" [Fisher and Pry, 1971]. Mathematicians and economists have tried to calculate the formula for the S curve, but none has achieved perfect prediction. While many people believe in the theory of the S curve, like the theory of supply and demand, no one believes that all curves can possibly be captured by one formula.

Another aspect of the S curve is that the y-axis can be many different things. In some cases, it is the parameter performance. However, in many cases, it is market penetration or level of sales. Such a graph would not strictly constitute technology forecasting but rather marketing forecasting or product sales. Businessmen know from experience that new products, like new technologies, are typically slow in their initial development and then experience a launching like a rocket (if indeed successful). All products, however, sooner or later reach some plateau of stability before declining. Therefore, the concept of the S curve is readily believable, but the actual S curve is virtually impossible to predict mathematically before the future yields data.

Philosophy of Forecasting

At this point, we need to make explicit some philosophical assumptions. One view is that the future will be a continuation of the past, so that trend extrapolations are valid. However, history teaches us that there are changes, some called "revolutions", that dramatically divert streams of events. Another view is that the future will be totally different from today. This is the science fiction approach to futurism. However, the reality is that the future is always some combination of continuity and change. Once an event has occurred, a historian can go back and reconstruct (at least in broad terms) what happened and what were the trends that converged. However, when looking forward, we cannot see but one certain outcome for the confusion of multiple and often-conflicting trends.

We do not literally study the past to know the future. Rather, we study the past to understand how we got to where we are today and to bound the uncertainties of the future. A long-term, stable trend is likely to continue into the future, if it is undisturbed. Change requires an infusion of energy. Where are the sources of change and how powerful are they? Do trends reinforce one another or cancel them out? These are questions that are too often assumed but never explained in two-dimensional analytical graphic trend analysis.

Another philosophical problem is that trend analysis too easily assumes that there is a future and that future is knowable when we have enough data. Too much technology forecasting requires too much data and too much mathematics. Trend analysis is deterministic. It may even assert predestination. To those who do not embrace predestination in their daily livelihood (such as investors and businessmen), technology forecasting may look entirely too mechanical.

Technology forecasting can also appear deceptively "objective". Forecasters kid themselves as well as others when they assert that "the data speak for themselves". All forecasting involving the selection of data and the building of relationship models requires expert judgment, which is human knowledge in action. There may be as much expert judgment in graphs as in scenarios, yet the graphs look more objective. The expert judgment, however, is hidden in the selection of the data and the building of the graphs, including the very selection of the most meaningful y axis.

As mentioned above, trend analysis, despite its limitations, has its proper place. However, technology and economic models both break down when there are many and diverse variables with relationships that change over time and when the forecast period is long term rather than short term.

The greatest shortcoming of technology forecasting, however, in a business context is the attitude that we are reactive to technology progress beyond our control. Businessmen want to be proactive in creating their own competitive advantages in the future rather than reactive to what the competition may do to them. Therefore, from a very practical point of view, technology forecasting is not so much about "knowing" the future as it is about seeing opportunities and making "good" futures happen. Technology futuring is more vigorously practiced by corporations than technology forecasting.

Concept of Futuring

Technology futuring, in the business context, is the rigorous thinking process of anticipating likely technology, product, and market conditions in some future date, recognizing emerging business opportunities and threats, and deriving robust strategies with necessary technology investments to achieve business objectives in forthcoming years. Its goal is to develop strategies and make decisions today that will influence future conditions. It relies heavily on expert judgment and scenarios analysis supported by trend analysis.

Futuring methods, like technology forecasting methods, fall into three major categories: (1) trend analysis, (2) expert judgment, and (3) alternative options analysis. Although often misunderstood and abused, trend analysis is a part of virtually all kinds of technology futuring. It is not a question of whether the past will provide a linear path to the future; but, given current momentum, it is a matter of the most likely future based on continuing trends and how might that trend rise or fall given other conditions. History does provide guideposts to the future.

The most common trend analyses, as mentioned above, include linear projections, regression analysis, and S curves. Patent trend analysis and bibliometrics are two other methods less well known. Patent trend analysis covers the number and type of patents applied for or published by year. Plotting numbers against years shows time lines. The method is more powerful when combined with portfolio analysis of company holdings. Bibliometrics is similar, but the yearly counts are for published articles rather than patents. A sudden rise in the number of published articles on a particular technology is an indicator of recent research and development and forthcoming product developments (which are better tracked in combination with patent trend analysis) [Millett and Honton, 1991].

We have already commented on both the philosophical assumptions and limitations of technology trend analysis. It should always be used in groups; many trends, with their possible intersections, are more useful than single trends considered in isolation.

All forms of forecasting include some element of expert judgment, whether implicitly or explicitly. Expert judgment has the advantages of capturing discontinuities (which trend analysis cannot do) and providing foresight into possibly effective strategies. There are qualitative strengths lacking in quantitative trend analysis. However, expert judgment has its methodological problems, too. No expert knows everything, and all experts have biases. Therefore, one should consult many experts for thoroughness and evenness.

Delphi surveys are rarely done today, and when done stand more as perception studies than as forecasts. Surveys of all types can be administered by e-mail as well as by mail or in person. Interviews are common. Expert focus groups, using idea generation or the Nominal Group Technique are power approaches to gathering much expert opinion within only a few hours and with very reasonable cost [Delbecq et al., 1975].

The most common form of alternative future methods is scenario analysis. The best known approach is the intuitive writing scenario method developed by Shell International and SRI International. The current method of four, quadrangular future scenarios practiced by Global Business Networks and others is an iteration on the earlier Shell approach. Battelle uses cross-impact analysis as the means to generate scenarios through the use of a personal computer software program called BASICS-PC. Typically, between two and five expert focus groups, using the Nominal Group Technique, are held to identify the most important issues concerning a topic question to be included in the scenario analysis. With 18 to 24 issues, we research each to identify current trends and alternative outcomes to the target year in the future. The issues cover economic, social, demographic, market, and competitive matters as well as technology. Cross-impact analysis allows us to see what trends accelerate or retard other trends. A computer-based software program sorts out the calculations and organizes the issue outcomes into consistent, but different, groupings, called scenarios [Huss and Honton, 1987].

The computer approach has two great strengths over the intuitive approach. It forces the scenario team to make explicit all judgments. The process is systematic, reviewable, and revisable — it is not an intuitive black box. The computer approach also allows us to play "what-if" simulations. The simulations give us guidance on what combination of trends are most likely to produce desirable outcomes. From these experiments, strategies can be formulated and technology investments made.

Conclusions

The state of the art in technology forecasting and futuring is in business rather than in academia. Theory and methodology fundaments have changed little over the last 2 decades. The activity has been more in business applications than in the further development of concepts. The most important applications have been in the so-called "fuzzy front end" of new product development.

Through many applications, however, old concepts need to be refined, even revised. The field of technology futuring is ripe for a new era of theories and advanced practices. One possible area is trend analysis. Perhaps with new computer power and neural nets, trend lines will acquire more predictiveness than in the past. Certainly there is a need for integrating multiple trends into broader synthesis. In addition, scenario analysis begs a new generation that combines the best features of intuitive scenario writing and computerized cross-impact analysis. Perhaps the highest-value improvement in futuring is not so much improved trend analysis or scenarios but rather greatly advanced tools to perform simulations of strategies and future outcomes, whether computer based or not.

The burden now rests on the industrial practitioners to more openly share their experiences with the academics, who in turn can apply new models and methods to addressing the age-old challenges of foresight.

Defining Terms

Technology forecasting: In the narrow sense, the expectation for achieving a performance parameter by a predicted date, typically as a trend line on a two-dimensional graph.

Futuring: The strategic thinking process about future conditions, their opportunities, and challenges and the making of decisions today to influence the conditions of tomorrow.

References

Ascher, W. *Forecasting. An Appraisal for Policy-Makers and Planners,* The Johns Hopkins University Press, Baltimore, 1978.

Bell, W. *Foundations of Futures Studies,* Vol. 1, Transaction Publishers, New Brunswick, NJ, 1997.

Delbecq, A. L., Van de Ven, A. H., and Gustafson, D. H. *Group Techniques for Program Planning. A Guide to Nominal Group and Delphi Processes,* Scott, Foresman and Company, Glenview, IL, 1975.

Fisher, J. C. and Pry, R. H. A simple substitution model of technological change, *Technol. Forecast. Soc. Change,* 3, 75–88, 1971.

Fowles, J., Ed. *Handbook of Futures Research.* Greenwood Press, Westport, CT, 1978.

Huss, W. R. and Honton, E. J. Scenario planning — what style should you use?, *Long Range Planning,* 20, 21–29, 1987.

Linstone, H. A. and Turoff, M., Eds. *The Delphi Method. Techniques and Applications,* Addison-Wesley, Reading, MA, 1975.

Martino, J. P. *Technological Forecasting for Decision Making, 2nd ed.,* North-Holland, New York, 1983.

Millett, S. M. and Honton, E. J. *A Manager's Guide to Technology Forecasting and Strategy Analysis Methods,* Battelle Press, Columbus, OH, 1991.

12.9   Learning Curves

Timothy L. Smunt

The **learning curve** concept has been used for over a half century to predict production costs and to determine pricing of complex, technical products. The learning curve was first discussed by Wright [1936] in his article that described the cost trends of the production of aircraft at the Curtiss-Wright Corporation. Wright found that, as the cumulative production experience doubled, the cost of production tended to decrease by a certain percentage. While other more-complicated learning curve models have been proposed and used since then, many companies still find that the simple "doubling" model is an accurate planning tool.

Learning curves can be used in a variety of ways in a company. Production costs for a new product are often estimated based upon past productivity trends of similar products previously produced by the company. Additionally, production costs for future runs of an ongoing manufacture of a product can be predicted based upon the actual learning rate of the product to date. Based upon the estimate of production costs for particular demand volume assumptions, pricing decisions can be made that maintain profitability while still keeping prices low enough to capture large market share. Texas Instruments' calculator pricing scheme was a prime example of how market share can be built by setting prices at a fair but low margin above their projected and decreasing production costs. The use of the learning curve was reported to be the central focus of their pricing decisions.

To control manufacturing costs and determine product pricing in this manner, a number of variables must be considered by managers. They include factors that can be manipulated both before and after the start of a new product's production. Further, these factors can affect either individual learning or organizational learning.

Factors Affecting Learning

In Table 12.3, a list of factors that affect both the learning rate and the initial costs of production are provided.

The factors shown above may be considered either before manufacturing begins or implemented sometime after the first set of units is completed. If more time and effort are placed on the preproduction activities, the first **unit cost** (sometimes referred to as T_1 cost, representing the "time" of unit 1) will probably be lower, but subsequent learning may be less steep. On the other hand, if most of the

TABLE 12.3   Individual vs. Organizational Learning Factors

Individual Learning	Organizational Learning
Previous experience of the worker	Degree of process rationalization
Methods design	Overall education level of employees
Individual incentive pay plans	Group incentive pay plans
Lighting and environmental conditions	Product design stability
Raw material and purchased part quality	Capital equipment specialization and automation
Worker training	Level of vertical integration
Workplace layout	Degree of planning and design efforts

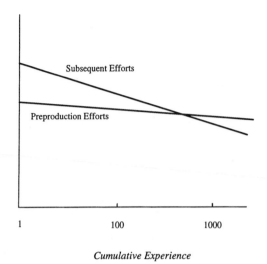

Cumulative Experience

FIGURE 12.15   Effect of preproduction vs. subsequent production management efforts.

management effort to increase production productivity is expended later in the process life cycle, then the earlier units may cost more to produce, although the later units of production will see higher rates of learning (see Fig. 12.15 for an illustration of this effect).

The optimal level of expenditures necessary to increase manufacturing productivity throughout the product life cycle will vary by company. In general, however, firms that compete early in the product life cycle will tend to place more effort on cost reduction after the first set of units during pilot production has been completed. Since these firms typically produce in small volumes, a job shop or batch production mode is often used. Thus, few changes to the general-purpose equipment utilized in these processes are possible before production starts. More often, method improvements, design changes for manufactura-bility, or worker training programs are implemented as production problems manifest themselves during the pilot production stage.

On the other hand, appliance manufacturers or other such firms that do not enter the market until the later parts of the growth stage (or early parts of the mature stage) of the product life cycle will likely find themselves expending a great deal of resources to determine the best product and process design before production begins.

Using Learning Curves in Cost Prediction and Price Analysis

There are a number of ways to model learning trends. For the purposes of this section, we will use the traditional model presented by Wright, which is still widely used today and forms the basis for most other models that have been devised. The basic assumption of the Wright learning curve model is that **cumulative average costs** decrease as the production experience doubles. The relationship between experience and cost is **log-linear** and can be modeled mathematically as follows:

$$Y(x) = T_1\left(x^{-b}\right) \tag{12.1}$$

where $Y(x)$ = cumulative average cost of the xth unit, T_1 = the cost of the first unit, and b = learning curve factor $= -\dfrac{\log(\text{learning curve }\%)}{\log(2.0)}$.

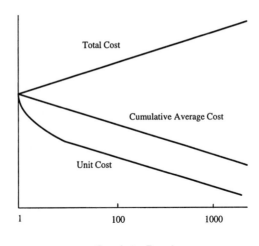

Cumulative Average Cost

FIGURE 12.16 Cumulative average learning curve model.

The total cost (TC) for the first x units is then

$$TC(x) = T_1(x^{-b})(x) \qquad (12.2)$$

or

$$TC(x) = T_1(x^{1-b}) \qquad (12.3)$$

The unit cost (U) is a discrete function based on the total cost through any given point in question less the cost of prior units:

$$U(x) = TC(x+1) - TC(x) \qquad (12.4)$$

Figure 12.16 graphically illustrates the cost curves represented by Eqs. 12.1, 12.3, and 12.4 for the Wright learning curve model.

In Table 12.4, the above formulas are applied to the situation where the first unit cost is $1000 and an 85% learning curve is assumed. All unit doublings from unit 1 to unit 16,384 are shown in order to illustrate the learning curve cost reductions expected in this model. Note that, at each doubling, the cumulative average cost for the latter unit number is equal to 85% of the one above. The use of the learning curve function in Eq. 12.1 provides the same cost calculations.

In Table 12.5, the cumulative average, total cost, and unit costs are shown for units 1 to 15 to illustrate the use of Eqs. 12.2 and 12.4.

As can be seen from Table 12.5, the rate of cost reduction for the unit costs is steeper than that for the cumulative average costs. Of course, this relationship is due to the assumptions of the Wright learning curve model, which uses the cumulative average cost curve as the log-linear function and is also illustrated in Fig. 12.16.

It is often useful to project the cost for a particular batch of production to one in the future, for example, for a new customer's order that might be produced after two orders committed to other

TABLE 12.4 Learning Curve Costs as Experience Doubles

Unit Number	Cumulative Average ($)	Total Cost ($)
1	1,000.00	1,000.00
2	850.00	1,700.00
4	722.50	2,890.00
8	614.13	4,913.00
16	522.01	8,352.10
32	443.71	14.198.57
64	377.15	24,137.57
128	320.58	41,033.87
256	272.49	69,757.57
512	231.62	118.587.88
1,024	196.87	201,599.39
2,048	167.34	342,718.96
4,096	142.24	582,622.24
8,192	120.91	990,457.80
16,384	102.77	1,683,778.27

T_1 = $1,000; learning curve = 85%; b = −0.2345.

TABLE 12.5 Unit Cost Calculations

Unit Number	Cumulative Average ($)	Total Cost ($)	Unit Cost ($)
1	1,000.00	1,000.00	1,000.00
2	850.00	1,700.00	700.00
3	772.91	2,318.74	618.74
4	722.50	2,890.00	571.26
5	685.67	3,428.36	538.36
6	656.98	3,941.87	513.51
7	633.66	4,435.60	493.73
8	614.13	4,913.00	477.40
9	597.40	5,376.58	463.58
10	582.82	5,828.20	451.63
11	569.94	6,269.35	441.14
12	558.43	6,701.17	431.82
13	548.05	7,124.63	423.46
14	538.61	7,540.51	415.88
15	529.97	7,949.48	408.97

T_1 = $1,000; learning curve = 85%; b = −0.2345.

customers. For example, assume that the new order will correspond to units 301 to 450 of production. The **batch cost** estimate (BATCH) is calculated as

$$\text{BATCH}\left(301 \text{ to } 450\right) = \text{TC}\left(450\right) - \text{TC}\left(300\right)$$

$$= \left(\$1000\right)450^{1.2345} - \left(\$1000\right)300^{1.2345} \tag{12.5}$$

$$= \$107,430.20 - \$78,763.03$$

$$= \$28,667.17$$

To determine the historical rate of learning from actual data, either a log-linear regression analysis can be performed using statistical software such as SAS, SPSS, JMP or SYSTAT, or a rough cut analysis

TABLE 12.6   Simulated Actual Data Points

Unit Number	Cumulative Average ($)
2	455.00
4	410.00
40	288.00
80	257.00
125	240.00
400	205.00
800	185.00

can be performed by calculating the learning rate factor, b, for various pairs of data points. The learning rate factor can be calculated in this manner:

$$b = \frac{\log y - \log y'}{\log x' - \log x} \tag{12.6}$$

where x and y = the first unit number of two different batches and x' and y' = the last unit number of two different batches.

In Table 12.6, a number of simulated actual data points are provided from which we can calculate the learning rate factor. Note, for example, that, by applying Eq. 12.6 to units 2 and 4, we get a **b** equal to 0.1502. Further, **b** can be calculated for Units 2 and 40 as 0.1527, for Units 80 and 125 as 0.1533, and for units 2 and 800 as 0.1502. Since the **b** factor is relatively consistent for each of these pairs, we can use the average of the four calculations in determining the learning curve percentage. The average **b** is equal to 0.1516. Converting to the learning curve percentage, we have

$$\begin{aligned} \text{Learning curve } \% &= 2^{-b} \\ &= 2^{-0.1516} \\ &= .9003 \\ &= 90.03\% \end{aligned} \tag{12.7}$$

If we found that **b** varied from each unit pair calculation, then it would be advisable to determine whether a change in learning curve rates occurred at a particular point in time due to a significant event, the addition of product lines to the factory, a change in the worker incentive system, or similar causes.

When Learning Curves Should be Used

The use of learning curves can be somewhat more time consuming than a standard cost analysis, which assumes that costs remain level. For more complex, technical products, however, it is likely that steady and fairly predictable productivity trends will exist. In this case the learning curve approach can provide a competitive advantage since it provides a tool to better understand the dynamic nature of costs and the potential for further improvements through specific investments for individual and organizational learning.

Smunt [1986a] has shown that a simpler approach to estimating productivity trends can be beneficial if the actual costs exhibit high degrees of variance. The use of a moving average can simplify trend analysis and provide quick updates to changes in the rate of improvement. On the other hand, when the cost data are fairly stable, explicit use of the learning curve function has been shown to be superior.

Additionally, cost data may be available for either detailed levels of production or for higher levels of product level aggregation. If the product design is stable, then the use of product level (end item) cost information is sufficient for learning curve analysis and pricing decisions. However, if portions of the product (subassemblies or components) are frequently redesigned, then it becomes necessary to estimate learning curves at more-detailed levels of production, especially when the learning rates for components vary for any reason. Aggregation issues as related to learning curve analysis are discussed in a number of articles, including Conway and Schultz [1959] and Smunt [1986b]. Conway and Schultz also discuss the impact of inflation on learning curve estimates. Inflation adjusted cost figures or labor hour data should always be used to determine the true learning curves.

Learning curve research on technical products continues today. Recent studies by Bohn [1995], Epple et al. [1996], and Adler [1990] are good examples of how learning curves can be used to better understand a wide variety of management issues in companies producing complex products. Other recent research has been published that integrates optimal lot sizing with learning curve analysis to provide methods to reduce total costs of production (see, for example, Hiller and Shapiro [1986], Mazzola and McCartle [1996], Pratsini et al. [1993], and Smunt and Morton [1985]).

In summary, learning curve analysis can be applied to a wide variety of cost estimating and pricing situations. Such application requires thorough understanding of the factors that contribute to the learning trends and the careful use of the learning curve formula.

Defining Terms

Batch cost: The sum of all the unit costs with a certain unit range.
Cumulative average cost: The average cost of all units produced up to a certain point (from unit 1).
Experience curve: The rate of productivity improvement for total costs, including overhead costs and other general expenses.
Learning curve: The rate of productivity improvement of direct costs as experience doubles, expressed as a percentage. For example, a 90% learning curve is one where the cumulative average cost of producing the fourth unit is 90% of the cumulative average cost of the second unit. The learning curve may also apply to unit costs.
Log-linear relationship: The straight-line relationship of the learning curve function when plotted on log-log coordinates.
Unit cost: The cost of producing a specific unit.

References

Adler, P. S. Shared learning, *Mgt. Sci.*, 36(8): 938–957, 1990.

Bohn, R. E. Noise and learning in semiconductor manufacturing, *Mgt. Sci.*, 41(1): 31–42, 1995.

Conway, R. W. and Schultz, A. Jr. The manufacturing progress function, *J. Ind. Eng.*, 10(1): 39–54, 1959.

Epple, D., Argote, L., and Murphy, K. An empirical investigation of the microstructure of knowledge acquisition and transfer through learning by doing, *Oper. Res.*, 44(1): 77–86, 1996.

Hiller, R. S. and Shapiro, J. F. Optimal capacity expansion planning when there are learning effects, *Mgt. Sci.*, 32(9): 1153–1163, 1986.

Mazzola, J. B. and McCardle, K. F. A Bayesian approach to managing learning-curve uncertainty, *Mgt. Sci.*, 42(5): 680–692, 1996.

Pratsini, E., Camm, J.D., and Raturi, A. S. Effect of process learning on manufacturing schedules, *Comput. Oper. Res.*, 20(1): 15–24, 1993.

Smunt, T. L. A comparison of learning curve analysis and moving average ratio analysis for detailed operational planning, *Decision Sci.*, 17(4): 475–495, 1986a.

Smunt, T. L. Incorporating learning curve analysis into medium-term capacity planning procedures: a simulation experiment, *Mgt. Sci.*, 17(4): 475–495, 1986b.

Smunt, T. L. and Morton, T. E. The effects of learning on optimal lot sizes: further developments on the single product case, *IIE Trans.*, 17(1): 3–37, 1985.

Wright, T. P. Factors affecting the cost of airplanes, *J. Aeronaut. Sci.*, 3(4): 122–128, 1936.

Further Information

Further examples of learning curve calculations and its uses can be found in *Learning Curve for Cost Control* by Jason Smith, Industrial Engineering and Management Press, Institute of Industrial Engineers, Norcross, GA, 1989.

An excellent review of the various types of learning curve models can be found in "The Learning Curve: Historical Review and Comprehensive Survey", *Decision Sci.*, 10: 302–328, 1979, by Louis E. Yelle.

12.10   Advertising High-Technology Products

Eitan Gerstner and Prasad Naik

Marketing involves activities that are traditionally classified according to the framework of the **4Ps**: product-related activities, pricing-related activities, place (or distribution)- related activities, and promotion (or communication)- related activities. Advertising is a component of promotions together with personal selling, sales promotions, and public relations, as illustrated in Fig. 12.17.

Advertising is defined as a paid communication to seek desired responses from customers. In this section we will focus on the role of advertising in the marketing of technology products.

Advertising Objectives and Tasks

Advertising can be viewed as an input to attain the following objectives.

1. Provide information about a product or service.
2. Generate leads for product inquiries or orders.
3. Build **brand equity** (goodwill).
4. Remind customers about the product features.
5. Persuade customers about the benefits of the product.

These objectives help move customers from stage to stage in the buying process shown in Fig. 12.18.

The buying process starts when customers recognize a certain problem or need that is not fulfilled, e.g., machine breakdowns, service failures, etc. To solve this problem, customers typically search for information. Because technology products are typically "new to the world" and complex, consumers need more information compared to simpler products such as consumer packaged goods. The information gathered helps customers identify alternatives to solve the problem. As customers evaluate these

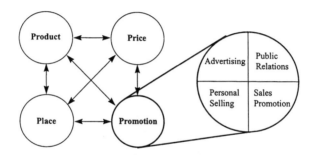

FIGURE 12.17   Placing advertising within the 4Ps of marketing.

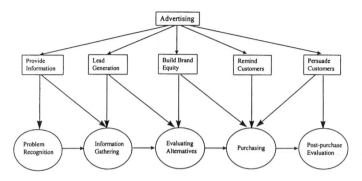

FIGURE 12.18   Objectives and tasks of advertising of technology products.

alternatives, they start forming their preferences, which lead to purchases of goods and services. After purchasing, customers use the product and evaluate its performance. Customer satisfaction is obtained if product performance is not below expectations.

This buying process is useful in understanding the role of advertising. First, sellers would like to identify potential customers in the problem-solving stage. Advertising provides preliminary information via magazines, trade shows, Web sites, etc. thereby inviting potential customers to request additional information. Second, advertising generates leads for prospective customers whose names and other information can be compiled into a database to be used for future marketing activities. Third, advertising creates a favorable image for the seller and its products and builds brand equity. This helps convince the prospective customers to include the firm's product in the set of alternatives to be considered. Fourth, advertising can also remind customers about the product benefits and its advantages relative to other brands and influence them to purchase it. Finally, advertising can reassure customers about the wisdom of their decision and help remove any doubts or second thoughts they may have after purchase.

Advertising Strategy: Push vs. Pull

Manufacturers typically sell products to customers through distributors. To stimulate demand, they can advertise their products directly to customers. This is called **pull advertising** because the manufacturer expects that customers will demand the product from the distributors as a result of advertising campaign. Alternatively, manufacturers can advertise to distributors or share advertising effort with them. This strategy is called **push advertising** because the manufacturers expect that advertising will motivate the distributors to push the products to customers (see Fig. 12.19).

Push strategy is more cost-effective than pull for hi-tech companies because the number of distributors is usually smaller than the number of customers. However, pull strategy can create strong name recognition with customers. Intel very successfully used pull strategy in combination with push in their "Intel Inside" campaign. They ran television ads directed to consumers and at the same time they shared advertising expenses with distributors who agreed to put the logo "Intel Inside" on their assembled computers, packaging materials, and promotions. U.S. Robotics is another hi-tech manufacturer that is pursuing such a push-pull advertising strategy.

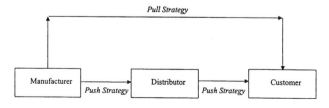

FIGURE 12.19   Push vs. Pull Advertising Strategies

TABLE 12.7   Percentage of Companies Using
Different Budgeting Methods

	Method	1982 (%)	1987 (%)
1	Objectives and tasks	40	50
2	Percentage of sales	38	25
3	Competitive parity	8	8
4	Affordable	49	50
5	Experimentation	14	20

Source: Lehmann, D. R. and Winer, R. S. *Product
Management*, p. 321, Richard D. Irwin, Chicago, 1997.

Once the advertising strategy has been chosen, the next step is to determine advertising expenditures to implement the strategy. Common methods of setting advertising budgets are discussed next.

Setting Advertising Budgets

Table 12.7 presents the most widely used methods for budgeting advertising expenditure.

In the objectives and tasks approach, the firm first determines the objectives that it wishes to attain (see Fig. 12.18). Then, advertising expenditure is obtained by computing the expenses involved in performing the various tasks to meet the objective.

In the percentage of sales method, advertising expenditures for the current year are determined as a percentage of sales in the previous years. The average percentage of sales in some industries can be as high as 18% (games and toys), while in others industries it can be below 1% (computer and office equipment).

In the competitive parity approach, a firm's spending on advertising is in proportion to the desired market share. For example, to obtain 20% market share, a firm will try to spend 20% of the total advertising in the industry. However, small companies spend more, while large companies spend less than the amount suggested by this method [Jones, 1990].

Small hi-tech companies typically spend most of their resources on technological research and product development, and, therefore, there is not much left to spend on advertising. Thus, advertising budget is basically a residual of what is left in the total expenditure budget. This is referred to as the affordable method.

Finally, in the experimentation approach, a company tries to determine the optimal advertising expenditures by varying the levels of spending and measuring its impact on sales. Eventually, the optimal expenditures is the one that generates the highest profit.

The next step after determining budget is to decide which specific media to select and what messages to communicate to the target audience.

Media Available and Creative Strategy

Media Available

Shaklin and Ryans [1987] conducted a survey to determine how different media are used by hi-tech marketers. The following list of media vehicles reflects the intensity of use in the order of importance:

- Specific industry trade journals
- Brochures and other promotional literature
- Direct mail
- Trade shows
- General business magazines
- Video tapes
- Newspapers
- Television
- Radio

A new and emerging medium that is growing at a very fast rate is the Internet, or World Wide Web. One advantage of advertising on the Internet is that it enables advertisers to target their audience efficiently because customers self-select the Web sites of their interest. In addition, unlike the conventional media, which permits only a fixed amount of space or time, the Internet via its interactivity provides the customer the option of obtaining as much information on a firm's products as desired by spending more or less time on that site.

Creative Strategy

In creating an advertising message there are several important things to keep in mind.

1. The typical customer is exposed to hundreds of messages every day through a variety of media. A good advertising message should break through this clutter by using imaginative ways to attract attention of the target customer.
2. A company should design messages that are consistent with other elements of its marketing mix (see Fig. 12.17). Therefore, the content of advertising copy should be carefully evaluated and tested before it is launched to the entire market.
3. The impact of advertising message should be tracked over time to detect any wearout due to repetition (boredom) or obsolescence (dated). The firm should consider replacing the advertised messages periodically to assure the effectiveness of its campaign.

Advertising agencies play an important role in media selection and creative strategy. Next, we discuss this issue.

Role of Advertising Agency

Most companies outsource advertising and media tasks to advertising agencies. There are a variety of advertising agencies. On one extreme, there are specialists who perform specific tasks such as producing jingles, creating graphic artwork, or buying and selling media space or time. On the other extreme, there are full-service advertising agencies that provide a complete range of services to the clients. These activities include copy design, artwork, production, media scheduling, media selection, marketing research, and consulting.

We recommend that companies should be highly involved with the design and monitoring of the campaign and not rely entirely on advertising agencies. Such involvement will ensure consistency between advertising campaign and overall marketing strategy.

Conclusions

Advertising is an important component of the marketing mix for any organization. It is important, in particular, for hi-tech companies because they bring new products to the market at a fast pace. To develop the market, hi-tech companies must quickly inform and educate consumers about the benefits of the new products or technology. Other forms of promotions such as word-of-mouth are important, but its impact is slower. Some of the most successful hi-tech companies such as Microsoft, Intel, and AT&T have achieved dominance in the marketplace not only because of their novel products but also because of their aggressive advertising campaigns.

Defining Terms

Brand equity: Additional value a brand name brings to a company because of its higher appeal and premium prices customers are willing to pay.

Pull advertising: Manufacturer's advertising aimed directly to consumers.

Push advertising: Manufacturer's advertising aimed directly to distributors.

4Ps of marketing: The four decision areas in marketing, which consist of product-related decisions, pricing decisions, distribution (place) decisions, and promotion decisions.

References

Lehmann, D. R. and Winer, R. S. *Product Management*, Richard D. Irwin, Chicao, 1997.

Jones, J. P. Ad spending: maintaining market share, *Harv. Bus. Rev.*, 68: 38–48, 1990.

Shanklin, W. L. and Ryans, J. K. *Essentials of Marketing High Technology*, Lexington Books, Lexington, MA, 1987.

Further Information

Association of National Advertisers, 155 East 44th Street, New York, N.Y. 10017; phone (212) 697-5950.

12.11 Efficient Direct Marketing for High-Technology Products

Peter LaPlaca

What is Direct Marketing

Direct marketing is a strategy for putting manufacturers of high-technology products directly in touch with their prospects and customers without the use of marketing intermediaries such as distributors, dealers, wholesalers, or retailers. Direct marketing of high-technology products can be used for consumer products as well as industrial or business products. Because the customers communicate directly with the manufacturer, response times, frequently expanded due to delays in communications between marketing middlemen and manufacturers, can be reduced. Direct marketing's purpose is to produce a customer reaction that generates an immediate or ultimate profit to the manufacturer. Its long-term goal is to build an effective relationship that continuously increases the lifetime value of the customer.

Direct marketing is fundamentally different from other types of marketing in that it places the customer in charge. Customers and prospects must respond to the invitation of the offering and this response comes directly to the manufacturer.[5] By coming directly to the manufacturer and bypassing the marketing middlemen, the typical margins of these marketing institutions can be eliminated.[6] This is extremely important for start-up and small high technology companies where manufacturing costs may be high and prospects for significant cost reductions due to experience curve factors are low. Typically, manufacturers like to price their products at four to five times the direct manufacturing cost. This may result in a satisfactory selling price (to the end customer) if it were purchased directly but a totally unsatisfactory price when marked up several times by a series of middlemen. For example, a high-tech product produced at a cost of $125 is factory priced at $500. This is purchased by an original equipment manufacturer and marked up 100% to $1000 and sold to a dealer or distributor who in turn adds a 25% markup on their cost so the product is now priced to the end customer at $1250. Clearly many more units could be sold to the final customer at $500 than at $1250. Direct marketing is one means of selling the product to the final customer closer to the factory price thereby increasing **customer value** and strengthening the manufacturer-customer relationship.

---

[5]For many consumer products where responses can come in the tens of thousands or more, manufacturers frequently employ a fulfillment house to help process customer orders and ship product. However, this is generally invisible to the customers who believe the manufacturer is completing the process without assistance.

[6]These margins (or markups) can range from a few percentage points for a very high volume item to markups of 100% or more for many low-volume, high-tech products. These margins are used to cover the operating costs and desired profits of the middlemen.

Types of Direct Marketing

Manufacturers can select from numerous types of direct marketing. These include direct mail, telemarketing, print media such as magazines, broadcast media, catalogs, or invitations to respond placed directly in or on products (to encourage repeat buying or the buying of related products and services.) The type of direct media used depends on the type of information to be conveyed, the prospects' and customers' familiarity with the product, and the media's ability to provide sufficient customer information to accurately target the direct marketing offering to appropriate customers.

Direct Mail

When thinking of direct marketing, many people think of direct mail. Unfortunately, many place this in the context of junk mail. Poorly designed and targeted direct mail **is** junk mail, but well-designed and targeted direct mail can not only be an effective generator of sales but can actually be welcomed and is quite useful to customers.

Direct mail has many advantages over other forms of advertising. With carefully developed databases containing information about customer purchase patterns, demographics, and other factors, you can obtain almost unlimited selectivity in targeting your mailing pieces. Direct mail also offers great flexibility in how you package the offering. For some products direct mail offers the opportunity to send product samples with personalized messages to high-probability prospects. Direct mail offerings can involve several media. For example, in addition to beautifully designed brochures, catalogs, or other print media, you can include videotapes, CD-ROM, audiotapes, floppy disks, and similar media to enhance the overall presentation to the customer or prospect.

While many untargeted mass mailings have response rates of 2 or 3%, carefully focused and prepared direct mailings for high-technology products that may have very limited customer bases can obtain responses from a much higher percentage of targeted prospects.

Catalogs

Catalogs began as a means of reaching widely dispersed customers that were not served by retail establishments. Today catalog sales represent 2% of our gross national product and 4% of retail sales. While many consider the overall catalog business a mature one, several sectors of this industry are experiencing rapid growth. These include technology products, business-to-business, and specialty niche-driven catalogs.

In reality, catalogs are specially focused magazines, and as such they are more than a mere listing of products offered for sale. They contain information for customers. High-technology catalogs can describe products and their uses. They can also provide useful reference material for prospects so that the prospect will retain and use the catalog over an extended period thereby increasing the number of times that they are exposed to your sales message. While many people view catalogs as product centered since they feature the company's products, effective catalogs are customer centered, beginning with the customer's needs and presenting products and information in such a way to maximize the transfer of information to the customer. Beyond the physical catalog, catalog operations must also be exceptional. Order entry, order processing, packaging, shipping and delivery, order tracking, billing, and customer follow-up must all contribute to a positive customer experience.

Magazines

During the past few decades, magazines have continued to become more focused on specific market niches, enhancing their potential as direct marketing vehicles. The difference between traditional advertising and direct marketing advertising is the way customers respond to the ads. Rather than merely presenting the product on the page, direct-response magazine ads require the reader to respond directly to the advertiser (the high-tech manufacturer) in the form of calling a toll-free number, circling a number on a bingo card that is processed and sent to the advertiser, or mailing in a business card or a request for more information on company letterhead. The message must grab the reader's attention, must interest the reader, must transfer information that the reader wants, and must stimulate the reader to take action.

Newsletters

A growing medium that can be very effective for direct marketing of high-technology products is newsletters. Newsletters can be very specialized and aimed at very specific markets. They are especially useful when used to maintain relationships with existing customers. In this case, every reader has already committed to the company by prior purchases, and the newsletters are very useful in launching new products or selling related products and services.

Naturally, the focus of the newsletter is the presentation of useful information to the customers. This information can be helpful hints on how to get more usefulness from the products already purchased or they can contain information that will help the customer operate more efficiently (for a business-to-business customer) or enjoy greater leisure time at home. Well-placed product ads or announcements can solicit readers to respond directly to the company to purchase a new upgrade or other item.

Newspapers

High technology products that serve a diverse consumer market may find newspaper preprints or inserts can be a effective direct response medium. These are usually on heavier stock and in full color. Products that are low to moderately priced and that appeal to a broad cross section of the market are best suited for this approach. Prospects can send in the reply card or call a toll-free number to either order the product or to obtain additional information. Due to the way newspapers are produced and the inherent currency of their content, newspaper direct response ads and inserts offer the most rapid response time of all print media. In most cases your direct response ad can appear in print within 72 hours; preprint materials will take longer due to production times.

Television

While any television ad that prompts the viewer to contact the advertiser direct can be termed a direct response ad, the specific type of television show that has the greatest potential for smaller high-technology manufacturers are the home shopping shows on cable television. Cable TV is much more focused than broadcast television, and cable operators can provide detailed information on the viewers of specific shows. They know who is hooked up and what they are watching! The two largest cable shopping operations are the Home Shopping Network (HSN) and QVC. These shows provide a longer format than normal television ads, so more information about the product can be included in the ad. Prospects call the toll-free number flashed on the screen and the order is taken. HSN or QVC then place the order directly with the manufacturer or the fulfillment company.

Another form of direct television advertising is the 30-minute sponsored sales show. These shows have the viewer's full attention for the duration of the show. Products are shown and described, users are interviewed, product benefits are explained, and, of course, operators are standing by for the full 30 minutes.

In preparing direct-response television advertisements, it is important to keep the following items in mind:

1. Television is a visual and action medium. Show the product in use and make sure the benefits are clearly the highlights of the visual treatment.
2. Make sure the viewer understands the offer. Exactly what will they get if they respond?
3. Create a sense of urgency by having the viewer call the number *now*!
4. Emphasize the fact that they cannot obtain the product except by calling the number. It is not available in their local store.
5. To facilitate rapid response indicate that the offer is good only for a limited time.
6. You must ask for the response many times during the commercial. Just once at the end will *not* do the job.

Radio

Radio is similar to television, but it offers a greater variety of program formats advertisers can select from and it is much less expensive than television. Radio also offers the opportunity for greater creativity. The advertiser is not bound by the limitations of what is possible visually, but can ask the listener to imagine things. Combinations of words, sounds, and music can provide a tremendous range of possibilities.

Co-Ops

Co-ops or cooperative promotions involve the combining of several manufacturers' direct-response ads into a common package for delivery. These cooperative efforts are typified by packages of coupons from a variety of manufacturers and packages of business reply postcards. The former are almost always for consumer products and services while the latter are commonly used for business-to-business products and services including high-tech products. The key to getting your co-op message read is to keep it simple with excellent graphics. Multiple sales pitches or complicated appearances get discarded.

Telemarketing

Telemarketing efforts can be either reactive or proactive. Reactive systems respond to customer initiated calls, usually on a toll-free number. Proactive systems are initiated by the seller calling the prospect. Consumers have a decidedly less-attractive attitude toward proactive telemarketing than reactive telemarketing.

Many high-tech firms have a reactive telemarketing effort, which is called inside sales. Without proper training, supervision, and control, many high-tech firms loose countless orders from customers who **want** to order but are stymied by the poor system. Excellent reactive systems, such as those at L.L. Bean and Lands End, provide callers with an attractive, easy-to-use system that is designed to minimize errors and to cross-sell callers on related or complementary products. High-tech firms could gain valuable knowledge by studying these two world-class systems.

While consumers in general do not respond well to proactive telemarketing systems, established customers tend to have a much better attitude toward this practice. This is especially true when the manufacturer-initiated call is not always perceived as a sales call but as relationship building. These calls are genuinely designed to see how the customer is doing, to inform the customer of new advances or hints in how they might use their existing purchase more effectively or efficiently. High-technology proactive telemarketing programs can be effectively coupled with customer newsletters and product or company announcements. The proactive call makes it easier for the customer to respond to information contained in the newsletter or announcement.

Internet

Due to its interactive nature, the internet is ideally suited to direct response marketing. Prospects interested in products actively seek out suppliers on the net using a variety of search engines. The user is then provided with a list of net sites that contain the information requested. High-tech companies can have Web pages that contain pictures of products, complete catalogs, technical manuals, and much more. Costs can range from less than $1000 to over $30,000 per year.[7] Once designed, the Web page must be maintained to contain the most current product information, customer hints, price data, and other items of interest to the site user.

The Internet can be a low-cost means of establishing instant global distribution. Prospects around the world can see your Web page and order directly through the site. Using credit or debit cards or other means of electronic funds transfer, the Web can fully automate your ordering process while minimizing billing procedures and collection problems. However, if you just let your Web page simply exist on the Web, it will get little use. You must advertise the Web address in all of your other promotional activities including print and broadcast advertising, product literature, product instructions, technical manuals,

---

[7]Unless you are a professional Web Page designer, it is highly recommended that you hire a professional to do this work. Poorly designed pages can cost you money with little if any response. Professionals will know how to structure the page for maximum impact, how to link it to other Web sites to maximize the "good hits", i.e., accesses by hot prospects and not just random Web surfers, and how to integrate the Web site with your customer database and information system.

product packaging, billing documents, business cards, letterhead and any other means you can. It must also be linked to other Web sites that will attract hot prospects with a high buying probability.

On-Product

A frequently overlooked area for direct response marketing efforts is the product itself. This is directly targeted to existing customers and can easily be used to build customer relationships and encourage repeat buying. High-tech manufacturers can enclose copies of their catalog in all product packages. Some firms enclose catalogs with all of their shipments to provide customers with the information on all products. You can also enclose materials to purchase extended warranty programs or long-term service contracts. On-product marketing is highly effective because you are dealing with customers who are already favorably inclined toward your product and your company.

Databases

The key ingredient in a successful direct marketing program is the quality and usefulness of your database. Databases contain the information used to target and focus direct marketing programs. Direct mail must focus on the right prospects to minimize wasted mailings. Analysis of the types of customers purchasing each product will enable you to fine tune product options. As the name implies, customer databases contain information about the customer. A single hierarchical customer database will contain all information about your customers, their purchase history, billing and payment history, etc. organized by specific customer. When the customer calls the company, the operator or customer service person can call up a complete history of the customer's relationship with the company. This information is useful in cross-selling other products and services, in deciding on what special treatment the customer deserves (high-volume customers deserve and should get special considerations), and in making sure the customer's needs are being met.

Relational databases are much simpler and provide the firm with more flexibility. Rather than being organized by customer record, the relationship approach links multiple databases together. For example, a simplified customer database (only containing demographic and purchase information) can be linked to a product database, to an accounting database, and to an inventory database. Use of relational databases provides maximum flexibility in the analysis of data to plan and target direct marketing activities.

Direct marketing databases must be maintained for maximum usefulness. Names and addresses must be correct and current. Duplications must be purged as must the names and addresses of people who have died or moved with no forwarding address. A current and accurate database is a valuable resource, and the firm must maintain security for the database. This prevents theft and unauthorized use as well as accidental or deliberate attempts to corrupt the database.

The Offer

The offer is what you present to the customer or prospect. It is the product and its features and benefits. It is the price. It is the promise of satisfaction guaranteed. It is what the prospect responds to. Of these items, the benefits that the customer derives from using the product are the most important. Promising benefits that customers don't want or deem important will not sell the product. Customer benefits, more than product features, are what sells the product. The offer must clearly present one or, at most, two customer benefits. Description of product features that reinforce the benefit are useful, but they cannot replace statements of customer benefit.

The price must also be clearly stated. Does it include shipping and handling? How much do you charge for special features or options? What kind of credit options does the customer have? What guarantees are offered?

For some high-technology products, the direct marketing program is *not* designed to generate sales. Rather it stimulates requests for more information, trial use, or other presale activities. Specific actions of the responder must clearly be described to avoid confusion. When evaluating these offerings, you must remember that sales are *not* the appropriate measure of effectiveness but rather the specific actions that are supposed to be stimulated by the offer.

Managing Direct Marketing Leads

All direct marketing programs generate leads, whether these are inquires for information or actual sales orders.[8] Successful lead management systems provide for rapid and complete follow-up. The system must provide for lead qualification and must direct qualified leads to the sales force (either field sales people, manufacturer's representatives, or the company's telemarketing department.) It is very important to manage the entire direct marketing effort so that the stream of leads is as uniform as possible. Periodic surges of leads will result in many not being followed-up, causing customer dissatisfaction. An insufficient number of leads or period shortages will result in the sales force losing faith in the lead-generating system.

Evaluating Direct Marketing Programs

As with any marketing effort, a direct marketing program must be continuously monitored and evaluated. Have the programs objectives (sales and otherwise) been met? Is the program (and its components) as effective and efficient as possible? Does the program change to keep pace with changes in customers desires? How does our direct marketing program compare to those of our competitors? How does it compare to world-class programs regardless of the industry?

The high-technology company must make certain that satisfactory answers to these and other questions can constantly be obtained. There is a tendency to let successful programs alone, but this is a mistake. Why is our program working? Are our goals too low? It's working, but how can it be even more efficient or effective? We live in a world of continuous improvement of products, services, and systems. Our direct marketing program must also strive for continuous advancement...someone may be gaining on you.

Defining Terms

Direct marketing: Direct marketing is a strategy for putting manufacturers of high-technology products directly in touch with their prospects and customers without the use of marketing intermediaries such as distributors, dealers, wholesalers, or retailers.

Customer response value: The total amount of sales received from a customer over a specified time period (year, multiyear, or lifetime.)

References

Kremer, J. *The Complete Direct Marketing Sourcebook,* John Wiley & Sons, New York, 1992.
Resnick, R. and Taylor, D. *The Internet Business Guide, 2nd ed.,* Sams.net Publishing, Indianapolis, IN, 1995.
Settles, C. *Cybermarketing: Essentials for Success,* Ziff-Davis Press, Emeryville, CA, 1995.
Stone, R. *Successful Direct Marketing Methods, 6th ed.,* NTC Business Books, Lincolnwood, IL, 1997.

Further Information

American Telemarketing Association, 5000 Van Nuys Boulevard, #400, Sherman Oaks, California 91403; phone (818) 995-7338.
Direct Marketing Association, 11 West 42nd Street, New York, NY 10036; phone (212) 768-7277.

---

[8]In general, the higher the cost of the product the more likely the lead will be a request for further information rather than an actual sales order. This is especially true of leads for high-technology products targeted to businesses rather than consumers.

12.12   Brand Equity

Kevin Lane Keller

According to the American Marketing Association, a **brand** is a "name, term, sign, symbol, or design, or a combination of them intended to identify the goods and services of one seller or group of sellers and to differentiate them from those of competition." A brand should be contrasted from a product. A **product** is anything that can be offered to a market for attention, acquisition, use, or consumption that might satisfy a need or want [Kotler, 1997]. A brand is a product then, but one that adds other dimensions that differentiate it in some way from other products designed to satisfy the same need or want. These differences may be rational and tangible — related to product performance of the brand — or more symbolic, emotional, and intangible — related to what the brand represents.

By creating perceived differences among products through branding and developing a loyal consumer franchises, marketers create value that can translate to financial profits for the firm. Brands are thus valuable intangible assets that need to be handled carefully. Accordingly, **strategic brand management** involves the design and implementation of marketing programs and activities to build, measure, and manage brand equity. In a general sense, most marketing observers agree that brand equity is defined in terms of the marketing effects uniquely attributable to the brand [Farquhar, 1989], that is, **brand equity** relates to the fact that different outcomes result in the marketing of a product or service because of its brand name or some other brand element, as compared to if that same product or service did not have that brand identification [Aaker, 1991, 1996; Keller, 1998].

A Conceptual Framework for Brand Equity

Recognizing the importance of the customer in the creation and management of brand equity, **customer-based brand equity** is defined as the differential effect that consumers' brand knowledge has on their response to the marketing of that brand [Keller, 1998]. A brand is said to have positive customer-based brand equity when customers react more favorably to a product and the way it is marketed when the brand is identified as compared to when it is not (e.g., when it is attributed to a fictitiously named or unnamed version of the product).

The basic premise behind customer-based brand equity is that the power of a brand lies in the minds of consumers and what they have experienced and learned about the brand over time. More formally, consumers' brand knowledge can be defined in terms of an associative network memory model as a network of "nodes" and "links" where the brand can be thought of as being a node in consumers' memory with a variety of different types of associations potentially linked to it. Brand knowledge can be characterized in terms of two components: brand awareness and brand image. **Brand awareness** is related to the strength of the brand node or trace in memory as reflected by consumers' ability to recall or recognize the brand under different conditions. **Brand image** is defined as consumer perceptions of and preferences for a brand, as reflected by the various types of brand associations held in consumers' memory.

Sources of Brand Equity

Customer-based brand equity occurs when the consumer has a high level of awareness and familiarity with the brand and holds some strong, favorable, and unique brand associations in memory. The latter consideration is often critical. For branding strategies to be successful and brand equity to be created, consumers must be convinced that there are meaningful differences among brands in the product or service category. In short, the key to branding is that consumers must *not* think that all brands in the category are the same.

In some cases, however, brand awareness alone is sufficient to result in more favorable consumer response, e.g., in low involvement decision settings where consumers lack motivation and/or ability and are willing to base their choices merely on familiar brands. In other cases, though, the strength, favorability, and uniqueness of brand associations play a critical role in determining the differential response making up the brand equity. These three critical dimensions of brand associations are determined by the following factors:

1. *Strength.* The strength of a brand association is a function of both the amount or quantity of processing that information initially receives and the nature or quality of that processing. The more deeply a person thinks about brand information and relates it to existing brand knowledge, the stronger are the resulting brand associations. Relatedly, two factors facilitating the strength of association to any piece of brand information is the personal relevance of the information and the consistency with which this information is presented over time.

2. *Favorability.* Favorable associations for a brand are those associations that are desirable to customers and are successfully delivered by the product and conveyed by the supporting marketing program for the brand. Associations may relate to the product or other intangible, non–product-related aspects (e.g., usage or user imagery). Not all brand associations, however, will be deemed important and viewed favorably by consumers nor will they be equally valued across different purchase or consumption situations.

3. *Uniqueness.* Finally, to create the differential response that leads to customer-based brand equity, it is important to associate unique, meaningful **points of difference** to the brand to provide a competitive advantage and a "reason why" consumers should buy it. For other brand associations, however, it may be sufficient that they are seen as comparable or roughly equal in favorability to competing brand associations. These associations function as **points of parity** in consumers' minds to establish category membership and negate potential points of difference for competitors. In other words, these associations are designed to provide "no reason why not" consumers should choose the brand.

Outcomes of Brand Equity

Assuming a positive brand image is created by marketing programs that are able to effectively register the brand in memory and link it to strong, favorable, and unique associations, a number of benefits for the brand may be realized, as follows:

1. Greater loyalty
2. Less vulnerability to competitive marketing actions
3. Less vulnerability to marketing crises
4. Larger margins
5. More inelastic consumer response to price increases
6. More elastic consumer response to price decreases
7. Greater trade cooperation and support
8. Increased marketing communication effectiveness
9. Possible licensing opportunities
10. Additional brand extension opportunities

Measuring Brand Equity

According to the definition of customer-based brand equity, brand equity can be measured indirectly, by measuring the potential sources of brand equity, or directly, by measuring the possible outcomes of brand equity. Measuring sources of brand equity involves profiling consumer knowledge structures in terms of breadth and depth of awareness and strength, favorability, and uniqueness of brand associations. Measuring outcomes of brand equity involves approximating the various benefits realized from creating these sources of brand equity. The two measurement approaches are complimentary and should be used together. In other words, for brand equity to provide a useful strategic function and guide marketing decisions, it is important for marketers to fully understand the sources of brand equity, how they affect outcomes of interest (e.g., sales), and how all of this information changes, if at all, over time.

Building Brand Equity

Building customer-based brand equity requires creating a brand that consumers are aware of and with which consumers have strong, favorable, and unique brand associations. In general, this knowledge building will depend on three factors:

1. The initial choices for the brand elements or identities making up the brand
2. The supporting marketing program and the manner by which the brand is integrated into it
3. Other associations indirectly transferred to the brand by linking it to some other entity.

Choosing Brand Elements

A number of options exist and a number of criteria are relevant for choosing brand elements. A **brand element** is visual or verbal information that serves to identify and differentiate a product. The most common brand elements are brand names, logos, symbols, characters, packaging, and slogans. Brand elements can be chosen to enhance brand awareness or facilitate the formation of strong, favorable, and unique brand associations. The test of the brand-building contribution of brand elements is what consumers would think about the product *if* they only knew about its brand name, associated logo, etc.

In terms of choosing and designing brand elements to build brand equity, five general criteria can be used:

1. *Memorability* — e.g., easily recognized and recalled.
2. *Meaningfulness* — e.g., credible and suggestiveness as well as fun, interesting, and rich in visual and verbal imagery.
3. *Transferability* — e.g., mobile both within and across product categories and across geographical boundaries and cultures.
4. *Adaptability* — e.g., flexible enough to be easily updated and made contemporary
5. *Protectability* — e.g., legally secure and competitively well guarded.

The first two criteria are more "offensive" considerations to create and build brand knowledge structures; the last three criteria are more "defensive" considerations to maximize and protect the value of those knowledge structures. Because some brand elements are more likely to satisfy certain criteria better than others, in most cases a subset or even all of the possible brand elements should be employed.

Integrating the Brand into the Supporting Marketing Program

Although the judicious choice of brand elements can make some contribution to customer-based brand equity, the primary input comes from the marketing activities related to the brand. Strong, favorable, and unique brand associations can be created by marketing programs in a variety of well-established ways reviewed elsewhere in this chapter. The following highlights some particularly important marketing guidelines for building brand equity:

1. Ensure a high level of perceived quality and create a rich brand image by linking tangible and intangible product-related and non–product-related associations to the brand [Park et al., 1986].
2. Adopt value-based pricing strategies to set prices and guide discount pricing policy over time that reflect consumers' perceptions of value and willingness to pay a premium.
3. Consider a range of direct and indirect distribution options and blend brand-building "push" strategies for retailers and other channel members with brand-building "pull" strategies for consumers.
4. "Mix" marketing communication options (e.g., advertising, promotion, public relations, sponsorship, etc.) by choosing a broad set of communication options based on their differential ability to impact brand awareness and create, maintain, or strengthen favorable and unique brand associations. "Match" marketing communication options by ensuring consistency and directly reinforcing some communication options with other communication options [Schultz et al., 1993].

Leveraging Secondary Associations

The third and final way to build brand equity is to leverage secondary associations. Brand associations may themselves be linked to other entities that have their own associations, creating "secondary" brand associations. In other words, a brand association may be created by linking the brand to another node or information in memory that conveys meaning to consumers. For example, the brand may be linked to certain source factors, such as the company (through branding strategies), countries or other geographical regions (through identification of product origin), and channels of distribution (through channel strategy) as well as to other brands (through ingredient or cobranding), characters (through licensing), spokespeople (through endorsements), sporting or cultural events (through sponsorship), or some other third-party sources (through awards or reviews). Because the brand becomes identified with another entity, even though this entity may not directly relate to product performance, consumers may *infer* that the brand shares associations with that entity, thus producing indirect or secondary associations for the brand. In essence, the marketer is borrowing or "leveraging" some other associations for the brand to create some associations of its own and thus help to build its brand equity.

Secondary brand associations may be quite important if existing brand associations are deficient in some way. In other words, secondary associations can be leveraged to create strong, favorable, and unique brand associations that otherwise may be lacking. These secondary associations may lead to a transfer of global associations such as attitude or credibility (i.e., expertise, trustworthiness, and likability). These secondary associations may also lead to a transfer of more specific associations related to the product meaning and the attributes or benefits of the brand.

Managing Brand Equity

Managing brand equity involves those activities designed to maintain and enhance the brand equity of all the products a firm sells over time. A key aspect of managing brand equity is the proper branding strategy for those products. Brand names of products typically do not consist of only one name but often consist of a combination of different brand names and other brand elements. A **branding strategy** for a firm identifies which brand elements a firm chooses to apply across the various products it sells. Two important tools to help formulate branding strategies are the brand-product matrix and the brand hierarchy. Combining these tools with customer, company, and competitive considerations can help a marketing manager formulate the optimal branding strategy.

The **brand-product matrix** is a graphical representation of all the brands and products sold by the firm. The matrix or grid has the brands for a firm as rows and the corresponding products as columns. The rows of the matrix represent brand-product relationships and capture the brand extension strategy of the firm with respect to a brand. A **brand extension** is when the firm uses an established name to introduce a new product. Potential extensions must be judged by how effectively they leverage existing brand equity to a new product as well as how effectively the extension, in turn, contributes to the equity of the existing parent brand [Aaker and Keller, 1990].

The columns of the matrix represent product-brand relationships and capture the brand portfolio strategy in terms of the number and nature of brands to be marketed in each category. A firm may offer multiple brands in a category to attract different — and potentially mutually exclusive — market segments. Brands also can take on very specialized roles in the portfolio — as flanker brands to protect more valuable brands, as low-end entry level brands to expand the customer franchise, as high-end prestige brands to enhance the worth of the entire brand line, or as cash cows to milk all potentially realizable profits.

A **brand hierarchy** reveals an explicit ordering of all brand names by displaying the number and nature of common and distinctive brand name elements across the firm's products [Farquhar et al., 1992]. By capturing the potential branding relationships among the different products sold by the firm, a brand hierarchy is a useful means to graphically portray a firm's branding strategy. One simple representation of possible brand elements and thus potential levels of a brand hierarchy is (from top to bottom):

1. Corporate or company brand (e.g., General Motors)
2. Family brand (e.g., Chevrolet)
3. Individual brand (e.g., Camaro)
4. Modifier (e.g., Z28)

The challenge in setting up the brand hierarchy and arriving at a branding strategy is (1) to design the proper brand hierarchy in terms of the number and nature of brand elements to use at each level and (2) to design the optimal supporting marketing program in terms of creating the desired amount of brand awareness and type of brand associations at each level.

Designing the Brand Hierarchy

First, the number of different levels of brands that will be employed and the relative emphasis or prominence that brands at different levels will receive when combined to brand any one product must be defined. In general, the number of levels employed typically are two or three. One common strategy to brand a new product is to create a sub-brand, where an existing company or family brand is combined with a new individual brand (e.g., as with *Microsoft Office* or *IBM Thinkpad*). When multiple brand elements are used, as with a sub-brand, the relative visibility of a brand element as compared to other brand elements determines its prominence. Brand visibility and prominence will depend on factors such as the order, size, color, and other aspects of physical appearance of the brand. To provide structure and content to the brand hierarchy, the specific means by which a brand is used across different products and, if different brands are used for different products, the relationship among those brands also must be made clear to consumers.

Designing the Supporting Marketing Program

Second, the desired awareness and image at each level of the brand hierarchy for each product must be defined. In a sub-branding situation, the desired awareness of a brand at any level will dictate the relative prominence of the brand and the extent to which associations linked to the brand will transfer to the product. In terms of building brand equity, determining which associations to link at any one level should be based on principles of (1) relevance and (2) differentiation. First, it is generally desirable to create associations that are *relevant* to as many brands nested at the level below and to distinguish any brands at the same level. Corporate or family brands can establish a number of valuable associations that can help to provide meaning to associated products such as common product attributes, benefits, or attitudes; people and relationships; programs and values; and corporate credibility. A corporate image will depend on a number of factors, such as (1) the products a company makes, (2) the actions it takes, and (3) the manner with which it communicates to consumers. Second, it is generally desirable to *differentiate* brands at the same level as much as possible. If two brands cannot be easily distinguished, then it may be difficult for retailers or other channel members to justify supporting both brands. It may also be confusing for consumers to make choices between them.

What Makes a Strong Brand?

Although a number of criteria are possible, the above discussion suggests some possible guidelines as to how to create a strong brand with much equity. Marketing managers must

1. Understand what brands mean to consumers and develop products that are appropriate to the brand and address the needs of the target market.
2. Properly position brands by achieving necessary and desired points of parity and points of difference.
3. Provide superior delivery of desired benefits all through the marketing program.
4. Maintain innovation in design, manufacturing, and marketing as well as relevance in brand personality and imagery.
5. Establish credibility and be seen as expert, trustworthy, and likable.
6. Communicate with a consistent voice at any one point in time and over time.

7. Employ a full range of complementary brand elements and supporting marketing activities.
8. Design and implement a brand hierarchy and brand portfolio that puts brands in the proper context with respect to other brands and other products sold by the firm.

Special Considerations for Technical Products

Branding is playing an increasingly important role in the marketing equation for technologically intensive or "high-tech" products [Pettis, 1995]. In many of these product markets, financial success is no longer driven by product innovation alone or by offering the "latest and greatest" product specifications and features — marketing skills can be crucial in the successful adoption of technical products. There are a number of distinguishing features of technical products — such as the complexity of the product, the technical sophistication (or lack thereof) of the target market, and the short product life cycles due to technological advances and research and development breakthroughs — that suggest several important branding guidelines:

1. *Create a corporate or family brand with strong credibility associations.* Because of the complex nature of high-tech products and the continual introduction of new products or modifications of existing products, consumer perceptions of the expertise and trustworthiness of the firm are particularly important. In a high-tech setting, trustworthiness also relates to consumers' perceptions of the firm's longevity and "staying power".
2. *Leverage secondary associations of quality.* Lacking ability to judge the quality of high-tech products, consumers may use brand reputation as a means to reduce risk. This lack of ability by consumers to judge quality also means that it may be necessary to leverage secondary associations to better communicate product quality. Third-party endorsements from top companies, leading consumer magazines, or industry experts may help to achieve the necessary perceptions of product quality. To be able to garner these endorsements, however, will typically necessitate demonstrable differences in product performance, suggesting the importance of innovative product development over time.
3. *Link nonproduct-related associations.* Non–product-related associations related to brand personality or other imagery can be important, especially in distinguishing near-parity products.
4. *Carefully design and update brand portfolios and hierarchies.* Several issues are relevant here. First, brand extensions are a common high-tech branding strategy. With new products continually emerging, it would be prohibitively expensive to brand them with new names in each case. Typically, names for new products are given modifiers from existing products, e.g., numerical (*Microsoft Word 6.0*), time based (*Microsoft Windows 95*), etc., *unless* they represent dramatic departures or marked product improvements for the brand, in which case a new brand name might be employed. Using a new name for a new product is a means to signal to consumers that this particular generation or version of a product is a major departure and significantly different from prior versions of the product. Second, family brands can be an important means of grouping products. Individual items or products within those brand families must be clearly distinguished, however, and brand migration strategies must be defined that reflect product-introduction strategies and consumer market trends. Other brand portfolio issues relate to the importance of retaining some brands. Too often, high-tech firms continually introduce new sub-brands, making it difficult for consumers to develop product or brand loyalty.

Branding, like much of marketing, is an art and a science. Adhering to these guidelines, and others in this section, however, provides guidance and structure and improves the odds for success.

Defining Terms

Brand: A name, term, sign, symbol, or design, or a combination of them intended to identify the goods and services of one seller or group of sellers and to differentiate them from those of competition.

Brand awareness: The strength of the brand node or trace in memory as reflected by consumers' ability to recall or recognize the brand under different conditions.

Brand elements: Trademarkable devices that serve to identify and differentiate the brand, e.g., brand names, logos, symbols, characters, slogans, jingles, and packages.

Brand equity: Marketing effects uniquely attributable to the brand, i.e., the fact that different outcomes result in the marketing of a product or service because of its brand name or some other brand element, as compared to if that same product or service did not have that brand identification.

Brand extension: When a firm uses an established brand to introduce a new product.

Brand hierarchy: An explicit ordering of all brand names by displaying the number and nature of common and distinctive brand name elements across the firm's products.

Brand image: Consumer perceptions of and preferences for a brand, as reflected by the various types of brand associations held in consumers' memory.

Branding strategy: The brand elements a firm chooses to apply across the various products it sells.

Brand-product matrix: A graphical representation of all the brands and products sold by the firm.

Category points of parity: Those associations that consumers view as being necessary to be a legitimate and credible product offering within a certain category.

Competitive points of parity: Those associations designed to negate competitors' points of difference.

Customer-based brand equity: The differential effect that consumers' brand knowledge has on their response to the marketing of that brand.

Points of difference: Those associations that are unique to the brand that are also strongly held and favorably evaluated by consumers.

Points of parity: Those associations that are not necessarily unique to the brand but are possibly shared with other brands.

Strategic brand management: The design and implementation of marketing programs and activities to build, measure, and manage brand equity.

References

Aaker, D. A. *Managing Brand Equity,* Free Press, New York, 1991.

Aaker, D. A. *Building Strong Brands,* Free Press, New York, 1996.

Aaker, D. A. and Keller, K. L. Consumer evaluations of brand extensions, *J. Market.,* 54(1): 27–41, 1990.

Farquhar, P. H., Han, J. Y., Herr, P. M., and Ijiri, Y. Strategies for leveraging master brands, *Market. Res.,* September: 32–43, 1992.

Keller, K. L. *Strategic Brand Management,* Prentice Hall, Upper Saddle River, NJ, 1998.

Kotler, P. *Marketing Management,* 9th ed., Prentice Hall, Upper Saddle River, NJ, 1997.

Park, C. W., Jaworski, B. J., and MacInnis, D. J. Strategic brand concept-image management, *J. Market.,* 50(4): 621–635, 1986.

Pettis, C. *Technobrands,* AMACOM, New York, 1995.

Schultz, D. E., Tannenbaum, S. I., and Lauterborn, R. F. *Integrated Marketing Communications,* NTC Business Books, Lincolnwood, IL, 1993.

Further Information

Kapferer, J. *Strategic Brand Management,* Kogan-Page, London, 1992.

Lehmann, D. and Winer, R. *Product Management,* Richard D. Irwin, Burr Ridge, IL, 1994.

12.13 Conjoint Analysis: Methods and Applications

Paul E. Green, Jerry Wind, and Vithala R. Rao

Conjoint analysis is one of many techniques for dealing with situations in which a decision maker has to choose among options that simultaneously vary across two or more attributes. The problem facing the decision maker is how to trade off the possibility that option X is better than option Y on attribute A but worse than option Y on attribute B, and so on. For over 40 years, researchers from a variety of disciplines — economics, operations research, psychology, statistics, marketing, and business — have studied aspects of the multiattribute choice problem.

Conjoint analysis is concerned with the day-to-day decisions of consumers — what brand of toothpaste, automobile, or photocopying machine to buy (or lease)? The marketing researcher may collect trade-off information for hundreds or even thousands of respondents. Data collection and processing techniques must be relatively simple and routinized to handle problems of this scope.

Following the theoretical work of Luce and Tukey [1964], conjoint analysis was introduced to the marketing research community in the early 1970s [Green and Rao, 1971]. Since that time, two extensive reviews of the field [Green and Srinivasan, 1978, 1990] have appeared. Conjoint has been one of the most documented methods in marketing research. Judging by the thousands of conjoint applications that have been conducted since 1970, it has become the most popular multiattribute choice model in marketing [Wittink and Cattin, 1989].

Conjoint analysis is both a trade-off measurement technique for analyzing preferences and intentions-to-buy responses and a method for simulating how consumers might react to changes in current product/services or the introduction of new products into an existing competitive array. Conjoint analysis has been applied to products and services (consumer and industrial) and to not-for-profit offerings as well.

Basic Ideas of Conjoint Analysis

To illustrate the basic concepts of conjoint analysis, assume that a marketer of credit cards wishes to examine the possibility of modifying its current line of services. One of the first steps in designing a conjoint study is to develop a set of attributes and levels that sufficiently characterize the competitive domain. Focus groups, in-depth consumer interviews, and internal corporate expertise are some of the sources used to structure the sets of attributes and levels that guide the rest of the study.

Table 12.8 shows an illustrative set of 12 attributes employed in an actual study of credit card suppliers. Note that the number of levels within attribute range from 2 to 5, for a total of 35 levels. However, the total number of possible combinations of levels is 186,624.

TABLE 12.8   Card Attribute Levels Used in Conjoint Survey

Annual price ($)
  0   10   20   80   100
Cash rebate (end-of-year, on total purchases)
  None   1/2%   1%
800 number for message forwarding
  None   9–5 p.m. weekdays   24 hours per day
Retail purchase insurance
  None   90 days' coverage
Common carrier insurance (death, injury)
  None $50,000 $200,000
Rental car insurance (fire, theft, collision, vandalism)
  None $30,000
Baggage insurance (covers both carry-on and checked)
  None $2,500 depreciated cost $2,500 replacement cost
Airport club admission (based on small entrance fee)
  No admission $5 per visit $2 per visit
Card acceptance
  Air, hotel, rental cars; AHC and most restaurants; AHCR and most general
    retailers (AHCRG); AHCR and department stores only (AHCRD)
24-hour medical/legal referral network
  No   Yes
Airport limousine to city destination
  Not offered   Available at 20% discount
800 number for emergency car service
  Not offered   Available at 20% discount

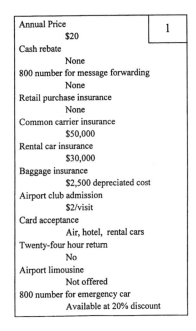

Annual Price	1
$20	
Cash rebate	
None	
800 number for message forwarding	
None	
Retail purchase insurance	
None	
Common carrier insurance	
$50,000	
Rental car insurance	
$30,000	
Baggage insurance	
$2,500 depreciated cost	
Airport club admission	
$2/visit	
Card acceptance	
Air, hotel, rental cars	
Twenty-four hour return	
No	
Airport limousine	
Not offered	
800 number for emergency car	
Available at 20% discount	

Annual Price	2
$50	
Cash rebate	
1/2%	
800 number for message forwarding	
24 hours per day	
Retail purchase insurance	
None	
Common carrier insurance	
None	
Rental car insurance	
$30,000	
Baggage insurance	
None	
Airport club admission	
$5/visit	
Card acceptance	
AHC and most restaurants	
Twenty-four hour return	
Yes	
Airport limousine	
Available at 20% discount	
800 number for emergency car	
Not offered	

FIGURE 12.20 Illustrative full-profile prop cards.

Conjoint analysts make extensive use of **orthogonal arrays** [Addelman, 1962] to reduce the number of stimulus descriptions to a small fraction of the total number of combinations. For example, in the preceding problem, an array of only 64 profiles (less than 0.1% of the total) is sufficient to estimate *all* attribute-level main effects on an uncorrelated basis. Since the study designers used a hybrid conjoint design [Green and Krieger, 1996], each respondent received only eight (balanced) profile descriptions, drawn from the 64 profiles.

Figure 12.20 shows two illustrative prop cards, prepared for the credit card study. After the respondent sorts the prop cards in terms of preference, each card is rated on a 0 to 100 likelihood-of-acquisition scale. In small conjoint studies (e.g., six or seven attributes, each at two or three levels), the respondent receives all of the full profiles, ranging in number from 16 to 32 prop cards. In these cases, the prop cards are sorted into four to eight ordered categories before likelihood-of-purchase ratings are obtained for each separate profile, within group.

Types of Conjoint Data Collection

There are four major types of data collection procedures that have been implemented for conjoint analysis:

1. *Full profile* techniques — each respondent sees a full set of prop cards, as illustrated in Fig. 12.20. After an initial sorting into ordered categories, each card is rated on a 0 to 100 likelihood-of-purchase scale.

2. *Compositional techniques*, such as the CASEMAP procedure [Srinivasan and Wyner, 1989] — preferences are collected by having each respondent rate the desirability of each set of attribute levels on a 0 to 100 scale. (This approach is also called self-explicated preference data collection.)

3. *Hybrid techniques* — each respondent receives both a self-explicated evaluation task and a set of full profiles for evaluation. The resulting utility function is a composite of data obtained from both tasks.

4. *Adaptive conjoint analysis* [Johnson, 1987] — this technique is also a type of hybrid model in which each respondent first receives the self-explication task followed by a set of partial profile descriptions, two at a time. The respondent evaluates each pair of partial profiles on a graded, paired comparisons scale. Both tasks are administered by computer.

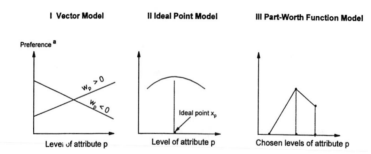

FIGURE 12.21   Alternative models of preference. Preference for different levels of attribute p while holding the values for the other attributes constant. (*Source:* Green, P. E. and Srinivasan, V. *J. Consum. Res.*, 5: 103–123, 1978.)

Conjoint Models

Most conjoint analysts fit what is known as the **part-worth** model to respondents' evaluative judgments, whether obtained by trade-off tables or full profile, self-explicated, or hybrid approaches. Let $p = 1, 2, \ldots$ and t denote the set of t attributes that are used in the study design. We let y_{jp} denote the level of the pth attribute for the jth stimulus; we first assume that y_{jp} is inherently continuous. The *vector* model assumes that the preference s_j for the jth stimulus is given by

$$s_j = \sum_{p=1}^{t} w_p \, y_{jp} \tag{12.8}$$

where w_p denotes a respondent's weight for each of the t attributes (see Fig. 12.21).

The *ideal point* model posits that preference s_j is negatively related to the weighted squared distance d_j^2 of the location y_{jp} of the jth stimulus from the individual's ideal point x_p, where d_j^2 is defined as

$$d_j^2 = \sum_{p=1}^{t} w_p \left(y_{jp} - x_p \right)^2 \tag{12.9}$$

The *part-worth* model assumes that

$$s_j = \sum_{p=1}^{t} f_p \left(y_{jp} \right) \tag{12.10}$$

where f_p is a function denoting the part-worth of different levels of y_{jp} for the pth attribute. In practice, $f_p(y_{jp})$ is estimated for a selected set of discrete levels of y_{jp}.

Figure 12.22 shows illustrative (averaged) part-worths for each of the attribute levels described in Table 12.9. As noted, part-worths are often scaled so that the lowest part-worth is zero, within each attribute. Strictly speaking, part-worth functions are evaluated at discrete levels for each attribute. However, in most applications, analysts interpolate between levels of continuous attributes, such as price (when the part-worths enter buyer choice simulators). Note that the scaling (vertical axis) is common across all attributes; this allows one to add up part-worths across all attributes to obtain the overall (product/service) utility of any profile composable from the basic attribute levels.

Stimulus Presentation

Conjoint data collection methods currently emphasize the full-profile and hybrid procedures. While some industry studies still employ paragraph descriptions of attribute levels, profile cards (with terse attribute-level

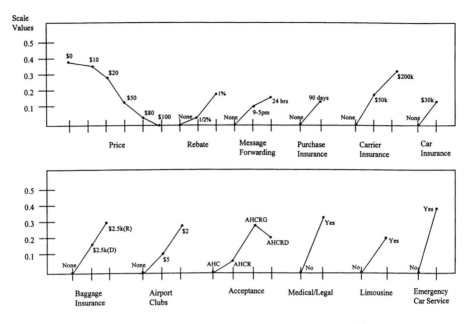

FIGURE 12.22 Average part-worths from conjoint model.

TABLE 12.9 Sample List of Conjoint Applications

Consumer nondurables	Other products
Bar soaps	Automotive styling
Hair shampoos	Automobile and truck tires
Carpet cleaners	Car batteries
Synthetic-fiber garments	Ethical drugs
Gasoline pricing	Toasters/ovens
Panty hose	Cameras
Lawn chemicals	Apartment design
Financial services	Other services
Branch bank services	Car rental agencies
Auto insurance policies	Telephone services and pricing
Health insurance policies	Employment agencies
Credit card features	Information-retrieval services
Consumer discount cards	Medical laboratories
Auto retailing facilities	Hotel design
High-tech maintenance service	Transportation
Industrial goods	Domestic airlines
Copying machines	Transcontinental airlines
Printing equipment	Passenger train operations
Facsimile transmissions	Freight train operations
Data transmission	International Air Transportation Association
Portable computer terminals	Electric car design
Personal computer design	

descriptions, as shown in Fig. 12.20) are, by far, the more popular stimulus presentation method. Increasing use of pictorial material is to be noted; these kinds of props not only make the task more interesting to the respondent but also provide easier and less ambiguous ways of conveying information. Moreover, conjoint methodology is increasingly being applied to the design of physical products (e.g., foods/beverages, fragrances, personal care products, and so on).

Part-Worth Estimation

With the swing toward behavior-intentions rating scales (as opposed to preference rankings), it is not surprising that several of the earlier *nonmetric regression* techniques for estimating part-worths have given way to OLS regression as the major parameter estimation technique. Attribute levels are simply recoded into dummy variables and entered into a standard multiple regression program. OLS regression does a respectable job in mimicking the output of nonmetric techniques, without the associated dangers of solution degeneracy or local optima associated with nonmetric methods. However, the Linmap nonmetric program [Srinivasan and Shocker, 1973] still continues to be popular.

Logit and probit models are also finding increasing application [Louviere and Woodworth, 1983] as well. In most instances these models have been applied to aggregate-level data, requiring the need for relatively large numbers of observations.

Recent Trends in Conjoint Analysis

Microcomputer Software Packages

Two principal conjoint software packages are currently available for applications researchers:

1. Bretton-Clark's package of Conjoint Designer, Conjoint Analyzer, Simgraf, Conjoint Linmap, and Bridger.
2. Sawtooth Software's Adaptive Conjoint Analysis and Price Value Analysis programs.

Bretton-Clark's package [Herman, 1988] is the more traditional of the two. It has a module for constructing orthogonal arrays (Conjoint Designer) and a module for estimating part-worths and for running new product simulations (Conjoint Analyzer). Newer additions to the package include nonmetric part-worth estimation (Conjoint Linmap), a more sophisticated simulator (Simgraf), and Bridger for dealing with large numbers of conjoint attributes. Data are collected by conventional pencil-paper questionnaire. However, part-worth estimation and buyer choice simulation are all done in the computer.

Sawtooth's Adaptive Conjoint Analysis package [Johnson 1987] utilizes the PC to collect conjoint data and uses a somewhat different procedure to estimate part-worths. Its Price Value Analysis package, however, also contains an option for preparing pencil-paper questionnaires. Otherwise, the Bretton-Clark and Sawtooth packages are quite similar in their components.

Applications of Conjoint Analysis

Over the past 20 years conjoint has been applied to virtually every industry sector, both in the United States and abroad. Every major marketing research supplier offers this service, and a few firms specialize in conjoint (and related) techniques. Table 12.9 lists a wide variety of conjoint applications, ranging from consumer nondurables to large complex industrial machinery.

Future Directions

Table 12.10 shows some of the directions that applications of conjoint appear to be taking. First, there appears to be an expansion in the diversity of applications and a scaling up in the use of conjoint for strategy formulation. Second, there is an increasing emphasis on improving the predictive power of conjoint forecasts and extending the models to include other marketing mix variables such as advertising, promotion, and distribution.

Much of the recent work in hybrid modeling and adaptive conjoint analysis has been motivated by practitioners' desires to expand the numbers of attributes and levels that can be accommodated. Conjoint studies are fast becoming more complex and more suited to higher-level decision strategies. In addition, researchers are also developing product design optimizers and dynamic simulators that take competitors' retaliatory strategies into consideration. In short, both the research and applications aspects of conjoint analysis continue to expand.

TABLE 12.10 Future Directions in the Practice of Conjoint Analysis

Extension of conjoint methodology to new application areas, such as litigation, employee benefits packages, conflict resolution (e.g., employer/employee negotiations), corporate strategy, and social/environmental trade-offs

Application of newly developed models for optimal product and product line design, including models that combine conjoint analysis with multidimensional scaling

Descriptive and normative studies for measuring customer satisfaction, perceived product, and service quality

Models and applications of conjoint analysis to simulated test marketing services that include ongoing prediction validation and the establishment of adjustment "norms" for converting survey responses to market forecasts

Extension of conjoint models and applications to include marketing mix strategy variables, such as advertising, promotion, and distribution

Models and applications that combine survey-based data (e.g., conjoint analysis) with single-source behavioral data obtained from scanning services and split-cable TV experiments

New computer packages that exploit recent developments in hybrid modeling, multivariate response analysis, and optimal product search.

Source: From Green, P. E. and Srinivasan, V. *J. Market.*, 54: 3–19, 1990.

Defining Terms

Compositional technique: A preference measurement procedure that collects respondent ratings of attribute-level desirabilities and attribute importances.

Full profile: A conjoint technique where OLS dummy variable regression is used to obtain derived part-worths (regression coefficients) from overall responses to designed profiles.

Hybrid conjoint technique: A combined conjoint model using both compositional data and a limited set of full-profile evaluations.

Nonmetric regression: A multiple regression technique where the dependent variable is assumed to be only ordinally scaled.

Orthogonal array: A highly fractionalized experimental design that enables the uncorrelated estimation of all main effects with maximum efficiency.

Part worth: The derived value that a respondent places on one level of one attribute.

References

Addelman, S. Orthogonal main-effect plans for asymmetrical factorial experiments, *Technometrics*, 4: 21–46, 1962.

Green, P. E. and Krieger, A. M. Individualized hybrid models for conjoint analysis, *Mgt. Sci.*, 42: 850–867, 1996.

Green, P. E. and Rao, V. R. Conjoint measurement for quantifying judgmental data, *J. Market. Res.*, 8: 355–363. 1971.

Green, P. E. and Srinivasan, V. Conjoint analysis in consumer research: issues and outlook, *J. Consum. Res.*, 5: 103–123, 1978.

Green, P. E. and Srinivasan, V. Conjoint analysis in marketing: new developments with implications for research and practice, *J. Market.*, 54: 3–19, 1990.

Herman, S. Software for full-profile conjoint analysis, Proceedings of the Sawtooth Software Conference on Perceptual Mapping, Conjoint Analysis, and Computer Interviewing, Sawtooth Software, Ketchum, ID, pp. 117–130, 1988.

Johnson, R. M. Adaptive conjoint analysis, Sawtooth Software Conference on Perceptual Mapping, Conjoint Analysis, and Computer Interviewing, Sawtooth Software, Ketchum, ID, pp. 253–266, 1987.

Louviere, J. and Woodworth, G. Design and analysis of simulated consumer choice or allocation experiments: an approach based on aggregate data, *J. Market. Res.*, 20: 350–367, 1983.

Luce, R. D. and Tukey, J. W. Simultaneous conjoint measurement: a new type of fundamental measurement, *J. Math. Psychol.*, 1: 1–127, 1964.

Srinivasan, V. and Shocker, A. D. Linear programming techniques for multidimensional analysis of preferences, *Psychometrika*, 38: 337–369, 1973.

Srinivasan, V. and Wyner, G. A. Casemap: computer-assisted self-explication of multi-attributed preferences, *New Product Development and Testing*, W. Henry, M. Menasco, and H. Takada, eds., Lexington Books, Lexington, MA, 1989.

Wittink, D. R. and Cattin, P. Commercial use of conjoint analysis: an update, *J. Market.*, 53: 91–96, 1989.

For Further Information

For a discussion of optimal product design models, see Green, P. E. and Krieger, A. M. An application of a product positioning model to pharmaceutical products, *Market. Sci.*, 11: 117–132, 1992.

For an example of dynamic product design, see Green, P. E. and Krieger, A. M. Using conjoint analysis to view competitive interaction through the customer's eyes, *Wharton on Dynamic Competitive Strategy*, G. S. Day and D. J. Reibstein, eds., pp. 343–367. John Wiley & Sons, New York, 1997.

For an elementary exposition of conjoint analysis, see Green, P. E. and Wind, Y. New way to measure consumers' judgments, *Harv. Bus. Rev.*, 53: 107–117, 1975.

12.14 Mass Customization

Barbara E. Kahn and Cynthia Huffman

Mass customization is a general term that refers to the process of combining the low-cost efficiencies of "mass marketing" (standardization) with the strategic effectiveness of customization (individualization). Whereas many industries specialized in standardized fare a decade ago, e.g., McDonald's mass-produced hamburgers or the standardized credit card, today in order to keep a competitive edge it is often necessary to provide a highly varied product line. Thus, we see varieties of hamburger/chicken/fish/taco sandwiches in fast-food restaurants and personalized credit cards guaranteeing individualized service.

Mass customization has become necessary as industries have become more competitive. Competition arises either because of more interfirm rivalry (as in the personal computer industry) or because buyers become more powerful (as in financial services where quick access to information gives investors more power) or because of deregulation (as in the telecommunications industry). In all of these cases, buyers are more demanding, requiring both a higher level of service and competitive prices. In order to respond to these demands better than the competition, firms have to create specialized products to meet individual needs. Frequently, the process of developing these customized products also creates a unique relationship between buyer and seller that can in and of itself generate long-term loyalty.

Cost-Effective Strategies

Mass customization lowers the costs of high-variety strategies through one of two ways. First, costs can be lowered through economies of scale, where standardized components are configured in a customized manner [Pine, 1993; Sawhney and Ramdas, 1997]. For example, paint stores now carry only standard basic colors and customized shades are mixed on demand using the basics. Bloated inventories carrying hundreds of shades of paint are no longer necessary, thus significantly reducing costs.

Another technique is called **component swapping**. Costs are reduced in this method by using a standardized platform, and then components are swapped on an as-needed basis to provide a customized product. For example, a children's book company allows parents the opportunity to customize a book for one's child by including his or her name throughout the book and by including a photograph of the child. Except for these components, the rest of the book is standardized.

The extreme example of standardized components is called modular architecture, where each component is exactly the same (thus significantly reducing costs of producing each one), and the overall product is configured uniquely from these standardized components. The best example of this is the Lego Amusement Park at the Mall of America in Minnesota. Here complicated sculptures are created using only Lego blocks.

The second way to lower the costs of higher-variety strategies is through economies of scope — or using a single process to achieve greater variety of products and/or services [Pine, 1993; Sawhney and Ramdas, 1997]. Process-based strategies include flexible manufacturing, where industrial robots can control manufacturing through software programming. Within a well-developed system and given a predetermined envelope of variety, there are no cost penalties for manufacturing any one part vs. another, yielding a manufacturing system that can quickly respond to changes in demand. Some of these systems can also be designed to switch quickly between runs of one and runs of thousands, thus providing the same production cost per unit regardless of the size of the run. Computer-aided manufacturing/design (CAD/CAM) systems are used to help lower costs and respond to individualized needs. Casio is able to make 5000 different watches, but manufacturing is completely automatic.

Degrees of Customization

Key strategic questions involved in implementing a mass-customization strategy include (1) how much customization should be offered, (2) which components should be customized, and (3) at what point in the delivery chain should the customization take place.

Deciding how much customization to provide is obviously a function of how competitive the marketplace is. If the marketplace is very competitive, then there will be a benefit in providing more customization and **micromarketing** to a finer and finer level. Presumably, by meeting the needs of the customer more precisely, price competition can be averted. In determining which components of the product should be standardized and which ones customized, it is necessary to take into consideration how the products are consumed at the customer level. Highly visible components that are relatively cheap to design are the most likely ones to be customized, whereas invisible components that are expensive to design are least likely to be customized [Sawhney and Ramdas, 1997]. For example, in an automobile, the braking system and chassis are usually standardized while the paint color or interior seat covers are frequently customized.

Customization can be incorporated at several different stages in the delivery chain. At one extreme, customization can be done by the customer. For example, individual needs in an office chair can be quite specific, depending upon the idiosyncracies of a person's physical shape and well-being. Here a manufacturer can make a standardized office chair that can be customized at the customer level by providing levers at various junctures. In this case, each individual chair is unique, but the customization occurs after the product is purchased.

At the other extreme, customization occurs at the manufacturing level where each product that leaves the factory is customized. To reduce costs when customization takes place at the factory, the cost strategies described above are utilized. In between these two extremes, customization can take place at the point of delivery site. For example, customized tee shirts are created at tee-shirt kiosks in the mall. Here the retailer is provided with standard tee shirts and the customization is done at purchase time.

Customization at the Customer Level

From the customer's point of view, a high-variety product line means only one thing: "do you have what I want?" The broadest product lines are meaningless if they don't answer the specific customer's needs. It's the marketer's job to determine what type of customized product to provide so that the customer can find what she or he wants or needs and is not confused or frustrated with the process of finding it.

Four distinct approaches to customization have been identified [Gilmore and Pine, 1997]: *adaptive, cosmetic, transparent,* and *collaborative.* Adaptive and cosmetic approaches both offer standardized products that do not require understanding specific customer needs, while transparent and collaborative approaches offer products that are differentiated and require knowledge of customer preferences.

Adaptive customization is where the customer adapts the product to fit his or her needs. The office chair example provided above fits in this category. This type of customization is appropriate if the customer is not confused by the customization process and perhaps even enjoys setting his or her own levels. Adaptive customization is *not* appropriate if the customer is not sophisticated enough to appreciate the capabilities of the product. For example, for some customers, programmable VCRs and customizable software are a nuisance even though they have the capability of fitting one's requirements.

Cosmetic customization is where the product is standardized but is presented in a "customized" way. At a basic level, cosmetic customization takes place when a bank puts the customer's name on a checkbook. Airlines and hotels provide more sophisticated cosmetic customization. In many hotels, the rooms are standardized, but "special" floors are created by providing service around the standard room, e.g., providing fancy mints or free drinks at cocktail hour. Similarly, although airline seats may be similar, the service can be customizable. Cosmetic customization is generally more inexpensive and, if tastes can be predicted adequately, can prove to be of real value to the customer. The problem with cosmetic customization occurs when preferences are predicted incorrectly. For example, assuming someone whose name is Robert goes by "Bob", sending customized stationary can backfire if that particular person hates nicknames.

When products are customized by the firm to meet the needs of an individualized customer, then each customer's preferences must be determined a priori. Gilmore and Pine [1997] suggest two ways to determine customers' preferences: *transparency* and *collaboration.* Discovering preferences *transparently* means that the marketer assumes the burden of learning a customer's needs without actually engaging the customer. For example, the Ritz Carlton Hotel makes a point of observing choices that their guests make and then recording them for the future. If a guest requests a *Wall Street Journal* one morning, then a *Wall Street Journal* will be provided every morning at any Ritz Carlton hotel that guest visits. Domino's Pizza keeps a record of their customers' phone numbers and addresses. When a customer calls, he is identified by the phone number, and the delivery address is automatically known. Transparent customization is not possible when preferences are not well defined or observable or when a series of interactions is not expected.

When the marketer learns customers' preferences in a *collaborative* fashion, then the marketer engages in a dialogue with customers to help them articulate their needs and identify the precise offering that fulfills those needs. Although ultimately this method can result in the ideal product, it can also be a terrible bother and take a great deal of time, as anyone who has designed a kitchen or a custom home knows. The goal of the marketer in learning preferences collaboratively is to determine a method that maximizes the learning without frustrating the customer. Further, if collaboration is done well, the marketer can amass a great deal of data about customer preferences that can be used to help sell other services and products. In the extreme, if the marketer does a good job of learning preferences, significant switching costs will be erected. The customer will depend on the marketer who already knows what s/he likes.

Forms of Collaborative Interaction

Once a marketer decides that some sort of collaboration in the customization process is desirable, the exact form of the interaction must be determined. There are several types of interaction possible that vary by the degree of effort and input required by the customer. Many of the interaction styles seen in practice may be some combination of the forms described below.

Choice

One way for the marketer to learn customer's preferences is through observation of a series of choices or evaluative judgments that the consumer makes. For example, people can determine what kind of clothing to buy as a gift for a friend by observing the choices of clothing that friend has made over time. **Choice-based conjoint analysis** is a formal modeling method that can be used to infer preferences through a series of choices that are made.

This type of interaction can be automated where the agent capturing preferences is in fact a computer program. Firefly, a music vendor on the World Wide Web, analyzes consumers' preferences by observing their selections. The program then matches those selections to a large database of consumers. People with like tastes are clustered, and recommendations are made based on other music that the matched set also enjoyed. This process is called **collaborative filtering.**

Usage

Another way for the marketer to learn customers' preferences is through a usage-style interaction. Here, the agent observes the consumer in the process of using the product and translates characteristics of the usage into the required product attributes. For example, Custom Club, a designer of custom golf clubs, takes a computer video of the customer swinging a golf club, along with measurements such as height and distance from ground to knuckles, and uses the information to create the clubs that are "perfect" for that person's natural swing. In other usage-style interactions, the consumer is asked to talk about intended use. For example, many customized sofa salespeople will inquire about the room in which the sofa is to be used: what activities generally occur in the room, whether it is used for entertaining, whether TV is watched in the room, etc.

Self-Talk

A variation of the usage-style interaction is for the consumer to talk about himself or herself, while the agent translates that information into product characteristics. For example, a marketer of face treatment products might ask consumers to talk about their skin, what they do to care for it, etc. From this information, the agent can characterize skin types, problems that need to be solved, and benefits that are desired. This method requires an agent who is highly trained not only in product attributes but also in personal communication techniques.

Attribute Preference Learning

Probably the method that requires the greatest input from the customer is asking him or her to learn about product attributes and to formulate preferences. A complete learning interaction would require that the consumer indicate (1) how important each attribute is relative to the others and (2) which attribute level is preferred. Customers unfamiliar with the product category may find this process too frustrating and difficult. A modified version of learning customer preferences may be more appropriate — where consumers are simply asked which attribute level they prefer and are not asked how important each attribute is [Huffman and Kahn, 1997].

The way the information is presented to the customer also matters. One way to elicit preferences is *attribute based*, where customers are asked what attribute level they prefer for each attribute. Dell Computer uses this method on the World Wide Web, asking consumers their preferences for memory, RAM amount, number of ports desired, etc. After the information is collected, the customized computer is developed and shown to the consumer for approval prior to ordering.

Another method of presenting information is *alternative based*, where the customer is shown several alternatives and asked to indicate which aspects are liked and disliked. For example, in customized sofa stores and kitchen design shops, consumers are often led through showrooms and encouraged to express their likes and dislikes of the various attributes based on the displayed exemplars.

Choosing a Method of Customer Collaboration

In determining which of the above methods of collaboration is best suited for learning customer needs, one needs to consider the type of product, whether the purchase is likely to be a single transaction or a series of transactions, and individual factors such as the level of product knowledge and motivation.

Generally speaking, the more complex the product category, the more important it is to have interaction with consumers to help them articulate preferences. Product-class complexity is in part a function of the number of possible alternatives. However, actual variety may be small even when the consumer perceives that a large set of options is being offered. For example, the typical Chinese menu seems to offer an enormous number of choices but can actually be reduced to four kinds of meats, four sauces, etc. Thus, the more crucial aspect of complexity is the number of attributes and attribute levels.

Experiments that we have run [Huffman and Kahn, 1997] lead us to believe that, for complex product categories, customers should learn their preferences for attribute levels rather than simply choosing without first systematically formulating those preferences. Satisfaction is increased when this kind of learning is facilitated because customers have more control over the ultimate choice and are more able to appreciate the product selection. For example, a customer is more likely to be satisfied with his customized sofa if the salesperson helps the customer learn the attributes of the sofa rather than leaving the customer to wander around on his own.

In addition, we found in complex product categories that the attribute-based method of preference elicitation is strongly preferred. Learning preferences from alternatives is perceived to be difficult. Attribute-based methods aid consumers in learning, facilitate readiness and willingness to make a choice, reduce the perceived complexity, and increase satisfaction with the interaction process.

The second factor to consider is the nature of the relationship with the customer. If the relationship is long term, and customers are likely to make more purchases in the future, an opportunity exists to use a choice-style interaction where the marketer observes the consumers' choices and induces preferences. A marketer may also wish to use a choice-style interaction when cross-selling is a possibility. USAA, for example, examines purchase records and uses those to inform the consumer of other products/services that might be of interest, based on that particular consumer's apparent needs and interests.

On the other hand, if the relationship with the customer is long term, it might also make sense to go through a more-involved collaboration where preferences are learned or usage styles are uncovered. The data can be kept in customer files and aid in future purchases. A side benefit for marketers who do a good job of learning preferences is that the customer feels known and valued. The "personal touch" possible in these types of interactions is the basis for a one-to-one relationship with each customer that ultimately increases long-term loyalty [Peppers and Rogers, 1993].

The third factor to consider is customer-specific characteristics. Some customers are very knowledgeable about the product category. These customers are likely to know the attributes and the available attribute levels. The marketer in this case can concentrate on making sure that full information is provided as needed. On the other extreme, when customers have low levels of knowledge, the marketer has to be especially careful not to frustrate or confuse. In this case transparent learning of preferences is generally appreciated, and recommendations from the marketer are usually welcomed.

Summary

As markets get more competitive, firms frequently look toward high-variety product lines to provide a differential advantage. Mass customization provides cost-efficient methods of creating the variety. Depending upon the customer, the product class, and the competition, marketers can customize their offering to varying degrees: the customization can take place after purchase by allowing the customers themselves to adjust the product to their needs, or the products can be customized before they are sold. If products are customized prior to sale, then the process of learning the customers' needs also should be carefully managed. Customer satisfaction with the preference learning process is intimately connected to reducing complexity, confusion, and information overload. An effective method of preference elicitation will ensure that the customers get what they need.

Defining Terms

Choice-based conjoint analysis: A tradeoff measurement technique for analyzing preferences and intentions to buy, where the dependent variable is choices rather than ratings.

Collaborative filtering: A technique where people answer questions about their likes and dislikes. These data are then matched to a large sample of consumers. People with similar tastes are clustered together. Recommendations can then be generated based on what the matched sample has liked in the past.

Component swapping: A cost-reducing method in which different components are paired with the same basic platform.

Micromarketing: The process of tailoring the marketing mix (product, place, price, and promotion) to the needs of a microsegment, typically down to the individual level.

References

Gilmore, J. H. and Pine, B. J., II. The four faces of mass customization, *Harv. Bus. Rev.*, January-February: 91–101, 1997.

Huffman, C. and Kahn, B. E. Variety for Sale: Mass Customization or Mass Confusion?, presentation for MSI Conference, Too Much or Too Little? Managing Product Assortment from Production to Point of Purchase, Scottsdale, AZ, March 13–14, 1997.

Peppers, D. and Rogers, M. *The One to One Future*, Doubleday, New York, 1993.

Pine, B. J., II. *Mass Customization: The New Frontier in Business Competition*, Harvard Business School Press, Boston, 1993.

Sawhney, M. and Ramdas, K. A Cross-Functional Approach to Evaluating Line Extensions for Assembled Products," presentation for MSI Conference, Too Much or Too Little? Managing Product Assortment from Production to Point of Purchase, Scottsdale, AZ, March 13–14, 1997.

For Further Information

Feitzinger, E. and Lee, H. L. Mass customization at Hewlett-Packard: the power of postponement, *Harv. Bus. Rev.*, January-February: 117–121, 1997.

Fisher, M. L., Hammond, J. H., Obermeyer, W. R., and Raman, A. Making supply meet demand in an uncertain world, *Harv. Bus. Rev.*, May-June: 83–93, 1994.

12.15 Focus Groups and Qualitative Research

John H. Mather

Focus groups are the most widely used information-gathering technique among qualitative research methodologies. A focus group is a discussion of a prepared topic outline by a moderator among 8 to 12 participants for about 90 minutes. Analysis of the discussion content provides valuable insight for numerous organization decisions. In addition, the ability to observe and to listen to focus groups often provides the only opportunity for interested individuals to interact indirectly with product users, communication receivers, customers, or unhappy consumers.

However, all is not fine with focus group research. Even though considered nearly the universal research methodology, focus groups are the most misused and abused research approach. Although these strong accusations and generally held perceptions are often ignored or not acknowledged because this is "informal" research, there is truth in the statement. The information value from focus groups is often compromised and organizations then do not benefit from the technique or information gained.

This section defines, explains, and provides directions on how to conduct and use focus group research. The technique is described within the context of qualitative research methodology and then discussed in terms of specific phases for realizing the full information benefits from focus groups. There is a significant material available about focus groups. The content of this section is a small sampling of that material, but enough to demonstrate the role and value of focus groups as well as to provide sufficient awareness of the correct procedures and mistakes for conducting reliable qualitative research.

Focus Group Definition

In the broadest definition offered by Jane Templeton [1994], a **focus group** is

> ...a small, temporary community, formed for the purpose of the collaborative enterprise of discovery. The assembly is based on some interest shared by the panel members, and the effort is reinforced because panelists are paid for the work.

Making the definition specifically relevant to a technology-based company, the focus group is composed of qualified individuals who are invited to a convenient location for the purpose of a discussing planned topics directed by a moderator. For participation, the group members receive a payment often referred to in the research industry as a cooperative fee.

Focus groups are a form of **qualitative research**, which means that the results are not hard numbers or quantitative and therefore cannot be projected to the population in general. Qualitative research is equated to exploratory research design and defined as the initial information-gathering methods to obtain better understanding of a problem or opportunity in terms of definition, ideas, relationships, insights, key variables, and semantics.

Exploratory research can help reduce complex problems to more manageable component areas for future investigation. Observing and hearing a focus group discussion about a new product concept often leads to significant improvements that were expressed based on anticipated usage within their respective user work environments and shared in the focus group experience As the product idea is developed, and prototypes emerge, subsequent focus groups among users can provide additional insight for the aesthetics and functional facets of the same new product. However, the major caution here is that, from a statistical perspective, the results from the focus group should be quantified using a larger sample from the designated audience, which then allows reliability estimates to be placed on all sample estimates. More information will be shared regarding the qualitative nature of focus groups and the methods employed to increase the value and confidence of the data obtained.

Returning to exploratory research approaches, focus groups are the most often used technique. In addition, there are literature searches and individual in-depth interviews. Two major purposes of qualitative or exploratory research are

- To gain familiarity with the problem by defining variables, relationships, and possible subproblems.
- To obtain initial reaction and direction concerning ideas or concepts involving products, services, communications, packaging, or pricing issues.

The following statement from Churchill's [1995] marketing research test summarizes concisely the strength of **exploratory research** and its popularity:

> Because knowledge is lacking when an inquiry is begun, exploratory studies are characterized by flexibility with respect to the methods used for gaining insight and developing hypotheses. "Formal design is conspicuous by its absence in exploratory studies." Exploratory studies rarely use detailed questionnaires or involve probability sampling plans. Rather, investigators frequently change the research procedure as the vaguely defined initial problem is transformed into one with more precise meaning. Investigators follow where their noses lead them in an exploratory study. Ingenuity, judgment, and good luck inevitably play a part in leading to the one or two key hypotheses that, it is hoped, will account for the phenomenon.

Thus, for obtaining initial reaction to product ideas or product concept in the very early stages of development, focus groups offer an enticing alternative. Many users of focus groups often characterize the focus group research method as being very seductive to the unwary and inexperienced individual. Although numerous focus groups are conducted in error, which renders the results even less useful, there are several steps that can and should be followed to maximize the value of the information. Often focus group users are deceived by the apparent simplicity of the focus group method. Consequently, in their haste to conduct the research, shortcuts are taken, attention to key elements of organization are overlooked, and the interpretive value of results is significantly lessened. The next section describes the critical procedures for organizing, conducting, and using focus groups.

Focus Group Procedures

For emphasis, it was stated that focus groups are qualitative because the results cannot be projected to the general population of interest. In other terms, the findings from the focus groups are not transferable to the population and assumed as given values within that population. However, the use of appropriate procedures provides a research environment for focus groups and yields strong results for decision makers. The procedures described here include

- Determining information objectives
- Recruiting participants
- Developing the discussion guide
- Selecting a moderator
- Analyzing and reporting results
- Planning the groups

There are many fine books available that cover in depth the previous focus group topics. The purpose in this short description is to emphasize the key points for each subject area, thereby generating initial familiarity with the technique.

Objective setting is first. Because the general purpose of qualitative research is to examine, explore, and understand the research topic so that definition emerges for a quantitative survey, focus groups allow the creation of these general qualitative guidelines for understanding within a group environment. Specifically, focus groups are employed to provide initial details of product or service experiences and the resulting opportunities for new products or product changes. The personal responses within a group setting generates discussion about specific product attitudes and usage topics. These topics can include

- Description of the problem, solution, and satisfaction
- Prioritization of key problem solution needs
- Presentation of new concept or product prototype
- Reaction to concept or product prototype
- Comparison to present solution and other solution alternatives
- Measurement of intent to try or to purchase the concept or product
- Investigation of purchase location, price, and information sources for new product

The previous outline is a rough topic sequence for a focus group discussion. Specific objectives can include the entire sequence — usually the case with new product development — or just two or three topics based on the information need as related to the complexity of the product.

The next important step is group *participant recruitment.* This phase is critical because the information obtained is only as good as the source. Wrong people yield wrong information. No organization has the unlimited resources to conduct incorrect focus group research. The basis of focus group recruitment is the definition of target user group in the population. This is best accomplished by reviewing Section 12.7. The market is defined and subsegments are identified. The correctness of the individuals in the subsegment targets represent the recruiting population for focus groups. It is through the use of focus groups that problems, solutions, new product reactions, and purchase intentions can be examined initially within one target segment or several on a comparative basis.

After respondent and group member qualifications are decided upon, a formal recruiting screening questionnaire should be drafted. This form is approved by the appropriate individual in the organization conducting the research. The screener is used for all group participants and ensures that the appropriate individuals will be members of the groups and the information obtained will be from the intended product user audience.

The group *discussion guide* is another important phase of focus group methodology. This is an outline of the topics that are to be covered in the group by the moderator. The finished document serves as a guideline for the discussion and can be considered an information contract among the interested individuals and organizations involved in the focus groups. The previous statement of a contract is meant to signify the importance of the discussion guide.

For the content of the discussion guide, the information objectives serve as the primary source of topics. The discussion guide must be developed several weeks prior to the first group session because the development requires conversation among the individuals requesting and using the information. As with participant recruitment, the discussion guide development is a formal procedure. It is not acceptable practice and is even dangerous to decide the night before or even an hour prior to the group's meeting to rough out a few subjects. The results will be questionable and the resources wasted.

Often, the development of the discussion guide requires careful consideration of the information needs and end use of the information. Prior experience, other research, or preliminary interviews among selected experts in the field (another qualitative research technique) provide the background for structuring the focus group discussion guide. The development is an iterative process so that the final draft represents the best topic content and flow for the group participants in the first focus group.

After the discussion guide is finalized prior to the first group, actual time segments are assigned to subjects according to priority of need and to gauge the total time requirements. An average focus group lasts 90 minutes. Even though serious time was taken to develop the discussion guide, be prepared for changes after the first focus groups are completed. No matter how much preparation was expended, the group participants can alter any part of the subject content or flow of the discussion guide.

The discussion guide is an expansion of the information objectives. Previously, general information objectives were listed for a new product focus group discussion. The discussion guide would emerge from the added topic details within each objective. Expansion of the first two objectives is described to illustrate the derivation process for the discussion guide.

- Discussion of the problem

    Most recent experience (job)
    Describe experience
    Key elements of the problem
    Most critical component
    Solution and alternatives evaluated
    Overall satisfaction of completed job
    Reason for satisfaction level
    Advice to friend encountering same problem or experience

- Solution needs

    Discussion of problem solution components
    Priority of key solution criteria or problem needs
    Reasons for top three criteria
    Criteria stable or changing
    Any new needs/criteria
    Best solution available now for criteria
    Reasons for best solution

Each topic then is expandable into several subareas of discussion that maximize the group members' participation and sharing of views. The development of the guide is an important process among the personnel involved in the focus group research project. A consensus should be reached regarding the content of the guide prior to conducting the first group session.

However, bear in mind the possibility that the discussion guide may not work because topics maybe out of order, respondents do not know the answers, the information obtained is not needed, or the topic ends up not being relevant to the research objectives. This is one of the advantages to focus groups because the discussion guide can be altered between groups and new topics added, order or questions can be changed, and sections can be expanded or deleted. The discussion guide is a critical step for successful focus groups; however, the person that uses the guide during the focus group is extremely important too.

The *moderator* with the discussion guide is a necessary and powerful combination in obtaining the desired information from the qualified individuals in the focus group. There are many individuals who claim to be focus group moderators and even more who believe they can do focus groups. Finding the right moderator is another big step for successful focus groups. The following list is a suggested profile for evaluating and deciding on the right moderator for focus group sessions [Krueger, 1994]:

- Able to exercise control over the group, but indirectly and according to acceptable guidelines
- Generate enthusiasm and interest for the topic based on self-curiosity
- Exhibit a sense of timing for pursuing topics or participant comments
- Show respect for the respondents as well as be polite in all group situations
- Have or acquire adequate background knowledge of the subject
- Be a strong communicator
- Be self-disciplined and refrain from expressing opinions as well as leading the responses from the participants
- Offer a friendly manner with humor so that respondents feel comfortable

Any moderator should be well experienced in group dynamics of small group interactions. Group moderation is a challenging and exhausting job. A professional moderator is often a worthwhile invest-ment to increase the value from focus groups. As stated earlier, the focus group approach can deceive potential users easily, and part of that deception is group moderation. Many persons believe that mod-eration is "just sitting down and talking with group members". That thinking yields worthless and even dangerous information.

One last thought for moderator selection. Being familiar with the subject is strategic, and familiarity often increases with the complexity of the subject, especially technology. Therefore, orientation sessions will be very helpful for the moderator so that all discussion in the group is relevant and therefore useful. Sometimes a pilot session can be conducted to give the moderator a preview of the subject to test the content and flow of the discussion guide and to clarify any subject areas related to the information objectives. Given a good job of group moderation and a workable discussion guide, the next phase is of analysis and interpretation of the group discussion content.

The *analysis and interpretation* of the group discussion is a very specialized research procedure. Because of the far-reaching effect of the information generated from the focus group methodology, the organi-zation, compilation, and presentation of the group discussion results should be systematic to maximize the credibility and application of the findings. The previous steps described are all critical elements of reliable and valued focus group sessions. The final preparation of the results is no exception.

Generally, the results produced for the focus group sessions are prepared by the moderator. This requirement does not preclude observers of the group from preparing summary reports either. Because the conclusions and applications from the groups can be influenced heavily by subjective interpretation, alternative analysis for comparison purposes can be beneficial to the end value of the qualitative research. At a minimum, it is always recommended that, after each group and, most importantly, the final group, debriefing sessions be held to review the findings from the discussion.

This procedure helps refine the group content structure and initiates the final observations from all the groups. In all debriefing meetings and the final analysis and reporting, a systematic approach should be employed. The guidelines for the systematic approach are [Krueger, 1994]

- Be prescribed and sequential process
- Be refinable and replicated by another individual using same group discussion data
- Be focused using the information objectives as the guide
- Be limited to the group discussion and content only

- Be practical and relevant to needed action
- Be timely, but not rushed and inclusive of all group discussion
- Reflect sequential learning from group sessions
- Be insightful for improved understanding of topics and respondents
- Be inclusive of alternative interpretations
- Be conducted by an individual having training, experience, and past performance qualifications

The analysis and interpretation is labor intensive. Notes from the sessions and review of the audiotapes provide the raw data that must be compiled, organized, and then interpreted based on the original information objectives. This is the last of the systematic steps taken to generate an objective, relevant, and useful document from the focus group process.

Conclusions

The previous component areas of focus group research are key to obtaining credible results from the group discussion sessions. The last area of focus group research concerns the *mechanics or logistics* of focus groups. This last section will provide initial familiarity with the scheduling, organizing, facilities, and costs of focus groups. The first arrangement for focus groups is locating a facility in the cities selected for the group sessions. The cities selected depend on the research objectives and respondents desired, but there are facilities across the country. Facilities should have a conference room, reception area, and client viewing area with a one-way mirror and soundproofing for observing the focus group sessions. A guide published for focus group facilities and research services is available for reference from the Marketing Research Association [Adams, 1997]. After the facility or facilities are selected in desired cities, the recruiting of the respondents proceeds with the screening questionnaire, which was described earlier. Generally, 15 respondents will be recruited in order for 10 to 12 to be present at the scheduled time. Usual times for focus groups are anytime during the day and at 6:00 and 8:00 in the evening. Because group sessions are increasing in demand, some cities are scheduling Saturday sessions. At least a month in advance for facility reservations is required in the larger urban locations.

Often, the facilities will provide refreshments for the group members. If the group is scheduled for a meal time, then additional food is provided. All facilities should generate audiotapes for each session. It is always a very good idea to check the recording system because this is the raw data from the group discussion. Videotapes are now available at most facilities but cost extra, and that cost varies considerably across location sites.

The facility provides the administrative support for focus group research. These are professionals and have decades of experience in recruiting and handling all the activities associated with focus group sessions. Included in the fees charged from the facility will be the respondent recruiting, receptionist for hosting the group, and disbursement of respondent fees. Today, an average base price for a focus group is $3000, which includes the four major costs: respondent recruitment, facility rental, respondent participation fee, and the moderator. The more specialized the respondent qualifications, the higher the cost will be for each focus group because of recruiting and the participation fee.

The applications of focus group research are numerous. In the most general way, focus groups can be used to obtain opinions, attitudes, ideas, and reactions from any audience that is critical to an organization's performance either internally or externally. New product development is an area that uses focus groups extensively. However, examination of problems, or opportunities in a product category, communications campaigns, packaging, purchase patterns, brand identity, product changes, customer service, product experiences, information sources, and direct mail catalogs represent a sample of focus group subject areas. The attraction of focus groups is the ability to interface with a qualified subset of the consumer target in any of the preceding topics.

Given that the guidelines presented in this section are followed, the results from the focus group research will be useful. Focus groups are not quantitative, and the opportunities to negate the value from the qualitative research method are ever present. Credible focus groups require a disciplined research approach that includes, at a minimum, careful planning and information objective setting, development of a respondent screening form, selection of a qualified moderator, systematic analysis and interpretation of the discussion results, and a professional facility that can assist and support the focus group process.

If commitment to the focus group research methodology is made, then the insights, perspectives, and ideas will be available to the users of this qualitative research approach that often leads to better products, programs, or communications in the marketplace.

Defining Terms

Exploratory research: Initial research conducted to define the problem and to examine variables as well as to observe possible variable relationships within the defined problem. Both secondary and primary research methods are used for exploratory research.

Focus groups: A qualitative research method used in an exploratory mode to structure a problem and to obtain initial insight into the problem prior to subsequent survey research.

Qualitative research: Research conducted from which the results cannot be projected to the general population under investigation.

References

Adams, A. *MRA 1997 Blue Book*, Marketing Research Association, Rocky Hill, CT, 1997.

Churchill, G. A., Jr. *Marketing Research, 6th ed.,* Dryden Press, Fort Worth, TX, 1995.

Krueger, R. A., *Focus Groups, 2nd ed.,* Sage, Thousand Oaks, CA, 1994.

Templeton, J. F. *The Focus Group, rev. ed.,* Richard D. Irwin, Burr Ridge, IL, 1994.

For Further Information

The *American Marketing Association* offers many professional development seminars at annual conferences. Focus group research is one of the development seminars. All facets of focus group research are covered by experts in the area. American Marketing Association, 250 South Wacker Drive, Chicago, IL 60606-5819; phone 1(800) 262-1150.

The Burke Institute offers a series of seminars on focus group methodology as well as qualitative research methods in general. The Burke Institute, 50 E. River Center Boulevard, Covington, KY, 41011; phone 1(800) 543–8635.

The Focus Group Kit, edited by David L. Morgan and Richard A. Krueger, Sage, 1997. This is a six-volume set that represents a complete and comprehensive guide for conducting focus groups. The material expands on the key points presented in this section.

12.16 Marketing Planning for High-Technology Products and Services

Peter LaPlaca

The key to success in the high-tech marketplace is to know as much about the market, the customers, and the competition as possible *before* you enter the market and to have this knowledge in a systematic and organized methodology. The old adage "Forewarned is forearmed" applies to the marketing of high-technology products and services.

In developing the marketing plan, there are five major components: foundations, strategies, goals and objectives, tactics, and evaluation and control. These will be examined in this section.

Foundations

The foundation for successful **planning** is knowledge of all of the factors that will impact the firm's operation in the marketplace. These include the customers and their environments, the technology, competition, and the external and internal environments faced by the firm. One of the biggest mistakes the high-tech firms make is their lack of planning to select their customers. Too many high-tech firms first develop their technology and then seek out possible applications. A better approach is to first identify those types of customers that you think might benefit from applications of your technology then research these customers to see exactly how applications of your core technology will benefit them, how they perceive the benefit, what is the value of this benefit to them, and how else they might receive the same benefit. By not first focusing on the target customers, firms risk spending limited resources on technology applications that provide little real benefit to customers. This results in limited sales and lower prices. Part of this foundation is developing a thorough understanding of how the customers view the technology and its use for their benefit. *This understanding can only be achieved by thoroughly researching customers' characteristics, needs and wants, behaviors, attitudes, and values.* It also gives the firm the opportunity to know how customers are currently meeting their needs for the targeted application of the technology, what they are paying for that method, and how much they would pay for the new alternative that the firm is attempting to market.

The Technology

It is important that the high-tech firm develops a complete understanding of the technology, from both the scientific and economic perspectives. From the technological perspective, the company must fully understand the technology, the rate and path of technological development, and the limits that the technology can achieve. This understanding must encompass both the company's core technology as well as the many derivative technologies and applications resulting from the core. From the economic perspective, the company must know the investments required to keep pace with the technology and the costs of developing new applications of the core technology. Economic and financial analysis will also help determine how much of the new technological application customers would be able to afford. Designing products that offer outstanding features but cost the customer many times what the benefit is worth is a surefire way to fail in the marketplace.

One of the fallacies believed by many high-tech firms is that they have no competitors. Theirs is a unique technology and customers must come to them. This is seldom if ever the case. Direct competitors are firms that are also working with the same core technology, *even though they may be developing different applications of that core technology.* With minimal investment these firms can, and frequently do, soon turn to producing similar products for the same customers. The result becomes an escalation of features (the "features war") combined with reduced prices and margins to gain market share. *It is imperative that high-tech firms maintain accurate and current information on all firms involved with the same core technologies as theirs.* This information should include their strengths and weaknesses, sales, market share, targeted markets, marketing and technology strategies, sales and marketing tactics, and any other relevant information on the firm's or it's principle employees.

In addition to direct competitors, attention should also be directed to indirect competitors. These are firms that employ a different technology to provide an alternative method to providing customers with the same inherent benefit. (For example, customers can travel between Chicago and New York via train, bus, auto, and plane. United Airlines' direct competitors are the other airlines that fly between these two cities. Indirect competitors would include any other means of traveling between the two cities. While the direct benefit [travel between the two cities] is the same for all modes of transportation, there are many differences in the secondary benefits such as time it takes, ability to view the scenery, ability to take side trips to see various sights, etc.) High-tech firms should develop profiles of these alternative technologies and their costs and value placed on the secondary benefits by the customers. This information is very useful in the development of marketing strategies and tactics (Table 12.11).

Strategies

The development of sound marketing **strategies** is one of the key elements to success in the marketplace. Indeed, one study found that, in 88% of the product failures studied, improper marketing contributed

TABLE 12.11   A Marketing Planning Checklist

Customers
  Who are they?
  What are their needs?
  How are their needs now being met?
  What is the value of meeting these needs?
Technology
  Define core technology
  Define specific applications
  Technology momentum
  Cost of keeping up
Competition
  Direct competitors and their strategies, strengths,
    and weaknesses
  Indirect competition
Strategies
  What possible strategies exist?
  What are their success requirements?
  What are their risks?
Goals and objectives
  Basic goals
  Operating goals
  Nonoperating goals
  Stretch goals
  Dollars and cents
Tactics
  Selling and sales management
  Advertising and promotions
  Pricing and price structure
  Product details
  Customer service
  Distribution and logistics
  Budgets
Evaluation and control
  Assessing effort
  Assessing performance
  Maintaining forward momentum

to the failure. Marketing strategy is a term applied to the broad approach that the company will follow. It begins with a definition of the customers being targeted and contains a broad description of how the high-tech product will be marketed. The type of marketing strategy planned for depends on the type of technology and the type of market. For technology aimed at industrial or business-to-business markets, a very focused strategy might be used. For example, there may be only a few dozen possible companies that would buy the new technology. In this case a marketing strategy of going directly to these limited customers would be appropriate. On the other hand, a new technology product aimed at residential household consumers that number in the millions would need a more indirect marketing approach using extensive advertising, multiple channels of distribution, and perhaps after-sales service centers. The strategic discussion section of the plan should contain a brief (one or two paragraphs at most) statement of each possible strategy that could be considered, including the known success requirements, upside and downside potentials, and possible risks and their consequences.

Goals and Objectives

The next part of the plan should be the company's statements of goals and objectives. This does not mean a simple statement of desired sales and profits, but rather a series of goals that impact current and future performance. In addition to an absolute dollar amount of sales revenues or profits, some of the goals that should be considered include

- Market penetration — This represents the percentage of possible customers that use the product. It does not imply that the customer uses only your product, but that your product is among those purchased.
- Market trials — This is similar to market penetration, but is used with new products and services. It represents the percentage of possible customers that try the product at least once.
- Market share — This is the percentage of total market sales that your sales represents. For new or recently introduced products, market share gains frequently have the highest priority, even higher than profits because achievement of a minimal market share provides the basis for *future* growth and profitability.
- Market position — This represents the rank your sales places within the industry. If your sales were the highest, your market position would be one. This could also be expressed as a change in position, for example, to move from fourth to second position within the next year.
- Change in sales — This is usually expressed as a percentage increase in current-year sales from the previous year as in a 10% increase in sales or a 10% revenue growth.
- Bookings — Bookings are similar to sales except that they represent future sales that are placed today (such as orders). For many high-technology items, particularly large capital goods, orders may be placed for delivery years into the future. The airline industry orders aircraft in this manner.
- Profitability — Profits are expressed in dollars and are measured after allowances for state, local, and federal taxes. Profitability is measured as the ration of net dollar profits divided by the total sales (e.g., 8% of sales.) Some firms also use gross margin or contribution margin as profitability goals.
- Unit sales — Rather than dollars, unit sales refers to the number of physical units sold during the planning period.
- Sales or profits per employee — This is simply the total sales (or profits) divided by the average number of people employed in the company for the year.

In addition to goals such as those above that reflect the firm's finances, many marketing goals are used that are not in terms of financial measures. For example, the firm might seek to achieve a specific reputation ("Most innovative in the industry", "Highest customer satisfaction", "Leader in X technology", etc.). As long as these goals can be quantified and objectively measured, they can serve to direct company efforts and measure company results. Other goals may involve aspects of the company's marketing efforts such as "Six new products introduced this year", "Distribution coverage in all fifty states", "Most recognized advertising", etc.

Regardless of which goals are used, all goals must have the ability to be objectively measured and must have a time constraint (usually a year). Some goals may conflict with other goals. For example, the goal of gaining market share frequently is counter to the goal of increased profits *during the same time period*. When goals conflict, employees must be given a clear signal as to which goal to pursue. For example, "During the next two fiscal years, the most important goal for Product Line X is to increase its market share by at least 10% annually. Following this time period the most important goal will be to maximize annual profit contribution." This sample goal statement also illustrates another important caveat for goals. They must be within the control of the entity being measured. For a single product or product line, a goal of increasing overall company profits is not useful since other aspects of the company's operation can influence overall profits. In these cases profit contribution (sales revenues minus costs of producing, managing, and selling the specific product) is a better goal.

Another class of goals to include are the stretch goals. These are goals that the firm *could* achieve if all things worked out. They are really beyond normal expectations, and failure to achieve them is not cause for concern. They are used to provide the company with additional targets to aim for when previous goals prove to be insufficiently optimistic.

Tactics

Tactics are the details of the firms marketing efforts and include selling and sales management, advertising and promotions, pricing and the overall price structure, product details, customer service, distribution and logistics, and the detailed budgets for each of these components of the marketing effort.

Selling and sales management decisions involve the type of sales effort that the firm will utilize. This can include a company-employed sales force that calls on potential customers at their place of business or residence (outside sales), use of independent manufacturers' representatives or sales agents, company personnel that respond to telephone inquires (inside sales), or an outside telemarketing company that calls on potential customers via telephone. The marketing plan should contain all details necessary to select, hire, train, and manage the sales force. Sales management information will include the compensation system, territory management, and sales support programs. The sales plan will also include individual and group sales targets (quotas) for all sales personnel.

Advertising and promotions are activities used by the company to persuade potential customers to purchase their products and services. The marketing plan will include details such as the overall promotional budget, proportion of the budget spent on trade and other media, planned expenditures of advertisement creation, trade shows and similar activities, brochures and other product literature, and similar activities. A good plan will include all of these items along a planned time line showing when each activity will be accomplished and which company personnel are responsible for task implementation. Coordination with outside agencies and consultants must also be included in the plan.

Pricing and the overall price structure involve the establishment of a detailed price list for all products, special features, options, services, and other items purchased by the customers. The firm must also include a detailed description of its pricing policy (allowable discounts, how to deal with competitive price pressures, minimum order sizes, inclusion of shipping charges, allowances and other details of company pricing.) For firms dealing with foreign markets, special attention must be given to currency for price quotes, collections, and exchanges.

The marketing plan must also include information about all product details such as special features, policy toward customization of products, packaging, and special handling arrangements. These must be as detailed as possible to avoid confusion and possible misrepresentation to customers. Similar attention must also be given to the entire area of customer service. How will customer orders be monitored? How will customer training and installation be accomplished? How will customer complaints and inquires be dealt with? Each aspect of customer service must have an accompanying budget and person responsible.

The area of distribution and logistics is concerned with the process of how the customer receives the products. Will the firm utilize dealers and distributors, retailers, or other entities as intermediaries between the customer and the firm? If so, how will they be selected, monitored, compensated, controlled, and evaluated? What types of contracts will be used to initiate relationships with these intermediaries? What power will they have on customer maintenance? Distribution is a critical but external asset of the company, and much care and attention must be used in developing a proper distribution network. This network will evolve and change over time, and the marketing plan must allow for this change *without causing confusion on the part of the customers.*

Evaluation and Control

Part of the marketing plan must also include control and evaluation. Control refers to maintaining the planned marketing efforts, and evaluation measures the results of the marketing program against the stated expectations.

Control looks at whether planned activities are being accomplished on schedule and within budgets. In many cases, underspending can result in as great a problem as overspending. If the plan has been carefully put together, anticipated expenditures should be carried forward to reach the plan's goals and objectives. Plan **evaluation** looks at the results achieved by the planned activities. How many of the goals are being achieved as the company goes forward with plan implementation? Were the goals established properly? Are they too pessimistic or too optimistic? What unforeseen events have happened that will require plan modifications?

For evaluation and control to make a positive contribution to the company, they must be undertaken simultaneously to plan implementation. Do not wait until the plan has been completed (the end of the year) to measure results. Do this continuously and make course corrections as the plan goes forward.

Defining Terms

Planning: A process whereby future company activities are coordinated to reflect the needs of the firm's environment as it seeks to achieve desired goals and objectives.

Strategy: A broad overview of the basic means the firm will employ to establish and strengthen its relationships with its customers.

Tactics: Detailed steps required to complete the company strategy.

References

Hiebing, R. G. and Cooper, S. W. *How to Write a Successful Marketing Plan, 2nd ed.*, NTC Business Books, Lincolnwood, IL, 1997.

McDonald, M. H. B. and Keegan, W. J. *Marketing Plans that Work: Targeting Growth and Profitability*, Butterworth-Heinemann, Woburn, MA, 1997.

Further Information

PlanWrite Marketing Plan Toolkit, Business Resource Software, Austin, TX.

QuickInsight: A Marketing Strategy Analysis Tool, Business Resource Software, Austin, TX.

12.17 Branding and Private Label Strategies

Eitan Gerstner and Prasad Naik[9]

The concept of branding is central to marketing. Branding transforms commodities such as shampoos and microprocessors into products such as Head & Shoulders shampoo and the Intel microprocessor. Branding involves not only advertising but also commitments on the part of the companies to deliver promises made and meet customer expectations. Branding creates value to customers that can be captured through premium prices, increased customer loyalty, and a larger market share. The additional company value created through branding is referred to as **brand equity**. Traditionally, branding was pursued mainly by manufacturers. Their brands are called "manufacturer brands" or **national brands**, because they are advertised and distributed nationally. National brands helped manufacturers to gain power over retailers, who were forced to carry these brands at their stores because of high customer demand.

Private labels are brands owned by distributors. Familiar examples are the Eight O'Clock coffee brand owned by A&P supermarket or Craftsman tools brand owned by Sears. One reason that pushed distributors to introduce private labels was the need to gain back power from manufacturers who used their popular brands to obtain concessions such as lower retail margin, unfavorable credit terms, and prime shelf space from the distributors. Another motivation for developing private labels was to increase customer loyalty to a distributor who competed with other distributors.

Private labels are very common for frequently purchased products such as those sold in supermarkets and drugstores; however, successful private labels have also been developed by distributors of technical products. For example, Sears has developed two successful private labels that included dozens of products: Kenmore and Craftsman. The opportunity for private labels, however, exists for other technical products

---

[9]We would like to thank Michal Gerstner for her help in collecting the data presented in this section.

such as computers, and their popularity is increasing as **direct marketing** channels such as catalogs and the Internet are growing quickly, as will be discussed below.

A Short History of Private Labels

The history of private labels can be traced back into the 19th century, when mail-order companies were first established in the United States. Two of the first private labels were Eight O'Clock and Our Own tea, both developed by A&P grocery chain. In fact, the chain managed these two brands so successfully that during the 1960s and 1970s they represented 35% of A&P's total store sales in more than 15,000 stores.

As part of their act to counter the power of A&P, a group of the Independent Grocers Alliance (IGA) was formed, and by the end of 1930s it included over 10,000 stores. Because IGA products were advertised in national radio stations, it was difficult to distinguish them from national brands. Private labels were also introduced by department stores such as J.C. Penney, and Sears Roebuck, which until the late 1980s had the majority of its sales in private labels. Only in the 1990s Sears has moved to increase the share of national brands in its stock. Private labels were also developed by manufacturers to be sold as alternative economy brands by retailers. For example, in 1964, 3M developed a line of private labels that captured 10% of the film market in 1993.

Two factors helped slow down the wide acceptance of private labels: inconsistencies in quality and the proliferation of brand names. The quality was inconsistent because products sold under the same private labels were procured from different manufacturers. The brand proliferation occurred when more and more retailers introduced private labels and competition intensified. As consumer confusion grew, brand loyalty declined.

The confusion with brand names brought to the market the nameless **generic brands**, which were no-frills products with simple packaging, no advertising, and often no mentioning of the producer. Although generic brands had the potential of providing a good value to consumers because of the cost savings in advertising, promotions, and packaging, they did not succeed because consumers were suspicious about their quality and safety. Today, generic brands are rare in consumer markets.

In the last 2 decades, private labels had a comeback, but this time with a greater emphasis on quality and consistency [Salmon and Cmar, 1987]. More attention is also given to packaging and promotions. Some companies, such as Loblaw in Canada, specialize in developing and licensing what is known as "premium private labels". Its private label, President's Choice, and other premium private labels such as Master's Choice are gaining popularity in grocery chains in the United States and abroad. A boost to private labels is also given by the growth of direct marketing channels as more and more customers buy products through catalogs and the Internet. Brands such as Dell and Gateway 2000 have become household names for personal computers distributed through mail-order operations.

In recent years, a major battle has developed between national brands and private labels of supermarket goods [Quelch and Harding, 1996]. Producers of national brands started to react to the threat of private labels by cutting prices aggressively [Giles, 1993]. In the PC business, manufacturers such as IBM had no choice but to enter the mail-order distribution business. It remains to be seen where this struggle will end.

Who Produces Private Labels?

In general, private labels are not produced by unknown and obscure manufacturers. Most often they are produced by the same manufacturers who produce the more-familiar national brands. For example, Whirlpool sells appliances under its own name, but its products are also sold by Sears under the retailer's private label. Admiral sold televisions under its own name and under Kmart's private label. To prevent channel conflicts, the manufacturers of national brands disguise the fact that they are also the producers of private labels by maintaining distinct packaging and labeling.

Why do these manufacturers choose to compete with themselves? First, large retailers who carry private labels provide the manufacturer access to very large markets. Second, if some manufacturers did not supply private label products, their competition would. Third, the production of private labels helps fill in downtime in production lines, use excess capacity in production facilities, and minimize advertising and other marketing costs.

Some manufacturers produce private labels only in categories in which they do not have market dominance. For example, Heinz is the marketing leader in the ketchup market and it produces no private label for this market. However, Heinz sells private labels in the soups product category because they find it difficult to compete directly with the national brands such as Campbell soups. In some cases, the private label market may become so attractive that the leading national brand manufacturer decides to enter it. For example, in 1991, Ralston Purina announced its entry into the private label business with up to 30 private label accounts [Fitzell, 1993].

Prices and Quality of Private Labels

Private labels are typically sold at a lower prices than the national brands. The reasons for this are related to costs and demand. On the cost side, private labels are promoted by distributors who sell thousands of products, and therefore promotion costs can be spread among many products. The manufacturers can pass the savings in advertising to the retailers and eventually to consumers in the form of lower prices. On the demand side, private labels give all sellers an opportunity to benefit from **price differentiation** in which slightly differentiated products are sold at two different prices to two different customer segments — at a higher price to customers who are willing to pay a high price for a well-recognized name brand and at a lower price to customers who are sensitive to prices.

An interesting case is that of the two private labels tools sold by Sears: Kenmore, and Craftsman. Table 12.12 gives relative quality rankings and prices of these brands in several product categories surveyed in the Buying Guide of Consumer Reports, 1996. For example, in the washing machines category, Kenmore model 25841 was ranked number 1 out of the 19 reviewed brands, and its price was $580 compared to the average price of only $406.

Table 12.12 shows that, for the product categories reported, Kenmore and Craftsman were typically ranked above average in quality performance (most of the brands ranked were national brands). Second, these brands are typically sold at a price close to or higher than the average national brand. Therefore, the conventional wisdom of selling the private labels at lower prices than national brands may not hold when a distributor consistently maintains good quality and service for its private labels.

There are two factors that need to be considered when pricing private labels relative to national brands. (1) the perceived quality gap between the private labels and national brands and (2) the extent to which selling the private label takes sales from a national brand, i.e., **brand cannibalization**. When national brands are perceived to be superior to private labels, the price of the national brand can be increased, but some customers would switch to the private label. How much brand cannibalization can be tolerated by the national brands? Gerstner and Holthausen [1986] showed within a theoretical framework that price differentiation can be profitable even when cannibalization is substantial. Recently, however, manufacturer of food products chose to roll back prices by more than 20% because brand cannibalization was perceived to be excessive [Giles, 1993].

TABLE 12.12   Relative Quality Rank and Prices of Sears Private Labels (Based on Consumer Reports, 1996)

Product Category	Private Label	Rank/No. of Brands	Price/Average Price ($)
Washing machines	Kenmore 25841	1/19	580/406
Midsize microwave ovens	Kenmore 89381	12/17	180/154
Electrical ranges	Kenmore 93541	3/10	500/473
Built-in dishwashers	Kenmore 16941	1/24	565/442
Room air conditioners	Kenmore 75089	4/12	450/438
Power blowers	Craftsman 79838	5/10	65/66
Gasoline lawn mowers	Craftsman 38274	7/26	220/170
Garage door openers	Craftsman 53425	6/17	160/198

Private Labels in Direct Marketing

Direct marketing channels have expanded dramatically in the last few decades, encompassing almost every product category. Using direct marketing, manufacturers can become distributors at a relatively low cost because there is no need to open many stores in highly desired locations.

Direct marketing allowed companies to develop brands that were associated with this distribution channel. In early 1900s, Sears started its direct marketing business by developing a comprehensive catalog with its own private labels. More recently, direct marketing has reached technical products such as the personal computer and its peripherals. Initially, it was difficult to imagine that customers would buy complex technical products through mail, but the conventional wisdom has proven wrong. A key to the development of this market was the establishment of reliable and reputable direct mail-order distributors such as Dell and Gateway.

Although these companies distribute computers directly to consumers, they also assemble the computers they sell. Therefore, one could consider their brands as either private labels or a combination of manufacturer and distributor brands. In contrast, some companies such as Compaq and Hewlett Packard (HP) still use traditional retailers to distribute their products and therefore should be considered as manufacturer brands. Table 12.13 shows that direct mail companies were successful in developing top-quality brands that are sold at prices well below those of the manufacturer brands.

Interestingly, Dell and Gateway 2000 are not only the top ranked brands but also the least-expensive ones. This excellent performance can be attributed to the efficiency of the direct channel relative to the traditional retailers. First, in this direct channel, orders are taken first and then computers are built to the buyer's specification. This guarantees a better fit between the customer and the product and requires a minimal inventory of finished products. Second, when a brand is manufactured as well as distributed by the same company, a greater control exists over customer satisfaction because the same company provides services that otherwise would be shared by the manufacturer and retailers. This control can be used to enhance brand reputation.

Penetration of Private Labels

The penetration (or market shares) of private labels varies across product categories and across countries. Table 12.14 shows the penetration percentage of private labels by country in 1991. As can be seen, Great Britain was leading the pack in this survey with 35% penetration, and the United States was placed in the eighth place with a 13% penetration. For the same year, Table 12.15 gives the penetration of private labels (in terms of category dollar shares) for several product categories in supermarket stores in the United States, and the percentage price differentials between them and the national brands.

A possible determinant of degree of penetration may be the perceived quality of private labels relative to national brands. Store brands of fresh foods such as cheesecakes may be perceived as good quality when consumers can inspect and try samples. Therefore, one would expect a higher penetration of private labels in such categories. On the other hand, health-related products such as antacid cannot be judged before purchase and product trial. The uncertainty concerning such "experience goods" probably encourages consumers to buy national brands, even at a significant premium (in particular, when health is involved). This may explain why the penetration of private labels in the health and beauty aid category is relatively low.

TABLE 12.13 Quality Rank and Prices of P/166 PC Systems (Based on Consumer Reports, September 1996)

Brand and Model	Quality Rank	Price ($)
Dell Dimension XPS P166s	1	2680
Gateway 2000 P5-166	2	2675
Compaq Presario 9660	5	3375
NEC Ready 9619	7	3500
HP Pavilion 7170	9	3550

TABLE 12.14   Worldwide Penetration of Private Label in 1991[2]

Country	Market Share (%)	Country	Market Share (%)
Great Britain	35	United States	13
Germany	24	Australia	11
Canada	19	Spain	8
Belgium	19	New Zealand	7
Austria	17	Italy	6
The Netherlands	17	Greece	3
France	15	Portugal	1

The information in the table was presented by Loblaw company at the June 1992 PLMA Consumerama meeting in Toronto.

TABLE 12.15   Penetration and Price Differential of Private Labels in 1991 (Compared to National Brands in U.S. Supermarkets)

Product Category	Category Dollar Share	% Price Differential
Milk	64.06	17.83
Cheesecake	56.52	17.44
Vitamins	30.16	27.30
Diapers	11.35	14.25
Razors	5.37	27.03
Antacids	4.34	21.90
Shampoo	2.02	23.24
Toothpaste	0.91	25.66

The penetration of private labels may also be a function of the price differential between them and the national brands. When manufacturers of national brands become too greedy, and the price differential increases, consumers are encouraged to take more risk and try the less-expensive private labels. If consumers discover that the quality of private labels is satisfactory, they will continue to buy them, and, consequently, the market share of private labels will increase (see, for example, Corts [1995]).

Finally, the penetration of private labels into different markets and countries may also depend on culture. In some cultures, it may be more embarrassing for people to use or serve private labels to friends and family because of the image it may convey. National brands usually are more prestigious than private labels because of the wide recognition they have due to heavy advertising.

Conclusions

Private labels are more common for packaged goods sold in supermarkets and drugstores than for technology products. However, the opportunity for using this branding strategy exists for technology products as well because of the cost savings in production, advertising, and distribution and because of the market power distributors of private labels can gain back from the manufacturers of national brands. With the growth of direct marketing channels and the Internet, one would expect that private labels for technology products will become more prevalent in the future.

Defining Terms

Brand cannibalization: Sales of one brand take a bite from the sales of another brand produced by the same manufacturer.

Brand equity: Additional value a brand name brings to a company because of its higher appeal and premium prices customers are willing to pay for it.

Direct marketing: Marketing to specific segments via direct mail, catalogues, and the Internet.

Generic brand:　Products that do not have a name brand (basically commodities).

National brands:　Brand names owned by manufacturers who promote them nationally.

Price differentiation:　Offering the same basic product (often under different brand names) at different prices to different customers.

Private label:　Brand names owned by distributors such as wholesaler and retailers.

References

Buying Guide 1996, *Consumer Reports*, Consumer Union, New York.

Corts, K. The Ready-to Eat Breakfast Cereal Industry in 1994, Case No. 9-795-191, *Harvard Business School*, 1995.

Fitzell, P. *Private Label Marketing in the 1990s*, Global Book Productions, New York, 1992.

Gerstner, E. and Holthausen, D. M., Jr., Profitable pricing when market segments overlap, *Market Sci.*, 5(1): 55–69, 1986.

Giles, M. The food industry, *Economist*, December 4th: 3–18, 1993.

Quelch, J. and Harding, D. Brands versus private label: fighting to win, *Harv. Bus. Rev.*, January-February: 99–109, 1996.

Salmon, W. and Cmar, K. Private labels are back in fashion, *Harv. Bus. Rev.*, May-June: 99–105, 1987.

Further Information

Private Label Association (PLMA), 369 Lexington Avenue, New York, NY 10017; phone (212) 972-3131.

12.18　Marketing to Original Equipment Manufacturers

David P. Paul, III and Earl D. Honeycutt, Jr.

Marketing is "the process of planning and executing the conception, pricing, promotion, and distribution of ideas, goods, and services to create exchanges that satisfy individual and organizational objectives" [Peter, 1995]. Central to the concept of marketing is the necessity that successful exchanges benefit both buyers and sellers. The mutual benefit of buyers and sellers is especially important if the relationship between the parties is to be a long-term one.

Original equipment manufacturers (OEMs) buy materials and/or components from suppliers to use in the process of creating the products and services they sell to others. These purchased goods and services may be specific to the ultimate end product or of general business use. For example, IBM buys specific electronic components such as circuit boards to use in the manufacture of its computers and also office equipment, supplies and financial/banking services necessary for the day-to-day conduct of its business. OEMs thus function as intermediaries between raw materials and finished products.

Kinds of Demand

While producers of consumer products are concerned with primary demand, OEMs are concerned with **derived demand**. Primary or consumer demand is affected by product, price, and availability and by consumers' personal tastes and discretionary income. On the other hand, demand for the products and services of OEMs is driven by, or derived from, the ultimate consumer demand. While ultimate consumer demand remains an important consideration for OEMs, other factors are quite important to them in marketing these products and services. Additional criteria important to the OEM's customers include the OEM's price, ability to meet specified quality specifications, ability to meet required delivery schedules, technical capacity, warranties and claim policies, past performance, and production facilities and capacity [Berkowitz et al., 1997]. These additional criteria relate to the fact that the direct customers of OEMs are themselves businesses, which rely on the OEMs to deliver certain products and services *on* *time* and *to specification* in order that production of the ultimate good/service is accomplished in a timely, efficient, and profitable manner.

Why should a supplier be concerned about derived demand? Because the relationship between supplier and ultimate consumer is relatively distant, what was a direct relationship between price and quantity demanded becomes an indirect and possibly reversed one [Reeder et al., 1991]. Consider the following scenario. A supplier to a large, quality-oriented OEM decreases the price of its product or service in order to increase its sales. If the OEM had doubts about the quality of the supplier's product or his ability to deliver that product as agreed, the decreased price may serve to confirm those doubts, leading to a decrease in the suppliers' sales rather than an increase.

Roles in the Buying Center

Who does the buying of the hundreds of billions of dollars worth of the goods and services sold by OEMs? In a business organization, typically more than one individual participates in the purchase decision. The decision-making unit of a buying organization is called the **buying center**, defined as "all those individuals and groups who participate in the purchasing decision-making process, who share some common goals and the risks arising from the decisions" [Webster and Wind, 1972]. Within the buying center, individual employees can play one or more of the following roles:

- **Initiators**: The first to communicate the need for a product or service, e.g., could be anyone in the organization.
- **Users**: People who actually use the product or service, e.g., production-line workers.
- **Influencers**: People who affect the buying decision, often by defining the specifications of what is to be purchased, e.g., quality assurance personnel.
- **Buyers**: Individuals with the formal authority and responsibility to select a supplier and negotiate the terms of purchase, e.g., purchasing agents, buyers.
- **Deciders**: Individuals with the formal or informal authority to approve the buying decision, e.g., Presidents, Vice Presidents, or plant managers.
- **Gatekeepers**: Parties that control or restrict the flow of information in the buying center, e.g., purchasing agents, secretaries, or receptionists.

Significance of Buying Center Roles

Why should OEMs be concerned about these various roles in their customers' buying centers? Because various employees with disparate roles may be active in different types of buying situations, it is important to determine exactly who their salespeople should contact, when the contact(s) should be made, and what sort of appeal is likely to be the most useful and persuasive. Unfortunately, answers to these questions tend to differ from customer to customer. However, some generalizations are possible. Presidents and vice presidents of small firms (defined as those with annual sales under $25 million) exert significantly more influence in all stages of the decision-making process than do purchasing agents or engineers, while the situation is reversed in larger companies [Bellizzi, 1981]. Customer buying centers are likely to involve a greater number of and more varied participants in the purchase of an expensive, technologically complex product or service than when the purchase involves a simpler or less-costly product [Jennings and Plank, 1995].

Information gained from past experience with each customer and from marketing research with present and potential customers is perhaps more important than generalizations. The *specific* wants and needs of each of the individuals in the buying center must be taken into consideration in marketing to OEMs. Although the buying center itself is rewarded for successful completion of its tasks, either individually or as a whole, personnel playing each of the buying center roles are rewarded according to different criteria. For example, users often are concerned with the technical specifications of the product or service — the ability of the product or service to "do the job", while buyers may be much more concerned with costs. Keeping how each individual in the buying center is ultimately rewarded in mind will help in deciding which features of the product or service should be emphasized in marketing to these individuals.

Types of Buying Situations

The number and type of buying center parties involved in the purchasing decision is influenced by the specific **buying situation**. Three types of buying situations have been identified, ranging from the routine reorder or **straight rebuy** to the completely new purchase or **new buy**. Intermediate between these extremes is the **modified rebuy**. In more detail,

- Straight rebuy: In this situation, someone (often a buyer or purchasing manager) reorders an existing product or service from a list of approved suppliers, perhaps without checking with the users or influencers of the buying center. Although price may be a consideration, conformance to specifications, delivery schedules, and other quality-control factors are also important influencers. Companies on the **approved vendors' list** try to maintain existing product and service quality and often suggest that automatic reordering systems (with themselves as the sole supplier) will save the purchasing agent time. Companies not on the approved vendors' list often attempt to make a small successful sale in order that their firm will be considered for future reorders. Examples include office supplies and maintenance services.
- Modified rebuy: Here the users, influencers, and/or deciders of the buying center require that something about an existing product or service be changed. Possibilities include the product specifications, price, delivery schedule, or packaging. Although the product or service itself is largely unchanged, individuals from outside the purchasing department (influencers) must be consulted.
- New buy: In this buying situation the organization is involved in the purchase of an entirely new product or service. Because the organization is basically unfamiliar with potential pitfalls involved with the selection and purchase of the entirely new product or service, there is increased risk of making a poor decision. The greater the cost and/or risk, the more decision participants and the more information are necessary before a decision can be reached. Thus, in this situation, individuals playing all of the roles in the buying center are normally involved.

Significance of Buying Situations on the Buying Center Roles

Individuals with different buying center roles may participate — and even exert different amounts of influence — at different stages in the decision process. For example, technical people — engineers and research and development staff — often exert great influence over a new buy decision, especially with respect to specifications and criteria that the new product must meet. On the other hand, the purchasing manager or agent often has the most influence in the selection among alternative suppliers. The makeup of the buying center also varies with the amount of experience the OEM has with the purchasing firm: the buying center tends to be smaller and the influence of the purchasing manager greater in the case of reordered products than with new buys. Therefore, a salesperson calling on OEMs must be able to discern the buying stage and the individual roles played within the buying center of his or her clients. A summary of how the buying situation can affect buying center behavior is presented in Table 12.16.

TABLE 12.16 How the Buying Situation Can Affect Buying Center Behavior

Decision Criteria	New Buy	Modified Rebuy	Straight Rebuy
People involved	Many	Intermediate	Few
Problem definition	Uncertain	Reasonably well defined	Well defined
Buying objective	Good solution	Improvement over current situation	Low-price supplier
Suppliers considered	New or present	New or present	Present
Decision time	Long	Intermediate	Short
Buying influence	Many (technical considerations may predominate)	Varies	Purchasing agent

Conclusions

Although the OEM supplier faces what appears at first glance to be a bewildering array of factors involved in its marketing decisions, it should not be disheartened. After all, the OEM supplier has a much smaller number of potential customers to satisfy than does the firm marketing directly to the ultimate consumer, whose wants and needs tend to change more quickly. In order to be successful, good market research is mandatory. Many firms use their sales force to collect useful information about their clients.

OEMs must keep firmly in mind that their customers are the producers of the ultimate product or service. Thus, market research regarding the wants and needs of consumers, while potentially important, should not be the central thrust of the OEMs' marketing research effort. Instead, OEMs' marketing research should concentrate on the wants and needs of their customers — the manufacturers. Keeping a firm grasp on the buy class of their products and services, OEMs should attempt to determine in detail the wants and needs of the various buying center members who exert the most significant impact on the manufacturers' purchase decisions.

Defining Terms

Approved vendors' list: A formalized list of vendors that has been certified by the purchasing department as suppliers for a particular product or service.

Buyer: An individual with the formal authority and responsibility to select a supplier and negotiate the terms of purchase.

Buying center: All the individuals and units that participate in the organizational buying decision process.

Buying situation: One of three buying classes faced by organizations: new buy, modified rebuy, and straight rebuy.

Decider: An individual with the formal or informal authority to approve the buying decision.

Derived demand: Organizational demand that ultimately comes from (is derived from) the demand for consumer goods.

Gatekeeper: An individual who controls or restricts the flow of information in the buying center.

Initiator: The first individual to communicate the need for a product or service.

Influencer: An individual who affects the buying decision, often by defining the specifications of what is to be purchased.

Modified rebuy: A buying situation in which the buyer wants to modify product or service specifications, prices, terms, or suppliers.

New buy: A buying situation in which the buyer purchases a product or service for the first time.

Straight rebuy: A buying situation in which the buyer routinely reorders something without modification.

User: An individual who actually uses the product or service.

References

Berkowitz, E. N., Kerin, R. A., Hartley, S. W., and Rudelius, W. *Marketing,* 5th ed., Richard D. Irwin, Chicago, 1997.

Bellizzi, J. A. Organizational size and buying influences, *Indust. Market. Mgt.,* 10: 17–21, 1981.

Jennings, R. G. and Plank, R. E. When the purchasing agent is a committee: implications for industrial marketing, *Indust. Market. Mgt.,* 24: 411–419, 1995.

Peter, D. B. *Dictionary of Marketing Terms,* 2nd ed., NTC Publishing Group, Lincolnwood, IL, 1995.

Reeder, R. R., Brierty, E. G., and Reeder, B. H. *Industrial Marketing: Analysis, Planning and Control,* Prentice Hall, Englewood Cliffs, NJ, 1991.

Webster, F. E., Jr. and Wind, Y. *Organizational Buying Behavior,* Prentice Hall, Englewood Cliffs, NJ, 1972.

12.19   Customer Adoption of Really New Products

Donald R. Lehmann

Forecasting the initial **adoption** of new products (technologies, services) is crucial for planning whether and how to launch them. For products that are not very new, forecasting is relatively straightforward. For really new products that create new categories, however, forecasting is quite difficult.

New Product Determinants of Adoption

Customer reactions generally fall into six basic categories [Rogers, 1983]:

1. Relative Advantage. A key, but far from the only, factor impacting adoption is whether the product or service is really better. Essentially this is the "better mousetrap" dimension, often described in terms of performance or economic characteristics relative to the currently used or best available alternative. Aspects of advantage include (1) quality, (2) convenience (easier and faster), and (3) avoidance of bad aspects of the alternatives (e.g., less expensive, dangerous, or polluting). Further, (4) difference often per se is an advantage, particularly in style/fashion goods. Obviously, the greater the advantage, the greater the likelihood of adoption.

2. Compatibility. The better a new product fits in with existing systems, the easier it is to adopt. Both hardware (e.g., physical connections and size) and software (e.g., operating procedures) are relevant, as are operator procedures. More specifically, incompatibility encompasses (1) physical space requirements, (2) current inventory of alternatives, (3) links to other products, and (4) behavioral changes required to use the product that differ from current use patterns. The QWERTY keyboard was developed to slow down typing in order to prevent mechanical typewriters from jamming but remains the standard because of users' familiarity with it. In the short run, incompatibility often blocks the adoption of seemingly better products.

3. Complexity. In general, the more complicated a device, the less likely it is to be adopted.

4. Communicability. When a new product has a clear and easily communicated advantage (e.g., capacity), then it is generally adopted sooner. In contrast, when the advantages of a product require extensive and sometimes taxing explanations, adoption is slower.

5. Trialability/divisibility. Products that are easy to try are more likely to be adopted. Hence low trial cost and reversibility (so if it doesn't work, you can quickly revert to the old approach) encourage trial.

6. Perceived risk. A sixth characteristic that impacts adoption is the perceived risk. This to some extent encompasses some of the earlier attributes, especially trialability. Risk has multiple components including (1) uncertainty over whether the product will perform as expected/promised, (2) likelihood and consequences of failure, (3) financial costs (price of purchase and upkeep), (4) physical/health concerns, and (5) psycho/social aspects of how using the product would be perceived by the user and relevant others. While it is easy to downplay psychological aspects, a remarkably large number of adoption decisions (positive and negative) can be better explained based on user self-image than on price-performance trade-offs. Unsurprisingly, the riskier the product, the less likely people are to adopt it.

Overall, it is generally sufficient to evaluate new products on three dimensions: relative advantage, compatibility, and risk [Holak and Lehmann, 1990].

Customer Determinants of Adoption

Adoption also depends heavily on characteristics that differ widely across potential customers. Essentially these combine to form an (implicit) estimate of the maximum they are willing to pay for a new product (the reservation price). These include

1. Perceived need for whatever the new product offers
2. Product currently used to satisfy the need (available substitutes)
3. General interest in the product category (involvement)
4. Innovativeness
5. Acquisitiveness
6. Ability to pay

General Issues

1. Patience. Rarely is anything new adopted as rapidly as its proponents think it should be. In many instances, years and even generations pass before anything resembling widespread acceptance occurs.
2. Expectations. Customers have come to expect technological products to systematically change over time. Specifically, they expect (1) quality to improve (features and reliability) and (2) price to decrease. These expectations lead many to delay adoption since the improved future product becomes the standard of comparison. In the case of generations to technology, customers often skip (leapfrog) generations entirely.
3. Relevant Customers. For a new product to succeed, a number of constituencies must all adopt it. These include (1) within company (managers, sales force, and functions/departments), (2) channel (suppliers, wholesalers, and retailers) and (3) outside publics (regulators and media), as well as (4) the actual user and their firm's or family's buying center (influential parties).

Modeling Trial/Adoption

Adoption normally starts slowly, driven by the (few) customers with high need for the product, ability to pay, and perceptions of its advantages (often called innovators or, in the case of electronic products, techies). Adoption then grows as more mainstream buyers enter the market, encouraged as the use by others reduces the risk, product quality increases, and price decreases. Toward the end of a product's growth, most of the potential buyers have been "captured", and decreasing potential leads to a decrease in adoptions. This so-called product life cycle (Fig. 12.23) has been found in numerous circumstances. Of course, other patterns do exist, such as the exponential decline seen in movie attendance at theaters for most releases where sales are highest the first week (encouraged by large promotional efforts), next highest the next week, and so forth. Fortunately, the model used to describe the life cycle also fits this pattern as well.

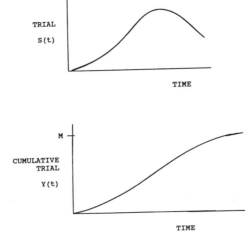

FIGURE 12.23   General Patterns.

The most popular model for describing the pattern of *initial* purchases of a new product in marketing is due to Bass [1969] and is often called a contagion model since it also describes the spread of a disease through a population. (Note the focus is on initial purchases and not upgrades, replacement, or multiple-unit sales, which later in the life cycle account for most of a product's sales.) The model describes sales over time as

$$S(t) = \text{sales in period } t = \left[p + q \left(Y(t) / M \right) \right] \left[M - Y(t) \right]$$

$$= pM + (q - p) Y(t) + q \left[Y(t) \right]^2 / M$$

(12.11)

where p = coefficient of innovation, q = coefficient of imitation, $Y(t)$ = number who have already bought before period t, and M = number who eventually will buy (**market potential**).

When $q = 0$, the model represents exponential decline. Otherwise, the model predicts sales as the product of (1) a probability of purchase (hazard rate) or $p + q(Y(t)/M)$ and (2) the potential adopters who have not yet adopted, $M - Y(t)$. In turn, the probability of purchase consists of two parts: the probability a person adopts without regard to the number of other adopters (hence, the term coefficient of innovation) and those who adopt because the product has been adopted by others (hence, the term coefficient of imitation). Variants of the model explicitly incorporate marketing variables such as price, advertising, and distribution or allow for a more flexible shape [Mahajan et al., 1990; Mahajan et al., 1995]. Still the model fits well without these refinements [Bass et al., 1994].

The model is best estimated by a nonlinear least-squares method [Srinivasan and Mason, 1986]. However, a simple regression approach works quite well in many cases. Basically, this involves running a regression of the form $S(t) = a + b\, Y(t) + c\, Y^2(t)$. One can then solve back to get p, q, and M from

$$p = a / M$$

$$q = -cM$$

(12.12)

$$M = \left(-b \pm \sqrt{b^2 - 4ac} \right) / 2c$$

Also

$$\text{Time of peak sales} = \left(1 / p + q \right) \ln \left(q / p \right)$$

$$\text{Peak sales} = \left(M (p + q)^2 \right) / 4q$$

(12.13)

For example, using sales in thousands for the first 8 years for electronic typewriters, we set up the data as:

Year	Sales	$Y(t)$	$Y^2(t)$
1978	30	0	0
1979	80	30	900
1980	110	110	12100
1981	180	220	48400
1982	330	400	160000
1983	550	730	532900
1984	840	1280	1638400
1985	1325	2120	4494400

Running a regression we get

$$\text{Sales} = 42.3 + 0.7\, Y(t) - 0.000046\, Y^2(t)$$

(12.14)

Solving back, we get $p = 0.002$, $q = 0.70$, and $M = 15,213$. (Note: another approach is to estimate market potential separately and then just estimate p and q from data.)

This model fits data well when several periods of data exist but is unstable when only the first three or four periods are available. Fortunately, p and q follow predictable patterns (average p about 0.02, q about 0.30, with p and q larger for industrial goods and subsequent country introductions.) There is also some evidence q, but not p, is increasing, probably due to enhanced information technology and media. A Bayesian approach, which combines the prior estimates (0.02 and 0.30) with limited data, is useful for early predictions [Sultan et al., 1990].

Forecasting the future requires putting $Y(t)$ and $Y^2(t)$ into the formula to forecast the next period $(T + 1)$. (To forecast sales $(t + 2)$, use $Y(t + 1) = Y(t) +$ forecast sales $(t + 1)$ and repeat the process.)

Summary

Forecasts of really new products can be generated by (1) conceptual analysis (managerial judgment) of the product's market potential and the time it will take to achieve it, (2) considering specific analogies or general patterns (averages) of similar past products, or (3) extrapolating data from either test situations or actual sales in early periods. In general, reliable estimates require a combination of past patterns and actual sales data.

Defining Terms

Adoption: The decision to use (buy) a new product or technology.
Market potential: The maximum sales achievable for a product.

References

Bass, F.M. A new product growth model for consumer durables, *Management Science.* 15: 215–227, 1969.

Bass, F.M., Krishnan, V., and Jain, D.C. Why the Bass model fits without decision variables, *Market. Sci.,* 13: 203–223, 1994.

Holak, S. and Lehmann, D. R. Purchase intentions and the dimensions of innovation: an exploratory model, *J. Prod. Innov. Mgt.,* 7: 59–73, 1990.

Mahajan, V., Muller, E., and Bass, F. New product diffusion models in marketing: a review and directions for future research, *J. Market.,* 54: 1–26, 1990.

Mahajan, V., Muller, E., and Bass, F. Diffusion of new products: empirical generalizations and managerial uses, *Market. Sci.,* 14(3): G79–G88, 1995.

Parker, P. M. Aggregate diffusion forecasting models in marketing: a critical review, *Int. J. Forecast.,* 10: 353–380, 1994.

Rogers, E. M. *Diffusion of Innovation,* Free Press, New York, 1983.

Srinivasan, V. and Mason, C. H. Non-linear least square estimation of new product diffusion models, *Market. Sci.,* 5: 169–178, 1986.

Sultan, F., Farley, J. U., and Lehmann, D. R. A meta-analysis of applications of diffusion models, *J. Market. Res.,* 27: 70–77, 1990.

Urban, G. L. and Hauser, J. R. *Design and Marketing of New Products,* 2nd ed., Prentice Hall, Englewood Cliffs, NJ, 1993.

Urban, G. L., Weinberg, B. D., and Hauser, J. R. Pre-market forecasting of really-new products, *J. Market.,* 60: 47–60. 1996.

For Further Information

The best summaries of the material on forecasting appear in Mahajan et al. [1990, 1995]. A more general discussion of new product adoption appears in Urban and Hauser [1993].

12.20   Assessing Marketing Capabilities

Peter LaPlaca

The greatest strength of high-technology companies is their technology; it is also their greatest weakness. How can this focus of the company help and hinder at the same time? The answer lies in the high levels of investment that many managers of high-tech companies place in their technology by reducing investment in learning more about their market, its customers, and competitive activities. Indeed, the largest cause of market failure for new products is lack of knowledge about the market and the customers. This section will take the reader through the components of a marketing **assessment** used with high-technology firms.

Assessing Marketing Capabilities

A marketing assessment starts with an analysis of the company's knowledge of costs and profits. Many high-tech firms fail to produce a complete set of financial statements (income statement, balance sheet, receivables' aging, cash flow analysis, etc.) and almost never know the profitability of specific products or product lines, individual models, individual jobs, or work orders. When financial data are available, they are often dated and lacking in sufficient detail to provide real information for decision making.

It is important to get a firm handle on the firm's marketing and sales costs. Selling, advertising, promotional, and trade show costs should be broken down into their components. The number, frequency, and costs of sales calls must be reported and anayyzed. Are there any reports that analyze the reasons for lost sales or not getting sales from specific prospects? What kind of cost data are used (actual, estimated, or standard costs) in developing prices or quotations?

One of the most critical decisions any firm can make is selection of a customer base. For high-tech companies, this targeting decision is frequently determined by seeing where the firm's specific technology has an immediate application. Key questions are "How does the company target market segments or selected market niches?" "Are there different sales and marketing strategies for different segments?"

Using Customer Information

High-technology firms know much about their customers, but few realize the potential that these data contain. An assessment will evaluate the customer database and information system. Does it exist and what does it contain? How often is it updated? How easy is it to use the information to develop marketing strategies and programs?

An assessment will look at how the firm identifies its most-valuable customers based on both sales and profits. Can a customer be too good or too valuable for the company's benefit? Does the firm focus on its most-valuable customers with specific programs to retain them year after year? A marketing assessment will investigate the firm's customer service and support function and their impact on customer satisfaction and whether the company also monitors customer reactions to its activities.

Sales and Marketing Objectives

Sales and marketing objectives are evaluated to determine their nature relative to market potential, value of the technology to the customers, competitive and regulatory actions, and other external and internal constraints. In addition to sales and revenue goals, a well-developed marketing strategy will contain additional goals that contribute to long-term growth and stability. These may include customer penetration (number of customers using our product), share of business they get from each of the most-valuable customers, customer satisfaction, product quality, market share, and similar measures.

Competitor and Market Information

A marketing assessment evaluates how the company gathers competitive information. Companies must know all major competitors, their strengths and weaknesses, the state of their technology relative to the firm's own technology, their marketing and technology plans and strategies, and their specific target markets.

Market Analysis

The marketing assessment investigates how the firm develops its segmentation strategy and the status of those segments selected for focused marketing efforts. Some questions that will be included in this part of the assessment are "How well does the firm describe its customers in terms of market segments or niches?" "Can they rank the selected niches in terms of profit potential?" "Do they know the size of each market segment or their share of market for each segment?" Not only must this information exist, but it must constantly updated.

Marketing Strategies and Marketing Functions

Beyond the foundations mentioned above, the assessment examines the company's sales and marketing functions in terms of detailed strategies required to develop a marketing plan. This portion of a marketing assessment will look at the company's planning process, the sales function, pricing strategy, promotional programs, the distribution function, export potentials, use of sales representatives, and ancillary activities such as the service and parts departments when appropriate.

General Strategy

While it is common for high-technology firms to know much about their core technology and even to have a *technology* plan to keep pace with the changes in this technology, far fewer high-tech firms have a written plan defining the company's marketing strategies and plans. Many of those that do have a plan fail to communicate the plan or the changes they make in these sales and marketing strategies to employees. A marketing assessment evaluates the detailed implementation program to achieve the plan's goals and objectives with specific individuals assigned the responsibility and authority for all phases of implementation and an appropriate budget and control system to monitor program implementation. High-tech firms with exemplary marketing have reporting systems to assess the impact of marketing efforts on sales.

Evaluating the Selling Function

High-technology companies employ a combination of methods to sell their products to their customers and to their distribution system. These are a company-employed outside sales force, company-employed inside sales department, manufacturers' representatives, and direct marketing techniques. A marketing assessment will investigate the efficiency and effectiveness of those methods employed by the company.

Outside Sales

Surprisingly the majority of sales departments is **not** organized along lines that relate to the marketing and sales tasks faced by the sales people. The assessment will look for an organizational setup that reflects the makeup of the customers (Fig. 12.24). This could be along SIC[10] segments, by geography, by customer demographics, by industry served, or by specific application. The assessment also determines if there are enough sales people for adequate face-to-face customer contact while keeping the cost of sales calls within industry norms. Sales force performance must also be evaluated.

The assessment should also look at how the company sets sales targets for their sales people and how these are used for evaluative purposes. Some common targets are gross sales revenue, gross margins, market share, target accounts, number of sales calls made, adherence to expense budgets, and similar items. Evaluations should be conducted several times a year. The marketing assessment will look at how well the qualifications of the sales force match the expectations of the target customers. Compensation of the sales force **and** sales management should be tied to the attainment of company sales goals (both sales volume and other sales goals).

---

[10]Standard industrial classification.

FIGURE 12.24

The inside sales function is evaluated in a manner very similar to the outside sales but reflects the limitations of inside sales not being able to be physically at customer locations. There are two types of inside sales activities: order taking and order getting. Order taking is a passive system where the inside sales people wait for customers to call them. They should attempt to cross sell to other products or services and inquire about additional customer needs.

Order getters are those inside sales people that actively call existing customers and prospects to sell them over the telephone. The marketing assessment will investigate how well this function is executed, the percentage of sales generated by these activities, growth in existing accounts, number of new accounts obtained, and similar measures. Company efforts for such items as selection and training of these inside sales people are also studied.

Manufacturers' Representatives

Manufacturers' reps (also called sales reps or manufacturers' agents) are independent sales organizations with whom the high-tech company contracts for selling services in a defined territory. They may represent several related, but noncompeting, companies and bring an assortment of products to their customers. The marketing assessment examines the coverage offered by the rep network and the system for selecting, contracting, controlling, and evaluating reps. The company is also evaluated for the support it gives its rep network. Other areas that are evaluated are industrial telemarketing, mail order, catalog sales, and internet marketing.

Advertising and Promotions

Advertising and promotions are active attempts at generating additional sales volume. Whether the company does its own advertising or secures the services of an outside advertising agency, the marketing assessment will evaluate all aspects of these efforts and their results. The assessment will evaluate how the advertising/promotion budget is determined and how well the company coordinates advertising and promotional efforts with its field sales people.

The Service Department

The service department provides physical work on the product either at customer locations or sent into your facilities. The marketing assessment looks at the contribution of the service department to company sales and profits. Using customer input, the marketing assessment evaluates the activities and services provided by the department. Companies that offer customers a toll-free telephone number on a 24-hour basis will be rated higher than firms that do not provide this service. It is imperative that the company measures customer satisfaction with the service department on a regular basis. Some measures used by high-tech companies include the number and seriousness of customer complaints, the length of emergency or down time situations, response time to customer requests for service, and comparing the company's service to that of their competitors.

The Parts Department

The parts department can provide valuable, high-margin sales for the company. The assessment examines the department's organization, the breadth of the parts lines stocked, supplying parts kits for frequently purchased parts, the existence and usefulness of a parts catalog, and the way the company measures customer satisfaction.

Pricing

Pricing of high-technology products is often considered an art rather than a science. The marketing assessment looks at the method the firm uses to establish prices, including the team of people that establishes the pricing strategy as well as prices of specific products and services. Companies with well-structured pricing usually have a cross-functional team [composed of senior people from sales, marketing, production, finance, engineering, and research and development] develop the pricing strategy and the specific prices. Prices should be reviewed on at least an annual basis. A marketing assessment looks at pricing implementation as well as pricing strategies. This includes the percentage of orders that are discounted in order to get the sale, periodic review of the firm's prices as compared to those of its competitors, and the establishment of specific pricing objectives (such as market penetration, building market share, or maximizing customer value). High-tech firms that excel in pricing analyze how much money its products either save the customer or how they impact customers' profits. All prices, procedures, and policies should be published and distributed to all appropriate people in the company (marketing, sales, accounting, invoicing, order entry, and similar functions) to ensure that they are followed.

Product Management

Daily management of each product or product line is important to the firm's success, profits, and growth. The marketing assessment examines how the company manages its products by determining target gross margins and seeing if they fall short, meet, or exceed the target; the company's knowledge of the need for products or product lines to be replaced or redesigned; the company's knowledge of where the products are on the product life cycle; whether the company's products are "me too" copies of competitive products or if they have a real advantage in the eyes of the customers; and how the company evaluates each product and eliminates those products with no or little profit potential.

New Product Development

It seems that high-technology companies are always introducing new products to the marketplace. Because the single, largest reason for marketplace failure of new products, particularly high-tech products, is a shortage (or complete absence) of market and customer information, a marketing assessment will devote significant attention to the firm's new product development process.

The assessment will look at how the firm modifies existing products or enhances current models to meet customer or market needs, the firm's track record of new product introductions, and where new product ideas originate. Unfortunately, the single, most-often cited source of ideas for high-tech companies is the firm itself (R&D, engineering, technical employees, and others). Too few ideas originate with customer requests or statements of customer problems that might be solved by application of the firm's base technology.

A sound new product development system will involve a standard procedure with specific steps and responsibilities to screen and evaluate new product and service ideas before funds are invested for developing prototypes. Market-driven high tech firms begin this system with a comprehensive and continuous monitoring of customer needs or the marketplace for new product or service ideas, and these firms evaluate the impact of new products or services in terms of their projected costs, sales, and profits before making substantial investments. This system will have a series of check points that are used to monitor the development process and make certain that the product's development is consistent with the latest customer/market/competitive feedback.

Channels of Distribution

To be successful, any firm must be able to effectively and efficiently deliver its products and services to its customers. A comprehensive marketing assessment will evaluate the firm's direct and indirect distribution system. Distribution is an essential function to have but difficult to develop and manage. This is because it is an *external* asset with independent management, goals, objectives, and ways of doing business.

The marketing assessment judges the sales coverage or overall performance of the firm's distribution system. It will evaluate the company's system for recruiting and hiring the best dealers and distributors to fit customers' needs and the marketing support and training received by these dealers' and distributors' companies. Successful high-tech companies use their dealers and distributors and strategic partners in serving the markets, and a well-developed assessment will utilize information from dealers and distributors, the company, and the customers in evaluating the company's overall distribution system.

Completing the Assessment

Most assessment reports are divided into two major components: findings of the assessment and recommendations for corrective actions. The initial part informs the company where marketing operations provide some measure of competitive advantage and where the company has a decided disadvantage. The best basis for comparison is for the assessor to compare the firm's marketing activities to those of firms that excel in marketing. This is true **benchmarking** and the true value of conducting a marketing assessment in the first place.

No program of marketing improvement should be undertaken without first conducting a comprehensive marketing assessment. The moneys invested in doing this will help focus future marketing investments so that the firm will achieve a true competitive advantage while establishing excellent relationships with its customers.

Defining Terms

Assessment: An objective evaluation of the way the firm performs specific functions as part of its normal business practices.

Benchmarking: Comparing the firm's various functions to similar functions performed by the acknowledged "best" performers in the firm's industry or in any industry in general.

References

Camp, R. C. *Benchmarking: The Search for Industry — Best Practices that Lead to Superior Performance,* Quality Resources, White Plains, NY, 1989.

Collins, M. *The Manufacturer's Guide to Business Marketing,* Irwin Professional Publishing, Chicago, 1995.

Davidow, W. H. *Marketing High Technology: An Insider's View,* The Free Press, New York, 1986.

Hlavacek, J. and Ames, B. C. *Market Driven Management, (rev. ed.,),* Irwin Professional Publishing, Chicago, 1997.

MacDonald, C. R. *The Marketing Audit Workbook,* Institute for Business Planning, Englewood Cliffs, NJ, 1982.

Rexroad, R. A. *High Technology Marketing Management,* Ronald Press, New York, 1983.

Shanklin, W. L. and Ryans, J. K. *Marketing High Technology,* Lexington Books, Lexington, MA, 1984.

Further Information

To find out more about the marketing assessment process, contact The Advantage Group, L.L.C., 24 Quarry Drive, Suite 201, Vernon, CT 06066-4917.

13

Production and Manufacturing

Richard J. Schonberger
University of Washington

Edward M. Knod, Jr.
Western Illinois University

Michael Pinedo
New York University

Sridhar Seshadri
New York University

Steven Nahmias
Santa Clara University

Matthew P. Stephens
Purdue University

Joseph F. Kmec
Purdue University

James P. Gilbert
Rollins College

William L. Chapman
Hughes Aircraft Company

A. Terry Bahill
University of Arizona

Layek Abdel-Malek
New Jersey Institute of Technology

Lucio Zavanella
Universita degli Studi di Brescia

Xiaoping Yang
Purdue University

C. Richard Liu
Purdue University

Hamid Noori
*Wilfrid Laurier University
and Hong Kong Polytechnique
University*

0-8493-8577-6/99/$0.00+$.50
© 1999 by CRC Press LLC

B. Michael Aucoin
Texas A&M University

Robert W. Hall
Indiana University and Purdue University

Kwasi Amoako-Gyampah
University of North Carolina

Mark Atlas
Carnegie Mellon University

Richard Florida
Carnegie Mellon University

13.1   Types of Manufacturing

Richard J. Schonberger

Although there are many ways to categorize manufacturing, three general categories stand out. These three (which probably have emerged from production planning and control lines of thought) are

1. **Job-shop production.** A job shop produces in small lots or batches.
2. **Mass production.** Mass production involves machines or assembly lines that manufacture discrete units repetitively.
3. **Continuous production.** The **process industries** produce in a continuous flow.

Primary differences among the three types center on output volume and variety and **process** flexibility. Table 13.1 matches these characteristics with the types of manufacturing and gives examples of each type. The following discussion begins by elaborating on Table 13.1. Next are comments on hybrid and uncertain types of manufacturing. Finally, five secondary characteristics of the three manufacturing types are presented.

TABLE 13.1   Types of Manufacturing — Characteristics and Examples

Volume	Very low	High	Highest
Variety	Highest	Low	Lowest
Flexibility	Highest	Low	Lowest
1. Job-shop production	Tool and die making Casting (foundry) Baking (bakery)		
2. Mass production		Auto assembly Bottling Apparel manufacturing	
3. Continuous production			Paper milling Refining Extrusion

Job-Shop and Batch Production

As Table 13.1 shows, job-shop manufacturing is very low in volume but is highest in output variety and process flexibility. In this mode, the processes — a set of resources including labor and equipment — are reset intermittently to make a variety of products. (Product variety requires flexibility to frequently reset the process.)

In tool and die making, the first example, the volume is generally one unit, for example, a single die set or mold. Since every job is different, output variety is at a maximum, and operators continually reset the equipment for the next job.

Casting in a foundry has the same characteristics, except that the volume is sometimes more than one, that is, a given job order may be to cast one, five, ten, or more pieces. The multipiece jobs are sometimes called lots or **batches**.

A bakery makes a variety of products, each requiring a new series of steps to set up the process, for example, mixing and baking a batch of sourdough bread, followed by a batch of cinnamon rolls.

Mass Production

Second in Table 13.1 is mass production. Output volume, in discrete units, is high. Product variety is low, entailing low flexibility to reset the process.

Mass production of automobiles is an example. A typical automobile plant will assemble two or three hundred thousand cars a year. In some plants, just one model is made per assembly line; variety is low (except for option packages). In other plants, assembly lines produce mixed models. Still, this is considered mass production since assembly continues without interruption for model changes.

In bottling, volumes are much higher, sometimes in the millions per year. Changing from one bottled product to another requires a line stoppage, but between **changeovers** production volumes are high (e.g., thousands). Flexibility, such as changing from small to large bottles, is low; more commonly, large and small bottles are filled on different lines.

Similarly, mass production of apparel can employ production lines, with stoppages for pattern changes. More conventionally, the industry has used a very different version of mass production: cutters, sewers, and others in separate departments each work independently, and material handlers move components from department to department to completion. Thus, existence of an assembly line or production line is not a necessary characteristic of mass production.

Continuous Production

Products that flow — liquids, gases, powders, grains, and slurries — are continuously produced, the third type in Table 13.1. In continuous process plants, product volumes are very high (relative to, for example, a job-shop method of making the same product). Because of designed-in process limitations (pumps, pipes, valves, etc.), product variety and process flexibility are very low.

In a paper mill, a meshed belt begins pulp on its journey though a high-speed multistage paper-making machine. The last stage puts the paper on reels holding thousands of linear meters. Since a major product changeover can take hours, plants often limit themselves to incremental product changes. Special-purpose equipment design also poses limitations. For example, a tissue machine cannot produce newsprint, and a newsprint machine cannot produce stationery. Thus, in paper making, flexibility and product variety for a given machine are very low.

Whereas a paper mill produces a solid product, a refinery keeps the substance in a liquid (or sometimes gaseous) state. Continuous refining of fats, for example, involves centrifuging to remove undesirable properties to yield industrial or food oils. As in paper making, specialized equipment design and lengthy product changeovers (including cleaning of pipes, tanks, and vessels) limit process flexibility; product volumes between changeovers are very high, sometimes filling multiple massive tanks in a tank farm.

Extrusion, the third example of continuous processing in Table 13.1, yields such products as polyvinyl chloride pipe, polyethylene film, and reels of wire. High process speeds produce high product volumes, such as multiple racks of pipe, rolls of film, or reels of wire per day. Stoppages for changing extrusion heads and many other adjustments limit process flexibility and lead to long production runs between changeovers. Equipment limitations (e.g., physical dimensions of equipment components) keep product variety low.

Mixtures and Gray Areas

Many plants contain a mixture of manufacturing types. A prominent example can be found in the process industries, where production usually is only partially continuous. Batch mixing of pulp, fats, or plastic granules precedes continuous paper making, refining of oils, and extrusion of pipe. Further processing may be in the job-shop mode: slitting and length-cutting paper to customer order, secondary mixing and drumming of basic oils to order, and length cutting and packing of pipe to order.

Mixed production also often occurs in mass production factories. An assembly line (e.g., assembling cars or trucks) may be fed by parts, such as axles, machined in the job-shop mode from castings that are also job-shop produced. Uniform mass-made products (e.g., molded plastic hard hats) may go to storage where they await a customer order for final finishing (e.g., decals) in the job-shop mode. An apparel plant may mass produce sportswear on one hand and produce custom uniforms for professional sports figures in the job-shop mode on the other.

More than one type of manufacturing in the same plant requires more than one type of production planning, scheduling, and production. The added complexity in management may be offset, however, by demand-side advantages of offering a fuller range of products.

Sometimes, a manufacturing process does not fit neatly into one of the three basic categories. One gray area occurs between mass production and continuous production. Some very small products — screws, nuts, paper clips, and toothpicks — are made in discrete units. However, because of small size, high volumes, and uniformity of output, production may be scheduled and controlled not in discrete units but by volume, thus approximating continuous manufacturing. Production of cookies, crackers, potato chips, and candy resembles continuous forming or extrusion of sheet stock on wide belts, except that the process includes die cutting or other separation into discrete units, like mass production. Link sausages are physically continuous, but links are countable in whole units.

Another common gray area is between mass and job-shop production. A notable example is high-volume production of highly configured products made to order. Products made for industrial uses — such as specialty motors, pumps, hydraulics, controllers, test equipment, and work tables — are usually made in volumes that would qualify as mass production, except that end-product variety is high, not low.

These types of manufacturing with unclear categories do not necessarily create extra complexity in production planning and control. The difficulty and ambiguity are mainly terminological.

Capital Investment, Automation, Advanced Technology, Skills, and Layout

The three characteristics used to categorize manufacturing — volume, variety, and flexibility — are dominant but not exhaustive. To some extent, the manufacturing categories also differ with respect to capital investment, technology, skills, and layout.

Typically, continuous production is highly capital intensive, whereas mass production is often labor intensive. The trend toward automated, robotic assembly, however, is more capital intensive and less labor intensive, which erodes the distinction. Job-shop production on conventional machines is intermediate as to capital investment and labor intensiveness. However, computer numerically controlled machines and related advanced technology in the job shop erodes this distinction as well.

As technology distinctions blur, so do skill levels of factory operatives. In conventional high-volume assembly, skill levels are relatively low, whereas those of machine operators in job shops — such as machinists and welders — tend to be high. In automated assembly, skill levels of employees tending the production lines elevate toward technician levels — more like machinists and welders. In continuous production skill levels range widely — from low-skilled carton handlers and magazine fillers to highly skilled process technicians and troubleshooters.

Layout of equipment and related resources is also becoming less of a distinction than it once was. The classical job shop is laid out by type of equipment: all milling machines in one area, all grinding machines in another. Mass and continuous production have been laid out by the way the product flows: serially and linearly. Many job shops, however, have been converted to cellular layouts — groupings of diverse machines that produce a family of similar products. In most work cells the flow pattern is serial from machine to machine, but the shape of the cell is not linear; it is U-shaped or, for some larger cells, serpentine. Compact U and serpentine shapes are thought to provide advantages in teamwork, material handling, and labor flexibility.

To some degree, such thinking has carried over to mass production, that is, the trend is to lay out assembly and production lines in U and serpentine shapes instead of straight lines, which was the nearly universal practice in the past. In continuous production of fluids, the tendency has always been toward compact facilities interconnected by serpentine networks of pipes. Continuous production of solid and semisolid products (wide sheets, extrusions, etc.), on the other hand, generally must move in straight lines, in view of the technical difficulties in making direction changes.

Defining Terms

Batch: A quantity (a lot) of a single item.

Changeover (setup): Setting up or resetting a process (equipment) for a new product or batch.

Continuous production: Perpetual production of goods that flow and are measured by area or volume; usually very high in product volume, very low in product variety, and very low in process flexibility.

Job-shop production: Intermittent production with frequent resetting of the process for a different product or batch; usually low in product volume, high in product variety, and high in process flexibility.

Mass production: Repetitive production of discrete units on an assembly line or production line; usually high in product volume, low in product variety, and low in process flexibility.

Process: A set of resources and procedures that produces a definable product (or service).

Process industry: Manufacturing sector involved in continuous production.

Further Information

Industrial Engineering. Published monthly by the Institute of Industrial Engineers.

Manufacturing Engineering. Published monthly by the Society of Manufacturing Engineers.

Schonberger, R. J. and Knod, E. M. *Operations Management: Customer-Focused Principles,* 6th ed., Richard D. Irwin, Burr Ridge, IL, 1997; see, especially, chapters 12, 13, 14, and 15.

13.2 Management and Scheduling

Edward M. Knod, Jr.

Prescriptions for how managers ought to establish and maintain world-class excellence in their organizations changed substantially during the 1980s and 1990s. Emerging markets, shifting patterns of global competitiveness and regional dominance in key industries, the spread of what might be called Japanese management and manufacturing technologies, and the philosophy and tools of the total quality movement are among the factors that combined to usher in a heightened focus on the overall capacity required to provide continuous improvement in meeting evolving customer needs. The effects of this contemporary management approach first appeared in manufacturing [Schonberger, 1982; Hall, 1983] and continue to have profound influence in that sector. Changes in the way managers approach scheduling serve to exemplify the new thinking.

This section begins with an overview of contemporary management, continues with a discussion of scheduling in various manufacturing environments, and concludes with references and suggested sources of additional information.

Management: Definition and Innovations

In a somewhat general-to-specific progression, **management** in contemporary competitive manufacturing organizations may be described by (1) duties and activities, (2) requisite skills and attributes, (3) trends and innovations, and (4) principles for managing operations.

Duties and Activities

Briefly, the goal of management is to ensure organizational success in the creation and delivery of goods and services. Popular definitions of management often employ lists of activities that describe what managers do. Each activity serves one or more of three general, and overlapping, duties: creating, implementing, and improving.

- Creating. Activities such as planning, designing, staffing, budgeting, and organizing accomplish the creativity required to build and maintain customer-serving capacity. Product and **process** design, facility planning and layout, work-force acquisition and training, and materials and component sourcing are among the tasks that have a substantial creative component.
- Implementing. When managers authorize, allocate, assign, schedule, or direct, emphasis shifts from creating to implementing — putting a plan into action. (A frequent observation is that the biggest obstacle to successful implementation is lack of commitment.) During implementation, managers also perform controlling activities, that is, they monitor performance and make necessary adjustments.
- Improving. Environmental changes (e.g., new or revised customer demands, challenges from competitors, and social and regulatory pressures) necessitate improvements in output goods and services. In response to — or better yet, in anticipation of — those changes, managers re-create, that is, they start the cycle again with revised plans, better designs, new budgets, and so forth.

Outcomes, desirable or otherwise, stem from these activities. A goal-oriented manager might describe the aim of the job as "increased market share" or "greater profitability", but he or she will try to attain that goal by creating, implementing, and improving.

Requisite Skills and Attributes

Exact requirements are difficult to pin down, but any skill or attribute that helps a manager make better decisions is desirable. Bateman and Zeithaml [1993] suggest that managers need technical skills, interpersonal and communications skills, and conceptual and decision skills. Extension of these broad categories into job-specific lists is perhaps unwarranted given current emphasis on cross-functional career migration and assignments to interdisciplinary project teams or product groups. Sound business acumen and personal traits such as demeanor, good time-management habits, and pleasant personality, however, are general attributes that serve managers in any job. More recently, emphasis on computer (and information system) literacy and knowledge of foreign languages and customs has increased as well.

Trends and Innovations

An array of publications, seminars, and other vehicles for disseminating "how and why" advice has bolstered the spread of contemporary management theory and research. Manufacturing managers constitute the primary target audience for much of this work. The information bounty can be reduced, tentatively, to a set of core trends and innovative approaches. Table 13.2 offers a short list of seven concept areas within which numerous interrelated trends or innovations have emerged. They are dominant themes in contemporary management literature and in that regard help define what today's managers are all about.

TABLE 13.2   Trends and Innovations in Management

Customers at center stage.   The customer is the next person or process — the destination of one's work. The provider-customer chain extends, process to process, on to final consumers. Whatever a firm produces, customers would like it to be better, faster, and cheaper; product managers therefore embrace procedures that provide *total quality, quick responses,* and *waste-free* (economical) *operations.* These three aims are mutually supportive and form the core rationale for many of the new principles that guide managers.

Focus on improvement.   Managers have a duty to embrace improvement. A central theme of the TQ movement is constant improvement in output goods and services and in the processes that provide them. Sweeping change over the short run, exemplified by business process reengineering [Hammer and Champy, 1993], anchors one end of the improvement continuum; the rationale is to discard unsalvageable processes and start over so as not to waste resources in fruitless repair efforts. The continuum's other end is described as incremental continuous improvement and is employed to fine tune already-sound processes for even better results.

Revised "laws" of economics.   Examples of the contemporary logic include the following. Quality costs less, not more. Costs should be allocated to the activities that cause their occurrence. Prevention (of errors) is more cost-effective than discovery and rework. Training is an investment rather than an expense. Value counts more than price (e.g., in purchasing). Desired market price should define (target) manufacturing cost, not the reverse.

Elimination of wastes.   Waste is anything that doesn't add value; it adds cost, however, and should be eliminated. Waste detection begins with two questions: "Are we doing the right things?" and "Are we doing those things in the right way?" Toyota identifies seven general categories of wastes [Suzaki, 1987], each with several subcategories. Schonberger [1990] adds opportunities for further waste reduction by broadening the targets to include nonobvious wastes. Simplification or elimination of indirect and support activities (e.g., production planning, scheduling, and control activities; inventory control; costing and reporting; etc.) is a prime arena for contemporary waste-reduction programs [Steudel and Desruelle, 1992].

Quick-response techniques.   Just-in-time (JIT) management, queue limiters, reduced setups, better maintenance, operator-led problem solving, and other procedures increase the velocity of material flows, reduce throughput times, and eliminate the need for many interdepartmental control transactions. Less tracking and reporting (which add no value) reduces overhead. Collectively, quick-response programs directly support faster customer service [Blackburn, 1991].

The process approach.   The process approach has several advantages [Schonberger and Knod, 1994]. Processes cut across functional departments; attention is drawn to overall or group results, ideally by a cross-functional team that may also include customers and suppliers. Processes are studied at the job or task level, or at the more detailed operations level. Automation can be beneficial *after* process waste is removed, and further reduction in variation is needed. Tools for measurement and data analysis, methods improvement, and team building are among those needed for successful process improvement.

Human resources management.   Increased reliance on self-directed teams (e.g., in cells or product groups) and/or on-line operators for assumption of a larger share of traditional management responsibilities is a product of the management revolution of the 1980s that had noticeable impact in the 1990s. Generally, line or shop employees have favored those changes; they get more control over their workplaces. There is a flip side: as employee empowerment shifts decision-making authority, as JIT reduces the need for many reporting and control activities, and as certain supervisory and support-staff jobs are judged to be non-value adding, many organizations downsize. Lower- and mid-level managers and support staff often bear the job-loss burden.

Principles for Managing Operations

The final and most detailed component of this general-to-specific look at contemporary management is a set of action-oriented, prescriptive principles for managing operations in any organization (see Table 13.3). The principles apply to managers at any level and define ways for increasing competitiveness in manufacturing organizations. Brief supporting rationale and techniques or procedures that exemplify each principle appear in the right-hand columns; Schonberger and Knod [1994] provide a more detailed discussion.

Scheduling

Basically, **scheduling** refers to the activities through which managers allocate capacity for the near future. It includes the assignment of work to resources, or vice versa, and the determination of timing for specific work elements and thus answers the questions "who will do what" and "when will they do it". In manufacturing, the scheduling time horizon is usually weekly, daily, or even hourly. In general, scheduling

TABLE 13.3   Principles for Managing Operations

Principle	Rationale and Examples
Get to know customers; team up to form partnerships and share process knowledge.	Providers are responsible for getting to know their customers' processes and operations. By so doing, they offer better and faster service, perhaps as a member of customers' teams.
Become dedicated to rapid and continual increases in quality, flexibility, and service and decreases in costs, response or lead time, and variation from target.	The logic of continuous improvement, or *kaizen* [Imai, 1986], rejects the "if it ain't broke..." philosophy; seeks discovery and then prevention of current potential problems; and anticipates new or next-level standards of excellence.
Achieve unified purpose through shared information and cross-functional teams for planning/design, implementation, and improvement efforts.	Information sharing keeps all parties informed. Early manufacturing/supplier involvement and concurrent or simultaneous product and process design are components of the general cross-functional team design concept.
Get to know the competition and world-class leaders.	Benchmarking [Camp, 1989] elevates the older notion of "reverse engineering" to a more formal yet efficient means of keeping up with technology and anticipating what competitors might do. Search for best practices.
Cut the number of products (e.g., types or models), components, or operations; reduce supplier base to a few good ones and form strong relationships with them.	Product line trimming removes nonperformers; component reduction cuts lead times by promoting simplification and streamlining. Supplier certifications and registrations (e.g., ISO 9000) lend confidence, allow closer partnering with few suppliers (e.g., via EDI), and reduce overall buying costs.
Organize resources into multiple chains of customers, each focused on a family of products or services; create cells, flow lines, plants in a plant.	Traditional functional organization by departments increases throughput times, inhibits information flow, and can lead to "turf battles". Flow lines and cells promote focus, aid scheduling, and employ cross-functional expertise.
Continually invest in human resources through cross-training for mastery of multiple skills, education, job and career path rotation, health and safety, and security.	Employee involvement programs, team-based activities, and decentralized decision responsibility depend on top-quality human resources. Cross-training and education are keys to competitiveness. Scheduling — indeed, all capacity management — is easier when the work force is flexible.
Maintain and improve present equipment and human work before acquiring new equipment, then automate incrementally when process variability cannot otherwise be reduced.	*TPM,* total productive maintenance [Nakajima, 1988], helps keep resources in a ready state and facilitates scheduling by decreasing unplanned downtime, thus increasing capacity. Also, process improvements must precede automation; get rid of wasteful steps or dubious processes first.
Look for simple, flexible, movable, and low-cost equipment that can be acquired in multiple copies — each assignable to a focused cell, flow line, or plant-in-a-plant.	Larger, faster, general-purpose equipment can detract from responsive customer service, especially over the longer run. A single fast process is not necessarily divisible across multiple customer needs. Simple, dedicated equipment is an economical solution; setup elimination is an added benefit.
Make is easier to make/provide goods and services without error or process variation.	The aim is to prevent problems or defects from occurring — the fail-safing (*pokayoke*) idea — rather than rely on elaborate control systems for error detection and the ensuing rework. Strive to do it right the first time, every time.
Cut cycle times, flow time, distance, and inventory all along the chain of customers.	Time compression provides competitive advantage [Blackburn, 1991]. Removal of excess distance and inventory aids quick response to customers. Less inventory also permits quicker detection and correction of process problems.
Cut setup, changeover, get-ready, and startup times.	Setup (or changeover) time had been the standard excuse for large-lot operations prior to directed attention at reduction of these time-consuming activities [Shingo, 1985]. Mixed-model processing demands quick changeovers.
Operate at the customer's rate of use (or a smoothed representation of it); decrease cycle interval and lot size.	Pull-mode operations put the customer in charge and help identify bottlenecks. Aim to synchronize production to meet period-by-period demand rather than rely on large lots and long cycle intervals.
Record and *own* quality, process, and problem data at the workplace.	When employees are empowered to make decisions and solve problems, they need appropriate tools and process data. Transfer of point-of-problem data away from operators and to back-office staff inhibits responsive, operator-centered cures.
Ensure that front-line associates get first chance at problem solving — before staff experts.	Ongoing process problems and on-the-spot emergencies are most effectively solved by teams of front-line associates; staff personnel are best used in advisory roles and for especially tough problems.

TABLE 13.3 (continued)	Principles for Managing Operations
Principle	Rationale and Examples
Cut transaction and reporting; control *causes*, not symptoms.	Transactions and reports often address problem symptoms (e.g., time or cost variances) and delay action. Quick-response teams, using data-driven logic, directly attack problem causes and eliminate the need for expensive reporting.

Source: Schonberger, R. J. and Knod, E. M., Jr. *Operations Management: Continuous Improvement,* 5th ed., chapter 1, Richard D. Irwin, Burr Ridge, IL, 1994. Adapted with permission.

(1) flows from and is related to planning, (2) is determined by manufacturing environment, and (3) can be simplified when appropriate management principles are followed.

Relationship to Planning

The planning activity also answers the "who", "what", and "when" questions, but in more general or aggregate terms for the longer time horizon — typically months, quarters, or years into the future. So, in the temporal sense, scheduling is short-term planning. However, planning involves more. With the other creative activities, planning also addresses the characteristics of what is to be done (e.g., designs), quantity and variety (e.g., the product mix), how work will be accomplished (e.g., methods and procedures), utilization of funds (e.g., budgeting), and so on — including how scheduling itself will be accomplished. When planning (or design) has been thorough and things go according to plan, little creativity should be required in scheduling; it ought to be mostly an implementation activity.

In manufacturing, aggregate demand forecasts are filtered by business plans and strategies — what the company wants to do — to arrive at aggregate capacity and production plans. The master schedule states what the company plans to provide, in what quantities, and when. To the extent that on-hand or previously scheduled inventories are insufficient to meet the commitments described by the master schedule, additional purchasing and production are required. Consequently, detailed planning and scheduling activities — for fabrication of components and subassemblies and for final assembly operations — come into play. Scheduling is among the production activity control duties that form the "back end" of the total manufacturing planning and control system [Vollmann et al., 1992]. Thus, it might be said that scheduling flows from planning.

Effects of Manufacturing Environment

The type of manufacturing environment determines the nature of scheduling, as shown in Table 13.4. As the table notes, project and repetitive/continuous environments present less extreme scheduling problems; the sources listed at the end of this section contain suitable discussion. For project scheduling, see Evans [1993], Kerzner [1989], and Schonberger and Knod [1994]. For scheduling in repetitive or continuous production, see Schniederjans [1993] and Schonberger and Knod [1994].

The variables inherent with traditional batch and job environments create the most complex scheduling (and control) problems. *Loading* (allocation of jobs to work centers), *sequencing* (determining job-processing order), *dispatching* (releasing jobs into work centers), *expediting* (rushing "hot" jobs along), and *reporting* (tracking job progress along routes) are among the specific activities. *Assignment models* can be of assistance in loading, and *priority rules* may be used for sequencing and dispatching. In batch operations, *lot splitting* and *overlapping* may help expedite urgent jobs. Unfortunately, however, throughput time in many job operations consists largely of queue time, setup time, move time, and other non–value-adding events. Perhaps even more unfortunate have been managers' attempts to "solve" the complexities of job scheduling and control by relying on more exotic scheduling and control tools.

Effects of Contemporary Management Practices

This last section closes the discussion of scheduling by appropriately returning to the topic of management. In the 1970s, North American managers allowed batch and job production scheduling and control

TABLE 13.4   Manufacturing Scheduling Overview

Manufacturing Environment	General Nature of Scheduling
Project	Activity scheduling and controlling (as well as overall project planning) may rely on program evaluation and review technique or critical path method. Large project complexity, cost, and uncertainties involved justify these tools. Smaller projects and single tasks may be scheduled with Gantt charts.
Job or Batch	Scheduling is time consuming due to low- or intermediate-volume output coupled with irregular production of any given item. Production typically occurs on flexible resources in functional or process layouts where considerable variation in products, routings, low sizes, and cycle times — along with the competition among jobs (customers) for resource allocation — adds to the scheduling burden. Rescheduling may be common.
Continuous or Repetitive	Regular — if not constant — production and equipment dedicated to one or a few products (e.g., lines or cells) combine to decrease the scheduling problem. In process flow systems, scheduling is minimal except for planned maintenance, equipment or product changeovers, and so forth. In repetitive production, line balancing may be used to assign work; JIT's pull system, regularized schedules, and mixed-model scheduling can closely synchronize output with demand and can accommodate demand variation.

systems to become too complicated, cumbersome, and costly. A more competitive approach lies in the simplification of the production environments themselves [Schonberger and Knod, 1994; Steudel and Desruelle, 1992; Schneiderjans, 1993]. Though necessary to some degree, scheduling itself adds no value to outputs; as such, it ought to be a target for elimination where possible and for streamlining when it must remain. Application of contemporary management practices, such as the principles detailed in Table 13.3, has been shown to improve scheduling, largely by removing the *causes* of the problems, that is, the factors that created a need for complicated and costly scheduling systems.

Steudel and Desruelle [1992] summarize such improvements for scheduling and related production control activities, especially in group-technology environments. Regarding scheduling, they note that manufacturing cells largely eliminate the scheduling problem. Also, sequencing is resolved at the (decentralized) cell level, and mixed-model assembly and kanban more easily handle demand variations. In similar fashion, manufacturing process simplifications foster improvements throughout the production planning and control sequence.

Just-in-time (JIT) management, for example, has been shown to greatly reduce the burden of these activities, especially scheduling and control [Vollmann et al., 1992]. Attempts to describe a "JIT scheduling system", however, are unnecessary; it is more productive to devote the effort to eliminating the need for scheduling at all. At this writing, it remains an oversimplification to suggest that the mere pull of customer demand is sufficient to *fully* schedule the factory, but that is a worthy aim. In sum, the less attention managers are required to devote to scheduling, the better.

Defining Terms

Management: Activities that have the goal of ensuring an organization's competitiveness by creating, implementing, and improving capacity required to provide goods and services that customers want.

Process: A particular combination of resource elements and conditions that collectively cause a given outcome or set of results.

Scheduling: Activities through which managers allocate capacity for the immediate and near-term future.

References

Bateman, T. S. and Zeithaml, C. P. *Management: Function and Strategy, 2nd ed.,* Richard D. Irwin, Burr Ridge, IL, 1993.

Blackburn, J. D. *Time-Based Competition,* Business One-Irwin, Homewood, IL, 1991.

Camp, R. C. *Benchmarking.* ASQ Quality Press, Milwaukee, WI, 1989.

Evans, J. R. *Applied Production and Operations Management, 4th ed.,* West, St. Paul, MN, 1995.

Hall, R. *Zero Inventories,* Dow Jones–Irwin, Homewood, IL, 1983.

Hammer, M. and Champy, J. *Reengineering the Corporation,* HarperCollins, New York, 1993.

Imai, M. *Kaizen: The Key to Japan's Competitive Success,* Random House, New York, 1989.

Kerzner, H. *Project Management: A Systems Approach to Planning, Scheduling, and Controlling,* 3rd ed., Van Nostrand Reinhold, New York, 1989.

Nakajima, S. *Introduction to TPM: Total Productive Maintenance,* Productivity Press, Cambridge, MA, 1988.

Schniederjans, M. J. *Topics in Just-In-Time Management,* Allyn & Bacon, Needham Heights, MA, 1993.

Schonberger, R. J. *Japanese Manufacturing Techniques: Nine Hidden Lessons in Simplicity,* Free Press, New York, 1982.

Schonberger, R. J. *Building a Chain of Customers,* Free Press, New York, 1990.

Schonberger, R. J. and Knod, E. M., Jr. *Operations Management: Continuous Improvement,* 5th ed., Richard D. Irwin, Burr Ridge, IL, 1994.

Shingo, S. *A Revolution in Manufacturing: The SMED [Single-Minute Exchange of Die] System,* Productivity Press, Cambridge, MA, 1985.

Steudel, H. J. and Desruelle, P. *Manufacturing in the Nineties,* Van Nostrand Reinhold, New York, 1992.

Suzaki, K. *The New Manufacturing Challenge: Techniques for Continuous Improvement,* Free Press, New York, 1987.

Vollmann, T. E., Berry, W. L., and Whybark, D. C., *Manufacturing Planning and Control Systems,* 3rd ed., Richard D. Irwin, Burr Ridge, IL, 1992.

Further Information

Periodicals

Industrial Engineering Solutions. Institute of Industrial Engineers.

Industrial Management. Society for Engineering and Management Systems, a society of the Institute of Industrial Engineers.

Journal of Operations Management. American Production and Inventory Control Society.

Production and Inventory Management Journal. American Production and Inventory Control Society.

Production and Operations Management. Production and Operations Management Society.

Quality Management Journal. American Society for Quality.

Books

Camp, R. C. *Business Process Benchmarking,* ASQC Quality Press, Milwaukee, WI, 1995.

Orlicky, *Materials Requirements Planning,* McGraw-Hill, New York, 1975.

Stonebraker, P. W. and Leong, G. K. *Operations Strategy.* Allyn & Bacon, Needham Heights, MA, 1994.

13.3 Capacity

Michael Pinedo and Sridhar Seshadri

The **capacity** of a system is defined as the maximum rate of production that can be sustained over a given period of time with a certain product mix. **Capacity management** focuses on the allocation and management of resources; it therefore affects almost every decision in a firm. It has an impact on lead times, inventories, quality, yield, and plant and maintenance costs.

Short-term capacity management contributes to meeting customer due dates, controlling inventories, and achieving high levels of labor and plant productivity. Medium-term capacity management deals with make or buy decisions, subcontracting, inventory levels, assignment of products to processes, and a variety of other decisions that affect production costs as well as customer service. Long-term capacity management deals with the planning of entire networks of facilities, determining facility locations, technology choices, and other economic factors. Long-term capacity expansion decisions have strategic implications for the firm and have to be taken under conditions of uncertainty in demand, technology, and competition.

Definition of Capacity and Measurement

In the first step of computing the capacity of a production facility, the appropriate groupings of machines that will form the basis for the analysis have to be determined. Each such grouping is called a **work center**. Dependent upon the purpose of the analysis, a work center may be an entire plant, a work area (such as turning or milling), or a single machine. The collection of all work centers, say M, will be referred to as the system. In the second step, the following data with regard to work center j, $j = 1, 2, ..., M$, have to be compiled:

- The product mix, $(x_1, x_2, ..., x_N)$, where x_i is the fraction of all products in the product mix, that are of type i.
- The mix of products that will use work center j, $(y_{1j}, y_{2j}, y_{3j}, ..., y_{Nj})$, where y_{ij} is the fraction of product i that will be produced at work center j.
- The average lot size, L_i, in which product i will be produced at work center j.
- The setup time, S_i, for a lot and the run time, r_i, for each unit in the lot.

Let q_i denote the processing time per unit of product i and p_j the average processing time for a unit of the product mix at work center j. Based on the data,

$$q_i = S_i / L_i + r_i$$

and

$$p_j = \sum_{i=1}^{N} \left(q_i \, x_i \, y_{ij} \right).$$

The **rated** (also called theoretical, nominal, or standing) capacity of work center j, denoted by C_j, is defined as $C_j = 1/p_j$.

Consider the following example: the raw materials for a total of 100 units are released into the system, and the desired product mix is $(x_1, x_2, ..., x_N)$. Then work center j will need to process $100 x_1 y_{1j}$ units of type 1, $100 x_2 y_{2j}$ units of type 2, and so forth. Therefore, $100 p_j$ is the time required to process all these units. The capacity is thus given by $1/p_j$. The capacity of the system, C_{system}, is defined as the smallest of all the C_js. The work center with the smallest capacity is called a bottleneck.

Typically, prior to these calculations, process analysis is used to determine the fractions y_{ij} (called the product routing decisions) and the lot sizes, L_i. The definition of C_j can easily be generalized to include multiple types of resources, such as machines, labor, space, material-handling equipment, and utilities. It is a matter of practical importance that the actual capacity need not be equal to C_j. With regard to work center j, let T_{av} denote the time period for which work is scheduled (such as two out of three shifts), T_{std} the standard hours of work produced, and T_{work} the hours actually worked. Based on these notions we can define the following concepts: (1) the utilization, $U_j = T_{work}/T_{av}$ (≤ 1) and (2) the efficiency, $E_j = T_{std}/T_{work}$.

The **capacity available** is said to equal $T_{av} \times C_j \times U_j \times E_j$. In this definition, the value of U_j reflects the fraction of time that work center j is unavailable due to breakdowns, repair, and preventive maintenance as well as the fraction of time it is not producing due to lack of work.

Factors that Affect Capacity

The major factors that affect capacity include

- Product mix
- Lot size
- Product yield, quality, rework, and rejection

- Routing and scheduling
- Input control and work regulation

The impact of the first three factors is clear from the formulas that define capacity. Product yield can be an important factor, especially in the microelectronics and process industries. If the product itself can be classified into several grades (such as the speed of a CPU chip), then quality too will play a major role in determining capacity. Rework reduces the capacity, especially when it is performed at the work center that was the cause of the rework. See also the section on the use of queueing networks for a further discussion on rework and yield. (The traditional method for dealing with rework and repair has been to include these times in the processing times themselves.)

Scheduling decisions are extremely important for ensuring twin goals: that resources are not idle for lack of work and that the buildup of work in progress inventory is kept minimal. While it clear that improper sequencing may lead to low levels of utilization, examples in practice as well as in theory show that excessive work in the system can actually lead to a reduction in throughput and thereby capacity. In recent years, it has also become more evident that proper timing of the release of work *into* the system can achieve the goals of minimizing inventory and maximizing capacity utilization.

Capacity Planning and Control Methods

The planning horizon chosen for capacity planning depends on the nature of the business. For example, capacity expansion plans could extend well over a decade for process industries but only 3 to 5 years for light engineering firms. The objective in capacity planning depends strongly on whether the planning is done for the short, medium, or long term.

Short-Term Capacity Planning

Short-term planning refers to the day-to-day management of work. It is well known that having the necessary resources for carrying out a plan in the aggregate (see the next section for an example of aggregate planning) is not sufficient to ensure that the plan can be executed in practice. Short-term planning can help bridge this gap between the aggregate plan and actual shop floor execution. With reference to the formula for capacity, an objective of short-term planning is to make the value of utilization (U) as close to unity as possible. The other objectives in short-term capacity planning include achieving the production target and meeting due dates, maintaining minimal inventory of work in progress (WIP), and controlling overtime. The short-term plan takes as the inputs higher-level decisions regarding the availability of resources, targets for finished goods and customer orders to be executed, along with due dates and priorities. The decisions that impact capacity in the short term include lot sizing, sequencing, controlling the release of work, routing of work, scheduling overtime, labor assignment, expediting, and ensuring that all the materials and resources required for carrying out the planned work are available at the right place and at the right time. These decisions are interrelated and complex. A recent innovation for short-term capacity management has been the introduction of manufacturing execution systems (MES). An advanced MES can provide support on a real-time basis for the scheduling, loading, and recording of work. These systems have become increasingly important, not just because of the complexity of the decision-making process but also because of the fact that traditional materials requirements planning (MRP) packages simply assumed unlimited capacity at all work centers. The MRP approach proved to be inadequate when firms were faced with shortening delivery lead times and growing product variety.

Medium-Term Capacity Planning

In the medium term, the objective of capacity planning is to minimize the cost of resources subject to meeting customer service levels. The decision variables normally include labor, overtime, the number of shifts in each period, the production volume in each period, the product mix, the average lot size, the level of subcontracting, the level of work in progress, and finished goods inventory. The inputs are the demand forecasts, the lead times for procuring materials, the lead times for production, resource requirement per unit of each product, and the service levels to be maintained over the planning horizon. The typical horizon can be as short as 2 weeks and as long as 1 year. Three different planning methods are discussed below: the mathematical programming approach, the capacity requirements planning method and the queueing network method.

Mathematical Programming

The medium-term capacity planning problem can be formulated either as a linear program or as a mixed integer program. The linear program ignores the integrality of labor and setups and cannot take into account explicitly a second or third shift option. In both formulations, it is convenient to group products into product families. The objective function may include wages, overtime costs, costs of hiring or laying off workers, setup costs, holding costs, and costs of subcontracting. The planning horizon is partitioned into periods (weeks or months). The constraints typically include

Inventory balance equations for each period:

$$\text{opening inventory} + \text{production} - \text{sales} = \text{closing inventory}$$

Resource requirement constraints for each period:

$$\text{setup plus run time needed for production} \leq \text{capacity of the resource} \\ + \text{subcontract contract hours}$$

Labor balance constraints for each period:

$$\text{labor at the beginning of the period} + \text{number of workers hired} \\ - \text{number of workers fired} = \text{labor at the end of the period}$$

Queueing Network Analysis

The 1980s saw the emergence of queueing theory as an approach for estimating and planning capacity. A production facility can be modeled as a network of queues, in which jobs of different types arrive randomly and, if accepted, enter the network and get processed in either a deterministic or random sequence (thereby catering to rework and product mix variations) and leave the system. It is in the modeling of the production facility as a queueing network that a planner can appreciate the impact of uncertainty and the close relationship between capacity and production lead times. For example, consider a simple single server queue whose utilization is U, in which arrivals come in according to a Poisson process, and job processing times are arbitrarily distributed. The average waiting time in this queue is proportional to the variability of the service time divided by $(1-U)$. The queueing approach allows the planner to establish the trade-offs between capacity and lead time, lot size and lead time, and sharing or not sharing of resources. The approach also allows the user to evaluate alternate routings, consider alternate configurations of the layout of resources, analyze the effect of overtime, experiment with different transfer lot sizes, and analyze the impact of breakdowns. As an example of this approach, the trade-off between lot size and the production lead time is shown in Fig. 13.1.

In this figure, the setup time per piece reduces as the lot size increases — thus leading to greater capacity. The greater capacity leads to smaller queueing delays for lots. However, the time spent in the system also includes the time spent waiting for a lot to build up at the work center, called the lot delay. The trade-off is due to these two effects and can be evaluated to choose an optimal lot size. Careful experimentation is often necessary to compute the jointly optimal lot sizes for several work centers. Similarly some experimentation is necessary when allocating labor to machines or processes because sharing a common resource can lead to interference (in the queueing literature referred to as the "machine interference problem").

A number of queueing network analysis software packages are available that employ rapid approximation techniques for evaluating the trade-offs discussed above. The queueing network approximation approach, however, *does not* allow the planner to capture the distribution of flow times through the

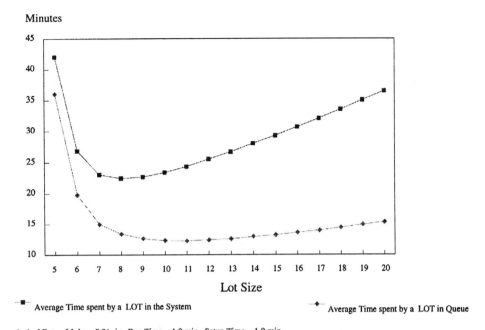

Minutes

Average Time spent by a LOT in the System
Average Time spent by a LOT in Queue

Arrival Rate of Jobs = 0.8/min., Run Time = 1.0 min., Setup Time = 1.0 min.

Arrival Process is Poisson, Service time is exponentially distributed, single machine (server)

FIGURE 13.1   Use of queueing theory for determining optimal lot size.

system, study the effect of priorities or expediting, or regulation of work into the system. Such finer details can be obtained only by constructing a simulation model of the system.

Capacity Requirements Planning

The manufacturing resources planning framework (MRPII) suggests four techniques for capacity planning, namely, **capacity planning using overall factors (CPOF), capacity bills, resource profiles, and capacity requirements planning (CRP)**. The choice of the technique depends on the industry and application.

CPOF uses data with regard to the direct labor hours required to make a product, the historical percentage of total hours used in each work center, and the production plan for each period. Using the production plan, the total hours required for the system are first calculated and then allocated to the work centers using the historical percentages. This method is appropriate when demand and product mix are steady and production lead-times are short.

Example

Products A and B have the processing requirements given in Table 13.5. There are two work centers, DEPT1 and DEPT2. Historically, 70% of the work load has gone to DEPT1 and the balance to DEPT2. This breakup of work load is based on the (given) historical product mix of 1:1. We are given the production plan for the two products, as shown Table 13.6. The capacity requirements using CPOF are also shown in Table 13.6.

The capacity bills (or labor bills) technique uses the same information as CPOF, but does not allocate the total hours to the work centers. Instead, the load for each work center due to each product is computed separately. This method is appropriate when demand is steady, product mix varies from period to period, and the production lead times are short. Capacity requirements computed using capacity bills for the same example are shown in Table 13.6.

The Resource Profiles technique uses the **operation set back chart** as additional input data. The operation set back chart is a Gantt chart that maps the capacity requirement for a product in two

TABLE 13.5 Processing Requirements

Part	Department	Operation (in sequence)	Lot Size	Run Time (min)	Setup Time (min)	Time per Part (min)	Planning Lead-time (pd)
A	2	1.00	20.00	0.05	1.00	0.10	1.00
	1	2.00	20.00	0.10	1.00	0.15	1.00
B	1	1.00	40.00	0.20	1.00	0.23	1.00
	2	2.00	40.00	0.05	0.50	0.06	1.00

TABLE 13.6 Production Plan and Capacity Requirements

		Period			
	(Backlog)	1	2	3	4
Production plan for A		20.00	30.00	25.00	15.00
Production plan for B		10.00	40.00	10.00	30.00
CPOF Calculations					
Total hours needed for A		5.00	7.50	6.25	3.75
Total hours needed for B		2.88	11.50	2.88	8.63
Total		7.88	19.00	9.13	12.38
Department 1 (70%)		5.51	13.30	6.39	8.66
Department 2 (30%)		2.36	5.70	2.74	3.71
Capacity Bill Calculations					
Department 1, product A		3.00	4.50	3.75	2.25
Department 1, product B		2.25	9.00	2.25	6.75
Total Department 1		5.25	13.50	6.00	9.00
Department 2, product A		2.00	3.00	2.50	1.50
Department 2, product B		0.63	2.50	0.63	1.88
Total Department 2		2.63	5.50	3.13	3.38
Resource Profile Calculations					
Department 1, product A		3.00	4.50	3.75	2.25
Department 1, product B	2.25	9.00	2.25	6.75	0.00
Total Department 1	2.25	12.00	6.75	10.50	2.25
Department 2, product A	2.00	3.00	2.50	1.50	0.00
Department 2, product B		0.63	2.50	0.63	1.88
Total Department 2	2.00	3.63	5.00	2.13	1.88

dimensions — work center on the y-axis vs. time on the x-axis. The production plan can then be converted to show the period in which the capacity will be required. This method is appropriate when demand and/or product mix variations are coupled with relatively long production lead times. For the previous example, if it is given that each operation takes one period to perform (called the planning lead time), then the capacity requirements will look as shown in Table 13.6.

The CRP technique is similar to the resource profiles technique, except that it uses the MRP data, accounts for work in progress, and additionally plans for service parts. This technique is more suitable for short-term planning purposes. The finite loading techniques embodied in MES will eventually replace the CRP method.

Planning for the Long Term

The objective of capacity planning in the long term is essentially strategic. The key decision variables include the location, size and technology of facilities, and the timing of the investment decisions for capacity expansion as well as equipment replacement. Other important decisions that bear upon capacity planning for the long term include partnerships for technology development, distribution system design, and the choice of the supply base and suppliers. Given the enormous scope and the uncertainties involved in long-term capacity planning, we shall only describe three approaches that have been used to model this problem. The approaches are dynamic programming, stochastic programming, and game theoretic models set in the industrial organization economics framework. Recent developments in the area of supply chain design are not covered in this list. Factors and techniques that are important for long-term capacity planning decisions (and that have not been covered) include tax incentives, government regulation, international laws, and investment analysis of projects under uncertainties.

The Dynamic Programming Approach

In this approach, demand is given either as a deterministic or as a random function of time, and plant costs are given as a function of size and time. Technological uncertainties can be built into the models, and a discount factor can be used to evaluate different expansion strategies. The decisions are when to invest and in what magnitude. The basic trade-off is between reduction in per unit cost of capacity (due to economies of scale) and having excess capacity. Closed-form solutions for the capacity expansion problem are available for simple demand and cost functions. The approach becomes computationally intensive with an increasing number of time periods, with a larger number of products as well as locations, and with the modeling of uncertainties in the technology.

Stochastic Programming

In this approach, the modeler evaluates expansion strategies under multiple scenarios. Consider, for example, a two-period, single resource capacity planning problem. The capacity required in the first period is 100. The capacity requirements in the second period can either be 100 in the low (L) demand scenario with probability p_L or 200 in the high (H) demand scenario with probability p_H. The demand in the second period will be known at the end of the first period. There are two technologies. Technology 1 (2) can be installed in capacity multiples of 100 (200) and each unit of 100 (200) costs \$1000 (\$1500). Let X_{i1}, $i = 1,2$ be the integer units of technology i installed in the first period, and X_{i2j}, $i = 1,2$; $j = L, H$, be the integer units of technology i installed under scenario j in the second period. The objective is to minimize the total expected cost in the two periods, and we are given that all demand has to be met. Then the optimization problem can be written as shown below.

$$\text{Min } 1000 \, X_{11} + 1500 \, X_{21} + p_L 1000 \, X_{12L} + p_H 1000 \, X_{12H} + p_L 1500 \, X_{22L} + p_H 1500 \, X_{22H}$$

subject to

First period demand: $\qquad\qquad\qquad 100 \, X_{11} + 200 \, X_{21} \qquad\qquad\qquad\qquad\qquad \geqq 100$

Second period low (L) demand: $\quad 100 \, X_{11} + 200 \, X_{21} + 100 \, X_{12L} + 200 \, X_{22L} \geqq 100$

Second period high (H) demand: $100 \, X_{11} + 200 \, X_{21} + 100 \, X_{12H} + 200 \, X_{22H} \geqq 200$

The optimal solution is to invest in one unit of the technology 1 in the first period if $p_L \leq 0.5$ and in one unit of technology 2 otherwise. There are a number of advantages in using this approach: the entire arsenal of mathematical programming is available for modeling the constraints as well as the dynamics and for choosing the objective function(s). The reaction of the competition can be taken into account in the model as well as taxes, subsidies, and regulation. This approach for capacity planning has been used in several industries, such as the electric utilities, oil exploration, the PVC industry, and several other process industries.

Game Theoretic Approach

Long-term capacity planning would need to include many additional factors such as preemptive behavior, competitive reaction to capacity changes, and the impact on market share. Game theoretical models can be used to understand these issues better. An introduction to the subject can be found in Lieberman [1987] and to the modeling ideas in Tirole [1990].

In closing this section on long-term capacity planning, it is important to mention the capacity strategy underlying just-in-time (JIT) implementations. In the JIT philosophy [Shingo, 1989], wasted labor time is considered one of the seven cardinal wastes — it is preferred that machines, instead of people, wait. While this strategy underscores excess capacity, it also goes hand in hand with the rest of the JIT strategy, such as building in-house expertise in the design and production of machines. More information about manufacturing strategy can be obtained from Hill [1993].

Defining Terms

Calculated capacity of a work center: Equals the rated capacity multiplied by the utilization, the efficiency, and the activation of that work center.

Capacity: The maximum rate of production that can be sustained over a given period of time with a certain product mix.

Capacity management: Allocation and management of resources.

Capacity planning using overall factors (CPOF), capacity bills, resource profiles, and **capacity requirements planning (CRP):** Four medium-term capacity planning techniques suggested in the manufacturing resource planning framework.

Operation set back chart: A Gantt chart that maps the capacity requirement for a product with the work center on the y-axis and time on the x-axis.

Rated (also called theoretical, nominal, or standing) **Capacity of a work center:** The inverse of the average processing time of a unit of work at that work center.

Work centers: The appropriate groupings of machines that form the basis for the analysis of capacity.

References

Bitran, G. R. and Morabito, R. Open queueing networks: performance evaluation models for discrete manufacturing systems, *Prod. Operat. Mgt.*, 5(2): 163–193, 1996.

Blackstone, J. H., Jr. *Capacity Management*, South-Western, Cincinnati, OH, 1989.

Buzacott, J. A. and Shanthikumar, J. G. *Stochastic Models of Manufacturing Systems*, Prentice Hall, Englewood Cliffs, NJ, 1993.

Hax, A. C. and Candea, D. *Production and Inventory Management*, Prentice Hall, Englewood Cliffs, NJ, 1984.

Hill, T. *Manufacturing Strategy: Text and Cases*, 2nd ed., Richard D. Irwin, Burr Ridge, IL, 1993.

Li, S. and Tirupati, D. Dynamic capacity expansion problem with multiple products: technology selection and timing of capacity additions, *Operat. Res.*, 42(5): 958–976, 1994.

Lieberman, M. B. Strategies for capacity expansion, *Sloan Mgt. Rev.*, Summer: 19–27, 1987.

Luss, H. Operations research and capacity expansion problems: a survey, *Operat. Res.*, 30(5): 907–945, 1982.

Malcolm, S. A. and Zenios, S. A. Robust optimization for power systems capacity expansion under uncertainty, *J. Operat. Res. Soc.*, 45(9): 1040–1049, 1994.

Ross, S. M. *A Course in Simulation*, Macmillan, New York, 1990.

Shingo, S. *A Study of the Toyota Production System*, Productivity Press, Cambridge, MA, 1989.

Tirole, J. *The Theory of Industrial Organization*, MIT Press, Cambridge, MA, 1990.

Further Information

For further details on capacity requirements planning and for other definitions of capacity see Blackstone [1989]. Details on mathematical programming applications for capacity planning can be found in Hax and Candea [1984]. For a comprehensive introduction to models employing the queueing network approach for capacity planning and control see Buzacott and Shanthikumar [1993]. An introduction to simulation modeling is given in Ross [1990]. The reader is referred to the survey article [Luss, 1982] for details on the use of dynamic programming for solving capacity expansion problems. For a mathematical programming approach for solving technology selection and capacity expansion problems, the reader is referred to Li and Tirupati [1994]. References and other details regarding stochastic programming applications for capacity expansion can be found in Malcolm and Zenios [1994].

13.4   Inventory

Steven Nahmias

Inventories are of concern at many levels of our economy. To be competitive retailers must maintain stocks of items demanded by consumers. Manufacturers require inventories of raw materials and work in process to keep production lines running. Inventories of spare parts must be available in repair centers for equipment maintenance and support. On a macro level, inventories are used to measure the health of the economy: larger inventories usually mean a slowdown of economic activity.

Why is inventory management so vital? Because the investment in inventories in the United States is enormous. As of April 1997, total business inventories in the United States was $1.02 *trillion* comprised of 44% manufacturing, 31% retail, and 25% wholesale. (Source: U.S. Department of Commerce Data.) Efficient management of inventories is clearly a top priority in our competitive economy. Inventory management is one of the most successful areas of application of operations research. For example, major weapons systems in the military, worth hundreds of billions of dollars, have been successfully managed using sophisticated mathematical **models** [Muckstadt, 1974]. Retailers employ large-scale information storage and retrieval systems with electronic data interchange to keep close tabs on inventory levels and consumer buying patterns.

This section is a brief overview of the models and methods for managing inventories when the following characteristics are present. There is a demand or need for the inventory, which may or may not be known in advance. There is an opportunity to replenish the inventory on an ongoing basis. Finally, there are accounting costs associated with various aspects of the inventory management process that can be measured or estimated.

Fundamentals

Let's start by reviewing the cornerstones of inventory modeling (here, the term model means a mathematical representation of a physical system). In the context of inventory control, the purpose of a model is to answer two questions: when should an order be placed (or production initiated) and how large should it be. Different control rules are appropriate in different circumstances, depending on several factors. These factors include the type of inventory, the motivation for holding the inventory, and the physical characteristics of the system.

Inventories can be classified in many different ways. One is based on increasing order of value added. This classification, appropriate in manufacturing contexts, gives the following categorization (listed in order of value added):

1. Raw materials
2. Components
3. Work in process
4. Finished goods

Other classification systems might be appropriate in other contexts. In order to understand inventory management, we must understand the underlying economic justification for holding inventories. Some of these are

1. *Economies of scale.* When substantial fixed costs accompany a replenishment, it is economical to produce in large lots. As we later see, the trade-off between fixed costs and **holding costs** forms the basis for the **EOQ** (economic order quantity) model, which itself is the basis for virtually all inventory modeling.

2. *Uncertainties.* Several uncertainties result in incentives to maintain inventories. The most important is uncertainty of the demand. In most contexts, demand cannot be predicted exactly, and inventories provide a buffer against demand uncertainty. Other relevant uncertainties include uncertainty of supply, uncertainty in replenishment **lead times**, uncertainties in costs, and uncertainties in the future value of the inventory.

3. *Speculation.* Holding inventory in anticipation of a rise in its value or a scarcity of supply is an example of the speculative motive. For example, silver is a requirement for production of photographic film. Major producers, such as Kodak, were at a competitive disadvantage when silver prices rose rapidly in the late 1970s.

4. *Transportation.* With the advent of the global economy, firms are producing and marketing products worldwide. One result is substantial in-transit or pipeline inventories. To reduce pipeline inventories, some companies choose to locate manufacturing facilities domestically even though labor costs may be higher in the U.S. than overseas.

5. *Smoothing.* Due to production capacity constraints, it makes sense to build inventories in anticipation of a sharp rise in demand. Many retailers do most of their business during the holiday season, for example. As a result, orders to manufacturers increase dramatically prior to year's end. Manufacturers must be prepared for this surge in demand.

Characteristics of Inventory Systems

The assumptions one makes about the underlying characteristics of the system determine the complexity of the resulting model. These characteristics include

- Demand
- Costs
- Review intervals
- Lead times
- Treatment of excess demand
- Changes over time
- Multiple echelons
- Item interactions

I will briefly review the most common cases treated.

Demand

The simplest case is when demand is known and constant. In other words, when it can be predicted exactly and the number of units consumed is the same every period. Known constant demand is rarely the case, but it can be a reasonable approximation. There are many ways to relax this assumption, but two are the most important. One is to assume that demand is known, but changing over time. This is known as nonstationary demand and is appropriate if there are significant seasonal variations, trends, or growth. The second is to allow for uncertainty of demand. In this case, demand is assumed to be random (or stochastic). Stochastic inventory models are based on an underlying probability distribution from which demand realizations are drawn. This distribution might be estimated from a past history of observations or from expert opinion. In most real-life environments, both nonstationarity and uncertainty are probably present to some extent. Little is known about dealing with these simultaneous sources or variation, however.

Costs

How costs are assessed also plays a major role in determining the complexity of the resulting model. Costs may be classified into the following broad categories: order costs, holding costs, and penalty costs. These categories incorporate most of the costs one encounters in practice. Let us consider each in turn.

Order costs are all costs that depend upon the quantity ordered or produced. The most common assumption is that there are both fixed and variable components. That is, the cost of ordering Q units, e.g., $C(Q)$, is

$$C(Q) = 0 \text{ if } Q = 0$$

$$C(Q) = K + cQ \text{ if } Q > 0$$

Here K is the fixed cost or setup cost and c is the marginal cost of each unit. More complex order-cost functions result when, for example, the supplier offers quantity discounts. Simpler models ignore fixed costs.

Holding costs, also known as carrying costs, are all costs that accrue as a result of holding inventory. Holding costs are composed of several components. These include the costs of:

1. Providing the physical space to store the item
2. Taxes and insurance
3. Breakage, spoilage, deterioration, pilferage, and obsolescence
4. Opportunity cost of alternative investment

In most cases, the last item is the most significant component of the holding cost. Money tied up in inventories could otherwise be invested elsewhere in the firm. This return could be quite substantial and relates to financial measures such as the cost of capital, the hurdle rate for new projects, and the internal rate of return. When inventory levels change continuously, holding costs can be difficult to calculate, since they are also changing continuously.

Penalty costs result from having insufficient stock to meet a demand when it occurs. Penalty costs include bookkeeping expenses when excess demand is backordered, foregone profit when excess demands are lost, and possible loss of customer goodwill. The last component is extremely important and difficult to measure. For example, loss of goodwill could affect the future demand stream of the product.

Review Intervals

The review interval corresponds to the times inventory levels are checked and reorder decisions made. Periodic review means that the opportunity to place orders occurs only at discrete points in time. Today, many systems are continuous review, which means that transactions are reported as they occur. This is the case, for example, with bar-code scanners that transmit information from point of sale to a centralized computer. Mathematical inventory models have been developed for both cases.

Lead Times

The lead time is the time that elapses from the point that an order is placed (or production is initiated) until the order arrives (or production is completed). The simplest assumption is that the lead time is zero. This may sound unrealistic, but is appropriate in some circumstances. When review periods are long and the replenishment lead time is less than a review period, assuming zero lead time is reasonable. In most environments, however, the lead time is substantial and must be included in the formulation of the model. In these cases, lead times are almost always assumed to be fixed and known. However, in many cases there is substantial uncertainty in the lead time. Mathematically, random lead times present many difficulties, so practical real-world applications rarely allow for them.

Treatment of Excess Demand

As noted earlier, excess demand may result in either backorders or lost sales. Another possibility is partial backordering, where some demands are backordered and some are lost. When excess demands are backordered, costs can be charged in several ways. The most common is a one-time cost for each unit of unfilled demand. In some circumstances, it is more appropriate to use a time-weighted backorder cost. This might be the case when the item is required in another process.

Changes over Time

Some inventories change over time. An example is perishable items with a known expiration date. Examples of perishables include processed foods, photographic film, and human blood. In these cases, the utility of an item is essentially constant until it expires, at which time the utility drops to zero. A somewhat different situation occurs when a fixed proportion of the inventory is lost each period due to spoilage, evaporation, or radioactive decay. This is known as exponential decay and may be a reasonable approximation of the more complex fixed-life perishable case. With the growth of the high-tech sector of the economy, obsolescence has become a more significant problem. In this case the inventory is not changing, but the environment is. The net result, that the items are no longer useful, is the same, however.

Multiple Echelons

In a large integrated inventory-control system, it is common for items to be stored at multiple locations. Retailers store inventory at regional distribution centers (DCs) before shipping to retail outlets. There may be several levels of intermediate storage locations between the producer and the consumer. Military applications sparked the original interest in these so-called **multiechelon** systems. The investment in spare parts in the military is huge, possibly as high as $1 trillion worldwide. The materiel support system includes as many as three levels or more. With the recent advent of EDI (electronic data interchange), interest in managing serial supply systems is greater than ever. These systems are known as "supply chains". The identification of unusual phenomenon such as the bullwhip effect (which corresponds to the increasing variance of orders as one moves from demand points to intermediate echelons to producing facilities), has contributed to the recent interest among academics and practitioners in supply chain management.

Item Interactions

Virtually all real-world inventory systems require simultaneous management of multiple items. It is not untypical to have hundreds of thousands of SKU's (stock keeping units) in one system. Often, interactions among the items arise that cannot be ignored. For example, if items compete for space or budget, explicit constraints expressing these limitations must be included into the analysis. In retailing, economic substitutes and complements are common. Hot dogs and hot dog buns are an example of complements, while hamburgers vs. hot dogs is an example of substitutes. Item interactions (such as these) are difficult to model.

The EOQ Model

The EOQ is the simplest mathematical inventory control model, yet is remarkably robust. The square root law that results appears frequently in more complex settings. The assumptions under which the EOQ model is exact are

- Demand is known and constant. The demand rate is λ units per unit time.
- Costs are assessed only against holding, at $\$h$ per unit held per unit time, and ordering at $K + cy$ per positive order of y units placed.
- The object is to find a policy that minimizes average costs per unit time.

As noted above, the objective of the analysis is to answer the questions (1) when should an order be placed? and (2) for how much? It turns out that the optimal policy is independent of the marginal ordering cost, c. Why is this so? Over an infinite horizon, all feasible policies must order exactly the demand. This will be paid for at the rate λc independent of the replenishment policy.

Assume for the time being that the order lead time is zero. In that case, one replenishes stock only when the level of on-hand inventory is zero. Any other strategy would clearly result in a higher holding cost. At the instant on-hand inventory hits zero, one places (and receives) an order. Suppose this order

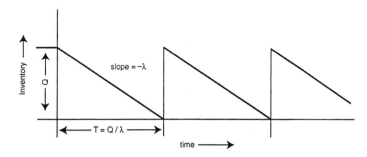

FIGURE 13.2   Inventory levels for EOQ model.

is for 5 units. This results in a sawtooth profile for the on-hand inventory as pictured in Fig. 13.2. Notice from Fig. 13.2 that the time between arrival of successive orders is Q/λ. Furthermore, the average on-hand inventory level is $Q/2$. From these observations alone, one can derive the EOQ formula.

The setup cost is incurred once each cycle. From Fig. 13.2 it is clear that the length of a cycle increases as Q increases, so that the average setup cost per unit time should be a decreasing function of Q. The average setup cost per unit time is K divided by the cycle length, Q/λ, giving $K\lambda/Q$.

The average holding cost per unit time is the average inventory level times the holding cost rate, which gives $hQ/2$. The optimal solution is found by differentiating $K\lambda/Q + hQ/2$ with respect to Q and set the resulting expression to zero to find the minimizing value of Q.

The result is the simple EOQ formula given by

$$Q = \sqrt{\frac{2K\lambda}{h}}\,.$$

This formula was discovered in 1915 by Ford Harris, an engineer with the Westinghouse Corporation. Amazingly, after more than 80 years and many thousands of technical papers on inventory theory, this formula remains the standard in many commercial inventory control systems. It is surprisingly robust. An error in the lot size or in the estimation of cost parameters results in significantly smaller penalties in the annual cost. For example, if one is using a lot size 50% larger than the optimal, the annual cost is only 8.33% higher than the optimal.

There are several relatively straightforward extensions to the basic model. These include (but are not limited to) (1) a positive order lead time, (2) a finite production rate, (3) quantity discounts, (4) constraints, and (5) backordering.

In this writer's opinion, the two most important generalizations of the basic EOQ model are to nonstationary demand and to random demand. A brief overview of the main results in each case follows.

Nonstationarity

Nonstationarity means that demands are changing over time. The easiest way to cast a non-stationary problem is as periodic review. That means that inventory levels are reviewed only at the beginning of discrete points in time, called periods. All demand is assumed to occur at one point within a period, and costs are assessed at one point in the period.

Suppose that demands over the next n periods are known constants. As with the EOQ model, assume that costs are charged against holding and setup only. The form of the optimal order policy is fundamentally different from the EOQ. Finding an optimal policy efficiently depends on the following result:

The optimal policy is an exact requirements policy, that is, in any period in which an order is placed, the size of that order is exactly the sum of the next k periods of demand where $1 \le k \le$ number of periods remaining in the horizon.

This result means that an optimal policy is completely specified by the periods in which ordering occurs. Forward or backward dynamic programming can be used to find optimal policies for even long planning horizons very efficiently. Nonstationary demand arises in several contexts. There are many situations where demands are relatively predictable (so randomness is not an issue), but there are anticipated peaks and valleys in the upcoming demand pattern. This would be the case, for example, for a highly seasonal retail item. Significant nonstationarities in demand also are common in MRP (materials requirements planning) systems. Even when end item demands are relatively smooth, the demand patterns for lower-level assemblies and components can be quite lumpy.

Randomness 1: The Newsvendor Model

Another fundamental extension of the EOQ model is to the case of demand uncertainty. Uncertainty in demand requires a fundamentally different way of looking at the system. Virtually all stochastic inventory models find policies to minimize expected costs. The use of the expected value as the appropriate operator can be justified by the law of large numbers. The law of large numbers says (roughly) that the arithmetic average of many draws from a fixed population eventually grows close to the expected value of the population. In the inventory context, that would suggest that the expected value is appropriate when the replenishment process is ongoing. By choosing a policy to minimize expected costs, one is guaranteed to minimize realized average costs over many planning periods. For one-shot replenishment decisions, the appropriateness of the expected value criterion is certainly not as clear.

As the EOQ model is the basis for all deterministic inventory models, so is the news vendor (originally "newsboy") model the basis for all stochastic inventory models. The situation is exactly that experienced by a news vendor. The product perishes quickly; that is, it can be used to satisfy demand for a single period only. A purchase decision must be made at the beginning of a planning period, and the demand during the period is a draw from a known probability distribution. Assuming a known probability distribution of demand is reasonable if there is a past history of demand observations during which the demand pattern has been stable. From this history, one can estimate both the shape and the parameters of the demand distribution.

There are several ways one might develop a cost structure for this system, but the easiest and most intuitive is the following. There are two ways one can err: by ordering too much or ordering too little. Ordering too much means that one pays for items that don't sell. Ordering too little means that excess demands are unmet, and profits foregone. Let us suppose that the cost of every unit purchased and not sold is c_o (for overage cost) and the cost of every unit of excess demand not filled is c_u (for underage cost). Furthermore, assume that the cumulative distribution function (cdf) of demand in a period is $F(t)$.

The solution to the news vendor problem turns out to be surprisingly simple. The order quantity, Q, which minimizes the expected costs of overage and underage for the period solves

$$F(Q) = \frac{c_u}{c_u + c_o}.$$

The right-hand side of this equation is known as the critical ratio. Note that, as long as both underage and overage costs are positive, the critical ratio must be between zero and one. How difficult this equation is to solve depends on the complexity of the demand distribution. However, since the cdf is a nondecreasing function, we know that this equation will always have a solution (as long as the distribution is continuous). Popular spreadsheet programs, such as Excel, include the inverses of the most common distributions, making this equation easy to solve in those cases. Under normally distributed demand (which is a very common assumption in practice), the optimal solution has the form

$$Q = \sigma z + \mu,$$

where μ and σ are, respectively, the population mean and standard deviation and z is the unit normal value corresponding to a left tail equal to the critical ratio.

Although the news vendor model is single period only, the form of the solution is exactly the same when stock can be carried from one period to the next and there are infinitely many periods remaining in the planning horizon. The only difference is that the overage and underage costs must be interpreted differently. As long as there is no fixed order cost, lead times can be incorporated into the analysis in a relatively straightforward manner as well. If the lead time is L periods, then the order up to level is found by inverting the $L + 1$-fold convolution of the single period demand distribution. When a salvage value is included equal to the purchase cost, the finite horizon solution can be found by solving a series of news vendor problems.

Randomness 2: The Lot Size Reorder Point Model

The periodic review multiperiod news vendor model is much more complex if one includes a fixed order cost. Since fixed costs are common in practice, a continuous review heuristic (i.e., approximate) model is much more popular. Continuous review means that the inventory level is known at all times. As long as demands do not occur in bulk, one has the opportunity to place an order when the inventory level hits a specified reorder level, say R.

Suppose there is a positive order lead time and the demand during the lead time is a random variable, D, with cumulative distribution function $F(t)$. As earlier, let λ be the expected demand rate. The following cost structure is assumed: ordering at $K + cQ$ per positive order of size Q, p per unit of backlogged demand, and h per unit held per unit time. The policy is when the inventory level hits R an order of size Q is placed. A heuristic analysis of this system leads to the optimal values of Q and R simultaneously solving the following two equations:

$$Q = \sqrt{\frac{2\lambda\left[\left(K + pn(R)\right)\right]}{h}}$$

$$1 - F(R) = \frac{Qh}{p\lambda}.$$

The term $n(R)$ is known as the loss integral. For normal demands, this function is tabled, but in general this term can be cumbersome to compute. Several researchers have recommended approximations that give good results and avoid the calculation of the loss integral. The simplest is to approximate Q by the EOQ, which gives fairly good results in most cases. If Q is given, finding R is equivalent to solving a news vendor problem.

Because it is often difficult to estimate the backorder cost, many users prefer to specify service levels instead. The service level can be defined in several ways, but the most common is the percentage of demands that can be met from stock. This is also known as the fill rate. Solving (Q,R) systems subject to a constraint on fill rate yields equations similar to those above. The reorder level is somewhat more difficult to find, however, even if one uses a simple approximation for 5. When a constraint on fill rate is specified, the reorder level R solves

$$n(R) = (1 - \beta)Q$$

where β is the fill rate. This equation requires inverting the loss function $n(R)$. The normal distribution is commonly assumed, and tables are used to perform the inversion. A recent application of (Q,R) models to a large-scale inventory system containing in excess of 30,000 parts is described in Hopp et al. [1997].

Historical Notes and Further Reading

As a subfield of operations research, inventory has long history. The EOQ model [Harris, 1915] predates most formal OR activities by almost 30 years. The origin of the news vendor model is unclear, but it appears to date to the late 1940s. Several important papers appeared in the early 1950s, which spawned the interest in this area among academics. These included the studies of Arrow et al. [1951] and Dvoretzsky et al. [1952a, 1952b]. The book by Whitin [1957] was important in linking inventory control to classical economics and was probably the origin of the (Q,R) model discussed here. Arrow et al. [1958] compiled a collection of highly technical papers, which served as the cornerstone of the mathematical theory of inventories. The results on the deterministic nonstationary problem are due to Wagner and Whitin [1958].

For those interested in further reading, both review articles and books provide good overviews of existing work. More comprehensive general reviews on inventory models can be found in Veinott [1966] and Nahmias [1978]. Several excellent and comprehensive review articles are contained in the book Graves et al. [1993]. There are also several reviews of particular subfields that might be of interest. These include a review of perishable inventory models [Nahmias, 1982], repairable inventory models [Nahmias, 1981], and inventory models for retailing [Nahmias and Smith, 1993]. Although out of print, Hadley and Whitin [1963] still provides an excellent summary of the basic models and theory. Other recommended texts include Brown [1967], Silver and Peterson [1975] (scheduled for a new edition soon), and Nahmias [1997]. Many technical journals include papers on inventory theory and practice. The following professional journals regularly carry articles on inventory theory: *Operations Research, Management Science, Industrial Engineering,* and *Naval Research Logistics,* to name a few.

Defining Terms

Model: A mathematical representation of a physical system.

Lead time: The elapsed time from the point an order is placed (or production is initiated) until the order arrives (or production is completed).

Multiechelon: An inventory system in which there are intermediate storage locations between producer and consumer.

Holding costs: Costs that result from physically carrying inventory. The main component is the opportunity cost of alternative investment.

EOQ: Economic order quantity given by the well-known square root formula.

References

Arrow, K. A., Harris, T. E., and Marschak, J. Optimal inventory policy, *Econometrica,* 19: 250–272, 1951.

Arrow, K. A., Karlin, S. A., and Scarf, H. E. *Studies in the Mathematical Theory of Inventory and Production,* Stanford University Press, Stanford, CA, 1958.

Dvoretzky, A., Kiefer, J., and Wolfowitz, J. The inventory problem I: case of known distributions of demand, *Econometrica,* 20: 187–222, 1952a.

Dvoretzky, A., Kiefer, J., and Wolfowitz, J. The inventory problem II: case of unknown distributions of demand, *Econometrica,* 20: 450–466, 1952b.

Harris, F. W. *Operations and Cost,* Factory Management Series, Shaw, Chicago, 1915.

Hopp, W. J., Spearman, M. L., and Zhang, R. Q. Easily implementable inventory control policies, *Operat. Res.,* 45: 327–340, 1997.

Muckstadt, J. A. A model for a multi-item, multi-echelon, multi-indenture inventory system, *Mgt. Sci.,* 20: 472–481, 1974.

Wagner, H. M. and Whitin, T. M. Dynamic version of the economic lot size formula. *Mgt. Sci.,* 5: 89–96, 1958.

Whitin, T. M. *The Theory of Inventory Management, rev. ed.,* Princeton University Press, Princeton, NJ, 1957.

Further Information

Brown, R. G. *Decision Rules for Inventory Management,* Dryden, Hinsdale, IL, 1967.

Graves, S. C., Rinnooy Kan, A. H. G., and Zipkin, P. H., Eds. *Handbooks in Operations Research and Management Science, Vol. 4,* Elsevier Science, Amsterdam, 1993.

Hadley, G. and Whitin, T. M. *Analysis of Inventory Systems,* Prentice Hall, Englewood Cliffs, NJ, 1963.

Nahmias, S. Inventory models, In *The Encyclopedia of Computer Science and Technology, Vol. 9,* Belzer, Holzman, and Kent, Eds., pp. 447–483, Marcel Dekker, New York, 1978.

Nahmias, S. Managing reparable item inventory systems: a review, In *Multilevel Production/Inventory Control Systems: Theory and Practice, Vol. 16,* L. Schwarz, Ed., TIMS Studies in the Management Sciences, pp. 253–277, North Holland, 1981.

Nahmias, S. Perishable inventory theory: a review, *Operat. Res.,* 30: 680–708, 1982.

Nahmias, S. and Smith, S. A. Mathematical models of retailer inventory systems: a review, In *Perspectives on Operations Management, Essays in Honor of Elwood S. Buffa,* Sarin, R., Ed., pp. 249–278, Kluwer Academic, Boston, 1993.

Nahmias, S. *Production and Operations Analysis, 3rd ed.,* Richard D. Irwin, Burr Ridge, IL, 1997.

Silver, E. A. and Peterson, R. *Decision Systems for Inventory Management and Production Planning, 2nd ed.,* John Wiley & Sons, New York, 1985.

Veinott, A. F. The status of mathematical inventory theory, *Mgt. Sci.,* 12: 745–777, 1966.

13.5 Quality

Matthew P. Stephens and Joseph F. Kmec

Although no universally accepted definition of **quality** exists, in its broadest sense quality has been described as "conformance to requirements", "freedom from deficiencies", or "the degree of excellence that a thing possesses". Taken within the context of the manufacturing enterprise, quality — or, more specifically, manufacturing quality — shall be defined as "conformance to requirements". This section focuses on the evaluation of product quality, with particular emphasis directed at statistical methods used in the measurement, control, and tolerances needed to achieve the desired quality. Factors that define product quality are ultimately determined by the customer and include such traits as reliability, affordability or cost, availability, user friendliness, and ease of repair and disposal. To ensure that quality goals are met, manufacturing firms have initiated a variety of measures that go beyond traditional product inspection and record keeping, which, by and large, were the mainstays of quality control departments for decades. One such initiative is total quality management (TQM) [Saylor, 1992], which focuses on the customer, both inside and outside the firm. It consists of a disciplined approach using quantitative methods to continuously improve all functions within an organization. Another initiative is registration under the ISO 9000 series [Lamprecht, 1993], which provides a basis for U.S. manufacturing firms to qualify their finished products and processes to specified requirements. More recently, the U.S. government has formally recognized outstanding firms through the coveted Malcolm Baldridge Award [ASQC, 1994] for top quality among U.S. manufacturing companies. One of the stipulations of the award is that recipient companies share information on successful quality strategies with their manufacturing counterparts.

Measurements

The inherent nature of the manufacturing process is variation. Variation is present due to any one or a combination of factors including materials, equipment, operators, or the environment. Controlling variation is an essential step in realizing product quality. To successfully control variation, manufacturing firms rely on the measurement of carefully chosen parameters. Because measurement of the entire population of products or components is seldom possible or desirable, **samples** from the **population** are

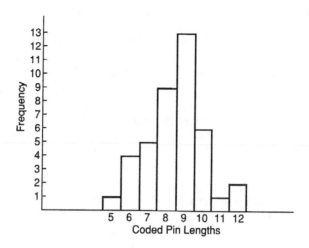

FIGURE 13.3   Coded pin lengths.

chosen. The extent to which sample data represent the population depends largely on such items as sample size, method of sampling, and time-dependent variations.

Measured data from samples taken during a manufacturing process can be plotted in order to determine the shape of the **frequency distribution**. The frequency distribution can give a visual clue to the process average and dispersion. The latter is referred to as **standard deviation**. Figure 13.3 shows a frequency distribution plot of 40 coded pin lengths expressed in thousands of an inch above 1 inch. Thus, the coded length 6 represents an actual length of 1.006 in. For the data shown, average coded pin length is 8.475 and standard deviation is 1.585.

Normal Distribution

Although there is an infinite variety of frequency distributions, the variation of measured parameters typically found in the manufacturing industry follows that of the normal curve. The normal distribution is a continuous bell-shaped plot of frequency vs. some parameter of interest and is an extension of a histogram whose basis is a large population of data points. Figure 13.4 shows a normal distribution plot superimposed on a histogram. Some important properties of the normal distribution curve are

1. The distribution is symmetrical about the population mean μ.
2. The curve can be described by a specific mathematical function of population mean μ and population standard deviation σ.

An important relationship exists between standard deviation and area under the normal distribution curve. Such a relationship is shown in Fig. 13.5 and may be interpreted as follows: 68.26% of the readings (or area under the curve) will be between $\pm 1\sigma$ limits, 95.46% of the readings will be between $\pm 2\sigma$ limits, and 99.73% of the readings will be between $\pm 3\sigma$ limits. The significance of this relationship is that the standard deviation can be used to calculate the percentage of the population that falls between any two given values in the distribution.

Statistical Quality Control

Statistical quality control (SQC) deals with collection, analysis, and interpretation of data to monitor a particular manufacturing or service process and ensure that the process remains within its capacity. In order to understand process capability, it is necessary to realize that variation is a natural phenomenon that will occur in any process. Parts will appear identical only due to the limitation of the inspection or measurement instrument. The sources of these variations may be the material, process, operator, time of the operation, or any other significant variables. When these factors are kept constant, the minor

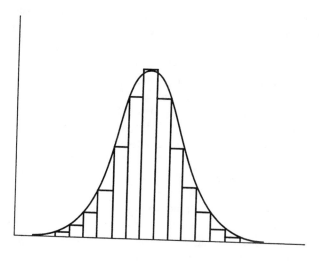

FIGURE 13.4 Normal distribution curve.

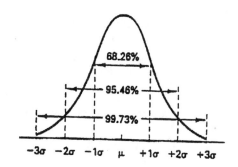

FIGURE 13.5 Percentages under the normal curve.

variations inherent in the process are called *natural* (or *chance*) *variations*, as opposed to variations due to **assignable causes**.

Control charts are utilized to determine when a given process variation is within the expected or natural limits. When the magnitude of variation exceeds these predetermined limits, the process is said to be *out of control*. The causes for out-of-control conditions are investigated and the process is brought back in control. Control charts or the control limits for the natural or chance-cause variations are constructed based on the relationship between the normal distribution and the standard deviation of the distribution. As stated earlier, since approximately 99.73% of a normal distribution is expected to fall between $\pm 3\sigma$ of the distribution, control limits are established at $\bar{X} \pm 3\sigma$ for the process. Therefore, any sample taken from the process is expected to fall between the control limits or the $\bar{X} \pm 3\sigma$ of the process 99.73% of the time. Any sample not within these limits is assumed to indicate an out-of-control condition for which an assignable cause is suspected.

Control charts can be divided into two major categories: control charts for variables (measureable quality characteristics, i.e., dimension, weight, hardness, etc.) and control charts for attributes (those quality characteristics not easily measurable and therefore classified as conforming or not conforming, good or bad, etc.).

Control Charts for Variables

The most common charts used for variables are the $\bar{X}$ and R charts. The charts are used as a pair for a given quality characteristic. In order to construct control charts for variables, the following steps may be followed:

1. Define the quality characteristic that is of interest. Control charts for variables deal with only one quality characteristic; therefore, if multiple properties of the product of the process are to be monitored, multiple charts should be constructed.
2. Determine the sample (also called the *subgroup*) size. When using control charts, individual measurements or observations are not plotted, but, rather, sample averages are utilized. One major reason is the nature of the statistics and their underlying assumptions. Normal statistics, as the term implies, assumes a normal distribution of the observations. Although many phenomena may be normally distributed, this is not true of all distributions. A major statistical theory called the *central limit theorem* states that the distribution of sample averages will tend toward normality as the sample size increases, regardless of the shape of the parent population. Therefore, plotting sample averages ensures a reasonable normal distribution so that the underlying assumption of normality of the applied statistics is met.

   The sample size (two or larger) is a function of cost and other considerations, such as ease of measurement, whether the test is destructive, and the required sensitivity of the control charts. As the sample size increases, the standard deviation decreases; therefore, the control limits will become tighter and more sensitive to process variation.
3. For each sample calculate the sample average, $\bar{X}$, and the sample **range**. For each sample record any unusual settings (e.g., new operator or problem with raw material) that may cause an out-of-control condition.
4. After about 20 to 30 subgroups have been collected, calculate

$$\bar{\bar{X}} = \frac{\Sigma \bar{X}}{g}; \quad \bar{R} = \frac{\Sigma R}{g}$$

   where $\bar{\bar{X}}$ is the average of averages, $\bar{R}$ is the average of range, and g is the number of samples or subgroups.
5. Trial upper and lower control limits for the $\bar{X}$ and R chart are calculated as follows:

$$\text{UCL}_{\bar{X}} = \bar{\bar{X}} + A_2 \bar{R}; \quad \text{UCL}_R = D_4 \bar{R}$$

$$\text{LCL}_{\bar{X}} = \bar{\bar{X}} - A_2 \bar{R}; \quad \text{LCL}_R = D_3 \bar{R}$$

   A_2, D_3, and D_4 are constants and are functions of sample sizes used. These constants are used to approximate process standard deviation from the range. Tables of these constants are provided in Banks [1989], De Vor et al. [1992], Grant and Leavenworth [1988], and Montgomery [1991].
6. Plot the sample averages and ranges on the $\bar{X}$ and the R chart, respectively. Any out-of-control point that has an assignable cause (new operator, etc.) is discarded.
7. Calculate the revised control limits as follows:

$$\bar{\bar{X}}_o = \frac{\Sigma \bar{X} - \Sigma \bar{X}_d}{g - g_d}; \quad R_o = \frac{\Sigma R - \Sigma R_d}{g - g_d}; \quad \sigma_o = \frac{R_o}{D_2}$$

$$\text{UCL}_{\bar{X}_o} = \bar{\bar{X}}_o + A \sigma_o \quad \text{UCL}_R = D_2 \sigma_o$$

$$\text{LCL}_{\bar{X}_o} = \bar{\bar{X}}_o - A \sigma_o \quad \text{LCL}_R = D_1 \sigma_o$$

The subscript *o* and *d* stand for revised and discarded terms, respectively. The revised control charts will be used for the next production period by taking samples of the same size and plotting the sample average and sample range on the appropriate chart. The control limits will remain in effect until one or

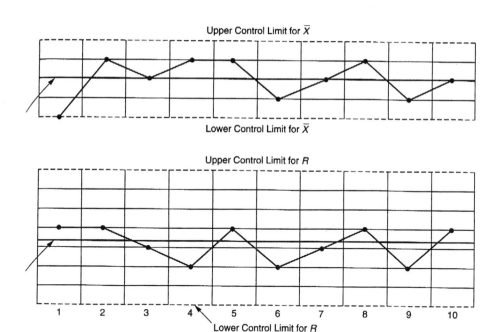

FIGURE 13.6 Control charts for $\overline{X}$ and R.

more factors in the process change. Figure 13.6 shows control charts of $\overline{X}$ and R values for ten subgroups. Each subgroup contained five observations because none of the ten data points lie outside of either upper and lower control limits; the process is designated "in control".

The control charts can be used to monitor the out-of-control conditions of the process. It is imperative to realize that patterns of variation as plotted on the charts should give clear indications to a process that is headed for an out-of-control condition or one that displays an abnormal pattern of variations. Whereas no point may actually fall out of the limits, variation patterns can often point to some unusual process behavior that requires careful study of the process.

Control Charts for Attributes

For those quality characteristics that are not easily measured — or in such cases where count of defects of defective items are involved or go-no-go gauges are used — control charts for attributes are used. These charts can be grouped into two major categories:

1. Charts for defectives or nonconforming items
2. Charts for defects or nonconformities

Charts for Nonconforming Items
The basic charts in this group are the fraction **nonconforming** chart (p chart), percent nonconforming chart ($100p$ chart), and the count of nonconforming chart (np chart). The procedure for the construction, revision, and the interpretation of control charts for attributes is similar to that for $\overline{X}$ and R charts. The following steps may be used to construct a p chart:

1. Once sample size has been established, fraction nonconforming, p, is determined for each sample by

$$p = \frac{np}{n}$$

where n is the sample size and np is the count of defectives or nonconforming items in the sample.

2. After roughly 20 to 30 subgroups have been collected, calculate $\bar{p}$, the value of the central line, or the average fraction defective.

$$\bar{p} = \frac{\sum np}{\sum n}$$

3. Trial control limits are calculated using:

$$\text{UCL} = \bar{p} + 3\sqrt{\frac{\bar{p}(1-\bar{p})}{n}}$$

$$\text{LCL} = \bar{p} - 3\sqrt{\frac{\bar{p}(1-\bar{p})}{n}}$$

4. Plot the fraction defective for each subgroup. The out-of-control subgroups that have assignable causes are discarded, and revised limits are calculated as follows:

$$P_o = \frac{\sum np - \sum np_d}{\sum n - n_d}$$

$$\text{UCL} = p_o + 3\sqrt{\frac{p_o(1-p_o)}{n}}$$

$$\text{LCL} = p_o - 3\sqrt{\frac{p_o(1-p_o)}{n}}$$

5. If the lower control limit is a negative number, it is set to zero. Sample points that fall above the upper limit indicate a process that is out of control. However, samples that fall below the lower limit, when the lower control limit is greater than zero, indicate a product that is better than expected. In other words, if a sample contains fewer nonconforming items than the process is capable of producing, the sample fraction defective will fall below the lower control limit. For this reason some practitioners may choose to set the lower limit of the attribute charts to zero. This practice, however, may mask other problems or potentials for process improvements.

Other charts for nonconforming items are simple variations of the *p* chart. In the case of the 100*p* chart, all values of the *p* chart are expressed as percentages. In the case of the *np* chart, instead of plotting fraction or percent defectives, actual counts of nonconforming or defective items are plotted. See Banks [1989], De Vor et al. [1992], Grant and Leavenworth [1988], and Montgomery [1991] for greater detail. The formulas for the central line and the control limits for an *np* chart are given below. It is assumed that the revised value for universe fraction defective, *p*, is known. If *p* is not known, then the procedure for the *p* chart must be carried out to determine the revised value for the universe fraction defective.

$$\text{Central line} = np_o$$

$$\text{Control limits} = np_o \pm 3\sqrt{np_o(1-p_o)}$$

where n is the sample size and p_o is the universe fraction defective.

Charts for Defects or Nonconformities

Whereas the charts for defective or nonconforming items are concerned with the overall quality of an item or sample, charts for defects look at each defect (i.e., blemish, scratch, etc.) in each item or sample. One may consider an item a nonconforming item based on its overall condition. A defect or **noncon-formity** is that condition that makes an item a nonconforming or defective item.

In this category are c charts and u charts. The basic difference between the two is the sample size. The sample size, n, for a c chart is equal to one. In this case the number of nonconformities or defects are counted per a single item. For a u chart, however, $n > 1$. See Banks [1989], De Vor et al. [1992], Grant and Leavenworth [1988], and Montgomery [1991] for the formulas and construction procedures.

Tolerances and Capability

As stated earlier, *process capability* refers to the range of process variation that is due to chance or natural process deviations. This was defined as $\bar{X} \pm 3\sigma$ (also referred to as 6σ), which is the expected or natural process variation. **Specifications** or **tolerances** are dictated by design engineering and are the maximum amount of acceptable variation. These specifications are often stated without regard to process spread. The relationships between the process spread or the natural process variation and the engineering specifications or requirements are the subject of process capability studies. Process capability can be expressed as

$$C_p = \frac{US - LS}{6\sigma}$$

where C_p = process capability index, US = upper engineering specification value, and LS = lower engineering specification value.

A companion index, C_{pk}, is also used to describe process capability, where

$$C_{pk} = \frac{US - \bar{X}}{3\sigma}$$

or

$$C_{pk} = \frac{\bar{X} - LS}{3\sigma}$$

The lesser of the two values indicates the process capability. The C_{pk} ratio is used to indicate whether a process is capable of meeting engineering tolerances and whether the process is centered around the target value $\bar{X}$. If the process is centered between the upper and the lower specifications, C_p and C_{pk} are equal. However, if the process is not centered, C_{pk} will be lower than C_p and is the true process capability index. See De Vor et al. [1992], Grant and Leavenworth [1988], and Montgomery [1991] for additional information.

A capability index less than one indicates that the specification limits are much tighter than the process spread. Hence, although the process may be in control, the parts may well be out of specification. Thus, the process does not meet engineering requirements. A capability index of one means that, as long as the process is in control, parts are also in spec. The most desirable situation is to have a process capability index greater than one. In such cases, not only are approximately 99.73% of the parts in spec when the process is in control, but, even if the process should go out of control, the product may still be within the engineering specifications. Process improvement efforts are often concerned with reducing the process spread and, therefore, increasing the process capability indices.

An extremely powerful tool for isolating and determining those factors that significantly contribute to process variation is statistical design and analysis of experiments. Referred to as "design of experiments", the methodology enables the researcher to examine the factors and determine how to control these factors in order to reduce process variation and therefore increase process capability index. For greater detail, see Box et al. [1978].

Defining Terms

Assignable causes: Any element that can cause a significant variation in a process.

Frequency distribution: Usually a graphical or tabular representation of data. When scores or measurements are arranged, usually in an ascending order, and the occurrence (frequency) of each score or measurement is also indicated, a frequency distribution results.

No. of Defectives	Frequency
0	10
1	8
2	7
3	8
4	6
5	4
6	2
7	1
8	1
9	0

The frequency distribution indicates that ten samples were found containing zero defectives.

Nonconforming: A condition in which a part does not meet all specifications or customer requirements. This term can be used interchangeably with *defective*.

Nonconformity: Any deviation from standards, specifications, or expectation; also called a *defect*. Defects or nonconformities are classified into three major categories: critical, major, and minor. A critical nonconformity renders a product inoperable or dangerous to operate. A major nonconformity may affect the operation of the unit, whereas a minor defect does not affect the operation of the product.

Population: An entire group of people, objects, or phenomena having at least one common characteristic. For example, all registered voters constitute a population.

Quality: Quality within the framework of manufacturing is defined as conformance to requirements.

Range: A measure of variability or spread in a data set. The range of a data set, R, is the difference between the highest and the lowest values in the set.

Sample: A small segment or subgroup taken from a complete population. Because of the large size of most populations, it is impossible or impractical to measure, examine, or test every member of a given population.

Specifications: Expected part dimensions as stated on engineering drawings.

Standard deviation: A measure of dispersion or variation in the data. Given a set of numbers, all of equal value, the standard deviation of the data set would be equal to zero.

Tolerances: Allowable variations in part dimension as stated on engineering drawings.

References

ASQC. *Malcolm Baldrige National Quality Award — 1994 Award Criteria*, American Society for Quality Control, Milwaukee, WI, 1994.

Banks, J. *Principles of Quality Control*, John Wiley & Sons, New York, 1989.

Box, G. E. P., Hunter, W. G., and Hunter, J. S. *Statistics for Experimenters*, John Wiley & Sons, New York, 1978.

De Vor, R. E., Chang, T. H., and Sutherland, J. W. *Statistical Quality Design and Control*, Macmillan, New York, 1992.

Grant, E. L. and Leavenworth, R. S. *Statistical Quality Control*, 6th ed., McGraw-Hill, New York, 1988.

Lamprecht, J. L. *Implementing the ISO 9000 Series*, Marcel Dekker, New York, 1993.

Montgomery, D. C. *Statistical Quality Control*, 2nd ed., John Wiley & Sons, New York, 1991.

Saylor, J. H. *TQM Field Manual*, McGraw-Hill, New York, 1992.

Further Information

Statistical Quality Control, by Eugene Grant and Richard Leavenworth, offers an in-depth discussion of various control charts and sampling plans.

Statistics for Experimenters, by George Box, William Hunter, and Stewart Hunter, offers an excellent and in-depth treatment of design and analysis of design of experiments for quality improvements.

Most textbooks on statistics offer detailed discussions of the central limit theorem. *Introduction to Probability and Statistics for Engineers and Scientists*, written by Sheldon Ross and published by John Wiley & Sons, is recommended.

American Society for Quality Control, P.O. Box 3005, Milwaukee, WI 53201-3005, phone: (800)248-1946, is an excellent source for reference material, including books and journals, on various aspects of quality.

13.6 Experience Curve

James P. Gilbert

Interest in organizational experienced-based learning is at an all-time high. The work of Peter Senge [1990] in his critically acclaimed book *The Fifth Discipline* has been read by many organizational leaders and found to be useful in improving their organizations. Senge invites us to use a holistic process of improvement that includes systems thinking, personal mastery, mental models, shared vision, and team learning.

The focus here is on a more narrow thrust of individual learning through personal experience. The **experience curve** is sometimes incorrectly used interchangeably with these terms: learning curve, progress curve, improvement curve, and startup curve. There is a hierarchy of concepts from the more global experience curve to the more operational level of learning curve. The concepts of experience and learning curves are very popular in business organizations and serve an important function in providing conviction and confidence to the vague notion of learning by doing [Hall and Howell, 1985].

The Experience Curve Logic

The terms experience curve and **learning curve** are occasionally taken to mean the same thing. These curves are both strategic and tactical tools useful in quantifying the rate at which cumulative experience allows a reduction in the amount of productive resources the firm must expend to complete its desired tasks [Melnyk and Denzler, 1996]. It is possible and indeed helpful to distinguish between them. The experience curve is broader in scope and refers to the reduction of costs that may occur over the life of a product, i.e., total costs [Hall and Howell, 1985]. The experience curve is based on the economic theory known as "economies of scale". The Boston Consulting Group empirically confirmed this effect with studies from many industries. Figure 13.7 illustrates the general shape for the experience curve where the decrease in average unit costs varies between 20 and 30% with each doubling of volume. Volume is a surrogate for accumulated experience and learning is a consequence [Starr, 1996]. With several iterations of doubling, the curve "flattens out" as most of the learning benefits have occurred. Thus, potential benefits for capacity, price, and market share implications are obtained in early iterations of productive doubling.

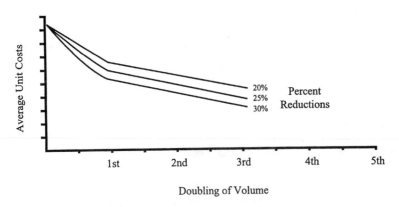

FIGURE 13.7   Average unit costs decrease with increasing experience. (Source: Starr, M. K. *Operations Management: A Systems Approach*, p. 431, Southwestern Publishing, Danvers, MA, 1996. With permission.)

The learning curve usually refers to a more micro concept employed in work analysis and cost control. Learning curve embodies the idea of learning by doing. These terms, experience curve and learning curve, differ in respect to the costs covered, the amount of productive output during the period of learning, and the causes of cost reduction [Hall and Howell, 1985].

The intuitive nature of the experience curve might mean that people frequently use this concept when repetitive actions improve results. For example, experience curve phenomena may have played a large role in the planning of the construction of the great pyramids of Egypt. The analytical use of the concept for business purposes first surfaced in airplane construction where Wright [1936] observed that, as the quantity of manufactured units doubled, the number of direct labor hours needed to produce each individual unit decreased at a uniform rate. This observation had strategic and operational importance for the development of aircraft. He illustrated the variation of labor cost with production quantity in the formula

$$F = \mathrm{Log}\ F/\mathrm{Log}\ N \tag{1}$$

where $F =$ a factor of cost variation proportional to the quantity N. The reciprocal of F then represents a direct percent variation of cost vs. quantity [Wright, 1936].

Wright shows that experience-based efficiencies in unit output are closely correlated with cumulative output and go beyond changes in design and tooling. This work presents evidence that as unit volume for a particular item increases there are predictable corresponding reductions in cost. These data then become central concepts for strategic and operational planning.

Experience Curve Principles

At the strategic planning level, the experience curve follows a profit enhancement desire via an increased market share strategy. It may be hoped that, by increasing market share while reducing costs, a detriment to market entrants will ensue [Lieberman, 1989]. Learning through experience becomes an important component of the increased market share strategy, hopefully leading to increased profits (see Fig. 13.8). This influence charts shows the underlying logic of the use of the experience curve. Quality learning is enhanced through the shared experience at the worker level and the organizational level. The critical aspect of quality products along with movement along the experience curve increases the time available for work, thus increasing productive efficiency. As the individual employees and the organization become more efficient, there should be a corresponding increase in productivity. More output for less input effectively increases capacity, which taken together with the increased efficiency and productivity should lead to a reduction in unit cost. The business is investing in a cost leadership posture based on the

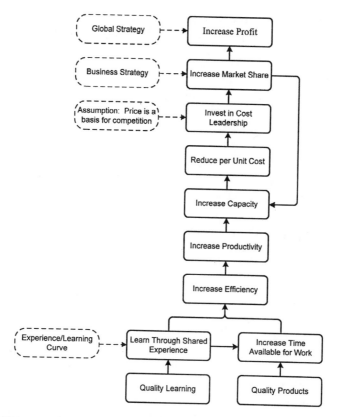

FIGURE 13.8   Influence chart of the experience curve strategy and application.

assumption that price is a basis for competition. If the firm is able to produce quality units and reduce market price, there is the opportunity for increased market share (the business strategy). Increased market share via a reduced price may lead to the global goal of improving profitability.

Generally, then, we see that the use of the cost leadership strategy using the experience curve is based on several assumptions [Amit, 1986]:

1. Price is a basis for competition.
2. If per unit cost is reduced, price may be reduced, which may lead to increased market share.
3. As cumulative output increases, the firm's average cost is reduced. Therefore, for any production rate, there is a reduction in the per-unit cost.
4. If market share is increased, profits will increase.

Note that another critical assumption of the experience curve is that learning can be kept within the organization [Lieberman, 1989]. Where industry-wide dissemination of process technology is rapid, the benefits of organizational learning through the experience curve may be short lived. The cost benefits, therefore, may not lead to increased market share even though industry costs are declining because all participants are learning at approximately the same rate.

Learning Curve Formulation

The formulation for the learning curve model is commonly shown in one of two ways: as a margin cost model and as a direct labor hour model. Both derivations are shown here for clarity, but the direct labor hours formulation may be more useful as hourly compensation typically changes over time and there may be inflation considerations as well. Also, direct labor hours may be easily converted into costs if necessary [Yelle, 1979]. By convention, we refer to experience curves by the complement of the improvement rate. For example, a 90% learning curve indicates a 10% decrease in per-unit (or mean) time or cost with each doubling of productive output. Experience and learning curves normally apply only to cost of direct labor hours.

Marginal Cost Model

The cumulative-average learning curve formulation is

$$y_{cx} = ax^{-b} \qquad (2)$$

where y_{cx} = the average cost of the first x units, a = the first unit cost, x = the cumulative unit number output, and b = the learning "elasticity", which defines the slope of the learning curve.

This learning curve model indicates that, as the quantity of units produced doubles, the average cost per unit decreases at a uniform rate.

Direct Labor Hours Model

$$Y = KX^n \qquad (3)$$

where Y = the number of direct labor hours required to produce the Xth unit, K = the number of direct labor hours required to produce the first unit, X = the cumulative unit number, $n = \log \phi / \log 2$, ϕ = the learning rate, and $1-\phi$ = the progress ratio.

These empirical models have been shown to fit many production situations very well. We must be cautious, however, as there are many other variables at work at the same time as the experience curve is being utilized. For example, production factors such as product design decisions, selection of tooling and equipment, methods analysis and layout, improved organizational and individual skills training, more effective production scheduling and control procedures, and improved motivation all play a role in decreasing cost and increasing capacity.

Applications and Uses

There are three general areas for the application and use of experience curves (see Table 13.7). As can be seen, there are numerous applications for the learning curve phenomenon. The usefulness depends on a number of factors: the frequency of product innovation, the amount of direct labor vs. machine paced output, and the amount of advanced planning of methods and tooling, all leading to a predictable rate of reduction in labor time.

Cautions and Criticisms

Abernathy and Wayne [1974] caution that the learning curve use needs to be managed in relation to firm innovation. This study of Ford Motor Company data illustrates that dependence on the learning curve phenomenon affects product innovation both in the nature of changes and the intensity of innovation activity. It should noted that the authors studied the model T Ford, which had a long life cycle. It is unlikely that we will see such long product lives today. The authors are not arguing that the use of the learning curve is inappropriate but rather that small innovative firms are having problems transitioning to repetitive, cost efficient, experienced-based manufacturing. The key point drawn from the study is that "there must be a theoretical limit to the amount by which costs can ultimately be reduced, a manufacturer reaches the practical limit first" [Abernathy and Wayne, 1974]. Caution, then, is necessary as management balances increasing competition, as competitors discover the process methods of the leader, and the firm's decreasing ability to be innovative, as dependence on cost reduction continues.

As with all data-driven decision tools, the selection of appropriate input data must be chosen wisely. The cautions described by Andress [1954] are still important today. Experience curve conclusions may be erroneous if the labor hour data are not correct. For example, if increased purchased components are

TABLE 13.7   Experience Curve: Applications and Uses

Strategic	Internal	External
Determine volume-cost changes	Develop labor standards	Estimate actual purchase costs
Estimate new product startup costs	Calculate rates of material supply required	Supplier scheduling
Determine capital needs	Manpower planning	Budgeting for cash flow
Pricing of new products	Production scheduling	Purchased goods planning
	Budgeting	
	Inventory planning	
	Evaluate new employees during training	
	Develop efficiency measures	
	Develop costs per unit	
	Plan delivery schedules	
	Make or buy decisions	
	Equipment planning	
	Work flow planning	
	Capacity planning	
	Establish wage incentive plans	

being used, this shift in labor hours to the supplier may seem to increase experience-based savings. This may be illusionary as the same amount of labor time is being expended and therefore no net gain has been achieved. Changes in labor mix, be they internal or external to the firm, may cause inappropriate data interpretations.

These situations may also exist if considerable capital has been spent on new tooling, engineering, and methods planning. We cannot separate direct labor hours from other cost elements. As the learning curve is only considering direct labor, changes in overhead may distort the picture. Management's use of the experience curve is an attempt to increase productive capacity; however, if the production rate is increased through sales volume, it may be possible to increase throughput by reorganizing direct labor as a function of this increased volume. Again, it may seem that experience is achieving cost savings when it may well be extraordinary efforts from marketing and sales.

Economic Implications

The experience curve continues to be a popular and useful tool for planning, budgeting, and controlling productive resources leading to increased profitability. Empirical evidence is mixed on the validity of the method. What appears to be unchallenged is that management continues to find the tool useful. This belief adds conviction and confidence to decision making and is certainly valuable in that light. With current thrusts for speed, flexibility, and ever-shortening product life cycles, it may be thought that experience curves are no longer necessary. However, the evidence to date is that early advantages accrue to those firms who take advantage of experience-based cost savings before the competitors' industry knowledge catches up. This time and cost advantage may lead to significant gains in market share, which will be difficult for lagging firms to overcome.

Defining Terms

Experience curve: An analytical tool designed to quantify the rate at which experience of accumulated output to date affects total lifetime costs. This term is broader than **learning curve** in respect to the costs covered, the range of output during which the reductions in costs take place, and the causes of cost reduction. (Adapted from Hall and Howell [1985] and Starr [1996]).

Learning curve: An analytical tool designed to quantify the rate at which cumulative experience of labor hours or cost allows an organization to reduce the amount of resources it must expend to accomplish a task. This function is usually described as with each doubling of productive unit output, the time per unit decreases at a predictable rate (Adapted from Melnyk and Denzler [1996]).

References

Abernathy, W. J. and Wayne, K. Limits of the learning curve, *Harv. Bus. Rev.*, 52: 109–119, 1974.

Amit, R. Cost leadership strategy and experience curves, *Strat. Mgt. J.*, 7: 281–292, 1986.

Andress, F. J. The learning curve as a production tool, *Harv. Bus. Rev.*, (Jan.-Feb.): 87–97, 1954.

Hall, G. and Howell, S. The experience curve from the economist's perspective, *Strat. Mgt. J.* 6: 197–212, 1985.

Lieberman, M. B. The learning curve, technology barriers to entry, and competitive survival in the chemical processing industries, *Strat. Mgt. J.* 10: 431–447, 1989.

Melnyk, S. A. and Denzler, D. R. *Operations Management: A Value-Driven Approach*, Richard D. Irwin, Chicago, 1996.

Senge, P. M. *The Fifth Discipline: The Art and Practice of the Learning Organization*, Doubleday Currency, New York, 1990.

Starr, M. K. *Operations Management: A Systems Approach*, Boyd & Fraser, Danvers, MA, 1996.

Wright, T. P. Factors affecting the cost of airplanes, *J. Aeronaut. Sci.*, 3(4): 122–128, 1936.

Yelle, L. E. The learning curve: historical review and comprehensive survey, *Dec. Sci.*, 10: 302–328, 1979.

For Further Information

For a nice overview of the development of the learning curve over time, the series published by the Harvard Business Review is excellent: *The Learning Curve as a Production Tool* (Jan.-Feb., 1954) by Frank J. Andress; *Profit from the Learning Curve* (Jan.-Feb., 1964) by Alfred B. Hirschmann, and *Limits of the Learning Curve* (Sept.-Oct., 1974) by William J. Abernathy and Kenneth Wayne.

13.7 Just-in-Time Purchasing

James P. Gilbert

Management of technology exists today within a business climate where manufacturers and service providers are driven by a passion for satisfying customers. We enhance the customers' lives by creating products and services that satisfy needs and wants as well as solve problems. Therefore, technology companies are looking for ever-increasing speed from the point of new product or service conceptualization to market availability for customers. Speed and flexibility of product offerings are aided by technology and drive new technology initiatives.

Product and service providers are increasingly becoming assemblers of items purchased from suppliers, as opposed to fabricators of significant numbers of parts. The make/buy decision is moving to the purchase side. We are buying the quality, process capability, and expertise of suppliers. Suppliers become critical to the success of our businesses and to the success of our efforts to deliver with speed and flexibility to our ultimate consumers. **Just-in-time (JIT) purchasing** develops the buyer/supplier relationships through true partnerships that aid both parties to increase profit performance and develop market share domination.

We are told by the National Academy of Engineering that there are eight major needs in the management of technology [McGaughey, 1989]:

- How to integrate technology into the overall strategic objectives of the firm.
- How to get into and out of technology faster and more effectively.
- How to assess/evaluate technology more effectively.
- How to best accomplish technology transfer.
- How to reduce new product development time.
- How to manage large, complex, and interdisciplinary or interorganizational projects.
- How to manage the organization's internal use of technology.
- How to leverage the effectiveness of technical professionals.

TABLE 13.8 Just-in-Time Purchasing Characteristics

Suppliers	Quantities
A few, nearby suppliers	Steady output rate (a desirable prerequisite)
Repeat business with same suppliers	Frequent deliveries in small lot quantities
Active use of analysis to enable desirable suppliers to become/stay price competitive	Delivery quantities variable from release to release but fixed for whole contract term
Clusters of remote suppliers	Long-term contract agreements
Competitive bidding mostly limited to new part numbers	Suppliers encouraged to reduce their production lot sizes
Buyer plant resists integration of supplier business	Little or no permissible overage of receipts
Suppliers encouraged to extend JIT buying to their suppliers	Suppliers encouraged to package in exact quantities
	Minimal release paperwork

Quality	Shipping
Minimal product specifications imposed on supplier	Use of company-owned or contract shipping, contract warehousing, and trailers for freight consolidation/storage where possible
Suppliers helped to meet quality specifications	Scheduling of inbound freight by buyer
Close relationships between buyers' and suppliers' quality assurance people	
Suppliers encouraged to use statistical process control instead of lot sampling inspection	

Source: Schonberger, R. J. and Gilbert, J. P. *Calif. Mgt. Rev.,* 26(1): 58, 1983.

JIT purchasing plays an important role in the accomplishment of these objectives by developing buyer/supplier relationships from the raw material suppliers, midlevel value-added suppliers, our firm's quality processes, and on to the final customer. Some have suggested that JIT purchasing is just a way of pushing inventory down the supply chain. Nothing could be further from the truth. The goal is to make the entire chain lean and efficient from supplier to customer to the next supplier and on to the ultimate customer so that speed and flexibility are integrated throughout the fabrication and service systems.

Fundamentals of JIT Purchasing

The major challenge facing purchasers in technology-driven companies is to proactively respond to multiple demands for (1) quality performance, (2) purchasing and inventory management through manufacturing resource planning and JIT production, (3) buyer-supplier partnerships, and (4) electronic data interchange [Farrell, 1990]. Purchasing professionals continue to apply the most sophisticated techniques in close cooperation with both the using departments and suppliers.

An emergence of purchasing's use of JIT was noticeable in the literature in the early 1980s. The term "JIT purchasing" was coined by Schonberger and Gilbert [1983] where the authors explained Japanese JIT purchasing practices and their benefits and showed that, despite obstacles, these practices could work in Western companies. Hahn [1983] provided an overview of the JIT approach (including purchasing) used by the Japanese and also highlighted the impact that JIT's implementation might have on the purchasing function in U.S. firms. Today, many Western firms have switched from traditional purchasing practices to the JIT purchasing concept [Lee and Ansari, 1986].

A number of characteristics of JIT purchasing are listed in Table 13.8. The JIT characteristics are interrelated and illustrated with four groups: suppliers, quantities, quality, and shipping.

Buyer/Supplier Partnership

The buyer/supplier partnership is the process and product of merging efforts to be responsive to each other's needs while conducting business in a way that provides maximum potential for growth and profit by both parties. This definition implies (1) communicating and planning as a team, (2) setting up and implementing a sourcing policy that makes sense to the buyer and the responsive suppliers, (3) fostering the building of long-term relationships, (4) emphasizing reduction of buyers' and suppliers' costs, not prices, and (5) the embodiment of JIT purchasing principles to improve continuously both parties' purchasing practices.

The primary underlying factors that describe a JIT partnership-oriented, buyer-supplier relationship are

1. Improved joint communication and planning efforts.
2. Improved responsiveness of receiving incoming purchased items benefiting both parties.
3. A sourcing policy beneficial to both, usually single or dual sourcing.
4. A more win-win approach to agreements and negotiations exchange whereby a mutually beneficial approach to the negotiations and the follow-up working relationship is enhanced.
5. Reducing costs for buyer and supplier — reduce both parties' costs and get better-quality processes and products moving through both systems.
6. Less concern with buyer's price — gaining the lowest price does not necessarily take the supplier's interests and needs into account.
7. Apply a concerted effort to use joint buyer/supplier efforts to improve relationships.
8. Become educated, involved in, and committed to increased use of JIT purchasing.

Improved Joint Communication and Planning Efforts

There is a need for improved joint communication and enhanced joint planning efforts between buyers and suppliers. Many buyers have recognized this and are experimenting with tighter integration with supplier planning and scheduling systems. Recently, electronic data interchange (EDI) has gathered momentum and has become the backbone of communications. This computer-to-computer communication interface eliminates dependence on the mail by allowing instant transmission of business documents. This has greatly enhanced many purchasing departments as EDI is the link between the company and its suppliers. Its role as the prime communication channel cannot be overemphasized.

Improved Responsiveness of Receiving

These aspects of the definition of the buyer/supplier partnership relationship deals with the buyer's response once purchased parts enter the receiving area. The importance of an integrated approach to the flow of both materials and information from the supplier to the buyer is crucial.

One of the first priorities of any plant should be to improve incoming product quality. Rather than attempting to inspect quality into a product, managers today must design and then purchase quality into the product. However, in order to meet the demand for more frequent delivery schedules and for zero-defect parts, and to decrease risk and uncertainty, companies must be willing to make long-term commitments, share engineering changes, supply the vendors with delivery schedules, and exchange product expertise. Improved quality benefits both buyer and supplier, as do reduced inspections. The quest for quality is a key ingredient in the sourcing policy of the buyer.

Beneficial Sourcing Policy

The first step in selecting suppliers is to identify all potential sources that appear capable of supplying the position. Deming [1986] believed that firms are better off with a single-source supplier. However, most manufacturers traditionally have preferred to work with multiple suppliers for fear that single sourcing could result in supply disruption.

While single sourcing offers some excellent opportunities for reducing costs and gaining control of purchases, there are some potential problem areas. These include the erosion of the supplier base for the buyer and, for the supplier, loss of technological thrust, excess control, and loss of supplier identity. Many manufacturers are reducing their number of suppliers so as to control quality, and they will give preference to those close to home. Small manufacturers feel less isolated because large companies are involved in their problems. A company can achieve enhanced performance from its suppliers by reducing their number and creating full partnerships that are supported by management commitment and trust on both sides. As a true partner, a buyer and supplier will attempt to reach more positive, jointly beneficial agreements.

Win-Win Approach to Agreements and Negotiations

One of the more prevalent attributes of a JIT purchasing partnership arrangement is a long-term relationship. With JIT purchasing, it is far more profitable and reliable to develop long-term relationships in which suppliers are partners, not victims. The result is a "win-win" situation for buyer and supplier. Long-term commitments providing exceptionally good communications also provide a strong under- standing of the business. Garnering those fewer but better suppliers is a crucial issue that some authors believe is vital to beginning a partnership arrangement with a supplier. Hahn et al. [1990] conclude that it is necessary to plan carefully and select competent suppliers in terms of quality, delivery, and cost capabilities. One essential task in setting up JIT is to integrate the suppliers into the overall strategy. Therefore, the process of vendor evaluation and selection is a time-consuming and critical aspect of establishing partnerships. A stronger partnership between manufacturers and suppliers is needed if purchased components are to be delivered to assembly lines ready for production, in exact quantities needed, and with zero defects (please see the section below on supplier certification).

The adversarial nature of face-to-face confrontations changes in a partnership-oriented relationship. Agreements based on competitive bidding using short-term contracts do little to fortify strong working relationships between the buyer and supplier. Furthermore, not attempting to gain the "upper hand" in negotiations is key to negotiating with a partner. This entails using joint problem solving and the desire to reach a win-win approach whereby every negotiation participant leaves the table with an agreement that both parties are comfortable with. One way to reach a win-win situation is through joint efforts to decrease buyer and supplier costs.

Reducing Costs for Buyer and Supplier

As buyers and suppliers work together in partnership to improve quality, they will inevitably find that prices and costs go down for both the buyer and the supplier. Executives believe that purchasing is no longer just a process of buying things at the best price. Rather, purchasing has evolved into a whole new management profession. Buying solely on the basis of the lowest price can often be one of the worst vendor selection decisions. Many buyers are choosing vendors on the least total cost method, which is based on their quality, on-time delivery, technical design coordination, sales representatives' assistance, product training workshops, and other factors (including price). In 1988, NCR Corporation was chosen as the sixth winner of *Purchasing's* Medal of Professional Excellence because of its enlightened approach to supplier relations and an emphasis on purchasing that shifted from price to value.

Less Concern with Buyer's Price

Many buyers are experimenting with various new approaches which are less "price" oriented, such as supplier base reduction and single or limited sourcing arrangements. Ishikawa [1986] claimed that at least 70% of the blame for defective purchased material lies with the purchasing organization driven by cost. Deming's [1986] view was that buyers have new responsibilities to fulfill, one of which is to end the practice of awarding business solely on price.

The lowest price embodies the philosophy of an adversarial-based approach to purchasing whereby the buyer wants the purchased items at the lowest price. Looking at this attribute from a supplier's standpoint, one would quickly conclude that to get the lowest price 100% of the time from the most qualified supplier may be difficult. The lowest price may not benefit both parties — it likely only serves the objectives of the buyer. This unilateral interest will likely erode any possibility of strong, long-term relationships.

Joint Buyer/Supplier Efforts to Improve Relationships

Much of the success of JIT is in the area of improved buyer/supplier relationships. Buyers benefit by having fewer suppliers to worry about and reduced labor in handling parts, while sellers benefit by concentrating on fewer and more productive sales calls. It is imperative to change the attitude that priorities and goals of the supplier are substantially different from those of the buyer, that buyers deal with many suppliers and change them frequently, and that buyer-seller transactions are random events with each transaction standing on its own.

Dobler et al. [1990] believe the most successful supplier management results when the buyer and the supplier view their relationship as a "partnership". Partnerships are based on mutual interdependency and respect. Respect permeates the partnership and replaces the adversarial attitudes present in too many buyer-supplier relationships.

Commitment is the most important factor in establishing good partnering agreements. However, because buyer-supplier relationships in the West have often been adversarial, we should be concerned that buyers will revert to former practices and drop suppliers after suppliers have made the necessary changes and investments to support JIT buyers.

The use of cooperative relationships with suppliers provides a method for purchasing managers to contribute positively to the strategic posture of their firms. They can do this by reducing the core supplier base to a few preferred suppliers and then by managing those suppliers accordingly. A close relationship with a supplier is possible if the company reduces the supplier base down to a manageable size. The real benefit of improved relationships with suppliers evolves from integrating them into the total business process of the company.

Education, Involvement, and Commitment to Increased Use of JIT Purchasing

Schonberger and Gilbert [1983] provide extensive lists of the positive impact that JIT purchasing can have upon an organization. For instance, Harley-Davidson reaped tremendous benefits after training its employees, management, and suppliers on JIT techniques. The employee is the key to the success of JIT purchasing implementation. Successful partnerships are built on a strong commitment to the importance of education in, and commitment to, the philosophy of JIT purchasing.

Supplier Certification

JIT purchasing partnership relationships often start with **supplier certification** programs. The processes of supplier certification are used to assure the customer that the supplier will consistently provide materials and processes that will meet and exceed all expectations and requirements. The strategy here is develop a win-win relationship between supplier and buyer which leads to increased market share, decreased costs, improved processes, quicker response times, increased reliability and conformance, and increased profits for both firms. Supplier certification is a joint effort of many skill areas within both firms. Often, the purchasing and quality assurance departments lead the effort, but engineering, marketing, and others are actively involved at various stages of the certification program.

The certification assessment process involves several in-depth, on-site investigations of the supplier's management expertise and practices as well as the overall stability of the supplier's business operations. The certification team will at a minimum review the supplier's: physical facilities, manufacturing capabilities, administrative systems, operating and order-processing systems, financial analysis, market data, customer satisfaction history, and technical support capabilities. Many detailed evaluations may occur during the certification process. Table 13.9 lists some of the areas of interest to the certifying firm.

The supplier certification team evaluates all data inputs relative to the decision and typically classifies suppliers into one of four categories: certified, conditionally certified, uncertified, and unevaluated. Supplier certification is an ongoing, never-ending process. Annual reviews are undertaken, as are periodic reviews as specific problems occur. These reviews must be satisfactory if the supplier is to maintain its certified status.

Conclusion

The rules, practices, and ideals are important to a JIT partnership among buyers and their suppliers. Without any one of them, the full potential may be more difficult to realize. Each purchasing organization should look inward before looking for help from its suppliers. If these eight underlying factors of a

TABLE 13.9 Supplier Certification Areas of Interest

Quality management processes	Quality history	Customer base	Quantity delivery history
Financial condition	Producibility	Cost control programs	Specification accuracy
On-time delivery record	Knowledgeable sales force	Education and training programs	Facilities and equipment
Labor conditions	Capacity availability and management	Market involvement	Organization of the firm
Research and development initiatives and capabilities	Prior and post-sales support	Tool tracking procedures	Preventative maintenance
Ethics	Subcontractor policies	Competitive pricing	Capability analysis
Management's commitment to the customer	Environmental programs	Policies and procedures	Percent of business partnership will represent
Housekeeping	Calibration history	Multiple plants	Geographical location

successful partnership are not being considered prior to asking the suppliers for substantial contributions, then the chances of a successful partnership will decrease. However, if the buyer is aware of the eight attributes, the buyer may be more willing to embrace a partnership-oriented relationship with the suppliers, and the chances for success in this area should improve.

Defining Terms

Supplier certification: Process used to identify those suppliers who the customer is assured will consistently provide materials and processes that conform to all expectations and requirements.

Just-in-time (JIT) purchasing: The process and product of merging buyer/supplier efforts to be responsive to each other's needs while conducting business in a way that provides maximum potential for growth and profit by both parties.

References

Deming, W. E. *Out of the Crisis,* MIT, Center for Advanced Engineering Study, Cambridge, MA, 1986.

Dobler, D. W., Lee, L., and Burt, D. N. *Purchasing and Materials Management: Text and Cases,* McGraw Hill, New York, 1990.

Ferrell, P. V. Purchasing into the '90s … and beyond, *Purchasing World,* 34(1): 27–29, 1990.

Hahn, C. K., Pinto, P. A., and Bragg, D. J. 'Just-in-time' production and purchasing, *J. Purch. Mater. Mgt.,* 26(7): 2–10, 1983.

Hahn, C. K., Watts, C. A., and Kim, K. Y. The supplier development program: a conceptual model, *J. Purch. Mater. Mgt.,* 26(2): 2–7, 1990.

Ishikawa, K. *Guide to Quality Control,* 2nd rev. ed., UNIPUB/Quality Resources, White Plains, NY, 1986.

Lee, S. M. and Ansari, A. Comparative analysis of Japanese just-in-time purchasing and traditional U.S. purchasing systems, *Int. J. Prod. Mgt.,* 5(4): 5–14, 1986.

McGaughey, N. W. Solving the technology puzzle, *Ind. Mgt.,* (July/August): 1989.

Schonberger, R. J. and Gilbert, J. P. Just-in-time purchasing: a challenge for U.S. industry. *Calif. Mgt. Rev.,* 26(1): 54–68, 1983.

For Further Information

An excellent source of information on purchasing in general, and JIT purchasing in particular, is the National Association of Purchasing Managers (NAPM). For information contact NAPM, P.O. Box 22160, Tempe, AZ 85285-2160, 800/888-6276 (http://www.napm.org.).

For nice summaries of the current state of JIT purchasing see Fawcett, S. E. and Birou, L. M. Just-in-time sourcing techniques: current state of adoption and performance benefits, *Prod. Inv. Mgt.,* first quarter, 1993; Goldhar, D. Y. and Stamm, C. L. Purchasing practices in manufacturing firms, *Prod. Inv. Mgt.,* third quarter, 1993.

13.8 Design, Modeling, and Prototyping

William L. Chapman and A. Terry Bahill

To create a product and the process that will be used to manufacture it, an engineer must follow a defined system design process. This process is an iterative one that requires refining the requirements, products, and processes of each successive design generation. These intermediate designs, before the final product is delivered, are called **models** or **prototypes**.

A model is an abstract representation of what the final system will be. As such, it can take on the form of a mathematical equation, such as $f = m \times a$. This is a deterministic model used to predict the expected force for a given mass and acceleration. This model only works for some systems and fails both at the atomic level, where quantum mechanics is used, and at the speed of light, where the theory of relativity is used. Models are developed and used within fixed boundaries.

Prototypes are physical implementations of the system design. They are not the final design, but are portions of the system built to validate a subset of the requirements. For example, the first version of a new car is created in a shop by technicians. This prototype can then be used to test for aerodynamic performance, fit, drivetrain performance, and so forth. Another example is airborne radar design. The prototype of the antenna, platform, and waveguide conforms closely to the final system; however, the prototype of the electronics needed to process the signal often comprises huge computers carried in the back of the test aircraft. Their packaging in no way reflects the final fit or form of the unit.

The System Design Process

The system design process consists of the following steps.

1. Specify the requirements provided by the customer and the producer.
2. Create alternative system design concepts that might satisfy these requirements.
3. Build, validate, and simulate a model of each system design concept.
4. Select the best concept by doing a trade-off analysis.
5. Update the customer requirements based on experience with the models.
6. Build and test a prototype of the system.
7. Update the customer requirements based on experience with the prototype.
8. Build and test a preproduction version of the system and validate the manufacturing processes.
9. Update the customer requirements based on experience with the preproduction analysis.
10. Build and test a production version of the system.
11. Deliver and support the product.

This can be depicted gradually on a spiral diagram as shown in Fig. 13.9.

The process always begins with defining and documenting the customer's needs. A useful tool for doing this is quality function deployment (QFD). QFD has been used by many Japanese and American corporations to document the voice of the customer. It consists of a chart called the "house of quality". On the left is listed what the customer wants. Across the top is how the product will be developed. These are often referred to as *quality characteristics*. The "whats" on the left are then related to the "hows" across the top, providing a means of determining which quality characteristics are the most important to the customer [Akao, 1990; Bahill and Chapman, 1993].

After the customer's needs are determined, the design goes through successive generations as the design cycle is repeated. The requirements are set, and a model or prototype is created. Each validation of a model or test of a prototype provides key information for refining the requirements.

For example, when designing and producing a new airborne missile, the initial task is to develop a model of the expected performance. Using this model, the systems engineers make initial estimates for the partition and allocation of system requirements. The next step is to build a demonstration unit of the most critical functions. This unit does not conform to the form and fit requirements but is used to

FIGURE 13.9   The system design process.

show that the system requirements are valid and that an actual missile can be produced. Requirements are again updated and modified, and the final partitioning is completed. The next version is called the *proof-of-design* missile. This is a fully functioning prototype. Its purpose is to demonstrate that the design is within specifications and meets all form, fit, and function requirements of the final product. This prototype is custom made and costs much more than the final production unit will cost. This unit is often built partly in the production factory and partly in the laboratory. Manufacturing capability is an issue and needs to be addressed before the design is complete. More changes are made and the manufacturing processes for full production are readied. The next version is the proof of manufacturing or the preproduction unit. This device will be a fully functioning missile. The goal is to prove the capability of the factory for full-rate production and to ensure that the manufacturing processes are efficient. If the factory cannot meet the quality or rate production requirements, more design changes are made before the drawings are released for full-rate production. Not only the designers but the entire design and production team must take responsibility for the design of the product and processes so that the customer's requirements are optimized [Chapman et al. 1992].

Most designs require a model upon which analysis can be done. The analysis should include a measure of all the characteristics the customer wants in the finished product. The concept selection will be based on the measurements done on the models. The models are created by first partitioning each conceptual design into functions. This decomposition often occurs at the same time that major physical components are selected. For example, when designing a new car, we could have mechanical or electronic ignition systems. These are two separate concepts. The top-level function — firing the spark plugs — is the same, but when the physical components are considered the functions break down differently. The firing of the spark plugs is directed by a microprocessor in one design and a camshaft in the other. Both perform the same function, but with different devices. Determining which is superior will be based on the requirements given by the customer, such as cost, performance, and reliability. These characteristics are measured based on the test criteria.

When the model is of exceptional quality a prototype can be skipped. The advances in computer-aided design (CAD) systems have made wire-wrapped prototype circuit boards obsolete. CAD models are so good at predicting the performance of the final device that no prototype is built. Simulation is repeated use of a model to predict the performance of a design. Any design can be modeled as a complex finite state machine. This is exactly what the CAD model of the circuit does. To truly validate the model, each state must be exercised. Selecting a minimum number of test scenarios that will maximize the number of states entered is the key to successful simulation. If the simulation is inexpensive, then multiple runs of this model should be done. The more iterations of the design process there are, the closer the final product will be to the customer's optimum requirements.

In the last two paragraphs, we mentioned functional decomposition and finite state machines as common design tools. There are many others and selection of the design tool is an important task [Bahill et al., 1998].

For systems where modeling works poorly, prototypes are better. Three-dimensional solids modeling CAD systems are a new development. Their ability to display the model is good, but their ability to manipulate and predict the results of fit, force, thermal stresses, and so forth is still weak. The CAD system has difficulty simulating the fit of multiple parts (such as a fender and car frame) because the complex surfaces are almost impossible to model mathematically. Therefore, fit is still a question that prototypes, rather than models, are best able to answer. A casting is usually used to verify that mechanical system requirements are met.

Computer-aided manufacturing (CAM) systems use the CAD database to create the tools needed for the manufacture of the product. Numerical control (NC) machine instructions can be simulated using these systems. Before a prototype is built, the system can be used to simulate the layout of the parts, the movement of the cutting tool, and the cut of the bar stock on a milling machine. This saves costly material and machine expenses.

Virtual reality models are the ultimate in modeling. Here the human is put into the loop to guide the model's progress. Aircraft simulators are the most common type of this product. Another example was demonstrated when the astronauts had to use the space shuttle to fix the mirrors on the Hubble telescope. The designers created a model to ensure that the new parts would properly fit with the existing design. They then manipulated the model interactively to try various repair techniques. The designers were able to verify fit with this model and catch several design errors early in the process. After this, the entire system was built into a prototype and the repair rehearsed in a water tank [Hancock, 1993].

Computer systems have also proved to be poor simulators of chemical processes. Most factories rely on design-of-experiments (DOE) techniques, rather than a mathematical model, to optimize chemical processes. DOE provides a means of selecting the best combination of possible parameters to alter when building the prototypes. Various chemical processes are used to create the prototypes that are then tested. The mathematical techniques of DOE are used to select the best parameters based on measurements of the prototypes. Models are used to hypothesize possible parameter settings, but the prototypes are necessary to optimize the process [Taguchi, 1976].

The progressive push is to replace prototypes with models because an accurate fully developed model is inexpensively simulated on a computer compared to the cost and time of developing a prototype. A classic example is the development of manned rockets. Figure 13.10 shows the number of test flights before manned use of the rockets.

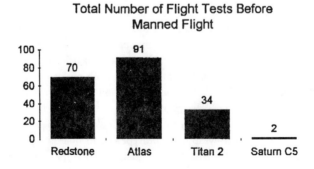

FIGURE 13.10   Flight tests for the manned space program rockets.

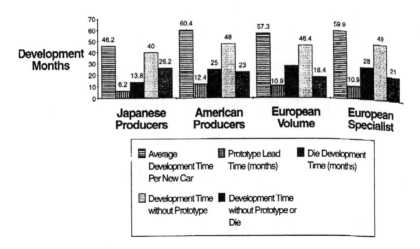

FIGURE 13.11 Comparison of Japanese, American, and European car producers. (*Source:* Womack, J. P., Jones, D. T., and Roos, D. *The Machine that Changed the World,* Rawson Associates, New York, 1990.)

The necessity for prototypes diminished rapidly as confidence in computer models developed. Initially, many of the rockets exploded in their first attempts at launch. As more was learned about rocketry, sophisticated models were developed that predicted performance of the rocket based on real-time measurements of valves, temperatures, fuel levels, and so forth. Using modern computers, the entire model could be evaluated in seconds and a launch decision made. This eliminated the need for many flight tests and reduced the cost of the entire Apollo moon-landing program.

Rapid Prototyping

Rapid prototyping is the key to reducing design time for parts and processes. Design is an iterative process. By creating prototypes quickly, the design can be completed faster. Japanese automobile manufacturers develop prototypes of their products in 6 months, whereas American companies take 13 months [Womack et al., 1990]. This advantage allows the Japanese companies to get to the market faster, or if they choose they can iterate their design one more time to improve their design's conformance to the customer's requirements. Japanese automakers' ability to create the prototype quickly is due in part to better coordination with their suppliers but also to the exacting use of design models to ensure that the design is producible.

As seen in Fig. 13.11, the lead in prototype development accounts for 44% of the advantage the Japanese producers have in product development time. The rest of the advantage if from rapid creation of the huge dies needed to stamp out the metal forms of the automobiles. The design of these important manufacturing tools is given as much attention as the final product. By creating flexible designs and ensuring that the teams that will produce the dies are involved in the design process, the die development time is cut from 25 months in the United States to 13.8 months in Japan. The rapid creation of the tooling is a key to fast market response.

Stereolithography is a new method of creating simple prototype castings. The stereolithography system creates a prototype by extracting the geometric coordinates from a CAD system and creating a plastic prototype. The solids model in the CAD system is extracted by "slicing" each layer in the z-axis into a plane. Each layer is imaged by a laser into a bath of liquid photopolymer resin that polymerizes with the energy from the laser. Each plane is added, one on top of the other, to build up the prototype. The final part is cured and polished to give a plastic representation of the solids model. It illustrates the exact shape of the part in the CAD system. This technique will not, however, verify the function of the final product because the plastic material used will not meet strength or thermal requirements [Jacobs, 1992].

Software developers also use rapid prototyping. This technique is used to get a fast though barely functional version of the product into the customer's hands early in the design cycle. The prototype is created using the easiest method to simulate functionality to the viewer. The customer comments on what is seen and the developers modify their design requirements. For example, when developing expert systems, models are almost never used. One of the driving rules is to show a prototype to the customer as soon as possible and afterwards throw it away! The purpose of building the prototype was to find out what the knowledge that needs to be represented is like so that the appropriate tool can be selected to build the product. If, as is usually the case, a nonoptimal tool was used for the prototype, then the prototype is thrown away and a new one is developed using better tools and based on better understanding of the customer's requirements. Beware, though — often a key function displayed in the prototype is forgotten when the prototype is abandoned. Be certain to get all the information from the prototype [Maude and Willis, 1991]. The fault with this technique is that the requirements are not written down in detail. They are incorporated into the code as the code is written, and they can be overlooked or omitted when transferred to a new system [Bahill et al., 1995].

When to Use Modeling and Prototyping

When should modeling vs. prototyping be used? The key difference is the value of the information obtained. Ultimately, the final product must be created. The prototype or model is used strictly to improve the final product. Costs associated with the prototype or model will be amortized over the number of units built. The major problem with models is the lack of confidence in the results. Sophisticated models are too complex for any single engineer to analyze. In fact, most models are now sold as proprietary software packages. The actual algorithms, precision, and number of iterations are rarely provided. The only way to validate the algorithm (not the model) is by repeated use and comparison to actual prototypes. It is easier to have confidence in prototypes. Actual parts can be measured and tested repeatedly, and the components and processes can be examined. Prototypes are used more often than models once the complexity of the device exceeds the ability of the computer to accurately reflect the part or process. During the initial design phases, models must be used because a prototype is meaningless until a concept has been more firmly defined. At the other extreme, modeling is of limited benefit to the factory until the configuration of the part is well known. A general rule is to build a model, then a prototype, then a production unit. Create even a simple mathematical model if possible so that the physics can be better understood. If the prototype is to be skipped, confidence in the model must be extremely high. If there is little confidence in the model, then a minimum of two or three prototype iterations will have to be done.

Defining Terms

Model: An abstract representation of what the final system will be. It is often a mathematical or simplified representation of a product.

Prototype: A physical representation of a product built to verify a subset of the system's requirements.

Stereolithography: A prototype-manufacturing technique used to rapidly produce three-dimensional polymer models of parts using a CAD database.

References

Akao, Y., Ed., *Quality Function Deployment: Integrating Customer Requirements into Product Design*, Productivity Press, Cambridge, MA, 1990.

Bahill, A. T. and Chapman, W. L., A tutorial on quality function deployment, *Eng. Mgt. J.*, 5(3): 24–35, 1993.

Bahill, A. T., Bharathan, K., and Curlee, R. F., How the testing techniques for a decision support system changed over nine years, *IEEE Trans. Systems, Man and Cybernetics*, 25: 1533–1542, 1995.

Bahill, A. T., Alford, M., Bharathan, K., Clymer, J., Dean, D. L., Duke, J., Hill, G., LaBudde, E., Taipale, E., and Wymore, A. W., The design-methods comparison project, *IEEE Trans. Systems, Man and Cybernetics*, Part C: Applications and Reviews, 28(1): 80–103, 1998. Simulations used in this project are available at http://www.sie.arizona.edu/sysengr/methods.

Chapman, W. L., Bahill, A. T., and Wymore, A. W., *Engineering Modeling and Design*, CRC Press, Boca Raton, FL, 1992.

Hancock, D., Prototyping the Hubble fix, *IEEE Spectrum*, 30(10): 34–39, 1993.

Jacobs, P. F., *Rapid Prototyping & Manufacturing: Fundamentals of StereoLithography*, McGraw-Hill, New York, 1992.

Maude, T. and Willis, G., *Rapid Prototyping: The Management of Software Risk*, Pitman, London, 1991.

Taguchi, G. *Experimental Designs, 3rd ed., vols. 1 and 2*, Maruzen, Tokyo, 1976.

Womack, J. P., Jones, D. T., and Roos, D. *The Machine that Changed the World*, Rawson Associates, New York, 1990.

Further Information

Pugh, S. *Total Design: Integrated Methods for Successful Product Engineering*, Addison-Wesley, London, 1990.

Suh, N. P. *The Principles of Design*, Oxford University Press, New York, 1990.

Wymore, A. W., *Model-Based Systems Engineering*, CRC Press, Boca Raton, FL, 1993.

13.9   Flexible Manufacturing

Layek Abdel-Malek and Lucio Zavanella

The birth of the global village poses a myriad of problems and challenges to manufacturers all over the world. Today's decerning consumers demand, in addition to customization and affordable prices, high-quality products as well as delivery in short lead times. As a result, *flexible manufacturing* has emerged as a strategic imperative for industry in order to succeed in the current competitive environment.

Many experts agree that the 20th century has witnessed the evolution of three inclusive strategic imperatives in manufacturing. The first one resulted from Fredrick Taylor, who had set forth the principles of scientific management [Taylor, 1911]. The implementation of its main principle, efficiency, in Ford's Motor company in 1913 led to a remarkable success. Soon after Ford's growth, many companies had focused on improving the efficiency of their plants. Later, in the 1950s, quality had become the second strategic imperative of the era. The consumers of that time had not only cared about low prices that resulted from high efficiency but they also valued the quality and the reliability of their goods.

In addition to efficiency and quality, and facilitated by advances in automation and computer technologies, **flexibility** and flexible manufacturing have lately emerged as the essential strategic imperative of the 1990s for manufacturers' viability.

Our taxonomy in this section is as follows: after this introduction, we define flexibility and its classes, present a historical background and definition of the current flexible manufacturing systems and their components, introduce the concept of telemanufacturing as a flexible paradigm, and conclude with a section for references and further readings.

Flexibility

Despite its importance in manufacturing environment and the numerous writings in this subject matter, flexibility means different things to different people. One of the widely accepted definitions of flexibility is the ability of the manufacturer to fulfill customers' demands in a timely fashion. The deliverables are expected to be customized products that enjoy high quality at affordable prices. Nevertheless, despite differences in its definition, practitioners and academicians seem to agree on flexibility major classes (attributes). Table 13.10 lists these classes and their brief descriptions. Of these listed flexibility classes, however, manufacturers appear to value more those of product mix, volume, process, and routing. It should be pointed out, though, that these flexibility classes are not necessarily independent of each other.

Evaluating a manufacturing system's flexibility is important for gauging a company's standing with respect to its competitors. Several methods and indices have been developed to measure flexibility and its attributes. The interested reader is referred to Abdel Malek et al. [1996], Benjaffar and Ramamrishnan

TABLE 13.10   Flexibility Classes

Flexibility Class	The ability to
Expansion	Add or reduce capacity easily and modularly
Machine	Change tools and fixtures to process a given part set
Mix	Absorb changes in product mix
Mix-change	Alter manufacturing processes to accomodate new part types
Volume	Operate economically at different output levels
Process	Interchange the ordering of operations for a given part type
Routing	Process a given part set on alternative machines
Programming	Alter basic operating parameters via control instructions
Communications	Transmit and receive information or instructions

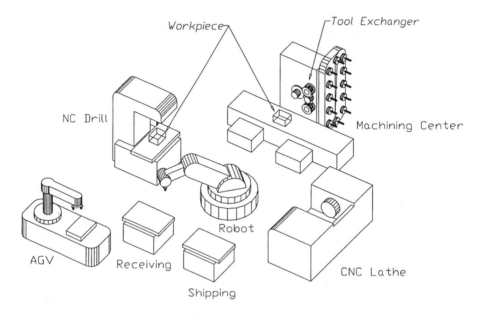

FIGURE 13.12   Example of a FMS.

[1996], Brill and Mandelbaum [1989], and Das [1996] for some of these measures. Also, further readings are provided in the reviews by Sethi and Sethi [1990] and Sarker et al. [1994].

The aforementioned discussion of flexibility and its classes is mainly pertinent to technology, where most of the published works have focused. However, it is equally important in attaining the agility of a manufacturing enterprise to have flexibility in its infrastructure as well as its labor force [Slack, 1987].

Flexible Manufacturing Systems

It is believed that the first **flexible manufacturing system (FMS)**, as it is known today, was designed in the early 1960s by Williamson and developed in 1965 by Sunstrand Corp. FMSs were conceived as a result of the ability to program the path of a cutting tool on a punched tape, introducing the concept of numerical control (NC). The off-line programmability of NC machine tools presented a significant evolution because of the ease of change between setups of different production runs. (This evolution in the concept of controlling machine tools was in contrast to that of the hard automation of manufacturing systems, which prevailed at that time, known as transfer lines. A transfer line consists of a set of special-purpose machines arranged in tandem, each carrying out single or multiple tasks. In a transfer line, the change between products requires a substantial effort and time because of the need to physically replace gadgets.)

Before defining the various types of FMSs, hereafter is a description of their most common components and subsystems (see also Fig. 13.12).

NC Machine

Developed at MIT in 1952, it is a machine tool guided by instructions that are punched on a tape and decoded by a reader interfaced with the tool positioning system.

Direct Numerical Control (DNC) of Machines

in the 1960s, technological advances in electronics allowed the control of several NC machines by one computer (or a hierarchy of them). This control system, referred to as DNC, stores in the central computer a database for the production cycles of the machines and monitors the plant's conditions. (Because of problems such as voltage fluctuations and possibility of fault of the mainframe computer, the use of this type of control system started to decline in the succeeding decade.)

CNC Machine (Computerized NC)

In the 1970s, the NC machine control was successfully implemented on an on-board computer. Because of the development of the powerful microcomputer, this controlling unit also attended to various machining functions (such as feed, depth of cut, etc.) and to tool paths, allowing the machine to perform various supervised trajectories around several axes at the same time.

Machining Center

It is an NC or CNC machine tool that is capable of executing different operations, such as drilling, boring, milling, etc. on different surfaces of the work piece. Several tools are arranged on a magazine. The machining center is usually equipped by an automatic tool and a workpiece exchanger.

Robot

Developed in the late 1950s, it is a reprogrammable multifunction manipulator that can perform a variety of tasks, such as material handling, welding, painting, drilling, machine loading and unloading, etc. Some of its most important characteristics are degrees of freedom, type of control (point-to-point, continuous path), payload, reach (envelope volume), repeatability, and accuracy. The type of use of the robot determines the importance of these characteristics.

Automated Guided Vehicles (AGVs)

Developed in the mid-1950s, they are computer-controlled carts for transporting jobs between stations on the factory floor. Their ability to select and alter routes and paths makes them suitable for transportation in a flexible environment. AGVs can be guided by wires or radio signals or optically by infrared sensors. These programmable transporters link different stations under computer control. Some AGVs are equipped with a robot arm.

Automated Storage/Retrieval System (AS/RS)

Its development is linked to that of minicomputer in the late 1960s. An AS/RS consists of several subsystems that operate under a central control system. Two of their most common types are the rack/container system and the carousel/bin system. In the first type, each of the items is stored in a set of containers or, alternatively, on pallets that are arranged in aisles and served by a shuttle. The shuttle automatically moves between aisles to bring or to pick up from the loading/unloading point the needed items. The other type is the carousel system. Usually, it is used for small and light products. Items are located in specified bins, and the whole carousel automatically rotates so that the needed bin is properly alligned in the proximity of the loading/unloading point to deliver or receive.

Different arrangements and combinations of the aforementioned resources result in various types of FMSs. Browne et al. [1984] categorize these systems into four types:

Type I, flexible machining cell, is a single CNC machine that interacts with material handling and storage systems.

Type II, flexible machining system, is a group of machining stations linked by a material handling system and connected by a flexible transport system. Part production and material flow are controlled online. Routes and scheduling are also controlled and adjusted according to the system's conditions.

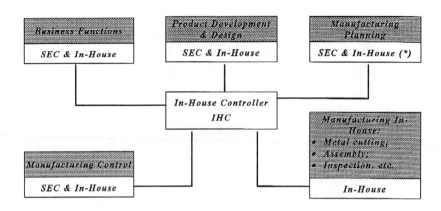

(*) *SEC & In-House:* acitivity could be performed In-House and/or by a SEC

FIGURE 13.13   The telemanufacturing enterprise.

Type III, flexible transfer lines, is a set of machines and workstations in tandem where each performs
   one operation under the control of a computer.
Type IV, flexible transfer multi-line, is the multiple case of the preceding category, where the various
   single flexible transfer lines are interconnected.

These types of systems have short setup times (due to the ability of off-line reprogramming and
preprogramming). This in turns allows the production of smaller batch sizes and large varieties of product
mix. Additionally, these systems offer capabilities for various processing sequences and machining alternatives of parts as well as the ability of producing a particular item via multiple routes. These characteristics, as mentioned before, are essential attributes of flexibility. Also, another positive characteristic of
these systems is the reduction of the amount of work in process. Nevertheless, these systems, especially
type II, require a large investment to acquire, and their flexibility is bound by the available technology
and budget at the time of installation. Moreover, the flexibility depends on the system designers' ability
to forecast the types and specifications of products that would be manufactured in the future. If market
conditions change from what was anticipated at the time of the design, and quite often they do, the
system acquired could suffer from lack of the needed flexibility. Upton [1995] reports incidences and
difficulties that were encountered where industries suffered because of the limitation of these FMSs.
   The leap in the information and computer technologies experienced in the mid-1990s is credited for
the renewed emphasize on flexible manufacturing and change in notion, that is, to achieve flexible
manufacturing, it is no longer necessary to have a highly automated environment on the factory floor
(as was the original notion in the early 1960s). The following section describes an emerging concept in
flexible manufacturing, which has been coined **telemanufacturing**.

Telemanufacturing

The premise of telemanufacturing is flexibility, adaptability, and affordability. The term telemanufacturing coins an infrastructure whereby a company can outsource several of its production and design
functions, particularly via information superhighways, rendering in real-time activities essential for
production of goods [Abdel Malek et al., 1996].
   As will be discussed later, this kind of structure avoids problems that are inherent in the types of FMSs
described before. Fig. 13.13 shows a schematic diagram for a telemanufacturing enterprise. In addition
to communication media (both internal and external), the telemanufacturing enterprise requires two
essential types of components: the in-house controller (IHC) and the other component is a set of service
providers such as houses specialized in design, production control, or customer billing. We designate
these houses as specialized expert centers (SECs). The following is a brief description.

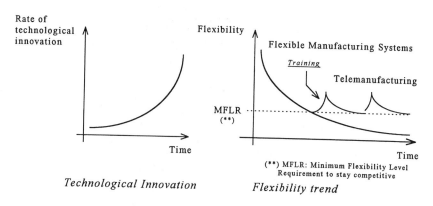

FIGURE 13.14 Telemanufacturing and flexibility level.

Specialized Expert Centers (SECs)

It is a center that possesses the state of the art in a certain field with various facets, whether it is software or hardware. Its human resources are well trained, enjoying advanced expertise in the respective fields, and have access to the latest pertinent information and development. Examples of the expert centers could be those research centers that currently exist in universities as well as those of EDS (electronic data systems). Today, many companies employ the latter type for their billing and payroll. To illustrate the former, consider university research centers that may be specialized in digital design or composite material; they can render their expertise to manufacturers that seek them. This provide the manufacturers with the option to select the center they wish to subscribe to. Also, manufacturers could choose the functions that they wish to perform themselves or outsource to SECs, according to their prevailing conditions and needs, consequently enhancing flexibility.

In-House Controller (IHC)

It is an essential control component that integrates the expert centers, manufacturing floor activities, and the other functions. Then it orchestrates the execution process according to predetermined protocols. The IHC consists of an in-house team and computers that have the necessary databases, software, and modules that interactively communicate with the different SECs and the manufacturing functions on the floor. It harmonizes the different decisions, and centralizes the final one, preceding execution. In essence, the IHC is the brain of the telemanufacturing enterprise.

Communication Media

There are two media necessary for the telemanufacturing enterprise: One for within the enterprise connections and the other for outside communication. The inside communication could be carried out using one of those commercially available for computer integrated manufacturing (CIM), such as local area network (LAN) or wide area network (WAN). Its basic function is to communicate between the IHC and the factory floor as well as between workstations.

As for the external media, the Internet could provide the enterprise with means to the outside connection, mainly to communicate between the IHC and the different SECs, or, depending on the frequency of usage, especially dedicated lines could be setup for the interaction.

The Logic of Telemanufacturing and its Relation to Flexibility Reenforcing

Telemanufacturing allows a company to remain flexible throughout. Since it permits a company to cosource some of its production functions to expert centers, consequently, as shown in Fig. 13.14, the flexibility level of the telemanufacturing enterprise is not dependent on the technological obsolescence of its systems' components. Its flexibility is mostly dependent on that of the SECs subscribed to. It is reasonable to assume

TABLE 13.11 A Comparison of Modern Manufacturing Concepts

	CIM Technology	Holonic Manufacturing	Telemanufacturing
Flexibility	Restricted by equipment	Restricted by equipment	Limited by state of the art
Adaptability	Workforce dependent	Workforce dependent	Semidependent on workforce
Investment	People, hardware, software	People, hardware, software	Subscription and low cost of software and hardware

that, in order for these SECs to survive, they must keep adopting new technologies; otherwise, subscribers can change to other service providers. Therefore, it can be conjectured that applying telemanufacturing sustains and reenforces flexibility of the enterprise. (Similar structures have been utilized with remarkable success. We cite the system developed by SAP and used in Boeing's aircraft programs.)

Additionally, it is plausible to assume that, because of technological innovation, a software or a machine that is very sophisticated (top of the line) today may not be the same tomorrow. Hence, if the company buys the equipment or the software, it will risk (depending on the equipment) its future obsolescence. Meanwhile, if it can use a SEC, in addition to avoiding training costs, initial investment, and time for acquisition, the problem of obsolescence will be that of the SEC (which has to maintain the upgrading of its equipment for survival). Table 13.11 provides a comparison between telemanufacturing and the various recent manufacturing structures vis a vis their flexibility, adaptability, and investment.

Acknowledgments

The first author appreciates the partial support provided by National Science Foundation (NSF) grant #DMI9525745 and the Italian Council of National Research (CNR).

Defining Terms

Flexibility: The ability of a manufacturer to efficiently respond to the volatile market needs in a timely fashion, delivering high-quality products at competitive prices.

Flexible manufacturing systems (FMS):   A set of reprogrammable machinery, automated material handling, storage and retrieval devices, and a host computer. This computer controls the various activities of the system.

Telemanufacturing: The infrastructure whereby a company can outsource several of its production and design functions, particularly via information superhighways, rendering in real-time activities essential for production of goods.

References

Abdel Malek, L., Wolf, C., and Guyot, P. Telemanufacturing: A Flexible Manufacturing Solution. *Preprints of 9th Int. Working Seminar on Production Economic,* on *Int. J. Prod. Econ.,* 3: 347–365, 1996.

Benjaffar, S. and Ramamrishnan, R. Modelling, measurement and evaluation of sequencing flexibility in manufacturing systems, *Int. J. Prod. Res.,* 34(5): 1195–1220, 1996.

Brill, P. H. and Mandelbaum, M. On measures of flexibility in machining systems, *Int. J. Prod. Res.,* 27(5): 5–23, 1989.

Browne, J., Dubois, D., Rathmill, K., Sethi, S., and Stecke, K. Classification of flexible manufacturing systems, *FMS,* 114–117, 1984.

Das, S. The measurement of flexibility in manufacturing systems, *Int. J. Flex. Manuf. Syst.,* 8: 67–93, 1996.

Sarker, B., Krshnamurthy, S., and Kuthethur, S. A survey and critical review of flexibility measures in manufacturing systems, *Int. J. Prod. Plan. Control,* 5(6): 512–523, 1994.

Sethi, A. K. and Sethi, S. P. Flexibility in manufacturing: a survey, *Int. J. Flex. Manuf. Syst.,* 2: 289–328, 1990.

Slack, N. The flexibility of manufacturing systems, *Int. J. Prod. Mgt.,* 1(4): 35–45, 1987.

Taylor, F. W. *Principles of Scientific Management,* Harper and Row, New York, 1911.

Upton, D. What really makes factories flexible?, *Harv. Bus. Rev.,* July-August: 74–84, 1995.

Further Information

Asfhal, C. R. *Robots and Manufacturing Automation,* John Wiley & Sons, New York, 1985.

Chang, T. C., Wang, H. P., and Wysk, R. *Computer-Aided Manufacturing,* Prentice Hall, Englewood Cliffs, NJ, 1991.

Groover, M. *Fundamentals of Modern Manufacturing Materials, Processes, and Systems.* Prentice Hall, Englewood Cliffs, NJ, 1996.

Gupta, D. and Buzacott, J. A. A framework for understanding flexibility of manufacturing systems, *J. Manuf. Syst.,* 8(2): 89–96, 1989.

Hitomi, K. *Manufacturing Systems Engineering,* Taylor & Francis, 1979.

Miller, R. K. *Automated Guided Vehicles and Automated Manufacturing,* SME, 1987.

Jamshidi, M., Lumia, R., Mullins, J., and Shahinpoor, M. *Robotics and Manufacturing: Recent Trends in Research, Education and Applications,* ASME Press, New York, 1992.

Kusiak, A. *Intelligent Manufacturing Systems,* Prentice Hall, Englewood Cliffs, NJ, 1990.

McAfee, A. and Upton, D. The real virtual factory, *Harv. Bus. Rev.,* July-August: 123–135, 1996.

Tompkins, J. A. and Smith, J. D. *The Warehouse Management Handbook,* McGraw-Hill, New York, 1988.

13.10   Design for Manufacturability

Xiaoping Yang and C. Richard Liu

In the launching of a new product, lead time, quality, and cost are the key factors that determine the competitiveness of the product. Consequently, shortening lead time, improving quality, and reducing cost are of paramount importance for any company to survive and excel. The significance of product manufacturability in achieving the forgoing goals has long been recognized and intensive efforts have been taken to improve product manufacturability. All the methods preceded **design for manufacturability (DFM)** such as **value analysis (VA)**, however, are *serial* in nature and do not enter consideration until the product design is completed. As a result, they incur high cost of design change (see Fig. 13.15). Moreover, they are not very thorough and cannot offset the poor decisions made during the early design stages. It has been recognized that design is the first step in product manufacture. Design decisions have greater influence on the cost of manufacturing than manufacturing decisions. Therefore, manufacturing issues should be considered as early as possible so as to make the right decisions the *first time.*

DFM is the practice that brings the consideration of manufacturing issues into early stages of product design process to insure that all issues are considered *concurrently* so that the product can be made in the least time with the desired level of quality and the least cost. DFM is no longer optional but *a must* for any company to be successful in today's competitive global market. DFM plays an important role in **computer-integrated manufacturing** and is an important tool for accomplishing **concurrent engineering (CE)**. It involves such procedures as ascertaining the customer's real needs, design for function, materials and processes selection, design for assembly (DFA), design for processes, design for reparability and maintainability, design for safety, and design for disposal and recycle.

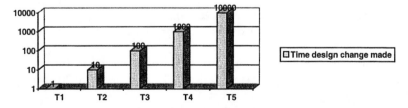

FIGURE 13.15   Comparative cost of an engineering change in different stages in the product cycle. T1: during design, T2: design testing, T3: process planning, T4: pilot production, and T5: final production. (*Source:* Shina, S. G. *Concurrent Engineering and Design for Manufacture of Electronics Products,* Van Nostrand Reinhold, New York, 1991.)

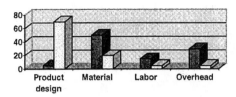
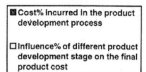

FIGURE 13.16   The product design has the greatest impact. (*Source:* Boothroyd, G. et al. *Product Design for Manufacture and Assembly,* Marcel Dekker, New York, 1994.)

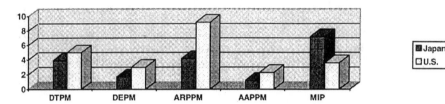

FIGURE 13.17   Comparison between Japanese and U.S. auto industry. DTPM: design time/model (years), DEPM: design effort/model (man hours), ARPPM: average replacement period/model (years), AAPPM: average annual production/model (×100,000), and MIP: models in production (×10). (*Source:* Shina, S. G. *Concurrent Engineering and Design for Manufacture of Electronics Products,* Van Nostrand Reinhold, New York, 1991.)

Why DFM and What Are the Benefits of Applying DFM?

Fragmenting a product design into subtasks in the traditional approach often leads to *conflicting decisions* made by different departments that are responsible only for their own duties and suboptimal design. Consequently, the product development cycle is long because iterations are needed to resolve decision conflicts between different departments, the quality is low, and the cost is high. In today's competitive world market, the company practicing the serial product design process finds itself losing profits and market share. It may even be wiped out the market. Therefore, DFM is *indispensable* for a company to survive and excel because, by transforming the *serial* process into a *parallel* one that facilitates the early resolving of conflicts among different departments and making decisions leading to the global optimization of the product design, DFM can shorten lead time, improve quality, and reduce cost.

It is emphasized that the DFM be considered in the early *conceptual design* stage to get the maximized benefits because 70% of the final product cost has been committed in product design. The later the engineering change, the higher the cost. Refer to Figs. 13.15 and 13.16. Figure 13.17 illustrates the main advantage of DFM. Detailed discussion is in Shina [1991].

DFM Guidelines

1. Design starts from ascertaining the *real* customer needs. It is important that the ascertained needs are real because unreal needs lead to increased complexity of the product and consequently result in increased manufacturing cost, decreased system reliability, and increased repair and maintenance cost. [Trappey and Liu, 1992].
2. Starting from the ascertained needs, the typical hierarchical structure of design includes conceptual design, configuration design, and parametric design. The design proceeds from one level to another in the hierarchy until the design is finished. In each level, design alternatives are generated and evaluated to find the best design candidate from the *global* point of view before it moves on to the next stage. See[Trappey and Liu [1992] for more information.
3. The best design is the *simplest* design that satisfies the needs. Design for function that aims at satisfying the minimum required function in the most effective manner can lead to simplified

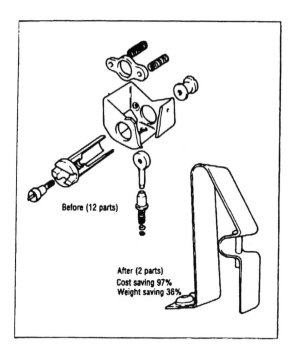

Before (12 parts)

After (2 parts)
Cost saving 97%
Weight saving 36%

FIGURE 13.18 Improved design of a locking mechanism by design for function. (*Source:* Jacobs, G. *Engineering,* February: 178–182, 1980.)

designs, the benefits of which are lower cost and improved quality and reliability. See Fig. 13.18 for an example.

4. It is emphasized that the interaction between materials and processes be considered in materials and processes selection because only certain combinations of manufacturing processes and materials are possible. The general considerations of materials selection are properties, cost, and availability. The general considerations for selecting manufacturing processes are materials, dimensional and surface finish, operational and manufacturing cost, and production volume. The selection process ends up with a compatible combination of materials and processes. When more than one materials/processes combination satisfies design requirements, all alternatives are ranked according to various criteria and the best selected. Boothroyd et al. [1994] summarize compatibility between processes and materials and shape generation capabilities of processes with DFA compatibility attributes. Cost factors for process selection are materials, direct labor, indirect labor (setup, etc.), special tooling, perishable tools and supplies, utility, invested capital, etc. Every effort should be made to minimize the *total* unit cost in the selection process. The setup cost and special tooling cost are production-volume dependent. For a product of low volume, the number of operations should be kept minimum because the setup cost accounts for a large percentage of the total product cost. See Bralla [1986] and Liu and Mittal [1996]. Tolerance plays an important role in product cost. The tighter the tolerance and surface finish, the higher the cost. Refer to Figs. 13.19 and 13.20.

5. The goal of DFA is to ease the assembly of the product. Boothroyd et al. propose a method for DFA that involves two principal steps:

   - Designing with as few number of parts as possible. This is accomplished by analyzing parts pair-wise to determine if the two parts can be created as a single piece rather than as an assembly.
   - Estimating the costs of handling and assembling each part using the appropriate assembly process to generate costs figures to analyze the cost savings through DFA.
   - In addition to the assembly cost reductions through DFA, there are reductions in part costs that are often more significant. Other benefits of DFA includes improved reliability and reduction

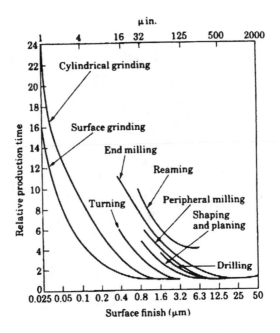

FIGURE 13.19   Relative production time as a function of surface finish produced by various manufacturing methods. (*Source*: Kalpakijian, S. *Manufacturing Processes for Engineering Materials, 3rd ed.*, Addison-Wesley, Menlo Park, CA, 1996.)

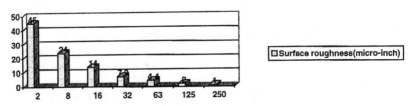

FIGURE 13.20   Relative cost corresponding to different surface roughness. (*Source*: Bralla, J. G. *Handbook of Product Design for Manufacturing*, McGraw-Hill, New York, 1986.)

in inventory and production control costs. Consequently, DFA should be applied regardless of the assembly cost and production volume. See Boothroyd et al. [1994] for more information.

6. Design for processes:

Design for machining: Machining is almost always needed if precision is required. The great flexibility is another advantage of machining, and it is particularly economical for products of low volumes. As there are always some materials wasted in machining, it becomes less favorable as the product volume increases. Avoid machining processes whenever possible. If machining has to be involved, then make every effort to minimize the materials wasted. Do not use tight dimensional tolerances and surface finish unless required by product function. Avoid expensive secondary operations such as grinding, reaming, and lapping if possible. Avoid features that need special cutters. Make sure that the features to be machined are accessible (see Fig. 13.21A). Design parts such that the number of machining operations are minimal (see Fig. 13.21B). Refer to Boothroyd et al. [1994] and Bralla [1986].

Design for metal forming: A major advantage of metal forming over machining is that there is little or no material waste in the processes. Materials savings become more significant as the raw materials become higher priced. However, the special tooling is usually a significant

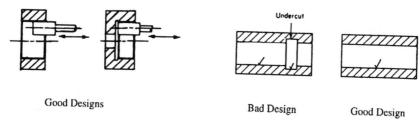

Good Designs   Bad Design   Good Design

FIGURE 13.21 Examples of good designs and bad designs for machining. (A) Internal grinding: make sure the feature to be machined is accessible. (*Source:* Boothroyd, G. et al. *Product Design for Manufacture and Assembly,* Marcel Dekker, New York, 1994.) (B) Compare the bad design with undercut and the good design without undercut. (*Source:* Bralla, J. G. *Handbook of Product Design for Manufacturing,* McGraw-Hill, New York, 1986.)

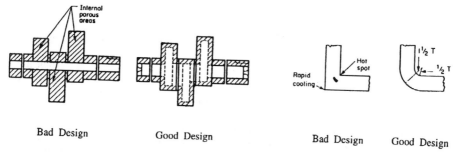

Bad Design   Good Design   Bad Design   Good Design

FIGURE 13.22 Examples of good designs and bad designs for casting. (A) Using uniform cross sections. (B) Rounding corners to avoid the hot spot. (*Source:* Bralla, J. G. *Handbook of Product Design for Manufacturing,* McGraw-Hill, New York, 1986.)

investment that can only be justified by a large volume of production. Because the materials flow plastically in the solid state, irregularities and intricate shapes should be limited in design and sharp corners avoided. Metal extrusion is particularly advantageous in producing long part with constant intricate cross-section. Because its relative low cost of tooling and its capability to incorporate irregular cross-sectional shapes in the extrusion dies such that machining could be eliminated, it could be justified for short runs. Avoid thin-walled extrusion and balance sections walls for extrusion. Stamping is only suitable for high-volume-production. Minimize bend stages and maximize stock utilization for stamping. A special benefit of forging process is that the grain structure can be controlled so as to improve physical properties of the metal. Aligning the parting line in one plane perpendicular to the axis of die motion if possible. See Boothroyd et al. [1994] and Bralla [1986].

Design for casting: Two major advantages of casting are (1) producing intricate shapes such as undercuts, complex contours, and reentrant angles and (2) allocating materials according to stress distributions of the parts. Depending on the actual mold used, its suitable production volume ranges from low to high. The major design considerations for casting are flow and heat transfer of the material. Try to design the parting line on a flat plane if possible, use uniform cross-section (Fig. 13.22A), and round corners to avoid hot spots (Fig. 13.22B). See Boothroyd et al. [1994] and Bralla [1986].

Design for Injection: For intricate parts that are not subject to high stresses, injection molding is the best candidate. It is a process for large-volume-production. Allow sufficient spacing for holes and making the main wall of uniform thickness. See Boothroyd et al. [1994] and Bralla [1986].

7. Reparability, maintainability, human factors, and safety issues are discussed in [Anderson, 1990]. See also Compton [1996] for general discussions on DFM.

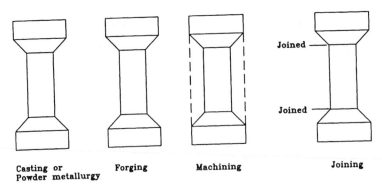

Casting or   Forging   Machining   Joining
Powder metallurgy

FIGURE 13.23   Different manufacturing processes for the same part.

TABLE 13.12   Comparison of Different Processes

Possible Manufacturing Processes	Casting, Powder Metallurgy	Forging	Machining	Joining (Welding, Brazing, Bonding)
Special tooling	Yes	Yes	No	No
Materials wasted	No	No	Yes	No
Comments	Net, near-net shape processing; good for large-volume production	Net-shape processing; improved strength; good for large-volume production	Economical for low-volume production; choose the materials such that materials wasted be minimal	Most appropriate choice if the tops and middle portion are made of different metals

Integration of CAD and DFM

The *greatest* benefits of DFM can be attained when DFM guidelines are applied in *early* design stages, while the current CAD system has been oriented to detail design. Research is needed to lay the foundation for the CAD system for the conceptual design so that DFM and CAD can be integrated successfully. Mukherjee and Liu [1995] proposed a method that is promising.

Application Examples

Process Selection

The product to be manufactured is shown in Fig. 13.23 and the materials are metal to satisfy the functional requirements. To manufacture a cost-competitive, high-quality product, it is important that the part be produced at net or near-net shape, which eliminates much secondary processing and reduces total manufacturing time and lower the cost. The process selection also depends on production volume and rate. See Table 13.12 for a comparison of different processes.

Reticle Assembly — The Amazing Improvement of the Product Design through DFA

On the left side of Fig. 13.24 was the original design of the reticle assembly by Texas Instrument and on the right side was the improved design. The comparison of the two design is shown in Fig. 13.25. See Boothroyd, et al. [1994] for detailed discussion.

Evaluation of Manufacturability for Stamped Part in Conceptual Design

Figure 13.26 is an example showing that the design with less bend stage is a better design for stamping. To represent parts for early manufacturability evaluation in computer is an active research area. Figure 13.26 is one way to achieve this goal. See Mukherjee and Liu [1997] for more information.

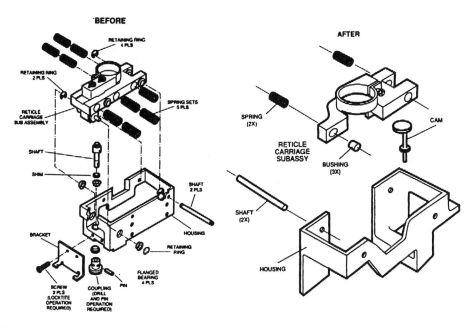

FIGURE 13.24 (*Source:* Boothroyd, G. et al. *Product Design for Manufacture and Assembly,* Marcel Dekker, New York, 1994.)

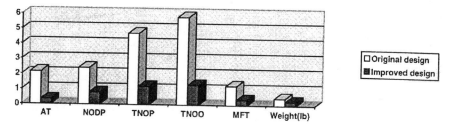

FIGURE 13.25 Comparison of original and new designs of the reticle assembly. AT: Assembly time (h), NODP: Number of different parts (×10), TNOP: Total numbers of parts (×10), TNOO: Total number of operations (×10), and MFT: Metal fabrication time (h). (*Source:* Boothroyd, G. et al. *Product Design for Manufacture and Assembly,* Marcel Dekker, New York, 1994.)

Defining Terms

Computer-integrated manufacturing: A methodology to integrate all aspects of design, planning, manufacturing, distribution, and management by a computer.

Concurrent engineering (CE): To integrate the overall company's knowledge, resources, and experience in design, development, marketing, manufacturing, and sales at the earliest possible stage so that the product developed can be brought to market with least time, high quality, and low cost, while meeting customer exceptions.

Design for manufacturability (DFM): The practice that brings the consideration of manufacturing issues into early stages of product design process to insure that all issues are considered concurrently so that the product can be made in the least time with the desired level of quality and the least development cost.

Value analysis (VA): A system of techniques that use creativity and the required knowledge to provide each function for lowest cost by identifying function and evaluating function.

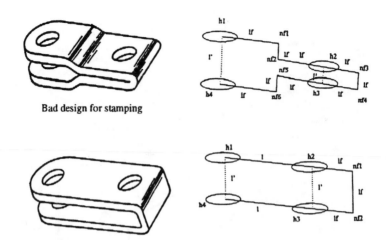

Bad design for stamping

FIGURE 13.26 (A) Examples of stamping parts. (B) Parts sketching abstraction that facilitates manufacturability evaluation in conceptual design. (*Source:* Mukherjee, A. and Liu, C. R. *Res. Eng. Des.,* 7: 253–269, 1995.)

References

Anderson, D. M. *Design for Manufacturability,* CIM Press, Lafayette, CA, 1990.

Boothroyd, G. et al. *Product Design for Manufacture and Assembly,* Marcel Dekker, New York, 1994.

Bralla, J. G. Handbook of Product Design for Manufacturing, McGraw-Hill Book Company, 1986.

Compton, W. D. *Engineering Management, Creating and Managing World-Class Operations,* pp. 264–307, Prentice-Hall, Upper Saddle River, NJ, 1997.

Jacobs, G. Designing for improved value, *Engineering,* February: 178–182, 1980.

Kalpakijian, S., *Manufacturing Processes for Engineering Materials, 3rd ed.,* Addison-Wesley, Menlo Park, CA, 1996.

Liu, C. R. and Mittal, S. Single-step superfinishing hard machining: feasibility and feasible cutting conditions, *Robot. Comput. Integr. Manuf.,* 12(1): 15–27, 1996.

Mukherjee, A. and Liu, C. R. Representation of function-form relationship for conceptual design of stamped metal parts, *Res. Eng. Des.,* 7: 253–269, 1995.

Mukherjee, A. and Liu, C. R. Conceptual design, manufacturability evaluation and preliminary process planning using function-form relationship in stamped metal parts, *Robot. Comput. Integr. Manuf.,* in press.

Shina, S. G. *Concurrent Engineering and Design for Manufacture of Electronics Products,* Van Nostrand Reinhold, New York, 1991.

Trappey, A. J. C. and Liu, C. R. An integrated system shell concept for design and planning, *J. Des. Manuf.,* 2: 1–17, 1992.

13.11 Computer-Aided Design and Manufacturing

Hamid Noori

Computer-aided design (CAD) and manufacturing (CAM) are defined by the use of computer software and hardware to assist in the manufacturing process. Inasmuch as the software is used to "drive" the production machinery in CAM, CAD is the use of software and hardware to assist in the design process.

CAM is not an entity unto itself, but simply the use of computers in the context of the manufacturing process. According to the *Automation Encyclopaedia,* CAM involves many disparate activities, including on-line planning, computer-numerical control (CNC), automated assembly, computer-aided process planning, scheduling, tool design and production, automated materials handling, and robotics.

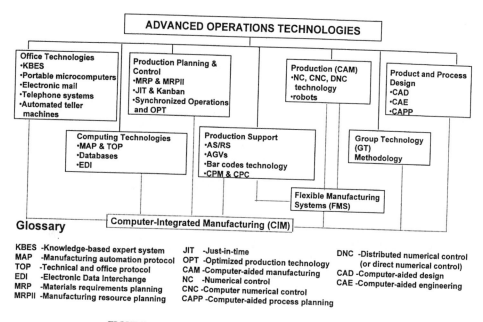

FIGURE 13.27   A list of advanced operations technologies.

CAD/CAM is a fundamental requirement for the implementation of a computer-integrated manufacturing (CIM) or, ultimately, an enterprise-wide manufacturing system (see Fig. 13.27).

A Brief History of Automated Production Processes

The use of automation in manufacturing can be traced back to the early 1800s, when Joseph Jacquard produced a fabric loom capable of running with a card system to produce patterned fabric. The mechanization of the factory proceeded throughout the 1800s and early 1900s. The introduction, in the early 20th century, of the assembly line created the perfect medium to accelerate the growth of controlled production processes.

The introduction of numerical-control (NC), in the 1920s and 1930s, expanded the application of card-directed production. By the late 1950s and early 1960s the use of a "computer-directed" manufacturing process had become fairly widespread. With the evolution of computers came the vastly expanded application of CNC production.

Moving into the late 1960s and early 1970s, as the cost of automation and computer-assisted applications dropped, the uses of CNC expanded rapidly. The wider availability of personal computers (PC) in office and manufacturing settings created the next wave of application adjustments to the automation of the production process.

Concurrently, the development of CAD in the late 1960s began what can be best described as the second prong of the CAD/CAM development process. The 1970s saw the increased integration of the two applications, so that they are now thought of as permanently linked.

What Is CAD?

CAD refers to the use of a computer to create or modify an engineering design. Traditionally, the designs, and supporting tooling for a product, were done on a drawing board. An engineer would prepare a blueprint manually for the total product, including the tooling to produce the product. Other engineers would provide the drawings for their specialized area, (i.e., electrical). These would be used throughout the product development process and updated or changed as the process continued. This process was

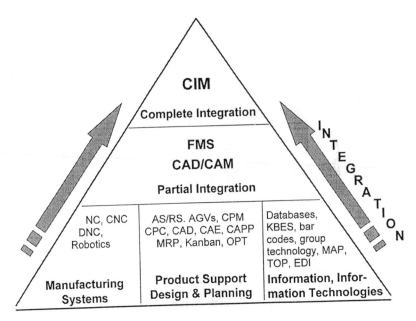

FIGURE 13.28 CIM.

very costly in terms of time and money. CAD has improved the process significantly and reduced dramatically the development costs and time.

What Is CAM?

CAM is a generic term used to describe the complete range of computer applications in direct manufacturing activities. At the heart of CAD/CAM is the linkage of the design process to the machine programs of the manufacturing facility. The effectiveness and efficiency of this linkage has been one of the prime reasons for enhancing the accuracy and reliability of the product development process, shortening the manufacturing process times, and hence shortening the product introduction time and improvement in overall productivity.

Computer-Integrated Manufacturing

The integration of CAD/CAM with other technologies such as computer-aided engineering (CAE), group technologies such as computer-aided process planning (CAPP), CNC, appropriate databases, and the like, when used concurrently, are commonly known as computer integrated manufacturing (CIM). Organizations today are in the process of automating the production process ranging from automation of specific functional areas to full integration (see Fig. 13.28).

Enterprise-Wide Automation Systems

The development and use of enterprise-wide automation systems are a phenomenon of the late 1980s and early 1990s. At the heart of these systems is the linkage of some, most, or all internal functional areas, often including upstream suppliers and downstream customers (see Fig. 13.29). For example, in the case of the **CATIA** system (computer-aided, three-dimensional, interactive application) used by Boeing in the development of the 777 airframe, the ability to develop, on-line, the various parts saved time and money. The 777 is the first Boeing airliner to be 100% designed using three-dimensional (3-D) solid modeling technology. The software enabled Boeing to design, model and adjust the design on-line. CATIA promotes concurrent engineering, on-line, in 3-D — linked to the appropriate databases. The

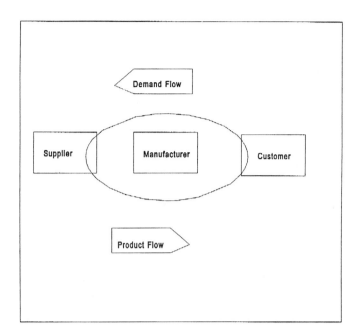

FIGURE 13.29 Enterprise-wide linkage through the value chain.

result of Boeing's use of the CATIA system was an aircraft that was developed in a much shorter time frame and at substantial cost savings. Boeing has since moved toward the integration of the entire organization into an on-line information technology system, ultimately creating the ability to track virtually any item.

Manufacturing Technology and Strategy

The search for ever tighter quality control and cost containment is creating the push for the use of manufacturing technologies. In essence, firms are no longer competing on cost or quality alone; they are competing on the total value provided. Whoever delivers the greatest possible value is the winner in the marketplace. It is difficult to imagine a firm being able to compete on multiple dimensions of competition (price, quality, flexibility, time, etc.) simultaneously without using some form of automated technologies (such as CAD/CAM). In fact, it is argued that adoption of new technology could have a major impact on the operational effectiveness of the company enabling it to compete more effectively on multiple dimensions of competition. At the same time, it is also important to note that any decision to adopt new technology must be embedded in the operational vision and overall strategy of the firm.

Strategy is the creation of a valuable position, involving a unique set of activities that determines how, where, when and for what type of customer a firm is going to compete, and what type of organizational structure the firm will require. In this context, choice of appropriate manufacturing technology is vital to the success of the firm. This implies that it is not enough for a firm to merely duplicate a competitor's manufacturing automation. Instead, it should base its decision on the fit between the technology and its strategy.

Investing in Advanced Manufacturing Technology

The decision to employ advanced manufacturing technologies is no simple task. Indeed, the costs, in terms of time, money, and commitment are considerable. However, the cost of *not* investing in these technologies may be greater.

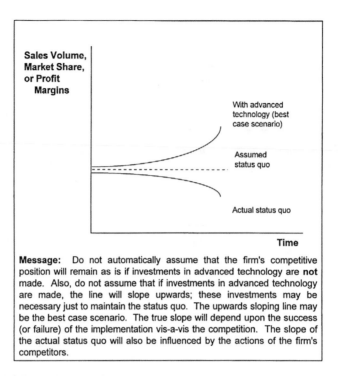

Sales Volume, Market Share, or Profit Margins

With advanced technology (best case scenario)

Assumed status quo

Actual status quo

Time

Message: Do not automatically assume that the firm's competitive position will remain as is if investments in advanced technology are **not** made. Also, do not assume that if investments in advanced technology are made, the line will slope upwards; these investments may be necessary just to maintain the status quo. The upwards sloping line may be the best case scenario. The true slope will depend upon the success (or failure) of the implementation vis-a-vis the competition. The slope of the actual status quo will also be influenced by the actions of the firm's competitors.

FIGURE 13.30   A Comparison of investment in advanced technology vis-a-vis maintaining the status quo. (*Source:* Adapted from Noori, H. and Radford. *Production and Operations Management: Total Quality and Responsiveness,* p. 287, McGraw-Hill, New York, 1995.)

Consider the situation whereby a competitor invests in new technologies that improve quality, cost, or both, effectively changing the value equation of the product. The options available to the target firm are limited to investing, not investing or attempting to redefine the value equation. In most industries, the options will be limited to, in the short run at least, investing or not investing. The decision then becomes: "Will the customers accept the product the firm currently delivers?" Once the market becomes aware of the changing value equation, the target firm will no doubt face erosion of the customer base. The firm, in actual fact, has no real option to maintaining the technology status quo (see Fig. 13.30).

Financial justification of new technology is usually a challenging job. This is so because many of the costs (such as software development and maintenance and productivity decline in transition) are hidden and some of the benefits (such as strategic flexibility) are not easily quantifiable. Attention to these hidden costs and deciding who should establish the (surrogate) criteria and specification to justify the acquisition are therefore important. In general, to justify new technologies, one can think of those benefits and costs that can be identified, quantified, and, finally, to which financial measures can be applied.

Concluding Remarks

The purpose of this section was to provide the practitioner with a snapshot of the issues involved with CAD/CAM and other manufacturing automation technologies. New technologies often provide increased flexibility and capability. For firms competing in dynamic, often chaotic environments, the creation and retention of a competitive advantage may lie in the utilization of advanced technologies. Flexibility or agility may be the deciding factor for which firms survive and prosper and those that do not in the future.

13.12 Total Quality Management

B. Michael Aucoin

Total quality management (TQM) is in many ways the heart of any endeavor, for it involves organizing for and delivering the inherent worth of a product or service. Continually improving the quality of a product or service is critical for the success of an organization and should be a key element of its business strategy. There are two major foundations to TQM. One is analytical and involves the management of variation from specification. The second is organizational and relates to the management of the value or worth of a product or service. The organizational aspect also involves working to continually improve the product or service and the environment in which it is produced. In brief, TQM is identifying what customers want and organizing to provide what they want.

Progress of Quality Movement

Anyone who has followed popular management over the last 15 years has seen much discussion and activity concerning TQM. During this period the quality movement reached its zenith with scores of books, conferences, and seminars on the subject. Recently, the level of activity has declined, signaling the rise and fall of faddish TQM and the firm establishment of genuine TQM.

For much of the 20th century little thought was given to product quality as customers were expected to buy whatever was offered. In large part this can be attributed to the dominance of U.S. goods in the world until the 1970s. With little competition, there was little reason to provide high quality. Where quality was considered important, it was typically approached as inspection for defects after production, based upon the belief that production workers had little interest in product improvement.

Two key developments led to the progression of the quality movement. First, after the devastation of World War II, Japan embarked on a commitment to quality manufacturing with the help of W. Edwards Deming, the father of the quality movement. With their rise and the rise of other nations with resulting competition for markets, and reduction of barriers to market entry, quality became a means for product differentiation. Ultimately, in the 1980s, the quality movement reached the United States as it finally recognized the need for instilling quality in products.

Second, in the development of management practice, it came to be recognized that all individuals in an organization can have a significant impact on quality and productivity. This understanding began with the landmark Hawthorne studies in the 1920s but did not reach fruition until the 1980s. This simple but fundamental shift in thinking became an important cornerstone of implementing quality in organizations, making quality the responsibility of everyone in an organization.

There is little argument that the quality movement has resulted in significant improvements in products and customer satisfaction. However, TQM as a movement has had mixed results. If so, why? The organizations that failed at TQM never really implemented it and instead approached it as faddish window dressing. TQM when sincerely implemented will result in demonstrable improvements in product or service quality, customer satisfaction, and organizational success.

Defining Quality and TQM

One can come up with various definitions for quality, but in fundamental terms quality involves satisfying a customer. Quality is whatever a customer says it is and may differ from one customer to another. While most people have a rough concept of quality, one can define it as appropriate functions, features and workmanship for a customer target cost. For this definition, quality does not necessarily imply the product with the most features or the highest cost, but the one that does the best job for a given price. The quality of a product or process is best approached by identifying characteristics that can be measured.

Simply speaking, TQM is the identification of what a customer wants and organizing to deliver what the customer wants. The identification of customer requirements is the result of research on customer preferences and is embodied in the product specification. Organizing for quality involves establishing the organizational systems for making quality a critical aspect of a product or service and providing the environment for delivery of quality.

At any given time the specification defines what the product should be as an understanding of what the customer wants. It defines in measurable quantities the features, function, and operation of the product under various constraints including price. The customer's expectation of the product is embodied in the specifications.

If the specification embodies customer preferences, then any variation of a product from specification can be considered a defect with an opportunity for improvement. This definition of quality and organizing to provide this aspect of quality involves analysis and feedback to production. By applying appropriate tools and techniques, variations in product quality can be readily identified and corrected in production.

The specification is best seen as a dynamic document, and, at any point in time, it is a snapshot of customer needs and the organization's best thoughts on how to meet them. However, customers, organizations, and technology are dynamic entities so there is always room for improvement. In this aspect of quality management, organizations strive to improve the inherent merit or worth of a product. Is the product what it should be? How can it be improved? In essence, this definition of quality occurs outside the specification and drives the specification as a feedback mechanism.

It can be seen that there are two ways of influencing or improving product quality: one is analytical and the other is behavioral or organizational. These two approaches will now be discussed.

Process Quality Management

Let us assume that we have a specification for a product and a process to produce it. We want the output of the process to conform to the specification. Output that does not comply with the specification within a given tolerance range is defective. The foundation of analytical quality management is taking data on a process to gain knowledge about the capabilities and limitations of the process. Based upon this knowledge we can ultimately control a process, that is, make it behave in the way we want it to behave, or in conformance to the specification. We want to have a process that is stable and predictable, so that it can always produce products in the same range of characteristics.

The tools of analytical quality management permit drawing conclusions about the quality of a process based upon sampling of the process. By implementing process quality management, we can make the transition from *detection* of defects to *prevention* of defects and thereby make dramatic improvements in quality (as conformance) at a reasonable cost. The act of measurement produces a feedback mechanism to the process that makes prevention possible. One of the other benefits of analytical quality management is that it enables the involvement of everyone associated with the process, not just the inspector.

One of the primary tools for analytical quality management is **statistical process control** (SPC), a methodology pioneered by Walter A. Shewhart in the 1920s, which involves a statistical sampling of measurable parameters of a process output such as products coming off a production line. A process that is operating properly produces outputs that exhibit some slight variations from one unit to another, but the overall average of a particular parameter should be stable. If there are excessive variations in the parameter, there may be defects that will result in lowered quality as perceived by the customer.

By examining the characteristics of variations, one can see whether the process is predictable in its behavior or under **statistical control**. A process is brought under control by eliminating or correcting all special causes of variation, such as operator error or power surges, which cause excessive variations. This procedure is accomplished by eliminating, one by one, special causes of excessive variation. The primary tools of SPC are the run chart, which depicts the measured values of a product characteristic over the course of a process run, and the control chart, which builds upon the run chart to permit one to visually determine if the process is under control.

Just because a process is under control does not mean it is necessarily producing the desired output. After bringing a process under control, the next step is to ensure that all variations are within tolerance. When this is achieved, the process is said to be **capable**. A process is made capable only through improvements in the process itself.

Even after a process is under control and capable, it is important to seek to improve the process. One principle of process improvement, initiated by Genichi Taguchi, holds that quality is not just conformance to specification within tolerance but is better described by the deviation from target value. One should seek to minimize the deviation from the target value because the quality of a product quickly degrades as a measured characteristic deviates further from the target value.

A powerful method for improving process quality is through **design of experiments** (DOE). This is a statistical method that interrogates the process to help optimize it. It involves making deliberate changes to inputs based upon statistically designed, orthogonal experiments to determine the effects of inputs on outputs. Through experimental design, one can screen the process to determine the factors that most affect the process and how they do so. Experimental design enables one to model the process and determine the sensitivity of the process to key input factors. Based on this information, one can optimize the process and make it more robust.

In addition to the numerical tools of SPC and DOE, there are many other valuable non-numerical techniques and tools for management of quality. Some of the other tools used for prevention of defects and quality management include

- Cause and effect (Ishikawa) diagram — a diagram often used in a group setting to identify causes and potential solutions of a problem.
- Flow chart — a chart that depicts activities and steps in a process to identify points at which improvement is needed.
- Pareto chart — a chart used to concentrate efforts on the few issues that contribute significantly to a problem rather than the many trivial issues that have little effect on a problem.
- Quality function deployment — a map of customer requirements into technical product characteristics, also called a "house of quality".

Continuous Improvement

It may be apparent by now that the pursuit of quality is an ongoing and never-ending mission. This aspect of quality can be described as the continuous improvement dimension of quality. For many products, there is some interaction or feedback from customer use to identify areas for improvement. Often, technological improvements are made and suggest application to the product. The best products are delivered in environments that solicit customer input, survey the environment for technology developments, and apply creative thinking, all in a mechanism that feeds back into the next round of product specification.

Organizational Aspects of Quality

Products and services are provided by people and often as part of the activity of an organization. It is therefore relevant to consider how individuals and groups affect quality. If TQM is determining what the customer wants and organizing to deliver what the customer wants, then the key word in TQM is *management*. The management of quality involves three key organizational dimensions: focus on the customer, the mission and model of the organization, and team commitment.

Focus on the Customer

The foundation of organizing to deliver what the customer wants begins and ends with the customer. The process of product and service delivery ultimately involves the satisfaction of a customer in a marketplace interaction. While there are many methods to research customer preference and satisfaction, organizations that do well at customer satisfaction go beyond these methods to get close to the customer and to experience the product or service as a customer would.

Consider the 1991 movie "The Doctor", which was directed by Randa Haines. In this movie, the main character is a physician who, while medically proficient, does little to interact with patients on an emotional level. When he becomes seriously ill, he is forced to become a patient and experience the impersonality and dehumanization he had delivered as a doctor. He recovers and returns to practice as a doctor who has learned to care about the patient as a human. Not only is the story moving, it is a dramatic lesson in learning how to deliver better service to a customer.

The best organizations live and breathe customer orientation, and they try to live the interaction from the perspective of the customer. There is still more to the story, however. The holy grail of customer satisfaction comes in delivering a product or service in such a way that is uniquely outstanding. It goes far beyond a feeling of satisfaction in the customer and results in the customer perceiving the transaction as exceptional. Tom Peters [1994] describes this as a "WOW" experience.

One example of this exceptional level of customer satisfaction is provided in *The E-Myth* by Gerber [1986]. He describes an accidental stay at a California inn that does an exceptional job at meeting and even anticipating his needs.

The staff watches when he leaves his room for dinner so that a lighted fire can be prepared for him before his return. He orders a brandy at dinner, and, without his asking, the staff has a second glass waiting next to his bed upon his return. In the morning his favorite coffee and newspaper are made available without his asking. Through observation and subtle questions, the staff learns how to tailor the lodging experience to each guest and provide key ingredients of the experience without being asked, *exactly when the guest anticipates wanting them.* The service is so exceptional that Gerber is awestruck. We are generally used to accepting mediocre customer service as the standard in our society. When we encounter exceptional service, it can transform us.

Organizing for Quality: Mission and Model

If we can determine what the customer wants, and we wish to provide an exceptional customer experience, how then can we organize to deliver it? One starts with the mission or core beliefs of the organization. Again, one example can be found in the story of the inn described in *The E-Myth.*

The manager of the inn explains that the goal of the staff is to provide an experience for the guest that replaces the home that most of us no longer have. The structure of the organization provides the lodging service in the context of a system, a game that is played to provide the sense of community and love that is present in what we would like to experience at home. The best quality organizations speak about the customer with a passion and organize systems, resources, and policies around this passion.

If having a quality organization is the vision, what are some ways of implementing this vision? Ultimately, there must be a singularity of purpose for the all members of the organization to strive to deliver quality products and processes to the customer. If the "soul" of the organization embodies a passion for customer happiness, the rest becomes easy. It is important for the organization leadership to exhibit this passion in behavior and attitude. For example, Herb Kelleher, CEO of Southwest Airlines, regularly does his part to load baggage and perform other tasks that bring him in direct contact with the customer experience.

Next, the leadership must implement a good model for the organization. A model for this purpose can be considered a set of process and operating principles and strategies that implement the core beliefs into the workings of the organization. To promote quality, the model for an organization should embody and strive to build an environment that values the delivery of quality to the customer, promotes honest communication, allows members to adapt the model to their specific situation, and provides sufficient resources to implement the model.

The organizational arena of quality provides the opportunity for success, but many organizations have failed in their efforts to implement TQM programs. If this is the case, what can go wrong with TQM?

The road to quality leads through humility or even pain because the pursuit of quality means always admitting that there is room for improvement. Many individuals and organizations are unwilling to admit their need for improvement, and, for this reason, they never seriously pursue quality. It is easier for them to talk about quality rather than do something about it. This is the essential difference between organizations that practice faddish quality and those which practice genuine quality: quality organizations don't just talk about quality, they live it.

It is relevant in this context to mention standards for quality, such as ISO 9000. These standards have been developed for systematically organizing, developing, and documenting an organization's processes and systems for delivering quality. These standards are valuable as guideposts, but they do not in themselves guarantee quality and do not substitute for the pursuit of genuine quality in an organization.

Team Commitment

For organizations that commit to genuine quality, the final organizational issue to address is obtaining the commitment of the members of the organization to implement the vision of quality. To be effective, quality management requires all members of an organization to contribute to continuous improvement. Such a commitment cannot be forced; it must be voluntary. The good news is that individuals in organizations desire meaning in what they do and want to be valued and successful. It is the responsibility of the leadership to build organizations and systems that promote such an environment. With this approach, individuals willingly commit to providing their best creative effort for the success of the endeavor in pleasing the customer. There is something worthy, almost transcendent, for individuals in following a good model.

There is a story about three bricklayers. One describes his work as laying bricks, another as constructing a building. The third explains that he is erecting a cathedral that will inspire many. It is in seeing one's work in such a model that leads to exceptional service delivery among individuals in an organization.

Conclusions

The pursuit of quality is a crucial core component of the mission of an organization. There are two major attributes of quality management: first, managing the quality of a process and, second, managing the continuous improvement of the behavior of the organization. TQM is implemented by continually identifying what customers want and organizing to deliver what they want. With the choices available to customers in the marketplace, organizations that do well at managing quality have a decided advantage over those that do not.

Defining Terms

Capable: A state of a process that is achieved when all variations in the process output are within tolerance.

Design of experiments: A statistical methodology for determining the effect of process inputs on outputs so one may improve or optimize the process.

Statistical process control: A statistical methodology for determining if a process is performing within acceptable parameters.

Statistical control: A state of a process that is achieved when all special causes of variation in the process output have been eliminated and the process is stable.

Total quality management: The identification of and organizing to deliver what a customer wants.

References

Gerber, M. *The E-Myth*, Harper Collins, New York, 1986.
Peters, T. *The Pursuit of WOW!*, Random House, New York, 1994.

Further Information

There is a multitude of reference materials available on quality management. Books that provide extensive coverage of the subject include *Total Quality Control* by Armand V. Feigenbaum and *Total Quality Management Handbook* by Jack Hradesky.

Those interested particularly in TQM in a production environment would likely benefit from H. G. Menon's *TQM in New Product Manufacturing* as well as *Improving Quality through Planned Experimentation* by Richard D. Moen, Thomas W. Nolan, and Lloyd L. Provost.

A study of quality should include review of work by W. Edwards Deming, such as his book *Out of the Crisis* and *The Deming Guide to Quality and Competitive Position* by Howard and Shelly Gitlow.

It is also helpful to engage in a critical examination of TQM successes and failures in *Why TQM Fails and What to Do About It* by Mark Graham Brown, Darcy E. Hitchcock, and Marsha L. Willard.

The organizational aspects and customer orientation of quality are addressed well in Michael Gerber's *The E-Myth* and *The Pursuit of WOW!* by Tom Peters.

The American Society for Quality Control is an excellent resource for information on quality through their monthly publication *Quality Progress*.

13.13   Process Improvement: The Just-in-Time Effect

Robert W. Hall

Just-in-time (JIT) production is the father of time-based management principles. JIT techniques are frequently rediscovered and applied to all kinds of processes under such names as concurrent engineering, one-stop customer service, reengineering, and others. Because of the tangible nature of machines and materials, the principles and techniques are easier to see in a production process.

One of the most unfortunate aspects of JIT production is its name, which implies that its objective is merely to make material arrive at each step of a process just in time to be used. That is only one observable result. The objective is to improve manufacturing quality, efficiency, and responsiveness to customers.

JIT prods continuous improvement in many forms and many places, not just the production floor. Through extension to a supply chain, the intent is to eventually cover a total process — from minerals in the ground to the end user. To prepare people for collaboration and continuous improvement, JIT development is usually accompanied by development of teamwork in some form. It's a mind set change.

Just-in-Time Principles

JIT is an unending improvement process that can be perceived in many ways. Its principles are flow, visibility, and compression of operations. Application of the principles depends upon the specifics of each case.

A first step in mind set change is to measure performance using time. In a plant, one measure is throughput time, the lead time from material starting production processes until it finishes. A broader measure is door-to-door time, from material entry to customer shipment. In most plants, these measures are approximated using days-on-hand inventory. In addition, order-to-ship time measures the customers' experience.

Another measure is value added ratio, the fraction of throughput time during which value is added (synthesis, fabrication, or assembly). Time consumed not doing value-added work correctly, or in doing anything else, is wasted. Activities that waste time are targeted for elimination. Simplify the manufacturing process into its essence, stripping away any activity that does not need to be done.

A plant that has never thought about JIT often has a value-added ratio of 5% or lower. A stretch goal could be to attain a value-added ratio of 50% or higher, depending on the potential inherent in the production process. Like snowflakes, many plants are similar, but none are identical.

Waste has been categorized in many ways: quality, idle inventory, poor equipment maintenance, unnecessary material handling, poor training, absence of information, or floods of it, and on and on. There is no formula for eliminating waste other than cutting lead times (inventories) to make problems so visible that they receive attention. Some problems are easily overcome. Others continue to foil solution after repeated attack, but the dictum of JIT is straightforward. Solve anything solvable, and never sink back into addiction to that drug of manufacturing, excess inventory.

Well-developed JIT production processes have little excess motion. Operations with a high value-added ratio have a minimum of conveyor belts, material handling, clutter, and confusion. Expediting isn't a crisis. Like good athletes, people using processes that add much value make it look easier than it really is.

People who mimic a few JIT techniques fail, or at least they never realize anything close to their potential. Those who succeed understand that the techniques support human development by stimulating problem seeing and problem solving. Those responsible for processes learn experientially. Even if new plants are laid out for fast flow using equipment designed it, operators must see their problems themselves, and largely overcome them themselves. You cannot successfully "hand" JIT to them.

Managements that have removed great waste have developed their people — the entire workforce, white collar, blue collar, or pink collar — to work together recognizing and solving problems. It's usually called teamwork and quality. There's no point cutting inventory to reveal problems if the organization isn't prepared to deal with them. Changing performance measurement, reward systems, and traditional roles of people — the "soft stuff" — is more difficult than adopting JIT techniques.

This prepares minds for JIT. Then simplify, simplify, simplify the processes themselves. Next, control operations by decentralized parallel processing (by people, not just computers) rather than by a centralized brain issuing orders.

JIT Techniques

The following brief review of techniques assumes that minds are prepared and that an existing manufacturing process is to be converted to JIT. The techniques still apply if one is designing a JIT process from a clean sheet of paper, no matter how manual or automated it may be. No process is perfectly designed from scratch. All need tuning, and, as requirements change, all need adaptation.

Workplace Organization, or 5S

The shorthand version of this is "A place for everything, and everything in its place, free from extraneous items." There are different versions of the five steps, but a common one is:

1. *Clearing and simplifying*: Remove everything unnecessary in the near future. Besides trash, that includes tools, materials, and instructions. Reduce each work place to the items essential to its functioning.
2. *Standardizing locations*: Think of surgical trays. Colocate everything used together, ready for use, and easily found by anyone (or any machine) that needs it: often-used items in standard spots for instant use, less frequently used ones at a greater distance, and rarely used ones stored away. Silhouette tool boards throughout a plant are a common marker of this practice. This discipline begins to develop *visibility* in the workplace.
3. *Cleaning*: Clean regularly so that dirt or accumulated offal never interferes with process functioning, whether it's in a class 10 clean room or a foundry.
4. *Discipline to maintain the system*: For workplace organization to be effective, everyone involved must learn the discipline of the system and follow it.
5. *Participation*: Everyone must understand and support the system, which means that no one — executives, engineers, etc. — messes up an organized workplace. Everyone cleans up after themselves.

Visibility

This is unspoken, often unwritten human communication. In a plant or a project office, anyone should be able to evaluate the situation at a glance: flow of material, inventory levels, machine status (by signal lights), goals, orders, schedule, and accomplishments. Disruption of an expected pattern of visibility is a cue that a problem exists and that it needs attention. A good visibility system displaces the need for managers and supervisors to intervene in routine operations. The shop floor runs itself.

The implications run deep. If a set of operations becomes a shared, readable environment to its participants, people are drawn to becoming students of the process and more willing to participate not only in its preservation, but its improvement.

Properly applied, computers, software, and sensors enhance visibility (and control processes) in ways that are impossible by humans alone. Poorly applied, they complicate the environment without removing waste. The need for human visibility suggests man-machine (and man-software) interface issues roughly analogous to those of pilots flying by computers and instruments rather than manually by visual rules.

Disciplined Pull Systems

A pull system of production control means that each production station signals its preceding stations what material or other items that it needs. If the pull system is disciplined, it also limits the volume of inventory in the pipeline between all pairs of work stations. Limiting the pipeline stock also limits the production throughput time. It also limits the time window within which each station must respond to the demands of its "customer" stations, which stimulates people to pay attention to eliminating waste in order to do it.

The signaling system is often called a kanban system, which can be set up in a variety of ways. In the simplest case, a part is simply transferred one at a time from one station to the next, by a person or by a device, sometimes a conveyor. If the storage space between stations is limited, the inventory is also limited, and knowing what part to send is elementary. It is the only one exiting the machine and going to the next.

One step up in complexity, a little larger space is marked into squares or compartments, one square being designated for each part number and limited in size. The supplying station simply fills open squares that have been emptied by the using station. This simple system is merely one aspect of an overall shop visibility system.

If a supplier station is out of sight from its customer stations, the signals may be conveyed using cards, or standard-sized, designated parts containers — empties returned from the customer. These signal the need to replace a fixed number of parts withdrawn from the pipeline. Thus, the range of the visibility system can be extended to customer sites many miles distant, but the inventory in the pipeline remains limited if the number of cards or containers in circulation for each part number is limited.

With the basic idea of a disciplined pull system in mind, imaginative people may concoct varying mechanisms for signals and inventory limits, including electronic ones, but the key point is to maintain pressure for process simplification rather than be diverted into fancy systems to manage waste.

An objection to this system is that it will not work unless process routings are standardized, and that is true. There are two response to this objection. First, if process routings could be standardized, but they have grown up in a tangle, that is itself a waste, and needs to be corrected.

Second, disciplined pull systems are only one aspect of visibility. Engineer-to-order job shops and process industries not having discrete parts are both subject to great improvement using all the other ideas of JIT, including simplify, simplify, simplify, visibility systems, and as much parallel processing as can be built in. This is more than a speculative concept; it's been done.

Setup Time Reduction

As long as setup times are long, lot sizes do not decrease much. Therefore, neither inventory or throughput times decrease very much. Setup processes are attacked the same way as any other, by eliminating the waste. First, eliminate any steps that are unnecessary, and apply 5S to setup processes. Then reduce the operations that must be done with the machine stopped to the barest minimum.

Without changing process technology or product designs, reorganizing the physical process of setup may cut setup times by 50 to 90 and even 100%, and all with a decrease of work required, not an increase. Sometimes maintenance and quality are issues. The objectives of quick setup are to make a quality piece the first time with no adjustments and to keep maintenance separate from setup. JIT stimulates plants to adopt preventive and predictive maintenance. The specifics of each setup must be worked through, but thousands of plants have now done it.

Common issues are broadening the training and responsibility of workers, therefore reducing the number of job classifications, and organizing workers into teams for setups. The human legacies are more difficult to overcome than the plant physics, but some of those are also challenges. Single-purpose, dedicated equipment and tooling does not become flexible easily.

Reducing setup times improves manufacturing flexibility as well as efficiency. For example, it may allow the scheduling of new tooling trials without disrupting production.

Attention to setup problems throughout a company stimulates other useful thinking, for example, design for manufacturing and design for assembly, simplicity of process with maximum use of common parts. Another line of thinking is to avoid tooling changes by making as much setup as possible depend only on a change of software — to cut a different path in a workpiece, for instance.

Moving in this direction leads to the idea of giving the customer maximum variety from his viewpoint, but from a standard flow production process. If both the engineering and execution of a customer solution can be done with minimal setup, a plant is in position to not only be JIT, but agile. JIT demands quality, and it improves efficiency and flexibility. Agility is the ability to respond to the unexpected.

Cellular Production

In many cases, one-piece-at-a time flow through a cell is ideal JIT. If all machines in a cell are capable of quick setup, a minute or less, and certainly less than 10 minutes, and if all have a similar routing, a cell can produce a family of parts with a very high value-added ratio.

There are many kinds of cells, which are machines for a sequence of operations grouped together. Some are highly automated; some minimally automated. Cellular manufacturing focuses attention on many issues associated with JIT.

The classic U-line is still the cell design used for reference. An operator moves parts from machine to machine in a walk-around, and he has good visibility and access to the process inside the cell. If the parts are nearly uniform, simple conveyances, such as chutes, can be used between the machines instead of an operator, and the total system can be automated very cheaply using limit switches and the like.

In some cases, an intelligent robot can do this task. However, beware of complexity and cost in programmed automation, particularly if all it does is material handling. Question whether the machine or the software adds value or contributes to waste. Elegant automation adds value. Believing that it must be programmed to handle a large variety of contingencies is evidence that the process has not yet been simplified enough to merit profitable automation. (Complex products have challenge enough without taking on the unnecessary ones.)

Overcoming the problems that discourage the organization of a production process into a cell is the point of JIT, and every case has a full measure of such problems. If machines have not been designed for cells, rearranging them into a cellular layout may be awkward. If people fear not having space to handle all manner of contingencies, machines will be spaced too far apart, which is usually done the first time one organizes a cell. The process needs simplification.

Accounting based on the assumption of totally independent machines and operations is a problem. An expensive machine may lose the volume thought necessary to economically recover its capital and overhead cost. (Keeping expensive machines busy is one of the chief sources of bottlenecks in a job shop. The consequent overproduction adds to waste.) For the same reason, accountants tend to be upset if a machine runs at less than full speed when it is matched with others in a cell. For a financial sanity check, estimate what a cell, plus all the simplification that it necessitates, will do to the total process, not a fraction of it.

However, the greatest challenge again is apt to be the human one. In some cells, it's easy for an operator to learn every operation, and learn preventive maintenance for every machine. In other cells, as for advanced precision machining, this is a stretch. People working in cells must broaden their skills and learn to work in teams, a tough transition made more difficult if historical work agreements and status systems have rewarded specialization.

Kaizen Improvement

Kaizen means seeing and overcoming problems. It takes place in many forms. Employee suggestions, made and adopted, are one form. Immediate corrective action is another form. Schedules in JIT plants often allow a few minutes of downtime each day to observe problems and make corrections, and, if needed, a line, cell, or machine will be stopped immediately while the problem is solved. Anyone observing a problem can stop the process. (The Japanese call this *jidoka*.)

In the United States, kaizen usually refers to a more intensive revision of an area or process for JIT or to make improvements in work flow. Everyone should participate in changing layout and making other revisions that drastically affect their own work.

All the principles and techniques of JIT may be employed in intensive kaizen. A concept very important to kaizen work design is takt time. A takt time is the amount of time allowed to complete all the work for one unit passing through a work station. For example, a line station in auto assembly running at a speed of 48 cars per hour would have 1.25 minutes, or 75 seconds, of takt time in which to properly accomplish work on each car passing through.

Takt time is a scheduled cycle time. Divide the scheduled daily work time available at a station by the number of units to be completed and one gets the takt time per unit. (If one divides the actual time worked by the actual number of units completed, the result is the actual cycle time per unit.)

The objective is to accomplish all the necessary value-added work within takt time. Devise or revise a station work process that eliminates all possible non–value-added work from the work cycle.

Who does this? The worker. When JIT is fully developed, the staff only does preliminary, approximate engineering of work for the workers. Each one finishes his own detailed industrial engineering for his own work station. That is the natural end point of workplace organization.

Again the biggest obstacle is human development, including the development of the managers. Kaizen presumes that almost everyone knows how to make improvement — how to reengineer his work. Problems can be of every kind, quality, time waste, safety, ergonomics, environment, and so on. Many of them cannot be solved by one person alone, or even by a work team in his area, but closely supporting workers in making the detailed improvements is a revolution in thinking.

Learning how to make improvement regularly is a tremendous experience for both individuals and organizations. One does not evolve from a command and control environment to responsibility for improvement quickly. Responsibility means having judgment about costs, so workers need access to appropriate cost databases.

Toyota is still the most advanced company in the practice of JIT. For many years, monthly schedules have come out with calculated takt times for operations, including those that feed assembly and those at suppliers. Using these takt times, the workers revise and improve their stations in preparation for next month's work, a regularly scheduled mini-kaizen. After documenting their plan on standards sheets, they work the plan, and adhering to it is very important.

Toyota calls this distributed production planning. As a consequence, a Toyota assembly plant can change line speed once a month — more often if necessary — a capability that auto companies immature in JIT cannot dream of doing. Only a few other companies have developed JIT this far.

Uniform Load Schedules

For the repetitive manufacturing case, a short description of a uniform load schedule is making a little bit of everything every day. If fabrication feeds final assembly, then the assembly schedule should plan for mixed model assembly, not long runs. Naturally, the assembly process must be physically developed for this also. The lot sizing and uniformity of a schedule should mirror the capability of the physical processes at the time.

Some assembly processes, such as automotive, must run a type of mixed-model sequence in order to maintain line balance. Different models may create overcycles and undercycles of work at various stations. Unless these are balanced out by the sequence spacing, the critical line stations are either starved of work or overworked.

Fortunately, mixed model assembly, or assembly of small-lot batches in a mixed sequence, also creates a relatively uniform demand for the parts that are fed to assembly. Therefore, the pull systems working down through fabrication processes tend to keep all of them working, but with a slightly varying workload. Most operations that have been developed for JIT are robust enough to accommodate swings of $\pm 10\%$ in daily loads and more than that in parts mix percentages. That kind of performance is needed if the takt times calculated from final assembly rates are to be meaningful. If machines or cells are assigned families of parts, it is easier to accomodate wider swings in part mixes.

There's more to uniform load scheduling than creating a mixed model assembly schedule and a disciplined pull system. If affects all processes, including order entry. One does not encourage sales by offering large lots at a big discount — not without careful planning so that the volume surge does not create waste. Rather, one offers everyday low prices, pointing out that the average cost over time is lower than if demand surges and dies using promotions, that is, having a uniform load schedule interacts with the marketing strategy and customer order flow process of a company.

The paradox is that a company with a uniform load schedule may be more responsive to customers. Many are. For example, 3M in making video cassettes for many years allowed 20% or so of each day's schedule to actually make phone-in orders not filled from stock. As long as the load was relatively uniform, the process easily accommodated major mix changes.

Suppliers

As a company develops its JIT capability, it soon becomes apparent that suppliers simply are part of the fabrication process not owned by the company. One would like to extend the system to them.

Partnerships with suppliers have many more considerations than JIT production, for example, participation in product design. However, one reason for limiting the number of suppliers is that it is preferable to give a supplier a uniform load covering a family of parts, just as is true of a work center in one's own shop. Where that becomes an obvious advantage, the partnership is worth seeking, and the value added by the partnership is more than just the supply of the parts. It's the flexibility of response.

To be a JIT supplier, a company must either develop its own people for JIT, or the customer must assist them. A few large companies develop suppliers. It's in their best interest, even if the suppliers also serve competitors. By so doing they are assured of having highly competent suppliers, at least in production.

References

Greif, M. *The Visual Factory*, Productivity Press, Portland, OR, 1991. Basic. Many pictures.

Hall, R. W. *Attaining Manufacturing Excellence*, Irwin-McGraw-Hill, Homewood, IL, 1987. An overview explaining the why's of JIT.

Hirano, H. *JIT Implementation Manual*, Productivity Press, Portland, OR, 1990. Expensive, but comprehensive and detailed. Includes analysis forms.

Ohno, T. *Toyota Production System*, Productivity Press, Portland, OR, 1988. How it began. Origin of the system by the Father of JIT.

Shingo, S. *The SMED System*, Productivity Press, Portland, OR, 1985. Reducing setup times. Loaded with how-to.

Shingo, S. *A Study of the Toyota Production System*, Productivity Press, Portland OR, 1989. A classic. Earlier version was the first book on JIT in English.

13.14   Lean Manufacturing

Kwasi Amoako-Gyampah

Increasing global competition is forcing companies to seek better and better ways of competing through manufacturing. Increasing global competition is also forcing companies to redefine how they compete. Whereas in the past companies could be successful by emphasizing either low-cost production, higher quality, delivery dependability, or higher flexibility, it has becoming increasing important for companies now to develop competencies in all four areas and to be able to compete in all areas. One approach to competing that has been used successfully by the firms in Japan and has found some acceptance in the United States (at least among automobile manufacturers) is lean manufacturing.

Lean manufacturing was developed as an alternative to traditional mass production. Mass production is based on producing large volumes of limited items at low cost in an environment where workers perform minute task in repetitive fashions and a separation of powers exists between management and labor. On the other hand, lean manufacturing is aimed at producing large varieties of high-quality items very quickly in a flexible and continuously learning organization with multiskilled workers at all levels of the organization.

Lean manufacturing is a system of manufacturing that seeks to achieve more with less resources. It is a manufacturing approach that focuses on total quality management, just-in-time production, waste elimination, continuous improvement, multifunctional teams, product design, and supplier partnerships. Lean manufacturing does not only focus on core production activities, but it is also aimed at product development, component procurement, and product distribution [Karlsson and Ahlstrom, 1996]. The ultimate goal of lean manufacturing is increased productivity, lower costs, increased quality of products, shortened lead times, faster and reliable delivery of products, and enhanced flexibility.

Although the tenets of lean production were developed by the Toyota Motor Corporation in the 1950s, the term "lean production" was actually coined by a researcher involved in the International Motor Vehicle Program at the Massachusetts Institute of Technology. The results of that program was published in the book *The Machine that Changed the World* [Womack et al., 1990] and this book has been largely credited with the development of the knowledge base on lean manufacturing.

Elements of Lean Manufacturing

This section examines the various elements of lean manufacturing as mentioned in the previous section. These descriptions are not in any order of importance since all the elements are necessary for the full implementation of lean manufacturing.

Total Quality Management

Lean manufacturing requires the adoption of total quality management (TQM) principles. Some aspects of TQM such as continuous improvement and employee empowerment will be discussed later. One aspect of TQM important in lean manufacturing is focus on the customer. Focus on the customer means that **value** has to be defined from the customer's perspective. By defining value as the features that a customer desires in a product that will be offered at the right time and at the right price, the company can then focus on eliminating the non value-adding items, that is, focus on **waste** elimination.

Another TQM element necessary for lean manufacturing is measurement. This implies the use of scientific measurement tools to reduce variability in product outcomes and to improve the quality of the products. Workers will be expected to be trained in the use of statistical quality control techniques in order to able to identify and solve quality-related problems so as to enhance value to the customer. In a lean manufacturing environment, workers will be empowered to stop production processes if they detect quality problems. The producers of a part are responsible for the ultimate quality of the part.

Just-in-Time Production

The implementation of lean manufacturing requires the use of just-in-time production (JIT) and delivery. JIT is described in greater detail elsewhere in this handbook. Briefly, JIT uses **pull manufacturing** to produce only the needed materials in the smallest quantities possible and at the latest possible time for onward delivery to the customer. Materials are pulled through the value-adding chain as opposed to the traditional **push manufacturing** where materials are pushed through the value-adding chain using preset lot sizes.

The production of goods in small lot sizes requires the reduction in setup or changeover times. Traditionally, the changeover time refers to the time required to change dies. If changeover times between batches are small, then companies do not have to produce large batches at a time in order to be cost competitive. Small batch production means an ability to respond faster to changing customer demands for the existing mix of products, reduction in inventory, and therefore less space requirements, and quality problems become more visible.

Waste Elimination

Another element of lean manufacturing is waste elimination. Waste is anything that does not add value to the customer, from product initiation to final delivery of the product to the customer. This includes waste of overproduction, raw materials, tools, equipment, labor, space requirements, and transportation requirements. All these components are related. Producing more than what is needed will mean using raw materials earlier than needed and therefore having to order more materials to replace those consumed. Tools and equipment will have to be used when not needed because of a decision to produce more, and more space will be required to store the extra inventory because of overproduction. The extra inventory hides quality and other production-related problems and increases lead times.

Continuous Improvement

The gains that can be achieved through the other elements of lean manufacturing require the implementation of a mentality of continuous improvement in the organization. The Japanese have developed a system known as **Kaizen** to focus on continuous improvement. Continuous improvement in a lean manufacturing environment requires the efforts of everyone in the organization involved in the manufacturing process — both management and labor. For example, an increase in problem identification skills and problem solving skills and a recognition that there is always a need to improve will be necessary. The goal is to be never satisfied but to constantly strive toward perfection. Emphasis is placed on small incremental improvements.

Several forums can be used in accomplishing the goals of continuous improvement. Among these are the use of quality circles, formal employee suggestion programs, multifunctional teams, and problem-solving teams. For the continuous improvement program to be successful there have to be mechanisms for implementing suggestions, providing feedback on employee suggestions, and an effective reward system for the employees. There also has to be a mechanism for monitoring the number of improvement suggestions for a given time period, the number implemented, reasons for not implementing others, and the outcomes of those implemented.

Another element that has to be present for continuous improvement to work is that it has to be embedded in a cooperative management-labor relationship [Klier, 1993]. Since most of the suggestions for continuous improvement will come from the shop floor workers, it will not work if workers have a fear that some of them will lose their jobs as a result of productivity gains made from those suggestions. Therefore, a trust relation between management and labor has to be developed, and management has to treat labor as an asset rather than cost. That commitment from management that employees will not lose their jobs as a result of a lean manufacturing project might not be easy to keep. However, with enhanced manufacturing capabilities and competitive strength arising from the adoption of lean manufacturing, the company is likely to win more business, leading to increased job opportunities [Day, 1994].

Supplier Relationships

The role of suppliers changes in a lean manufacturing environment. Suppliers are selected right at the onset of product development. The basis for supplier selection is not the typical low-cost supplier, but rather supplier selection is based on long-term relationships that are likely to have developed from proven abilities to supply other components in the past [Womack et al., 1990]. The suppliers for a manufacturer or an assembler are usually divided into "tiers". A first-tier supplier is usually given a complete component or subsystem to produce, such as a brake system for an auto assembler. The first-tier supplier might have other suppliers known as "second tiers" who might supply components or subsystems to the first tier and so on.

The number of suppliers (first tier) used by a lean manufacturer is typically low. The total number of suppliers might be large; however, only the first-tier suppliers deal directly with the assembler/manufacturer. The first-tier suppliers, because of large volumes and also because they deliver large components or complete systems, typically will locate close to their customers. This is facilitated by the longer-term contracts that the assemblers offer those suppliers. The lower tiers on the other hand will make their location decisions mostly based on cost because they typically will be supplying low–value-added parts.

Suppliers in a lean manufacturing environment are expected to make frequent deliveries and deliver only the quantities desired, that is, delivery quantities are typically small and ideally are expected to occur daily. The delivered components will go straight from the delivery vehicles to the production floors and bypass warehouses completely. This approach means that the quality of the materials has to meet the expectations of the assembler since there is no time for inspection. The approach is also based on trust and frequent communication between the suppliers and the assemblers.

Multifunctional Teams

A critical element of lean manufacturing is the use of multifunctional teams. One should expect to find the use of multifunctional teams to increase in an organization that implements lean manufacturing. In a lean manufacturing environment, workers in a team are cross-trained to perform many different tasks. Some of these tasks might involve preventive maintenance, quality control checks, process control checks, housekeeping, setups, and other production activities.

Cross-training of workers ensures that the required workforce will be available to meet the fluctuations in demand that might result from changing customer demands for various products. Workers can be moved from low demand areas to areas where the demand has picked up. For the multifunctional teams to perform well, the employees must be empowered to set their work schedules, determine breaks, make work assignments, and sometimes even to decide on new members for the teams.

Product Design

The ability to make quick changeovers during production of goods, the ability to meet ever-changing customer requirements with regard to product features, and the ability to respond quickly and flexibly to all these demands depend also on having effective product designs. Lean manufacturing requires that design for manufacturing (DFM) be built into the product right at the design stage. DFM entails that the product can be easily manufactured, will have the right quality, be reliable, and will be easily serviced once in use. Lean manufacturing and DFM require the use of cross-disciplinary teams in product development and engineering. This requires teams of professionals who are cross-trained, possess multi-skills, and are willing to be team players. In this approach, design and manufacturing are not viewed as separate sequential activities but are as integrated synchronous activities. Complete unification of the design and manufacturing activities is not required. What is needed is effective coordination and coop-eration of the team members.

Under lean manufacturing, product development also requires the input of suppliers. Suppliers are relied on more and more to develop component items and subsystems. This requires more research and development effort on the part of the suppliers since they no longer simply produce items according to the manufacturer's specifications [Klier, 1993].

What Can You Expect from Adopting Lean Manufacturing?

One of the best documented examples of the benefits that can be expected from the adoption and implementation of lean manufacturing techniques is that of NUMMI, the GM-Toyota joint venture in Fremont, CA. At this plant, the use of lean manufacturing principles led to dramatic improvements in quality, productivity, inventory reduction, and low absenteeism using essentially the same workforce and facilities that previously had developed a reputation for bad quality, low productivity, and high absen-teeism in the assembly of cars [Adler, 1993]. The implementation of lean manufacturing techniques by the Frendenberg-NOK General Partnership led to production lead times being cut in half, an increase in productivity of 52%, a 78% reduction in cycle time, and a reduction of 63% in move times [Day, 1994].

The results of a survey conducted among 24 companies with sales ranging from $30 million to more than $1.5 billion that had implemented lean manufacturing techniques showed the same benefits of lower costs, increased productivity, higher quality, shortened lead times, and higher inventory turnover. In addition, an increase in worker empowerment was reported by the respondents [Struebing, 1995]. Klier [1993] reports of two companies, Luk Inc. and Eaton Corp., who were able to achieve higher than average productivity gains in their operations as a result of the use of lean manufacturing techniques. These few examples demonstrate the benefits that can be obtained with the implementation of lean manufacturing.

How to Make Lean Manufacturing Successful

Even though many companies outside the automotive industry have generally accepted the principles of lean manufacturing, very few have actually taken the necessary steps to incorporate the concepts into a coherent business system [Womack and Jones, 1996]. Some executives fear that costs might increase, that the process might take too long, and that the organization might not be able to handle the change in culture required. However, it has to be emphasized that the costs for not adopting lean manufacturing can be also be tremendous.

To achieve lean manufacturing success requires long-term commitment and a recognition that there are bound to be failures and setbacks along the way. It has been estimated that moving from mass production to lean manufacturing can take at least 5 years [Bergstrom, 1995]. This might require a "champion" in the organization who will be committed to the program.

Lean manufacturing success might require that the organization be reorganized structurally. This reorganization not only has to incorporate the use of teams but also allow for the development of multiple career paths within the organization. Lean manufacturing requires an environment where workers are empowered, teamwork is encouraged, creativity is fostered, and the complete involvement of all employees is nurtured [Day, 1994]. The entire supply chain also has to be adequately managed. Whereas the quality of the products of first-tier suppliers might be adequate, the quality of items supplied by lower-tier suppliers might not and unless that is properly managed the desired results of lean manufacturing might not be achieved.

Lean manufacturing can result in increased stress. The stress results from the synchronized nature of operations, the fact that there are no built-in slacks in the operations, and the emphasis on continuous improvement whereby sometimes problems are intentionally injected into the system in order to foster improvement. However, high stress level of employees does not mean they will not accept lean manufacturing. It has been demonstrated that employees' acceptance of lean manufacturing principles is more likely to depend on their commitment to the company and their motivation [Shadur et al., 1995]. Thus, management has the responsibility to put in place the processes necessary to ensure employee commitment to the lean manufacturing program.

Summary

This section has examined some of the basic elements of lean manufacturing. Under lean manufacturing, small batches matched to customers' needs are produced just in time using teams of workers. This contrasts with traditional manufacturing where parts are produced in large volumes by workers performing only few functions with several non–value-adding stages in the process. Lean manufacturing emphasizes team work, employee empowerment, continuous improvement, and waste elimination. Lean manufacturing requires reliance on a few core suppliers to produce large proportions of a company's output. These suppliers are also very often involved in product design and development.

Companies that have implemented aspects of lean manufacturing have reported several benefits. Some of these benefits include increased worker productivity, decreased work in process, decreased move times on the plant floor, reduced cycle times, decreased worker compensation costs, reduced product introduction times, higher quality, greater employee empowerment, reduced absenteeism, and an increase in business volume.

To be successful, the top management in the organization must be committed to the lean manufacturing program, and the presence of a lean manufacturing champion in the organization will help.

Management will have to relinquish some of the decision making to lower levels. Workers will have to be cross-trained and be prepared to work in teams. Supplier cooperation will have to be developed. An environment of continuous improvement will have to be fostered. Managers considering lean manufacturing should be prepared for difficulties and failures along the way, but the benefits of increased competitive strength should make the effort worthwhile.

Defining Terms

Kaizen: A term used to characterize the Japanese approach to continuous improvement

Pull manufacturing: A manufacturing approach in which item production at any stage of the process is dictated by the demand at the next downstream operation.

Push manufacturing: A manufacturing approach in which item production at any stage is based on predetermined batch sizes regardless of the demand for the item at the next downstream operation.

Value: A customer's subjective evaluation of a product on how a product meets his or her expectations taking into consideration the product's cost.

Waste: Anything that does not add value to the customer.

References

Adler, P. S. Time-and-motion regained, *Harv. Bus. Rev.,* 71(1): 97–108, 1993.

Bergstrom, R. Y. Toward lean success, *Production,* 107(2): 58–61, 1995.

Day, J. C. The lean-production imperative, *Ind. Week,* 243(15): 70, 1994.

Karlsson, C. and Ahlstrom, P. Assessing changes towards lean production, *Int. J. Prod. Operat. Mgt.,* 16(2): 24–42, 1996.

Klier, T. How lean manufacturing changes the way we understand manufacturing sector, *Econ. Perspec.,* 17(3): 2–10, 1993.

Shadur, M. A., Rodwell, J. J., and Bamber, G. J. Factors predicting employees' approval of lean production, *Hum. Relat.,* 48(12): 403–421, 1995.

Struebing, L. Survey finds lean production yields more than reduced costs for U.S. companies, *Qual. Progr.,* 28(11): 16–18, 1995.

Womack, J. P. and Jones, D. T. Beyond Toyota: how to root out waste and pursue perfection, *Harv. Bus. Rev.,* 74(5): 140–158, 1996.

Womack, J. P., Jones, D. T., and Roos, D. *The Machine that Changed the World,* Rawson Associates, New York, 1990.

Further Information

A good comprehensive material on lean manufacturing is the book *The Machine that Changed the World* (see reference list above). Even though it focuses on lean manufacturing in the automotive industry, the principles are applicable universally.

The paper "Beyond Toyota: How to Root Out Waste and Pursue Perfection" also by James P. Womack and Daniel T. Jones, *Harvard Business Review,* September-October, 1996, discusses important steps needed in implementing a comprehensive lean manufacturing system.

For companies interested in implementing lean manufacturing in a global supply chain environment, the article "Lean Production in an International Supply Chain" by David Levy, *Sloan Management Review,* Winter, 1997, provides valuable insights.

Managers interested in understanding how the adoption of lean manufacturing might affect the way they deal with their suppliers should read the article "The Impact of Lean Manufacturing on Sourcing Relationships" by Thomas Klier, *Economic Perspectives* [Federal Reserve Bank of Chicago], 18(4): 8-18, 1994.

13.15   Green Manufacturing

Mark Atlas and Richard Florida

There are many ways that industrial facilities can implement technologies and workplace practices to improve the environmental outcomes of their production processes (i.e., **green manufacturing**) and many motivations for doing so. Green manufacturing can lead to lower raw material costs (e.g., recycling wastes, rather than purchasing virgin materials), production efficiency gains (e.g., less energy and water usage), reduced environmental and occupational safety expenses (e.g., smaller regulatory compliance costs and potential liabilities), and an improved corporate image (e.g., decreasing perceived environmental impacts on the public) [Porter and van der Linde, 1995].

In general, green manufacturing involves production processes that use inputs with relatively low environmental impacts, that are highly efficient, and that generate little or no waste or pollution. Green manufacturing encompasses **source reduction** (also known as waste or pollution minimization or prevention), **recycling**, and green product design. Source reduction is broadly defined to include any actions reducing the waste initially generated. Recycling includes using or reusing wastes as ingredients in a process or as an effective substitute for a commercial product or returning the waste to the original process that generated it as a substitute for raw material feedstock. Green product design involves creating products whose design, composition, and usage minimizes their environmental impacts throughout their lifecycle.

Source reduction and recycling activities already have been widely adopted by industrial facilities. According to 1993 U.S. Environmental Protection Agency (EPA) Biennial Reporting System (BRS) data, which cover facilities that generate over 95% of the country's hazardous waste, 57% and 43% of these facilities had begun, expanded, or previously implemented source reduction and recycling, respectively. According to a 1995 survey of over 200 U.S. manufacturers, 90% of them cited source reduction and 86% cited recycling as main elements in their pollution prevention plans [Florida, 1996].

Organizing for Green Manufacturing

Green manufacturing provides many opportunities for cost reduction, meeting environmental standards, and contributing to an improved corporate image. However, finding and exploiting these opportunities frequently involve more than solving technological issues. The ten most frequently cited hazardous waste minimization actions are listed in Table 13.13.

As the data show, only a small portion of these actions involves new or modified technology. Most involve improving operating practices or controls or fairly basic ideas — such as waste segregation or

TABLE 13.13   Most Frequently Cited Hazardous Waste Minimization Actions

Percent of All Actions	Waste Minimization Action
8.9	Improved maintenance schedule, recordkeeping, or procedures
8.0	Other changes in operating practices (not involving equipment changes)
7.1	Substituted raw materials
6.5	Unspecified source reduction activity
5.1	Stopped combining hazardous and nonhazardous wastes
4.8	Modified equipment, layout, or piping
4.6	Other process modifications
4.4	Instituted better controls on operating conditions
4.1	Ensured that materials not in inventory past shelf life
4.0	Changed to aqueous cleaners

$N = 81,547$ waste minimization actions.

Source: Tabulations from 1989, 1991, 1993, and 1995 EPA BRS databases.

raw material changes — that production workers can suggest and implement. Thus, it is first necessary to organize production operations, management functions, and personnel for green manufacturing to facilitate the identification and development of both technical and common-sense waste minimization ideas [Dillon and Fischer, 1992].

There are several important prerequisites for this process. First, it is critical to have an accounting of inputs, wastes, and their associated costs at each point in the production process. According to 1994 EPA data, 31% of all reported source reduction actions were first identified through pollution prevention opportunity or materials balance audits [EPA, 1996]. The normal financial incentives to reduce costs can be highly efficient within such an accounting system, but the actual efficiency greatly depends on the extent to which true costs are accounted for. The pinpointing of costs, particularly tracking them back to specific production processes, and the projection of future costs are challenging [Florida and Atlas, 1997; Todd, 1994]. Second, the facility must thoroughly know the environmental laws with which it must comply now and in the foreseeable future. This includes environmental permits specifically applicable to it. The facility also must assess the legal implications of possible changes in its operations (e.g., the need for permits if certain changes are made or any restrictions on using particular chemicals).

Third, green manufacturing must be a central concern of the facility's top management [Florida and Atlas, 1997; Hunt and Auster, 1990]. This is usually helped by outside pressure (from government or environmentalists) or by the convincing demonstration of its benefits (e.g., reduced production costs) [Lawrence and Morell, 1995]. Fourth, it is typically very helpful to involve production workers in green manufacturing [Florida and Atlas, 1997; Makower, 1993]. When they are involved in the environmental implications of their activities, they often make substantial contributions, especially improvements in industrial housekeeping, internal recycling, and limited changes in production processes. According to 1994 EPA data, 42% of all reported source reduction activities were first identified through management or employee recommendations [EPA, 1996].

Fifth, green manufacturing will greatly benefit from the easy availability of technical and environmental information about cleaner technology options. Both in-house technical and environmental experts and outside consultants can be useful. It also can be desirable to involve the facility's suppliers and customers in the effort [Georg et al., 1992]. Often they can provide solutions not easily perceived by the facility involved in the actual production. Finally, setting challenging objectives and monitoring the facility's progress toward achieving them can help in creating effective green manufacturing [Florida and Atlas, 1997]. The targets may be financial (e.g., cost reduction), physical (e.g., input and/or discharge reduction), legal (e.g., lowering emissions to avoid the need for an environmental permit), and personnel (e.g., fewer injuries).

Choosing Green Manufacturing Options

Once the proper organizational approach is established, the first step in choosing options for green manufacturing is making an inventory by production operation of the inputs used (e.g., energy, raw materials, and water) and the wastes generated. These wastes include off-specification products, inputs returned to their suppliers, solid wastes, and other nonproduct outputs sent to treatment or disposal facilities or discharged into the environment. The second step is selecting the most important nonproduct outputs or waste streams to focus upon. Their relative importance could depend upon the costs involved, environmental and occupational safety impacts, legal requirements, public pressures, or a combination thereof.

The third step is generating options to reduce these nonproduct outputs at their origin. These options fall into five general categories: product changes, process changes, input changes, increased internal reuse of wastes, and better housekeeping. The fourth step is to pragmatically evaluate the options for their environmental advantage, technical feasibility, economic sufficiency, and employee acceptability. With respect to economic sufficiency, calculating the payback period is usually adequate.

This evaluation usually leads to a number of options, especially in better housekeeping and input changes, which are environmentally advantageous, easy to implement, and financially desirable. Thus, the fifth step is to rapidly implement such options. There typically are other options that take longer to evaluate but that also usually lead to a substantial number that are worth implementing.

Potential Green Manufacturing Options

As noted earlier, the options for green manufacturing can be divided into five major areas: product changes, production process changes, changes of inputs in the production process, internal reuse of wastes, and better housekeeping. The following discussion focuses on the physical nature of changes that can be implemented (excluding green product changes, which are discussed elsewhere).

Changes in Production Processes

Many major production process changes fall into the following categories: (1) changing dependence on human intervention, (2) use of a **continuous** instead of a **batch** process, (3) changing the nature of the steps in the production process, (4) eliminating steps in the production process, and (5) changing cleaning processes.

Production that is dependent on active human intervention has a significant failure rate. This may lead to various problems, ranging from off-specification products to major accidents. A strategy that can reduce the dependence of production processes on active human intervention is having machines take over parts of what humans used to do. Automated process control, robots used for welding purposes, and numerically controlled cutting tools all may reduce wastes.

With respect to using a continuous, rather than batch, process, the former consistently causes less environmental impact than the latter. This is due to the reduction of residuals in the production machinery and thus the decreased need for cleaning, and better opportunities for process control, allowing for improved resource and energy efficiency and reducing off-specification products. There are, however, opportunities for environmentally improved technology in batch processes. For chemical batch processes, for instance, the main waste prevention methods are (1) eliminate or minimize unwanted by-products, possibly by changing reactants, processes, or equipment, (2) recycle the solvents used in reactions and extractions, and (3) recycle excess reactants. Furthermore, careful design and well-planned use can also minimize residuals to be cleaned away when batch processes are involved.

Changing the nature of steps in a production process — whether physical, chemical, or biological — can considerably affect its environmental impact. Such changes may involve switching from one chemical process to another or from a chemical to a physical or biological process or vice versa. In general, using a more selective production route — such as through inorganic catalysts and enzymes — will be environmentally beneficial by reducing inputs and their associated wastes. Switching from a chemical to a physical production process also may be beneficial. For example, the banning of chlorofluorocarbons led to other ways of producing flexible polyurethane foams. One resulting process was based on the controlled use of variable pressure, where carbon dioxide and water blow the foam, with the size of the foam cells depending upon the pressure applied. An example of an environmentally beneficial change in the physical nature of a process is using electrodynamics in spraying. A major problem of spraying processes is that a significant amount of sprayed material misses its target. In such cases, waste may be greatly reduced by giving the target and the sprayed material opposite electrical charges.

Eliminating steps in the production process may prevent wastes because each step typically creates wastes. For example, facilities have developed processes that eliminated several painting steps. These cut costs and reduce the paint used and thus emissions and waste. In the chemical industry, there is a trend to eliminate neutralization steps that generate waste salts as by-products. This is mainly achieved by using a more selective type of synthesis.

Cleaning is the source of considerable environmental impacts from production processes. These impacts can be partly reduced by changing inputs in the cleaning process (e.g., using water-based cleaners rather than solvents). Also, production processes can be changed so that the need for cleaning is reduced or eliminated, such as in the microelectronics industry, where improved production techniques have sharply reduced the need for cleaning with organic solvents. Sometimes, by careful consideration of production sequences, the need for cleaning can be eliminated, such as in textile printing, where good planning of printing sequences may eliminate the need for cleaning away residual pigments. In other processes, reduced cleaning is achieved by minimizing carryover from one process step to the next. The switch from batch to continuous processes will also usually reduce the need for cleaning.

Changes of Inputs in the Production Process

Changes in inputs is an important tool in green manufacturing. Both major and minor product ingredients and inputs that contribute to production, without being incorporated in the end product, may be worth changing. An example where changing a minor input in production may substantially reduce its environmental impact is the use of paints in the production of cars and airplanes. The introduction of powder-based and high solids paints substantially reduces the emission of volatile organic compounds. Also, substituting water-based for solvent-based coatings may lessen environmental impacts.

Internal Reuse

The potential for internal reuse is often substantial, with many possibilities for the reuse of water, energy, and some chemicals and metals. Washing, heating, and cooling in a countercurrent process will facilitate the internal reuse of energy and water. Closed-loop process water recycling that replaces single-pass systems is usually economically attractive, with both water and chemicals potentially being recycled. In some production processes there may be possibilities for **cascade-type reuse**, in which water used in one process step is used in another process step where quality requirements are less stringent. Similarly, energy may be used in a cascade-type way where waste heat from high-temperature processes is used to meet demand for lower-temperature heat.

Better Housekeeping

Good housekeeping refers to generally simple, routinized, nonresource-intensive measures that keep a facility in good working and environmental order. It includes segregating wastes, minimizing chemical and waste inventories, installing overflow alarms and automatic shutoff valves, eliminating leaks and drips, placing collecting devices where spills may occur, frequent inspections aimed at identifying environmental concerns and potential malfunctionings of the production process, instituting better controls on operating conditions (e.g., flow rate, temperature, and pressure), regular fine tuning of machinery, and optimizing maintenance schedules. These types of actions often offer relatively quick, easy, and inexpensive ways to reduce chemical use and wastes.

Defining Terms

Batch process: A process that is not in continuous or mass production and in which operations are carried out with discrete quantities of material or a limited number of items.

Cascade-type reuse: Input used in one process step is used in another process step where quality requirements are less stringent.

Continuous process: A process that operates on a continuous flow (e.g., materials or time) basis, in contrast to batch, intermittent, or sequential operations.

Green manufacturing: Production processes that use inputs with relatively low environmental impacts, that are highly efficient, and that generate little or no waste or pollution.

Recycling: Using or reusing wastes as ingredients in a process or as an effective substitute for a commercial product or returning the waste to the original process that generated it as a substitute for raw material feedstock.

Source reduction: Any actions reducing the waste initially generated.

References

Dillon, P. and Fischer, K. *Environmental Management in Corporations: Methods and Motivations,* Tufts University Press, Medford, MA, 1992.

Florida, R. Lean and green: the move to environmentally conscious manufacturing, *Cal. Mgt. Rev.,* 39: 80–105, 1996.

Florida, R. and Atlas, M. *Report of Field Research on Environmentally-Conscious Manufacturing in the United States,* Carnegie Mellon University, Pittsburgh, PA, 1997.

Georg, S., Ropke, I., and Jorgensen, U. Clean technology — innovation and environmental regulation, *Environ. Resource Econ.* 2: 533–550, 1992.

Hunt, C. and Auster, E. Proactive environmental management: avoiding the toxic trap, *Sloan Mgt. Rev.,* Winter: 7–18, 1990.

Lawrence, A. and Morell, D. Leading-edge environmental management: motivation, opportunity, resources, and process, In *Research in Corporate Social Performance and Policy, Supplement 1,* J. Post, D. Collins, and M. Starik, eds., pp. 99–126, JAI Press, Greenwich, CT, 1995.

Makower, J. *The e Factor: The Bottom-Line Approach to Environmentally Responsible Business,* Times Books, New York, 1993.

Porter, M. E. and van der Linde, C. Green and competitive: ending the stalemate, *Harvard Bus. Rev.,* 73: 120–134, 1995.

Todd, R. Zero-loss environmental accounting systems, In *The Greening of Industrial Ecosystems,* B. Allenby and D. Richards, ed., pp. 191–200, National Academy Press, Washington, DC, 1994.

U.S. Environmental Protection Agency. *1994 Toxics Release Inventory Public Data Release,* Office of Pollution Prevention and Toxics, Washington, DC, 1996.

Further Information

The Academy of Management has an Organizations and the Natural Environment section for members interested in the organizational management aspects of green manufacturing. For membership forms, contact The Academy of Management Business Office, Pace University, P.O. Box 3020, Briarcliff Manor, NY 10510-8020; phone (914) 923-2607. Also, many of the Web sites cited below lead to other organizations with particular interests in green manufacturing.

The quarterly *Journal of Industrial Ecology* provides research and case studies concerning green manufacturing. For subscription information, contact MIT Press Journals, 55 Hayward Street, Cambridge, MA 02142; phone (617) 253-2889.

There are numerous Web sites with green manufacturing-related information, including the following: http://www.epa.gov/epaoswer/non-hw/reduce/wstewise/index.htm; http://es.inel.gov/; http://www.epa.gov/greenlights.html; http://www.hazard.uiuc.edu/wmrc/greatl/clearinghouse.html; and http://www.turi.org/P2GEMS/.

14

Product Management

John Heskett
Illinois Institute of Technology

Susan Walsh Sanderson
Rensselaer Polytechnic Institute

Mastafa Usumeri
Auburn University

Robert G. Cooper
McMaster University

Robert P. Smith
University of Washington

Catherine Banbury
Saint Mary's College

Marc H. Meyer
Northeastern University

Roger J. Best
University of Oregon

Liora Salter
York University

Robert J. Thomas
Georgetown University

Stanley Slater
University of Washington

Ajay Menon
Colorado State University

Denis Lambert
IDX Corporation

Robert B. Handfield
Michigan State University

Michael R. Hagerty
University of California

0-8493-8577-6/99/$0.00+$.50
© 1999 by CRC Press LLC

Noellette Conway-
Schempf
Carnegie Mellon University

Lester Lave
Carnegie Mellon University

Kim B. Clark
Harvard Business School

Steven C. Wheelwright
Harvard Business School

14.1   Design Function of a Product

John Heskett

Reactions against technology are often fed by applications being framed without adequate consideration of their meaning for people — the interface problems of VCRs is a notorious example. Avoiding this is possible by managing the design function of a product from its earliest origins on the basis of user-centered methods and processes, which can ensure that technology is adapted to people. This section will describe how design methods can be applied in development processes to ensure innovative products are developed rapidly and cost-effectively, in terms specifically appropriate for users.

The term "design" has many applications. In this section, it will signify industrial design. If engineering design is mainly concerned with the successful technical function of a product and process technology, industrial design is concerned with how the technology encapsulated in products is made accessible, applicable, and appropriate for users. Both approaches to design are crucial and cooperation between them is clearly essential.

Developing concepts based on what users say they want has considerable limitations. Few users can suggest substantially new ideas and the result can often be products barely distinguishable other than in incremental detail from those of competitors. On the other hand, with innovative concepts there is a risk they may be too far ahead of what users are prepared to accept. The aim if one wishes to be competitively innovative is to give customers products they never knew they wanted. This can be achieved if, in the earliest stages of development processes before major costs are incurred, designers apply **early prototyping** to frame innovative product concepts that allow rapid **behavioral testing** of users' reactions. This can also enable products to be brought to market faster, embodying higher value, and capable of systemic development to ensure a longer life cycle.

User-Centered Design

Basically, four generic design functions of a product, each emphasizing a dominant attribute, can be identified. Designs can be

1. Technology centered
2. Marketing centered
3. Image centered
4. User centered

The precise combination of these will vary according to product, but all should be closely integrated, though this is not always the case. Divisions often exist between technology-centered aspects of design in products and manufacturing, marketing demands, and ideas about "styling". All too often, user needs are subordinated to available technology or attempts to create a superficial image for purposes of differentiation. This "push" approach to production and marketing has obvious limitations; one research program revealed that, in over half the projects studied, products needed partly redesigning by the time final prototype testing took place, since customer requirements were found not to match those originally planned after concept evaluation [Bonnet, 1986].

The current competitive situation in many product sectors is now more oriented to a "pull" approach, requiring products to be more specifically adapted to consumer needs. This makes it imperative, however, that user needs be placed in clearer perspective, identified with greater specificity, and integrated into development processes at an early stage. Yet there are still problems in achieving this, even in a company as well known for its design standards as Apple. The company's head of industrial design has been quoted as saying (*The Sunday Times*, 1996), "There's an incredible tendency to focus on value that you can easily measure. Things like chip speed are easy to talk about and people feel more comfortable with that sort of thing. When you get into these emotive sort of areas which are one of the reasons our customers love the product, it becomes sometimes downright embarrassing to talk about in a corporate environment."

Implicit in this comment is the point that, although there are limitations to quantitative methods, the affective areas of user behavior are not easily dealt with. Clearly, the information and values regarded as significant largely determine the product outcome. Therefore, if the needs that technology can satisfy for users are to be integral to any development process, they need to be acknowledged from the outset, and as far as possible, based on understanding the experience of users, in all their levels and diversity.

A corollary of the above point is that how industrial design is utilized in developing successful products will depend to a great extent upon how its significance is viewed by managers of the product development process. If it considered to be only concerned with the superficial details of style, it will essentially function as a decorative afterthought. If employed, however, in terms of user-centred approaches as a generator of superior product value and a contributor to the overall efficiency of production, then it will be best utilized early in the product development process.

In terms of its essential contribution to product development, three major functions of industrial design can be identified:

- It gives a product concept tangibility, which is a vital stage in translating from abstract idea to an actual form as perceived by users, thus enabling decisions on the feasibility of ideas to be more firmly grounded.
- The form of a design has important implications for manufacturing feasibility and therefore cost. Identifying any incompatibilities, or need for new equipment, at an early stage can be a vital element in decision making on a project.
- The reality of a design and its value as perceived by users is the ultimate determinant of market success and should be the core focus of any development process.

Making a Concept Tangible

Any product development process depends initially on the generation of ideas representing a range of possibilities. Many can be eliminated in the early stages as unsuitable for various reasons. The time and cost of generating ideas and their preliminary weeding out is relatively low. Those considered feasible can be further investigated in terms of potential, but this will involve increasing levels of cost and resources. Without a clear idea of what products might be in real terms and real time, much time and cost can be wasted since they cannot be effectively evaluated. Until any concept is realized in tangible form, it will be difficult to clearly establish whether full development should proceed or not — decisions will be based on instincts or assumptions. Getting into tangible form fast and testing it is an excellent way of improving decision making at an early stage. It can be seen as a way of managing risk — be prolific with concepts, test them early and rapidly, and ditch them if they don't work before further time and money are invested.

Design Determines Cost

The role of industrial design as a primary determinant of the tangible reality of product concepts automatically has in-built implications for cost at every subsequent stage of development. Yet the relatively low cost of design decision making up front in the development process, in relation to the high proportion of development costs committed as a result of these, should make it easier to ensure that decisions taken early do not create unnecessary complications downstream. An increase in expenditure in the early stages can therefore be a sound investment.

The actual rate of increase in cost spent, and the parallel level of future costs committed, as development progresses will be affected by factors such as type of product and the complexity or simplicity of its technology, but in all cases the benefits of problems being more specifically defined and accurately solved in the early stages of development should be obvious. If errors can be eliminated when costs are low, the possibility of them being compounded, and thus more complex and difficult to solve later in the process, will be reduced, with significant cost savings.

Satisfying Customers

What, however, is meant by getting ideas right? Design processes may be cost effective and may generate product ideas compatible with such factors as available or foreseeable resources that still fail in the market. Extensive studies of failure rates for new product introductions [Crawford, 1979, 1987] conclude that the average rate is around 35%. Another ongoing research project [Cooper, 1986] concludes that only one out of every four new products that are fully developed becomes a commercial success. These figures can be explained as an inevitable Darwinian process in which high wastage rates are a necessary aspect of the struggle to produce successful products, a commercial survival of the fittest. Undoubtedly, many innovations inevitably involve considerable risk. There is nothing inevitable about such failure rates, however, and, although some unsuccessful products can perhaps be attributed to unforeseeable changes in market conditions, inadequacies in managing product development are also a major cause. What is most frequently overlooked in this respect is the second general function of design referred to above: the need to ensure, at the earliest possible stage, that a product concept is appropriate for its intended users.

Competitive advantage in any respect is barely conceivable unless it is manifested in the product. It is ultimately the embodiment of values and the generator of reputations in the marketplace — it is a company or brand to a user. An essential focus of industrial design is the factors that create uniqueness and superior value, the prime determinants of economic value in a product. The specification of the tangible form of the product therefore defines its reality, as it will be seen, heard, felt, and understood by every purchaser and user.

Three primary functions at different points of a product's life cycle can be identified as key points at which the contributions of design in product development can be maximized to give vital competitive advantages, summed up as faster, higher, and stronger:

Faster — Design concepts and methods are available that can significantly speed up development processes while giving them more exact focus and definition at an early stage, which can create greater stability in more costly and complex phases of development.

Higher — One of the most frequent claims for design is that it adds value, which overlooks a much more powerful role in creating value. If this can be sustained, the means of adding and creating value should be capable of closer definition and precise application to ensure more rapid acceptance of products in their intended market and a faster return on investment.

Stronger — In an age when the turnover in product concepts can be extremely rapid and volatile, and in some cases so short that return on investment is minimal, methods of extending product market lives can considerably enhance profitability.

Early Prototyping and Speeding the Development Process

The need to speed product development processes results from many trends, among them changes in production technology, markets, and shortening product life cycles. In addition to factors such as cost or quality, therefore, companies can compete by rapidly developing products that create new markets, by identifying and responding to emerging needs, redefining niches, or stimulating and being sensitive to changes in existing stable markets. This means an ability to develop products that are truly innovative.

As noted earlier, this implies risk. It is therefore imperative to develop prototypes that enable crucial decisions to be made about the viability of product concepts before time and resources are expended on them. Many concepts of prototyping do not help in this stage, being too detailed and expensive. Two approaches can be used to give early focus to the product development process: **conceptual prototypes** and **behavioral prototypes**.

Conceptual prototypes are simple, preliminary concepts in two or three dimensions that embody earliest ideas about a product concept, developed on the basis of behavioral observation, which can discussed for feasibility, and then subjected to behavioral testing. The aim at this stage is not verisimilitude in the prototype but a tangible concept that can be rapidly and iteratively tested and adapted.

It is important to emphasize that the origins and testing of conceptual prototypes are in behavioral terms, which means aiming at replicating how users actually behave — not what they say they want or how they say they would behave — in relation to the proposed design. Simple simulations of user conditions and responses enable conceptual prototypes to be tested, recorded using observation techniques such as photography and video-ethnography, and rapidly changed to acknowledge every problem users encounter. Concepts developed in the light of users' behavior enable the summary of this testing, a behavioral prototype, to be evolved, resulting in a product concept that is defined in the essential terms in which the product must succeed in the market before moving to more detailed and expensive prototypes. It is this tighter focus on user needs at the earliest stage that can deliver substantial benefits later: in speed of process, elimination of errors in concept and specification that are exceedingly costly to rectify, giving greater assuredness and stability in the later and vastly more expensive stages of development, and being fast to market.

Moreover, by involving all major players in this early prototyping phase, it is possible to build a team approach before any group establishes turf rights over an idea. The prototype establishes a common concept and vocabulary that is directly focused on users, whoever they might be. Indeed it becomes possible to involve suppliers and customers in ensuring the viability of a concept before it reaches the more expensive stage of development.

Adding Higher Value to Products

Speed in development can also be highly advantageous, but only if the products emerging have high added value as perceived by customers. To achieve this, in addition to techniques of behavioral testing, **Human factors** methodologies can more precisely determine user needs and concepts of value. It has four subcategories:

Physical human factors — These consists of studies of human dimensions and capabilities, e.g., ergonomics and anthropometrics.

Cognitive human factors — Studies of human processes of cognition and perception.

Social human factors — How people work in groups and social contexts.

Cultural human factors — How values and habits differ between groups and societies.

The purpose of the methodologies in all these categories is to make industrial design less reliant on personal inclination and more capable of applying specific tools to highlight actual behavior that can be observed, analyzed, understood, and tested.

Systemic Development

Generating one-off product ideas, however innovative they may be, is becoming increasingly difficult to justify in many product sectors and is being replaced by more systemic concepts of design. There are two levels on which systemic development in industrial design is important. Innumerable products cannot be understood as objects in themselves but exist as part of a wider environment. An ATM machine, for example, has no meaning outside the context of the system in which it functions. Many products incorporate both hardware and software and the design of interfaces has become integral with products. These linkages also need to be incorporated at the earliest possible stage. Secondly, if a product is understood to be not an end in itself, but the starting point for a wide variety of possibilities for further incremental development, or transfer to other markets and products, it can give a concept longer market life through systemic development of its full potential and possibilities.

Speeding development, creating higher value, and systemically developing products are ways in which industrial design, when managed as an integral element in development processes, can make powerful contributions to market competitiveness through the potential to make processes more efficient and products superior.

Defining Terms

Behavioral testing: Tests that simulate a user scenario and seek to identify exactly how users behave in the context of actual use.

Behavioral prototypes: The outcome of behavioral testing, a product concept based on user reactions in practice.

Conceptual prototypes: Simple, preliminary, two- or three-dimensional forms embodying the earliest ideas about product concepts, used for behavioral testing.

Early prototyping: The use of conceptual prototypes for behavioral testing at the earliest stage of product development. The outcome is a behavioral prototype.

Human factors: A series of methodologies to specifically identify user needs at various levels of action, reaction and values.

References

Bonnet, D. Nature of the R&D/marketing co-operation in the design of technology advanced new industrial products, *R&D Mgt.*, 16: 121, 1986.

Cooper, R. G. *Winning at New Products*. Addison-Wesley, New York. p.12, 1986.

Crawford, C. M. New product failure rates-facts and fallacies, *Res. Mgt.*, September: 9–13, 1979; New product failure rates: a reprise, *Res. Mgt.*, July-August: 20–24, 1987.

The Sunday Times (London). Design Leads Apple Revival, December 8, 1996.

Further Information

Lorenz, C. *The Design Dimension: The New Competitive Weapon for Product Strategy and Global Marketing*, Blackwell, Cambridge, MA, 1990.

Walsh, V., Roy, R., Bruce, M., and Potter, S. *Winning by Design: Technology, Product Design and International Competititiveness*, Blackwell, Cambridge, MA, 1992.

14.2   A Framework for Product Model and Family Evolution[1]

Susan Walsh Sanderson and Mastafa Usumeri

The emergence of global markets has fundamentally altered competition as many firms have known it. Perhaps the most prominent, and disconcerting, features of these markets to firms accustomed to relatively stable customer bases and known competitors are the speed with which they grow and change and the increasing incidence of incursions by former noncompetitors.

Market dynamics are forcing the compression of product development times and the expansion of model variety. The challenge that manufacturing firms face today, simply put, is that global markets demand frequent product innovation and extensive product diversification *simultaneously.* When to prune, and when to plant, a new family tree is rapidly becoming *the* critical decision for manufacturers. Helpful clues lie in the forces that drive and shape patterns of product model and family evolution.

The Bases of Analysis

In the following sections we present a framework for competition among product models and that reflects growing acknowledgment by management theory that (1) product variety is important and (2) product designs are becoming obsolete much more rapidly than ever before. We define units of analysis employed in the framework and analyze the key descriptive measures, product variety, and rate of design change, at two levels of analysis, for product models and for product families. Empirical examples illustrate the application of these measures in a number of product categories. Finally, we identify, change, and integrate into a life cycle model of product competition the forces that drive product variety and rate of change.

Product models and product families, long and widely accepted in practice as basic units of analysis, have recently begun to receive attention in academic circles. Researchers have variously advanced examples of "design families" and "design variants" among automobiles, jet engines, aircraft, and hovercraft; Roy and Gardiner [1988] demonstrated effective management of incredible varieties of product models that target customer needs in distinct international markets; Sanderson and Uzumeri [1992] suggested as a structural rationale a "design hierarchy" that includes a "core concept" and solutions to various "sub-problems"; Clark [1985] developed an integrative typology for design projects that includes "platform" as well as enhancement, hybrid, and derivative projects [Wheelwright and Clark, 1992].

Model distinctions and family relationships are fairly easily established for sophisticated technical products, particularly assembled systems. Few would quarrel, for example, with Boeing's determination that the 747-200, 747-300, 747-400, and 747-SP are individual models within the same product family [Rothwell and Gardiner, 1988]. Similarly, the definition of models in different passenger car product families tends to be fairly consistent, typically combining basic styles such as two- and four-door sedans, two-door coupes, three- and five-door hatchbacks, and, perhaps, a wagon with a few, well defined "packages" that, collectively, cover the full spectrum of features and price.

However, as one moves away from sophisticated assemblies, reliance on technical criteria increasingly gives way to subjective perceptions of subtler characteristics. Boundaries are particularly fuzzy for product families, such as house paint, that are based on recipes. What constitutes a model — each basic color or every mixture a customer takes home? Is each grade of paint a product model or a family?

Some contend that such subjectivity imperils any attempt at classification or categorization [Zeithaml, 1988], others that the threat is overdrawn, that knowledgeable observers generally make consistent interpretations [Tornatsky and Klein,1982]. For most commercially manufactured products, the many published suppliers catalogs and industry buyers' guides reflect industry consensus regarding product model and family designations. We thus define a **model** to be *a product design that differs sufficiently from other*

---

[1]*Source:* Sanderson, S. W. and Uzumeri, M., *Managing Product Families*, Irwin/McGraw-Hill, 1997.

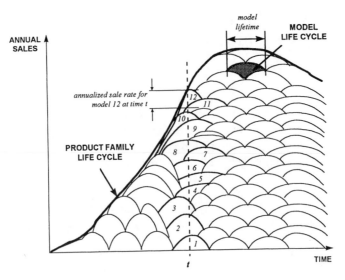

FIGURE 14.1   The product family life cycle.

designs that the manufacturer assigns it a distinctive commercial designation and a **product family** to be *a set of models that a given manufacturer makes and considers to be related.*

Industry-derived definitions, albeit easy to apply, are not without problems, however. It may, for example, serve the marketing interests of a given manufacturer to fudge the definitions, either to inflate variety, or to oversell change. It is thus important to validate manufacturer's definition against a model's technical specifications.

The Product Family Life Cycle

Having established product models and families as our units of analysis, we analyze each in terms of variety and rate of change in order to arrive at an integrated picture of a firm's *pattern* of product competition. Note that in Fig. 14.1, which presents the sales history for a family of related models, each model has its own *model life cycle*. A firm's *product family life cycle* is the aggregation of these model cycles.

The model variety available to customers at t, for example, is 12. We can estimate model variety thus only if the boundary of the product family is known, which reinforces the need to attend to industry criteria for family membership.

We observed earlier that *model change* has two components, *rate* and *type*, the former reflecting the frequency, the latter the degree of change. We can identify two types of change. *Additive* change occurs when existing models are augmented by one or more new designs, yielding (provided these designs do not supplant existing models) a concomitant increase in model variety. With *replacement*, or *serial*, change, which suggests the discontinuation of earlier models, model variety is reduced or remains unchanged.

Model lifetimes — easily identified after the fact and, to the extent that a manufacturer has good product planning and forecasting, reasonably predictable — can be used to generate a rough estimate of the *rate of serial model change*, that is, the rate at which a model is being replaced. This can be approximated by the reciprocal of the model's lifetime. Models with 2-year lifetimes, for example, will be replaced at a rate of 0.5 models per year if existing model variety is to be preserved.

Sales volume (reflected in the height of the curves in Fig. 14.1) for both individual models and a product family are central concerns of any manager. Some models (e.g., 7 and 11 in Fig. 14.1) have lower sales volumes than others. A complete analysis of this product family would require detailed sales histories for each model. Manufacturers can readily assemble this information for their own products, but it is difficult to obtain such data for competing firms. Product introductions tend to be well publicized, and

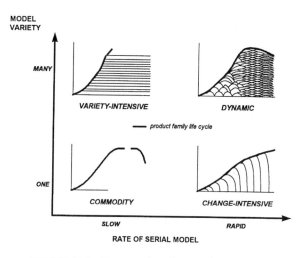

FIGURE 14.2   Patterns of product model evolution.

the withdrawal of models is usually well known through the industry grapevine. As companies produce more models targeted to distinct customer needs, sales volumes of individual models tend to become smaller.

The Model Variety-Change Framework

The shape of the product family life cycle depicted in Fig. 14.2 will vary greatly from one product family to another. Juxtaposing model variety and rate of model change, as in Fig. 14.1, yields four distinct patterns that reflect a firm's competitive reaction to internal drives and external forces, particularly those exerted by markets, the introduction of new technologies, and the actions of competitors.

The pattern of model evolution for *commodities* — eggs, carbon black, soda ash, chemical feedstocks, and other products that do not, or cannot, exhibit significant design variety — is depicted in the lower left quadrant of Fig. 14.2. Design variants for these products are not offered, either because no one can envision an alternative or because customers are unwilling to accept a change or substitute. That most early published descriptions of the product life cycle resemble this pattern reflects the slow rates of model replacement consistent with mass production.

Products such as hand tools, light bulbs, and door hardware exhibit a *variety-intensive* pattern of model evolution (upper left quadrant of Fig. 14.2) characterized by additive model change. Stanley Works' hand tools, for example, are offered in thousands of models that tend to change very slowly, a consequence, as one Stanley manager put it, of their market lives being "measured in centuries.[2]" Packaging changes are frequent, new models much less so; fewer than 3% of the more than 1500 products listed in the 1990 Stanley hand tool catalog were identified as new and most of those models represented only cosmetic changes. Clearly, even slow rates of change can yield extraordinary model variety in a firm that remains in business for 150 years.

Products characterized by replacement change, e.g., software, exhibit a *change-intensive* pattern of model evolution (lower right quadrant of Fig. 14.2). Firms constrained to put all their effort into maintaining the pace of competitive redesign, to develop follow-on models even as they introduce new models, strive to minimize model variety. Microsoft's DOS, for example, has since 1981 been marketed in versions 1.1, 2.0, 2.1, 2.2, 3.0, 3.1, 4.0, 4.1, 5.0, and 6.0, each of which effectively replaced its predecessor as far as new sales were concerned.

A complex and volatile mix of additive and replacement change yields a *dynamic* pattern of model evolution (upper right quadrant of Fig. 14.2). A growing number of globally competitive manufactured

---

[2]Personal communication, Tim Walsh, VP Marketing, Stanley-Proto Tools, Norcross GA, November 1989.

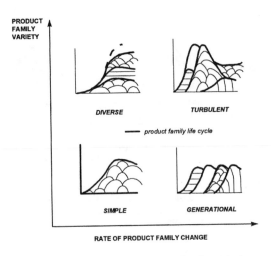

FIGURE 14.3 Patterns of product family evolution.

products — computers, machine tools, automobiles, specialized semiconductors, and medical equipment, among others — exhibit this pattern.

Product Family Variety and Rate of Change

Beyond the competitive patterns within individual product families, companies have a larger opportunity to develop multiple product families simultaneously. These, too, can be plotted in terms of variety and rate of change (see Fig. 14.3).

For simple product families, the product model and family life cycles are essentially the same (lower left quadrant of Fig. 14.3). For most other products, patterns of family evolution are independent of, and can be quite different from, patterns of model evolution. Product families that replace one another in rapid, serial fashion exhibit a *generational* pattern of evolution, as illustrated by the lower right quadrant of Fig. 14.3. One might find in successive generations of a product family, for example, any or all of the four patterns of product model evolution. Of all the patterns in product model and family evolution, this is the most confusing, the difficulty being to distinguish generational replacement from change-intensive model evolution. The logic advanced thus far suggests three criteria for generational product evolution:

- Powerful and persistent market demand for continuous improvement (without which major change will not occur).
- More than one technological way to satisfy the market need (given only one technological approach, generational model evolution will dominate).
- Strong market resistance to the simultaneous existence of more than one product family (without such resistance, turbulent product family competition will result).

Products that satisfy all three criteria are difficult to find. The shift from records to audiotape cassettes to compact discs, and, finally, to digital audiotape is perhaps the best example of generational product family change, but even here product family generations overlap. Are the different media in fact serving distinct niches, with analog audiotape the inexpensive solution, compact disc the choice for durability, and digital audiotape the new standard for fidelity and convenience?

Concurrent as opposed to successive product families in which diverse patterns of model evolution coexist and prosper, linked by a common technology or marketing insight that encourages observers to view them as a group, are characterized as having a turbulent life cycle (turbulent patterns of product family evolution can vary widely, as suggested by the two upper quadrants in Fig. 14.3. Again, Stanley Works provides a helpful example. Its hand tool product family represents only a fraction of the company's

TABLE 14.1   Evolution of Stanley's Hand Tool Product Families

Operating Unit	Year Acquired	# SKUs[a]	New Models	1989 Sales[b]	Representative Products
Stanley Tools	1843	1500+	50–75/yr	$560M	Woodworking and construction hand tools
Fastening systems					
Bostitch, Hartco, Sutton-Landis, Halstead, Spenax, Parker Tools	1986-89	4000+	10–20/yr	$320M	Fasteners and fastening systems
Mechanics Tools					
MAC Tools, Proto Tools, Peugeot Tools, National Hand Tool and Beach Industries	1980-87	11,000+	300–500/yr	$400M	Auto mechanics tools, industrial hand tools, auto mechanics tools, and toolboxes

[a] SKU = stock keeping unit.
[b] *Source:* estimated, Shearson Lehman Hutton, March 8, 1989.

overall product variety. During the 1980s, thousands of new models were added as Stanley Works expanded beyond the maturing carpentry hand tool market into product families for industrial hand tools, auto mechanic's tools, hydraulic and pneumatic power tools, glue guns, power fastening devices and fasteners, tool boxes, and a number of other hardware-related products (see Table 14.1).

Both generational and turbulent product family evolution present major challenges to manufacturers, inasmuch as each new product family represents a threat to, or opportunity for, every competitor. Each must manage its own complex pattern of product family evolution while contending with competition from the similarly volatile product families of its competitors. The generational and turbulent characterizations of product family evolution are consistent with anecdotal reports of intensified product competition among globally marketed products. Competitive pressures contend some observers, by forcing its compression are driving the product family life cycle into generational and turbulent patterns [von Braun, 1991]. An example can be found in the office electronics marketplace, which has evidenced an increase in product family variety over the past 20 years.[3] Equally striking is the decline in relative sales revenue for some product families, notably portable clocks and calculators.

The proliferation of product families, particularly when many are based on fundamentally different technologies, is capable of generating complex evolutionary dynamics [Young, 1991]. New families can complement or replace existing families, be introduced individually or in groups, exit quickly, or remain viable for years. Given market uncertainties, firms are seldom able to predict the effect of a particular product family change. Persistent incremental changes at the model level, though they may not determine the long range outcomes of product family evolution, may play a major role in the short-term battle for market share and revenue that enables a firm to remain competitive.

The Structural Implications of Variety and Change

The capacity of our framework to illuminate the relationships among product variety and rate of change and the internal structure and operation of firms — which it does by characterizing, across a broad spectrum of industries and levels of analysis, patterns of variety and rate of change for both groups of models and groups of product families — enables us to describe succinctly competitive behavior that is not accounted for in previous frameworks and models.

---

[3] The sales data were obtained from the survey of electronic product sales that merchandising and its successor magazine, *Dealerscope Merchandising*, has published annually throughout the study period. The definitions of the product families and of the overall industry of office electronics were borrowed directly from the publication's classification system. Because the publications are aimed at chain store executives, they focus on sales revenues through retail outlets and miss large corporate purchases. Hence, the industry might be more accurately referred to as "office electronics for home and small office use".

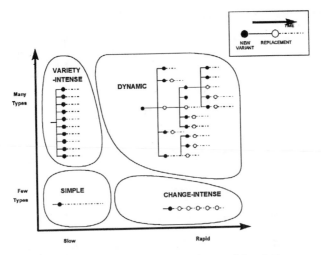

FIGURE 14.4   A framework of product model evolution.

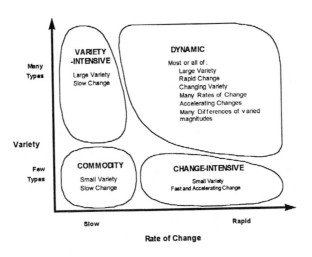

FIGURE 14.5   Sources of product complexity.

Variety-change patterns do not, of themselves, indicate competitive success or strategic causality. They do discriminate among the behaviors of firms competing in the same product, even though the firms' approaches may appear quite similar by other criteria.

The manner in which sequences of individual designs relate to one another in each of the four regions of model evolution is depicted in Fig. 14.4. The inherent complexity of dynamic product model evolution, which is distinguished by the *interactions* between product variety and rapid change is clearly evident in the upper right quadrant.

This dynamic region of product model evolution is qualitatively different from the others inasmuch as it involves dealing with different rates of design change, which individually might be accelerating or decelerating and greater disparity in the magnitudes of the differences between models and over time.

Implications of Patterns of Product Model Evolution

In general, our framework is likely to have important *operational* implications that differ markedly between the extremes. Moreover, because product competition is so central to its operations, these implications are likely to be felt throughout the firm. Below, we consider specifically implications of the

TABLE 14.2 A Framework for Design, Operations, and Organization

Variety Intensive	Dynamic
Firms search for economies of scope	Firms search for cost-effective flexibility
Firms invest in manufacturing technologies that can be shared by many product designs; flexible manufacturing processes are exploited to shift easily among multiple designs in a stable universe of products	Firms pursue all forms of manufacturing flexibility; try to build "core competencies" that can be applied widely and flexibly; are early adopters of "lean production systems" and "concurrent engineering"
Marketing develops sophisticated models to forecast sales volumes and continually debates the wisdom of adding new variants and dropping old ones	Marketing listens intently to customers in order to anticipate pending shifts in needs
Product designers may work on many different incremental designs at the same time	Product designers work on a wide range of design projects, from minor cosmetic changes to major, multiyear technological innovations
Organizational learning occurs mainly through the sharing of technical ideas among the designers and makers of similar products	There is constant tension between the desire for rapid, focused learning to master change and the need to formalize lessons and apply them across related designs

Commodity	Change Intensive
Firms search for economies of scale	Firms search for economies of scale, with full knowledge of the short life span of their capital investments
Manufacturing processes are relatively inflexible and firms emphasize process learning curve and capital process improvements	Firms place great store on the learning curve and invest in design and engineering technologies (e.g., CAD/CAM, simulators, simultaneous engineering) and in improved manufacturing process improvement
Firms concentrate on selling all of their plant capacity	Marketing concentrates on selling product as rapidly as possible and on guessing the timing of competitors' new product introductions
Innovation activities are limited to process technologies that yield costsavings and quality benefits	Product and process designers work very intensely on one project at a time
Organizational learning occurs slowly and exerts only a modest impact	Organizational learning occurs in focused groups and concentrates on immediate application of findings to the next design generation

several patterns of product model evolution for: customer needs, design, operations and organizations, strategic management, and raw materials supply.

A Framework for Design, Operations, and Organization

Table 14.2 summarizes ways in which firms that can identify the patterns of evolution that characterize their product models can adjust their design, operational, and organizational approaches to be compatible with those patterns. These are elaborated below for each of the quadrants depicted in the framework.

Commodity Quadrant

With little incentive, in the absence of variety and change, to invest in flexibility makers of commodity products might be expected to focus on capital process improvements that accelerate movement down the process learning curve. Economies of scale in production, where technically possible, will be most relevant to products in the commodity region. Capital investment in large plants to exploit the potential for high-volume sales will have ample time to pay off.

Variety-Intensive Quadrant

The need to share design, production, distribution, and marketing assets among many products with relatively low sales volumes leads firms to stress modular product and process design and invest heavily in focused factories or flexible manufacturing facilities that facilitate switching among models in a well-defined universe of product designs. Economies of scope are likely to be emphasized by such firms.

Change-Intensive Quadrant

Rapid change in product families necessitates investment in the development of new component technologies as well as in technologies that facilitate new product design (e.g., computer-aided design) that speed new designs into production (e.g., computer-aided manufacturing) in machines that can be rapidly retooled and in simultaneous engineering programs. Sizable investments in fixed manufacturing are likely to be avoided, particularly if manufacturing requirements are subject to change from one design generation to another. Opportunities for new generation designs pose a dilemma: they exert pressure to produce at high volumes, increasing the potential for economies of scale, but the relatively short life of each design generation renders the requisite capital investments difficult to recoup. This is increasingly a problem for the approximately 15 active manufacturers of dynamic random access memory chips (DRAMs): a new generation of DRAM emerges every 2 to 3 years — entailing a capital investment in manufacturing capability on the order of $1 billion per facility — to satisfy a worldwide market valued at roughly $6 to 7 billion per year.

Dynamic Quadrant

To accommodate the production of many related designs each selling at a different rate, firms must exploit economies of both scope and scale, aware that volatile sales rates increase the risk of fixed capital investment. If Toshiba's DRAM division suddenly traded its management team for that of Stanley Works, chaos would almost certainly result for both organizations. Acquisition strategies and decisions that were successful at Stanley Works would fail utterly at Toshiba. By the same token, large, focused investments in new product technology of the scale that is necessary in the semiconductor industry would be wasted at Stanley Works. The difference between a variety-intensive and change-intensive product is so profound that it is even difficult to see how a firm could effectively manage both types of product under the same roof.

Summary

In highlighting differences in product model and family evolution, our framework sheds light on a number of new and unresolved competitive issues. However, although it makes a substantial contribution to the body of knowledge on innovation, it nevertheless leaves many questions unanswered. For example, we have applied the framework at only two levels of analysis, variety and rate of change, whereas reality offers dozens of levels of analysis. Taking product models and families as the units of analysis, moreover, glosses over many important competitive distinctions, among them, components, subassemblies, platform designs, brands, divisional product lines, business units, industry collaborations, and multinational cartels. Aspects of product competition play out at all these levels.

Nor is our framework always easy to apply. The case of DRAMs illustrates the need to be clear about assumptions when assigning product families or companies to particular quadrants in the framework. For more than a decade, DRAMs exhibited a persistent pattern of serial product and process innovation, with major new models appearing approximately every 3 years.[4] To keep up this pace, DRAM manufacturers have had to introduce new models within a short time of the market leader, with the result that rates of innovation have necessarily been quite similar. DRAMs would seem to be an ideal example of generational product family evolution. Yet many semiconductor engineers consider the attendant improvements in processing technology to be merely incremental, not radical. To accept their interpretation would — the enormous expense associated with the new process technology — notwithstanding recast DRAMs as an example of change-intensive model evolution.

---

[4] That DRAM generations until very recently exhibited little model variety is a function of the nature of DRAM manufacturing. Manufacturers rely on learning curve effects to achieve the finer tolerances, higher yields, and increased profitability needed to reach the next density level. To divert factory production from high-volume, learning-intensive products to marginally different designs that contribute little knowledge useful for process improvement can impair a firm's ability to achieve that next plateau. DRAM design variety increased dramatically in 1989, largely for technological reasons. Previously, each generation of DRAMs had outdone its predecessor across all relevant dimensions of performance. This became more difficult with the much denser chips that began to emerge, attended by trade-offs between, for example, information density and access time and voltage. The consequence was a proliferation of chip designs.

Defining Terms

Model: A product design that differs sufficiently from other designs that the manufacturer assigns it a distinctive commercial designation.

Product family: A set of models that a given manufacturer makes and considers to be related.

References

Clark, K. B. The interaction of design hierarchies and market concepts in technological evolution, *Res. Policy*, 4: 235–251, 1985.

Rothwell, R. and Gardiner, P. Re-innovation and robust designs: producer and user benefits, *J. Market. Mgt.*, 3(3): 372–387, 1988.

Sanderson, S. W. and Uzumeri, M. Industrial design: the leading edge of product development for world markets, *Des. Mgt. J.*, Summer: 28, 1992.

Tornasky L. G. and Klein, K. J. Innovation characteristics and innovation adoption-implementation: a meta-analysis of findings, *IEEE Trans. Eng. Mgt.*, EM-29: 28–45, 1982.

Young, L., June 17. Product development in Japan: evolution vs. revolution, *Electron. Bus.*, 75–77, 1991.

von Braun, C.F. The acceleration trap, *Sloan Mgt. Rev.*, Fall: 49-58, 1990; The acceleration trap in the real world, *Sloan Mgt. Rev.*, Summer: 43–52, 1991.

Wheelwright, S. C. and Clark, K. B. Creating project plans to focus product development, *Harv. Bus. Rev.*, 70: 70–82, 1992.

Zeithaml, V. A. Consumer perceptions of price, quality and value: a means-end model and synthesis of evidence, *J. Market.*, 52: 2–22, 1988.

Further Information

Meyer, M. H. and Lehnerd, A. P. *The Power of Product Platforms*, The Free Press, New York, 1997.

Utterback, J. P. *Mastering the Dynamics of Innovation*, The Harvard Business Press, Boston, 1996.

14.3 Attributes of Successful New Products and Projects

Robert G. Cooper

An understanding of the attributes of successful new products and projects is central to effective new product management. It provides insights for executing new product projects (for example, are certain best practices strongly linked to success?) and yields clues to new product selection (what is the profile of a winner?). These success factors can be approximately divided into two groups:

1. Process attributes — those factors that capture the nature of the new product process and how the project is undertaken. These are often *controllable* factors ... doing *projects right.*
2. Selection attributes — those factors that describe the new product project and its situation. These tend to be *outside of the control* of the project leader and team but are useful in *project selection ...* doing the *right projects* (see Table 14.3).

Of the two sets, process factors, presented first, have by far the strongest impact on success.

The Critical Success Factors — Process Related

In Search of the Critical Success Factors

The keys to new product success outlined in this section are based on numerous research studies into why new products succeed, why they fail, and comparisons of winners and losers. Many of these investigations have been reported over the years in various journals and books. The most revealing of these studies have been the large sample, quantitative studies of successful vs. unsuccessful new products. They

TABLE 14.3 Impacts of Success Factors on New Product Profitability and Timeliness

Factor	Correlation with Profitability* (Rank Ordered)	Correlation with Timeliness**
Unique, superior product — differentiated	0.557	—
Strong market and customer orientation	0.440	0.406
Sharp, early product definition	0.413	0.242
Proficient up-front, predevelopment homework	0.366	0.408
Cross-functional team	0.351	0.483
Synergy (leveraging core competencies)	0.316	—
Market potential (size, growth, and need)	0.312	0.215
Effective market launch	0.312	(0.205)
Familiarity	0.272	—
Technical proficiency	0.269	0.316
Competitive situation	—	—

* Profitability: degree to which project's profits met or exceeded the acceptable profit level for this type of investment [Cooper, 1995].

** Timeliness: on-time performance (launch date vs. scheduled date) and time efficiency (time to market vs. fastest possible).

Correlations are Pearson product-moment correlations, significant at the 0.05 level. Parenthesis indicates not quite significant.

began with Project SAPPHO in the early 1970s, followed by the NewProd series of studies, the Stanford Innovation Project, and more recently studies in countries outside of North America and Europe. This long tradition of research has enabled us to pinpoint the critical success factors — those factors that separate winners from losers — that are outlined in this section. See Cooper [1993, 1995], Cooper and Kleinschmidt [1986, 1996], Maidique and Zirger [1984], Montoya-Weiss and Calantone [1994], Rothwell et al. [1974], and Sanchez and Elola [1991].

A Unique Superior Product

A superior and differentiated product is the number one driver of success and profitability, with success rates reported to be three to five times higher than for reactive products. However, repeated studies show that *reactive products* and *technically driven products that lack customer benefits* are the rule rather than the exception, and the majority fail! Winning products

- Are superior to competing products in terms of meeting users' needs, offer unique features not available on competitive products, or solve a problem the customer has with a competitive product.
- Provide excellent relative product quality, relative to competitors' products, and in terms of how the user measures quality.
- Feature good value for money for the customer, reduce the customer's total costs (high value in use), and boast excellent price/performance characteristics.
- Offer product benefits or attributes easily perceived as useful by the customer and benefits that are highly visible.

A point of distinction: *benefits* are what customers or users value and pay money for; by contrast, *attributes* are product features, functionality, and performance — the things that engineers and designers build into products.

A Strong Market Orientation — Market Driven, Customer Focused

A thorough understanding of customers' needs and wants, the competitive situation, and the nature of the market are an essential component of new product success. This tenet is supported by virtually every study of product success factors. Conversely, a failure to adopt a strong market orientation in product innovation, an unwillingness to undertake the needed market assessments, and leaving the customer out of product development spells disaster. Even in the case of technology-driven new products (where the idea comes from a technical or laboratory source), the likelihood of success is greatly enhanced if customer and marketplace inputs are built into the project soon after its inception. Market-oriented project teams build in the market and customer from idea through to launch, for example, idea generation with lead users, market research to determine the customer's "wish list" and the ingredients of a superior product, constant and iterative testing with customers all the way through the development phase, rigorous field trials and market tests to ensure positive product performance and purchase intent, and finally a carefully crafted, well-resourced market launch.

A strong market orientation is missing in the majority of firms' new product projects, however. Detailed market studies are frequently omitted — in more than 75% of projects, according to one investigation [Cooper and Kleinschmidt, 1986]. Further, marketing activities are the weakest rated activities of the entire new product process, with relatively few resources and little money spent on the marketing actions (less than 20% of total project expenditures).

Predevelopment Work — The Homework

The steps that precede the design and development of the product make the difference between winning and losing. Successful businesses spend about twice as much time and money on these vital up-front or predevelopment activities:

- Initial screening — the first decision to get into the project (the idea screen).
- Preliminary market assessment — the first and quick market study.
- Preliminary technical assessment — the first and quick technical appraisal of the project.
- Detailed market studies or marketing research — user needs and wants studies, competitive analysis, and concept tests.
- Business and financial analysis just before the decision to "go to development" (building the business case).

Most businesses confess to serious weaknesses in the up-front or predevelopment steps of their new product process. Small amounts of time and money are devoted to these critical steps: only about 7% of the expenditure and 16% of the effort. Far from adding extra time to the project, research reveals that homework pays for itself in reduced development times, the result of sharper and more stable product definition, and fewer surprises (and time wasters) later in the project.

Sharp and Early Product Definition

Sharp, early product definition is one result of solid up-front homework. How well the product is defined prior to entering the development stage is a major success factor, impacting positively on both profitability and reduced time to market. This definition includes

- Specification of the target market: exactly who the intended users are.
- Description of the product concept and the benefits to be delivered.
- Delineation of the positioning strategy.
- A list of the product's features, attributes, requirements, and specifications.

Unless the these four items are clearly defined, written down and agreed to by all parties prior to entering the development stage, the odds of failure increase by a factor of three.

Quality of Execution

Certain key activities — how well they are executed and whether they are done at all — are strongly tied profitability and reduction in time to market. Particularly pivotal activities include the vital homework actions outlined above and market-related activities. However, proficiency of most activities in the new product process impacts on outcomes, with successful project teams consistently doing a better quality job across many tasks.

There is a *quality crisis* in product innovation, however. Investigations reveal that the typical new product project is characterized by serious errors of omission and commission:

- Pivotal activities, often cited as central to success, are omitted all together. For example, more than half of all projects typically leave out detailed market studies and a test market (trial sell).
- Quality of execution ratings of important activities are also typically low. In post-mortems on projects, teams typically rate themselves as "mediocre" in terms of how good a job they did on these vital activities.

New product success is thus very much within the hands of the men and women leading and working on projects. To improve quality of execution, the solution that some firms have adopted is to treat *product innovation as a process.* They have adopted a formal stage-and-gate product delivery process; they build in quality assurance approaches, such as check points and metrics that focus on quality of execution; and they design quality in by making mandatory certain vital actions that are often omitted yet are central to success.

The Correct Organizational Structure and Climate

Product innovation is very much a team effort! Do a post-mortem on any bungled new product project and invariably you'll find each functional area doing its own piece of the project, with very little communication between players and functions — a silo mentality, and no real commitment of players to the project, that is, inadequate people resources devoted to the project, with players having numerous other functional tasks underway at the same time. Many studies concur that good organizational design is strongly linked to success and reduced time to market.

Product development must be run as a multidisciplinary, cross-functional effort. Good organizational design means projects that are

- Organized as a *cross-functional team* with members from research and development (R&D), engineering, marketing and sales, operations, and so on (as opposed to each function doing its own part independently).
- Where the team is dedicated and focused (i.e., devotes a large percentage of its time to this project, as opposed to spread over many projects).
- Where the team members are in constant contact with each other, via frequent but short meetings, interactions, project updates, and even colocation.
- Where the team is accountable for the entire project from beginning to end (as opposed to accountability for only one stage of the project).
- Where there is a strong project leader who leads and drives the project.

While the ingredients of good organizational design should be familiar, surprisingly many firms have yet to get the message [Cooper and Kleinschmidt, 1996].

A second organizational success ingredient is climate and culture. A positive climate is one that supports and encourages intrapreneurs and risk-taking behavior, where new product successes are rewarded and recognized (and failures not punished), where team efforts are recognized, rather than individuals, where senior managers refrain from "micromanaging" projects and second guessing the team members, and where resources and time are made available for creative people to work on their own "unofficial projects". Idea submission schemes (where employees are encouraged to submit new product ideas) and open project review meetings (where the entire project team participates) are other facets of a positive climate.

Focus and Sharp Project Selection Decisions

Most companies suffer from too many projects and not enough resources to mount an effective or timely effort on each — a lack of focus. This stems from inadequate project evaluation and poor prioritization. Project evaluations are consistently cited as weakly handled or nonexistent: decisions involve the wrong people from the wrong functions (no functional alignment), no consistent criteria are used to screen or rank projects, or there is simply no will to kill projects at all.

The desire to weed out bad projects coupled with the need to focus limited resources on the best projects means that tough go or kill and prioritization decisions must be made. Some companies have built *funnels* into their new product process via decision points in the form of tough gates. At gate reviews, senior management rigorously scrutinizes projects and makes go or kill and prioritization decisions.

TABLE 14.4 Typical Go/Kill Gate Criteria

Strategic fit and importance
Reasonable likelihood of technical feasibility
Product differentiation, superiority, and good value for money
Attractive market (size, growth, potential, and competitive situation)
Ability to leverage the businesses's core competencies or strengths (marketing, technological, operations)
Adequate reward vs. risk (financial and other)

Effective gate decisions rely on visible go/kill criteria (see Table 14.4 for sample go/kill criteria). Progressive businesses are also moving to portfolio management, which attempts to select the right set of new product projects in order to maximize the value of the portfolio, achieve the right balance of projects, and yield a portfolio of projects that supports the business's strategy [Roussel et al., 1991; Cooper et al., 1997].

Planning and Resourcing the Launch

Not only must the product be superior, but it must also be launched, marketed, and supported in a proficient manner. A quality launch is strongly linked to new product profitability. Note that competitive advantage gained via *elements other than product superiority* has an impact on new product success. These elements include brand name or company reputation, superior marketing communications (advertising and promotion), a superb sales force or distribution channel, superior technical support and tech service, or simply product availability. The limited evidence available, however, suggests that the impact of nonproduct advantage pales in comparison to the impact of product advantage — less than half the effect.

The Role of Top Management

Top management's role in product development is as a facilitator — to set the stage — and not be an actor front and center. One important role of senior management is to articulate a new product strategy for the business, something that is often missing. An effective new product strategy means defined new product goals (e.g., percentage of the business's sales to be derived from new products), delineated arenas of focus (e.g., product types, markets, technologies, and technology platforms where the business unit intends to concentrate its development efforts), and strategies with both a longer-term orientation and that are visible to everyone in the business. Management must also deploy the necessary product development resources and keep the commitment. These two factors — an articulated new product strategy for the business and adequate resources for development — are two of the three strongest drivers of new product performance at the business unit level, according to a recent benchmarking study; the third driver is a high-quality new product process [Cooper and Kleinschmidt, 1996].

Speed — But Not at the Expense of Quality of Execution

Speed yields competitive advantage (the first on the market), less likelihood that the market situation has changed, and a quicker realization of profits. Therefore, the goal of reducing the development cycle time is admirable. A word of caution here: speed is only an interim objective; the ultimate goal is profitability. While studies reveal that speed and profitability are connected, the relationship is anything but one to one. Further, often the methods used to reduce development time yield precisely the opposite effect and in many cases are very costly. The objective remains successful products, not a series of fast failures! Additionally, an overemphasis on speed has led to trivialization of product development in some businesses — too many product modifications and line extensions and not enough real new products and new platforms [Crawford, 1992].

Some of the ways that project teams have reduced time to market have been highlighted above (see also Table 14.3). Other methods include

- *Parallel processing:* Activities are undertaken in parallel (rather than sequentially) with the team members constantly interacting with each other. New product rugby with time compression is the result.
- *Flowcharting:* Here the team maps out its entire project from beginning to end and focuses on reducing the time of each element or task in the process.

- *A time-line and discipline:* Most project teams use computer software to plan their projects in a critical path or Gantt chart format. The rules are simple: practice disciple, the time line is sacred, and resources can be added but deadlines never relaxed.

A Multistage, Disciplined New Product Process

A systematic new product process — a *Stage-Gate™ process* — is the solution that many companies have turned to in order to overcome the deficiencies that plague their new product programs [Cooper, 1993]. *Stage-Gate* processes are simply roadmaps for driving new products from idea to launch; successfully and efficiently. These systems break the innovation process into stages, each stage comprising multiple, concurrent, cross-functional activities or actions (a team approach is mandatory). Gates are the quality control checkpoints in the process, opening the door for the project to proceed to the next stage. Gates are where the tough go/kill decisions are made: gates typically specify deliverables (what must be delivered by the team to a given gate), criteria for go (upon which the go/kill and prioritization decisions are based), and outputs (an action plan for the next stage and resources approved).

The payoffs of such processes have been frequently reported: improved teamwork, less recycling and rework, improved success rates, earlier detection of failures, a better launch, and even shorter cycle times (by about 30%).

Critical Success Factors: Selection Related

The next five factors describe the new product project and its setting. Unlike the ones above, which are process related, the factors below are less controllable by the project team. They tend to be more useful as project selection criteria.

Market Potential

New products targeted at the following kinds of markets tend to do better, yielding higher success rates and profitabilties:

- Large markets, where the product type is essential for the customer
- High growth markets and noncyclical markets
- Markets with a positive economic climate
- Markets where potential customers are innovative adopters, profitable, and relatively price insensitive

Few of these characteristics on their own are highly predictive of success, but, when taken together, they are more strongly linked to performance.

Competitive Situation

The *competitive situation* is a second facet of the market environment. Numerous studies reveal that the competitive situation has surprisingly little impact on new product outcomes: new products are not that much less successful when aimed at markets characterized by many and strong competitors, competitors who would defend their positions, aggressive competitors and who compete on the basis of price, low barriers to entry, and customers who are loyal to competitors.

Product Life Cycle

A final market descriptor is the *stage of the product life cycle* of the product market at which the new product is aimed. New products on average do best in product markets that are in the early growth and growth phases of the life cycle and worst at either end of the life cycle — in mature markets and in markets in the introductory phase (the two growth phases feature almost double the success rates vs. mature and introductory phases).

Synergy or Leveraging Core Competencies

Synergy means having a strong fit between the needs of the new product project and the business's resources — the ability to leverage core competencies and strengths. Results are better when synergy exists in terms of R&D resources; marketing, selling (sales force) and distribution (channel) resources; manufacturing or operations capabilities and resources; technical support and customer service resources; market research and intelligence resources; and management capabilities. These six synergy ingredients become obvious checklist items in a scoring or rating model to help prioritize new product projects.

Familiarity

Familiarity is a parallel concept to synergy. Some new product projects take the business into unfamiliar territory: a product category new to the firm; new customers and unfamiliar needs served; unfamiliar technology; new sales force, channels, and servicing requirements; or an unfamiliar manufacturing process. Here the business often pays the price: step-out projects tend to fail, so beware the unknown.

Conclusion

New products are vital to the growth and prosperity of the modern corporation. New products fail at an alarming rate, however. An understanding of what successful products and projects have in common helps both in project management as well as project selection.

References

Cooper, R. G. *Winning at New Products: Accelerating the Process from Idea to Launch,* Addison-Wesley, Reading, MA, 1993.

Cooper, R. G. Developing new products on time, in time, *Res. Technol. Mgt.,* 38(5): 49–57, 1995.

Cooper, R. G., Edgett, S. J., and Kleinschmidt, E. J. *Portfolio Management for New Products,* McMaster University, Hamilton, Ontario, Canada, 1997.

Cooper, R. G. and Kleinschmidt E. J. An investigation into the new product process: steps, deficiencies and impact, *J. Prod. Innov. Mgt.,* 3(2):71–85, 1986.

Cooper, R. G. and Kleinschmidt, E. J. Winning businesses in product development: critical success factors, *Res. Technol. Mgt.,* 39(4): 18–29, 1996.

Crawford, C. M. The hidden costs of accelerated product development, *J. Prod. Innov. Mgt.,* 9 (3): 188–199, 1992.

Maidique, M. A. and Zirger, B. J. A study of success and failure in product innovation: the case of the U.S. electronics industry, *IEEE Trans. Eng. Mgt.,* EM-31: 192–203, 1984.

Montoya-Weiss, M. M. and Calantone, R. Determinants of new product performance: a review and meta-analysis, *J. Prod. Innov. Mgt.,* 11(5): 397–417, 1994.

Rothwell, R., Freeman, C., Horseley, A., Jervis, V. T. P., Robertson A. B., and Townsend, J. SAPPHO updated — project SAPPHO Phase II, *Res. Policy,* 3: 258–91, 1974.

Roussel, P., Saad, K. N., and Erickson, T. J. *Third Generation R&D, Managing the Link to Corporate Strategy,* Harvard Business School Press and Arthur D. Little, Inc., Boston, MA, 1991.

Sanchez, A. M. and Elola, L. N. Product innovation management in Spain, *J. Prod. Innov. Mgt.,* 8: 49–56, 1991.

14.4 Design Quality

Robert P. Smith

What Is Quality?

As managers and technical professionals go about fulfilling market needs, they need to have measures for how well the design artifact fulfills those needs. One important dimension for determining success in the customer's view is *quality.*

Quality is a difficult-to-define term. A customer, a purchasing agent, a manufacturing manager, and a design engineer may all have different conceptions of what quality means, and all of them are useful.

Their measures of quality are also all strongly influenced by decisions made during the design process. The focus of this section is to examine those varied definitions of quality, and to determine how design decisions affect those conceptions.

Conformance and Performance

Many of our conceptions of quality were developed in the manufacturing world; let us examine manufacturing before we think about design. The two primary meanings of quality are conformance and performance. **Conformance** is the absence of error, that the artifact has the intended attributes. In the manufacturing world this means that the produced product is within the specified tolerance range, that no defect has been introduced. **Performance** is the degree to which the needs of the customer are fulfilled. Every designer must recognize that there are customer needs that a product will not be able fulfill at a price that the market is willing to pay. (This section discusses design in terms of product, but the ideas apply equally well to the design of a process or the design of software.)

Consider two bicycle manufacturers, one high-volume mass-market and one a specialty racing producer. The racing bicycle producer can legitimately claim that it is the higher-quality producer because it better fulfills the needs of the customer, and it can therefore charge a premium for its product. They have greater *performance*, although it is possible that because of handcrafted parts that some fraction of produced bicycles contains one or more defects.

The high-volume producer, by contrast, can potentially claim higher *conformance* quality. They may exercise greater control over the processes that produce and assemble parts, have more complete tracking of production activities, and have fewer bicycles that fail inspection steps. Their production system is introducing fewer errors into the product, and they can therefore rightly claim high conformance.

Both conformance and performance have important implications to examining design quality. Below we will examine how these concepts map on to design activities and examine the implications of these considerations.

Conformance

The conformance aspects of quality have several relevancies for design. Since conformance is concerned with the absence of error in the finished product, conformance concerns in design must account for how errors may arise. Errors may be introduced by misspecification of the design goals and specifications, through designers' failure to meet the stated design goals, through an error in technical analysis or testing that causes the product to perform in an unintended way, or through errors introduced in production due to difficulty in fulfilling the stated level of tolerance. Let us consider each of these sources of error separately.

Misspecification of Goals

Customer needs are generally only poorly understood, by the designer, by the marketing organization, and by the customer. One of the most important roles of the marketing professional is to transform and solidify these poorly understood needs into a form that can be used by the designer. One important tool to accomplishing this goal is the House of Quality (see Section 12.6). Correctly and completely administered, this process will minimize the introduction of error into the specification generation process. Since the marketing function and the House of Quality are covered elsewhere in this volume, we will not consider them further despite their important role in design quality.

Design Errors

The next two categories of conformance problems involve what we will consider design errors. Design errors are designers' failure to fulfill stated specifications or failure to detect departure from the specifications. Design is the creation of a complex, novel technical artifact for an unknown situation under conditions of strong time and resource pressures. It is not surprising that errors do arise under these circumstances. One goal of the design process is to minimize the number of errors that remain in the product through avoidance, detection, and remedy.

All design processes are iterative; designers repeat tasks. One reason that iteration occurs is that errors are discovered in previously completed design work. Often these discoveries are made through the examination or testing of a prototype of all or part of the system being designed. The prototype may be a full working version, a scale version with only a few functions, or a software-based simulation of the product. All of these methods of testing prototypes have value, and all are appropriate for discovering errors in design decisions, depending on the technology under consideration.

Discovering these errors and fixing them is an important part of the design process. Often, finding the errors is difficult and costly, as is redoing the work that may be required to create a remedy. Leaving the errors in the product is often even more costly, as the errors will adversely affect the customer. (Design errors where safety and/or liability are involved are treated more diligently, see Section 20.2.) Addressing quality in the design process includes the prevention, detection and correction of technical errors in an efficient and timely manner.

Robust Design

Another aspect of design quality is the degree to which a product fulfills the needs of the customer despite any other changes or unforeseen events that occur. It is necessary to design the product so that it can allow for other things to go wrong. A product that is relatively insensitive to changes is robust, and the process of identifying how product design decisions can achieve this goal is **robust design**.

Performance variation can be reduced by exploiting the nonlinear effects between a product's parameters and the product's desired performance characteristics. Consider a thermocouple, which has an output voltage that is a nonlinear function of the ambient temperature as well as of its material properties. It is desirable to match the material properties and the expected operating range of the device so that small variations in the material properties do not cause large errors in temperature estimation. Typically, for thermocouples this is done by choosing from a catalog of known products with known operating conditions. For design of new products, it is necessary to determine what the effects are and to choose parameters and operating ranges so that output variation is small.

Parameter design is the determination of specific nominal settings for product parameters. Some outcome performance variables are strongly a function of some product parameters over only a portion of the possible range. Much of robust design activity is focused on the parameter design phase and on identifying these ranges of relative insensitivity.

There are several sources of variation in performance. Environmental sources of variation are those that change within the user environment: users treat the product in different ways, the product may be used in different operating conditions (humidity, vibration, heat, radiation, and so on). Each of these changes in the user environment has the potential of affecting the performance of the product.

There is product deterioration. Each product, as it ages, has its performance change in subtle ways. There may be more play within a gear box, the insulation on an electric cable may lessen, or the control surface operated by the user may become worn. Each of these changes may affect the performance of the product.

Additionally, each product is subject to manufacturing variation. Try as they may to control their processes, production operations are inherently variable, and no two products will be identical in all details. To some extent the topic of manufacturing variation is the realm of good manufacturing quality processes, but good products are able to operate correctly for the user even in the presence of manufacturing variation.

A good design anticipates all of these sources of variation and produces a product that is relatively insensitive to all of them. Product parameter settings that reduce performance variation can be identified with experimentation. Since the product parameters combine nonlinearly to produce outcome performance, we must study the various levels in combination. The techniques of robust design have been developed to address these needs.

There are several stages to the robust design process. First, the set of dependent quality-related outcome variables is selected. Second, the parameters and noise variables that may affect the outcome variables are identified. Third, design parameter levels are chosen. Design parameter variables can be set at a variety of levels; often these can be reduced to two or three likely candidates for each important parameter. Fourth, noise variable settings are chosen. These settings are chosen to cover the expected range of manufacturing and operating characteristics. Fifth, a thoughtfully designed set of statistical experiments is conducted. Because there are, in general, sufficiently many variables that not all combinations of design parameter and noise factor can be examined, statistical theory is used to test an efficient set. Fifth, the outcome of the experiments is analyzed in order to identify which product parameter settings that are best able to reduce performance variation.

These techniques are applied to design a product able to operate in a variety of conditions while fulfilling its intended purpose. These techniques are valuable improvements to the performance quality of the design processes.

Improving Design Quality

In summary, there are a number of important interactions between quality and the design process. It is valuable to structure the design process so that it is of high quality both in terms of performance and conformance.

It is necessary to understand the customer and to design the product so that it fulfills their needs correctly. The House of Quality is a useful tool to insure that this is done correctly. The customer need identification process has both important conformance and performance relevance.

It is necessary to structure the design process so that few mistakes are made. Equip designers with the tools necessary to perform complete and accurate analysis, involve designers who have the necessary levels of experience and expertise, and be thorough and rigorous in investigation. All that being said, errors will occur. It is necessary to correct these errors as soon as feasible and whenever possible before the product is delivered to the customer.

It is necessary to design the product so that variation in the manufacturing process or in the user environment does not adversely affect performance. This process of robust design is vital to the design of a superior product.

Quality is the lifeblood of design success. Inattention to the various performance factors that lead to quality success is inimical to well-managed design processes.

Defining Terms

Conformance: The degree to which an article conforms to its specifications (is free of defects). One of the two main measures of quality.

Performance: The degree to which an article fulfills the needs of the customer. The other main measure of quality.

Quality function deployment: A technique for identifying customer needs and translating them into useful engineering specifications. This approach is appropriate for achieving performance quality. Also known as the House of Quality.

Robust design: A technique for designing an artifact so that its performance is not unduly affected by variations in the manufacturing process or in the user environment. This technique is appropriate for achieving conformance quality.

Further Information

There are several professional associations that are relevant to design quality and related design management topics. Each of these associations runs conferences and symposia, publishes journals, maintains

Web-based information, and provides other opportunities to learn more about these topics. The associations include the Product Development Management Association (http://www.pdma.org/), the Design Management Institute (http://www.designmgt.org/), and the International Association for the Management of Technology (http://www.iamot.org/iamot).

Total Quality Development: A Step-by-Step Guide to World Class Concurrent Engineering, by Don Clausing (ASME Press, 1994) is a comprehensive guide to the application of quality function deployment and robust design (as well as other product development tools) in a coherent, systematic way.

Tools and Methods for the Improvement of Quality by Howard Gitlow and others (Irwin, 1989) is an excellent book that describes the theoretical basis behind many quality tools used in engineering work.

Product Design and Development by Karl Ulrich and Steven Eppinger (McGraw-Hill, 1995) is a useful introductory textbook to product development well informed by the latest thinking on the design process.

Quality Function Deployment: Integrating Customer Requirements into Product Design by Yoji Aoki (Productivity Press, 1990) is the definitive and complete description about quality function deployment and the House of Quality. Well worth further study for any serious student of the relationship between the customer and the engineer.

Quality Engineering using Robust Design by Madhav S. Phadke (Prentice Hall, 1989) is a useful introductory text that presents the elements of robust design.

Designing Engineers by Louis L. Bucciarelli (MIT Press, 1994) is a very insightful and entertaining book about what it is that engineers actually do when engaged in the process of design.

14.5   Life Cycle

Catherine Banbury

The concept of the product life cycle helps us understand how innovations in a product technology evolve over time and how these changes are linked to firm strategy, structure, and processes as well as core product technologies. The links between products and organization and products and technologies provide us with useful information regarding the competitive dynamics of what have come to be known as the various stages in the life of a product. Theoretical and empirical work on the product life cycle has established that innovation in product technologies tends to follow a predictable pattern over time [Utterback and Abernathy, 1975; Sahal, 1981, 1986; More and Tushman, 1982; Butler, 1985]. At industry inception, initial innovative efforts usually emphasize product performance, followed by product variety, and toward the end of a product's life cycle the emphasis shifts to product standardization and cost efficiencies.

For example, according to Anderson and Tushman [1990], a new product technology will usually undergo an era of fermentation often culminating in a standard product design. A period of technological progress will follow in which improvements are generally considered to be incremental. This incremental stage may be interrupted by a subsequent technological discontinuity. The cycle — commercialization, fermentation, dominant design, incremental change, and discontinuity — will then repeat. These concepts and the competitive implications will be addressed in the following sections.

Before proceeding with the literature review, however, it is important to note the ubiquity of the terms product and life cycle. "Life cycle" is used frequently in the organizational sciences with reference to an organization's, a technology's as well as a product's life cycle, for example. The term "product" tends to vary in scope depending on the study cited and the industrial sector studied. This makes the identification/definition of the relevant unit of analysis an important one when reviewing the literature on this topic. For these reasons, different definitions for the product life cycle will be referenced where relevant in the follow discussion.

The Product Life Cycle

In this section, we look at the product life cycle as presented by Utterback and Abernathy [1979] and as extended by Moore and Tushman [1982] to include its relationship to firm strategies, structures and processes, and market dynamics. We then go on the explore more fully the links between product changes over time and technological evolution. This leads us to the use of a number of concepts such as design, periods of ferment and of incremental change, and technological discontinuities that are explored more fully in preceding chapters Finally, we conclude with a discussion of the limitations of current knowledge and prospects for future research.

Organizational Connections

The key dynamic to the product life cycle concept is that the rates of product and process innovation vary over time and vary inversely to each other. Product changes are frequent early on, while process innovations increase as market demand increases and production becomes more specialized. In the later stages of the life cycle, a reduction in the number of both product and process innovations is usually observed.

According to Utterback and Abernathy [1975], the unit of analysis is the productive unit, the product line, and its associated production process, since the product produced and the manufacturing process used become so interdependent. They concluded that firms can be identified as being in one of three stages of the product life cycle: the uncoordinated, the segmental or the systemic. The uncoordinated stage is characterized by high rates of product innovation and product diversity between competitors and a lack of a standardized means of production, generally referred to as process technology. By definition the dominant source of innovation is the market. The segmental stage is characterized by increased price competition driving increases in production efficiency. During this stage sufficient sales volume exists to permit standardization of both product and process, which leads to a few stable product designs and increased emphasis on process innovation and product variety. The dominant source of innovation tends to be technology during this stage. During the third stage of the product life cycle, the systemic stage, the levels of investment required to generate innovation in both the product and the process increases significantly and market demand slows. The potential for product obsolescence is generally greatest during this stage of the product life cycle.

When Moore and Tushman refer to the product life cycle, they use the term to encompass the product line or class, and those parts of the firm that have "some control over marketing, manufacturing, and research, development and engineering" [1982] the product line. Thus they are focusing on the life cycle of microprocessors and automobiles, as opposed to the 486 and Model T. By broadening the definition of the product life cycle in this way, it makes it easier for the authors to link the various stages in a product's life cycle to firm strategies, organizational structures and processes, and roles of managers.

Moore and Tushman [1982] identify three stages in the product life cycle: the introductory, growth, and maturity. These stages closely parallel the uncoordinated, segmental, and systemic stages of Utterbach and Abernathy's model. In the introductory stage, the basis of competition is usually product performance, production volumes are low, and the organization's structure tends to be free form. This experimental stage ends when a dominant design emerges and firms begin to increase production volume. The next stage, the growth stage, then is characterized by increased product standardization and rationalization of the production processes. The organization then takes on a functional organic form on its way to a project/matrix-based structure. During the mature stage, there is generally little growth in sales, innovation in the product and process technologies are minimal, as the product becomes commodity like and the production processes increasingly capital intensive, automated, and efficient. Employees' roles become more specialized and rule bound and the organizational structure becomes increasingly bureaucratic. As the organization evolves through these stages, the role of management and communication networks also becomes increasingly formal.

In summary, the literature on the links between product variations over time and organizational characteristics concur on a number of important points, namely, that innovations in product technologies tend to evolve through three general yet distinct phases, within which firms, innovations, and market demand demonstrate a range of definable characteristics and that is usually during the third stage that new and/or revolutionary innovations tend to occur.

Links to Technological Evolution

There is precedence in the design literature and strategic management literature for making a distinction between the product as a whole and its component parts, where each component is itself a technology embodying a core design concept [Henderson and Clark, 1990]. Anderson and Tushman [1990] suggest that it may be helpful to conceptualize products as systems of core technologies and associated linking technologies each of which have their own technological trajectories. Closer examination then, of the interdependencies between products and their component technologies provides us with finer-grained insight into the patterns of innovation that tend to occur over a product's life cycle.

New products and new industries sometimes arise as a result of a technological change. Similarly, technological change may result in product obsolescence. These two "end" points are generally characterized by radical/paradigmatic/revolutionary innovation. The relationship between product life cycles and technological paradigms do not completely overlap. They are close enough, however, that they provide us with additional and more detailed insight into the variety of product innovations that are most likely to occur over the product's life cycle. Within a product's life cycle (as governed by the technological paradigm governing the product's core technology), innovations in the product technology tend to follow a reasonably predictable pattern. Paradigms act as a means of channeling a firm's innovative efforts within a certain limited set of possibilities, generally referred to as technological trajectories. In this sense then innovative activities, particularly those that occur during the incremental, are strongly selective, often cumulative activities finalized in rather precise directions [Dosi, 1982]. The concept of a technological paradigm enables us to delineate the boundaries of technological change cycles (paradigms) and the expected direction of changes (trajectories) over the product's life cycle.

The early stages of commercialization (referred to as the introductory stage in the previous section) are characterized by competition between a variety of alternative core technologies and product configurations, that is, initially competition for market share exists between competing technological paradigms. For example, during the early stages of the motor vehicle industry three different designs based on three different methods of propulsion (gas, steam, and electricity) coexisted and competed for market dominance. Even when the battle between paradigms has been resolved, firms may still compete on the basis of fiddering product configurations, for example, beta vs. VHS video formats. This early commercialization period is often referred to as the era of fermentation in the technology literature. What is important about the period of fermentation is that market adoption of a new product technology or a switch from the old to the new product technology is not necessarily smooth and that in both cases this period usually culminates in industry adoption of a standard.

In most competitive industries a number of product designs emerge that enable a certain amount of product standardization and market growth. The design standards/parameters then delineate the boundaries of and progression of future innovations in the product technology and ushers in the era of incremental change (referred to as the growth stage earlier). Competition during this stage of the life cycle is primarily driven by incremental improvements in the dominant product design. This pattern will persist until the industry faces design obsolescence.

The existence of dominant product design provides firms with a framework within which technological change is most likely to occur in the product. This framework limits the set of possible changes in the product technology that are likely to persist in the industry until such time as the present dominant design is superseded. Once a product design is adopted by the market and dominates sales, a body of shared knowledge develops regarding the technological problems that need to be solved to improve product performance. As demand for a specific design increases, firms producing the popular design will benefit from learning associated with increased production.

Increased adoption of a certain product design enable firms and buyers in the industry to accumulate knowledge along a specific technological trajectory. In effect, a dominant design narrows the boundaries of the search for future product innovation. A firm's search for innovation is usually in areas closely associated with its current technology and aimed at improving its current technology. In general, firms have a tendency to stay close to home technologically speaking; future innovations build on past innovations and related applications of the core technology.

Further, within a technological paradigm, and more specifically the growth and mature stages of the product life cycle, the knowledge acquired from one advance is often a necessary input to the next, each subsequent change building on prior changes. Ongoing competition between firms in an industry, then, will tend to revolve around the development of these path dependent innovations and the speed with which they are introduced to the market and imitated. What this means is that engineers, technicians, and other professionals on both the demand and supply side of an industry share a common and public body of knowledge and search for an develop innovations within the boundaries of the current technical paradigm. The competitive race between firms during the incremental era will be tight, and the firm's search for innovation will most likely be internally driven with an eye on competitor actions. Another consequence of these path-dependent changes is that each subsequent innovation will tend to improve on the preceding models and in most cases cannibalize existing product models and generations through leaving much of user learning intact, for example, Intel's 286, 386, 486, and Pentium chip technology.

In summary, the history of a product over its life cycle is contextual to the history of its core technology and to a lesser extent its associated component technologies. Although there is scope for much more work to be done in this area, initial attempts to look at how technologies and products coevolve, though complete (since the relationships are not strictly one to one), show great promise of further explicating both phenomena.

Missing Pieces

There are two major limitations in the current rendition of the product life cycle concept. The first has to do with our limited understanding of the growing interdependency between formerly distinct products and markets, and the second has to with our lack of understanding of the processes by which radical innovations arise that render current product technologies obsolete. With respect to the first point, the product life cycle concept loses descriptive power, theoretically and empirically, as industry boundaries blur with the increase in connections between formerly distinct product technologies. For example, with the growing convergence of the telecommunications, entertainment, consumer electronics, and computer industries, it is becoming increasingly difficult for firms in any one or a combination of these competitive spaces to identify emerging product markets, let along identify potential competitors. These linkages between product technologies are creating new industrial ecologies or networks, and new conceptualizations of the term "product".

With respect tot he second point, although we know much regarding the organizational effects of product obsolescence, we know little about how companies might survive the transition between the old and the new, let alone how to "foresee" a product's likely successor.

Summary

Finally, the concept of the product life cycle offers practitioners and academics a useful framework for anticipating the likely generic types of changes a product technology will undergo over time. It removes much of the mystery surrounding product-based innovations, their likely timing as an industry evolves, and their organizational and competitive implications. In conclusion, however, since not all incremental innovations in a product technology are equivalent, they will vary with respect to the easy with which they may be implemented or imitated, and they will vary with respect to their competitive necessity and competitive effects. Therefore, much scope remains for bringing work on technological evolution to bear on our understanding of the types of incremental innovations that occur over a product's life cycle and of their effects on competitive dynamics.

Defining Terms

Incremental innovations: Incremental product innovations are defined as refinements and extensions of established designs that result in substantial price or functional benefits to users.

References

Anderson, P. and Tushman, M. L. Technological discontinuities and dominant designs. A cyclical model of technological change, *Admin. Sci. Q.*, 35(4): 604–633, 1990.

Butler, J. E. Theories of technological innovation as useful tools for corporate strategy, *Strat. Mgt. J.*, 9: 15–29, 1988.

Dosi, G. Technological paradigms and technological trajectories, *Res. Policy*, 11: 147–162, 1982.

Henderson, R. M. and Clark, K. B. Architectural innovation: the reconfiguration of existing product technologies and the failure of established firms, *Admin. Sci. Q.*, 35: 9–30, 1990.

Moore, W. L. and Tushman, M. L. Managing innovation over the product life cycle, In *Readings in the Management of Innovation*, M. L. Tushman and W. L. Moore, eds., pp. 131–150, Ballinger, Cambridge, MA, 1982.

Sahal, D. *Patterns of Technological Innovation*, Addison Wesley, Reading, MA, 1981.

Sahal, D. Technological guideposts and innovation avenues, *Res. Policy*, 14: 61–82, 1986.

Utterback, J. M. and Abernathy, W. J. A dynamic model of process and product innovation, *Omega, Int. J. Mgt. Sci.*, 3(5): 639–656, 1975.

For Further Information

An excellent chart describing how strategic and organizational factors vary over the product life cycle is given on page 143 of Moore and Tushman [1982], referenced above.

For an exposition of some of the theoretical underpinnings of the concept of the product life cycle please refer to Sahal's work [1981,1986], referenced above.

In his 1982 paper, Dosi gives a great exposition of the determinants and directions of technical change.

14.6 Managing Cycle Time in New Product Development

Marc H. Meyer

Managing the cycle time for the development of new products and services should not be viewed in isolation from the larger issues and concerns confronting the business. The long-term success of an enterprise depends on a stream of new products — some replacing older ones, others pioneering new markets, and all satisfying customer needs. It is this stream of new products, exploiting advances in both product technologies and technologies used to manufacture, distribute, and provide support that provide the fuel for corporate growth and renewal.

Unless a firm's new products provide value to the customer, it really doesn't matter how fast they are developed. Time does serve as a constraint on development cycles — timely innovation is as important as efficacious innovation. One cannot rush poorly conceived and executed products to market — for whatever reason; yet, one cannot let the perfect get in the way of the good. The essence of managing cycle time in new product development is to find an appropriate balance between these two drivers that fits the needs of the firm in the context of its industry and customers.

Traditional Perspectives on Cycle Time can Be Limiting

The statements above should convey to the reader our own concerns about traditional perspectives on cycle time. Management systems typically consider the design and development of new products *one product at a time*: the faster any given product shoots through the development funnel, the better. Approval gates swing open and closed as single development projects move forward. "Slip rate" — the gap between expected and actual values as applied to project time and project budget — becomes the singularly important management metric.

The truth be known, slip rate is really a better measure of the manager's ability to estimate a project's schedule and budget. Predictability in product development is not and should not be equated with effectiveness or excellence. Emphasis on slip rates undoubtedly encourages schedule and budget "padding" by project managers. Further, we have seen companies prosper from late and overbudget projects — projects that would not measure up on the slip rate yardstick. A study of a successful pharmaceutical company, for example, showed that the firm experienced overruns on 80% of its new product efforts, with an average slip of 1.78 times planned costs, and 1.61 times planned schedules. Yet, this firm consistently prospered from these late and overbudget products [Mansfield et al., 1971]. Another study of an electronics manufacturers that the author conducted found a similar result: no correlation between development cycles times and commercial success [Meyer and Utterback, 1995].

Single products, particularly in competitive, technology-intensive industries, are important only in their cumulative impact. In fast-paced industries, single products hardly matter — it is the continuous stream of products that counts most in a firm's ability to build and maintain share.

A single-minded focus on cycle times can also have a pernicious effect on the firm's ability to survive and prosper over the long term. Slip rate measures drive managers away from more innovative new product efforts and toward incrementalism instead, less ambitious efforts whose costs and development schedules are more predictable. Working on the trivial to the exclusion of fundamental innovation is precisely what can make a firm's products and services obsolete and prone to attack.

Projects that contain high levels of technological and market familiarity can and should be pursued rapidly. On the other hand, projects having higher technological and market uncertainty — the hallmarks of new product platforms or next generation products — require greater patience. Demanding tight schedules for these projects may be self-defeating [Utterback et al., 1992]. It is far better to start such initiatives early and give them time to ripen and flourish.

Managing Cycle Time Expectations

Engineering managers must provide the air cover required for engineers to do good work. This means managing expectations on the part of senior business management. The more ambitious the initiative, the more carefully expectations need to be managed.

We believe that the best way set appropriate expectations is to elevate the time cycle issue above the single product treadmill and associated stage-gate processes of management. We have argued that the most important indicator of success in terms of cycle time is not the schedule to slip rate of any single product but the ability of the firm to introduce a stream of exciting, value-rich products over time. Many managements have recognized this, measuring the percentage of revenues from products introduced in the past few years to supplement slip rates and time-to-market measures. Research has shown a strong correlation between industry leadership and a high percentage of revenues generated from recently introduced products. For example, one recent study found that Hewlett-Packard gained 60% of its annual revenue from products introduced within the prior 5-year period, and companies such as Gillette, 3M, Corning, and Johnson & Johnson, between 25 and 35%.[5]

Successful product developing companies create robust product families. Product families do not have to emerge one product at a time. In fact, they can plan so that a number of derivative products can be efficiently created from the foundation of common core technology. We call this foundation of core technology the "product platform", which is *a set of subsystems and interfaces that form a common architecture from which a stream of derivative products can be efficiently developed and produced.* A platform approach to product development dramatically reduces manufacturing costs and provides significant economies in the procurement of components and materials because so many of these are shared between individual products. Perhaps as important, the building blocks of product platforms can be integrated with new components to rapidly address new market opportunities.

In our book, *The Power of Product Platforms,* the reader will find numerous examples of strong product families whose respective individual members share architecture, a substantial amount of subsystem technology, and many downstream processes for manufacturing, distribution, and customer support

---

[5]Product Development and Management Association survey of 200 companies, 1991.

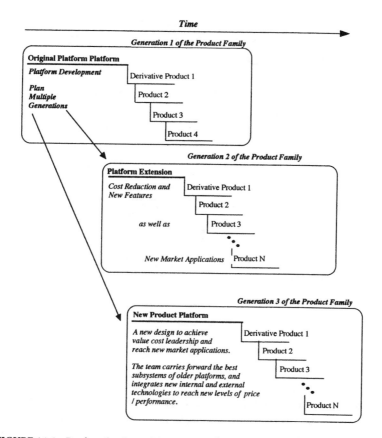

FIGURE 14.6 Product family evolution, platform renewal, and new product creation.

[Meyer and Lehnerd, 1997]. Each example — be it Hewlett-Packard's inkjet printers, Black and Decker's consumer power tools, EMC's large scale storage systems, or Boeing's 777 airplanes — is representative of a more generalized model of product development that is shown in Fig. 14.6. The figure represents a single **product family** starting with the initial development of a product platform, followed by successive major enhancements to the core product and process technology of that platform, with derivative product developments within each generation. Successful companies continuously renew their platform architectures and their manufacturing processes by integrating advances in core product and process technologies. These advances may be created internally or externally. By consistently obsoleting its own products with better ones, the company keeps the heat on its competitors and insures the perpetuity of the enterprise.

Add to this figure greater depth and you end up with the framework shown in Fig. 14.7. At the top of the figure are the market applications of a company's technology. The market for the product family is defined in a traditional way through a matrix of market segments that identifies particular user groups and product price and/or performance characteristics. The market applications of a product family take the form of derivative products based on product platforms. Most corporations tend to view their market segments in isolation from one another. Simply placing these segments on one page may allow management to then consider how product technology and manufacturing processes can be shared or made common across product lines serving different market segments.

In the middle tier are the company's product platforms as defined earlier. Every company must determine precisely the structure of the product platforms suitable for its business, e.g., those subsystems and interfaces that are the essence of the stream of products or services it provides. Product platforms capable of accommodating new component technologies and variations make it possible for firms to create derivative products at incremental cost relative to initial investments in the platform itself. This is possible because the fundamental subsystems and interfaces of each new derivative are carried forward.

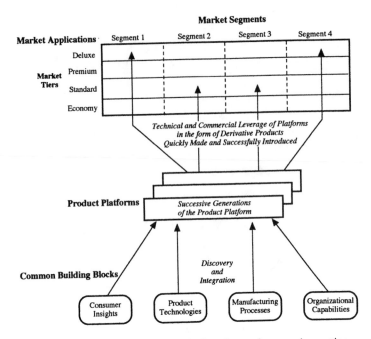

FIGURE 14.7 An integrative model of product and process innovation.

Since the costs associated with the carried forward elements are essentially sunk costs, only the incremental costs of creating variations to them accrue to the derivatives. Typically, these incremental costs are a small fraction of the cost of developing the original product platform, leading to what may be called **"platform leverage"**. Product platforms can also improve development cycle times of derivative products by facilitating a more streamlined development process and more frequent model changes [Clark and Fujimoto, 1991].

At the bottom tier of Fig. 14.7 lies the heart of all product development activity: those core technologies and competencies in product and process arenas that are brought together to form a current generation product platform. We think of technology as the implementation of knowledge with the potential to be incorporated into a product or service. Product technology takes many forms: chemistries, programming languages and algorithms, hardware or logic design, and so forth. One level up from these basic technologies are actual implementations of proprietary knowledge and skills in the form, for example, of chemicals, or materials, software modules, or chips. The building blocks are the essential components within the subsystems of product platforms. Product technologies also include subsystem interfaces, be they proprietary connections or those based on de facto or regulatory imposed standards.

Rapid platform development and renewal is facilitated by ongoing research that advances the state of basic technologies. These advances can then be integrated into new platform designs. "Integration" is the operative word. Just as derivative products should be rapidly developed through incremental improvements to existing product platforms, new platforms should themselves be created by integrating complete component technologies, either from the firm itself or from external suppliers. To make platform renewal work, senior management must invest in the basic research and development of technology building blocks as well as specific product development.

Consider 3M's Post It. The long-term research was on the backside adhesive, the "platform" development was on the manufacturing processes for coating paper with that adhesive, and incremental refinements have come in the different size and color Post Its. All three categories of technical work were required for success. Essential product technologies may come from sources external to the firm.

Management must encourage engineers to aggressively search the world for breakthroughs in core technologies that may prove useful for their own product designs and manufacturing processes.

The cycle time implications of these framework is fairly straightforward. The development of proprietary technology building blocks will ordinarily take the longest time. New measurement techniques, new materials, or new production processes can easily consume over 5 years of basic research. New platform development, on the other hand, can proceed far more rapidly, particularly if the approach is one of integrating (as opposed to building from the ground up) together existing internal and external technologies into next-generation product architectures. In the electronics, consumer products, and software industries (the author's own strongest experience base), new product platform development efforts typically span from 1 to 3 years. Products created from existing platforms, on the other hand, can be produced in much faster time cycles — given that the platforms themselves are robust and flexible in both product technology and process or production technology. Quarterly development schedules are not unusual for companies that operate this way.

Implications for Cycle Time Measurement and Management

We can now consider how to measure R&D beyond individual project slip rate. What are appropriate cycle times for core technologies, new product platforms, and derivative products in your company? First look at the historical evolution of the mainstream product families in your own company. Differentiate between periods of success and failure during that history. How long did core technology vs. new architecture vs. incremental product development take during those respective periods? How do these experiences benchmark against similar activities by industrial partners or competitors?

The potential power of your platform architectures may be shown by computing a simple cycle time efficiency measure. Using the start and end of R&D for product platforms and derivative products, an elapsed time cycle measure of platform efficiency for any single product can be expressed as

$$\text{Cycle time efficiency} = \frac{\text{elapsed time to develop a derivative product}}{\text{elapsed time to develop the product platform}}$$

Similarly, an average cycle time efficiency value for a generation of products based on a particular version of a product platform can be computed as

$$\text{Average cycle time efficiency} = \frac{\text{average}\left(\text{elapsed time to develop derivative product}\right)}{\text{elapsed time to develop the product platform}}$$

Simply plotting the development times for platforms vs. those of derivative products based on those platforms can be highly instructive. We can illustrate this with a product family of medical equipment. Figure 14.8 shows the elapsed calendar time in years for the development of two distinct platform versions of this product family. It also shows the average cycle times in years for the derivative products in each generation of the product family.

The cycle time for the initial product platform was about 5 years and, for successive extensions of that platform, about 2 years. Derivative products were developed on average in about 2 years. The development of the second product platform was conducted in 2.5 years, which from discussions with team members, was far too rushed and released too earlier due to competitive pressures. Subsequent platform extensions to this second platform took longer to complete, and you will notice that the third platform extension took almost 7 years to finish. Similarly, development times for derivatives products based on

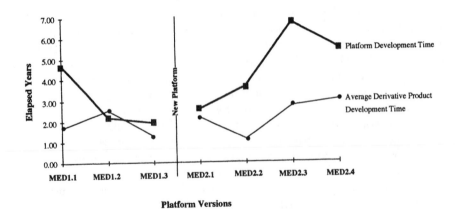

Platform Versions

FIGURE 14.8   Platform and derivative product cycle time efficiency.

this second platform were longer than those of the first platform. The team was trying to add increasingly complex clinical functionality to the product line and was finding the underlying platform architecture a barrier to rapid development.

Most managers sense when the underlying architectures of their major product lines are running out of gas. Their responsibility, then, is to fund new platform development projects to supersede them. When senior management treats new platform development efforts as if they were derivative product developments, both in terms of time and resource allocation, the tangible results of these new platform efforts will probably be disappointing. New platform developments, those in which multidisciplinary teams must look deeply into technical and marketing alternatives, takes time and effort. Such efforts must be started early. It also stands to reason that, if new platform development efforts take longer to complete than derivative product efforts, R&D aimed at platform renewal should be pursued concurrent within derivative product developments on existing platforms. This insures a continuous stream of products embodying competitive technology.

Introduce External Perspectives to Cycle Time Measurement

Companies have found it very useful to track the degree to which a firm has beaten its competitors to the marketplace with new features or capabilities in its products or services. We call the measure lead-lag competitive responsiveness. It readily lends itself to systematic study in the context of the evolving product family.

To perform the analysis, start by constructing a product family map or vintage chart. In addition to noting the various platform versions and derivative products within the family, ask the team members to identify the important new features or capabilities that have appeared over time in the stream of products. Then ask: when did our competitors introduce similar capabilities? Plot the comparative data in the format shown in Fig. 14.9.

Figure 14.9 shows the lead-lag competitive response for a rather substantial consumer products family over 20 years. While the company started off as a leading innovator, over time the company lost its position as the innovation leader. The company's engineers either couldn't innovate fast enough, or, when they did, the manufacturing organization impeded rapid deployment. The old slip-rate measure of success also contributed to incremental in innovation.

Using Composite Design to Establish Focus and Reduce Cycle Time

Composite design is a process for breaking down product platform architectures into major subsystems and interfaces and then systematically determining for which subsystems the firm can best add proprietary value and for which others external technologies might be best integrated into the overall design. This

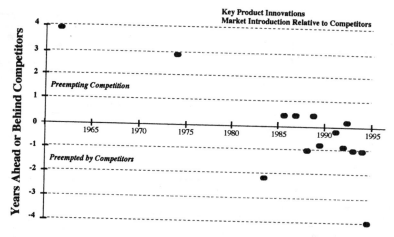

FIGURE 14.9 Lead lag competitive response.

process is fully described in *The Power of Product Platforms*. The implications of composite design for cycle time in new product development boils down to one very important goal: focusing engineers on technologies where the firm has distinctive competence and integrating in those other technologies required to complete the design.

A company must break down its products and those of competitors into logical subsystems. The goal of this activity is to identify which products can be built at the lowest cost, which use the fewest parts and types of parts, which are easiest to manufacture, and which provide superior reliability and performance. Tear-down analysis or reverse engineering of competing products, of course, is nothing new. However, examining competitor's products in terms of their systems and subsystems tends to be the exception.

The next task is to rate, or index, present designs against those of the keenest competitors in terms of both functionality and cost. Once each individual subsystem of a design is indexed, you can then aggregate these indices into an overall index of function and cost for the design as a whole.

Figure 14.10 shows a hypothetical indexing for a product platform. The platform design has six major subsystems. For each subsystem, indices would be created on various dimensions of function and cost. Function is typically some key aspect of performance that can be clearly measured. Cost can be considered in terms of materials cost, yield, or manufactured cost — whatever makes most sense the particular subsystem and your business. Index values are determined for function and cost by studying the products of lead competitors, tearing them down, and studying their own subsystems in the areas of function, materials, and cost. The best in class subsystem would then be scored as "1.00" and all other competitors subsystems (including your own) judged relative to it.

From this process, the firm achieves a clear understanding of the best subsystems from across the industry, both in terms of function and cost. Its own technology will probably excel in some subsystem areas and be at parity with competitors in others. In fact, *the firm need not be superior in every single subsystem*. The key is to determine those subsystems within the platform design where the company can truly excel and that will place competitors at a clear disadvantage. In Fig. 14.10, for example, Subsystem C is presented as a major new breakthrough — a discontinuity — for which competitors have no immediate response.

The next step is to create composite designs. As shown in Fig. 14., the composite design is represented in the central integration block. Since all subsystems may not be of equal importance in the overall design, you may wish to assign weights to each of the particular subsystems. These weights can be factored into the calculation of overall function and cost.

The composite designs will have multiple iterations. The first set of these should be a simple computation of the total index scores for function and cost for the company's present products and those of

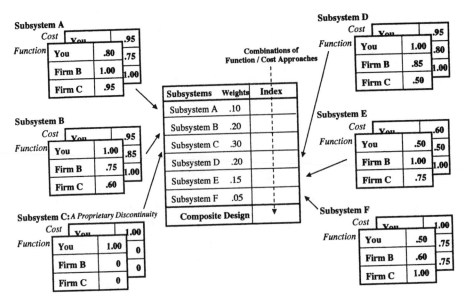

FIGURE 14.10 Composite design: indexing on function and cost.

key competitors. You may find, for example, that the current designs lead in total function but lag in cost, or lead in cost but fall short in total function. The goal, of course, is to create a new composite design that is the combined value-cost leader in the industry.

Perhaps most important in this process from a cycle time perspective is that the firm identify those subsystems that are most important to the overall architecture — the drivers of the platform design. Then, it should do whatever it takes to achieve or acquire best in class cost performance in those critical subsystems and then standardize on that solution across the product line. This will create commonality across products, and further focus solution development teams will not reinvent the wheel every time they go about the business of creating a new derivative product for a market niche. The clearest example is Gillette's razors — or more accurately — its blades. You will find the same blade technologies — not different ones — used in the men's and women's shavers. That is effective platform subsystem management.

Make Teams a Reality: Collocate Different Functional Disciplines

Effective use of teams also has clear cycle time implications. The idea of attacking product development problems through teams is not new. Over a decade ago, Takeuchi and Nonaka [1986] reported on the effectiveness of teams in Japanese companies, contrasting them to traditional approaches in product development. More recently, Katzenbach and Smith (management consultants) observe that "teams are the primary unit of performance in an increasing numbers of organizations . . [and] teams naturally integrate performance and learning [Katzenbach and Smith, 1993]."

Despite these endorsements, many companies remain burdened by slow, functionally based bureau-cracies that impede breakthrough technological innovations, new product platforms, and future streams of exciting new products. Even when teams are formed, few are given with the autonomy required for true success. We never ceased to be surprised when we facilitate a product line meeting for a corporation only to find the marketing vice president "missing in action."

Commercial innovation requires the collaboration of engineering, production, and marketing. Collo-cation of individuals from these functions is essential. A clear charter signed by company executives is also essential. A product family plan that defines both current generation architecture and derivatives products — "locked and loaded" for several or more years and not second guessed by corporate planners every 6 months — is also essential. Consistency on the part of senior management is essential.

Just bringing team members together into one physical place has been shown to improve communication and information sharing. There, small bits of knowledge and information that by themselves mean nothing can be pieced together with other bits to form meaningful insights. Team collocation also fosters bonding between individual members and the commitment needed for focused, fast, high-risk projects. Academic research has measured the communications aspects of collocation. After extensively documenting the specific communications between engineers and engineering managers of a large manufacturer, Allen [1977] found that, if two individuals reporting to different managers were located just 30 feet apart from one another, their probability of communicating dropped to a mere 5%! If they had the same boss, that probability was only about 15%. Considering that R&D, manufacturing, and sales departments are often located in different buildings (if not in different countries), one can easily imagine the difficulty of achieving the amount and quality of communications required for simultaneous design and renewal of product lines.

Full-time team members should have offices or desks adjacent to one another. They must also have a large, open team room where many types of information can be displayed on flip charts or on notes on the walls. Competitors' products should also be on display, both assembled and dissembled, with notes indicating their suppliers and estimated material and labor costs. These team rooms create the cohesiveness and cross-fertilization essential to team building, even within large corporations.

Be Wary of Processes that Seek to Micromanage the Work of Teams

Surely one of the greatest impediments to the realization of fast cycle times are the management processes in which committees of senior managers examine the particular details of individual projects on a frequent basis. The insights and energies of senior management are far better placed into the functionality and markets for next generation product platforms — and not on reviewing the specifics of individual derivative products based on existing platforms. Excessive review leads to too much second guessing, to changes in projects that are already well underway. This interferes with the productivity of the corporation.

A number of observers have noted that "the fuzzy front end" of product development is the greatest consumer of resource and time. We ask, "is that because corporations do not know how to perform effective market research?" In some case certainly yes; but in most, no. We suspect that the fuzzy front end is attributable as much to the second guessing of senior management over design particulars as it is to anything else, such as dysfunctional relationships between marketing and engineering and manufacturing.

Like great products, management processes should be elegant in their simplicity and effectiveness, with every element serving an important purpose. Overly complex processes are unmanageable. Traditional product generation processes tend to be cumbersome, inflexible, and force people to spend more time doing paper work and making presentations than designing and building products. This results in long cycle times and added costs. One manufacturer that studied the overhead imposed by its existing development process of checklists and frequent reviews discovered that it added 18 months to the new product cycle time! Not surprisingly, the real innovations in this company came from "guerrilla" campaigns operating entirely outside the system.

For large organizations, processes are instruments intended to bring order to chaos. Too much order and control, however, discourages innovation and renewal. A certain level of creative chaos is needed to move any organization, large or small, beyond its current technology and habits. This is one of the lessons we have learned from entrepreneurial startups. These enterprises are almost devoid of management process, yet they provide the economy with many of its innovations. Overcontrol by senior management simply will grind down that creative chaos and undermine the initiative and speed of development teams.

Defining Terms

Development cycle time: The time period between the start and end of development for a given product or system, measured separatedly on both an elapsed or person-year basis.

Full cycle time: The time period between the start of development and the commercial shipping date for a given product, measured separately on an elapsed or person-year basis.

Platform cost leverage: The engineering costs incurred in developing a derivative product, divided by the engineering costs incurred in developing the version of the platform upon which the products are derived.

Platform time leverage: The time period between the start and end of R&D for derivative products, divided by the time period between the start and end of R&D for the version of the product platform upon which the products are derived, computed on either an elapsed or person-year basis.

Product platform: A set of subsystems and itnerfaces that form a common structure from which a stream of derivative products can be efficiently developed and produced.

Product family: A set of individual products that share common technology and address a related set of market applications.

References

Allen, J. *Managing the Flow of Technology,* MIT Press, Cambridge, MA, 1977.

Clark, K. and Fujimoto, T. *New Product Development Performance,* Harvard Business School Press, Boston, Smith, P. and Reinertsen, D. *Developing Products in Half the Time,* Van Nostrand Reinhold, New York, 1991.

Katzenbach, J. and Smith, D. *The Wisdom of Teams,* Harvard Business School Press, Boston, 1993.

Mansfield, E., Rapoport, J., Schnee, J., Wagner, S., and Hamburger, M. *Research and Innovation in the Modern Corporation,* W. W. Norton, New York, 1971.

Meyer, M. H. and Lehnerd, A. P. *The Power of Product Platforms,* Free Press, New York, 1997.

Meyer, M. H. and Utterback, J. Product development cycle time and commercial success, *IEEE Trans. Eng. Mgt.,* 42(4): 1–8, 1995.

Utterback, J. M., Meyer, M. H., Tuff, T., and Richardson, L. When speeding concepts to market can be a mistake, *Interfaces,* 22(4): 24–37, 1992.

Takeuchi, H. and Nonaka, I., The new product development game, *Harvard Business Review,* Jan.-Feb., 1986.

Further Information

Visit the author's Web sit at www.cba.neu.edu/~nmeyer.

14.7   Product Positioning

Roger J. Best

The combination of unique product benefits at a specific price creates a particular product position relative to competing products. However, being distinct and different in price and product benefits is not sufficient for success. To be successful, a business must create, communicate, and deliver a product position and value proposition that is appealing to target customers and differentially superior to competing alternatives [Best, 1997]. The purpose of this section is to present the product positioning process and the role technology management plays in successful product positioning.

The Product Positioning Process

The product positioning process is diagramed in Fig. 14.11 and consists of the following steps:

   Inputs — Customer needs and use situations, the positioning of competing alternatives, and product and process technologies.

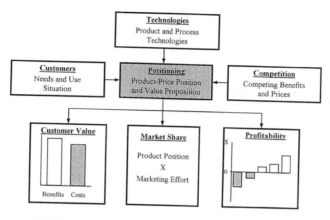

FIGURE 14.11   Process inputs and positioning outputs.

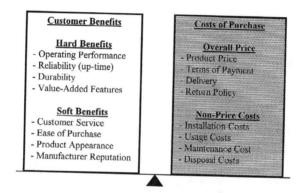

FIGURE 14.12   Customer benefits vs. costs of purchase.

Product positioning — A distinct product-price position and value proposition that are attractive to target customers and differentially superior to competing alternatives.

Outputs — A customer value that is superior to competing products, a desired level of market share, and an acceptable level of profitability.

Inputs to Product Positioning

Customer Inputs

This is where the product positioning process has to start. A technologically superior product that does not meet customer needs will result in limited customer demand and economic failure. To be successful, product positioning must deliver an attractive combination of customer benefits at an acceptable cost or an attractive cost with an acceptable level of customer benefits [Yeager, 1985]. As shown in Fig. 14.12, customer benefits can include *hard benefits* such as operating performance, reliability, durability, and value-added features. For a desktop computer, hard benefits would include memory, speed, disk drive capacity, screen size, multimedia capability, etc. However, product positioning can also include *soft benefits*. With respect to desktop computers, soft benefits could include local buying and customer service, accompanying software, on-line technical support, product warranty, and manufacturer reputation.

Balanced against any combination of hard and soft benefits is the total cost of purchase, which includes the overall price of the product and other, non-price costs of purchase. The overall price of the product goes beyond the purchase price to include the terms of payment, delivery charges, price discounts, and warranty. Nonprice costs are associated with installation, usage, maintenance, and disposal of the product

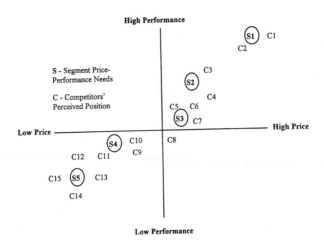

FIGURE 14.13   Product positioning map.

[Forbis and Mehta, 1981]. These nonprice costs of purchase are often overlooked by both customers and businesses. As a result, they provide important product positioning opportunities that have the potential to create greater customer value.

Competitor Inputs

No product operates in isolation from competition. Whether existing products with comparable benefits, substitute products with different benefits, or emerging products with new or improved benefits, each offers target customers an alternative solution to the problem they are wanting to solve. To better understand various competing alternatives relative to customer needs, a product positioning map of competing product positions can be developed [Urban and Star, 1991]. Figure 14.13 illustrates a product positioning map in which there are 5 distinct customer segments and 15 competing alternatives. The position of each competitor is based upon how customers *perceive* each alternative as a competing substitute [Devine and Morton, 1984]. The distance from a segment's needs to a specific competitor represents the *relative attractiveness* of that competitor's product position.

For example, consider the product positions of competitors C1 and C2 in Fig. 14.13. They have the most attractive product positions for customers in segment S1, but we would expect more customers to prefer competitor C2, since it is closer to the needs of this segment. This might lead one to ask why any customer in segment S1 would buy from competitor C1. There are several potential explanations. First, some customers in segment S1 might require a specific performance level that only competitor C1 can provide. Second, many customers in segment S1 may be unaware of alternative C2. Third, it may be that customers in segment S1 prefer competitor C2's product but cannot purchase it due to limited availability, the result of insufficient production and/or distribution. Thus, a product positioning map provides important competitor input into a business's product positioning based on customer perceptions of the business's product position relative to all competing alternatives.

Technology Inputs

A common error in technology-driven businesses is a singular focus on technology. Developing a superior product position requires the right combination of the many competing product and process technologies [Best and Angelmar, 1989]. For example, after reviewing earthmoving equipment customer needs (both fulfilled and unfulfilled) and competitors' product positions, Komatsu determined the product and price requirements needed to achieve a superior product position. This created a challenge for Komatsu engineers, since current technologies could not satisfy these product positioning requirements. This motivated Komatsu to further develop existing technologies and seek new technologies in order to achieve a desired product-price position.

The technology inputs used in developing a product position often impact both product benefits and cost. For example, the development of disk drives for desktop computers has increased average disk drive capacity from 120 megabytes in 1992 to 2.5 gigabytes in 1997, and the average is projected to increase to 12 gigabytes by the year 2000. This will deliver improved desktop computer performance and user productivity. In addition, process technologies can significantly lower the cost of a product. For example, Intel continues to develop its wafer process technology that will allow a single 0.25-μm micron silicon wafer to be made into two. This will both increase the production capacity of a chip plant and lower per unit cost.

Product Positioning

Based on the positioning inputs outlined in Fig. 14.11, a business must develop a product-price position that is either differentially superior in product performance or lower in cost [Porter, 1980]. A product position built around superior product benefits can often be attractive to target customers even at a price premium. Conversely, a product position built around a lower price can attract customers when product benefits match that of competing alternatives.

Cost vs. Benefits

A variety of product positions can be built with combinations of hard and soft benefits and price and nonprice costs. However, to be attractive, a product position needs to be focused on a specific set of customer benefits and customer costs with a knowledge of how competitors currently serve these target customers.

In the desktop computer market, Packard Bell has sought to position itself with a low price while meeting minimum customer requirements for hard and soft benefits. Compaq Computer, on the other hand, adds more hard benefits to its product position but seeks to have a slightly lower price than competitors such as IBM, which leverages the soft benefit that comes with the IBM reputation and customer support. However, Dell and Gateway have positioned themselves with hard benefits similar to Compaq and IBM but with fewer soft benefits (local purchase and customer service) in exchange for a lower price. Toshiba also competes with these same companies but has positioned itself with a greater array of value-added hard benefits and slightly higher prices. Each of these desktop computer manufacturers has developed a slightly different product-price position that is attractive to different desktop computer customer needs.

Value Proposition

A value proposition is a short statement designed to communicate a business's product position to target customers. It highlights key costs and benefits and how the customer should derive increased value from their product [Band, 1995]. If cost is the value driver, then low price or lower nonprice costs of purchase need to be built into a product's position and communicated in its value proposition. On the other hand, a product position could be built around specific hard or soft benefits or various combinations of customer benefits and costs. Whichever the case, a value proposition must be a clear statement of how a target customer will derive increased value from the purchase of the product.

To illustrate, Loctite Corporation developed Quick Metal, a chemical compound that enables factory maintenance workers to repair worn or cracked machine parts and drastically reduces machine downtime and the cost of carrying spare parts. Based on these value drivers, the value proposition developed for Quick Metal product position was, "*Keep the machine running until the new part arrives.*" This value proposition explicitly communicates the key benefit of increased machine operating time and the implicit cost savings of greater machine up-time and lower cost of parts inventory.

Outputs of Product Positioning

Customer Value

To attract customers, a product position must deliver a superior value relative to competing alternatives. This means that the overall benefits derived from the product must exceed the total cost of purchase, as illustrated in Fig. 14.11. The greater the positive difference between overall benefits and cost, the greater the customer value and the more attractive the product position [Gale, 1994]. Of course, an attractive customer value could be the result of superior benefits, lower cost of purchase, or some combination of the two.

Market Share

A potentially attractive product position will fail if target customers are unaware of a product's value proposition or cannot obtain the product. Achieving market penetration requires an attractive product position *and* marketing effort [Best, 1997]. As expressed in Eq. 1, a business's market share is a function of its product position *times* its marketing effort relative to the product of these two factors for all competing alternatives (including the business). For example, if there were three competitors with equally attractive product positions and comparable marketing efforts, all would achieve market shares of roughly 33%. To obtain a higher market share, a business must have a differentially superior product position and a marketing effort comparable to competitors or a comparable product position and a superior marketing effort. Of course, having a superior product position *and* a superior marketing effort greatly increases the opportunity to grow market share.

$$\begin{matrix} \text{Market share} \\ \text{(business)} \end{matrix} = \frac{\text{Business's} \left(\text{product position} \times \text{marketing effort} \right)}{\sum_{i=1}^{n} \text{Competitor}_i \left(\text{product position} \times \text{marketing effort} \right)} \tag{1}$$

Profit Impact

While an attractive customer value and market share penetration are key outputs of a successful product position, to achieve overall success, the product position must also deliver a desired level of profitability. This means that, in addition to obtaining a sufficient product volume (market share times market demand), the product position must yield an adequate unit margin (price less variable costs) so that the total contribution (volume times margin) will exceed direct operating expenses, as illustrated in Eq. 2 through 4. When the net product contribution is positive, it contributes to indirect overhead expenses and the net profits. If the net product contribution is relatively small or negative, the product position is failing, in that it is not making a sufficient contribution to indirect overhead and profits.

$$\begin{matrix} \text{Net product} \\ \text{contribution} \end{matrix} = \left(\begin{matrix} \text{market} \\ \text{demand} \end{matrix} \times \begin{matrix} \text{market} \\ \text{share} \end{matrix} \right) \times \left(\begin{matrix} \text{price} \\ \text{per unit} \end{matrix} - \begin{matrix} \text{variable cost} \\ \text{per unit} \end{matrix} \right) - \begin{matrix} \text{direct operating} \\ \text{expenses} \end{matrix} \tag{2}$$

$$= \begin{matrix} \text{product} \\ \text{volume} \end{matrix} \times \begin{matrix} \text{contribution} \\ \text{margin per unit} \end{matrix} - \begin{matrix} \text{direct operating} \\ \text{expenses} \end{matrix} \tag{3}$$

$$= \text{total contribution} - \text{direct operating expenses} \tag{4}$$

Summary

The overall net profit of a business is the sum of the net product contributions for *all* of its products less the overhead expenses needed to support the business and, indirectly, its products, as shown in Eq. 5, where n equals the number of products in a given business.

$$\text{Net profits} = \sum_{i=1}^{n} \text{net product contribution}_i - \text{indirect overhead expenses} \tag{5}$$

Thus, each product must be positioned in such a manner that it is attractive to target customers, achieves meaningful market share, and delivers a desired level of net product contribution that contributes to the overall net profits of a business.

References

Band, W. Customer-accelerated change, *Market. Mgt.*, Winter: 19–33, 1995.

Best, R. J. *Market-Based Management: Strategies for Growing Customer Value and Profitability*, Chapter 7, Prentice Hall, Upper Saddle River, NJ, 1997.

Best, R. and Angelmar, R. *Strategies for Leveraging Technology Advantage. Handbook on Business Strategies*, Vol. 2, pp. 1–10, Warren, Gorham and Lamont, New York, 1989.

Devine, H., Jr. and Morton, J. How does the market really see your product?, *Bus. Market.*, July: 70–79, 1984.

Forbis, J. and Mehta, N. Value-based strategies for industrial products, *Bus. Horizons*, May: 32–42, 1981.

Gale, B. *Managing Customer Value*, Free Press. New York, 1994.

Porter, M. *Competitive Strategy*, Free Press. New York, 1980.

Urban, G. and Star, S. *Advanced Marketing Strategy*, p. 144, Prentice Hall, Upper Saddle River, NJ, 1991.

Yeager, R. Customers don't buy technologies; they buy solutions: here's how five advanced technology marketers saw the light and avoided becoming high-tech commodities, *Bus. Market.*, November: 61–76, 1985.

14.8   Selecting and Ending New Product Efforts

Robert J. Thomas

The only thing more difficult than selecting the right **new product** effort is terminating a once-promising project that has been championed by a fervent new product team. Consequently, the decisions to select and end new product efforts require a managerial discipline that follows an organization's strategic direction rather than a project-by-project evaluation. This discipline should include the following major activities:

1. Clarify the organization's strategic objectives
2. Define evaluative criteria based on strategic objectives
3. Define a decision model that assesses the value of the new product option
4. Implement the decision model in a portfolio framework
5. Monitor and track performance

Creating a managerial discipline to select and end new product efforts demands careful thought, attention to data collection, respect for judgment modeling, tolerance for mistakes, and continuous learning. The five major activities are described in the following sections.

Clarify the Organization's Strategic Objectives

New product efforts make most sense in the context of business strategy, where the goal is to improve an organization's long-run performance and increase the prospects for survival in competitive business environments. The latter is especially critical for start-up firms that often invest their future in a single new product effort. Whatever the type of organization, there are at least eight strategic reasons (or objectives) motivating new product development [Thomas, 1993]:

- *Establishing long-run competitive advantage:* New products can provide potential buyers with need-satisfying capabilities that don't exist in competing products.
- *Reinforcing or changing strategic direction:* New products can signal a firm's direction in its primary markets or become the basis for new strategic imperatives.
- *Enhancing corporate image:* New products can symbolize a firm's position to its primary stakeholders.
- *Improving financial return:* New products developed in the past 3 to 5 years provide an increasing proportion of a firm's sales and profits.

- *Increasing research and development (R&D) effectiveness:* New products enable firms to capitalize on technology.
- *Improving utilization of production and operations:* New products can enhance capacity utilization.
- *Leveraging marketing effectiveness:* New products can extend a firm's brand equity and create marketing economies of scale.
- *Effectively utilizing human resources:* New products can create jobs and enhance career opportunities.

Of course there are other strategic objectives for new products, which are specific to a firm or its industry and which emerge from a firm's situation analysis of its business. In any case, the organization should clarify the primary reasons and objectives for any new product effort. Not only does this activity give the new product effort a clear purpose, but, more importantly it provides a strategic foundation for selecting criteria to evaluate specific efforts.

Define Evaluative Criteria Based on Strategic Goals

The selection and termination of new product efforts require clear operational criteria (or factors) to make these decisions. Experience shows that managers must decide which criteria they will use to evaluate new products, though research suggests these kinds of factors can be generic across firms and industries [de Brentani, 1986]. The approach suggested here is to be aware of generic factors; however, the specific criteria selected should follow directly from stated strategic objectives. The criteria can be a simple listing, or they can be more usefully organized into relevant categories for portfolio analysis.

In the simplest case a two-dimensional multifactor **portfolio approach** can be utilized to support decision making. The two most-often used dimensions organize the criteria by considerations that are external and internal to the firm. External criteria can be defined as the expected market attractiveness of the new product effort, and internal criteria as the organization's expected business position with respect to the new product effort. Typically, external market attractiveness criteria are linked to the first strategic objective of establishing long-run competitive advantage. These include criteria such as

- Market opportunity (What is the size, growth, etc. of the market for the new product?)
- Market acceptance (Will potential buyers adopt it in a timely fashion?)
- Competitiveness (Does it have a competitive advantage?)
- Environmental fit (Does it fit environmental trends?)
- Stakeholder fit (Does it meet concerns of major stakeholders such as suppliers, retailers, regulators, etc.?)

Internal business position criteria are linked to the other seven strategic objectives, respectively:

- Strategic fit (Does it fit our business strategy?)
- Corporate image fit (Does it fit our corporate image?)
- Financial fit (Will it make money for us?)
- R&D fit (Do we have the technology to design it and do we have a development process to translate the idea into a concept, a prototype, a product, and a launch marketing program?)
- Manufacturability/production processing fit (Can we make it?)
- Marketing effectiveness relative to competition (Can we sell it?)
- Organizational human resources fit (Do we have a team and a product champion to make it happen?)

Each criterion can be directly operationalized and measured, or subfactors can be identified that provide more specific meaning. For example, the "market opportunity" factor above can be characterized by the eight items listed below, with possible measures in parentheses:

1. Size/volume of market (units, dollars)
2. Growth of market (average annual growth rate of units, dollars)
3. Price trend (declining, stable, increasing)

4. Concentration of industry (few large buyers vs. many small buyers)
5. Seasonality of demand (volatile vs. stable)
6. Cyclicality of demand (volatile vs. stable)
7. Geographic location (highly dispersed vs. concentrated)
8. Newness of market to firm (first time in market vs. experience with market)

Measures on such criteria can be point estimates or forecasts (e.g., sales, profits, growth rates), rating scales (e.g., degree of industry concentration measured on a tenpoint interval scale), or meaningful categories (e.g., industry life cycle measured as growth, maturity, or decline). The selection of criteria (and measures) should reflect strategic objectives derived from a situation analysis of the business and should be carried out by managers involved in the new product effort. The management team should meet to discuss the criteria and decide which should be used to evaluate various new product efforts and how they will be measured.

Define a Decision Model that Assesses the Value of the New Product Option

The basic decision involved in selecting or ending new product efforts is the new product **Go/NoGo decision**. The Go/NoGo decision has four basic outcomes:

- *Go:* Continue to invest resources into the development of this new product.
- *No/Go:* End or terminate the new product; it has no future.
- *Hold:* The new product cannot be adequately evaluated with available data. Hold the decision until necessary data are acquired, then evaluate Go/NoGo.
- *Shelve:* The idea has promise but the market is not ready for it or it cannot be made with current technologies and resources. Evaluate Go/NoGo when taken off the shelf.

Making these decisions typically involves managerial review of the selected criteria (including experience and any data available on each option) followed by a review of the status of the new product effort. After discussing various aspects of the project, managers make the Go/NoGo decision. The decision can be based on a relative comparison of a specific new product effort to others under evaluation (selecting the project which best meets the criteria) or by comparison to a benchmark (typically based on past experience with similar new product efforts).

Even though all Go/NoGo decisions are eventually based on this type of informed judgment, the preferences of a dominant executive or other managerial biases [Hogarth and Makridakis, 1981] compel the consideration of decision models that provide more systematic and objective support [Baker and Albaum, 1986]. Three such models are briefly considered here: simple hurdle rate, index of attractiveness, and weighted average.

Hurdle Rate Models

Hurdle rate models involve three fundamental steps: (1) managers select only essential criteria for evaluation (such as the 12 major factors identified above or a subset of them), (2) they then set a hurdle rate for each criterion used or a measured value that the criterion must meet for the new product effort to continue, and (3) they then establish Go/NoGo decision rules based on the hurdle rates. There are numerous variations on the decision rules, the selection of which depends on a combination of judgment and benchmarks established from prior new product experience. For example, managers can

- Assign equal importance to all criteria then decide Go only if 100% of the criteria hurdle rates are satisfactorily met; otherwise, decide NoGo;
- Assign equal importance to all criteria then decide Go only if some set proportion (e.g., 60%) of the criteria hurdle rates are satisfactorily met; otherwise, decide NoGo;
- Assign ordered importance to the criteria then decide Go only if some set proportion or number (e.g., top five) of the hurdle rates on the top ranked criteria are satisfactorily met; otherwise, decide NoGo.

Illustrative Ratings for the Weighted Average Model

Evaluative criteria	Importance of criteria (I)*	New product option 1 rating (R_1)**	New product option 2 rating (R_2)**
Market opportunity	.40	.6	.5
Market acceptance	.10	.9	.6
Strategic fit	.20	.4	.9
Financials	.30	.5	.8
	1.00		

* Based on a constant sum scale of importance, such that the importance of criteria sum to 1.00.
** Based on a 0.00 to 1.00 rating, where 1.00 is the highest score.

Index of Attractiveness Model

The **index of attractiveness** (IA) model recognizes four major criteria in evaluating whether to select or end a new product effort: technical development, marketability, profit, and development costs [Urban and Hauser, 1993]. The index is essentially a measure of expected return on development costs, weighted by the probabilities of technical and commercial success. More specifically it is defined as

$$IA = \left(T \times C \times P\right)/D \tag{1}$$

where IA = index of attractiveness, T = probability of successful technical development, C = probability of commercial success (given it is technically successful), P = estimated profit if successful, and D = estimated cost of development.

Although the above formula can be used to estimate an index of attractiveness for a single new product effort, more insight into the value of the effort can be gained when placed in the context of other new products. Given an index, a product would be selected or ended according to a benchmark, usually based on experience from past new product ideas. Clearly the accuracy of cost and sales estimates, as well as the probabilities, are critical to the reliability of such a model.

Weighted Average Model

The **weighted average** model provides an expected value for the new product effort based on the importance of the evaluative criteria and the rating of each new product effort on each criterion. The model takes the following form:

$$W_k = \Sigma\left(I_j\right)\left(R_{jk}\right) \tag{2}$$

where W = computed value of the weighted rating scores for k product options, I = importance of each of the j criteria, and R = rating of the extent to which each of the k new product options meets the j criteria.

The table above illustrates one application of the weighted average model using four criteria to evaluate two new product efforts.

Using Eq. 2, the calculations for new product options one and two are shown below:

$$W_1 = (0.4)(0.6)+(0.1)(0.9)+(0.2)(0.4)+(0.3)(0.5) = 0.56$$

$$W_2 = (0.4)(0.5)+(0.1)(0.6)+(0.2)(0.9)+(0.3)(0.8) = 0.68$$

These two summary values indicate that the second new product option receives a higher score than the first one. Without relative comparisons, benchmarks must be established to implement a decision rule.

In selecting and using such models it is important to note that the latter two models (index of attractiveness and weighted average) are compensatory, in that high scores on one criterion can compensate low scores on another. It is also important to note that the outcome of the decision can vary with the criteria used, the measurements employed, the quality of managerial judgments in making

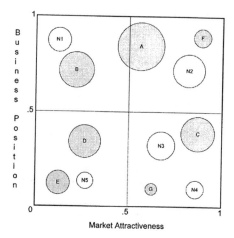

FIGURE 14.14   Portfolio of new and existing (shaded) products.

estimates, and the benchmarks set in the decision rules. Because of this potential for variability, it is recommended that firms experiment with multiple models (using a variety of criteria and measures) and track their performance. Ultimately, a specific model or a composite approach based on multiple models will become favored through experience.

Implement the Decision Model in a Portfolio Framework

Though not essential for evaluating a specific new product effort, it is often helpful to visualize one or more new products in the context of an existing portfolio of products for purposes of resource allocation. The premise of the portfolio concept applied to new product decisions is that the strategic value of the new product and existing products can be portrayed in a two-dimensional space. By showing the array of new and existing products that define the business, key strategic decisions can be made (such as whether to invest in the product, maintain it, harvest it, or divest it). As noted above, these two dimensions are often characterized as market attractiveness vs. business position and can be defined by multiple factors for each.

To illustrate the approach, the "attractiveness" (A) and "position" (P) coordinates in the two-dimensional space can be computed from the weighted average model described above. Rather than one summary score for all criteria included in the model, the criteria are separated into two sets (attractiveness vs. business position), and a score for each is estimated. The new product can then be plotted in the two-dimensional space with the respective (A,P) coordinates. By computing the estimated scores for existing products, comparisons with possible new product efforts are possible. Further, the (A,P) coordinates can be plotted as the center of a circle (or "bubble") with a size (diameter) based on a key measure. The size is computed relative to other new and existing products in the portfolio (e.g., calculating the proportion of sales from each product). Available software programs facilitate computational aspects of these tasks.

The portfolio illustrated in Fig. 14.14 reveals that the metric used to compute the summary coordinates (A,P) involved a zero-to-one scale, with the value of one being the highest business position and market attractiveness. In this illustration, the shaded circles represent existing products. Of the five new products shown (N1 to N5), N2 appears to have the best position on the two dimensions and the largest diameter (e.g., representing forecasted profit). If a firm could only invest in one new product, then N2 would appear to be a good decision (assuming a valid and reliable scoring model). The portfolio also suggests that N5 may be a candidate for termination — given its relatively weak position and size. The same could be said for existing product E, which may also be a candidate for deletion. If product E were divested, resources saved (or gained from its sale) could be invested in other efforts. For example, closer scrutiny of N3 may be merited. If the possibility of increased investment can improve the product's business position in the market, then it too may receive a Go decision.

Monitor and Track Performance

The first four activities described above should be applied to products as they move through various "forms" of development: from ideas to concepts to prototypes to viable products and marketing programs for them. The approach can be used to strategically screen new product ideas and subsequently be updated as selected ideas evolve into concepts and products. New data from market research, costs, and business case analyses enhance the original estimates, such that once attractive ideas may become candidates for termination. In this way a product's performance can be tracked throughout its development, recognizing the need to revisit often the Go/NoGo decision.

As suggested earlier, this continuous monitoring and tracking provides the necessary "discipline" to select or end new product efforts with some degree of objectivity and confidence. Monitoring and tracking also involves developing and maintaining an ongoing data base of information about the new product evaluation process (criteria used, measures, data sources, etc.). This becomes essential to validate the performance of various decision models suggested above and to establish benchmarks.

Conclusions

Selecting and ending new product efforts involve decision processes that are central to new product development. There are a variety of criteria, measures, and decision models to consider as support in making these decisions. Some have been briefly reviewed in this section. Because judgment is central to all new product decision making, it is subject to human biases. It is therefore important to define a fairly systematic approach and follow it to provide the necessary intellectual discipline to improve new product decision making.

However, several considerations for using such an approach are recommended:

- Pay special attention to the selection of evaluative criteria, their measures, and the decision model used. They seriously influence the outcome. To the extent possible use group consensus methods to help make such selections.
- Collect as much data as possible for the various criteria, given time and other resource constraints, even using experts as often as possible to provide important insights.
- Recognize the imprecision of such modeling approaches and be willing to make decisions under uncertainty. Models only help structure the problem, not make decisions.
- This is as much a process of learning as it is decision making. Maintain records of decisions and track the performance of various models to validate their effectiveness.

Finally, recognize that initial experience with the process will be difficult and cumbersome; however, as experience is gained in selecting and ending new product efforts, it becomes more tractable. Ultimately, it becomes a necessity and builds a culture that sustains profitable innovation.

Defining Terms

Go/NoGo decision: Because new product development processes carry potential risk and return, the decision of whether to continue or not with the project must be continually considered. The decision has four basic outcomes: go, or continue to invest resources in the new product; no/go, or terminate the investment of resources; hold, or suspend the project until additional data are available to assess its potential; and shelve, or indefinitely hold the new product's development until internal organizational capabilities or external business conditions change.

Hurdle rate: A measured value (or set of values) that must be met on a specific criterion (or set of criteria) to make a decision.

Index of attractiveness: An index computed from selected criteria and estimates of measures on those criteria. The index is typically a multiplicative or additive model that includes variables, which, when multiplied or added, produce a summary index.

New product: A multidimensional concept with need-satisfying capabilities not previously experienced by the stakeholders interested in it.

Portfolio approach: An approach for arraying alternative options in a two-dimensional space for visual understanding and analysis. Each dimension can summarize information from multiple factors. The options of interest here are comparing the current and expected performance of existing and new products, respectively.

Weighted average: A method for computing a summary measure or expected value for several options, with each option characterized by selected criteria. Estimates of the importance of the criteria are used to weight ratings taken on each criterion for each option.

References

Baker, K. G. and Albaum, G. S. Modeling new product screening decisions, *J. Prod. Innov. Mgt.*, 3: 32–39, 1986.

de Brentani, U. Do firms need a custom-designed new product screening model?, *J. Prod. Innov. Mgt.*, 3: 108–119, 1986.

Hogarth, R. M. and Makridakis, S. Forecasting and planning: an evaluation, *Mgt. Sci.*, 27: 115–138, 1981.

Thomas, R. J. *New Product Development: Managing and Forecasting for Strategic Success*, John Wiley & Sons, New York, 1993.

Urban, G. L. and Hauser, J. R. *Design and Marketing of New Products*, Prentice Hall, Englewood Cliffs, NJ, 1993.

Further Information

The *Journal of Product Innovation Management* provides regular articles and research on the various aspects of assessing new product development projects. JPIM on-line can be reached at http://www.elsevier.com/locate/jpimonline.

The *Product Development and Management Association* and the *Commercial Development Association* provide numerous seminars and annual meetings that often feature the experience of organizations in assessing the performance of new product efforts. The PDMA can be reached at http://www.pdma.org.

New Product Development: Managing and Forecasting for Strategic Success (in the Portable MBA Series) by Robert J. Thomas describes a system of forecast-based spreadsheets that help to assess the performance of a new product development effort on an ongoing basis.

The *Design and Marketing of New Products*, by Glen L. Urban and John R. Hauser, provides a compendium of research techniques and models that support decision making throughout the new product development process.

14.9 Time to Market: The Benefits, Facilitating Conditions, and Consequences of Accelerated Product Development

Stanley Slater, Ajay Menon, and Denis Lambert

In a time of shrinking product life cycles, a strategy for reducing time to market seems to be essential, e.g., Slater [1993]. A time-based competitive strategy focuses on the speed with which a company can take a product from the concept stage to market introduction. A McKinsey & Company economic model suggested that high-technology products that reach the market 6 months late without having exceeded budget earn 33% less profit over a 5-year period than products that reach the market on time. In contrast, products that are on time and as much as 50% over budget lose only 4% in profitability [Reinertsen, 1983].

During the 1980s and 1990s, several European, American, and Japanese corporations have responded to the demands of a rapidly changing marketplace and unanticipated economic conditions by developing accelerated product development strategies to sharpen their competitive edge. For example, in 1992 Motorola set a goal of reducing product development cycle time to 1/10 by 1997. According to a study by Pittiglio, Rabin, Todd, and McGrath, the high-tech industry's average product development cycle-time was reduced by almost 10% between 1992 and 1994, with an average improvement goal of over 20% between 1994 and 1996. Demonstrating the results of rapid product development, over 60% of Hewlett-Packard's 1996 revenues came from products introduced in the 2 previous years.

Recognition of the growing importance of accelerated product development is reflected in its status as a central element in the strategic business plan and in organizational endeavors to streamline operations and processes and to improve efficiency. In the following sections, we discuss the benefits of accelerated product development, organizational conditions that facilitate accelerated product development, and a number of pitfalls that can more than offset the benefits of an accelerated product development process.

Benefits of Accelerated Product Development

Accelerated product development is particularly important to businesses that are committed to pioneering. Pioneers are the first entrants into a market with a new product or a new generation of product. As the first entrants, they have a monopoly position until a rival emerges. This position may lead to a leadership reputation and enable premium pricing and may lead to the pioneering product establishing itself as the de facto standard. For products with high **switching costs**, pioneers can secure their position by creating a large installed base before significant rivals emerge. The experience gained from the pioneer's lead may also translate into a cost advantage or a sustainable lead in technology development. Finally, the first entrant often has first access to actual experiential customer feedback. Having the ability to rapidly translate this new market knowledge into the next generation of product means that the pioneer is more likely to continue to satisfy market needs.

Early followers can avoid the pioneering risks while still competing effectively in fast-moving markets if they can respond quickly and proactively to the pioneer's move. Rather than trying to establish a monopoly position in the market, early followers attempt to understand shortcomings in the pioneer's product or its positioning and rapidly develop a superior alternative. Due to the uncertainty of market and product development, pioneers often "leave something on the table". Followers can take advantage of this opportunity, generally at a lower cost and with lower risk, if they are able to define and seize the opportunity rapidly. Thus, while accelerated product development may look different in an early follower business than it does in a pioneer business, as a general capability it may be just as important.

Another potential category of benefit is the opportunity for reducing the cost of the product development process. A common misconception is that accelerated product development must be more expensive since it will require more engineers, manufacturing people, and marketers working over a shorter period of time to achieve their objective. While some leaders in time-based competition have achieved their goals by committing increased amounts of resources to drive down cycle times, this approach often does not produce competitive advantage. Accelerated product development should be about **reengineering** the product development process to improve speed, effectiveness, and efficiency. If this is the approach, it will mean the same number of people, or fewer, working for a shorter period of time to bring the product to market.

If accelerated product development leads to any of these outcomes, the result should be improved financial performance for the business. The next section will introduce some of the organizational principles that will facilitate the achievement of this result.

How Organizations Facilitate Accelerated Product Development

In this section, we discuss seven characteristics of businesses that have achieved relatively short product development cycles. These organizational characteristics should reinforce each other so that the result is a speedy organization. These characteristics have been identified in a number of recent studies, e.g., Cooper and Kleinschmidt [1994] and Zirger and Hartley [1996].

A Market-Oriented Product Definition Process

The objective of accelerated product development is to develop the right product for the market before the competition can fill that competitive space. A market orientation is a set of values that permeate the entire organization and encourage continuous learning about how buyers' needs are evolving in the market [Slater and Narver, 1995]. Learning occurs as organizations acquire market information by working with lead customers, through experiences with existing customers, or from traditional market research techniques. This information is then disseminated broadly so that a shared interpretation of its implications for product development is achieved. This shared interpretation provides a clear objective and strong motivation for the entire organization. Market-oriented learning continues throughout the product development process so that new information can be incorporated into the product's design.

Dedicated, Cross-Functional Project Teams

Empowered, dedicated, accountable cross-functional teams have been extensively promoted as a key to accelerated product development. Product development is characterized by high levels of interdependence among the research and development, manufacturing, and marketing functions, as well as others. A team-based product development organization facilitates the rapid transfer of information and other resources among team members. This type of structure is most appropriate for the development of innovative new products where uncertainty and task difficulty are the greatest.

Predevelopment Planning

Another critical success factor is the step of laying the foundation for the development project before actually undertaking it [Cooper and Kleinschmidt, 1994]. The predevelopment planning includes performing the necessary market research, identifying both the resources and capabilities that are required for product success, specifying clear objectives, developing a realistic but aggressive schedule for completing the project, and conducting a financial analysis of the project. Good, up-front planning will help to determine the critical path for the project. Thus, the process of acquiring critical resources or skills can take place far before they are actually needed, allowing the process to progress in a smooth manner. Planning also provides the opportunity to identify project risks and to develop contingency plans for dealing with them. It is important to recognize that the planning activities will be valuable only if progress against the objectives and milestones in the plan is monitored regularly and appropriate adjustments are made. Finally, the financial analysis should be viewed only as one more piece of information. Many development projects produce less than stellar initial results. However, they may provide a platform for future development projects whose value cannot be quantified and incorporated into the financial analysis.

Overlapping Development Phases

A traditional product development project proceeds through a sequence of discrete phases such as market investigation, product design, prototype development, manufacturing process engineering, and product introduction. Accelerated product development requires that the phases overlap [Zirger and Hartley, 1996]. Thus, product engineering might commence before a prototype is completed. This approach has the benefits of increased speed and flexibility as well as stimulating communication and a shared sense of responsibility among members of the development team.

Unfortunately, this approach has the potential to create additional tension and conflict in the group. It also complicates the communication process, coordination with suppliers, and the preparation of contingency plans. This may explain the need for cross-functional teams and colocation on accelerated product development projects. However, this could be the most crucial part of reengineering the product development process for reducing time to market.

Focus on Core Competencies

The development of technical proficiency in a new technology is often very time consuming. Accelerated product development requires proficiency in the core technology [Cooper and Kleinschmidt, 1994] so that the development team is not burdened with having to acquire and disseminate knowledge about new technologies or with having to integrate multiple core technologies [Meyer and Utterback, 1995] during development. A focus on core competencies often reduces the breadth of the development portfolio and the number of projects in it. This helps to insure that the resources required for timely project completion will be available and that bottlenecks will not occur [Adler et al., 1995].

Incremental Product Development Based on Reuse and Leverage

One key to being able to accelerate product development without an associated increase in staffing and investment is a system where components and subsystems can be leveraged and reused. Extensive **reuse and leverage** enable the product development organization to rapidly evolve a product through successive generations by concentrating on significantly improving only a few components or subsystems at a time. Instead of trying to hit the infrequent home run, you can regularly deliver base hits that produce the same, or higher, score. This strategy reduces risk, generates positive cash flows more rapidly, and enables access to early market feedback for midcourse corrections.

Accessible Organizational Memory

Product development leaders are learning organizations. Learning organizations continuously create or discover knowledge by seeking and acquiring information, sharing and challenging it, taking action based on it, and analyzing the results of those actions [Slater and Narver, 1995]. The most effective organizations store the knowledge accumulated through discovery and analysis and reuse it or share it throughout the organization.

This characteristic is particularly important to organizations that are attempting to accelerate product development. These organizations will try many things, some of which will succeed and some of which will fail. In order to keep from making the same mistakes over and over, or to enable other development teams to build on specific successes, it is essential to have an accessible organizational memory. This may take the form of shared databases, cross-functional teams, rotating team memberships, or high-quality, published project audits [Hurwitz, 1996]. This is a crucial step in the organizational learning process.

Accelerated Product Development: The Bottom Line

Does accelerated product development deliver the promised benefits? The answer to this question has been the subject of important, recent research. Surprisingly, the findings do not strongly support the existence of a competitive advantage based either on accelerated product development or on pioneering advantage. As explained earlier, it is generally believed that pioneers achieve higher than average profits and greater market share than followers. The most thorough and well-controlled study of pioneering contradicts this assumption. By tracking the market share of the main competitors in 36 product categories, starting with the initial market entry, Golder and Tellis [1993] show that the average long-term market share for pioneers is approximately 10%. Market pioneers were the current leaders in only 4 of the 36 categories. The median period of market leadership for the pioneers was approximately 5 years. However, followers that entered the market approximately 5 years after the pioneer averaged a 28% market share.

Focusing specifically on the benefits of accelerated product development, in-depth research in one firm showed no correlation between development time and expected commercial success and that forcing accelerated development when market and technological uncertainties are high increases the probability of failure [Meyer and Utterback, 1995]. In a broad-based study, Cooper and Kleinschmidt [1994] found a positive but relatively weak relationship between timeliness and financial performance, and it turns out that the oft-cited McKinsey study is based on a simplistic economic model and no empirical research. Furthermore, the conclusion of that paper [Reinertsen, 1983] was that, "The new-product race does not always go to the swift. Speed is sometimes secondary and, if unduly emphasized, can lead to disaster. The key to winning is a company's flexibility in adapting a product's development to its market and strategy."

Conclusion

The pitfalls associated with pursuing a time-based strategy are numerous and potentially serious. For example, incremental product improvement can drive out longer-term innovations; the organization's development efforts may become too dependent on existing competencies; key process steps can be skipped or shortchanged in the interest of speed; the development process may become too internally focused and may not be able to adequately respond to market-driven changes; customers may be frustrated by the speed with which their purchases seem to become obsolete; product development pressures can force out necessary long-term infrastructure, process, tool, and product development; and the organization may end up competing more with itself than its competitors, prematurely ending the life of viable products, e.g., Crawford [1993].

Accelerated product development does have benefits, but only when it is focused on satisfying market needs and when the process is thoughtfully conducted. The steps for accelerating product development that are outlined in this section are based on good theory and on strong evidence. They should provide a reasonable framework for an accelerated product development program. However, every business faces a unique set of circumstances. Your product development program should reflect your business's overall strategic orientation, the market pressures it faces, and its capabilities and resources.

Defining Terms

Reengineering: Defining the business in terms of customer-value adding processes and eliminating work that does not add value.

Reuse and leverage: Identifying hardware and software components that can be reused from one product generation to the next or leveraged across multiple products.

Switching costs: Costs (e.g., steep learning curves, large buyer investments, infrastructure dependencies, etc.) that tend to tie buyers to one supplier who is then protected from raids by competitors.

References

Adler, P., Mandelbaum, A., Nguyen, V., and Schwerer, E. Getting the most out of your product development process, *Harv. Bus. Rev.,* 74(2): 134–152, 1996.

Cooper, R. and Kleinschmidt, E. Determinants of timeliness in product development, *J. Prod. Innov. Mgt.,* 11(5): 381–396, 1994.

Crawford, C. M. The hidden costs of accelerated product development, *Eng. Mgt. Rev.,* Summer: 21–28, 1993.

Golder, P., and Tellis, G. Pioneer advantage: marketing logic or marketing legend, *J. Market. Res.,* 30(2): 158–170, 1993.

Hurwitz, D. Getting to market first, *Mach. Des.,* 68(9): 186, 1996.

Meyer, M. and Utterback J. Product development cycle time and commercial success, *IEEE Trans. Eng. Mgt.,* 42(4): 297–304, 1995.

Reinertsen, D. Whodunit? The search for new-product killers, *McKinsey Q.,* July: 35–37, 1983.

Slater, S. Competing in high velocity markets, *Ind. Market. Mgt.,* 22: 255–263, 1993.

Slater, S. and Narver, J. Market orientation and the learning organization, *J. Market.,* 59(3): 63–74, 1995.

Zirger, B. and Hartley, J. The effect of acceleration techniques on product development time, *IEEE Trans. Eng. Mgt.,* 43(2): 143–152, 1996.

Further Information

Deschamps, J-P. and Nayak, P. R. *Product Juggernauts,* Harvard Business School Press, Boston, 1995.

Smith, P. and Reinertsen, D. *Developing Products in Half the Time,* Van Nostrand Reinhold, New York, 1995.

Wheelwright, S. and Clark, K. *Revolutionizing Product Development,* Free Press, New York, 1992.

14.10 Concurrent Engineering

Robert B. Handfield

Product development lead time is recognized as a critical determinant of successful innovation and market competitiveness. The benefits of reducing development lead time include increased market share, heightened barriers to new competitors, and higher sales further into the future. Shorter development times also allow firms to delay the release of new products in order to introduce a better targeted product at the same time as slower competitors. Companies moving quickly in new product development can enjoy high profit margins through progression down the manufacturing learning curve and exploitation of technical opportunities. Research has also shown that the inability to develop new products quickly can provide a window of opportunity for competitors to move from follower positions and seize industry leadership in less than 10 years [Stalk and Hout, 1990]. The incurred costs in the early conception stage of product development are low, yet, in real terms, 40 to 60% of the costs are committed to in the form of material and process choice decisions. Changes made to the product are very inexpensive at this stage, as no resources other than manpower have been committed. Changes made during final production are estimated to be 10,000-fold more expensive. At this stage, resources have been committed, processes established, and any changes that occur require long man hours and expensive retooling to implement.

As part of an initiative to become fast to market, many time-based competitors are embracing a new forms of product development known as "**concurrent engineering**" (see Defining Terms). Concurrent engineering is typically associated with a number of other product development strategies, including planned obsolescence in the product life cycle, parallel scheduling activities, a "time-based" corporate culture, greater reliance on external supplier design capabilities, improved manufacturing techniques, and greater reliance on automated design technologies. Recent studies suggest that concurrent engineering is best applied in conjunction with well-developed product strategies aimed at incremental innovation [Handfield, 1994]. In this section, we compare the sequential and concurrent engineering approaches to product development and note the advantages and disadvantages of each product development strategy.

The Sequential Product Development Process

The product development process varies across different industries and types of organizations. The majority of American firms rely on **sequential development** of new products, in which functions "hand off" the product after completion of their stage to the next functional group responsible in the process. The metaphor in this case is a relay, in which runners pass the project on to the next stage in the development process. The manner in which activities are carried out in product development directly impacts the transition between development and the supply chain processes, which occur later in the product's life cycle. Each set of activities performed in development is typically evaluated at a "gate" to determine whether a set of formal criteria were met. At each of the four gates shown in Fig. 14.15, the project is measured against a number of clearly defined attributes that are "signed off" by the functional personnel involved. The decision to pass a gate is based on marketing studies, engineering reports, and manufacturing, testing, and cost assessments. If there exists any substantive reason for termination at any gate, the project is aborted and formally classified as a failure. Although a gate is occasionally attempted again, it is often difficult to pass the second time.

In the first stage of new product development, marketing studies provide commercial specifications of the attributes that the future product should possess. Since the source of successful innovations are typically customer driven, marketing's role is to tap customers as a source of innovation and transform these assessed needs into a potential product. The response or instrument that results at this gate is a product specification agreement with the engineering/design function. The role of engineering and design is then to develop a product agreement based on the level of available and projected technology as well as the level of expertise in the firm. At this "breadboard" stage, the principal task is that of examining the feasibility of such a product. The interaction between marketing and engineering departments at this

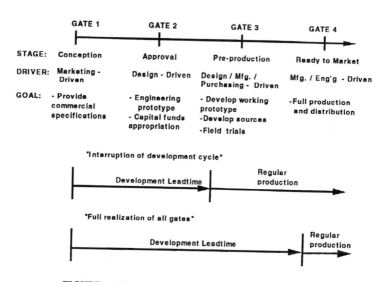

FIGURE 14.15 Gates in the product development cycle.

stage is critical in obtaining an accurate portrayal of available technology vs. market needs. However, the criteria for passing gate 1 are primarily marketing driven.

The second gate is a design-driven effort. The goal of design engineers is to design a product that will meet or surpass the attributes determined by marketing in the conception stage, yet not exceed cost estimates. The design task involves the choice between alternative or existing product technologies which will ensure market success. The result of these activities is the development of an engineering prototype. This prototype embodies the type of "package" to be used as well as the required components and materials involved. The use of computer-aided design technologies have made great strides in reducing lead time within this process. Capital funds are approved for the project once this gate is passed, which signifies that the product is formally approved for development.

Between gates 2 and 3, the first manufactured prototypes are run on existing or procured equipment. Purchasing departments play an important role in the make or buy decision of materials and may facilitate information exchanges between the supplier and the buying firm. Suppliers often prove to be invaluable sources of information in determining simpler and more manufacturable designs. In some cases, suppliers perform the design function for many components and parts, contributing significantly to reduced development lead times. For those parts produced in-house, manufacturing personnel are instrumental in matching process capability and design tolerances. In cases where the available process technology is not available, capital funds are allocated to either procure the necessary equipment or develop an external source.

Tests occurring in-house and through field trials determine the performance characteristics of the manufacturing prototypes. Processes and tooling are tested, and the first preproduction runs occur. Employees go through extensive training and progress down the learning curve, and the capabilities of chosen suppliers are observed for the first time in a large-scale production environment. These runs provide an assessment of the match between materials, product, and process. In this sense, the primary locus of decision making at gate 3 is manufacturing and purchasing oriented.

Customers may also play a key role in passing gate 3, in testing manufactured prototypes in the field. Based on field trials, customers will either reject, reject with conditions, or accept the product in its existing form. Since large customers are frequently experimenting with competing alternatives to new products, a failure to accept may indicate an alternative has been found, thereby "shutting the window of opportunity".

The final gate is passed when the product is ready to market. At this stage all testing has been approved and completed, the customer's suggestions have been integrated into the product, and all process capability

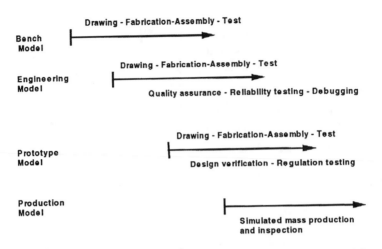

FIGURE 14.16   The concurrent ("Sashimi") overlapping development schedule.

studies have been formally agreed upon. If all is in order, the product proceeds easily into full production. At this point, engineering efforts focus on cost reductions through alternative materials, value analysis, process improvements, and progression down the learning curve.

Although product development is not usually considered complete until "first realization" of the product, production and shipments to customers sometimes occur as soon as a single working prototype has been developed (before gate 3 is passed). In cases when the product has been properly designed, this is not a problem. However, in cases when the designed product has not be fully tested, the product is shipped to customers with existing defects and "bugs" that have not yet been discovered. The result is that technical problems must be worked out on an order-by-order basis, increasing the delivery lead time to customers. This suggests that the "full realization" of the serial development cycle may result in a product that is more "manufacturable" with a shorter delivery lead time. That is, the integration of learning early in the development cycle can prevent manufacturing problems from occurring later in production. By completing all of the activities in the cycle, a product that is better focused to its target market with fewer undiscovered defects will be shipped.

Concurrent Engineering Approaches to Product Development

Many firms continue to employ a "sequential approach" to product development. This approach is typified by the serially phased program planning as shown in Fig. 14.15. These types of systems were developed by NASA in working on large projects for the space program in the 1950s and 1960s. The disadvantage to these systems, however, is that a bottleneck in any one phase can slow down the entire development product, thereby increasing development lead time. Within the last 10 years, many Japanese companies such as Fuji, Canon, and Honda began using a new type of system that uses a system in which activities take place concurrently.

As shown in Fig. 14.16, concurrent engineering involves having events in the product development cycle scheduled in parallel at the interface of different phases. The benefits of concurrent engineering at the Fuji corporation included faster speed of development, increased flexibility, and greater sharing of information [Imai et al., 1989]. This system (which has been referred to as "sashimi" development by Fuji, so called because of the manner in which slices of raw fish overlap one another on a plate) also increases shared responsibility of project members. Members of sales, research and development (R&D), manufacturing, and suppliers all work together as a team to solve the problems occurring in new product design. Some of the disadvantages of the sashimi system include the amplification of ambiguity, tension, and conflict in the group.

If we compare the "sashimi" and "sequential" approaches to product design, it is apparent that the former involves the compression of development lead times. In order to design a product without defects, all of the stages in product development must be carried out, and both systems achieve this goal if not interrupted. Cooper [1990] also found that the more activities left out, the higher the likelihood of product failure. While the total time (in man hours) spent on development are equal using both approaches, the sashimi system compresses development time through overtime and parallel development activities. In addition, such approaches often require that the organization be able to compress learning into a shorter time period. For instance, descriptions of the process used at Fuji often involved monthly overtime of 60 to 100 hours for project members [Imai et al., 1989]. This means that product development team members must learn to work more as a team rather than as functions separated by "walls".

North American firms have begun to use concurrent engineering processes more in the last 5 to 10 years. The primary attribute associated with concurrent engineering is scheduling parallel activities, in which functions transfer information in parallel rather than serially (analogous to a "rugby" game as compared to a relay race), as shown in Fig. 14.16. For instance, the "Work flow Analysis" program at General Electric moves a majority of noncritical support work off-line or in parallel so that only necessary actions are performed on the main line resulting in a shortened total time cycle [Hughes, 1990]. In describing this evolution at General Electric, a manager noted that all aspects of the product, process, information systems, and organizational capabilities were addressed at the same time. Parallel scheduling typically involves the sharing of partial information at critical design interfaces. Opportunities for over-lapping can be created by considering the types of data that must be transmitted from one activity to another to support the sequence of design decisions. By sharing "imperfect" information at various nodes, activities which follow can be started before preceding ones are fully complete. Profound changes in decision-making styles based on imperfect data are required. Managers must learn how to fully use all available information, consider all of the possibilities, and make decisions on shorter notice. A prerequisite for this to occur is a shift in power structures. Moreover, greater levels of empowerment are needed to assign this level of responsibility to individuals who will be making decisions, which may have a tremen-dous impact on future capital investments in product and process technology.

Tradeoffs between Concurrent and Sequential Engineering

Disadvantages of concurrent engineering include a greater investment in up-front activities such as extra man hours, engineering expenses and contracting fees, and the amplification of ambiguity, tension, and conflict in the group. Nevertheless, the total time required for development can be decreased substantially through concurrent engineering. This suggests the trade-off that exists between development and delivery lead time in the case of sequential development, which can be overcome through concurrent engineering processes.

To illustrate this, consider the production frontier curve L1 shown in Fig. 14.17. This curve represents the theoretical set of trade-offs that exists for a firm using sequential product development. Product managers faced with the task of determining reasonable goals to achieve in terms of time to market vs. delivery time can use such curves as an aid in decision making. The curves can be developed using data from previous product introductions, or data on competitor's product introductions, and applying simple or polynomial regression. This data can also be useful in developing economic life cycle models, which are used in making decisions regarding allocations of funds for compression of development time. Such models should be simple and developed by teams of engineering, manufacturing, marketing, and finance managers, and should consider the implications of trade-offs in development speed, product cost, product performance, and development expense. Although such models do not necessarily address how to design the product, a properly developed market analysis should include when the product needs to be released as well as providing reasonable delivery time expectations in the future. Decisions made early in the development of the product will subsequently affect the choice of suppliers, technology, process, and training involved.

This framework can also be useful in considering whether concurrent engineering should be used. In the example shown in Fig. 14.17, a firm with production frontier L1 for a new product must decide on timing the market release at time "B", with resulting delivery lead time of "D", or of running through the

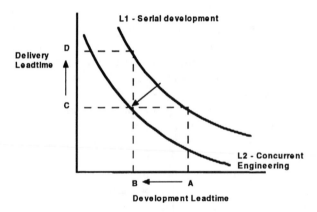

FIGURE 14.17   Trade-offs in development and delivery time.

entire sequence of development activities requiring time "A", with a resulting shorter delivery lead time of "C". In many cases, there is some evidence to show that development lead time is a higher priority. However, if the firm decides to compress development time through the use of concurrent engineering, a transition from L1 to a whole new production frontier L2 occurs. This represents a major evolution within the design/manufacturing interface. Firms making this transition are able to simultaneously benefit from lower development and delivery lead times at coordinate (B,C), while laggards remain at coordinates (B,D). This progression from L1 to L2 constitutes a major organizational evolution. Such changes take time and vision, and are not achieved overnight. Neither cross-functional involvement nor quality programs can explain the phenomenon entirely. The prevalent structures in many organizations make the implementation of such techniques extremely difficult. Further, compressed development also involves a greater investment upfront such as extra manhours, engineering expenses, and contracting fees.

It is also important to note that the concurrent engineering does not eliminate any of the stages in product design; it focuses on concurrent consideration of all stages. Sequential development often involves expensive iterations of the design process, due to the low degree of communication between different stages. Concurrent engineering does not change the technological precedence relationship between any of the design tasks, but benefits accrue because of the increased communication among all the departments, reduction of iterations, and improved learning processes.

In comparing the "concurrent engineering" and "sequential" approaches to product design, it is possible that in the former case the compression of activities may lead to mistakes or oversights. In order to design a product without defects, all of the stages in product development must be well executed. In compressing product development cycles, the likelihood of mistakes is increased, as project members are forced to make critical decisions on short notice. This happens because of the use of imperfect information at critical decision nodes; there is less technical R&D/engineering information available, a minimum of concept testing, and little if any product use testing. Because the process is accelerated, there is often no attempt to study options at critical decision points. The result is that products designed using concurrent engineering may be less "manufacturable" and may consequently have a greater number of quality problems, due to sloppy or underresourced execution of quality testing and preproduction activities. Another danger is that critical technologies may be overlooked, resulting in a product which does not meet customers' expectations. Innovation becomes an end in itself, and the company may launch more varieties of product but may fail to invest the time and energy to look for new ways to do business with customers that will take full advantage of newly enhanced capabilities or technologies.

The implication of this is that not all products justify acceleration. For instance, large complex projects and those for which basic technical discoveries are needed may not be suitable for concurrent engineering. In addition to the potential for "burnout" of committed people, rapid development makes extra demands on resources and requires extra managerial attention to move a project through development quickly.

Defining Terms

Concurrent engineering: Defined by the Institute for Defense Analysis (IDA) as "... the systematic approach to the integrated concurrent design of products and related processes including manufacture and support. This approach is to cause the developers, from the outset, to consider all the elements of product life-cycle from conception through disposal including quality, cost, schedule and user requirements" [Kusiak and Belhe, 1992].

Sequential product development: A process wherein functions "hand off" the product after completion of their stage to the next functional group responsible in the process. At each major gate (concept, engineering prototype, manuacturing prototype, and full production), each function "signs off" on the project, indicating their agreement with the decisions made up to that point.

References

Cooper, R. G. and Kleinschmidt, E. J. What makes a new product a winner: success factors at the project level, *R&D Mgt.,* 17: 175–189, 1987.

Cooper, R. G. Stage-gate systems: a new tool for managing new products, *Bus. Horizons,* May-June: 44–54, 1990.

Handfield, R. B. Effects of concurrent engineering on make-to-order products, *IEEE Trans. Eng. Mgt.,* 41(4): 1–11, 1994.

Hughes, J. N. Approaches to product-process development management, In *Managing the Design-Manufacturing Process,* J. Ettlie and H. W. Stoll, eds., p. 159–185, McGraw-Hill, New York, 1990.

Imai, K., Nonaka, I., and Takeuchi, H. Managing the New Product Development Process: How Japanese Companies Learn and Unlearn, Colloquium on Productivity and Technology, Harvard Business School, March 27–29, 1989.

Kusiak, A. and Belhe, U. Concurrent engineering: a design process perspective, *Proc. Am. Soc. Mech. Eng.,* PED- 59: 387–401, 1992.

Stalk, G. S. and Hout, T. M. *Competing Against Time,* The Free Press, New York, 1990.

Further Information

A good reference for internal benchmarking of concurrent engineering practices is presented in *Reengineering for Time-based Competition,* by R. Handfield (Quorum Press, Westport, CT, 1995). The author provides field data to benchmark product development performance.

A discussion of specific best practices in new product development teaming is *Supplier Integration into New Product Development* by R. Monczka, G. Ragatz, R. Handfield, and R. Trent (ASQC Press, 1998).

14.11 Launch Strategies for Next-Generation Products

Michael R. Hagerty and Prasad Naik

Firms are often urged to "improve cycle time" and launch the next-generation product or technology quickly. Yet there are many situations where launching the next generation will *not* be profitable for a firm. For example, next-generation technology may not provide enough new benefits to customers, as in the Apple Newton, or the next generation may make obsolete the firm's own current products, as with IBM's next generation of mainframes. Or the firm may not be able to protect its innovations from competitors. If *all* firms improve their cycle time, then *no* firm will achieve relative advantage. Therefore, this section considers the alternative new product strategies a firm might pursue to maximize its own relative advantage in a market.

Some firms can gain advantage by investing in marketing or cost reduction instead of investing heavily in next-generation technology. These strategies are (1) invest in product line extensions using the existing

TABLE 14.5 Conditions Favoring Four New Product Strategies

	Technology Follower	Technology Leader
Broad Product Lines	Customer needs are diverse and are feature sensitive Product variants have low substituitability, so cannibalization is low Niche competitors require flanking Technology permits economies of scope R&D costs are low Firm has flexible manufacturing capabilities	Some customers incur costs to switch to new generations while others don't Product is an integral part of a system that is replaced only periodically Competitors, suppliers, and buyers cooperate to set new generational standards Technology leads to cost-effective flexibility R&D investment is high and is a barrier to other firms Firm has high market share or deep skills in R&D
Narrow Product Lines	Customer preferences are stable Product specifications conform to the requirements of downstream buyers Competitors can imitate product designs or technology Technology offers economies of scale and improves production process to achieve cost leadership Negligible R&D costs Firm does not have flexible manufacturing capability	Customers want the latest and the best design Product demand is uncertain Competitive intensity if low Technology permits economies of scale R&D investment is high and is a barrier to other firms Firm has high market share or deep skills in R&D

technology or (2) become a technology follower instead of leader. In contrast, some firms can gain advantage by investing aggressively in next generation technology, obsoleting the current products. Other strategies are (3) develop the next-generation upgrade or (4) invest in coevolution of a new system with partners. We summarize conditions under which each strategy may be followed and outline market research methods that help to make the decision.

Four New Product Strategies

Table 14.5 summarizes four major types of new product strategy that can give a firm competitive advantage. The strategies vary along the rows by whether the firm chooses to invest in a broad or a narrow product line. The strategies along the columns differ by whether the firm chooses to invest in the next-generation technology (technology leader) or invest in other comparative advantages, such as marketing or cost-reduction (technology follower). This two-by-two table results in the four possible product development strategies proposed by Sanderson and Uzumeri [1997], which we label as follows: product line extensions, cost reduction, generational upgrade, and coevolution strategies.

In *product line extensions*, firms invest in existing technology and expand product features and uses. Product line extensions are of two types: brand extensions and line extensions. In brand extensions, firms leverage their existing brand names to enter new product categories (e.g., Sony markets computer monitors). In line extensions, firms leverage the current technology to enter new market segments (e.g., IBM markets home computers). Several researchers have developed models and methods that provide guidance to firms pursuing product line extension strategies. Wilson and Norton [1989] suggest that the timing of introduction will depend on cannibalization rate, relative margins, and length of planning horizon. They show that product line extensions should occur *early* in the life cycle of the current technology or never at all.

The *generational upgrade* strategy in Table 14.5 involves replacing the current technology with the new one. For example, consider Intel's microprocessor chips 286, 386, 486, and Pentium or Microsoft's DOS versions 1 through 6. Such firms set the pace of competitive redesign by developing technological upgrades concurrent with marketing the older generation. For a firm that chooses the strategy of generational upgrade, there is some research that highlights the costs and benefits of various strategic options (see Saunders and Jobber [1994]).

A mix of the above two strategies in Table 14.5 is termed as *coevolution*, i.e., evolution of current product lines *as well as* new generations of technology [Moore, 1993]. Multinational corporations often follow this strategy. Consider, for example, the global line of Walkman products manufactured by Sony

Electronics, Inc. In the United States alone they offered 24 new variants in a single year — far outstripping their competitors [Sanderson and Uzumeri 1997] — which exhibits a product line extension strategy. Concurrently, Sony maintained the leadership in inventing new major "platforms" or generational upgrades. Finally, the last strategy in Table 14.5, cost reduction, pursues none of the above strategies and the firm concentrates its efforts on managing the existing product portfolio.

Conditions that Favor Each Type of New Product Strategy

Which of these strategies should a firm choose to attain competitive advantage? Table 14.5 shows how this choice depends on six characteristics of customers, products, competitors, technology, research and development (R&D) costs, and the firm itself. Boulding and Staelin [1995] provide strong support for the importance of competitor and firm characteristics from analysis of 2000 businesses in the United States. They show that investment in product R&D causes higher returns for firms that either face low competitive intensity or have high market share, but not both. This is because the firm creates a sustainable advantage in R&D only when it has both ability and motivation to do so. They define "ability" as high market share, which allows high cash flow to fund R&D activities. However, clearly other firm characteristics contribute to ability, especially deep skills in product R&D. For example, biotech firms such as Genentech or Calgene have developed expertise in translating scientific knowledge into commercial products that give them competitive advantage in pursuing technology leadership. Hence Table 14.5 shows that strategies requiring high investment in product-related R&D (coevolution and generational upgrade) require high firm ability (market share and deep R&D skills) or low competitive intensity to achieve high returns.

In contrast, the strategies of Product Line Extension or Cost Reduction can be pursued even with small market shares, low R& D skills, or higher competitive intensity. Instead of investing in product-related R&D, a successful firm with these strategies would invest in cost-reduction efforts or in building marketing expertise to find and serve many customer segments (product line extension). Similarly, firms with technology that is not patent protected and is easy to copy typically finds it difficult to sustain technological leadership. For example, technological leadership in small kitchen appliances is difficult to sustain: less than 6 months after the first electric knife was introduced, *ten competitors* had a similar knife on the market. In such situations, firms would be better off reducing costs or finding creative uses or new users.

Table 14.5 also shows other factors — customer diversity and cannibalization — that drive success of new product strategy. If customer needs are diverse and cannibalization of sales by product line extension is low, then the firm will realize higher returns under product line extension or coevolution strategies.

Economic Forecasts of Next-Generation Success

If a firm selects a strategy of generational upgrade from Table 14.5, they must still decide *when* to launch the next generation. The present section reviews methods to forecast profitability for particular new product launch decisions. Figure 14.18 shows the six drivers of next-generation profitability and how they enter the economic forecast of launch success. The drivers are classified into two factors: those external and internal (to the firm). The five economic variables are shown in the center column and come from the basic equation for product profitability:

$$\text{Profits} = (\text{new margin}) \times (\text{potential}) \times (\text{market share})$$
$$- (\text{old margin}) \times (\text{cannibalized sales}) - (\text{fixed costs}) \tag{1}$$

where new margin is the gross margin received on the product with new technology, potential is the expected sales of the new generation, market share is the proportion of sales achieved by the firm, old margin is the gross margin received on the older generation product (usually lower), cannibalized sales is the volume of the firm's older-generation sales displaced by the new-generation product, and fixed costs are expenditures on R&D and launch.

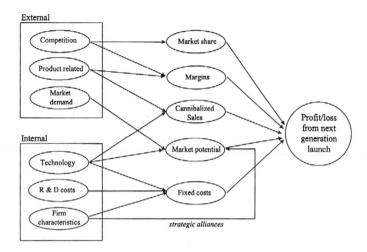

FIGURE 14.18   Relations between six drivers and five economic variables that impact profitability of a next-generation launch.

The arrows in Fig. 14.18 show that market share is determined primarily by competition (e.g., the number of firms, timing of entry, and nature of products introduced). Margins are influenced by competition and product-related factors. Cannibalized sales are driven by product-related factors and technology. Potential is influenced by the demand for the products and the innovativeness of the technology. Fixed costs are driven by the technology, the nature of R&D activities and the characteristics of the firm. All of these factors influence the forecast of ultimate profitability of the next generation. Though no unified model has been developed that includes all of these factors, we will next review models for forecasting market potential and market share, which are usually known with least certainty.

Several models have been developed to predict market potential. Norton and Bass [1987] estimate a diffusion model of new-generation sales growth over time. They assume that information about and ability to use an innovation increase with the level of sales, while actual sales decrease as the pool of potential adopters is depleted. Hence product sales grow initially, reach a peak, and then decline over time. Figure 14.19 shows results of fitting such a model to multiple generations of DRAM memory chips. They later proposed a Law of Capture, which states that the next-generation technology will eventually "capture" all the users of previous technology. This model was shown to hold across different product categories. They also observe that, for the categories studied, demand for the earlier generation continues to grow after launch of the next generation before declining to zero.

An alternative approach to forecasting market potential has been taken by Bemmaor [1996]. Here, market research studies are conducted in which potential customers examine prototypes of the new product and then rate their likelihood of buying on intention-to-buy scales. After making the adjustments for bias, these ratings may be used to obtain an unbiased forecast of sales in the next year along with its confidence interval.

Which of these methods should be used to estimate market potential for the new generation? While Norton and Bass' diffusion model fits data on historical sales of earlier generations quite well, it may not forecast future sales of the next generation accurately. This is because, as they point out, the forecasts are very unstable when only the first few periods of sales are known for the new generation. Their model requires more precise estimates of eventual total sales of the new generation in order to make better forecasts. We suggest using Bemmaor's method to obtain estimates for eventual total sales. This combined method should yield a better forecast of sales growth and decline over time.

The other economic component in Eq. 1 associated with great uncertainty is market share — the percent of potential sales accruing to the firm relative to other competing firms. Since market share depends on competitors, it is important to know *who* and *how many* competitors the firm will face, *when* they will enter, and the eventual *market share* they will achieve. Unfortunately, research to date has tended

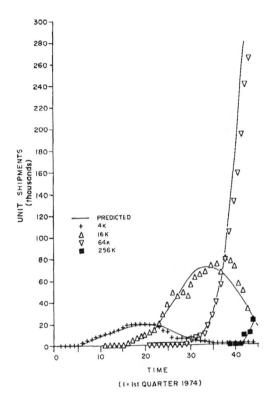

FIGURE 14.19 Backcast of three generations of DRAM chips. (*Source:* Norton, J. A. and Bass, F. M. *Mgt. Sci.,* 33: 1069–1087, 1987.)

to separate these issues into separate streams, so that separate models must be used to predict each aspect of competition. The most research has been done on the last aspect: eventual market share that competitors will achieve. Kalyanaram et al. [1995] review results of over 400 consumer goods industries and 1200 industrial goods industries and show how expected market share varies with price, quality, advertising, as well as the *order of entry.* Interestingly, on average, market share was found to be proportional to $1/\sqrt{N}$, where N is the order of entry. Thus, according to their model, a monopolist would have 100% share; when the second firm enters, the ratio of market shares of the two firms will be 1: $1/\sqrt{2}$ (or 59% vs. 41%); when the third firm enters, the ratios of market shares of the three firms will be in the proportions 1: $1/\sqrt{2}$: $1/\sqrt{3}$.

Besides forecasting market share, the *number* and *timing* of competitors entering a market must be forecast. In some well-defined industries, firms may know that only their existing competitors will enter but do not know when. In others, competitors may enter from other industries as well, thereby creating more uncertainty. These issues needs to be resolved in the future.

Conclusion

We reviewed current research to answer the question "When should the next-generation product be launched?" Our answer requires both a strategic choice as well as specific product forecasting. First, the firm should choose a general new-product strategy from Table 14.5, based on competitive advantages of the firm. If the firm chooses to be a technology leader, it should launch the next generation as soon as a prototype has been developed that passes the profit hurdle specified in Eq. 1. On the other hand, if the firm chooses a strategy of technology follower, the firm should invest less in prototype development and should not be the first to launch new technology, regardless of the profitability predictions of Eq. 1. Instead, they should concentrate resources on marketing and cost reduction, waiting for others to launch the next generation. Only then should they aggressively pursue next-generation technology, using reverse engineering and cost reduction to develop candidates that can be evaluated by Eq. 1.

Note that the decision rule we propose gives primacy to the firm's strategic stance and only secondary status to market research. This is because the market research required by Eq. 1 at present contains too much error to base the launch decision exclusively on it. Urban and Hauser [1996] point out that, even after a good new product idea is identified, the conditional probability that it will go on to be a market success is only 26%. This is a situation where it is dangerous to rely solely on the data. Thus, besides the market and product development research described in much of this handbook, firms must make careful strategic choices that will yield comparative advantage in the market.

Defining Terms

Cannibalization: Occurs when a firm introduces a new product that steals market share from the firm's older products in higher proportion than it steals from competitors' products.

Launch strategies: A set of decisions on whether and when to develop and introduce a new product.

Next-generation product: Refers to a new product or service that will displace sales of an older product or service because it offers improved benefits or lower cost to the consumer.

Technology leader: The first firm to introduce a new generation or technology in an industry.

References

Bemmaor, A. C. Predicting behavior from intention-to-buy measures: the parametric case, *J. Market. Res.,* 32: 176–191, 1995.

Boulding, W. and Staelin, R. Identifying generalizable effects of strategic actions on firm performance: the case of demand-side returns to R&D spending, *Market. Sci.,* 14: G222–G236, 1995.

Kalyanaram, G., Robinson, W. T. and Urban, G. L. Order of market entry: established empirical generalizations, emerging generalizations, and future research, *Market. Sci.,* 14: G212–G221, 1995.

Moore, J. F. *The Death of Competition,* Harper Collins, New York, 1996.

Norton, J. A., Bass, F. M. A diffusion theory model of adoption and substitution for successive generations of high-technology products, *Mgt. Sci.,* 33: 1069–1087, 1987.

Sanderson, S. W. and Uzumeri, M. *Managing Product Families,* Richard D. Irwin, Chicago, 1997.

Saunders, J. and Jobber, D. Product replacement: strategies for simultaneous product deletion and launch, *J. Prod. Innov. Mgt.,* 11(5): 433–450, 1994.

Urban, G. L. and Hauser, J. R. *Design and Marketing of New Products, 2nd ed.,* Prentice Hall, Englewood Cliffs, NJ, 1996.

Wilson, L. O. and Norton, J. A. Optimal Entry Timing For A Product Line Extension, *Market. Sci.,* 8: 1–17, 1989.

Further Information

For software tools and research in this area, see the *Special Issue* on Innovation and New Products, *Journal of Marketing Research,* February 1997.

For information on professional activities, see the Web site of the Product Development and Management Association (http://www.pdma.org/).

14.12   Green Products

Noellette Conway-Schempf and Lester Lave

A green product can be defined rather broadly as one that (1) uses less materials and energy in its production, use, and disposal than other products of similar function (particularly less nonrenewable resources) and (2) uses fewer toxic materials or results in lower discharges of hazardous materials than other products. These environmental improvements may be the result of dematerialization (less material usage), material substitution, processing changes, or increased recycling and reuse of materials, components, and products. Substituting green products for less environmentally conscious products is a step toward preserving the environment and giving future generations the same opportunities we enjoy: an ideal known as **sustainable economic development**.

Green Products

A green product is not defined in any absolute sense, but only in comparison with other products of similar function. For example, a product could be entirely made of renewable materials, use renewable energy, and decay completely at the end of its life. However, this product would not be green if, for example, a substitute product uses less resources during production and use, or results in the release of fewer hazardous materials. Other things being equal, a car that gets 50 miles per gallon is greener than one that gets 30 miles per gallon — unless the owner family cannot fit into the more fuel-efficient, car necessitating two trips. A fully loaded bus is greener than either car, but a bus with one passenger is not at all green. Still more important than green products is the concept of green systems. A 50-mile-per-gallon car whose components can be recycled easily is relatively green. However, as a means of getting from one place to another, it is much less green than a bicycle, even if the bicycle is made of materials that cannot be reused or recycled. A community where people can walk to work, school, and shopping is inherently greener than cities such as Los Angeles and Dallas where an automobile is required to get to work, shopping districts, schools, and recreation areas. Thus, while one product is greener than another, the more important attribute has to do with the system in which each product is used.

Rarely is one product greener in every dimension (resource and energy use, emissions, reyclability, etc.) than other products; there are usually trade-offs among characteristics. For example, making cars more fuel efficient requires making them lighter. This can be accomplished by substituting aluminum or plastic for steel. Both aluminum and plastic require more energy during production than does steel. How should we compare the energy required during production with the energy required for operation? One approach is to calculate how many miles the car must go to "pay back" the energy required during production. Another example is that some new materials and composites, such as carbon fibers, have many advantages but cannot be recycled. Which is more important, the ability to recycle or the lighter weight and strength? Finally, if petroleum is in extremely short supply, saving gasoline may be desirable even if the result is increasing total energy use, e.g., cars powered by methanol from coal.

In the sections that follow, we discuss the importance of focusing on the design phase when we desire green products, the principal tools for assessing the greenness of a product, comparing the environmental impacts of different materials, approaches to making products more recyclable, the importance of selecting the proper materials, and a tool for allowing companies and consumers to make green choices without having to become environmental experts. We end with a brief look at regulations designed to promote green products and the social benefits that flow from making the economy greener.

Designing Green Products

The largest potential for green products is to start with the design phase minimizing the environmental impact of a product, through all its life stages — from resource extraction to manufacturing and processing to disposal by designing out negative environmental attributes (Fig. 14.20). Product design is an ideal point to address environmental problems [Hendrickson and McMichael, 1992]. It is at the design stage that manufacturing decisions, resource requirements, toxic materials usage, energy use, waste disposal options, etc. are made. An oft-quoted NRC study [NRC, 1991] estimates that over 70% of the environmental costs of product development, manufacture, and use are determined during the design stage — this certainly holds true for environmental costs and impacts.

Design is a complicated messy process. Designers are faced with numerous competing requirements and constraints. For example, a personal computer must be fast and powerful and cheap. To be green it

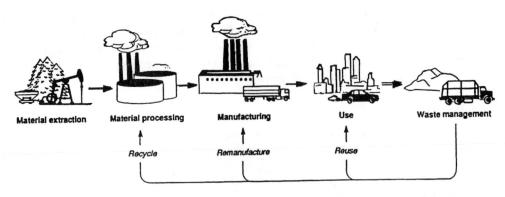

FIGURE 14.20   Stages of a product life cycle. (*Source:* Green Products by Design: Choices for a Cleaner Environment, U.S. Congress, Office of Technology Assessment, OTA-E-541.)

must also be energy efficient, and easily recycled. Designers have struggled to achieve the first set of attributes. Achieving environmental goals makes the task more difficult. Particularly as for most consumers, energy efficiency and recyclability are less important product attributes, which means that designers cannot compromise other product attributes in becoming green.

Designing and manufacturing green products requires appropriate knowledge, tools, production methods, and incentives. Aids for green design must be easy and quick to use and understand. Ideally, these design tools will help identify design changes that have lower costs while improving materials use and recyclability, for example, using snap fits rather than screws may have little additional cost burden at the design stage and may significantly increase recycling potential.

Green product design tools generally fit into the categories described in the following sections.

Life Cycle Assessment

Life cycle assessment (LCA) is a method of examining the environmental effects of a product or process throughout its life [EPA, 1993]. A traditional LCA consists of (1) defining a system boundary, (2) carrying out an inventory of all the materials and energy used and all the environmental discharges resulting from the product's manufacture, use, and disposal within the defined boundary, (3) carrying out an assessment of the environmental implications resulting from the discharges and materials use identified in the inventory, and, finally, (4) carrying out an assessment of the opportunities for improvement. Unfortunately, the most commonly used tools for LCA have major flaws. They are generally expensive and time consuming and require data on environmental impacts that are not available. Most LCAs involve the use of boundaries that limit the analysis and generate controversy. A new approach solves many of these difficulties, although it has problems in carrying out detailed product or materials analysis [Lave et al., 1995]. The obvious need and demand for LCA data and tools, and the possibility of the development of international standards for LCA, should result in the availability of widely accepted tools within the next decade.

Environmental Impact or "Green" Indices

How can an analyst compare a pound of organic matter dumped into the environment with a pound of dioxin? Green indices or ranking systems attempt to summarize various environmental impacts into a simple scale. The designer or decision maker can then compare the green score of alternatives (materials, processes, etc.) and chose the one with the minimal environmental impact. This would contribute to products with reduced environmental impacts. Examples of green indices currently being used include Volvo's Environmental Priority Strategies system (EPS — this involves calculating "environmental load units" of alternatives [Graedel and Allenby, 1995]) and Carnegie Mellon University's CMU-ET toxicity weighting system (this uses US EPA toxicity release inventory data and worker exposure safety data to compare alternatives [Horvath et al., 1995]. Although none of these tools is yet capable of incorporating many different types of environmental impact, all provide at least rudimentary guidance to the designer in choosing materials, components, or process alternatives that have reduced environmental impacts.

Design for Disassembly and Recycling Aids

Design for disassembly and recycling (DFD/R) means making products that can be taken apart easily for subsequent recycling and parts reuse. For example, Kodak's "disposable" cameras snap apart, allowing 87% of the parts (by weight) to be reused or recycled. Unfortunately, the economic costs associated with physically taking apart products to get at valuable components and materials often exceeds the value of the materials. Reducing the time (and thus cost) of disassembly might reverse this balance. Thus, DFD/R acts as a driver for recycling and reuse. DFD/R software tools generally calculate potential disassembly pathways, point out the fastest pathway, and reveal obstacles to disassembly that can be "designed out".

Material Selection and Label Advisors

Any of several materials can produce a particular quality component or product. However, they have different environmental implications. Material selection guidelines attempt to guide designers toward the environmentally preferred material. For example, Graedel and Allenby [1995] present the following guiding principles for materials selection:

- Choose abundant, nontoxic materials where possible.
- Choose materials familiar to nature (e.g., cellulose), rather than man-made materials (e.g., chlorinated aromatics).
- Minimize the number of materials used in a product or process.
- Try to use materials which have an existing recycling infrastructure.
- Use recycled materials where possible.

In addition to these generic guidelines, companies such as IBM and Chrysler have been developing and using specific materials selection guidelines for environmentally sound product development, which describe in detail the materials that should be used for specific applications.

Label advisors are generally marks on materials or products that reveal information about the material content relevant to materials handling and waste management. For example, the plastic bottles used in many consumer products usually have a plastics identification symbol, which can be used in plastics resorting and recycling efforts.

Full Cost Accounting Methodologies

Many corporations and consumers want to support green products and sustainability but do not know how to make greener decisions. Designers and plant managers are specialists who cannot be expected to be environmental experts capable of estimating the environmental and sustainability implications of their decisions. As a result, a company often incurs high costs from using a material or process that creates environmental problems when an environmentally benign material or process exists. Often consumers purchase products that create environmental problems because they do not know about green alternatives.

Companies need management information systems that reveal the cost to the company of decisions about materials, products, and manufacturing processes. This sort of system is called a "full cost accounting" system. For example, when an engineer is choosing between protecting a bolt from corrosion by plating it with cadmium vs. choosing a stainless steel bolt, a full cost accounting system could provide information about the purchase price of the two bolts *and* the additional costs to the company of choosing a toxic material such as cadmium.

In many cases, the choices that a designer or consumer makes also impose costs on society. For example, choosing a cadmium coating increases the possibility of human exposure to a toxic substance. The designer and consumer might be informed by showing them this social cost of the cadmium, i.e., the cost of preventing the exposure and the potential health risks of exposure. This information might be communicated by having a "social" cost listed on the price tag. A still stronger step would be to actually charge the designer and consumer for the social costs of environmentally damaging materials or products.

Thus, the government might add an environmental tax or effluent fee that would account for the social damage.

A number of companies are beginning to launch preliminary full cost accounting efforts in order to spotlight products which result in relatively large environmental costs.

The Benefits of Green Products

Developing and marketing green products is a concrete step toward sensible resource use and environmental protection and toward sustainable economic development. Green products imply more efficient resource use, reduced emissions, and reduced waste, lowering the social cost of pollution control and environmental protection.

Greener products promise greater profits to companies by reducing costs (reduced material requirements, reduced disposal fees, and reduced environmental cleanup fees) and raising revenues through greater sales and exports. Designing green products offers much to the current generation as well as providing future generations with a planet that will enable them to survive and prosper.

Governments, particularly in Europe, have been providing incentives for greener product development. As examples, Germany's **product takeback laws** have prompted European and U.S. industries to reduce packaging and start designing products with disassembly and recycling in mind. France and The Netherlands have special government agencies to foster clean technologies. In the United States, many government purchasing criteria specify the use of recycled materials; the federal government has ordered its employees to look for recycled products and has initiated a number of energy efficiency programs for products and buildings.

Significant progress has already been made as companies see that they can lower costs and increase revenues by making green products. Consumers have been slower to respond to green products. In the end, progress is limited by what consumers are willing to purchase.

Defining Terms

Product takeback laws: Laws in some European nations that transfer the responsibility of packaging and product disposal from the consumer to the producer. For example, in Germany, manufacturers are responsible for ensuring that packaging is recycled appropriately, and recent laws also require manufacturers of certain products to recover and recycle their products when the consumer no longer needs the product.

Sustainable economic development: Defined by WCED [1987] as "meeting the needs of the present without compromising the ability of future generations to meet their needs". For example, a vehicle powered by solar radiation would be sustainable at least in terms of energy for operation.

References

Conway-Schempf, N. and Lave, L. Pollution prevention through green design, *Pollut. Prev. Rev.*, Winter: 11–20, 1996.

Environmental Protection Agency, 1993. Life cycle assessment: inventory guidelines and principles, EPA/600/R-92/245, Washington, DC, 1993.

Graedel, T. E. and Allenby, B. R., *Industrial Ecology*, Prentice Hall, Englewood Cliffs, NJ, 1995.

Hendrickson, C. T. and McMichael, F. C., Product design for the environment, *Environ. Sci. Technol.*, Vol. 26: 844, 1992.

Horvath, A., Hendrickson, C., Lave, L., McMichael, F., and Wu, T.-S., Toxics emissions indices for green design and inventory, *Environ. Sci. Technol.*, Vol. 29: 86–90, 1995.

Lave, L., Cobas-Flores, E., Hendrickson, C., and McMichael, F. C., Using input-output analysis to estimate economy-wide discharges, *Environ. Sci. Technol.*, Vol. 29: 420A–426A, 1995.

National Research Council. *Improving Engineering Design: Designing for Competitive Advantage,* National
University Press, Washington, DC, 1991.
Office of Technology Assessment. *Green Products by Design,* OTA-E-541, US GPO, 1992.
World Commission on Environment and Development (WCED). *Our Common Future,* Oxford University
Press, New York, 1987.

Further Information

http://www.ce.cmu.edu/greenDesign/.

14.13   Product Cost Management

Roger J. Best

Product cost management is generally approached from a strictly internal point of view. From this perspective, the management of costs is typically limited in scope to materials, labor, production, research, distribution, and other such costs of producing and distributing a product [Shank and Govindarajan, 1989]. While the management of these internal costs is certainly critical to achieving a desired level of profitability, it is also important to consider the total cost of a product to the customer. Customer costs include, in addition to the purchase price, the cost of acquiring and installing a product as well as costs related to using, maintaining, and disposing of a product. Consideration of these postpurchase customer costs, along with internal product costs, can provide a more complete approach to product cost management.

Product Cost Management Process

Illustrated in Fig. 14.21 is a cost management process that includes three inputs — technology, production, and systems. Each has a meaningful effect on the three major areas of product costs — customer costs, variable costs, and fixed costs. Effective management of these inputs and areas of cost has the potential to lower customer costs, which can translate into increased customer demand, and indirectly lower a business's variable and fixed costs, which, in turn, contributes to greater profitability. The purpose of this section is to explain the product cost management process, as presented in Fig. 14.21.

The first step in the product cost management process is to understand how technology, production, and systems inputs affect the three major areas of product cost. To achieve a meaningful reduction in any of these areas requires management of one or more of these inputs.

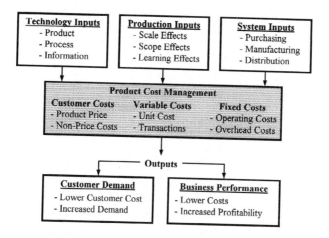

FIGURE 14.21   Product cost management process.

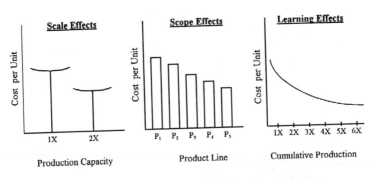

FIGURE 14.22 Production volume and unit cost reduction.

Technology Inputs

The first of the three product cost management inputs is technology. Product, process, and information technologies each have the potential to lower product costs and customer nonprice costs. Gains in any of these broad areas of technology can reduce variable costs with lower unit manufacturing, distribution, and inventory costs. Technology inputs can also be used to lower fixed costs associated with product development time, marketing, and product administration. These technology inputs also have the potential to reduce customer nonprice costs through product designs/redesigns that lower the customer's costs of acquiring, installing, using, maintaining, and/or disposing of a product.

Production Inputs

There are three production inputs that have the potential to lower product costs [Best, 1997]. The first of these is economies of scale. A production capacity of 2X will achieve a lower unit cost than a production capacity of 1X, as illustrated in Fig. 14.22. However, to achieve a lower unit cost, a business has to operate near full-production capacity. A business with a 2X production capacity that operates appreciably below that capacity could actually have a higher per unit cost than a competitor with 1X capacity that operates near full capacity.

Product line scope is another production input into product cost management, as shown in Fig. 14.21. As more products using similar product designs or processes are added to a business's product line, the unit manufacturing cost of each product is likely to be reduced, as illustrated in Fig. 14.22. This is due to the purchase and manufacture of common components or subassemblies in larger quantities as well as providing a higher utilization of capital equipment. Thus, product line scope has the potential to lower unit costs, which could impact prices (a customer cost) and margins (a profitability component).

Learning through production experience also lowers the unit cost of a product. As a business learns more from the experience of producing a product, there is a tendency for the unit cost of a product to decrease exponentially, as illustrated in Fig. 14.22. For example, most microchip production plants are on learning curves near 70%. This means that, every time the cumulative production experience for a specific product doubles, the unit cost decreases by 30%. Thus, production experience (learning) also lowers unit cost, which could lower customer costs (through lower prices) and/or improve profitability (with higher margins).

Systems Inputs

Systems inputs can be derived from three major areas — manufacturing, purchasing, and distribution. Systems such as just-in-time inventory management, materials resource planning, CAD/CAM design systems, outsourcing, and shared production each offer product cost management opportunities that could lower customer costs and business costs. For example, a more efficient inventory system could lower customer acquisition costs, since customers would be able to reduce inventories of a purchased product. Likewise, a more efficient distribution system could lower transportation costs, which could lower customer acquisition costs as well as a business's variable transactions costs. CAD/CAM design

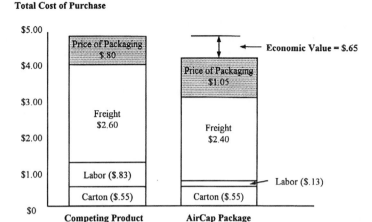

FIGURE 14.23   Total cost of purchase and economic value.

systems and outsourcing have also been shown to shorten product development time and the fixed cost of product development, which could benefit customers with earlier access to better or lower-cost products as well as benefit a business with lower fixed operating expenses.

Product Cost Management

As shown in Fig. 14.21, product cost management focuses on three major areas of cost — customer costs, variable costs, and fixed costs. Each of these specific areas of product cost is central to building customer demand and business profitability.

Customer Costs

While managing internal product costs is a natural point of focus in product cost management, we are going to start the discussion of the process of product cost management from an external point of view. How can technology, production, and systems inputs be managed to lower the customer's total cost of purchase? Before we answer that question, we have to understand how the inputs to cost management might impact customer costs.

First, it is important to recognize that there are two types of customer costs — price and nonprice costs. Price includes the cost of the product and the cost of acquiring it which could include transportation charges, discounts, and terms of payment, as well as the cost of returns, service agreements, or inventory when availability and delivery are not reliable. Nonprice costs include postpurchase life cycle costs. These customer costs include the cost of startup (installation and training), usage (performance, product life, and ease of use), maintenance (reliability and serviceability), and disposal (salvage value and cost of disposal) of a product. Each of these costs can be significant and, when managed to lower levels, has the potential to lower customer costs and enhance customer attractiveness and product demand [Forbis and Mehta, 1981].

To illustrate this dimension of product cost management, consider the example presented in Fig. 14.23. Sealed Air Corporation's AirCap uses a superior product technology to manufacture a packaging product that actually costs a customer more per package ($1.05) than a competing product ($.80). However, the use of AirCap significantly lowers labor cost (a nonprice use cost) and slightly reduces freight costs (a nonprice transaction cost). The net result of this customer cost management effort is an economic value, or customer savings, of $.65 per package [Best, 1997]. It is also important to note in this example that AirCap lowered the customers total cost of purchase while still obtaining a price premium relative to a competing product.

Variable Costs

There are two major areas of variable costs to be managed in product cost management as depicted in Fig. 14.21. Each varies with each unit sold and can be managed to lower levels with technology, production, and systems inputs, which provides an opportunity to improve customer attractiveness and profitability.

Unit Costs

This is the manufacturing cost per unit. The unit cost of a product includes material, direct labor, and use of manufacturing capital that is amortized over a given level of production volume. Product design, engineering redesign, and process improvements are technology inputs that can lower unit cost. Production inputs such as economies of scale, scope, and learning can each contribute to lower unit costs as discussed earlier. Each of these inputs plays a key role in the management of unit cost.

Intel, for instance, continues to develop its wafer process technology and thereby reduce the size of silicon wafers. Each reduction in size reduces the unit cost of a silicon wafer chip, which improves the margin per unit but also increases production capacity, which further impacts unit cost and the ability to serve a greater number of customers.

Transaction Costs

This is a sales and distribution cost that varies with each unit sold. Sales commissions, distributor discounts, shipping charges, rebates, and cash discounts are examples of transactions costs that occur with each unit sold. For example, a business may pay its sales force a 5% sales commission, which lowers the net price, or offer customers a 2% price discount if they pay their bill within 10 days.

Alternative selling and distribution systems can shift the impact of variable transactions. For example, a business that uses a commissioned sales force to sell to distributors is likely to incur both sales commissions and distributor discounts, which are transactions costs that lower margins. By switching to a salaried company sales force and shipping directly to end user customers, the business could eliminate the variable transactions costs related to sales commissions and distributor discounts. However, the business would incur the increased cost of a salaried sales force and the cost of direct distribution to end user customers.

Fixed Costs

There are two types of fixed costs that occur over a given planning horizon. These costs include direct (fixed) operating costs and indirect (fixed) allocated overhead costs. Each impacts performance and profitability and needs to be managed in order to achieve a specific target profit.

Direct Fixed Operating Costs

These include various fixed product management and marketing costs associated with a given positioning strategy. Both of these are direct product costs that would disappear if the product were to be discon-tinued. These fixed product costs are often affected by systems inputs and management practices. For example, a business may outsource certain engineering, production, or customer services in order to lower the fixed operating costs for a specific product.

Marketing costs are also a fixed operating cost related to the cost of product communications (advertising, trade shows, product literature, etc.) and the cost of sales and distribution. Marketing expenses can be shifted from a fixed cost to a variable cost with different marketing strategies. For example, a business using a direct sales and distribution system incurs a certain level of fixed marketing expense. If the business switched to the use of distributors and manufacturers' representatives, these fixed marketing expenses would disappear as they are shifted to variable costs in the form of distributor discounts and manufacturers' rep sales commissions.

Indirect Fixed Costs

These costs are not related to specific products but are critical to running the overall business. These indirect costs are overhead costs that traditional accounting practices generally allocate to products based on sales, production volume, or other related activities. Although these overhead costs need to be covered

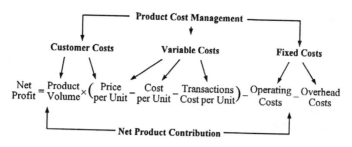

FIGURE 14.24 Product cost management and profit impact.

in order for the business to be profitable, the danger is that they may significantly distort the process of a product cost management when they are allocated to products [Shank and Govindarajan, 1988]. For example, a telecommunications business allocated its overhead costs on the basis of sales to various direct operating costs, variable transactions costs (including sales commissions), and unit costs. This greatly distorted true product costs and led to product elimination decisions which actually reduced overall profitability. Thus, in product cost management, it is important to isolate fixed overhead costs as a separate cost element.

Outputs of Product Cost Management

Product cost management includes management of customer costs, variable costs, and fixed costs. Each plays a role in shaping customer demand and business performance, as illustrated in Fig. 14.21.

Customer Demand

As the total cost of purchase of a product goes down relative to competing products of comparable quality and benefits, the demand for that product will go up. Effective product cost management will allow a business to reduce the unit cost of a product and often lower prices, which enhances product attractiveness and customer demand. Reductions in nonprice costs can also lower the overall cost of purchase to the customer and increase product attractiveness and customer demand for a product. As in the case of AirCap, decreasing nonprice costs to the customer may even allow a business to charge a higher price for a product while still lowering the total cost of purchase to the customer.

Business Performance

Decreases in variable unit costs and transactions costs contribute to increases in unit margins and to higher levels of net profit. Reductions in fixed operating costs also directly affect net profits. In addition, proper treatment of allocated overhead costs allows management to better understand the true cost and profitability of a product. As illustrated in Fig. 14.24, net profit is determined by the level of net product contribution that is produced to cover indirect overhead costs and contribute to profits. While the same net profit would occur if the indirect overhead costs were allocated to various product costs, our ability to manage, measure, and track these costs and how they impact customer attractiveness and profits is greatly diminished when indirect overhead is spread across various areas of product costs.

References

Best, R. J. *Market-Based Management: Strategies for Growing Customer Value and Profitability,* Prentice Hall, Upper Saddle River, NJ, 1997.

Forbis, J. and Mehta, N. Value-based strategies for industrial products, *Bus. Horizons,* May: 32–42, 1981.

Shank, J. and Govindarajan, V. The perils of cost allocation based on production volumes, *Account. Horizons,* 4: 71–79, 1988.

Shank, J. and Govindarajan, V. *Strategic Cost Analysis,* pp. 99–112, Richard D. Irwin, New York, 1989.

14.14   Standards

Liora Salter

Standards are technical documents reflecting agreed upon qualities, capabilities, or by-products of products, industrial processes, systems, or networks. However developed or adequate they are, standards serve as points of reference, their influence being determined by the number of people who abide by them.

There are standards for virtually every aspect of technology, and many different standards organizations, all with their own acronyms. As such, it has proven useful to classify standards. One method is by their point of impact on industrial processes. Thus, there are quality standards, process standards, and compatibility standards. **Quality standards** pertain to the level of performance expected, while **process standards** deal with how products are manufactured, or with the design and possible outputs of systems or networks. **Compatibility standards** concern the interconnection or interworking of various products or systems.

Another common method is by how standards are developed. In this case, a distinction made is between *de jure* and *de facto* standards. *De jure* **standards** are developed by standards development organizations. *De facto* **standards** are outcomes of market decisions. Most often, *de facto* standards originate with a single firm that is able to command allegiance for its technology, products, or process by virtue of its dominant market position. *De facto* standards have no official status, but they sometimes serve as the basis of *de jure* standards.

As well, standards can be classified by their legal status. Some standards are proprietary, while others are made available through licensing arrangements and others yet are distributed without cost or sold as publications by standards organizations. *De jure* standards are not normally proprietary. They can be either voluntary or regulatory. Voluntary standards are guidelines. **Regulatory or mandatory standards** are set or adopted into law by governments, enforceable though courts or agencies. Regulatory standards can be either **prescriptive** or **performance** based, the former requiring compliance with predetermined levels of product or system capability, etc., the latter focused on results or outputs or on mitigating potential harm. Regulating drugs, pesticides, or the height of smokestacks involves prescriptive standards, while emission standards are an example of regulatory performance standards.

In fact, the distinction between regulatory and voluntary standards is less clear than it appears, although firms clearly prefer voluntary standards. Many regulatory standards were originally developed as voluntary standards, only later to be adopted (sometimes as revised) and referenced as regulations [Salter, 1985]. Further, many regulatory standards are not enforced strictly, so that they actually operate as guidelines not regulations. As well, voluntary standards (and especially *de facto*) standards can be so widely accepted that their impact in the market is coercive.

Finally, standards can be classified by the type of organizations involved in developing them. Some standards are developed by intergovernmental organizations operating under the auspices of the United Nations, for example, the International Telecommunications Union. Even when these standards remain voluntary (i.e., when they are not eventually adopted as national regulations), they are considered to reflect the best achievable technical options and are thus very influential, especially in countries lacking technical or financial resources to develop standards on their own. No less influential, however, are the voluntary standards under the auspices of the nongovernmental international organizations, for example, the International Organization of Standardization (ISO). International bodies have been complemented today by regional ones, operating in the Asia Pacific, the Americas, and Europe in particular. Today increasingly, emphasis is being given to promoting international standards, but national standards remain important. National standards bodies organize industry and government contributions to the international organizations, and they recognize and disseminate international standards, as well as set standards of their own.

There are many kinds of national standards bodies, including government departments or agencies, (private, nonprofit) accredited standards organizations, accrediting bodies, testing and certification organizations, and also diverse industry, research, and professional organizations. Government agencies and accredited standards organizations both must ensure that public input is sought. Accredited organizations normally also require that a consensus-based process be used. Industry, research, and professional organizations are tailored to the needs of their members and seldom involve either public input or consensus. However, their standards can be accepted by governments as substitutes for the standards otherwise set by governments or accredited standards organizations. Recently, a new source of standards has emerged: forums and consortia. Many of these are international in scope but organized around a single technology or company product or standard. Some forums restrict their membership, but others are open to participation (usually for a fee) although a consensus process is rarely used.

Classifications of standardization are important for more than academic reasons. Standards are an economic and political phenomenon as well as technical documents. The trade implications of standards are easily identified. National standards (including both *de jure* and *de facto* standards) can operate as nontariff barriers to trade, often to the benefit of firms, while liberalization and globalization, also benefiting firms, require standards compatibility and harmonization. Deliberations about harmonization are fraught with geopolitical consequences, often producing conflicts about whether particular national standards are discriminatory. Standards are thus frequently the subject of bilateral or multilateral negotiations. The World Trade Organization (WTO) and the regional trade bodies (for example, the North American Free Trade Agreement) promote (or, in some instances, require) harmonization and play key roles assessing or adjudicating disputes.

As such, standardization also often has political implications. These pertain to the interstate conflicts arising from countries promoting globalization (with or without relying on standards) while relying upon (but never admitting to using) standards as important nontariff barriers to trade. Questions about economic development (especially in developing countries) and national sovereignty are often addressed in the WTO and the international (and especially intergovernmental) standards organizations. Debates about whether standards should be voluntary or regulatory, prescriptive or performance, are also highly political at the national level, where national style and public pressures play important roles.

Everyone agrees that standards have economic implications, but disagreement exists about their impact. The argument advanced by standards participants (and in OECD or the G 7) is that standards are essential to the globalized economy where international *de jure* standards organizations have important roles to play. It is also suggested that the economic risks associated with emerging technologies are so great that coordination among the major players is beneficial. Because, thus far, standardization has not fallen within the ambit of antitrust provisions in any country (although some standards and organizations have occasioned legal action), standardization provides as excellent venue for cooperation. The contrary argument, offered mainly in the economics literature, is that both regulatory and *de jure* standards stifle innovation by committing everyone to a single approach, vendor or established technology. [Berg, 1989; Besen and Farrell, 1994; David and Greenstein, 1990]. Consensus-based standardization is slow, cumbersome, and costly, regulatory standardization even more so. Standards are said to reflect only what the market has already determined, making standardization not worth the considerable investment of time and resources involved. Significant reforms have been undertaken by all *de jure* standards organizations to deal with these negative arguments, but they remain persuasive to some analysts and firms.

In fact, firms have many options: whether or not to participate in which types of organizations, for which of their many products or processes. Firms also exert influence to prevent standards from becoming regulatory or to shape regulations. They must decide whether their standards will be proprietary, when to enter licensing agreements, and whether to make their standards available to the *de jure* organizations where their competitors are also participants. Very large firms often choose to be involved in many *de jure* standards organizations, locally, regionally, and internationally, but will also establish their own standards. Smaller firms are likely to be standards followers, participating, if at all, in *de jure* standardization to keep informed about market and new technical developments.

As noted, standards are technical documents, developed by engineers working in committees where the priority is achieving agreements about the fundamentals of technical design (definitions, basic performance characteristics) or the best achievable solution to complex technical problems. However, firms sponsor participation in these committees, and pay for the background research supporting standards. For many years, technical issues took precedence in determining actual standards in the *de jure* standards organizations. Today everyone agrees that standardization should be industry led and responsive to market needs primarily. This has meant restructuring all the major *de jure* standards development organizations to include users and user groups in setting priorities, greater reliance upon industry bodies as "feeder organizations", more emphasis being given to precompetitive standards, fast track decision making even in consensus standards organizations, limiting de jure standardization to essential requirements of new technologies or systems, and, last, the use of the forums as a resource for standardization. The typical *de jure* standard now requires less than 2 years (often as little as 1 year) to be developed, as compared with the previous norm of 8 years. Regulatory standards take longer to develop, but, in high technology fields, their influence is diminishing.

In sum, technology managers deal with standards primarily in terms of their technical quality, but the trade, economic and political implications of standardization cannot be ignored. Senior management has long been involved in deciding whether, when, and where to fund standardization, which organization(s) to support, and how to influence governments. It is now becoming involved at the level of policy making in the major standards organizations and in forums and consortia as well.

Defining Terms

Compatibility standards: Concern the interconnection or interworking of various products or systems
De facto standards: Are the outcome of market decisions.
De jure standards: Are developed by standards development organizations.
Performance standards: Are focused on results or outputs or on mitigating potential harm.
Prescription standards: Require (or promote) compliance with predetermined levels of product or system capability, etc.
Process standards: Deal with how products are to be manufactured or with the design and possible outputs of systems or networks.
Quality standards: Pertain to the level of performance expected.
Regulatory or **mandatory standards:**   Are *de jure* standards set or adopted into law by government departments or agencies, enforceable though courts or agencies.
Standards: Are technical documents reflecting agreed upon qualities, capabilities, or by-products of products, industrial processes, systems, or networks.

References

Berg, S. The production of compatibility: technical standards as collective goods, *Kyklos*, 42(3): 361–383, 1989.

Besen, S. M. and Farrell, J. Choosing how to compete: strategies and tactics in standardization, *J. Econ. Perspect.*, 8(2): 117–132, 1994.

David, P. A. and Greenstein, S. The economics of compatibility standards: an introduction to recent research, *Econ. Innov. New Technol.*, 1(1/2): 3–41, 1990.

Salter, L. *Mandated Science: Science and Scientists in the Making of Standards*, Kluwer, Dordrecht, The Netherlands, 1985.

Further Information

Cargill, C. F. *Open Systems Standardization: A Systems Approach*, Prentice Hall, Upper Saddle River, NJ, 1997.

Commission of the European Communities. *Commission Green Paper on the Development of European Standardization: Action for Faster Technological Integration in Europe*, COM 90, 456 final, Brussels, 1990.

Gabel, H. L., Ed. *Standardization and Competitive Strategy,* Elsevier/North Holland, Amsterdam, 1987.

Hawkins, R., Mansell, R., and Skea, J. *Standards Innovation and Competitiveness,* Edward Elgar, Aldershot, England, 1995.

Hemenway, D. *Industry Wide Voluntary Product Standards,* Ballinger, Cambridge, MA, 1975.

Salter, L. *Mandated Science: Science and Scientists in the Making of Standards,* Kluwer, Dordrecht, The Netherlands, 1985.

U.S. Congress, Office of Technology Assessment. *Global Standards: Building Blocks for the Future,* 1992.

14.15 Organizing and Leading "Heavyweight" Development Teams[6]

Kim B. Clark and Steven C. Wheelwright

Effective product and process development requires the integration of specialized capabilities. Integrating is difficult in most circumstances, but is particularly challenging in large, mature firms with strong functional groups, extensive specialization, large numbers of people, and multiple, ongoing operating pressures. In such firms, development projects are the exception rather than the primary focus of attention. Even for people working on development projects, years of experience and the established systems — covering everything from career paths to performance evaluation, and from reporting relationships to breadth of job definitions — create both physical and organizational distance from other people in the organization. The functions themselves are organized in a way that creates further complications: the marketing organization is based on product families and market segments; engineering around functional disciplines and technical focus; and manufacturing on a mix between functional and product market structures. The result is that in large, mature firms, organizing and leading an effective development effort is a major undertaking. This is especially true for organizations whose traditionally stable markets and competitive environment are threatened by new entrants, new technologies, and rapidly changing customer demands.

This section zeros in on one type of team structure — "heavyweight" project teams — that seems particularly promising in today's fast-paced world yet is strikingly absent in many mature companies. Our research shows that when managed effectively, heavyweight teams offer improved communication, stronger identification with and commitment to a project, and a focus on cross-functional problem solving. Our research also reveals, however, that these teams are not so easily managed and contain unique issues and challenges.

Heavyweight project teams are one of four types of team structures. We begin by describing each of them briefly. We then explore heavyweight teams in detail, compare them with the alternative forms, and point out specific challenges and their solutions in managing the heavyweight organization. We conclude with an example of the changes necessary in individual behavior for heavyweight teams to be effective. Although heavyweight teams are a different way of organizing, they are more than a new structure; they represent a fundamentally different way of working. To the extent that both the team members and the surrounding organization recognize that phenomenon, the heavyweight team begins to realize its full potential.

Types of Development Project Teams

Figure 14.25 illustrates the four dominant team structures we have observed in our studies of development projects: functional, lightweight, heavyweight, and autonomous (or tiger). These forms are described

[6]*Source:* California Management Review, Spring 1992, Vol. 34, #3, adapted from Chapter 8 of *Revolutionizing Product Development: Quantum Leaps in Speed, Efficiency and Quality* by Steven C. Wheelwright and Kim B. Clark. Copyright © 1992 by Steven C. Wheelwright and Kim B. Clark. Reprinted with permission of the *California Management Review.*

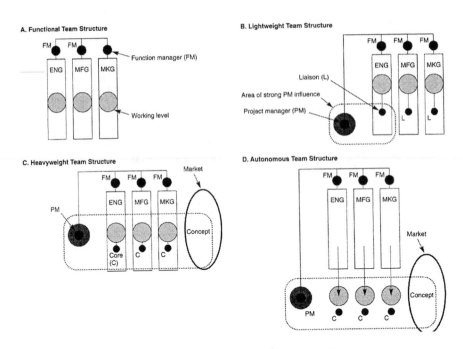

FIGURE 14.25   Types of development teams.

below, along with their associated project leadership roles, strengths, and weaknesses. Heavyweight teams are examined in detail in the subsequent section.

Functional Team Structure

In the traditional functional organization found in larger, more mature firms, people are grouped principally by discipline, each working under the direction of a specialized subfunction manager and a senior functional manager. The different subfunctions and functions coordinate ideas through detailed specifications all parties agree to at the outset, and through occasional meetings where issues that cut across groups are discussed. Over time, primary responsibility for the project passes sequentially — although often not smoothly — from one function to the next, a transfer frequently termed *throwing it over the wall.*

The functional team structure has several advantages, and associated disadvantages. One strength is that those managers who control the project's resources also control task performance in their functional area; thus, responsibility and authority are usually aligned. However, tasks must be subdivided at the project's outset (i.e., the entire development process is decomposed into separable, somewhat indepen- dent activities). But, on most development efforts, not all required tasks are known at the outset, nor can they all be easily and realistically subdivided. Coordination and integration can suffer as a result.

Another major strength of this approach is that, because most career paths are functional in nature until a general management level is reached, the work done on a project is judged, evaluated, and rewarded by the same subfunction and functional managers who make the decisions about career paths. The associated disadvantage is that individual contributions to a development project tend to be judged largely independently of overall project success. The traditional tenet cited is that individuals cannot be evaluated fairly on outcomes over which they have little or no control. But, as a practical matter, that often means that no one directly involved in the details of the project is responsible for the results finally achieved.

Finally, the functional project organization brings specialized expertise to bear on the key technical issues. The same person or small group of people may be responsible for the design of a particular component or subsystem over a wide range of development efforts. Thus the functions and subfunctions capture the benefits of prior experience and become the keepers of the organization's depth of knowledge while ensuring that it is systematically applied over time and across projects. The disadvantage is that every development project differs in its objectives and performance requirements, and it is unlikely that specialists developing a single component will do so very differently on one project than on another. The "best" component or subsystem is defined by technical parameters in the areas of their expertise rather than by overall system characteristics or specific customer requirements dictated by the unique market the development effort aims for.

Lightweight Team Structure

Like the functional structure, those assigned to the lightweight team reside physically in their functional areas, but each functional organization designates a liaison person to "represent" it on a project coordinating committee. These liaison representatives work with a "lightweight project manager", usually a design engineer or product marketing manager, who coordinates different functions' activities. This approach usually figures as an add-on to a traditional functional organization, with the functional liaison person having that role added to his or her other duties. The overall coordination assignment of lightweight project manager, however, tends not to be present in the traditional functional team structure.

The project manager is a "lightweight" in two important aspects. First, he or she is generally a middle- or junior-level person who, despite considerable expertise, usually has little status or influence in the organization. Such people have spent a handful of years in a function, and this assignment is seen as a "broadening experience", a chance for them to move out of that function. Second, although they are responsible for informing and coordinating the activities of the functional organizations, the key resources (including engineers on the project) remain under the control of their respective functional managers. The lightweight project manager does not have power to reassign people or reallocate resources and instead confirms schedules, updates time lines, and expedites across groups. Typically, such project leaders spend no more than 25% of their time on a single project.

The primary strengths and weaknesses of the lightweight project team are those of the functional project structure. But, now at least one person over the course of the project looks across functions and seeks to ensure that individual tasks — especially those on the critical path — get done in a timely fashion, and that everyone is kept aware of potential cross-functional issues and what is going on elsewhere on this particular project.

Thus, improved communication and coordination are what an organization expects when moving from a functional to a lightweight team structure. Yet, because power still resides with the subfunction and functional managers, hopes for improved efficiency, speed, and project quality are seldom realized. Moreover, lightweight project leaders find themselves tolerated at best, and often ignored and even preempted. This can easily become a "no-win" situation for the individual thus assigned.

Heavyweight Team Structure

In contrast to the lightweight setup, the heavyweight project manager has direct access to and responsibility for the work of all those involved in the project. Such leaders are "heavyweights" in two respects. First, they are senior managers within the organization; they may even outrank the functional managers. Hence, in addition to having expertise and experience, they also wield significant organizational clout. Second, heavyweight leaders have primary influence over the people working on the development effort and supervise their work directly through key functional people on the core teams. Often the core group of people are dedicated and physically colocated with the heavyweight project leader. However, the longer-term career development of individual contributors continues to rest not with the project leader — although that heavyweight leader makes significant input to individual performance evaluations — but with the functional manager, because members are not assigned to a project team on a permanent basis.

The heavyweight team structure has a number of advantages and strengths, along with associated weaknesses. Because this team structure is observed much less frequently in practice and yet seems to have tremendous potential for a wide range of organizations, it will be discussed in detail in the next section.

Autonomous Team Structure

With the autonomous team structure, often called the "tiger team," individuals from the different functional areas are formally assigned, dedicated, and colocated to the project team. The project leader, a "heavyweight" in the organization, is given full control over the resources contributed by the different functional groups. Furthermore, that project leader becomes the sole evaluator of the contribution made by individual team members.

In essence, the autonomous team is given a "clean sheet of paper"; it is not required to follow existing organizational practices and procedures, but allowed to create its own. This includes establishing incentives and rewards as well as norms for behavior. However, the team will be held fully accountable for the final results of the project: success or failure is its responsibility and no one else's.

The fundamental strength of the autonomous team structure is focus. Everything the individual team members and the team leader do is concentrated on making the project successful. Thus, tiger teams can excel at rapid, efficient new product and new process development. They handle cross-functional integration in a particularly effective manner, possibly because they attract and select team participants much more freely than the other project structures.

Tiger teams, however, take little or nothing as "given"; they are likely to expand the bounds of their project definition and tackle redesign of the entire product, its components, and subassemblies, rather than looking for opportunities to utilize existing materials, designs, and organizational relationships. Their solution may be unique, making it more difficult to fold the resulting product and process — and, in many cases, the team members themselves — back into the traditional organization upon project completion. As a consequence, tiger teams often become the birthplace of new business units or they experience unusually high turnover following project completion.

Senior managers often become nervous at the prospects of a tiger team because they are asked to delegate much more responsibility and control to the team and its project leader than under any of the other organization structures. Unless clear guidelines have been established in advance, it is extremely difficult during the project for senior managers to make midcourse corrections or exercise substantial influence without destroying the team. More than one team has "gotten away" from senior management and created major problems.

The Heavyweight Team Structure

The best way to begin understanding the potential of heavyweight teams is to consider an example of their success, in this case, Motorola's experience in developing its Bandit line of pagers.

The Bandit Pager Heavyweight Team

This development team within the Motorola Communications Sector was given a project charter to develop an automated, on-shore, profitable production operation for its high-volume Bravo pager line. (This is the belt-worn pager that Motorola sold from the mid-1980s into the early 1990s). The core team consisted of a heavyweight project leader and a handful of dedicated and colocated individuals, who represented industrial engineering, robotics, process engineering, procurement, and product design/CIM. The need for these functions was dictated by the Bandit platform automation project and its focus on manufacturing technology with a minimal change in product technology. In addition, human resource and accounting/finance representatives were part of the core team. The human resource person was particularly active early on as subteam positions were defined and jobs posted throughout Motorola's Communications Sector, and played an important subsequent role in training and development of operating support people. The accounting/finance person was invaluable in "costing out" different options and performing detailed analyses of options and choices identified during the course of the project.

An eighth member of the core team was a Hewlett-Packard (HP) employee. HP was chosen as the vendor for the "software backplane", providing an HP 3000 computer and the integrated software communication network that linked individual automated workstations, downloaded controls and instructions during production operations, and captured quality and other operating performance data. Because HP support was vital to the project's success, it was felt essential they be represented on the core team.

The core team was housed in a corner of the Motorola Telecommunications engineering/manufacturing facility. The team chose to enclose in glass the area where the automated production line was to be set up so that others in the factory could track the progress, offer suggestions, and adopt the lessons learned from it in their own production and engineering environments. The team called their project Bandit to indicate a willingness to "take" ideas from literally anywhere.

The heavyweight project leader, Scott Shamlin, who was described by team members as "a crusader", "a renegade", and "a workaholic", became the champion for the Bandit effort. A hands-on manager who played a major role in stimulating and facilitating communication across functions, he helped to articulate a vision of the Bandit line and to infuse it into the detailed work of the project team. His goal was to make sure the new manufacturing process worked for the pager line, but would provide real insight for many other production lines in Motorola's Communications Sector.

The Bandit core team started by creating a contract book that established the blueprint and work plan for the team's efforts and its performance expectations; all core team members and senior management signed on to the document. Initially, the team's executive sponsor — although not formally identified as such — was George Fisher, the sector executive. He made the original investment proposal to the board of directors and was an early champion and supporter, as well as direct supervisor in selecting the project leader and helping get the team underway. Subsequently, the vice president and general manager of the Paging Products division filled the role of executive sponsor.

Throughout the project, the heavyweight team took responsibility for the substance of its work, the means by which it was accomplished, and its results. The project was completed in 18 months as per the contract book, which represented about half the time of a normal project of such magnitude. Further, the automated production operation was up and running with process tolerances of five sigma (i.e., the degree of precision achieved by the manufacturing processes) at the end of 18 months. Ongoing production verified that the cost objectives (substantially reduced direct costs and improved profit margins) had indeed been met, and product reliability was even higher than the standards already achieved on the off-shore versions of the Bravo product. Finally, a variety of lessons were successfully transferred to other parts of the sector's operations, and additional heavyweight teams have proven the viability and robustness of the approach in Motorola's business and further refined its effectiveness throughout the corporation.

The Challenge of Heavyweight Teams

Motorola's experience underscores heavyweight teams' potential power, but it also makes clear that creating an effective heavyweight team capability is more than merely selecting a leader and forming a team. By their very nature — being product (or process) focused, and needing strong, independent leadership, broad skills and cross-functional perspective, and clear missions — heavyweight teams may conflict with the functional organization and raise questions about senior management's influence and control. And even the advantages of the team approach bring with them potential disadvantages that may hurt development performance if not recognized and averted.

Take, for example, the advantages of ownership and commitment, one of the most striking advantages of the heavyweight team. Identifying with the product and creating a sense of esprit de corps motivate core team members to extend themselves and do what needs to be done to help the team succeed. However, such teams sometimes expand the definition of their role and the scope of the project, and they get carried away with themselves and their abilities. We have seen heavyweight teams turn into autonomous tiger teams and go off on a tangent because senior executives gave insufficient direction and the bounds of the team were only vaguely specified at the outset. And even if the team stays focused, the rest of the organization may see themselves as "second class". Although the core team may not make that distinction explicit, it happens because the team has responsibilities and authority beyond those commonly given to functional team members. Thus, such projects inadvertently can become the "haves" and other, smaller projects the "have-nots" with regard to key resources and management attention.

Support activities are particularly vulnerable to an excess of ownership and commitment. Often the heavyweight team will want the same control over secondary support activities as it has over the primary tasks performed by dedicated team members. When waiting for prototypes to be constructed, analytical tests to be performed, or quality assurance procedures to be conduced, the team's natural response is to "demand" top priority from the support organization or to be allowed to go outside and subcontract to independent groups. While these may sometimes be the appropriate choices, senior management should establish make-buy guidelines and clear priorities applicable to all projects — perhaps changing service levels provided by support groups (rather than maintaining the traditional emphasis on resource utilization) — or have support groups provide capacity and advisory technical services but let team members do more of the actual task work in those support areas. Whatever actions the organization takes, the challenge is to achieve a balance between the needs of the individual project and the needs of the broader organization.

Another advantage the heavyweight team brings is the integration and integrity it provides through a system solution to a set of customer needs. Getting all the components and subsystems to complement one another and to address effectively the fundamental requirements of the core customer segment can result in a winning platform product and/or process. The team achieves an effective system design by using generalist skills applied by broadly trained team members, with fewer specialists and, on occasion, less depth in individual component solutions and technical problem solving.

The extent of these implications is aptly illustrated by the nature of the teams Clark and Fujimoto studied in the auto industry.[7] They found that for U.S. auto firms in the mid-1980s, typical platform projects — organized under a traditional functional or lightweight team structure — entailed full-time work for several months by approximately 1500 engineers. In contrast, a handful of Japanese platform projects — carried out by heavyweight teams — utilized only 250 engineers working full-time for several months. The implications of 250 vs. 1,500 full-time equivalents (FTEs) with regard to breadth of tasks, degree of specialization, and need for coordination are significant and help explain the differences in project results as measured by product integrity, development cycle time, and engineering resource utilization.

However, that lack of depth may disclose a disadvantage. Some individual components or subassemblies may not attain the same level of technical excellence they would under a more traditional functional team structure. For instance, generalists may develop a windshield wiper system that is complementary with and integrated into the total car system and its core concept. However, they also may embed in their design some potential weaknesses or flaws that might have been caught by a functional team of specialists who had designed a long series of windshield wipers. To counter this potential disadvantage, many organizations order more testing of completed units to discover such possible flaws and have components and subassemblies reviewed by expert specialists. In some cases, the quality assurance function has expanded its role to make sure sufficient technical specialists review designs at appropriate points so that such weaknesses can be minimized.

Managing the Challenges of Heavyweight Teams

Problems with depth in technical solutions and allocations of support resources suggest the tension that exists between heavyweight teams and the functional groups where much of the work gets done. The problem with the teams exceeding their bounds reflects in part how teams manage themselves, in part how boundaries are set, and in part the ongoing relationship between the team and senior management. Dealing with these issues requires mechanisms and practices that reinforce the team's basic thrust — ownership, focus, system architecture, integrity — and yet improve its ability to take advantage of the strengths of the supporting functional organization — technical depth, consistency across projects, senior management direction. We have grouped the mechanisms and problems into six categories of management action: the project charter, the contract, staffing, leadership, team responsibility, and the executive sponsor.

---

[7]See Kim B. Clark and Takahiro Fujimoto, *Product Development Performance* (Boston, MA: Harvard Business School Press, 1991).

TABLE 14.6   Heavyweight Team,
Contract Book — Major Sections

Executive summary
Business plan and purposes
Development plan
Schedule
Materials
Resources
Product design plan
Quality plan
Manufacturing plan
Project deliverables
Performance measurement and incentives

The Project Charter

A heavyweight project team needs a clear mission. A way to capture that mission concisely is in an explicit, measurable project charter that sets broad performance objectives and usually is articulated even before the core team is selected. Thus, joining the core team includes accepting the charter established by senior management. A typical charter for a heavyweight project would be the following:

> The resulting product will be selected and ramped by Company X during Quarter 4 of calendar year 1991, at a minimum of a 20% gross margin.

This charter is representative of an industrial products firm whose product goes into a system sold by its customers. Company X is the leading customer for a certain family of products, and this project is dedicated to developing the next generation platform offering in that family. If the heavyweight program results in that platform product being chosen by the leading customer in the segment by a certain date and at a certain gross margin, it will have demonstrated that the next generation platform is not only viable, but likely to be very successful over the next 3 to 5 years. Industries and settings where such a charter might be found would include a microprocessor being developed for a new computer system, a diesel engine for the heavy equipment industry, or a certain type of slitting and folding piece of equipment for the newspaper printing press industry. Even in a medical diagnostics business with hundreds of customers, a goal of "capturing 30% of market purchases in the second 12 months during which the product is offered" sets a clear charter for the team.

The Contract Book

Whereas a charter lays out the mission in broad terms, the contract book defines, in detail, the basic plan to achieve the stated goal. A contract book is created as soon as the core team and heavyweight project leader have been designated and given the charter by senior management. Basically, the team develops its own detailed work plan for conducting the project, estimates the resources required, and outlines the results to be achieved and against which it is willing to be evaluated. (The table of contents of a typical heavyweight team contract book is shown in Table 14.6). Such documents range from 25 to 100 pages, depending on the complexity of the project and level of detail desired by the team and senior management before proceeding. A common practice following negotiation and acceptance of this contract is for the individuals from the team and senior management to sign the contract book as an indication of their commitment to honor the plan and achieve those results.

The core team may take anywhere from a long week to a few months to create and complete the contract book; Motorola, for example, after several years of experience, has decided that a maximum of seven days should be allowed for this activity. Having watched other heavyweight teams — particularly in organizations with no prior experience in using such a structure — take up to several months, we can appreciate why Motorola has nicknamed this the "blitz phase" and decided that the time allowed should be kept to a minimum.

Staffing

As suggested in Fig. 14.25, a heavyweight team includes a group of core cross-functional team members who are dedicated (and usually physically colocated) for the duration of the development effort. Typically there is one core team member from each primary function of the organization; for instance, in several electronics firms we have observed core teams consisting of six functional participants — design engineering, marketing, quality assurance, manufacturing, finance, and human resources. (Occasionally, design will be represented by two core team members, one each for hardware and software engineering.) Individually, core team members represent their functions and provide leadership for their functions' inputs to the project. Collectively, they constitute a management team that works under the direction of the heavyweight project manager and takes responsibility for managing the overall development effort.

While other participants — especially from design engineering early on and manufacturing later on — may frequently be dedicated to a heavyweight team for several months, they usually are not made part of the core team though they may well be colocated and, over time, develop the same level of ownership and commitment to the project as core team members. The primary difference is that the core team manages the total project and the coordination and integration of individual functional efforts, whereas other dedicated team members work primarily within a single function or subfunction.

Whether these temporarily dedicated team members are actually part of the core team is an issue firms handle in different ways, but those with considerable experience tend to distinguish between core and other dedicated (and often colocated) team members. The difference is one of management responsibility for the core group that is not shared equally by the others. Also, it is primarily the half a dozen members of the core groups who will be dedicated throughout the project, with other contributors having a portion of their time reassigned before this heavyweight project is completed.

Whether physical colocation is essential is likewise questioned in such teams. We have seen it work both ways. Given the complexity of development projects, and especially the uncertainty and ambiguity often associated with those assigned to heavyweight teams, physical colocation is preferable to even the best of on-line communication approaches. Problems that arise in real time are much more likely to be addressed effectively with all of the functions represented and present than when they are separate and must either wait for a periodic meeting or use remote communication links to open up cross-functional discussions.

A final issue is whether an individual can be a core team member on more than one heavyweight team simultaneously. If the rule for a core team member is that 70% or more of his or her time must be spent on the heavyweight project, then the answer to this question is no. Frequently, however, a choice must be made between someone being on two core teams — for example, from the finance or human resource function — or putting a different individual on one of those teams who has neither the experience nor stature to be a full peer with the other core team members. Most experienced organizations we have seen opt to put the same person on two teams to ensure the peer relationship and level of contribution required, even though it means having one person on two teams and with two desks. They then work diligently to develop other people in the function so that multiple team assignments will not be necessary in the future.

Sometimes multiple assignments will also be justified on the basis that a function such as finance does not need a full-time person on a project. In most instances, however, a variety of potential value-adding tasks exist that are broader than finance's traditional contribution. A person largely dedicated to the core team will search for those opportunities and the project will be better because of it. The risk of allowing core team members to be assigned to multiple projects is that they are neither available when their inputs are most needed nor as committed to project success as their peers. They become secondary core team members, and the full potential of the heavyweight team structure fails to be realized.

Project Leadership

Heavyweight teams require a distinctive style of leadership. A number of differences between lightweight and heavyweight project managers are highlighted in Table 14.7. Three of those are particularly distinctive. First, a heavyweight leader manages, leads, and evaluates other members of the core team and is also the

TABLE 14.7 Project Manager Profile

	Lightweight (limited)	Heavyweight (extensive)
Span of coordination responsibilities	|————————————————|
Duration of responsibilities	|————————————————|
Responsible for specs, cost, layout, components	|————————————————|
Working level contact with engineers	|————————————————|
Direct contact with customers	|————————————————|
Multilingual/multidisciplined skills	|————————————————|
Role in conflict resolution	|————————————————|
Marketing imagination/concept champion	|————————————————|
Influence in		
Engineering	|————————————————|
Marketing	|————————————————|
Manufacturing	|————————————————|

TABLE 14.8 The Heavyweight Project Manager

Role	Description
Direct market interpreter	First-hand information, dealer visits, auto shows, has own marketing budget, market study team, direct contact and discussion with customers
Multilingual translator	Fluency in language of customers, engineers, marketers, stylists; translator between customer experience/requirements and engineering specifications
Direct engineering manager	Direct contact, orchestra conductor, evangelist of conceptual integrity and coordinator of component development; direct eye-to-eye discussions with working level engineers; shows up in drafting room, looks over engineers' shoulders
Program manager "in motion"	Out of the office, not too many meetings, not too much paperwork, face-to-face communication, conflict resolution manager
Concept infuser	Concept guardian, confronts conflicts, not only reacts but implements own philosophy, ultimate decision maker, coordination of details and creation of harmony

person to whom the core team reports throughout the project's duration. Another characteristic is that rather than being either neutral or a facilitator with regard to problem solving and conflict resolution, these leaders see themselves as championing the basic concept around which the platform product and/or process is being shaped. They make sure that those who work on subtasks of the project understand the concept. Thus they play a central role in ensuring the system integrity of the final product and/or process.

Finally, the heavyweight project manager carries out his or her role in a very different fashion than the lightweight project manager. Most lightweights spend the bulk of their time working at a desk, with paper. They revise schedules, get frequent updates, and encourage people to meet previously agreed upon deadlines. The heavyweight project manager spends little time at a desk, is out talking to project contributors, and makes sure that decisions are made and implemented whenever and wherever needed. Some of the ways in which the heavyweight project manager achieves project results are highlighted by the five roles illustrated in Table 14.8 for a heavyweight project manager on a platform development project in the auto history.

The *first role* of the heavyweight project manager is to provide for the team a direct interpretation of the market and customer needs. This involves gathering market data directly from customers, dealers, and industry shows, as well as through systematic study and contact with the firm's marketing organization. A *second role* is to become a multilingual translator, not just taking marketing information to the various functions involved in the project, but being fluent in the language of each of those functions and making sure the translation and communication going on among the functions — particularly between customer needs and product specifications — are done effectively.

TABLE 14.9   Responsibilities of Heavyweight Core Team Members

Functional Hat Accountabilities
 Ensuring functional expertise on the project
 Representing the functional perspective on the project
 Ensuring that subobjectives are met that depend on their function
 Ensuring that functional issues impacting the team are raised proactively within the team
Team Hat Accountabilities
 Sharing responsibility for team results
 Reconstituting tasks and content
 Establishing reporting and other organizational relationships
 Participating in monitoring and improving team performance
 Sharing responsibility for ensuring effective team processes
 Examining issues from an executive point of view (answering the question, "Is this the appropriate business response
  for the company?)
 Understanding, recognizing, and responsibly challenging the boundaries of the project and team process

A *third role* is the direct engineering manager, orchestrating, directing, and coordinating the various engineering subfunctions. Given the size of many development programs and the number of types of engineering disciplines involved, the project manager must be able to work directly with each engineering subfunction on a day-to-day basis and ensure that their work will indeed integrate and support that of others, so the chosen product concept can be effectively executed.

A *fourth role* is best described as staying in motion: out of the office conducting face-to-face sessions, and highlighting and resolving potential conflicts as soon as possible. Part of this role entails energizing and pacing the overall effort and its key subparts.

A *final role* is that of concept champion. Here the heavyweight project manager becomes the guardian of the concept and not only reacts and responds to the interests of others, but also sees that the choices made are consistent and in harmony with the basic concept. This requires a careful blend of communication and teaching skills so that individual contributors and their groups understand the core concept and have sufficient conflict resolution skills to ensure that any tough issues are addressed in a timely fashion.

It should be apparent from this description that heavyweight project managers earn the respect and right to carry out these roles based on prior experience, carefully developed skills, and status earned over time, rather than simply being designated "leader" by senior management. A qualified heavyweight project manager is a prerequisite to an effective heavyweight team structure.

Team Member Responsibilities
Heavyweight team members have responsibilities beyond their usual functional assignment. As illustrated in Table 14.9, these are of two primary types. Functional hat responsibilities are those accepted by the individual core team member as representative of his or her function. For example, the core team member from marketing is responsible for ensuring that appropriate marketing expertise is brought to the project, that a marketing perspective is provided on all key issues, that project subobjectives dependent on the marketing function are met in a timely fashion, and that marketing issues that impact other functions are raised proactively within the team.

However, each core team member also wears a team hat. In addition to representing a function, each member shares responsibility with the heavyweight project manager for the procedures followed by the team, and for the overall results that those procedures deliver. The core team is accountable for the success of the project, and it can blame no one but itself if it fails to manage the project, execute the tasks, and deliver the performance agreed upon at the outset.

Finally, beyond being accountable for tasks in their own function, core team members are responsible for how those tasks are subdivided, organized, and accomplished. Unlike the traditional functional development structure, which takes as given the subdivision of tasks and the means by which those tasks will be conducted and completed, the core heavyweight team is given the power and responsibility to change the substance of those tasks to improve the performance of the project. Since this is a role that core team members do not play under a lightweight or functional team structure, it is often the most difficult for them to accept fully and learn to apply. It is essential, however, if the heavyweight team is to realize its full potential.

The Executive Sponsor

With so much more accountability delegated to the project team, establishing effective relationships with senior management requires special mechanisms. Senior management needs to retain the ability to guide the project and its leader while empowering the team to lead and act, a responsibility usually taken by an executive sponsor — typically the vice president of engineering, marketing, or manufacturing for the business unit. This sponsor becomes the coach and mentor for the heavyweight project leader and core team and seeks to maintain close, ongoing contact with the team's efforts. In addition, the executive sponsor serves as a liaison. If other members of senior management — including the functional heads — have concerns or inputs to voice, or need current information on project status, these are communicated through the executive sponsor. This reduces the number of mixed signals received by the team and clarifies for the organization the reporting and evaluation relationship between the team and senior management. It also encourages the executive sponsor to set appropriate limits and bounds on the team so that organizational surprises are avoided.

Often the executive sponsor and core team identify those areas where the team clearly has decision-making power and control, and they distinguish them from areas requiring review. An electronics firm that has used heavyweight teams for some time dedicates one meeting early on between the executive sponsor and the core team to generating a list of areas where the executive sponsor expects to provide oversight and be consulted; these areas are of great concern to the entire executive staff and team actions may well raise policy issues for the larger organization. In this firm, the executive staff wants to maintain some control over:

- Resource commitment — head count, fixed costs, and major expenses outside the approved contract book plan.
- Pricing for major customers and major accounts.
- Potential slips in major milestone dates (the executive sponsor wants early warning and recovery plans).
- Plans for transitioning from development project to operating status.
- Thorough reviews at major milestones or every three months, whichever occurs sooner.
- Review of incentive rewards that have companywide implications for consistency and equity.
- Cross-project issues such as resource optimization, prioritization, and balance.

Identifying such areas at the outset can help the executive sponsor and the core team better carry out their assigned responsibilities. It also helps other executives feel more comfortable working through the executive sponsor, since they know these "boundary issues" have been articulated and are jointly understood.

The Necessity of Fundamental Change

Compared to a traditional functional organization, creating a team that is "heavy" — one with effective leadership, strong problem-solving skills, and the ability to integrate across functions — requires basic changes in the way development works. However, it also requires change in the fundamental behavior of engineers, designers, manufacturers, and marketers in their day-to-day work. An episode in a computer company with no previous experience with heavyweight teams illustrates the depth of change required to realize fully these teams' power.[8]

Two teams, A and B, were charged with development of a small computer system and had market introduction targets within the next 12 months. While each core team was colocated and held regular meetings, there was one overlapping core team member (from finance/accounting). Each team was

---

[8]Adapted from a description provided by Dr. Christopher Meyer, Strategic Alignment Group, Los Altos, CA.

charged with developing a new computer system for their individual target markets but by chance, both products were to use an identical, custom-designed microprocessor chip in addition to other unique and standard chips.

The challenge of changing behavior in creating an effective heavyweight team structure was highlighted when each team sent this identical, custom-designed chip — the "supercontroller" — to the vendor for pilot production. The vendor quoted a 20-week turnaround to both teams. At that time, the supercontroller chip was already on the critical path for Team B, with a planned turnaround of 11 weeks. Thus, every week saved on that chip would save one week in the overall project schedule, and Team B already suspected that it would be late in meeting its initial market introduction target date. When the 20-week vendor lead time issue first came up in a Team B meeting, Jim, the core team member from engineering, responded very much as he had on prior, functionally structured development efforts: because initial prototypes were engineering's responsibility, he reported that they were working on accelerating the delivery date, but that the vendor was a large company, with which the computer manufacturer did substantial business, and known for its slowness. Suggestions from other core team members on how to accelerate the delivery were politely rebuffed, including one to have a senior executive contact his or her counterpart at the vendor. Jim knew the traditional approach to such issues and did not perceive a need, responsibility, or authority to alter it significantly.

For Team A, the original quote of 20-week turnaround still left a little slack, and thus initially the supercontroller chip was not on the critical path. Within a couple of weeks, however, it was, given other changes in the activities and schedule, and the issue was immediately raised at the team's weekly meeting. Fred, the core team member from manufacturing (who historically would not have been involved in an early engineering prototype), stated that he thought the turnaround time quoted was too long and that he would try to reduce it. At the next meeting, Fred brought some good news: through discussions with the vendor, he had been able to get a commitment that pulled in the delivery of the supercontroller chip by 11 weeks! Furthermore, Fred thought that the quote might be reduced even further by a phone call from one of the computer manufacturer's senior executives to a contact of his at the vendor.

Two days later, at a regular Team B meeting, the supercontroller chip again came up during the status review, and no change from the original schedule was identified. Since the finance person, Ann, served on both teams and had been present at Team A's meeting, she described Team A's success in reducing the cycle time. Jim responded that he was aware that Team A had made such efforts, but that the information was not correct, and the original 20-week delivery date still held. Furthermore, Jim indicated that Fred's efforts (from Team A) had caused some uncertainty and disruption internally, and in the future it was important that Team A not take such initiatives before coordinating with Team B. Jim stated that this was particularly true when an outside vendor was involved, and he closed the topic by saying that a meeting to clear up the situation would be held that afternoon with Fred from Team A and Team B's engineering and purchasing people.

The next afternoon, at his Team A meeting, Fred confirmed the accelerated delivery schedule for the supercontroller chip. Eleven weeks had indeed been clipped out of the schedule to the benefit of both Teams A and B. Subsequently, Jim confirmed the revised schedule would apply to his team as well, although he was displeased that Fred had abrogated "standard operating procedure" to achieve it. Curious about the differences in perspective, Ann decided to learn more about why Team A had identified an obstacle and removed it from its path, yet Team B had identified an identical obstacle and failed to move it at all.

As Fred pointed out, Jim was the engineering manager responsible for development of the supercontroller chip; he knew the chip's technical requirements, but had little experience dealing with chip vendors and their production processes. (He had long been a specialist.) Without the experience, he had a hard time pushing back against the vendor's "standard line". However, Fred's manufacturing experience with several chip vendors enabled him to calibrate the vendor's dates against his best-case experience and understand what the vendor needed to do to meet a substantially earlier commitment.

Moreover, because Fred had bought into a clear team charter, whose path the delayed chip would block, and because he had relevant experience, it did not make sense to live with the vendor's initial commitment, and thus he sought to change it. In contrast, Jim — who had worked in the traditional functional organization for many years — saw vendor relations on a pilot build as part of his functional job, but did not believe that contravening standard practices to get the vendor to shorten the cycle time was his responsibility, within the range of his authority, or even in the best long-term interest of his function. He was more concerned with avoiding conflict and not roiling the water than with achieving the overarching goal of the team.

It is interesting to note that in Team B, engineering raised the issue and, while unwilling to take aggressive steps to resolve it, also blocked others' attempts. In Team A, however, while the issue came up initially through engineering, Fred in manufacturing proactively went after it. In the case of Team B, getting a prototype chip returned from a vendor was still being treated as an "engineering responsibility", whereas in the case of Team A, it was treated as a "team responsibility". Since Fred was the person best qualified to attack that issue, he did so.

Both Team A and Team B had a charter, a contract, a colocated core team staffed with generalists, a project leader, articulated responsibilities, and an executive sponsor. Yet Jim's and Fred's understanding of what these things meant for them personally and for the team at the detailed, working level was quite different. While the teams had been through similar training and team-startup processes, Jim apparently saw the new approach as a different organizational framework within which work would get done as before. In contrast, Fred seemed to see it as an opportunity to work in a different way — to take responsibility for reconfiguring tasks, drawing on new skills, and reallocating resources, where required, for getting the job done in the best way possible.

Although both teams were "heavyweight" in theory, Fred's team was much "heavier" in its operation and impact. Our research suggests that heaviness is not just a matter of structure and mechanism, but of attitudes and behavior. Firms that try to create heavyweight teams without making the deep changes needed to realize the power in the team's structure will find this team approach problematic. Those intent on using teams for platform projects and willing to make the basic changes we have discussed here can enjoy substantial advantages of focus, integration, and effectiveness.

15

Project Management

Hans J. Thamhain
Bentley College

John L. Richards
University of Pittsburgh

Jeffrey K. Pinto
The Pennsylvania State University

Howard Eisner
The George Washington University

E. Lile Murphree, Jr.
The George Washington University

15.1   Project Evaluation and Selection

Hans J. Thamhain

For most organizations, resources are limited. The ability to select and fund the best projects with the highest probability of success is crucial to an organization's ability to survive and prosper in today's highly competitive environment. Project selections, necessary in virtually every business area, cover activities ranging from product developments to organizational improvements, customer contracts, R&D activities, and bid proposals. Evaluation and selection methods support two principal types of decisions:

1. Judging the chances of success for one proposed project
2. Choosing the best project among available alternatives

Although most decision processes evaluate projects in terms of cost, time, risks, and benefits, such as shown in Table 15.1, it is often extremely difficult, if not impossible, to define a meaningful aggregate measure for ranking projects regarding business success, technical risks, or profit. Managers can use traditional, purely rational selection processes toward "right", "successful", and "best" only for a limited number of business situations. Many of today's complex project evaluations require the integration of both analytical and judgmental techniques to be meaningful.

Although the literature offers a great variety of project selection procedures, each organization has its own special methods. Approaches fall into one of three principal classes:

TABLE 15.1   Typical Criteria for Project Evaluation and Selection

The criteria relevant to the evaluation and selection of a particular project depend on the specific project type and business situation such as project development, custom project, process development, industry and market. Typically, evaluation procedures include the following criteria:

Development cost
Development time
Technical complexity
Risk
Return on investment
Cost benefit
Product life cycle
Sales volume
Market share
Project business follow-on
Organizational readiness and strength
Consistency with business plan
Resource availability
Cash flow, revenue, and profit
Impact on other business activities

Each criterion is based on a complex set of parameters and variables.

1. Primarily quantitative and rational approaches
2. Primarily qualitative and intuitive approaches
3. Mixed approaches, combining both quantitative and qualitative methods

Quantitative Approaches to Project Evaluation and Selection

Quantitative approaches are often favored to support project evaluation and selections if the decisions require economic justification. Supported by numeric measures for simple and effective comparison, ranking, and selection, they help to establish quantifiable norms and standards and lead to repeatable processes. However, the ultimate usefulness of these methods depends on the assumption that the decision parameters — such as cash flow, risks, and the underlying economic, social, political, and market factors — can actually be quantified. Typically, these quantitative techniques are effective decision support tools if meaningful estimates of capital expenditures and future revenues can be obtained and converted into net present values for comparison. Table 15.2 shows the cash flow of four project options to be used for illustrating the quantitative methods described in this section.

Net Present Value (NPV) Comparison

This method uses discounted cash flow as the basis for comparing the relative merit of alternative project opportunities. It assumes that all investment costs and revenues are known and that economic analysis is a valid singular basis for project selection.

We can determine the *net present value* (NPV) of a single revenue, stream of revenues, and/or costs expected in the future.

Present worth of a single revenue or cost (often called annuity, A) occurring at the end of period n and subject to an effective interest rate i (sometimes referred to as discount rate or **minimum attractive rate of return, MARR**) can be calculated as

$$PW\left(A\middle|i,n\right) = A\frac{1}{\left(1+i\right)^{n}} = PW_{n}$$

TABLE 15.2 Cash Flow of Four Project Options or Proposals*

End of Year	Do-Nothing Option P1	Project Option P2	Project Option P3	Project Option P4
0	0	−1000	−2000	−5000
1	0	200	1500	1000
2	0	200	1000	1500
3	0	200	800	2000
4	0	200	900	3000
5	0	200	1200	4000
Net cash flow	0	0	+3400	+7500
$\text{NPV}\|N = 5$	0	−242	+2153	+3192
$\text{NPV}\|N = \infty$	0	+1000	+9904	+28030
$\text{ROI}\|N = 5$	0	20%	54%	46%
$\text{CB} = \text{ROI}_{\text{NPV}}\|n = 5$	0	76%	108%	164%
$N_{\text{PBP}}\|i = 0$	0	5	1.5	3.3
$N_{\text{NPV}}\|i$	0	7.3	5	3.8

*Assuming an MARR of $i = 10\%$
Note: The first line of negative numbers represents the initial investment at the beginning of the life cycle.

Net present value of a series of revenues or costs, A_n, over N periods of time is as follows:

$$\text{NPV}\left(A_n \middle| i, N\right) = \sum_{n=1}^{N} A_n \frac{1}{\left(1+i\right)^n} = \sum_{n=1}^{N} PW_n$$

Table 15.2 applied these formulas to four project alternatives, showing the most favorable 5-year net present value of $3192 for project option P4. (There are three special cases of net present value: (1) for a uniform series of revenues or costs over N periods, $\text{NPV}(A_n|i,N) = A\{(1 + i)^N − 1\}/i(1 + i)^N$; (2) for an annuity or interest rate i approaching zero, $\text{NPV} = A \times N$; and (3) for the revenue or cost series to continue forever, $\text{NPV} = A/i$.)

Return-on-Investment Comparison

Perhaps one of the most popular measures for project evaluation is the *return on investment* (ROI):

$$\text{ROI} = \frac{\text{Revenue}\left(R\right) - \text{Cost}\left(C\right)}{\text{Investment}\left(I\right)}$$

It calculates the ratio of net revenue over investment. In its simplest form it entails the revenue on a year-by-year basis relative to the initial investment (for example, project option 2 in Table 15.2 would produce a 20% ROI). Although this is a popular measure, it does not permit a comparative evaluation of alternative projects with fluctuating costs and revenues. In a more sophisticated way we can calculate the average ROI per year

$$\overline{\text{ROI}}\left(A_n, I_n \middle| N\right) = \left[\sum_{n=1}^{N} \frac{A_n}{I_n}\right] \middle/ N$$

and compare it to the minimum attractive rate of return. All three project options, P2, P3 and P4, compare favorably to the MARR of 10%, with project P3 yielding the highest average return on investment of 54%. Or we can calculate the net present value of the total ROI over the project lifecycle, also known as *cost-benefit* (CB). This is an effective measure of comparison, especially for fluctuating cash flows. (Table 15.2 shows project 3 with the highest 5-year ROI_{NPV} of 108%.)

$$\text{ROI}_{\text{NPV}}\left(A_n, I_n \middle| i, N\right) = \left[\sum_{n=1}^{N} \text{NPV}\left(A_n \middle| i, N\right)\right] \middle/ \left[\sum_{n=1}^{N} \text{NPV}\left(I_n \middle| i, N\right)\right]$$

Pay-Back Period (PBP) Comparison

Another popular figure of merit for comparing project alternatives is the *payback period* (PBP). It indicates the time period of net revenues required to return the capital investment made on the project. For simplicity, *undiscounted* cash flows are often used to calculate a quick figure for comparison, which is quite meaningful if we deal with an initial investment and a steady stream of net revenue. However, for fluctuating revenue and/or cost streams, the net present value must be calculated for each period individually and cumulatively added up to the "break-even point" in time, N_{PBP}, when the net present value of revenue equals the investment.

$$\sum_{n=1}^{N} \text{NPV}\left(A_n \middle| i\right) \geq \sum_{n=1}^{N} \text{NPV}\left(I_n \middle| i\right)$$

Pacifico and Sobelman Project Ratings

The previously discussed methods of evaluating projects rely heavily on the assumption that technical and commercial success is ensured and all costs and revenues are predicable. Because of these limitations, many companies have developed their own special procedures for comparing project alternatives. Examples are the *project rating factor* (PR), developed by Carl Pacifico for assessing chemical products, and the *project value factor* (z), developed by Sidney Sobelman for new product selections:

$$\text{PR} = \frac{pT \times pC \times R}{TC} \qquad z = \left(P \times T_{LC}\right) - \left(C \times T_D\right)$$

Pacifico's formula is in essence an ROI calculation adjusted for risk. It includes probability of technical success [$.1 < pT < 1.0$], probability of commercial success [$.1 < pC < 1.0$], total net revenue over project life cycle [R], and total capital investment for product development, manufacturing setup, marketing, and related overheads [TC]. The Sobelman formula is a modified cost-benefit measure. It takes into account both the development time and the commercial life cycle of the product. It includes average profit per year [P], estimated product life cycle [T_{LC}], average development cost per year [C], and years of development [T_D].

Limitations of Quantitative Methods

Although quantitative methods of project evaluation have the benefit of producing relatively quickly a measure of merit for simple comparison and ranking, they also have many limitations, as summarized in Table 15.3.

Qualitative Approaches to Project Evaluation and Selection

Especially for project evaluations involving complex sets of business criteria, the narrowly focused quantitative methods must often be supplemented by broad scanning, intuitive processes, and collective,

TABLE 15.3 Comparison of Quantitative and Qualitative Approaches to Project Evaluation

Quantitative Methods	Qualitative Methods
Benefits	*Benefits*
Simple comparison, ranking, selection	Search for meaningful evaluation metrics
Repeatable process	Broad-based organizational involvement
Encourages data gathering and measurability	Understanding of problems, benefits, opportunities
Benchmarking opportunities	Problem solving as part of selection process
Programmable	Broad knowledge base
Input to sensitivity analysis and simulation	Multiple solutions and alternatives
	Multifunctional involvement leads to buy-in
	Risk sharing
Limitations	*Limitations*
Many success factors are nonquantifiable	Complex, time-consuming process
Probabilities and weights change	Biases via power and politics
True measures do not exist	Difficult to proceduralize or repeat
Analysis and conclusions are often misleading	Conflict- and energy-intensive
Methods mask unique problems and opportunities	Do not fit conventional decision processes
Stifle innovative decision making	Intuition and emotion dominates over facts
Lack people involvement, buy-in, commitment	Justify wants over needs
Do not deal well with multifunctional issues and dynamic situations	Lead to more fact finding than decision making
Pressure to act prematurely	

multifunctional decision making, such as Delphi, nominal group technology, brainstorming, focus groups, sensitivity analysis, and benchmarking. Each of these techniques can either be used by itself to determine the best, most successful, or most valuable option, or be integrated into an analytical framework for *collective multifunctional decision making*, which is discussed in the next section.

Collective Multifunctional Evaluations

Collective multifunctional evaluations rely on subject experts from various functional areas for collectively defining and evaluating broad **project success** criteria, employing both quantitative and qualitative methods. The first step is to define the specific organizational areas critical to project success and to assign expert evaluators. For a typical product development project, these organizations may include R&D, engineering, testing, manufacturing, marketing, product assurance, and customer services. These function experts should be given the time necessary for the evaluation. They also should have the commitment from senior management for full organizational support. Ideally, these evaluators should have the responsibility for ultimate project implementation, should the project be selected.

The next step is for the evaluation team to define the factors that appear critical to the ultimate success of the projects under evaluation and arrange them into a concise list that includes both quantitative and qualitative factors. A mutually acceptable scale must be worked out for scoring the evaluation criteria. Studies of collective multifunctional assessment practices show that simplicity of scales is crucial to a workable team solution. Three types of scale have produced most favorable results in field studies: (1) 10-point scale, ranging from +5 = most favorable to −5 = most unfavorable; (2) 3-point scale, +1 = favorable, 0 = neutral or can't judge, −1 = unfavorable; and (3) 5-point scale, A = highly favorable, B = favorable, C = marginally favorable, D = most likely unfavorable, F = definitely favorable. **Weighing of criteria** is not recommended for most applications, since it complicates and often distorts the collective evaluation.

Evaluators score first individually all of the factors that they feel qualified to make an expert judgment on. Collective discussions follow. Initial discussions of project alternatives, their markets, business opportunities, and technologies involved are usually beneficial but not necessary for the first round of the evaluation process. The objective of this first round of expert judgments is to get calibrated on the opportunities and challenges presented. Further, each evaluator has the opportunity to recommend

(1) actions needed for better assessment of project, (2) additional data needed, and (3) suggestions that would enhance project success and the evaluation score. Before meeting at the next group session, agreed-on action items and activities for improving the decision process should be completed. With each iteration the function-expert meetings are enhanced with more refined project data. Typically, between three and five iterations are required before a project selection can be finalized.

Recommendations for Effective Project Evaluation and Selection

Effective evaluation and selection of project opportunities is critical to overall project success. With increasing complexities and dynamics of the business environment, most situations are too complex to use simple economic models as the sole basis for decision making. To be effective, project evaluation procedures should include a broad spectrum of variables for defining the project value to the organization.

Structure, discipline, and manageability can be designed into the selection process by grouping the evaluation variables into four categories: (1) consistency and strength of the project with the business mission, strategy, and plan; (2) multifunctional ability to produce the project results, including technical, cost, and time factors; (3) success in the customer environment; and (4) economics, including profitability. Modern **phase management** and **stage-gate processes** provide managers with the tools for organizing and conducting project evaluations effectively. Table 15.4 summarizes suggestions that may help managers to effectively evaluate projects for successful implementation.

A Final Note

Effective project evaluation and selection requires a broad-scanning process that can deal with the risks, uncertainties, ambiguities, and imperfections of data available at the beginning of a project cycle. It also requires managerial leadership and skills in planning, organizing, and communicating. Above all, evaluation team leaders must be social architects in unifying the multifunctional process and its people. They must share risks and foster an environment that is professionally stimulating and strongly linked with the support organizations eventually needed for project implementation. This is an environment that is conducive to **cross-functional** communication, cooperation, and integration of the intricate variables needed for effective project evaluation and selection.

Defining Terms

Cross-functional: Actions that span organizational boundaries.

Minimum attractive rate of return (MARR): The annual net revenue produced on average by projects in an organization as a percentage of their investments. Sometimes MARR is calculated as company earnings over assets.

Net worth: Discounted present value of a future revenue or cost.

Phase management: Projects are broken into natural implementation phases, such as development, production, and marketing, as a basis for project planning, integration, and control. Phase management also provides the framework for *concurrent engineering* and *stage-gate processes*.

Project success: A comprehensive measure, defined in both quantitative and qualitative terms, that includes economic, market, and strategic objectives.

Stage-gate process: Framework for executing projects within predefined stages (see also **phase management**) with measurable deliverables (*gate*) at the end of each stage. The gates provide the review metrics for ensuring successful transition and integration of the project into the next stage.

Weighing of criteria: A multiplier associated with specific evaluation criteria.

TABLE 15.4 Suggestions for Effective Project Evaluation and Selection

1. **Seek out relevant information.** Meaningful project evaluations require relevant quality information. The four categories of variables can provide a framework for establishing the proper metrics and detailed data gathering.
2. **Take top-down look; detail comes later.** Detail is less important than information relevancy and evaluator expertise. Don't get hung up on lack of data during the early phases of the project evaluation. Evaluation processes should iterate. It does not make sense to spend a lot of time and resources gathering perfect data to justify a "no-go" decision.
3. **Select and match the right people.** Whether the project evaluation consists of a simple economic analysis or a complex multifunctional assessment, competent people from those functions critical to the overall success of the project(s) should be involved.
4. **Success criteria must be defined.** Deciding on a single project or choosing among alternatives, evaluation criteria must be defined. They can be quantitative, such as ROI, or qualitative, such as the probability of winning a contract. In either case, these evaluation criteria should cover the true spectrum of factors affecting success and failure of the project(s). Only functional experts, discussed in point 3, are qualified to identify these success criteria. Often, people from outside the company, such as vendors, subcontractors, or customers, must be included in this expert group.
5. **Strictly quantitative criteria can be misleading.** Be aware of evaluation procedures based only on quantitative criteria (ROI, cost, market share, MARR, etc.). The input data used to calculate these criteria are likely based on rough estimates and are often unreliable. Evaluations based on predominately quantitative criteria should at least be augmented with some expert judgment as a "sanity check".
6. **Condense criteria list.** Combine evaluation criteria, especially among the judgmental categories, to keep the list manageable. As a goal, try to stay within 12 criteria.
7. **Communicate.** Facilitate communications among evaluators and functional support groups. Define the process for organizing the team and conducting the evaluation and selection process.
8. **Ensure cross-functional cooperation.** People on the evaluation team must share a strategic vision across organizational lines. They also must sense the desire of their host organizations to support the project if selected for implantation. The purpose, goals, and objectives of the project should be clear, along with the relationship to the business mission.
9. **Don't lose the big picture.** As discussions go into detail during the evaluation, the team should maintain a broad perspective. Two global judgment factors can help to focus on the big picture of project success: (1) overall benefit-to-cost perception and (2) overall risk-of-failure perception. These factors can be recorded on a ten-point scale: -5 to $+5$. This also leads to an effective two-dimensional graphic display of competing project proposals.
10. **Do your homework between iterations.** As project evaluations are most likely conducted progressively, action items for more information, clarification, and further analysis surface. These action items should be properly assigned and followed up, thereby enhancing the quality of the evaluation with each consecutive iteration.
11. **Stimulate innovation.** Senior management should foster an innovative ambience for the evaluation team. Evaluating complex project situations for potential success or failure involves intricate sets of variables, linked among organization, technology, and business environment. It also involves dealing with risks and uncertainty. Innovative approaches are required to evaluate the true potential of success for these projects. Risk sharing by senior management, recognition, visibility, and a favorable image in terms of high priority, interesting work, and importance of the project to the organization have been found strong drivers toward attracting and holding quality people to the evaluation team and gaining their active and innovative participation in the process.
12. **Manage and lead.** The evaluation team should be chaired by someone who has the trust, respect, and leadership credibility with the team members. Further, management can positively influence the work environment and the process by providing some procedural guidelines, charters, visibility, resources, and active support to the project evaluation team.

References

Brenner, M. Practical R&D project prioritization, *Res. Technol. Mgt.*, 37(5): 38–42, 1994.

Bulick, W. J. Project evaluation procedures, *Cost Eng.*, 35(10): 27–32, 1993.

Menke, M. M. Improving R&D decisions and execution, *Res. Technol. Mgt.*, 37(5): 25–32, 1994.

Obradovitch, M. M. and Stephanou, S. E. *Project Management: Risk and Productivity*, Daniel Spencer, Bend, OR, 1990.

Remer, D. S., Stokdyk, S. B., and Van Driel, M., Survey of project evaluation techniques currently used in industry, *Int. J. Prod. Econ.*, 32(1): 103–115, 1993.

Schmidt, R. L. A model for R&D project selection, *IEEE Trans. EM*, 40(4): 403–410, 1993.

Shtub, A., Bard, J. F., and Globerson, S. *Project Management: Engineering, Technology, and Implementation*, Prentice Hall, Englewood Cliffs, NJ, 1994.

Skelton, M. T. and Thamhain, H. J. Concurrent project management: a tool for technology transfer, *Proj. Mgt. J.*, 26(4): 41–48, 1993.

Ward, T. J. Which product is BEST?, *Chem. Eng.*, 101(1): 102–107, 1994.

Further Information

The following journals are good sources of further information: *Engineering Management Journal* (ASEM), *Engineering Management Review* (IEEE), *Industrial Management* (IIE), *Journal of Engineering and Technology Management*, *Project Management Journal* (PMI), and *Transactions on Engineering Management* (IEEE).

The following professional societies present annual conferences and specialty publications that include discussions on project evaluation and selection: American Society for Engineering Management (ASEM), Rolla, MO 65401, (314) 341-2101; Institute of Electrical and Electronic Engineers (IEEE), East 47 St., New York, NY 10017-2394; and Project Management Institute (PMI), Upper Darby, PA 19082, (610)734-3330.

15.2 Critical Path Method

John L. Richards

The purpose of this section is to describe the three-step, iterative decision-making process of planning, scheduling, and controlling with the **critical path method (CPM)**. CPM is a network-based analytical tool that models a project's activities and their predecessor/successor interrelationships. **Planning** is the development of a **work breakdown structure (WBS)** of the project's activities. **Scheduling** is the calculation of **activity parameters** by doing a forward and a backward pass through the network. **Controlling** is the monitoring of the schedule during project execution by **updating** and **upgrading**, as well as the modifying of the schedule to achieve feasibility and optimality using cost duration analysis and critical resource analysis.

Planning the Project

Project planning requires the development of a work breakdown structure, which then becomes the basis for a network model of the project. This model can then be used to evaluate the project by comparing regular measures of performance.

Project Performance Measures

The three common performance measures in project management are *time, cost,* and *quality*. The overall objective is to accomplish the project in the least time, at the least cost, with the highest quality. Individually, these objectives conflict with each other. Thus, the manager must seek an overall solution by trading off among them. Further, since the overall project is defined by the activities that must be done, the overall project duration, cost, and quality will be determined by the individual activity times, cost, and quality levels.

Activity Time-Cost Trade-off

For a specified quality level for a given activity, the manager initially selects that combination of resources (labor, equipment, and material) that accomplishes that particular activity at the least cost. This is the *normal* duration on an *activity time-cost trade-off curve* (Fig. 15.1). Thus, since each activity is to be done at its least cost, the overall project will be done at the least total cost. However, in order to reduce an activity's duration, the activity cost must increase. For example, one can work overtime at premium rates or use more expensive equipment, which increases cost, in order to reduce an activity's duration. *Crash*

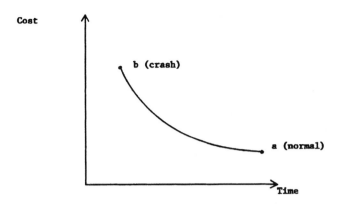

FIGURE 15.1 Activity time-cost trade-off curve. The normal time (point a) is the least cost/longest activity duration. The crash time (point b) is the least activity duration/highest cost.

is the shortest possible activity duration, no matter how high the cost. The inverse relationship between time and cost yields curves with negative slopes.

Activity Interrelationships

There are two possible relationships between a pair of activities in a project network: (1) one must immediately precede the other (*predecessor*), or (2) one must immediately follow the other (*successor*). If there is no predecessor/successor relationship, the activities may be done simultaneously. These predecessor/successor relationships are derived from absolute constraints such as physical/technological, safety, and legal factors; or imposed constraints such as the selection of resources, methods, and financing. The manager should initially incorporate relationships derived only from absolute constraints. Relationships derived from imposed constraints should be added only as necessary to achieve feasibility. This approach to predecessor/successor relationships yields the least constrained project network initially.

The basic predecessor/successor relationship is finish to start with no lead or lag. However, more sophisticated models allow three other types: start to finish, finish to finish, and start to start. In addition, each of the four could have a lead time or a lag time. Thus there are twelve possible ways to describe a particular predecessor/successor relationship.

Work Breakdown Structure

The work breakdown structure (WBS) of a project is the listing of all the individual activities that make up the project, their durations, and their predecessor/successor relationships. It should be the least costly and least constrained method of executing the project, that is, normal durations and absolute constraints only. Therefore, if the schedule resulting from this initial WBS is feasible, then it is also optimal. If, however, the schedule is infeasible because of time and/or resource considerations, then the manager would want to achieve feasibility with the least additional cost. (Scheduling and feasibility/optimality are discussed in later sections.)

There are three approaches to developing a WBS: (1) by physical components, (2) by process components, and (3) by spatial components. *Physical components* model a constructed product (e.g., build wall). *Process components* model a construction process (e.g., mix concrete). *Spatial components* model a use of time or space (e.g., order steel or cure concrete). No matter which of the three approaches is used to define a particular activity in a WBS, each should be described by an action verb to distinguish an activity from an event. (Note: There can be special activities in a WBS that involve time only and no cost, such as curing concrete. There can also be dummy activities for logic only that have no time or cost.)

A project's WBS must be developed to an appropriate level of detail. this means that activities must be broken down sufficiently to model interrelationships among them. Also, a standard time period (hour, shift, day, week, etc.) must be chosen for all activities. An appropriate WBS will have a reasonable number of activities and reasonable activity durations.

TABLE 15.5 WBS of Swimming Pool Construction

Activity ID	Duration	Description	Immediate Predecessors
A101	10	Order and deliver filtration equipment	—
A202	5	Order and deliver liner/piping	—
B301	4	Excavate for pool	—
B202	3	Install liner/piping	A202, B301
B102	2	Install filtration equipment	A101
C301	2	Fill pool	B202
B401	5	Construct deck	B202
C302	2	Connect and test system	C301, B102
B501	3	Landscape area	B401

Example — Work Breakdown Structure. Table 15.5 shows a WBS for constructing a backyard in-ground swimming pool with vinyl liner. Note that a time period of days was selected for all activities.

CPM Network Models

There are two types of network models: activity oriented and event oriented. Both types have nodes connected by arrows that model events (points in time) and activities (processes over time).

Activity-Oriented Diagram

The activity-oriented diagram is also called an *arrow diagram* (ADM) or *activity on arrow* (AOA). The activities are the arrows, and the events are the nodes. Dummy activities (depicted as dashed arrows) may be required to correctly model the project for logic only. Activity identification is by node pairs (node *i* to node *j*). This was the original diagramming method. It is easy to visualize but difficult to draw.

Event-Oriented Diagram

The event-oriented diagram is also called a *precedence diagram* (PDM), *activity on node* (AON), or *circle network*. The activities are the nodes, and the events are the ends of arrows. There are no dummies. All arrows are for logic only, which also allows for modeling the twelve types of activity interrelationships discussed earlier. This diagramming method is easier to draw and well suited to computer use. Although developed after AOA, the AON has become the preferred method.

CPM Network Calculations

There are three steps in the manual analysis of a CPM network that determine the activity parameters. The forward pass determines the *earliest start time* (EST) of each activity. The backward pass determines the *latest finish time* (LFT) of each activity. The other times, the *earliest finish time* (EFT) and the *latest start time* (LST), and the **floats**, *total float* (TF) and *free float* (FF), are then determined from a table. The calculation process is the same for either type of network (ADM or PDM). Before beginning the process, one needs to establish a time convention — beginning of time period or end of time period. The example in this chapter uses beginning of time period, thus the first day is day one. (The end-of-time-period convention would begin with day zero.)

Forward Pass

To determine the EST of an activity, one compares all the incoming arrows — that is, the *heads* of arrows — choosing the *largest* earliest event time. The comparison is made among all the immediately preceding activities by adding their ESTs and respective durations. The process begins at the start node (the EST being zero or one) and proceeds to the end node, taking care that all incoming arrows get evaluated. At the completion of the forward pass, one has determined the overall project duration. The ESTs can be placed on diagrams as shown in Fig. 15.2.

Backward Pass

The backward pass begins at the end node with the overall project duration (which is the LFT) and proceeds to the start node. This process determines the LFT for each activity by comparing the outgoing

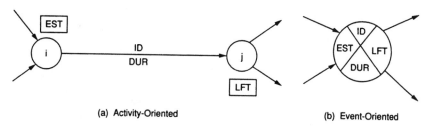

(a) Activity-Oriented   (b) Event-Oriented

FIGURE 15.2 Activity graphical models. the activity identification (ID) and duration (DUR) are from the WBS. the EST and LFT are calculated from the forward and backward passes, respectively.

arrows — that is, the *tails* of arrows — choosing the *smallest* latest event time. The comparison is made among all the immediately succeeding activities by subtracting their durations from their respective LFTs. At the completion of the backward pass one should calculate the original project start time. The LFTs can be placed on diagrams as shown in Fig. 15.2.

Floats

The other two times (EFT and LST) and floats (TF and FF) are determined in a tabular format using the following relationships: (1) EFT = EST + duration, (2) LST = LFT − duration, (3) TF = LFT − EFT or TF = LST − EST, and (4) FF = EST (of following activities) − EFT. Activities with a TF = 0 are on a **critical path**. There may be more than one critical path in a network. If the duration of any critical path activity is increased, the overall project duration will increase. Activities with TF > 0 may be increased without affecting the overall project duration. On the other hand, free float is that amount of total float that can be used by the activity without affecting any other activities. If TF equals 0, then FF equals 0, necessarily. Free float may be some, all, or no portion of total float.

Example — CPM Scheduling. This example continues with the project introduced in Table 15.5. The CPM calculations for the forward and backward passes are shown in Fig. 15.3 for both network types. Table 15.6 lists all the times and floats.

Controlling the Project

CPM-based project management provides the tools to control time and money in a dynamic and hierarchical project environment during project execution.

Managing Time and Money

Computerized project management systems can provide a variety of informational outputs for use in managing a project. These include network diagrams showing activity interrelationships, bar charts showing activity durations, tabular listings showing activity parameters, and profiles showing cash flows or resource utilization during the project.

Hierarchical Management

Project management generally occurs in a multiproject environment with multiple time parameters and multiple managerial levels. Multiple project models integrate cash flow and resource profiles of several projects. Multiple calendar models allow activities to be done at different time frames (for example, some may be done 5 days per week, others 7 days per week). Project management information systems can provide summary information for upper-level management and detailed information for workers in the field.

Managing the Schedule

The project schedule is a dynamic managerial tool that changes during project execution.

Updating the Schedule

As the project proceeds the schedule should be **updated** at periodic intervals to reflect actual activity progress. Such updates can incorporate percent complete, remaining durations, and actual start and end

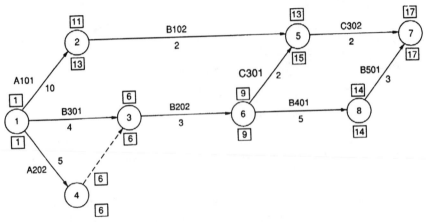

(a) Activity-Oriented Network (ADM or AOA)
Note that activity 4-3 is a dummy and is also on the critical path.

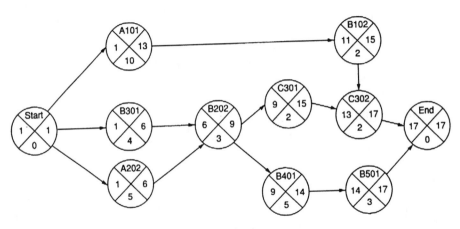

(b) Event-Oriented Network (PDM or AON)
Note that implied start and end nodes are shown with zero durations.

FIGURE 15.3 CPM network calculations (forward and backward passes). The overall project duration was determined to be 16 days ($17 - 1 = 16$).

TABLE 15.6 Activity Times and Floats

Activity ID	Duration	EST	EFT	LST	LFT	TF	FF
A101	10	1	11	3	13	2	0
A202*	5	1	6	1	6	0	0
B102	2	11	13	13	15	2	0
B202*	3	6	9	6	9	0	0
B301	4	1	5	2	6	1	1
B401*	5	9	14	9	14	0	0
B501*	3	14	17	14	17	0	0
C301	2	9	11	13	15	4	2
C302	2	13	15	15	17	2	2

*Activities on critical path.

dates for each activity. The updates can be the basis for evaluating overall project objectives (time and money) and for making progress payments. After any update, a new schedule calculation must be done to determine new times and floats.

Upgrading the Schedule

Any change to an existing schedule either by changing a planned duration or an activity relationship is a schedule upgrade. A schedule upgrade can occur either prior to start of the project, or any time during the project, based on new information. After any upgrade a new schedule calculation must be done to determine new times and floats.

Managing the Floats

A negative total float indicates that the project will overrun the stated completion date. A positive (>0) total float indicates that the project will be completed earlier than the stated completion date.

A particular network will generally have many different total float paths, including negative ones if it is behind schedule, and no zero ones if it is ahead of schedule.

Free float indicates that amount of time an activity can be manipulated without affecting any other activity (and therefore the project as a whole). When managing the floats, one would want to use free float before total float. Once total float is used, the activity becomes part of a new critical path.

The use of the floats can be a difficult contractual issue among the parties. It is a good idea to decide beforehand how the floats can be used. Otherwise the issue may be part of a delay claim at the completion of the project.

Example — Updating and Upgrading.   This example continues with the project introduced in Table 15.5 and originally scheduled in Table 15.6.

1. *Updating.* Assume that it is the beginning of day 6 and that activity B301 is done, and A101 has 9 days left and A202 has 2 days left. These updated durations are used in Fig. 15.3 to recalculate activity times and floats as shown in Table 15.7. Notice that a second critical path has developed and that the overall project duration has been extended two days.
2. *Upgrading.* Now assume that immediately after updating, the duration for activity C302 gets changed to three days and B401 must precede it. These upgrades could be incorporated into a revised Fig. 15.3 (not shown) to recalculate another set of activity times and floats as shown in Table 15.8. Notice that there is now only one critical path and the overall project duration has been extended another day to day 20.

Modifying the Project Schedule

Project planning, scheduling, and controlling is an iterative decision-making process. It is highly unlikely for an initial schedule to be both feasible and optimal in the first iteration. Likewise, it is highly unlikely that the actual project execution will match the original project plan exactly. Therefore, one must know

TABLE 15.7   Activity Times and Floats (Update)

Activity ID	Duration	EST	EFT	LST	LFT	TF	FF
A101*	9	6	15	6	15	0	0
A202*	2	6	8	6	8	0	0
B102*	2	15	17	15	17	0	0
B202*	3	8	11	8	11	0	0
B301	0	—	—	—	—	—	—
B401*	5	11	16	11	16	0	0
B501*	3	16	19	16	19	0	0
C301	2	11	13	15	17	4	4
C302*	2	17	19	17	19	0	0

*Activities on critical path.

TABLE 15.8 Activity Times and Floats (Upgrade)

Activity ID	Duration	EST	EFT	LST	LFT	TF	FF
A101*	9	6	15	6	15	0	0
A202	2	6	8	7	9	1	0
B102*	2	15	17	15	17	0	0
B202	3	8	11	9	12	1	0
B301	0	—	—	—	—	—	—
B401	5	11	16	12	17	1	0
B501	3	16	19	17	20	1	1
C301	2	11	13	15	17	4	4
C302*	3	17	20	17	20	0	0

*Activities on critical path.

how to modify the project schedule in order to achieve feasibility and optimality. The modification process involves either changing activity duration, changing activity relationships, or both.

Cost Duration Analysis

Cost duration analysis (CDA) utilizes activity time/cost trade-off curves (discussed earlier) in order to compress the overall project schedule. The objective is to buy back each time unit in the cheapest possible manner until the desired completion date is reached (feasibility). Only activities on a critical path need be reduced. The others with positive float simply have the float reduced. The problem can become very complex in a large network with multiple critical paths where the incremental additional costs for the activities are different.

Critical Resource Analysis

The approach to **critical resource analysis (CRA)** is different from CDA in that it seeks to extend the overall project duration the least among in order to resolve resource conflicts (i.e., achieve feasibility). CRA can be viewed from one of two perspectives: (1) constrained resources — staying below a specified limit or (2) resource leveling — selecting a constant limit. The solution approach for either is the same. The problem is one of ordering (predecessor/successor relationships) those activities that have resource conflicts during the same time period. The pairwise comparison of all such activities in a large network with many critical resources presents a huge combinatorial problem. The only viable solution approaches are based upon heuristic decision rules. (A simple rule could be that the predecessor activity should be the one with the smaller LST.)

Combined CDA and CRA

Combining CDA and CRA to achieve a feasible and optimal schedule is virtually impossible for all but the simplest networks. Although the CDA problem does have rigorous mathematical solutions, they are not incorporated in most commercial software. On the other hand, the software generally does incorporate heuristic-based solutions for the CRA problem. Therefore, one should use the software in an interactive decision-making manner.

Example — CDA and CRA.   Assume that after the upgrade as shown in Table 15.8, it is decided that the desired completion date is day 19, and also that activities B102 and B401 cannot be done simultaneously because of insufficient labor. Further, assume that B401 can be reduced from 5 days to 3 days at an additional cost of $200 per day and the C302 can be reduced from 3 days to 2 days at an additional cost of $400.

Solution.   The solution approach to this problem is to work two cases: (1) B102 precedes B401, and compress B401 and/or C302 if they lie on a critical path, and (2) B401 precedes B102, and again compress B401 and/or C302. Case 1 would yield an overall project duration of 25 days, and one can readily see that it is impossible to reduce it 6 days (one can only gain a total of 3 days from activities B401 and C302). Case 2 (B401 precedes B102) yields an overall project duration of 21 days, with both B401 and

C302 on the same critical path. One should choose the least expensive method to gain one day — that is, change B401 to 4 days for $200. This yields a project duration of 20 days and an additional critical path. Therefore, reducing B401 another day to 3 days does not get the project to 19 days. Instead, the more expensive activity (C302) must be reduced from 3 to 2 days for $400. The answer to the problem, then, is B401 goes from 5 to 4 days for $200, and C302 goes from 3 to 2 days for $400. Thus, the overall project duration is compressed from 21 to 19 days from a total additional cost of $600.

Project Management Using CPM

CPM was first developed in the late 1950s by the Remington Rand Corporation and the DuPont Chemical Company. Since then, many software manufacturers have developed sophisticated computer-based management information systems using CPM. In addition to performing the CPM calculations discussed in this chapter, such systems can provide data for creating the historical file of an ongoing project, for developing estimating information for a future project, and for performance evaluation of both the project and the participating managers. CPM has even become a well-accepted means for analyzing and resolving construction disputes.

The successful use of CPM as a managerial tool involves not only the analytical aspects discussed in this chapter, but also the attitude that is displayed by those using it in actual practice. If CPM is used improperly as a weapon, rather than as a tool, there will be project management failure. Therefore, successful project management must include positive team building among all project participants, along with the proper application of the critical path method.

Defining Terms

Activity parameters: The activity times (EST, EFT, LFT, and LST) and activity floats (TF and FF) calculated in the scheduling step.

Controlling: The third step in the interactive decision-making process, which monitors the accomplishments of the project by updating and upgrading and seeks feasibility and optimality by cost duration analysis and critical resource analysis.

Cost duration analysis (CDA): Reducing durations of selected activities in the least costly manner in order to achieve a predetermined project completion date.

Critical path (CP): String of activities from start to finish that have zero total float. There may be more than one CP, and the CPs may change after an update or upgrade.

Critical resource analysis (CRA): Sequencing selected activities in such a manner as to minimize the increase in project duration in order to resolve resource conflicts among competing activities.

Planning: The first step in the interactive decision-making process, which determines the work breakdown structure.

Scheduling: The second step in the interactive decision-making process, which determines the activity parameters by a forward and a backward pass through a network.

Update: Changing remaining activity durations due to progress only, then rescheduling.

Upgrade: Changing activity durations and interrelationships due to new information only, then rescheduling.

Work breakdown structure (WBS): Listing of the individual activities that make up the project, their durations, and their predecessor/successor relationships.

References

Antill, J. M. and Woodhead, R. W. *Critical Path Methods in Construction Practise,* 4th ed., John Wiley & Sons, New York, 1990.

Hendrickson, C. and Au, T. *Project Management for Construction,* Prentice Hall, Englewood Cliffs, NJ, 1989.

Moder, J. J., Philips, C. R., and Davis, E. W. *Project Management with CPM, PERT and Precedence Diagramming,* 3rd ed., Van Nostrand Reinhold, New York, 1983.

Further Information

Journal of Management in Engineering and Journal of Construction Engineering and Management, published by the American Society of Civil Engineers.

Project Management Journal, published by the Project Management Institute.

The Construction Specifier, published by the Construction Specifications Institute.

Journal of Industrial Engineering, published by the American Institute of Industrial Engineers.

15.3 Fundamentals of Project Planning

Jeffrey K. Pinto

The key to successful project management is comprehensive planning. While the topic of project planning is wide and far ranging, this section is intended as an introduction to some of the rudiments in the project planning process. Its purpose is to familiarize project managers with some common project planning tools: **scope** management, work breakdown structures, scheduling, and risk management. The section describes the basic elements and principles of utilizing these planning techniques.

Sound project planning forms the cornerstone of any successful implementation effort. When we speak of *planning activities*, we are generally referring to the process of creating a road map that will guide us from the start of our project journey to its intended conclusion. Planning takes into account all aspects of the upcoming implementation: ensuring the availability of needed personnel, acquiring and distributing resources to logistically support the project, and taking the seemingly insurmountable overall project goal and breaking it down into a series of logical and mutually supporting sub-goals, all of which, taken together, serve to achieve the objectives. Planning also serves the purpose of bringing the project team, top management, and the project's clients together early in the process in order to exchange information, learn about each others' priorities, needs, and expectations, and develop the beginnings of a supportive working atmosphere. Finally, project planning encompasses creating a workable project schedule — the practical conversion of project plans into an operational timetable. Through the process of scheduling, project managers are able to attach specific sequences and times to a series of action steps that will bring the project to fruition. As such, scheduling forms the basis for any monitoring and feedback by members of the project team.

There are a number of useful developments that arise out of early and comprehensive project planning and scheduling:

- The interrelatedness of all subprocesses in a project are clearly illustrated.
- The creation of a working team utilizing individuals from different functional departments who realize that a project's success can only result from their mutual cooperation.
- A sufficiently informed completion date for implementing the project.
- Departmental conflicts are minimized through emphasizing task interrelatedness.
- Tasks are scheduled to achieve maximum efficiency.
- "Critical tasks" are identified which could potentially delay the installation and adoption of the project.

Project planning is a catch-all term that comprises a number of different elements. These elements include the development of project scope management, **work breakdown structures (WBS)**, project activity scheduling, and **project risk** analysis. It is through the systematic integration of each of these component parts that project planning is effectively performed. The balance of this section will examine, in a necessarily broad manner, each of these facets of project planning.

Project Scope Management

The first step in initiating a project is documenting and approving the requirements and parameters necessary to complete the project. Project scope management, one of the most important steps in successful project implementation, includes project definition. Project definition requires all relevant project stakeholders (e.g., project team, clients, etc.) to identify, as clearly as possible, the purpose behind the project, including its specific goals and outcomes. Without effective project definition and scope management, different project stakeholders may misinterpret their roles and assignments, even to the point of working at cross-purposes. Scope management has six elements, consisting of (1) conceptual development, (2) scope statement, (3) work authorization, (4) scope reporting, (5) control systems, and (6) project closeout.

1. Conceptual development entails identifying, defining, and documenting the project's objectives to meet the goals of the project's sponsors [PMBOK, 1987]. Effective conceptual development should have the following characteristics as its end product: expected results, time frame to completion, cost estimates, benefit/cost analysis, and identifiable constraints. It should also include preliminary schedules and work breakdown structures.

2. The scope statement serves as the core of future project work. Until the formal scope statement has been ratified by top management and relevant stakeholders both inside and outside the organization, it is impossible to proceed to operationalizing these goals into action steps.

3. Work authorization entails examining the project scope definition, planning documents and contracts in light of project objectives and ensuring that they complement each other [PMBOK, 1987]. Work authorization gives formal approval to the commencement of project work.

4. Scope reporting requires the project manager and team to agree on methods for recording, accumulating, and disseminating project information. Essentially, the reporting component consists of a priori agreement on what information will be collected, how often it will be collected, and how these data will be reported to the project team and other interested parties.

5. Determining the type of and approach to project control systems is a necessary element of scope management. Control systems are the project teams' device for ensuring that plans are regularly updated, corrections noted, and exceptions reported. They serve to upgrade project plans over the development cycle.

6. Project closeout is the process of using historical records on similar past projects and archiving current records for future use. The closeout phase is an opportunity for the project team to learn appropriate lessons, reflect on experiences, and develop action plans for future project efforts.

Work Breakdown Structures

The WBS defines the actual steps that must be taken to complete the project. Its purpose is to subdivide in some hierarchical manner the activities and tasks of the implementation effort. WBS is used as a coding scheme, based on the hierarchy of task relationships, to create a set of "work packages", each of which has a relatively limited time frame and series of objectives. The benefit of WBS is that through using a common terminology regarding the tasks and objectives, members of the project team as well as the organization's controllers have a basis for communicating to each other on the project's status. Put together, they serve to offer a full picture of the catalog of tasks, activities, and **events** that are necessary in order for a project to be successfully implemented. Creating a numerical code to identify each project activity or subroutine is critical in helping cost accountants follow and record project expenditures.

The three steps in the WBS process are

1. Construction of a project activity chart along with codings and specific responsibilities and time frames.
2. Reporting of progress and expenditures to date.
3. Reconciliation of the progress reports (milestones and budget targets) against original projections.

The WBS is a useful tool for managing the implementation process because it serves to illustrate how each piece of the project is tied to the others in terms of technical performance, budgets, schedules, and personal responsiblity. The following steps offer a simplified process for developing a WBS for project management:

1. Break down the project into a series of interrelated (but distinct) tasks. Further, break each of these tasks areas down into subtasks, applying successively finer levels of detail until the project manager and team are satisfied that they have identified all relevant and distinct elements in the project. To determine whether tasks have been sufficiently broken down ask: can each step in the process can be assigned a budget, completion date, and team member responsible for its completion?
2. For each task or subroutine:

   • Develop a task statement that shows the necessary inputs, action steps, responsibilities, and expected end results of the activity.
   • List any additional organizational members who may need to be involved, or whose expertise will need to be tapped.
   • Coordinate with cost accountant to establish code budgets and account numbers.
   • Identify all relevant resource needs: additional personnel, computer time, support funds, facilities, equipment, and supplies.
   • List the personnel (or departments) responsible for addressing each task. It is sometimes helpful to develop a model showing the responsiblity matrix of all personnel on the team.

3. The WBS is reviewed by both upper management and the personnel responsible for performing each activity. This step is important because it accomplishes two goals. First, it assesses whether projections are accurate and achievable. Second, the WBS helps encourage the commitment of the personnel who have been selected to perform these tasks. By asking their approval of the WBS and soliciting their suggestions for its improvement, the project manager can develop a team of individuals who are confident in the planning that has been done and are committed to fulfilling their specific tasks.
4. As the project is carried out, the project manager can conduct periodic updates of the WBS to assess actual progress against the projections. Their comparisons help managers identify problems, solidify estimates of costs or schedule, and begin to take corrective actions that will bring the implementation process back onto course.

Project Scheduling

Project scheduling typically consists of establishing time estimates for individual activities that are then transferred to a master schedule activity network. Probably the most common methods for scheduling are program evaluation and review technique (**PERT**) and the **critical path method (CPM)**. The techniques are quite similar and are often referred to together as PERT/CPM. While both techniques are used for activity scheduling, PERT is chiefly concerned with the time element of projects and makes use of probabilistic estimates to determine the likelihood of each of the project's activities being completed by a specific date. The CPM, on the other hand, uses deterministic activity time estimates and is employed to control both time and cost aspects of a project. This trait is important because it gives project managers the ability to understand time/cost trade-offs. Using CPM, certain activities can be "crashed" (expedited) at a calculatable extra cost in order to speed up the implementation process.

The benefit of both techniques is that they allow the project manager to determine the *critical path* — activities that cannot be delayed while identifying those activities with *slack time* that can be delayed without exerting pressure on the projected completion date. **Slack** time refers to additional time built into the schedule along the noncritical paths. Obviously, the critical path has *zero slack time* associated with it. However, PERT, as a scheduling technique, is no longer used to the degree it was once because far more elaborate and precise methods are currently available. For example, the enhanced power and speed of personal computers now make it possible to employ monte carlo simulation techniques to time estimation, in which project managers can simulate the project schedule, altering variables in order to determine optimal scheduling. The advent of the personal computer has been an immeasureable help to project managers in terms of time estimation and has consequently rendered many techniques such as PERT increasingly obsolete.

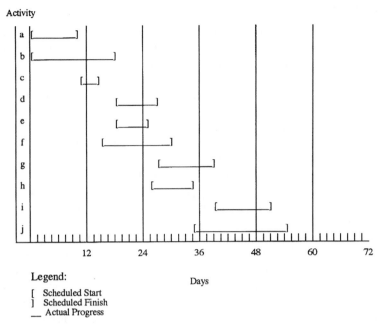

FIGURE 15.4 Sample Gantt chart.

Another scheduling technique is the *Gantt Chart* (Fig. 15.4). Gantt charts are one of the oldest and still most useful of scheduling tools. Originally developed in 1917 by Henry Gantt, the Gantt chart shows planned and actual progress for a number of project tasks displayed on a horizontal time scale. One of the strengths of the Gantt chart is that it is easy to read and quite accessible to all types of managers — those with considerable training to novices. Gantt charts give managers the opportunity to compare actual progress against the planned schedule in a manner that PERT/CPM does not allow. As a result, project managers can look at the chart at any point in time to see if there exist any major discrepancies between actual and planned activity completion times. Where these problems do occur, managers are able to reallocate resources, alter sequencing, or delay various activities until the other events have occurred.

There are some important advantages in using Gantt charts. First, they are easy to read and interpret. They are laid out in a manner that makes them highly accessible to those interested in determining the status of the implementation at any point in time. Please note, however, that, once major resources reallocations or major changes are made to the activity schedules, the chart must be reconfigured. A second advantage is that they are easy to construct. Project managers can quickly master the logic behind Gantt charts. Finally, Gantt charts are easy to update. They use a "floating date" point so that those reviewing the project's progress are able to immediately assess where the implementation stands at any point in time.

One point that bears mention is that it was not my intention to include both PERT/CPM and Gantt chart methodologies as competing scheduling techniques. Rather, these tools are most effective when they are combined and applied in a complementary fashion. PERT/CPM has its strengths: it details the critical path, activities and events that can be accomplished in parallel (simultaneous) fashion, identifies sources of slack time, and the appropriate sequencing of activities. Likewise, Gantt charts have many advantages: they are easy to read, provide good graphic detail concerning activities that are being completed on time, and illustrate the impact of the activities not being accomplished in a timely manner. Used in tandem, they are powerful techniques for implementation scheduling and control.

Risk Management

The final element in developing comprehensive project plans is to engage in risk management. Project risk management requires project managers to be proactive and prepare for possible events in advance, rather than reacting to them after they have occurred. Risk management involves three elements: (1) risk identification, (2) impact analysis, and (3) response planning. Risk identification must be sub-divided into two distinct forms of risk: business risk, which is the likelihood of profit from any business endeavour, and insurable risk [PMBOK, 1987]. Insurable risk differs from business risk in that it is only negative in nature; that is, there is no chance for profit from insurable risk, only loss. In assessing the probability of both types of risk, it is important to engage in environmental scanning, seeking to determine as early as possible the relevant risk factors that can derail the project.

The next step in risk management consists of impact analysis. While it is appropriate to determine the potential sources of risk in the environment, their discovery is only half the battle. Equally important is the effort to assess the impact that each intervening event could have on project success. Impact analysis, in its simplest form, typically consists of breaking down each project activity or external intervening event into categories of "high risk", "moderate risk", and "low risk". Further, it determines the likelihood of each source of risk to develop a clearer picture of the status, or riskiness of the project.

The final component of risk management is response planning. Associated with each form of project risk should be complementary contingency plans to address problems if and when they are discovered. Research has demonstrated that the biggest single cause of project failure is the absence of adequate contingency plans to bring a project back on track after it has experienced difficulties [Pinto and Mantel, 1990]. Response planning is the project management equivilant of an "ounce of prevention" through early corrective action rather than a pound of belated cure.

In summary, risk management requires project managers to consider answers to the following questions:

1. What is likely to happen (the probability and impact)?
2. What can be done to minimize the probability or impact of these events?
3. What cues will signal the need for such action (i.e., what clues should I be actively looking for)?
4. What are the likely outcomes of these problems and my anticipated reactions?

Summary

Clearly, project planning is a monumental undertaking, so great, in fact, that this section only covers the introductory material of project management. The planning techniques discussed are only good if they are used effectively. In this section I have offered the basics of project planning. I urge any manager to thoroughly read up on these procedures. For that purpose, I have included a list of recommended readings on planning and scheduling at the end of this section. These bibliographic selections offer more in-depth discussion of these techniques, often with sample problems for practice.

Defining Terms[1]

Activity: A task or series of tasks performed over a period of time.

CPM (critical path method):   A scheduling technique that employs deterministic time estimates for project activities.

Critical path: The series of interdependent activities of a project, connected end to end, that determines the shortest total length of the project.

Event: An event is an identifiable single point in time on a project.

---

[1]Unless otherwise identified, all terms are defined according to the *Project Management Institute Body of Knowledge*, PMI, Upper Darby, PA, 1987.

Network diagram: A schematic display of the sequential and logical relationships of the activities that comprise the project.

Path: The continuous, linear series of connected activities through a network.

PERT: Program evaluation and review technique. A scheduling technique that employs probablistic time estimates for project activities.

Project risk: The cumulative effect of the chances of uncertain occurrences which will adversely affect project objectives. It is the degree of exposure to negative events and their probable consequences. Project risk is characterized by three factors: risk event, risk probability, and the amount at stake.

Scope: The work content and products of a project or component of a project. Scope is fully described by naming all activities performed, the resources consumed and the end products which result, including quality standards.

Slack (or Float): The extra time associated with noncritical activities on a network activity chart. Slack time is associated with activies not on the critical path.

Work Breakdown Structure (WBS): A task-oriented "family tree" of activities that organizes, defines, and graphically displays the total work to be accomplished in order to achieve the final objectives of the project.

References

Meredith, J. R. and Mantel, S. J., Jr. *Project Management: A Managerial Approach, 2nd ed.*, John Wiley & Sons, New York, 1989.

Pinto, J. K. and Mantel, S. J., Jr. The causes of project failure, *IEEE Trans. Eng. Mgt.*, EM-37: 269–276, 1990.

PMBOK (Project Management Body of Knowledge), PMI, Upper Darby, PA, 1987.

Further Information

Cleland, D. I. and Kocaoglu, D. F. *Engineering Management*, McGraw-Hill, New York, 1981.

Lavold, G. D. Developing and using the work breakdown structure, In *Project Management Handbook, 2nd ed.*, D. I. Cleland and W. R. King, Eds., 302–323, Van Nostrand Reinhold, New York, 1988.

Moder, J. J. Network techniques in project management, In *Project Management Handbook, 2nd ed.*, D. I. Cleland and W. R. King, Eds., Van Nostrand Reinhold, New York, 324–373, 1988.

Moder, J. J., Phillips, C. R., and Davis, E. W. *Project Management with CPM, PERT, and Precedence Diagramming, 3rd ed.*, Van Nostrand Reinhold, New York, 1983.

Weist, J. D. and Levy, F. K. *A Management Guide to PERT/CPM, 2nd ed.*, Prentice-Hall, Englewood Cliffs, NJ, 1977.

Woolsey, G. The fifth column: Production scheduling as it really is, *Interfaces*, 12: 115–118, 1982.

15.4 Controlling Cost and Schedule

Howard Eisner

The capability to **control** both cost and schedule is a critical aspect of successful project management. The matter of project control is often viewed as one of the four elements of project management, namely, (1) planning, (2) organizing, (3) directing, and (4) controlling [Kerzner, 1989]. At times, control is considered attainable by appropriate monitoring and directing so that a second view of the key elements of project management consists of planning, organizing, directing, and monitoring [Eisner, 1997]. Cost and schedule problems can occur in related or unrelated ways, and at times such problems are consequences of shortfalls in performance. Whatever the source, the project manager must be able to exercise the necessary control so as to satisfy overall cost budgets and time allocations that almost always serve as constraints for a project.

Cost Control

Many of the notions of cost control have roots that go back to U.S. government programs and systems and are related to planning, programming, and budgeting systems (PPBS), systems analysis methods, "should cost" and "design to cost" principles, the formulation of **cost-estimating relationships (CERs)**, and even the use of commercial-off-the-shelf (COTS) systems and components. In the latter case, the presumption is that the use of readily available commercial products will lead to reduced costs. Other actions, such as competitive subcontracting, are generally believed to have similar effects. Thus, cost control is often thought of as synonomous with ways and means of reducing costs.

Budgeting

The first step of cost control of a project is that of budgeting. Every project should have a budget that is the maximum amount of funding that is allocated to the execution and completion of the project. The project budget contains several parts, sometimes called line items. The conventional elements of project cost include the cost of direct labor, fringe benefit costs, overhead (OH) costs, other direct costs (ODCs) such as materials and supplies, consultants and subcontractors, as well as general and adminstrative (G & A) costs. Dollars reserved for profit are normally not considered in the category of cost elements. Most of these costs can be broken down into subordinate costs. For example, fringe benefits include sick leave, holidays, vacation, insurance, taxes, pension, and possibly other costs. In the same vein, both OH and G & A costs have component parts [Eisner, 1997]. From the point of view of cost control, fringe benefits, OH, and G & A are not normally under the jurisdiction of the project manager. They are broader costs borne by the enterprise, and thus cost control is exercisable at the officer (vice president or president) levels. The project manager therefore tends to focus on control of the direct labor and the ODCs.

Generic Cost Control

One might consider five fundamental steps in exercising project cost control [Kezsbom et al., 1989]. The first requires a cost measurement system at an appropriate level of detail. Second, the system must actually make the measurements of actual costs and compare them to planned costs. Third, this information must then be presented in an appropriate form to several project personnel, including the project manager. Fourth, the situation must be analyzed, and new forecasts must be made, e.g., costs to complete. In this step it is usually important to consider how to reduce future costs. Finally, the fifth step is to actually implement corrective action so that overall costs goals and budgets are met. In general terms, these same steps apply to controlling schedule.

Tracking and Controlling Project Expenditures

Control of project costs is most often achieved by tracking these costs every month, and at times more frequently, and comparing recent as well as cumulative expenditures to planned expenditures. One can then set up a spreadsheet in which the rows are the elements of cost or expenditure, and the columns list, as a mimimum, the following aspects of the costs: (1) budgeted cost, cumulative to date, (2) actual costs, cumulative to date, (3) difference between budgeted and actual, cumulative, (4) budgeted cost, this month, (5) actual cost, this month, (6) difference between budgeted and actual, this month, (7) budgeted cost to complete, (8) current estimate of cost to complete, (9) difference between budgeted and current estimate of cost to complete, (9) budgeted cost of entire project, and (10) current estimate of cost of entire project. Wherever actual expenditures exceed planned or budgeted expenditures, the project manager must look very closely at how to bring the project costs back into line.

The cost elements listed as the rows of the cost expenditure spreadsheet are often broken down further by task and subtask, or by entries in a work breakdown structure (WBS) [Frame, 1987]. The WBS enumerates all the work to be done in the project and should correlate with the tasks and subtasks, if indeed they are different. This additional breakdown and reporting of costs give the project manager insight into where the problems might lie, especially for a large project.

Earned Value Analysis (EVA)

Earned value analysis (EVA) is a well-recognized method of analyzing both cost and schedule variances [Eisner, 1997; Kezsbom et al., 1989]. EVA is based upon developing and working with cost curves defined as follows: (1) budgeted cost of work scheduled (BCWS), (2) budgeted cost for work performed (BCWP), and (3) actual cost of work performed (ACWP). These are curves since they are normally plotted as a function of time.

The cost variance (CV) is then defined as $CV = BCWP - ACWP$, and the schedule variance may be calculated as $SV = BCWP - BCWS$. With respect to the cost variance, if $ACWP > BCWP$, then the project is clearly overexpended, and the cost variance will be calculated as a negative number. In a similar fashion, if the $BCWS > BCWP$, the schedule variance will be determined to be a negative number, implying that the project is behind in schedule. The EVA formalism further extrapolates linearly to new costs (ECAC) and times (ETAC) at completion, using the following relationships: $ECAC = (ACWP)(BAC)/(BCWP)$ and $ETAC = (BCWS)(TAC)/(BCWP)$ where BAC is the original budgeted cost and TAC is the original time at completion. This linear extrapolation has the benefit of being a simple calculation but has its distinct drawbacks. An alternative approach is to reestimate cost and schedule to complete by obtaining new inputs from the project staff. This will often reveal definite nonlinearities in both cost and schedule.

The Authority Matrix

A practical and often-used method of cost control is the Authority Matrix. This controls cost by setting limits on the costs that can be expended by different people on a project or in an organization. Depending upon position and level of responsibility, the project establishes maximum values of cost that can be spent, taking into account project needs and plans as well as schedule. In such a situation, the lead software engineer on a project, for example, may not be able to spend more than a prescribed amount for a given purpose, in a given month, unless approved by the engineer's boss, and so on up the chain of command.

Business Process Reengineering

Another subject that is related to cost control is **business process reengineering (BPR)**. This subject and process came upon the scene in 1993 [Hammer and Champy, 1993] and promised to be a reliable method for improvements in cost, time, and performance. With respect to cost, the rationale is that, by reengineering a given process, within or outside the domain of project management, it is likely that significant cost reductions can be achieved, if such is the main focus for the reengineering. A similar argument applies as well to schedule reduction and control. Cost reductions by downsizing have been attributed, at times, to the adoption of BPR.

Some of the principles espoused by the originators of BPR may be identified as follows: (1) reengineering requires a fundamental rethinking and radical redesign of processes, (2) it focuses on dramatic rather than marginal or incremental improvements, (3) workers are empowered to make important decisions and changes, (4) steps in a redesigned process are performed in a more natural order, (5) processes may have several versions, (6) work is carried out where it makes the most sense, (7) checks and controls are reduced in favor of an improved process, and (8) performance measures change from looking at activities to focusing on results. These principles of BPR may be considered at both the project and the organizational levels within an enterprise.

Cost Risk Reduction

Reductions in the risk associated with possible cost overruns may be achieved by a systematic examination of such factors as [Eisner, 1988] (1) key cost drivers within a project or systems, (2) focusing on low-cost design alternatives in distinction to "gold plating", (3) proving performance through the use of simulation and modeling techniques, (4) maximum use of commercial-off-the-shelf (COTS) equipment, (5) early breadboarding, prototyping, and testing, (6) utilizing design-to-cost approaches, and (7) obtaining multiple bids from subcontractors.

Schedule Control

Schedule control involves all analysis and actions that are taken in order to assure that a project is carried out within a prescribed period of time. Generic schedule control follows approximately the same steps listed above with respect to cost control, i.e., a measurement system, the measurement of actual times vs. planned times, appropriate presentation, analysis and forecasting, and the implementation of corrective action.

Schedule formats that are widely used include the Gantt Chart and the **PERT (program evaluation review technique)** or CPM (critical path method) diagram, with the latter two being very similar (see "Critical Path Method" in this handbook and Cleland and King [1983]. The time required to complete each designated activity is estimated, and the relationship between an activity and a task or subtask or work breakdown element is precisely defined. Activities have beginning and end points that are usually called events. Events must also be defined with precision so that one knows whether or not they have occurred. Under the PERT or CPM formalism, the dependencies between all activities are established, leading to a network representation. This can occur either before or after the individual time estimates are made. Activities in a network that precede and lead into other activities must be carried out first. This type of network dependency is normally not part of the Gantt chart procedure and is considered to be a central feature of the PERT and CPM approaches.

Estimating Activity Times

Although the PERT or CPM method appears to be straightforward, problems often arise in the matter of estimating the times for each of the activities. In these formalisms, there is an opportunity to estimate optimistic, pessimistic, and most likely times, leading to the computations of expected values and variances for these activities [Eisner, 1997]. However, times are usually estimated with some implicit assumption as to the number and type of people available to carry out the activities. If such an assumption turns out to be incorrect (the number and type of people are not available when required), activity time estimates can be grossly inaccurate. Thus, the first opportunity to control a schedule is to try to assure that the input data to the schedule is as accurate as possible. This can mean taking at least the following two steps: (1) verifying the implicit people assumptions cited above and (2) having several people make the original estimates, independent of one another. The former is likely to reveal difficulties with the original assumptions, leading to their reconciliation. The latter may show disparate results between the independent estimates, also leading to discussions and a final reconciliation.

The Critical Path in Schedule Networks

The critical path is the longest path through a PERT or CPM network schedule. This means that, if one or more activities on the critical path actually take more time than the original estimate, a slippage in the overall schedule for the project will occur. Such is not necessarily the case for paths off the critical path since there may be sufficient slack in these paths to make up for such a discrepancy. This attribute of the critical path makes it the focal point for exercising schedule control. Once the critical path is calculated and established, special efforts need to be made to verify the original time estimates for all activities on that path. This special attention continues as the project moves forward, always comparing actual times vs. estimated times. Immediate corrective action is normally resorted to at the first signs of a discrepancy in the above comparison. A framework for examining a possible "schedule variance" was discussed above under the topic of earned value analysis (EVA), including the notion of extrapolating to a new estimated time to complete the project.

Probabilistic Schedule Estimates

As alluded to above, the use of three input time estimates for each network activity creates the opportunity to look at the critical and near-critical paths from the point of view of quantifying uncertainty. Using basic principles of the addition of means and variances for sums of independent random variables, it is possible to calculate the mean and variance of the estimated project end date. Using these estimates, it is further possible to estimate the likelihood (probability) that the overall project will exceed some arbitrary date. If the Central Limit Theorem is invoked, then the distribution associated with the project completion time may be assumed to be Gaussian [Parzen, 1960], in which case a table look-up, together with the estimate of the mean and standard deviation, will reveal the desired probability estimate. This procedure is normally usable only with very large networks involving hundreds of activities.

Schedule Risk Reduction

A variety of actions may be considered in order to try to reduce the risk associated with meeting schedule constraints. These include [Eisner, 1988] (1) utilizing the most productive personnel on time-critical paths and activities, (2) overtime work with related incentives, (3) shifting to parallel instead of serial activities wherever possible, (4) having parallel teams working on critical activities with the expectation that at least one will be able to meet schedule, and (5) reducing performance goals and expectations where such an approach is feasible. The latter is a countermeasure against the inappropriate tendency, at times, to "gold plate" a product.

Relationship between Cost and Schedule

It should also be recognized that there is normally a relationship between cost and schedule, leading often to the consideration of both when control alternatives are being examined. For example, cost reductions may lead to time increases, and time reductions may be achievable only with the additional expenditure of costs. The former may occur when, in an attempt to reduce cost, personnel are taken off the project and reassigned elsewhere. In the latter case, personnel may indeed be added to a project in an attempt to compress timelines. In addition to this type of relationship, project direct labor costs are generally derivable from the initial consideration of activity time estimates together with the eventual assignment of personnel to the activities. Person months or person weeks are directly convertible into direct labor costs.

Using Teams for Cost and Schedule Control

A method that is perhaps not used widely enough in cost and schedule control is that of utilizing a team approach to the issues and problems. This has at least two dimensions: (1) using independent teams to develop initial estimates of time and cost and (2) using teams to deliberate about problems of excessive costs and schedule overruns. The former addresses the matter of critical input data upon which later analyses depend. The latter suggests that a top-level investigation of problem areas by one or more teams is likely to yield more effective results. Team problem solving by interconnected groups of people has been shown to be especially productive. This can be further explored under such topics as integrated product teams, situation analysis, concurrent engineering, and integrative management [Eisner, 1997]. Group processes have also been assisted through the use of group decision support systems (GDSSs).

Reserves

The project manager, as well as bosses up the chain of command, often find it desirable to set aside reserves in both the cost and time dimensions to try to assure that the target budget and schedule will be achieved. This has proven to be a useful procedure on particularly important projects.

Software for Cost and Schedule Control

Cost and schedule control can be greatly facilitated through the use of rather powerful, and relatively inexpensive, project management software tools. Features of such tools include such items as (1) Gantt charting, (2) PERT networking, (3) the work breakdown structure, (4) a task responsibility matrix, (5) personnel assignments and loading profiles, (6) resource leveling, (7) cost slices, comparisons and aggregations, and (8) various types of standard and user-defined output reports [Eisner, 1997]. Software tools provided by different vendors differ on how each of the above is implemented, as is the case with respect to such aspects as ease of use, speed of operation, hardware requirements, multiproject handling, capacity, product support, and the cost of the product itself. The latter has generally run, over a number of years, from a low end of about $100 to a high end of the order of $3500. Low-end (in cost) packages include Project Scheduler, Microsoft Project, Time Line, Harvard Project Manager, Superproject, and others. At the high end, vendor offerings have included Primavera, Promis, Qwiknet, Open Plan, Viewpoint, and others. The government has played a role with respect to project management support tools by evaluating them and publishing results both widely and inexpensively [Berk et al., 1992]. The low cost and significant capability represented by these tools make their use virtually mandatory for even relatively small projects.

Defining Terms

Business process reengineering (BPR): A process improvement perspective and set of methods, introduced by Hammer and Champy in 1993 (see References) that gained wide acceptance in business and government.

Control: A combination of monitoring and directed action that leads to changes in behavior, one of the four elements of project management, also referred to as an aspect of cybernetics, as in control and communication in the animal and machine (defined by cyberneticist Norbert Wiener).

Cost estimating relationship (CER): A mathematical relationship that is derived from empirical data whereby one or more independent variables are related to cost, the dependent variable.

Earned value analysis (EVA): A formal project management method of calculating cost and schedule variances and extrapolating to estimate cost and time at completion of the project.

PERT (program evaluation and review technique): A network-based method of scheduling, introduced in a key Navy program in the 1950s, that has become a standard tool for managing large, complex projects, contained in most project management software packages, a foundation for certain types of stochastic network formalisms.

References

Berk, K., Barrow, D., and Steadman, T. *Project Management Tools Report*, Software Technology Support Center (STSC), Hill Air Force Base, UT 84056, March 1992.

Cleland, D. I. and King, W. R. *Systems Analysis and Project Management*, McGraw-Hill, New York, 1983.

Eisner, H. *Essentials of Project and Systems Engineering Management*, John Wiley & Sons, New York, 1997.

Eisner, H. *Computer-Aided Systems Engineering*, Prentice-Hall, Englewood Cliffs, NJ, 1988.

Frame, J. D. *Managing Projects in Organizations*, Jossey-Bass, San Francisco, 1987.

Hammer, M. and Champy, J. *Reengineering the Corporation*, HarperCollins, New York, 1993.

Kerzner, H. *Project Management — A Systems Approach to Planning, Scheduling, and Controlling*, Van Nostrand Reinhold, New York, 1989.

Kezsbom, D. S., Schilling, D. L., and Edward, K. A. *Dynamic Project Management*, John Wiley & Sons, New York, 1989.

Parzen, E. *Modern Probability Theory and Its Applications*, John Wiley & Sons, New York, 1960.

Further Information

The *IEEE Transactions on Engineering Management* and the *IEEE Engineering Management Review*, both provided by the Institute of Electrical and Electronics Engineers (IEEE), with Headquarters at 345 East 47th Street, New York, NY 10017.

The *Engineering Management Journal*, produced by the American Society for Engineering Management (ASEM); Member Support Department and Journal can be contacted at P. O. Box 820, Rolla, MO 65402.

Harvard Business Review, a distinctive journal on a wide variety of management issues; 60 Harvard Way, Boston, MA 02163.

The American Management Associations (AMA) and publishing arm division AMACOM, 135 West 50th Street, New York, NY 10020.

15.5 Feasibility Studies

E. Lile Murphree, Jr.

The term *feasibility studies* comprises all the *objective and subjective analyses*, including economic and financial projections, legality of the proposed project, its environmental impact, including biological and social impacts, and all other aspects of the project's life cycle, including its ultimate disposal or recycling, that impinge on the decision that the project, as proposed, is *feasible*. The distinction between *feasibility* and *desirability* is made, as is the point that a purely *feasible* project may, by dint of its undesirability from, say, a social point of view, be made practically *infeasible*.

The present discussion focuses primarily on the economic viability of a project, and touches only lightly on legal and environmental issues, both areas being outside the scope of this section. References are given for guidance on preparing environmental impact analyses; legal issues are not addressed in any depth at all here, and the reader is urged to seek competent legal advice on any proposed project that may be subject to regulation by government at any level, such as many construction activities.

At its essence, the feasibility study is the process of measuring the proposed project against those appropriate criteria that have the potential of influencing the *go-no-go* decision for the project. The procedures presented in detail in this section are those commonly used to assess the economic viability of a project that requires a capital investment in one or more phases and returns a stream of income over the expected life of the project, followed by a capital cost, or a lump sum income, at disposal at the end of the project's life cycle.

Every project, prior to the final decision regarding implementation, should be subjected to careful scrutiny and measured against all appropriate measures of advisability of proceeding: Is the project as proposed doable, i.e., is it possible, given the state of all the requisite resources, including know-how, to actually do the project? In view of the primary and secondary effects the project will have on others, including the physical, biological, and social environment, is it desirable to do the project? Finally, assuming the proposed project has passed the foregoing hurdles of possibility and desirability, is the project as proposed economically/financially feasible, when measured against some appropriate standard? These aspects of project assessment are considered in this section.

Definitions

In order for us to make sense of a set of complex concepts relating to the viability of projects, we need some definitions of terms and agreement on how the concepts relate to projects and to each other. Within this section, the following definitions hold:

Possible: There exist realizable scenarios in which this project would be feasible, but not necessarily those that have been considered in the present study.

Impossible: There are no known realizable scenarios in which the project would be feasible.

Desirable: The project has subjective (e.g., social, humane, artistic, altruistic) features that make it desirable, independent of the scenario chosen for its implementation, and independent of its possibility under any realizable scenario.

Undesirable: The project is undesirable under some scenario of implementation and undesirable independently of its possibility.

Feasible: Reasonably likely for the project as proposed to succeed.

Infeasible: Not reasonably likely for the project as proposed to succeed.

Success:	The project will likely meet the feasibility standards set.
Failure:	The project will likely not meet the feasibility standards set.
Standards:	Net value is positive with
	time horizon chosen
	discount rate chosen
	inflation rate chosen
	that is, the mathematical model, as constructed, indicates a positive discounted cash flow, and there are no other factors considered that would render the project infeasible.
Scope:	The scope of the analysis is a single project; no alternative project is compared with the project under consideration.

Possibility, Desirability, and Feasibility

Under the definitions stated above, a proposed project may be *possible* or *impossible*, *desirable* or *undesirable*, *feasible* or *infeasible*, depending on the scenario in which it is considered. A project that is impossible cannot be feasible, although it may be desirable. An example is a project to eradicate cancer from humanity. This is clearly impossible, since we do not know how to do it, and therefore infeasible, though its desirability is undeniable. A project to turn the Washington, DC, waterfront into a deep water industrial seaport is technically possible, although surely undesirable from society's point of view, and probably financially infeasible as well, considering the high cost of land in downtown Washington, DC, and the competition for revenues from existing ports nearby, e.g., Baltimore. A project to build low-cost housing for the urban poor in Chicago may be possible and desirable, but infeasible in some, but not necessarily all, financial scenarios.

Among the factors that influence desirability is the future legality of the project. The possibility and feasibility of a proposed casino are attenuated if the local government is petitioned by citizens' groups to outlaw the operation of gambling establishments. Political instability and social policy can influence the desirability of a project: a pipeline across a politically unstable country on the brink of civil war is undesirable whatever its present feasibility. Predictability of continuing management of the project in the future, its ownership, and control of its operation and assets are vital to maintaining desirability of the project. Actual feasibility fades in view of these overriding considerations.

While we are here primarily concerned with the objective feasibility of a proposed project, experience has shown that it is prudent first to determine a project's possibility and its desirability before proceeding to the relatively more expensive exercise of predicting its feasibility. To many, it is both possible and desirable to replace the aging Woodrow Wilson Bridge over the Potomac River south of Washington, DC. Determining its feasibility is not a trivial undertaking, the outcome of which will depend on the scenarios presented for analysis. For any possible and desirable project, there are many facets to the question of feasibility. Some of the more obvious are discussed below.

Feasibility Factors

It is useful to consider a feasibility study an exercise in determining *in*feasibility, for, as soon as we have established that a proposed project is infeasible, we need look at that particular implementation scenario no further; that analysis is complete, with a negative report. Following is a structured approach to determining feasibility in steps, each of which, if resulting in the negative, allows us to abandon the study with an infeasible verdict.

Technical Feasibility

In the context of this section, it is useful first to determine the technological feasibility of the proposed project: Is the project realizable with the technology available now or firmly expected at the time of planned implementation? Many software projects depend upon the existence of other software products; for example, certain implementations of popular word processing programs cannot function outside one of the versions of Microsoft's Windows operating systems. Indeed, an important feature of the software industry is the availability of incomplete operating systems to software developers months in advance of the final, consumer versions, allowing the developers to comfortably anticipate a technological environment for their products long in advance of its actual, full completion. Without this assurance of future technology, the software industry would be forced to wait until shipment of enabling systems before they could with confidence begin work on applications software, thus attenuating the pace at which the industry now grows and changes.

Legal Feasibility

Any realizable project must be legal under the laws of the relevant jurisdictions, federal, state, and local. Otherwise, the project, as conceived, cannot be feasible, even though it may otherwise be possible and even desirable, to some constituency. A project to grow marijuana for recreational purposes is illegal in the United States, and therefore not legally feasible. The same project, with the crop destined for research, may be legal in some jurisdictions, and therefore legally feasible. Legal infeasibility terminates the study with a negative result.

Environmental Feasibility

Environmental feasibility is established by means of environmental impact studies, the details of which are beyond the scope of the present section. By law, any project whose implementation will disrupt the biological, physical, or social environment beyond established thresholds must be the subject of environmental impact studies, summarized in an environmental impact statement. The federal Environmental Protection Agency and its state-level counterparts have set standards that control the go-no-go decisions with respect to environmental impacts of proposed projects. Impact studies are nontrivial undertakings, often costing hundreds of thousands of dollars, and virtually any major construction project such as a highway segment, airport runway addition, or bridge replacement would have to pass this hurdle on the way to ultimate approval. Many major infrastructure projects have been derailed permanently by a finding of adverse impact to the environment. Notable among these is the dam on the Tennessee River aborted because of its potential impact on the welfare of the endangered snail darter, a small, rare fish found at the proposed site. Environmental impact studies have evolved into a high art; references to enable the reader to delve deeper into this area of feasibility studies can be found at the end of this article [Jain et al., 1981, 1993].

Environmental impacts are not limited to biological. Major freeway construction has been halted in Boston and San Francisco by organized resistance by political action groups angered and dismayed by the destruction of the social fabric of old, established neighborhoods by the arteries of concrete. Construction is routinely halted when evidence is unexpectedly unearthed of ancient cultures, until archeologists have had time to excavate and remove artifacts and human remains. Political action has put on hold for years the destruction of giant sequoias in California and clear cutting of timber in Oregon. Runoff and removal of topsoil from home sites is rigidly controlled in Virginia and Maryland, to minimize physical changes to the land and to lower the levels of sediment in waterways.

If biological, social, and physical impacts to the environment do not scuttle the project, then we can begin to assemble data for economic and financial studies. Feasibility studies for any project that will exist over a substantial duration of time present the analyst a host of problems with data, most notably the prediction of appropriate values for the future. Equally challenging is the prediction of technological change and societal changes, both of which impact the future viability of projects, which may, in today's world, appear entirely feasible. Future technological developments, which cannot be quantified or even known with any degree of certainty, have the potential of making obsolete any project we may implement today. Obsolescence can render economically noncompetitive a process or a plant based on a technology that has been superseded by new developments. For example, a computer chip fabricating facility based on optical semiconductor lithography will surely be obsolesced by X-ray lithography 5 years from now. Societal changes, in expectations of technology users, made long-play, vinyl records unsalable in the face of compact disks. The introduction of automobile air conditioning did not so much make obsolete something that already existed, but rather raised the expectations of the motoring public, so that un–air-conditioned automobiles are seldom sold now in the United States and the buyer runs the risk of difficulties in selling such an automobile at trade-in time.

Basic Feasibility Analysis

One basic criterion for deciding if a proposed project is feasible financially is whether or not its net present value (NPV) is positive [Steiner, 1989]; that is, if the discounted value of the incomes over the life of the project is greater than that of the investments and costs over time, the project is considered feasible. This criterion is only one of several possible criteria, such as internal rate of return (IRR), and relates to a *go–no-go* decision based on a particular scenario. In other words, a project that is infeasible under one financial scenario may very well be feasible under an alternate scenario. Here, we present the mechanics of performing the analysis on a *given* scenario, but will not attempt to address the creative act of generating scenarios. We will also ignore the possibility of a net present worth equaling zero; in such a case, the analysis is indifferent to the decision and therefore inconclusive.

The analysis of financial feasibility requires that the analyst estimate the relevant costs and the expected incomes from the project over the expected economic life of the project. Estimates of costs for construction projects in the United States are often taken from reference books of current cost data from recently completed projects, weighted for geographical region [R. S. Means, 1997]. Even with such data, cost estimating is an art more than a science, and estimates before the fact are often off the mark. Assumptions must be made about the availability of labor with requisite skills and reliability, wage rates over time, which are usually the subject of labor union contracts, and productivity [R. S. Means, 1997]. The effects on both costs and returns can be affected in important ways by taxes and depreciation, which can be estimated fairly well since rates are relatively stable over time, and inflation, which is unknowable for future times, but can be estimated from past experience. Taxes are location specific, and are therefore not considered in the following. Depreciation is knowable in advance, but is irrelevant in the absence of a consideration of taxes. Inflation can be included in the analysis, if assumptions are made about the rate over the life of the project. We ignore the effects of inflation in the immediately following mathematical analysis, but explicitly include it later on in this article.

Mathematically, the financial feasibility analysis and criterion are expressed by

$$-P+\sum_{j=1}^{N}\left(B_j-C_j\right)\left(P/F,i,j\right)>0 \tag{15.1}$$

where P is the initial investment in the project, at $j = 0$, B_j is the value of the benefits, or income, at the end of period j, C_j is the sum of the costs at the end of period j, and $(P/F,i, j)$ is the single payment present value factor for period j, i.e.,

$$\left(\frac{1}{1+i}\right)^{j} \tag{15.2}$$

for financial feasibility to be indicated.

Toll Road Example 1

An example is useful. Let us suppose that we plan to build a toll road in eastern Kentucky. The initial cost is estimated to be $6 million, annual operation and maintenance of the road is $.5 million, and income from the road will come from tolls from cars ($0.25 each) and trucks ($1.00 each). The life of the road as a toll road is expected to be 10 years, after which it will be turned over to the state, at no cost. Expected traffic and income from cars and trucks over the 10 year life of the project are shown in Table 15.9. Discount rate is assumed to be 8.00% per annum.

TABLE 15.9 Data for Toll Road Example 1

						Annual operation = $500,000				
1	2	3	4	5	6	7	8	9	10	11
year	cars/day	cars/yr	$,cars	tk/day	tk/yr	$,tk	$,total	$,net/yr	factor	NPV, $
1	2000	730000	182500	2000	730000	730000	912500	412500	0.925926	381944
2	2400	876000	219000	2200	803000	803000	1022000	522000	0.857339	447531
3	2880	1051200	262800	2420	883300	883300	1146100	646100	0.793832	512895
4	3456	1261440	315360	2662	971630	971630	1286990	786990	0.735030	578461
5	4147	1513728	378432	2928	1068793	1068793	1447225	947225	0.680583	644665
6	4977	1816474	454118	3221	1175672	1175672	1629791	1129791	0.630170	711960
7	5972	2179768	544942	3543	1293240	1293240	1838182	1338182	0.583490	780816
8	7166	2615722	653930	3897	1422563	1422563	2076494	1576494	0.540269	851731
9	8600	3138866	784717	4287	1564820	1564820	2349536	1849536	0.500249	925229
10	10320	3766640	941660	4716	1721302	1721302	2662962	2162962	0.463193	1001870
						Present value of operations and income				6,837,102
						Investment				6,000,000
						Net Present Value				837,102

At the beginning of this project, at $j = 0$, the initial investment to build the toll road was $6,000,000. Therefore, the NPV for this project is the algebraic sum of the initial investment (–$6,000,000) plus the sum of the annual NPVs (+$6,837,102), which equals +$837,102. Since this is a positive number, the project is considered feasible *under this scenario*. The reader should note that, under difference assumptions, say, a different discount rate, the project may show a negative overall NPV, and be considered financially infeasible.

In many cases, $B_j - C_j$ is constant for all *j*s except for when $j = 0$, i.e., the initial investment in the project. In such a case inequality (1) becomes

$$-P+\left(B_N - C_N\right)\left(P/A, i, N\right) > 0 \qquad (15.3)$$

where $(B_N - C_N)$ = the constant net income, N = the last period, and $(P/A, i, N)$ is the uniform series present value factor, expressed by

$$\frac{\left(1+i\right)^N - 1}{i\left(1+i\right)^N} \qquad (15.4)$$

Toll Road Example 2

In the foregoing toll road example, if we revise the annual usage by cars to 4000 per day and trucks to 2500 per day, at the same tolls as above, $0.25 per car and $1.00 per truck, we have an annual income for car tolls of $365,000 and for truck tolls $912,500, for a total toll income for each year of the 10-year life of the toll road project, B_N, of $1,277,500. The annual operation and maintenance for each year, C_N, remains at $500,000, and the initial cost of the project, P, remains at $6,000,000. Keeping the discount rate at the same level as in the earlier example, 8% per annum, and the life of the project, N, at 10 years, we substitute these numbers into Eq. 15.4 to get $(P/A, i, N) = 6.71008$. Then, Eq. 15.3 becomes

$$-\$6,000,000 + (\$1,277,500 - \$500,000)(6.71008) = -\$782,913$$

which violates the criterion of positivity. Therefore, under this scenario, the project is not financially feasible.

Toll Road Example 3

Let us now suppose that, rather than turn over the toll road to the state at the end of the 10-year period, we sell the road to the state for $2,000,000. Inequality (15.3) now becomes

$$-P+\left(B_N-C_N\right)\left(P/A,i,N\right)+S_N\left(P/F,i,N\right)>0 \tag{15.5}$$

where S_N = the salvage (sales) value at the end of period N and $(P/F, i, N)$, as above, is the single payment present value factor for period $j = N$, as shown in expression (15.2).

We have, then, evaluating inequality (15.5),

$$-\$6,000,000 + (\$1,277,500 - \$500,000)(6.71008) + \$2,000,000(0.46319) = \$143,467.$$

This analysis, which results in a positive number for inequality (15.5), indicates that this scenario is financially feasible.

Toll Road Example 4, with Inflation

Finally, let us suppose that the annual projected costs and income of example 3 are subjected to inflation of 3% per year. In cases of uniform inflation of $f\%$ per year, we must adjust the uninflated discount rate u by the following expression to get the actual discount rate i

$$i=\left(\frac{u-f}{1+f}\right) \tag{15.6}$$

Substituting in Eq. 15.6 0.03 for f and 0.08 for u, we have

$$i=\left(.08-.03\right)\big/\left(1+.03\right)$$

$$i=0.04854$$

and $(P/A, i, N)$, the uniform series present value factor, becomes

$$\frac{\left(1+0.04854\right)^{10}-1}{0.04854\left(1+0.04854\right)^{10}}=7.77678 \tag{15.7}$$

Similarly, with $i = 0.04854$, the single payment present value factor becomes

$$\left(\frac{1}{1+0.04854}\right)^{10}=0.62252 \tag{15.8}$$

Evaluating again inequality (15.5), with the new uniform series present value factor to account for inflation during the 10 years and the new single payment present value factor to account for the inflated sales price at the end of year 10, we have

$$-\$6,000,000 + (\$1,277,500 - \$500,000)(7.77678) + \$2,000,000(0.62252) = \$1,291,487,$$

and we see that, again, despite inflation, the project is financially feasible.

Conclusions

We have discussed the general process of determining if a proposed project is possible, desirable, or feasible under a given scenario of implementation. There is much to consider that we have been constrained by space to exclude, but the patterns of analysis given here represent those that must be used to prune away those projects from further consideration that are either impossible or undesirable, and to concentrate on objective analyses of those that might be feasible, legally, environmentally, and financially. Following are some recommended references for the interested reader who wishes to dig deeper into this important subject.

References

Jain, R. K. et al., *Environmental Assessment*, McGraw-Hill, New York, 1993.

Jain, R. K., Urban, L. V., and Stacey, G. S., *Environmental Impact Analysis: A New Dimension in Decision Making*, Van Nostrand Reinhold, New York, 1981.

Steiner, H. M., *Basic Engineering Economy, rev. ed.*, Books Associates, Glen Echo, MD, 1989.

R. S. Means Company, Inc., Means Building Construction Cost Data 1997, Kingston, ME, 1997.

R. S. Means Company, Inc., Means Heavy Construction Cost Data 1997, Kingston, ME, 1997.

R. S. Means Company, Inc., Means Labor Rates for the Construction Industry 1997, Kingston, ME, 1997.

Stanford graduates Dave Packard and Bill Hewlett first started to work together in 1938. They started off in a garage in Palo Alto with $538 in working capital. Their first product was an audio oscillator (HP200A), an electronic instrument used to test sound equipment. And one of their first clients was Walt Disney, who ordered eight oscillators (HP 200B) for the production of the movie *Fantasia*.

The partnership was formed on January 1, 1939, and a coin toss decided the company name. Bill Hewlett won so his name was first in the company name.

It wasn't until 1943 that HP entered the microwave field with signal generators developed for the Naval Research Laboratory and a radar-jamming device. A complete line of microwave test products followed World War II and HP became the acknowledged leader in signal generators.

- Incorporated August 18, 1947.
- The first public stock offering was November 6, 1957.

Growth in revenue

1940	$34,000
1951	$5.5 million
1958	$30 million
1969	$365 million
1980	$3 billion
1991	$14.5 billion
1992	$16.4 billion
1993	$20.3 billion
1994	$25.0 billion
1995	$31.5 billion
1996	$38.4 billion

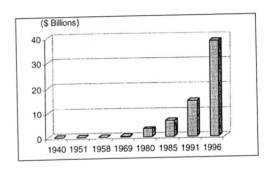

Innovations:

1938 — Audio oscillator (HP 200A-B)
1943 — Radar-jamming device for the Navy
1951 — High-speed frequency counter (HP 524A)
1958 — High-quality graphic recorders, the beginning of printers
1964 — A cesium-beam standard instrument (HP 5060A)
1966 — HP's first computer (HP 2116A)
1972 — Scientific hand-held calculator (HP-35)
1980 — HP's first personal computer (HP-85)
1984 — Inkjet printer, laser printer
1991 — Color scanner
1994 — Color laser printer

Strategic and Management Issues:

- Strong commitment to research and development in electronics and computer technology.
- "Management by Wandering Around", an informal HP practice which involves keeping up to date with individuals and activities through informal or structured communication. Trust and respect for individuals are apparent when MBWA is used to recognize employees' concerns and ideas.

V

Strategy
of the Firm

16

Business Strategy

Robert A. Burgelman
Stanford University

Richard S. Rosenbloom
Harvard Business School

Richard Goodman
*University of California,
Los Angeles*

Robert D. Hisrich
Case Western Reserve University

James C. Collins
University of Virginia

Jerry I. Porras
Stanford University

Nick Oliver
University of Cambridge

Dorothy Leonard
Harvard University

16.1   Design and Implementation of Technology Strategy: An Evolutionary Perspective[1]

Robert A. Burgelman and Richard S. Rosenbloom

Technology is a resource that, like financial and human resources, is pervasively important in organizations. Managing technology is a basic business function. This implies the need to develop a technology strategy, analogous to financial and human resource strategies. Technology strategy serves as the basis for fundamental business strategy decisions. It helps answer questions such as

1. Which distinctive technological competences and capabilities are necessary to establish and maintain competitive advantage?

---

[1]This paper is an elaboration and extension of Burgelman, R. A. and Rosenbloom, R. S. Technology Strategy: An Evolutionary Process Perspective, in Rosenbloom, R.S. and Burgelman, R.A., eds., *Research on Technological Innovation, Management and Policy, Vol. 4*, 1989.

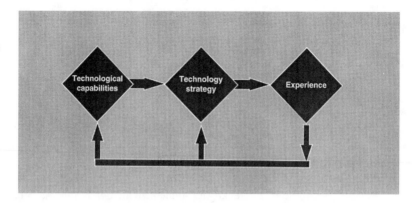

FIGURE 16.1   A capabilities-based organizational learning framework of technology strategy.

2. Which technologies should be used to implement core product design concepts and how should these technologies be embodied in products?
3. What should be the investment level in technology development?
4. How should various technologies be sourced — internally or externally?
5. When and how should new technology be introduced to the market?
6. How should technology and innovation be organized and managed?

Technology strategy encompasses, but extends beyond, research and development (R&D) strategy [Mitchell, 1986; Adler, 1989].

Technology Strategy: An Evolutionary Organizational Learning Process

Strategy making concerning technology can be conceptualized as an evolutionary organizational learning process. This is shown in Fig. 16.1.

Figure 16.1 shows the linkages between technical competencies and capabilities, technology strategy, and experience. Technology strategy is a function of the quantity and quality of technical capabilities and competences. Experience obtained from enacting technology strategy feeds back to technical capabilities and technology strategy.

The remainder of this section examines the three interrelated elements of technology strategy making. The next section examines technological competencies and capabilities. The third section discusses the substance of technology strategy: the theoretical dimensions in which technology strategy can be expressed. The fourth section presents an overview of the internal and external forces that shape the evolution of a firm's technology strategy. The fifth section discusses experience through the enactment of technology strategy: the various key tasks through which the firm's technology strategy is actually implemented and experience accumulated.

Technological Competence and Capability

Over time organizations develop distinctive competences that are closely associated with their ability to cope with environmental demands [Selznick, 1957]. McKelvey and Aldrich [1983] view distinctive competence as "… the combined workplace (technological) and organizational knowledge and skills … that together are most salient in determining the ability of an organization to survive." Nelson and Winter [1982], in similar vein, use the concept of "routines", which they consider to play a role similar to genes in biological evolution. It is important to note that research has revealed that distinctive competences can become a competence trap [Levitt and March, 1988] or core rigidity [Leonard-Barton, 1992]. Other research has found that there are strong inertial forces associated with distinctive technological competences, but that strong technological competences are also likely to generate innovations [Burgelman, 1994].

In general, a firm's distinctive competence involves the differentiated skills, complementary assets, and routines used to create sustainable competitive advantage [Selznick, 1957; Andrews, 1981; Teece et al., 1990]. Prahalad and Hamel [1990], building on the work of Selznick and Andrews, define *core compe-tencies* as ".the collective learning in the organization, especially how to coordinate diverse production skills and integrate multiple streams of technologies." These authors also provide criteria for identifying a firm's core competence: A core competence should

1. Provide potential access to a wide variety of markets.
2. Make a significant contribution to the perceived customer benefits of the end product.
3. Be difficult for competitors to imitate.

In this section, we are primarily concerned with the subset of technological competences of the firm, but the interrelationships with competences in other key areas, such as marketing, are always to be considered as well.

Stalk et al. [1992] distinguish core competence from a firm's strategic capabilities: "...whereas core competence emphasizes technological and production expertise at specific points along the value chain, capabilities are more broadly based, encompassing the entire value chain. They define a *capability* as ".a set of business processes strategically understood. ...The key is to connect them to real customer needs." Thus, technological competences and capabilities are complementary concepts, and value chain analysis (e.g., Porter [1985]) provides a useful tool for examining their interrelationships.

Substance of Technology Strategy

Technology strategy can be discussed in terms of (1) the deployment of technology in the firm's product-market strategy to position itself in terms of differentiation (perceived value or quality) and delivered cost and to gain technology-based competitive advantage, (2) the use of technology, more broadly, in the various activities comprised by the firm's value chain, (3) the firm's resource commitment to various areas of technology, and (4) the firm's use of organization design and management techniques to manage the technology function. These constitute four substantive dimensions of technology strategy [Burgelman and Rosenbloom, 1989; Hampson, 1993].

Competitive Strategy Stance

Technology strategy is an instrument of more comprehensive business and corporate strategies. As part of these broader strategies, a business defines the role that technology should play in increasing the differentiation and/or reducing the costs of its products and services [Porter, 1983, 1985]. From a competitive strategy point of view, technology can be used defensively to sustain achieved advantage in product differentiation or cost or offensively as an instrument to create new advantage in established lines of business or to develop new products and markets.

Technology Choice

Recent work on the distinction between design concepts and their physical implementation [Clark, 1985] and on the distinction between components and architecture in product design and development [Clark, 1987; Henderson and Clark, 1990] is useful to establish a framework for technology choice. Henderson and Clark [1990] offer the example of a room fan, which is a system for moving air in a room. The major components of a room fan include the blade, the motor that drives it, the blade guard, the control system, and the mechanical housing. A component is defined as ". a physically distinct portion of the product that embodies a core design concept [Clark, 1985] and performs a well-defined function." Core design concepts correspond to the various functions that the product design needs to embody so that the manufactured product will be able to serve the purposes of its user. For instance, the need for the fan to move corresponds to a core design concept. Core design concepts can be implemented in various ways to become components. For instance, movement of the fan could be achieved through using manual power or electrical motors. Each of these implementations, in turn, refers to an underlying technological knowledge base. For instance, designing and building electrical motors requires knowledge of electrical and mechanical engineering. Each of the core concepts of a product thus entails technology choices. In addition to components, a product also has an architecture that determines how its components fit and work together. For instance, the room fan's overall architecture lays out how its various components will work together. Product architectures usually become stable with the emergence of a "dominant design" [Abernathy and Utterback, 1978] in the industry. Product architecture also affects technology choice.

Technology choices require careful assessments of technical as well as market factors and identify an array of targets for technology development. The relative irreversibility of investments in technology makes technology choice and targets for technology development an especially salient dimension of technology strategy. Targeted technology development may range from minor improvements in a mature process to the employment of an emerging technology in the first new product in a new market [Rosenbloom, 1985].

Technology Leadership

The implications of technological leadership have been explored in earlier writings on technology and strategy (e.g., Ansoff and Stewart [1967]; Maidique and Patch [1979]). Discussions of technological leadership are often in terms of the timing (relative to rivals) of commercial use of new technology, that is, in terms of product market strategy. A broader strategic definition views technological leadership in terms of relative advantage in the command of a body of technological competencies and capabilities. This sort of leadership results from commitment to a "pioneering" role in the development of a technology [Rosenbloom and Cusumano, 1987] as opposed to a more passive "monitoring" role. Technological leaders thus have the capacity to be first movers, but may elect not to do so.

A firm's competitive advantage is more likely to arise from the unique aspects of its technology strategy than from characteristics it shares with others. Companies that are successful over long periods of time develop technological competences and capabilities that are distinct from those of their competitors and not easily replicable. Crown Cork and Seal, Marks and Spencer, and Banc One are examples briefly discussed below. Canon [Prahalad and Hamel, 1990] and Walmart [Stalk et al., 1992] are other examples. The capabilities-based strategies of such companies cannot easily be classified simply in terms of differentiation or cost leadership; they combine both. The ability to maintain uniqueness that is salient in the marketplace implies continuous alertness to what competitors are doing and should not be confused with insulation and an inward-looking orientation.

The competencies and capabilities-based view of technological leadership draws attention to the importance of accumulation of capabilities, e.g., Barney [1986] and Itami [1987]. Technological leadership cannot be bought easily in the market or quickly plugged into an organization. A firm must understand the strategic importance of different competencies and capabilities and be willing to patiently and persistently build them, even though it may sometimes seem cheaper or more efficient in the short term to rely on outsiders for their procurement.

Thinking strategically about technology means raising the question of how a particular technical competence or capability may affect a firm's future degrees of freedom and the control over its fate. This involves identifying and tracking key technical parameters, considering the impact on speed and flexibility of product and process development as technologies move through their life cycles. It also requires distinguishing carefully between technologies that are common to all players in the industry and have little impact on competitive advantage and those that are proprietary and likely to have a major impact on competitive advantage. Furthermore, it requires paying attention to new technologies that are beginning to manifest their potential for competitive advantage and those that are as yet only beginning to emerge [Arthur D. Little, 1981].

Technology Entry Timing

The timing of bringing technology to market, of course, remains a key strategic issue. Porter [1985] identifies conditions under which pioneering is likely to be rewarded in terms of lasting first mover advantages. While noting the various potential advantages accruing to first movers, Porter also identifies the disadvantages that may ensue, highlighting the significance of managerial choice of timing and the importance of situational analysis to determine the likely consequences along the path chosen.

Teece [1986] extends the analysis by identifying the importance of appropriability regimes and control of specialized assets. *Appropriability regimes* concern the first mover's ability to protect proprietary technological advantage. This usually depends on patents, proprietary know-how, and/or trade secrets. The legal battle between Intel and Advanced Micro Devices about access to Intel's microcode for microprocessor development is an example of the importance of appropriability regimes [Steere and Burgelman, 1994]. An important consideration here is the cost of defending one's proprietary technological position. For instance, the prospect of rapidly escalating legal costs and/or claims on scarce top management time may sometimes make it difficult for smaller firms to decide to go to court to protect their proprietary position unless violations are very clear. *Control of specialized assets* concerns the fact that, in many cases, the first mover may need access to complementary specialized assets owned by others. Gaining access to those assets — through acquisitions or strategic alliances — may absorb a large part of the rent stream coming from the innovation. The alliances between startup and established firms in the biotechnology industry are an example of the importance of complementary assets (e.g., Pisano and Teece [1989]). Unless companies command strong positions in terms of appropriability regimes and complementary assets, their capacity to exploit potential first mover advantages remains doubtful.

Technology Licensing

Sometimes firms need to decide whether they will bring technologies to market by themselves or also offer other firms the opportunity to market them through licensing arrangements, e.g., Shepard [1987]. Ford and Ryan [1981] identify several reasons companies may not be able to fully exploit its technologies through their own product sales alone. First, not all technologies generated by a firm's R&D efforts fit into their lines of business and corporate strategy (see also Pavitt et al. [1989]). Second, companies may need to consider licensing their technology to maximize the returns on their R&D investments because patents provide only limited protection against imitation by competitors. Licensing may be a strategic tool in discouraging imitation by competitors or in preempting competitors with alternative technologies. Third, smaller firms may be unable to exploit their technologies on their own because they lack the necessary cash and/or complementary assets (e.g., manufacturing). Fourth, international market development for the technology may require licensing local firms because of local government regulation. Fifth, antitrust legislation may sometimes prevent a company from fully exploiting its technological advantage on its own (Kodak, Xerox, and IBM are fairly recent examples). Technology-rich companies should therefore consider developing a special capability for marketing their technologies beyond embodiment in their own products.

Value Chain Stance

Technology pervades the value chain. A competencies and capabilities-based view goes beyond the strategic use of technology in products and services and takes a competitive stance toward its use in all the value chain activities [Porter, 1985].

Scope of Technology Strategy

Considering technology strategy in relation to the value chain defines its scope: the set of technological capabilities that the firm decides to develop internally. This set of technologies can be called the *core technology*. Other technologies, then, are peripheral. Of course, in a dynamic world, peripheral technologies today may become core technologies in the future and vice versa. Core technologies are the areas in which the firm needs to assess its distinctive technological competences and to decide whether to be a technological leader or follower and when to bring them to market. The scope of technology strategy is especially important in relation to the threat of new entrants in the firm's industry. All else equal, firms with a broader set of core technologies would seem to be less vulnerable to attacks from new entrants attempting to gain position through producing and delivering new types of technology-based customer value. However, resource constraints will put a limit on how many technologies the firm can opt to develop internally. Thus, it is important to limit the scope of technology strategy to the set of technologies considered by the firm to have a material impact on its competitive advantage.

The scope of a firm's technology strategy may be determined to a significant extent by its scale and business focus. Businesses built around large, complex systems such as aircraft, automobiles, or telecommunication switches demand the ability to apply and integrate numerous distinct types of expertise creating *economies of scale, scope, and learning* (synergies). General Electric, for instance, was reportedly able to bring to bear high-powered mathematical analysis, used in several divisions concerned with military research on submarine warfare, to the development of computerized tomography (CT) products in its medical equipment division [Rutenberg, 1986]. Other fields may actually contain diseconomies of scale, giving rise to the popularity of "skunkworks" (e.g., Rich and Janos [1994]) and discussion of the "mythical man-month" [Brooks, 1975]. The emergence of a new technology raises issues concerning the scope of the technology strategy. This, in turn, may impact the delineation of the set of core technologies of the firm and the boundaries between the business units that it comprises [Prahalad et al., 1989].

Resource Commitment Stance

The third dimension of the substance of technology strategy concerns the intensity of its resource commitment to technology. The variation among manufacturing firms is pronounced: many firms do not spend on R&D; a few commit as much as 10% of revenues to it. While interindustry differences can explain a large share of this variation, substantial differences in R&D intensity still remain between rivals.

Depth of Technology Strategy

Resource commitments determine the *depth* of the firm's technology strategy: its prowess within the various core technologies. Depth of technology strategy can be expressed in terms of the number of technological options that the firm has available.

Depth of technology is likely to be correlated with the firm's capacity to anticipate technological developments in particular areas early on. Greater technological depth may offer the benefit of increased flexibility and ability to respond to new demands from customers/users. It provides the basis for acting in a timely way.

Management Stance

Recent research [Hampton, 1993] based on Burgelman and Rosenbloom's [1989] framework suggests that the substance of technology strategy also encompasses a management stance: the choice of a management approach and organization design that are consistent with the stances taken on the other substantive dimensions.

Organizational Fit

Firms that can organize themselves to meet the organizational requirements flowing from their competitive, value chain, and resource commitment stances are more likely to have an effective technology strategy. For instance, a science-based firm that has decided to be a technology leader for the long run probably needs to create a central R&D organization. One that is satisfied with commercially exploiting existing technologies to the fullest may be able to decentralize all R&D activity to its major businesses. Imai et al. [1985] describe how Japanese firms' use of multiple layers of contractors and subcontractors in an external network to foster extreme forms of specialization in particular skills which, in turn, provide them with flexibility, speed in response, and the potential for cost savings since the highly specialized subcontractors operate on an experience curve even at the level of prototypes.

Evolutionary Forces Shaping Technology Strategy

An evolutionary process perspective raises the question of how a firm's technology strategy actually comes about and changes over time. Evolutionary theory applied to social systems focuses on variation-selection-retention mechanisms for explaining dynamic behavior over time (e.g., Campbell [1969], Aldrich [1979], Weick [1979], Burgelman [1983, 1991], and Van de Ven and Garud [1989]). It recognizes the importance of history, irreversibilities, invariance and inertia in explaining the behavior of organizations. However, it also considers the effects of individual and social learning processes, e.g., Burgelman [1988]. An evolutionary perspective is useful for integrating extant literatures on technology. The study of technological

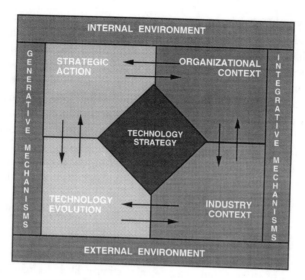

FIGURE 16.2 Determinants of technology strategy.

development, for instance, contains many elements that seem compatible with the variation-selection-retention structure of evolutionary theory (e.g., Rosenberg [1979], Krantzberg and Kelly [1978], Abernathy [1978], Clark [1985], Henderson and Clark [1990], and Burgelman [1994]). While cautioning against the fallacy of unwarranted analogy in applying concepts from biological evolution to cultural (organizational) evolution, Gould's [1987] interpretation of the establishment of QWERTY [David, 1985] as the dominant, if inferior, approach to laying out keys on typewriter keyboards shows the power of evolutionary reasoning to identify and elucidate interesting phenomena concerning technological evolution.

The evolutionary factors shaping technology strategy making comprise internal and external generative and integrative forces. In this section, we explore a simple framework to conceptualize these forces. The framework is presented in Fig. 16.2.

The ideas expressed in Fig. 16.2 are that technology strategy is shaped by the generative forces of the firm's strategic action and the evolution of technology and by the integrative, or selective, forces of the firm's organizational context and the industry context.

Technology Evolution

A firm's technology strategy is rooted in the evolution of its technical capabilities. However, the dynamics of these capabilities, and hence the technology strategy, are not completely endogenous. A firm's technical capabilities are affected in significant ways by the evolution of the broader areas of technology of which they are part and which evolve largely independently of the firm. Different aspects of technological evolution have been discussed in the literature: (1) the evolution of technologies along S-curve trajectories (e.g., Twiss [1982] and Dosi [1982]), (2) the interplay between product and process technology development within design configurations over the course of a particular technological trajectory [Abernathy, 1978], (3) the emergence of new technologies and their trajectories (S-curves) [Foster, 1986], (4) the competence enhancing or destroying consequences of new technologies [Astley, 1985; Abernathy et al., 1983; Tushman and Anderson [1986], (5) dematurity: renewed technological innovation in the context of well-established markets, high production volumes, and well-established organizational arrangements [Abernathy et al., 1983], and (6) the organizational determinants of technological change [Tushman and Rosenkopf, 1992]. These are some of the major evolutionary forces associated with technological development, which transcend the strategic actions of any given firm.

Industry Context

Important aspects of industry context are (1) the industry structure that can be understood in terms of five major forces [Porter, 1980], all of which can be affected in important ways by technology [Porter,

1983], but the interplay of which, in turn, determines the technological competences that can form the basis of competitive advantage [Burgelman, 1994], (2) the appropriability regime associated with a technological innovation [Teece, 1986], (3) the complementary assets needed to commercialize a new technology [Teece, 1986], (4) the emergence of dominant designs [Utterback and Abernathy, 1975; Abernathy, 1978], (5) increasing returns to adoption for particular technologies [Arthur, 1988; David, 1985], (6) the emergence of industry standards [Farrell and Saloner, 1985; Metcalfe and Gibbons, 1989], (7) the social systems aspects of industry development [Van de Ven and Garud, 1989], and (8) the competitive effects of the interplay of social systems characteristics and technological change [Barnett, 1990]. These various factors and their interplays affect the likely distribution of profits generated by a technological innovation among the different parties involved as well as the strategic choices concerning the optimal boundaries of the innovating firm. They also affect the expected locus of technological innovations [von Hippel, 1988].

Strategic Action

A firm's strategy captures organizational learning about the basis of its past and current success [Burgelman, 1983, 1991; Donaldson and Lorsch, 1983; Weick, 1987]. Strategic action is, to a large extent, induced by the prevailing concept of strategy. Induced strategic action is likely to manifest a degree of inertia relative to the cumulative changes in the external environment [Hannan and Freeman, 1984]. Cooper and Schendel [1978] found that established firms, when confronted with the threat of radically new technologies, were likely to increase their efforts to improve existing technology rather than switch to new technology even after the latter had passed the threshold of viability. Abernathy et al. [1983] point out that "core design concepts" may emerge that "'lock in' or stabilize the domain of relevant technical effort [Clark, 1985]. Henderson and Clark [1990] found that firms faced with architectural innovations were often unable to adapt their product development efforts. Burgelman [1994] found that inertial forces associated with a firm's distinctive competence impeded adaptation to changes in the basis of competition as a product moved from specialty to commodity.

Firms also exhibit some amount of autonomous strategic action aimed at getting the firm into new areas of business (e.g., Penrose [1968] and Burgelman [1983, 1991]). These initiatives are often rooted in technology development efforts. In the course of their work, for example, technical people may serendipitously discover results that provide the basis for redirection or replacement of major technologies of the firm. The existence of a corporate R&D capability often provides a substratum for the emergence of such new technical possibilities that extend beyond the scope of the firm's corporate strategy [Rosenbloom and Kantrow, 1982; Burgelman and Sayles, 1986; Pavitt et al., 1989]. Participants engaging in autonomous strategic action explore the boundaries of a firm's capabilities and corresponding opportunities sets [Burgelman, 1983]. As Itami [1983] has observed, "in reality, many firms do not have … complete knowledge and discover the full potential of their ability only after the fact."

Organizational Context

The industry context exerts strong external selection pressures on the incumbent firms and their strategies. However, a key feature of established firms is that they have an organizational context that allows them, to some extent, to substitute internal for external selection. Organizational context thus serves as an internal selection environment [Burgelman, 1991]. Continued survival may very well depend on the effectiveness of the firm's internal selection environment. It affects the capacity of the firm to deal with major strategic management challenges: (1) the ability to exploit opportunities associated with the current strategy (induced process), (2) the ability to take advantage of opportunities that emerge spontaneously outside the scope of the current strategy (autonomous process), and (3) the ability to balance challenges (1) and (2) at different times in the firm's development [Burgelman and Sayles, 1986].

Organizational context takes shape over time and reflects the administrative approaches and dominant culture of the firm. The dominant culture as it relates to technology may be different, depending on whether the firm's distinctive competencies are rooted in science (e.g., pharmaceutical firms), engineering (e.g., semiconductor firms), or manufacturing (e.g., Japanese firms); whether the product development process has been driven by technology push, need pull, or a more balanced approach; and so on. Some research (e.g., Boeker [1989]) suggests that the background and management approaches of the founders have lasting impact on the firm's organizational context. Other research [Henderson and Clark, 1990] has shown that product architecture becomes reflected in organization structure and culture and greatly affects communication channels and filters. This, in turn, makes it difficult for organizations to adapt to architectural innovations that change the way in which the components of a product are linked together while leaving the core design concepts (and thus the basic technological knowledge underlying the components) intact. Still other research [Burgelman et al., 1997] has found evolving links between a firm's technology strategy and substantive and generic corporate strategies. While, initially, technology strategy drives substantive as well as generic corporate strategies, over time these relationships become reciprocal. Generic corporate strategy, however, may be a more enduring driver than substantive strategy because it becomes embedded in the firm's organizational context (internal selection environment).

Applying the Framework: Research and Practice

Several examples can be offered to indicate how the framework may help situate new research findings in the field of strategic management of technology, as well as gain insight in the practice of firms' strategic management of technology.

Research

Recent empirical work on "disruptive technologies" by Clayton Christensen and collaborators (e.g., Christensen and Bower [1996]) allows to illustrate the usefulness of the framework to integrate new research findings as they come along. Disruptive technologies have the potential to cannibalize existing technologies and to undermine the strategic position of incumbent firms in an industry. Christensen's research shows this phenomenon in the magnetic disk drive industry, where the emergence of a new generation of smaller form factor (smaller disk) materially affected the success of the prominent firms in the preceding generation. Other examples of disruptive technologies, among many, are the transistor, electronic fuel injection, and the personal computer. Christensen distinguishes disruptive from sustaining technologies. The latter leverage the existing technologies and may reinforce the strategic position of incumbents. (In the terminology of economics, disruptive technologies may be viewed as substitutes; sustaining technologies as complements.)

Christensen's research shows that disruptive technologies often emerge first in incumbent firms where technical people engage in autonomous *strategic action* to champion the newly discovered technological opportunity. The existing *organizational context*, however, is usually not receptive to these initiatives. This is so, in part, because the firm's sales force experiences a lack of interest on the part of existing customers. This, in turn, implies unpromising returns to investment in the foreseeable future. The existing customers are typically interested in improvements of performance along the technology dimensions that are relevant to them and their own current product development efforts and not in the improvement along different technology dimensions promised by the new technology. This is, therefore, very much a rational source of resistance. Hence, initially, the *industry context* represented by existing customers and the organizational context reinforce each other in resisting the new technology.

Often, however, in the face of organizational rejection or recalcitrance, the initiators of the new technology will leave the incumbent firm and start their own to pursue the technological opportunity. The new entrepreneurial firm will have to find and pursue new customers who are interested in the improvements offered by the new technology along the different dimensions. For instance, personal computer makers may be more interested in the size of the disk than in its price/performance ratio. As new users adopt the technology, *technological evolution* is likely to lead to improvements along the technology's other dimensions (e.g., price/performance). As a result, the new technology may become attractive to the customers of the old technology. This may precipitate a major shift in the market (*industry context*) and propel the new companies to prominence in the industry. [Note that the performance of the old technology usually continues to advance as well (*technological evolution*) but in ways that exceed performance levels that are valued by the old customers and therefore becoming irrelevant.]

Practice

Crown Cork and Seal was able to do very well over a 30-year period as a relatively small player in a mature industry. Technology strategy seems to have contributed significantly to Crown's success. When taking over the company in 1957, CEO John Connelly recognized the existence of Crown's strong skills in metal formation and built on these to specialize in "hard-to- hold" applications for tin cans (*competencies/capabilities*). Crown did not have an R&D department, but developed strong links between a highly competent technical sales force and an applications-oriented engineering group to be able to provide complete technical solutions for customers' "filling needs" (*competencies/capabilities*). In the face of a major external technological innovation — the two-piece can — initiated by an aluminum company (*technology evolution*) Crown was able to mobilize its own capabilities as well as those of its steel suppliers to quickly adapt the innovation for use with steel. Over several decades, CC&S continued to stick to what it could do best (*strategic action*) — manufacturing and selling metal cans — driven by a strong customer-oriented culture (*organizational context*) while its competitors were directing their attention to diversification and gradually lost interest in the metal can industry (*industry context*). CC&S has continued to refine the skill set (*experience*) that made it the only remaining independent metal can company of the four original major players.

The success of Marks and Spencer (M & S), a British retailer with a worldwide reputation for quality, is based on a consistent strategy founded on an unswerving commitment to giving the customer "good value for money". The genesis of its technology strategy was the transformation, in 1936, of a small textile testing department into a "Merchandise Development Department", designed to work closely with vendors to bring about improvements in quality. Thirty years later, the M & S technical staffs, then numbering more than 200 persons working on food technology as well as textiles and home goods, allowed M&S, quite literally, to control the cost structure of its suppliers. The development of the technical capability itself was driven by the strong value of excellent supplier relationships held and continuously reinforced by top management.

Banc One Corporation is a Midwest banking group that consistently ranks among the most profitable U.S. banking operations. In 1958, the new CEO of City National Bank of Columbus Ohio (CNB), John G. McCoy, persuaded his Board to invest 3% of profits each year to support a "research and development" activity. Over the net two decades, CNB, which became the lead bank of Banc One Corporation, developed capabilities that made it a national leader in the application of electronic information-processing technologies to retail banking. It was the first bank to install automatic teller machines and a pioneer in the development of bank credit cards, point-of-sale transaction processing, and home banking. While not all of its innovative ventures succeeded, each contributed to the cumulative development of a deep and powerful technical capability that remains a distinctive element of the bank's highly successful competitive strategy.

The three companies cited above are notable for the consistency of their strategic behavior over several decades. The following example illustrates the problems that can arise in a time of changing technology and industry context when a fundamental change in strategy is not matched by corresponding adaptation of the organizational context.

The National Cash Register Company (NCR) built a dominant position in worldwide markets for cash registers and accounting machines on the basis of superior technology and an outstanding sales force created by the legendary John H. Patterson. By 1911, NCR had a 95% share in cash register sales. Scale economies in manufacturing, sales, and service presented formidable barriers to entry to its markets (*industry context*), preserving its dominance for another 60 years. Highly developed skills in the design and fabrication of complex low-cost machines (*competencies/capabilities*) not only supported the strategy, they also shaped the culture of management, centered on Dayton where a vast complex housed engineering and fabrication for the traditional product line (*organizational context*). In the 1950s, management began to build new capabilities in electronics (*a revolution, not just evolution in technology*) and entered the emerging market for electronic data processing (EDP). A new strategic concept tried to position traditional products (registers and accounting machines) as "data-entry" terminals in EDP systems (*changing strategic behavior*). However, a sales force designed to sell stand-alone products of moderate unit cost proved ineffective in selling high-priced "total systems". At the same time, the micro-electronics revolution destroyed the barriers inherent in NCR's scale and experience in fabricating mechanical equipment. A swarm of new entrants found receptive customers for their new electronic registers (*changing industry context*). As market share tumbled and red ink washed over the P&L in 1972, the chief executive was forced out. His successor was an experienced senior NCR manager who had built his career entirely outside of Dayton. He moved swiftly to transform the ranks of top management, decentralize manufacturing (reducing employment in Dayton by 85%), and restructure the sales force along new lines. The medicine was bitter, but it worked; within 2 years, NCR had regained leadership in its main markets and was more profitable in the late 1970s than it had been at any point in the 1960s.

Experience through Enactment of Technology Strategy

A Note on Performance (Enactment) as Experience

The conventional view of performance in the strategy literature is in terms of outcomes such as ROE, P/E, market share, and growth. Typically, strategy researchers have tried to establish statistical associations between such outcomes and strategic variables, for instance, the association between profitability measures and measures of product quality. Establishing such associations is useful, but little insight is usually provided in how exactly outcomes come about (e.g., how quality is achieved and how it influences buyer behavior). In the framework represented in Fig. 16.1, experience is viewed in terms of actually performing (enacting) the different tasks involved in carrying out the strategy. This view of performance is akin to that used in sports. For instance, to help a swimmer reach his/her highest possible performance, it is not enough to measure the time needed to swim a certain distance and communicate that outcome to the swimmer, or to establish that, on average, a certain swimming style will be associated with better times. Sophisticated analysis of each and every movement of the swimmer traversing the water provides clues on how this particular swimmer may be able to achieve a better time. Similarly, experience derived from using technology in strategy provides feedback concerning the quantity and quality of the firm's technical competences and capabilities and the effectiveness of its strategy. The learning and unlearning potentially resulting from this, in turn, serve to leverage, augment and/or alter the firm's capabilities, the strategy or both (e.g., Maidique and Zirger [1984], Imai [Nonaka [and Takeuchi [1985], Itami [1987], and Burgelman [1994]).

Technology strategy is realized in practice through enactment of several key tasks: (1) internal and external technology sourcing, (2) deploying technology in product and process development, and (3) using technology in technical support activities. Performing these activities, in turn, provides valuable experience that serves to augment and change the firm's technical competencies and capabilities and to reconsider certain substantive aspects of the technology strategy.

Technology Sourcing

Since the sources of technology are inherently varied, so must be the mechanisms employed to make it accessible within the firm. It is useful to distinguish between internal R&D activity and acquisitive functions which import technology originating outside the firm.

Internal Sourcing

Internal sourcing depends on the firm's R&D capability. Each firm's technology strategy finds partial expression in the way it funds, structures, and directs the R&D activities whose mission is to create new pathways for technology. Relatively few firms — primarily the largest ones in the R&D intensive industries — are able to support the kind of science-based R&D that can lead to important new technologies. A recent example is the emergence of high-temperature superconductivity from IBM's research laboratories. Most established technology-based firms emphasize applied R&D in support of existing and emerging businesses. Cohen and Levinthal [1990] found that an internal R&D capability is also an important determinant of the firm's "absorptive capacity," that is, the firm's ".ability to recognize the value of new, external information, assimilate it, and apply it to commercial ends." This indicates a close link between internal sourcing of technology and the capacity to use external sources of technology.

External Sourcing

Many important technologies used in the value chain are outside of the technological capabilities of the firm. While internal sourcing seems necessary for most of the firm's core technologies, some may need to be sourced externally through exclusive or preferential licensing contracts or through strategic alliances. Every firm finds that it must structure ways to acquire certain technologies from others. The choices made in carrying out those tasks can tell us a great deal about the underlying technology strategy. To what extent does the firm rely on ongoing alliances, as opposed to discrete transactions for the acquisition of technology [Hamilton, 1986]? Is the acquisition structured to create the capability for future advances to be made internally, or will it merely reinforce dependence [Doz et al., 1989]?

Continuous concern with improvement in all aspects of the value creation and delivery process may guard the firm against quirky moves in external technology sourcing that could endanger the firm's competitive product market position. Viewing the issue of external sourcing from this perspective highlights the importance of managing interdependencies with external providers of capabilities. One requirement of this is a continuous concern for gaining as much learning as possible from the relationship in terms of capabilities and skills rather than a concern solely with price. To the extent that a firm engages in strategic alliances, it seems necessary to establish the requisite capabilities for managing the relationships. For instance, to develop unique and valuable supplier relationships, a company may need a strong technical staff to manage these relationships. If the strategic alliance was necessary because the firm is behind in an important area of technology, it may need to invest in plant and equipment to apply what is learned from the partner and begin building the technological capability in-house. For example, when Japanese electronics firms acquired technology from abroad during the 1950s and 1960s, most of them structured it in ways that made it possible for them to become the leaders in pushing the frontiers of those same technologies in the 1970s and 1980s [Doz et al., 1989].

Product and Process Development

Technology strategy is also enacted by deploying technology to develop products and processes. Product and process development activities embody important aspects of the dimensions of technology strategy. An understanding of what the strategy is can be gained from considering the level of resources committed, the way they are deployed, and how they are directed in product and process development. For instance, how does the organization strike the delicate balance between letting technology drive product development and allowing product development and/or market development to drive technology? The availability of integrated circuit technology drove product development in many areas of consumer electronics. New product development efforts, on the other hand, sometimes stimulate the development of new technologies. Notebook computers, for instance, drove the development of new disk drive technology and semiconductor "flash" memory. The mammoth personal computer industry was founded when the young engineers of Apple sought to exploit the potential inherent in the microprocessor, at a time when corporate managers in Hewlett-Packard, IBM, and others were disdainful of the commercial opportunity. A decade later, however, it was clear that market needs were now the primary force shaping efforts to advance the constituent technologies.

Wheelwright and Clark [1992] suggest three potential benefits associated with product and process development: (1) market position, (2) resource utilization, and (3) organizational renewal and enhancement. They point out that in most firms, the potential is seldom fully realized because they lack a development strategy framework that helps them consistently integrate technology strategy with product-market strategy. Recent work on technology integration [Iansiti, 1997] goes one step further, focusing on the role of technology evaluation and selection processes that precede actual product development processes. These technology integration processes are different from the various types of "advanced development projects" identified by Wheelwright and Clark. They are concerned with how choices of new technological possibilities (deriving from fundamental research) in relation to the existing application context (represented by the current product, manufacturing, and customer/user systems) affect the speed and productivity of the product development processes at the project level.

Substance	Enactment (Modes of Experience)				
	External Technology Sourcing	Internal Technology Sourcing	Product Development	Process Development	Technical Support
Competitive strategy stance (Choice/leadership/ entry timing/licensing					
Value chain stance (scope)					
Resource commitment stance (depth)					
Management stance (organizational fit)					

FIGURE 16.3 Substance and enactment in technology strategy making.

Technical Support

The function commonly termed "field service" creates the interface between the firm's technical function and the users of its products or services. Experience in use provides important feedback to enhance the firm's technological capabilities [Rosenberg, 1982]. Airline operations, for example, are an essential source of information about jet engine technology. In some industries, such as electronic instrumentation, important innovations often originate with the users [von Hippel, 1977]. The technology strategy of a firm, then, finds important expression in the way it carries out this important link to users. Two-way flows of information are relevant: expert knowledge from product developers can enhance the effectiveness of field operations, while feedback from the field informs future development.

Enactment Reveals Substance of Technology Strategy

Studying the processes involved in performing the key tasks sheds light on how technology strategy relates technical capabilities to competitive advantage and how organizational learning and unlearning actually come about. In other words, enactment reveals the substance of technology strategy. The matrix illustrated in Fig. 16.3 presents a framework for mapping the interactions among the substance and enactment in technology strategy making.

Two Conjectures

Implicit in the foregoing discussion are two normative conjectures about technology strategy. The first is that the substance of technology strategy should be comprehensive, that is, technology strategy, as it is enacted through the various tasks of acquisition, development, and technical support, should address the four substantive dimensions and do so in ways that are consistent across the dimensions. Second, we suggest that technology strategy should be integrated, that is, each of the key tasks should be informed by the positions taken on the four substantive dimensions in ways that create consistency across the various tasks.

Conclusion

This section argues that an evolutionary process perspective provides a useful framework for thinking about technology strategy and about its role in the broader competitive strategy of a firm. The essence of this perspective is that technology strategy is built on technical competencies and capabilities and tempered by experience. These three main constructs — technical competencies and capabilities, strategy, and experience — are tightly interwoven in reality. Technical competencies and capabilities give technology strategy its force; technology strategy enacted creates experience that modifies technical competencies and capabilities. Central to this idea is the notion that the reality of a strategy lies in its enactment, not in those pronouncements that appear to assert it. In other words, the substance of technology strategy can be found in its enactment of the various modes by which technology is acquired and deployed — sourcing, development, and support activities. The ways in which these tasks are actually performed and the ways in which their performance contributes, cumulatively, to the augmentation and deepening of competencies and capabilities convey the substance of technology strategy in practice.

A second central idea in this paper is that the ongoing interactions of technical capabilities-technology strategy-experience occur within a matrix of generative and integrative mechanisms that shape strategy. These mechanisms (sketched in Fig. 16.2) are both internal and external to the firm. Anecdotal evidence suggests that successful firms operate within some sort of harmonious equilibrium of these forces. Major change in one, as in the emergence of a technological discontinuity, ordinarily must be matched by adaptation in the others. Which leads to the final conjecture of this paper, namely, that it is advantageous to attain a state in which technology strategy is both comprehensive and integrated. By comprehensive we mean that it embodies consistent answers to the issues posed by all four substantive dimensions. By integrated we mean that each of the various modes of performance is informed by the technology strategy.

References

Adler, P. Technology strategy: a review of the literatures, In *Research on Technological Innovation, Management and Policy, Vol. 4*, R. S. Rosenbloom and R. A. Burgelman, eds., pp. 25–152, JAI Press, Greenwich, CT, 1989.

Andrews, K. *The Concept of Corporate Strategy*, Irwin, Homewood, IL, 1981.

Abernathy, W. J. *The Productivity Dilemma: Roadblock to Innovation in the Automobile Industry*, Johns Hopkins University Press, Baltimore, 1978.

Abernathy, W. J., Clark, K., and Kantrow, A. M. *Industrial Renaissance*, Basic Books, New York, 1983.

Abernathy, W. J. and Utterback, J. Patterns of industrial innovation, *Technol. Rev.*, 1978.

Aldrich, H. E. *Organizations and Environments*, Prentice Hall, Englewood Cliffs, NJ, 1979.

Ansoff, H. L. and J. M. Stewart, Strategies for a technology-based business, *Harv. Bus. Rev.*, November-December, 1967.

Arthur, W. B. Competing technologies: an overview, In *Technical Change and Economic Theory*, G. Dosi et al., eds., Columbia University Press, New York, 1988.

Arthur D. Little. The strategic management of technology, *Eur. Mgt. Forum*, 1981.

Astley, W. G. The two ecologies: population and community perspectives on organizational evolution, *Admin. Sci. Q.*, 30: 224–241, 1985.

Barnett, W. P. The organizational ecology of a technological system, *Admin. Sci. Q.*, 35, 1990.

Barney, J. Strategic factor markets: expectations, luck, and business strategy, *Mgt. Sci.*, 32, 1986.

Boeker, W. Strategic change: the effects of founding and history, *Acad Mgt. J.*, 32: 489–515, 1989.

Brooks, F. P. *The Mythical Man-Month*, Addison-Wesley, Reading, MA, 1975.

Burgelman, R. A. Corporate entrepreneurship and strategic management: insights from a process study, *Mgt. Sci.*, 29: 1349–1364, 1983.

Burgelman, R. A. Strategy-making and evolutionary theory: towards a capabilities-based perspective, In *Technological Innovation and Business Strategy*, Tsuchiya, M. ed., Nihon Keizai Shinbunsha, Tokyo, 1986.

Burgelman, R. A. Strategy-making as a social learning process: the case of internal corporate venturing, *Interfaces*, 18: 74–85, 1988.

Burgelman, R. A. "Intraorganizational ecology of strategy making and organizational adaptation: theory and field research, *Org. Sci.*, 2: 239–262, 1991.

Burgelman, R. A. "Fading memories: a process theory of strategic business exit in dynamic environments, *Admin. Sci. Q.*, to be published.

Burgelman, R. A., Cogan, G. W., and Graham, B. K. "Strategic business exit and corporate transformation: evolving links of technology strategy and substantive and generic corporate strategies, In *Research on Technological Innovation, Management and Policy,* Vol. 6, R. A. Burgelman and R. S. Rosenbloom, eds., oo, 89–153, 1997.

Burgelman, R. A. and Rosenbloom, R. S. Technology strategy: an evolutionary process perspective, In *Research on Technological Innovation, Management and Policy, Vol 4,* Rosenbloom, R.S. and Burgelman, R.A. eds., pp. 1–23. JAI Press, Greenwich, CT, 1989.

Burgelman, R. A. and Sayles, L. R. *Inside Corporate Innovation,* Free Press, New York, 1986.

Campbell, D. T. Variation and selective retention in sociocultural evolution, *Gen. Syst.,* 14: 69–8S, 1969.

Christensen, C. M. and Bower, J. L. Customer power, strategic investment, and the failure of leading firms, *Strat. Mgt. J.,* 17: 197–218, 1996.

Clark, K. B. The interaction of design hierarchies and market concepts in technological evolution, *Res. Policy,* 14: 235–251, 1985.

Clark, K. B. Managing technology in international competition: the case of product development in response to foreign entry, In *International Competitiveness,* M. Spence and H. Hazard, eds., pp. 27–74, Cambridge, MA, 1987.

Cohen, W. M. and Levinthal, D. A. Absorptive capacity: a new perspective on learning and innovation, *Admin. Sci. Q.,* 35: 128–152, 1990.

Cooper, A. C. and Schendel, D. Strategic responses to technological threats, *Bus. Horizons,* February: 61–63, 1976.

David, P. A. Clio and the economics of QWERTY, *Am. Econ. Rev.,* 75: 332–337, 1985.

Dosi, G. Technological paradigms and technological trajectories: a suggested interpretation of the determinants and directions of technical change, *Res. Policy,* 11: 147–162, 1982.

Doz, I. L., Hamel, G., and Prahalad, C. K. Collaborate with Your Competitors — and Win, *Harv. Bus. Rev.,* January-February, 133–139, 1989.

Farrell, F. and Saloner, G. Competition, compatibility and standards: the economics of horses, penguins and lemmings, In *Product Standardization and Competitive Strategy,* G. Landis, ed., pp. 1–21, Elsevier, New York, 1987.

Ford, D. and Ryan, C. Taking technology to market, *Harv. Bus. Rev.,* March-April: 1981.

Foster, R. N. *Innovation: The Attacker's Advantage,* Summit, New York, 1986.

Gould, S. L. The Panda's thumb of technology, *Nat. Hist.,* January: 14–23, 1987.

Hamilton, W. F. Corporate strategies for managing emerging technologies, *Technol. Soc.,* 7: 197–212, 1985.

Hampson, K. D. *Technology Strategy and Competitive Performance: A Study of Bridge Construction,* doctoral dissertation, Stanford University, Stanford, CA, 1993.

Hannan, H. T. and Freeman, J. H. Structural inertia and organizational change, *Am. Soc. Rev.,* 43: 149–164, 1984.

Henderson, R. M. and Clark, K. B. Architectural innovation: the reconfiguration of existing product technologies and the failure of established firms, *Admin. Sci. Q.,* 35: 9–30, 1990.

Iansiti, M. *Technology Integration,* Harvard Business School Press, Boston, 1997.

Imai, K., L. Nonaka, L., and Takeuchi, H. Managing the new product development process: how Japanese learn and unlearn, In *The Uneasy Alliance: Managing the Productivity-Technology Dilemma,* K. B. Clark, R. H. Hayes, and C. Lorenz, eds., Harvard Business School Press, Boston, 1985.

Itami, H. The Case for Unbalanced Growth of the Firm, Research Paper Series #681, Graduate School of Business, Stanford University, Stanford, CA, 1983.

Itami, H. *Mobilizing Invisible Assets,* Harvard University Press, Cambridge, MA, 1987.

Kelly, P. and Kranzberg, M., Eds. *Technological Innovation: A Critical Review of Current Knowledge,* San Francisco Press, San Francisco, 1978.

Leonard-Barton, D. Core capabilities and core rigidities: a paradox in new product development, *Strat. Mgt. J.,* 13: 111–126, 1992.

Levitt, B. and March, J. G. Organizational learning, *Annu. Rev. Sociol.,* 14: 1988.

Maidique, M. A. and Patch, P. Corporate strategy and technological policy, Harvard Business School, Case #9-679-033, rev. 3/80, 1978.

Maidique, M. A. and Zirger, B. J. The new product learning cycle, *Res. Policy,* 1–40, 1984.

McKelvey, B. and Aldrich, H. E. Populations, organizations, and applied organizational science, *Admin. Sci. Q.,* 28: 101–128, 1983.

Metcalfe, J. S. and Gibbons, M. Technology, variety and organization: a systematic perspective on the competitive process, In *Research in Technological Innovation, Management and Policy, Vol. 4,* R. S. Rosenbloom and R. A. Burgelman, eds., pp. 153–194, JAI Press, Greenwich, CT, 1989.

Mitchell, G. R. New approaches for the strategic management of technology, *Technol. Sci.,* 7: 132–144, 1986.

Nelson, R. R. and Winter, S. G. *An Evolutionary Theory of Economic Change,* Harvard University Press, Cambridge, MA, 1982.

Pavitt, K. L. R., Robson, M. J., and Townsend, J. F. Technological accumulation, diversification, and organization of U.K. companies, 1945–83, *Mgt. Sci.,* 35: 91–99, 1989.

Penrose, E. T. *The Theory of the Growth of the Firm,* M. E. Sharpe, White Plains, NY, 1980.

Pisano, G. and Teece, D. J. Collaborative arrangements and global technology strategy: some evidence from the telecommunications equipment industry, In *Research on Technological Innovation, Management and Policy, Vol. 4,* R. S. Rosenbloom and R. A. Burgelman, eds., pp. 227–256. JAI Press, Greenwich, CT, 1989.

Porter, M. E. *Competitive Strategy,* Free Press, New York, 1980.

Porter, M. E. The technological dimension of competitive strategy, In *Research on Technological Innovation, Management and Policy, Vol. 1,* R. S. Rosenbloom, ed., pp. 1–33, 1983.

Porter, M. E. *Competitive Advantage,* Free Press, New York, 1985.

Prahalad, C. K., Doz, I. L., and Angelmar, R. Assessing the scope of innovation: a dilemma for top management, In *Research on Technological Innovation, Management and Policy, Vol. 4,* R. S. Rosenbloom and R. A. Burgelman, eds., pp. 257–281, JAI Press, Greenwich, CT, 1989.

Prahalad, C. K. and Hamel, G. The core competence of the corporation, *Harv. Bus. Rev.,* May-June: 79–91, 1990.

Rich, B. R. and Janos, L. *Skunk Works: A Personal Memoir of My Years at Lockheed,* Little Brown, Boston, 1994.

Rosenberg, N. *Inside the Black Box,* Cambridge University Press, Cambridge, 1982.

Rosenbloom, R. S. Technological innovation in firms and industries: an assessment of the state of the art, In *Technological Innovation: A Critical Review of Current Knowledge,* P. Kelly and M. Kranzberg, eds., San Francisco Press, San Francisco, 1978.

Rosenbloom, R. S. Managing technology for the longer term: a managerial perspective, In *The Uneasy Alliance: Managing the Productivity-Technology Dilemma,* K. B. Clark, R. H. Hayes, and C. Lorenz, eds., Harvard Business School Press, Boston, 1985.

Rosenbloom, R. S. and Cusumano, M. A. Technological pioneering: the birth of the VCR industry, *Calif. Mgt. Rev.,* XXIX: 5176, 1987.

Rosenbloom, R. S. and Kantrow, A. M. The nurturing of corporate research, *Harv. Bus. Rev.,* January-February: 115–123, 1982.

Rutenberg, David. Umbrella Pricing, working paper, Queens University, 1986.

Selznick, P. *Leadership in Administration,* Harper and Row, New York, 1957.

Shepard, A. Licensing to enhance demand for new technologies, *Rand. J. Econ.,* 21: 147–160, 1967.

Stalk, G., Evans, P., and Shulman, L. E. Competing on capabilities: the new rules of corporate strategy, *Harv. Bus. Rev.,* March-April: 57–69, 1992.

Steere, D. and Burgelman, R. A. Intel Corporation (D): Microprocessors at the Crossroads, case BP-256D, Stanford Business School, Stanford, CA, 1994.

Teece, D. J. Profiting from technological innovation: implications for integration, collaboration, licensing and public policy, *Res. Policy,* 15: 285–305, 1986.

Teece, D. J., Pisano, G., and Shuen, A. Firm Capabilities, Resources, and the Concept of Strategy, Working paper #90-9, University of California at Berkeley, Center for Research in Management, 1990.

Tushman, M. L. and Rosenkopf, L. Organizational determinants of technological change, In *Research in Organizational Behavior, Vol. 14,* B. Staw and L. Cummings, eds., pp. 311–325, JAI Press, Greenwich, CT, 19??.

Tushman, M. L. and Anderson, P. Technological and organizational environments, *Admin. Sci. Q.,* 31: 439–465, 1986.

Twiss, B. *Managing Technological Innovation,* Longman, London, 1980.

Van de Ven, A. H. and Garud, R. A framework for understanding the emergence of new industries, In *Research on Technological Innovation, Management and Policy, Vol. 4,* R. S. Rosenbloom and R. A. Burgelman, eds., pp. 195–226, JAI Press, Greenwich, CT, 1989.

von Hippel, E. *The Sources of Innovation,* Oxford University Press, New York, 1988.

von Hippel, E. A. Has a customer already developed your next product?, *Sloan Mgt. Rev.,* Winter: 1978.

Weick, K. *The Social Psychology of Organizing,* Addison-Wesley, Reading, MA, 1979.

Wheelwright, S. C., and Clark, K. B. *Revolutionizing Product Development,* Free Press, New York, 1992.

16.2   The Dual Mission of Technology in Strategic Affairs

Richard A. Goodman

Excellence in technology strategy does not imply the finest products imaginable. Rather, it implies the corporation's ability to **balance** technological potential against the needs of operational effectiveness, the market viability of the results, and the nature of the barriers to competition. A well-defined and executed technology program provides the firm with defensible strategic advantage in the marketplace. Such a technology program requires strategic resource allocations over considerable time periods and effective coordination throughout the organization — from finance and marketing to production and distribution.

The potential impact of technology on the firm's strategic affairs is clear. What is not so clear is the various role's that technology can play in strategic affairs. In some cases, technology programs provide direct strategic advantage and in other cases they provide a platform for corporate strategic moves. Sorting out and understanding the differences in mission requires working through several nested conceptualizations.

Alternative Technology Roles in Creating Advantage

At a broad level, it is possible to logically identify seven technologically based actions that can be linked with the several sources of barriers to competition. For instance, an investment in productivity drives costs down and permits price reductions, which in turn lowers the cost to the customer and requires the competitors to respond by making a similar investment thus increasing their costs. Alternatively, the increased margin can be employed by the firm to differentiate its product through a variety of advertising and/or service actions, which also raises the cost to the competitor.

An investment in capacity normally has a cost reduction effect and a potential preemptive effect. The cost reduction benefits are the same as those above. A preemptive effect occurs when the new capacity brings down the unit cost while simultaneously creating a capacity or overcapacity condition in the industry. Under these circumstances, the relative cost to competitors is driven up.

Investment in flexibility reduces costs to the customer when high change rate or niche strategies are called for and, of course, drives up costs to competitors who would need to make the same investment in order to compete (Table 16.1).

The three product research and development (R&D) strategies all create informational barriers to the competitors by developing proprietary knowledge, creating higher costs for the competitors who have to respond in kind. Experience curve benefits can also be associated with these developments as the firm becomes increasingly sophisticated and better understands the technologies involved in product development.

The final category deals with benefits of formal, contractual, or informal technological, vertical integration. By definition, these actions create contractual or regulatory barriers to competitors as they formally preempt certain markets or market segments.

TABLE 16.1   Technology Actions and Competitive Barriers

Action	Type of Barrier Created
Invest in productivity	Cost/price or cost/reputation
Invest in capacity	Cost/market size
Invest in flexibility	Cost/market size
R&D for new general products	Information barriers
R&D for new niche markets	Information barriers
R&D for hierarchical design	Information barriers
Negotiate hierarchical governance	Contractual/regulatory barriers

The Duality of Technology Programs

The most important framework for understanding is the "dual mission" of technology programs. One fundamental mission of technology is to be *directly supportive of corporate strategic initiatives*. The other fundamental mission of technology is to *contribute to the operational effectiveness* of the corporation. Corporate strategic initiatives derive from analysis but rely on choice and judgment. Operational effectiveness on the other hand requires technology programs that anticipate technological change.

Many elements of the firms competencies can be assessed and arrayed. The competitive situation and market opportunities can be evaluated. Alternative actions can be identified to build or strengthen the firms' advantage. While at the bottom of all this analysis, it is up to the executive team to exercise judgment and decide what to risk and what to protect. This involves the decision to rapidly push frontiers or to smoothly evolve the business from its current strengths. The technological programs that support a "push" strategy are often very different from those that support a "build" strategy.

Operationally directed technology programs anticipate technological change and serve to create learning curve benefits in advance of the need to insert new technologies into the firms' business operations. The mission of technology programs here is to understand what technological changes are going to be required and to begin building intellectual capital about the nature of new technologies. Thus, when a new product or process needs to be introduced, the technological risk has been reduced.

The dual mission of technology programs is to develop intellectual capital in support of the firm strategic objectives and in support of operational effectiveness. Both are proactive stances. This is best represented by considering the development of intellectual capital as the primary objective of the firm and product or process innovations as by-products. This, of course, is the opposite of the traditional way of thinking about technology. This frame helps remind management that intellectual capital is the basis of strategic and operational leverage and that innovations are the by-products of the state-of-technological capital at the time that the development is initiated (Fig. 16.4).

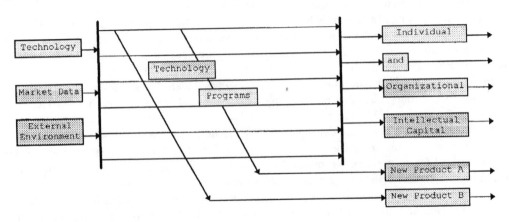

FIGURE 16.4   Technology programs and intellectual capital.

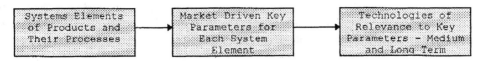

FIGURE 16.5   Technology mapping.

TABLE 16.2   Technology Program Investment Levels

Base technologies	Necessary but not advantage	Little investment
Key technologies	Basis of firm's advantage	Systematically built
Pacing technologies	Basis of future advantage	Selective development
Emerging technologies	Long-term changes to advantage	Monitoring

The ability to simultaneously build for the future while satisfying present needs can be represented by a practical measurement metaphor. When evaluating "new product A", three questions can be addressed. Does the project *meet specifications?* This suggests that earlier technology program choices were on the mark. Does the product receive *broad customer acceptance?* This indicates that the technology programs supported the short run tactical/strategic initiatives of the firm. Does the product contribute to *downstream payoff* for next-generation products and processes? This indicates that information on successes and failures along the way has been accumulated into the firm's intellectual capital for later technostrategic deployment.

A technology mapping exercise in each of the firms product and process areas can identify the key parameters that will have leverage in the marketplace. These key parameters can then be associated with the diverse technological approaches that hold promise and an investment plan can be created. Technological mapping represents the second level of nested framework required for understanding the potential role of technology in strategy (Fig. 16.5).

Given the "key parameters — technologies of relevance" assessment, an additioanl framework can be employed. In this framework the technological potential is sorted into base, key, pacing, and emerging technologies. **Base** technologies represent the technologies that all firms must and do have to participate in the product marketplace. These are necessary technologies but do not differentiate one firm from another. Little investment should be directed in these arenas with the exception of some training resources for update and maintenance of personnel. **Key** technologies are those technologies that provide the firm with current advantage (Table 16.2). These technologies should be the focus of systematic building.

Pacing technologies are those technologies that will alter the bases of competition. These technologies require serious investment and need to be selectively developed. Here we can see that this investment has a near- to medium-term strategic implication. The technologies themselves provide or maintain a competitive position in the marketplace. The systematic building of these technologies means that the deployment process will be smoother or more efficient since many of the technological uncertainties will have been resolved by the investment. The dual mission of technology programs is easily observed here. **Emerging** technologies represent medium- to long-term changes to the bases of competition. From an investment perspective, emerging technologies need to be monitored. This again is consistent with the idea of technology programs focusing upon strategic and operational objectives. The monitoring of emerging technologies allows the firm to anticipate opportunities and threats and to prepare to integrate new technologies into their strategic thinking.

In Summary

The fundamental mission of technology in the corporation is to support the corporate strategy by enabling the firm to create defensible competitive advantage *and* to contribute to the operational efficiency of the firm. This latter effort is accomplished by being far enough down the learning curve on new technologies that the adoption of new technological approaches in products, services, or processes can proceed with reasonable assurance of the operational outcomes. This statement can best by understood when the main objectives of the firm are seen as developing technological intellectual capital and the by-products of the firm are technological innovations. The technology programming approach turns the traditional view of the corporation on its side but expressed somewhat eloquently the basic issues of technological capital management.

The future can be studied by employing key parameter analysis to describe the technologies of relevance in the future and by using an investment schema based upon the idea of base, key, pacing and emerging categories. Different investments create different opportunities for defensible advantage.

Defining Terms

Key parameters: The dimensions of a future system, or a subsystem, that will drive purchasing behavior. They should be the focus of the next generation development.

Technologies of relevance: The various technologies that might be employed to solve the key parameters needs of the customer.

Technology mission: Supports corporate strategy by employing the evolving technologies of relevance to create defensible competitive advantage.

Technology programs: The corporate investments int he develoment of underlying technologies, i.e., the basic technological approaches of subsystems. This does not refer to project development programs — but new product developments can contribute to a technology program through furthering the understanding of one or another technological approach.

References

Abernathy, W. J. and Goodman, R. A. Strategies for development projects: an empirical study, *R&D Mgt.*, 2: 125–129, 1972.

Arthur D. Little, Inc. The Strategic Management of Technology, 1981.

Balderson, J., Birnbaum, P., Goodman, R. A., and Stahl, M., *Modern Management Techniques in Engineering and R&D*, Van Nostrand Reinhold, New York, 1984.

Goodman, R. A. and Lawless, M. W. *Technology Strategy: A Conceptual and Diagnostic Approach*, Oxford University Press, New York, 1994.

Furthur Information

There are two central books for this arena and one major reference volume plus several professional associations. Aside from the Goodman and Lawless *Technology and Strategy* text, good background can be found in Burgelman Maidique, and Wheelwright's *Strategic Management of Technology*, 2nd ed. (Chicago: Irwin, 1996), Tushman and Anderson's *Managing Strategic Innovation and Change* (New York: Oxford, 1997), and the Clarke and Reavley *Science and Technology Management Bibliography* (Ottawa: Stargate, 1993).

For professional enrichment, further perspective is available through the Strategic Management Society and the Society of Competitive Intelligence Professionals. This can be supplemented academically at the Technology Innovation Management Division and/or the Business Policy and Strategy Division of the Academy of Management or the College of R&D and/or the College of Planning of the Institute of Management Sciences.

16.3   The Business Plan

Robert D. Hisrich

In the hypercompetitive environment of today, there is perhaps nothing more important than planning and specifically developing a business plan. In any organization, there are many different types of plans — financial, human resource, marketing, production, and sales. These plans may be short term or long term, strategic or operational and may vary greatly in scope. In spite of the differences in scope and coverage, each plan has a common purpose: to provide guidance and structure on a continuing basis for managing the organization in a rapidly changing hypercompetitive environment.

Purpose of a Business Plan

In this era of planning and hypercompetition, a business plan is an integral part of strategically managing an organization. By describing all of the relevant external and internal elements involved in managing the organization, the business plan integrates the functional plans such as finance, human resources, marketing, production, and sales, providing a road map for the future of the organization.

A business plan is usually read by a variety of stakeholders, and often has many different purposes. It needs to be comprehensive enough to address the issues and concerns of advisors, bankers, consultants, customers, employees, investors, and venture capitalists. Generally, the purpose of a business plan is to obtain financial resources, obtain other resources, develop strategic alliances, and/or provide direction and guidance for the organization. While a business plan can serve several purposes, its most frequent use is to obtain financial resources. Bankers, investors, and venture capitalists will not take an investment possibility seriously without a business plan. Some will not even meet with an organization without first reviewing the organization's business plan. A well-developed business plan then is important as it (1) provides guidance to the entrepreneur and managers in decision making and organizing the direction of the company, (2) indicates the viability of an organization in the designated market, and (3) serves as the vehicle in obtaining financing.

Aspects of a Business Plan

Given the importance and purpose of a business plan, it is imperative that it be comprehensive and cover in depth all aspects of the organization. The plan will be read by a variety of individuals, each of whom is looking for a certain level of detail. As is indicated in Table 16.3, the business plan can be divided into three primary areas, each of which has several sections.

The first primary area, while the shortest, is perhaps the most significant, particularly when the purpose is to secure financing. This area consists of the title page, table of contents, and executive summary. The title page should contain the following information: (1) the name, address, telephone, fax, and e-mail numbers of the organization, (2) the name and position of the principle individuals in the organization, (3) three to four sentences briefly describing the nature of the organization and the purpose of the business plan, and (4) a statement of confidentiality such as "this is confidential business plan number 3, which cannot be reproduced without permission." This statement is important as each numbered business plan needs to be accounted for by recording the person and organization of the individual receiving it and the date of receipt. When trying to obtain financing, this is particularly important as follow-up can be scheduled at the appropriate time, which is about 45 days from the receipt date assuming the individual has not already initiated contact. As one venture capitalist commented, "One way I get a feel for the hunger and drive of the entrepreneur is by waiting to see if he/she initiates follow-up at the appropriate time."

The table of contents is perhaps the easiest part of the business plan to develop. It should follow the standard format with major sections and appendixes (exhibits) indicated along with the appropriate page numbers.

The final part of the first primary area of the business plan — the executive summary — is the most important and most difficult to develop, particularly when the purpose of the business plan is to secure financing or other resources. This no more than three-page summary is frequently used by upper-level investors, venture capitalists, and bankers to determine if the entire business plan is worth reading and analyzing. The executive summary then becomes the screen or hurdle that determines if more detailed attention will be given to the plan. Imagine a typical venture capitalist who receives about 80 business plans/month of about 200 pages each. He or she needs to employ some mechanism for screening this large number down to perhaps 10 to 15 for more focused initial attention.

TABLE 16.3   Business Plan Outline

Title Page	*Primary*
Table of Contents	*Area*
Executive Summary	*One*

1.0 Description of business
    Product(s)
    Services
    Size of business
    Office equipment and personnel
    Background of entrepreneurs
2.0 Description of technology
    Nature of technology
    Related products
    Patent information
3.0 Description of industry
    Future outlook and trends
    Analysis of competitors
    Market segment
    Industry and market forecasts
4.0 Marketing plan
    Pricing distribution
    Promotion
    Product
    Service needed
5.0 Financial plan   *Primary*
    Pro forma income statement   *Area*
    Cash flow projections   *Two*
    Pro forma balance sheet
    Break-even analysis
    Sources and applications of funds statement
6.0 Production plan
    Manufacturing process (amount subcontracted)
    Physical plant and layout
    Machinery and equipment
    Names of suppliers of raw materials
7.0 Organization plan
    Form of ownership
    Identification of partners or principal shareholders
    Authority of principals
    Management team background
    Roles and responsibilities of members of organization
8.0 Operational plan
    Description of company's operation
    Flow of orders and goods
9.0 Summary   *Primary*
Appendices (exhibits)   *Area*
    A. Resumes of principals   *Three*
    B. Market statistics
    C. Market research data
    D. Competitive brochures and price lists
    E. Patent material
    F. Leases and contracts
    G. Supplier price lists

Source: Adapted from Hisrich, R. D. and Peters, M. P. *Entrepreneurship: Starting, Developing and Managing a New Venture,* p. 120, Irwin, Burr Ridge, IL, 1993.

Given its importance, the executive summary should be written last and be written and rewritten until it highlights the organization in a concise, convincing manner covering the key points in the business plan. The executive summary should emphasize the three most critical areas for the success of the organization. In order of importance, these are the characteristics, capabilities, and experience of the management team; the nature and degree of innovativeness of the product or service and its market size and characteristics; and the expected results in terms of sales and profits over the next 3 years.

The second primary area of the business plan is the essence of the plan, which contains the following sections: description of the business, description of the technology, description of the industry marketing plan, financial plan, productions plan, organization plan, operations plan, and summary (see Table 16.3). This area should be self-contained, flowing smoothly from the description of the business section to the summary section. Part one (title page, table of contents, and executive summary) precedes this area and is also self contained. Part three (appendices or exhibits) is also self-contained, providing supportive material for the various sections in the second primary area.

The first section in this second primary area — the description of the business — describes in detail the past, present, and future of the organization. Answering the questions indicated in Table 16.4 will provide the information needed for the person reading the business plan to understand the history of the organization, its present size and scope, and its future over the next 3 years.

The second section contains a description of the technology. Business plans for nontechnology-based products or services, of course, would not contain this section. For technologically-based products, this section is extremely important as it provides information on the nature of the technology, the unique differential advantage the technology has over rival technology, and the degree that the technology is protectable, particularly by patents, copyrights, or trade secrets. In many cases, it is this section alone that makes someone interested in the organization.

Following the technology section comes the description of the industry. This section is important as it puts the organization in its proper context and competitive position. The critical issues needing to be addressed in this section are indicated in Table 16.5. This section gives a historical overview of the industry, its present situation in terms of seize and offerings, and its future outlook. A particularly important part of this section is the competitive analysis, which should describe the strengths and weaknesses of each major competitor with respect to the corresponding strengths and weaknesses of the organization. This section should conclude with a forecast of the size and future outlook of the industry.

TABLE 16.4 Description of the Business

1. What are your product(s) and/or services?
2. Describe the product(s) and/or service(s), including patent, copyright, or trademark status
3. Where will the business be located?
4. Is your building new? Old? In need of renovations? (If renovation needed, state cost)
5. Is the building leased or owned? (State the term)
6. Why is this building and location right for your location?
7. What additional skills or personnel will be needed to operate the business?
8. What office equipment will be needed?
9. Will equipment be purchased or leased?
10. What is your business background?
11. What management experience do you have?
12. Describe personal data such as education, age, special abilities, and interests
13. What are your reasons for going into business?
14. Why will you be successful in this venture?
15. What work has been completed to date?

Source: Hisrich, R. D. and Peters, M. P. *Entrepreneurship: Starting, Developing and Managing a New Venture,* p. 123, Irwin, Burr Ridge, IL, 1993.

TABLE 16.5 Critical Issues for an Industry Description

1. What are total industry sales over the past 5 years?
2. What is anticipated growth in this industry?
3. How many new firms have entered this industry in the past 3 years?
4. What new products have been recently introduced in this industry?
5. Who are the nearest competitors?
6. How will your business operation be better than this?
7. Are each of your major competitors' sales growing, declining, or steady?
8. What are the strengths and weaknesses of each of your competitors?
9. What is the profile of your customers?
10. How does your customer profile differ from that of your competition?

Source: Hisrich, R. D. and Peters, M. P. *Entrepreneurship: Starting, Developing and Managing a New Venture,* p. 122, Irwin, Burr Ridge, IL, 1993.

The fourth section begins the four parts dealing with the important functional aspects of the organization. The first, the marketing plan (Section 4.0), is an important section as it describes the nature of the product or service and how it will be distributed, priced, and promoted to achieve the amount of sales indicated each year for the next 3 years. Since everyone realizes that marketing is involved in achieving the necessary sales, each marketing aspect should be discussed in detail in as comprehensive terms as possible. The sales forecasted each month in the first year and each quarter the second and third years should reflect the activities described in this section and be as accurate as possible as the sales figures are necessary for accurate profitability forecasting.

Given the close relationship, the financial plan follows logically as the next section (Section 5.0) in the second primary area of the business plan. This section evolves around the preparation of four basic statements: the sources and uses of funds statement, the pro forma income statements for at least the next 3 years, the pro forma cash flow statements for at leas the next 3 years, and the pro forma balance sheets for at least the next 3 years. If the company has already been in operation, then past income statements and balance sheets should be in the appendix and discussed in the financial plan section. The first statement — the sources and uses of funds statement — indicates the amount of money needed by use such as working capital, inventory raw materials and supplies, administration expenses, and operating expenses and the sources of this money, such as self, relatives and friends, banks, venture capitalists, and other investors. The second statement — the pro forma income statement — indicates the sales costs and profits or losses for at least the first 3 years. For high-technology companies with large research and development costs and a longer time to break even, a 5-year plan may be needed. The pro forma income statement is a forecasted statement that contains the same items as a regular income statement such as sales, cost of goods sold, operating expenses, taxes, and net profit (or loss). These are forecasted by month for the first year and by quarter for each year after. The third statement — the cash flow statement — indicates the amount of cash the company has available. It's very similar to the pro forma income statement with the first year by month and each statement thereafter by quarter. The final statements in the financial plan are the pro forma balance sheets. These indicate the assets and liabilities of the company and are done at the end of each year for the next 3 years or the number of years of the pro forma income statements.

Following the financial plan is the sixth section — the production plan. As was the case with the description of the technology section, this section will not be present in business plans for service companies where no manufacturing occurs. The production plan completely describes the manufacturing process by describing the process itself, physical plant layout, machines and equipment, raw materials and suppliers names, and the costs of manufacturing. All information that the reader needs to understand the manufacturing operation should be supplied.

The seventh section — the operational plan — describes overall how the company will operate. This section should focus on the flow of orders and goods enabling the reader to understand the process that occurs from the time an order is received until the product or resource is delivered. This provides an understanding of the overall operation of an organization.

Primary area two concludes with a summary. This short section merely summarize the preceding sections by highlighting the most important points and concludes with any requests of the reader such as providing the financing or resources needed. This summary finalizes primary area two, making the area self-contained.

Following this core of the business plan comes primary area three — the appendices or exhibits. These provide supporting and additional information amplifying the material presented in primary area two. Typical appendices include: resumes of principals, markets statistics, market research data, competitors brochures and price lists, leases and contracts, and supplier price lists (see Table 16.3).

Using The Plan

Given the amount of time and effort needed to develop a good business plan, it is important that the plan be carefully implemented and used to provide guidance for the firm in all areas of its operation. The business plan will be most effective when controls are simultaneously implemented and the progress toward the established objectives are reviewed on a regular basis. Since the business is operating in a hypercompetitive changing environment, it is important that the organization be sensitive to changes in the company, industry, and market and make the appropriate changes in the business plan as needed. This will allow the business plan to be most effective in successfully guiding the organization.

16.4 Building Your Company's Vision[2]

James C. Collins and Jerry I. Porras

> We shall not cease from exploration
> And the end of all our exploring
> Will be to arrive where we started
> And know the place for the first time.

> T.S. Eliot, Four Quartets

Companies that enjoy enduring success have core values and a core purpose that remain fixed while their business strategies and practices endlessly adapt to a changing world. The dynamic of preserving the core while stimulating progress is the reason that companies such as Hewlett-Packard (HP), 3M, Johnson & Johnson, Procter & Gamble, Merck, Sony, Motorola, and Nordstrom became elite institutions able to renew themselves and achieve superior long-term performance. Hewlett-Packard employees have long known that radical change in operating practices, cultural norms, and business strategies does not mean losing the spirit of the HP way — the company's core principles. Johnson & Johnson continually questions its structure and revamps its processes while preserving the ideals embodied in its credo. In 1996, 3M sold off several of its large mature businesses — a dramatic move that surprised the business press — to refocus on its enduring core purpose of solving unsolved problems innovatively. We studied companies such as these in our research for *Built to Last: Successful Habits of Visionary Companies* and found that they have outperformed the general stock market by a factor of 12 since 1925.

Truly great companies understand the difference between what should never change and what should be open for change, between what is genuinely sacred and what is not. This rare ability to manage continuity and change — requiring a consciously practiced discipline — is closely linked to the ability to develop a vision. Vision provides guidance about what core to preserve and what future to stimulate progress toward. However, *vision* has become one of the most overused and least understood words in the language, conjuring up different images for different people: of deeply held values, outstanding achievement, societal bonds, exhilarating goals, motivating forces, or raisons d'être. We recommend a conceptual framework to define vision, add clarity and rigor to the vague and fuzzy concepts swirling

---

[2]From Harvard Business Review, Sept.-Oct. 1996. With permission.

FIGURE 16.6   Articulating a vision.

around that trendy term, and give practical guidance for articulating a coherent vision within an organization. It is a prescriptive framework rooted in 6 years of research and refined and tested by our ongoing work with executives from a great variety of organizations around the world.

A well-conceived vision consists of two major components: core ideology and envisioned future (see Fig. 16.6). Core ideology, the yin in our scheme, defines what we stand for and why we exist. Yin is unchanging and complements yang, the envisioned future. The envisioned future is what we aspire to become, to achieve, to create — something that will require significant change and progress to attain.

Core Ideology

Core ideology defines the enduring character of an organization — a consistent identity that transcends product or market life cycles, technological breakthroughs, management fads, and individual leaders. In fact, the most lasting and significant contribution of those who build visionary companies is the core ideology. As Bill Hewlett said about his longtime friend and business partner David Packard upon Packard's death not long ago, "As far as the company is concerned, the greatest thing he left behind him was a code of ethics known as the HP Way." HP's core ideology, which has guided the company since its inception more than 50 years ago, includes a deep respect for the individual, a dedication to affordable quality and reliability, a commitment to community responsibility (Packard himself bequeathed his $4.3 billion of HP stock to a charitable foundation), and a view that the company exists to make technical contributions for the advancement and welfare of humanity. Company builders such as David Packard, Masaru Ibuka of Sony, George Merck of Merck, William McKnight of 3M, and Paul Galvin of Motorola understood that it is more important to know who you are than where you are going, for where you are going will change as the world around you changes. Leaders die, products become obsolete, markets change, new technologies emerge, and management fads come and go, but core ideology in a great company endures as a source of guidance and inspiration.

Core ideology provides the glue that holds an organization together as it grows, decentralizes, diversifies, expands globally, and develops workplace diversity. Think of it as analogous to the principles of Judaism that held the Jewish people together for centuries without a homeland, even as they spread throughout the Diaspora. Or think of the truths held to be self-evident in the Declaration of Independence, or the enduring ideals and principles of the scientific community that bond scientists from every nationality together in the common purpose of advancing human knowledge. Any effective vision must embody the core ideology of the organization, which in turn consists of two distinct parts: core values, a system of guiding principles and tenets; and core purpose, the organization's most fundamental reason for existence.

TABLE 16.6 Core Values Are a Company's Essential Tenets

Merck	Encouraging individual initiative
Corporate social responsibility	Opportunity based on merit, no one is entitled to anything
Unequivocal excellence in all aspects of the company	Hard work and continuous self-improvement
Science-based innovation	Sony
Honesty and integrity	Elevation of the Japanese culture and national status
Profit, but profit from work that benefits humanity	Being a pioneer — not following others; doing the
Nordstrom	impossible
Service to the customer above all else	Encouraging individual ability and creativity
Hard work and individual productivity	Walt Disney
Never being satisfied	No cynicism
Excellence in reputation, being part of something	Nurturing and promulgation of "wholesome American
special	values"
Philip Morris	Creativity, dreams, and imagination
The right to freedom of choice	Fanatical attention to consistency and detail
Winning — beating others in a good fight	Preservation and control of the Disney magic

Core Values

Core values are the essential and enduring tenets of an organization. A small set of timeless guiding principles, core values require no external justification; they have *intrinsic* value and importance to those inside the organization. The Walt Disney Company's core values of imagination and wholesomeness stem not from market requirements but from the founder's inner belief that imagination and wholesomeness should be nurtured for their own sake. William Procter and James Gamble didn't instill in P&G's culture a focus on product excellence merely as a strategy for success but as an almost religious tent. And that value has been passed down for more than 15 decades by P&G people. Service to the customer — even to the point of subservience — is a way of life at Nordstrom that traces its roots back to 1901, 8 decades before customer service programs became stylish. For Bill Hewlett and David Packard, respect for the individual was first and foremost a deep personal value; they didn't get it from a book or hear it from a management guru. And Ralph S. Larsen, CEO of Johnson & Johnson, puts it this way: "The core values embodied in our credo might be a competitive advantage, but that is now *why* we have them. We have them because they define for us what we stand for, and we would hold them even if they became a competitive *dis*advantage in certain situations."

The point is that a great company decides for itself what values it holds to be core, largely independent of the current environment, competitive requirements, or management fads. Clearly, then, there is no universally right set of core values. A company need not have as its core value customer service (Sony doesn't) or respect for the individual (Disney doesn't) or quality (Wal-Mart Stores doesn't) or market focus (HP doesn't) or teamwork (Nordstrom doesn't). A company might have operating practices and business strategies around those qualities without having them at the essence of its being. Furthermore, great companies need not have likable or humanistic core values, although many do. The key is not *what* core values an organization has but that it has core values at all.

Companies tend to have only a few core values, usually between three and five. In fact, we found that none of the visionary companies we studied in our book had more than five: most had only three or four (see Table 16.6). Indeed, we should expect that. Only a few values can be truly *core*, that is, so fundamental and deeply held that they will change seldom, if ever.

To identify the core values of your own organization, push with relentless honesty to define what values are truly central. If you articulate more than five or six, chances are that you are confusing core values (which do not change) with operating practices, business strategies, or cultural norms (which should be open to change). Remember, the values must stand the test of time. After you've drafted a preliminary list of the core values, ask about each one. If the circumstances changed and *penalized* us for holding this core value, would we still keep it? If you can't honestly answer yes, then the value is not core and should be dropped from consideration.

A high-technology company wondered whether it should put quality on its list of core values. The CEO asked, "Suppose in ten years quality doesn't make a hoot of difference in our markets. Suppose the only thing that matters is sheer speed and horsepower but not quality. Would we still want to put quality on our list of core values?" The members of the management team looked around at one another and finally said no. Quality stayed in the *strategy* of the company, and quality-improvement programs remained in place as a mechanism for stimulating progress; but quality did not make the list of core values.

The same group of executives then wrestled with leading-edge innovation as a core value. The CEO asked, "Would we keep innovation on the list as a core value, no matter how the world around us changed?" This time, the management team gave a resounding yes. The managers' outlook might be summarized as, "We always want to do leading-edge innovation. That's who we are. It's really important to us and always will be. No matter what. And if our current markets don't value it, we will find markets that do." Leading-edge innovation went on the list and will stay there. A company should not change its core values in response to market changes; rather, it should change markets, if necessary, to remain true to its core values.

Who should be involved in articulating the core values varies with the size, age, and geographic dispersion of the company, but in many situations we have recommended what we call a *Mars Group*. It works like this: Imagine that you've been asked to re-create the very best attributes of your organization on another planet but you have seats on the rocket ship for only five to seven people. Whom should you send? Most likely, you'll choose the people who have a gut-level understanding of your core values, the highest level of credibility with their peers, and the highest levels of competence. We'll often ask people brought together to work on core values to nominate a Mars Group of five to seven individuals (not necessarily all from the assembled group). Invariably, they end up selecting highly credible representatives who do a super job of articulating the core values precisely because they are exemplars of those values — a representative slice of the company's genetic code.

Even global organizations composed of people from widely diverse cultures can identify a set of shared core values. The secret is to work from the individual to the organization. People involved in articulating the core values need to answer several questions: What core values do you personally bring to your work? (These should be so fundamental that you would hold them regardless of whether or not they were rewarded.) What would you tell your children are the core values that you hold at work and that you hope *they* will hold when they become working adults? If you awoke tomorrow morning with enough money to retire for the rest of your life, would you continue to live those core values? Can you envision them being as valid for you 100 years from now as they are today? Would you want to hold those core values, even if at some point one or more of them became a competitive *dis*advantage? If you were to start a new organization tomorrow in a different line of work, what core values would you build into the new organization regardless of its industry? The last three questions are particularly important because they make the crucial distinction between enduring core values that should not change and practices and strategies that should be changing all the time.

Core Purpose

Core purpose, the second part of core ideology, is the organization's reason for being. An effective purpose reflects people's idealistic motivations for doing the company's work. It doesn't just describe the organization's output or target customers; it captures the soul of the organization (see Table 16.7). Purpose, as illustrated by a speech David Packard gave to HP employees in 1960, gets at the deeper reasons for an organization's existence beyond just making money. Packard said,

> I want to discuss why a company exists in the first place. In other words, why are we here? I think many people assume, wrongly, that a company exists simply to make money. While this is an important result of a company's existence, we have to go deeper and find the real reasons for our being. As we investigate this, we inevitably come to the conclusion that a group of people get together and exist as an institution that we call a company so they are able to accomplish something collectively that they could not accomplish separately — they make a contribution to society, a phrase which sounds trite but is fundamental.... You can look around (in the general business world and) see people who are

TABLE 16.7   Core Purpose Is a Company's Reason for Being

3M: To solve unsolved problems innovatively	**McKinsey & Company:** To help leading corporations and governments be more successful
Cargill: To improve the standard of living around the world	**Merck:** To preserve and improve human life
Fannie Mae: To strengthen the social fabric by continually democratizing hoe ownership	**Nike:** To experience the emotion of competition, winning, and crushing competitors
Hewlett-Packard: To make technical contributions for the advancement and welfare of humanity	**Sony:** To experience the joy of advancing and applying technology for the benefit of the public
Lost Arrow Corporation: To be a role model and a tool for social change	**Telecare Corporation:** To help people with mental impairments realize their full potential
Pacific Theatres: To provide a place for people to flourish and to enhance the community	**Wal-Mart:** To give ordinary folk the chance to buy the same things as rich people
Mary Kay Cosmetics: To give unlimited opportunity to women	**Walt Disney:** To make people happy

interested in money and nothing else, but the underlying drives come largely from a desire to do something else: to make a product, to give a service — generally to do something which is of value.[3]

Purpose (which should last at least 100 years) should not be confused with specific goals or business strategies (which should change many times in 100 years). Whereas you might achieve a goal or complete a strategy, you cannot fulfill a purpose; it is like a guiding star on the horizon — forever pursued but never reached. Yet although purpose itself does not change, it does inspire change. The very fact that purpose can never be fully realized means that an organization can never stop stimulating change and progress.

In identifying purpose, some companies make the mistake of simply describing their current product lines or customer segments. We do not consider the following statement to reflect an effective purpose: "We exist to fulfill our government charter and participate in the secondary mortgage market by packaging mortgages into investment securities." The statement is merely descriptive. A far more effective statement of purpose would be that expressed by the executives of the Federal National Mortgage Association, Fannie Mae: "To strengthen the social fabric by continually democratizing home ownership." The secondary mortgage market as we know it might not even exist in 100 years, but strengthening the social fabric by continually democratizing home ownership can be an enduring purpose, no matter how much the world changes. Guided and inspired by this purpose, Fannie Mae launched in the early 1990s a series of bold initiatives, including a program to develop new systems for reducing mortgage underwriting costs by 40% in five years; programs to eliminate discrimination in the lending process (backed by $5 billion in underwriting experiments); and an audacious goal to provide, by the year 2000, $1 trillion targeted at 10 million families that had traditionally been shut out of home ownership — minorities, immigrants, and low-income groups.

Similarly, 3M defines its purpose not in terms of adhesives and abrasives but as the perpetual quest to solve unsolved problems innovatively — a purpose that is always leading 3M into new fields. McKinsey & Company's purpose is not to do management consulting but to help corporations and governments be more successful: in 100 years, it might involve methods other than consulting. HP doesn't exist to make electronic test and measurement equipment but to make technical contributions that improve people's lives — a purpose that has led the company far afield from its origins in electronic instruments. Imagine if Walt Disney had conceived of his company's purpose as to make cartoons, rather than to make people happy; we probably wouldn't have Mickey Mouse, Disneyland, EPCOT Center, or the Anaheim Might Ducks Hockey Team.

One powerful method for getting at purpose is the five whys. Start with the descriptive statement We make X products or We deliver X services, and then ask, Why is that important? five times. After a few whys, you'll find that you're getting down to the fundamental purpose of the organization. We used this

---

[3]David Packard, speech given to Hewlett-Packard's training group on March 8, 1960; courtesy of Hewlett-Packard Archives.

method to deepen and enrich a discussion about purpose when we worked with a certain market-research company. The executive team first met for several hours and generated the following statement of purpose for their organization: To provide the best market research data available. We then asked the following question: Why is it important to provide the best market-research data available? After some discussion, the executives answered in a way that reflected a deeper sense of their organization's purpose: To provide the best market-research data available so that our customers will understand their markets better than they could otherwise. A further discussion let team members realize that their sense of self-worth came not just from helping customers understand their markets better but also from making a *contribution* to their customers' success. This introspection eventually led the company to identify its purpose as: To contribute to our customers' success by helping them understand their markets. With this purpose in mind, the company now frames its product decisions not with the question Will it sell? but with the question Will it make a contribution to our customers' success?

The five whys can help companies in any industry frame their work in a more meaningful way. An asphalt and gravel company might begin by saying, We make gravel and asphalt products. After a few whys, it could conclude that making asphalt and gravel is important because the quality of the infrastructure plays a vital role in people's safety and experience; because driving on a pitted road is annoying and dangerous; because 747s cannot land safely on runways built with poor workmanship or inferior concrete; because buildings with substandard materials weaken with time and crumble in earthquakes. From such introspection may emerge this purpose: To make people's lives better by improving the quality of man-made structures. With a sense of purpose very much along those lines, Granite Rock Company of Watsonville, California, won the Malcolm Baldrige National Quality Award — not an easy feat for a small rock quarry and asphalt company. And Granite Rock has gone on to be one of the most progressive and exciting companies we've encountered in *any* industry.

Notice that none of the core purposes fall into the category "maximize shareholder wealth." A primary role of core purpose is to guide and inspire. Maximizing shareholder wealth does not inspire people at all levels of an organization, and it provides precious little guidance. Maximizing shareholder wealth is the standard off-the-shelf purpose for those organizations that have not yet identified their true core purpose. It is a substitute — and a weak one at that.

When people in great organizations talk about their achievements, they say very little about earnings per share. Motorola people talk about impressive quality improvements and the effect of the products they create on the world. HP people talk about their technical contributions to the marketplace. Nordstrom people talk about heroic customer service and remarkable individual performance by star salespeople. When a Boeing engineer talks about launching an exciting and revolutionary new aircraft, she does not say, "I put my heart and soul into this project because it would add 37 cents to our earnings per share."

One way to get at the purpose that lies beyond merely maximizing shareholder wealth is to play the "Random Corporate Serial Killer" game. It works like this: Suppose you could sell the company to someone who would pay a price that everyone inside and outside the company agrees is more than fair (even with a very generous set of assumptions about the expected future cash flows of the company). Suppose further that this buyer would guarantee stable employment for all employees at the same pay scale after the purchase but with no guarantee that those jobs would be in the same industry. Finally, suppose the buyer plans to kill the company after the purchase — its products or services would be discontinued, its operations would be shut down, its brand names would be shelved forever, and so on. The company would utterly and completely cease to exist. Would you accept the offer? Why or why not? What would be lost if the company ceased to exist? Why is it important that the company continue to exist? We've found this exercise to be very powerful for helping hard-nosed, financially focused executive reflect on their organization's deeper reasons for being.

Another approach is to ask each member of the Mars Group, How could we frame the purpose of this organization so that if you woke up tomorrow morning with enough money in the bank to retire, you would nevertheless keep working here? What deeper sense of purpose would motivate you to continue to dedicate your precious creative energies to this company's efforts?

As they move into the twenty-first century, companies will need to draw on the full creative energy and talent of their people. But why should people give full measure? As Peter Drucker has pointed out, the best and most dedicated people are ultimately volunteers, for they have the opportunity to do something else with their lives. Confronted with an increasingly mobile society, cynicism about corporate life, and an expanding entrepreneurial segment of the economy, companies more than ever need to have a clear understanding of their purpose in order to make work meaningful and thereby attract, motivate, and retain outstanding people.

Discovering Core Ideology

You do not create or set core ideology. You *discover* core ideology. You do not deduce it by looking at the external environment. You understand it by looking inside. Ideology has to be authentic. You cannot fake it. Discovering core ideology is not an intellectual exercise. Do not ask, What core values should we hold? Ask instead, What core values do we truly and passionately hold? You should not confuse values that you think the organization ought to have — but does not — with authentic core values. To do so would create cynicism throughout the organization. ("Who're they trying to kid? We all know that isn't a core value around here!") Aspirations are more appropriate as part of your envisioned future or as part of your strategy, not as part of the core ideology. However, authentic core values that have weakened over time can be considered a legitimate part of the core ideology — as long as you acknowledge to the organization that you must work hard to revive them.

Also be clear that the role of core ideology is to guide and inspire, not to differentiate. Two companies can have the same core values or purpose. Many companies could have the purpose to make technical contributions, but few live it as passionately as HP. Many companies could have the purpose to preserve and improve human life, but few hold it as deeply as Merck. Many companies could have the core value of heroic customer service, but few create as intense a culture around that value as Nordstrom. Many companies could have the core value of innovation, but few create the powerful alignment mechanisms that stimulate the innovation we see at 3M. The authenticity, the discipline, and the consistency with which the ideology is lived — not the content of the ideology — differentiate visionary companies from the rest of the pack.

Core ideology needs to be meaningful and inspirational only to people inside the organization; it need not be exciting to outsiders. Why not? Because it is the people inside the organization who need to commit to the organizational ideology over the long term. Core ideology can also play a role in determining who *is* inside and who is not. A clear and well-articulated ideology attracts to the company people whose personal values are compatible with the company's core values; conversely, it repels those whose personal values are incompatible. You cannot impose new core values or purpose on people. Nor are core values and purpose things people can buy into. Executives often ask, How do we get people to share our core ideology? You don't. You can't. Instead, find people who are predisposed to share your core values and purpose; attract and retain those people; and let those who do not share your core values go elsewhere. Indeed, the very process of articulating core ideology may cause some people to leave when they realize that they are not personally compatible with the organization's core. Welcome that outcome. It is certainly desirable to retain within the core ideology a diversity of people and viewpoints. People who share the same core values and purpose do not necessarily all think or look the same.

Don't confuse core ideology itself with core-ideology statements. A company can have a very strong core ideology without a formal statement. For example, Nike has not (to our knowledge) formally articulated a statement of its core purpose. Yet, according to our observations, Nike has a powerful core purpose that permeates the entire organization: to experience the emotion of competition, winning, and crushing competitors. Nike has a campus that seems more like a shrine to the competitive spirit than a corporate office complex. Giant photos of Nike heroes cover the walls, bronze plaque of Nike athletes stand alongside the running track that rings the campus, and buildings honor champions such as Olympic marathoner Joan Benoit, basketball superstar Michael Jordan, and tennis pro John McEnroe. Nike people who do not feel stimulated by the competitive spirit and the urge to be ferocious simply do not last long in the culture. Even the company's name reflects a sense of competition: Nike is the Greek goddess of victory. Thus, although Nike has not formally articulated its purpose, it clearly has a strong one.

Identifying core values and purpose is therefore not an exercise in wordsmithery. Indeed, an organization will generate a variety of statements over time to describe the core ideology. In HP's archives, we found more than half a dozen distinct versions of the HP Way, drafted by David Packard between 1956 and 1972. All versions stated the same principles, but the words used varied depending on the era and the circumstances. Similarly, Sony's core ideology has been stated many different ways over the company's history. At its founding, Masaru Ibuka described two key elements of Sony's ideology: "We shall welcome technical difficulties and focus on highly sophisticated technical products that have great usefulness for society regardless of the quantity involved; we shall place our main emphasis on ability, performance, and personal character so that each individual can show the best in ability and skill."[4] Four decades later, this same concept appeared in a statement of core ideology called Sony Pioneer Spirit: "Sony is a pioneer and never intends to follow others. Through progress, Sony wants to serve the whole world. It shall be always a seeker of the unknown.... Sony has a principle of respecting and encouraging one's ability...and always tries to bring out the best in a person. This is the vital force of Sony."[5] Same core values, different words.

You should therefore focus on getting the content right — on capturing the essence of the core values and purpose. The point is not to create a perfect statement but to gain a deep understanding of your organization's core values and purpose, which can then be expressed in a multitude of ways. In fact, we often suggest that once the core has been identified, managers should generate their own statements of the core values and purpose to share with their groups.

Finally, don't confuse core ideology with the concept of core competence. Core competence is a strategic concept that defines your organization's capabilities — what you are particularly good at — whereas core ideology captures what you stand for and why you exist. Core competencies should be well aligned with a company's core ideology and are often rooted in it; but they are not the same thing. For example, Sony has a core competence of miniaturization — a strength that can be strategically applied to a wide array of products and markets. But it does not have a core *ideology* of miniaturization. Sony might not even have miniaturization as part of its strategy in 100 years, but to remain a great company, it will still have the same core values described in the Sony Pioneer Spirit and the same fundamental reason for being — namely, to advance technology for the benefit of the general public. In a visionary company like Sony, core competencies change over the decades, whereas core ideology does not.

Once you are clear about the core ideology, you should feel free to change absolutely *anything* that is not part of it. From then on, whenever someone says something should not change because "it's part of our culture" or "we've always done it that way" or any such excuse, mention this simple rule: If it's not core, it's up for change. The strong version of the rule is, *If it's not core, change it!* Articulating core ideology is just a starting point, however. You also must determine what type of progress you want to stimulate.

Envisioned Future

The second primary component of the vision framework is *envisioned future*. It consists of two parts: a 10- to 30-year audacious goal plus vivid descriptions of what it will be like to achieve the goal. We recognize that the phrase *envisioned future* is somewhat paradoxical. On the one hand, it conveys concreteness — something visible, vivid, and real. On the other hand, it involves a time yet unrealized — with its dreams, hopes, and aspirations.

Vision-level BHAG

We found in our research that visionary companies often use bold missions — or what we prefer to call BHAGs (pronounced BEE-hags and shorthand for Big, Hairy, Audacious Goals) — as a powerful way to

---

[4]See Nick Lyons, *The Sony Vision* (New York: Crown Publishers, 1976). We also used a translation by our Japanese student Tsuneto Ikeda.

[5]Akio Morita, *Made in Japan* (New York: E. P. Dutton, 1986), p. 147.

TABLE 16.8 Big, Hairy, Audacious Goals Aid Long-Term Vision

Target BHAGs can be quantitative or qualitative	Role-model BHAGs suit up-and-coming organizations
Become a $125 billion company by the year 2000 (Wal-Mart, 1990)	Become the Nike of the cycling industry (Giro Sport Design, 1986)
Democratize the automobile (Ford Motor Company, early 1900s)	Become as respected in 20 years as Hewlett-Packard is today (Watkins-Johnson, 1996)
Become the company most known for changing the worldwide poor-quality image of Japanese products (Sony, early 1950s)	Become the Harvard of the West (Stanford University, 1940s)
Become the most powerful, the most serviceable, the most far-reaching world financial institution that has ever been (City Bank, predecessor to Citicorp, 1915)	Internal-transformation BHAGs suit large, established organizations
Become the dominant play in commercial aircraft and bring the world into the jet age (Boeing, 1950)	Become number one or number two in every market we serve and revolutionize this company to have the strengths of a big company combined with the leanness and agility of a small company (General Electric Company, 1980s)
Common-enemy BHAGs involve	Transform this company from a defense contractor into the best diversified high-technology company in the world (Rockwell, 1995)
David-vs.-Goliath thinking	
Knock off RJR as the number one tobacco company in the world (Philip Morris, 1950s)	Transform this division from a poorly respected internal products supplier to one of the most respected, exciting, and sought-after divisions in the company (Components Support Division of a computer products company, 1989)
Crush Adidas (Nike, 1960s)	
Yamaha wo tsubusu! We will destroy Yamaha! (Honda, 1970s)	

stimulate progress. All companies have goals. But there is a difference between merely having a goal and becoming committed to a huge, daunting challenge — such as climbing Mount Everest. A true BHAG is clear and compelling, serves as a unifying focal point of effort, and acts as a catalyst for team spirit. It has a clear finish line, so the organization can know when it has achieved the goal; people like to shoot for finish lines. A BHAG engages people — it reaches out and grabs them. It is tangible, energizing, highly focused. People get it right away; it takes little or no explanation. For example, NASA's 1960s moon mission didn't need a committee of wordsmiths to spend endless hours turning the goal into a verbose, impossible-to-remember mission statement. The goal itself was so easy to grasp — so compelling in its own rights — that it could be said 100 different ways yet be easily understood by everyone. Most corporate statements we've seen do little to spur forward movement because they do not contain the powerful mechanism of a BHAG.

Although organizations may have many BHAGs at different levels operating at the same time, vision requires a special type of BHAG — a vision-level BHAG that applies to the entire organization and requires 10 to 30 years of effort to complete. Setting the BHAG that far into the future requires thinking beyond the current capabilities of the organization and the current environment. Indeed, inventing such a goal forces an executive team to be visionary, rather than just strategic or tactical. A BHAG should not be a sure bet — it will have perhaps only a 50% to 70% probability of success — but the organization must believe that it can reach the goal anyway. A BHAG should require extraordinary effort and perhaps a little luck. We have helped companies create a vision-level BHAG by advising them to think in terms of four broad categories: target BHAGs, common-enemy BHAGs, role-model BHAGs, and internal-transformation BHAGs (see Table 16.8).

Vivid Description

In addition to vision-level BHAGs, an envisioned future needs what we call *vivid description*, that is, a vibrant, engaging, and specific description of what it will be like to achieve the BHAG. Think of it as translating the vision from words into pictures, of creating an image that people can carry around in their heads. It is a question of painting a picture with your words. Picture painting is essential for making the 10- to 30-year BHAG tangible in people's minds.

For example, Henry Ford brought to life the goal of democratizing the automobile with this vivid description: "I will build a motor car for the great multitude.... It will be so low in price that no man making a good salary will be unable to own one and enjoy with this family the blessing of hours of pleasure in God's great open spaces.... When I'm through, everybody will be able to afford one, and everyone will have one. The horse will have disappeared from our highways, the automobile will be taken for granted...(and we will) give a large number of men employment at good wages."

The components-support division of a computer products company had a general manager who was able to describe vividly the goal of becoming one of the most sought-after divisions in the company: "We will be respected and admired by our peers.... Our solutions will be actively sought by the end-product divisions, who will achieve significant product 'hits' in the marketplace largely because of our technical contribution.... We will have pride in ourselves.... The best up-and-coming people in the company will seek to work in our division.... People will give unsolicited feedback that they love what they are doing.... (Our own) people will walk on the balls of their feet.... (They) will willingly work hard because they want to.... Both employees and customers will feel that our division has contributed to their life in a positive way."

In the 1930s, Merck had the BHAG to transform itself from a chemical manufacturer into one f the preeminent drug-making companies in the world, with a research capability to rival any major university. In describing this envisioned future, George Merck said at the opening of Merck's research facility in 1933, "We believe that research work carried on with patience and persistence will bring to industry and commerce new life; and we have fait that in this new laboratory, with the tools we have supplied, science will be advanced, knowledge increased, and human life win ever a greater freedom from suffering and disease;. We pledge our every aid that this enterprise shall merit the faith we have in it. Let your light so shine — that those who seek the Truth, that those who toil that this world may be a better place to live in, that those who hold aloft that torch of science and knowledge through these social and economic dark ages, shall take new courage and feel their hands supported."

Passion, emotion, and conviction are essential parts of the vivid description. Some managers are uncomfortable expressing emotion about their dreams, but that's what motivates others. Churchill understood that when he described the BHAG facing Great Britain in 1940. He did not just say, "Beat Hitler." He said, "Hitler knows he will have to break us on this island or lose the war. If we can stand up to him, all Europe may be free, and the life of the world may move forward into broad, sunlit uplands. But if we fail, the whole world, including the United States, including all we have known and cared for, will sink into the abyss of a new Dark Age, made more sinister and perhaps more protracted by the lights of perverted science. Let us therefore brace ourselves to our duties and so bear ourselves that if the British Empire and its Commonwealth last for a thousand years, men will still say, 'This was their finest hour.'"

A Few Key Points

Don't confuse core ideology and envisioned future. In particular, don't confuse core purpose and BHAGs. Managers often exchange one for the other, mixing the two together or failing to articulate both as distinct items. Core purpose — not some specific goal — is the reason why the organization exists. A BHAG is a clearly articulated goal. Core purpose can never be completed, whereas the BHAG is reachable in 10 to 30 years. Think of the core purpose as the star on the horizon to be chased forever; the BHAG is the mountain to be climbed. Once you have reached its summit, you move on to other mountains.

Identifying core ideology is a discovery process, but setting the envisioned future is a creative process. We find that executives often have a great deal of difficulty coming up with an exciting BHAG. They want to analyze their way into the future. We have found, therefore, that some executives make more progress by starting first with the vivid description and backing from there into the BHAG. This approach involves starting with questions such as, We're sitting here in 20 years; what would we love to see? What should this company look like? What should it feel like to employees? What should it have achieved? If someone writes an article for a major business magazine about this company in 20 years, what will it say? One biotechnology company we worked with had trouble envisioning its future. Said one member of the executive team, "Every time we come up with something for the entire company, it is just too generic to be exciting — something banal like 'advance biotechnology worldwide.'" Asked to paint a picture of the company in 20 years, the executives mentioned such things as "on the cover of *Business*

TABLE 16.9 Putting It All Together: Sony in the 1950s

Core Ideology	Envisioned Future

Core Values
   Elevation of the Japanese culture and national status
   Being a pioneer — not following others, doing the impossible
   Encouraging individual ability and creativity
 	
Purpose
   To experience the sheer joy of innovation and the application of technology for the benefit and pleasure of the general public
	BHAG
	   Become the company most known for changing the worldwide poor-quality image of Japanese products
	
	Vivid Description
	   We will create products that become pervasive around the world.... We will be the first Japanese company to go into the U.S. market and distribute directly.... We will succeed with innovations that U.S. companies have failed at — such as the transistor radio.... Fifty years from now, our brand name will be as well known as any in the world...and will signify innovation and quality that rival the most innovative companies anywhere.... "Made in Japan" will mean something fine, not something shoddy

Week as a model success story...the *Fortune* most admired top-ten list...the best science and business graduates want to work here...people on airplanes rave about one of our products to seatmaters... 20 consecutive years of profitable growth...an entrepreneurial culture that has spawned half a dozen new divisions from within...management gurus use us as an example of excellent management and progressive thinking," and so on. From this, they were able to set the goal of becoming as well respected as Merck or as Johnson & Johnson in biotechnology.

It makes no sense to analyze whether an envisioned future is the right one. With a creation — and the task is creation of a future, not prediction — there can be no right answer. Did Beethoven create the right Ninth Symphony? Did Shakespeare create the right *Hamlet*? We can't answer these questions; they're nonsense. The envisioned future involves such essential questions as Does it get our juices flowing? Do we find it stimulating? Does it spur forward momentum? Does it get people going? The envisioned future should be so exciting in its own right that it would continue to keep the organization motivated even if the leaders who set the goal disappeared. City Bank, the predecessor of Citicorp, had the BHAG "to become the most powerful, the most serviceable, the most far-reaching world financial institution that has ever been" — a goal that generated excitement through multiple generations until it was achieved. Similarly, the NASA moon mission continued to galvanize people even though President John F. Kennedy (the leader associated with setting the goal) died years before its completion.

To create an effective envisioned future requires a certain level of unreasonable confidence and commitment. Keep in mind that a BHAG is not just a goal; it is a Big, Hairy, Audacious Goal. It's not reasonable for a small regional bank to set the goal of becoming "the most powerful, the most serviceable, the most far-reaching world financial institution that has ever been," as City Bank did in 1915. It's not a tepid claim that "we will democratize the automobile," as Henry Ford said. It was almost laughable for Philip Morris — as the sixth-place player with 9% market share in the 1950s — to take on the goal of defeating Goliath RJ Reynolds Tobacco Company and becoming number one. It was hardly modest for Sony, as a small, cash-strapped venture, to proclaim the goal of changing the poor-quality image of Japanese products around the world (see Table 16.9). Of course, it's not only the audacity of the goal but also the level of commitment to the goal that counts. Boeing didn't just envision a future dominated by its commercial jets; it bet the company on the 707 and, later, on the 747. Nike's people didn't just talk about the idea of crushing Adidas; they went on a crusade to fulfill the dream. Indeed, the envisioned future should produce a bit of the "gulp factor": when it dawns on people what it will take to achieve the goal, there should be an almost audible gulp.

But what about failure to realize the envisioned future? In our research, we found that the visionary companies displayed a remarkable ability to achieve even their most audacious goals. Ford did democratize the automobile; Citicorp did become the most far-reaching bank in the world; Philip Morris did rise from sixth to first and beat RJ Reynolds worldwide; Boeing did become the dominant commercial aircraft company; and it looks like Wal-Mart will achieve its $125 billion goal, even without Sam Walton.

In contrast, the comparison companies in our research frequently did not achieve their BHAGs, if they set them at all. The difference does not lie in setting easier goals; the visionary companies tended to have even more audacious ambitions. The difference does not lie in charismatic, visionary leadership: the visionary companies often achieved their BHAGs without such larger-than-life leaders at the helm. Nor does the difference lie in better strategy: the visionary companies often realized their goals more by an organic process of "let's try a lot of stuff and keep what works" than by well-laid strategic plans. Rather, their success lies in building the strength of their organization as their primary way of creating the future.

Why did Merck become the preeminent drug maker in the world? Because Merck's architects built the best pharmaceutical research and development organization in the world. Why did Boeing become the dominant commercial aircraft company in the world? Because of its superb engineering and marketing organization, which had the ability to make projects like the 747 a reality. When asked to name the most important decisions that have contributed to the growth and success of HP, David Packard answered entirely in terms of decisions to build the strength of the organization and its people.

Finally, in thinking about the envisioned future, beware of the We've Arrived Syndrome — a complacent lethargy that arises once an organization has achieved one BHAG and fails to replace it with another. NASA suffered from that syndrome after the successful moon landings. After you've landed on the moon, what do you do for an encore? Ford suffered from the syndrome when, after it succeeded in democratizing the automobile, it failed to set a new goal of equal significance and gave General Motors the opportunity to jump ahead in the 1930s. Apple Computer suffered from the syndrome after achieving the goal of creating a computer that nontechies could use. Start-up companies frequently suffer from the We've Arrived Syndrome after going public or after reaching a stage in which survival no longer seems in question. An envisioned future helps an organization only as long as it hasn't yet been achieved. In our work with companies, we frequently hear executives say, "It's just not as exciting around here as it used to be; we seem to have lost our momentum." Usually, that kind of remark signals that the organization has climbed one mountain and not yet picked a new one to climb.

Many executives thrash about with mission statements and vision statements. Unfortunately, most of those statements turn out to be a muddled stew of values, goals, purposes, philosophies, beliefs, aspirations, norms, strategies, practices, and descriptions. They are usually a boring, confusing, structurally unsound stream of words that evoke the response "True, but who cares?" Even more problematic, seldom do these statements have a direct link to the fundamental dynamic of visionary companies: preserve the core and stimulate progress. That dynamic, not vision or mission statements, is the primary engine of enduring companies. Vision simply provides the context for bringing this dynamic to life. Building a visionary company requires 1% vision and 99% alignment. When you have superb alignment, a visitor could drop in from outer space and infer your vision from the operations and activities of the company without ever reading it on paper or meeting a single senior executive.

Creating alignment may be your most important work. But the first step will always be to recast your vision or mission into an effective context for building a visionary company. If you do it right, you shouldn't have to do it again for at least a decade.

16.5 Benchmarking

Nick Oliver

Ensuring that products and processes are competitive is a perennial problem for managers from all disciplines, but especially technology managers. Although the market ultimately provides feedback on product and service attributes such as quality, cost, and delivery, such feedback is imprecise and rarely permits useful operational interventions. Moreover, a product may succeed or fail in the marketplace for reasons that are unconnected to the internal workings of the organization that produces it — for example, being on the right side of the industry standards, enjoying a position of market dominance, marketing muscle, and tacit or explicit government intervention may all be significant.

Consequently, on many occasions it will be necessary to gather specific information on how an organization is performing. Often, this occurs through formal performance measurement systems, but these typically focus on the internal operations of an individual organization. It can also be useful for an organization to look *outside* its own boundaries and to compare its performance and internal processes with those of other organizations. Benchmarking is about gathering this external comparative information and then taking action based upon it.

The term "benchmark" was originally a surveyor's term and referred to a mark used as a sighting point. It denotes a yardstick or reference point against which measures can be taken. With the advent of computing in the latter part of the 20th century, "benchmark" computer programs developed. These programs comprised standard sets of tasks that could be run on different machines to test their relative performance. Benchmarking entered the management vocabulary during the latter part of the 1980s as a label to describe comparisons in business and process performance. The activity was popularized by the Xerox Corporation, which compared its U.S. operations with its counterparts in Fuji-Xerox in Japan and then used these comparisons to drive improvement activities.

Benchmarking Defined

The term benchmarking covers many approaches to process and performance comparison. Strictly speaking, benchmarking refers to true apples to apples comparisons between companies, plants, functions, or other specific operations. Over time the term benchmarking has become diluted, and today many so-called benchmarking programs represent little more than industrial tourism. Where this is the case, comparisons tend to be casual at best, and benchmarking refers to little more than the borrowing of ideas from other organizations. This ambiguity over terminology is troublesome, and demonstrates how the parties embarking on a benchmarking program need to establish a shared view of what they mean by benchmarking (and agree the level of precision to which they intend to work) at an early stage of the benchmarking process.

There are many definition of benchmarking. These include

The continuous process of measuring products, services, and practices against the toughest competitors or those companies recognized as industry leaders [Camp 1989].
A continuous search for and application of significantly better practices that lead to superior performance [Watson 1993].

Within these broad definitions there are a number of different types of benchmarking, but broadly these fall into four main types:

- Performance
- Process
- Strategic
- Product benchmarking

"Performance" benchmarking typically covers company-wide measures such as sales per employee or inventory turns. "Process" benchmarking refers to comparisons based on discrete business or other processes. For example, manufacturing processes may be compared in terms of yield rates, throughput times, and direct labour productivity. Warehouse operations may be compared according to criteria such as pick rates and the number of product lines they can handle. Indirect functions such as human resource functions may make comparisons of absence figures, human resource staff costs per employee, and so on. Such comparisons may be made between units within the same organization or between different organizations.

The term "strategic" benchmarking frequently appears in the literature, though purists balk at describing this activity as true benchmarking. Strategic benchmarking compares the strategies that companies pursue in order to compete in their particular markets. Of necessity, benchmarking exercises in this area are broad brush and concerned with general patterns rather than precise, systematic comparisons.

Product benchmarking compares the features and performance of actual products. Many companies routinely carry out tear-down analyses of competitor products to see how they compare in terms of design, manufacturability, and other features. Tests may also be conducted on product performance features, such as noise levels, durability, reliability, robustness, power consumption, etc. Such comparisons may form the basis of reverse engineering exercises. Independent product test data may be available from public sources (e.g., trade or consumer magazines) although rarely at a sufficient level of sophistication to permit detailed analysis. Such data, usually at a much higher level of detail, may also be available for a fee from closed networks.

Functions of Benchmarking

Benchmarking can perform several functions, and the approach adopted should clearly reflect the intended function of an individual study. Typical functions include

1. General "awareness-raising" ("What's going on out there?")
2. Identification of good practices, for assimilation ("What we can learn about how to do x better?")
3. Verification of a company's own position ("Are we as good as we think we are?")
4. Legitimization of change ("Look at how much we must improve")

Although the awareness-raising function smacks of industrial tourism, it cannot be denied that many companies have made breakthroughs based on what they have seen in other companies, even if these observations were not part of a systematic process of comparison. Bogan and English [1994] report that Henry Ford had the idea of a moving assembly line after seeing carcasses being moved around a Chicago slaughterhouse via hooks hanging from an overhead monorail. Toyota derived its inspiration for what later evolved into the Toyota Production System from Eliji Toyoda's observations of the factories of the U.S. automakers and of U.S. supermarkets, whose efficiency in restocking their shelves caught his attention. Such inspiration can often stem from unlikely places rather than from observations of operations closely comparable to one's own. The implication of this is that the intelligence-gathering net must be spread sufficiently widely, so that new ways of doing things are encountered, not just marginally better ways of doing the same things.

As one moves through functions 2, 3, and 4, the need for precision in the comparison process increases greatly, and considerable focus and discipline in the benchmarking process are required. This is particularly so if the real purpose of a benchmarking study is to drive through change, in which case the benchmarking data need to provide valid, inarguable evidence of the need to do things differently — typically by demonstrating a substantial performance shortfall. Such data will only have the desired effect if they are accepted as legitimate by those who will be required to change their working routines in order to close the gap. In their discussion of the diffusion of lean production methods Womack et al. [1990] talk about the importance of local demonstrations of the methods to break down the attitude of "It can't be done here". Benchmarking data perform a similar function, in lending credibility to improvement targets that might otherwise be regarded as little more than management wish lists.

Whatever function benchmarking is intended to perform, a key dilemma in designing a benchmarking study is the balance between breadth and depth of the comparison. The most powerful comparisons are those that reveal substantial performance differences between operations that deliver highly comparable products or services. However, the number of directly comparable organizations is likely to very small in many industries, even in large economies such as the United States or Japan. This problem means that many benchmarking studies are forced to include a more heterogeneous mix of units than is ideal in terms of strict comparability. The price of this is paid when the time comes to interpret performance differences. Are such differences due to different practices, from which much can be learned? Or are they due to differences in product, market, or sectoral characteristics?

A number of alternative approaches have evolved to address this problem of comparability of outputs. One approach is to construct a stereotype of "best practice"; the benchmarking exercise then tests for the presence or absence of the stereotype rather than gathering actual performance data. Several industry award exercises utilize variants of this method. Thus, rather than gathering actual data on units per hour or defect rates from a manufacturing company, this approach uses indicators of the practices a "good" company ought to exhibit (e.g., just-in-time production, kanban, quality improvement teams) and then tests for conformity to this model. Scoring systems are often used, companies being given points according to the characteristics that they report.

This approach to benchmarking carries a number of drawbacks. First, stereotypes of best practice in many areas are well articulated in the management literature, and consequently respondents may claim that best practices are being used to a much greater extent than is the case in reality. Second, because this approach tends to avoid actual measures of outcomes, the link between practices and performance is assumed rather than tested. Such links may not be proven but mediated by local conditions. For example, running a manufacturing system on a just-in-time basis (a typical example of manufacturing best practice) may yield many benefits in a high-volume, repetitive, manufacturing environment where process equipment and suppliers are reliable and demand is predictable. However, where these conditions do not apply, the same practice of low inventories may risk unfulfilled orders and hence reduced performance. Without regular validation of the practice/performance linkage, there is a danger that best practice becomes a set of self-perpetuating myths. Many national and international awards (e.g., the Baldrige award and the European Quality Award) are derived from assessments based on these principles.

A second alternative to real performance benchmarking is to compare specific processes across organizations whose overall missions and operations may be quite different. For example, Camp [1989] describes how Xerox compared its warehouse performance with that of a footwear manufacturer on measures such as orders per person per day, product lines per person day, and so on. This certainly creates opportunities for the cross-industry learning discussed earlier, but it also risks denial of the validity of the comparison.

Benchmarking Studies

Examples of benchmarking can be found in many areas of business. Benchmarking studies of manufacturing operations have received extensive attention, especially when these have involved international comparisons. One of the best known studies is the 1985-1990 International Motor Vehicle Program (IMVP), which was written up by Womack et al. in *The Machine that Changed the World* [1990]. This study alleged an enormous superiority on the part of Japanese auto plants compared to their non-Japanese counterparts in labour hours and defects per vehicle (approximately 2:1 on both measures). These differences were ascribed to the practices the Japanese producers, in particular Toyota, employed in their factories, practices that were summarized by the phrase "lean production".

The IMVP study illustrates the impact of a well-executed benchmarking study. The study provided a powerful cocktail of shocking statistics (the best Japanese plants produce vehicles using half the labor hours and with half the defects compared to Western plants) and recipe for success in the form of lean production principles. Hundreds of thousands of copies of *The Machine that Changed the World* have been sold, and many managers have used it — often rather uncritically — as a blueprint for high-performance manufacturing. This example demonstrates how benchmarking functions. First, the shock of the comparison creates a receptivity to new ideas and leads to a recognition of a need for change. Second, the exercise can reveal alternative, and in this case apparently superior, methods to address this need.

Benchmarking studies are by no means restricted to manufacturing operations, nor to the private sector. There have been benchmarking studies of new product development processes, software development, warehousing and distribution operations, human resource functions, information system functions, health services, and rescue services, to name but a few. In all such studies, the dynamics are essentially the same: compare, learn, and then improve.

TABLE 16.10 The Process of Benchmarking

Activity	Specific Actions
Planning	Ascertain the main purpose of the benchmarking exercise
	Identify the processes that is both feasible and desirable to benchmark
	Identify units to be included in the study
	Establish sources of data and collection methods
	Ascertain likely levels of cooperation (if collecting original data)
Data collection	Gather data (ensuring validity and consistency)
	Identify any special conditions to be taken into account in the comparison
Analysis	Collate data from all units, checking for consistency of approach; eliminating or correcting suspect data
Communication	Prepare dissemination materials, aiming for simplicity and transparency
	Place emphasis on learning, not evaluation
Action	Concentrate on a handful of key performance shortfalls identified by the exercise and establish root causes
	Specify programs (with realistic time scales) to address these gaps and establish priorities.
	Identify methods to capture learning
Monitoring	Track progress over time
	Repeat periodically — but not too frequently

The Benchmarking Process

Virtually every book on benchmarking offers a multistage representation of the benchmarking process — the Xerox process consists of 12 steps, AT&T's 9 steps, Motorola's process has 5 steps. Table 16.10 presents a summary of the key stages in a generic benchmarking process. In the interests of simplicity, Table 16.10 is produced as if the reader is conducting the benchmarking exercise. However, before the benchmaking process commences, there is the question of whether to conduct the study oneself to or commission another party to carry it out. If the exercise involves comparison between direct competitor organizations, it is likely that an independent third party may be needed. However, it is much more difficult to capture learning in general, and tacit knowledge in particular, if the exercise is outsourced.

Many crucial elements of a benchmarking exercise are determined during the planning phase, and it is extremely important that time and effort are invested at this stage. Clarity about the main purpose of the exercise is vital. If the exercise is being used as a lever for change, the results may be greeted with scepticism at best and denial at worst. In such cases, it is important that all the obvious possibilities to explain away the results are closed off — consequently, the emphasis in the study is likely to be on rigor and comparability. If the purpose is a more general awareness-raising exercise, a less stringent approach will be acceptable.

The choice of process or operations to be included is a second key decision during the planning phase. Issues of comparability should be given thorough consideration at this time, if necessary trading off the breadth of a study against the precision of the comparison. In the IMVP, for example, only certain activities that were common to all auto assembly plants (body construction, painting, and final assembly) were included in the process comparisons. The reason for this was simple — comparing a car plant that includes, for example, engine manufacture, with one that does not clearly tips the balance in favor of the less integrated plant on measures of hours per vehicle — simply because less work is undertaken there. It is often not necessary or desirable to compare units on every activity in which they engage. Genuinely comparable studies of a subset of standard activities may be much more powerful than vague, ill-defined comparisons of larger entities.

The availability of appropriate units willing and able to take part in the exercise should be investigated carefully. This in itself is typically the major constraint to benchmarking. Although it will not be feasible to get all possible partners to buy-in to a benchmarking exercise from day one, there should be a minimum critical mass on board from an early stage — three units in a study where the position of individual units is not to be revealed.

Options for the collection of original data include postal questionnaires, site visits and telephone surveys or a combination of these. In general, it is difficult to obtain satisfactory data of any complexity using any means other than a combination of a site visit and a structured, preplanned schedule of questions. Site visits also yield valuable contextual information that greatly aids the interpretation of quantitative data.

The discussion so far has been based on the assumption that the exercise will involve the collection of original data, but this may not be necessary if secondary sources can be used. There are a number of options available in this respect, including

- Existing databases. These may be run by dedicated benchmarking associations, management consultancies, industry associations, government bodies, or academic organizations.
- Publicly available data, such as company accounts.

There are clear advantages to using secondary data, particularly in terms of the speed and cost with which the exercise can be conducted. However, the precision of the comparison may be sacrificed, and it is likely to be a case of accepting what is already on offer rather than crafting a study ideally suited to one's needs. In addition, many existing databases include a wide range companies and operations, further diluting the comparison process.

Even with extensive and careful planning, unforeseen issues inevitably emerge during data collection, often in the course of site visits. Even an apparently simple question about, for example, defect rates may have hidden complexities (What about returns where there were no faults found? If our customers find one defect, they return the whole batch. Do we count the whole batch or just the single defective item?). In benchmarking, consistency of measurement is vital. The way such issues are dealt with in each case should be logged and a common response agreed among those conducting the data collection. Small, focused data collection teams find it easier to maintain consistency of approach and to spot and rectify anomalies that do larger teams of part-timers.

Anomalies in measurement often do not emerge until the data are collated and the position of each company is seen alongside that of all the others. Time and resources should be built into the program so that these can be investigated and the database debugged before data analysis and the release of the results. If there are significant doubts about the reliability or validity of particular items, it is best to eliminate them at this stage, as such items may undermine the legitimacy of the whole exercise.

Once data analysis is complete, materials can be prepared for presentation and dissemination. The principles concerning constructive and destructive feedback apply to benchmarking as much as elsewhere. Two common problems at this stage are

1. The desire to present every piece of information that has been gathered in the course of the exercise.
2. The temptation to enter premature evaluation of the findings.

Ideally, the initial presentation of the findings should serve as a springboard to a process of enquiry and not as a judicial review of who was responsible for the situation revealed by the data.

In presenting the findings of a benchmarking study, more often equals less. On most measures, simple maximum, minimum, and averages are sufficient to permit a company to work out if its position is good, bad, or indifferent. Measures that require a substantial interpretation, or that reveal inadequacies of the approach adopted, are best dropped or relegated to an appendix. Selectivity (though not with the intent of distorting the messages), simplicity, and transparency are the keys to the effective presentation of benchmarking data.

The emphasis should be on *learning* from the benchmarking data rather than using them as a vehicle to distribute blame or glory. However, the dominant culture of the organization is likely to assert itself at this point, and it is here that many of the gains of benchmarking can be won or lost. Unlike traditional (and ongoing) performance measurement systems, which have personal assessments and promotions attached to their outcomes, benchmarking exercises are generally one-offs. As such the process may be less politicized than other methods of performance measurement. Framing the follow-up to a benchmarking exercise as a process of enquiry or problem solving will yield much more productive results than if the exercise is seen as a witch hunt. A corollary of this is that benchmarking is not necessarily an exercise that should be conducted too frequently, lest it becomes just another aspect of an organization's system of performance measurement and evaluation.

The challenge, as with many initiatives, is to capitalize on the spark of energy and enthusiasm for improvement that can be engendered by a well-designed and executed benchmarking program. Some companies are really able to extract value and learning from the comparison process. Sadly, for others, benchmarking seems to be like watching a horror movie — the experience is shocking, stimulating, but essentially transitory. Much of this is clearly due to the wider organizational context, such as cycles of firefighting and defensive organizational routines that squeeze out development and improvement activities. Related to this is the problem of knowledge capture; as improvements are made, how is knowledge about these institutionalized and built upon, thereby avoiding the repeated reinventing of the wheel?

Summary

Benchmarking represents a powerful tool, which can serve a number of functions — as a reality check on organizational strategy and performance, as a method for learning from others, and as a driver of change and improvement. As with many management tools, this means that benchmarking has both a technical dimension (to do with the design of the study, the collection of valid, appropriate data, and appropriate methods of analysis) and a social or political dimension. The latter dimension concerns the role of benchmarking in driving and legitimating change. A benchmarking study whose messages are lost due to poor presentation, or denied because of methodological or other inadequacies, has clearly failed in many important respects. Benchmarking studies do not occur in a vacuum. The learning and improvement activities that should follow from a well-executed study can be profoundly mediated by the wider organizational context. Ensuring that the environment is conducive to organizational learning is vital if the true benefits of benchmarking are to be realized.

References

Bogan, C. E. and English, M. J. *Benchmarking for Best Practices — Winning Through Innovative Adaptation,* McGraw-Hill, New York, 1994.

Camp, R. C. *Benchmarking: The Search for Industry Best Practices that Lead to Superior Performance,* ASQC Quality Press, Milwaukee, WI, 1989.

Clark, K. B. and Fujimoto, T. *Product Development Performance,* Harvard Business School Press, Boston, 1991.

Delbridge, R., Lowe, J., and Oliver, N. The process of benchmarking: a study from the automotive industry, *Int. J. Prod. Oper. Mgt.,* 15(4): 50–62, 1995.

IBM Consulting Group. *Made in Europe: A Four Nations Best Practice Study,* IBM Consulting Group, London, 1994.

Oliver, N, Gardiner, G., and Mills, J. Benchmarking the design and development process, *Des. Mgt. J.,* 8(2): 72–77, 1997.

Watson, G. *Strategic Benchmarking: How to Rate Your Company's performance Against the World's Best,* John Wiley & Sons, New York, 1993.

Womack, J. P., Jones D. T., and Roos, D. *The Machine that Changed the World: The Triumph of Lean Production,* Rawson Macmillan, New York, 1990.

Further Information

International Benchmarking Clearinghouse, Houston, TX.

The European Foundation for Quality Management, Eindhoven, The Netherlands.

The American Society for Quality Control, The Malcolm Balrige Quality Award Criteria, National Institute of Standards and Technology, U.S. Department of Commerce, Gaithersburg, MD.

16.6   Core Capabilities

Dorothy Leonard

Core capabilities are interdependent systems of knowledge (both content and process) that have built up over time, are not easily imitated or transferred, provide a basis for multiple product or service lines, and convey a strategic advantage to the organization possessing them. They are frequently, but not always, technology-based. Core capabilities may be distinguished from supplemental capabilities (which are beneficial but are not strategically essential) and from enabling capabilities (which are important as a minimum basis for competition in the industry but which convey no particular competitive advantage). So, for example, a supplemental capability might be brand recognition or eye-catching graphics in a company icon or advertising. An example of an enabling capability is excellent service or world-class manufacturing.

History

The concept underlying the term **core capabilities** is not new. Researchers and management theorists have written for years about resources and competencies [Rumelt, 1974, 1986] using a number of different descriptive terms: distinctive or organizational competencies [Hayes et al., 1988], firm-specific competence, resource deployments, invisible assets [Itami and Roehl, 1987] and core competencies [Prahalad and Hamel, 1990]. Core capabilities are, by their nature, difficult to identify, isolate, and measure, and there has been much debate over whether they should be defined at the level of a business unit or for the total corporation. Although core capabilities must by definition provide a strategic advantage to the organization, researchers have found it difficult to prove that advantage empirically. Economists have conducted industry-level analyses to support the argument that the existence of capabilities explains the differential performance of firms [Rumelt, 1991]; management theorists have studied individual firms at depth to suggest that their superiority is based on their core competencies [Collis, 1991].

Dimensions of a Core Capability

Capabilities comprise four interdependent dimensions:

1. *Employee knowledge and skills.* Some portion of such knowledge is tacit, i.e., held in the heads of individuals, not codified or captured in a transferable form. Some theorists argue that this tacit knowledge is in fact the most important asset of the firm [Nonaka and Takeuchi, 1995]. This know-how may relate to any of the activities of the firm, from research to service, and may be diffused among many people or concentrated in the minds and hands of a few. For example, in a minimill such as Chaparral Steel in Texas, even operators on the line are highly skilled in experimentation and have high degrees of knowledge about metallurgy. The firm also boasts one of the industry's top experts in mold design.
2. *Physical systems.* Knowledge built up over time becomes embodied in software, hardware, and even in infrastructure. Database systems or simulation software are obvious repositories of information generated by people who have left the organization, but there are many perhaps less obvious ways in which employee, customer, and vendor knowledge has been captured over time. Proprietary or greatly adapted machinery, old product prototypes, even the arrangement of facilities, may represent knowledge contributing to a core capability. Chaparral Steel has a patented near-net-shape casting process; Kodak has proprietary machinery for manufacturing film; Ford has sophisticated software simulations that contain years of data on automobile crashes.
3. *Managerial systems.* The accumulation of corporate knowledge is guided by the organization's system of education, rewards, incentives, and modes of interaction with the market. Although such systems are not universally recognized as part of core capabilities, some authors have referred to "routines" [Teece and Pisano, 1994] or "organizational assets," including culture [Barney, 1986] that contribute to a sustainable strategic advantage. Chaparral Steel has some highly unusual

apprenticeship programs and extensive arrangements for travel and education through which the firm builds competitively advantageous knowledge. Hewlett-Packard has constructed a complex management matrix that enables divisional entrepreneurship and yet enough central control to allow cross-divisional product development. Proctor & Gamble is known for its highly sophisticated market research techniques.

4. *Values and Norms.* Attitudes and accepted behavior also affect capabilities. "These determine what kinds of knowledge are sought and nurtured, what kinds of knowledge-building activities are tolerated and encouraged. There are systems of caste and status, rituals of behavior, and passionate beliefs associated with various kinds of technological knowledge that are as rigid and complex as those associated with religion. Therefore, values serve as knowledge-screening and -control mechanisms" [Leonard-Barton, 1995]. Microsoft is able to attract some of the very brightest college graduates because of the very high value placed in the company on intelligence and potential rather than on experience. Top chemical engineers have traditionally migrated toward companies such as Dow or Kodak, whereas marketers head for Proctor & Gamble and financial analysts go to Wall Street companies. By attracting the people who best embody the kinds of knowledge highly valued in their organization, managers enhance the company's core capabilities.

While all capabilities have some portion of each of these four dimensions, the relative importance of each dimension varies according to the basis of competition within the industry. Thus, a company that competes through excellence in logistics (e.g., Federal Express) may consider the physical systems of software that control information flow to be relatively critical, whereas a management consulting company will emphasize the importance of employee skills and knowledge.

Issues in the Management of Core Capabilities

Because core capabilities are deeply ingrained systems of knowledge, they are very difficult to manage. In fact, they contain paradoxical characteristics. As noted above, much of the knowledge implicit in core capabilities may be tacit and therefore difficult to identify and evaluate. In manufacturing, codification of knowledge is highly desirable, and tacit knowledge is usually viewed as an indication of an immature, primitive process. However, tacit knowledge as part of a core capability is recognized as sophisticated, complex, and often advantageous [Nonaka and Takeuchi, 1995] The more that such knowledge is captured and codified in some transferable form, the more easily the capability can be imitated and therefore the less strategic advantage it offers. Consequently managers must decide what kinds of tacit knowledge can and should be codified and diffused.

One of the most difficult characteristics of core capabilities is that they are simultaneously **core rigidities** [Leonard-Barton, 1992]. The core capabilities that led to superior performance in the past may be inadequate or inappropriate for current and future competition. Thus, the very systems of knowledge that have led to an organization's success make two important activities difficult: exploration and renewal [Doz, 1994]. The systems that comprise a core capability within a firm become over time increasingly efficient. Routines are established; employees know how to import, process, and use information in support of core capabilities. Human resource departments can sort through resumes quickly to hire more people that fit the profile of the firm; operations run with increasing smoothness. When the whole system is therefore optimized for certain kinds of knowledge flows, it is difficult to experiment, just as it is difficult to interrupt an in-line manufacturing process to prototype a new product.

For the same reason, it is difficult to renew the core capabilities. Any renewal will challenge some part of the system, and, as we have seen, capabilities are not simple, static knowledge repositories with discrete modules that can be replaced or modified in isolation. Rather, any change to one of the four dimensions mentioned above inevitably affects the other three interdependent dimensions. Even the introduction of new physical systems (e.g., machinery or new software) is likely to affect skills, reward systems, and norms [Leonard-Barton, 1995].

The more extensive is the renewal, the more difficult is the change, of course. Therefore, companies that move from one technological base to another entirely different one (e.g., Kodak moving from silver halide to electronics) often experience great trauma. However, research suggests that a technical change of this nature is not as great a challenge to core capabilities as is the occurrence of a discontinuity that shifts the basis on which customers judge a given line of products [Bower and Christensen, 1995], that is, well-established companies lead the industry in developing even radical technologies when those technologies address existing customers' needs — but find it almost impossible to shift to a new technology when it is attractive only to an emerging market. Yet, the technologies initially used in emerging markets often become so advantageous that entrant firms overtake the incumbents. For example, IBM dominated the mainframe segment of the computer market, but missed the minicomputer market. Digital Equipment and Data General were successful in the minicomputer market, but were overtaken by Apple, Commodore, and Tandy (and later, by IBM) in the desktop computer markets. Apollo and Sun Microsystems were new players in the industry when they introduced engineering workstations.

These examples suggest why it is important to think of core capabilities as more than technology-based competencies. The companies that missed the opportunities offered by technological discontinuities were able to create radical and sophisticated technical innovations — so long as these were in response to the urgent needs of current customers. What they could not do was to shift all their supplier, vendor, engineering, and marketing foci to a different set of customers — because that shift would have affected their core capabilities, not just technical innovation in their product development. All their physical systems, skill sets, managerial systems, and norms were targeted at a given market; their core capabilities functioned as core rigidities by inhibiting their ability to address a totally different value chain.

Defining Terms

Core capabilities: Interdependent systems of knowledge that have built up over time, are not easily imitated or transferred, provide a basis for multiple product or service lines, and convey a strategic advantage to the organization possessing them.

Core rigidities: The flip side of core capabilities; they have all the same characteristics except that they convey no strategic advantage.

Dimensions of capabilities: Employee knowledge and skills; physical systems (e.g., software or hardware); managerial systems (rewards, evaluation, and training systems); values and norms (attitudes and accepted behaviors that influence the way that knowledge is imported, evaluated, and assimilated into the organization).

References

Barney, J. B. Organizational culture: Can it be a source of sustained competitive advantage?, *Ac. Mgmt. Rev.*, 11: 656–665, 1986.

Bower, J. and Christensen, C. M. Disruptive technologies: catching the wave, *Harv. Bus. Rev.*, Reprint #95103, January-February: 1995.

Collis, D. A resource-based analysis of global competition: the case of the bearings industry, *Strat. Mgt. J.*, 12: 49–68, 1991.

Doz, Y. Managing Core Competency for Corporate Renewal: Towards a Managerial Theory of Core Competencies, INSEAD Working Paper 94/23/SM (rev. 17/05/94), reprinted in Campbell, A. and Luchs, K. S. *Core Competency-Based Strategy*, International Thomson Business Press, London, 1997.

Hayes, R., Wheelwright, S., and Clark, K. *Dynamic Manufacturing: Creating the Learning Organization*, Free Press, New York, 1988.

Hitt, M. and Ireland, R. D. Corporate distinctive competence, strategy, industry and performance, *Strat. Mgt. J.*, 6: 273–293, 1985.

Hofer, C. W. and Schendel, D. *Strategy Formulation: Analytical Concepts*, West, St. Paul, MN, 1978.

Itami, H. and Roehl, T. *Mobilizing Invisible Assets*, Harvard University Press, Cambridge, MA, 1987.

Leonard-Barton, D. Core capabilities and core rigidities: a paradox in managing new product development, *Strat. Mgt. J.*, 13: 111–125, 1992.

Leonard-Barton, D. *Wellsprings of Knowledge: Building and Sustaining the Sources of Innovation,* Harvard Business School Press, Boston, 1995.

Nonaka, I. and Takeuchi, H. *The Knowledge-Creating Company,* Oxford University Press, New York, 1995.

Pavitt, K. Key characteristics of the large innovating firm, *Br. J. Mgt.,* 2: 41–50, 1991.

Prahalad, C. K. and Hamel, G. The core competence of the corporation, *Harv. Bus. Rev.,* 68 (3): 79–91, 1990.

Rumelt, R. P. *Strategy, Structure and Economic Performance,* Harvard Business School Press, Boston, 1974, 1986.

Teece, D. and Pisano, G. The dynamic capabilities of firms: an introduction, In *Industrial and Corporate Change, Vol. 3,* pp. 537–556, Oxford University Press, New York, 1994.

Further Information

Leonard-Barton (1995, see references) provides the most in-depth *operational* examination of capabilities. Most of the examples in the book are drawn from manufacturing firms.

For an understanding of how capabilities are regarded by *strategists,* see Prahalad and Hamel, 1990, (see references) or Gary Hamel and C. K. Prahalad, *Competing for the Future,* Harvard Business School Press, 1994.

The Strategic Management Society, comprised of both academics and practitioners, has recently focused more attention on the resource-based view of the firm, and therefore papers on core capabilities are sometimes presented at their annual conferences.

17

Strategic Action

Mark L. Sirower
New York University

Nikhil P. Varaiya
San Diego State University

Kathryn Rudie Harrigan
Columbia University

Donald D. Myers
University of Missouri-Rolla

Robert Simons
Harvard University

Gerardo A. Okhuysen
The University of Texas

Kathleen M. Eisenhardt
Stanford University

Robert B. Handfield
Michigan State University

17.1 Diversification through Acquisition

Mark L. Sirower and Nikhil P. Varaiya

Corporations exist as intermediaries between investors around the world and the "projects" in which those corporations will make investments. The managerial challenge for a corporation is to earn a return on invested capital at least as high as could be earned by investors on investments of similar risk (the cost of capital). The 1980s and 1990s represent a dramatic period in industrial history of managers of firms choosing to grow and diversify through acquisitions of other existing companies. This method of diversification and implementation of strategic change is unique in that the owners of acquiring firms could have essentially made these investment decisions on their own by buying shares in these **target companies** directly.

Unfortunately, the vast majority of these acquisition decisions may have actually destroyed value for the owners of the acquiring firms [Varaiya, 1986; Porter, 1987; Sirower, 1997]. The purpose of this section is to summarize the strategic and financial analysis so essential to making informed acquisition decisions and creating value for the shareholders of acquiring firms.

Chandler [1977] provides a thorough historical analysis of the factors that lead firms to diversify from their original core business: (1) firms begin as single product line businesses supplying a local market, (2) improvements in methods of transportation and communication allow firms to serve wider regional and even national markets, (3) firms expand by vertical integration, for example, as firms grew to compete nationally they integrated forward into marketing and distribution systems or backward into the raw materials needed for production, and (4) excess capacity in production, marketing, and distribution systems causes firms to diversify their product ranges. In practice, we can classify firm diversification based on whether it is directly related to the **value chains** of the core businesses, vertically related either backward or forward in the value chain of operations, or unrelated to the value chains of the core businesses [Rumelt, 1974].

Regardless of the type of diversification firms pursue, it is most often facilitated through acquisitions and the strategic and financial conditions for value-creating acquisitions are consistent across diversification type. Acquisitions are in fact a very unique investment decision. First, acquisitions normally require the payment of a control premium for the shares of the firm being acquired (the target). The **control premium** is the amount by which the offer price per share of the target exceeds its preacquisition share price, expressed as a percent, and has averaged between 40 to 50% over the past 2 decades. Paying an acquisition premium creates an additional business problem for acquirers — managing new performance gains that never existed and were not expected (called **synergies**). Thus acquirers must effectively maintain stand-alone values in addition to managing the new business problem created by the premium.

Second, unlike virtually any other capital investment decision, the *total* price — stand-alone preacquisition value of the target plus the premium — is paid up front before the acquirer gains control of the assets of the target. The up-front payment of a premium for some uncertain stream of additional benefits in the future has been called the acquisition game [Sirower, 1997]. Existing stock market prices are formed on expectations of the future. It is these preexisting performance expectations that create the base performance case. Paying a control premium necessitates performance gains above those that are already expected. This forms the competitive challenge. Synergy must imply gains in competitive advantage, that is, competing better than was previously expected. Simply diversifying through acquisition will not create value because shareholders can buy shares in other companies *without* paying a premium [Alberts, 1974].

Assuming the acquirer can at the minimum maintain the stand-alone value of itself and the target, the value to the shareholders of the acquirer as a result of the acquisition is the difference between the value of real performance improvements and the control premium required to acquire the target. These two distinct parts of the acquisition game require careful analysis. In the majority of acquisitions, the performance requirements implicit in the premium far outweigh the value of likely synergies — even where the target's businesses are closely related to those of the acquirer [Alberts and Varaiya, 1989; Sirower, 1997]. We first consider the strategic conditions that must be met in order to create additional value and justify the payment of *any* premium.

Strategic Analysis for Value-Creating Acquisitions

When evaluating acquisition strategies, acquirers must be able to identify where performance improvements can be achieved as a result of acquiring the target. Extending Alberts [1974], improvements in target operating performance can come about if acquisition of the target results in some combination of the following bargain opportunities:

1. *Position bargain:* In this case acquirer management can improve the target's performance by *further* differentiating its offerings (by enhancing existing attributes and/or adding new ones), *further* improving its relative efficiency or both as a stand-alone company. There are no plans to integrate the target, but there are opportunities for the target to be better managed with regard to either product or cost positions. If the target was competitively disadvantaged along its value chain relative to rivals, or its stand-alone economic profits were negative rather than positive, the prospect

of being able to improve the target's competitive position would make the opportunity a turnaround or restructuring bargain. Leveraged buyout firms commonly seek these type of bargains often by replacing the current management or creating improved incentives for current management.

2. *Expansion bargain:* In this case acquirer management will be capable of selling the product profitably in other geographical markets that the target's *present* management cadre has chosen not to enter (perhaps because of human resource and capital constraints).

3. *Internal synergy bargain:* In this case acquirer management: (1) has the capability of integrating the target's positioning strategies and the positioning strategies of one or more of the acquirer's other units and (2) by doing so could bring about further differentiation of the target's offerings (e.g., by putting the acquirer's brand on the offering), a further increase in the target's efficiency (e.g., by obtaining from the excess capacity of the acquirer a component or service at a lower price than the one the target has been paying another supplier), or both. The prefix "internal" means that the improvement in operating performance comes as a result of the ability of the acquirer to integrate the value chains and supporting organization systems, structures, and procedures of the target's businesses with those of the acquirer. This integration will involve the effective transfer of strategic assets, capabilities, and/or specialized knowledge between the acquirer and the target.

4. *Market power bargain:* In this case, acquiring management by acquiring the target can influence the industry structure of either its own industry or that of the target. In doing so, the acquirer can prevent operating margin deterioration by lessening the power of large buyers or suppliers or by decreasing the threat of substitute products.

5. *Undervaluation/divestment bargain:* In this case the present stand-alone market value of the target does not correctly represent the intrinsic value of the free cash flow generation power of the assets or the existence of so-called hidden assets. This might also include a more optimistic forecast of demand for an existing or potential product than the market consensus. Further, upon acquisition, the acquirer may divest value destroying businesses that target management was unwilling or unable to fix or sell.

6. *Leverage/internal tax bargain:* In this case, benchmarking by acquirer management will reveal that the target is underleveraged relative to its rivals (its ratio of permanent debt to invested capital is significantly lower than the average ratio), the implication of which is that, by matching these rivals, the target's spread between return on capital and the cost of capital can be increased by using the tax shields provided by a more appropriate capital structure. The acquirer may also choose to structure the acquisition transaction in a way that affects the cash tax rate that the acquirer or the target must pay under the current tax code. Further, the target may have unused tax shields in the form of loss carryforwards from past operations.

The preceding bargain opportunities can only be realized when the management cadre who will oversee the target have sufficient knowledge and capabilities to identify, plan, and implement the potential sources of target performance improvement. Where acquirer management does not possess the required capabilities and knowledge for implementation, these bargain opportunities may represent failed potential.

It is also crucial to note that the benefits derived from any of the preceding bargains must be net of the potentially substantial costs required to effectively implement a given acquisition strategy. These costs include advisory fees, severance and relocation expenses for employees, shutdown and renovation costs, and any transition costs related to the acquisition.

We now specify the improvements in financial performance that are required to create value when the acquisition of the target necessitates the payment of a control premium. These performance improvements must be driven by the ability to exploit the bargain opportunities described in the preceding section.

Premium Recapture Analysis for Value-Creating Acquisitions

Over the period from 1976 to 1990 premiums in large U.S. industrial acquisitions averaged around 50% and ranged up to 185%. For any acquisition to create value there must be, at the minimum, a sufficiently large improvement in the target's postacquisition financial performance to fully recapture the control premium [Alberts and Varaiya, 1989; Sirower, 1997]. Alberts and Varaiya [1989] develop an operational premium recapture model in which required improvements in the target's financial performance are characterized as a combination of required improvements in expected future profitability (denoted the spread, and measured as the difference between expected future return on equity and the cost of equity capital) and expected future earnings growth rate to fully recapture the offer premium. Increasing this spread between the returns on capital and the cost of capital of a business is central to the objective of a value-creating diversification strategy.

If M_1 denotes the preacquisition equity (market) value of the target, then M_1 is the present value of expected equity cash flows and will be given for an indefinitely long horizon by the equation [Miller and Modigliani, 1961]

$$M_1 = \sum_{t=1}^{t=\infty} \frac{B_t R_t \left[1 - r_t\right]}{\left(1 + k_e\right)^t} \tag{17.1}$$

where B_t = book equity at the beginning of year t, R_t = the rate of return on B_t in year t, r_t = the fraction of earnings reinvested at the end of year t, and k_e = the appropriate cost of equity capital for the target. The numerator of Eq. 17.1 represents the equity cash flow for the target in year t; $B_t R_t$ is the target's earnings in year t.

If the R_t and r_t vectors were thought to be trendless and stable over time, the target's earnings growth rate (as well as its growth rate of capital and equity cash flow) — the product of R_t and r_t — also would be trendless and stable. As a consequence, Eq. 17.1 can be simplified to

$$M_1 = B_1 + B_1 \left[\frac{\bar{\theta}}{k_e - g} \right], \quad \bar{\theta} = \bar{R} - k_e, \tag{17.2}$$

where $\bar{g}$ (which is less than k_e) designates the target's average rate of growth of equity cash flows forecasted by investors in determining M_1 and $\bar{R}$ is the deduced specific uniform or sustainable rate of return on book equity which satisfies Eq. 17.2. Then, $\bar{\theta}$ is designated the **uniform annual equivalent spread** between $\bar{R}$ and k_e. The latter equation tells us that, if $\bar{\theta}$ equals zero, M_1 will equal B_1; if $\bar{\theta}$ is positive, M_1 will be greater than B_1; and if $\bar{\theta}$ is negative, M_1 will be less than B_1.

If the target is acquired at a price M so that $M = M_1(1+p)$, and p is the control premium expressed as a fraction, then, if the acquisition is to create value through full premium recapture the acquirer, will want the target's postacquisition equity value to be *no less* than a magnitude we denote M^*:

$$M^* = M_1 \left(1 + p\right)\left(1 + v\right) \tag{17.3}$$

where v is the acquirer's minimum acceptable value increment, in addition to the control premium, also expressed as a fraction. Thus, the acquirer's challenge would be to improve the target's return-growth performance sufficiently to transform it from a unit worth M_1 to a unit worth M^*.

Since M_1 can be explained by Eq. 17.2, M^* can be explained similarly with the equation

$$M^* = B_1 + B_1 \left[\frac{\theta^*}{k_e - g} \right], \quad \theta^* = R - k_e \tag{17.4}$$

where k_e is the target's cost of equity capital, R^* and g^* are the target's postacquisition expected return-growth performance parameters that imply a magnitude of the right-hand side of Eq. 17.4 equal to M^*, and R^* is calculated from an initial equity base of B_1. The differences $\bar{\theta}^* - \bar{\theta}$ and $g^* - \bar{g}$ can be described as the sustainable performance improvements necessary to increase the target's equity value to M^*.

If we assume that the target's postacquisition growth rate will not change materially as a consequence of the acquisition, g^* will equal $\bar{g}$, and the postacquisition target spread, θ^*, required to recapture "fully" the premium (that is, to recapture p *and* add the value increment v) is then given by the equation

$$\theta^* = \left[\frac{M^*}{B_1} - 1\right]\left[k_e - g^*\right]; \ M^* = M_1\left(1+p\right)\left(1+v\right)$$

(17.5)

Consider that the median premium for a large sample of acquisitions of U.S. industrial firms for the period 1976 to 1990 is 56%. To fully recapture this premium, the target's postacquisition spread (using the inputs to Eq. 17.5) would have to be at least 11%, assuming that the target's postacquisition growth rate of earnings did not deviate significantly from its preacquisition growth rate.

By comparison for *FORBES* industrials the median spread over the period 1977 to 1991 was −2.9% and for the 90th percentile industrial it was 9.4%. This suggests that premium recapture of the median offer premium requires that the target's postacquisition performance be better than the 90th percentile *FORBES* industrial, a truly "heroic" performance.

Managerial Pitfalls and Lessons for Value-Creating Acquisitions

Integrating the strategic and financial fundamentals for creating value when diversifying through acquisition yields the following managerial pitfalls and resulting lessons [Sirower 1997]:

Prior Expectations and Additional Resource Requirements

Before the acquisition, the stock market price of the target company will, in most cases, have substantial expected improvements already built in. What might appear to be postacquisition performance gains may have nothing to do with synergy because the expectation was already priced in the stand-alone values. Additional costs or investments such as increased research and development, new executives, golden parachutes, new plants, or increased advertising can negate any additional benefits in addition to damaging prior expectations. Finally, acquisitions can divert important managerial resources away from the acquirer's other businesses.

> *Lesson 1:* Pay close attention to what is required to maintain value in the stand-alone businesses. This is the base case. When acquirers make organizational or strategic changes to gain the value they paid for in the premium, they run the risk of destroying the growth and/or value that may already be priced by the markets. Additional investments in the businesses are like additions to the premium and must be considered as such if maintaining value is the objective.

Competitors

If the proposed changes in strategy following an acquisition are easily replicated by competitors, performance improvements will be negated. The result is that acquirers will likely need to make major additional commitments if the changes have a chance of improving competitive advantage. On the other hand, acquirers must question whether integration moves will cause inflexibility such that the moves of their competitors become difficult to contest. For example, workforce cuts might help efficiency but slow competitive response in the next round of competition. Finally, the longer the acquirer delays in implementing a postacquisition strategy, the more time competitors have to learn what an acquirer is attempting to do. They will find ways to challenge the acquirer's anticipated moves before "improvements" even begin.

> *Lesson 2:* Synergies will be the result of competitive gains and must be viewed in this context. An acquisition strategy will not create synergy with only a vision of why it might be a good thing to do. Unless acquirers carefully consider where changes will actually improve performance along the value chain, whether they have the capabilities to affect these changes and the additional resources that will be needed to put them in place, financial performance improvements will not occur. The transition to the postacquisition integration phase must be done quickly and decisively.

Otherwise, transition management becomes a resource drain and can make the acquirer even more vulnerable to competitors.

Time, Value, and the Premium

The premium translates into required performance improvements that only grow with time, so improvements need to begin immediately. However, many improvements can come only from changes that take significant time to plan and implement (distribution, product development, new plant locations, and executive succession), and rushing them may prove to be a disaster to stand-alone value.

> *Lesson 3:* The market has already valued expected future performance of the target company as a stand-alone. The premium must represent improvements above this. Thus, acquirers need to value the *improvements* when they are reasonably expected to occur. The premium must be driven by the ability of the acquirer to plan and implement the bargains available from a strategic analysis, not by the premiums that other acquirers have paid for similar acquisition targets. There is no credible way to enter negotiations on price if these issues are not clearly considered. These are the rules of the game.

Defining Terms

Control premium: The amount paid by an acquirer for a target above the stand-alone market value of the target.

Position: Includes the product/service offerings of a company and its costs to compete in those offerings.

Synergy: The increase in performance of the combined firm above what is already expected for the stand-alone firms to achieve independently.

Target: The company that is acquired in the acquisition process.

Uniform annual equivalent spread: The sustainable spread between the return on equity capital and the cost of equity capital required to justify a given equity market value.

Value chain: The sequence of activities — inputs, processes, and outputs — in which a firm is involved in the course of doing business.

References

Alberts, W. W. The profitability of growth by merger, In *The Corporate Merger*, W. W. Alberts and J. E. Segall, eds., University of Chicago Press, Chicago, 1974.

Alberts, W. W. and Varaiya, N. P. Assessing the profitability of growth by acquisition: a 'premium recapture' approach, *Int. J. Ind. Org.*, 7: 133–149, 1989.

Chandler, A. D. *The Visible Hand: The Managerial Revolution in American Business*, Harvard University Press, Cambridge, MA, 1977.

Miller, M. H. and Modigliani, F. Dividend policy, growth and the valuation of shares, *J. Bus.*, 34: 411–433, 1961.

Porter, M. E. From competitive advantage to corporate strategy, *Harv. Bus. Rev.*, May/June: 43–59, 1987.

Rumelt, R. P. *Strategy, Structure and Economic Performance*, Harvard University Press, Boston, 1974.

Sirower, M. L. *The Synergy Trap: How Companies Lose the Acquisition Game*, Free Press, New York, 1997.

Varaiya, N . P. The returns to bidding firms and the gains from corporate takeovers: a reexamination. In *Research in Finance, Vol. 6*, A. Chen, Ed., pp. 149–178, JAI Press, Greenwich, CT, 1986.

Further Information

Goold, M., Campbell, A., and Alexander, M. *Corporate-Level Strategy: Creating Value in the Multibusiness Company*, John Wiley & Sons, New York, 1994. This is an excellent book on the value that can be created and destroyed through diversification strategies.

17.2 Joint Venture

Kathryn Rudie Harrigan

A **joint venture** is the organizational form of cooperation between sponsoring firms that creates a separate entity. It is often called the *equity joint venture* because partners that sponsor this form of enterprise own it, each according to their respective *pro rata* shares. Although ownership splits vary by culture, national law, and tax incentives, the most common division of equity shares is the **50%-50% joint venture**, in which two partners own a venture equally. Although this essay will illustrate the many areas of conflict between partners, joint ventures are an especially comfortable vehicle for cooperation between companies that do not know each other very well and hence do not yet trust each other (or the respective laws of each firm's domicile). Because the joint venture creates an entity with **property rights**, managers feel that this alternative adds a greater comfort level in the face of uncertainty.

Terms of joint ventures are defined by an **ownership agreement contract** as well as a series of **ancillary contracts**, which specify with greater precision partners' obligations and expectations concerning specific operating-level transactions that may occur in the daily course of business between the joint venture and each of its sponsoring firms. Although the ownership agreement, which captures the general **purpose** for cooperating (and other terms explained below) may remain relatively unchanged over time, the specifics of sponsoring partners' contributions to (and benefits derived from) their joint venture may evolve. Change is often needed in the face of changing competitive conditions as well as changes in the situations of the parties themselves.

Beyond the codified meeting of minds concerning day-to-day operations within joint ventures (and the organizations which form liaisons with them), which is reflected by this plethora of contracts, is the *reality* of sponsoring firms' goals and aspirations. Contracts cannot ensure that organizational learning occurs. Nor can due diligence performed during the precontract phase ensure that **potential synergies** from working together will be realized.

Over time, joint ventures develop their own organizational goals, particularly if technological successes have been enjoyed for which personnel within sponsoring firms made few contributions (other than funding). To the extent that sponsoring firms desire continuing opportunities for collaboration with each other on future projects of mutual interest, it will be important for them to develop managerial systems for maintain harmony between personnel within sponsor and venture as well as between each other.

The Contract: Typical Negotiated Terms

Discussions leading to joint venture formation typically include agreements regarding the following points, among others: (1) purpose, (2) resources to be provided by each sponsor (and means of valuing each contribution), (3) benefits to be received by each sponsor (and a methodology for valuing what was taken), (4) identification of activities where close coordination with one (or more) sponsoring firm is required (and a system for achieving appropriate coordination), (5) specification of activities where substantial joint venture autonomy will be permitted (including guidelines for swift resolution of unclear situations), (6) scope of activities where partner exclusivity is required, (7) expected duration of joint venture activity (as well as duration of negotiated restrictions), (8) specification of milestones, timelines, and other forms of owner-imposed controls for assessing joint venture performance, and (9) terms by which joint venture will be terminated ("divorce clause"). Because the single most important predictor of joint venture success is choosing the right partner, the negotiation process should be used as an opportunity to learn about the appropriateness of potential partners and whether the firm seeking such alliance partners can work with them. Key operating managers — those who will be most involved in day-to-day liaisons with the venture's activities — should be included in some discussions of aspects of joint venture to ensure that the desired outcomes of cooperation are feasible and likely to be achieved by personnel who will be working jointly to achieve them.

Purpose for Cooperating

Discussions of purpose should include agreement concerning the joint venture's mission as well as a statement of its strategy. Specificity regarding which **products** will be encompassed by the agreement to be sold to which **markets** (customers) using which problem-solving methodologies (**technologies**) can clarify the scope of activity that is contemplated. Joint ventures should support and complement sponsoring firms' existing strategies, respectively. They should not be used to give sponsors strategic direction.

Statements of joint venture strategy should include goals and tangible performance measures. Sponsoring firms must be clear about what is most important to achieve through cooperation and be prepared to accept short-term operating losses to support long-term strategic goals. Sometimes changes in the asset positions (or operating successes) of partners have changed so dramatically that renegotiation of some terms among partners may be required. Throughout these joint venture restructurings, the strategy statement is a touchstone that reminds each sponsoring firm of their motives for cooperating with their partner.

Resources to be Contributed

Agreements concerning which sponsoring firm contributes what resources toward the operations of their joint venture often parallel firms' motives for taking partners; sponsors seek cooperation in areas where they lack resources that would be too costly (or time consuming) to develop effectively in-house. In descending order of importance, these resources may include (1) distribution access to the most attractive customers comprising key markets, (2) access to each firm's most appropriate technologies and other intellectual properties as needed for entering critical lines of business, (3) access to each firm's most valuable personnel or "knowledge workers" (and the organizational systems within which they are most effective) as needed for seamless implementation of their path-breaking gambles, (4) use of physical assets (machines, networks, and other crucial infrastructure) that a particular firm lacks, (5) use of partners' trademarks and other intangible properties that facilitate market acceptance of products, (6) cash or access to patient providers of capital, and (7) access to other needed resources. Fungible financial resources are valued less highly than other, unique resources that may be critical to operating success. Possession of such resources gives potential partners great bargaining power over firms needing access to the opportunities such resources provide.

Partners should agree on a means of valuing each firm's contributions to the joint venture as they negotiate its terms, although sponsors may assign higher values to contribution of some types of resources than to others. In cases where asymmetrical valuations have been assigned to specific resources, sponsoring firms may find that they can acquire inexpensive access to resources that their partners take for granted (hence value less highly) if they are astute learning organizations. Although not central to their purpose for cooperating, the bleedthrough of knowledge that occurs when firms work together could provide invaluable benefits to receptive corporations that have organized their **liaison activities** to capture such learning opportunities. Their willingness to respect and absorb ideas and practices from their partners (and evaluate these critically for useful adaptation within their own organization) allows such firms to receive more benefit than they may have bargained for.

Benefits to be Taken

The mirror image of negotiations concerning which *access* to resources partners will provide to each other pertains to the role each partner will play in the other's value-creating system of activities. Because partners may be using each other's resources in lieu of the joint venture's, these areas of cooperation constitute a type of outsourcing arrangement whereby firms lacking competitive resources can supplement their internal assets and capabilities with those of partners. In descending order, the most desirable benefits of cooperating include (1) market access to partner's customers who do not yet know the sponsoring firm's products and reputation, (2) use of partner's technology, patents, know-how, trade secrets, and other knowledge needed for competitive success, (3) access to a reliable network of suppliers for raw materials, services, and other resources that sponsoring firm lacks, (4) access to partner's highly trained and experienced personnel ("knowledge workers") and management systems and/or practices as needed to solve problems and implement programs, (5) use of partner firm's physical assets, infrastructure, and logistical systems for distribution, (6) use of trademarks and other intangibles that sponsoring firm lacks, (7) access to favorable financing, and (8) use of other resources as needed for competitive success. Firms that depend upon partners for crucial steps in their value-adding systems have lesser bargaining power than firms which possess scarce and valuable resources, unless they offer something of comparable worth in return.

A methodology should be negotiated (as part of the ownership agreement contract) for maintaining fairness within the partnership — a way to keep partners mutually attracted to each other's participation. The *quid-pro-quo* philosophy of valuing what each sponsoring firm takes from their joint venture distinguishes a balanced partnership from various forms of sourcing arrangements where cash is exchanged for the benefits one sponsoring firm can provide. Changes in the balance between need for and control of resources often force partners to reevaluate the viability of their joint venture, sometimes prompting renegotiations.

Coordination of Activities between Sponsor and Venture

Firms' expectations regarding where (and by whom) particular value-adding tasks will be performed are articulated in the part of the ownership agreement that identifies activities where close coordination with one (or more) sponsoring firm is required. Precautions — such as domain definition (turf identification) — are necessary to avoid redundancy and internecine rivalries that could develop where joint ventures evolve into competitors of their owners. Sponsoring firms typically specify conditions pertaining to the joint venture's (1) use of their respective distribution systems (or not), (2) freedom to sell its products and services to third parties (or not), (3) autonomy in procuring raw materials and supplies of other needed resources (or not), (4) obligations to use their respective technology, patents, know-how, trade secrets and other knowledge that sponsoring firms would consider proprietary (rather than using outsiders' technology), (5) access to critically skilled "knowledge workers" within parent companies (rather than hiring outside experts), (6) requirements to utilize sponsoring firms' plants, machines, infrastructure, and other physical assets to absorb excess capacity (or not), (7) permission to use sponsoring firms' brand equity, trademarks, and other marketing intangibles (or not), (8) reliance upon owners to raise additional capital as needed (or not), and (9) close coordination of other germane activities with those of sponsoring firms. Some ownership agreements also define metrics for auditing joint venture coordination to ensure that the benefits of shared activity that partners envisioned are indeed realized.

Although sponsoring firms will typically require close coordination between their joint venture's activities and those of wholly owned business units in areas where the competencies, capabilities, resources, and knowledge workers to be shared are core to their respective strategies, use of outsiders (or the choice to do so) may be specified for other activities where substantial joint venture autonomy will likely benefit sponsoring firms. In such cases, the joint venture may be empowered to (1) develop its own (or use others') distribution channels distinct from those of sponsoring firms, (2) create a marketing image separate from that of parent firms, (3) purchase resources and services from outside vendors, (4) sell its products and services to outside customers, (5) use technical standards, designs, technologies, patents, know-how, and other intangibles supplied by outside firms, (6) recruit outside personnel for critically skilled jobs, (7) build its own plants and other physical facilities, (8) invest in other assets for the joint venture's unique purpose (including acquisitions), (9) create its own market image (by developing its own trademarks and brand equity), (10) raise capital from outside sources, as needed, and (11) take other actions autonomously.

Negotiation of the ownership agreement contract should include some guidelines for swiftly resolving unforeseen situations where the joint venture's desire to operate autonomously of its sponsors' activities — although not provided for — does not appear to sacrifice the likely realization of sponsoring firms' strategic goals or any germane scale, scope, vertical integration, or experience curve economies that may be available by coordinating parent's and child's activities closely. The ownership agreement contract should also acknowledge (and provide for) the need to change coordination requirements as conditions affecting sponsoring firms (and their joint venture itself) evolve over time.

Exclusivity between Partners

Negotiations between potential partners should establish an policy regarding the scope of joint ventures (or other forms of strategic alliance) that will be permitted with third parties. In environments of rapidly changing technological standards where several alternative routes to problem solving could become commercially successful, sponsoring firms often reserve the option of partnering with other firms for commercialization of alternative technologies by forming parallel joint ventures. Such practices pit each respective technological (or market) approach against the others to see which alternative is most successful.

A firm possessing strong bargaining power over potential partners may form a **spider's web** of joint ventures, with itself at the hub of communications linkages between dyads of partners. Firms possessing asymmetrical **bargaining power** over partner firms frequently exert such power to prevent them from forming **parallel joint ventures** (or alliances) with competitors. Where **bilateral bargaining power** exists, each sponsoring firm can preside over its own respective spider's web of alternative joint ventures (or strategic alliances), each testing the use of different valuable resources in various markets.

Negotiations among sponsoring firms should address the contingency of adding new partners to the extant joint venture. Whether partners favor such changes in their relationship (or not), sponsoring firms should recognize that their resources and capabilities may prove to be inadequate to compete effectively without infusions of assets and skills controlled by third parties who may demand ownership equity as the *quid pro quo* for sharing these **bottleneck** resources.

Concerns raised by adding new parties to the joint venture extend beyond issues to fairness to those of intellectual property rights associated with the joint venture's products. Since the likelihood of **unintended bleedthrough** of proprietary information is often reduced by reducing the number of contacts between authorized personnel within the sponsor and venture, respectively, adding parties to the deal increases the difficulty of controlling information geometrically. Coordination problems frequently increase as the number of partners rises. Issues concerning joint ownership of intellectual properties are also compounded by the presence of multiple parents.

Expected Duration of Joint Venture Activity

Ownership agreement contracts typically specify an expected lifespan for the joint venture, especially to avoid antitrust concerns, where very powerful competitors will be temporarily joined in partnership to pursue a specific project to its commercial outcome. Life span issues may also address the expected duration of negotiated restrictions (regarding exclusivity, access to resources, territorial agreements, or other terms that could be otherwise construed as being anticompetitive) and such duration may be shorter than that of the joint venture itself. For example, partners frequently agree that their joint venture shall have an 18-month "head start" in commercializing technology, during which time no sponsoring firm can sell products using the joint venture's technology in competition with it (or license said technology to outsiders).

The purpose of such exclusivity agreements is to give the joint venture (or its sponsoring firms) an opportunity to become established in the marketplace before competitors also possess the critical resource covered by such agreements. Granting such temporary protection represents a trade-off among alternatives whereby sponsoring firms seek to extract value from their investments in the marketplace. Since joint ventures frequently evolve into stand-alone entities that develop their own market presence (distinct from that of their sponsors), property rights are at the heart of negotiations covering duration and exclusivity. Partners which seek to avoid such dilemmas frequently choose another form of cooperative strategy that does not create a separate entity.

Measures of Performance and Owner-Imposed Controls

Because joint ventures are (at best) a transitory form of organization — a stepping-stone approach to organizational evolution — sponsoring firms must continually assess whether it is time to move on to another form of joint activity (or go it alone). Carefully specified milestones, timelines, and other forms of owner-imposed controls for assessing joint venture performance assist sponsors in deciding whether to reopen negotiations (to obtain compensating **side payments**) or terminate the partnership altogether.

Short-term performance measures should support the venture's mission and frequently include nonfinancial activity measures. Because many joint ventures have required more time to pay back sponsors' sunk costs than parent firms are willing to invest, performance evaluations involving financial measures alone can undervalue the capabilities that partnering organizations have been exposed to, especially where negotiations regarding the joint venture's termination are acrimonious.

Because joint ventures possess many characteristics like those of a start-up enterprise, sponsors should be wary of overburdening the fledgling firm with onerous measurement and reporting systems requiring gratuitous data collection, maintenance of parallel systems (to serve each sponsor's dissimilar information requirements), and other governance provisions that encumber the venture's competitiveness. At a minimum, negotiations that create a parental control system should specify who will serve as **operator** in the joint venture's day-to-day decision-making process and identify decisions requiring sponsor approval. Briefly, in addition to capital expenditures requiring board approval, partners that do not assume day-to-day operating responsibility for a particular activity of interest to them may specify certain decisions for which they must be consulted, provided their requests are not unduly cumbersome. Specification of responsibilities for each type of value-creating activity depend, in part, on which firm possesses the greatest relative competencies in each area, as well as which firm has the greatest strategic need for possession of that type of competency. **Assimilation of competency** cannot occur without organizational participation in the associated activity, and suboptimal performance may occur if one partner values its organizational learning more highly than the venture's efficiency.

Joint venture managers (who are frequently neutral outsiders who were recruited specifically for the joint undertaking) should ensure that personnel within sponsoring firms comprehend the implications of actions that are taken to ensure that they can implement necessary steps quickly. Ignorant sponsors often drag their feet in providing necessary resources that support the venture's well-being because they fear that crucial information has previously been withheld from them. In arenas where timing is critical to competitive success, partners must move quickly to support their venture's activities rather than take destructive actions. Systems must be created beforehand to manage this implementation process flawlessly to compensate for speed disadvantages created by the venture's ownership form.

Divorce Clauses

Negotiations to create joint ventures devote disproportionate time and contractual language to specifying what happens when it terminates. Although partners may develop formal systems for reviewing their joint venture's performance at regular intervals (as well as formulas for evaluating value creation and dividing the jointly created spoils), it is difficult to write a contract incorporating all future contingencies describing the joint venture's evolution. Specifying a fair process for adjudicating disagreements at the time of termination is easier than trying to anticipate every way in which the partnership could evolve while negotiating its birth.

Nevertheless, negotiations may be helped by developing and discussing scenarios that may be faced when the project has ended and the partnership dissolves. Although it may be painful to confront a partner's worst fears about working with outsiders, articulating them helps managers sharpen their expectations regarding what the joint venture must accomplish as well as foresee areas of conflict with proposed partners. Discussions during the formation phase regarding how to manage eventual resource ownership, property rights, seconded employees, personnel absorption (or outplacement), **nonsurviving ownership compensation,** and other changes may suggest cooperation alternatives that managers had not previously contemplated. Conversations that illuminate the values and priorities of potential partners are useful an assessing whether an appropriate chemistry is likely to exist between firms (and the personnel that must work together to achieve the joint venture's purpose).

Mutatis Mutandis

In spite of all of the care that sponsoring firms lavish on creating a well-specified ownership agreement contract, joint ventures are an inherently unstable form of organization, which is very difficult to operate as negotiated over time. Because competitive conditions necessitate adaption in the joint venture's activities, it may be necessary to change key parameters of the carefully negotiated agreement if partners intend to stay together. Every time partners return to the bargaining table to resolve a difficulty, other irritating issues can permeate their discussions and undo a Pandora's Box of compromises. Assuming that each sponsoring firm's organization has been learning about the true value of their partners' resources and practices during the time when they worked together, partners may value each other's contributions and needs differently than when their bargain was first struck.

In spite of the many operational difficulties that **vertically integrated** joint ventures will create for sponsoring firms, they have become the "lesser among several evils" in competitive environments where firms must work with outsiders to keep up with the pack. Since cooperating firms should expect to "go to the well" repeatedly to augment their resource bases, managers should begin *early on* to identify compatible partners that can work on a stream of collaborative projects of mutual interest to each other. Since the field of **Prince Charmings** shrinks with every consolidating merger between competitors, managers should devise systems for interfirm cooperation that make their firms more highly preferred as partners than others in the field.

The introspective process of developing a strategy for working jointly with outsiders is both time consuming and devastating to organizations that resist change. Learning organizations are humble in the face of their many inadequacies of knowledge. *Hubris* makes managers unwilling to acknowledge that their firm got the short end of a bargain. Managers must devise strategies to cope with both organizational extremes because joint ventures (and other forms of strategic alliance) are part of the arsenal of competitive weapons that firms must use adroitly from now on.

Defining Terms

Ancillary contracts: Specific agreement regarding provision (or purchase) of resource that a partner controls outside of the jointly owned venture and its partners desire the use of.

Assimilation of competency: Attempt to appropriate valuable knowledge from partner by working closely with its personnel within joint venture's activities. Such attempts are usually unsuccessful unless firm's personnel rotate between joint venture and parent organization in formal program of knowledge diffusion (and absorption).

Bargaining power: Manifestation of ownership (or control) of valuable resources, competencies, capabilities, and knowledge workers that other firms value highly enough to acquiesce to controlling firm's demands.

Bilateral bargaining power: Each firm possesses (or controls) resources, competencies, capabilities, and knowledge workers of comparable worth to those of its counterpart in negotiations.

Bottleneck: Control of a scarce resource for which there is no comparable worth counterpart, hence, owner extracts tolls from all requiring its use.

50%-50% Joint venture: Partners own equity equally (shares of proceeds may vary, according to whatever profit split was negotiated).

Liaison activities: Coordination between personnel employed by sponsoring firms and those employed by joint venture. Information sharing could occur in content of governance, auditing of performance milestones, working jointly on activities, or other opportunities for communication between representatives of sponsor and venture.

Market: Customers willing to pay rents for use of products.

Nonsurviving ownership compensation: Value of sponsoring firm's ownership interest in property rights created through joint activity (or contributed by said nonsurviving partner to joint venture); assumes nonsurviving partner relinquishes all claims to said property rights for negotiated compensation.

Operator: Firm taking day-to-day responsibility for most decisions associated with implementation of joint venture's purpose.

Ownership agreement contract: Master contract delineating structure and terms of joint venture. The "contract" is actually a series of contracts — each delineating the specific performance of what a particular partner brings to the party.

Parallel joint ventures: Joint ventures with identical purposes, but different firms each in partnership with same powerful firm.

Potential synergies: Performance improvements realized by using extant resources, competencies, capabilities, and people differently by virtue of access to previously unavailable resources competencies, capabilities, and people that are controlled by partners.

Prince Charmings: Ideal joint venture partner for purpose under consideration. Frequently located via process of trial and error, as in "You must kiss a lot of frogs to find your Prince Charming."

Products: Tangible or intangible manifestation of firm's control of valuable resources from which it can extract rents.

Property rights: If each partner contributes tangibles (intangibles), he wishes to protect his intellectual property rights which may comprise its basis for competitive advantage. Short of selling each partner's contribution into the venture in a "black box" configuration, specializing on a part of the activity seems to be the cleanest way to protect the value of each partner's contributions. Delineating ownership becomes even more difficult if the jointly owned venture creates a new resource that no sponsoring firm possesses.

Purpose: Unifying justification for creating and operating joint venture. All terms of ownership agreement contract should be consistent with this basic motivation.

Side payments: Bribes paid by one partner to another to keep said firm involved in joint venture without formally re-negotiating terms of ownership agreement contract

Spider's web: Powerful firm negotiates several joint ventures (or other forms of strategic alliance) with competitors, each having the same purpose in a slightly different market, product, or technology context.

Technologies: Know-how, practices, methodologies, and other valuable knowledge that can be applied to sell products to markets.

Unintended bleedthrough: Proprietary information becomes known to unauthorized outsiders because of carelessness in guarding firm's property rights.

Vertically integrated: Joint venture buys from (or supplies to) sponsoring firm under contractual arrangements, which may not permit sales to (purchases from) outsiders without subsequent renegotiations between partners.

References

Beamish, P. and Killing, P. *Cooperative Strategies*, Jossey-Bass, San Francisco, 1997.

Friedman, P., Berg, S., and Duncan, J. *Joint Venture Strategies and Corporate Innovation*, Oelgeschlager, Gunn & Hain, 1982.

Gomes-Cassares, B. *The Alliance Revolution: The New Shape of Business Rivalry*, Harvard University Press, Boston, 1996.

Gray, B. and Chatterjee, K. *International Joint Ventures: Economic and Organizational Perspectives*, Kluwer Academic, Dordrecht, The Netherlands, 1996.

Harrigan, K. R. *Managing for Joint Venture Success*, Free Press, New York, 1986.

Harrigan, K. R. *Strategies for Joint Ventures*, Free Press, New York, 1986.

Inkpen, A. *The Management of International Joint Ventures: An Organizational Learning Perspective*, Routledge, New York, 1995.

Lewis, J. *Partnerships for Profit: Structuring and Managing Strategic Alliances*, Free Press, New York, 1990.

Newman, W. H. *Birth of a Successful Joint Venture*, University Press of America, .

Rangan, U. S. and Yoshino, M. Y. *Strategic Alliances: An Entrepreneurial Approach to Globalization*, Harvard Business School Publishing, 1996.

Starr, M. K. *Global Corporate Alliances and the Competitive Edge: Strategies and Tactics for Management*, Greenwood, 1991.

17.3   Strategic Technology Transfer

Donald D. Myers

The management of strategic technology transfer is an important consideration for any organization for which technology is one of it's strategic competitive advantages. In a broad sense, technology transfer should permeate the technology-intensive organization in its strategy and tactics, i.e, Intel, and in the units of other organizations where its technology differentiates competitiveness, i.e., WalMart's inventory control. Technology is dynamic and therefore to be competitive for the long term must constantly be evolving and/or transitioning to a new paradigm.

The term *technology transfer* is a relatively new term, only coming into existence in the 1960s. Despite varying definitions with resulting imprecision of its meaning, when used in a broad sense, it has critical implications for the technological organization and, also, is important for all other organizations dependent upon technology in any aspect of its activities. For this treatise, technology transfer refers to the *process by which technology contained within one organizational setting is brought into use within another organization setting.*

Technology includes knowledge concerned with the making and doing of useful things. Accordingly, this particular knowledge may be embodied in various forms including personnel, hardware, software, processes, drawings, books, reports, and patents. In the case of personnel, technology transfer may occur by transferring research and development (R&D) employees, having developed a new process, to operations that are about to proceed with installation or licensing a patent to an independent organization that utilizes the technology taught in the patent. Likewise, the technology embodied in software is transferred when the software is used within another organization setting.

Science is separate from, and generally precedes, technology development. Science involves the understanding of principles of nature, whereas technology is concerned with application of this knowledge for making and doing useful things. Although technology may be developed without a thorough understanding of the related science, the technology can best be refined and optimized only with this knowledge. Accordingly, technology's dependency on science makes the separation an academic exercise when contemplating the practical considerations of technology transfer. All knowledge useful to the receiver should be acquired.

The consequences of the above definition of technology transfer is that it should be contemplated in all stages of the technological innovation process and in all functional areas with technology responsibilities. At the earliest innovation stages where R&D is performing applied research, plans should be made for the greatest utilization of the research results. This may require transferring the technology internally and/or externally according to the utilization purposes. Sometimes having an effective transfer process is as important as successful R&D.

In addition to technology being transferred from R&D to engineering for embodiment in the product (i.e., schematics, flow diagrams, etc.), engineering should also be considering external technology for embodiment in the product through performing its design responsibilities (i.e., improved electronic packaging materials, higher performance pumps, etc.). Further, external technology for improving the process to perform the design process (i.e., computer automated design system, rapid prototyping, etc.) should be of interest. Accordingly, technology may flow into engineering from three different directions. Similar multiple technology flows may be occurring into all the functional areas participating in the product development process.

Since this treatise is devoted to *strategic* technology transfer, further clarification of this distinction is useful. Strategic technologies are those technologies in the value chain that differentiate competitiveness among competitors of the organization. This distinguishes it from base technologies (or enabling technologies) that are common to other competitors (computer automated design, CAD, systems that can be purchased by any of the competitors) or provide only a minor impact on the achievement of organization goals and do not provide a competitive differentiation among competitors [Burgelman et al., 1996]. The technologies embodied in the product may be strategic as well as the technology for performing the engineering function (unique CAD software that provides a sustainable competitive advantage that is not available to competitors). It is vital that an organization fully understand its technologies that provide a strategic advantage. Priority by higher-level management responsible for technology must be on its strategic technology needs and advancement.

Strategic Technology Transfer Fundamentals

Categorizing technology transfer by identifying whether the **source** and **user** of technology is internal or external of the organization is useful for discussion of strategic considerations. This includes three possible scenarios: (1) external source transferring to an internal user, (2) internal source transferring to an internal user, and (3) internal source transferring to an external user.

The organization must insure the advancement of existing strategic technologies and identification of the organization's strategic technology needs. If the technologies are strategic, then someone in the organization should be assigned the task. In many organizations, this would be the **chief technology officer** (CTO). This person should possess the **technological overview** for the organization. Through organizational technological audits, competitor technology benchmarking, technological forecasting techniques, and technology mapping, organizational strategic technology needs, and their timing availability needs [Betz, 1993] that are supportive of the overall organizational strategic plans can be identified. These strategic technologies may be developed and continuously improved internally to the organization. However, the concept of limited resources would dictate that consideration be given to the strategic value of acquiring the technologies externally.

External Source to Internal User

Considerations for whether to look externally for strategic technologies include (1) availability of the technology externally, (2) timing needs, (3) ability to develop the technology internally, (4) ability to obtain a proprietary position and protect it, (5) ability and motivation to assimilate the technology, (6) comparative economics, and (7) strategic need to build internal capabilities. If the desired technology is available externally and can be assimilated faster or less expensively than if developed internally, then efforts should be made to acquire the technology. However, for strategic technology to have a long-term sustainable competitive advantage, there must be the ability to protect it from competitors. Generally it is imperative, by definition, that a proprietary position that excludes competitors be obtained if the technology is going to be the basis for the sustainable competitive advantage. This can be accomplished by buying the patent rights, getting an **exclusive license**, or even buying the company in certain cases. One possible exception to the generalization of the need for a proprietary position, for example, could exist in a process-oriented industry where a nonproprietary technology is acquired and then enhanced. If this enhanced technology provides a strategic advantage that is protectible, then it's not necessary to have obtained a proprietary position for the acquired technology. The ability and motivation to enforce contractual rights and patent infringement must be considered. This can be very expensive and time consuming for the organization. Further, consideration must be given to the **absorptive capacity** of the organization to exploit the technology [Cohen and Levinthal, 1990].

The stages of technology transfer from an external source may be categorized as including (1) identification of available technology, (2) evaluation of the technology, (3) acquiring the technology rights if a positive evaluation, (4) implementing the technology, including enhancing when appropriate, and (5) **diffusion** of the technology. These categories are supportive of the broad definition of technology transfer above and are inclusive of much more than merely acquiring technology for successful technology transfer.

Strategic technology transfer and non-strategic technology transfer would not have these same stages for technology transfer. There would, however, be different criteria for the identification and evaluation stages. These questions would relate to the strategic fit to the list of technology needs identified by the CTO. Likewise, the mechanisms for acquisition and implementation may vary because of considerations such as obtaining a proprietary position of the strategic technology and the absorptive capabilities of the organization. It should be expected that diffusion efforts would be resisted by competitors to prevent the strategic technology from becoming an industry standard.

Rubenstein [1989] suggest that the more common methods by which industrial organizations acquire external technology includes (1) licensing, (2) joint ventures, (3) limited R&D partnerships, (4) minority interests in firms with R&D programs, (5) contracts for R&D to other companies and research institutes, (6) university contracts, grants and consortia, (7) bilateral cooperative technology arrangements, (8) hiring individual (or teams of) specialists, (9) stepping up technical intelligence activities, (10) buying technology imbedded in products, materials, equipment, or processes, (11) increasing pressure on suppliers to innovate, and (12) acquiring small high-technology companies. Other methods not specifically mentioned include education, obtaining technical publications, and attending research conferences.

Internal Source to External User

The stages for transfer of internal technology to an external source by an organization are the same as for the transfer from the external source to the internal user with only the roles reversed. Rogers [1983] designated these stages as (1) knowledge (make potential user aware of the existence of the technology), (2) persuasion (for positive evaluation by potential user), (3) decision (to acquire by potential user), (4) implementation (by user), and (5) confirmation (through diffusion of the technology beyond the user). Rather than needing to identify sources of technology, the role is now to identify potential users of the organization's technology. This suggests that the CTO needs to develop a list of internal technologies the organization is willing to transfer. Generally, there would be no interest in transferring strategic technologies since a sustainable competitive advantage should be very carefully protected from potential competitors. Consequently, the CTO needs to insure that strategic technologies are not included in the list. Exceptions may include where the production/market capabilities cannot be timely developed, a customer requires a second source, an exchange for another organization's strategic technologies, or for use in a noncompeting market.

The reason, generally, for transferring technologies externally is for strategic leveraging of the technology assets. Many companies have such programs that make significant contributions to the "bottom line". Such an effort should be a consideration by all CTOs in managing the organization's technology assets. Managing for maximum success includes timely identifying internal technologies that are valuable to others but are no longer strategic to the sourcing organization. As Apple Computer has learned, identifying the right time to allow others to use the technology is not an easy task.

There are some organizations that are in the business of developing technologies and selling them. Their core competencies would be in developing these technologies. For success they must insure that the technologies are successfully transferred. If that process is flawed, success will be limited.

Internal Source to Internal User

Internal technology transfer to other internal organizational units permeates all technology-intensive organizations. The primary occurrence for this category of technology transfer in this organization will be related to the product (or process or service) the organization is producing for sale. It is an integral part of managing the product development process and deserves attention throughout the organization. The driving force for improving this process includes the shorter product cycles and increasing competition. The market leader is often the first to market. An organization with a first-to-market strategy would need to emphasize techniques that would permit product technology transfer from one unit to another unit quickly through means such as concurrent engineering. Even if the strategy is quick to follow the leader, it may be even more imperative to have such capabilities to remain competitive when technologies change. The process for implementation of such strategies will not be addressed in this treatise and will be left to others.

Certainly there are internal-to-internal technology transfers that flow separately from the progress of the product development process. There may be, "off-line" from product development, an internal effort to improve the hardware and/or software for CAD. If one of the organization's core competencies is design and this technology represents something more than a insignificant advancement in that competency, then transferring this technology to the product designer is strategic and should receive priority from top management. If an organization's strategy is "me-too", i.e., copy the competitor's product, then efforts to advance its core competency, i.e., for design, manufacture, or whatever it may be, becomes of higher priority than product technology. By definition of this strategy, the technology for the product is not strategic but rather the ability to copy may be a core competency. The ability to compete successfully, in addition to perhaps being able to reverse engineer the leader's product, may be from having lower costs because of design and/or manufacturing capabilities.

Organization Factors

If technology is considered strategic to an organization, it is essential that the organizational structure be considered for enhancement of technology transfer. Johnson and Tornatzaky [1981] suggest that a general framework for shaping interorganizational relationships would consider three major factors: (1) goal congruity and capability, (2) boundary-spanning structures, and (3) organizational incentives. Similar factors that exist also for intraoganizational relationships, however, would likely be much less of a mismatch. Differences of goals, capability, and timing needs will decrease the effectiveness of technology transfer. Providing incentives and removing barriers will do much to improve congruity of goals of the individuals and their organization with the other organizations.

In consideration of external technology transfer, it must be recognized that organizations, and organization units, are separated by boundaries including space, time, and culture. These boundaries are going to be much more severe for interorganizational relationships but may be severe even for intraorganizational relationships for large, multidivision international organizations. Designating a capable office or individual in an organization as the scout for strategic technology is one approach to lowering the boundary barriers for interorganization relationships. Universities and the federal laboratories have done much to address these factors to improve technology transfer with industrial organizations. Virtually all research universities have offices assigned the responsibilities for licensing university owned patents. In addition, they have in place many other programs designed to transfer technology [Schoenecker et al., 1989]. Federal laboratories have legislative designated Office of Research and Technology Applications (ORTAs). Many corporations have an office specifically designated to identify and coordinate acquisition of external strategic technologies as well as license technology to other organizations.

Finally, technology transfer is a "contact" activity. It is about the transfer of technical knowledge and is highly dependent upon person-to-person contact. Technology transfer should be the responsibility of every person in the organization with technical responsibilities. How to achieve this if technology is to be a sustainable competitive advantage, particularly in a highly dynamic technology-intensive industry, is an issue to be considered by top management. Technology transfer is about having capable personnel that are motivated to learn and teaching this knowledge to others that have a "need to know". Providing a culture that is supportive of learning and the necessary incentives will result in successful technology transfer [Hayes et al., 1988].

Defining Terms

Absorptive capacity: The ability to assimilate and exploit technology. It is depended not only upon the organization's direct interface with the external environment but also upon transfer of technology across and within subunits that may be removed from the original point of entry.

Chief technology officer: A top-level executive that provides a technology perspective on strategic issues and promotes the organization's technological commitment throughout the organization. The key tasks are that of a technical business person deeply involved in shaping and implementing overall corporate strategy.

Diffusion: The pattern by which new technologies spread through a population of potential adopters following its initial adoption.

Exclusive license: A contract giving the sole right to make, use, and/or sell the technology. Others are excluded from the rights granted.

Source: An organizational unit that has possession of technology and possess the capability and right to transfer it.

Technological overview: An understanding of the areas of rapidly changing technology, areas of defensive technology, levels of technological effort, major competitors' technology strategy, and major university/industry/government research consortia.

User: An organizational unit, other than the source of the technology, that will apply the technology in some practical way.

References

Betz, F. *Strategic Technology Management*, McGraw-Hill, New York, 1993.

Burgelman, R. A., Maidique, M. A., and Wheelwright, S. C. *Strategic Management of Technology and Innovation, 2nd ed.*, Irwin, Chicago, 1996.

Cohen, W. M. And Levinthal, D. A. Absorptive capacity: a new perspective on learning and innovation, *Admin. Sci. Q.*, 35: 128–152, 1990

Hayes, R. H., Wheelwright, S. C., and Clark, K. B. *Dynamic Manufacturing: Creating the Learning Organization*, Free Press, New York, 1988.

Johnson, E. C. and Tornatzky, L. G. *Academia and industrial innovation*, In *New Directions for Experiential Learning: Business and Higher Education — Toward New Alliances*, G. Gold, ed., Jossey-Bass, San Francisco, 1981.

Rogers, E. M. *Diffusion of Innovations*, Free Press, New York, 1983.

Rubenstein, A. H. *Managing Technology in the Decentralized Firm*, John Wiley & Sons, New York, 1989.

Schoenecker, T. S., Myers, D. D., and Schmidt, P. Technology transfer at land-grant universities, *J. Tech. Transfer*, 14(2): 28–32, 1989.

Further Information

DeBruin, J. P. and Corey, J. D. Technology Transfer: Transferring Federal R&D Results for Domestic Commercial Utilization, Sandia Report SAND88-1716-UC-29, Albuquerque, NM, 1988.

National Technology Transfer Center Home Page @ http://www.nttc.edu/.

Northwestern University Infrastructure Technology Institute Technology Transfer Bibliography Home Page @ http://iti.acns.nwu.edu/clear/tech/jen12.html.

Smith, P, G., and Reinertsen, D. G. *Developing Products in Half the Time*, Van Nostrand Reinhold, New York, 1991.

The Journal of Technology Transfer, Technology Transfer Society, Chicago.

17.4   Corporate Performance

Robert Simons

Corporate strategy is concerned with decisions about how to maximize the value of resources controlled by the corporation. These decisions focus primarily on the allocation of resources inside the corporation. In single business firms, all resources are devoted to only one business. In multibusiness firms — firms that are organized to compete in more than one product market — decisions must be made about how to allocate scarce resources across business units to maximize value creation.

Managers of individual business units formulate and implement business strategies to create value for both customers and corporate owners. **Business strategy** — how a company creates value for customers and differentiates itself from competitors — determines the goals and measures that managers use to coordinate and control internal business processes and activities.

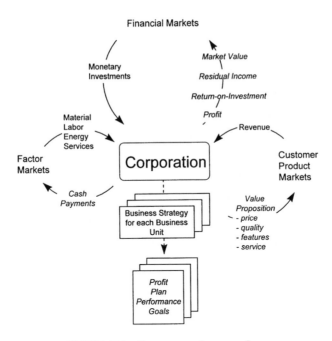

FIGURE 17.1 Corporate performance flows.

The success of corporate strategy is reflected in **corporate performance**. *Corporate performance refers to a firm's level of achievement in creating value for market constituents.* Ultimately, corporate performance is determined by the achievement of business goals across the different business units of the firm. High-performing companies create value over time; low performing companies do not. However, different constituents seek different types of value. Thus, the creation of value — and corporate performance — can be assessed only from the perspective of major market constituents.

Figure 17.1 presents the flows that must be considered in assessing corporate performance. The key constituents of value creation from a corporate performance perspective are customers, owners, and suppliers. All three groups transact with the firm through markets. That is, customers buy (and sometimes resell) goods and services through product markets; owners buy and sell ownership claims, i.e., shares of stock in the company, through financial markets; suppliers sell products and services to the firm through factor markets.

Corporate owners invest in a company in the expectation of receiving economic gains over time. The financial resources provided by owners are used by the firm's managers to acquire productive assets to be used in the production of goods and services for sale. Customers trade money — recognized as revenue to the firm — for goods and services of desired price, quality, and functionality. The firm uses the revenue received from customers to pay suppliers for the inputs they have provided — materials, parts, labor, energy, information, and support services. After all expenses have been paid, any excess revenue is recognized as profit, which can be reinvested in the business and/or used to pay financial dividends to corporate owners.

Corporate Performance from the Perspective of Customer Markets

Customers are interested in the value and performance that is offered by the business's goods and services. However, it is important to remember that customers deal with businesses as corporate entities. In addition to the purchase of any product or service, they are buying the brand name, value, and reputation of the corporation. In buying a computer from Compaq, a refrigerator from General Electric, or an automobile from Ford, a customer is transacting with a corporate entity of known reputation and quality.

Customers in most competitive product markets face choices: they can typically choose among several different product or service offerings to meet their needs. Managers of each firm that competes in a defined product market will attempt to develop a unique *value proposition* to attract customers and generate profitable sales. A value proposition refers to the mix of product and service attributes that a firm offers to customers in terms of price, product features, quality, availability, and after-sales warranty and service.

Customers and potential customers must perceive that a business's value proposition offers superior value and performance or they will choose to buy from a competitor.

Customer Value Measures

To ensure an adequate flow of revenue and profit, managers must ensure that (1) their value proposition is sufficiently differentiated from competitors and (2) their firm's products and services are meeting customer needs and expectations. Managers monitor their value proposition in the marketplace by focusing on key customer value measures. Customer value measures can be either financial — expressed in monetary units such as dollars — or nonfinancial — expressed in units, counts, or quantities.

Financial Measures

- Revenue or revenue growth — This measure indicates customer willingness to purchase a firm's goods and services.
- Gross profit margin — Gross profit margin (sales revenue minus the direct and indirect costs of producing those goods or services) reflects the willingness of customers to pay premium prices in return for perceived value in the firm's products and services.
- Warranty expenses and/or product returns — These measures provide insight into product quality and the extent to which products meet customer expectations concerning features and attributes.

Nonfinancial Measures

- Market share or market share growth — Market share is a measure of customer acceptance relative to competitive offerings in the market place. Market share is calculated as the sales of the firm (i.e., revenue) divided by total sales of all competitors in a defined market.
- Customer satisfaction — These measures are typically collected through survey techniques administered by telephone or mail following purchase of goods and services. Data are collected on customer perceptions of value and the extent to which products or services have met expectations.
- Referrals — This measure of customer loyalty is calculated by gathering data on the source of new business and maintaining a tally of new business generated by referrals.

Corporate Performance from the Perspective of Financial Markets

From a stockholder's perspective, corporate performance is reflected in increases in the monetary value and financial return of their investment. For publicly traded companies, this value can be measured in daily changes in stock price of the firm. For privately held companies, increases in value can only be assessed with certainty at the time when shares of the company change hands.

In competitive financial markets, there are always alternatives for investment funds. Thus, the economic performance of a firm — as reflected in its stock price — must be sufficient to attract new investment and induce existing stockholders to maintain their ownership position. Failure to generate adequate economic returns can lead to the replacement of existing management.

Financial Value Measures

Owners invest their personal monetary resources in the business. In return, they expect financial returns through dividends and/or appreciation in the market value of their investment. Managers must ensure that the financial returns created by the sustained profitability of the business are meeting the expectations of owners.

To assess value creation from an owner's perspective, managers commonly monitor four financial measures of increasing aggregation which focus on corporate performance: profit, return-on-investment (ROI), residual income, and market value.

Profit, as disclosed on a firm's income statement, is the cornerstone of business performance from an investor's perspective. Profit is the residual amount that is retained by the business after subtracting all expenses from the revenues earned during that accounting period.

$$\text{Accounting profit} = (\text{revenues for the period}) - (\text{expenses for the period})$$

Profit is a measure of how much of the revenue received from customers for goods and services is available for reinvestment in the business or distribution to owners. Note, however, that profit as a stand-alone measure does not take into account the level of investment needed to generate that profit. Thus, it is difficult to evaluate the economic performance of a business that earns $100 profit without knowing whether the investment needed to generate that profit was $500 or $1000.

ROI remedies this problem by considering explicitly the level of financial investment. ROI for any period is calculated as the ratio of accounting profit divided by the investment (i.e., assets) that was needed to create that income.

$$\text{ROI} = (\text{accounting profit})/(\text{investment in business})$$

Since ROI takes account of the investment made by owners to support profits, higher levels of profit for a given level of investment can be expected to yield higher financial returns for investors and increased market values. Common variations of this measure used within firms are return on assets, or return on capital employed.

Residual income is a measure of value creation that goes one step further by considering how much profit investors *expect* to earn from their capital. Residual income is a measure of how much additional profit remains for (1) investment in the business or (2) distribution to owners after allowing for normal (i.e., expected) returns on investment. It is calculated by subtracting the normal cost of capital used in the business, calculated at current market rates, from accounting profit.

$$\text{Residual income} = (\text{accounting profit}) - (\text{charges for capital used to generate profit})$$

$$= (\text{accounting profit}) - (\text{value of assets used to generate profit} \times \text{expected financial returns on those assets})$$

Some firms refer to residual income as *economic value added*. Positive residual income correlates with increases in the market value of the firm since positive residual income indicates that a business is accumulating net resources at a greater rate than is needed to satisfy the providers of capital. The firm should, therefore, be in a position to grow and increase future cash flows (or pay out an abnormally high level of dividends to owners).

Market value represents the highest, most aggregate, measure of value creation since it represents the value of ownership claims in the business as priced by financial markets. Market value is the price at which shares in the company trade on the open market. For publicly traded companies, market value is priced daily on a per share basis and reported in the financial press. The total market value of a company, or *market capitalization*, is calculated by multiplying the total number of ownership shares outstanding times the price per share.

$$\text{Total market value} = (\text{number of ownership shares outstanding}) \times (\text{price per share})$$

Market value fluctuates with investor perceptions of the level and timing of expected future cash flows of the business. Market value can be expected to increase for companies in which investors believe that future cash flow growth will be positive.

Corporate Performance from the Perspective of Factor Markets

Any corporate strategy, i.e., attempts to maximize the value of resources controlled by the corporation, must rely on the resources provided by factor markets. These suppliers provide critical resources such as labor, materials, energy, leased land and buildings, support services, etc.

Suppliers think of value in a very different way than customers and owners. They are not investing in the firm or making a purchase decision that may have long-term consequences. Instead, they are selling their own goods and services to the firm in exchange for the promise of a cash payment under defined terms (e.g., net 30 days). Thus, the primary performance measure for suppliers is the promptness and reliability of payment for goods and services received.

In terms of supplier value, the applicable measures that managers monitor relate to liquidity — cash flow and days outstanding in accounts payable. Managers of any business must project and manage cash balances carefully to ensure that cash on hand is sufficient to meet obligations as they become due.

Linking External Markets and Internal Operations

The measurement of corporate performance is an assessment of the quality and/or quantity of outputs created by the various businesses of the firm. These outputs encompass products and services, profit, and cash flow. Of course, it is internal business processes that create these outputs of value to customers and financial returns for owners. Therefore, understanding the linkage between internal business processes and value from a markets perspective is key to corporate performance.

To measure internal business performance, managers must set standards against which to calibrate outputs. Standards can be establishes by reference to some preset benchmark (e.g., an effectiveness criterion based on cost, quality, customer acceptance, etc.) or by reference to the amount of inputs required to produce a given level of outputs (an efficiency criterion). For example, the effectiveness in developing a new product can be compared to previous product development efforts or best industry practice. Order processing can be calibrated against efficiency standards based on the performance of previous time periods. The success of parts manufacturing can be assessed by reference to quality standards and throughput benchmarks.

A Profit Plan Prices Strategy

Inside the firm, business strategy determines the business goals that are set for each separate operating unit and for individual managers. These *internal performance goals* — reflecting performance variables such as cost, quality, speed, market share growth, etc. — are a reflection of the specific business strategy that is being implemented by managers as they attempt to create value for customers, differentiate their products from competitors, and generate satisfactory returns for investors. Thus, internal performance goals based on individual business strategies represent the critical linkage to corporate performance from the perspective of external markets.

Profit plans are the principal tools that managers use to price their business and operating plans, make trade-offs between different courses of action, set performance and accountability goals, and evaluate the extent to which business performance is likely to meet the expectations of different constituents. Managers use profit plans to communicate — both inside and outside the firm — the strategic choices that have been made and level of aspirations that have been set as goals.

Profit plans are used to coordinate and control the internal processes that link the firm's activities with factor markets (suppliers of inputs), customer markets (purchasers of goods and services), and financial markets (suppliers of capital). Profit plans show the quantity and pricing of factor inputs, the revenue to be derived from the expected sale of goods and services, and the amount of profits that will be left over for reinvestment or the payment of dividends.

Examples of the types of strategic decisions that are reflected in a firm's profit plan include the following:

Revenue

1. Number of products in product portfolio
2. Mix and type of products
3. Price points of products (a function of features, quality, and competitive products)
4. Changes in any of the above, including

    a. New product introductions
    b. Product deletions

Cost of goods sold

1. Cost of features
2. Cost of quality
3. Efficiency of internal processes

    - Production scale and batch sizes
    - Economies of purchasing
    - Economies of distribution
    - Capacity

4. Customization
5. Investment in research and development
6. Investment in plant and equipment (through depreciation)

Gross margin

1. Sustainability of business
2. Success of pricing strategy
3. Market acceptance of product differentiation strategy

Selling, general, and administrative expenses

1. Level of support services
2. Outsourcing

Profit

1. Attractiveness of business for future investment
2. Willingness of stockholders to invest resources

Defining Terms

Business strategy: How a business in a defined product market creates value for customers and differentiates itself from competitors.

Corporate performance: A corporation's level of achievement in creating value for market constituents.

Corporate strategy: How a corporation maximizes the value of resources controlled by the corporation.

Profit plan: A projected income statement for a defined future period that estimates flows related to revenues, expenses, and profit.

Further Information

Kaplan, R. S. and Norton, D. P. *The Balanced Scorecard,* Harvard Business School Press, Boston, 1996.

Palepu, K. G., Bernard, V. L., and Healy, P. M. *Business Analysis and Valuation,* South-Western College Publishing, Cincinnati, OH, 1995.

Simons, R. *Levers of Control,* Harvard Business School Press, Boston, 1995.

17.5   How Teams Make Smart Decisions in Technology-Based Firms

Gerardo A. Okhuysen and Kathleen M. Eisenhardt

Success in the high-technology arena depends on management teams making wise strategic decisions: decisions about which markets to enter, which products to make, how to allocate scarce resources in research and development, and so on. In technology-based firms, top management teams make choices with critical consequences under conditions of fast pace and high uncertainty. These firms exist in high-velocity markets, markets where technology and consumer preferences can change at lightning speed. In this kind of environment, managers have difficulty predicting the future, which makes planning impossible. At the same time, the competition is intense and ruthless. This makes even small mistakes potentially costly. Competitors are often poised to steal customers by offering the same goods or services for less money and with better quality.

It is in this uncertain and fast-moving marketplace that top managers in technology based firms make strategic decisions. In addition to making the right decisions, these strategic choices must be made quickly. To make the challenge more difficult, the team must also be prepared to make important decisions again and again. How do effective management teams do it? How can a team function effectively to make decisions and meet the demands of today's high-velocity environment?

In this section, we discuss the approaches that effective management teams use to make strategic decisions in technology-based companies. A management team must overcome three challenges to be successful: the team must be able to make decisions in a marketplace that does not allow for planning, the team must make decisions without delay, and the team cannot make big mistakes. We will explore three strategies that can transform ineffective teams into forward thinking and proactive teams that make very fast and high quality decisions. These strategies center around the critical structures and processes of the top management team. First, we discuss how to build an effective decision-making team. Second, we describe how effective teams make fast decisions. Third, we focus on how effective teams are able to make good decisions again and again.

Step One: Building an Effective Decision-Making Team

Effective strategic decision making begins by building an effective team. The key is to create a team that has diverse viewpoints. We have found two strategies for creating such a team: building team diversity and building team roles.

Building Team Diversity

The first step in building an effective team is creating a diverse team. The goal of diversity is to provide multiple perspectives during decision making. Teams that have members of different ages, academic background, gender, and ethnicity are more likely to have different points of view and are therefore more likely to contribute different ideas in the decision making process. For example, in a successful computer equipment manufacturer the CEO was in his mid-forties and had spent his entire career in sales at a major electronics corporation. Also on the team was the vice president of research and development, a computer prodigy in his late twenties with a background primarily in academics. A third executive, in his fifties, had worked in a variety of other firms before this one. The team was complemented by several other members who were around forty years old, occupying positions in other departments, such as marketing and finance.

Effective teams typically value diversity in team membership. They recognize, often from experience, the need for different opinions in the team, and the need for conflict to ensure that the quality of the decision-making process is high. In these teams, speaking one's mind is not only allowed, but is required by colleagues on the team.

Building Roles

In effective decision-making teams, demographic diversity is not enough. In addition, these teams increase the range of perspectives by using roles. These executives fill these roles in an almost caricature-like manner. There may be a Ms. Action, who is a "doer", a "go-getter", "real-time oriented", and "impatient". Ms. Action is interested in pushing the team forward, in moving the project along at almost any cost. As a complement to Ms. Action, effective teams also include a Mr. Steady. Mr. Steady is the "Rock of Gibraltar" in the team. Mr. Steady's moderating influence advocates for caution in action. Mr. Steady is responsible for moderating the enthusiasm of other team members by expressing skepticism openly. A third common role is that of a futurist, someone who is exceptionally proficient on technical and industry matters and who has ideas, sometimes crazy ones, about where the company should go.

When different members of the team take on different roles, they are able to express points of view that compete with and at the same time complement one another. The differing points of view each of these roles embodies are all helpful in developing multiple options for the team. This process of role building helps keep a broader range of corporate concerns in mind throughout the decision-making process. For example, futurists are typically not concerned with day-to-day operations and instead focus on the long-term direction of the firm. By focusing on tomorrow, they can help the group avoid being trapped in today's circumstances. Futurists can be very strategy oriented, and their concerns within the team reflect this orientation.

How are differing points of view used by effective teams? This is another critical difference between effective and ineffective teams. Ineffective teams interpret divergence of opinions and attitudes as negative to the team. In these teams, consensus and unanimity are valued, at any cost. This attitude stifles the exchange of opinions between team members and prevents the team from capitalizing on the different points of view that members have. Effective teams, on the other hand, not only value differences of opinion, but actively use them to enhance the quality of their interaction and their decisions.

Effective teams use diversity in membership to highlight different points of view, which provides the team with flexibility and speed. Different points of view serve as a starting point for careful analysis, by forcing members to assess the advantages and the disadvantages of different alternatives. The pros and cons of each alternative are hashed out in detail among supporters and detractors. This in-depth understanding of the choices improves the quality of the decision when it is eventually made and helps the group make good decisions.

Step Two: Making Fast Decisions

How do top management teams make fast decisions? High-performing organizations have three characteristics in their fast decision-making process: they excel at gathering real-time data, they develop multiple choices for each decision, and they interact frequently.

Gathering Information

Why is real-time data gathering important? Teams that are effective in dealing with the future are first effective in dealing with the present. Effective teams recognize that the present does not predict the future, but at the same time they understand that the present is the springboard to the future and that it must be tended to carefully. As a consequence, effective decision-making teams have well-structured processes to gather and analyze real-time company performance data.

Effective decision making teams have very well-developed processes to gather information, processes that focus on up-to-date information for which there is little or no time lag. Routinely monitoring the performance of the company is critical to accelerate the decision-making process. When a decision has to be made, having all the available information ready means that there is less background analysis to be done. Instead, the team can focus on the impact of the choices that must be made and move quickly to implement them. This information is usually gathered from different functional areas in the company and the marketplace by people responsible for each functional area. Effective teams focus on day-to-day measures such as orders, inventory, or hourly tracking of production from the manufacturing floor. Ineffective teams, on the other hand, focus on time-lagged measures such as sales and profit and use this obsolete information in their decision-making process.

Developing Multiple Alternatives

Using different points of view, as we mentioned earlier, is important in making the right decision. Making decisions fast also depends on the development of multiple alternatives. Since effective teams are comfortable with handling multiple options at the same time, they become comfortable with ambiguity. Ineffective teams, on the other hand, often engage in an incomplete search. As a result, the number of options considered is insufficient and the quality of the options is deficient.

In the computer industry, the reputation of an organization is critical for success. One ineffective team had the task of fashioning a new image for the company. As part of this process, the management team worked on a name change for the company, using three consulting firms to explore different alternatives. However, the first consulting firm was dismissed, and the third was never consulted because the second seemed to fit the needs of the firm. Once the decision was made and implemented, the team realized through the consequences that they had not developed sufficient alternatives. Limiting the search for options virtually guaranteed that the decision that was eventually made was not the best one.

Developing multiple alternatives for a decision in parallel has other benefits. By viewing a problem through multiple lenses, a team can become more flexible. While one strategic direction might be useful for the company under one set of conditions, a different strategic direction might be appropriate if the conditions in the market change. The development of multiple options implicitly includes the development of fallback positions for the firm. Through the use of these fallback positions, firms remain flexible in turbulent markets.

One effective team headed a startup firm in the computer industry. The issue was the need for new financial backing. These managers initially considered multiple options, such as bank financing, going public, or finding a strategic partner. When their preferred partner for a strategic alliance backed out, the team was ready to sign a deal with their second choice for a partnership. If that had failed, the team had ensured a line of credit and was ready to go public on short notice. The development of multiple options in the initial phases of decision making allowed this firm to maintain its flexibility as circumstances changed.

Interacting Frequently

One last tool used to speed up decision making is the use of frequent interactions among team members. Effective decision-making teams have very frequent interactions and go out of their way to talk to one another. One team, for example, sets aside one day a week as a "no-travel" day. This means everyone is available at least once a week for a full team meeting. At the same time, it is not unusual for members of effective teams to have multiple meetings with one another fixed at different times during the week.

By providing many opportunities for team members to talk to one another, they get to know each other and learn each other's opinions. Through these interactions, two important things happen. First, information is constantly shared, and any change in the market or the company that one team member discovers is quickly communicated to the rest of the team. The other members can then use the new information when they consider the challenges and choices for the firm. Second, the knowledge that comes from frequent interactions translates into a solid understanding of each others' concerns. As a consequence, these concerns are constantly on people's minds as alternative courses of action are discussed.

Step Three: Making Good Decisions Fast, Again and Again

Once teams develop skills to make good decisions, and after they master making fast decisions, one challenge remains: they must make good decisions fast, and they must do it again and again. Maintaining a high pace of activity in the face of constant environmental and competitive changes is not easy, and the members of the team cannot afford to alienate one another. Instead, they must devote time and attention to getting along for the long term.

Establishing a Balance of Power

One of the most common challenges to maintaining long-term effectiveness in teams is politics, politics that mask underlying conflict. Politics appear in a top management team for a variety of reasons, such as conflicting goals or poor firm performance. Effective decision-making teams, though, avoid significant amount of politicking. In contrast, ineffective teams are typically hobbled by them. Politics slow down the decision-making process and cause dissension in the team. But, how do effective teams eliminate politics in the decision-making process? The simplest way of eliminating politics is by establishing a balance of power.

Effective decision-making teams have a balance of power. When power is not balanced, the decision-making process is controlled by only one (or a few) team member. Typically this is the leader. In these teams, the other members feel their only alternative is to use politics to make their case. Members of these teams complain about their inability to do things independently, and they complain about the centralization of power. These teams have much behind-the-scenes wheeling and dealing, reflected in "outlaw" meetings and routine withholding of information. In order to eliminate politics under these conditions, members of the team need independence of thought and action. Members of effective teams recognize and value this independence, and at the same time they recognize and accept the scope of their authority as well as the scope of their responsibility. To complement this, they also understand the scope of authority and responsibility of other team members. This explicit recognition of the bounds of authority means team members do not need to use politics to voice their concerns.

To maintain a balance of power, a second element is needed. All members of the team must be involved in the decision-making process. Without this participation, team members are often dissatisfied both with the decision-making process and with the outcome. Members of ineffective teams often complain about how little participation they have in the decision-making process. Again, this is usually the case in teams controlled by one individual, and, again, this situation often leads to politics. By contrast, effective decision-making teams involve all of the members in the process. Through the constant sharing of information, members of effective teams feel that their opinions are considered in the decision-making process. This way, members develop a buy-in for the decisions that are made by the team.

Developing Consensus with Qualification

One of the challenges to the long-term viability of decision-making teams is resentment that builds up on the part of some team members toward others. Resentment appears when one or more members of the team are consistently seen as losers in the strategic decision-making process.

Effective teams avoid this problem by using consensus with qualification. When it is time to make a decision, the team tries to arrive at consensus on the strategy the company will follow. In effective teams, the final decision is never put to a vote. A vote, by its nature, classifies members of the team as either winners (the majority) or losers (the minority). Instead of taking a vote, effective teams focus on trying to convince dissenters to have a change of heart. To support this effort, an effective team does not penalize individual members for changing their mind and accepts that as a normal part of the process. By allowing for changes of opinion, members are less likely to blindly support a particular choice or position to save face, even after it has been discredited.

However, in effective teams, the consensus building process is never carried to the extreme. Unanimity is not required to make a decision. Instead, if a broad consensus is achieved in the team, the opinion of the holdouts is considered, but a decision is then made by the CEO or the president. This way, ultimate responsibility for the decision rests with the leader of the team. Ineffective teams, by contrast, typically have a naive view that consensus is always possible. This approach is ineffective because the search for consensus, which is sometimes not possible, can extend the decision-making process. This search for consensus often involves endless haggling and often ends in frustration for all. A blind search for consensus means that every individual in the team has veto power. As a consequence, decisions can be delayed indefinetely at great cost. Developing consensus with qualification distinguishes effective teams from ineffective ones.

Keeping a Sense of Humor

So far, we have discussed structures and processes that aid the decision-making process. Sometimes, however, simple tactics make an important contribution to help groups get along. Teams that get along over the long term have a well-developed sense of humor. In some teams, humor is articulated as one of the necessary goals of the decision-making team. Team members place a premium on making others laugh, and on getting along with one another. Laughter is commonplace, and practical jokes abound during April Fools' and Halloween. Effective teams use humor to broach difficult subjects while defusing serious interpersonal challenges. Effective teams engage in the use of humor as a way of maintaining conflict focused on task-oriented issues rather than interpersonal ones. By contrast, humor is unthinkable in ineffective decision-making teams. Outside of the mandatory holiday party, there are no social activities in these teams, and there are certainly no attempts at humor.

Conclusion

In this section, we have discussed strategies that distinguish effective teams from ineffective ones, strategies that affect critical aspects of the decision-making process. First, we discussed how to build an effective team to make good decisions. These teams capitalize on the team's diversity and the use of roles to fully exploit different approaches to each problem. Second, we focused on how effective teams make fast decisions through intensive data gathering, the development of multiple alternatives, and frequent interactions among team members. Last, we described how effective teams are able to make decisions again and again. To do this, they establish a balance of power, they develop consensus with qualification to make decisions, and they maintain a sense of humor in their interactions.

In the introduction, we indicated that technology-based firms are unique because they face fast-moving and uncertain markets. This makes planning impossible. The competition is often intense and ruthless. Mistakes can be fatal. By following the strategies presented in this section, top management teams can develop their decision-making skills, and they can effectively overcome the challenges they face in today's competitive environment. The task for decision makers is at the same time fairly straightforward to describe but complex and difficult to achieve. Effective teams are those that are willing to try and succeed.

References

Bourgeois, L. J., III and Eisenhardt, K. M. Strategic decision processes in high velocity environments: four cases in the microcomputer industry, *Mgt. Sci.,* 34: 816–835, 1988.

Eisenhardt, K. M. and Bourgeois, L. J., III. Politics of strategic decision making in high-velocity environments: toward a midrange theory, *Acad. Mgt. J.,* 31: 737–770, 1988.

Eisenhardt, K. M., Kahwajy, J. L., Bourgeois, L. J., III. Conflict and strategic choice: how top management teams disagree, *Calif. Mgt. Rev.,* 39: 42–62, 1997.

Further Information

Eisenhardt, K. M. Speed and strategic choice: how managers accelerate decision making, *Calif. Mgt. Rev.,* 32: 39–54, 1990.

Eisenhardt, K. M., Kahwajy, J. L., Bourgeois, L. J., III. How management teams can have a good fight, *Harv. Bus. Rev.,* 75: 77–85, 1997.

17.6   Insourcing/Outsourcing

Robert B. Handfield

One of the most complex and important business decisions facing business today is whether to produce or provide a component, assembly, or service internally ("**insourcing**") or whether to purchase that same component, assembly, or service from an outside supplier ("**outsourcing**"). The impact of such decisions is often felt for many years. For example, the decision by U.S. electronics firms to outsource radio transmitter components in the early 1950s to Japanese suppliers helped to establish the electronics industry in Japan and Hong Kong. The transfer of technology that took place led to these same suppliers evolving into major competitors who eventually captured many U.S. markets.

Insourcing/outsourcing ("make or buy") decisions also can be very controversial. American workers have increasingly become concerned over the fact that organizations have been outsourcing manufacturing tasks to suppliers in developing countries with lower wage rates. These sentiments became especially evident during the North American Free Trade Agreement discussions in 1994. In March 1996, these issues came to a head, as members of the United Auto Workers Union within General Motors (GM) went on strike as a protest against increased outsourcing of components by GM.

Because of the importance of insourcing/outsourcing to organizational competitiveness, a number of variables and factors must be systematically considered. These factors include a firm's competency and specific costs, quality, delivery, technology, responsiveness, and continuous improvement. Cross-functional teams are increasingly being used to make the insourcing/outsourcing decisions. In most cases, these teams are made up of key representatives of critical functions such as manufacturing, engineering, finance, and procurement.

The professional manager must bring a wide variety of knowledge and technical skills to bear on the insourcing/outsourcing decision, ranging from strategic thinking to in-depth cost analysis so as to provide the greatest insight about the external supply base capabilities. This section will provide an introduction to the insourcing/outsourcing decision. The decision to outsource or insource is complex and often requires extensive analysis. Each alternative involves a different set of facts, which must be thoroughly evaluated.

Core Competence

Insourcing involves keeping the product of a product or service within the firm. This option also has its pros and cons. The tremendous growth in the number of highly capable global competitors facing most organizations in the United States and North America markets has forced them to dramatically improve overall competitiveness by concentrating on their **core competence** [Prahalad and Hamel,1990]. A core competence refers to a skill, process, or resource that distinguishes a company and makes them "stand out from the rest". Core competencies often provide global competitors with a unique competitive advantage, in terms of cost, technology, flexibility, or overall capability.

Insourcing/outsourcing decisions are strategic in nature because they determine where a firm allocates its manufacturing or service resources. These decisions reflect where management believes the company possesses a strong level of core competence. Unfortunately, many firms have locked themselves into a "make" position through years of tradition, (i.e., "we have always made this item, so we are not about to stop now!"). The concept of core competence dictates that a firm should concentrate on developing or enhancing those activities that it does better than the competition. For example, Hewlett-Packard probably will benefit more by developing new technologies and products than focusing its attention on manufacturing printed circuit boards, which is a mature technology.

How does one define a core competence? In general, core competence is the firm's long-run, strategic ability to build a dominant set of technologies and/or skills that enable the firm to adapt to quickly changing marketplace opportunities. In their classic *Harvard Business Review* article, Pralahad and Hamel define core competence as "the collective learning in the organization, especially how to coordinate diverse production skills and integrate multiple streams of technologies." A manager or team tasked with making an insourcing/outsourcing decision must develop a true sense of what the core competence of the organization is and whether the product or service under sourcing consideration is an integral part of that core competence. A key product or service that is closely interrelated with the firm's core competence would more likely be reflected in a favorable insourcing ("make") decision rather than an outsourcing ("buy") decision. If a firm errs and mistakenly outsources a core competence, they may lose a strong a competitive advantage.

There are three key principles related to the concept of core competence that may guide the professional manager in effectively participating in an insourcing/outsourcing decision.

1. The firm faced with the insourcing/outsourcing decision should concentrate on providing those components, assemblies, systems, or services that are critical to the end product and that the firm has a distinctive advantage in providing.
2. The firm should consider outsourcing components, assemblies, systems, or services when suppliers have a particular advantage over the firm in providing those goods or services. Advantages may occur because of economies of scale, higher quality, familiarity with a technology, economies of scope, favorable cost structure, or greater performance incentives.
3. The firm can use strategic outsourcing as a way of enhancing employee motivation and commitment to improve the firm's manufacturing or service performance in order to save internal jobs.

An example of an effective outsourcing decision is the case of a mountain bike manufacturer, which was interested in producing bicycles with titanium frames. Titanium, because it is so light and strong, has many advantages over aluminum and alloy frames. However, the material has many properties that make it extremely difficult to work with, and the bicycle producer did not have the equipment or the workforce skills required to produce the frames. The manufacturer carried out an insourcing/outsourcing analysis and found a former arms manufacturer in Russia that possessed excellent skills in working with titanium. With the arms market in Russia essentially gone, the arms producer was looking for new business. This proved to be an excellent match, and the bicycle manufacturer now outsources all of its titanium frames from the former arms producer, who is also very happy with the arrangement!

Time-Based Competition

In addition to global competitiveness, another significant factor impacting the insource/outsource decision is **time-based competition**. In particular, time is very important in the new product development process, one of the key decision points where a detailed insourcing/outsourcing analysis is appropriate. For example, Japanese automakers are consistently able to design and build a new automobile in less than 30 months. Until very recently, the Big Three automotive manufacturers required from 48 to 60 months to accomplish the same set of tasks [Handfield, 1995]. It is easy to see that, after a relatively short period of time, the Japanese automakers will have steadily leapfrogged their American and European counterparts by several generations of models, achieving a significant technological and marketing advantage in terms of quality, design, and performance.

Reliance on Suppliers

In the last several years, there appears to be a greater trend toward outsourcing. A recent study polled a number of purchasing executives, who indicated that they perceived the potential contributions of suppliers to be increasing in importance. Furthermore, these executives believed that they would rely increasingly on external suppliers for product and process technology [Monczka et al., 1992]. In addition, despite the increased reliance on suppliers, the study of executives also found supplier performance lags in meeting future buyer expectations. Supplier performance does not appear to be increasing at a level to fully meet perceived future requirements of buying companies. These data suggest that the insourcing/outsourcing decision will become ever more difficult as managers must carefully decide which suppliers to outsource to.

To summarize, there are a number of forces that will influence the insourcing/outsourcing decision process.

1. There will continue to be severe pressures to reduce costs as firms attempt to utilize their productive resources more efficiently. This is particularly true in companies that are facing increased global competition.

2. Firms will continue to become more highly specialized in products and production technology. As a result, cost differentials between making and buying will increase as firms are unable to specialize in everything!
3. Firms will increasingly focus on what they do best (core competencies) while externally sourcing areas in which they do not possess expertise.
4. The need to respond rapidly and effectively to customer requirements will encourage greater levels of outsourcing and less vertical integration in industries with rapidly changing technologies and marketplaces.
5. Improved tools such as computer simulation and forecasting software will also help firms to perform insourcing/outsourcing analyses with greater precision and allow the easy use of sensitivity analysis to compare different possibilities.
6. While firms will probably outsource more in the future, managers must carefully evaluate and in some cases develop the capabilities of suppliers with whom they intend to outsource.

Initiating the Insourcing/Outsourcing Decision

There are several situations when an insourcing/outsourcing decision is appropriate. These include the new buy situation and re-consideration of making to buying and from buying to making. These situations occur at any number of times within a product's life cycle. Similar analysis of the various insourcing/outsourcing factors is necessary for each situation.

New Product Development

The first point of the insourcing/outsourcing decision process occurs during the new product development cycle, at which point an initial insourcing/outsourcing analysis is conducted. Because the product, service, and/or components have not yet been provided, there may be minimal information available to guide the analyst in the decision-making process. The components, assemblies, systems, or services under consideration for insourcing or outsourcing may represent unfamiliar new technology or different processes. This fact, coupled with the high levels of uncertainty associated with any new product introduction, dictate that an outsource decision be made initially unless the parts under appraisal are considered core competencies. The company will also consider the stability of the technology, the possible duration of the product life cycle, and the availability of reliable sources.

Sourcing Strategy

The second initiator during the product life cycle when an insourcing/outsourcing analysis might be appropriate is related to an organization's sourcing strategy. During the business strategy development process, top-level executives may decide that a change in sourcing patterns is necessary. This decision will be supported by a detailed insourcing/outsourcing analysis, using both cost and noncost factors as inputs to the decision. As a result, a firm may decide to change its particular strategic focus regarding the business or markets it competes in and the core competencies required to become and remain competitive in those businesses and markets. For example, many companies such as Spring, Union Pacific, Tenneco, Anheuser-Busch, and ITT have "spun-off" noncore businesses and are choosing to outsource some of these components. The CEO of General Electric, Jack Welch, formally implemented a policy that "spun-off" any division that was not considered to be a "core competence" (even though many of them were profitable!).

Supplier Performance

A third initiator of insourcing/outsourcing decisions is related to the failure of the current supplier's performance. If a supplier demonstrates an inability or unwillingness to manufacture a particular part or provide a key service (or shows an unwillingness to continuously improve), then the purchasing firm must decide whether to bring back manufacture of the part or component in-house or to develop another capable external source. Likewise, if internal manufacturing performance has proven itself to be unsatisfactory, then the firm is faced with a decision to outsource with a capable supplier or spend time and resources to improve its internal capabilities.

Demand Patterns

The fourth key decision point occurs when there are significant shifts in the marketplace stemming from such conditions as changing sales demand or changing market economics from technological innovation. If demand decreases dramatically, then there might be an impetus to shift production from internal to external sources to make more effective use of the firm's physical assets and skills. Likewise, if demand increases significantly, then the firm might consider insourcing the part or component instead of continued outsourcing to garner economies of scale or scope.

Technology Life Cycles

Changes in the technology used to produce a particular part or component may also favor a decision to insource or to outsource. **Technology life cycles** refer to the duration of a particular technology before it becomes outdated. An example here is the memory chip: the 386 was replaced by the 486, which was replaced by the Pentium, and so on. If a product technology is relatively mature or stable, then the technology life cycle will probably last a good long while. In such cases, there is some reasonable assurance that investment in capital equipment to produce that technology will have a longer payback period. On the other hand, if the technology is changing rapidly, outsourcing can help shift the risk to an external source.

Careful analysis of past life cycles of components and materials will provide insights into potential technology life cycles. In many technology-oriented firms, designers also keep very close tabs on component technology and are usually in the forefront of anticipated change. Regular feedback and contact with designers can serve to provide information on the potential life cycle of materials. The purchasing trade literature and the trade literature of various technology-oriented industries will also provide information on technology life cycles of various materials. Regular new product introductions are often announced in the trade literature as suppliers push new products for customer purchase. Procurement specialists must be careful in analyzing trade literature since the suppliers of new technology are often overly optimistic with regard to availability. Finally, the purchasing professional can often gain insights into technology life cycles through regular contact and interaction with component suppliers, who can provide information on current component product development and expected timing for release of new technologies [Handfield and Pannesi, 1994].

It is important for a firm to understand the general conditions under which an outsourcing relationship should be modified or terminated before the outsourcing relationship is initiated or consummated. This evaluation calls for weighing the firm's intentions for the relationship, the potential supplier's intentions for the relationship, and the likely progression of technology. These factors are very difficult to measure and analyze prior to the relationship, but it is important to make the effort to include these factors so as to arrive at an appropriate decision. In addition, such analysis should be continued over time as appropriate.

Advantages/Disadvantages of Insourcing

There are a number of advantages to vertically integrating (i.e., insourcing) the product or service. First, the degree of control the buyer wishes to exert over the transfer of technology should be considered. If a high degree of control is desired so that proprietary designs or processes can be protected from unauthorized use through, then vertical integration may be preferred over outsourcing. Other alternatives to maintaining control over technologies include confidentiality and **nondisclosure agreements**. Vertically integrating can also help a firm to prevent a currently competitive market from deteriorating into a noncompetitive market over time. Once the supplier is acquired, the buying firm can stipulate both price and unrestricted availability to key resources over an indefinite period of time. This type of backward vertical integration allows firms to gain greater control over their supply of scarce inputs and can help to secure a competitive advantage. Firms can control key supply markets, preventing both current or future competitors from having sufficient access to critical key inputs. A vertically integrated firm increases its visibility over each step of the process by having more of the factors of production available under its daily control. A dedicated facility can also result in lower per unit costs when economies of scale or scope provide the firm with higher efficiencies.

The disadvantages of vertical integration are related to the level of investment typically required when the insourcing decision is made. A high level of investment is required when new plant and equipment is bought. The firm must ensure that adequate volume is present to sufficiently pay back the plant and equipment bought to manufacture the product internally. Second, if the investment is made in dedicated plant and equipment that cannot be utilized for other types of products in the future, the risk associated with the insourcing alternative increases. A good example is the semiconductor industry. In 1995 alone, at least a dozen new semiconductor plants were under construction, including three by Intel and two by Motorola. The average cost of a chip plant is currently about $1.5 billion, but is expected to rise to $3 billion by 1999. The life of these plants is often as little as 6 months, before the chips are replaced by newer technologies. Plant expansions are being made on the premise that investments in new capacity can produce rapid market-share gains. On the risk side, however, analysts are worried that the chip business may gyrate in future years, and the increasing cost of wafer-fabrication plants could soar beyond the reach of all but a few companies [Business Week, 1995].

Another disadvantage might be the lack of flexibility encountered if a firm tries to change or alter the product in accordance with market needs or demand. Matching demand to requirements in the various parts of the supply chain is an intricate process. Unbalanced output levels at one location can result in waiting times, while inventories build up in another linkage. Finally, more than one firm has expected its supplier's performance to improve once it was acquired through investment. In many cases, the supplier actually becomes more complacent than before, and, since there is no incentive to improve performance, it actually deteriorates!

Advantages/Disadvantages of Outsourcing

Outsourcing generally refers to the process of sourcing a particular product or service from an outside organization. Other terms that have been used to describe the sourcing process include (1) contracting, (2) procurement, (3) external manufacturing, and (4) purchasing. The common element in each of these is that they all indicate that the good or service sought is produced from outside the firm where it is ultimately used.

The major advantage to outsourcing is that it allows the buying firm a greater degree of flexibility. As market demand levels change, the firm can more easily make changes in its product or service offerings in response. Because there are lower levels of investment in specific assets, it is easier for the firm to make unexpected changes in its own production resources. There is a minimal investment risk for the buyer as the supplier assumes the uncertainty inherent in plant and equipment investment. Ideally, both the buying and the supplying firms should concentrate on their own distinct core competencies while outsourcing other products and services that are not considered areas of expertise, yet are necessary for competitive advantage. Also, outsourcing allows for improved cash flow because there is less up-front investment in plant and equipment. A firm may achieve reductions in labor costs by transferring production to an outsource location that pays lower wages or has higher efficiencies, resulting in lower per unit costs of manufacture. Especially in recent years as the high cost of providing retirement and medical benefits continues to escalate, firms are increasingly avoiding hiring more full-time personnel whenever possible.

Conversely, there is great risk if the firm chooses the wrong supplier or suppliers to provide the product or service being outsourced. The supplier's capabilities may have been misstated, the process technology may have become obsolete, or the supplier's performance may not meet the buying firm's expectations or requirements. In one case, a manufacturer trusted a supplier to develop a component, based on the supplier's claim that it had mastered the required process technology. By the time they realized that the supplier was incapable of producing the product, the market for the final product had already been captured by a competitor! There is also the issue of loss of control. The buying firm may perceive that it has lost the ability to effectively monitor and regulate the quality, availability, or performance of the goods or services being bought because it is not produced under the firm's direct supervision. This may lead to concern over product/service performance. Ultimately, the buying firm will create additional safeguards to prevent poor performance by changing specifications, increasing inspection, or periodic audits to ensure that the supplying firm is meeting expectations. One example of this behavior is the use of "guardbanding". In such cases, the buying company's engineers will design products with specifications that are "tighter" than normal, with the expectation that supplier's will never be able to produce to these specifications. In fact, this extra "safety margin" often creates additional problems, as suppliers may charge a higher price in attempting to meet the specifications on their equipment. In order to prevent this situation from occurring, improved communication among engineers, buyers, and suppliers is important.

A final disadvantage of outsourcing is the potential for losing key skills and technology that may weaken a company's future competitive position. The phenomena of North American firms outsourcing their manufacturing to low-cost suppliers who later become global competitors has been described as the "hollowing out of the corporation". Indeed, some would argue that many U.S. firms are now only empty "shells" that no longer produce anything, but simply act as distribution and sales networks. While there is probably some truth to this statement, the outsourcing decision has to be balanced off with the need to remain globally competitive and to outsource those tasks over which the firm no longer has a competitive advantage.

Defining Terms

Core competence: A firm's long-run, strategic ability to build a dominant set of technologies and/or skills that enable the firm to adapt to quickly changing marketplace opportunities. Core competencies often provide global competitors with a unique competitive advantage, in terms of cost, technology, flexibility, or overall capability.

Insourcing: The decision to develop a product/component and process internally and subsequently manufacture it. This decision typically occurs when that product or process is considered to be an organizational "core competence".

Nondisclosure agreement: An agreement signed between two contracting parties that stipulates that a given technology will not be shared with other external parties.

Outsourcing: The decision to subcontract the development of a product/component and process to an external supplier. This process typically involves an extensive comparison of internal capabilities vs. suppliers' costs. Once the decision is reached, a contract is negotiated with the supplier for a fixed period.

Technology life cycles: The duration of a particular technology before it becomes obsolete.

Time-based competition: A firm's ability to reduce product development time, customer order cycle time, and/or response time. This ability will often give organizations a significant market advantage over their competitors and is especially important in product markets with short life cycles and customized customer requirements.

References

The Great Silicon Rush of '95, *Business Week,* October 2: 134–136, 1995.

Handfield, R. B. *Re-Engineering for Time-based Competition,* Quorum Press, Westport, CT, 1995, 1995.

Monczka, R. M., Trent, R. J., and Callahan, T. J. "Supply base strategies to maximize supplier performance, *Int. J. Phys. Distrib. Log. Mgt.*, 23(4): 42–54, 1993.

Pralahad, C. K. and Hamel, G. The core competence of the corporation, *Harv. Bus. Rev.*, May-June: 79–91, 1990.

Further Information

A good textbook that covers many details of the outsourcing decision is *Purchasing and Supply Chain Management*, by R. Monczka, R. Trent, and R. Handfield (Southwestern College Publishing, Cincinnati, OH, 1997).

A book that describes in detail the specific processes to be used in the outsourcing decision in new product development is *Supplier Integration into New Product Development*, by R. Monczka, G. Ragatz, R. Handfield, and R. Trent, (ASQC Press, 1998).

An article describing one organization's experience with insourcing outsourcing is provided by Ravi Venkatesan, "Strategic Sourcing: To Make or Not to Make," Harvard Business Review, November-December: 98–107, 1992.

IBM started in 1911, in Auburn New York, with the creation of the Computing-Tabulating-Recording Company (C-T-R). It wasn't until 1924 that CTR adopted the name International Business Machines Co. Ltd. Eventually the company was called IBM.

Growth in revenue

1914	$4 million
1920	$12 million
1937	$30 million
1948	$156.4 million
1955	$563 million
1969	$6.9 billion
1975	$14.4 billion
1980	$26.2 billion
1984	$40 billion

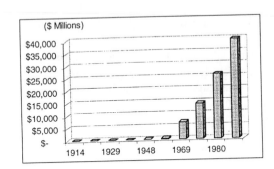

Innovations:

1930 — Accounting machines using punching cards
1952 — First production computer (IBM 701)
1955 — 608 Transistor calculator, high-speed printers
1961 — Dictation system and portable dictating unit
1967 — Binary synchronous communications
1973 — Thin film memory that operates at a speed of 100 billionths of a second
1980 — Single-element typewriter, designed to meet the Japanese Katakana alphabet requirements
1983 — IBM Personal Computer XT

Social and Educational Actions:

1933 — New educational laboratories
1937 — Paying employees 6 holidays
1940 — Military service benefits
1946 — Hospitalization plan for employees
1947 — Permanent disability income plan
1952 — IBM family major medical plan
1958 — All workers previously paid on hourly basis become salaried
1959 — Matching grants program

Strategic and Management Issues:

- The first private company to implement social benefits.
- Big scale operations — 1936 making accounting machines for the Social Security Program and NASA.
- Marketed the IBM personal computer in a mass market.

VI

Core Relationships for the Technology Manager

18

Human Resources

Mary Ann Von Glinow
Florida International University

Terri A. Scandura
University of Miami

Angelo S. DeNisi
Texas A&M University

Adrienne J. Colella
Texas A&M University

Dana M. Johnson
Wayne State University

E. Brian Peach
The University of West Florida

Ralph Roberts
The University of West Florida

Karol I. Pelc
Michigan Technological University

18.1   Recruitment of High-Technology Workers: Today's Real Challenges and Keys to Success

Mary Ann Von Glinow and Terri A. Scandura

There is an increasing demand for workers that have high-technology engineering and computer skills. It is expected that a shortage of engineers will pose a threat to the future economic survival of the United States [Forbes and Edosomwan, 1990]. Due to the emergence of **high-technology industrial** growth, both in the manufacturing and service sectors, the need for scientists has been projected to increase by 36% between 1986 and 2000. At the same time, the phasing out of the post-war "baby boom" has resulted in a college-age population of fewer than 24 million in the year 2000 compared with 30 million in 1986. The resulting population of college-age individuals has been referred to as the "baby bust" generation or "generation X". Persons in this group grew up with computers in their classrooms from grades K to 12 and are increasingly demanding interesting, challenging, and well-paying jobs of today's employers. Recruiting and attracting these new high-technology workers, now in short supply, is a major challenge for today's organizations [Edwards, 1986].

Recruiting High-Tech Workers: Today's Real Challenges

The Need for Management Competence

The shortage of high-technology workers has created a number of challenges for the organization attempting to ensure their technological advantage by having employees with necessary technology savvy to meet the demands of today's competitive technological environment. As noted by Von Glinow [1988], the need for management competence in the ranks of the high-technology workers is still very much the case. McCarthy et al. [1987] conclude that management competence is "… an essential factor that distinguishes successful high-technology companies from those that make little or no progress." Managers of high-tech firms must be able to find and retain the rare, highly skilled people who also have excellent managerial skills needed in today's high technology businesses. One recent survey conducted by the University of California found that companies will be hiring 17% more "techno-MBAs" in 1997 than in 1995, and the projected hirings of traditional MBAs will be down by 13% [Pepe, 1996]. As noted by Warner [1986], the need for managers with **hybrid skills** and the need for skill integration in the high-technology workplace has resulted in recruitment being even more selective. Hybrid skills are defined as the ability to integrate the needs of markets, technology, and products.

Mergers and Acquisitions

The current trend in high-technology firms of mergers and acquisitions has also created challenges for recruiting and retaining high-tech workers. Harwood and Briscoe [1987] report a case study of firms that had been acquired by a larger firm, resulting in an exacerbation of its inefficient hiring practices. The acquiring firm brought different belief systems regarding personnel selection issues. Review of the recruitment practices in this firm revealed that (1) managers and supervisors had no training in interviewing skills, (2) too many people were involved in the recruitment process, (3) the amount of information on prospective employees collected was uneven (either too much or too little), (4) meetings to review job candidates were vague and leaderless, (5) managers and interviewers failed to coordinate efforts, and (6) managers had significant differences in values and expectations, which affected recruiting criteria. A new approach was instituted in this firm, including mandatory interviewing skills workshops to bring greater consistency to the process. Restructuring processes can turn the spotlight on recruitment practices as the type of employees hired into the acquired or merged firms will be placed under scrutiny following mergers and acquisitions.

Cultural Diversity of High-Technology Workers

As noted by Lewis [1987], many high-technology companies have a high percentage of foreign-born employees. This cultural variety has added even more complexity to managing already complex industries. For example, California's high-technology firms in Silicon Valley are highly dependent on Chinese engineering talent (there are an estimated 10,000 Chinese engineers working in Silicon Valley [Johnstone, 1989]). A case study of a Silicon Valley semiconductor company illustrates some of the issues that may arise from an increasingly diverse high-technology workforce [Lewis, 1987]. When communicating with others from different cultural backgrounds, people tend to make false assumptions about the other by **projecting cognitive similarity**. Projecting cognitive similarity is defined as expecting another person to respond in the same manner to situations and to hold the same values, attitudes, and goals. Obviously, persons of differing cultural backgrounds may not be cognitively similar, and this may give rise to numerous misunderstandings and miscommunications in high technology environments. Many Chinese want to control their own destinies and work in an environment in which they can gain promotion or are free to venture out on their own [Johnstone, 1989]. Also, it has been projected that the shortage of engineers will be met by increases in women, minorities, and the disabled [Forbes and Edosomwan, 1990]. Hence, it appears that the trend toward increasing diversity in the high-technology workforce will continue.

Given the challenges in recruiting high-technology workers today, our review revealed several "keys to success" for recruiting in high-technology firms. With the goal of attracting the highest levels of high-tech talent available, numerous approaches to facing these issues are being implemented.

Keys to Success

Clearly, in identifying successful from unsuccessful high-technology companies, the recruitment of management competence is central — among which is technical know-how and market savvy [McCarthy et al., 1987; Von Glinow, 1988]. Recruiting for these factors in an increasingly shrinking domestic market demands a long-term strategic outlook, paired with quick response time, all delivered in a global environment. Recruiting nowadays demands sourcing of candidates from across borders and, most certainly, across cultures. Recruiters must be able to locate, hire, motivate, and retain these highly skilled people. Recruiting the right people isn't the only ingredient in the successful high-technology company; however, it is an awfully important one that links personnel practices with technology. There are now research studies that conclude that successful high-technology companies are the ones that give the personnel/human resource function greater recognition [Yuen, 1991]. This speaks to the partnering of human resources (HR) with the various technology departments. This is where recruiters can appeal to their highly skilled job candidates by emphasizing that, in addition to superior technological systems, there will also be superior personnel practices in place, thus enhancing satisfaction and quite possibly performance.

Recruiting techniques that enhance that fit between people and technology include attention to the skilled workers' career and personal development. Many companies offer company-sponsored personal and skill-specific development programs as well as the more common full tuition reimbursement programs [Chen, 1990]. In addition, recruiting initiatives that stress flexibility appear superior. This includes flex-time, flex-work, flex-pay, and flex-benefits. High-tech firms that appear to enjoy the "preferred employer" status also recruit by emphasizing esprit de corps, identifiable career paths, a quality environment, and other employee assistance programs such as on-site child care, fitness facilities and housing assistance. While more difficult in recruiting to the international post, these elements are stressed in the recruiting interview as well. One well-known high-technology company in China actually provides not only housing assistance; it allows home purchase under the umbrella of its benefit plan.

Successful recruiting initiatives also stress a good performance management system and acknowledges and rewards creative and excellent performance. This seems somewhat obvious, but in practice few high-technology firms have found the right mix of the reward/performance equation. This is true because most of our existing knowledge on how to reward employees is at high variance with current empirical research on how best to reward high-technology and professional workers.

In culling through the research on successful recruiting techniques used, there are a number of critical issues that have been identified. For organizations seeking top-notch talent, care should be taken so as not to turn the candidate off with the idiosyncrasies of the recruiting process. Management has often neglected to pinpoint specifically the nature of the job before the interview process or tries to oversell the candidate on the job or falls prey to other predictable problems in recruiting. To alleviate these issues, care should be taken to have a reasonable job description (or a set of skills for the tasks to be accomplished), identify the candidate's real needs, lead with honesty by providing a realistic preview of the job as well as the organization's culture, and by emphasizing a balanced career and compensation package.

As we enter the new millennium, employee recruitment is increasingly a challenge, forcing even the most resourceful of companies to explore alternative recruiting options. As one frame of reference, we might use the following phrase in recruiting the best and the brightest: to buy, borrow, or build talent.

In buying talent, the firm must concentrate on the perks associated with the compensation package as well as the fit with the company culture. In borrowing talent, the prescription has more to do with a "shared services" model, wherein the appeal to the job candidate might reside with the opportunity to work on many exciting projects. In building talent, the firm recruits to the training, educational, and developmental programs that may appeal to job candidates. Nontraditional recruitment methods include sourcing candidates from outside the United States. Given that organizations have taken on a "boundaryless" quality, recruiting has now become "borderless" and through electronic means. As the cost of overdoing the "virtual organization theme", there may be some merit in recruiting the "virtual worker." The virtual worker is a skilled specialist operating anywhere, on anything, in a highly team-based, information-rich environment. Thus, an employee may literally work hundreds of "jobs" while never changing positions, quite akin to telecommuting, however, in the context of the global organization. Such employees, coupled with sophisticated and complicated technological changes to hardware and software, tax older and often inadequate existing HR systems and practices. Thus, high-technology companies that have a balance scorecard for their HR function tend to anticipate these recruiting challenges, factor them in *a priori*, and fare quite well overall in the recruiting game.

Acknowledgment

The authors would like to thank Linda Clarke and Ethyln Williams for their assistance in compiling information for this work.

Defining Terms

High-technology industry: The American Electronics Association (AEA) uses 45 standard industrial classification (SIC) codes to define the high-technology industry. They fall into three broad categories — high-technology manufacturing, software- and computer-related services, and communications services. The 45 codes do not comprehensively cover the entire high-technology industry, as the structure of the SIC system is limited. In an effort to produce solid statistics, AEA does not include broad categories of the high-technology portion because it does not represent a clear majority.

Projected cognitive similarity: Expecting another person to respond in the same manner to situations and to hold the same values, attitudes, and goals.

Hybrid skills: The ability to integrate the needs of markets, technology, and products.

References

Chen, C. Recruitment: TRW S&D strives to be the preferred employer, *Person. J.,* 69(7): 70–73, 1990.

Edwards, C. Aggressive recruitment: the lessons of high-tech hiring, *Person. J.,* 65(1): 40–48, 1986.

Forbes, L. H. and Endosomwan, J. A. The role of women, minorities and the handicapped in the engineering sector, *Ind. Eng.,* 22(9): 47–49, 1990.

Harwood, S. and Briscoe, D. R. Improving the interview process: a case study, *Personnel,* 64(9): 48–50, 1987.

Johnstone, B. California's brain gain, *Far East. Econ. Rev.,* 146(42): 55, 1989.

Lewis, J. C. Issues concerning high technology managers from Multiple cultural backgrounds, *J. Mgt. Dev.,* 6(3): 73–85, 1987.

McCarthy, D. J., Spital, F. C., and Laurenstein, M. C. Managing the growth at high-technology companies: a view from the top, *Acad. Mgt. Exec.,* 1(4): 313–323, 1987.

Pepe, M. 'Techno-MBAs' gaining prominence, *Comput. Resell. News,* (715): 167, 1996.

Von Glinow, M. A. *The New Professionals,* Ballinger, Cambridge, MA, 1988.

Warner, M. Human-resources implications of new technology, *Hum. Syst. Mgt.,* 6(4): 279–287, 1986.

Yuen, E. Human resource management in high- and medium-technology companies, *Int. J. Manpower,* 12(6): 10–20, 1991.

Further Information

http://careers.computerworld.com
http://www.year2000.com
http://www.aeanet.org

18.2 Human Resources: Selection

Angelo S. DeNisi and Adrienne J. Colella

Selection is the process by which organizations choose, from among a group of applicants, whom they wish to hire. Virtually every organization needs employees, and so they must engage in selection. Once a need is identified, some form of job analysis is typically employed to help understand what the job involves and to estimate the knowledges, skills, abilities, and other requirements (or KSAOs) required to be successful on the job. Organizations recruit a pool of qualified applicants, and selection takes place when there are two or more qualified applicants for an opening. In this section, we will discuss basic techniques for selection, and the continuing legal issues surrounding selection, and some of the more recent developments in this area. We will review some of the more common selection techniques, again, pointing out some more recent developments, and end with a definition of terms.

Selection Techniques

Traditionally, once organizations identified requisite KSAOs, they implemented some selection technique to determine the levels of these KSAOs possessed by applicants and then selected the applicant who possessed the highest levels. More recently, some organizations have begun selecting individuals, not because they had the highest levels of KSAOs, but because they "fit" better in the organization (more on this below), but, in either case, the process involves some formal selection techniques.

There is no one best way to select among applicants, and the specific techniques used should depend upon the KSAOs that need to be assessed (or upon how fit is to be assessed). Nonetheless, the most popular selection technique in use is the interview. Interviews involve face to face (typically) sessions where interviewers ask applicants a series of questions in order to decide if they should be hired. Interviews are seen as especially effective for determining interpersonal skills and oral communication skills, and interviews are also seen as an excellent tool for determining fit. In fact, there is evidence that, in all interview settings, interviewers tend to select persons who are similar to them. In cases where we are trying to select on the basis of KSAOs this might be a problem, but, in the case of selecting on the basis of fit, this might be exactly what is needed (see review by Dipboye [1994]). Whenever possible, interviews should be structured, that is, interviews are given a standardized list of questions to ask all applicants before beginning the process. In fact, unstructured interviews are much less likely to be valid predictors of job performance.

The next most common selection technique is the paper and pencil test. Although these tests are sometimes administered on computers, they still refer to series of items for which there are typically right and wrong answers (although there is an exception that we will discuss later). Test are available that assess cognitive skills, such as verbal ability, numerical ability, and spatial relations as well as psychomotor skills. Care must be taken to select tests that have been professionally developed and for which there is information about reliability, validity, and potential adverse impact, regardless of the type of test used. As a result, tests should always be purchased from reputable test publishers, and validity information collected, since the organization bears all the legal responsibility for selection systems if there is a problem.

One special and intriguing area for testing is intelligence. General intelligence (referred to as *g*), refers to basic cognitive reasoning skills, and these tests are popularly known as IQ tests. These tests have been found to be predictive of performance in a wide variety of settings (e.g., Ree et al. [1994]), but, probably due to differential access to educational opportunities, non-whites tend to score lower on these tests than do whites. The potential legal problems that this state of affairs gives rise to are beyond the scope of the present section, but the technology manager needs to exercise caution when using intelligence tests in selection because there are potential legal issues.

A somewhat different type of test is a personality test. These tests typically do not have right and wrong answers, but try to assess applicants on series of personality constructs. Recently the "Big Five" personality traits (neuroticism, extroversion, openness, agreeableness, and conscientiousness) have been found to be successful in the prediction of job performance (e.g., Tett et al. [1991]) but, in any case, the assumption here is that persons who have certain personality traits (e.g., conscientiousness) are more likely to be successful on a given job than people who have other personality traits. Personality tests designed to identify aberrant behavior patterns (i.e., psychoses) should probably not be sued in selection settings, except where it might be important to screen out certain types of individuals who might do harm to others, and, even here, these tests often require specialized training for effective administration and interpretation of scores. One final note concerning personality tests is that they may be useful in assessing fit (e.g., Kristof [1996]), that is, these tests may be able to tell us more about whether a person has the kind of personality that is required to fit in with this organization, although, in some organizations, fit is assessed relative to work or life values rather than personality.

Drug tests and honesty tests are two other types of tests that are growing in popularity. Drug tests are self-explanatory and, although there are some issues of invasion of privacy, many large organizations routinely screen applicants to insure that they do not use illegal drugs. Honesty, or integrity, tests, on the other hand, are typically paper and pencil measures designed to predict who might steal from the organization. These tests are especially popular in the retail sector, and, although they are somewhat controversial, their use is growing.

Legal Issues

Probably the biggest challenge facing managers making selection decision stem from the legal system. The Civil Rights Act (CRA) of 1964 (and the more recent Civil Rights Act of 1991) and the Americans with Disabilities Act (ADA) of 1990 stand as the two most serious legal constraints on the selection system. Title VII of the CRA deals with employment settings and states that it is illegal to discriminate against anyone, in an employment setting, on the basis of race, gender, color, religion, or national origin. Furthermore, the CRA established certain groups as being "protected" and requiring special consideration under the law and established the Equal Employment Opportunity Commission to oversee this Act.

All Americans are protected by Title VII, but the assumption was made that some groups have been the target of discrimination in the past and so deserve special protection. Basically, these protected groups include anyone who is not a white, Anglo-Saxon, Protestant male, but most concerns have focused on women, African Americans, Hispanics, Asians, and Native Americans. Under the law, any selection practice (or any employment practice for that matter) that adversely affects members of a protected class will be considered illegal, unless the organization can prove otherwise. Therefore, if a test or other selection procedure results in a disproportionate rejection rate among members of a protected class, the company will be considered to be guilty of discrimination. The organization can only defend itself by demonstrating that the test systematically predicts performance for members of both the protected and the non-protected groups (i.e., that the test is valid for all applicants).

Recently, this process of alleging and responding to charges of discrimination has brought several issues to the forefront of concerns. One issue that has emerged involves the use of different cutoff scores for selection. Once a relationship between tests scores and performance has been established, the firm still needs to select a cutoff score such that persons scoring above this score will be hired while persons scoring below this point will be rejected. In the past, some scholars and professionals have advocated the use of separate cutoff scores for different groups, but, since the Civil Rights Act of 1991 essentially banned their use, other solutions are needed.

One such proposal involved the use of "banding". Although there are variations on banding schemes, and they tend to get complex, the central notion here is that, in almost all cases, there are ranges or

RANK	TEST SCORE	PREDICTED PERFORMANCE RATING	
1	100	5	
2	98	5	
3	97	4.6	
4	96	4.5	BAND
5	94	4.3	
6	90	4.0	
.	.	.	
.	.	.	
10	85	3.5	
11	80	3.0	
12	78	2.9	
13	76	2.8	BAND
14	75	2.6	
15	73	2.5	

FIGURE 18.1 Illustration of banding. In this example, applicants are ranked according to test scores, and predicted performance scores (predicted ratings on a five-point scale) are presented as well. The first ten candidates here are shown as being in a single band, since the differences among their predicted performance ratings are not meaningful to the company. As a result, within this band, the company could select on the basis of some other criterion, such as increased diversity. If there were a need to select more applicants, a second band is illustrated as well. The alternative model would be to start with the top-ranked candidate and work down until all positions were filled.

"bands" of test scores within which job performance in not really predicted to be different, that is, as illustrated in Fig. 18.1, although someone scoring 100 on a hypothetical test might be expected to perform better than someone scoring 70, the difference between someone scoring 100 and someone scoring 90 might be rather small. In such cases, it might be possible to select on the basis of something other than test scores (e.g., increased workforce diversity). Perhaps not surprisingly, this approach is controversial, and some suggest it is illegal.

The final issue related to the CRA that we will discuss involves the attempt to select people based on their "fit" with the organization rather than whether they can perform a given job. This is not to say that unqualified persons would be hired. Instead, the assumption is that most KSAOs can be learned as long as the applicant has some basic levels and so differences in KSAOs beyond these basic levels will not influence ultimate performance on the job. Instead, what will influence performance is whether the person shares the personality or values of the organizations or generally fits in with the other people in the organization. In these cases, selection is made on the basis of evaluations of the applicants' values and personality and whether they are consistent with the organization's values. Although there is some suggestion that this procedure results in higher levels of performance and employee satisfaction, it has also been suggested that these approaches can result in discrimination, since there is a suspicion that, in many organizations, people who do not "look" like the current employees will not "fit" in as well (see review by Kristof [1996]). We present this approach without endorsing or rejecting it. Instead, our purpose is to make the manager aware of the fit model and to note that, as traditional notions of jobs disappear and employee responsibilities change rapidly, the fit approach to selection will likely become more popular.

The ADA also has implications for selection. The ADA outlaws discrimination against persons with disabilities in all employment decisions and defines disabilities broadly enough (see notes) that many applicants and employees are affected. In practice, in cases where papers and pencil tests are used in the selection process, since these tests require reading, persons with certain learning disabilities might be at a disadvantage when taking such a test. Under the ADA, the applicant could choose *not* to identify the fact that he or she had this learning disability and simply take the tests. The reluctance to self-identify might be based on a belief that some organizations discriminate against persons with disabilities and the applicant preferred to simply try to pass the test. Later, if the person were rejected, he or she could *then* identify the fact that had this leaning disability and request an accommodation. Specifically, they could request that the test be given in a different format or with different time limits. The firm would be required to comply with this request. Thus, the ADA, in addition to raising issues concerning the need for accommodations in general also has clear implications for selection and for the mode in which selection devices are administered (see also Stone and Colella [1996]).

The Effectiveness of Selection Systems

How do we know if a selection system is effective and whether a firm is selecting the best people? Traditionally, the outcome measure of interest has been individual performance. Selection systems have been considered successful if the average individual performance in a groups selected by the systems being evaluated is significantly greater than that in a group selected by some other means (this is another way of stating that the system is valid). If the increase in average individual performance exceeded the costs involved in the implementation of the new selection system, the system was considered successful. In fact, establishing this relationship is refereed do as performing a utility analysis (for a fuller discussion of the issues of utility analysis, see Cascio [1993]).

However, more recently, some have argued for other ways of evaluating selection systems. Of course, there have always been calls for considering success from the point of view of the individual selected, and so focusing on employee satisfaction as a criterion, but recent calls have been for a different approach. Several scholars have begun discussing the impact of any HR practice on firm performance. In this view of the world, HR practices should be evaluated in terms of whether they contribute to the success of the enterprise (i.e., the "bottom line"). Thus, HR practices have been related to profitability and stock prices and have generally found that using carefully developed and validated selection procedures do have "value added". Even here, though, there is disagreement. Some have argued that there is no one best way to select people, but the selection systems used and the people selected should depend upon the firm's strategy (e.g., Delery and Doty [1996]), while others have argued for "best practices" or "high-performance work systems", which should be effective regardless of the firm's strategy (e.g., Huselid [1995]). At this point, it is not clear which view is correct, but it is clear that selection systems can affect firm performance.

Finally, as noted above, models already challenge the view that a person should be selected because he or she possesses high levels of the KSAOs needed for a specific job, but the new emphasis on work teams and team performance further challenges the traditional view. As organizations move more toward the implementation of teams, it becomes important, not only for individuals to be able to perform a given job, but also that they can function as part of a team. There have been a few specific recommendations about how to best select persons for teams, but it is clear that team work requires additional KSAOs as well as new approaches to selection (see Stevens and Campion [1994]).

Conclusions

The need to select individuals to fill jobs is virtually universal. Therefore, all managers need to understand the challenges facing them when then try to select individuals as well as the means available to do them to carry out this task. The selection process should begin with an identification of skills or abilities needed to carry out the job, and the managers can then use any of a number of means for assessing where applicants fall on these skills and abilities. At that point, the manager and the organization must decide if they want to select the individuals who scored the highest on the requisite skills or those individuals who will best fit in with the organization. Whatever the basis for the selection decision, this decision is complicated by laws outlawing discrimination in hiring practices. These laws impact upon the specific selection techniques used as well as on how they are used and their impact upon applicants, and they are especially problematic when we consider the use of intelligence tests for selection. However, whatever selection techniques are used, and whatever basis is chosen for making selection decisions, it is important for the technology manager to understand that these selection decisions will not only affect the individual effectiveness of the persons hired, but they can also have a direct impact upon the organization's bottom line.

Defining Terms

Disability: Any condition that substantially limits any major life activity, the history of such a condition, or the perception of such a condition.

Reliability: The extent to which scores on a selection technique are free from random errors of measurement.

Validity: The extent to which a test measures what it was intended to measure. In selection, this is often demonstrated by a significant relationship between scores on a selection technique and performance on the job, indicating that test scores can predict on the job performance.

References

Cascio, W. F. Assessing the utility of selection decisions: theoretical and practical considerations, In *Personnel Selection in Organizations,* pp. 310–340, N. Schmitt and W. Borman, eds., Jossey-Bass, San Francisco, 1993.

Delery, J. E. and Doty, D. H. Modes of theorizing in strategic human resource management, tests of universalistic, contingency, and configurational performance predictions, *Acad. Mgt. J.,* 39: 802–835, 1996.

Dipboye, R. L. Structured and unstructured selection interviews: beyond the job fit model, *Res. Pers. Hum. Res. Mgt.,* 12: 79–123, 1994.

Huselid, M. A. The impact of human resource management practices on turnover, productivity, and corporate financial performance, *Acad. Mgt. J.,* 38: 635-670, 1995.

Kristof, A. L. Person-organization fit: an integrative review of its conceptualizations, measurement, and implications, *Pers. Psychol.,* 49: 1–49, 1996.

Ree, M. J., Earles, J. A., and Teachout, M. S. Predicting job performance: not much more than g, *J. Appl. Psychol.,* 79: 518–524, 1994.

Stevens, M. J. and Campion, M. A. The knowledge, skill, and ability requirements for teamwork: implications for human resource management, *J. Mgt.,* 20: 503–530, 1994.

Stone, D. L. and Colella, A. A model of factors affecting the treatment of disabled individuals in organizations, *Acad. Mgt. Rev.,* 21: 352–401, 1996.

Tett, R. E., Jackson, D. N., and Rothstein, M. Personality measures as predictors of job performance: a meta-analytic review, *Pers. Psychol.,* 44: 703–742, 1991.

Further Reading

Borman, W. C., Hanson, M. A., and Hedge, J. W. Personnel selection, *Annu. Rev. Psychol.,* 48: 299-337, 1997.

Jackson, S. E. and Schuler, R. S. Understanding human resource management in the context of organizations and their environments, *Annu. Rev. Psychol.,* 46: 237–264, 1995.

18.3   Training Virtual Associates

Dana M. Johnson

As the number of virtual corporations or organizations increases, the need for virtual associates increases. A **virtual associate** is defined as a technical or support professional that performs work on a project or as needed basis for a virtual corporation. The virtual associate will develop a relationship with one or more virtual corporations depending on the type of skills that he or she has to offer. In today's world, the virtual associate may be referred to as an independent contractor. The real difference between the virtual associate and the independent contractor is that the virtual associate may never meet face to face with the organization that provides technical assistance.

Training and Skill Maintenance

Training and maintenance of skills becomes an issue for virtual corporations and virtual associates. No longer is the organization responsible for providing training because the traditional employment relationship no longer exists. It now becomes the virtual associate's responsibility to keep their skills current to continue to remain as a virtual associate.

There will no longer be a training department or a human resource director ensuring that training is taking place.

There are serious concerns from an organization perspective.

1. Whose responsibility is it to ensure that virtual associates maintain and upgrade their skills?
2. How does the virtual organization ensure that the necessary pool of resources are available for staffing projects and programs?
3. How will the skills be validated to ensure they meet the virtual organization's quality standards?
4. If a shortage of qualified virtual associates occurs, how will this impact the virtual organization in achieving its goals and objectives?

These questions and many others will be discussed.

Continual Learning

The process of continual learning is important to the "just-in-time" workforce. The virtual associates are utilized when they are needed. It becomes the responsibility of the virtual associate to ensure they possess the following:

- Technical skills
- Behavioral attributes
- Leadership
- Business skills

These skills are required to maintain continuing relationships with one or more virtual organizations. Failure to maintain and upgrade all the aforementioned skills will result in the inability to obtain work.

A base of knowledge can be obtained from college and universities. Academic institutions will need to provide more hands-on training and laboratories at a much earlier stage of the educational process. Those who work in virtual environments must be highly competent in all dimensions. These are highly trained performance people [Grenier and Metes, 1995]. Often these virtual associates are referred to as knowledge workers, possessing specialize skills to meet the needs of a virtual enterprise [Bracco, 1996].

The continual learning process will include continual self-assessment of skills by the virtual associate. The virtual associate will need to have a self-awareness of their learning style. The virtual associate is responsible for accessing information and knowledge suited to the situation and the form that suits their learning style [Grenier and Metes, 1995].

Methods of Learning

Technological methods of **distance learning** have been used by many companies in the automotive industry to allow for education at the undergraduate and graduate level. Satellite transmission, delay broadcasting, and videotape serve as the media for learning. The major disadvantage of this method has been the lack of interaction with a faculty member or slow or no response to questions posed to the lead faculty member. This causes frustration and hinders the learning process.

Interactive approaches may be feasible where the computer is the medium of exchange and there is immediate or timely response to questions posed by the learner. Interactive computer training is being used now for some programs but can only provide responses to preprogrammed information. This is also a limiting factor in the effectiveness of training.

The "just-in-time" skill acquisition is an important element to consider. If a virtual associate has a relationship established with a virtual organization and the associate becomes aware of technology that is needed to support additional projects or programs, how do they obtain the skills? The Internet may prove to be a valuable tool in the future to provide access to alternative training programs. The virtual associate will be responsible for determining the "tools of the trade" and having the capability to efficiently use the technology [Bracco, 1996].

Experiential learning is another approach. Experiential learning is a collective term for human resource development (HRD) activities based on participants' reactions to the practical activities during an exercise, as opposed to passive learning. The experiential learning design model portrays the process as a circle of five revolving steps: experiencing, publishing, processing, generalizing, and applying [Reynolds, 1993].

The virtual associate is responsible for keeping a database of all the education, training, and experience that they have obtained over the years. It is the virtual associate's responsibility to develop a training/continuing education/skill enhancement plan, which becomes a part of information and documentation submitted to potential virtual organizations that they would support.

Ability to Obtain Necessary Resources

The virtual organization is faced with the task of obtaining the necessary resources. Some skill sets that are required may be in short supply. A shortage may require that the virtual organization takes on some responsibility of training or they will be faced with the inability to meet their customer's requirements. If the virtual organization takes on too much responsibility in training, then the virtual associate will begin to rely on the organization. The virtual associates are compensated based on their skills. The virtual organization has indirectly paid the cost of training to the virtual associate in the form of compensation for services provided. The virtual associate will determine their fees based on work to be performed, business overhead, and costs and expenses, including continual learning. The virtual organization will be incurring an additional cost and should possibly request a reduction in fees from the virtual associate. By reducing the fees, the virtual organization runs the risk of destroying their relationship with the virtual associate and is in no better position.

Validating Skills

The **validation** process is a means of ensuring that a learning activity, instructional product, measurement instrument, or system is capable of achieving its intended aims or functions [Reynolds, 1993]. A transformation of this definition can be developed to validate the skills of the virtual associate, which means that they have the appropriate skills to be able to adequately and consistently perform the functions required to complete the work for the virtual organization.

To validate the skills, the virtual organization will need to develop a test that is an accurate measure of the virtual associate's ability to perform the function. For some functions, standardized tests may be available. For example, if a virtual associate needs to possess blueprint and geometric dimensioning and tolerancing skills, the associate can take a test to validate these skills.

In some cases, standardized tests may not be appropriate because the skill may be too specialized, too new, or very limited in application to justify development of a standardized tool. In this situation, a specialized test would need to be developed to confirm the virtual associate's ability to perform a specific function. The newer the skill, the more difficult it will be to test the ability to perform the function.

Achieving Virtual Organizational Goals and Objectives: Human Resource Perspective

Since most virtual organizations will no longer rely on employees for most of their needs, there is an increasing role of the human resource (HR) function. The HR function will determine the necessary skills that are required by virtual associates and ensure who they contract with has these skills to perform the necessary technical functions. The HR function will usually be combined with another functional area and will no longer stand alone.

The HR function will work closely with all the technical professionals to ensure an understanding of the technical skills required of the virtual associate. Most HR professionals will possess a technical degree as well as skills in the human resource area.

The HR professional will play a significant role in the company strategic planning process to ensure a clear understanding of the overall virtual organization's goals and objectives.

The HR professional will be responsible for ensuring that the virtual associates used by the organization possess basic and key skills and knowledge to fill the need for technical support. The way that this can be accomplished will include verification of university education, just-in-time training, seminar and continuing education training, and any other relevant activities that the virtual associate was involved.

Conclusion

As a company shifts to become a virtual organization having limited number of employees and operate on an ad hoc basis, the responsibility of training shifts to the virtual associate. This does not mean that the virtual organization no longer has responsibility for training. This is especially true when there is a shortage of technical virtual associates available to support a program or project.

In the case of a shortage, the virtual organization has to make decisions of how to obtain the necessary resources at a minimum cost. This may mean making an investment and offering training to a virtual associate that the organization has developed a relationship.

There is the risk of this talent defecting if better offers come along and with limited and scarce resources; this is always a possibility.

The role of the HR professional has changed as it relates to the training function. They now play a more active role in contracting with virtual associates because it becomes their responsibility to validate skills.

Companies will need to manage the change to ensure that a pool of technical professionals are available and to minimize their training investment.

Defining Terms

Distance learning: An instructional method in which the instructor or facilitator is geographically separated from the learner.

Experiential learning: A collective term for HRD activities based on participants' reactions to the practical activities during an exercise, as opposed to passive learning. The experiential learning design model portrays the process as a circle of five revolving steps: experiencing, publishing, processing, generalizing, and applying [Reynolds, 1993].

Validation process: A means of ensuring that a learning activity, instructional product, measurement instrument, or system is capable of achieving its intended aims or functions [Reynolds, 1993].

Virtual associate: A technical or support professional that performs work on a project or as need basis for a virtual corporation.

References

Bjerklie, D. and Cole, P. E. Age of the road warrior, *Time,* Spring: 1995.

Bleeker, S. E. The virtual organization, *Futurist,* March/April: 1994.

Bracco, D. M. Profile of the Virtual Employee and Their Office, *IEMC 96 Managing Virtual Enterprises: A Convergence of Communications, Computing, and Energy Technologies Proceedings,* August 1996.

Byrne, J. A. The virtual corporation, *Bus. Week,* February 8, 1993.

Davidow, W. H. and Malone, M. S. *The Virtual Corporation,* Harper Business, New York, 1992.

Grenier, R. and Metes, G. *Going Virtual: Moving Your Organization Into the 21st Century,* Prentice-Hall, Upper Saddle River, NJ, 1995.

Handy, C. Trust and the virtual organization, *Harv. Bus. Rev.,* May/June: 1995.

Mannes, G. The virtual office, *Pop. Mech.,* March: 1995.

Reynolds, A. *The Trainer's Dictionary,* Human Resource Development Press, Amherst, MA, 1993.

18.4 Performance Appraisal (Evaluation)

E. Brian Peach and Ralph Roberts

Performance appraisal is the systematic evaluation of an individual's job performance to evaluate adequacy of work [Yoder and Heneman, 1974]. Because it directly affects an individual's career, it is one of the most painful aspects of personnel management for both the appraiser and the appraised. Some people contend that performance appraisal does more harm than good. W. Edwards Deming [1986], the world-famous quality expert, calls performance appraisal one of American management's seven deadly diseases. Research consistently indicates that the appraisal process is generally imperfect, yet organizations almost universally maintain performance appraisal as an important assessment tool [Cleveland et al., 1989]. Because employee performance will inevitably be judged in some manner, proper understanding and use of formal appraisal systems will enhance worker performance and organizational effectiveness.

Purpose of Performance Appraisal

The major objectives of performance appraisal are to provide feedback about employee strengths and weaknesses for development purposes and to distinguish among employees for rewards such as promotions, raises, and bonuses. Other common uses include documenting personnel actions such as transfers or dismissals, retention or dismissal, identifying training needs, and assigning work. Performance appraisal provides both employees and management with information about employee job performance.

Because most organizations link rewards to performance appraisals, accurate appraisals are critical in identifying high performers. One study found that the difference in productivity between average and high performance ranged from 15% for blue collar workers to 46% for professionals to 97% for some sales people [Hunter et al., 1990]. Accurate performance appraisal, indicating needed improvements, is also necessary to provide appropriate developmental guidance. Research by Zedeck and Cascio [1982] indicates, however, that very different results are obtained from developmental as opposed to reward appraisals, especially when peers and subordinates are involved in the appraisal process. Ideally, therefore, organizations should use separate developmental and reward systems. In practice most organizations use a single appraisal system.

Types of Performance Measures

There are two general types of performance appraisal: objective and subjective. Many organizations focus on the objective aspect, believing that strict mechanical technique should be substituted for supervisor judgment. Others favor subjective approaches relying on human assessment. Both approaches have advantages and disadvantages, and the best performance appraisal systems combine them to incorporate human judgment with objective techniques to minimize human errors. **Objective performance measures** are usually based on measurable results. Objective measures include quantity of output, quality of output, timeliness, and resource usage. These are intuitively appealing, especially to technically oriented people who like to count tangible and identifiable events. Unfortunately, objective performance measures have many theoretical and practical limitations. One limitation occurs when performance is influenced by situational characteristics beyond the control of the individual. A second limitation arises when nonobjective measures critical to organizational success are left out of performance criteria because individuals typically emphasize what they are measured on before what is important to the organization. A third limitation is that results based measures do not specify *how to achieve* results, thus leaving the possibility open for illegal, unethical, or other suboptimal actions. A fourth limitation occurs for jobs where the output cannot be determined in the short term, such as for research and development and strategic planning.

Subjective performance measures are usually behaviorally based, describing behaviors believed to lead to job success. Subjective measures may be either relative (compared to other employees) or absolute (measured against a specified written description). Subjective measures have the advantage of defining desired activities but they suffer from a number of drawbacks. First, they depend on human judgment and are therefore prone to biases that are described in a later section. Second, a job situation may have a number of possible behaviors leading to job success, but specifying behaviors limits employee flexibility, which may suboptimize organizational performance. **Trait** criteria are personality characteristics such as "friendly", "pleasant personality", and "creative". Subjective performance measures based on traits should be avoided as they are ambiguous, unreliable, and have little to do with job performance.

In establishing appraisal criteria, whether using objective or subjective measures, the list should be necessary and sufficient. *Necessary* means that each criterion measures a job related performance and that each criterion is not redundant to other criteria. *Sufficient* means that enough measures are included to ensure that meeting them all will result in a completely successful job performance.

Problems and Biases in Performance Appraisal

A number of human biases may distort the appraisal so potential appraiser bias should be recognized up-front in the process. *Leniency/severity* bias may occur because each rater has unique internal standards, assumptions, and expectations. This means that a rater may provide consistently higher or lower ratings than other raters. *Recency* bias may occur when a rater allows recent performance to overshadow earlier performance in the rating period. Employee absences or significant error shortly before an appraisal may negatively affect ratings. Similarly, a spectacular performance on a recent task may blur mediocre performance over the entire period. *Central tendency* bias causes raters to avoid high or low evaluations. Some raters perceive everybody as "average"; others are concerned that low appraisals will cause employees to do less work or to create negative consequences for the organization. *Halo* bias occurs when raters assign specific item ratings based on an overall impression of the ratee. A rater who has an extremely positive (or negative) overall impression or who has substantial knowledge about performance on a specific task may allow this to affect other ratings where there is less performance information. *Political* bias refers to personal actions and preferences that are not based on employee performance. Raters may deliberately or unconsciously enhance or worsen an appraisal for political reasons or personal self-interest.

There are many other factors that affect subjective appraisals: sex, race, age, education level, self-confidence, intelligence, job experience, leadership style, organizational position, rater knowledge of ratee/job, prior expectations, and stress. These may play a part in distorting the rater's perception and the accuracy of the consequent performance appraisal.

Approaches to Performance Appraisals

A wide variety of appraisal instruments exist, ranging from relatively simple to fairly complex. More complex systems attempt to remove biases and errors from the appraisal process to make appraisals more consistent across raters and more accurate in capturing performance. There are two major approaches to appraisal systems: relative and absolute. Relative systems compare employees to each other and include ranking, paired comparisons, and forced distribution. Absolute systems attempt to link ratings to predefined scales and include essays, behavior checklists, critical incidents, and graphic rating scales.

Simple ranking requires that the rater rank all employees from highest to lowest. Thus, raters must compare each employee with all other employees. A more complex variation is *paired comparison,* in which the rater pairs each employee with every other employee. The individual pairs are then summarized, and the highest rated employee is the one ranked higher in a pair the most times. Ranking approaches have some drawbacks. One is whether to rank employees once on an overall criterion or multiple times for several criteria. Also, it is difficult to compare ranked individuals across groups, divisions, or locations. Consistency of rank order is another problem as raters often provide different results at different times. Ranking also penalizes members of a superior group and rewards members of an inferior group. *Forced distribution* systems attempt to compensate for inflation bias — the bias of rating all employees highly. These systems use a bell curve to allocate employees to performance levels. A typical distribution might limit top scores to 10% of all employees within a group, division, or organization. Above-average scores might be limited to 20%, and average scores to 50%, meaning that some percentage of employees must be designated as below average. This approach clearly has problems when raters (or employees) believe more than 10% of employees are excellent and/or no employee is deficient. There is no evidence that bell curves are appropriate for all groups because in fact most rating systems result in 60 to 70% of employees being highly rated. Forced distribution also penalizes members of superior groups and rewards members of inferior groups.

Absolute rating systems describe a ratee without comparison to other employees and includes narrative methods such as essays and critical incidents. *Essays* are written descriptions of an employee's strengths and weaknesses in great detail. They are useful for developmental feedback, but their unstructured nature makes them less useful for comparisons between employees. *Critical incidents* involve knowledgeable supervisors recording employee actions that were effective or ineffective in job performance. Critical incidents should include actions that are crucial in job performance, but in practice they vary by supervisor. In addition, recording incidents is time consuming and requires diligence by the supervisor, and employees may become fearful of their supervisor writing in a book.

Category rating systems include graphic rating scales, checklists, and forced choice. *Graphic rating scales* are the oldest and most widely used because they are simple to construct and easy to use. Evaluators rate performance along a descriptive scale ranging from low to high performance. Scales for selected performance measures use a subdivided line or series of boxes labeled with performance **anchors** so that the supervisor can indicate levels of performance (see Fig. 18.2).

Scales vary in the type of anchors used for evaluation. At the simplest level only the ends may be anchored using words such as unsatisfactory and outstanding. Other approaches may include descriptive statements for each point along the scale. Problems occur when inappropriate or distasteful anchors such as average are used. No one wants to be average. Anchors such as "meets standard" are more acceptable. Problems also occur when an anchor means different things to different raters. Trying to avoid this can lead to increasingly elaborate anchors, which may still be misinterpreted. *Behaviorally anchored rating scales* (BARS) is an attempt to improve graphic rating scales that tries to provide a system that is both appealing to raters and valid in performance appraisal. BARS' two goals are to use anchors that accurately discriminate between levels of performance and that are interpreted similarly by diverse raters. To construct a BARS instrument requires careful and extensive work by employees, supervisors, and psychometricians (professional test developers). Unfortunately, a major research effort by Bernardin and Beatty [1984] found little evidence of superiority for BARS over other rating methods. *Behavioral checklists* use descriptive statements of job-related activities with the rater checking which statements apply. Analysis of the good and bad statements that are checked provides an evaluation of performance. *Forced choice* attempts to compensate for human biases. Pairs of seemingly equal statements are presented to the rater, with only one actually measuring performance on the criterion. The rater's choice thus depends on actual performance and is not subject to rater biases. Forced choice systems give more accurate and consistent results but they are complex to construct and they are disliked by raters.

(1) Quality of Work Low _____ High

(2) Quality of Work Low _____ High
                                          1   2   3   4   5

(3) Quality of Work _____
                         Unsatisfactory   contains   meets   exceeds   is very high
                                           numerous   standard   standard   quality
                                           errors

(4) Quality of Work
     Accuracy   X
     Consistency   X
     Timeliness   X
     Quantity   X
                    _____
          Low 1 2 3 4 5 6 7 8 9 10 11 12 13 14 15 16 17 18 19 20 21 22 23 24 25 26 27 28 High

(5) Quality of Work _____
                         Poor   Below   Average   Above   Excellent
                                           Average   Average

(6) Quality of Work ‖ | | ‖ | | | ‖ | | ‖ | | ‖ | | ‖
                         Poor   Below   Average   Above   Excellent
                                           Average   Average

(7) Quality of Work 1 2 3 4 5 6 7 8 9 10 11 12 13 14 15 16 17 18 19 20 21 22 23 24 25 26 27 28
                         Frequent   Occasional   Work   Few   Errors
                         errors   errors   acceptable   errors   rare

(8) Quality of Work
     Accuracy   __Superior __ Excellent __ Acceptable __Needs Improvement __ N/A
     Consistency   __Superior __ Excellent __ Acceptable __Needs Improvement __ N/A
     Timeliness   __Superior __ Excellent __ Acceptable __Needs Improvement __ N/A
     Quantity   __Superior __ Excellent __ Acceptable __Needs Improvement __ N/A

FIGURE 18.2   Examples of graphic rating scales.

Management by objectives involves the rater and ratee mutually agreeing at the beginning of an evaluation period on performance objectives. Challenging and mutually agreed goals have the added benefit of motivating effort. MBO suffers from the earlier mentioned requirements of sufficiency in that, if all desired tasks are not specified, employees are less likely to accomplish all necessary tasks. This can happen because some tasks are difficult to specify as clear and measurable objectives or potentially from manipulation by the ratee in setting easy goals.

Feedback of 360° uses ratings from peers and subordinates as well as supervisors. The intended advantage is that multiple views of an individual's performance are obtained, providing a clearer and more complete picture of employee performance. In many cases, peers or subordinates may be more aware of an individual's actual performance than a supervisor. In practice, if peers or subordinates are competing for the same rewards or promotions as the person being rated, political bias may adversely affect their appraisal. Table 18.1 provides a comparison of the various ratings systems.

Improving Performance Appraisal

In the final analysis, ratings involve human judgment. Although this is a problem regardless of the format used, the process can be improved in several ways. Providing structure helps limit rater discretion by clearly specifying standards and criteria. Reducing ambiguity in the rating scales helps the rater match perceived performance with appraisal ratings and achieve consistency between rating periods and between raters. Rater training about potential errors and appraisal objectives may improve observational skills and reduce judgmental biases.

Communicating appraisal results through performance feedback is often one of the more disliked and poorly accomplished aspects of the appraisal process. Training raters about feedback can help to transmit results of the appraisal in a manner that is positive for both the organization and the ratee.

TABLE 18.1 Evaluations of Performance Appraisal Techniques

Technique	Providing Feedback and Counseling	Allocating Rewards and Opportunities	Minimizing Costs	Avoiding Rating Erros
Management by objectives (MBO)	Excellent Specific problems, deficiencies, and plans are identified	Poor Nonstandard objectives across employees and units makes comparisons difficult	Poor Expensive to develop; time consuming to use	Good Tied to observations, reflects job content, low errors
Checklist	Average General problems identified, but little specific guidance for improvement	Good-average Comparative scores available, and dimensions can be weighted	Average Expensive development, but inexpensive to use	Good Techniques available to increase job relatedness and reduce errors
Graphic rating scale	Average Identifies problem areas, and some information on behaviors/outcomes needing improvement	Average Comparative scores available but not easily documented and defended	Good Inexpensive to develop and use	Average Substantial opportunity for erors, though they can be linked to specific dimensions
Behaviorally anchored rating scales (BARS)	Good Identifies specific behaviors leading to problems	Good Scores available, documented, and behavior based	Average Expensive development, but inexpensive to use	Good Based on job behaviors, can reduce errors
Essay	Unknown Depends on essay topics chosen by evaluators	Poor No overall score available, not comparable across employees	Average Inexpensive development, but expensive to use	Unknown Good observation can reduce errors, but lack of structure poses a danger
Comparing individuals (ranking, forced distribution)	Poor Based on general factors, with few specifics	Poor-average Overall score available, but difficult to defend	Good Inexpensive to develop and to use	Average Usually consistent, but subject to halo error and artificiality

(*Source:* Milkovich, G. T. and Boudreau, J. W. *Human Resource Management,* Irwin, Burr Ridge, IL, 1994.)

Performance Feedback

Whether for developmental purposes or reward allocations, employees need regular feedback about their performance. The manner of feedback can influence whether the employee feels encouraged and motivated or offended and resentful. Appraisal should be a continuous process of feedback on performance with no surprises at the formal interview. Frequently supervisors are reluctant to transmit criticisms until forced to at a formal appraisal interview. There they dump their pent-up perceptions, often to the surprise of the employee. Feedback is better received when it is perceived as timely, constructive, and well intentioned.

Ratees should also be encouraged to prepare for their appraisal interviews by preparing their own feedback for their supervisor. They should be encouraged, perhaps formally, to assess their performance, identifying both their strengths and weaknesses. During the interview, the rater should encourage participation, strive to judge performance not personality, and offer specific examples to support evaluative statements. The rater should listen actively and avoid destructive criticism. If the rater and ratee can agree on the appraisal, this is an optimum outcome. However, the rater does control the process and must provide an accurate, but considerate, appraisal for the benefit of the organization and the employee. If the appraisal is developmental, then a process of personal improvement should be initiated. If the appraisal is for rewards, then any rewards must be clearly linked to the evaluations. Any manager or supervisor who is unfamiliar with the organization's appraisal process should plan a lengthy visit with the personnel office or with seasoned peers to better understand the specifics of the organization's appraisal system and its rater-ratee process.

Defining Terms

Anchor: Descriptive statement intended to provide meaningful distinction between levels of performance along a performance dimension continuum.

Objective performance measures: Quantifiable measures of output, such as sales production, quality level, or typing errors per page.

Performance appraisal: A systematic assessment and description of job-relevant strengths and weaknesses demonstrated by an individual in the performance of assigned duties.

Subjective performance measures: Specify job behaviors the organization considers important to properly accomplish the job.

Trait performance measures: Personality characteristics such as "strong leader" or "self-starter" that have been found ambiguous and unreliable.

References

Bernardin, H. J. and Beatty, R. W. *Performance Appraisal: Assessing Human Behavior at Work*, Kent, Boston, 1984.

Cleveland, J. N., Murphy, K. R., and Williams, R. E. Multiple uses of performance appraisal: prevalence and correlates, *J. Appl. Psychol.*, 74: 130–135, 1989.

Deming, W. E. *Out Of The Crisis*, MIT Institute for Advanced Engineering, Cambridge, MA, 1986.

Hunter, J. E., Schmidt, F. L., and Judiesch, M. K. Individual differences in output variability as a function of job complexity, *J. Appl. Psychol.*, 75: 28–42, 1990.

Yoder, D. and Heneman, H. G. *Staffing Policies and Strategies. American Society for Personnel Administration Handbook of Personnel and Industrial Relations*, Vol 1, The Bureau of National Affairs, Washington, DC, 1974.

Zedeck, S. and Cascio, W. F. Performance appraisal decisions as a function of rater training and purpose of the appraisal, *J. Appl. Psychol.*, 67: 752–758, 1982.

Further Information

More detailed discussions of appraisals can be found in college textbooks on human resources. A good example is Milkovich, G. T., and Boudreau, 1994, *Human Resource Management*, 7th ed., Irwin, Boston, A more advanced reference is Cascio, W. F., 1991, *Applied Psychology in Personnel Management*, 4th ed., Prentice Hall, Englewood Cliffs, NJ.

The Academy of Management Human Resources division maintains a noncommercial electronic bulletin board HRNet for the exchange of HR-related ideas and questions. All HR topics including appraisal are discussed and past answers are available in a searchable database. The membership is varied including HR professionals, academics, and managers with HR interests. To subscribe, send an e-mail message to listserv@cornell.edu. In the body of message put: SUB HRNET firstname lastname

A number of computer-based appraisal systems are available that assist in preparing clear and consistent appraisals. One example is *Performance Now!* from HR Pro at 1-888-464-7776.

A leading resource for personnel issues is the Society for Human Resource Management (SHRM), which has numerous offerings for members, including a magazine, information center, and MemberNet, which shares solutions. More information can be found on the Web at http://www.ahrm.org/shrm/shrm.htm or obtained by contacting SHRM at 606 North Washington Street, Alexandria, VA 22314-1997; phone (703) 548-3440; Fax (703) 836-0367.

18.5   Consulting

Ralph Roberts and E. Brian Peach

Organizations hire consultants for their special skills and expertise. Usually "consultant" means an expert hired from outside the organization, although some larger organizations have employees who function in an internal consulting capacity. Outside consultants will be discussed here, since that is the usual consulting arrangement. The consulting process may be viewed from either the client or consultant perspective. Of course, what is important from one perspective also is important from the other. The perspective of a client organization will be emphasized here, but with some mention of the consultant's interests as well. References include both perspectives. This presentation will offer a comprehensive overview of the client-consultant-project process suitable for major undertakings. Smaller projects, use of known consultants, quick action needs, exploratory situations, and other factors may allow a stream-lined process that can be extracted from this presentation.

Why a Consultant?

The first question in the client-consultant process is why bother: what is the point? Typically, the reason for retaining a consultant is that a problem, issue, or need exists that cannot be solved feasibly in-house. In-house expertise may be unavailable or fully committed to other tasks. Temporary assistance may be needed, speed may be required, fresh, unbiased, frank opinions may be important, a direct focus or catalyst on a specific issue may be needed, and perhaps political or other entrenched organizational problems need to be either dealt with or avoided by employing outsiders. Another important use of consultants is for overview or second opinions about past and future management actions. These may be available only from an outside source. Also, while consultant fees exceed internal pay rates on a per hour basis, it may be less expensive on a total cost basis to hire external assistance (the make or buy decision).

Participation and Problem Definition

At the beginning of the consulting process, it is critical that the client organization make two initial efforts. First, the client organization should use a very participatory approach in the consulting process. Wide participation will greatly enhance the likely success of most consulting engagements and should be used unless there are clear reasons for limiting information and participation. Some strategic planning and personnel issues may fall in this limited category. Extremely important to the process is top management participation up to the level that will approve, fund, implement, and be responsible for the results of the consulting engagement. Consultants with successful records will insist that top management be included in the consulting process because inclusion will greatly enhance the likelihood of project success. Furthermore, the hiring and managing of the consultant should be a serious interest of management; consultants are expensive, the results are important, and failure is debilitating to all involved. A consultant should beware if the appropriate top management does not clearly indicate strong support for the project.

Second, the organization must very carefully, as best possible, define the need or problem for the consultant to meet or solve. What is the gap or deficiency between desirable and actual conditions? What is the issue? Is it a management issue such as personnel or strategic planning? Is it a technical issue such as inventory control or information systems? Is it an unbiased review of some element of the organization? Next, define the expected outcomes, objectives, goals, and results; specifically, define how the changed situation will be measured. If the desired results cannot be specified (even with revision during client-consultant negotiations), then the consultant has no direction! Be wary of a consultant who is not extremely interested in what will be considered a successful outcome.

After the needs and results are as well specified as is possible in the initial stage of the project, the members of the organization must consider their willingness or reluctance to work with a consultant and to institute recommended changes. Will the organization enthusiastically support the consultant? Is the problem not that important? Will bad news be rejected? Are political interests and entrenched power brokers unlikely to change? What if a high-level manager is the focus of the solution? Will he or she adopt and adapt — or bury the project with reservations and inaction?

What Kind of a Consultant?

After defining needs and results, attaining full support, and confirming readiness for change, the organization can consider the type of consultant which is needed. The consultant may have to offer generalist skills, to be able to make a diagnostic view of a general organizational problem. Problem areas might include management functions such as strategic planning, communications, or control, or the consultant may have to offer technical skills to provide in-depth assistance with complex details in such areas as computer systems, inventory management, logistics, or legal advice. Consultants may bring expert knowledge and advice, gathering evidence and presenting conclusions, or they may bring an action-research process so the client can self- analyze and self-generate conclusions.

At the same time the scope of the consultant's duties must be considered. Is it to provide information, to make recommendations, or to implement a solution? Also, does the client expect to be empowered by the consultant, to grow out of needing the consultant? Or is the consulting to be a continuous engagement? The former requires the consultant to educate the client, while the latter requires a continuing contract, frequently on a retainer basis.

Contacting Potential Consultants

With the previous steps, the client organization has a basis to seek appropriate consultants. A steering committee or individual should be designated to develop a list of potential consultants. An excellent source is business associates who may be contacted for recommendations. Deans of appropriate colleges may be called, and also trade associations, trade journals and editors of trade journals should be queried for leads. Previous consultants may be asked for recommendations, but use caution as they may believe they can do anything and inappropriately recommend themselves for consideration. **Gale Research Company** (see references) provides an encyclopedia of both associations and consultants that will be helpful in the search process. The references also list useful Internet addresses.

Exploratory telephone conversations about needs and objectives with possible consultants will generate a list of five to ten who are worthy of further review. They should be asked to send information describing their competencies and experience. From this list the steering committee should select about five good candidates for prearranged telephone interviews. The steering committee may operate via speaker telephone and should follow a preplanned interview structure with questions to the consultant including who are you, what do you do, here is our project, what do you think of it, how would we proceed, what is your fee structure, what resources would you use, and so forth. Remember, the potential consultants will be evaluating the client as well, so they should be provided frank and complete answers to their questions.

Evaluating Potential Consultants

From these conversations, which are low cost to all parties, the organization should select two or three potential consultants for in-depth on-site interviews if feasible. The client should coordinate the time and expense of consultant selection with the scope of the organizational project. Small projects may require selecting only one candidate and making arrangements for a confirmatory visit. The client organization must be sensitive to the financial strength of the consultant as interviews are arranged. If a university professor from 500 miles away is to be considered, then an offer of paying for direct expenses is appropriate; a Big Six accounting firm may be expected to bear the expenses of a visit.

As the consultants visit and discuss the project, they must be treated as partners in the mutual determination of a satisfactory relationship. Each one should be asked a similar list of questions, in depth. Each consultant should enjoy an ending interview with the steering committee, individual interviews with top managers, and group meetings, which are open to all involved in the consulting project. Meetings should inquire deeply into the consultant's resources and capabilities for the project. Not that a relatively small, new consultant organization should be rejected, but its resources should be particularly investigated. Throughout the interview process the consultants should be asked to comment on the proposed project: is it necessary, is there a better way, is the focus correct, is it too premature (other groundwork needs to be accomplished), what resources will the host organization provide, and what about learning experiences (if appropriate) for the organization, i.e., how can the organization grow out of dependency on the consultant? Also, mutual explorations about results are important: what will constitute success and how will success be measured.

Detailed discussion, late in the interview process, should include fees and expenses; no surprises is the goal here. If the fees quoted seem high, remember the consultants have considerable overhead (including the free time provided in the engagement process). Also the fees of several consultants may be compared to determine if there is an out-of-range proposal, either high or low. If so, and if the organization really prefers that consultant, then the disparity should be frankly presented and discussed.

One final result of interviewing several candidates is that the organization can revisit and rewrite the scope and expectations of the consulting project. This will be furnished to the short list of candidates as a basis for their written proposals of work, results, and costs.

One warning about the interview process is in order. The client organization should be wary of rainmakers. Rainmakers are usually senior partners of large consulting firms who have excellent sales skills. They are smooth, knowledgeable, reassuring, and quick witted, but they may have little to do with the project once it is sold. The client must pursue a very level-headed selection process — make no decisions before their time. Also, the client should get clear confirmation about the specific individuals, by name and qualifications, who will execute the project. It is very appropriate for the client organization to request interviews with the proposed on-site leaders. Positive chemistry between client and consultant leaders is important and should be evaluated. As part of avoiding rainmaking problems, the client organization should be wary of consultants offering prepackaged hammers who define all problems as nails. A standard solution or approach offers proven expertise and cost-effectiveness if it is appropriate to the problem. However, if the problem is not a nail, then the client must look for tailored work, not a hammer. The more technical and specific a problem is, the more likely a packaged solution will fit the need.

After the on-site visits, the client should obtain references from its short list of perhaps three potential consultants. The client-consultant relationship is a serious partnership, so serious background checks are well in order. Consultants should be asked for references from all similar engagements in the last year. If this is not at least five, then the consultant should be asked for additional references. Having few similar engagements is not grounds for immediate rejection, but experience should weigh in the decision. Also, the consultant should be asked to furnish the last two or three reports that were written for similar engagements. Be wary if the consultant says confidentiality precludes giving references or reports. If necessary, ask the consultant to get permission from previous clients to provide their names as references and to sanitize their reports so they may be provided without violating confidentiality. It is important to call, or even visit, references and to expand the reference list beyond that supplied by the consultant. It is appropriate also for a potential consultant to ask permission to speak to consultants used previously by the client organization.

There may remain some uncertainties toward the end of the consultant evaluation process; perhaps the project seems too big or expensive or perhaps there are questions about the capabilities of the consultant. If there are uncertainties, one approach is to unbundle the project and break it into smaller parts. A beginning may be a feasibility study that allows the consultant a thorough investigation of the problem or issue, with a new proposal for a solution. Sometimes an abbreviated feasibility study is offered at no charge. However, if it is used in the organization without retaining the consultant, then payment should be made for at least the direct expenses of the study.

The Client-Consultant Contract

When the consultant is finally identified, the client organization must take a last, realistic, business view of the likely outcomes of the project against its costs. Costs include actual consultant's charges, organizational time, and organizational upheaval. The decision to retain the consultant requires a positive answer to "is this worth it?" Is there an appropriate return-on-investment explicit or implicit in the project? If the answer is yes, a very thorough and careful contract should be written that mirrors the size and complexity of the consulting project. Here the objective is predictability from both parties. There should be no surprises for either side because the client and consultant are now a team in a mutually working partnership.

The contract will cover many of the matters discussed previously, including problem statement, expected results in as much detail as possible, how the results will be evaluated, time lines, progress reports, resources committed by both consultant and client, an inclusive list of fees and expenses and how they are derived, and payment expectations. The consultant may ask for a beginning payment before proceeding, and the client may ask for a similiar withholding until the project is fully completed. Other matters for written understandings include the termination process if either side is unsatisfactory, the process of settling disputes (binding arbitration is recommended), the main client decision maker, and the main consultant decision maker. Both client and consultant should operate under agreements for returning calls, participation in meetings, and access/relationships to members of the opposite organization. If there are learning experiences expected by the client (even a client group to work with the consultant), these should be identified. How the consultant will operate, organizational access, and working hours may also be addressed.

Structuring the Engagement

The project may grow after it is initiated: the client may request additions and the consultant may suggest new client needs. Extra free services should not be requested by the client. The consultant may agree, but cut corners elsewhere, or the mutual working relationship may be threatened. Additional fees are appropriate if the original scope of work does not reasonably include the additional activity. On the other hand, the consultant may accept or offer small additions as a goodwill gesture. New, discrete, project proposals should receive a cautious look and perhaps a deferral by the client organization until the consultant can be fully evaluated on the existing project.

If a committee is planned to make decisions on either side, this may auger conflicted and slow decision making, so both client and consultant would be wise to press for a single decision making person for project management. This person, of course, may be advised by a committee, but all parties are aware of the single place where the client's and consultant's "buck" stops.

The final decision maker may not be the day-to-day liaison/project manager for each side. Project managers are the information point for their organization. They must be specifically designated and must establish a close and trustworthy relationship between themselves. Project managers handle problems as they arise, monitor progress and budgets, and schedule/facilitate/administer progress meetings. To be effective, project managers must have direct and quick access to the final decision makers as well as their unqualified support.

As the contract is established, organizational entry by the consultant must be planned. An announcement should be made by client management stating the advantages of the consulting arrangement and asking for full cooperation. Everyone in the client organization who is affected by the project should be given some opportunity for comments and discussion. Small efforts here will overcome much of the normal doubt, fear, and resistance that occurs with organizational change. Departure and closure need consideration as well. When and how is it over? Will there be a last meeting, a circulated report, a celebration, and so forth. This is the point when both parties agree that the required deliverables have been fulfilled, the agreed-upon fees have been earned, and neither party has additional responsibility to the other.

Managing the Consulting Process

After the contract is signed, there should be no sense of relief in turning the problem over to the consultants. The client-consult process must be managed on both sides for optimum results. The culture of the working relationship should be established before the contract is signed and then nurtured thereafter. Regular progress meetings with specified senior people are crucial. The consultant's senior managing partner, the organization's top manager, the project managers, and other highly involved individuals must have scheduled times to meet, perhaps weekly, monthly, or whenever it is necessary for timely feedback and problem solving. Productive individual and group project management meetings require frankness on both sides. Client or consultant misunderstandings or progress impediments must be discussed immediately. Problems will then become routine opportunities for solutions. Also, positive feedback is helpful to both sides; each side needs to know what is going well — for continuance and motivation. Be careful of establishing friendships, at least until the project is completed. The consultancy is a business relationship with results expected on both sides. This is best attained with an arms-length working relationship.

At the closure stage, it is useful to evaluate the entire project both in-house and with the consultant. The object here is learning about consulting/project management, primarily by the client, but perhaps by the consultant as well. In-house questions include the following: were our expected results achieved, was it cost effective, what did we do right or wrong in the entire consulting process, what were the surprises? How did we work effectively/ineffectively with the consultant? Would we use this consultant again?

The client also will have frank discussions with the consultant about the entire consulting process. How did we do as a client, could we have assisted more, were there surprises for you as consultant? How can we improve our client-consultant process in the future? The consultant also will be interested in an unbiased review of the work, so the feedback and debriefing activity will be beneficial to both parties.

Conclusion

In conclusion, good clients enable good consultants and good consultants enable good clients. The relationship is not adversarial; it is mutually beneficial. The benefits of success to clients are specified and apparent, but consultants also need success because they depend primarily on reputation and word of mouth for future opportunities. The consultant cannot be a know-it-all, and the client cannot sigh with relief while passing the problem completely to the consultant. Selecting consultants and managing the consultancy is critical to the success of the project. Top decision makers on both sides must be committed and must be sure that resources, management arrangement, and results are clearly defined for the project. A well-planned partnership process will almost guarantee a successful, cost-effective engagement for both client and consultant.

References

Golightly, H. O. *Consultants: Selecting, Using, and Evaluating Business Consultants,* Franklin Watts, New York, 1985.

Grupe, F. H. Selecting and effectively using computer consultants, *Small Bus. Forum,* Fall: 43–65, 1994.

Shenson, H. L. *How to Select and Manage Consultants,* D.C. Heath, Lexington, 1990.

Ucko, T. J. *Selecting and Working with Consultants,* Crisp Publications, Los Altos, 1990.

Further Information

More discussion of the consulting process from the consultant's viewpoint can be found in *Personal Selling Strategies for Consultants and Professionals* by Richard K. Carlson (1993, John Wiley & Sons, New York) and *Shenson on Consulting,* by Howard L. Shenson (1994, John Wiley & Sons, New York).

Listings of consultants and associations are provided by Gale Research Company in Detroit: *Consultants and Consulting Organizations Directory* and *The Encyclopedia of Associations.*

The Association of Management Consulting Firms (ACME) provides consulting information and a referral service at Web site http://www.acmeworld.org. Another Web site by Expert Marketplace at http://www.expert-market.com provides posting opportunities for consultants and a massive listing of consulting firms.

18.6 Virtual Networks

Karol I. Pelc

Under competitive pressure and due to rapid growth of technological knowledge, many companies must concentrate their resources around carefully selected *core competencies*. As a consequence, it becomes necessary that they establish collaborative structures and alliances with external partners. The collaborative structures are based on complementarity of competencies and expertise of the partners involved. Several projects, tasks, products, and services are subject of *outsourcing*. This tendency coincides with current availability of *information technology* and computer networking systems that support distributed projects and activities. Collaborative structures and distributed activities constitute probably the most important modes of operation in modern industry. They provide high efficiencies in almost all types of companies. In high-technology industries they play the most critical role. At the same time, they create new working conditions and impose new requirements for engineering and managerial staff. More and more frequently, they have to operate as partners in **virtual networks** or **virtual teams**. Strategic orientation toward outsourcing and advances in communication technology, including computer networks and multimedia, made it possible to establish **virtual organization** as a new type of business structure and new form of organization [Chesbrough and Teece, 1996; Davidow and Malone, 1992; Dorf, 1996; Nohria and Berkley, 1994].

Virtual networks, virtual organizations, and virtual teams became important concepts in management of technology. They are also directly related with other functional areas of management. For instance, virtual networks establish new forms of dissemination of marketing information, create very close links between customers and suppliers, and organize commercial transactions within so-called *virtual communities* [Hagel and Armstrong, 1997].

In this section, we focus attention on such virtual networks which are either directly related to development of technology or are affecting the technical functions of a company, i.e., research and development (R&D), new product development, and manufacturing.

Virtual Networks for Technology Development

Cooperative arrangements between companies, in order to support new technology development, are expanding rapidly and include joint exploratory research activities, new product development projects, and outsourced production or service operations. Each case involves some amount of knowledge sharing via new communication systems and takes advantage of high speed of information processing and transfer in computers networks. New forms and protocols are created to allow for fast and direct communication between individuals located in geographically distant places. Virtual networks involve both the organizations and the individuals.

In R&D, scientists and engineers discuss their ongoing tasks with each other to obtain advice or to share results. In competitive situation, several limitations exist for that kind of collaboration among R&D laboratories of different companies. However, there are three specific situations, and corresponding strategic goals, when it is recommended that a company establish an R&D-oriented virtual network with other partners [Harris et al., 1996].

1. Monitoring and development of an *emerging technology*. If the company's own capability in given domain is relatively weak or moderate, then the R&D laboratory should collaborate with a selected partner in monitoring and developing an emerging technology. Risks related with investing in such technology may be shared by the partners.

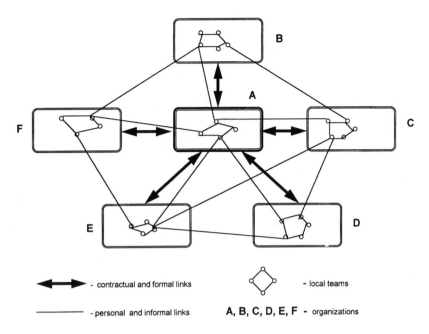

FIGURE 18.3 Institutional and personal links in a virtual network of multiple partners.

2. *Outsourcing of R&D* project or task. If the company's own capability is weak, then outsourcing of given technological project or its part may be the only viable option.
3. *Leveraging capabilities* in basic technology. If the company's own capability in basic technology is moderate, then it may be leveraged through partnership in R&D activities.

In new product development projects, joint efforts may be organized as a virtual network either at the level of cooperating teams or at the level of individuals belonging to the *joint virtual team.* Responsibility and task allocation between partners, in a joint development project, may be made according to *phases of the project* or according to *modules of product design.* In the first case, each team (or author) works on a different phase of the same project (e.g., design, engineering, testing, etc.) and shares information on progress, difficulties, and results with respective partners. In the second case, each team (or author) works on a different module (part) through all phases and shares information on current state of the project with partners responsible for other modules or subassemblies and with those responsible for total product design (including corrections and adjustments of design as necessary to guarantee compatibility between the modules) [Carmel and Zettl-Schaffer, 1996]. Strength of ties and intensity of information exchange among partners in such virtual network-based projects depend on the project category and on the stage of project. Typically, the nonroutine projects are associated with large *weak tie networks* while routine projects are associated with smaller *strong tie networks* [Gales et al., 1995]. At the same time, links between partners have dual character. Contractual relationships between companies create framework for formal information exchange, transactions and cooperative links. The latter are frequently paralleled by personal (and informal) links between members of project teams of respective companies. This dual character of information flows in a virtual network is illustrated in Fig. 18.3.

Recent advances in information technology, related with facilitation of file transfers and with the graphic user interfaces (GUI), allow to exchange large amounts of data, including engineering drawings and three-dimensional pictures, as needed for visualization of new product parts and assemblies. Frequency of those file transfers among partners within the virtual network is one of new measures of the project progress. It usually achieves peak values in the neighborhood of project milestones [Hameri and Nihtila, 1997].

TABLE 18.2 Properties of Virtual Networks in Technology Development

Business/Management Perspective	Technical Perspective
Just-in-time expertise	Shared laboratory notes and research records
Expertise may be obtained from the networked partners or subcontractors at any time; it allows the company to reduce cost of maintaining its own technological expertise	Computer recorded communications among project members allow for up-to-date progress reporting and data sharing
Flexibility of the R&D system	Shared databases
Project portfolio doesn't need to be restricted by the internal set of technological competencies	Access is possible to one another's databases, which constitute a common reference for projects
Concurrent development projects	Interactive project evaluation
Projects can be coordinated asynchronously and in real time across the geographically dispersed organizations (also in different time zones)	Difficulties or problems arising during the lifetime of a project and project outcomes may be evaluated on current basis by all partners
Customer-oriented projects	Virtual colocation
Individual tasks may be easily consulted with and adjusted to customer needs through direct networking with them	Individuals and teams participate in joint projects independently of their geographical location

In summary, the virtual networks provide several advantages and support effectively the technology development projects. Their properties are presented in Table 18.2, as viewed from the business and technical perspectives [Pelc, 1996].

Design of Virtual Teams

Based on the Internet and the World Wide Web, several forms of virtual teams are emerging. Many of them, such as electronic bulletin boards, lists, and news groups, operate on principle of spontaneous exchange of information and opinions. Membership in those groups is voluntary, on ad hoc basis, and is usually temporary. Those unstable virtual teams are valuable only in the initial exploration of information background for future R&D projects.

Another form of virtual team is needed for working on a specific project. Such team may be organized by a company that operates with own geographically dispersed units or collaborates with other companies. In this case, responsibilities need to be assigned and mutual expectations need to be declared among the virtual team members.

Virtual team imposes a number of special requirements on its members [Grenier and Metes, 1995] to assure efficiency when working on a project:

1. *Competency* of each member must be known and recognized by other members of the team.
2. Members need to have capability and willingness for *communicating freely* and sharing their ideas and opinions within the team, without the face-to-face contacts.
3. Members need to develop *trust* within the team; it becomes essential when the project is developed under conditions of stress or when difficulties arise.
4. The team needs to have a *leader who is facilitating communication* among members, monitoring their interaction and focusing the team attention on the project as a whole.
5. Sometimes, the team members have to overcome interpersonal barriers emerging due to *cultural differences*, which may be substantial when the virtual team is international.
6. *Intellectual flexibility and learning capabilities* of members are critical factors of the virtual team success.

Information technology infrastructure for the project-oriented virtual teams is frequently based on the internal computer networks of the companies involved in a joint project. Protocols and rules of communication need to be determined in advance for each project in such a way that they are compatible with properties of the information systems of all parties involved. Assurance of that *compatibility* is one of the critical responsibilities of the team leader.

Conclusion

Virtual networks have both managerial and technical importance in contemporary business environment. They provide support for development and management of technology whenever it involves collaborative efforts of several dispersed organizations, teams, or individuals. In those networks, the human resources, and competencies they represent, are merged and may be used efficiently to solve problems and to perform complex tasks. In this way, the virtual networks may contribute to technological and business successes, almost independently of physical location of partners involved.

Defining Terms

Virtual network: A set of dispersed actors (individuals, groups, or organizations), linked by an information technology system, e.g., via computer network, to share knowledge, to interact, or to perform joint tasks.

Virtual organization: A network of dispersed companies, or their units, linked by an information technology system, e.g., via computer network, to cooperate in business activities. Virtual organization usually includes suppliers, subcontractors, and customers acting as partners in outsourced and dispersed activities.

Virtual team: A team of dispersed cooperating authors linked by an information technology system, e.g., via computer network, and involved in a joint activity or project. Members of a virtual team share skills, access common databases, and exchange information, sometimes informally and/or across the organizational boundaries.

References

Carmel, E. and Zettl-Schaffer, K. Globally dispersed software development teams: a definition and framework, In *Management of Technology V: Technology Management in a Changing World*, R. M. Mason, L. A. Lefebre, and T. M. Khalil, eds., pp. 649–650, Elsevier Advanced Technology, Oxford, England, 1996

Chesbrough, H. W. and Teece, D. J. When is virtual virtuous? Organizing for innovation, *Harv. Bus. Rev.*, Jan.–Feb.: 61–69, 1996.

Davidow, W. H. and Malone, M. S. *The Virtual Corporation*, Harper Collins, New York, 1992.

Dorf, R. C. Designing the virtual enterprise, In *IEMC'96 Proceedings: Managing Virtual Enterprises*, Vancouver, B. C., August 18-20, 1996, pp. 139–141, IEEE, Piscataway, NJ, 1996.

Gales, L. M., Tierney, P., and Boynton, A. C. The nature of information ties and the development of technology, In *Advances in Global High-Technology Management*, Vol. 5, L. R. Gomez-Mejia and M. W. Lawless, eds., JAI Press, Greenwich, CT, 1995.

Grenier, R. and Metes, G. *Going Virtual: Moving Your Organization Into the 21st Century*, Prentice-Hall, Upper Saddle River, N J, 1995.

Hagel, J., III and Armstrong, A. G. *Net Gain*, Harvard Business School Press, Boston, 1997.

Hameri, A. P. and Nihtila, J. Distributed new product development project based on Internet and World-Wide Web: a case study, *J. Prod. Innov. Mgt.*, 14: 77–87, 1997.

Harris, R. C., Insinga, R. C., Morone, J., and Werle, M. J. The virtual R&D laboratory, *Res. Technol. Mgt.*, 39(2): 32–36, 1996.

Nohria, N. and Berkley, J. D. The virtual organization: bureaucracy, technology, and the implosion of control, In *The Post-Bureaucratic Organization*, C. Heckscher and A. Donnellon, eds., pp. 108–127, Sage, Thousand Oaks, CA, 1994.

Pelc, K. I. Maps of virtual structures in R&D, In *IEMC'96 Proceedings: Managing Virtual Enterprises*, Vancouver, B. C., August 18-20, 1996, pp. 459–463, IEEE, Piscataway, NJ, 1996.

Further Information

The International Conference on Engineering and Technology Management, IEMC'96, which was held in Vancouver, British Columbia, August 18 to 20, 1996, reviewed several topics related to virtual networks. Proceedings of that conference (over 130 papers) have been published by the IEEE Engineering Management Society: *IEMC'96 Proceedings Managing Virtual Enterprises: A Convergence of Communications, Computing, and Energy Technologies,* 1996, The Institute of Electrical and Electronics Engineers, Piscataway, NJ.

18.7   Compensation

E. Brian Peach and Ralph Roberts

Compensation refers to all things of tangible value received by employees, both direct and indirect. It consists of all forms of financial returns and services as well as benefits employees receive as part of an employment relationship [Milkovich and Boudreau, 1996]. **Direct compensation** consists of financial returns such as wages, incentives, bonuses, and merit pay plus special payments such as cost of living adjustments. **Indirect compensation** consists of benefits such as paid time off, pensions, medical plans, counseling services, etc. Formal compensation does not include **alternative rewards**, such as recognition parties, job enrichment, or personal satisfaction for a job well done. Alternative rewards are a very important part of an organization's total reward system, but are beyond the scope of this section.

There are three primary **compensation objectives**: to attract qualified employees, to retain desirable employees, and to motivate employees to perform. Achieving each of these objectives requires different compensation elements, and often objectives are in conflict. In addition, achieving these objectives must be accomplished within constraints: the organization's ability to pay, applicable laws and legislative mandates, labor contract requirements, limitations, and internal and external labor markets. Although the design and implementation of a compensation system is complex, compensation systems are the primary tool for influencing employee behavior in support of organizational goals and thus are an absolutely critical aspect of the overall human resources program. To be most effective, compensation systems and strategies should derive from and support overall organizational, divisional, and departmental objectives.

Theories of Job Worth

A common perception, at least in Western economies, is that employees are paid in accordance with their contribution to the organization. In fact, there are many possible bases for paying people (see Wallace and Fay [1988]). Historically, employees have been paid on the basis of a "just wage", which based the value of labor on social status or on the basis of need, meaning employees supporting families should be paid more than single employees. Economic theories such as marginal revenue product contend an employee's worth equals the net increase in revenues due to the employee. Labor scarcity theories using supply and demand claim that employees are paid more if their services are scarce, regardless of their contribution to the firm. Human capital models contend employees should be paid based upon what they know and *can* do rather than their current work assignment. Other models place value on position in the organizational hierarchy — higher-ranking managers should be paid more than supervisors, etc.

Theories of job worth are important because they guide a firm's compensation system (either explicitly or more often implicitly) and thus signal to employees what the firm values. Most compensation experts contend that a straight hourly wage signals employees that the compensation objective is *attendance*. Thus, the components of a compensation system must be carefully structured to support critical organizational goals and expectations.

Balancing Equity

A major challenge for compensation system managers is the balance of three perceived fairness concerns of employees: internal, external, and individual equity. Perceptions of inequity can create dissatisfaction and poor performance in an organization. **Internal equity** considers whether jobs within an organization of equal worth are paid similarly and whether jobs of unequal worth have appropriate pay differentials. Employees compare their salaries with other employees in similar jobs and different jobs to determine if they are being paid fairly. **External equity** considers whether jobs within an organization are paid comparably to similar jobs in other organizations. If employees perceive their compensation is externally inequitable, they may leave for a higher-paying position. **Individual equity** concerns employees comparing their current compensation or their compensation over time with their personal feelings of value.

Achieving equity goals almost always requires trade-offs. A compensation manager may believe two positions are similar and deserve equal compensation, but external labor market shortages may force higher pay for one of the positions in order to attract and retain employees. Paying more than market for only just the position with a labor shortage creates internal inequity and possible employee dissatisfaction. Paying more for both creates external inequity, and the organization's labor costs are higher than its competitors.

Assessing Work

Internal job worth is established through job analysis and job evaluation. External job worth is established through salary surveys and market comparisons. **Job analysis** involves assessing the content of an organization's jobs based on specified criteria. These criteria must be work related and should reflect what the organization values. Typical criteria include actual work performed and employee characteristics, including behaviors and abilities. A job analysis starts with collecting data. Each specific job is identified, including its title, department, location, number of incumbents, and whether it is hourly or managerial (exempt from Fair Labor Standards Act). Each job is then defined, including its purpose, what it accomplishes for the organization, and how it fits with other jobs. The job is then described, covering work performed (task data) and employee characteristics (behavioral and abilities data).

Salary surveys are conducted to determine what other firms are paying for similar jobs. Different jobs will have different **relevant labor markets**. Some jobs, such as secretaries and maintenance workers, are filled locally. Some professional or executive positions may be advertised nationally or internationally. To save on survey costs, an organization may choose to survey only a few jobs, using internal job analysis data for remaining positions. Commercial surveys are available, but they may be misleading if they include firms not part of the relevant labor market. Also, critics of market surveys contend that these tend to continue inequitable practices such as paying less for historically female jobs such as nurses and librarians.

Designing the Pay Structure

Another critical but extremely challenging part of compensation is **job evaluation**: systematically establishing the relative value of jobs within an organization as a basis for establishing differential pay levels. There are four common approaches to job evaluation: ranking, classification, factor comparison, and the point method. Whichever approach is used, its implementation must be defensible in a court of law as fair, equitable, and work related. *Ranking* orders jobs from highest to lowest on some value scale and is the easiest to use and explain but is in disfavor as the criteria are typically vague and not clearly work related. *Classification* establishes a series of classes of increasing value to the organization, placing each job into a class. It is more complex and works best for stable jobs and is used primarily by the Civil Service for general schedule (GS) ratings. *Factor comparison* defines a universal set of factors for establishing job value such as mental or physical requirements or responsibility and then ranks a few key jobs on these factors. These key jobs usually have easily established external market pay levels. Other jobs are ranked and compared to the key jobs to establish pay levels. Factor comparison can be very complex; it is difficult to explain to employees, who often are reluctant to accept the results.

The most widely used approach is the *point method* of job evaluation. Although complex, it can be understood by employees and is defensible to courts if properly done. Compensable factors, which are specific tasks/behaviors the organization values, are established and then scaled with relative points assigned for increasing levels of that factor. For example, five levels of communication skills could be established, ranging from ten points for a job that requires the ability to converse and write at a high school level to fifty points for a job that requires interpreting and communicating complex topics to diverse audiences. The compensable factors are weighted according to the importance of the factor to the job, and the weighted points across all factors are summed. These sums are then compared across jobs to determine relative value of each job to the organization. Once the relative value of jobs within the organization is established, pay for key jobs is established and relative pay for all other jobs derive from their relative value to the key jobs. The pay for key jobs can be based either upon internal pay practices or market surveys or a combination of the two.

Once firms have assigned relative value to jobs, they are grouped in *pay grades*, which are clusters of similarly paid jobs. Each pay grade then has a *pay range* assigned to it, which defines the minimum and maximum amounts the firm will pay for that pay grade. Once an employee reaches the maximum for a pay grade, no further raises (other than cost of living) will be given until the employee is promoted to a higher pay grade. The midpoints of each pay grade establish the firm's *pay line*, which reflects the firm's overall pay structure relative to external competitors. Some firms prefer to *lead* competitors by establishing a pay line above the market average, believing it gives them best employees. Others *follow* the market and pay the average. Some prefer to *lag* the market and pay below average believing they save on labor costs without a significant penalty in employee performance. The proper strategy depends on a number of factors, such as a firm's competitive situation, general economic conditions, and relative labor scarcity. *Broadbanding* refers to having a few pay grades with wide pay ranges. Broadbands have the advantage of flexibility in defining job responsibilities but shift pay decisions to assessing individuals rather than jobs. Broadbands also tend to be more expensive as there are fewer controls to upward pay migration.

The overriding concern throughout the pay structure process is balancing internal and external equity concerns. Firms that primarily, if not exclusively, rely on external comparisons (market price pay) forgo the ability to link pay with corporate strategies and objectives. Firms that ignore external realities may lose employees or experience higher compensation costs.

Pay for Performance

Pay for performance (PFP) is based on the belief that higher pay is directly associated with higher performance because higher pay will motivate employees to work harder and more effectively. *Need theories* contend people are motivated by internal needs and pay is motivating to the extent it satisfies a need. *Expectancy theory* postulates that motivation depends on an individual's personal belief in having the ability to perform job requirements, that higher performance will lead to the reward, and the reward must be valued by the employee more than rewards for other possible activities. *Equity theory* predicts that, if employees perceive others are paid more for the same or lesser work, they will react negatively and reduce output to correct the equity imbalance. *Agency theory* predicts that pay directs and motivates employee performance. Thus, pay must be directly linked to desired behaviors.

The conclusion is that PFP systems will motivate higher levels of performance, if certain conditions exist: (1) employees must be able to control their levels of performance, (2) employees must believe they can achieve the desired performance, (3) performance must be measurable, (4) performance expectations must be perceived as fair, (5) employees must value the potential reward, and (6) employees must believe successful performance will result in receiving the reward. Compensation focuses on tangible pay and benefits as motivational rewards. Other types of motivational rewards are possible and may, given specific circumstances, be more motivating than performance based financial rewards. These include security, status, recognition, promotion, social interaction, work content and variety, autonomy, performance feedback, and development opportunity.

There are two primary types of PFP plans: merit and incentive. **Merit pay** provides increases in pay based on past performance, where some amount or percentage is available for high-performing employees. The theory is that, by rewarding high levels of past performance, employees will be motivated to continue their high performance. Problems with merit pay are many. Usually merit pay increases are small relative to total pay — in the range of 3 to 5%. Research indicates that performance is not significantly affected until at least 20 to 30% of an employee's compensation is affected. In practice, merit pay plans provide almost all employees some increase. Thus, the top performer may get a 5% increase, and an average performer receives a 3% increase. High performers perceive this as inequitable and may perform at lower levels in subsequent years, and average performers see no need to improve. Another problem is that merit pay increases are based on supervisory evaluations, which are often perceived as unfair by employees. Thus, typical merit pay systems violate several of the requirements for a PFP system.

Incentive pay systems offer a reward for some objective, predetermined level of performance. Incentive pay systems can be either individual or group based, and there are many approaches for each one. *Individual plans* can be based on piece work or standard hours. Individual plans work best where work is simple and repetitive, the pace is under control of the employee, and no interaction with or dependence on others is required. Common *group-based* PFP plans include gain sharing, profit sharing, and employee stock option plans (ESOPs). *Profit sharing* divides a portion of a firm's profits among employees but is criticized because employees often do not see the link between their effort and overall organizational performance. This is especially true if industry or economic conditions hurt firm profitability despite increased employee efforts. *Gain sharing* attempts to compensate for this by setting increased productivity goals and rewarding employees for meeting these even if overall firm performance is not strong.

A major problem in establishing a PFP system is defining the desired and measurable performance. For many jobs, it is very difficult to quantify all of the specific desired performances and link them to rewards. Those performances that are not rewarded in the PFP plan will receive less employee effort. If the PFP plan is group based, some group members may not work as hard as others, yet receive the same reward, creating perceptions of inequity. Another difficulty arises when an organization tries to modify the performance requirements necessary for rewards. Even if labor saving equipment is installed, employees are typically very suspicious of any attempts by management to increase the performance requirements.

PFP systems address a basic belief within most people — that those who do more should be paid more. However, designing and maintaining an effective PFP system that is accepted as fair by employees and also is supportive of specific organizational goals can be expensive and complicated.

Benefits

Employee benefits are an increasingly important part of compensation systems for two reasons. First, they are becoming an ever-larger portion of the total compensation expense. Second, they are being recognized as an important compensation tool for supporting the three objectives of compensation: attracting, retaining, and motivating employees. Complicating the role benefits play in compensation is the view by many employees that benefits are an entitlement, rather than a component of compensation and a reward for doing a good job. Thus, for benefits to achieve the objectives of compensation, an effective communication program has to be established that makes employees aware of the nature and value of benefits received. Many employees are not aware of the sizeable expense a firm sustains in providing benefits. Memos, bulletins, training, and/or informational sessions can help to inform employees of the value received through benefit programs.

Some benefits are *legally required* such as social security, worker's compensation, unemployment insurance, up to 12 weeks unpaid leave for specified family situations, and extended health coverage for employees leaving a firm. In addition, there are many possible types of discretionary benefits, including health, dental, and life insurance, dependent care, vacation and other paid time off, and retirement plans.

Retirement plans can be defined benefit or defined contribution plans. In a *defined benefit plan*, the employer agrees to provide a specific retirement amount, in either dollars or a percentage of salary. A *defined contribution plan* specifies contribution amounts by the employer in dollars or percentage of salary but makes no promise about the amount to be received by the employee at retirement. Actual retirement benefits depend on the investment success of the retirement plan. *Vesting* refers to the length of time before an employee is entitled to all of an employer's contribution to a retirement plan. Recent legislation requires full vesting after no more than 7 years. *Portability* refers to the ability of an employee to transfer retirement funds from an old employer to a new employer's plan.

Benefits can be contributory or noncontributory. Under *noncontributory plans*, all expenses are paid by the employer, and all employees are covered. *Contributory plans* involve payments by both the employer and employee. This provides greater flexibility to the employee and induces some financial responsibility in using the benefits. *Cafeteria* or *flexible benefits* plans allow employees to select benefits, allowing a closer fit between benefits received and employee preferences. Younger employees will typically favor medical and dental benefits while older employees will favor retirement benefits. Combining flexible and contributory plans helps contain costs, but possible drawbacks are increased administrative costs and that employees will make poor choices, leaving them uncovered for some emergencies.

There are many laws that have been passed regulating benefits. The Fair Labor Standards Act of 1938 linked pay to some benefits (social security). The Health Maintenance Act of 1973 requires employers to offer alternative health coverage. The Employee Retirement Income Security Act (ERISA) of 1974 established pension guidelines. The Tax Reform Act of 1986 affects individual retirement accounts (IRAs) and 401K plans. Various antidiscrimination legislation proscribes assigning benefits based on a variety of grounds. The Consolidated Omnibus Budget Reconciliation Act (COBRA) of 1984 extends benefit rights after an employee leaves work. Recent family acts and other legislation continues to extend the rights of employees to benefits both during and after terminating employment. Benefits are an increasingly complex field with many possible legal and competitive risks, and as a consequence many small- to medium-sized firms are contracting out the administration of benefits to specialists.

Conclusion

In the final analysis, all firms face the same compensation challenges. First is the level of compensation sufficient to attract, retain, and motivate employees. Second is the level of compensation perceived as fair by other employees, stockholders, and the community. Third, but very important, is the level and nature of compensation such that the firm remains competitive on labor costs. A compensation system that achieves the three objectives, is perceived as fair, and is cost competitive is an effective system.

Defining Terms

Alternative compensation: All rewards other than tangible financial rewards that are part of the formal compensation system, such as recognition, promotions, and job enrichment.

Compensation: Includes direct and indirect returns to employees and consists of all forms of financial returns as well as all tangible services and benefits received by employees in return for services to an organization.

Compensation objectives: To attract qualified people to join the organization, to retain desirable employees, and to motivate employee performance.

Direct compensation: Consists of financial returns such as wages, incentives, bonuses, and merit pay plus special category payments such as cost of living adjustments.

External equity: One of the three fairness concerns of employees and exists when jobs within the organization are paid the same as in similar jobs in other organizations.

Indirect compensation: Consists of returns to employees that have tangible value but are in forms other than direct payment of funds, such as services and benefits provided to employees. These include paid time off, pensions, medical plans, counseling services, etc.

Internal equity: One of the three fairness concerns of employees and exists when similar jobs are perceived as being paid similarly; different jobs have pay differentials proportionate to the difference in job requirements.

Job analysis: A systematic process of collecting information using work-related criteria about the task and employee requirements of a specific job and presenting it in a structured manner.

Job evaluation: Identifies the relative complexity and requirements of jobs within an organization to establish the relative worth to the organization of those jobs as an aid in establishing pay differentials between jobs.

References

Adams, J. S. Toward an understanding of inequity, *J. Abnorm. Soc. Psychol.*, 67: 422–436, 1963.

Milkovich, G. T. and Newman, J. M. *Compensation, 5th ed.,* Irwin, Boston, 1996.

Further Information

Reasonably straightforward discussions of compensation systems may be found in any good college textbook. Good examples are *Compensation, 5th ed.,* by George Milkovich and Jerry Newman, published by Irwin, and *Compensation Decision Making, 2nd ed.,* by Fred Hills, Tom Bergmann, and Vida Scarpello, published by Dryden.

An advanced discussion of the basis for pay and theories of job worth can be found in *Compensation Theory and Practice, 2nd ed.,* by Mark Wallace and Charles Fay. This was published in 1988 but provides an authoritatively technical, yet still readable, exposition of compensation theory.

Job analysis is both a critical and complex undertaking. The U.S. Department of Labor, Manpower Administration, 1972, *Handbook for Analyzing Jobs*, U.S. Government Printing Office, is a basic reference. For a more detailed description of the process, see Robert J. Harvey, 1991, Job Analysis, in the *Handbook of Industrial and Organizational Psychology, Vol. 2,* edited by Marvin Dunnette and L. Hough, published by the Consulting Psychologists Press in Palo Alto, CA. The original *Position Analysis Questionnaire* (PAQ) can be obtained from the Purdue University Bookstore, 360 West State St., West Lafayette, IN 47906. However, the best source for PAQ information is PAQ Services, 1625 North 1000 East, Logan UT 84321, phone (801) 752-5698. PAQ Services provides extensive support for the PAQ, including a newsletter.

Further information on job evaluation is in the *Handbook of Wage and Salary Administration,* edited by Milton Rock and published by McGraw-Hill in 1984. For more information on the Hay approach, see *The Guide Chart Profile of Job Evaluation,* published by Hay Associates.

A professional organization comprised of consultants, academics, and practitioners devoted to this topic is the American Compensation Association, 14040 N. Northsight Boulevard, Scottsdale, AZ 85260, phone (602) 951-9191 or visit them at http://www.aca.online.org/.

19

Personal Issues

Ralph Katz
Northeastern University

Hyrum W. Smith
Franklin Covey Company

Lenny Ralphs
Franklin Covey Company

Richard O. Mason
Southern Methodist University

Florence M. Mason
F. Mason and Associates

Terry Pearce
Leadership Communication

Mike Markel
Boise State University

Mary Munter
Dartmouth College

David Birchall
Henley Management College

19.1 Careers of Technical Professionals

Ralph Katz

Organizations employing many professional specialists, engineers, and scientists in particular face the dilemma of establishing **career** opportunities that are both stimulating to the professional and productive for the organization. This problem stems, in part, from the notion that research and development (R&D) professionals bring to the organization a set of attitudes and career aspirations that are in conflict with the organization's pressures and work requirements. As argued by Kerr and Von Glinow [1977] as well as Katz et al. [1995], many technical professionals are socialized into occupations with values and definitions of success that differ significantly from those prevailing in the traditional managerial setting. In the typical work setting, for example, management expects authority to be discharged according to the **hierarchical** principle, delegated downward through a progressive series of well-ordered job positions. In sharp contrast, however, many technical professionals, especially some of the more creative and independent contributors, would strongly prefer to have the freedom to pursue their own technical ideas and interests, to feel responsible for making judgments in their particular areas of knowledge and expertise, and to be able to exercise control and make decisions through collegial discussions and reviews rather than through managers in formal job positions or through formal structures and procedures.

Careers have typically been viewed as a set of successive stages relating to the individual's work experiences, job activities, and organizational contributions. However, if technical professionals and their employing organizations differ significantly with respect to their individual needs and expectations, then how are the career orientations and interests of engineers and scientists being met by their work assignments and promotional opportunities at different career stages? The primary focus of this section, therefore, will be to examine several career stage **models** of technical professionals and to see how well an organization's reward and promotional practices can satisfy the needs of its technical work force as they change over the course of these career stages.

Career Stage Models

Dalton and Thompson's Four-Stage Model

Empirical research by Dalton and Thompson [1985] has identified four distinct stages through which the more successful, or at least the more highly rated, technical professionals pass. Each of the four stages differs from the others in the tasks that individuals must perform, in the types of relationships that they form, and in the psychological adjustments that they need to make. Determining one's career stage is crucial to assessing past success and making future decisions. In the first stage, scientific or engineering technologists are new and relatively inexperienced, typically working under the supervision and direction of more senior professionals within their field. For the most part, their work is not entirely their own since the assignments they are given are usually connected to a larger program or project activity. Professionals in stage one are expected to willingly accept supervision and direction, to do more of the detailed and routine project work, to exercise incremental creativity and initiative, and to learn how to perform well and complete tasks under pressure and within budgeted costs and schedule. Since many younger scientists and engineers are eager to undertake more responsibility and challenging jobs, they may find this initial work situation frustrating if they remain in it too long.

Independence is the primary prerequisite for the second stage according to Thompson and Dalton. Technical professionals make the transition to this stage by developing a reputation for being technically competent and having the ability to work independently to produce significant results. Most engineers and scientists progress to stage two by becoming specialists within certain fields or areas of expertise. Depth of knowledge and specialized skills that can be applied to a wide variety of problems in a given area are likely to gain an individual increased visibility and opportunity within a large organization. A successful transition to this second career stage also involves a change of attitude and behavior. The R&D professional is now expected to assume responsibility for key portions of the project's activities by developing more of his or her own resources to solve problems, relying less on one's supervisor or **mentor** for answers. Doing well in the second stage is vital for increasing one's confidence and credibility within the organization.

While in the second stage, scientists and engineers developed their own problem-solving abilities and learned to take care of themselves. In the third stage, however, they learn to take care of others, leading and motivating them in coordinated work activities. Professionals in stage three develop greater technical breadth and business perspective and they begin to assume increased responsibility for directing and developing other professional employees in their work. They also learn to play multiple roles, including informal mentoring to younger professionals, idea leading in project work activities, and assuming a more formal supervisory position. Successful transition to stage three requires a willingness to assume responsibility for others' performances and a capacity for dealing with the tensions and uncertainties that result from bridging the worlds of management and professional disciplines.

In progressing to stage four, the technical professional is now providing direction to the organization by highlighting opportunities and problems, by focusing strategic activities, and by managing the processes by which critical decisions are made. Senior-level managers who are usually in the fourth stage of their careers formulate policies, make decisions, and sponsor programs by bringing resources, money, and people together to pursue new ideas and business ventures. Not only do they need to make fast and good operating decisions, they must also learn to adopt a longer range perspective, keeping the big picture in mind. A crucial adjustment for success in this stage is becoming accustomed to using managerial power and influence, fighting for projects and programs in which they firmly believe and forming the alliances necessary for getting them done. Scientists and engineers in this stage come to understand just how important it is to insert themselves into the politics of the decision-making process if they truly hope to lead the organization and influence its strategic direction and focus.

Dalton and Thompson's research has shown that, as technical professionals get older, they tend to maintain higher performance ratings only if they can move to at least stage three. This notion that performance is linked to progress through career stages is crucial to establishing reward policies, training and professional development programs, and job assignments that are appropriate for R&D professionals in all of the stages. An organization, therefore, should examine the distribution of its technical work force not only by age, salary, or educational degree, but also by career stage. The four stages can then be used to determine those programs, policies, and job assignments that are more suitable for stimulating and developing an employee's career at various points in his or her professional development. Professionals in stage one, for example, are generally concerned about their initial job assignments. They want assignments that will allow them to make a significant contribution and learn new skills. Even though the amount of challenge in first assignments can have a strong positive impact on an employee's entire career [Katz, 1997], all too often organizations do not give many of their new technical recruits initial task assignments that are sufficiently meaningful and challenging.

Schein's Career Anchor Model

A person's career anchor, according to Schein [1996], is his or her self-concept, consisting of (1) self-perceived talents and abilities, (2) basic values, and, most importantly, (3) the individual's evolved sense of motives and needs. Career anchors evolve as one gains occupational and life experiences. Once they are formed, however, they function as a stabilizing force, making certain job and career changes either acceptable or unacceptable to the individual. Schein's research has shown that most people's self-concepts revolve around eight categories reflecting basic values, motives, and needs, and, as careers and lives evolve, most people discover that one of these eight categories is the anchor, the thing they will *not* give up, even though most careers permit the fulfilling of several of the needs that underlie the different anchors.

1. *The managerial anchor.* Individuals who found themselves in this category made it clear that their fundamental motivation was to become competent in the complex set of activities comprising general management. These are those professionals who want to rise in the organization to higher levels of responsibility and authority and there are three component areas of competence that they need to develop. First, they must have the analytical competence to identify and solve conceptual problems under conditions of uncertainty and incomplete information, and, second, they must have the interpersonal competence to influence and lead people toward the more effective achievement of organizational goals. It is also critically important that professionals in general management positions develop emotional competence, that is, the ability to remain strong and make tough decisions in highly pressured and stressful situations without feeling debilitated or paralyzed.

2. *The technical-functional anchor.* Many professionals indicate that their careers are motivated by the technical challenges of the actual work they do. Their anchor lies in using the knowledge and skills of their technical fields or disciplinary areas, not the managerial process itself. If members of this grouping hold supervisory responsibility, they are usually most comfortable supervising others who are like themselves or who are working in the same disciplinary area. Unlike those in the managerial anchor category, it is the nature of the technical work and not the supervising that excites them. They have a strong preference not to be promoted to job positions that place them outside their technical area, although many admit to having taken positions of greater managerial responsibility in the hope of garnering more influence and rewards.

3. *The security-stability anchor.* A third group of individuals tries to tie its careers to particular organizations. These professionals seem to have an underlying need for security in that they seek to stabilize their careers by linking them to particular organizations. The implication is that a professional with this anchor is more willing than other anchor types to accept the organization's definition of his or her career. Regardless of their personal aspirations or areas of technical competence, these individuals come to rely on the organization to recognize their needs and provide them the best possible options and opportunities. Professionals anchored in security/stability may have considerable difficulty in higher levels of managerial responsibility where emotional competence is the prime requisite for effective performance. They may also experience a great deal of stress as organizations eschew policies of no layoffs and guaranteed employment, thereby forcing many professionals to shift their career dependence from the organization to themselves.

4. *The entrepreneurial creativity anchor.* Some professionals have a strong need to create something of their own. This is the fundamental need operating in an entrepreneur, and it manifests itself in the desire to commercialize a new product or service or in some way to create something that can be clearly identified with the individual. Although these individuals often express a strong desire to be on their own and away from the constraints of established organizations, the decisive factor for leaving their prior organizations was not to achieve autonomy or to make a great deal of money [Roberts, 1991]. Instead, technical entrepreneurs typically start their own companies and businesses because they really believe in a given product or service, and the organization in which they had been working would not allow them to move forward with their idea. Schein [1987] also reports that entrepreneurially anchored people are often obsessed with the need to create and can easily become bored or restless with the demands and routines of running a business. Other studies have drawn a strong distinction between professionals who may be good at generating new creative ideas and those professionals who have the strong desire and capability to grab good ideas and persevere with them until they have been commercialized in the marketplace [Roberts and Fusfeld, 1997].

5. *The autonomy-independence anchor.* Some professionals discover that they strongly prefer not to be bound by the kind of rules, policies, procedures, dress codes, working hours, and other behavioral norms that are present in almost any traditional organization. These individuals are primarily concerned about their own sense of freedom and autonomy. They have an overriding need to do things their own way, at their own pace, and along their own standards. They have found organizational life to be restrictive, somewhat irrational, or intrusive into their own private lives. As a result, they prefer to pursue more independent careers on their own terms, often seeking work as consultants, professors, independent contractors, or researchers in R&D-type environments. Professionals anchored in autonomy are often highly educated and self-reliant, and they will usually decline much better work and job offers if such work significantly impinges on their independence.

6. *The lifestyle anchor.* Professionals in this anchor category are focusing on careers that can be integrated with their total lifestyle. They value careers that permit them to stabilize their life patterns by settling into a given region without having to move and relocate. These individuals differ from the security-stability anchor professionals in that they do not define their careers in terms of economic security but rather see their careers as part of a larger life system. Most are in **dual career** situations and so they seek to work in organizational settings that will provide them the flexibility and kinds of options they need to integrate two careers and two sets of person and family concerns into a coherent overall pattern. Such options might include working at home, flexible schedules, part-time work when needed, leaves of absence or sabbaticals, paternity and maternity leaves, day-care coverage, etc. Although the size and strength of this anchor grouping is probably tied to various social trends within our society, what these professionals want from their employing organizations is an understanding attitude and genuine respect for their lifestyle pressures and needs.

7 and 8. *The service-dedication* and *pure challenge anchors:* A growing number of individuals, according to Schein [1996], report that they are feeling the need not only to maintain an adequate income but also to do something meaningful in a larger context. As the information technology explosion has made many of the world's problems more visible, professionals anchored in service are devoting their careers to organizations and activities that allow them to contribute to these issues in a significant way. Finally, there has always been a small group of people who define their careers in terms of overcoming impossible odds, solving the unsolved problem, or winning out over competitiors. These professionals seek jobs in which they face tougher challenges or more difficult kinds of strategic problems, but, in contrast to the technical-functional anchored group, they are less concerned about the particular kind of problem or technology that is involved.

Since individual career anchors vary so widely, Schein argues that constructive career management is impossible unless the individual knows his or her own needs and biases. It is therefore critically important that professionals gain more self-insight by analyzing their own career anchors and then managing more actively their own career courses. Organizations also need to create more flexible career paths and incentive systems in order to meet this wide range of individual needs. If individuals are given a more accurate picture of career patterns and the work that needs to be done, then they will be better able to set a constructive direction for themselves.

Katz Job Longevity Model

Over the course of their careers, technical professionals try to structure their work activities to reduce stress and ensure a level of certainty [Katz, 1997]. People do not do well with uncertainty; they would like to know as much as possible about what will happen next or how they will be affected. Based on this notion, as technical professionals continue to work in a given job environment, they are likely to develop standard work patterns that are both familiar and comfortable, patterns in which predictable routines play a relatively large part. Management, however, needs to create a work setting in which professionals are both comfortable with their job requirements but also challenged and excited by them.

In the course of working at one's job over a long period of time, Katz argues that a technical professional passes through three broad stages: *socialization, innovation,* and *stabilization*. During the socialization period, professionals are primarily concerned with understanding and coming to terms with their new and unfamiliar task and social environments. They must learn the customary **norms** of behavior within their group settings, how reward systems really operate, the expectations of their supervisors and/or managers, and a host of other considerations that are necessary for them to funtion effectively. While this *breaking-in* period is critical for new hires, it should also be understood that veteran employees assigned to new jobs or new teams must also *resocialize* themselves since they too must now deal with unfamiliar tasks and colleagues. It is in this period that professionals learn not only the technical requirements of their new job assignments but also the behaviors and attitudes that are necessary for becoming an accepted and contributing member of the organization.

As individuals become increasingly familiar with their new jobs, they are freer to devote their energies and concerns less toward socialization and more toward performance and accomplishment. It is this innovation stage of a job that individuals become increasingly capable of acting in a responsive and undistracted manner. The movement from socialization to innovation implies that professionals no longer require much assistance in deciphering their new job and organizational surroundings. Instead, they can divert their attention from *getting on board* and *fitting-in* to concerns for achievement and influence. Opportunities to change the status quo and to respond to new challenging demands within their job settings become progressively more pertinent to individuals in this job longevity stage.

As the length of time spent in the same job environment stretches out, however, professionals gradually enter the stabilization stage, a period in which the person has slowly shifted away from active involvement and receptivity to the challenges in his or her job and toward a greater degree of unresponsiveness to these challenges. In time, even the most engaging job assignments and responsibilities can become less exciting, little more than habit, to technical professionals who have mastered and become accustomed to their everyday task requirements. With prolonged job longevity and stability, professionals' perceptions of their present conditions and future possibilities become increasingly impoverished. If individuals cannot maintain, redefine, or expand their jobs for continued growth and change, then their work enthusiasm will deteriorate. If opportunities for development are continued, however, then the stabilization period may be held off indefinitely.

In moving from innovation to stabilization, technologists who continue to work in the same overall job situation for long periods gradually adapt to such steadfast employment by becoming increasingly indifferent to the challenging aspects of their assignments, and, as individuals come to care less about the intrinsic nature of their work, their absorption in contextual features such as benefits, vacations, compatible coworkers, and friendly supervisors tends to increase. Interestingly, entry into the stabilization period does not necessarily imply a reduced level of job satisfaction. On the contrary, they have adapted to their jobs by becoming very satisfied with the comfortableness and predictability of their work situations. As pointed out by Manners et al. [1997], "fat, happy rats never run mazes." With stability comes greater loyalty to precedent and to the established patterns of behavior. Stabilized professionals become increasingly content with the customary ways of doing things and reject those challenges that require exceptional effort and/or innovative change. The differences between the innovation and stabilization stages are indicative of the distinctions between creative and routine performance, between excitement and complacency. What they are asked to do, they do well, but that spark is missing; the willingness to stretch one's efforts, to go beyond what is expected and try something new is just not there [Organ, 1990].

With ongoing job changes and promotions, individuals can easily cycle between socialization and innovation or they can slowly proceed from innovation to stabilization over time. Direct movement from stabilization back to innovation, however, is very unlikely without the individual first going through a new socialization (or resocialization) opportunity. Thus, rotation per se is *not* the solution to rejuvenation; instead, it is rotation coupled with a new socialization experience that provides the professional with fresh perspective and a new chance to regain responsiveness to new task challenges. The intensity of the resocialization experience, moreover, must match the strength of the prior stabilization period. All too often, organizations expend much time and effort planning the movements and reassignments of individuals but very little effort in managing the socialization process that transpires *after* rotation. This is very unfortunate, for it is the experiences and interactions that occur after reorganizations that are so important for influencing and framing an individual's disposition and eventual responsiveness.

Career Orientations and Dual-Ladder Reward Systems

For many years now, much has been written about how professional rewards and incentives clash with those options normally available in an organization. Supposedly, many technical professionals are motivated by a desire to contribute to their fields of knowledge and to be known for their expertise within their technical disciplines. They are more *cosmopolitan* in their career orientations, developing strong commitments to their specialized skills and outside professional reference groups, which often leads to less organizational loyalty. Managers, on the other hand, are seen as individuals whose career orientations are more *locally* defined within their organization. They do this by focusing more of their efforts on the achievement of company goals in the hope of being rewarded with greater advancement along the organization's managerial hierarchy. In short, the argument is that technical professionals acquire status and define success from the perspectives of their technical colleagues as they practice their technical specialty while managers build these same attributes from the perspectives of their organizational superiors.

Despite these purported differences in professional and managerial career orientations, the highest rewards in most organizations are conferred on those who assume additional managerial responsibilities. Promotion or advancement up the managerial ladder secures increases in status, recognition, salary, influence, and power. For many technical professionals, movement into management becomes the most viable career strategy simply because their opportunities to achieve success in other ways are very limited. As a result, many of the most productive engineers and scientists become frustrated as they feel pressured to take on managerial and administrative roles they really do not want in order to attain higher salary and more prestige. Organizations may even be compounding this problem if they fail to provide alternative rewards for those technical professionals who either do not aspire to or who show no aptitude for management. The migration of many of the organization's most competent engineers and scientists into management can seriously deplete the company's pool of experienced, creative technical talent, and, of course, not all high-performing technologists have the interpersonal, communication, and leadership skills necessary to make the successful transition from technical specialist to manager. In such cases, not only does the organization lose their technical abilities, it also encounters additional problems and discontent as their effectiveness in their new managerial role diminshes.

The dual-ladder system of career advancement is an organizational arrangement that was developed to solve these individual and organizational dilemmas by providing meaningful rewards and alternative career paths for technical professionals working in organizations. The dual-ladder approach is the formalization of promotions along two parallel hierarchies. One hierarchy provides a managerial career path while the other provides advancement as a technical or individual contributor. Promising equal status and rewards at equivalent levels of the two hierarchies, the dual ladder was established many years ago to reward engineers and scientists for their technical performance and contributions without having to remove them from their professional work.

Although dual ladders have been in use for some time, their success is often the focus of much debate and a whole host of problems has been identified. For example, most cultures associate prestige with managerial advancement. Titles of department head and vice president convey images of success while titles of senior researcher and lead engineer are considerably more ambiguous. Frequently too, the organization does a poor job of publicizing technical ladder promotions, and little observable or real change takes place either in work activities or technical opportunities after such promotions. Further problems even arise because technical promotions are typically discussed and justified in terms of past contributions while managerial promotions are framed more positively in terms of future promise and potential.

Another set of problems concerns the incentives associated with each ladder. Advancement up the managerial ladder usually leads to positions of increased influence and power. Since the number of individuals under a manager usually increases with promotion, resources can be mobilized more easily to carry out the manager's needs and demands. Contrastingly, advancement up the technical ladder usually leads to increased autonomy in the pursuit of one's technical interests but often at the expense of organizational influence and power. Neither the number of subordinates nor any visible means of power increase, fostering perceptions that the technical ladder might really be less important. The risk is that the technical ladder becomes a *parking lot* for bright technologists whose abilities to generate ideas and advances easily outstrips the capability of the organization for dealing with them. Rewards of freedom and independence can sometimes bring with them feelings of rejection and disconnection.

In addition to these problematic issues, there is also the tendency to *pollute* the technical side of the dual ladder when unsuccessful or failed managers are moved onto it. Even more unfortunate, the criteria for technical promotion often become corrupted over time to gradually encompass not just excellent technical performance but also organizational loyalty, rewarding those professionals who have been *passed over* for managerial promotions. Any of these or similar kinds of practices can transform the technical ladder into a consolation prize, with technical professionals interpreting technical promotions not as a reward but simply as a signal that they aren't good enough to be a manager. Such misuses certainly undermine the integrity of the dual-ladder reward system.

Fortunately, much has been done over the past few years to improve the formal structures of dual-ladder systems in an attempt to alleviate these kinds of problems. Using internal and external peer reviews, organizations have begun *policing* their technical ladders to protect their promotion criteria and prevent the *dumping ground* and other aforementioned abuses. Organizations have tried to strengthen their commitments to the technical side through increased publicity, recognition, career counseling, and information dissemination; through making the ladders more comparable in numbers of people and equivalent perquisites; and through clearer job descriptions, qualifications, responsibilities, performance standards, and reporting relationships. However, most importantly, organizations have taken important steps by encouraging and even training professionals on the technical ladder not only to become involved in the organization's decision-making process but also to influence the organization's technical strategies and overall business direction.

Defining Terms

Career: A set of job activities and positions within a profession for which one trains and is educated.
Dual career: A family situation in which both spouses are actively working.
Hierarchy: A governing body of individuals in job positions organized into orders or ranks, each subordinate to the one above it.
Mentor: A trusted counselor who provides coaching and guidance to a more novice individual.
Model: An abstract representation of the relationships among a set of variables and concepts.
Norms: A set of beliefs and principles that guide and control acceptable behaviors within a group.

References

Dalton, G. W. and Thompson, P. *Novations: Strategies for Career Management*, Scott Foresman, Glenview, IL, 1985.

Katz, R. Organizational socialization and the reduction of uncertainty, In *The Human Side of Managing Technological Innovation*, R. Katz, ed., pp. 25–38, Oxford University Press, New York, 1997.

Katz, R., Tushman, M., and Allen, T. The influence of supervisory promotion and network location on subordinate careers in a dual ladder RD&E setting, *Mgt. Sci.*, 41(5): 848–863, 1995.

Kerr, S. and Von Glinow. M. Issues in the study of professionals in organizations: the case of scientists and engineers, *Org. Beh. Human Perf.*, 18: 329–345, 1977.

Manners, G., Steger, J., and Zimmerer, T. Motivating your R&D staff, In *The Human Side of Managing Technological Innovation*, R. Katz, ed., pp. 3–10, Oxford University Press, New York, 1997.

Organ, D. The subtle significance of job satisfaction, *Clin. Lab. Mgt. Rev.*, 25(4): 94–98, 1990.

Roberts, E. *Entrepreneurs and High Technology: Lessons From MIT and Beyond*, Oxford University Press, New York, 1991.

Roberts, E. and Fusfeld, A. Critical functions: needed roles in the innovation process, In *The Human Side of Managing Technological Innovation*, R. Katz, ed., pp. 273–287, Oxford University Press, New York, 1997.

Schein, E. Individuals and careers, In *Handbook of Organizational Behavior*, J. Lorsch, ed., pp. 155–171. Prentice-Hall, Englewood Cliffs, NJ, 1987.

Schein, E. Career anchors revisited: implications for career development in the 21st century, *Acad. Mgt. Exec.*, 10(4): 80–88, 1996.

Further Information

Badawy, M. *Management as a New Technology*, McGraw-Hill, New York, 1993.

Katz, R. and Allen, T. Managing dual ladder systems in RD&E settings, In *The Human Side of Managing Technological Innovation*, R. Katz, ed., pp. 472–486, Oxford University Press, New York, 1997.

McCall, M. and Lombardo, M. *The Lessons of Experience: How Successful Executives Develop on the Job*, D.C. Heath, Lexington, MA, 1989.

Raelin, J. *Clash of Cultures*, Harvard Business School Press, Boston, 1985.

Von Glinow, M. *The New Professionals: Managing Today's High Tech Employees*, Ballinger, Cambridge, MA, 1988.

19.2 Time Management

Hyrum W. Smith and Lenny Ralphs

Why is it that one technology manager gets more done than another technology manager in the same amount of time? The answer may be found in how well they manage their time. Time is the most important resource we have. **Time management** is the art and science of determining how best to use one's time. Getting things done is "efficient"; getting the right thing done is "**effective**". Ideally, we are both **efficient** and effective with our time.

There are two common myths about time management. The first myth is we believe that we can get more time. We cannot get more time; we have all the time there is. Each of us has only 24 hours a day, no more, no less. Regardless of our status in life or our positions in an organization, we all have the same amount of time in a day. Second, we believe we can save time. We cannot save time, every minute that goes by is gone forever. We have exactly 1440 minutes each day. Once each minute is passed, we can never recover that minute. It is a misnomer to say that we can save time. What we really mean is that some people may be able to make wiser decisions and accomplish more than others in the same amount of time, but no one can save time. We start out each day with 1440 minutes, and at the end of the day it's gone. We cannot store it or accumulate it.

Values, Goals, and Priorities

Ask yourself two questions: (1) What are my highest priorities? (2) What am I doing about them? Simply by answering these two questions you will have a good idea of what you **value** most. What you value most are typically your highest priorities. If you are serious about managing your time most effectively, the place to start is to identify what you value most. Why do this? Because what you value most is the driving force behind every **goal**. Goal setting is the key to achieving productivity on a long-term basis. Breaking down those goals into bite-size steps is a critical process on the way to achieving specific actions on a daily basis. The process of seeing your values acted upon on a daily basis is the key to successful time management. After you've identified what you value most, identify one or two goals in each value area you want to achieve. Once you've identified a goal, you may want to put that through the "SMART" test. SMART is an acronym for "Specific, measurable, action oriented, realistic, and timely". In the book, *The Ten Natural Laws of Successful Time and Life Management*, by Hyrum W. Smith, he defines SMART goals as follows:

Specific — Make sure your goals are specific. If your goal is to increase business, that's too general. You need to be specific about how much you want the business to grow and when the growth will be reached.

Measurable — Goals need to have specific completion dates and criteria for how well they will be achieved. For example, if your goal is to teach the customer service seminars three times by October 24, receiving an average evaluation of 9.0 or better on a 10 point scale; the completion date is October 24, and the specific criteria are three seminars and 9.0 evaluations.

Action-oriented — Some action needs to be taken. Goals are not passive; they are proactive and dynamic. Every well-written goal should have an action verb. In the example above, the action verb is "teach". As a rule of thumb, the best action verbs are those that allow you to picture someone accomplishing the action.

Realistic — Goals need to be realistic. The goals should be attainable, given the resources and constraints in the situation at hand.

Timely — The time allowed for accomplishing the goal needs to be reasonable. If you don't have the time to take on another goal right now, or if you set the completion date too far in the future, you may feel frustrated and ultimately fail.

You are more likely to achieve a goal if you have written down the goal. There is a physiological process our mind and bodies go through that directs us to be more successful in achieving goals when we have written them down. Take time to write down your goals.

Organizing: Desk, Paper, and Files

Managing your Desk

An effective way to organize your desk is to use some type of prioritization approach, such as A, B, C. To start with, identify what you would consider to be the "A" space on the top of your desk. Typically, that's the space right in front of you. Make sure nothing invades your "A" space except for that critical project you're working on at that moment in time. The more you keep things out of that area, the better you focus all of your energy on the task at hand. When you are focusing all of your energy on the task at hand, you will be more effective with your time.

Identify a "B" space, where you put those things that may be needed for achieving your goal but not at that moment. Your "B" space may be behind you on a credenza or some other place in the room. Finally, anything that is peripheral, such as your file cabinets, would be put in your "C" space. The "C" space may be something you need, but only when you work on that critical project in your "A" space.

You may want to use the same idea with setting up the drawers in your desk or your file cabinets. Set aside critical drawers as your "A" drawers, the less critical drawers as your "B" drawers, and the least critical drawers as your "C" drawers.

Paperwork

According to some industry experts, we may have as much as 35 to 42 hours of paperwork within arms reach of our desk and have only 90 minutes to spend on it. We need to make sure that we are much more ruthless in our decision-making process before we spend a lot of time on needless paperwork.

You have probably heard the saying that we should only pick up a piece of paper once. Generally speaking, that's a good rule of thumb; however, oftentimes it's unrealistic. Ideally, if you can make a decision when you pick up a document the first time and take care of it, that is clearly the best way to go. If you can't get it done the first time around, then make sure you don't fall into the trap of continually looking at documents over and over again without making any kind of a decision. When you pick up a document, determine what kind of action needs to take place.

There are four decisions you can make each time you come into contact with any document: (1) Can I act on this right now? (2) Can I **delegate** this to someone else? (3) Can I discard this because it's not worth doing? (4) Can I do this better at a later point in time? Ideally, you want to take action immediately and get it out of your way. Unfortunately, in some situations it's not the best time to take care of it and you must postpone it for another time.

In those situations, make sure that you schedule time to take action, rather than letting it go off into "never-never land". For some things, decide they are not worth doing, they are not a good return on your investment of time. For still other things, just decide you are not responsible for the task and delegate it to someone else.

Files

It's important to have deadlines on files. Too much space is wasted by people putting documents into files they never look at again. When you put something into a file, make sure that you make a decision "before" you put it in the file cabinet that you will use this item again on an ongoing basis. If you are not going to use the item on an ongoing basis, then determine a date and discard it some time in the future. Unfortunately, most of what we put into file cabinets does not get reviewed again. We waste a lot of time and space by putting things into file cabinets because we think we are going to get around to reviewing it later when in fact most of the time we don't. A good rule of thumb is to decide if someone else has the original document. Let them be the keeper of the document so that it doesn't have to occupy your work space as well.

Do your filing on Friday afternoon, just before a weekend or vacation. This will force you to go quicker and be much less tolerant of much of the paperwork that you might have a tendency to keep around.

Planners and Organizers

Using a paper-based **planner** or **organizer** it is extremely helpful in organizing time. One of the most useful things about a planner is that you can capture ideas and thoughts on paper before you forget them. Many people start the day by analyzing their day and making up a list. Once you create a list of things you need to accomplish in a day, that act alone will cause the reticular activating system in your brain to want to achieve those things. Prioritize your list once you have written it down by identifying those things that are most important and the work on those things first. Unfortunately, most planners or organizers are only as effective as the people who use them. In fact, you'll remember the term "GIGO" — garbage in, garbage out. Any planner will only be useful if we put useful information into it so we can, in turn, get useful information.

Delegation

One of the most critical things that we can do in managing our time effectively is to delegate. There's nothing that will burden you down more than to think that you have to do all the tasks yourself. An effective manager is one that learns to delegate. Always be asking yourself the question, "Is there some way that I can delegate this to someone else?" It is particularly important to delegate those things that you determine to be "B" or "C" tasks if possible. Delegating these items frees you up to spend more time on the "A" priorities, the things that are most important to you.

There are two things that you must do in order to delegate effectively: (1) make sure that you clearly communicate, exactly what you expect to have done. Like a hand-off at a football game, if that hand-off is not made in an effective manner, the ball will be dropped. (2) Make sure that you follow-up with the person to whom the task has been delegated. Have an appointed time when he will report back to you so that you ensure the task is accomplished in a timely manner.

Make a commitment to yourself to begin to delegate more, and you will find a benefit of having more time for yourself to work on the things that matter most. It is satisfying to lift some of those burdens off your back.

Time and Technology

One of the better books on using technology to help people be more effective with their time is by Jeffrey J. Mayer called *Winning the Fight between You and Your Desk*. In that book, he tells of a number of things that you can do to interface with the technology of the day and become more effective. For example, he talks about personal information managers (PIMS), computers, and other pieces of technological equipment that have entered into our lives in the last couple of decades. The following are a few suggestions that might be helpful to you regarding technology.

E-Mail/Voice Mail/Fax

You can save much time sending out messages to a group of people by using e-mails/voice mails, and group distribution lists. For example, it's not uncommon to be working with groups of 10 to 30 people where you can very quickly type one e-mail or give one voice mail message that can be distributed within a matter of minutes to all of them. The same principle applies to faxing documents, particularly if you have a program similar to Goldfax, which allows for one document to be sent to a number of machines at once.

Personal Information Managers

There are a number of PIMs on the market today, and new ones are coming out all the time. We highly recommend that you get a PIM that is compatible with your personal computer. The PIMs have the convenience of being portable and yet you can download from them to your personal computer when you are back in your office. In our own company, we are associated with the U.S. Robotics Pilot, which has the Ascend software program we produce loaded directly into the Pilot. Use whichever PIM meets your needs.

Distance Learning/Conferencing

One way to save on travel costs is to use video and teleconferencing. There are locations, such as Kinkos, where you can access these. In the near future, this same technology should be available through Internet and intranet networks.

Internet/Intranets

Probably the most important technology of this decade, perhaps this century, is the Internet. Nothing is growing as fast or having more of an impact as quickly as the Internet. Not only has the concept caught on between individuals and groups, now organizations are formalizing their own intranets to interface with all employees.

Using the Internet may not necessarily be a time saver. In many cases, it can become a time waster because we spend too much time focusing on things that are not important to us. On the other hand, the Internet can be a tremendous time saver because we can access so many new areas of information in a research mode that we were never able to access before. The world is, literally, at our fingertips as a result of this new technology. The key is to focus on what is most important so that you don't waste much time on the Internet focusing on those things that you don't need.

Focus on what you value, whether you are on the Internet, using your computer, through teleconferencing or sending an e-mail. Make sure you have your priorities in balance so that you're most effectively utilizing your time.

Defining Terms

Delegate: To assign a task or project to someone else.
Effective: Doing the right things.
Efficient: Doing things right.
Goal: Something you want to achieve in a certain period of time.
Planner/organizer: A tool, either paper based or electronic, to help one organize time.
Time management: The art and science of determining how best to use one's time.
Value: Something that matters most to you, a high priority in your life.

References

Alesandrini, K. *Survive Information Overload,* Richard D. Irwin, Homewood, IL, 1992.
Mayer, J. J. *Winning the Fight Between You and Your Desk,* HarperCollins, New York, 1993.
Phillips, S. R. The new time management, *Train. Dev. J.,* April: 73–77, 1988.
Smith, H. W. *The 10 Natural Laws of Successful Time & Life Management,* Warner Books, New York, 1994.
Winston, S. *The Organized Executive,* Warner Books, New York, 1983.

Further Information

Time management education

- www.FranklinCovey.com
- www.Daytimer.com
- www.Trainingnet.com

Associations

Association & Society Management, Inc.
1033 La Posada Drive
Austin, Texas 78752-3038

National Association of Professional Organizers
C/O Association & Society Management, Inc.
(512) 206-0151
WWW.CCSI.com/~ASMI/GROUPS/NAPO.html
E-mail NAPO@ASSNMGNT.com

Check your local library for further references.

19.3 Ethics

Richard O. Mason and Florence M. Mason

Ethics is the study and evaluation of human conduct in the light of moral principles. It pertains to all situations in which a moral agent expresses free will and makes choices. Managers, as moral agents, make decisions in order to achieve goals for an organization. These decisions, in turn, materially affect one or more stakeholders — employees, users, customers, etc. — ability to achieve their goals. The affected parties may either be helped or harmed (ethics works for the good as well as for bad or evil), but they are impacted nevertheless. The moral context of a decision is engaged when an agent addresses two central questions: "What action should I take?" and "What kind or organization do I want to create?" These questions are eternal; they deal with the good, the right, or the just. The context in which they must be answered, however, differs with changes in social organization and is affected by science, technology, and other forces. Ethical decision making is the process of identifying a problem, generating alternatives, and choosing the course of action from among the alternatives that realizes ethical as well as organizational values. These basic values include accountability, citizenship, caring, excellence, fairness, honesty, integrity, loyalty, promise keeping, respect for others, and truth telling. Ethical decision making is guided by two great traditions. The duty or **deontological** tradition recognizes that interpersonal relationships create obligations, also called "**prima facie duties**" [Ross 1930], on the part of the actor or agent. Prima facia duties are binding at first sight and include beneficence, nonmalefeasance (do no harm), fidelity, reparation, gratitude, justice, self improvement, and a respect for the autonomy of others. The **utilitarian** tradition holds that a decision should yield the greatest utility, good, or happiness for the greatest number of stakeholders.

Managers and professionals are guided by **codes of ethics**, which are used to help define expected standards of behavior beyond those legally required. Many codes of ethics exist. Businesses such as Johnson & Johnson, Marriott Corporation, and Texas Instruments have codes. Professional organizations also have codes such as the Association of American Association of Engineering Societies, Association for Computing Machinery (ACM), and the American Library Association. Governmental agencies and organizations also use codes including the Ethics in Government Act and the International City Managers Association Code of Ethics. Codes of ethics typically express a set of broad guiding principles and then provide specifics or applications. Most codes provide guidance and address these areas: (1) defining key responsible professional behaviors, (2) exhorting competence in the execution of duties, (3) adhering to moral and legal standards, (4) making public statements, (5) preserving confidentiality, (6) acting in the interest of the customer, (7) developing and maintaining professional knowledge, and (8) prohibiting certain kinds of behavior such as bribery or deception.

Historical Forces on Ethics

As human society has evolved, ethical thinking has responded to several major societal transformations. There was the transition from older, hierarchical Egyptian and Greek societies to the more democratic polis of 5th century BC Athens. This stimulated the reflections of Plato and Aristotle, who redefined the meaning of "the good" and of "virtue". During the 18th century, the polis and agrarian society gave way to the industrialized society. The new focus on mechanization resulted in the rapid substitution of machine-based energy for human labor. Different views of human productivity, automation, and management emerged at this time as did ethical challenges related to managing the human workforce in the mechanized workplace. The immediate post-World War II era saw the emergence of an explicit scientific/technically based society, which valued scientific objectivity and progress as important social goals. This period has also witnessed the emergence of a knowledge- and information-based society, founded on computers and communications technologies. The increasingly pervasive use of computers and information technology has created new organizational forms, flatter structures, more highly networked, with intensive external relationships.

Managerial Ethics

Ethical dilemmas arise from managers' ability to influence or control other individuals or organizations, that is, from the exercise of managerial power that engenders responsibility. The control of organizational information by management, including the capability to produce and handle it, is a fundamental source of managerial power and also a generator of responsibility. For instance, the principle intent of the strategic and marketing use of information technology is to enhance organizational power. Wielding this power, however, may result in considerable help or harm to others. In industry, for example, capturing vital information sources can result in monopolistic power, which, in a free market economy, raises serious questions for managers and for government as to how this power is to be channeled, allocated, and used responsibly. Technically oriented managers assume ethical responsibilities in three roles: (1) as managers, (2) as designers and users of science and technology, and (3) as information handlers and processors. These roles are played out on several levels, individual, professional, organizational, and social. In the human relations area, managerial responsibility falls into three broad areas. Managers are accountable for the activities and the behavior of their employees, must be able to assist employees in identifying ethical implications of their decisions, and serve as role models for their employees. Overall, managers are responsible for the mobilization and allocation of human resources in the organization. This includes decisions on (1) acquisition of personnel — providing for the recruitment, selecting, hiring, and socialization of new members of the organization (ethical hiring practices are based on the applicant's talent and ability, not on prejudice, nepotism, and perpetuation of stereotypes); (2) development — providing for the education, training, and development of personnel; (3) allocation — assigning personnel to jobs and roles; (4) utilization — determining the organizational structure, information and communications flow patterns, leadership style, and motivation scheme for using personnel; (5) evaluation — measuring an individual's contribution to the organization (ethical managers set standards for expected behaviors which are understood by employees); (6) maintenance and compensation — determining the appropriate economic and noneconomic rewards such as salaries, wages, benefits, promotions, disciplining, job conditions, status, recognition, and social incentives necessary to maintain each individual's contribution (and that they apply consistent discipline for similar types of infractions). Ethical disciplinary practices mean seeking constructive methods of applying discipline, conducting disciplinary activities in private, and giving employees input into the disciplinary process; and (7) termination — determining when to fire, lay off, or retire personnel (termination should be fair: based on facts, conducted under conditions of due process, and protective of the dignity of the individual). Each of these decisions requires managers to balance the rights and dignity of each individual employee against the rights of the organization, teams, or groups of employees in which he or she works. Other key ethical areas include creating a more diverse workforce by more open acquisition practices and managing that diversity, defining and abating sexual harassment, managing family and personnel issues in the workplace such as child care, and dealing with employee problems caused by chemical dependencies and substance abuse.

Technology and Scientific Ethics

Ethical dilemmas also arise from the deployment of technology in organizations and from its effect on human processes and social progress. There are two basic views of technology: that technological change is inevitable and may not be under human control — technologically determinism — or that technology is a tool that can be used to accomplish socially desired ends and means — technological optimism. Technological optimism (or pessimism) raises major ethical issues. The implementation of technology is guided typically by human goals and is intended to secure economic and social gains. In the process the flow of benefits and burdens to the organization's stakeholders is changed. This creates two types of ethical issues.

One is to safeguard human concerns in light of advances in technology. Ethical thinking about technology must consider its ability to alter humankind, either to harm human beings, to demoralize them, or, alternatively, to enable them to more fully develop their talents and capabilities. This is the lesson of Mary Shelley's *Frankenstein*. It involves a basic ethical dilemma: exhorting, on the one hand, moral agents to do good for human beings (beneficence), while, on the other hand, avoiding doing harm to them — the principal of nonmalfeasance. Technology ethics includes the use of sociotechnical design principles for designing systems — including jobs — to simultaneously accomplish both technological ends and human needs fulfillment. Important ethical themes include giving full consideration to the appropriate economic costs of pursing technological ends, environment damage as the result of use of technologies, and the assessment of dangers inherent in technology itself.

A second growing area of ethical concern is user-valued and user-centered technological design. A number of writers, including Donald Norman, have established the need for respect for user's dignity, which calls for designs that meet several criteria: are easy and graceful to use, are fault tolerant, and draw heavily on the knowledge already commonly available in the world and in the user's head. A key requirement is to avoid designs that are error ridden, inflexible, lacking in imagination, and insensitive to human needs and desires. Henry Petroski [1994] writes about the importance of understanding failure as it relates to the design process, noting that, "The concept of failure is central to the design process, and it is by thinking in terms of obviating failure that successful designs are achieved."

Finally, an overarching issue is the question of who should have responsibility for technology decisions. During the industrial era, the people highest in the organizational hierarchy made these decisions. In the modern era, however, managers must ensure that these decisions are made under circumstances whereby not only high-level managers and technologists but also stakeholders, including clients, customers, and users, are involved.

Information and Computer Ethics

The most recent area of managerial concern to emerge focuses on the need for ethical decisions related to acquiring, processing, storing, disseminating, and using information. Increasingly, these decisions can affect workers' quality of life. As computer-based work becomes more pervasive and affects customers and users, crucial issues such as privacy, information access, and protection of intellectual property arise. As organizations, and life in general, become more dependent upon computational and informational processes for control, and the quality of human life is greatly influenced by the use of all forms of information technology. Modern technology also processes considerable information about people and their behavior. Most workplaces today are computer oriented and depend heavily upon information manipulation as a business process. Ethical issues include (1) the misuse of computers, including mischief — for example, playing computer games on company time, (2) misappropriation — for example, using the company computer for conducting personal business, (3) destruction of property — "hacking" and trespassing into systems, including introducing software viruses, worms, logic bombs, or Trojan Horses, and (4) computer crime including using systems for theft and fraud. Key ethical issues include the requirement for individual privacy and the confidentiality of information. As information technology is used more heavily, more information is desired by decision makers in business, government, and other organizations. Some of this information may be gathered at the ethical cost of invading individuals' privacy. Significant ethical concerns at the organizational and employee levels arise related to the demand for privacy of individuals from intrusion in the workplace and in their private lives, the right to protection for their intellectual products and labor, and protection from unauthorized use of personal information. Sensitive, sometimes quite intimate information, is capable of being revealed about people to individuals who do not have a legitimate need to know or who are not authorized by the subject party to have it. Managers must seek a balance between their temptation to acquire this data and their obligation to respect the privacy and autonomy of others. Relevant U.S. legislation on this topic includes the Freedom of Information Act of 1966, the Fair Credit Reporting Act of 1970, the Privacy Act of 1974, and the Privacy Protection Act of 1980. This legislation is based on the principle of informed consent and includes requirements that no information about an individual should be acquired on a secret basis, an individual should be able to discover what personal information is being kept about him or her, an individual should be able to correct errors in records about themselves, and an individual who gives consent for the collection of information for one purpose should be able to prohibit its use for any other purposes. If organizations collect and handle personal information, they are expected to assure its accuracy and reliability and to take reasonable precautions to prevent its misuse. The use of new communication technologies such as fax, electronic mail, and the Internet raises new issues about preserving confidentiality of information in the workplace. Questions have arisen as to whether employers have dominion over this technology since they are installed and maintained at the employer's cost, or, whether employees have certain rights to confidentiality and a latitude to use these employer-owned systems as they see fit, in a manner similar to that of the telephone and personal mail communications, which have been legislated as private and confidential. Recent legislation has tended to be somewhat ambivalent in this area. For example, the issue of obscene and pornographic content remains contentious. What individuals' rights to their intellectual property are has also been challenged. "Digital plasticity" — the capability to readily change digital images, text, and forms — is increasingly making all types of digital manipulation possible. Central to the ethical and legal argument is the view that intellectual property belongs to its creator. In the 17th century, John Locke argued that people earn the right to make their works their property by virtue of their physical and intellectual labor. Property is something that can be possessed, controlled, or owned while excluding others from these privileges. Overall, the U.S. copyright laws have upheld that one's intellectual capability and knowledge are protected under law. Now, however, due to the unique attributes of information — it is intangible and mental, symbolic, readily reproducible, transmittable, and easily sharable — it is difficult for the law to effectively control widely practiced human behavior that runs contrary to it. Although one's intellectual property is a source of personal wealth and value, other people are motivated, tempted, and, frequently, able to take intellectual material without compensating its owner. Common practices such as using "clip art" and "clip" sources, scanning correspondence, copying reports, reformatting electronic documents for compatibility purposes, and making electronic tags for retrieval purposes are readily accomplished through the digitization of information. Another issue related to digital plasticity is the ease with which information can be rearranged, edited, and transformed using software tools. This permits images to be readily manipulated into new forms or shapes to produce derivative works. Managers must steward and safeguard their organization's intellectual property and ensure that their organizations and employees respect the property of others. This includes issues such as software piracy, fraud and theft in electronic funds transfers and accounts, and copyright infringements of all types. Relevant U.S. legislation includes the Copyright Act of 1976, the Electronic Funds Transfer Act of 1980, the Semiconductor Chip Protection Act of 1984, the Computer Fraud and Abuse Act of 1986, and proposed computer virus legislation. The accessibility and reliability of available information is another critical ethical issue. It is an ethical responsibility of management to ensure that their organization's information systems are both reliable and accurate. People rely on information to make decisions, decisions that materially affect their lives and the lives of others. People depend on computers, communication devices, and other technologies to provide this information. Information errors can result in bad decisions, personal trauma, and significant harm to other, often innocent, parties. Accordingly, employees, users, and customers are entitled to receive information that is accurate, reliable, valid, and of high quality (at least, adequate for the purpose for which it was produced). Safeguarding these requirement, however, entails a significant opportunity cost. Error-free, high-quality information can be approximated only if substantial resources are allocated to the processes by which it is produced. Consequently, managers must make an ethical trade-off between the benefits received from other allocations of these resources and the degree of accuracy in information they are obligated to provide to meet the legitimate needs of their users. In any case, a certain minimal, socially acceptable level of accuracy is required of all information and information systems.

The combined forces of the technological "push" and demand "pull" of information work to exacerbate ethical issues for managers. The dynamics are as follows: increased use of information technology results in more information being made available. In turn, this increased availability creates its own demand due to the perceived benefits of its use. As a result, more parties rush in to use the information in order to secure its benefits for themselves and are tempted to intrude on individuals' privacy, capture their intellectual property, hoard information, or be careless with its accuracy and use. Thus, information and computer ethics will continue to be central topics for managers in the foreseeable future.

Defining Terms

Codes of ethics: A list of moral principles or duties for an organization or preofssional society that describes the acceptable and expected behavior of its members. It also holds the organization or profession as a whole and its members accountable to the public.

Deontology: The Greek word "Deon" means duty or to bend. Deontological ethical theories center on the "act" taken by the agent and the duties, rights, privileges, or responsibilities that pertain to that act.

Prima facie duties: Duties that are binding "at first sight", before closer inspection, because they are intuitively self-evident. Typical duties include beneficence, nonmalfeasance, justice, gratitude, reparation, fidelity, truth telling, loyalty, and self-improvement.

Utilitarianism: An ethical theory that focuses on "the consequences" of an "act". Attributed to Jeremy Bentham and John Stuart Mill, this theory evaluates consequences in terms of their "utility", or the ability to produce (1) benefits, advantages, good, pleasure, or happiness (2) costs, mischief, pain, evil, or unhappiness. Often summarized as the principle of the greatest net good for the greatest number.

References

Mason, R. O., Mason, F., and Culnan, M. *Ethics of Information Management,* Sage, Thousand Oaks, CA, 1995.

Norman, D. *The Design of Everyday Things,* Doubleday, New York, 1990.

Petroski, H. *Design Paradigms: Case Histories of Error and Judgment in Engineering,* Cambridge University Press, Cambridge, England, 1994.

Ross, W. D. *The Right and the Good,* Oxford University Press, New York, 1930.

Further Information

Hosmer, L. T. *The Ethics of Management,* 3rd ed., Irwin, Chicago, 1996.

Teich, A. *Technology and the Future,* St. Martin's Press, New York, 1993.

19.4 The Human Voice in the Age of Technology

Terry Pearce

The predominate use of technology has redefined communication at its core, causing those of us involved in the field to rethink the distinctions that used to be so easy to make. Now, we have a vast array of choices for message delivery or for dialogue, and it makes the choice of media a taxing decision. Trade-offs between efficiency and effectiveness are even more important to think through, when the cost of e-mail is half that of a telephone call, and a fraction of the cost of a plane ticket to that overseas location.

My field is leadership communication…the kind of communication that inspires others to take action to effect change. Most of the time, I coach people in leadership positions to communicate in a way that generates commitment rather than mere compliance. My clients naturally want to deploy themselves in the most efficient way, travel as little as possible, and use e-mail, voice-mail, and video to their maximum advantage. Lately, however, the proliferation of easy information flow has demonstrated the value of straightforward authentic oral communication. Perhaps the most compelling supportive examples are like the ones reported in a recent *Time* article. Some managers, it seems, are using e-mail to apprise employees of their performance…oh yes, and to criticize them with CAPITAL LETTERS. (i.e., "THIS SIMPLY MUST, REPEAT MUST, STOP"). Clearly such managers are using mechanical means to avoid the real work of the manager, to coach and help the employee become better.

Many studies have reported recently on the number of messages received and sent daily via voice-mail and e-mail; but, whether you believe it is 17 or 70 per person (the range of estimates), there seems to be general agreement that some messages just do not lend themselves to technology. These media may not be well suited to the kind of communication that makes members of your staff want to work that extra hour or come up with that new idea. When trust is needed to enhance performance, there is no substitute for real-time and live human voice.

In this section, I will explore why that is true and why, particularly in the high-tech environment, we need to rededicate ourselves to live human interaction.

What We Hear

Albert Mehrabian of UCLA did the definitive, and oft-cited, research in the 1970s on how people draw conclusions about what someone is saying. Mehrabian showed that we rely on words for only 7% of our judgment, compared to 38% for voice quality, tone, and inflection and 55% from other physical cues. While this research spawned an entire consultant industry (the use and interpretation of body language), the data also suggest that we use a very small percentage of our faculty for making judgments when we are limited to the words (as in the case of e-mail) or even to the voice and the words (as in the case of voice mail). Accordingly, when we confine ourselves to the use of these technologies, we limit the possibilities of deeper and even different understanding with others. Most of us have been guilty of deleting an e-mail or voice mail after the first few words…"yeah, yeah, I know!"

In other words, we frequently confuse hearing the *content* of the message with listening to the *people* sending it.

A fellow consultant was recently putting a leadership client through a "360" review, a process in which the executive is given performance feedback from direct reports, peers, and his or her boss. One of the categories was "listening", and my friend asked her client how he thought his direct reports would rate him. He replied, "I think they will say that I'm a bad listener, but, if you press them, they would have to admit that I always get their content." He was right, my friend confirms. All of his direct reports said that he was a terrible listener, but they all admitted that he generally heard the content of what they were saying.

This case suggests, of course, something we all know to be true. We don't listen to content. We *comprehend* content, but we *listen* to people. When content is the issue, a memo will suffice. Much of oral communication is about *being heard*, not about understanding content.

In the age of the knowledge worker, this distinction might seem irrelevant to some. After all, aren't we really trying to share and transform information as a competitive advantage?

Let's look at the business environment to see what's true.

Loyalty at Work

First, depending on which statistics you read, the unemployment rate for technology workers is anywhere from −15% to −30%. One of every three or four jobs is unfilled at any given time. Further, and more significantly, there is at least a perception that these highly skilled workers are moving from company to company, portraying themselves as bundles of skill for sale to the highest bidder.

Loyalty to a company, it is said, is a thing of the past. In fact, in *The 500-Year Delta*, authors Jim Taylor and Watts Wacker, both bona fide futurists, suggest that loyalty takes the fun out of work. "Work can be fun," they suggest, if we can only "shed the notion that any loyalty is to be given or received in a business relationship, realize that you are a freelancer moving from deal to deal even when you are in someone else's employ, and understand that there is only one person you are working for: yourself. You're the boss…this is freedom" [Taylor and Wacker, 1997].

In this scenario, which is becoming more and more popular with business theorists, independent people move around in parallel play, never really connecting with one another around the business as a whole, but rather connecting like bumper cars in a giant amusement park called the workplace. These players are object oriented, moving in and out of groups to accomplish specific tasks, and then moving on. Business, in this scenario, is merely a holding tank for accomplishment of objectives, and, in a world like this, information is enough. There is a premium on preciseness of input and clarity of task. "Just the facts, please."

However, there is more. It is also true that there has never been more need for loyalty in a company. The competitive advantage based on product innovation is fleeting at best. What used to be called *models* that changed annually are now called *versions* that change daily. The only product advantage that is sustainable is a major leap. Competitors can copy your product or even enhance it in a heartbeat, so the premium is on fast upgrades and quantum breakthroughs.

This kind of production and application of knowledge requires that people work well together, and share knowledge and a sense of urgency. Businesses need commitment, not merely compliance. Everyone has to participate. Ownership is not just a financial concept; it needs to be a psychic reality as well. In order to generate the kind of effort and results that are needed in today's environment, employees need to "own" the business.

So, in a world in which there is little loyalty, loyalty itself becomes a competitive advantage. One of the central questions of business is how to generate that environment of commitment. The question has become so pervasive that Gary Hamel, arguably the defining authority on business strategy today, suggests that the central role of the strategist is not to do the strategic plan or even to try to forecast the products that will be successful. It is to find ways of generating an environment where "new voices can have new conversations, forming new perspectives, to generate new passion and conviction, to foster the raft of new ideas" that will be needed [Hamel and Scholes, 1997]. Innovation itself is the central competitive advantage.

How do we generate the kind of environment where people display conviction and passion?

The Spirit of the Leader

Enter leadership, and enter oral communication.

Like it or not, commitment and passion are spiritual words. They are not generated from the body or the mind. We do not merely "figure out" commitment; it is an integrated phenomena, encompassing our entire selves, mind and heart. Commitment gives rise to passion, the driving and exciting force in which meaning reveals itself. Who has not felt the surge of excitement that comes with knowing that you can actually make a difference in the world? We need only look to our volunteer activities to discover that we are committed to what we believe in, to the causes that create meaning in our lives. Is this too much to ask of our company? Not at all. In fact, great ones provide lots of opportunity for such expression, right in the workplace.

How can this meaning be generated? Can compelling vision be communicated via memo and brochure? Apparently not, even in the best companies. In *Built to Last*, Collins and Porras examined this very point. "Some managers are uncomfortable with expressing emotion about their dreams, but it's the passion and emotion that will attract and motivate others" [Collins and Porras, 1994].

Charles Schwab is one of the most successful growth companies in the financial services business. The company has had a strong vision since its inception: "to provide the most ethical and useful brokerage services in America." Charles Schwab, the man, is an ICON of consumer advocacy, and has always operated the company by the values of fairness, empathy, responsiveness, striving for excellence, team-work, and trustworthiness. Not only are the vision and values written down, they are inscribed on a wall hanging that every San Francisco employee sees when he or she comes to work every day.

In 1994, while the company was growing at a compound rate of more than 20%, it conducted its first employee opinion survey. The results showed three rather startling findings. Many of Schwab's 5000 people didn't know the direction of the company, were not sure that they trusted senior management, and did not think that their voices were being heard.

After the initial shock, David Pottruck, then president and chief operating officer, initiated a 10-month program of oral communication. Every member of Schwab's senior management team was asked to contribute to the renewal of the vision and values. In a series of off-site meetings, every word was discussed, scrubbed, and rescrubbed. This group of more than 80 people then developed a list of ten strategic priorities that were most critical to meeting the new vision and toasted their commitment to these ideals and priorities in a ritual at the Sheraton Hotel in San Francisco.

This was only the beginning. In the ensuing 2 months, Chuck Schwab and Dave Pottruck spoke to every single employee in the company in town hall meetings and through two-way satellite hookups, explaining the vision, values, and strategic priorities and asking each employee to examine his own tasks in the company for alignment. Each employee was given the opportunity to ask questions, probe for sincerity, and find in these two men the authenticity necessary for commitment.

In the next survey taken a few months later, the responses to all three issues — direction of the company, having a voice, and trusting senior management — improved by more than thirty points. The company results speak for themselves. It has since continued its compound rate of growth and it enjoys a rich reputation as trustworthy in an generally selfish industry. Both its employees and its customers are loyal to the firm.

The voice is as distinctive as our fingerprint. The voice carries us with it, the unedited version of what is inside of us. It is traditionally, "like the eyes and the face, a window to the soul" [Whyte, 1994]. It conveys more than the words, it conveys the spirit of the person speaking, and therefore carries with it the meaning in the words being said.

In a world of faster and faster transfer of information, meaning becomes a fleeting luxury. Leaders can only convey the real meaning of their work with the voice.

Trusting One Another

If compelling vision is the fuel for an environment of commitment, real teamwork is the day-to-day grease. William Miller reported on the research of the leader of the laboratories at Hewlett-Packard, Frank Carrubba. When he took over in 1987, Carrubba conducted a study of the most significant differences among teams that failed to achieve their objectives, teams that were successful, and teams that produced extraordinary results. When he compared marginally successful teams with consistently superior teams, he concluded that they were reasonably equal in terms of talent, motivation, and the ability to create realistic goals. As to the difference: 'the teams that really stood out…had a relationship with others that was personal…that allowed people to be themselves, without struggling to represent themselves as being something that they weren't."

Can such relationships be generated with memos? No. Miller sums up Carrubba's findings "…the difference between 'plain' success and extraordinary results was found in communicating with integrity and authenticity" [Miller, 1993].

Taylor and Wacker [1997, p. 113] reach the same conclusion. "Trust counts," they say, as one of the fundamental coins of the new corporate realm. They contend that both clarity and depth are required for trust to develop. They conclude, as I do, that trust is not established through statistics alone, but rather through myth, through live storytelling.

Telling stories, like poetry and other oral traditions, allows us to suspend judgment and listen to an experience as a whole. When was the last time you listened to a storyteller, and then remarked, "I disagree?" Of course, this is nonsensical. Stories are about our human experience. Essentially, they form the foundation of what we trust.

Kouzes and Posner [1995] report some research conducted by organizational sociologists Joanne Martin and Melanie Powers to explore the power of stories.

They compared the persuasiveness of four different methods of convincing a group of MBA students in their study that a particular company practiced a policy of avoiding layoffs. In one situation, Martin and Powers used only a story to persuade people. In the second, they presented statistical data that showed that the company had significantly less involuntary turnover than its competitors. In the third, they used the statistics and the story. And in the fourth, they used a straightforward policy statement written by an executive of the company. The students in the groups that were given the story believed the claim about the policy more than any of the other groups. Other research studies also demonstrate that information is more quickly and accurately remembered when it's first presented in the form of an example or story.

Stories, only once removed from experience, will become more and more important as sources of meaning. As information itself becomes more plentiful and as facts are used to prove and disprove reality, it is the story and the telling of it by a real human being that will become the arbiter of truth.

As we continue down this road of technology, paradox, speed, and chaos, we will meet, head-on, the issues of our own spirit, the way in which we express ourselves as human beings at work. Will we jump from place to place, taking the symbolic rewards, ignoring meaning and the power of relationships? I don't think so.

Personal authentic oral communication will actually be a hallmark of people and companies who lead and thrive in these conditions. Information is a commodity and too plentiful. However, technology will ultimately help us decide which information is really essential to help us with the decisions that we need to make. The real frontier, then, is not what to do with the next and fastest chip, but how to manage the freedom it gives us in an obviously interdependent world. As we are testing the breadth of the application of technology, we will also be plumbing the depths of the human spirit and trying to bring more of our best selves into play in our chosen work. The voice, the ultimate expression of that spirit, will be the central access point.

References

Taylor and Wacker, *The 500-Year Detla*, pp. 207–208, Harper Business, New York, 1997.
Hamel and Scholes, The quest for new wealth. In *Leader to Leader,* Drucker Foundation, 1997.
Collins and Porras, *Built to Last,* p. 234, Harper Business, New York, 1994.
Whyte, *The Heart Aroused,* p. 124, Currency Doubleday, New York, 1994.
Miller, *Quantum Quality,* p. 24, AMA, 1993.
Taylor and Wacker, *The 500-Year Delta,* p. 113, Harper Business, New York, 1997.
Kouzes and Posner, *The Leadership Challenge,* p. 226, Jossey-Bass, San Francisco, 1995.

19.5   Written Communication

Mike Markel

Engineers and managers are writers, whether they want to be and whether they trained to be. Every important idea, proposal, or project has to described in words and graphics, even if only for the record. Most of the time, however, writing is not merely a thing, an artifact at the end of some process. It is the process itself, the means by which people work in an organization. Writing is way people create their ideas, test them, and communicate them to other people. Without writing skills, engineers and managers have nothing to say.

Surveys routinely show the amount of writing engineers and managers do (usually it is from one third to more than half of the job), the importance of writing in the professional's career (usually it is near or at the very top of the required job skills), and the value professionals place on their own communication skills (usually it is very high). (See Beer and McMurray [1997] for a review of some of this literature.) One simple way to get a feel for the importance of writing and other communication skills is to read job ads; rare is the notice that does not state that a candidate needs excellent communication skills.

In this section, I provide an overview of the writing process: the steps you can follow to **plan, draft, and revise** any kind of document. If you think of writing as a process consisting of steps to follow, rather than as the act of waiting for inspiration, you will find that it is a lot closer to the other technical tasks you do every day. Once you realize that workplace writing is a way to think about your job systematically, not an art form, you have cleared the first and most important hurdle.

Let me start with three brief comments about topics I will not be addressing:

- I do not get into any detail on how to write the different kinds of documents: reports, manuals, letters, and so forth. There is no space here. In addition, there are plenty of excellent books that describe these formats (see Further Information). Finally, the best way to learn what the different formats should look like in your organization is to study documents in the files.
- I do not go into any detail on writing collaboratively. We know that most professionals write collaboratively at least sometimes [Ede and Lunsford, 1990]. For suggestions on how to structure writing teams and work together productively, see Markel [1998].
- I do not go into any detail on graphics. Many engineers and scientists think visually rather than verbally, and much of their writing is in fact a series of annotated graphics. If these graphics are well done, that is fine; there is no reason to value words more than pictures. An effective graphic has a purpose in the communication; it is not merely decorative. It is simple and uncluttered, with clear labels and legends. And it is placed in an appropriate location in the document, introduced and (if appropriate) explained clearly in words. One caution about spreadsheet graphics: the people who design them sometimes seem more interested in showing off what the software can do than in communicating clearly to a reader. The result is three-dimensional bar graphs that are almost impossible to read. Stick with the simplest presentations; unless you have a good reason to present a graphic in three dimensions, stay with two. See the text by Tufte listed in Further Information for an excellent discussion of graphics.

Planning

Many writers like to get started right away. They want to get some words on paper, for few things are as disconcerting as blank sheet of paper or screen. However, I recommend that you stop. *Don't* start writing. Start thinking about what you are trying to do. A good portion of the total time you devote to the document — perhaps a quarter or a third — should be devoted to thinking about the direction the document should take. Here are five suggestions for planning the document:

1. Analyze your audience: who they are, what they already know about the topic, what they already think about it, how they are going to read the document, and so forth. Until you can get a clear picture of who is going to be using your document, you cannot make even the most basic decisions, such as what kind of document to write, how long it should be, how it should be structured, and how technical the information in it should be.
2. Analyze your purpose. It's critical at this point to distinguish between the technical purpose of the project and the rhetorical purpose of the document. The technical purpose of the project is a statement of what the project is intended to accomplish. For instance, the technical purpose of the project might be to determine the best system for inventory control at a manufacturing plant. The rhetorical purpose, on the other hand, concerns the document itself: what the document is intended to accomplish. The rhetorical purpose of the document might be to describe three major approaches to designing the system for inventory control and recommend methods for determining

which is best for our company. Notice how this statement of rhetorical purpose contains the embryonic outline of the document: the problem, the three major approaches, the recommendation. You will know you have a good understanding of your rhetorical purpose when you can finish the following sentence: "The purpose of this report [manual, memo, proposal, etc.] is to…" Think in terms of words such as *describe, explain, summarize, analyze, recommend,* and *forecast.*

3. Figure out what you're going to say before you start making sentences and paragraphs. **Brainstorming** involves listing, as quickly as you can, ideas that might belong in the document. By *ideas* I don't mean clearly articulated sentences; rather, I mean mere phrases, such as "software needs", "who has to approve it?", and "down time". You are trying merely to jot down possible topics. Don't censor yourself by worrying about whether the topic will be included in the document or where you will discuss it. Just write it down. One fact about brainstorming is that two or three people brainstorming together create a synergy and generate much more material than one person sitting alone in front of the monitor. A second technique for generating material is to talk with another person. The act of putting your ideas into words forces you to think of things that might belong in the document, and the other person's simple questions — such as "What's wrong with the approach we're using now?" — help you realize important topics you might otherwise overlook because they are too obvious to you.

4. Arrange the material. This step is really two different steps: grouping related topics and figuring out the right sequence. Grouping related material involves going through your list of possible topics and linking related ones. For example, some topics relate to the problem you are investigating, some to personnel needs, some to safety concerns, and so forth. Some writers like to use **sketches** at this point. One common structure for a sketch looks like a tree, with a main trunk branching off to smaller and smaller limbs. Another structure involves writing a main idea in the middle of a page, then adding smaller ideas around it, like planets circling a sun. Of course, each planet can have its own satellites. The advantage of these different sketches is that they are hierarchical but nonsequential: you don't have to determine which topic to discuss first. Determining the right sequence is a question of your audience and purpose. Perhaps your audience expects an executive summary first. Maybe they expect to see the technical details integrated in the body of the document; perhaps they expect to see them at the end, in appendices.

5. Run it past the boss. After you work out the basic contents and structure for an important document, it's a good idea to make sure your supervisor understands and approves of your plans. Write a brief memo describing your audience and purpose and listing the major topics you plan to cover. This way, if your boss has a very different idea of how you should do the document, you'll know it early in the process, when it's easier to change course.

Drafting

Many writers spend the bulk of their time drafting, that is, they craft each sentence carefully. They stop to look up the spelling of words, to check their grammar, to verify an equation or number. In short, they try to write their final draft as they write their first draft. This is a mistake. The purpose of drafting is to turn your outline into words, sentences, and paragraphs. They might not be very good words, sentences, and paragraphs, but you can fix that later, as you revise. In drafting, you just want to see whether you understand what you are trying to say, whether you have a general idea of the points you are trying to get across. Leave the polishing until later.

There are two main reasons to draft quickly. First, at this point in the process, you want to stay focused on the big picture. If you stop every few seconds to look something up or fix something, you will lose your train of thought. You will end up on side streets when you want to be on the highway. Second, you probably write better when you move along at a conversational pace. One of the big problems in workplace writing is that it doesn't flow naturally. If you turn out a sentence every 2 minutes, of course, it's not going to flow. It might be a wonderful sentence — full of information and grammatically correct — but it won't link up with the sentence before it and the one after it. However, if you draft quickly, the sentences will mesh better, the rhythms will be more conversational, and your word choice will be more natural. You will have less editing to do later than you would if you had labored over every sentence.

How do you draft quickly? Here are five suggestions:

- Draft for a certain period of time, without stopping. About an hour or two is the longest most people can stand. However, if you really do draft that long, you can easily turn out a couple of thousand words.
- Don't start at the beginning. Psychology is a large part of writing, and we are very good at psyching ourselves out. Why start at the beginning, where you have to introduce the detailed discussion, which you haven't figured out yet? Instead, begin with a section you know well, and draft it quickly. Move your cursor around from spot to spot on the outline.
- Turn down your monitor. If you are the kind of writer who can't stand a misspelled word, a clumsy sentence, or a missing statistic, get rid of the distractions by turning the contrast knob on your monitor so the screen is harder to see. This way, you'll be less tempted to stop and revise when you should be drafting.
- Use abbreviations. One of the wonderful things about using a computer is that you can do a search and replace. If you are writing about potentiometers, there is no need to keep typing out that long word. Instead, type something like p*. Then, when you revise later, change every p* to potentiometer; you'll have to spell it correctly only once.
- When it's time to stop drafting, stop in the middle of an idea. This can help prevent writer's block when you start drafting again. When you call up the file, you will jump right in and finish the idea you were working on. The mere act of making your fingers move gets your brain moving, too, and you will be less likely to sit and stare than you would be if you had quit drafting at a logical juncture.

Revising

Revising is not the same as proofreading. Too many people think that revising involves reading through the draft, hoping that problems will leap off the page. Some problems will, especially smaller points of grammar and spelling. However, it is a mistake to think that you can read the draft once and do a thorough job revising. To revise effectively, you have to budget much time — maybe a third of the total time you devote to the document.

The following discussion covers ten basic principles of revising:

- Let it sit. Wait as long as you can after drafting before you start revising. Try to let the document sit at least overnight. Your goal is to try to forget what you have written, so that when you look at it again you will be more like a reader and less like its author.
- Get help. No one, no matter how talented, can revise a document as effectively working alone as working with someone else. If possible, have someone read through the draft, then talk to that person about it. A careful reading by someone who is similar in background to the document's intended readers can give you a tremendous amount of information about what works and what doesn't.
- Look for different kinds of problems as you revise. Go through the document several times. Start with the bigger issues. Have you included all the information from the outline? Does the overall point come across clearly? Is the structure appropriate, allowing different kinds of readers to find the kinds of information they need? Is the emphasis appropriate, or do some minor points get too much attention and some major points not enough?
- Use headings. You don't want your document to consist of a succession of paragraphs. Instead, try for two or three headings or subheadings per page. Headings provide the cues that readers need to help them understand where you're going in the paragraphs that follow. Headings help you, too, for they force you to stay on track and they make it unnecessary for you to write awkward sentences such as "Now let's turn to the topic of rapid prototyping."

- Use lists. This section is full of paragraph lists, such as the one you are in now. Lists increase the reader's comprehension because they give a clear and informative visual design to the information they convey. A list is simply a collection of related items, which the eye sees as visually related.
- Use **topic sentences**. Engineers are trained to withhold judgments or conclusions until they have all the facts. This is good engineering, but it can make for bad writing. To help your readers understand the main point of the paragraph, state it up front in a topic sentence. Don't think that the topic sentence has to be self-supporting. It doesn't; it just has to point the direction for the rest of the paragraph, as in the following sentence: "In the new century, new tools for technology management will abound, but they will never replace the manager who has good common sense." Given this topic sentence, the reader expects to see a brief discussion of some of the tools that will be used, followed by an explanation of the continuing need for the wise manager.
- Revise sentences. There is not enough space here to discuss sentence-level matters, of course, but, in general, the most basic sentence structures work best. Try to write sentences so that the subject — the object, person, or idea that you are talking about — comes at the start and the main action is communicated in a clear, simple verb. Instead of writing, "An improvement in our inventory management capabilities is needed," write "We need to improve the way we manage inventory."
- Choose the simplest and most common word. Companies that make vocabulary-building books and tapes want you to think that, if you don't use fancy words, people won't respect you. Do you respect people just because they use big words? Probably not. You respect people who get their points across clearly and quickly. What impresses smart people is good ideas expressed simply. Instead of writing "It is necessary that state-of-the-art communication modalities be utilized," write, "We need to use the most modern communication tools."
- Be careful with style checkers and other revision tools. Your word-processing software includes a number of tools intended to help you improve your writing, including style checkers and a thesaurus. Style checkers are generally more trouble than they are worth, for they produce many false-positives: they accuse you of making errors or using faulty constructions when you probably haven't. Style checkers cannot know whom you are writing to or how you are using the word or construction: they work by recognizing words or constructions that are sometimes troublesome. For instance, style checkers alert you when you start a sentence with the word "but". Some sentences that begin with "But" are awkward or incorrect. But some are perfectly fine. When you use a thesaurus, make sure you understand the connotations of a word that is listed as a synonym. One popular program lists "infamous" as a synonym for "famous." Einstein was famous; Hitler was infamous.
- Be careful with spell checkers. The spell checker tells you whether you have typed a word correctly; it doesn't tell you whether you have typed the correct word. You still have to proofread your document to see that you have spelled the right word correctly.

Finally, a qualifier. The best way to learn how to write well is to write often and to have the writing edited by a competent editor. If your organization has a writing group or a technical publications group, take advantage of its expertise, or try to find a training course or good college course in technical communication. Reading about how to write is useful, but, as with most other skills, you have to practice it.

Defining Terms

Brainstorming: A means of listing topics for a document. Brainstorming is the process of quickly jotting down ideas without worrying about whether they are appropriate or where they will go.

Drafting: The process of turning your outline or notes into rough sentences and paragraphs.

Planning: The process of determining your audience and purpose and figuring out what information will go into your document.

Revising: The process of turning your draft into a polished document. Revising involves studying your document numerous times, looking for different kinds of problems each time.

Sketches: Visual representations of the scope and content of a document. Sketching is a hierarchical representation of document.

Topic sentence: Usually the first sentence in a paragraph; the topic sentence states or forecasts the main point that will be supported in the paragraph.

References

Beer, D. and McMurrey, D. *A Guide to Writing as an Engineer,* John Wiley & Sons, New York, 1997.

Ede, L. and Lunsford, A. *Singular Texts/Plural Authors: Perspectives on Collaborative Writing,* Southern Illinois University Press, Carbondale, IL, 1990.

Markel, M. *Technical Communication: Situations and Strategies,* 5th ed., St. Martin's Press, New York, 1998.

Further Information

Beer, D. *Writing and Speaking in the Technology Professions: A Practical Guide,* IEEE Press, New York, 1991.

Markel, M. *Writing in the Technical Fields,* IEEE Press, New York, 1994.

Robertson, B. *How to Draw Charts and Diagrams,* North Light Books, Cincinnati, OH, 1988.

Tufte, E. R. *The Visual Display of Quantitative Information,* Graphics Press, Cheshire, CT, 1983.

Nagle, J. G. *Handbook for Preparing Engineering Documents,* IEEE Press, Piscataway, NJ, 1996.

Rubens, P. *Science and Technical Writing: A Manual of Style,* Henry Holt, New York, 1992.

19.6   Effective Meeting Management

Mary Munter

EMS. CSCW. CRT vs. **LCD. Meeting** technology seems to change daily. However, you can only benefit from these technologies if you use them wisely. So, the first part of this section will provide the necessary techniques and skills — "the human side of meeting management" — that will enable you to utilize the technologies discussed in the second half of the section — "the technical side of meeting management".

The Human Side of Meeting Management

The next time you are thinking about calling a meeting, always ask yourself, "Is the meeting necessary?" Perhaps the single most prevalent complaint about meetings is that they are called unnecessarily. A good way to start is to state your meeting objective specifically: "As a result of this meeting, we will accomplish _____." Meetings should be reserved for situations in which you need group idea generation, discussion, feedback, and decision making and/or when you are trying to build group identity, relationships, and trust. Meetings should not be for routine announcements or for presenting your own finalized ideas.

Once you have established why you are meeting, here are some guidelines to help you reach your objective: (1) preparation before the meeting, (2) participation during the meeting, and (3) decision making and follow-up after the meeting.

Preparation before the Meeting

Before the meeting, think carefully about your agenda and about delegating meeting roles.

Set the Agenda

Since the whole purpose of a meeting to is elicit information from other people, prepare your agenda carefully and in advance, so that participants can think of ideas in advance.

- *State the objective:* State the meeting objective or meeting impetus as specifically as possible, so that participants will have no doubt about its purpose.
- *Think about timing:* Productivity tends to drop after about 2 hours, or if you have too many topics to cover. Schedule a series of short meetings if the agenda requires more time.

- *Clarify group preparation:* Let participants know how they should prepare, how are they will be expected to contribute. Don't put people on the spot in the meeting. Instead, let them know in advance what will be expected of them — for example, "Think about the pros and cons of this proposal" or "List five ideas before the meeting."

Delegate Roles

Decide which role(s) you are going to perform yourself, and which you will delegate to someone else.

- *Facilitator:* If you have strong feelings about the subject at hand or want to participate actively, you should consider asking someone else to facilitate the discussion. If you choose to facilitate yourself, you must refrain from dominating the discussion: it's difficult to listen to others' point of view when you are trying to convince them of your own.
- *Scribe:* Most meetings need a method to record participant comments during the meeting and to provide a summary at the end of the meeting. In nonelectronic meetings, you may wish to appoint someone else to serve as the scribe to avoid having to talk and write at the same time, to improve legibility, and to save time because you can go on to discuss the next point while the scribe is still recording the previous point.
- *Timer:* You may also wish to appoint someone else to serve as timekeeper because it's difficult to concentrate on the discussion and keep you mind on the time all at once. Going over the time limit, running off on tangents, and losing control of time can be big problems; conversely, controlling the flow too much or cutting people off can also be problems. Think about how you are going to deal with time issues, and, within reason, stick to your decision or group contract on timing.

Participation during the Meeting

Facilitating a meeting demands firmness, flexibility, and a willingness not to dominate the discussion. You need encourage people to participate, yet make sure they don't dominate; you need to record ideas so you can follow-up effectively, yet not get so caught up in your writing that you break the flow of the discussion. Here are some techniques to help you manage this complex set of skills.

Opening the Meeting

At the beginning of the meeting,

- *Set the tone:* Start on time. Get people interested, involved, and enthusiastic by giving a short introduction and then involving them early. The earlier you can get participants involved in some way, the more likely they are to participate.
- *Explain the agenda:* Make sure everyone understands and agrees on the meeting's purpose, impetus, agenda, and decision-making technique. Sometimes, you may wish open the meeting by modifying the agenda or adding discussion items with the group.
- *Get people to agree on ground rules:* Meetings will run much more effectively if everyone agrees explicitly on the ground rules at the outset. If you wait until someone has erred before you clarify the rules, the person will feel humiliated. If, however, you have the rules clear from the start, a brief reminder will usually work. You can either work together with the group to set up ground rules or you can list the ground rules in advance yourself. Examples of ground rules include

  We will start and stop on time.
  We will not interrupt.
  We will stick to the agenda.
  We will show respect for one another and not engage in personal attacks.
  We will treat all information as confidential.

During the Meeting: What Gets Said

Throughout the meeting, use the following skills to encourage everybody's participation:

- *Use good listening skills.* Model good listening skills, especially (1) asking open-ended questions that cannot be answered "yes" or "no", such as "What are your reactions to this proposal?", (2) paraphrasing responses by restating what each person says verbally and by recording all comments, and (3) in nonelectronic meetings, using effective body language, such as a relaxed, open, involved-looking posture, eye contact to show interest, and eliminating distractions such as doodling, tapping your paper, or fidgeting.
- *Show support for every person's right to speak.* Support does not necessarily mean agreement. Instead, showing support means you hear and acknowledge each idea. In fact, you may very well end up hearing contradictory ideas. That is perfectly appropriate in a meeting; you can go back and evaluate the ideas after they are all out on the table. Responses to show support include

  "That idea shows a lot of thought."
  "What do the rest of you think?"
  "Let's consider what Chris has just recommended."

Responses that do not show support include "I disagree" or "That's wrong because…" or "That won't work, because…" Finally, encourage discussion of ideas, not of personalities.

- *Don't talk too much.* It is difficult to avoid dominating a meeting you are running. To control yourself, avoid interrupting, arguing, criticizing, or overdefending; don't talk for more than a couple of minutes; ask other people to contribute; ask someone else to present background information; ask questions instead of answering them; talk with them, not to them. You have decided to call a meeting; therefore, don't deliver a lecture or presentation.
- *State disagreements carefully.* Remember to disagree with ideas — not with the individual person. Say, for example, "That project may be very time-consuming" instead of "The project Tom has proposed will take too long!" Disagree by saying "I'm not comfortable with" or "What concerns me about your ideas…" instead of "I disagree with" or "That won't work because."
- *Avoid dominance by any one person or subgroup.* Here are some techniques to avoid letting any one person or group dominate the discussion. (1) Give each person a chance to speak to the issue, saying something along the lines of "Let's hear from those who haven't spoken yet," but don't put them on the spot by calling on them by name. (2) Use a firm but tactful reminder of the ground rules, such as "Wait, Elizabeth. Remember, we agreed on no interrupting" or "Excuse me, Darcy, but we need to keep our remarks brief so everyone has the chance to talk." (3) Talk to disruptive, very verbal, or high-status people privately before or after the meeting avoid a direct confrontation in front of the group. Try to understand them and enlist their help in making the next meeting more productive.

During the Meeting: What Gets Written

You should almost always record ideas during a meeting. Why?

1. *For accuracy:* Recording ideas publicly ensures accuracy and a record of what was said.
2. *For morale:* Recording ideas publicly also makes people feel heard and appreciated, even if their ideas are rejected later.
3. *For timing control:* Besides making everyone feel included and heard, writing on visuals may be used to control timing. If people talk too long or repetitively, recording their points may assure them of feeling heard, so you can move on.
4. *For a permanent record:* From what you recorded, you can come up with a permanent record after the meeting — either electronically or by writing minutes.

In nonelectronic meetings, writing to record participants' ideas is not easy. Here are four suggestions to improve your effectiveness:

- *Record accurately:* To check the accuracy of what you record, either check with the speaker verbally (e.g., "OK?" or "Is this what you mean?") or nonverbally (e.g., questioning eye contact).

- *Record essential phrases only:* Make sure you record essential words and phrases only. Don't bore your audience by laboriously writing long, complex sentences.
- *Include everybody's comments:* Write down all comments — not just some people's. You don't want to appear to be ignoring people or leaving people out. Recording their ideas does not necessarily mean you agree with them; instead, it means you hear and acknowledge them. You may very well end up recording contradictory ideas; you can go back later and evaluate the comments.
- *Make the notes readable:* Keep them in full view of the group, so everyone can see what's going on. Make sure your writing is large enough to be read by all. Write neatly or use a scribe.

Decision Making and Follow-Up After the Meeting

Don't waste all the valuable ideas you have gained from the meeting participants: use the following techniques to make a decision and to follow up.

Decision Making

Some items on your agenda may require a decision. Choose, or have the group choose, how decisions will be made — and make it clear to the participants in advance which method you plan to use. Keep in mind that decision-making methods vary in different organizations and cultures.

- *By one person or majority vote:* These two methods are quite fast; they are effective when the decision is not particularly important to everyone in the group or when there are severe time constraints. The disadvantage of these methods, however, is that some people may feel left out, ignored, or defeated — and these people may later sabotage the implementation. On the other hand, most people do not mind serving in an advisory board capacity only, as you make it clear to them in advance.
- *By consensus:* Consensus means reaching a compromise that may not be everybody's first choice but that each person is willing to agree on and implement. Consensus involves hearing all points of view and incorporating these viewpoints into the solution, so it is time consuming and requires group commitment to the process. Unlike majority rule, consensus is reached by discussion, not by a vote. For example, the facilitator might ask "Do you all feel comfortable with this solution?" or "Seems to me we've reached consensus around this idea. Am I right?" Consensus does not mean unanimity; every participant does not have veto power. If one person seems to be the lone holdout for a position, say something like "Well, Fran, we understand your point clearly, but the rest of us don't want that solution. Can you live with this one instead?"

Follow-Up

How you end the meeting can be the most crucial key to success: all the time and effort spent on the meeting itself will be wasted if the ideas are not acted upon.

- *Permanent record:* Most meetings should be documented with a permanent record of some kind, usually called the "minutes", to record what occurred and to communicate those results. The nature of the minutes will vary, given the meeting's purpose; ineffective minutes are either too detailed or too general; effective minutes include the issues discussed, the alternatives considered, the decisions reached, and the action plan. You can write the minutes yourself, appoint someone else at the beginning of the meeting, or use the electronic methods discussed later. Eventually, participants should receive hard or electronic copy of the minutes and the action plan.
- *Action plan:* The group should agree to an action plan, to include (1) what actions are to be taken, (2) who is responsible for each action, (3) the time frame for each action, and (4) how the action will be reported back to the group. A good way to start your next meeting might be to present an update on the previous meeting action plan.

The Technical Side of Meeting Management

Only after you have thought through the human issues described above are you ready to tackle the technical issues described in the following section. Think about the advantages and disadvantages of each option discussed below to make an informed decision or to capitalize on their advantages and overcome their disadvantages.

When you are choosing technologies, however, always consider four general questions first.

1. *Group expectations:* What are the expectations of your particular group, organization, and culture? Some groups expect and always use a certain kind of equipment. For some groups, certain kinds of equipment would be perceived as too slick, flashy, or technical. For other groups, not using the latest technology would be perceived as too old-fashioned.
2. *Timing and location issues:* Do you need people to participate at the same or different times? Are they in the same or multiple locations?
3. *Group size:* Consider also the number of people in your audience. For example, a group of 3 might be easier to convene face to face; a group of 15 might be easier to convene electronically.
4. *Resource availability:* Finally, be realistic about the equipment and resources you have available.

Then, consider the following meeting options: face-to-face meetings (same site) or **groupware** meetings (teleconferences, e-mail meetings, or electronic meetings).

The Low-Tech Option: Face-To-Face Meetings

Use the lowest-tech option — the face-to-face meeting — when nonverbal, human contact is especially important.

- *Advantages of face-to-face meetings:* Face-to-face meetings are appropriate (1) when you need the richest nonverbal cues, including body, voice, proximity, and touch, (2) when the issues are especially sensitive, (3) when the people don't know one another, (4) when establishing group rapport and relationships are crucial, and (5) when the participants can be in the same place at the same time. Because they are less technologically complex, same-site systems are easier to use, less likely to crash, and less likely to have compatibility problems.
- *Disadvantages of face-to-face meetings:* Such meetings (1) do not allow the possibility of simultaneous participation by people in multiple locations, (2) can delay meeting follow-up activities because decisions and action items must be written up after the meeting, and (3) may be dominated by overly vocal, quick-to-speak, and higher-status participants.

When you are meeting face to face, choose from among various technologies — flip charts, traditional boards, copy boards, electronic boards, or projectors.

Option 1:   Using Flip Charts and Nonelectronic Boards

Flip charts include large standing charts and small desktop charts; nonelectronic boards include blackboards and whiteboards. Charts and boards often elicit a great deal of group discussion because they are so low-tech that they are nonthreatening and unintimidating to certain kinds of participants and because they work in a brightly lit room. Additional advantages of flip charts are that they can be taped up around the room for reference during the session and taken with you for a permanent record at the end of the session. On the downside, they may appear unprofessional and too low-tech for certain groups; they cannot show complex images; and they are too small to be used for large groups. Additional drawbacks of nonelectronic boards are that you must erase to regain free space and they do not provide hard copy.

Option 2:   Using Electronic Boards

Electronic boards come in two varieties. Electronic "copy" boards look like traditional white boards, but they can provide a hard copy of what was written on them — unlike nonelectronic boards, which need to be erased when they are full. Electronic "live" boards provide digitized hard copy of documents, computer files, and other two-dimensional objects, and can be annotated real time.

Option 3:   Using Still Projectors

You can also annotate on two kinds of still projectors, although some people associate such projectors with noninteractive presentations, so they may be less likely to talk. One option, overhead projectors, uses acetate sheets and special marking pens (which are not the same as flip-chart markers). A second option, document cameras, also known as electronic overheads or visualizers is more convenient because it uses regular paper (or any two-dimensional or three-dimensional object) instead of acetate slides, but the resolution is not as high as that of an overhead projector.

Option 4:   Using Multimedia Large-Screen Projectors
Multimedia projectors (for computer or video images) come in three varieties. (1) "Three-beam projectors" have the best quality and resolution, but are not portable. (2) Self-illuminating LCD projectors have moderate quality and resolution (although they are improving); some are medium weight and compact enough to move. (3) LCD projectors that use an overhead projector as a light source have lower-quality resolution than the others, but they are the easiest to move.

Groupware Meetings (Usually in Multiple Sites)

Unlike the face-to-face meetings discussed so far, the following higher-tech kinds of meetings use different kinds of "groupware" — a broad term for a group of related technologies that mediate group collaboration through technology — that may include any combination of collaborative software or intraware, electronic and voice-mail systems, electronic meeting systems, phone systems, video systems, electronic boards, and group document handling and annotation systems — variously known teleconferencing, computer conferencing, document conferencing, screen conferencing, GDSS ("group decision support systems"), or CSCW ("computer-supported cooperative work"). Groupware participants may communicate with one another through telephones or through personal computers with either individual screens or a shared screen.

- *Advantages of groupware:* Groupware meetings are especially useful (1) for working with geographically dispersed groups — because they give you the choice of meeting at different places/same time, or different places/different times, or the same place/same time, or same place/different times and (2) for speeding up meeting follow-up activities because decisions and action items may be recorded electronically.
- *Disadvantages of groupware:* They (1) lack richest nonverbal cues of body, voice, proximity, and touch simultaneously, (2) are not as effective when establishing new group rapport and relationships that are crucial, and (3) may be more difficult to use and more likely to crash than low-tech equipment, such as flip charts and overheads.

Groupware Option 1:   Teleconferencing
Teleconferencing may take place on a large screen in a dedicated conference room with a dedicated circuit, on television screens in multiple rooms, on desktop computers with a small camera installed, using a shared resource such as the Internet, or on telephones.

Audioconference participants hear one another's voices. Videoconference participants see and hear one another on video. In addition, they may also be able to (1) view documents and objects via document cameras and (2) see notes taken on electronic live boards (some of which can be annotated in only one location, but viewed in multiple locations; others of which can be annotated and viewed in multiple locations, and may have application-sharing capabilities as well).

- *Advantages of teleconferencing:* Teleconferences are useful (1) when the participants are in different places, but you want to communicate with them all at the same time, (2) when you want to save on travel time and expenses, (3) when you want to inform, explain, or train (as opposed to persuade or sell), (4) when you do not need the richest nonverbal cues, such as proximity and touch, and (5) in the case of audioconferencing, when vocal cues are sufficient to meet your needs, you do not need the richer nonverbal cues of body language, or you need a quick response without much set-up time.
- *Drawbacks of teleconferencing:* (1) They are usually not as effective as face-to-face meetings when you need to persuade or to establish a personal relationships, (2) fewer people tend to speak and they speak in longer bursts than in other kinds of meetings, and (3) in the case of audioconferencing, they do not include visual communication or body language.

Groupware Option 2: E-mail Meetings

E-mail meetings differ from other kinds of meetings in several ways. They differ (1) *from electronic meetings* because participants respond at different times and their responses are not coordinated through a facilitator, (2) *from teleconferencing* because responses usually consist of words and numbers only, with no nonverbal cues — except those provided by "emoticons" such as : -) and : - (or unless participants have the ability to scan in documents or animation, (3) *from face-to-face meetings* because participants may be in different places at different times and because they lack the richest nonverbal cues of body, voice, proximity, and touch.

- *Advantages of e-mail meetings:* At its best, e-mail can (1) increase participation because it overcomes dominance by overly vocal and quick-to-speak participants, (2) increase communication across hierarchical boundaries, (3) decrease writing inhibitions — using more simple sentences, brief paragraphs, conversational style with active verbs and less pompous stiff language than in traditional writing, (4) decrease transmission time when you are circulating documents, and (5) speed up meeting follow-up activities because all decisions and action items are recorded electronically and can be distributed electronically.
- *Disadvantages of e-mail:* At its worst, email can (1) decrease attention to the person receiving the message and to social context and regulation, (2) be inappropriately informal, (3) consist of "quick and dirty" messages, with typos and grammatical errors, and, more importantly, lack of logical frameworks for readers, such as headings and transitions, (4) increase use of excessive language and other irresponsible and destructive behavior — including blistering messages sent without thinking, known as "flaming", and (5) overload receivers with trivia and unnecessary information. Two final problems with e-mail are that the sender cannot control if and when a receiver chooses to read a message and that, because the message is electronic, it is less private than hard copy.

Groupware Option 3: Electronic Meetings

Unlike e-mail meetings, electronic meetings utilize a trained technical facilitator who must be proficient at using EMS and at managing group control. Electronic meetings automate all of the traditional meeting techniques — such as brainstorming (sometimes called "brainwriting" when performed on-line), recording ideas on flip charts and tacking them to the wall, organizing ideas, ranking ideas, examining alternatives, making suggestions, casting votes, planning for implementation, and writing meeting minutes — all electronically.

Participants "meet" using a keyboard and screen, usually from their individual workstations, with information appearing and becoming updated real-time on all participants' screens simultaneously. However, in addition to the possibility of being at their workstations (different places/same time) participants can be in a dedicated room with a large screen (same place/same time), or at either place participating at their convenience (same place/different time or different place/different time).

Participants communicate by viewing and manipulating words, numbers, documents, spreadsheets, and shared electronic files and sometimes computer-generated graphics, presentations, or images.

- *Advantages of EMS:* Electronic meetings, (1) like all the options except face-to-face meetings, are useful when the participants are geographically dispersed or when you want to save on travel time and expenses, (2) unlike e-mail, are mediated and usually data are entered and viewed by all participants simultaneously, but can be used by participants at different times at their convenience, (3) can maximize audience participation and in-depth discussion because everyone can "speak" simultaneously, so shy members and are more likely to participate and the "vocal few" are less likely to dominate the discussion, (4) unlike any other option, allow the possibility of anonymous input, which may lead to more candid and truthful replies, equalize participants' status, and increase participation among hierarchical levels (with some programs offering the choice to self-identify or not and others making it impossible for even the facilitator to tell each participant's identity), (5) can lead to better agenda management and keeping on task, (6) can generate more ideas and alternatives quicker than with a traditional note taker, and (7) can provide immediate documentation when the meeting is finished.

- *Disadvantages of EMS:* EMS (1) cannot replace face-to-face contact, especially when group efforts are just beginning and when you are trying to build group values, trust, and emotional ties, (2) may exacerbate dysfunctional group dynamics, and increased honesty may lead to increased conflict, (3) may make it harder to reach consensus, because more ideas are generated and because it may be more difficult to interpret the strength of other members' commitment to their proposals, (4) depend on all participants having excellent keyboarding skills to engage in rapid-fire, in-depth discussion (at least until voice recognition technology, converting dictation into typewritten form, is more advanced), and (5) demand a good deal of preparation time and training on the part of the facilitator.

From the past, when people met in caves, to the future, when holograms will meet in collaboratories, the latest technological changes are only as effective as the people using them. So, use the skills from the first section of this chapter to enable you to use technologies from the second section most effectively and productively.

Defining Terms

EMS: "Electronic meeting systems", a type of groupware that allows you to perform meeting tasks electronically.

Groupware: A broad term for a group of related technologies that mediate group collaboration through technology.

Meeting: A situation in which you want to (1) elicit group ideas, discussion, feedback, or decision making and/or (2) build group identity, relationships, and emotional ties — not to present your own finalized ideas.

References

Aiken, M. and Chrestman, M. Increasing quality and participation through electronic meeting systems, *J. Qual. Partic.*, 18: 98–101, 1995.

Chisholm, J. Electronic meeting systems, *Unix Rev.*, 12: 13–18, 1994.

Coleman, D. and Raman, K. *Groupware: Technology and Applications*, Prentice Hall, Englewood Cliffs, NJ, 1995.

Dumville, J. and Breneman, G. When to use videoconferencing, *Inform. Strat. Exec. J.*, 13(1): 22–25, 1996.

Kettelhub, M. How to avoid misusing electronic meeting support, *Plan. Rev.*, 22: 34–38, 1994.

Langham, B. Mediated meetings, *Success. Meet.*, 44(1): 75–76, 1995.

Levasseur, R. *Breakthrough Business Meetings: Shared Leadership in Action*, Bob Adams, Holbrook, MA, 1994.

Munter, M. *Guide to Managerial Communication*, 4th ed., Prentice Hall, Upper Saddle River, NJ, 1997.

Price, R. An analysis of stylistic variables in electronic mail, *J. Bus. Tech. Commun.*, 9(4): 5–23. 1997.

Snyder, J. Sizing up electronic meeting systems, *Network World*, 12: 29–32. 1995.

Further Information

Association for Business Communication. Robert J. Myers, Executive Director, Baruch College, Department of Speech, G1326, 17 Lexington Avenue, New York, NY 10010.

Journal of Business Communication, Journal of Business and Technical Communication, Management Communication Quarterly.

Lloyd, P. *Groupware in the 21st Century*, Praeger, Westport, CT, 1994.

19.7   Lifelong Learning within the Learning Organization

David Birchall

Two expressions, lifelong learning and the learning organisation, have entered our everyday language during the 1990s. They are both a response to perceived business needs as we progress into the knowledge economy.

The rate of change of technology is an obvious stimulus for personal development as employees strive to maintain technical competence. However, the changing nature of business is requiring technologists at all levels in organizations to exhibit a broader range of skills to operate effectively.

Here we explore some of the changes taking place and their implications for personal development.

The Need for Lifelong Learning

Human beings have an infinite capacity for learning and the ability to keep on applying newly acquired knowledge in novel ways. So, why now the interest in the concept of lifelong learning?

There are many estimates of the explosion of knowledge. Whether we only know $1/35$th of the knowledge that will be known by the year 2030 or whether it is $1/20$th or $1/50$th is of little significance here. What is obvious is that what is known and understood about the universe in which we live is increasing, possibly exponentially. Business is one of the drivers of this expansion of knowledge as it searches for new and cheaper products to meet evermore-demanding customers who have a wider choice on which to spend their increasing disposable income.

One consequence is the increasing speed at which knowledge becomes redundant. In some areas of computing the half-life of knowledge may already be as low as 4 to 6 months. In work environments such as this, it is clear that those who wish to remain employed in these industries must themselves be not only acquiring new knowledge to keep up with the change but also, as the knowledge base explodes, be acquiring knowledge in greater magnitude and depth.

Allee [1997] recently wrote "security lies in what you know how to do, what you can learn to do, and how well you can access knowledge through collaboration with others." She was referring in this statement to job security, which she, along with many others, sees as something now consigned to our past.

So, it seems highly likely that people will be engaged by organizations for the currency of their knowledge and ability to apply that knowledge for the benefit of their employer. Organizations are less and less likely to keep people employed as a reward for loyalty and service. Competition faced by business is driving out those whose contribution is marginal.

In order to enjoy a prosperous lifestyle in the next decade, there is no doubt that people will have to be proactive in searching out new learning opportunities, will have to learn how to make good use of this learning in the workplace, and will have to ensure that they make their employers aware of their particular knowledge and expertise and their potential contribution. As we indicated earlier, the human race has a built-in capacity for learning, but what now appears necessary is the acceleration of the process. Learning how best to learn seems an essential aspect of this. Learning the most effective ways of using knowledge to personal advantage is also important for career success.

This process of more systematic learning is the basis of lifelong learning. Focusing the learning so as to open up new avenues for self-fulfilment and personal reward underpins the concept.

Changing Workplace Demands

Those involved in the field of technology management are well conversed with many of the issues facing our organizations within the rapidly changing competitive environment. Few organizations can now claim anywhere near complete internal mastery of the technologies needed to deliver their products and services and to develop new offering. Daily we read about the creation of strategic alliances, joint ventures, partnerships and other interorganization arrangements to enable organizations to gain access to the technologies needed for the creation of new or upgraded products or services. A company such as IBM has as many as 800 strategic alliances. This is despite the inherent difficulties in creating and managing such ventures.

At the same time consumers, in particular, are increasingly exercising their rights to redress where products or services fail to meet specification. This is requiring more sophisticated systems and procedures to assure quality. For many who are used to more laissez-faire methods of working, this represents an increased bureaucracy.

Increasing educational levels, as well as impacting on the business relationship with the consumer and business customers, is impacting on the labor force. Employees expect organizations to behave differently than would have been the case even 20 years ago. Moves to "empower" front line and other staff reflect — on the one hand — the rising expectations of staff but — on the other — the need to continuously reduce costs and be more responsive to customers.

What do the changes highlighted here mean for the nature of work? New styles of working, accompanied by new competencies, are required by all levels in organizations.

Also, as "value" is stressed by organizations, individuals will be expected to understand how their contribution "adds value" to the process and be personally focused on increasing that value added. This implies a general understanding of management principles and practice. Currently, this is possibly no more apparent than within corporate research and development (R&D) where organizations are pressing for clearer value for money resulting from investment. Project evaluation and management with a clear customer focus has been driven through many business corporate R&D divisions. Many have become profit centers; many others are set up as separate companies, stating as a principal aim the introduction of a clear business culture.

It is clear that much more emphasis will be placed on the individual's capability to contribute not only technical expertise but to the broader organizational need. This is likely to involve the management of relationships as more and more we recognize a customer for our own activity and also we become a customer for someone higher up the process or value chain. So everyone is likely to be required to demonstrate some elements of management competence. The author has provided more details of these competencies elsewhere [Birchall, 1996].

The Changing Organization

It is interesting to attempt predictions of what organizations will look like in, say, 5 years' time. It is also interesting to look back at changes that have taken place over the last 5 years. As an example, the financial services industry is seeing the impact of deregulation, technology, and global competition leading to new entrants into the territory formerly controlled by banks and insurance companies. Supermarkets, oil companies, airlines, utilities, media companies, retail stores, and others offer competitive services such as credit cards, insurance over the phone, off-the-shelf pensions, Internet banking, and the like. Many other industries are facing similar change.

Prusak [1996] recently wrote "the only thing that gives an organization a competitive edge — the only thing that is sustainable — is what it knows, how it uses that it knows and how fast it can know something new." Many organizations in recognizing this have also questioned why they need many of their fixed assets and overhead items. Many have recognized that their business is not really real estate management or the running of computer systems or recruitment of personnel, hence, the tendency to downsize and the rapid growth in outsourcing. Many organizations have set about identifying their core business and core competencies. They have then sought to maximize the return on investment from these strengths. This has often taken them into new business areas, e.g., Lockhead Martin applies logistics systems designed for military application in running postal services, supermarkets use the database generated from the use of loyalty cards to target new products and services to specific groups of customers, and the automanufacturers are heavily involved in financial services arranging customer finance, a business probably more profitable than manufacture.

As organizations seek to improve their return to shareholders, they are introducing more flexible operations. For the employee this may mean flexible hours of work, flexibility in the location of work (hot desking and less regular use of an office, teleworking), and more flexible contracts (annualized hours rather than fixed days, weeks, or months). It may result in more work being subcontracted and outsourced. The network of suppliers may become partners with their numbers reduced (see, for example, Lamming's [1993] account of the automotive industry). The organization may be moving towards a more flexible form so as to be more responsive to customer need (see Birchall and Lyons [1995] for an explanation of the network organization).

These changes are, in part, designed to make the organization more responsive to customer needs. However, in addition to a major rethink about the structural aspects of organization, many businesses are striving to become what Senge [1990] in his seminal work described as "learning organizations". This is an organization that has strong competence in creating, acquiring, and transferring knowledge as well as changing its behavior to reflect new knowledge and understanding. This form of organization is capable of continuous adaptation at the level of individuals, teams, and the organization itself as well as across its supply network. An important aspect of this is to understand the assumptions being made within the organization — "we do it this way because...." and challenging them. Also important is a culture of openness, risk taking, and reflection to learn from the actions taken.

Within the learning organization there is a strong emphasis on personal development and a commitment to lifelong learning for all. However, there is also a clear corporate vision and direction guiding the learning processes.

The Individual and Lifelong Learning

The last section began by suggesting that predicting organizational futures is an interesting exercise. For those committing to lifelong learning who wish to remain employed within fast-moving sectors of the economy, this is a good first step in defining the areas for personal development. However, a more fundamental question is "what do I want to be doing in 5 to 10 years time?" Identifying personal development needs is a logical step in the process, but this does depend upon a personal audit of competencies against an ideal for a particular or future role.

Competence frameworks can help in this process (see for example, Birchall [1996]), but these often tend to focus on the skills necessary to operate in today's business environment rather than looking ahead. They may need some interpretation.

Having identified areas for development, choices have to be made. The opportunities available for personal development range from the more familiar forms of educational program (classroom based) to distance learning (with or without electronic delivery and support) to schemes of work-based learning. This includes much distance learning material free of charge on the World Wide Web. However, it is necessary to recognize that personal competencies are more than knowledge acquisition. Competence implies an ability to use that knowledge at the workplace. This involves a range of personal competencies (interpersonal skills, negotiations, problem solving, etc.) in addition to knowledge of the technical aspects of the task. Much "real" learning takes place through attempts to apply knowledge and then learning from mistakes. Openness to feedback and review is an important element in this process.

Gaining insights into how one learns can also aid the process significantly. Some people learn from reading and reflection, some from debate, and some from active participation in a project followed by review. Understanding how effective learning is achieved can accelerate the process. Planning and organization, an external commitment to personal goals and regular review, are found essential by many learners as a way of bringing discipline to the process.

Another important consideration for the individual lifelong learner is being awarded credit for the learning achieved. In an era of low job security, transferable qualifications become more important. Additionally, the reputation and standing of the awarding body becomes more of an issue as competition for the higher-paid posts becomes more intense.

Conclusions

The case for adopting lifelong learning is indisputable. The need for organizations to become learning organizations is also widely recognized. However, just as organizations are struggling to interpret the concepts and apply them appropriately, for the individual, the adjustment in thinking and behavior is also in many cases a major challenge.

It requires the adoption of a more long-term view and a more strategic approach to personal development. Understanding the learning process and one's own preferred style can assist. However, to be successful, it clearly requires personal investment in accepting the challenge and devoting time. However, in an increasingly competitive business environment, the personal curriculum vitae and track record on recent projects will be key to gaining future work. Performance is underpinned by personal competence and subject mastery. Therefore, lifelong learning has to become a natural part of our lives.

References

Allee, V. *The Knowledge Evolution — Expanding Organizational Intelligence,* Butterworth Heinemann, Oxford, 1997.

Birchall, D. W. *The New Flexi Manager,* International Thomson Business Press, London, 1996.

Birchall, D. W. and Lyons, L. *Creating Tomorrow's Organization — Unlocking the Benefits of Future Work,* FT Pitman, London, 1995.

Lamming, R. *Beyond Partnership — Strategies for Innovation and Lean Supply,* Prentice Hall, Hemel Hempstead, 1993.

Prusak, L. The knowledge advantage, *Strat. Leader.,* 24: 1996.

Senge, P. *The Fifth Discipline — the Art and Practice of the Learning Organization,* Doubleday, New York, 1990.

20

Reliability and Maintainability

Rama Ramakumar
Oklahoma State University

B. S. Dhillon
University of Ottawa

Anil. B. Jambekar
Michigan Technological University

Karol I. Pelc
Michigan Technological University

20.1   System Reliability

Rama Ramakumar

Application of system **reliability** evaluation techniques is gaining importance because of its effectiveness in the detection, prevention, and correction of failures in the design, manufacturing, and operational phases of a product. Increasing emphasis on the reliability and quality of products and systems, coupled with pressures to minimize cost, further emphasize the need to study and quantify reliability and arrive at innovative designs.

Reliability engineering has grown significantly during the past 5 decades (since World War II) to encompass many subareas, such as reliability analysis, failure theory and modeling, reliability allocation and optimization, reliability growth and modeling, reliability testing (including accelerated testing), data analysis and plotting, quality control and acceptance sampling, maintenance engineering, software reliability, system safety analysis, Bayesian analysis, reliability management, simulation, Monte Carlo techniques, and economic aspects of reliability.

The objects of this section are to introduce the reader to the fundamentals and applications of classical reliability concepts and bring out the importance and benefits of reliability considerations.

Catastrophic Failure Models

Catastrophic failure refers to the case in which repair of the component is not possible, not available, or of no value to the successful completion of the mission originally planned. Modeling such failures is typically based on life test results. We can consider the "lifetime" or "time to failure" T as a continuous random variable. Then,

$$P\left(\text{survival up to time } t\right) = P\left(T > t\right) \equiv R\left(t\right) \tag{20.1}$$

where $R(t)$ is the *reliability* function. Obviously, as $t \to \infty$, $R(t) \to 0$ because the probability of failure increases with time of operation. Moreover,

$$P\left(\text{failure at } t\right) = P\left(T \leq t\right) \equiv Q\left(t\right) \tag{20.2}$$

where $Q(t)$ is the unreliability function. From the definition of the distribution function of a continuous random variable, it is clear that $Q(t)$ is indeed the distribution function for T. Therefore, the failure density function $f(t)$ can be obtained as

$$f\left(t\right) = \frac{d}{dt}Q\left(t\right) \tag{20.3}$$

The **hazard rate function** $\lambda(t)$ is defined as

$$\lambda\left(t\right) \equiv \lim_{\Delta t \to 0}\frac{1}{\Delta t}\left[\begin{array}{l}\text{probability of failure in }\left(t,\ t+\Delta t\right),\\ \text{given survival up to } t\end{array}\right] \tag{20.4}$$

It can be shown that

$$\lambda\left(t\right) = \frac{f\left(t\right)}{R\left(t\right)} \tag{20.5}$$

The four functions $f(t)$, $Q(t)$ $R(t)$, and $\lambda(t)$ constitute the set of functions used in basic reliability analysis. The relationships between these functions are given in Table 20.1.

The Bathtub Curve

Of the four functions discussed, the hazard rate function $\lambda(t)$ displays the different stages during the lifetime of a component most clearly. In fact, typical $\lambda(t)$ plots have the general shape of a **bathtub curve**

TABLE 20.1 Relationships between Different Reliability Functions

	$f(t)$	$\lambda(t)$	$Q(t)$
$f(t)=$	$f(t)$	$\lambda(t)\exp\left[-\int_0^t \lambda(\xi)d\xi\right]$	$\dfrac{d}{dt}Q(t)$ $-\dfrac{d}{dt}R(t)$
$\lambda(t)=\dfrac{f(t)}{1-\int_0^t f(\xi)d\xi}$	$\lambda(t)$	$\dfrac{1}{1-Q(t)}\dfrac{d}{dt}(Q(t))$ $-\dfrac{d}{dt}\left[\ln R(t)\right]$	
$Q(t)=\int_0^t f(\xi)d\xi$	$1-\exp\left[-\int_0^t \lambda(\xi)d\xi\right]$	$Q(t)$	$1-R(t)$
$R(t)=1-\int_0^t f(\xi)d\xi$	$\exp\left[-\int_0^t \lambda(\xi)d\xi\right]$	$1-Q(t)$	$R(t)$

Source: Ramakumar, R. *Engineering Reliability: Fundamentals and Applications.* Prentice Hall, Englewood Cliffs, NJ, 1993. With permission.

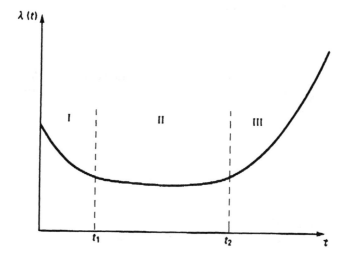

FIGURE 20.1 Bathtub-shaped hazard function. (*Source:* Ramakumar, R. *Engineering Reliability: Fundamentals and Applications.* Prentice Hall, Englewood Cliffs, NJ, 1993. With permission.)

as shown in Fig. 20.1. The first region corresponds to *wear-in* (infant mortality) or early failures during debugging. The hazard rate goes down as debugging continues. The second region corresponds to an essentially constant and low failure rate — failures can be considered to be nearly random. This is the useful lifetime of the component. The third region corresponds to the *wear-out* or fatigue phase with a sharply increased hazard rate.

    Burn-in refers to the practice of subjecting components to an initial operating period of t_1 (see Fig. 20.1) before delivering them to the customer. This eliminates all the initial failures from occurring after delivery to customers requiring high-reliability components. Moreover, it is prudent to replace a component as it approaches the wear-out region (i.e., after an operating period of $(t_2 - t_1)$). Electronic components tend to have a long useful life (constant hazard) period. The wear-out region tends to dominate in the case of mechanical components.

Mean Time to Failure

The mean or expected value of the continuous random variable *time to failure* is the **mean time to failure** (MTTF). This is a very useful parameter which is often used to assess the suitability of components. It can be obtained using either the failure density function $f(t)$ or the reliability function $R(t)$ as follows:

$$\text{MTTF} = \int_0^\infty t f(t) dt \quad \text{or} \quad \int_0^\infty R(t) dt \qquad (20.6)$$

In the case of repairable components, the repair time can also be considered as a continuous random variable with an expected value of MTTR. The mean time between failures, MTBF, is the sum of MTTF and MTTR. For well-designed components, MTTR $\ll$ MTTF. Thus, MTBF and MTTF are often used interchangeably.

Average Failure Rate

The average failure rate over the time interval 0 to T is defined as

$$AFR(0,T) \equiv AFR(T) = -\frac{\ln R(T)}{T} \qquad (20.7)$$

A Posteriori Failure Probability

When components are subjected to a burn-in (or wear-in) period of duration T, and if the component survives during $(0, T)$, the probability of failure during $(T, T + t)$ is called the *a posteriori failure probability* $Q_c(t)$. It can be found using

$$Q_c(t) = \frac{\int_T^{T+t} f(\xi) d\xi}{\int_T^\infty f(\xi) d\xi} \qquad (20.8)$$

The probability of survival during $(T, T + t)$ is

$$R(t|T) = 1 - Q_c(t) = \frac{\int_{T+t}^\infty f(\xi) d\xi}{\int_T^\infty f(\xi) d\xi} \qquad (20.9)$$

$$= \frac{R(T+t)}{R(T)} = \exp\left[-\int_T^{T+t} \lambda(\xi) d\xi\right]$$

Units for Failure Rates

Several units are used to express failure rates. In addition to $\lambda(t)$, which is usually in number per hour, $\%/K$ is used to denote failure rates in percent per thousand hours, and PPM/K is used to express failure rate in parts per million per thousand hours. The last unit is also known as FIT for "fails in time." The relationships between these units are given in Table 20.2.

TABLE 20.2 Relationships between Different Failure Rate Units

	λ (#/hr)	%/K	PPM/K (FIT)
$\lambda =$	λ	10^{-5} (%/K)	10^{-9} (PPM/K)
%/K =	$10^{5}\,\lambda$	%/K	10^{-4} (PPM/K)
PPM/K (FIT) =	$10^{9}\,\lambda$	10^{4} (%/K)	PPM/K

Source: Ramakumar, R. *Engineering Reliability: Fundamentals and Applications.* Prentice Hall, Englewood Cliffs, NJ, 1993. With permission.

Application of the Binomial Distribution

In an experiment consisting of n identical independent trials, with each trial resulting in success or failure with probabilities of p and q, the probability P_r of r successes and $(n - r)$ failures is

$$P_r = {}_nC_r p^r \left(1 - p\right)^{n-r}$$

(20.10)

If X denotes the number of successes in n trials, then it is a discrete random variable with a mean value of (np) and variance of (npq).

In a system consisting of a collection of n identical components with the probability p that a component is defective, the probability of finding r defects out of n is given by the P_r in Eq. 20.10. If p is the probability of success of one component and if at least r of them must be good for system success, then the system reliability (probability of system success) is given by

$$R = \sum_{k=r}^{n} {}_nC_k p^k \left(1 - p\right)^{n-k}$$

(20.11)

For systems with **redundancy**, $r < n$.

Application of the Poisson Distribution

For events that occur *in time* at an average rate of λ occurrences per unit of time, the probability $P_x(t)$ of exactly x occurrences during the time interval $(0, t)$ is given by

$$P_x(t) = \frac{\left(\lambda t\right)^x e^{-\lambda t}}{x!}$$

(20.12)

The number of occurrences X in $(0, t)$ is a discrete random variable with a mean value of μ of (λt), and a standard deviation σ of $\sqrt{\lambda t}$. By setting $X = 0$ in Eq. 20.12, we obtain the probability of no occurrence in $(0, t)$ as $e^{-\lambda t}$. If the event is failure, then no occurrence means success and $e^{-\lambda t}$ is the probability of success or system reliability. This is the well-known and often used exponential distribution, also known as the constant hazard model.

The Exponential Distribution

A constant hazard rate (constant λ) corresponding to the useful lifetime of components leads to the single parameter exponential distribution. The functions of interest associated with a constant λ are

$$f(t) = \lambda e^{-\lambda t}, \ t > 0 \tag{20.13}$$

$$R(t) = e^{-\lambda t} \tag{20.14}$$

$$Q(t) = Q_c(t) = 1 - e^{-\lambda t} \tag{20.15}$$

The a posteriori failure probability $Q_c(t)$ is independent of the prior operating time T, indicating that the component does not degrade no matter how long it operates. Obviously, such a scenario is valid only during the useful lifetime (horizontal portion of the bathtub curve) of the component.

The mean and standard deviation of the random variable *lifetime* are

$$\mu \equiv \mathrm{MTTF} = \frac{1}{\lambda} \quad \text{and} \quad \sigma = \frac{1}{\lambda} \tag{20.16}$$

The Weibull Distribution

The Weibull distribution has two parameters — a scale parameter α and a shape parameter β. By adjusting these two parameters, a wide range of experimental data can be modeled in system reliability studies. The associated functions are

$$\lambda(t) = \frac{\beta t^{\beta-1}}{\alpha^{\beta}}; \quad \alpha > 0, \beta > 0, t \geq 0 \tag{20.17}$$

$$f(t) = \frac{\beta t^{\beta-1}}{\alpha^{\beta}} \exp\left[-\left(\frac{t}{\alpha}\right)^{\beta}\right] \tag{20.18}$$

$$R(t) = \exp\left[-\left(\frac{t}{\alpha}\right)^{\beta}\right] \tag{20.19}$$

With $\beta = 1$, the Weibull distribution reduces to the constant hazard model with $\lambda = (1/\alpha)$. With $\beta = 2$, the Weibull distribution reduces to the Rayleigh distribution.

The associated MTTF is

$$\mathrm{MTTF} = \mu = \alpha \Gamma\left(1 + \frac{1}{\beta}\right) \tag{20.20}$$

where Γ denotes the gamma function.

Combinatorial Aspects

Analysis of complex systems is facilitated by decomposition into functional entities consisting of subsystems or units and by the application of combinatorial considerations and network modeling techniques.

FIGURE 20.2 Series or chain structure. (*Source:* Ramakumar, R. *Engineering Reliability: Fundamentals and Applications.* Prentice Hall, Englewood Cliffs, NJ, 1993. With permission.)

A **series structure** (or chain structure) consisting of n units is shown in Fig. 20.2. From the reliability point of view, the system will succeed only if all the units succeed. The units may or may not be physically in series. If R_i is the probability of success of the ith unit, then the series system reliability R_s is given as

$$R_s = \prod_{i=1}^{n} R_i \tag{20.21}$$

if the units do not interact with each other. If they do, then the conditional probabilities must be carefully evaluated.

If each of the units has a constant hazard, then

$$R_s(t) = \prod_{i=1}^{n} \exp(-\lambda_i t) \tag{20.22}$$

where λ_i is the constant failure rate for the ith unit or component. This enables us to replace the n components in series by an equivalent component with a constant hazard λ_s where

$$\lambda_s = \sum_{i=1}^{n} \lambda_i \tag{20.23}$$

If the components are identical, then $\lambda_s = n\lambda$ and the MTTF for the equivalent component is $(1/n)$ of the MTTF of one component.

A **parallel structure** consisting of n units is shown in Fig. 20.3. From the reliability point of view, the system will succeed if any one of the n units succeeds. Once again, the units may or may not be physically or topologically in parallel. If Q_i is the probability of failure of the ith unit, then the parallel system reliability R_p is given as

$$R_p = 1 - \prod_{i=1}^{n} Q_i \tag{20.24}$$

if the units do not interact with each other (i.e., are independent).

If each of the units has a constant hazard, then

$$R_p(t) = 1 - \prod_{i=1}^{n} \left[1 - \exp(-\lambda_i t) \right] \tag{20.25}$$

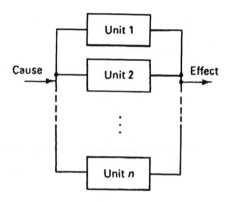

FIGURE 20.3 Parallel structure. (*Source:* Ramakumar, R. *Engineering Reliability: Fundamentals and Applications.* Prentice Hall, Englewood Cliffs, NJ, 1993. With permission.)

and we do not have the luxury of being able to replace the parallel system by an equivalent component with a constant hazard. The parallel system does not exhibit constant hazard even though each of the units has constant hazard.

The MTTF of the parallel system can be obtained by using Eq. 20.25 in Eq. 20.6. The results for the case of components with identical hazards λ are $(1.5/\lambda)$, $(1.833/\lambda)$, and $(2.083/\lambda)$ for $n = 2$, 3, and 4, respectively. The largest gain in MTTF is obtained by going from one component to two components in parallel. It is uncommon to have more than two or three components in a truly parallel configuration because of the cost involved. For two nonidentical components in parallel with hazard rates λ_1 and λ_2, the MTTF is given as

$$\text{MTTF} = \frac{1}{\lambda_1} + \frac{1}{\lambda_2} - \frac{1}{\lambda_1 + \lambda_2} \tag{20.26}$$

An r-out-of-n structure, also known as a partially redundant system, can be evaluated using Eq. 20.11. If all the components are identical, independent, and have a constant hazard λ, then the system reliability can be expressed as

$$R(t) = \sum_{k=r}^{n} {}_nC_k e^{-k\lambda t} \left(1 - e^{-\lambda t}\right)^{n-k} \tag{20.27}$$

For $r = 1$, the structure becomes a parallel system. For $r = n$, it becomes a series system.

Series-parallel systems are evaluated by repeated application of the expressions derived for series and parallel configurations by employing the well-known network reduction techniques.

Several general techniques are available for evaluating the reliability of complex structures that do not come under purely series or parallel or series-parallel. They range from inspection to cutset and tieset methods and connection matrix techniques that are amenable to computer programming.

Modeling Maintenance

Maintenance of a component could be a scheduled (preventative) one or a forced (corrective) one. The latter follows in-service failures and can be handled using Markov models discussed later. Scheduled maintenance is conducted at fixed intervals of time, irrespective of the system continuing to operate satisfactorily.

Scheduled maintenance, under ideal conditions, takes very little time (compared to the time between maintenances) and the component is restored to an "as new" condition. Even if the component is irreparable, scheduled maintenance postpones failure and prolongs the life of the component. Scheduled

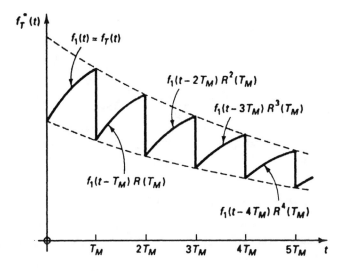

FIGURE 20.4 Density function with ideal scheduled maintenance incorporated. (*Source:* Ramakumar, R. *Engineering Reliability: Fundamentals and Applications.* Prentice Hall, Englewood Cliffs, NJ, 1993. With permission.)

maintenance makes sense only for those components with increasing hazard rates. Most mechanical systems come under this category. It can be shown that the density function $f_T^*(t)$, with scheduled maintenance included, can be expressed as

$$f_T^*\left(t\right)=\sum_{k=0}^{\infty}f_1\left(t-kT_M\right)R^k\left(T_M\right)$$

(20.28)

where

$$f_1\left(t\right)=\begin{cases}f_T\left(t\right) & \text{for } 0<t\le T_M \\ 0 & \text{otherwise}\end{cases}$$

(20.29)

and $R(t)$ = component reliability function, T_M = time between maintenance, constant, and $f_T(t)$ = original failure density function. In Eq. 20.28, $k = 0$ is used only between $t = 0$ and $t = T_M$, and $k = 1$ is used only between $t = T_M$ and $t = 2T_M$ and so on.

A typical $f_T^*(t)$ is shown in Fig. 20.4. The time scale is divided into equal intervals of T_M each. The function in each segment is a scaled-down version of the one in the previous segment, the scaling factor being equal to $R(T_M)$. Irrespective of the nature of the original failure density function, scheduled maintenance gives it an exponential tendency. This is another justification for the widespread use of exponential distribution in system reliability evaluations.

Markov Models

Of the different Markov models available, the discrete-state, continuous-time Markov process has found many applications in system reliability evaluation, including the modeling of reparable systems. The model consists of a set of discrete states (called the state space) in which the system can reside and a set of transition rates between appropriate states. Using these, a set of first-order differential equations is derived in the standard vector-matrix form the time-dependent probabilities of the various states. Solution of these equations incorporating proper initial conditions gives the probabilities of the system residing in different states as functions of time. Several useful results can be gleaned from these functions.

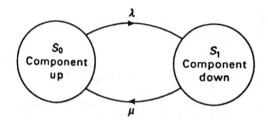

FIGURE 20.5   State space diagram for a single reparable component. (*Source:* Ramakumar, R. *Engineering Reliability: Fundamentals and Applications.* Prentice Hall, Englewood Cliffs, NJ, 1993. With permission.)

Binary Model for a Reparable Component

The binary model for a reparable component assumes that the component can exist in one of two states — the *up* state or the *down* state. The transition rates between these two states, S_0 and S_1, are assumed to be constant and equal to λ and μ. These transition rates are the constant failure and repair rates implied in the modeling process and their reciprocals are the MTTF and MTTR, respectively. Figure 20.5 illustrates the binary model.

The associated Markov differential equations are

$$\begin{bmatrix} P_0'(t) \\ P_1'(t) \end{bmatrix} = \begin{bmatrix} -\lambda & \mu \\ \lambda & -\mu \end{bmatrix} \begin{bmatrix} P_0(t) \\ P_1(t) \end{bmatrix} \tag{20.30}$$

with the initial conditions

$$\begin{bmatrix} P_0(0) \\ P_1(0) \end{bmatrix} = \begin{bmatrix} 1 \\ 0 \end{bmatrix} \tag{20.31}$$

The coefficient matrix of Markov differential equations, namely

$$\begin{bmatrix} -\lambda & \mu \\ \lambda & -\mu \end{bmatrix}$$

is obtained by transposing the matrix of rates of departures

$$\begin{bmatrix} 0 & \lambda \\ \mu & 0 \end{bmatrix}$$

and replacing the diagonal entries by the negative of the sum of all the other entries in their respective columns. Solution of Eq. 20.30 with initial conditions as given by Eq. 20.31 yields

$$P_0(t) = \frac{\mu}{\lambda + \mu} + \frac{\lambda}{\lambda + \mu} e^{-(\lambda + \mu)t} \tag{20.32}$$

$$P_1(t) = \frac{\lambda}{\lambda + \mu} \left[1 - e^{-(\lambda + \mu)t} \right] \tag{20.33}$$

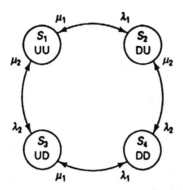

FIGURE 20.6 State space diagram for two dissimilar reparable components. (*Source:* Ramakumar, R. *Engineering Reliability: Fundamentals and Applications*. Prentice Hall, Englewood Cliffs, NJ, 1993. With permission.)

The limiting, or steady state, probabilities are found by letting $t \to \infty$. They are also known as limiting **availability** A and limiting unavailability U and they are

$$P_0 = \frac{\mu}{\lambda + \mu} \equiv A \quad \text{and} \quad P_1 = \frac{\lambda}{\lambda + \mu} \equiv U \tag{20.34}$$

The time-dependent $A(t)$ and $U(t)$ are simply $P_0(t)$ and $P_1(t)$, respectively.

Referring back to Eq. 20.14 for a constant hazard component and comparing it with Eq. 20.32 which incorporates repair, the difference between $R(t)$ and $A(t)$ becomes obvious. Availability $A(t)$ is the probability that the component is up at time t, and reliability $R(t)$ is the probability that the system has continuously operated from 0 to t. Thus, $R(t)$ is much more stringent than $A(t)$. While both $R(0)$ and $A(0)$ are unity, $R(t)$ drops off rapidly as compared to $A(t)$ as time progresses. With a small value of MTTR (or large value of μ), it is possible to realize a very high availability for a reparable component.

Two Dissimilar Reparable Components

Irrespective of whether the two components are in series or in parallel, the state space consists of four possible states: S_1 (1 up, 2 up), S_2 (1 down, 2 up), S_3 (1 up, 2 down), and S_4 (1 down, 2 down). The actual system configuration will determine which of these four states correspond to system success and failure. The associated state space diagram is shown in Fig. 20.6. Analysis of this system results in the following steady state probabilities:

$$P_1 = \frac{\mu_1 \mu_2}{\text{Denom}}; \quad P_2 = \frac{\lambda_1 \mu_2}{\text{Denom}}; \quad P_3 = \frac{\lambda_2 \mu_1}{\text{Denom}}; \quad P_4 = \frac{\lambda_1 \lambda_2}{\text{Denom}} \tag{20.35}$$

where

$$\text{Denom} \equiv \left(\lambda_1 + \mu_1 \right)\left(\lambda_2 + \mu_2 \right) \tag{20.36}$$

For components in series, $A = P_1$, $U = (P_2 + P_3 + P_4)$, and the two components can be replaced by an equivalent component with a failure rate of $\lambda_s = (\lambda_1 + \lambda_2)$ and a mean repair duration of r_s, where

$$r_s \equiv \frac{\lambda_1 r_1 + \lambda_2 r_2}{\lambda_s} \tag{20.37}$$

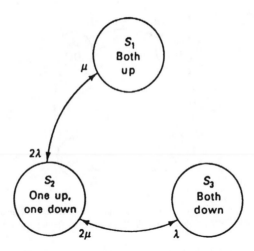

FIGURE 20.7   State space diagram for two identical reparable components. (*Source:* Ramakumar, R. *Engineering Reliability: Fundamentals and Applications.* Prentice Hall, Englewood Cliffs, NJ, 1993. With permission.)

Extending this to n components in series, the equivalent system will have

$$\lambda_s = \sum_{i=1}^{n} \lambda_i \quad \text{and} \quad r_s \cong \frac{1}{\lambda_s} \sum_{i=1}^{n} \lambda_i r_i \tag{20.38}$$

and system unavailability $= U_s \cong \lambda_s r_s = \sum_{i=1}^{n} \lambda_i r_i$ \hfill (20.39)

For components in parallel, $A = (P_1 + P_2 + P_3)$, $U = P_4$, and the two components can be replaced by an equivalent component with

$$\lambda_p \cong \lambda_1 \left(\lambda_2 r_1 \right) + \lambda_2 \left(\lambda_1 r_2 \right) \quad \text{and} \quad \mu_p = \mu_1 + \mu_2 \tag{20.40}$$

and system unavailability $= U_p = \lambda_p \left(1/\mu_p \right)$ \hfill (20.41)

Extension to more than two components in parallel follows similar lines. For three components in parallel,

$$\mu_p = \left(\mu_1 + \mu_2 + \mu_3 \right) \quad \text{and} \quad U_p = \lambda_1 \lambda_2 \lambda_3 \, r_1 \, r_2 \, r_3 \tag{20.42}$$

Two Identical Reparable Components

In this case, only three states are needed to complete the state space: S_1 (both up), S_2 (one up and one down), and S_3 (both down). The corresponding state space diagram is shown in Fig. 20.7. Analysis of this system results in the following steady state probabilities:

$$P_1 = \left(\frac{\mu}{\lambda + \mu} \right)^2; \quad P_2 = \frac{2\lambda}{\mu} \left(\frac{\mu}{\lambda + \mu} \right)^2; \quad P_3 = \left(\frac{\lambda}{\lambda + \mu} \right)^2 \tag{20.43}$$

Frequency and Duration Techniques

The expected residence time in a state is the mean value of the passage time from the state in question to any other state. Cycle time is the time required to complete an *in* and *not in* cycle for that state. Frequency of occurrence (or encounter) for a state is the reciprocal of its cycle time. It can be shown that the frequency of occurrence of a state is equal to the steady state probability of being in that state multiplied by the total rate of departure from it. Also, the expected value of the residence time is equal to the reciprocal of the total rate of departure from that state.

Under steady state conditions, the expected frequency of entering a state must be equal to the expected frequency of leaving the state (this assumes that the system is *ergodic*, which will not be elaborated for lack of space). Using this principle, frequency balance equations can be easily written (one for each state) and solved in conjunction with the fact that the sum of the steady state probabilities of all the states must be equal to unity to obtain the steady state probabilities. This procedure is much simpler than solving the Markov differential equations and letting $t \to \infty$.

Applications of Markov Process

Once the different states are identified and a state space diagram is developed, Markov analysis can proceed systematically (probably with the help of a computer in the case of large systems) to yield a wealth of results used in system reliability evaluation. Inclusion of installation time after repair, maintenance, spare, and stand-by systems, and limitations imposed by restricted repair facilities are some of the many problems that can be studied.

Some Useful Approximations

For an *r*-out-of-*n* structure with failure and repair rates of λ and μ for each, the equivalent MTTR and MTTF can be approximated as

$$\text{MTTR}_{eq} = \frac{\text{MTTR of one component}}{n-r+1} \tag{20.44}$$

$$\text{MTTF}_{eq} = \left(\frac{\text{MTTF}}{\text{of one component}} \right) \left(\frac{\text{MTTF}}{\text{MTTR}} \right)^{n-r} \left[\frac{(n-r)!(r-1)!}{n!} \right] \tag{20.45}$$

The influence of weather must be considered for components operating in an outdoor environment. If λ and λ' are the normal weather and stormy weather failure rates, λ' will be much greater than λ, and the average failure rate λ_f can be approximated as

$$\lambda_f \cong \left(\frac{N}{N+S} \right) \lambda + \left(\frac{S}{N+S} \right) \lambda' \tag{20.46}$$

where N and S are the expected durations of normal and stormy weather. For well-designed, high-reliability components, the failure rate λ will be very small and $\lambda t \ll 1$. Then, for a single component,

$$R(t) \cong 1 - \lambda t \quad \text{and} \quad Q(t) \cong \lambda t \tag{20.47}$$

and for *n* dissimilar components in series,

$$R(t) \cong 1 - \sum_{i=1}^{n} \lambda_i t \quad \text{and} \quad Q(t) \cong \sum_{i=1}^{n} \lambda_i t \qquad (20.48)$$

For the case of n identical components in parallel,

$$R(t) \cong 1 - (\lambda t)^n \quad \text{and} \quad Q(t) \cong (\lambda t)^n \qquad (20.49)$$

For the case of an r-out-of-n configuration,

$$Q(t) \cong {}_{n}C_{(n-r+1)} (\lambda t)^{n-r+1} \qquad (20.50)$$

Equations 20.47 to 20.50 are called rare-event approximations.

Defining Terms

Availability: The availability $A(t)$ is the probability that a system is performing its required function successfully at time t. The steady state availability A is the fraction of time that an item, system, or component is able to perform its specified or required function.

Bathtub curve: For most physical components and living entities, the plot of failure (or hazard) rate versus time has the shape of the longitudinal cross section of a bathtub.

Hazard rate function: The plot of instantaneous failure rate vs. time is called the hazard function. It clearly and distinctly exhibits the different life cycles of the component.

Mean time to failure: The mean time to failure (MTTF) is the mean or expected value of time to failure.

Parallel structure: Also known as a completely redundant system, it describes a system that can succeed when at least one of two or more components succeeds.

Redundancy: Refers to the existence of more than one means, identical or otherwise, for accomplishing a task or mission.

Reliability: The reliability $R(t)$ of an item or system is the probability that it has performed successfully over the time interval from 0 to t. In the case of irreparable systems, $R(t) = A(t)$. With repair, $R(t) \leq A(t)$.

Series structure: Also known as a chain structure or nonredundant system, it describes a system whose success depends on the success of all of its components.

References

Billinton, R. and Allan, R. N. *Reliability Evaluation of Engineering Systems: Concepts and Techniques*, 2nd ed., Plenum, New York, 1992.

Lewis, E. E. *Introduction to Reliability Engineering*, John Wiley & Sons, New York, 1987.

Ramakumar, R. *Engineering Reliability: Fundamentals and Applications*, Prentice Hall, Englewood Cliffs, NJ, 1993.

Shooman, M. L. *Probabilistic Reliability: An Engineering Approach*, 2nd ed., R. E. Krieger, Malabar, FL, 1990.

Further Information

Green, A. E. and Bourne, A. J. *Reliability Technology*, Wiley-Interscience, New York, 1972.

Henley, E. J. and Kumamoto, H. *Probabilistic Risk Assessment — Reliability Engineering, Design, and Analysis*, IEEE Press, New York, 1991.

IEEE Transactions on Reliability, Institute of Electrical and Electronics Engineers, New York.

O'Connor, P. D. T. *Practical Reliability Engineering, 3rd ed.,* John Wiley & Sons, New York, 1985.

Proceedings: Annual Reliability and Maintainability Symposium, Institute of Electrical and Electronics Engineers, New York.

Siewiorek, D. P. and Swarz, R. S. *The Theory and Practice of Reliable System Design,* Digital Press, Bedford, MA, 1982.

Trivedi, K. S. *Probability and Statistics with Reliability, Queuing, and Computer Science Applications,* Prentice Hall, Englewood Cliffs, NJ, 1982.

Villemeur, A. *Reliability, Availability, Maintainability and Safety Assessment, Vol. 1 and 2,* John Wiley & Sons, New York, 1992.

20.2 Design for Reliability

B.S. Dhillon

Reliability is an important consideration in the design of engineering systems because it helps ensure the success of such systems. Reliability may be described as the probability that an item will perform its defined function satisfactorily for the given period when used according to the designed conditions.

The history of reliability goes back to the 1930s, when probability concepts for the first time were applied to electric power generation [Dhillon, 1983]. However, generally World War II is regarded as the real beginning of the reliability discipline, when the Germans applied the basic reliability concept to increase the reliability of their V1 and V2 rockets. In 1950, the U.S. Department of Defense established an Ad Hoc Group on Reliability of Electronic Equipment, which later was referred to as the Advisory Group on Reliability of Electronic Equipment (AGREE). Some of the recommendations of the Ad Hoc Group, in 1952, included collecting better data on equipment and component **failures** from the field, establishing quantitative requirements for the reliability of equipment and components, developing better components, and establishing AGREE. In 1957, AGREE published a report entitled "Reliability of Military Electronic Equipment" [Cappola, 1984].

Several important events in the history of reliability took place in the 1950s: development of an important statistical distribution [Weibull, 1951], publication of *IEEE Transactions on Reliability* [1953], National Symposium on Reliability and Quality Control [1954], and a commercially available book entitled *Reliability Factors for Ground Electronic Equipment* [Henny, et al., 1956]. Since the 1950s, the field of reliability has branched out into many areas: mechanical reliability, software reliability, human reliability, etc., and a comprehensive list of publications on many such areas is available in [Dhillon, 1992].

Factors for Considering Reliability in Product Design

There could be many reasons for considering reliability during the design phase, i.e., competition, public demand, insertion of reliability related clauses in design specification, product complexity, etc. In fact, the increase in the product complexity is one of the most important reasons for the increased emphasis on reliability during the product/system design phase. Three examples of increase in the product complexity with respect to parts alone are a Boeing 747 jumbo jet airplane is made up of approximately 4.5 million parts including fasteners, in 1935 a farm tractor was composed of around 1200 critical parts but in 1990 the total tractor critical parts increased roughly 2½-fold, and the 1964 Mariner Spacecraft was composed of approximately 138,000 parts. Furthermore, various studies performed over the years indicate that the most important factor for profit contributions is the reliability professionals' involvement with designers. In fact, the past experience has shown that, if it costs $1 to rectify a design fault before the initial drafting release, the cost would increase by 10-fold after the final release, 100-fold at the prototype stage, 1000-fold at the preproduction stage, and 10,000-fold at the production stage.

In addition, many past studies indicate that, generally, in the case of electronic equipment, the greatest causes for the product failures are the design-related problems. In particularly, a study performed by the

U.S. Navy on electronic equipment provided the breakdowns for the causes of failure: design: 43% (circuit misapplication — 12%, circuit and component deficiencies — 11%, inadequate components — 10%, etc.), operation and maintenance: 30% (abnormal or accidental condition — 12%, manhandling — 10%, and faulty maintenance — 8%), manufacturing: 20% (faulty workmanship, inadequate inspection, and process control — 18%, defective raw materials — 2%), and miscellaneous: 7% [Niebel, 1994].

The Design for Reliability (DFR) Process and DFR Activities

Markets for newly designed engineering products have changed forever. A successful product must satisfy three conditions: better, cheaper, and faster. "Better" means that the products must have higher performance and reliability, and on the other hand "cheaper" means the products must have lower factory cost and prices. The meanings of the word "faster" is that the products must get to market more quickly.

In the past, high product reliability and low cost traditionally have been achieved through evolutionary changes rather than revolutionary ones, but in the today's environment customer expectations and technology are changing at a very fast pace for the traditional approach to be successful. Under such condition, there is a definite need to adopt a design process that maximizes the use of the most effective approaches and techniques and reduces the ineffective ones. One such process is called design for reliability. This process is the result of efforts made by a large U.S. company that became alarmed by a slow increase in warranty costs and decrease in its market share. Consequently, the chief executive officer (CEO) of the company stated "The goal we chose was a 10-fold reduction in the failure rates of our products in the 1980s" [Wall Street Journal, 1983]. This firm established a task force to study the problem and in turn the task force made several recommendations including survey the reliability improvement methods used company wide and developing a program to spread the most useful methods to all company departments/divisions [Grant Ireson et al., 1996]. The survey methodology helped to identify 37 activities to improve hardware reliability, and 8 activities were considered to correlate well with reliability improvement. Those activities were as follows:

- Worst-case analysis
- Supplier qualification testing (component qualification)
- Thermal design and measurement
- Supplier process audits
- Component stress derating (technology selection and establishing design rules)
- Goals high priority
- Failure modes and effect analysis and fault tree analysis
- DFR training for designers

The above activities are known as the **DFR** activities. The electronic firm's experience indicated the departments/divisions that practiced these eight DFR activities about 30% of the time had approximately threefold higher failure rates than the ones that utilized them 90% of the time. It is to be noted that there are basically seven important DFR activities and an eighth or the last being concerned with training in the other seven activities.

Definition, Design, and Release to Manufacturing Phase DFR Activities

During the definition phase, a fairly complete description of the proposed product/system is developed in addition to product market investigation, proposing schedules and budgets, and estimation of investment required to bring the proposed product to market. Three DFR activities are executed during this phase:

- Establishing reliability improvement goals
- Technology selection and establishing design rules
- Providing DFR training for designers

Developing reliability improvement goals is an important task of the definition phase with respect to product reliability improvement. Usually, the management set aggressive goals for reliability improvement in terms of general rates of improvement, e.g., a factor of three, fivefold in 5 years. The technical people break down such goals into a reliability budget and allocate it, subsequently. There are many different ways of expressing reliability goals: mean time between failures (MTBF), failure rate, etc.

Selecting technology and establishing design rules is another DFR activity of the definition phase. The most appropriate technology for each part of the proposed product is selected on the basis of extensive knowledge of the physics of failure of each alternative rather than its mere price and performance. Also, there are various design rules that must be established and acted upon. Such rules are because of the manufacturing processes, and some of the examples include layout rules for printed wiring boards and temperature and chemical resistance the parts/components must possess to survive through the soldering and mounting processes. It must be remembered that the majority of the defects discovered in factory test are due to damage done to components/parts by the manufacturing processes.

The third DFR activity, i.e., providing DFR training for designers, occurs at or prior to the start of the definition phase. Under the ideal condition, the people involved with design would attend a series of classes/seminars on reliability related topics. Training is one of the key elements for the success of DFR. Examples of the training topics are design qualification testing, worst-case analysis, thermal design and measurement, and failure modes and effect analysis (**FMEA**).

During the design phase, the product under consideration is created and the following three DFR activities occur:

- Worst-case analysis
- Thermal design and measurement
- Failure modes and effects analysis and fault tree analysis

The worst-case analysis activity interacts with thermal design since maximum input/output will normally lead to maximum rise in temperature. Other worst-case analysis components involve factors such as maximum variations in power supply voltage, component parameter tolerances, and timing or signal frequency. It is to be remembered that a robust design that can sustain a wide range of variations and still functions effectively is very crucial for high reliability.

Thermal design may simply be described as the practice of designing an item so that each and every part experiences minimum rise in temperature and never by passes a reasonable percentage of its maximum temperature rating. Thus, the thermal design approach helps to increase the item reliability as usually high temperature is the greatest enemy of reliability. Thermal measurement is performed in many different ways including indirect means (i.e., by measuring the infrared emission) and direct contact (i.e., by thermocouples, thermistors, or chemicals that experience change in appearance with temperature). Nonetheless, whichever approach is practiced, measurements are crucial to calibrate the simulations to be certain the calculated temperatures are correct.

FMEA was developed in the early 1950s and is used to evaluate design from the reliability aspect. The approach demands listing of potential failure modes of each and every component on paper and its effects on the listed subsystems. Another important design evaluation approach from the reliability standpoint is known as the fault tree analysis (**FTA**). The fault tree method was developed at Bell Laboratories in the early 1960s to evaluate the design of the Minuteman launch control system. The important difference between FMEA and the FTA is that, while FMEA is a "bottom-up" approach, the FTA is "top down". Both these techniques, i.e., FMEA and FTA, are described in more detail, subsequently.

During the release to manufacturing phase, the new product design nears completion and the two DFR activities play an important role: component qualification and supplier process evaluation and monitoring.

The component qualification DFR activity is very important as approximately 70% of all hardware warranty claims results in components/parts being replaced; in turn, the component supplier's reliability problem becomes product manufacturer's reliability problem. The primary step in qualifying a part/component usually involves in performing of testing of samples in product manufacturer's application or a test fixture. The supplier is asked to perform analysis of any failures that occur during testing, and the product manufacturer conducts its own parallel analysis on repeat failures.

The supplier process evaluation and monitoring activity is very useful because it is much more beneficial to determine which supplier has the best processes than exhaustively inspecting and testing the product, hoping to detect defects/change in processes as prior to parts/components' installation in the product being developed. After the selection of the supplier, it is a good practice to continue monitoring performance of its processes.

Additional DFR Activities

In addition to the above DFR activities, there are many more DFR activities that may help to improve product reliability. These include design reviews, design qualification testing, failure and root cause analysis, and statistical data analysis.

Design reviews form a crucial component of modern industrial practice, and their primary purpose is to insure that correct design principles are being applied as well as to examine design from many different aspects: reliability, producibility, human factors, maintainability, and standardization. Some of the reliability-related items that could be reviewed during design reviews include results of FMEA, reliability allocations, reliability predictions, and reliability test results.

Design qualification testing of the newly designed product is as important as testing of supplier's parts. Several types of tests are associated with the design qualification testing: margin test, accelerated life testing, etc. The purpose of margin tests is to determine if the newly designed product will operate 5 or 10% beyond its ratings. Accelerated life testing is performed to produce failures for purposes such as to find failure mechanisms that can be eradicated or minimized and to allow statistical analysis and reliability prediction.

The failure and root cause analysis DFR activity is based on the premise that "the key to improving product quality and reliability is discover, eradicate, or minimize the root causes of failures." This is accomplished in two steps: first to force or obtain failures and then conduct failure analysis to determine the cause.

The main purpose of statistical analysis is to identify the major or frequently occurring failures, whose eradication will lead to greatest improvement in reliability. There are many techniques available to perform statistical failure data analysis [Dhillon, 1988].

Design Reliability Analysis Methods

The two most widely used techniques to perform analysis during product design with respect to reliability are FMEA and FTA. It may be stated that both these approaches are the means of highlighting potential failure modes and their related causes.

The FMEA approach may simply be described as a systematic class of activities intended to recognize and evaluate all possible potential failures associated with a product/process and their effects, highlight measures that could eradicate or lower the probability of the potential failure occurrence, and document the process. The major steps associated with performing FMEA include define system boundaries and detailed requirements, list all components and subsystems in a system, list necessary failure modes, describe and identify the component under consideration, assign failure rates to each component failure mode, list each failure mode effect or effects on subsystem and plant, enter remarks for each failure mode in question, and review each critical failure mode and take corrective measures.

Some of the benefits of the FMEA approach are that it uses a systematic approach to classify hardware failures, serves as each possible failure mode occurrence detection method, identifies all possible failure modes and their effects on mission, personnel, and system, provides input data for test planning, is useful for comparing designs, and is easy to understand. The primary difference between FMEA and FTA is that the former is failure oriented and the latter event oriented. The cost is one of the most important factors in deciding whether to use the FMEA or FTA approach in design analysis.

FTA begins by identifying an undesirable event, called top event, associated with a system/product. The failure events that could cause the undesired event are generated and connected by logic operators OR, AND, etc. The OR gate provides a TRUE output if one or more inputs are TRUE (failures). On the other hand the AND gate provides a TRUE (failed) output only if all the inputs are TRUE (failures). The fault tree construction proceeds by generation of events in a successive manner until the events need not be developed further. More specifically, the generation of fault trees involves successively asking the question, "How could this event occur?" The fault tree itself is the logic structure relating the top event to the primary events and it makes use of symbols to represent gates and events [Dhillon and Singh, 1981].

The benefits of FTA include providing options to perform quantitative or qualitative reliability analysis, ferreting out failures deductively, requiring the reliability analyst to understand the system/product thoroughly and deal specifically with one particular failure at a time, providing insight into the system behavior, providing a visibility tool to designers, users, and management to justify design changes and trade-off studies, and handling complex systems more easily.

Reliability Data Sources

Availability of failure data on parts/components/products is very crucial to improve reliability of new products. There are numerous failure data banks throughout the world. Some of the sources of obtaining failure data are as follows:

- GIDEP Data Bank. This is a computerized data bank and the term GIDEP stands for Government Industry Data Exchange Program (GIDEP). The data bank is managed by the GIDEP Operations Center, Fleet Missile Systems, Analysis and Evaluation Group, Department of Navy, Corona, CA.
- NPRD Data. These data are in the form of reports containing failure probabilities of nonelectronic parts used by the U.S. military. The term NPRD stands for Non-Electronic Parts Reliability Data (NPRD) and the NPRD reports are periodically released by the Rome Air Development Center, Griffis Air Force Base, Rome, NY.
- NERC Data. These data are made available by the National Electric Reliability Council (NERC), New York. The Council publishes data collected from U.S. power generation plants annually.

Defining Terms

DFR: This stands for design for reliability (DFR).
Failure: The inability of an item to function within initially defined guidelines.
FMEA: This stands for failure modes and effects analysis (FMEA) and is a systematic class of activities intended to recognize and evaluate all possible potential failures associated with a product/process and their effects, highlight measures that could eradicate or lower the probability of the potential failure occurrence, and document the process.
FTA: This is an important reliability analysis method used to analyze product design. The term FTA stands for fault tree analysis (FTA).
Reliability: This is the probability that an item will perform its specified function satisfactorily for the specified period when used according to designed conditions.

References

Cappola, A. Reliability engineering of electronic equipment: a historical perspective, *IEEE Trans. Reliabil.*, 33(1): 29–35, 1984.

Dhillon, B. S. *Power System Reliability, Safety and Management*, Ann Arbor Science, Ann Arbor, MI, 1983.

Dhillon, B. S. *Mechanical Reliability: Theory, Models and Applications*, American Institute of Aeronautics and Astronautics, Washington, DC, 1988.

Dhillon, B. S. *Reliability and Quality Control: Bibliography on General and Specialized Areas*, Beta Publishers, Gloucester, Ontario, Canada, 1992.

Dhillon, B. S. and Singh, C. *Engineering Reliability: New Techniques and Applications*, John Wiley & Sons, New York, 1981.

Grant Ireson, W., Coombs, C. F., and Moss, R. Y. *Handbook of Reliability Engineering and Management*, McGraw-Hill, New York, 1996.

Henny, K., Lopatin, I., Zimmer, E. T., Adler, L. K., and Naresky, J. J. *Reliability Factors for Ground Electronic Equipment*, McGraw-Hill, New York, 1956.

Niebel, B. W. *Engineering Maintenance Management*, Marcel Dekker, New York, 1994.

Wall Street Journal, July 25, 1983, Manager's Journal, New York.

Weibull, W. A Statistical Distribution Function of Wide Applicability, *J. Appl. Mech.*, 18: 293–297, 1951.

Further Information

Dhillon, B. S. *Reliability Engineering in Systems Design and Operation*, Van Nostrand Reinhold, New York, 1983.

Kapur, K. C. and Lamberson, L. R. *Reliability in Engineering Design*, John Wiley & Sons, New York, 1977.

Smith, C. O. *Introduction to Reliability in Design*, McGraw-Hill, New York, 1976.

20.3   Maintainability: A Process Perspective

Anil B. Jambekar and Karol I. Pelc

Providing customers with products that maintain their operational capability for a long time, and under changing conditions, constitutes one of the most important responsibilities of a technology manager. He or she needs assurance of proper performance not only at the time of delivery but also through a whole lifecycle of the product. Potential direct and/or indirect losses due to failure of product may be very detrimental to reputation of a manufacturer and may influence the future of a company. Special attention must be paid to prevention or reduction of downtime and to minimization of repair costs. To meet this long-term challenge, a technology manager needs to develop a *process perspective* on three interrelated features of a product in its life-cycle: *reliability, maintainability, and availability*. Those features may be analyzed at different levels of product complexity and evaluated in several dimensions.

Maintainability, Reliability, and Availability

As **reliability** is centered around the frequency of breakdowns, **maintainability** is focused on the time of breakdown. **Availability** is viewed as being the consequence of reliability and maintainability. It is measured by proportion of time during which a product is effectively available for operational use. The equation can be written as

$$A = \frac{MTBF}{MTBF + MTTR}$$

where A = availability, $MTBF$ = mean time between failures (function of reliability), and $MTTR$ = **mean time to repair** (function of maintainability).

The equation sets up logic of trading off reliability vs. maintainability in order to attain a desired availability, which is what the user of the system or the product is asking for. These three features are sometimes used (e.g., in defense sector) as an integrated requirement, under an acronym of RAM (reliability-availability-maintainability), imposed on acquisition systems. The RAM requirement has to

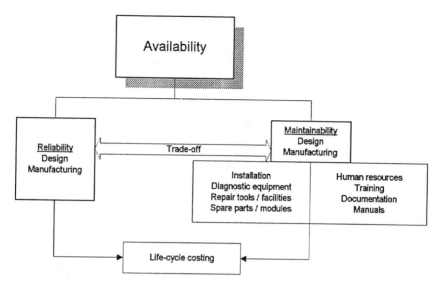

FIGURE 20.8   Relations among maintainability, reliability, availability, and life-cycle costing.

TABLE 20.3   Time- and Cost-Related Maintainability Parameters

Time-Related Parameters	Cost-Related Parameters
Mean time to repair	**Maintenance labor-hours per year**
Median corrective maintenance time	Maintenance labor-hours per operating cycle
Mean preventive maintenance time	Maintenance labor-hours per operate-hour
Mean active maintenance time	
Mean down time	
Mean logistic down time	

insure that the systems are ready for use when needed, will successfully perform assigned functions, and can be operated and maintained economically [Department of Defense, 1991]. Figure 20.8 shows a model of availability and life-cycle costing [Cunningham and Cox, 1972; Raheja, 1991; Nakajima, 1988].

Dimensions of Maintainability

Maintainability may be evaluated by parameters that are either time related or cost related. A list of basic parameters in these two categories is shown in Table 20.3. Definitions of individual parameters are given at the end of this section [Kowalski, 1996]. The parameters refer to two different types of activities: **preventive maintenance** and **corrective maintenance**. In addition to those parameters, several other factors and indexes may be used to describe maintainability of particular types of products, e.g., machines, instruments, equipment, systems, etc. or to determine more specific design requirements [Blanchard and Lowery, 1969; Gotoh, 1991; Locks, 1995; Raheja, 1991].

A Process Perspective of Maintainability

This section describes maintainability as an ongoing process that directly impacts effectiveness of a product or system. The process is circular and is shown in Fig. 20.9. A technology manager is generally concerned about product or system effectiveness during its operations.

The system or product effectiveness integrates three concepts: availability, **dependability**, and **capability**. Preventive and corrective maintenance actions restore the product or system effectiveness.

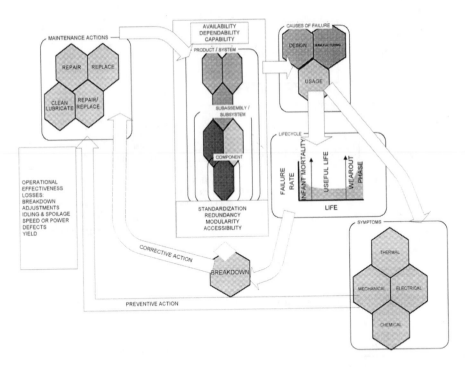

FIGURE 20.9   A process perspective of maintainability program.

Although the maintainability requirements are built into the product or system during design and manufacturing stage, the causes of failure are realized during operations or use of the product or system. During usage the product or system generates much information that can be systematically collected and analyzed to initiate required maintenance actions. A reliability bathtub model of failure rates offers useful insights [Nakajima, 1988]. During the startup phase, the failure rates are due to design and manufacturing errors. Trial runs and startup control procedures can be used as countermeasures, and the processes should be documented in operational manuals accompanying the system. During the useful life phase, the failures are due to chance and mostly to operational errors. Training of operators and prescribed operations under normal conditions can reduce the chance errors and, hence, accidental failures. Wear-out failures are due to limited natural life span of components or subsystems. The life can be extended through preventive maintenance.

Ideally, breakdowns can be eliminated entirely through the adoption of maintenance-free design or through preventive maintenance. The conditions for most systems or products are far from ideal and, hence, breakdowns are expected. By carefully monitoring symptoms that are specific to a product or a system, the hidden defects can be exposed and eliminated before they become stronger causing both function-reduction (loss of a standard function, longer and difficult setup/adjustment, frequent idling and minor stoppages, and reduction in power, speed, and cycle time) and function-failure breakdowns. The symptoms can be visible, such as dirt, or can be some combination of mechanical, thermal, chemical, and electrical causes. Tracking signals based on symptoms that can offer leading indicators of the presence of hidden problems are a critical part of any preventive maintenance program, which by its name implies exposing hidden problems and treating the product or system before it breaks down. Table 20.4 presents the relationship between breakdown countermeasures and skills required in each case [Nakajima, 1988].

Maintenance actions are (1) repair encountering down time, (2) replace using a spare backup and throw away the part, or (3) replace using a spare backup and (4) repair the deteriorated part. The actions are taken either at component level, subsystem level, or at the product or the system level. The nature

TABLE 20.4 Breakdown Countermeasures and Human Resource Skills

Countermeasures	Operations skill improvements	Maintenance skill improvements
Maintain basic conditions	***	
Adhere to operating procedures	***	***
Discover and predict deterioration	***	***
Establish repair methods		***
Correct design/manufacturing errors		***
Prevent operational errors	***	***
Prevent repair errors		***

*** Indicates the required human resource skills.

of the system or the product will dictate the human resource skills and maintenance logistical requirements. Logistical delay modifies the availability equation to

$$A = \frac{MTBF}{MTBF + MTTR + MLDT}$$

where $MLDT$ = mean logistical down time.

Conclusion

Maintainability is one of the most important features of a product or system. This feature is initially determined by design and manufacturing, but it becomes critical after delivery, during installation, and in the process of the product or system operational use. The level of maintainability influences strongly the future behavior of a customer/user and it may constitute a condition for trust to the manufacturer. Methods of prediction, assurance, and support of maintenance depend to some extent on specific types of products/systems. Remarks presented in this section, emphasizing a process perspective, are intended as a framework for analysis, selection and application of those methods.

Defining Terms

Availability: A measure of the degree to which an item is in the operable and committal state at the start of a mission when the mission is called for at an unknown (random) time [Department of Defense, 1991].

Capability: Measure of performance during system or product use. May include speed, power, range, accuracy, etc.

Corrective maintenance: All actions necessary to restore an item to satisfactory operational condition after a malfunction has caused degradation of the item below the specified performance [Blanchard and Lowery, 1969].

Dependability: Measure of operational level system or product condition. May include such measures as reparability, safety, adaptability, survivability, etc.

Maintainability: (1) The ability of an item (product, machine, system) to be retained in or restored to specified conditions when maintenance is performed by personnel having specified skills, using prescribed procedures and resources at each prescribed level of maintenance or repair [Department of Defense, 1991]. (2) Knowledge of eliminating or reducing the need for maintenance and minimizing the down time if maintenance is necessary [Raheja, 1991].

Maintenance labor-hours per year: The sum of the preventive and corrective maintenance labor-hours expended in a year.

Mean active maintenance time: The mean time to perform a corrective or preventive maintenance action, excluding supply delay and administrative delay times [Kowalski, 1996].

Mean down time: The mean total time of corrective or preventive maintenance including the average supply delay time and the average administrative delay time.

Mean logistics down time: The mean time to bring together necessary spare components or subsystems and skilled maintenance human resource to begin maintenance actions.

Mean preventive maintenance time: The mean time of preventive maintenance action weighted by the annual frequency of those actions.

Mean time to repair: (1) The mean of the times to repair an item weighted by the probability of occurrence. (2) The total corrective maintenance time divided by the number of corrective maintenance actions performed [Kowalski, 1996].

Median corrective maintenance time: The time within which half of the maintenance actions can be accomplished.

Preventive maintenance: All actions necessary for retaining an item in satisfactory operational condition including inspection, detection, and prevention of failures.

Reliability: The probability that a product or system will perform its intended function for a given time period under a given set of conditions [Raheja, 1991].

References

Blanchard, S. B., Jr. and Lowery, E. E. *Maintainability: Principles and Practices*, McGraw-Hill, New York, 1969.

Carruba, E. R. and Gordon, R. D. *Product Assurance Principles: Integrating Design Assurance and Quality Assurance*, McGraw-Hill, New York, 1988.

Cunningham, C. E. and Cox, W. *Applied Maintainability Engineering*, John Wiley & Sons, New York, 1972.

Gotoh, F. *Equipment Planning for TPM*, Productivity Press, Cambridge, MA, 1991.

Kowalski, R. *Maintainability and Reliability*, In *Handbook of Reliability Engineering and Management*, W. G. Ireson, C. F. Coombs, Jr., and R. Y. Moss, eds., pp. 15.1–15.29, McGraw-Hill, New York, 1996.

Locks, M. O. *Reliability, Maintainability, and Availability Assessment*, ASQC Quality Press, Milwaukee, WI, 1995.

Nakajima, S. *Total Productive Maintenance*, Productivity Press, Cambridge, MA, 1988.

Raheja, D. G. *Assurance Technologies: Principles and Practices*, McGraw-Hill, New York, 1991.

Department of Defense. Glossary, In *Defense Acquisition Acronyms & Terms*, 5th ed., 1991. Department of Defense, Defense Systems Management College, Acquisition Policy Department, Fort Belvoir, VA.

Further Information

The journal *IEEE Transactions on Reliability* published by the IEEE Reliability Society reports advances in maintenance and maintainability engineering and management, usually in a special section titled: Maintenance and Maintainability.

Proceedings and tutorial notes of *The Annual Reliability and Maintainability Symposium* include materials presented at the annual symposia. Those publications are available from AR&MS, c/o Evans Associates, 804 Vickers Ave., Durham, NC 27701.

Detailed definitions, specifications, procedures, and programs for maintainability assurance are presented in the following military standards and documents: MIL-STD-470, MIL-STD-472, MIL-STD-721, MIL-STD-741, and RADC-TR-83-29. The list of military and other standards pertaining to maintainability is presented in the following publication: McGill, A. A. Reliability standards and specifications, In *Handbook of Reliability Engineering and Management*, W. G. Ireson, C. F. Coombs, Jr., and R. Y. Moss, eds., pp. C.1– C.12, McGraw Hill, New York, 1996.

20.4   Design for Maintainability

B. S. Dhillon

Maintainability is a characteristic of system/product design and is concerned with system/product attributes including displays, accessibility, **safety**, test points, controls, and test equipment.

In the modern era, the documented concerns about maintainability/maintenance may be traced back to 1901 in the contract between the Army Signal Corps and Wright Brothers Concerning the development of an airplane. This document specifically stated that the airplane be "simple to operate and maintain" [AMCP-133, 1976]. A series of articles that appeared in *Machine Design* in 1956 covered maintainability areas such as designing electronic equipment for maintainability, designing for installation, recommendations for designing maintenance access in electronic equipment, and design of maintenance controls [Retterer and Kowalski, 1984]. In 1959, the first U.S. Department of Defense document on maintainability specification (i.e., MIL-M-26512) was released. Subsequently, the Defense Department prepared many other documents on maintainability: MIL-STD-778 (definitions of maintainability terms), MIL-STD-470 (maintainability program requirements), MIL-STD-471 (maintainability demonstration), MIL-HDBK-472 (maintainability prediction), etc.

In 1969, a book on maintainability principles and practices was published [Blanchard and Lowery, 1969]. This was one of the first books on the subject in the English language. Since the late 1960s, many publications on the subject have appeared. A comprehensive list of publications on maintainability is given in Dhillon [1983].

Maintainability Importance and Objectives, Expectations, and Questions for Clarifying Maintainability Requirements during the Design Phase

Two important factors for the need of product maintainability are the alarmingly high operating and support costs because of failures and subsequent maintenance. One study conducted by the U.S. Air Force in the 1950s indicated that one third of all Air Force personnel was engaged in maintenance and one third of all Air Force operating cost was for maintenance.

Some of the objectives of applying maintainability engineering principles include reducing projected maintenance time and cost through design modifications directed at maintenance simplification, utilizing maintainability data to estimate product **availability**/unavailability, determining labor hours and other related resources required to perform the projected maintenance, and determining projected maintenance down time and then making comparisons with the allowable down time for the purpose of examining if redundancy is justified to provide satisfactory level of a needed vital function. Nonetheless to which (equipment) maintainability engineering principles have been effectively applied can be expected to achieve the following factors:

- Reduced down time resulting in higher availability or operational readiness.
- The capability of being kept in an operational readiness state by inhibiting the occurrence of those types of failures that come from item's age or wear out (preventive maintenance).
- The capability of being repaired efficiently to operating condition when the down time is the result of random failures (corrective maintenance).

Occasionally, during the design phase, the maintainability professionals find that the stated (maintainability) requirements are incomplete or unclear. In order to overcome this difficulty, the maintainability professionals must seek answers on various areas, including reason for designing the system, place, and how the product under consideration to be supported, maintenance objectives, nature of personnel required to support the product under consideration, and regulations concerning environments [Dhillon, 1996].

Maintainability and Availability Analyses Purpose during Product's Design and Development Phases and Maintainability Design Characteristics

There are certain purposes for performing maintainability and availability analyses throughout a product's design and development phases. Some of those include establishing the most effective ways and means to reduce the requirement for maintenance as well as to achieve specified maintainability performance, quantifying maintainability requirements at the design level, evaluating the design for its performance with respect to both qualitative and quantitative requirements, and generating maintainability-related data for application in maintenance planning and in performing logistics support analysis [Grant Ireson et al., 1996].

During the product design phase, there are many maintainability-related basic characteristics that should be carefully considered. The maintainability design characteristics incorporate those factors and features that help to decrease product unavailability and down time. Examples of such factors are preventive and corrective maintenance tasks, maintenance ease, and support cost reduction. Some of the specific features that affect the maintainability aspects of engineering design are described below.

- Interchangeability. Basically, what it is means is that a given component can be replaced by any similar part, and the replacing part must be capable of performing the specified functions of the replaced part effectively. Some of the factors to consider in determining interchangeability needs include field conditions and manufacture and inspection cost-effectiveness. In order to assure maximum interchangeability, the design professionals must take into consideration factors such as providing satisfactory level of information in the task instructions, non-existence of physical interchangeability in places where functional interchangeability is not expected, and physical similarities including shape, size, mounting, etc.
- Standardization. This is concerned with imposing limitations on the variety of items to be used to meet the product requirements. Some of the advantages of standardization are improvement in item reliability and maintainability, reduction in item life-cycle acquisition and support costs, etc. The goals of standardization include maximizing the use of interchangeable parts, reducing the use of different types of parts, using the maximum number of common parts in different products, and so on.
- Equipment packaging. This is concerned with the manner in which equipment is packaged, for example, item layout, access, ease of parts removal, part mounting, and so on. Nonetheless, a careful consideration must be given to factors such as requirements for standardization and modularization, accessibility needs, manufacturing requirements, reliability and safety factors, environmental factors, and quality peculiar to each part. Two important factors associated with equipment packaging are accessibility and modularization. Accessibility may be described as the relative ease with which an item can be reached for such actions as replacement, inspection, or repair. There are many factors that affect accessibility: type of maintenance task under consideration, types of tools required, frequency of access usage, and distance to be reached to perform the required action. On the other hand, modularization is concerned with the division of a product into separate physical and functional units to assist removal and replacement. The benefits of modularization include easily maintainable product, reduction in maintenance time and cost, simplification of new design, and requirement of relatively low skill levels and fewer tools.
- **Human factors**. These are crucial because maintainability depends on both the operator and the maintainer. The designers concerned with maintainability must take environment into account carefully since it can vary quite significantly from one application to another. Some of the classifications of these environments are physical, human, and operational. The physical environment include factors such as noise, radiation, temperature, vibration, pressure, and dust. The factors belonging to the human environment category are physical, psychological, physiological, and human limitations. The components of the operational environment are illumination, acoustics, ventilation, work duration, and maintenance work space arrangement.

• Safety. This is an important maintainability design factor because people performing maintenance may be exposed to various kinds of hazards and accident situations. These could be the result of inadequate attention given to safety during the product design. Some of the human safety guidelines include installing appropriate fail-safe devices, providing adequate amount of tracks, guides, and stops to facilitate equipment handling, installing items requiring maintenance such that hazard in accessing them is minimized, fitting all access openings with appropriate fillets and rounded edges, carefully examining potential sources of injury by electric shock, and providing appropriate emergency doors and other emergency exits for maximum accessibility.

Design Reviews

Design reviews are the integral part of equipment development and allow people involved in the development process to assess factors such as adequacy of program planning, maturity of the design effort, and the contractual adequacy of the program's efforts. Two important design reviews during the design review process are known as the preliminary design review (PDR) and critical design review (CDR). The PDR may simply be described as a formal review of the primary design approach for an equipment or a functionally related group of lower-level elements, and the review is usually performed after the availability of the hardware development specification [Grant Ireson et al., 1996]. During the PDR, some of the maintainability issues that may be addressed, include maintainability characteristics of the system, quantitative maintainability, requirements, plans for maintainability demonstrations, maintainability prediction methods, repair-rate data sources, conformance to the maintainability ground rules and design criteria, measures to be followed if current predictions fail to satisfy requirements, and preventive-maintenance schedules.

The CDR is a formal review for an equipment or functionally related class of lower-level elements before proceeding to fabrication or production to assess if the proposed design solutions meet the requirements established in the design specification. Some of the maintainability issues addressed during the CDR are quantitative maintainability requirements and a comparison with available preliminary predictions, conformance to the maintainability ground rules and design criteria, preventive-maintenance schedules for compatibility with system needs, plans for maintainability demonstrations, approaches for automatic/semiautomatic/manual recovery from failure, and system conformance to the planned-maintenance concept.

General Maintainability Design Guidelines and Common Design Errors Affecting Maintainability

Over the years, design professionals involved with maintainability have developed many guidelines, including those that provide for visual inspection, label units, design for minimum maintenance skills, use color coding, design for safety, design for minimum tools, use standard interchangeable parts, provide for trouble shooting techniques, use captive-type chassis fasteners, design for minimum adjustment, group subsystems, provide test points, use plug-in rather than solder-in modules, and provide handles on heavy components for easy handling [Pecht, 1995].

The professionals working in the maintainability field have highlighted many common design errors that affect product maintainability, including placing an adjustment out of arm's reach, using access doors with numerous small screws, omitting handles, placing removable items such that they cannot be dismantled without taking the entire unit from its case, placing a low-reliability part underneath other parts, placing a screwdriver adjustment beneath a module, providing insufficient space for the maintenance person to get a gloved hand into the unit to perform necessary adjustment, locating an adjustable screw in a place cumbersome for the maintenance person to discover, using chassis/cover plates that drop when the last screw is taken out, installing inside parts on the same screws, placing adjustable screws close to a hot component/an exposed power-supply terminal, installing unreliable built-in test equipment that lead to a false reporting a failure, and locating fragile parts just inside the bottom edge of the chassis [Pecht, 1995].

Design Maintainability-Related Parameters

Quantitative maintainability specifications are based on desired limiting conditions imposed on product/equipment down time, maintenance labor hours, etc. Thus, during the design phase, the estimation of various maintainability parameters is important for an effective end product maintainability. Some of those parameters follows

Mean Time to Repair

This is probably the most widely used maintainability parameter and is expressed by

$$\text{MTTR} = \left(\sum_{i=1}^{k} \lambda_i T_{ri} \right) \Big/ \sum_{i=1}^{k} \lambda_i \qquad (20.51)$$

where MTTR is the mean time to repair, k is the total number of units/items, λ_i is the **failure rate** of unit/item i, and T_{ri} is the repair time of unit/item i.

It is to be noted that MTTR is the average of the number of times to repair weighted by the probability of occurrence.

Median Corrective Maintenance Time

This is used to determine the time within which half of all corrective maintenance can be accomplished. Its estimation depends upon the probability density function that represents the times to repair, e.g., exponential, lognormal. For an exponential distribution, the median corrective maintenance time (MCMT) is defined by

$$\text{MCMT} = 0.69 \big/ \mu \qquad (20.52)$$

where μ is the constant repair rate.

Similarly, for a lognormal distribution, MCMT is expressed by

$$\text{MCMT} = \text{antilog} \left[\sum_{i=1}^{k} \log T_n \big/ k \right] \qquad (20.53)$$

Mean Preventive Maintenance Time

Usually, the preventive maintenance demands are estimated on an annual basis by establishing an annual frequency for each preventive-maintenance task and then utilizing these frequencies to weigh the time to perform each task. Thus, mean preventive maintenance time (MPMT) is given by

$$\text{MPMT} = \left(\sum_{i=1}^{k} f_i D_i \right) \Big/ \sum_{i=1}^{k} f_i \qquad (20.54)$$

where k is the total number of tasks, f_i is the frequency of the ith task, D_i is the duration of the ith task.

Median Preventive Maintenance Time

This is an important parameter used for determining the time within which half of all preventive maintenance tasks are to be conducted. For the lognormally distributed preventive maintenance times, the median preventive maintenance time (MPMT) is expressed by [Blanchard, 1981].

$$MPMT = \text{antilog}\left[\left(\sum_{i=1}^{k} \log T_{pi}\right)\Big/ k\right] \qquad (20.55)$$

where $\log T_{pi}$ is the logarithm of the preventive maintenance time of unit/item i and k is the total number of units/items.

Defining Terms

Maintainability: This is a characteristic of system/product design or the probability that a failed unit, will be repaired to its satisfactory operating condition.

Failure rate: This is the number of failures of an item per unit measure of life.

Safety: This is a conservation of human life and its effectiveness and the prevention of damage to items as per mission requirements.

Availability: This is the probability that a system/equipment is available for use at the moment of need.

Human factors: This is the body of knowledge related to human abilities, limitations, etc.

References

AMCP-133. *Engineering Design Handbook: Maintainability Engineering Theory and Practice*, Department of the Army, Alexandria, VA, 1976.

Blanchard, B. S. *Logistics Engineering and Management*, Prentice-Hall, Englewood Cliffs, NJ, 1981.

Blanchard, B. S. and Lowery, E. E. *Maintainability Principles and Practices*, McGraw-Hill, New York, 1969.

Dhillon, B. S. *Reliability Engineering in Systems Design and Operation*, Van Nostrand Reinhold, New York, 1983.

Dhillon, B. S. *Engineering Design: A Modern Approach*, Richard D. Irwin, Chicago, 1996.

Grant Ireson, W., Coombs, C. F., and Moss, R. Y. *Handbook of Reliability Engineering and Management*, McGraw-Hill, New York, 1996.

Pecht, M., Ed. *Product Reliability, Maintainability and Supportability Handbook*, CRC Press, Boca Raton, FL, 1995.

Retterer, B. L. and Kowalski, R. A. Maintainability: a historical perspective, *IEEE Trans. Reliabil.*, 33(1): 56–61, 1984.

Further Information

Blanchard, B. S., Verma, D., and Peterson, E. L. *Maintainability*, John Wiley & Sons, New York, 1995.

Cunningham, C. E. and Cox, W. *Applied Maintainability Engineering*, John Wiley & Sons, New York, 1972.

Patton, J. D. *Maintainability and Maintenance Management*, Instrument Society of America, Research Triangle Park, NC, 1988.

21

Safety and Human Factors

Mansour Rahimi
University of Southern California

Waldemar Karwowski
University of Louisville

21.1 Managing Sources of Hazards in Engineered Systems

Mansour Rahimi

The management of technological systems will become increasingly complex. Engineers, scientists, and other technical personnel are being called upon to engage in technical as well as managerial activities. These technology mangers are required to direct technical project groups, cross-disciplinary work teams and task forces, evaluate and compensate group effectiveness, and solve the problems associated with complex work environments. A major issue facing today's **engineered systems** is how to reduce the risks associated with **hazards** of products, work processes, industrial and occupational settings, and work environments.

The overall costs of occupational accidents in the United States is more than $150 billion [Goetsch, 1996]. These costs include direct losses such as medical expenses, equipment and material, insurance, lost wages, property damages, legal fees, regulatory fines, and indirect losses such as lost time of production, equipment obsolescence, loss of public confidence, loss of prestige, degradation of employee morale, loss of market share, and loss of company reputation for quality and safety.

Generally speaking, engineered systems are often designed without an extensive evaluation for their potential hazards. For example, last year, there were over 10,400 work-related deaths in the United States (see Table 21.1 for a classification of these fatalities). The multidisciplinary field of safety and health contains a number of tools and techniques suitable for hazard identification and analysis of products, tools/equipment, vehicles, and technological subsystems. Unfortunately, product and process design teams do not make use of "safety" resources commonly recommended by the safety community nor are they fully familiar with hazard identification and analysis techniques [Main and Frantz, 1994]. One reason for this lack of attention may be that little standardization exists for using these techniques across the science and engineering fields. Moreover, as the system under study becomes more technologically

TABLE 21.1   A Classification of the Causes and
Percentages of Work-Related Deaths in the United States

Cause	Percentage
Surface transport	37.2
Trips, slips, and falls	12.5
Electric current	3.7
Drowning	3.2
Fire	3.1
Air transportation	3.0
Poisons (solid and liquid substances)	2.7
Water transport	1.6
Poisons (gas and vapor substances)	1.4
Other	31.6

Source: Goetsch, D. L. *Occupational Safety and Health: In
the Age of High Technology for Technologists, Engineers, and
Managers, 2nd ed.,* Prentice Hall, Englewood Cliffs, NJ, 1996.

complex in terms of hardware, software, human, environmental, and organizational variables, the need to employ a more comprehensive hazard evaluation and control methodology becomes more critical. In addition, the recent developments in environmental-conscious design and manufacturing require a deeper understanding of hazard consequences within and across the **product life-cycle** span (i.e., design, development, production, operation, deployment, use, and disposal). Therefore, successful technology managers are expected to employ comprehensive safety programs with support from all levels of management within a safety-cultured organizational structure [Hansen, 1993].

Hazard (Risk) Control

As it is impossible to design an absolutely "safe" system, it is suggested to manage and mitigate the critical hazards identified in a system and reduce their risks to an acceptable level. The effectiveness of these **risk control** measures is diminished as we move away from the source of the hazards. It is also important to mention that the principles and techniques outlined in this section work best when applied early in the system design process. Based on these notions, the engineering control of hazards can be classified as

1. Identifying and eliminating the source of hazardous energy, for example, to eliminate the use of high-voltage electricity, or considering the use of noncombustible and nontoxic material in environments with fire potentials. This approach is not always practical or cost-effective.
2. Reducing the degree of hazardous energy, for example, using low-voltage solid-state devices to reduce heat buildup in areas with explosion hazards. This approach is practical in some cases, yet costly in some other design applications.
3. Isolate the source of hazard. Provide barriers of distance, energy shields, and personal protective equipment to limit the harmful transfer of the hazardous agents. To control the sequence of events in time and space, a lockout/tagout procedure is recommended.
4. Failure minimization. Include constant monitoring of critical safety parameters (e.g., gas concentrations or radiation levels). The monitoring system should detect, measure, understand, and integrate the readings and respond properly.
5. Warnings. Similar to consumer product warnings, all system components should be equipped with warning and proper communication systems. Operators (and general public) should be warned of the type of hazard present and the means by which information can be obtained in case of an accident. However, too many warning signs and display indicators may confuse the operator.
6. Safe procedures. A common cause of accidents is the inadequacy of procedures and the failure to follow proper procedures.

7. Backout and recovery. In case of an accident, this defensive step is taken to reduce the extent of injury and damage. This step incorporates one or more of the following actions: (1) normal sequence restoration in which a corrective action must be taken to correct the faulty operation, (2) inactivating only malfunctioning equipment that is applied to redundant components or temporarily substituting a component, (3) stopping the entire operation to prevent further injury and damage, and (4) suppressing the hazard (e.g., spill containment of highly hazardous substances).

In some instances, hazard exposure cannot be reduced through the above technical or engineering controls. In these cases, an effort should be made to limit the employee's exposure through administrative controls such as (1) rearranging work schedules and (2) transferring employees who have reached their upper personal exposure limits.

Sources of Hazards

In order to study a system for its risk exposure, one needs to identify the sources of hazards as potential risk agents. One way of classifying the sources of hazards is by the type of energy within the system. This dominant **energy source** may be released (intentionally or unintentionally) and become the source of system failure and/or injury. From a hazard analysis point of view, an injury or illness is traced back to the root cause of the accident using the sources of energy as a guiding mechanism. In this section, the following hazards are considered: physical, chemical, airborne contaminants, noise, and fire. Other hazards are extensively reviewed in the reference material, and ergonomic hazards are discussed in the next section. Due to the advent of intelligent transportation systems (ITS), the nature and treatment of transportation-related hazards deserve a different section for discussion. Also, this section presents a brief discussion for a number of widely used hazard analysis techniques. Within the context of an engineering approach to hazard reduction, the objective is to control losses (e.g., injuries and damages) by identifying and controlling the degree of exposure to these sources of energy.

Physical Hazards

Physical hazards are usually in the form of kinematic force of impact by or against an object.

Human Impact Injuries

The severity of an injury depends on the velocity of impact, magnitude of deceleration, and body size, orientation, and position. The kinetic energy formula used to describe the impact injury is

$$E_{ft-lb} = \left(Wv^2\right)/2g$$

where W is the weight of an object or part of the body (lb), v is the velocity (ft/sec), and g is gravity (ft/sec²). However, if the impacting surface is soft, the kinetic energy for the impact is

$$E_{ft-lb} = \left[W(2sA)\right]/2g$$

where s is the stopping distance (ft) and A is the deceleration (ft/sec²). For example, for both of the above cases, the human skull fracture occurs at 50 ft-lb of kinetic energy. Hard hats are expected to prevent the transfer of this energy to the human skull.

Trip, Slip, and Fall

These injuries comprise 17% of all work-related injuries. Falls are the second largest source of accidental death in the United States (after motor vehicle injuries). Jobs related to manufacturing, construction, and retail and wholesale activities are the most susceptible to these types of hazards, comprising about

27% of all worker compensation claims in the United States. These hazards include slipping (on level ground or on a ladder), falling from a higher level to a lower one (or ground), falling due to the collapse of a piece of floor or equipment, and failure of a structural support or walkway. Principles of tribology are being used to study the control mechanisms for these accidents. Tribology is the science that deals with the design and analysis of friction, wear, and lubrication of interacting surfaces in relative motion. The basic measure of concern is the coefficient of friction (COF), which in its simplest form is the horizontal force divided by the vertical force at the point of relative motion. COF greater than 0.5 appears to provide sufficient traction for normal floor surfaces. However, a number of other conditions make this hazard evaluation difficult: unexpectedness of surface friction change vs. human gait progression, foreign objects and debris on a path, walkway depression, raised projections (more than 0.25 inch), change in surface slope, wet surfaces, improper carpeting, insufficient lighting, improper stair and ramp design, improper use of ladders, guardrails, and handrails, and human visual deficiencies (color weakness, lack of depth perception and field of view, inattention, and distraction).

Mechanical Injuries

The U.S. Occupational Safety and Health Act (OSHA) specifically states that one or more methods of machine guarding shall be provided the operator and other employees to protect them from **mechanical injury** hazards such as those created by point of operation, ingoing nip points, rotating parts, flying chips, and sparks. Other hazards in this category are cutting by sharp edges, sharp points, poor surface finishes, splinters from wood and metal parts; shearing by one part of a machine moving across a fixed part (e.g., paper cutters or metal shearers); crushing of a skin or a tissue caught between two moving parts (e.g., gears, belts, cables on drums); and straining of a muscle (overexertion) in a manual lifting, pushing, twisting, or repetitive motions.

Another important category is the hazards caused by pressure vessels and explosion hazards. These hazards are generally divided into two types: boilers, which are used to generate heat and steam, and unfired pressure vessels, which are used to contain a process fluid, liquid, or gas without direct contact of burning fuel. The American Society of Mechanical Engineers (ASME) covers all facets of the design, manufacture, installation, and testing of most boilers and process pressure vessels in the Boiler and Pressure Vessel Code (total of 11 volumes). The primary safety considerations relate to the presence of emergency relief devices or valves to reduce the possibility of overpressurization or explosion. Explosions can be classified on the basis of the length-to-diameter ratio (L/D) of the container. If the container has an L/D of about 1, the rise in pressure is relatively slow, and the overpressurization will cause the container to rupture. In containers with a large L/D ratio, such as gas transfer pipes and long cylinders, the initial flame propagates, creating turbulence in front of it. This turbulence improves mixing and expansion of the flame area and the speed of travel along the vessel, increasing the pressure by as much as 20 times very rapidly. Other types of explosions are caused by airborne dust particles, boiling liquid-expanding vapor, vessels containing nonreactive materials, deflagration of mists, and runaway chemical reactions. Explosions may have three types of effects on a human body: the blast wave effect, which carries kinetic energy in a medium (usually air). Decaying with distance, it can knock a person down (overpressure of 1.0 lb/in^2) or reach a threshold of lung collapse (overpressure of 11 lb/in^2). The second type of effect is thermal, usually resulting from the fire. The amount of heat radiated is related to the size of the fireball and its duration of dispersion. The radiant energy is reduced according to the inverse distance-squared law. Most explosive fireballs reach temperatures about 2400°F at their centers. The third effect is the scattering of the material fragments. All three of these injury effects are multifactorial and they are very difficult to predict.

Chemical Hazards

A Union Carbide plant in Bhopal, India, accidentally leaked methyl isocyanate gas from its chemical process. It left over 2500 dead and about 20,000 injured. There are over 3 million chemical compounds, and an estimated 1000 new compounds are introduced every year.

Hazardous/Toxic Substances

The health hazards of chemicals can be classified into acute or chronic. The acute ones are corrosives, irritants, sensitizers, toxic, and highly **toxic substances**. The chronic ones are carcinogens, liver, kidney and lung toxins, bloodborne pathogens, nervous system damages, and reproductive hazards. For a reference listing of maximum exposures to these substances, refer to the technical committees of the American Conference of Governmental Industrial Hygienists (ACGIH) and American Industrial Hygiene Association (AIHA). Rules and standards related to manufacture, use, and transportation of these chemicals are promulgated and their **compliance** is enforced by governmental agencies such as the Occupational Safety and Health Administration (OSHA), the Environmental Protection Agency (EPA), and the Department of Transportation (DOT).

Routes of Entry

There are four ways by which toxic substances can enter into the human body, causing external or internal injuries or diseases [Shell and Simmons, 1990].

1. Cutaneous (on or through the skin)

   - Corrosives: damage skin by chemical reaction
   - Dermatitis: irritants such as strong acids; sensitizers, such as gasoline, naptha, and some polyethylene compounds
   - Absorbed through skin, but effecting other organs

2. Ocular (into or through the eyes)

   - Corneal burns due to acids or alkali
   - Irritation due to abrasion or chemical reaction

3. Respiratory inhalation (explained later in Airborne Contaminants)
4. Ingestion

   - Toxic substances may be ingested with contaminated food
   - Fingers, etc. contaminated with toxic chemicals may be placed in mouth
   - Particles in the respiratory system are swallowed with mucus
   - Mechanisms of injury.

Toxic agents cause injury by one or a combination of the following mechanisms:

1. Asphixiants

   - Asphixia refers to a lack of oxygen in the bloodstream or tissues with a high level of carbon dioxide present in the blood or alveoli
   - Gas asphyxiants (e.g., carbon dioxide, nitrogen, methane, hydrogen) that dilute the air, decreasing oxygen concentration
   - Chemical asphyxiants make the hemoglobin incapable of carrying oxygen (carbon monoxide) or keep the body's tissues from utilizing oxygen from the bloodstream (hydrogen cyanide)

2. Irritants that can cause inflammation (heat, swelling, and pain)

   - Mild irritants cause hyperemia (capillaries dilate)
   - Strong irritants produce blisters
   - Respiratory irritants can produce pulmonary edema
   - Secondary irritants can be absorbed and act as systemic poisons

3. Systemic poisons

   - Poisons may injure the visceral organs, such as the kidney (nephrotoxic agents) or liver (hepatotoxic agents)
   - Poisons may injure the bone marrow and spleen, interrupting the production of blood (benzene, naphthalene, lead, propyl alchohol)

- Poisons may affect the nervous system, causing inflammation of the nerves, neuritis, pain, paralysis, and blindness (methyl alcohol, mercury)
- Poisons may enter the bloodstream and affect organs, bones, and blood throughout the body (usually with prolonged exposure)

4. Anesthetics

- May cause loss of sensation
- May interfere with involuntary muscle actions causing respiratory failure (halogenated hydrocarbons)

5. Neurotoxins

- Neurotics affect the central nervous system
- Depressants cause drowsiness and lethargy (alcohol)
- Stimulants cause hyperactivity
- Hypnotics which are sleep-inducing agents (barbituates, chloral hydrate)

6. Carcinogens

- Cancers of the skin at points of contact (tar, bitumen)
- Cancers of internal organs and systems have numerous known and suspected causes (labeled as suspected carcinogens)

7. Teratogenic effects. A substance that may cause physical defects in the developing embryo or fetus when a pregnant female is exposed to the substance for a period of time.

Airborne Contaminants

The greatest hazard exists when these contaminants have sizes smaller than 0.5 μm where they can be directly introduced into the bloodstream through alveolar sacs (e.g., zinc oxide, silver iodide). Particles larger than half micron are entrapped by the upper respiratory tract of trachea and bronchial tubes (e.g., insecticide dust, cement and foundry dust, sulfuric acid mist). There are two main forms of airborne contaminants [Brauer, 1990]: particulates: dusts, fumes, smoke, aerosols and mists, and gases or vapors.

Dusts are airborne solids, typically ranging in size from 0.1 to 25 μm, generated by handling, crushing, grinding, impact, detonation, etc. Dusts larger than 5 μm settle out in relatively still air due to the force of gravity.

Fumes are fine solid particles less than 1 μm, generated by the condensation of vapors. For example, heating of lead (in smelters) vaporizes some lead material that quickly condenses to small, solid particles.

Smokes are carbon or soot particles less than 0.1 micron resulting from incomplete combustion of carbonaceous material.

Mists are fine liquid droplets generated by condensation from the gaseous to liquid state or by the dispersion of same by splashing, foaming, or atomizing.

Gases are normally formless fluids that occupy space and that can be changed to liquid or solid by a change in pressure and temperature.

Vapors are the gaseous form of substances that are normally in a liquid or solid state.

The measures for toxicity of the above substances are given in part per million or ppm (for gases and vapors) and milligram per meter cubed, mg/m^3 (for other airborne contaminants). The criteria for the degree of toxicity are the threshold limit values, TLVs, based on review of past research and monitory experience (early OSHA standards listed permissible exposure limit, PEL). TLVs are airborne concentrations of substances that are believed to represent conditions to which nearly all workers may be repeatedly exposed, 8 hours a day, for lifetime employment, without any adverse effects. For acute toxins, short-term exposure levels (STELs) are indicated for a maximum of 15 minutes of exposure, not more than four times a day. Since exposures vary with time, a time-weighted average (TWA) is adopted for calculating TLVs:

$$X = \left(C_1/T_1\right) + \left(C_2/T_2\right) + \cdots + \left(C_n/T_n\right)$$

where E_i is the exposure to the substance at concentration level i and T_i is the amount of time for E_i exposure in an 8-hour shift.

In many environments, there are several airborne substances present at the same time. If the effects of these substances are additive and there exist no synergistic reactions, the following formula can be used for the combination of TLVs

$$\text{TLV}\left(\text{TWA}\right) = \Sigma\left(E_i T_i/8\right)$$

where C_i is the atmospheric concentration of a substance and T_i is the TLV for that substance. If $X < 1$, the mixture does not exceed the total TLV; if $X \geq 1$, the mixture exceeds the total TLV.

Noise

Noise-induced hearing loss has been identified as one of the top ten occupational hazards by the National Institute for Occupational Safety and Health (NIOSH). In addition to hearing loss, exposure to excessive amounts of noise can increase worker stress levels, interfere with communication, disrupt concentration, reduce learning potential, adversely affect job performance, and increase accident potential [Mansdorf, 1993]. Among many types of hearing loss, sensorineural hearing loss is the most common form in occupational environments. Sensorineural hearing loss is usually caused by the loss of ability of the inner ear (cochlea nerve endings) to receive and transmit noise vibrations to the brain. In this case, the middle ear (the bone structures of maleus, incus, and stapes) and the outer ear (ear drum, ear canal, and ear lobe) may be intact. A comprehensive and effective hearing conservation program can reduce the potential for employee hearing loss, reduce workers compensation costs due to hearing loss claims, and lessen the financial burden of noncompliance with government standards. Current OSHA standards require personal noise dosimetry measurements in areas with high noise levels. Noise dosimeters are instruments that integrate (measure and record) the sound levels over an entire work shift. Noise intensities are measured by the dBA, scale which most closely resembles human hearing sensitivity. For continuous noise levels, OSHA's permissible noise exposure is 90 dBA for an 8-hour shift. If the noise levels are variable, a TWA is computed. For noise levels exceeding the limit values, an employee hearing conservation program must be administered.

Fire Hazards

Fire is defined as the rapid oxidation of material during which heat and light are emitted. The National Fire Protection Association (NFPA) reported that large fire losses in the United States for the year 1991 exceeded $2.6 billion. The most frequent causes of industrial fires are electrical (23%), smoking materials (18%), friction surfaces (10%), overheated materials (8%), hot surfaces (7%), and burner flames (7%). The process of combustion is best explained by the existence of four elements: fuel, oxidizer (O_2), heat, and chain reaction. A material with a flash point below 100°F (vapor pressure < 40 lb/in²) is considered flammable, and higher than 100°F is combustible. In order to extinguish a fire, one or a combination of the following must be performed: the flammable/combustible material is consumed or removed, the oxidant is depleted or below the necessary amount for combustion, heat is removed or prevented from reaching the combustible material not allowing for fuel vaporization, or the flames are chemically inhibited or cooled to stop the oxidation reaction. Fire extinguishers are classified according to the type of fire present: class A involves solids that produce glowing embers or char (e.g., wood or paper), class B involves gases and liquids that must be vaporized for combustion to occur, class C include class A and B fires involving electrical sources of ignition, and class D involves oxidized metals (e.g., magnesium, aluminum, and titanium). In addition to heat, hazardous gases and fumes are the most dangerous

TABLE 21.2   A Sample for the Application of Preliminary Hazard Analysis to the Design of Metal Organic Chemical Vapor Deposition (only Two of the Hazards are Listed)

Hazard	Cause	Main Effects	Preventive Control[a]
Toxic gas release	Leak in storage cylinder	Potential for injury fatality from large release	Develop purge system to remove gas to another tank Minimize on-site storage Provide warning system Develop procedure for tank inspection and maintenance Develop emergency response system
Explosion, Fire	Overheat in reactor tube	Potential for fatalities due to toxic release and fire Potential for injuries, fatalities due to flying debris	Design control system to detect overheat and disconnect heater Provide warning system for temperature fluctuation, evacuate reaction tube, shut off input valves, activate cooling system

[a]This column is simplified to show the major categories of hazard control techniques.
Source: Kavianian, H. R. and Wentz, C. A. *Occupational and Environmental Safety Engineering and Management,* Van Norstrand Reinhold, NY, 1990.

by-products of fires, such as acrolein formed by the smoldering of cellulosic materials and pyrolysis of polyethyne, CO and CO_2, phosgene ($COCl_2$) produced from chlorinated hydrocarbons, sulfur dioxide, oxides of nitrogen (NO_x) resulting from wood products, ammonia (NH_3) when compounds containing nitrogen and hydrogen burn in air, and metal fumes from electronic equipment. For a complete reference to facility design requirements and fire-exiting requirements refer to NFPA Code 101.

Methods of Hazard Analysis

As mentioned earlier, knowledge about different types and sources of hazards in technological systems helps the analyst to identify ways by which these hazards can be controlled. A number of **hazard analysis techniques** have been developed to aid the analyst in making descriptive and comparative studies. This section presents some basic qualitative techniques for hazard evaluation associated with a technological system [for a detailed review, see Guidelines for Hazard Evaluation Procedures, American Institute of Chemical Engineers, 1985; Gressel and Gideon, 1991; Vincoli, 1993]. These techniques vary in terms of their hazard evaluation approaches and the degree to which the system under study lends itself to quantitative analysis. A precursor to using any of these techniques is the system hazard analysis. System hazard analysis includes the division of the system into its small and manageable components, the analysis for causes and consequences of any subsystem operation, and then the synthesis of hazard effects on the entire system. Five hazard analysis techniques are briefly presented here.

Preliminary Hazard Analysis

PHA is the foundation for effective systems hazard analysis. It should begin with an initial collection of raw data dealing with the design, production, and operation of the system. The purpose of this procedure is to identify any possible hazards inherent in the system. One example is energy as the source of hazard to explore the multitude of circumstances by which an accident can occur in a system. Table 21.2 demonstrates an actual use of PHA in the design phase of metal chemical vapor deposition. The four main categories of this table are the hazards, causes, main effects, and prevention controls. The hazard effects and corrective/preventative measures are only tentative indicators of potential hazards and their possible solutions.

Failure Mode, Effects and Criticality Analysis

The previously mentioned PHA studies hazards within the overall system operation, whereas FMECA analyzes the components of the system and all of the possible failures that can occur at different points in time. This form of analysis identifies components of a system that have potential for hazardous consequences. In this analysis, each item's function must be determined. Once this is done, the failure

TABLE 21.3 A Sample for the Application of Failure Mode, Effects, and Criticality Analysis to the Metal Organic Chemical Vapor Deposition Process (only Three System Components are Listed)

System Component	Failure mode	Effects	Criticality Ranking[a]
Reactor tube	Rupture	Release of pyrophoric gas causing fire and release of toxic gases	III
Control on reactor heater	Sensor fails; response control system fails	Reactor overheating beyond design specification	II
Refrigeration equipment	Failure to operate	Increase in vapor pressure; cylinder rupture	IV

[a]Criticality ranks are based on a scale from I to IV.

Source: Kavianian, H. R. and Wentz, C. A. *Occupational and Environmental Safety Engineering and Management,* Van Norstrand Reinhold, NY, 1990.

TABLE 21.4 Hazard Risk Assessment Matrix

Frequency of Occurrence	Hazard Category			
	I Catastrophic	II Critical	III Marginal	IV Negligible
Frequent	1	3	7	13
Probable	2	5	9	16
Occasional	4	6	11	18
Remote	8	10	14	19
Improbable	12	15	17	20

Hazard Risk Index	Suggested Criteria
1–5	Unacceptable
6–9	Undesirable
10–17	Acceptable with review
18–20	Acceptable without review

cause and effects of the components are indicated. Then, the criticality factor for each failure is determined, and a quantified severity rating is given to this factor. Table 21.3 shows an example for FMECA. Since the frequency of each potential occurrence is also an important factor, a risk assessment matrix (depicted in Table 21.4) can be used to codify the risk assignment. A design team (including the safety engineer) can use this FMECA to redesign components or parts of the system to reduce the criticality rating to a predetermined acceptable region (preferably a hazard risk index between 1 and 5).

Hazard and Operability Study

Hazard and operability study (HAZOP) is one of the most thorough forms of hazard analysis. It identifies potentially complex and interactive hazards in a system. HAZOP examines a combination of every part of the system and analyzes the collected data to locate potentially hazardous areas. The first step is to define the system and all of its subsystems, from which data will be collected. This will help in deciding the nature and the expertise needed in a HAZOP team. The team usually consists of experts in several fields of engineering, human behavior, and hygiene organization and management and other personnel who may have operational expertise related to the specific system being analyzed. Once this is done, an intensive information gathering process begins. Every aspect of the systems operation and its human interfaces is documented. The information is then broken down into small information nodes. Each node contains information on the procedure or specific machine being used in the system. Each node is then interconnected logically with other nodes in the system. Each node is also given "guide words", which help identify its conditions. Table 21.5 gives an example of guide words. Each guide word is a functional representation of subsystem hazard. The team can analyze and determine the criticality/likelihood

TABLE 21.5   HAZOP Data Table for Vacuum Air Vent Node and Reverse Flow Guide Word (Design Intention: to Vent Air into the Sterilize Following a Vacuum Stage)

Guide Word	Cause	Consequence	Recommendation
Reverse flow	Control valve leakage	Ethylene oxide leak into utility room	Air vent should not pick up air from utility room but from exhaust duct to reduce risk from reverse flow leakage of ethylene oxide

Source: Hazard and Operability Study of an Ethylene Oxide Sterilizer, National Institute for Occupational Safety and Health, NTIS Publication No. PB-90-168-980, Springfield, Virginia, 1989.

that this node could produce a hazard. At this point, the HAZOP team will need to determine what course of action to take in order to alleviate the potential hazard. This procedure is one of the most comprehensive yet time consuming of the analysis tools. It is widely used in large and complex systems with critical safety components such as utility and petrochemical facilities.

Fault Tree Analysis

Fault tree analysis (FTA) is one the most powerful tools for a deductive analysis of system hazards. FTA uses deductive reasoning to quantitatively (and qualitatively) depict possible hazards that occur due to failure of the relationships between the system components (e.g., equipment or plant procedures). FTA uses a pyramid-style tree analysis to start from a top undesired event (e.g., accident or injury) down to the initial causes of the hazard (e.g., machine failure due to a joint separation under vibrating forces or erroneous operating procedure). There are four main types of event symbols: fault event, basic event, undeveloped event, and normal event. A fault event is considered to be an in-between event and never the end event. A basic event is considered to be the final event. An undeveloped event is an event that requires more investigation due to its complexity or lack of analytical data. A normal event is an event that may or may not occur. Each one of these events is joined in the tree by a logic symbol. These gates explain the logic relationship between each of the events. For example, a main event is followed in the tree by two possible basic events. Either event can produce the main event. The logic gate in this case would be an "or" gate, but if both events could have contributed to the main event an "and " gate would be used. FTA's ability to combine causal events together to prevent or investigate accidents makes it one of the most desirable accident investigation and analysis tools.

Event Tree Analysis

Event tree analysis (ETA) is similar to FTA with the exception that ETA uses inductive reasoning to determine the undesired events that are caused by an earlier event. ETA uses the same pyramid structure as the previous analysis. However, rather than working from top to bottom, ETA works from left to right dividing the possibility of each event into two outcomes: true (event happening) or false (event not happening). By taking an initial failure of an event, the tree designer tries to incorporate all possible desired and undesired results of the event. The advantage to this system is that it helps predict failures in a step-by-step procedure. This helps the analyst to provide a solution or a countermeasure at each analysis node under consideration. ETA's main weakness lies with its inability to incorporate occurrence of multiple events at the same time.

Other techniques such as management oversight and risk tree, "what if" analysis, software hazard analysis, and sneak circuit analysis are discussed in the publications listed in the Further Information section.

Defining Terms

Compliance: The minimum set of requirements by which an environment conforms to the local, state, and federal rules, regulations, and standards. According to the seriousness of the violation, a workplace may be cited by OSHA or EPA (e.g., OSHA fines of up to $70,000 per violation) or other civil and criminal charges may be brought against responsible supervisors and managers.

Energy source: The point of production, maintenance, or transfer of energy within a system or subsystem.

Engineered systems: An aggregation or assembly of human-made objects or components united by some form of regular interaction or interdependence to function, operate, or move in unison as a whole.

Hazard: A set of system's potential and inherent characteristics, conditions, or activities that can produce adverse or harmful consequences including injury, illness, or property damage (antonym to "safety").

Hazard analysis techniques: A number of analytical methods by which the nature and causes of hazards in a product or a system are identified. These methods are generally designed to evaluate the effects of hazards and offer corrective measures or countermeasures.

Mechanical injuries: A type of physical injury caused by excessive forces applied to human body components such as cutting, crushing, and straining. Some ergonomic hazards are also included in this category.

Product life-cycle: Existence of a product from cradle to grave. From a hazard point of view the most important phase of a product life cycle is the design process, where most hazards can be eliminated or mitigated with minimal cost.

Risk control: The process by which the probability, severity and exposure to hazards (per mission and unit of time) are considered to reduce the potential loss of lives and property.

Toxic substances: Those substances that may, under specific circumstances, cause injury to persons or damage to property because of reactivity, instability, spontaneous decomposition, flammability, or volatility (including those compounds that are explosive, corrosive, or have destructive effects on human body cells and tissues).

References

Brauer, R. L. *Safety and Health for Engineers,* Van Nostrand Reinhold, New York, 1990.

Occupational Hazards, p. 6, Editorial, November, 1993.

Goetsch, D. L. *Occupational Safety and Health: In the Age of High Technology for Technologists, Engineers, and Managers, 2nd ed.,* Prentice Hall, Englewood Cliffs, NJ, 1996.

Gressel, M. G. and Gideon, J. A. An overview of process hazard evaluation techniques, *Am. Ind. Hyg. Assoc. J.,* 52(4): 158–163, 1991.

Guidelines for Hazard Evaluation Procedures, American Institute of Chemical Engineers, New York.

Hansen, L. Safety management: a call for (r)evolution, *Prof. Safety,* November: 16–21, 1993.

Main, B. W. and Frantz, J. P. How design engineers address safety: what the safety community should know, *Prof. Safety,* February: 33–37, 1994.

Mansdorf, S. Z. *Complete Manual of Industrial Safety,* Prentice Hall, Englewood Cliffs, NJ, 1993.

Shell, R. L. and Simmons, R. J. *An Engineering Approach to Occupational Safety and Health in Business and Industry,* Institute of Industrial Engineers, Norcross, GA, 1990.

Vincoli, J. W. *Basic Guide to System Safety,* Van Norstrand Reinhold, New York, 1993.

Further Information

This section only refers to some to the technical (e.g., engineering) aspects of the field of safety science. Since this field is broad and multidisciplinary, the reader is encouraged to study other aspects of safety such as management and administration for a more comprehensive coverage of the field. For further reading please see:

Accident Prevention Manual for Business and Industry: Engineering and Technology, 10th ed., National Safety Council, Itasca, IL, 1992.

Bass, L. *Products Liability: Design and Manufacturing Defects,* Shepard's/McGraw-Hill, Colorado Springs, CO, 1986.

Bird, F. E., and Germain, G. L. *Practical Loss Control Leadership: The Conservation of People, Property, Process, and Profits,* International Loss Control Institute, Loganville, GA, 1986.

Hammer, W. *Product Safety Management and Engineering, 2nd ed.,* American Society of Safety Engineers, Des Plaines, IL, 1993.

Hansen, D. J., Ed. *The Work Environment: Occupational Health Fundamentals,* Lewis, Chelsea, MI, 1991.

Hoover, S. R. *Fire Protection for Industry,* Van Nostrand Reinhold, New York, 1991.

Kavianian, H. R. and Wentz, C. A. *Occupational and Environmental Safety Engineering and Management,* Van Norstrand Reinhold, New York, 1990.

21.2 Ergonomics and Human Factors

Waldemar Karwowski

The science of **ergonomics** originated in 1857 when Wojciech Jastrzebowski of Poland defined the term by combining two Greek words: *eron* (work) + *nomos* (laws). This new science signified then the human work, play, thinking, and devotion as reflected in the manner of optimizing the use of four distinct human characteristics: (1) motor (physical); (2) sensory (aesthetic), (3) mental (intellectual); and (4) spiritual or moral [Karwowski, 1991]. The term *ergonomics* was independently reinvented by K. E. H. Murrell in 1949. Contemporary **human factors** (the parallel term for this new scientific discipline adopted in the United States), "discovers and applies information about human behavior, abilities, limitations, and other characteristics to the design of tools, machines, systems, tasks, jobs, and environments for productive, safe, comfortable, and effective human use" [Sanders and McCormick, 1993]. For example, human factors/ergonomics deals with a broad scope of problems relevant to the design and evaluation of work systems, consumer products, and working environments, whereas human-machine interactions affect human performance and product usability. the wide scope of issues addressed by ergonomics is presented in Table 21.6.

Human factors design and engineering aims to optimize the design and functioning of human-machine systems with respect to complex characteristics of people and the relationships between system users, machines, and outside environments. According to the Board of Certification in Professional Ergonomics (BCPE) a practitioner of ergonomics is a person who (1) has a mastery of a body of ergonomics knowledge, (2) has a command of the methodologies used by ergonomists in applying that knowledge to the design of a product, system, job, or environment, and (3) has applied his or her knowledge to the analysis, design testing, and evaluation of products, systems and environments. The areas of current practice in the field can be best described by examining the focus of technical groups of the Human Factors and Ergonomics Society, as illustrated in Table 21.7.

The Concept of Human-Machine Systems

A human-machine system can be broadly defined as "an organization of man and woman and the machines they operate and maintain in order to perform assigned jobs that implement the purpose for which the system was developed" [Meister, 1987]. The human functioning in such a system can be described in terms of perception, information processing, decision making, memory, attention, feedback, and human response processes. Furthermore, the human work taxonomy can be used to describe five distinct levels of human functioning, ranging from primarily physical tasks to cognitive tasks [Karwowski, 1992]. These basic but universal human activities are (1) tasks that produce force (primarily muscular work), (2) tasks of continuously coordinating sensory-monitor functions (e.g., assembling or tracking tasks), (3) tasks of converting information into motor actions (e.g., inspection tasks), (4) tasks of converting information into output information (e.g., required control tasks), and (5) tasks of producing information (primarily creative work).

Any task in a human-machine system requires processing of information that is gathered based on perceived and interpreted relationships between system elements. The processed information may need to be stored by either a human or a machine for later use. One of the important concepts for ergonomic

TABLE 21.6   Classification Scheme for Human Factors/Ergonomics

---

1. General

Human Characteristics

2. Psychological aspects
3. Physiological and anatomical aspects
4. Group factors
5. Individual differences
6. Psychophysiological state variables
7. Task-related factors

Information Presentation and Communication

8. Visual communication
9. Auditory and other communication modalities
10. Choice of communication media
11. Person-machine dialogue mode
12. System feedback
13. Error prevention and recovery
14. Design of documents and procedures
15. User control features
16. Language design
17. Database organization and data retrieval
18. Programming, debugging, editing, and programming aids
19. Software performance and evaluation
20. Software design, maintenance, and reliability

Display and Control Design

21. Input devices and control
22. Visual displays
23. Auditory displays
24. Other modality displays
25. Display and control characteristics

Workplace and Equipment Design

26. General workplace design and buildings
27. Workstation design
28. Equipment design
    Environment
29. Illumination
30. Noise
31. Vibration
32. Whole body movement
33. Climate
34. Atmosphere
35. Altitude, depth, and space
36. Other environmental issues

System Characteristics

37. General system features
38. Total system design and evaluation
39. Hours of work
40. Job attitudes and job satisfaction
41. Job design
42. Payment systems
43. Selection and screening
44. Training
45. Supervision
46. Use of support
47. Technological and ergonomic change

TABLE 21.6 (continued)   Classification Scheme for Human Factors/Ergonomics

Health and Safety

48. General health and safety
49. Etiology
50. Injuries and illnesses
51. Prevention

Social and Economic Impact of the System

52. Trade unions
53. Employment, job security, and job sharing
54. Productivity
55. Women and work
56. Organizational design
57. Education
58. Law
59. Privacy
60. Family and home life
61. Quality of working life
62. Political comment and ethical
63. Approaches and methods

Source: Ergonomics Abstracts, published by Taylor & Francis, London.

design of human-machine systems is the paradigm of the stimulus-response compatibility [Wickens, 1987]. This paradigm relates to the physical relationship (compatibility) between a set of stimuli and a set of responses, as this relationship affects the speed of human response. The spatial relations between arrangements of signals and response devices in human-machine systems with respect to direction of movement and adjustments are often ambiguous, with a high degree of uncertainty regarding the effects of intended control actions. It should be noted that the information displayed to the human operator can be arranged along a continuum that defines the degree to which that information is spatial-analog (i.e., information about relative locations, transformations or continuous motion), linguistic-symbolic, or verbal (i.e., a set of instructions, alphanumeric codes, directions, or logical operations). The scope of ergonomic factors that need to be considered in design, testing, and evaluation of any human-machine system is shown in Table 21.8 in the form of an exemplary ergonomic checklist.

Ergonomics in Industry

The knowledge and expertise offered by ergonomics as applied to industrial environments can be used to (1) provide engineering guidelines regarding redesign of tools, machines, and work layouts, (2) evaluate the demands placed on the workers by the current jobs, (3) simulate alternative work methods and determine potential for reducing physical job demands if new methods are implemented, and (4) provide a basis for employee selection and placement procedures.

The basic foundations for ergonomics design are based on two components of industrial ergonomics: engineering anthropometry and biomechanics. **Occupational biomechanics** can be defined as the application of mechanics to the study of the human body in motion or at rest [Chaffin and Anderson, 1993]. Occupational biomechanics provides the criteria for application of anthropometric data to the problems of workplace design. **Engineering anthropometry** is an empirical science branching from physical anthropology that (1) deals with physical measurements for the human body (such as body size, form (shape), and body composition), including, for example, the location and distribution of center of mass, weights, body links, or range of joint motions, and (2) applies these measures to develop specific engineering design requirements.

The recommendations for workplace design with respect to anthropometric criteria can be established by the principle of *design for the extreme,* also known as the *method of limits* [Pheasant, 1986]. The basic idea behind this concept is to establish specific boundary conditions (percentile value of the relevant

TABLE 21.7 Subject Interests of Technical Groups of the Human Factors and Ergonomics Society

Technical Group	Description/Areas of Concerns
I. Aerospace systems	Applications of human factors to the development, design, operation, and maintenance of human-machine systems in aviation and space environments (both civilian and military).
II. Aging	Human factors applications appropriate to meeting the emerging needs of older people and special populations in a wide variety of life settings.
III. Communications	All aspects of human-to-human communication, with an emphasis on communication mediated by telecommunications technology, including multimedia and collaborative communications, information services, and interactive broadband applications. Design and evaluation of both enabling technologies and infrastructure technologies in education, medicine, business productivity, and personal quality of life.
IV. Computer systems	Human factors aspects of (1) interactive computer systems, especially user interface design issues; (2) the data-processing environment, including personnel selection, training, and procedures; and (3) software development.
V. Consumer products	Development of consumer products that are useful, usable, safe, and desirable. Application of the principles and methods of human factors, consumer research, and industrial design to ensure market success.
VI. Educators' professional	Education and training of human factors and ergonomics specialists in academia, industry, and government. Focus on both degree-oriented and continuing education needs of those seeking to increase their knowledge and or skills in this area, accreditation of graduate human factors programs, and professional certification.
VII. Environmental design	Human factors aspects of the constructed physical environment, including architectural and interior design aspects of home, office, and industrial settings. Promotion of the use of human factors principles in environmental design.
VIII. Forensics professional	Application of human factors knowledge and technique to "standards of care" and accountability established within legislative, regulatory, and judicial systems. The emphasis on providing a scientific basis to issues being interpreted by legal theory.
IX. Industrial ergonomics	Application of ergonomics data and principles for improving safety, productivity, and quality of work in industry. Concentration on service and manufacturing process, operations, and environments.
X. Medical systems and functionally impaired populations	All aspects of the application of human factors principles and techniques toward the improvement of medical systems, medical devices, and the quality of life for functionally impaired user populations.
XI. Organizational design	Improving productivity and the quality of life by an integration of psychosocial, cultural, and technological factors and with user interface factors (performance, acceptance, needs, limitations) in design of jobs, workstations, and related management systems.
XII. Personality and individual differences in human performance	The range of personality and individual difference variables that are believed to mediate performance.
XIII. Safety	Research and applications concerning human factors in safety and injury control in all settings and attendant populations, including transportation, industry, military, office, public building, recreation, and home improvements.
XIV. System development	Concerned with research and exchange of information for integrating human factors into the development of systems. Integration of human factors activities into system development processes in order to provide systems that meet user requirements.
XV. Test and evaluation	A forum for test and evaluation practitioners and developers from all areas of human factors and ergonomics. Concerned with methodologies and techniques that have been developed in their respective areas.
XVI. Training	Fosters information and interchange among people interested in the fields of training and training research.
XVII. Visual performance	The relationship between vision and human performance, including (1) the nature, content, and quantification of visual information and the context in which it is displayed; (2) the physics and psychophysics of information display; (3) perceptual and cognitive representation and interpretation of displayed information; (4) assessment of workload using visual tasks; and (5) actions and behaviors that are consequences of visually displayed information.

TABLE 21.8   Examples of Factors to Be Used in the Ergonomics Checklists

1. Anthropometric, biochemical, and physiological factors
  1.   Are the differences in human body size accounted for by the design?
  2.   Have the right anthropometric tables been used for specific populations?
  3.   Are the body joints close to neutral positions?
  4.   Is the manual work performed close to the body?
  5.   Are there any forward-bending or twisted trunk postures involved?
  6.   Are sudden movements and force exertion present?
  7.   Is there a variation in worker postures and movements?
  8.   Is the duration of any continuous muscular effort limited?
  9.   Are the breaks of sufficient length and spread over the duration of the task?
 10.   Is the energy consumption for each manual task limited?

II. Factors related to posture (sitting and standing)
  1.   Is sitting/standing alternated with standing/sitting and walking?
  2.   Is the work height dependent on the task?
  3.   Is the height of the work table adjustable?
  4.   Are the height of the seat and backrest of the chair adjustable?
  5.   Is the number of chair adjustment possibilities limited?
  6.   Have good seating instructions been provided?
  7.   Is a footrest used where the work height is fixed?
  8.   Has the work above shoulder or with hands behind the body been avoided?
  9.   Are excessive reaches avoided?
 10.   Is there enough room for the legs and feet?
 11.   Is there a sloping work surface for reading tasks?
 12.   Have the combined sit-stand workplaces been introduced?
 13.   Are handles of tools bent to allow for working with the straight wrists?

III. Factors related to manual materials handling (lifting, carrying, pushing and pulling loads)
  1.   Have tasks involving manual displacement of loads been limited?
  2.   Have optimum lifting conditions been achieved?
  3.   Is anybody required to lift more than 23 kg?
  4.   Have lifting tasks been assessed using the NIOSH (1991) method?
  5.   Are handgrips fitted to the loads to be lifted?
  6.   Is more than one person involved in lifting or carrying tasks?
  7.   Are there mechanical aids for lifting or carrying available and used?
  8.   Is the weight of the load carried limited according to the recognized guidelines?
  9.   Is the load held as close to the body as possible?
 10.   Are pulling and pushing forces limited?
 11.   Are trolleys fitted with appropriate handles and handgrips?

IV. Factors related to the design of tasks and jobs
  1.   Does the job consist of more than one task?
  2.   Has a decision been made about allocating tasks between people and machines?
  3.   Do workers performing the tasks contribute to problem solving?
  4.   Are the difficult and easy tasks performed interchangeably?
  5.   Can workers decide independently on how the tasks are carried out?
  6.   Are there sufficient possibilities for communication between workers?
  7.   Is there sufficient information provided to control the assigned tasks?
  8.   Can the group take part in management decisions?
  9.   Are the shift workers given enough opportunities to recover?

V. Factors related to information and control tasks
Information
  1.   Has an appropriate method of displaying information been selected?
  2.   Is the information presentation as simple as possible?
  3.   Has the potential confusion between characters been avoided?
  4.   Has the correct character/letter size been chosen?
  5.   Have tests with capital letters only been avoided?
  6.   Have familiar typefaces been chosen?
  7.   Is the text/background contrast good?
  8.   Are the diagrams easy to understand?
  9.   Have the pictograms been properly used?
 10.   Are sound signals reserved for warning purposes?

TABLE 21.8 (continued)   Examples of Factors to Be Used in the Ergonomics Checklists

Controls

1. Is the sense of touch used for feedback from controls?
2. Are differences between controls distinguishable by touch?
3. Is the location of controls consistent and is sufficient spacing provided?
4. Have the requirements for the control-display compatibility been considered?
5. Is the type of cursor control suitable for the intended task?
6. Is the direction of control movements consistent with human expectations?
7. Are the control objectives clear from the position of the controls?
8. Are controls within easy reach of female workers?
9. Are labels or symbols identifying controls properly used?
10. Is the use of color in controls design limited?

Human-computer interaction

1. Is the human-computer dialogue suitable for the intended task?
2. Is the dialogue self-descriptive and easy to control by the user?
3. Does the dialogue conform to the expectations on e part of the user?
4. Is the dialogue error-tolerant and suitable for user learning?
5. Has command language been restricted to experienced users?
6. Have detailed menus been used for users with little knowledge and experience?
7. Is the type of help menu fitted to the level of user's ability?
8. Has the QWERTY layout been selected for the keyboard?
9. Has a logical layout been chosen for the numerical keypad?
10. Is the number of function keys limited?
11. Have the limitations of speech in human-computer dialogue been considered?
12. Are touch screens used to facilitate operation by inexperienced users?

VI. Environmental factors

Noise and vibration

1. Is the noise level at work below 80 dBA?
2. Is there an adequate separation between workers and source of noise?
3. Is the ceiling used for noise absorption?
4. Are the acoustic screens used?
5. Are hearing conservation measures fitted to the user?
6. Is personal monitoring to noise/vibration used?
7. Are the sources of uncomfortable and damaging body vibration recognized?
8. Is the vibration problem being solved at the source?
9. Are machines regularly maintained?
10. Is the transmission of vibration prevented?

Illumination

1. Is the light intensity for normal activities in the range of 200 to 800 lux?
2. Are large brightness differences in the visual field avoided?
3. Are the brightness differences between task area, close surroundings, and wider surroundings limited?
4. Is the information easily legible?
5. Is ambient lighting combined with localized lighting?
6. Are light sources properly screened?
7. Can the light reflections, shadows, or flicker from the fluorescent tubes be prevented?

Climate

1. Are workers able to control the climate themselves?
2. Is the air temperature suited to the physical demands of the task?
3. Is the air prevented from becoming either too dry or too humid?
4. Are draughts prevented?
5. Are the materials/surfaces that have to be touched neither too cold nor too hot?
6. Are the physical demands of the task adjusted to the external climate?
7. Are undesirable hot and cold radiation prevented?
8. Is the time spent in hot or cold environments limited?
9. Is special clothing used when spending long periods in hot or cold environments?

Chemical substances

1. Is the concentration of recognized hazardous chemical substances in the air subject to continuous monitoring and limitation?
2. Is the exposure to carcinogenic substances avoided or limited?

TABLE 21.8 (continued)   Examples of Factors to Be Used in the Ergonomics Checklists

3.	Does the labeling on packages of chemicals provide information on the nature of any hazard due to their contents?
4.	Can the source of chemical hazards be removed, isolated, or their releases from the source reduced?
5.	Are there adequate exhaust and ventilation systems in use?
6.	Are protective equipment and clothing — including gas and dust masks for emergencies and gloves — available at any time if necessary?

Based on Dul, J. and Weerdmeester, B. *Ergonomics for Beginners: A Quick Reference Guide.* Taylor & Francis, London, 1993.

human characteristic), which, if satisfied, will also accommodate the rest of the expected user population. The main anthropometric criteria for workplace design are clearance, reach, and posture. Typically, clearance problems involve the design of space needed for the legs or safe passageways around and between equipment. If the clearance problems are disregarded, they may lead to poor working postures and hazardous work layouts. Consideration of clearance requires designing for the largest user, typically by adapting the 95th percentile values of the relevant characteristics for male workers. Typical reach problems in industry include consideration of the location of controls and accessibility of control panels in the workplace. The procedure for solving the reach problems is usually based upon the fifth percentile value of the relevant characteristic for female workers (smaller members of the population). When anthropometric requirements of the workplace are not met, biomechanical stresses that manifest themselves in postural discomfort, lower-back pain, and overexertion injury are likely to occur.

The Role of Ergonomics in Prevention of Occupational Musculoskeletal Injury

Lack of attention to ergonomic design principles at work and its consequences have been linked to occupational musculoskeletal injuries and disorders. Musculoskeletal work-related injuries, such as cumulative trauma to the upper extremity and lower-back disorders (LBDs), affect several million workers each year, with total costs exceeding $100 billion annually. For example, the upper-extremity **cumulative trauma disorders (CTDs)** account today for about 11% of all occupational injuries reported in the U.S. and have resulted in a prevalence of work-related disability in a wide range of occupations.

The dramatic increase of musculoskeletal disorders over the last 10 to 15 years can be linked to several factors, including the increased production rates leading to thousands of repetitive movements every day, widespread use of computer keyboards, higher percentage of women and older workers in the workforce, and better record keeping of employers as a result of a crackdown on industry reporting procedures by the Occupational Safety and Health Administration (OSHA). Other factors include greater employee awareness of these disorders and their relation to the working conditions, as well as a marked shift in the social policy regarding the recognition of compensation for work-related disorders. Given the epidemic proportions of the reported work-related musculoskeletal injuries, the federal government increases its efforts to introduce minimum standards and regulations aimed to reduce the frequency and severity of these disorders. For more information about these efforts, see Waters et al. [1993]; for NIOSH guidelines for manual lifting, see the ANSI [1994] draft document on control of CTDs.

The current state of knowledge about CTDs indicate that the chronic muscle, tendon, and nerve disorders may have multiple work-related and non–work-related causes. Therefore, CTDs are not classified as occupational diseases but rather as *work-related disorders,* where a number of factors may contribute significantly to the disorder, including work environment and human performance at work. The frequently cited risk factors of CTDs are (1) repetitive exertions; (2) posture — shoulder (elbow above mid-torso reaching down and behind), forearm (inward or outward rotation with a bent wrist), wrist (palmar flexion or full extensions), and hand (pinching); (3) mechanical stress concentrations over the base of palm, on the palmar surface of the fingers, and on the sides of the fingers; (4) vibration;

(5) cold, and (6) use of gloves [Putz-Anderson, 1988]. A risk factor is defined here as an attribute or exposure that increases the probability of the disease or disorder. Cumulative trauma disorders at work are typically associated with repetitive manual tasks that impose repeated stresses to the upper body, that is, the muscles, tendons, ligaments, nerves, tissues, and neurovascular structures. For example, the three main types of disorders to the arm are (1) tendon disorders (e.g., tendonitis), (2) nerve disorders (e.g., carpal tunnel syndrome), and (3) neurovascular disorders (e.g., thoracic outlet syndrome or vibration-Raynaud's syndrome).

From the occupational safety and health perspective, the current state of ergonomics knowledge allows for management of CTDs in order to minimize human suffering, potential for disability, and the related worker's compensation costs. Ergonomics can help to (1) identify working conditions under which the CTDs might occur, (2) develop engineering design measures aimed at elimination or reduction of the known job risk factors, and (3) identify the affected worker population and target it for early medical and work intervention efforts. The ergonomic intervention should allow management to (1) perform a thorough job analysis to determine the nature of specific problems, (2) evaluate and select the most appropriate intervention(s), (3) develop and apply conservative treatment (implement the intervention), on a limited scale if possible, (4) monitor progress, and (5) adjust or refine the intervention as needed.

Most of the current guidelines for control of the CTDs at work aim to (1) reduce the extent of movements at the joints, (2) reduce excessive force levels, and (3) reduce exposure to highly repetitive and stereotyped movements. Workplace design to prevent the onset of CTDs should be directed toward fulfilling the following recommendations: (1) permit several different working postures; (2) place controls, tools, and materials between waist and shoulder heights for ease of reach and operation; (3) use jigs and fixtures for holding purposes; (4) resequence jobs to reduce the repetition; (5) automate highly repetitive operations; (6) allow self-pacing of work whenever feasible; and (7) allow frequent (voluntary and mandatory) rest breaks. For example, some of the common methods to control the wrist posture are (1) altering the geometry of tool or controls (e.g., bending the tool or handle), (2) changing the location/positioning of the part, or (3) changing the position of the worker in relation to the work object. In order to control the extent of force required to perform a task, one can (1) reduce the force required through tool and fixture redesign, (2) distribute the application of the force, or (3) increase the mechanical advantage of the (muscle) lever system.

Fitting the Work Environment to the Workers

Ergonomic job redesign focuses on fitting the jobs and tasks to capabilities of workers — for example, "designing out" unnatural postures at work, reducing excessive strength requirements, improving work layout, introducing appropriate designs of hand tools, and addressing the problem of work/rest requirements. As widely recognized in Europe, ergonomics must be seen as a vital component of the value-adding activities of the company. Even in strictly financial terms, the costs of an ergonomics management program will be far outweighed by the costs of not having one. A company must be prepared to accept a participative culture and to utilize participative techniques. The ergonomics-related problems and consequent intervention should go beyond engineering solutions and must include design for manufacturability, total quality management, and work organization alongside workplace redesign or worker training.

An important component of the management efforts to control musculoskeletal disorders in industry is the development of a well-structured and comprehensive ergonomics program. The basic components of such a program should include the following: (1) health and risk factor surveillance, (2) job analysis and improvement, (3) medical management, (4) training, and (5) program evaluation. Such a program must include participation of all levels of management; medical, safety, and health personnel; labor unions; engineering; facility planners; and workers.

The expected benefits of ergonomically designed jobs, equipment, products, and workplaces include the improved quality and productivity with reductions in errors, enhanced safety and health performance, heightened employee morale, and accommodation of people with disabilities to meet the recommendations of the Americans with Disabilities Act and affirmative action. However, the recommendations offered by ergonomics can be successfully implemented in practice only with full understanding of the production processes, plant layouts, quality requirements, and total commitment from all management levels and workers in the company. Furthermore, these efforts can only be effective through participatory cooperation between management and labor through development of in-plant ergonomics committees and programs. Ergonomics must be treated at the same level of attention and significance as other business functions of the plant — for example, the quality management control — and should be accepted as a cost of doing business, rather than an add-on activity calling for action only when the problems arise.

Defining Terms

Cumulative trauma disorders (CTDs): CTDs at work are typically associated with repetitive manual tasks with forceful exertions, performed with fixed body postures that deviate from neutral, such as those at assembly lines and those using hand tools, computer keyboards, mice, and other devices. These tasks impose repeated stresses to the soft tissues of the arm, shoulder, and back, including the muscles, tendons, ligament, nerve tissues, and neurovascular structures, which may lead to tendon and/or joint inflammation, discomfort, pain, and potential work disability.

Engineering anthropometry: An empirical science branching from physical anthropology that deals with physical measurements for the human body — such as body size, form (shape), and body composition — and application of such measures to the design problems.

Ergonomics/human factors: The scientific discipline concerned with the application of the relevant information about human behavior, abilities, limitations, and other characteristics to the design, testing, and evaluation of tools, machines, systems, tasks, jobs, and environments for productive, safe, comfortable, and effective human use.

Occupational biomechanics: The application of mechanics to the study of the human body in motion or at rest, focusing on the physical interaction of workers with their tools, machines, and materials in order to optimize human performance and minimize the risk of musculoskeletal disorders.

References

Chaffin, D. B. and Anderson, G. B. J. *Occupational Biomechanics, 2nd ed.*, John Wiley & Sons, New York, 1993.

Dul, J. and Weerdmeester, B. *Ergonomics for Beginners: A Quick Reference Guide*, Taylor & Francis, London, 1993.

Karwowski, W. Complexity, fuzziness and ergonomic incompatibility issues in the control of dynamic work environments, *Ergonomics*, 34(6): 671–686, 1991.

Karwowski, W. Occupational biomechanics, In *Handbook of Industrial Engineering*, G. Salvendy, ed., pp. 1005–1046, John Wiley & Sons, New York, 1992.

Meister, D. Systems design, development and testing, In *Handbook of Human Factors*, G. Salvendy, ed., pp. 17–42, John Wiley & Sons, New York, 1987.

Pheasant, S. *Bodyspace: Anthropometry, Ergonomics and Design*, Taylor & Francis, London, 1986.

Putz-Anderson, V., Ed. *Cumulative Trauma Disorders: A Manual for Musculoskeletal Diseases of the Upper Limbs*, Taylor & Francis, London, 1988.

Sanders, M. S and McCormick, E. J. *Human Factors in Engineering and Design, 6th ed.*, McGraw-Hill, New York, 1993.

Waters, T. R., Putz-Anderson, V., Garg, A., and Fine, L. J. Revised NIOSH equation for the design and evaluation of manual lifting tasks, *Ergonomics*, 36(7): 749–776, 1993.

Wickens, C. D. Information processing, decision-making, and cognition, In *Handbook of Human Factors*, G. Salvendy, ed., pp. 72–107, John Wiley & Sons, New York, 1987.

Further Information

ANSI. *Control of Cumulative Trauma Disorders,* ANSI Z-365 Draft, National Safety Council, Itasca, IL, 1994.

Clark, T. S. and Corlett, E. N. Eds. *The Ergonomics of Workspaces and Machines: A Design Manual,* Taylor & Francis, London, 1984.

Eastman Kodak Company. *Ergonomic Design for People at Work, Vol. 1,* Lifetime Learning, Belmont, CA, 1989.

Eastman Kodak Company, *Ergonomic Design for People at Work, Vol. 2,* Van Nostrand Reinhold, New York, 1986.

Grandjean, E. *Fitting the Task to the Man, 4th ed.,* Taylor & Francis, London, 1988.

Helander, M., Ed. *Handbook of Human-Computer Interaction,* North-Holland, Amsterdam, 1988.

Kroemer, K. H. E., Kroemer, H. B., and Kroemer-Elbert, K. E. *Ergonomics: How to Design for Ease and Efficiency,* Prentice Hall, Englewood Cliffs, NJ, 1994.

Salvendy, G., Ed. *Handbook of Human Factors,* John Wiley & Sons, New York, 1987.

Salvendy, G. and Karwowski, W., Eds. *Design of Work and Development of Personnel in Advanced Manufacturing,* John Wiley & Sons, New York, 1994.

Wilson, J. R. and Corlett, E. N., Eds. *Evaluation of Human Work: A Practical Methodology,* Taylor & Francis, London, 1990.

Woodson, W. E. *Human Factors Design Handbook,* McGraw-Hill, New York, 1981.

Ergonomics Information Sources and Professional Societies

International Ergonomics Association (IEA)
Poste Restante
Human Factors and Ergonomics Society
P.O. Box 1369
Santa Monica, CA
Phone: (310) 394-1811/9793. Fax: (310) 394-2410

Crew System Ergonomics Information Analysis Center
AL/CFH/CSERIAC
Wright Patterson AFB
Dayton, OH 45433-6573
Phone: (513) 255-4842. Fax: (513) 255-4823

Ergonomics Information Analysis Centre (EIAC)
School of Manufacturing and Mechanical Engineering
The University of Birmingham
Birmingham B15 2TT
England
Phone +44-21-414-4239. Fax: +44-21-414-3476

Journals

Ergonomics Abstracts. Published by Taylor & Francis, London.

Human Factors. Published by the Human Factors and Ergonomics Society, Santa Monica, CA.

International Journal of Human Factors in Manufacturing. Published by John Wiley & Sons, New York.

Applied Ergonomics. Published by Butterworth-Heinemann, Oxford, England.

International Journal of Human-Computer Interaction. Published by Ablex, Norwood, NJ.

International Journal of Industrial Ergonomics. Published by Elsevier, Amsterdam.

International Journal of Occupational Safety and Ergonomics. Published by Ablex, Norwood, NJ.

| **MICROSOFT** | **Headquarters** — Redmond WA 98052 | **Web Site** — http://www.microsoft.com |

In 1975 Paul Allen and Bill Gates started setting out to adapt Basic for a new machine — MITS Altair, the first personal computer. Allen flew to Albuquerque to demonstrate the language and to everyone's surprise and relief it worked perfectly the very first time. Allen soon accepted a position with MITS as Director of Software Development, and Gates followed him later that year to form an informal partnership called Micro-soft, complete with hyphen. Microsoft moved to Seattle in 1978 with 13 employees and revenue of $1.4 million.

Growth in revenue

1975	$16	thousand
1980	$7.52	million
1985	$140	million
1990	$1,183	million
1995	$5,940	million
1996	$8,671	million

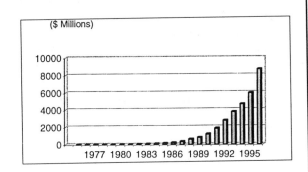

Innovations:

Even though Microsoft is the fastest growing company in the last two decades, much of its growth cannot be attributed to innovation but rather to excellent business management and software design management. The fast growing Microsoft and its CEO Bill Gates engulfed the industry with new ideas.

1975 — Basic
1980 — Pascal language and XENIX (enhanced version of Unix)
1981 — Disc Operating System (MS- DOS), Basic, Cobol, and Pascal for PC
1985 — Windows
1995 — Windows 95

Strategic and Management Issues:

- Gates speaks out for the first time against software piracy in February 1976: "If you are STILL using Altair BASIC 1.1, you have a copy that was stolen in March 1975!"
- Aggressive software design and marketing.
- Development of products for the Internet.

VII

Global Business Management

22

International Business

Douglas C. West
Henley Management College

Walter Kuemmerle
Harvard University

Lan Xue
The George Washington University

Chung-Shing Lee
The George Washington University

Hildy Teegen
The George Washington University

22.1 Global Marketing

Douglas C. West

At the root of **global marketing** is the premise that products can be standardized to an extent that the key elements of the marketing mix need not be significantly modified across different countries. A global industry, therefore, is one in which the actions of competitors in local markets are determined by their respective global positions. The management of marketing, namely, market research, product and channel, pricing and promotion, remain the same for global marketers as international advertisers. However, the *implementation* is different. International advertisers allow greater local control of strategy and will customize all or several elements of the marketing mix. By contrast, global marketers prefer a more centralized and standardized approach to the mix [Kashani, 1989]. Global marketing is commonly used as a term to include international marketing as well. This section will examine the reasons for standardizing international marketing programs and provide an assessment of the key components in the global marketing mix.

Standardized vs. Customized Marketing

Levitt [1983] first noted the move toward global marketing. Interestingly, given the subject of this handbook, Levitt emphasized the importance of technology. According to Levitt, "a powerful force drives the world toward a converging commonality, and that force is technology…The result is a new commercial reality — the emergence of global markets for standardized consumer products on a previously unimagined scale of magnitude." Furthermore, companies such as Volvo and Saab have demonstrated that you not need be a large firm to adopt global marketing. Ohmae [1987] cites a long list of products including

Kodak, Nike, and Revlon that are identical in Japan, Europe, and the United States. High-technology companies are especially poised well to take advantage of global marketing as their customers have extremely similar requirements internationally. By taking control and standardizing the marketing mix internationally, global marketers can benefit from:

- Economies of scale in marketing and production such as single brand names, advertising, packaging, product specifications, and so on.
- Exploiting excellent marketing strategies.
- The market of around 200 million international travelers.
- Erecting barriers of entry by defending their local market as well as attacking global and international rivals and preempting potential local rivals.

Despite the many benefits of being a global marketer and the enormous number of global products, there are many problems in implementing truly global marketing strategies. With standardization there is a risk of mediocrity. Standardized global marketing may lead to a strategy that "almost" fits every market but may not be "exactly" right anywhere. Even companies held up to be global marketers, such as Pepsi and Coca-Cola, modify their products in sweetness and carbonation, pricing, distribution, and advertising according to the regional/local environment. Companies often adapt and modify their marketing in different countries because of local variances (e.g., McDonald's replacing ketchup with chili sauce in Mexico) and/or differences in the stages in the international product life cycle (e.g., National Cash Register sells crank-operated cash registers in Africa). The current trend is for increasing **regionalization**. For example, in the case of advertising, Ogilvy & Mather shot a Pepsi commercial featuring Tina Turner performing with six different pop stars. Depending on where you lived in the world you saw a different partner for Turner in the commercial (e.g., David Bowie in the U.K.). Production costs were minimized with one shoot and the adapted commercials were shown in over 30 countries.

Global Marketing Programs

Marketing Research

The global research process follows the same pattern as the local one but differs in implementation. Global marketing research starts with the setting of the objectives and determination of the required information. However, the method of gathering the information may need to be customized depending on local circumstances. Secondary research might begin with an assessment of business libraries, local government, local embassy, associations, other firms in the market, and own company data. Very little secondary research may be available in less developed markets. Primary research is normally conducted in the country concerned (but immigrants, students, or a company's own staff may also be used for primary research by proxy). Business-to-business companies might use conferences at home or abroad to gather information but not necessarily in the country concerned. The major problems arising in primary research are often cultural (e.g., finding out influences on household purchasing in countries where woman cannot be interviewed), and sometimes there are legal hurdles (e.g., data protection laws).

Product

Straight product extension [Keegan, 1995] is the ideal product strategy for global marketers. It has proved successful among a range of consumer products such as foodstuffs, consumer electronics, and clothing, but it is not without problems. For example, there is a tiny market for American refrigerators, stoves, and top-load washers in Europe because such appliances are too big for the smaller kitchens. However, much greater success for straight extension has occurred in the global business-to-business market where many products such as banking, insurance, machine tools, office equipment, and retailing have been standardized to a great degree. If a standardized product is not feasible, then a company will need to adapt it to suit regional or local preferences. One final point is that global marketers may exploit international market opportunities. For example, Honda has been extremely successful marketing lawn mowers in North America despite the lack of grass in Japanese gardens.

Promotion

Global marketers aim to standardize all their promotions internationally. However, the international marketplace is such to make the working of promotions less predictable, and so this aim is not often achievable. One of the key issues in global promotion is translation error. Perhaps the most oft-quoted blunder is General Motors strapline "Body by Fisher". The Flemish translation became "Corpse by Fisher". However, most blunders occurred over 20 years ago, and it is rare in today's sophisticated marketing environment for companies to make such translation errors. The practice of **backtranslating** and seeking the advice of local advertising agencies and/or distributors has made translation errors uncommon. Meaning rather than translation has become more of an area of concern. International advertisers have sometimes encountered problems when their appeals have been counter to local culture, e.g., the Gulda beer campaign where Nigerians failed to respond because the person in the advertisement always drank alone, whereas Nigerians see drinking as a social activity.

A major local difference that has forced many advertisers to abandon global promotions has been regulation. Different countries often require the modification of appeals. For example, the Germans prohibit product claims of superiority, and Scandinavian countries have extremely strict laws on product claims of all kinds. Many Moslem countries outlaw appeals involving women revealing their bodies (causing particular problems for deodorant/perspirant advertisers). Every country has strict codes relating to advertising to children. Also, certain product categories are strictly regulated around the world (especially cigarettes and alcohol). It is the norm for international advertisers to seek local professional advice if the legality of an appeal or product advertising is in doubt or to seek advice from the local medium or regulatory body.

International trade fairs are one of the best means for a company to promote abroad. About 2000 trade fairs are held a year in 70 countries. Many are booked up several years ahead. The main functions of trade fairs are to

- Meet potential customers.
- Develop mailing lists.
- Make distributor contacts
- Demonstrate products.
- See what your competitors are doing.
- Save time and money in making contacts.

There are several **horizontal trade fairs** around the world. The biggest is the Hanover Trade Fair held in April, which attracts around 500 to 550,000 industrial buyers and sellers from around the world. Others worth mentioning are in Leipzig, Canton, and Milan. **Vertical trade fairs** are held all over the world several times a year. These are often more technical in nature and often do not allow entrance to the public.

Channel Organization

Distribution is extremely difficult for global marketers to standardize, given the variations in such countries as France, Japan, Poland, and India. There are enormous differences in the number and type of intermediaries in international markets through and including stores, hypermarkets, supermarkets, local shops, open markets, and individual sellers and a multitude of intermediaries in between. Companies planning to or already involved in international marketing face a number of options in distribution. The first decision is whether the market is big enough to justify the expense of setting-up a local international division or to simply decide to export (exporting is part of the global marketing mindset given the economies of scale). Exporting involves the use of services either located in the domestic market or abroad to reach the foreign market. Brokers are useful when a firm is looking for a short-term sales relationship with buyers as brokers do not take title to the products traded. Manufacturer's export agents (MEAs) are more specialized than brokers and offer a more long-term service. As with the broker, MEAs do not take title to the goods. Merchant middlemen take title to products through purchase for resale within the foreign market. As they absorb most of the risk in selling, their margins can be high.

Generally, a global marketer will seek to establish its own international division to increase its control and bypass domestic export services.

In international marketing, a firm may use **licensing** to gain a market presence or further a technological lead without having to commit resources (e.g., Philips licensing of cassette recorders) and whereby risky economic or political investments can be avoided. The downside is that the firm will reduce its share of the profits from the venture compared to the alternative of exporting or setting up a local production facility. Another difficulty is that, once the license has run out, the former licensee may turn into a competitor in both the domestic and foreign market. Licensing is particularly common in the drinks industry. Franchising is a kind of licensing that is more global in nature. Here the franchiser offers the franchisee a total marketing program from production to promotion. Fast food chains are proponents (Burger King, McDonald's, and KFC) as well as services such as car rental (e.g., Hertz) and hotels (e.g., Holiday Inns).

One way of bypassing channel problems is **direct marketing**. Direct mail was pioneered in the United States by leading catalog houses such as Sears. It is now widely used in the developed western world by an enormous range of consumer and business-to-business companies. Direct mail tends to be more local than global in character as it is less concerned with branding. However, items of a general nature can be translated. Furthermore, direct mail can offer a company the ability to do business in another country without establishing a presence in that country. Telemarketing is another growth area of direct marketing. Its growth has been spectacular in Britain, Canada, Denmark, France, Norway, Sweden, United States, and Germany in particular. Selling by telephone to consumers and businesses can be considerably cheaper than personal selling. Its effectiveness depends on the relevance of lists and the sensitivity of the operation. International telemarketing is viable if can be used effectively and the potential sales can outweigh the costs. This tends to limit its use to business-to-business prospects of relatively small numbers.

Price

Global marketers generally seek to establish similar relative prices in their international markets according to their global strategies, e.g., Volkswagen tries to sell its cars globally at a similar price to its domestic competitors. A firm with a technological edge or with a specialized product has greater pricing flexibility than others. In such cases, there tend to be fewer competitors and no local production (competitors are also global and face similar international costs), and transportation costs are relatively less important to high-technology products compared to bulkier low-technology ones.

Factors to take account of in global pricing include market demand and competition, given differing income levels, spending power, and demand elasticities. Culture may also play a role. For example, as negotiation is normally a prerequisite in the Middle East, many companies use a higher price list there to allow for discretion. Where rapid inflation is common (e.g., Brazil and Israel) companies tend to price in a stable currency and translate prices into local currencies on a monthly to daily basis. Foreign exchange is probably the single most important variable in price changes in global marketing.

Defining Terms

Backtranslating: Translate from English into the foreign language and then a different person translates it back into English.

Direct marketing: Any method of marketing to individual customers, but principally direct mail and telemarketing.

Global marketing: The coordination and performance of marketing across national boundaries based upon a set strategic position.

Horizontal trade fair: Nonspecific product categories, e.g., machine tools and cement.

Licensing: Selling either a trademark or a patent (protecting a technology) or both to another firm.

Regionalization: Marketing strategy is adapted or modified regionally (e.g., Western Europe or Australasia).

Straight product extension: Marketing the product without any modification.

Vertical trade fair: Specific product category, e.g., Paris Air Show or Earls Court (London) Toy Fair.

References

Levitt, T. The globalization of markets, *Harv. Bus. Rev.*, May-June: 92–102, 1983.
Keegan, W. J. *Multinational Marketing Management*. Prentice Hall, Englewood Cliffs, NJ, 1995.
Kashani, K. Beware the pitfalls of global marketing, *Harv. Bus. Rev.*, September-October: 91–97, 1989.
Ohmae, K. The triad world view, *J. Bus. Strat.*, Spring: 8–16, 1987.

Further Information

Hamel, G. and Prahalad, C. K. Do you really have a global strategy?, *Harv. Bus. Rev.*, July-August: 139–148, 1985.
International Financial Statistics (published by the International Monetary Fund).
Kanter, R. M. Thriving locally in the global economy, *Harv. Bus. Rev.*, September-October: 151–160, 1995.
Maruca, R. G. The right way to go global: an interview with Whirlpool CEO David Whitwam, *Harv. Bus. Rev.*, March-April: 135–145, 1994.
United Nations Statistical Yearbook.

22.2   Multinational Corporations

Walter Kuemmerle

Multinational corporations (MNCs) generally have a home base in one country but operate wholly or partially owned subsidiaries in other countries. The economic rationale for the existence of MNCs is either cheaper access to existing and new markets or cheaper access to resources. MNCs carry out cross-border transactions internally that would otherwise be carried out through cross-border contracts with third parties, such as sales agents or original equipment manufacturers. Over the last decades, the importance of MNCs in the world economy has increased significantly. In some countries, foreign firms are now responsible for close to 50% of manufacturing output (see Fig. 22.1).

The Evolution of Multinational Corporations

Firms generally start their economic activity in one location, where they perform administrative tasks and the development of new products, manufacturing, and marketing. As firms grow larger they tend

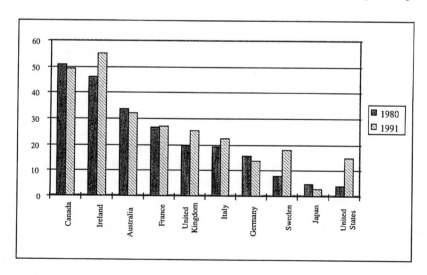

FIGURE 22.1   Share of foreign enterprises in manufacturing production. (*Source:* OECD, *Globalisation of Industry*, 1996.)

to move some (or all) of their economic activity to different locations, including locations abroad. Economic theory argues that firms will relocate activities if the costs of relocation are outweighed by the benefits of cheaper local inputs or better access to local markets in the new location [Weber, 1929].

Since national boundaries rather than mere geographical distances generally represent discrete changes in costs of local inputs and/or access to local markets, foreign countries often represent more attractive locations for the establishment of new sites of economic activity than distant locations in the same country. For example, if a company located in Massachusetts seeks to reduce the cost of labor content of its products, it will probably take a closer look at Mexico as a potential site for a factory rather than the state of Arizona; labor costs in Arizona are not that much lower than in Massachusetts, yet Arizona is almost as far from Massachusetts as is Mexico. Thus, the geographical dispersion of a growing firm's economic activity generally occurs *across* rather than *within* national boundaries.

The location where a firm is founded generally remains the firm's home base. Home base can be defined as the location where top management resides, where key strategic decisions get made, and where a considerable part of the firm's pool of firm-specific knowledge resides. Firm-specific knowledge is knowledge about a firm's particular technological capabilities and about how these capabilities can be combined with market opportunities in a way that maximizes the firm's profit. In other words, a firm's home base is the location of those employees and managers who know what their firm can achieve at any given point in time and how things can get done.

The dominant model described in the literature argues that, as firms grow, they will start their geographical dispersion by locating sales and marketing subsidiaries away from the firm's home base in order to increase sales through closer proximity to customers [Caves, 1996; Vernon, 1966].

As a next step, firms will often locate manufacturing facilities in countries with lower labor costs. Depending on the industry and the type of product, this sequence might be reversed, and firms will first locate manufacturing facilities abroad and then follow up with sales subsidiaries. For example, a manufacturer of automobile heating systems that sells its products only to one or two car manufacturers might not need an additional sales office but might invest in a new manufacturing location in a lower-wage country.

As a third step of geographical expansion, managers might decide to locate research and development (R&D) facilities abroad (see also third section of this article). Just as with the second step (establishment of manufacturing facilities), a reversal in the order of steps of geographical expansion can occur, depending on a firm's specific situation. Figure 22.2 gives an idea of the typical geographical evolution of the MNC.

An Organizational Model for Multinational Corporations

MNCs seek to carry out a complex set of tasks within the firm's boundaries. The decision to carry out these tasks internally rather than to outsource them is based on senior management's assumption that in the long run the internally provided activities will be cheaper or of higher quality than externally provided ones [Porter, 1986]. This competitive advantage is either based on economies of scale or economies of scope or monopolistic competition or a combination of these three factors. Firms need to choose an organizational design that is appropriate for the realization of competitive advantage. The appropriate organizational design of the MNC depends on a number of factors. Firms that carry out only a small part of their economic activity abroad often exercise tight control over all cross-border activities from the firm's home base. As the importance of foreign activities grows, however, units abroad require more local decision autonomy. Senior management at the firm's home base is often unfamiliar with the specific local requirements and therefore cedes authority to local managers. The willingness of top management to give local managers power increases with the cultural and geographic distance of the local country from the firm's home base. The upside of increased local autonomy are shorter decision routines and a quicker response to market needs. The potential downside is a disintegration of the MNCs network structure. However, it is exactly this network of informal and formal links between geographically

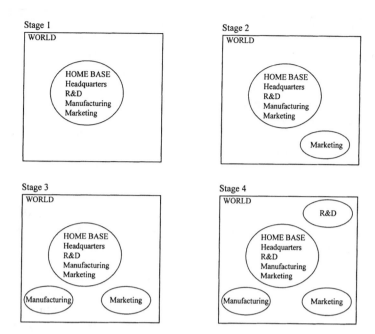

FIGURE 22.2 A model of a multinational firm's geographical expansion over time. The expansion of marketing, manufacturing, and R&D can also occur in reverse order.

dispersed firm sites that leads to economies of scope and scale within the MNC. Therefore, managers in MNCs have to make sure that their choice of organizational structure protects information channels and decision links between different units of their firm as the firm grows in size and geographical coverage. The more countries an MNC operates in and the more subsidiaries it runs, the more complex the task of intrafirm coordination becomes.

Bartlett and Ghoshal [1989] have argued that the challenges of managing a large MNC require a flexible management solution that makes use of specific strengths of overseas operations without hampering the network of primarily informal links between subsidiaries in different countries. Bartlett and Ghoshal call this solution the "**transnational management model**". In many ways the transnational model represents a compromise between local autonomy and centralized decision making. Under the transnational model, an MNC's assets and capabilities are dispersed according to the most beneficial location for a specific activity. At the same time, overseas operations are interdependent, and knowledge is developed jointly and shared worldwide.

ABB, a firm that was created through a merger of a Swedish and a Swiss company, can be regarded as a successful implementation of the transnational management model. Within ABB, designated subsidiaries are in charge of product categories on a worldwide level. All major activities are coordinated from a headquarters in Switzerland that ensures that information about technology and markets flows not only within but also across subsidiaries and across geographical distances. The case of ABB is also interesting since this MNC has two home bases, Sweden and Switzerland. While most firms still have one nation as a home base, there is a trend toward cross-national mergers, and an increasing number of firms will have to cope with multiple home bases in the future.

Technology Management in Multinational Corporations

The management of knowledge creation and knowledge flows is a key challenge for successful management of MNCs. The management of R&D across national boundaries is arguably the most difficult part of knowledge management within MNCs. Most MNCs have been successful in foreign markets because of the technological capabilities of R&D operations at their home base. However, over the last decade many firms in industrialized countries have realized the need to establish R&D activities away from their home base. The underlying reason for firms' growing propensity to locate R&D facilities abroad is twofold.

First, there seems to be an increasing number of "knowledge-creating clusters" around the world that firms deem important for their own survival and growth. While in the past the knowledge created by universities and competitors at the firm's home base was sufficient, now there are more clusters away from a firm's home base that offer knowledge on relevant technical and scientific topics. In addition, more clusters in *adjacent* fields of science and technology seem to be important for innovating and creating new products. In the pharmaceutical industry, for example, knowledge from the biotechnology industry is increasingly important (although not sufficient) for the creation and introduction of new drugs. Pharmaceutical firms located in Europe, Japan, and the United States have acquired or established research sites close to leading universities in Boston and in the San Francisco Bay area. In the electronics industry, a number of foreign computer hardware manufacturers have established R&D sites in Silicon Valley, primarily to stay up to date on recent developments in the software industry and to understand how software impacts the future of computer hardware manufacturing.

Second, as demand by local customers in foreign markets grows increasingly sophisticated and as MNCs operate more sophisticated manufacturing facilities abroad, there is a strong need for local R&D capabilities. Local R&D sites represent a transfer mechanism for knowledge that is created at the firm's home base and is designated to be exploited locally. A number of U.S. MNCs in the electronics industry have established R&D sites close to factories in Southeast Asia. These R&D sites make sure that the latest knowledge from the firm's home base R&D sites is applied effectively and speedily in manufacturing operations.

Results from an empirical study [Kuemmerle, 1997] of all R&D facilities of 32 MNCs in the electronics and pharmaceutical industries show that these firms had 156 dedicated R&D sites abroad, more than 60% of them established after 1984. The same study also showed that the two most important drivers for FDI in R&D are, on the one hand, firms seeking spillovers from universities' and competitors' R&D sites and, on the other hand, firms seeking spillovers within their own boundaries.

In the first case, firms primarily seek to learn new knowledge and skills from the local environment. This knowledge subsequently gets transferred back to the firm's home base where it is combined with the firm's existing knowledge and transformed into new products. This motive can be called ***home-base-augmenting*** investment in R&D since this type of investment helps firms to augment their knowledge base at home.

In the second case, firms primarily seek to exploit their existing pool of knowledge by transferring knowledge from the firm's home base to new markets and manufacturing facilities that are already located abroad. This motive can be called ***home-base-exploiting*** since firms use these R&D sites to exploit their home base.

The distinction between home-base-augmenting and home-base-exploiting R&D sites is essentially based on the economics of knowledge acquisition and knowledge transfer. Figure 22.3 gives an idea of the direction of knowledge flows within MNCs. It becomes clear from this figure that home-base-augmenting and home-base-exploiting R&D sites fulfill very different roles within the MNC. Distinguishing R&D sites according to their main mission can be a very useful tool for technology management in MNCs. Home-base-augmenting sites have to be well connected to the MNC's local environment in order to capture relevant knowledge. They should generally be staffed with a respected local scientist as a leader. Home-base-exploiting R&D sites, on the other hand, have to be particularly well connected to the firm's home base. They should generally be staffed with a scientist who has working experience at the MNC's home base R&D site and who can identify knowledge that could be transferred from the home base to the local R&D site.

As MNCs proliferate, cross-national management skills in general and cross-national technology management skills in particular will be in increasing demand. Particularly medium-sized firms that are on the threshold of becoming an MNC would be well advised to prepare for these managerial challenges early.

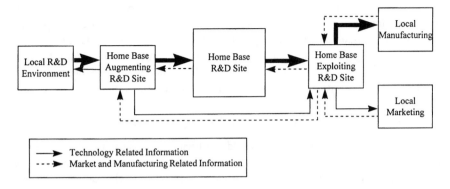

FIGURE 22.3 Direction of R&D information flows in multinational corporations.

Defining Terms

Home-base-augmenting R&D: R&D activities that are carried out abroad primarily in order to augment an MNC's stock of knowledge. Home base augmenting R&D activities are generally located close to leading universities or close to main competitors.

Home-base-exploiting R&D: R&D activities that are carried out abroad primarily in order to exploit an MNC's existing stock of knowledge. Home base exploiting R&D activities are generally located close to a firm's manufacturing sites or close to major customers.

Multinational corporation (MNC):   A firm that has a home base in one country but operates wholly or partially owned subsidiaries in other countries.

Transnational management model: A management model that acknowledges the complex challenges of managing an MNC. The transnational management model argues that an MNC should be managed as a flexible network of units. One of top management's main functions is the facilitation of information flows between units.

References

Bartlett, C. and Ghoshal, S. *Managing across Borders: The Transnational Solution*, Harvard Business School Press, Boston, 1989.

Caves, R. E. *Multinational Enterprise and Economic Analysis*, Cambridge University Press, Cambridge, 1996.

Kuemmerle, W. Building effective R&D capabilities abroad, *Harv. Bus. Rev.*, (March/April): 61–70, 1997.

Porter, M. E. Competition in Global Industries, Harvard Business School Press, Boston, 1986.

Vernon, R. International investment and international trade in the product cycle, *Q. J. Econ.*, LXXX (May): 190–207, 1966.

Weber, A. *The Theory of Location of Industries*, University of Chicago Press, Chicago, 1929.

Further Information

OECD. *Globalisation of Industry*, OECD, Paris, 1996.

22.3 Overseas Production Facilities

Lan Xue and Chung-Shing Lee

Traditionally, motivations for doing business overseas include securing key supplies, seeking foreign markets, and accessing low-cost factors of production. In recent years, however, some new forces have begun to emerge, which include increasing scale economies, reducing research and development (R&D) costs, and shortening product life cycles. These new forces, joined by the old ones, have made global scale of operations not a matter of choice but an essential prerequisite for companies to survive and succeed in many industries. The increased uncertainty and complexity from the expanded activities overseas present a major management challenge. Managing overseas production facilities is no exception. The following discussion will focus on three sets of issues in managing overseas production facilities: (1) the integration of the decision to produce overseas with a company's global strategy, (2) site selection for overseas production facilities, and (3) management of overseas operations.

Global Strategy and Overseas Manufacturing Facilities

Corporations operating in many parts of the world must balance pressures for global integration (e.g., global standardization in product manufacturing and marketing) against pressures for local responsiveness (e.g., consumer tastes, political, economic, sociocultural, and technological environment). A multinational corporation (MNC) follows a **multidomestic** strategy when each of its multiple overseas operations act as a completely distinctive place. Alternatively, MNCs can follow a **global** strategy that standardizes or tightly integrates operations under a single unifying strategy from country to country. However, a hybrid or **transnational** strategy [Bartlett and Ghosal, 1992] that balances the two extreme approaches was often adopted by many MNCs. The particular strategy employed by an organization has strong strategic implications for its establishment and expansion of overseas production facilities.

International Manufacturing Strategy

The decision to establish and expand overseas production facilities is determined by a company's manufacturing strategy. In general, an MNC's international **manufacturing strategy** involves the following interrelated decision categories [Wheelwright, 1984; Buffa, 1984; Flaherty, 1986]:

- Capacity: Amount, timing, and type.
- Facilities: Location, build or buy, size, and specialization.
- Technology: Equipment, product, process, and system.
- Vertical integration: Direction, extent, and balance.
- Workforce and job design: Skill, work rules, wage, labor-management relationship, and safety and security.
- Quality assurance: Defect prevention, monitoring, intervention, and TQM.
- Production planning and materials control: Manufacturing engineering, computerization, centralization, standards and decision rules.
- Organization: Structure, global linkages, management reporting system, and support groups.

Overall, companies should integrate their manufacturing strategy with other functional strategies (e.g., marketing, logistics, R&D, and finance) and systematically evaluate various manufacturing alternatives on the basis of their congruence with national environments and economic conditions.

Global Configuration and Coordination of Manufacturing Facilities

Having established manufacturing capacity overseas, the MNC has to decide upon the roles and relationships each plant will play in the company's international manufacturing strategy as well as the overall **configuration** and **coordination** of its geographically dispersed activities. Numerous strategies to manage global operation have been prescribed by many scholars. *Product standardization* [Levitt, 1983; Hout et al., 1982] can be adopted to exploit **economies of scale** through global volume and managed interdependently to achieve synergies across different activities while a broad *product portfolio* is better suited to share investments in technologies and distribution channels and *cross-subsidization* across products and markets [Hamel and Prahalad, 1985]. *Strategic flexibility* can create options by using multiple sourcing, production shifting to benefit from changing factor costs and exchange rates, and arbitrage to exploit imperfections in financial and information markets [Kogut, 1985]. Ultimately, decisions on sourcing and production location as well as organization structure have to be based on careful examination of the sources of competitive advantage, which come from national differences in input and output

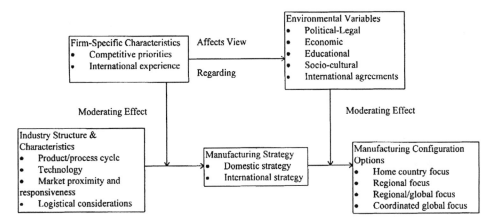

FIGURE 22.4 Major factors taht may influence a firm's international manufacturing configuration decision. (*Source:* DuBois, F. L., Toyne, B., and Oliff, M. D. *J. Int. Bus. Studies,* 2nd quarter: 307–333, 1993.)

markets, economies of scale in production, and **economies of scope** or synergies across activities [Ghoshal, 1987].

Figure 22.4 identifies the major elements of the internal and external environments that were found to have an influence on an organization's manufacturing strategy and its *international manufacturing configuration* patterns. More attention needs to be placed on the role that management decisions with regard to product quality, process flexibility, delivery dependability, and manufacturing costs play in decisions to increase or decrease a firm's international involvement [DuBois et al., 1993].

Location Selection for Overseas Production Facilities

Location selection for overseas production facilities is an important decision with significant consequences. Minor errors in location selection could result in major losses in the future because of the sunk cost involved. While cost savings may have been the driving factor in location selection in the past, rapid changes in societies, business practices, and technologies have introduced many new considerations that were unthought of previously.

Location Analysis

The first step in locating production facilities overseas is to analyze relevant factors that might influence location decisions. Table 22.1 gives a list of these factors, which include business climate and risks, infrastructure, cost, human resources, and site-specific issues. These factors can be relevant at the national, regional/community, and site levels [Mullin, 1996; Chase and Aquilano, 1995; Tong and Walter, 1980].

Business Climate and Risks
In comparing business climate and risks in different countries, one is concerned with the political stability, government regulations, and business barriers at the country level. While the end of Cold War presents many exciting new opportunities, the social and political transformations many countries are undergoing also bring great political uncertainties one should never underestimate. Relations between the country of location and the host country must also be taken into account. Among various government regulations, environmental regulations is particularly important since it will not only have measurable cost implications but also influence the relationship with the local community. Business barriers include both tangible ones, such as those imposed by trade blocs on nonmember companies, and intangibles ones, such as those related to business practice and culture. At the community/regional level, companies should take advantages of hospitable environment provided in free trade zones, export processing zones, or special economic zones, which were created in many developing countries to attract foreign investment.

TABLE 22.1   Factors for Location Analysis

Factors	Country	Regional/community	Site
Business climate and risks			
Political stability	x		
Government regulations	x	x	
Business barriers	x	x	
Community environment		x	x
Infrastructure			
Transportation	x	x	x
Telecommunication	x	x	x
Water, sewers, and utilities			x
Business services		x	x
Cost			
Total cost	x	x	x
Tax rates	x	x	
Human resources			
Availability and quality of labor	x	x	
Availability and quality of management and technical personnel	x	x	
Site-specific issues			
Size of the land			x
Adjoining land			x
Zoning			x

Infrastructure

Infrastructure requirements today include not only adequate road, rail, air, and sea transportation but also telecommunication facilities that are vital for communication between the overseas facilities and their headquarters. Water, sewers, utilities, and business services are also important factors in selecting a particular community or a site.

Cost

The bottom line in selecting one location over another boils down to the total cost of locating and operating at the selected site. Components of this total cost include labor, transportation, land, construction, energy, and others. A hidden cost often being ignored is the loss of customer responsiveness arising from locating away from the main customer base. Of course, tax rates at the national and local level are always important.

Human Resources

Considerations in this respect include not only the adequate supply of labor but also level of education, wage rate, skill level, work ethics, and the degree of unionization. Availability of management and technical talents is also extremely important.

Site-Specific Issues

These relate to the size of the land, adjoining land, zoning, drainage, soil, and others concerns specific to the selected site.

The analysis of these factors requires a great amount of information that is often available from government agencies, business associations, banks, and consulting companies specializing in site selection.

Facility Location Methods

In addition to qualitative analysis of the factors relevant to location selections, quantitative methods are also developed to help managers to evaluate alternative regions and sites. Some of these techniques are more appropriate to the regional selection and others to the site selection; all of them can be used for any stage. A detailed cost analysis should also accompany each of these methods.

Weighted score model [Meredith, 1992] is probably the most widely used technique because it provides a mechanism to combine diverse factors in an easy-to-understand format. The first step to use this method is to identify all important location factors, which has been discussed previously. The next step is to assign a weight to each factor depending on its importance to the manager. The more important a factor is, the more weight it receives. In the third step, a score is assigned to each of the location alternatives on each factor, again, with higher scores representing better results. In the last step, get the sum of the product of the factor weights and the factor scores for each location alternative. The location with the largest weighted score is considered the best.

The **center of gravity method** [Chase and Aquilano, 1995; Meredith, 1992] is often used to locate single facilities that consider the existing facilities, the distance between them, and the volumes of goods to be shipped. In its simplest form, this method assumes that inbound and outbound transportation costs are equal, and it does not include special shipping costs for less-than-full loads. It begins by placing the existing locations on an arbitrary coordinate system. The purpose is to establish relative distances between locations.

There are some other popular methods being used for location selection, such as the **rate-volume-distance transportation cost model** [Meredith, 1992], the **analytic delphi model** [Azani and Khorram-shahgol, 1990], and the **breakeven model** [Meredith, 1992]. Interested readers can get more information from the relevant references.

Management of Overseas Production Facility

In addition to the typical tasks of managing a production facility, operating production facilities overseas presents new managerial challenges; among them, how to strike an appropriate balance between central control and local flexibility is probably the most critical one.

Traditional approach has focused on the division of responsibilities between center and overseas facilities and how to match this division with proper organizational structures. However, international managers have long found that there is no single magic formula that can effectively deal with all the complex issues faced by managers of overseas facilities. There are activities for which central control is absolutely necessary in order to protect certain core competencies of the company and to achieve economies of scale. However, there are times when control of these same activities must be delegated to the local management to respond to the changing local market environments. The critical issue here is to create complementary mechanisms to generate the kind of flexibility needed to balance the biases (either toward center or local) embodied in the existing organizational structure. Based on the experience of Japanese and European MNCs, Bartlett and Ghoshal [1992] find two sets of mechanisms that help to create conditions that facilitate better integration of local and central activities.

Mechanisms that will help to make central activities more effective and local activities more efficient include (1) creating multiple linkages to gain subsidiary inputs in corporate decision-making processes, particularly for issues related to local market such as product specifications and design, (2) empowering local management through the dispersal of organizational assets and resources and decentralization of authority, (3) establishing "internal" market mechanisms for centralized activities such as product R&D, which creates intense competitions for projects among research groups and allow overseas managers to act as customers to select the services they need provided by the center, and (4) building integrative mechanisms for linking technical, marketing, and other functions within the headquarters and each overseas facility through personnel flow and other means.

Another critical issue is related to human resource management. Managerial positions in the parent company and foreign subsidiaries have to be occupied by suitably qualified individuals. The MNC has to devise policies covering the nationality composition of staff, the selection and management of expatriates, international compensation, as well as transnational collective bargaining.

Staffing Policy

Staffing policy of foreign subsidiaries is determined by the strategic disposition of the MNC, the characteristics of the subsidiary and the host country, as well as the related costs (i.e., the economic costs of recruiting host nationals as opposed to expatriates). In general, MNCs face four options when staffing their foreign subsidiaries [Heenen and Perlmutter, 1979]: (1) appointment of home-country nationals, which often occurs at the initial stages of internationalization, (2) appointment of host-country nationals to gain the benefit of management with local market knowledge, (3) appointment of the best people for the job irrespective of nationality, and (4) a variation of the third option, which may be necessary when regional expertise is essential. In essence, nationality per se should not determine the staffing policy.

Expatriate Policy

Failure or ineffectiveness of overseas personnel assignments is costly both to the MNC and in terms of the individuals' career advancements. The high failure rate of expatriate assignments is largely due to difficulties in cross-cultural adjustment. However, failure to adjust applies not so much to the expatriate manager as to the expatriate's spouse and family whom MNC often fail to consider in the selection process [Black and Gregersen, 1991]. The MNC should ascertain the opinion of the spouse on the proposed assignment, provide spouses as well as employees with predeparture training, and assist spouses to overcome social isolation and help them develop networks with host-country nationals, all of which are essential to reduce the probability of failure.

Other key issues related to international human resource management include selection of an appropriate collective bargaining and union policy for each country (i.e., anti-union, multi-union, or single-union agreement) and adaptation to changes in the legal environment of the host countries concerning issues such as antidiscrimination legislation, health and safety regulation, and the appointment and promotion of women and minority groups.

Defining Terms

Configuration: The extent to which various of the company's activities are concentrated in one location from which the MNC's international network is served.

Coordination: The way that similar activities performed in different countries are coordinated with each other.

Economy of scale: Reductions in unit costs resulting from increased size of operations.

Economy of scope: Cost of the joint production of two or more products can be less than the cost of producing them separately.

References

Azani, H. and Khorramshahgol, R. *Eng. Costs Prod. Econ.*, 20(1), 23–28, 1990.

Bartlett, C. A. and Ghoshal, S. *Transnational Management: Text, Cases, and Readings in Cross-Boarder Management*, Richard D. Irwin, Chicago, 1992.

Black, S. J. and Gregersen, H. B. When Yankee comes home: factors related to expatriate and spouse repatriation adjustment, *J. Int. Bus. Studies*, 22(4): 671–694, 1991.

Buffa, E. S. Making American manufacturing competitive, *Calif. Mgt. Rev.*, Spring: 29–46, 1984.

Chase, R. B. and Aquilano, N. J. *Production and Operations Management: Manufacturing and Services*, 7th ed., Richard D. Irwin, Chicago, 1995.

DuBois, F. L., Toyne, B., and Oliff, M. D. International manufacturing strategies of U.S. multinationals: a conceptual framework based on a four-industry study, *J. Int. Bus. Studies*, 2nd quarter: 307–333, 1993.

Flaherty, M. T. Coordinating international manufacturing and technology, In *Competition in Global Industries*, M. E. Porter, ed., pp. 83–109, Harvard Business School Press, Cambridge, MA, 1986.

Ghoshal, S. Global strategy: An organizing framework, *Strat. Mgt. J.*, 8(5): 425–440, 1987.

Hamel, G. and Prahalad, C. K. Do you really have a global strategy?, *Harv. Bus. Rev.*, July-August: 139–148, 1985.

Heenen, D. A. and Perlmutter, H. V. *Multinational Organization Development*, Addison-Wesley, Reading, MA, 1979.

Hout, T., Porter, M. E., and Rudden, E. How global companies win out, *Harv. Bus. Rev.*, September-October: 98–108, 1982.

Kogut, B. Designing global strategies: profiting from operational flexibility, *Sloan Mgt. Rev.*, Fall: 27–38, 1985.

Levitt, T. The globalization of markets, *Harv. Bus. Rev.*, May-June: 92–102, 1983.

Meredith, J. R. *The Management of Operations: A Conceptual Emphasis, 4th ed.*, John Wiley & Sons, New York, 1992.

Mullin, R. Tactical site selection, *J. Bus. Strat.*, May/June: 26–42, 1996.

Tong, H.-M. and Walter, C. K. An empirical study of plant location decisions of foreign manufacturing investors in the United States, *Columbia J. World Bus.*, Spring: 66–73, 1980.

Wheelwright, S. C. Manufacturing strategy: defining the missing link, *Strategic Mgt. J.*, 5: 77–91, 1984.

Further Information

Columbia Journal of World Business and *Journal of International Business Studies* are good sources of advances in international management. *The Operations Management Review* and *Journal of Operations Management* reports latest development in the operations management.

22.4 International Strategic Alliances

Hildy Teegen

Strategic alliances are partnerships between two or more independent organizations. These partnerships can be formed between firms or governmental bodies. They are distinct from single transactions that occur in the market that are between independent firms for a single sale or purchase in that they endure over time. They are also distinct from activities between firms that have merged, have been acquired, or are subsidiaries of a common business entity. Thus, to be a strategic alliance, business activities take place between (among) *independent* organizations. **Transaction cost economics** [Williamson, 1975] is useful in explaining why firms choose to develop, manage, and commercialize their technology using a given organizational structure such as a strategic alliance. The structure chosen will be one that will minimize the (direct) costs of transacting with a counterpart in a commercial exchange.

The types of strategic alliances relevant to technology issues in firms range from research and development (R&D) consortia to simple licensing of a single product technology to another firm and include various forms in between. These collaborative ventures allow domestic firms to partner with other firms in order to enhance their competitive ability in foreign markets.

Global trends such as economic integration that creates large markets, shortened product and technology life cycles, and enhanced communications and transport all provide incentives to firms to partner in their international business dealings. The combining of each firm's unique resources and capabilities allows the partners to succeed in markets where it would be too costly or risky to participate otherwise. Accessing new markets is often the key to a technology's commercial viability; strategic alliances afford technology firms such access in a rapid, relatively inexpensive manner.

The experience of existing international technology-related alliances offers insights into guidelines for successful international strategic alliances. Behavioral, structural and process-related dimensions of alliances all significantly impact the potential success of a technology firm's alliance(s) internationally.

Markets, Hierarchies, and Alliances

Transaction cost economics [Williamson, 1975] indicates how firms will best organize their necessary transactions (sourcing, distribution, etc.) by examining which structure minimizes the costs associated with performing and monitoring the transactions. **Market transactions** between independent firms are characterized by information asymmetry between buyers and sellers, bounded rationality, which limits a decision maker's ability to access and/or consider all information relevant to a decision, and small numbers bargaining (a limited number of firms/organizations with which a firm may transact). These conditions allow for *opportunism* by the counterpart in the transaction, that is, the counterpart may take actions that negatively impact the firm without (much) risk of retribution. The single-transaction nature of many market encounters exacerbates this tendency. So, for a firm to limit the risk and impact of such opportunism, it must invest (money, management time, etc.) in *monitoring the transaction* to ensure that the counterpart is not behaving opportunistically. These monitoring costs are the transaction costs associated with a market transaction.

Many firms deal with the opportunism problem associated with market transactions by internalizing the activities necessary to their firm's functioning; they form internal **hierarchies** that perform the firm's activities. Vertical integration is a form of internalization whereby the firm itself becomes responsible for its own sourcing, production, marketing, etc. instead of relying on independent firms for the performance of these activities. Since firm members perform the activities, the risk of opportunism is reduced. However, large, integrated firms suffer a distinct type of transaction cost — the bureaucratic costs of managing and motivating members of a large firm. Shirking and weak incentives for members of the firm tend to bog down these internal transactions, adding to their cost.

Strategic alliances are sought by firms seeking to mitigate the heavy transaction costs of the market and of hierarchies: opportunism monitoring and bureaucratic costs, respectively. Alliances reduce opportunism by linking the partners' fates across time. Thus, a counterpart is given less incentive to "cheat" on its partner as it will continue to transact with this partner in the future. However, since the partner firms still maintain their independence, there is incentive for them to perform their activities well or else their counterpart will have reason to leave the alliance. In this way, well-founded, structured, and managed alliances can result in reduced transaction costs for the participating firms.

Types of Strategic Alliances for Technology Firms

Strategic alliances take many forms, which vary according to the degree of interaction between the partners and how partners are compensated for their resource contributions [Contractor and Lorange, 1988].

Technology providers may form alliances for **technical training** for the purchasers of their products. These alliances tend to be fee based in nature, but often endure longer that a single market transaction might. In many sectors, these technical training alliances may extend to provide complete turn-key, operation **startup** services whereby the counterpart is provided with all the equipment, product, and process know-how necessary to become viable in a given business venture.

Given distinctions in factor endowments of countries, it is often advantageous to seek a partner in a low-wage country to produce using labor-intensive technology developed in a higher-wage country. Also, given threshold levels of optimal production capacity, certain firms' client bases and/or national markets are of insufficient size to allow for optimal production runs. In both of these cases, alliances with partners in **production agreements** can be useful in reducing costs of production, allowing for greater margins to the partnering firms.

Buyback arrangements are a special sort of alliance that links production with technology transfers. Where purchasers of technology are concerned about the efficacy of the transfer itself and/or where hard currency to pay for the acquired technology is scarce, a buyback arrangement may be formed [Lecraw, 1988]. In these arrangements, the seller of the technology receives all or part of its compensation for the transferred technology in the form of products produced with that technology. Thus, the seller has a strong incentive to ensure that the technology is well transferred or else its payment may suffer (poor product quality, which would limit the firm's ability to liquidate the product, converting its compensation to a cash basis).

Basic technology transfers typically are undertaken under **license** or through the sale of a **patent**. In these cases, the transferring firm cedes full or partial rights to the usage of the technology. Both process and product technology may be transferred under these arrangements. They may involve a lump sum payment or a royalty fee based upon sales or some other measure of performance. Where only partial rights have been transferred (e.g., use for a given territory), the transferring firm will tend to be more involved in the ongoing alliance than in the case where all rights have been ceded. This is because the transferring firm must ensure the proper usage of the technology by the partner; incorrect or inappropriate usage may otherwise degrade the value of the retained technology rights held by the transferring firm. This is particularly important for transfers to partners in jurisdictions that do not provide adequate protection of intellectual property rights. Although the World Trade Organization is working explicitly on the issue of convergence in intellectual property rights protection internationally, significant differences in protection still exist. Savvy firms will seek partners with this risk in mind.

An increasingly important alliance vehicle for transferring process technology is the **franchise**. Franchise alliances allow the franchiser to transfer process technology to the franchisee. As in the case of a license, lump sums and royalty fees are typical compensation forms. Over time, franchise relationships serve as reciprocal technology transfers as adaptations (internationally) to the original process technology are often suggested by on-site franchisees in the markets where the franchise operates.

Of increasing importance to technology firms is the **R&D consortium** [Olk, 1997]. These tend to be alliances of multiple firms and/or governments that work together to produce next-generation technologies. The Sematech Alliance in Austin, TX, is an example of a research consortium of independent firms with the goal of jointly produced innovations. The incidence of alliances of partners from various countries is growing steadily.

Motivations for International Strategic Alliances

Firms partner internationally to reduce the risks and costs of doing business overseas. Additionally, they can access new and/or complementary technologies and access new markets via their partner's position in the market or via resources that enable the alliance to enter the market [Contractor and Lorange, 1988].

The risk of doing business internationally can be largely offset by participating in a strategic alliance of one form or another. By partnering with firms in different product or national markets, a firm can diversify itself through the partnership, insulating it from negative downturns in its domestic market or industry. Also, by sharing the investment and fixed maintenance costs of a business venture with a partner, a smaller portion of the firm's capital is invested, reducing the potential loss of a venture. Also, by accessing a partner's expertise and capital, markets can be entered more quickly, and the payback to the initial investment is faster.

Besides sharing costs with a partner, actual costs can also decrease through alliances. By producing at optimal production levels, or by accessing a larger market with a partner, economies of scale can be achieved, reducing unit costs. Also, each partner to an alliance provides unique resources, and each can specialize in activities for which they have a comparative advantage in the partnership. In this way, the costs of performing those activities diminishes as each partner can move more quickly along the learning curve, further reducing total costs for the alliance.

Accessing technology, either through that provided by a partner (**substitution mode**) or through that jointly developed by the alliance partners (**fusion mode**) or through that created by the unique combination of the alliance partners' specific technologies (**complementary mode**) is a strong motivation for international alliance formation [Afriyie, 1988]. Given the high costs of internal development of technology, the short life cycles of technology, and the potential for transferring technology proven in one market that could be viable in other national markets, these technology access concerns are particularly important for international alliances.

The risk, cost, and technology access motivators for alliance formation are as relevant domestically as they are internationally. For international business, market access for technology firms is a critical impetus for alliance formation. Permission to enter a market and relationships with or knowledge about key customers are often impenetrable entry barriers for firms operating independently in seeking overseas markets.

Many foreign markets are protected in favor of local concerns. Forms of protection range from tariff barriers to licensing requirements and local content laws. By partnering with a local concern, a firm can acquire local firm status and thus be permitted to operate in the market. Technology-providing firms tend to be more welcome entrants, particularly in the case of developing countries that promote economic development via enhancements in the local technology base. Even where a firm is welcomed into a foreign market, the use of a foreign partner that is familiar with rules and regulations can greatly speed the time to initiate business in the market and improve the firm's likelihood of success there.

Perhaps of greater concern to firms entering foreign markets is lack of knowledge about the market and its consumers (industrial or final). A local partner that has existing commercial relationships with viable customers/clients is invaluable for quickly tapping those opportunities. In addition, a local partner can provide useful information on technological adaptations to improve the diffusion rate of the technology in the market. In many cases, these adaptations are then transferred back to the home country where they may enhance the utility of the technology there as well.

Guidelines for International Alliance Success

Alliances, like marriages, are collaborations between organizations — groups of human members. Given the human dimension of alliances, many behavioral aspects of alliances must be considered by collaborating firms. Recall in the discussion of transaction costs that in market transactions firms are at risk of opportunistic behavior by their partners. Although not as great in alliances, this risk still is present. *Trust* between alliance partners, however, serves as an important mitigator of opportunistic behavior by firms [Aulakh et al., 1997]. Partnering firms that share *organizational culture values* and that come *from culturally congruent countries* are those that find it easiest to trust one another. Of particular importance is the cultural norm of *flexibility*. Firms that are flexible in their interactions with their partner are better able to adapt to changing international environments and thus will have better performing alliances.

The structure of alliances can also impact the alliance's success. Management of the alliance's activities and the establishment of rights and responsibilities for each alliance member directly impact alliance performance. Control over alliance activities is a concern to allying firms. Some firms stipulate control via equity ownership in the alliance — an independent organization is formed by the partners who own portions of the newly created **joint venture**. In addition to ownership, areas of responsibility may be agreed upon to ensure that each partner "controls" the areas of the alliance most critical to them. These areas tend to be associated with each firm's area of comparative competence. Related to the notion of control in alliances is that of *dependence*. Since many alliances are formed to allow partners to specialize in specific activities, a risk of alliances is a firm's becoming dependent upon its partner for the activity. Successful alliances tend to have balanced dependence/interdependence, whereby the partners are relatively equal in terms of their dependence on their partner.

The structure of successful alliances accurately reflects partners' perceptions of *fair exchange* in the alliance. Alliances that provide outputs to each partner that are commensurate with their respective inputs are deemed fair alliances. Where partners view an alliance as fair, they will have incentive to work for the partnership's (joint) success.

The process of allying itself provides perhaps the greatest opportunity as well as the greatest challenge to collaborating firms. Transaction cost economics is useful in identifying the best organizational option among market, hierarchy, and alliances. This decision is based upon the direct short-term costs of transacting. The alliance process also entails long-term opportunity costs and benefits, though. These longer-term factors are particularly relevant when considering *organizational learning* through alliances.

By sharing resources with a partner, a firm can access vast sources of information on how to better produce, innovate, and commercialize — in short, to better compete in a given product or national market. Thus, learning from the partner is a great contribution of an alliance. Unfortunately, this benefit is a double-edged sword. Where a firm can learn from its partner, its partner too can learn from the firm. The greatest risk from allying is the potential for competitor creation through the partnership. To best ensure against this risk, core competencies of the firm should be actively guarded through control mechanisms in the structuring of the alliance. Where a firm provides core resources in the form of technology, the intellectual property protection provided in the jurisdiction of the alliance is particularly important for assessing the risk of allying with a given partner.

Conclusion

Globalization of markets has prompted many firms to seek collaborative partnerships — strategic alliances with domestic and international firms in order to better compete in foreign markets. The alliance structure provides benefits over market transactions and integrated, hierarchical firms in doing business internationally. The forms of alliances available to technology firms are varied, and they provide clear benefits in risk and cost reduction as well as technology and market access to allying firms. Well-structured alliances that consider the behavioral elements relevant to interacting firms and that promote learning without concomitant core capability leakage to a partner will produce favorable results to technology firms active internationally.

Defining Terms

Buyback arrangement: Technology transfers that are compensated wholly or partially with products produced using the transferred technology.

Complementary mode: Technology transfer mode that utilizes the technologies of partner firms as separate but compatible elements of the joint technology package.

Franchise: An alliance where a franchiser sells/licenses product/process technology and trademarks concerning an operating system to a franchisee.

Fusion mode: Technology transfer mode that combines the technologies of partner firms thereby producing a unique, new technology that is completely distinct from the initial technology contributions of the partners.

Hierarchy: Organizational mode that internalizes the firm's activities under the firm's direct ownership and control. Vertical integration is a hierarchical approach to organization.

Joint venture: A form of strategic alliance where the partner firms create and share equity in a new, separate, third entity: the joint venture.

License: A prescribed right of usage of a technology that typically requires payment of a lump sum or royalty fee as compensation to the transferor.

Market: Where transactions occur between legally separate firms with no expectation of future interaction.

Patent: Legally protected proprietary rights over intellectual property.

Production agreement: Strategic alliance where partners share production in a facility or where one partner produces on behalf of the other partner.

R&D consortium: Strategic alliance of (typically) multiple partners from the private and/or public sector established to jointly innovate technologies.

Startup accord: Strategic alliance where a firm establishes an enterprise for its partner and provides the requisite training to ensure the establishment's effective operation under the partner firm's guidance.

Substitution mode: Technology transfer mode whereby one partner's technology is used by the partnership to the exclusion of the other firm's technology.

Technical training agreement: Strategic alliance where a firm provides product and/or process technical training to the partner firm.

References

Afriyie, K. A technology-transfer methodology for developing joint production strategies in varying technological systems, In *Cooperative Strategies in International Business,* F. J. Contractor and P. Lorange, eds., pp. 3–30, Lexington Books, Lexington, MA, 1988.

Aulakh, P., Kotabe, M., and Sahay, A. Trust and performance in cross-border marketing partnerships: a behavioral approach, *J. Int. Bus. Studies,* 27(5): 1005–1032, 1997.

Contractor, F. J. and Lorange, F. Why should firms cooperate? The strategy and economics basis for cooperative ventures, In *Cooperative Strategies in International Business,* F. J. Contractor and P. Lorange, eds., pp. 3–30, Lexington Books, Lexington, MA, 1988.

Lecraw, D. J. Countertrade: a form of cooperative international business arrangement, In *Cooperative Strategies in International Business*, F. J. Contractor and P. Lorange, eds., pp. 3–30, Lexington Books, Lexington, MA, 1988.

Olk, P. The effect of partner differences on the performance of R&D consortia, In *Cooperative Strategies: North American Perspectives*, P. W. Beamish and J. P. Killing, eds., pp. 133–162, The New Lexington Press, San Francisco, 1997.

Williamson, O. *Markets and Hierarchies*, Free Press, New York, 1975.

Further Information

The *Cooperative Strategies Series* of the New Lexington Press contains the most up-to-date work available on international alliances — it is a three-volume set, which describes North American, European, and Asian perspectives on alliances, P. W. Beamish and J. P. Killing, eds., 1997.

Appendixes

Appendix A
Glossary of Business Terms

Terms	Definitions
Account	A running record of transactions between two transactors. In accounting, a formal record of a transaction.
Accounts payable	Amounts of money owed to others such as vendors.
Accounts receivable	Money owed to a business by customers who purchased goods or services on credit.
Accrual basis	Method of accounting where revenues are recorded as they are earned, regardless of when the money is received.
Acid test	See **Quick ratio**. A measure of liquidity.
Advertising	Public messages sent via any medium that are intended to attract and influence consumers.
Agent	An intermediary company or individual with the authority to carry out transactions with third parties on behalf of the principal.
Amortize	To write off assets gradually over a specified number of time periods.
Arbitrage	The simultaneous purchase and sale of substantially identical assets in order to profit from a price difference in two markets.
Asset	Something of monetary value that is owned by a firm or an individual.
Attribute	A feature or characteristic of a product or service.
Audit	An examination of a firm's financial records.
Automation	Using specialized equipment to carry out operational tasks.
Balance sheet	Record of a company's finances at a given time.
Barrier to entry	An obstacle to entry to an industry or market.
Brand	The attribution to a product of a name, trademark, or symbol.
Break-even volume	Fixed costs divided by the selling price minus the variable cost per unit.
Budget	A forecast of revenue and expenditure for a forthcoming period.
Bureaucracy, economic theory of	Assumes that governmental agencies will act as budget maximizers.
Business cycle	A fluctuation in the level of economic activity which forms an expansion followed by a contraction.
Capacity	The amount that a plant or enterprise can produce in a given period.
Capital	Financial assets.
Capital asset pricing model	Describes the tradeoff of risk and return for a portfolio of assets.

Capitalism	A political and economic system in which property and capital assets are owned and controlled primarily by private persons or companies.
Capitalization	The amounts and types of long-term finance used by an enterprise.
Cartel	A group of companies with the express purpose of reducing competition between them by fixing prices or other means.
Cash flow	Amount of net cash flowing into a company during a specified period. The sum of retained earnings and depreciation provision.
Cash cow	A company or product that generates significant cash without great effort or investment.
Centralization	The process of concentrating control of an organization at its center or the top of its hierarchy.
Chief executive officer	The person directing the enterprise and responsible to the firm's board of directors.
Commodity	A generic, relatively unprocessed, good that can be reprocessed and sold.
Comparative advantage	Principle that states that the world economy is advantaged if each nation concentrates on producing what it does best.
Competitive advantage	Strengths that a company holds as an advantage over its competitors.
Conglomerate	A holding company that owns companies in a wide range of different businesses or industries.
Contract	An agreement in which one party agrees to provide certain goods or services in return for a consideration or money.
Contribution margin	Sales revenues minus the variable cost of a product or service.
Core competence	The particular activity or know-how at which an enterprise is significantly good.
Corporate culture	The style, personality, or way of working of an organization.
Cost of capital	The weighted average of the cost of a firm's debt and equity capital.
Cost-benefit analysis	A framework for project evaluation.
Current assets	Assets that can quickly be converted into cash.
Current ratio	A measure of liquidity equal to the current assets divided by the current liabilities.
Debt	An obligation arising from borrowing or purchasing on credit.
Demand	The amount that people are ready to purchase at the prevailing price.
Depreciation	The reduction in value of an asset due to wear and use.
Diffusion process	The pattern of consumer acceptance of a new good or service. The rate at which an innovation spreads.
Diminishing returns	See **Law of diminishing returns**.
Distribution channel	The route by which a product travels from manufacturer to customer.
Distributor	A firm that takes title to goods and then sells them.
Diversification	A strategy for reducing risk by being active in a variety of markets.
Economies of scale	The decrease in the unit cost of production as a firm increases its volume of production.
Economies of scope	The decrease in the unit cost of production as a firm increases its range of related products.

Efficient market	A market in which prices reflect all available information and adjust rapidly to any new information.
Elasticity (price)	Responsiveness of the quantity purchased to a change in the item's price.
Empowerment	An increase in the power of individuals to act on their own initiative.
Entrepreneur	A person who takes the risk of establishing a new business and who will accrue the wealth or other benefits of success.
Equity	An ownership interest in a company, often held as shares in the capital of the company.
Ex ante	The planned or intended future outcome.
Ex post	The actual or realized outcome.
Exchange rate	The price for which one country's currency can be exchanged for another.
Externality	An outcome or impact upon a firm that occurs from causes not directly involved in the activity that causes the outcome.
Factoring	Selling your firm's accounts receivable to another firm for collection by them.
Finance	The form of funds obtained as capital to undertake an expenditure.
First mover	The first entrant into a market.
Fixed asset	An asset that remains in the business over time such as buildings and land.
Fixed cost	A cost that does not vary with the quantity of goods produced over the short run (near term).
Focus group	A group of people who are brought together to discuss a product or service.
Forecasting	A systematic method of obtaining an estimate of the fixture value of one or more variables, such as revenues or number of units sold.
Functional organization	A company structured into units by the type of work accomplished, such as engineering or finance.
Futures	Any transaction that involves a contract to buy or sell a commodity or security at a fixed date at an agreed price.
Generally accepted accounting principles (GAAP)	The official financial reporting criteria for United States firms.
Generic product	Unbranded, common product.
Global business	Activities that take advantage of worldwide opportunities and resources.
Goodwill	The amount that an acquiring company pays for the acquired company over and above the value of the acquired company's assets.
Gross margin	The difference between revenues and the cost of goods sold.
Hedging	An action taken by a buyer or seller to protect the income from a change in prices in the future.
Income statement	A summary of a firm's revenues and experiences for a specified period.
Increasing returns	See **Law of increasing returns**.
Inflation	A sustained rise in the price of goods and services.
Initial public offering	The initial sale of common stock to individuals in the public market.

Innovation	The act of introducing a new process or product into a market.
Intangible asset	A non-physical asset such as patents, copyrights, and business secrets.
Internal rate of return	The discount rate on an investment that equates the net present value of the cash outflows with the net present value of the cash inflows.
Inventory	Finished goods held for use or sale by a business firm.
Joint venture	A business enterprise officially formed by two or more unrelated firms.
Just in time	Receiving goods from a supplier when they are needed.
Keuretsu	Japanese business coalitions.
Law of diminishing returns	The principle that as increasing incremental resources are put into an activity, the less are the resulting incremental increases in outputs.
Law of increasing returns	The principle that for certain industries, such as software, that as increasing incremental resources are put into an activity, the incremental increases in output will grow.
Leadership	The ability of a person through force of personality and skills to influence others toward a goal.
Learning curve	A record of the increase in productivity caused by experience in a set of operations.
Learning organization	A firm that uses its corporate experiences to improve and adapt.
Lease	An agreement permitting the lessee to use property owned by the lessor.
Leverage	The use of borrowed funds to increase the return on an investment. A firm's debt-to-equity ratio is a measure of its leverage.
Liabilities	Any financial claims on the firm.
Liquidity	Assets that are cash or can be easily converted to money.
Management	The staff within a firm who control the resources and financial decisions of the firm.
Management by objectives	A system using specific performance goals for each employee.
Marginal cost	The incremental cost of producing an additional unit of output.
Market	The potential buyers and sellers of a good or service aggregated for purposes of exchange.
Market share	The proportion of total market sales accounted for by one firm.
Marketing	The planning and execution of the activity that identifies and satisfies customer requirements in a market. This activity may include sales promotion, advertising, and market research.
Matrix structure	An organizational arrangement for managing cross-functional activities.
Mission statement	A brief statement of a firm's core products, markets, and goals.
Money	A commonly accepted means of payment or means of exchange in a market.
Monopoly	A market served by only one firm acting as seller. The total absence of competition.
Multinational	A large firm with a home base in one country by operating subsidiaries in other countries.
Net present value (NPV)	The current value of future amounts of money, discounted at an agreed rate (discount rate).

Oligopoly	A market served by only a few firms acting as sellers.
Opportunity cost	The value of the next best alternative that is foregone by choosing the best alternative.
Option	A contract allowing another party to buy or sell a given item within a given time period at an agreed price.
Original equipment manufacturer (OEM)	A firm that assembles its product with modules made by suppliers.
Outsourcing	A firm's purchase of services or goods from suppliers rather than doing them within the firm.
Overhead	The indirect cost of doing business that cannot be directly attributed to a unit sold, such as the cost of company insurance.
Payback period	The period of time that it takes a firm to recover its investment in a product or activity.
Price elasticity	See **Elasticity (price)**.
Pricing	The process of determining the price to charge customers.
Private label	A product sold using the retailer's brand but produced for the retailer by a supplier or manufacturer.
Product line	A series of products that have similar characteristics but vary in one or two attributes.
Process	A sequence of activities used to obtain an output given a set of inputs.
Productivity	The ratio of the output obtained to the inputs utilized for a given process.
Quick assets ratio	The ratio of liquid assets to current liabilities.
Rate of return	The net income as a percent of investment.
Reengineering	The radical redesign of a firm's processes to achieve significant productivity improvement.
Research and development	A unit or group that uses or generates technical research information to generate new products or processes.
Risk	The potential for a loss of revenues or profits due to the occurrence of an event or externality.
Stakeholder	All individuals that have a vested interest in the achievements of an organization.
Strategic business unit	A business entity within a larger firm. It is normally organized around a product line or a market.
Strategy	A plan of action to achieve long-run outcomes.
Sunk costs	Costs that cannot be recovered if a firm leaves an industry or decides not to market a product it has developed.
Supply	The amount of a good or service that firms are ready to sell in a given market.
Technology transfer	The exchange among entities, such as firms or nations, of technological knowledge.
Technological progress	The involvement of economic growth which enables more output to be produced for unchanged inputs of labor and capital. This improvement is attributed to the development and use of new technologies.
Trademark	A name, symbol, or mark a firm uses to designate its product.

Value added	The value of a process output minus the value of its inputs.
Value chain	The sequence of steps or subprocesses that a firm uses to produce its product or service.
Variable costs	The costs that vary with the number of units produced.
Vertical integration	The extension of a firm's activities over more than the one stage in which it is currently active. Backward integration is when the firm extends its activities into a previous stage. Forward integration is when it extends into a succeeding stage of activity.
Virtual corporation	A firm that only works on activities based on its core competency and outsources most other tasks.
Work in process	Materials that are being developed into a product but are not finished goods.
Working capital	The firm's current assets which are financed from long term sources of finance.

Appendix B
The Bookshelf
of Great Books

Part I. The Technology Manager and the Modern Context

Chapter 1.　Entrepreneurship and New Ventures
Bygrave, W. P., *The Portable MBA in Entrepreneurship,* John Wiley & Sons, 1997.

Chapter 3.　Innovation and Change
Kotter, J. P., *Leading Change,* Harvard Business School Press, 1996.
Stewart, T., *Intellectual Capital,* Doubleday, 1997.
Tushman, M. and Anderson, P., *Managing Innovation and Change,* Oxford University Press, 1997.

Part II. Knowledge for the Technology Manager

Chapter 4.　Economics
Becker, G. S. and Becker, G. N., *The Economics of Life,* McGraw-Hill, 1996.
Landsburg, S. E., *The Armchair Economist,* Free Press, 1993.
Samuelson, P. and Nordhaus, W., *Economics,* McGraw-Hill, 1998.

Chapter 5.　Statistics
Keller, G. and Warrack, B., *Statistics for Management and Economics,* Wadsworth Publishing, 1997.

Chapter 6.　Accounting
Murray, D. and Neumann, B., *Using Financial Accounting,* Southwestern Publishing, 1997.
Stickney, C. and Weil, R., *Financial Accounting,* Dryden Press, 1998.

Chapter 7.　Organizations
Gibson, J. and Ivancevich, J., *Organizations,* Irwin-McGraw-Hill, 1991.

Part III. Tools for the Technology Manager

Chapter 8.　Finance
Higgins, R. C., *Analysis for Financial Management,* Irwin-McGraw-Hill, 1995.
Spiro, H. T., *Finance for the Nonfinancial Manager,* John Wiley & Sons, 1996.

Part IV. Managing the Business Function

Chapter 12.　Marketing of Technical Products
Best, R. J., *Market Based Management,* Prentice-Hall, 1997.
Kotter, P., *Marketing Management,* Prentice-Hall, 1997.
Moore, G., *Crossing the Chasm,* HarperCollins, 1991.

Chapter 13.   Production and Manufacturing
Goldratt, E., *The Goal,* North River Press, 1992.
Schonberger, R. and Knod, E., *Operations Management,* Irwin-McGraw-Hill, 1996.
Smith, G. M., *Statistical Process Control and Quality Improvement,* Prentice-Hall, 1997.

Chapter 14.   Product Management
Urban, G. and Hauser, J., *Design and Marketing of New Products,* Prentice-Hall, 1993.

Chapter 15.   Project Management
Meredith, J. and Mantel, S., *Project Management,* John Wiley & Sons, 1995.

Part V.   Strategy of the Firm

Chapter 16.   Business Strategy
Pearce, J. and Robinson, R., *Strategic Management,* Irwin-McGraw-Hill, 1998.

Part VI.   Core Relationships for the Technology Manager
Yate, M., *Career Smarts,* Ballantine Books, 1997.

Part VII.   Global Business Management

Chapter 22.   International Business
Buzzell, R. and Quelch, J., *Global Business Management,* Addison-Wesley, 1995.
Deans, C. and Dakin, S., *The Thunderbird Guide to International Business Resources on the World Wide Web,* John Wiley & Sons, 1996.

Appendix C
Twelve Great Innovations

1. **The Cotton Gin**, 1974. Eli Whitney devised a machine that removed the seeds from the cotton fibers, permitting the growth of the cotton and cloth industry.
2. **The Sewing Machine**, 1846. Elias Howe and Issac Singer invented the machine that industrialized clothes making and built the garment industry.
3. **Barbed Wire**, 1874. Joseph Glidden's wire enabled ranchers to fence the plains. By 1887, 173,000 tons a year were being sold and the frontier was being widely fenced.
4. **The Telephone**, 1876. Alexander Graham Bell developed and patented the telephone, which opened voice communications to all people.
5. **The Light Bulb**, 1880. Thomas Edison and his research team developed the light bulb and by 1882 built a system to provide electric power to enable lighting systems.
6. **The Automobile**, 1895. George Selden, Karl Benz, and Gottlieb Daimler developed the automobile and Henry Ford built the industry with the founding of Ford Motor Co. in 1903.
7. **The Radio**, 1901. Guglielmo Marconi demonstrated radio transmission and reception. By 1920, KDKA, Pittsburgh became the first commercial station.
8. **The Airplane**, 1906. Orville and Wilbur Wright demonstrated the airplane in 1903 and Glen Curtiss built early models for commercial use in 1908.
9. **Xerography**, 1942. Chester Carlson worked out the process in a makeshift laboratory. Fourteen years later in 1956, the first commercial copier was demonstrated.
10. **The Transistor**, 1950. John Bardeen, Walter Brattin, and William Shockley invented the device at AT&T Bell Laboratories. By 1956 the transistor was rapidly replacing the vacuum tube.
11. **The Integrated Circuit**, 1958. Jack Kilby at Texas Instruments fabricated an integrated circuit containing the transistor, a resistor, and a capacitor. Independently Robert Noyce and Gordon Moore of Intel developed an integrated circuit in 1959. Integrated circuits were commercial by 1961.
12. **The Internet (World Wide Web)**, 1990. Tim Berners-Lee invented the World Wide Web (WWW), an internet-based hypermedia developed for global information sharing. Marc Andreesen developed a graphical user interface to the WWW in 1993. Netscape and Microsoft commercialized a WWW graphical user interface in 1995.

0-8493-8577-6/99/$0.00+$.50
© 1999 by CRC Press LLC

Appendix D
Associations and Government Organizations

Name	Phone	Fax
Institute of Industrial Engineers	(770) 449-0461	(770) 263-8532
American Accounting Association	(941) 921-7747	(941) 923-4093
American Management Association	(212) 586-8100	(518) 891-0368
American Association of Advertising Agencies	(212) 682-2500	(212) 682-8391
American Economics Association	(615) 322-2595	
American Marketing Association	(312) 648-0536	
American Society for Training and Development	(703) 683-8100	
Date Processing Management Association	(709) 825-8124	
Conference Board	(212) 759-0900	(212) 980-7014
American Society for Quality Control	(414) 272-8575	(414) 272-1734
American Society of Mechanical Engineering	(212) 705-7722	(212) 705-7674
MIT Enterprise Forum	(617) 253-8240	(617) 258-7264
National Association of Manufacturers	(202) 637-3000	(202) 637-3182
National Technical Information Sources	(703) 487-4600	
Small Business Administration	(800) 827-5722	
Society of Competitive Intelligence Professionals	(202) 223-5885	(202) 223-5884
Society of Manufacturing Engineering	(313) 271-1500	(313) 271-2861
United States Trademark Association	(212) 986-5880	(212) 687-8267

Appendix E
Magazines and Journals

Name	Phone	Fax
Bureau of Economics Analysis, U.S. Gov't.	(202) 898-2450	
Business Week	(212) 512-2511	(212) 512-6458
California Management Review	(510) 642-7159	(510) 642-1318
Commerce Clearing House	(312) 583-8500	
Economist Magazine	(212) 460-0600	(212) 995-8837
Forbes Magazine	(212) 620-2200	(212) 620-2417
Fortune Magazine	(800) 541-1000	
Harvard Business Review	(617) 496-1449	(617) 495-6985
IEEE Transactions on Engineering Management	(908) 981-0060	
Sloan Management Review (MIT)	(617) 496-1449	(617) 495-6985
Standard and Poors	(212) 208-8000	(212) 412-0241
Wall Street Journal	(800) 568-7625/	(212) 416-2658
World Bank	(202) 477-1234	

Index